LIQUID CRYSTALS AND THEIR COMPUTER SIMULATIONS

Standing as the first unified textbook on the subject, *Liquid Crystals and Their Computer Simulations*, provides a comprehensive and up-to-date treatment of liquid crystals and of their Monte Carlo and Molecular Dynamics computer simulations. Liquid crystals have a complex physical nature, and, therefore, computer simulations are a key element of research in this field. This modern text develops a uniform formalism for addressing various spectroscopic techniques and other experimental methods for studying phase transitions of liquid crystals and emphasizes the links between their molecular organization and observable static and dynamic properties. Aided by the inclusion of a set of Appendices containing detailed mathematical background and derivations, this book is accessible to a broad and multidisciplinary audience. Primarily intended for graduate students and academic researchers, it is also an invaluable reference for industrial researchers working on the development of liquid crystal display technology.

CLAUDIO ZANNONI obtained his PhD in Chemical Physics from Southampton University in 1975 and has been Professor of Physical Chemistry (now Emeritus) at the University of Bologna since 1987. He has extensive experience in the field, having published some 300 papers and delivered over 350 lectures worldwide on computer simulations and molecular theories of liquid crystals. He is also past president (2012–2016) of the International Liquid Crystal Society. In 1998 he founded, and since then directs, the International School of Liquid Crystals in Erice.

LIQUID CRYSTALS AND THEIR COMPUTER SIMULATIONS

CLAUDIO ZANNONI

CAMBRIDGE
UNIVERSITY PRESS

CAMBRIDGE
UNIVERSITY PRESS

University Printing House, Cambridge CB2 8BS, United Kingdom

One Liberty Plaza, 20th Floor, New York, NY 10006, USA

477 Williamstown Road, Port Melbourne, VIC 3207, Australia

314–321, 3rd Floor, Plot 3, Splendor Forum, Jasola District Centre,
New Delhi – 110025, India

103 Penang Road, #05–06/07, Visioncrest Commercial, Singapore 238467

Cambridge University Press is part of the University of Cambridge.

It furthers the University's mission by disseminating knowledge in the pursuit of education, learning, and research at the highest international levels of excellence.

www.cambridge.org
Information on this title: www.cambridge.org/9781108424059
DOI: 10.1017/9781108539630

First published 2022

A catalogue record for this publication is available from the British Library.

Library of Congress Cataloging-in-Publication Data
Names: Zannoni, Claudio, author.
Title: Liquid crystals and their computer simulations / Claudio Zannoni.
Description: New York : Cambridge University Press, 2022. |
Includes bibliographical references and index.
Identifiers: LCCN 2021058085 (print) | LCCN 2021058086 (ebook) |
ISBN 9781108424059 (hardback) | ISBN 9781108539630 (epub)
Subjects: LCSH: Liquid crystals. | Liquid crystals–Computer simulation. |
BISAC: SCIENCE / Physics / General
Classification: LCC QD923 .Z36 2022 (print) | LCC QD923 (ebook) |
DDC 530.4/29–dc23/eng20220208
LC record available at https://lccn.loc.gov/2021058085
LC ebook record available at https://lccn.loc.gov/2021058086

ISBN 978-1-108-42405-9 Hardback

To my wife, Nicoletta, for her support over all these years and for believing that I would finish this project even when I doubted it.

Contents

Preface

There are two main approaches to the theoretical study of liquid crystals: continuum and molecular.

The first, well covered in various good books [Chandrasekhar, 1992; de Gennes and Prost, 1993; Virga, 1994; Kleman and Lavrentovich, 2003; Stewart, 2004; Oswald and Pieranski, 2005, 2006; Barbero and Evangelista, 2006], considers anisotropic systems at macroscopic level and typically deals with optical and elastic properties as well as with many practical electro-optical applications of liquid crystals. At the continuum level, liquid crystals are assumed to exist and their properties (e.g. elastic constants and viscosities) to be known, insofar as they are needed to parameterize the relevant equations. Molecules, phase transitions and spectroscopic properties are not normally taken into consideration. In this line of work computer simulations typically refer to a determination of the preferred orientation (director) or of the ordering tensor field that minimize the elastic free energy under a variety of boundary conditions, while dynamics is normally related to the solution of hydrodynamics equations for anisotropic fluids.

The other main line of investigation deals with the molecular organization of liquid crystals and how their macroscopic behaviour can be understood in terms of constituent molecules (or colloidal particles, as appropriate) and their interactions, particularly with the help of computer simulation techniques. It is definitely this microscopic approach that we shall follow in this book, discussing in some detail the main types of liquid crystal phases as well as theoretical and computer simulation approaches. I believe that such a book does not exist at the moment and that it might be useful to have one. On one hand, books dealing with liquid crystals [de Gennes, 1974; Chandrasekhar, 1992; Chaikin and Lubensky, 1995; Collings and Hird, 1997; Khoo, 2007; Blinov, 2011] hardly talk of computer simulations, since they are focussed on other aspects or, possibly, because their development is relatively recent. On the other, good textbooks on computer simulations also exist [Frenkel and Smit, 2002; Berendsen, 2007; Allen and Tildesley, 2017], but none deals specifically with liquid crystals. This is a major problem, since computer simulations of liquid crystals need to go beyond the standard calculations of thermodynamics properties or radial distributions and should relate to relevant experiments in the field. In particular, this requires developing appropriate methodologies to calculate the anisotropic, tensorial, observables, order parameters, space and time correlation functions, director field and defects, that are characteristic features of liquid crystals, and to make contact with what is actually measurable, e.g. from spectroscopic or diffraction experiments. Some of these aspects have been addressed in multi-author books [Luckhurst and Gray, 1979; Pasini and Zannoni, 2000; Lavrentovich et al., 2001; Pasini et al., 2005b], some of which I have co-edited. However, these books are now at least 15 years old, while very many new applications, e.g. all the predictive atomistic simulations of liquid crystals, have been developed more recently. It is also worth stressing that liquid crystals are an intrinsically interdisciplinary topic and many of the background tools needed for their understanding are drawn from different curricula, especially physics and chemistry, but also mathematics, biology, etc. A similar problem arises even within a single discipline when we wish to treat different anisotropic materials like low-molar-mass liquid crystals, polymers and membranes. At the moment, these topics are presented separately in reviews or book chapters. While these have the advantage of a detailed treatment of specific advanced topics, we aim here at a consistent approach that tries to amalgamate the various topics. For example, much of the background required to understand the application to liquid crystals of different spectroscopic techniques, such as Nuclear Magnetic Resonance (NMR), Fluorescence Depolarization (FD), Dielectric Relaxation (DR), X-ray, etc., is largely similar, even though the different fields have developed independently and often with a different jargon and notation for the same quantities, so that a unified treatment should now be timely. Such an approach, in terms of order parameters and correlation functions, is also key to predicting observables from computer simulations and comparing with experimental results. The book provides the basic conceptual and technical tools needed by a student towards the end of an undergraduate curriculum or at the beginning of a postgraduate course (in physics, chemistry, material sciences, engineering or mathematics), or more generally by someone starting research in liquid crystals. The book has grown from undergraduate and graduate courses that I have taught for a number of years at Bologna University as well as from lectures that I have given at a number of summer schools and at universities around the world, from Southampton to Kuala Lumpur. On the basis of this experience, I have made an effort to put together some of the contents useful for a fairly gentle introduction to liquid crystals at molecular level.

In summary, the organization of the book is as follows. The first part of the book introduces the various kinds of mesophases and their phase transitions from the thermodynamic point of view (Chapters 1 and 2) as well as in terms of order parameters (Chapter 3). The essentials of how various experimental techniques (Linear Dichroism (LD), FD, NMR, etc.) can be employed to determine order parameters are introduced. Pair correlations and their relation to various experimental quantities (elastic constants, X-ray scattering) are presented in Chapter 4, while the reorientational dynamics of molecules in liquid crystals is described in Chapter 6, with a detailed discussion of orientational correlation functions and of their properties. The calculation of these time dependent correlation functions using stochastic models (rotational diffusion in particular) is also presented. Connection with experiments providing information on dynamic properties is introduced with Linear Response Theory and some important cases (DR, ionic transport, thermal conductivity, viscosities) examined in some detail. Given the huge variety of liquid crystal phases, the systems are chosen with modelling and simulations in mind. Simulations are also viewed as a set of ‘computer experimental' techniques able to generate 'configurations', i.e. snapshots of the positions and orientations of a sample of N molecules at equilibrium. The availability of these sets of configurations or of trajectories, i.e. of their time evolution, will, perhaps unconventionally, be assumed to be available, at least in principle. even in the first part of the book, so as to connect the various concepts introduced to characterize the liquid crystal phases to simulations. However, the details of how to perform the simulations will only be given in the second part of the book. Intermolecular and more generally particle–particle interactions are introduced in Chapter 5 and Molecular Field and Onsager theories, the most important approximate statistical mechanical approaches currently used, are discussed in Chapter 7. We then turn to computer simulation techniques. Both Monte Carlo (MC) and Molecular Dynamics (MD) methodologies are introduced in Chapter 8 and in Chapter 9, respectively, with special attention given to the calculation of anisotropic properties. The following chapters are devoted to the application of computer simulation techniques to liquid crystals at multiple length scale: Lattice (Chapter 10), Off Lattice Molecular (Chapter 11) and fully Atomistic models (Chapter 12). Most of the required mathematics is covered in a series of Appendices, hopefully making the book fairly self contained. Thus, spherical tensors, Wigner matrices, quaternions and other tools useful for dealing with rotations, which have normally to be extracted from books on angular momentum and quantum mechanics, are treated here with our applications in mind. Even simpler topics, like orthogonal basis sets, Dirac delta functions and Fourier transforms, typically treated in a physics curriculum, but not always in chemistry courses, are covered, with an eye to the practical user. The majority of chapters also have a detailed treatment of some 'simple' but relevant cases (sections) that can be read independently from the rest and could be used, e.g. for undergraduate courses. If the huge increase in computer performance and resources continues (it has been of a factor of the order of 10^5 in the last 20 years), the vision is that computer simulations will become very widespread and used more and more by industry and by non-specialists in the field. Knowing the basic ingredients of computer simulations thus seem important even for potential users, rather than just for developers, even when dealing with materials as complex as liquid crystals.

In closing I wish to thank the many friends, students and colleagues that have helped providing advice and support. I am particularly grateful to Lara Querciagrossa, also for much essential help with the figures, and to Sergio Cataliotti who have both carefully read and corrected all the chapters. I am indebted to Andy Emerson, Alessandro Porreca and Riccardo Tarroni for some figures and to Matteo Babbi, Gianni Bendazzoli, Roberto Berardi, Martin Čopič, Raffaele della Valle, Juho Lintuvuori, Luca Muccioli, Silvia Orlandi, Guido Raos, Matteo Ricci, Lorenzo Soprani, Marco Mazza and Francesco Spinozzi for reading, correcting and commenting on some parts of the draft. All remaining errors are of course my responsibility. I am also very grateful to Oleg Lavrentovich for the beautiful image of a liquid crystal texture used for the cover and last, but certainly not least, to Roberto Berardi (unfortunately now prematurely deceased) and to Geoffrey Luckhurst for many essential discussions over the last few decades.

Part of this book was written at the Isaac Newton Institute, Cambridge, UK, and I am extremely grateful for the hospitality and for the stimulating atmosphere and the discussions with many colleagues that I thoroughly enjoyed there.

1

Phases and Mesophases

Any process which is not forbidden by a conservation law actually does take place with appreciable probability.

M. Gell-Mann, Il Nuovo Cimento 4, 1956

1.1 Introduction

It takes two antithetic words to indicate *liquid crystals* (LCs) and this gives immediately a hint of the complexity and the fascination of the state of matter that we are about to investigate. Despite this, liquid crystals are not necessarily exotic in their composition or rare in their occurrence, and indeed tens of thousands of liquid-crystalline compounds have been described already.

To start from the very beginning, here we define an equilibrium *phase of matter* as a molecular organization stable within a certain range of thermodynamic variables, e.g. in a certain temperature interval. We are all familiar with the crystalline solid, liquid and gas phases and with their macroscopic properties. For example, we know that crystals have a particular shape that they maintain over time and that they typically have different properties along different directions. Thus, if we measure some optical property of a crystal by sending a beam of light along its different axes we typically find different values. Such a material is accordingly called *anisotropic*. At the other extreme, liquids can flow and take the shape of their container and their physical properties are the same in any direction, thus liquids are *isotropic*. The gaseous state too is isotropic, like a liquid. As we shall see the gas state, except for the density, is indeed very similar to the liquid state, to the point of not being fundamentally distinct from it. On a microscopic level we can imagine an ideal crystal as formed by its constituent particles (molecules or atoms, ions, nanoparticles, ...) with positions regularly arranged on a lattice and, as long as they are non-spherical, with orientations parallel, or however very precisely organized, as shown schematically in Fig. 1.1a. A structure like this is said to possess both *positional* and *orientational order*. In a liquid, molecular positions and orientations are instead disordered overall, as sketched in Fig. 1.1b. We can expect a certain amount of correlations in the positions and orientations of nearby molecules, since each of them will have to adjust to its neighbours to avoid occupying the same space and to optimize attractive interactions. However, this local correlation will rapidly disappear as the separation between molecules increases, so that in

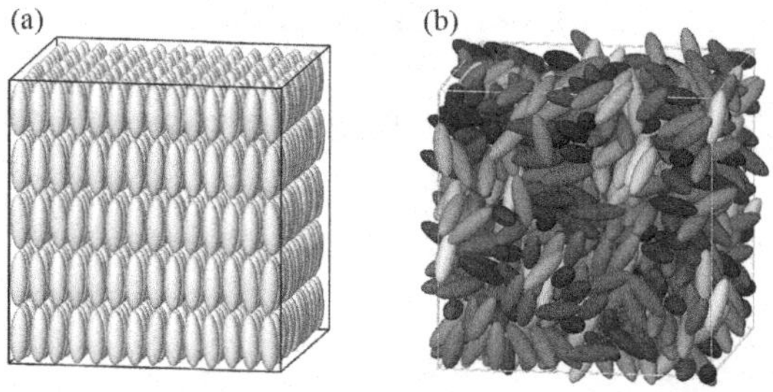

Figure 1.1 (a) A sketch of the molecular organization of a crystal and (b) of an isotropic fluid formed by elongated particles. The grey shade indicates the particles orientation.

an ordinary liquid (or gas) we have no *long range order*. It is apparent that there is a relation between the order at molecular level and the macroscopic properties of a system, and much of this book will be devoted to trying to establish and analyze this connection. It is also worth realizing that there is no rule of nature that forbids the existence of states of matter with long-range order intermediate between that of crystals (three-dimensional positional and orientational) and that of liquids (no positional and no orientational). Since all that is not forbidden can take place in nature or be artificially prepared, we do indeed have a variety of intermediate phases, some examples of which are schematically shown in Table 1.1, with order decreasing from top to bottom. It is reasonable to expect that phases like the *plastic crystal* that have positional, but not orientational, order will be formed by molecules of globular shape that can reorient without disrupting the structure. In practice, tetrahedral (e.g. tetrachloromethane, neopentane), octahedral (e.g. tetramethylbutane), cyclic (cyclobutane), bridged (camphor, adamantane) molecules, etc., give rise to plastic crystals. These rather special crystals have isotropic optical properties and usually they can be easily cut or extruded. Some, e.g. perfluoro cyclohexane, can even flow under their own weight [Kovshev et al., 1977].

When molecules significantly deviate from a globular form, e.g. when they are elongated or discoidal, we have the possibility of phases with orientational order and with a reduced or altogether absent positional order intervening between the crystal and the liquid phases. These intermediate phases or *mesophases* are called *liquid crystals* [de Gennes, 1974; Chandrasekhar, 1992; de Gennes and Prost, 1993; Chaikin and Lubensky, 1995; Collings and Hird, 1997; Khoo, 2007; Blinov, 2011]. Liquid crystals can be obtained from the isotropic liquid by cooling, or from the crystal by heating, and these materials are called *thermotropic*. However, liquid crystal phases can also be formed by mixing a liquid with one or more components formed by anisotropic particles in a suitable concentration range (lyotropics, colloidal suspensions, ...). In the next few sections we shall briefly describe the properties of both families, starting from thermotropics. In these systems the phase transformations can be reversible, with or without hysteresis, and in this case, they are called *enantiotropic* or, as found in a number of materials, the phase transformations take place only in one direction, e.g. upon cooling, and these are called *monotropic*.

Table 1.1. *A sketch of the molecular organization of various phases of matter displaying a combination of positional and orientational order*

	Phase	Positional order	Orientational order
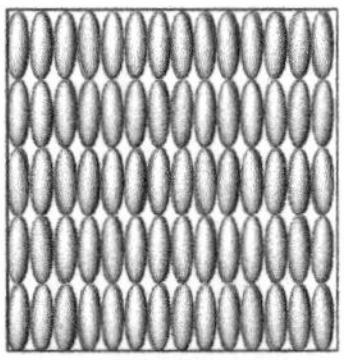	Crystal	Yes, 3D	Yes
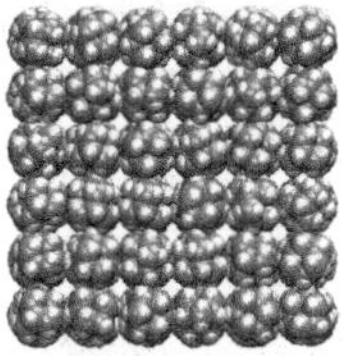	Plastic crystal	Yes, 3D	No
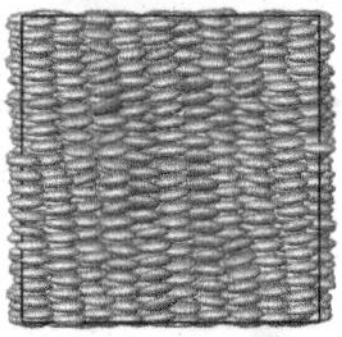	Columnar LC	Yes, 2D	Yes
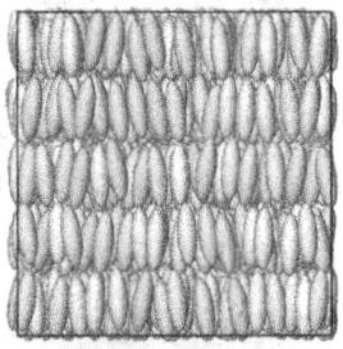	Smectic LC	Yes, 1D	Yes
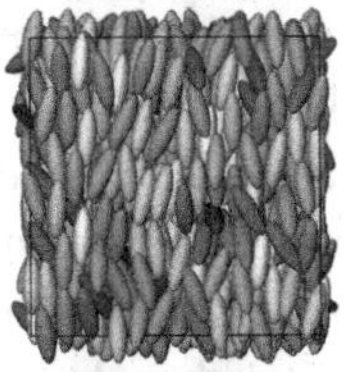	Nematic LC	No	Yes
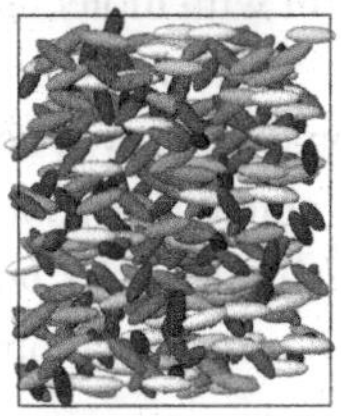	Liquid	No	No
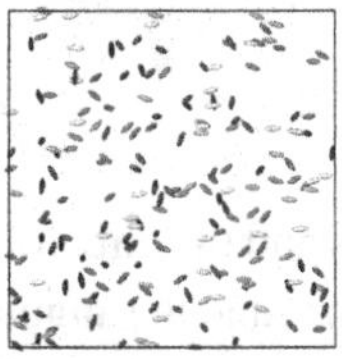	Gas	No	No

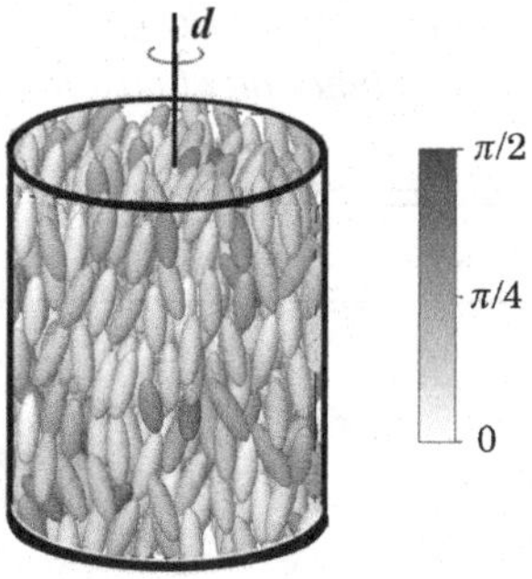

Figure 1.2 Microscopic representation of a nematic monodomain. The molecules tend to be aligned parallel to a common director $\boldsymbol{d}$, here along the vertical direction. Their orientation with respect to the director is indicated by the grey level, from white when parallel, to black when perpendicular.

1.2 Nematics

The characteristic property of the molecular organization of nematic liquid crystals is that their molecules tend, on average, to be parallel to one another and to a preferred direction $\boldsymbol{d}$, called the *director*. The director at a certain position in space $\boldsymbol{r}$ can have a different orientation from that at another position $\boldsymbol{r}'$, but a nematic can be easily aligned by relatively weak external fields of various kinds: magnetic fields of the order of a few tenths of Tesla, electric fields of the order of Volts per micron [de Gennes and Prost, 1993] or even by surfaces, at least for sufficiently thin layers [Jérôme, 1991], yielding a monodomain sample with a uniform director $\boldsymbol{d}$. The resulting aligned nematic, schematically shown in Fig. 1.2, has normally *uniaxial* symmetry around the director, in the sense that its physical properties do not change if we rotate of an arbitrary angle around this direction and here, unless explicitly stated, we shall always assume this to be the case. The properties of nematics are also invariant when we turn a sample upside down, so that $\boldsymbol{d}$ and $-\boldsymbol{d}$ are equivalent and, if we consider $\boldsymbol{d}$ as a unit vector, only its direction will be important. The symmetry of a monodomain nematic can thus be taken to be equivalent to that of a cylinder or, using group theory terminology (see, e.g., [Lax, 1974]), $D_{\infty h}$. This is consistent with molecules forming mesophases (*mesogens*) being apolar or, as is normally the case, being distributed with the same probability along $\pm\boldsymbol{d}$. The formal description of orientational order will be discussed in detail in Chapter 3, but the tendency of molecules yielding nematics (*nematogens*) to be parallel to $\boldsymbol{d}$ can be quantified as a first approximation by a simple *order parameter S* first introduced by Tsvetkov [1939]. Consider each mesogen to be a uniaxial object whose orientation is given by a unit vector $\boldsymbol{u}$

$$S \equiv \langle P_2 \rangle = \frac{3}{2}\langle (\boldsymbol{u} \cdot \boldsymbol{d})^2 \rangle - \frac{1}{2}, \tag{1.1}$$

where $\boldsymbol{u} \cdot \boldsymbol{d} \equiv \cos\beta$, with β the angle between the molecular axis and the director, $P_2(\cos\beta)$ is the second Legendre polynomial [Abramowitz and Stegun, 1965] and the angular brackets indicate an average over all the molecules in the system. It is easy to see that $\langle P_2 \rangle$ is a scalar

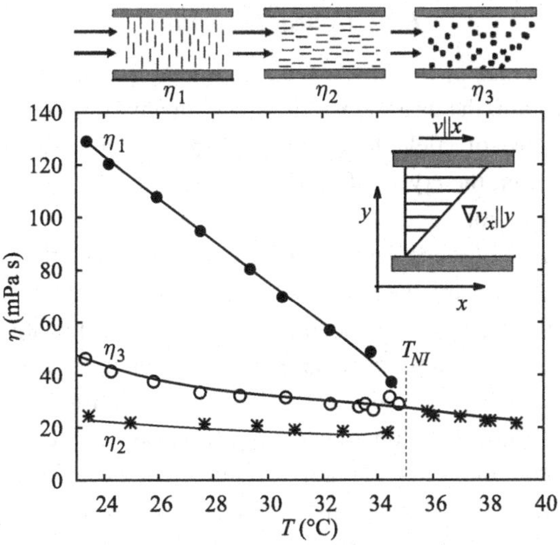

Figure 1.3 The temperature dependence of the Miesowicz shear viscosities η_1, η_2, η_3 for 5CB [Chmielewski, 1986]. In the inset we show a sketch illustrating the definition of these three viscosities [Orr and Pethrick, 2011] for the nematic flowing in a channel with a certain velocity $\boldsymbol{v}$ along $\boldsymbol{x}$ ($\boldsymbol{v}||\boldsymbol{x}$), and a flow velocity gradient across the channel ($\nabla v_x||\boldsymbol{y}$). Each viscosity is measured aligning the director with an external magnetic field $\boldsymbol{H}$ oriented in different directions: $\boldsymbol{H}||\boldsymbol{d}||\boldsymbol{y}$ for η_1, $\boldsymbol{H}||\boldsymbol{d}||\boldsymbol{x}$ for η_2 and $\boldsymbol{H}||\boldsymbol{d}||\boldsymbol{z}$ for η_3.

and that it has the properties that we would intuitively expect an order parameter to possess. For a system of molecules perfectly aligned with respect to $\boldsymbol{d}$, that we take to define our Z laboratory axis, $\beta = 0$ or $\beta = \pi$ for every molecule and $\langle P_2 \rangle = 1$. At the other extreme, for a completely disordered system, such as an ordinary isotropic fluid, we have

$$\langle \cos^2 \beta \rangle = \langle u_Z^2 \rangle = \langle u_X^2 \rangle = \langle u_Y^2 \rangle = \frac{1}{3}, \tag{1.2}$$

since in an isotropic system there will be no preference for any of the three axes and also $u_Z^2 + u_X^2 + u_Y^2 = 1$. Therefore, for a disordered system we have $\langle P_2 \rangle = 0$.

Nematics have long-range orientational order but not long-range positional order and their molecules can move and reorient quite easily, like in a liquid. Indeed the viscosities and the densities of materials in their nematic or isotropic liquid phases are quite similar, typically differing less than 5% on both sides of the transition [Ibrahim and Haase, 1976; Dunmur et al., 2001; Würflinger and Sandmann, 2001]. The orientational order gives, however, different optical, dielectric, diamagnetic and rheological properties in different directions with respect to the director, i.e. a nematic liquid crystal is *anisotropic*. The viscosity itself is different for different relative orientations of the flow velocity $\boldsymbol{v}$, of the director and of the flow velocity gradient [Miesowicz, 1946], as sketched in Fig. 1.3, where we also plot the three Miesowicz viscosities (η_1, η_2, η_3) for 4-n-pentyl-4′-cyano-biphenyl (5CB), showing that the lowest one corresponds to flow along the director (η_2) [Chmielewski, 1986; Orr and Pethrick, 2011].

1.2.1 Optical Properties

Even though the molecules of a nematic tend to arrange parallel to each other, defining a local preferred direction, the director will not point in any specific direction on a macroscopic scale in the absence of a field, for example when we obtain a nematic by cooling from the liquid or melting from the crystal. Rather, its direction will vary continuously so as to maintain on the one hand local uniaxial and on the other macroscopic isotropic symmetry. We can think of the molecules as being aligned within local domains, but with the domains being themselves randomly oriented one with respect to the other. This inhomogeneity gives rise to the scattering of visible light by an unoriented nematic (see Fig. 1.4) and to its characteristic turbid, milky appearance, which disappears at the nematic-isotropic transition, for this reason also called the *clearing* point. In turn, the strong scattering of visible light indicates that local domains leading to birefringence inhomogeneities have dimensions ξ of the order of the wavelength of visible light, i.e. a few hundred nanometres, corresponding to a number of spontaneously correlated molecules of the order of 10^8. The strong correlation between individual molecules indicated by this huge size is at the origin of the aforementioned easy alignment of nematics under an external field. Indeed a uniform alignment can be obtained by magnetic fields of the order of 0.1T or, e.g., by electric fields of the order of 1 V/μm. The free energy contribution corresponding to application of an electric field $\boldsymbol{E}$ is (see, e.g., [Khoo, 2007])

$$\mathcal{G}_{\mathrm{E}} = -\frac{1}{2}\Delta\epsilon\,(\boldsymbol{d}\cdot\boldsymbol{E})^2. \tag{1.3}$$

The alignment will thus tend to be in the direction of the applied electric field, $\boldsymbol{E}$, or perpendicular to it, if the *dielectric susceptivity* anisotropy of the material, $\Delta\epsilon = \epsilon_{\parallel} - \epsilon_{\perp}$,

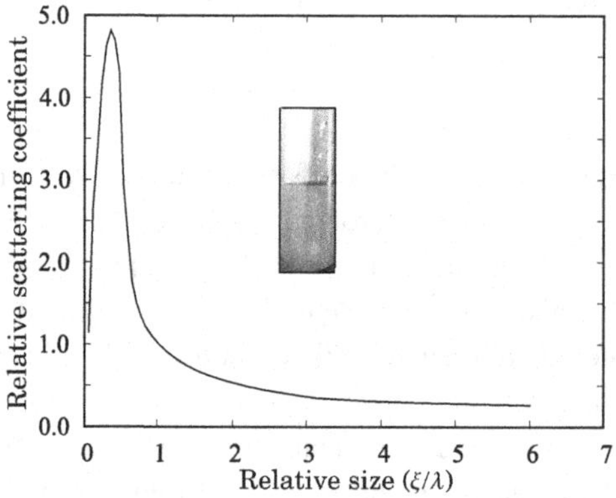

Figure 1.4 Relative scattering coefficient as a function of the ratio between size of the inhomogeneous domains, ξ, and incoming light wavelength λ [Tilley, 2000]. The milky appearance of nematics (inset) indicates that ξ is of the order of a few hundred nanometres, the visible light wavelength.

is positive or negative, respectively. A good alignment will be achieved if this free energy overcomes the thermal disordering energy. In the same way for an applied magnetic field $\boldsymbol{H}$

$$\mathcal{G}_B = -\frac{1}{2}\Delta\chi(\boldsymbol{d}\cdot\boldsymbol{H})^2, \tag{1.4}$$

where $\Delta\chi$ is the *diamagnetic susceptivity* anisotropy $\Delta\chi = \chi_\parallel - \chi_\perp$.

An aligned nematic has the optical properties of a uniaxial crystal, like calcite or quartz, with the director representing the optical axis, and shows *birefringence* [Jenkins and White, 2001]. A beam of light, propagating through the sample at an angle with respect to the optical axis, is split in two beams with parallel and perpendicular polarization, corresponding to the refractive indices parallel and perpendicular to the director, $n_\parallel$, $n_\perp$. Thus, in a liquid crystal the refractive index, **n**, is a tensor (see Appendix B). In a laboratory fixed system with the z-axis parallel to $\boldsymbol{d}$, the 3×3 matrix representing the refractive index tensor will be diagonal, with components $(n_\perp, n_\perp, n_\parallel)$. A simple experimental setup for measuring the birefringence, $\Delta n = n_\parallel - n_\perp$, i.e. the difference between the parallel (or *extraordinary*) and perpendicular (or *ordinary*) components of the refractive index tensor, is shown in Fig. 1.5a. The intensity of a beam of light with wavelength λ emerging through

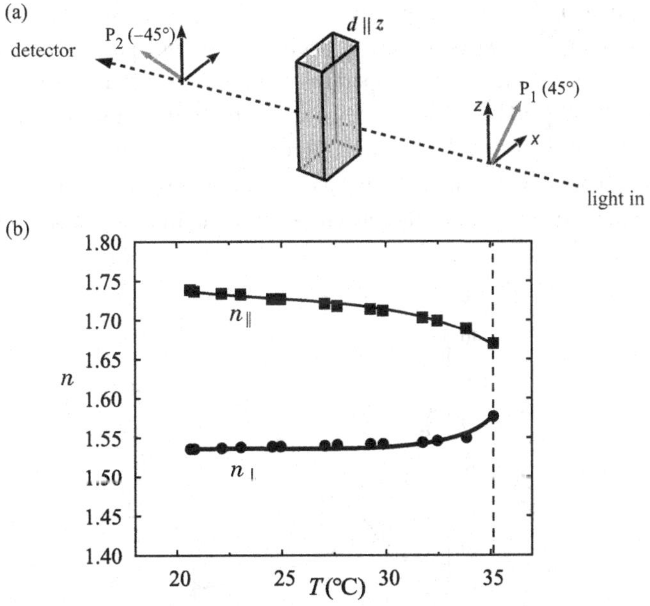

Figure 1.5 (a) Sketch of an experimental setup to measure the birefringence of an aligned nematic from the intensity of light transmitted through the two polarizers P_1 and P_2 at $\pm 45°$ with respect to the director $\boldsymbol{d}$. (b) Refractive indices $n_\parallel$ and $n_\perp$ for the nematic 5CB in the visible ($\lambda = 546.1$ nm) as a function of temperature [Karat and Madhusudana, 1976]. The vertical dashed line indicates the transition to isotropic.

the two crossed polarizers set at $\pm 45^\circ$ from the director of a vertically aligned sample of thickness δ is (see Appendix L)

$$I = I_0 \sin^2(\pi \delta \Delta n/\lambda), \tag{1.5}$$

where I_0 is the input intensity. The birefringence $\Delta n = n_\parallel - n_\perp$ typically decreases with increasing temperature, as shown in Fig. 1.5b, indicating an increasing disorder in the molecular organization. At a well-defined temperature T_{NI}, the nematic-isotropic transition temperature, the anisotropy vanishes abruptly and the material becomes an ordinary isotropic liquid. We note, however, that even above the transition the isotropic phase has some short-range ordering, with ordered clusters of molecules of a typical size (*coherence length*) ξ_I, that grows larger on approaching the nematic transition from above. This *pretransitional effect* is observed in a relatively large range of temperatures above T_{NI} (some 10–20 degrees) and is demonstrated by the anomalously large susceptivity to an applied electric field (*Kerr effect*) or magnetic field (*Cotton–Mouton effect*) measured, e.g., by the induced birefringence Δn. Thus (see, e.g., [Haynes et al., 2014]),

$$\Delta n = \lambda \mathscr{K} E^2, \tag{1.6}$$

where λ is the wavelength of the probe light, $\mathscr{K}$ is called the Kerr constant and E is the electric field applied. The Kerr susceptivity of nematic liquid crystals is linked to the size of the oriented domains, as we shall see in Section 4.11, and increases on cooling from the isotropic phase, diverging as the temperature approaches a characteristic temperature, $T_{NI^\star}$, which is typically $\approx$1K below the nematic-isotropic transition temperature T_{NI}, as we see in Fig. 1.6, although it can vary for different nematics [Blachnik et al., 2000]. The easy alignment of a nematic when applying an external field is even more strikingly shown by surface alignment. A thin (a few microns thick) nematic film on a glass or polymer slab

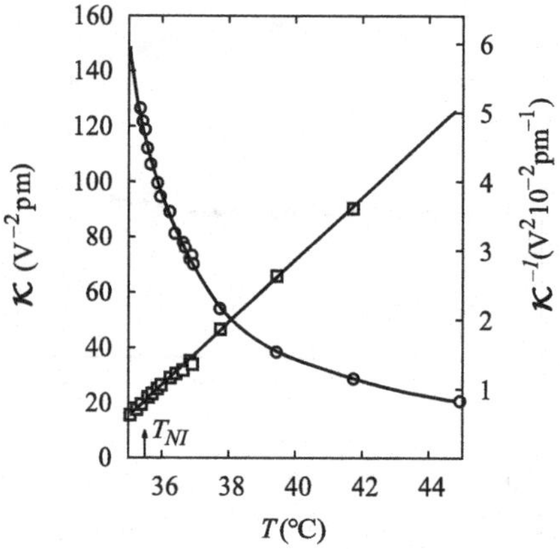

Figure 1.6 Kerr constant $\mathscr{K}$ (○) and its inverse $\mathscr{K}^{-1}$ (□) as a function of temperature for 5CB (obtained from the birefringence Δn at a wavelength $\lambda = 441.6$ nm). The divergence temperature $T_{NI^*} = 33.8^\circ$ C is 1.33 degrees below T_{NI} [Coles and Jennings, 1978]. By comparison, $\mathscr{K}$ of 5CB in a dilute solution of CCl_4 is only 2.8 (V^{-2} pm).

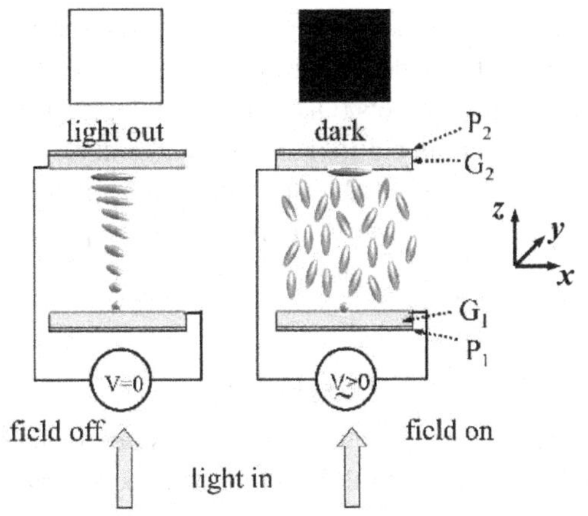

Figure 1.7 Sketch of a twisted nematic (TN) display pixel in electric field off and on states.

can be aligned by surface interactions. An alignment of the director along a certain *easy axis* can be achieved by simply rubbing or lapping the support surface with a soft tissue.

The possibility of changing between the director orientation established by surface forces in a small, micron-size region (a *pixel*) and that obtained by switching on and off an electric field, with the resulting change in the pixel optical properties, is at the heart of the many applications of these mesophases in the electro-optic display industry, where liquid crystal displays (LCDs) have become dominant for applications ranging from mobile phones, TV screens, etc. (see, e.g., [den Boer, 2005; Semenza, 2007; Kim and Song, 2009]). In Fig. 1.7 we see a sketch of the director configurations at rest and with the electric field on for one of the simplest and most widespread LC display types: the twisted nematic (TN). In a TN LCD, a few microns thick film of a nematic with $\Delta\epsilon > 0$ is confined between two glass slides G_1, G_2, each treated so as to induce a uniform alignment along a certain direction of the surface: $\boldsymbol{u}_1$, $\boldsymbol{u}_2$, $\boldsymbol{u}_1 \perp \boldsymbol{u}_2$. Going across the film from one surface to the other, the director changes in a helical way between the two perpendicular boundary directions. This chiral structure will be able to rotate the plane of the back illumination light, linearly polarized along $\boldsymbol{u}_1$ by a first polarizer P_1, so that at rest (off) light will be able to emerge through the second polarizer, P_2, set along $\boldsymbol{u}_2$. However, an electric field can be applied to the pixel, printed with a transparent conducting ink (typically, Indium-Tin Oxide, ITO) and connected to the device circuitry. Applying to the pixel a suitable voltage across the two surfaces, the small nematic volume (*voxel*) subjected to the field aligns along the field direction, blocking transmission through the pixel. The pixel then operates as an optical switch: it will appear black when the helix is completely unwound, or partially transparent, according to a number of grey levels n_G (typically $n_G = 256$) established by changing the applied field. When the field is switched off again, the LC interaction with the surfaces will re-establish the original situation of transparency. Colours are obtained by additive synthesis having, instead of a single pixel a set of three (*sub-pixels*) so close that the eye does not spatially resolve them. Illuminating with red, green and blue (RGB) light respectively, n_G^3 colours can be obtained. Each of the three lights is typically obtained from a white back-light by a colour filter. A more recent and better approach is to illuminate with UV light a film containing semiconductor quantum dots (QDs) [Reed, 1993; Ness and Niehaus, 2011], for instance, CdSe nanoparticles of three different sizes that emit respectively rather pure red, green and blue lights, more easily filtered than the original white light. Even more simply a blue light-emitting diode (LED) can be used to illuminate the polymer film containing two QDs emitting in the red and the green [Luo et al., 2014].

1.2.2 Defects

When a thin film of nematic on an untreated glass slide is observed between crossed polarizers, the birefringence coupled to the distribution of director orientations in the sample yields a typical texture, shown in Fig. 1.8a, called *schlieren* [de Gennes, 1974; Brochard, 1977]. The black threads correspond to regions where the director is in the plane parallel to one of the crossed polarizers or where the system is locally isotropic. The points or lines where these differently oriented directors meet, represent singularities of the director field corresponding to topological *defects* [Frank, 1958; Mermin, 1979; Kleman, 1982; Trebin, 1982; Lavrentovich et al., 2001; Muševič, 2017]. For nematics we note singularities (*noyeaux*) with two and four brushes (see Fig. 1.8) that correspond to defects with strength, or *winding number*, $s = \frac{1}{2}$ and $s = 1$, respectively (Fig. 1.8b). The value of the winding number, s, can be assigned, assuming for simplicity that in the thin film the director distribution is two-dimensional, by drawing a closed circuit around the defect and observing the total angle of rotation, α_d, of the director upon returning to the same point. Clearly, α_d is a multiple of π and $s = \alpha_d/(2\pi)$, and the s is just $1/4$ of the number of brushes observed. The sign of the defect can be obtained following the movement of the brushes as the polarizers are rotated: the sign is taken to be positive if the brushes rotate in the same direction as that of the crossed polaroids and negative if the brushes rotate in the opposite direction.

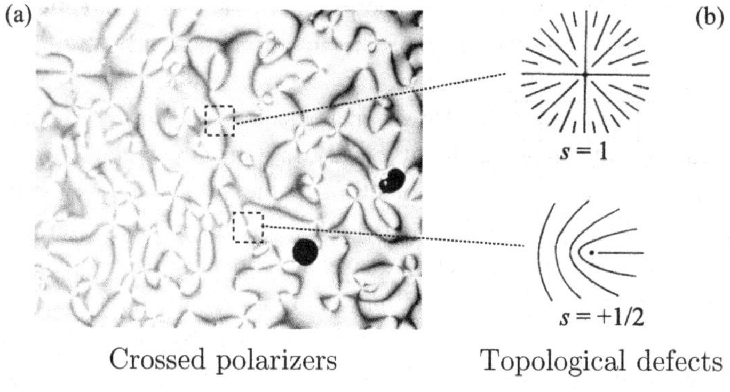

Figure 1.8 The schlieren texture of nematics between (a) crossed polarizers and (b) two topological defects showing the origin of the black threads. The two deep black regions correspond to hotspots, where the nematic has turned isotropic.

The procedure to classify defects can be generalized to director distributions in three dimensions considering a closed sphere surrounding the singular point and applying the methods of algebraic topology, in particular so-called homotopy groups, beyond the scope of this book and well treated elsewhere, e.g. by Kurik and Lavrentovich [1988] and Mermin [1990]. An important point is that, while in two dimensions an infinity of different strengths: $s = \pm 1/2, \pm 1, \pm 3/2, \ldots$ can in principle occur, in 3D only defects of strength $1/2$ can exist. The polarizing optical microscope (POM) textures of thin films of liquid crystals do not depend on the specific chemical nature of the compound, but, as we shall see, are characteristic, given a certain surface treatment of the film, of the different categories of liquid crystals. Indeed, this constitutes an important practical tool in assigning liquid crystal types [de Gennes, 1974; Gray and Goodby, 1984; Neubert, 2001a].

1.2.3 Elastic Constants

Nematics also present an elastic response: when their director is deformed by applying some weak external field, they return to the original molecular organization as the stimulus is released, somehow similarly to what happens to crystals but not to isotropic liquids [Frank, 1958; Nehring and Saupe, 1971; Stephen and Straley, 1974; Clark, 1976; Crawford and Žumer, 1995; Kleman and Lavrentovich, 2003]. To see how this effect can be observed, let us imagine we align a liquid crystal film with a uniform director in any point of the material and consider this is the state at rest. We can then try to deform the material and write the elastic free energy density $\mathcal{G}_{\mathrm{el}}$ for the slightly distorted nematic, following Frank [1958], as an expansion in powers of the (small) gradient components of $\boldsymbol{d}$. In the various notations commonly used these are $d_{ij} \equiv \nabla_i d_j \equiv [\nabla \boldsymbol{d}]_{ij} \equiv [\nabla \otimes \boldsymbol{d}]_{ij}$. The free energy density can be written as a sum of a bulk term $\mathcal{G}_{\mathrm{el}}^b$, and a surface term $\mathcal{G}_{\mathrm{el}}^s$:

$$\mathcal{G}_{\mathrm{el}} \equiv \mathcal{G}_{\mathrm{el}}(\boldsymbol{d}, \nabla \boldsymbol{d}) = \mathcal{G}_{\mathrm{el}}^b + \mathcal{G}_{\mathrm{el}}^s, \tag{1.7}$$

where $\mathcal{G}_{\mathrm{el}}(\boldsymbol{d}, \nabla \boldsymbol{d})$ should be invariant for arbitrary sample rotations and changes of sign of $\boldsymbol{d}$. Retaining only combinations allowed by symmetry and considering only quadratic terms like in the classical Hooke's law of elasticity gives the classical Frank–Oseen expression [Oseen, 1933; Frank, 1958; Stewart, 2004]:

$$\mathcal{G}_{\mathrm{el}}^b = \frac{K_{11}}{2}(\nabla \cdot \boldsymbol{d})^2 + \frac{1}{2}K_{22}[\boldsymbol{d} \cdot (\nabla \times \boldsymbol{d})]^2 + \frac{1}{2}K_{33}|\boldsymbol{d} \times (\nabla \times \boldsymbol{d})|^2. \tag{1.8}$$

The last bulk term can also be written as $+\frac{K_{33}}{2}\,[\boldsymbol{d} \cdot (\nabla\, \boldsymbol{d})]^2$ and the surface term contains divergence terms [Kleman and Lavrentovich, 2003]:

$$\mathcal{G}_{\mathrm{el}}^s = -K_{24}\nabla \cdot (\boldsymbol{d} \times \nabla \times \boldsymbol{d} + \boldsymbol{d}\nabla \cdot \boldsymbol{d}) + K_{13}\nabla \cdot (\boldsymbol{d}\nabla \cdot \boldsymbol{d}). \tag{1.9}$$

The total elastic free energy is obtained integrating over all the sample volume: $\mathcal{G}_{\mathrm{el}}^{tot} = \int_V \mathrm{d}\boldsymbol{r}\, \mathcal{G}_{\mathrm{el}}\,(\boldsymbol{d}(\boldsymbol{r}), \nabla \boldsymbol{d}(\boldsymbol{r}))$. We see that the expression for $\mathcal{G}_{\mathrm{el}}^b$ contains only three essential modes of deformation of the director: $(\nabla \cdot \boldsymbol{d})$, $[\boldsymbol{d} \cdot (\nabla \times \boldsymbol{d})]$, $|\boldsymbol{d} \times (\nabla \times \boldsymbol{d})|$, called *splay*, *twist* and *bend*, shown in Fig. 1.9. The corresponding *splay*, K_{11}, *twist*, K_{22}, and the *bend*, K_{33},

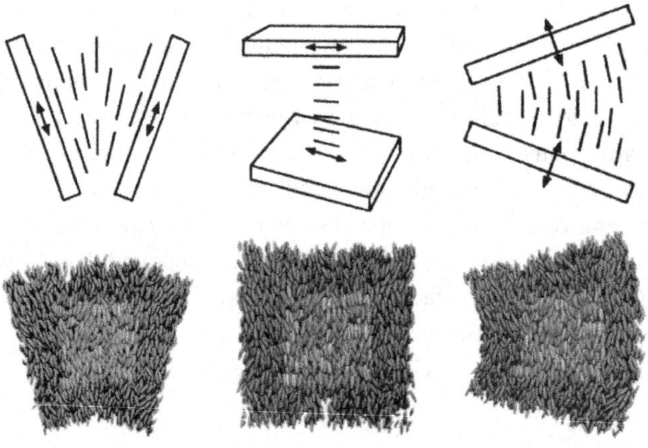

Figure 1.9 A sketch of the splay, twist and bend deformations of a liquid crystal corresponding to the Frank elastic constants K_{11}, K_{22} and K_{33}.

elastic constants [Frank, 1958] express the resistance of the material to these deformations. When calculating the total bulk free energy by volume integration only the first term, $\mathcal{G}^b_{\text{el}}$, survives, while the second, $\mathcal{G}^s_{\text{el}}$, involving the *saddle-splay* constant K_{24} and the mixed *splay-bend* constant K_{13}, averages to 0. However, these contributions should be considered when the system is confined or in any case a surface [Crawford and Žumer, 1995].

The elastic constants (or *moduli*) are positive, different from one another as they correspond to physically different distortions, they change with temperature and pressure and are fairly small, with typical values of the order of piconewtons (see, e.g., Table 1.3). For low-molar-mass nematic liquid crystals the differences between the three elastic constants are normally relatively small [Stannarius, 1998a; Dunmur, 2001] and, accordingly, the approximation of equal elastic constants in nematics is often made in theoretical work. Within this assumption of $K_{11} = K_{22} = K_{33} = K$, $K_{24} = 0$ and the identity Eq. A.27, the integrand becomes the scalar contraction (Eq. A.22) of the director gradient [Ball, 2017]: $\mathcal{G}^{tot}_{\text{el}} = \frac{1}{2} K \int_V \mathrm{d}\boldsymbol{r}\, ||\nabla \boldsymbol{d}||^2$. However, it is worth mentioning that in various important cases we may expect the elastic constants to differ significantly. For instance, de Gennes [1977] observed that splay distortions should be unlikely to occur, and, correspondingly, K_{11} should be very large for systems of long rods, like those of polymeric LC (see Section 1.3) because of the crowding of rods at one end caused by splay. This was experimentally observed for a main-chain polyester [Zhengmin and Kleman, 1984] where $K_{11} \approx 0.1$ pN $\approx 10K_{33}$. For another polyester, Martins et al. [1983] found $K_{11} \approx 2-3\, K_{33}$. In other LC types, like the twist-bend nematic phase found in certain dimers (see Section 1.5), the bend elastic constant K_{33} is much smaller than the other two [Borshch et al., 2013].

In a great variety of instances, the equilibrium distribution of the director can be obtained by setting up and minimizing the elastic free energy subject to appropriate boundary

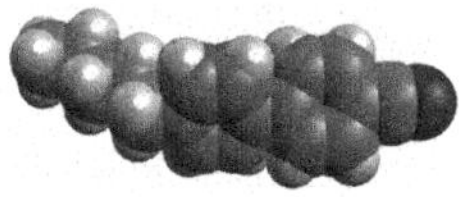

Figure 1.10 A space filling atomistic model of the structure of the nematogen 5CB.

conditions. As discussed by Ericksen [1966], the constants should obey the inequalities [Kleman and Lavrentovich, 2003]:

$$K_{11} \geq 0,\ K_{22} \geq |K_{24}| \geq 0,\ K_{33} \geq 0,\ 2K_{11} - K_{22} - K_{24} \geq 0. \tag{1.10}$$

1.2.4 Nematogenic Molecules

Nematics are formed of anisotropic molecules, that is, by molecules, typically non-spherical, which have different properties in different directions. The simplest kind of nematogen molecules, and for a long time the only ones, have elongated (also called *calamitic*) structures. In Fig. 1.10 we show, as an example, the nematogen molecule 5CB. As we can see from a space filling model, where each atom is represented by a sphere of characteristic radius, this molecule is relatively elongated, consisting of a fairly rigid core and of a flexible alkyl chain, although the *aspect ratio* (length to width ratio) is not by itself the only factor determining the existence or absence of a mesophase.

In *thermotropic* materials, nematic phases can be obtained by simply melting a crystal or cooling from an ordinary isotropic phase. In general, the nematicity range is from a few degrees to a few tens of degrees Celsius; then another transition to a solid or to another liquid crystal phase takes place. In Table 1.2 we list the nematic ranges for a few common liquid crystals, taking also the opportunity to introduce their code names. We note that when dealing with nematics we are nearly always in the neighbourhood of a phase transition. In particular, as we see from the data in Table 1.2, the temperature range of nematicity for the pure materials listed here $(T_{KN} - T_{NI})/(273.15 + T_{NI})$ is at most 30% from the nematic-isotropic transition temperature and is not much larger even for mixtures like E63 or Phase 5 especially designed to broaden the interval. The code names in Table 1.2 conform to their common usage and will be used throughout the book. They correspond to the following chemical compositions:

- E63: a commercial mixture (from BDH) of six cyano-biphenyls and terphenyl with a particularly wide nematic range.
- Phase 5: a eutectic mixture with the approximate percent content: 4-butyl, 4′-methoxy azoxybenzene (25 wt%), 4-methoxy, 4′-butyl azoxybenzene (40 wt%), 4-ethyl, 4′-methoxy azoxybenzene (12 wt%) and 4-methoxy, 4′-ethyl azoxybenzene (23 wt%).
- E7: a commercial liquid crystal mixture (from Merck) containing approximately [Lee et al., 2008] 45.53 wt% 5CB, 28.74 wt% 7CB, 16.28 wt% 8OCB ([1,1′-biphenyl], 4-carbonitrile, 4′-octyloxy) and 9.46 wt% 5CT ([1,1′,4′-1″-terphenyl], 4-carbonitrile, 4″-pentyl).

Table 1.2. *The melting, T_{KN}, and clearing, T_{NI}, temperatures for some mixtures and for pure nematics with their structural chemical formula*

Nematic	T_{KN}(°C)	T_{NI}(°C)	Formula
E63	−20.1	87.4	Mixture
Phase 5	−5.1	74.9	Mixture
E7	< -30	58.3	Mixture
E9	6.9	82.4	Mixture
MBBA	20.2	45.9	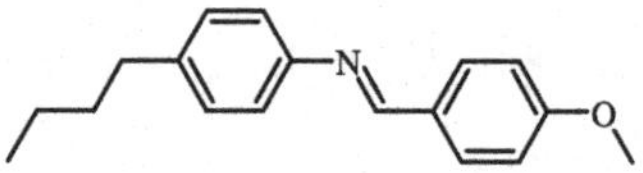
5CB	22.5	35.0	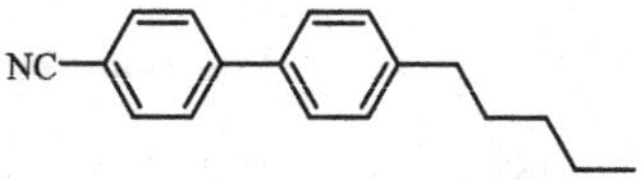
NAD	22.9	48.9	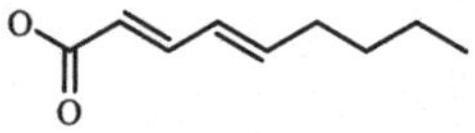
I52	23.9	103.6	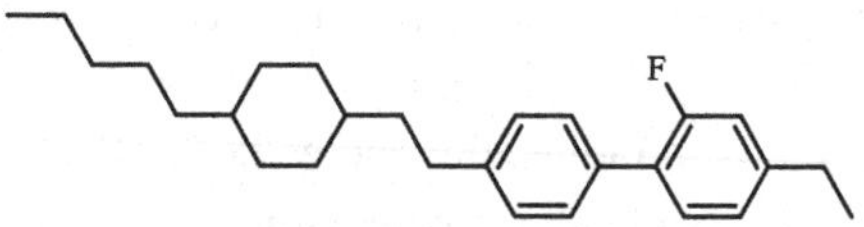
PCH5	29.9	55.1	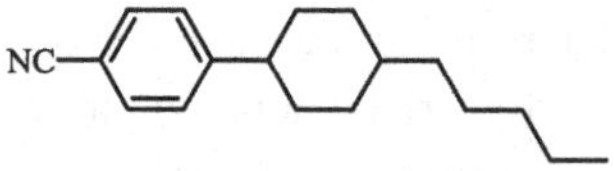
HAB	40.0	47.0	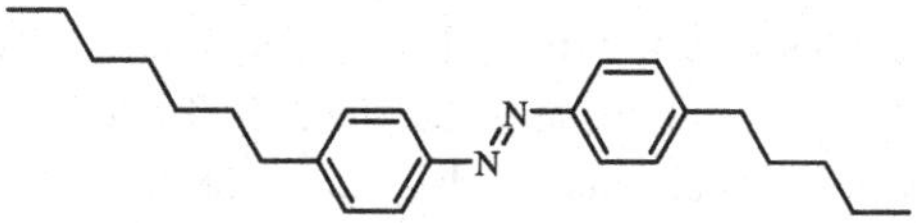
5FTP	60.0	120.0	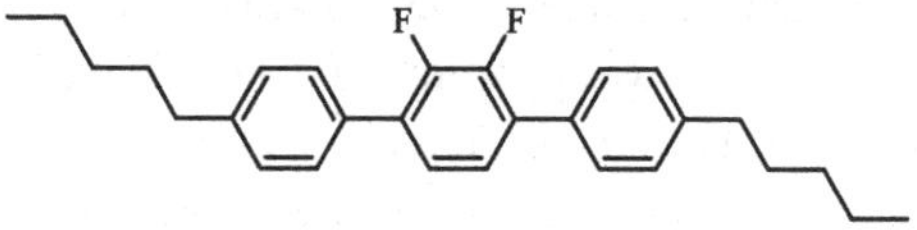
PAA	118.1	135.3	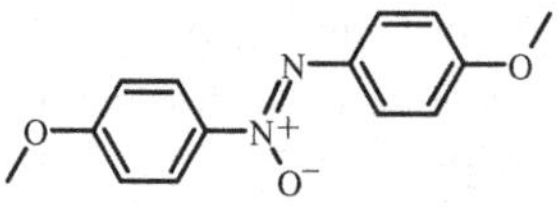
AAD	168.9	182.9	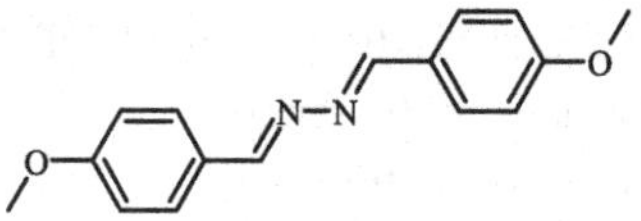
P5	400.9	444.9	

- E9: a mixture of 15% 4-propyl, 4′-cyano-biphenyl, 38% 4-pentyl, 4′-cyano-biphenyl, 38% 4-heptyl, 4′-cyano-biphenyl and 9% 4-pentyl, 4′-cyano-terphenyl.
- MBBA: (4-methoxy benzylidene)-4′-*n*-butylaniline (a Schiff base).
- 5CB: 4-*n*-pentyl-4′-cyano-biphenyl.

Table 1.3. *Some physical properties of the common nematics PCH5, 5CB and MBBA. PCH5 melts at* $T_{KN} = 30°C$ *and becomes isotropic at* $T_{NI} = 54.9°C$*, while for 5CB* $T_{KN} = 22.5°C$ *and* $T_{NI} = 35.0°C$ *and for MBBA* $T_{KN} = 22.0°C$ *and* $T_{NI} = 48.0°C$

Property	PCH5[a] 30.3°C	PCH5[a] 46.7°C	5CB 25.0°C	MBBA 25.0°C
Density, ρ(g cm^{-3})	0.9630	0.9496	1.022	1.042[d]
Shear (dynamic) viscosity, η (mPa s)	13.96	8.45	28[d]	23[d]
Rotational viscosity, γ_1(mPa s)	83.2	32.6		
Surface tension, γ (mN/ m)			30	
Refractive index, $n_\parallel$ at $\lambda = 589$ nm	1.6040	1.5849	1.71[c]	1.764[d]
Refractive index, $n_\perp$ at $\lambda = 589$ nm	1.4875	1.4860	1.53[c]	1.549[d]
Dielectric constant, $\varepsilon_\parallel$ at 1 kHz	17.1	15.9	18.5[c]	4.7[f]
Dielectric constant, $\varepsilon_\perp$ at 1 kHz	5.0	5.7	7.0[c]	5.4[f]
Elastic constant, K_{11}(pN)	8.5	5.9	6.2[b]	6.0[f]
Elastic constant, K_{22}(pN)	5.1	3.9	3.9[b]	4.0[f]
Elastic constant, K_{33}(pN)	16.2	9.9	8.2[b]	7.5[f]
Diamagnetic anisotropy, $\Delta\chi(10^{-8}\text{m}^3\,\text{kg}^{-1})$	3.9	3.7	1.13[e]	9.7[f]

[a][Finkenzeller et al., 1989], [b][Dunmur, 2001], [c][Cummins et al., 1975], [d, e][Pestov and Vill, 2005]: [d]@ 30°C, [e](@ 25.6°C, [f][Priestley et al., 1975].

- NAD: 2,4-nonadienic acid.
- I52: 4-ethyl-2-fluoro-4′-[2-(*trans*-4-*n*-pentyl-cyclohexyl)-ethyl]-biphenyl *n*-propyl-cyclohexyl-ethyl-6-fluoro-*n*-butyl-biphenyl.
- PCH5: 4-*n*-pentyl-4′-cyano-phenylcyclohexyl.
- HAB: 4,4′-diheptyl-azobenzene.
- 5FTP: 4,4″-pentyl-2′,3′-difluoro-terphenyl [Gray et al., 1989].
- PAA: 4,4′-dimethoxy-azoxybenzene.
- AAD: anisaldazine.
- P5: *p*-quinquephenyl.

These are just a few common examples of nematics. In Table 1.3 we report, as an example, a few important physical properties for the nematics PCH5, 5CB, and MBBA.

There are now tens of thousands of known compounds giving nematic phases with extremely different chemical structures [Gray, 1962; Demus, 1989; Kaszynski et al., 2001]. In particular, a variety of aromatic, aliphatic, polar and non-polar compounds have been found to yield nematics.

A very interesting class of compounds is also that of mesogens incorporating a metal, or *metallomesogens* (see, e.g., Fig. 1.11), that somehow combine the characteristics of liquid crystals with those of metal coordination complexes [Hudson and Maitlis, 1993; Donnio and Bruce, 1999]. In particular the introduction of metal atoms with a certain hybridization adds enormous possibilities of precisely directing ligands in space, beyond what is allowed by the linear, triangular or tetrahedral geometry of carbon single, double and triple bonds (sp, sp^2 and sp^3 hybridization) with the effect of varying the shape of a mesogen molecule

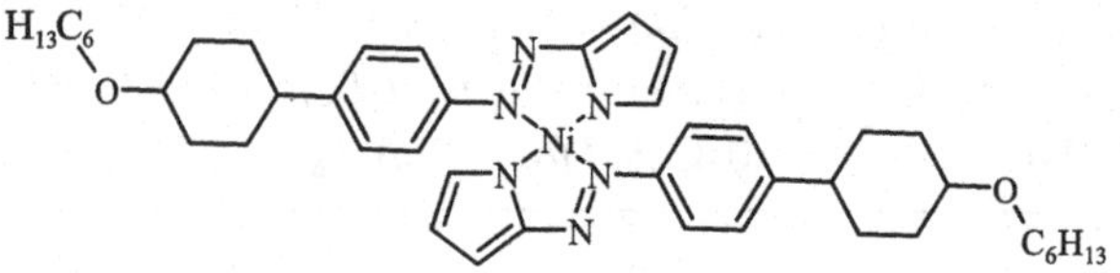

$$K \overset{173.7^\circ C}{\longleftrightarrow} N \overset{200.7^\circ C}{\longleftrightarrow} I$$

Figure 1.11 An example of metallomesogen: a tetracoordinated nickel (II) complex [Serrano and Sierra, 1996].

in a controlled way. The relation between molecular structure and mesogenic behaviour is far from obvious, also because many factors like shape or polarizability anisotropy that might intuitively favour nematicity could favour crystallization too, and possibly to a greater extent. Whether a liquid-crystalline phase is observed or not is a result, not just of one single overwhelming factor, but of a quantitative competition between different possible candidate organizations. However, synthetic chemists have mastered the art of making good candidate mesogenic molecules, developing effective sets of practical rules based on the type of fragments (e.g. rigid or non-rigid cores and chains of some chemical nature, ...) forming the candidate molecule as well as the way they are assembled together [Gray, 1979; Hird, 2001; Neubert, 2001b].

1.3 Polymeric Nematics

We have implicitly assumed that the liquid crystals we have introduced are 'low molar mass', certainly a rather ill-defined term that, just to fix ideas, we could apply to molecules containing up to one hundred atoms or so. However, a large and important class of materials of high molecular weight: liquid crystal polymers (LCPs) [Blumstein, 1985; Khokhlov, 1991; Donald and Windle, 1992; Shibaev and Bobrovsky, 2017], can be obtained if the nematogens are also *reactive monomers* (RM), i.e. if they contain suitable reactive groups that lead to polymerization. LCPs have orientational order similar to ordinary nematics, but are typically much more viscous (100 times or more). The fundamental requirement for a molecule to act as a monomer M is the ability to connect to two or more other molecules in a repetitive fashion via polymerization. There are many routes to this process [Flory, 1953; Finkelmann, 1982] and discussing them is clearly beyond the scope of this book. However, it is useful to consider at least one example: the category of addition polymerization via free radical reaction kinetics [Broer et al., 1988; Thiem et al., 2005]. The basic steps are:

(i) Initiation step. Some external stimulus (usually heat or light of suitable wavelength) is used by an added initiator, R_2 (see Fig. 1.12) to form reactive activated species, for instance by fragmentation in two free radicals $R^\bullet$ (or more generally a $R^\bullet_A$ and $R^\bullet_B$). Each $R^\bullet$ can react with a monomer M to form a new activated species $RM^\bullet$ completing the initiation process:

$$\begin{aligned} R_2 &\overset{h\nu}{\longrightarrow} 2R^\bullet \\ R^\bullet + M &\longrightarrow RM^\bullet. \end{aligned} \tag{1.11}$$

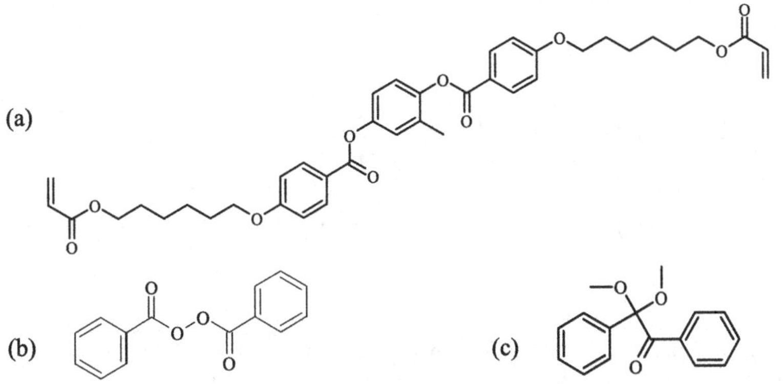

Figure 1.12 (a) A mesogenic monomer: methyl substituted 1,4-phenylene-bis{4-[6-(acryloyloxy) hexyloxy] benzoate} with end reactive acrylates [Broer et al., 1989; Thiem et al., 2005]. On the bottom are (b) the photoinitiators benzoyl peroxide and (c) 2,2-dimethoxy-1,2-diphenylethan-1-one (Irgacure 651© from Ciba©).

(ii) Propagation stage. Reaction of the activated species with monomers causes rapid chain growth:

$$R(M)_k M^\bullet + M \longrightarrow R(M)_{k+1} M^\bullet. \tag{1.12}$$

(iii) Termination step ends the reaction. The polymer chain length grows until, after a certain number of steps, the process is terminated by a quencher species Q which can react with the activated species $M^\bullet$ but cannot participate in further reactions:

$$R(M)_n M^\bullet + Q \longrightarrow R(M)_{n+1} Q\,. \tag{1.13}$$

A practical example is that where the monomer M is a reactive mesogen, with two active acrylates groups at the end of the molecule (see Fig. 1.12). The acrylate functional group is particularly reactive, due to the vinyl double bond. Initiation starts by irradiation of the mixture of monomer with photoinitiator (even in very small concentration, $\ll$1%) at a wavelength where the host reactive mesogens do not absorb light, so that the photopolymerization can proceed homogeneously through the whole sample. When the active groups are at the end of the elongated monomer, a *main-chain* polymer consisting of a string of monomer units is formed. However, with an alternative strategy mesogenic substituents can be attached to the chain backbone, forming *side-chain* polymers [Finkelmann, 1982; Finkelmann and Rehage, 1984; McArdle, 1989]. In Fig. 1.13 we see schematic examples of main-chain and side-chain polymers [Blumstein, 1978; Finkelmann, 1982; Finkelmann and Rehage, 1984; Samulski, 1985; Gleim and Finkelmann, 1989]. Alignment of LCP chains can be achieved by mechanical stretching of the material in various ways [Xue et al., 2015]. Some interesting classes of composite materials can be obtained performing polymerization of reactive monomers and nematics. Thus, for immiscible liquid crystals (roughly 30%–50%) and partially polymerized but still liquid RMs an emulsion can be formed, which results, by further polymerization (*curing*) of the RMs, in a phase separation producing a dispersion of micron size droplets of LC in a solid polymer. Films of these

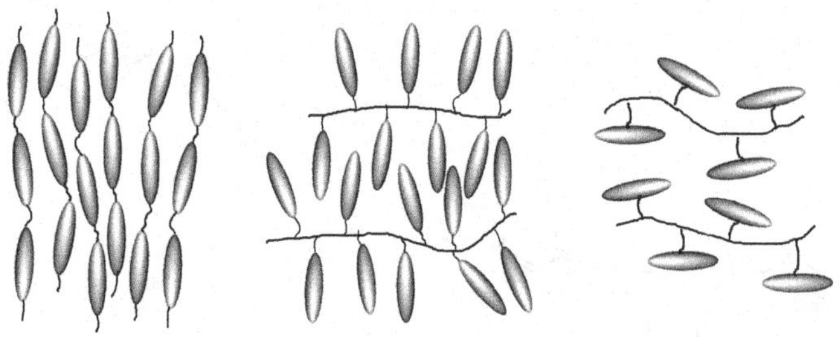

Figure 1.13 Schematics of (a) main-chain, (b) side-chain end-on and (c) side-chain side-on nematic LC polymers showing how monomers are connected to the backbone.

polymer dispersed liquid crystals (PDLCs) [Doane, 1990] scatter visible light giving a non-transparent state that, if the nematic trapped inside the droplets has positive dielectric anisotropy, can be switched to transparent subjecting them to an electric field across the film, with applications, e.g. for smart windows (even if the off opaque state may be non-ideal in many cases).

Increasing the polymer concentration to 60%–80% together with a fast curing with stronger light intensity gives nanosize droplets. A film of such a composite is transparent in the visible, even though it can still be switched, giving fast modulation of light useful in photonics applications. In particular, holographic gratings for switchable transmissive and reflective diffractive optics can be fabricated by the coherent interference of laser radiation of the reactive mixture, hence the name H-PDLC for these materials [Bunning et al., 2000].

A third type of polymer–nematic composites is obtained for low (1%–10%) polymer content, cured in a way that yields a uniform dispersion of the polymer chains (polymer stabilized LC, PSLC) or a polymer network in the nematic. The polymer provides some memory of the original orientation of the polymer chains to the system, facilitating and speeding up switching. A polymer network LC (PNLC) film with *homeotropic* (i.e. perpendicular to the surface) polymer chains orientations loaded with a nematic with negative dielectric anisotropy can be converted from transparent to opaque when a field applied across the film is switched on.

1.4 Chiral Nematics

Chiral[1] nematics, N*, also called *cholesterics*, are a naturally twisted variety of nematics, where the director assumes a helical configuration. They can be produced by chiral mesogens or by adding a chiral solute to a nematic [Chilaya and Lisetski, 1986; Oswald and Pieranski, 2005], see Fig. 1.14. Curiously, cholesterol esters were the first liquid crystals of any kind discovered (by the botanist Reinitzer in 1888) [Sluckin et al., 2004]. Maybe this is less surprising, when thinking that chiral liquid-crystalline structures abound in plant and

[1] Any geometrical figure, or a group of points or a particle in a d-dimensional space ($d = 1, 2, 3$), is said to be chiral if it cannot be superimposed to its mirror image by any rotation in the same space. The two mirror images or *enantiomers* can be called Left (L) or Right (R).

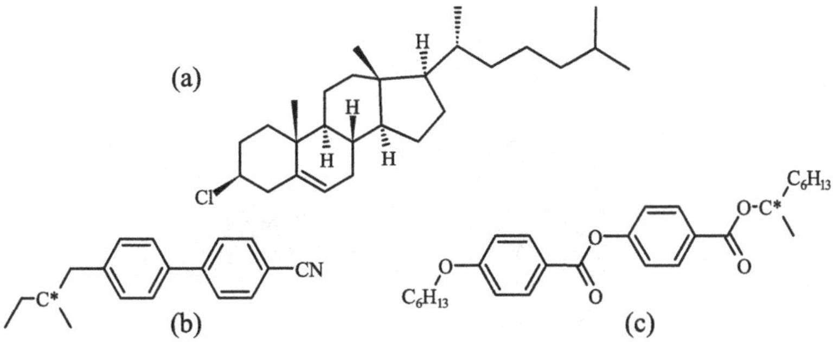

Figure 1.14 A chiral mesogen: (a) cholesteryl chloride and two chiral dopants: (b) CB15 and (c) S811 [Ko et al., 2009].

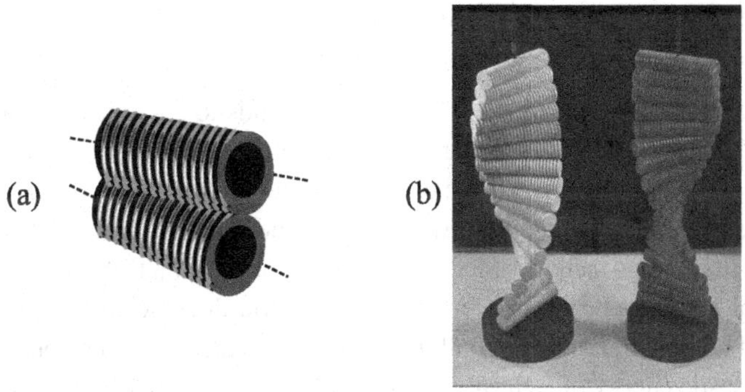

Figure 1.15 (a) Chiral threaded rods in contact cannot be placed exactly parallel to each other, but naturally adopt a twisted mutual orientation when trying to align. (b) They then form a helical organization. Changing the chirality of the rod changes the handedness of the resulting helix.

animal systems [Mitov, 2017]. To see how a chiral mesogen can induce the formation of a twisted structure, a simple hard particle model made of threaded rods can help (Fig. 1.15). The molecular organization in cholesterics is normally represented as in Fig. 1.16, with nematic-like planes twisted one with respect to the next. This model is compatible with the optical properties of cholesterics, but in reality, the system has a uniform distribution of centres of mass, and no true layer structure is present. The repeat distance or *pitch*, p, i.e. the distance over which the local director of the cholesteric helix rotates of 360°, is typically of a few hundred nanometres, so that on a local, molecular scale, very little difference exists between cholesterics and nematics. The environment around a molecule and the ordering of the molecule with respect to the local director is thus quite similar, with the difference that the director remains perpendicular to the helix axis while progressively rotating around it. If the director $\boldsymbol{d}$ describes a right-handed screw along the laboratory z-axis, in a right-handed coordinate system, its explicit form can be written as:

$$\boldsymbol{d}(z) = \big(\cos[q_0 z + \phi], \sin[q_0 z + \phi], 0\big), \tag{1.14}$$

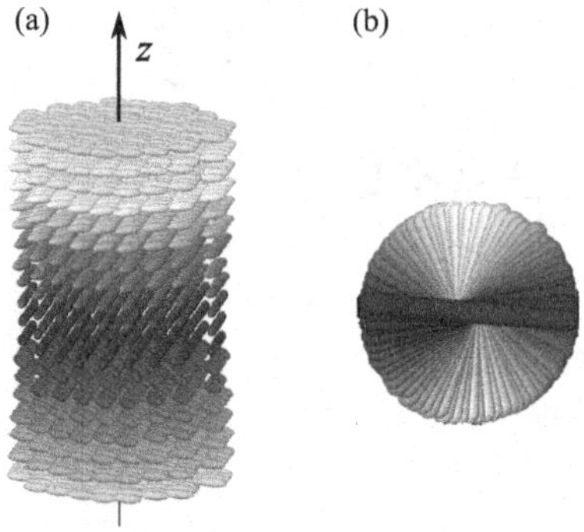

Figure 1.16 A simplified representation of the structure of a cholesteric in terms of nematic regions ('pseudo-layers') (a) twisted one with respect to the other, and (b) a view from top, showing the effective uniaxiality around the helix axis.

where ϕ is an arbitrary phase and $q_0 = 2\pi/p$ corresponds to the repeat distance along the column (pitch) p, with q_0 positive or negative for a right- or left-handed material. The sense of the cholesteric helix is reversed if the chirality of the mesogen is reversed. However, no general way exists of relating the left or right twist of the helix to the absolute conformation or the sign of the optical activity of the mesogen. The fact that chirality is essential to produce optical activity and to give rise to a cholesteric does not mean that the two are directly related but just that they are both not forbidden under the same conditions.

The Frank elastic energy of a chiral nematic, $\mathcal{G}_{\text{el}}^{ch}$, is very similar to the nematic one, but contains the wave vector q_0:

$$\mathcal{G}_{\text{el}}^{ch} = \frac{K_{11}}{2}(\nabla \cdot \boldsymbol{d})^2 + \frac{K_{22}}{2}[\boldsymbol{d} \cdot (\nabla \times \boldsymbol{d}) + q_0]^2 + \frac{K_{33}}{2}|\boldsymbol{d} \times (\nabla \times \boldsymbol{d})|^2 . \tag{1.15}$$

The helical structure implies that the plane of polarization of a beam of light propagating along the helix axis will be rotated by an amount proportional to the number of helix windings and thus to the sample thickness in an ideally oriented sample. A one-millimetre-thick sample of uniformly aligned cholesteric can give an optical rotation of the order of 10^4–10^5 degrees, a huge amount compared with that caused by single-molecule optical activity or to optical rotation in crystals. By means of comparison, the plane of polarization is rotated by 0.665 degrees upon passing through 1 mm of a water solution containing 10 g/l of sucrose, a simple chiral molecule. The pitch in cholesterics can vary a lot with the nature of the material and for cholesterol esters is in the range 300–500 nm. For a given material the pitch can be easily changed with a variety of perturbation agents: temperature, pressure, impurity concentration, etc.

Cholesteric phases can be also induced by the addition of a small quantity of an anisotropic chiral dopant to a nematic (see Fig. 1.14). The pitch of the induced cholesteric phase, p, is inversely proportional to the chiral dopant concentration C^*, but it is usually longer than that of pure materials, with reflection bands in the infrared (IR) rather than in the visible region. The inverse of the proportionality constant is called *helical twisting power*: $HTP \propto 1/(p\,C^*)$. The two chiral dopants CB15 and S811 shown in Fig. 1.14(b, c) are an example of solutes that have large and opposite helical twisting power: $HTP \approx 7\,\mu\text{m}^{-1}$ and $HTP \approx -11\,\mu\text{m}^{-1}$ respectively, when dissolved in the nematic mixture E7 [Ko et al., 2009].

1.4.1 Selective Wavelength Reflection

It is important to note that the periodic structure of cholesterics can act as a diffraction grid for light with wavelength of the same order of magnitude of its repeat distance. According to classical Bragg law, if we have a beam of white light incident at an angle θ with respect to the helix axis on a transparent cholesteric with an idealized perfectly planar surface, to have constructive interference we need the difference in optical path (the geometric difference of path times the refractive index) to be a multiple of the wavelength $n(\theta)p\sin\theta = m\lambda$, with m an integer, $n(\theta)$ the *effective refractive index* and p being the helix pitch. Assuming nearly normal incidence, $\theta \approx 90°$, this implies selective reflection of light (cf. Fig. 1.17a) at the wavelength $\lambda = p(n_\parallel + n_\perp)/2$ with the bandwidth $\Delta\lambda = p(n_\parallel - n_\perp)$, where $n_\parallel$ and $n_\perp$ are the extraordinary and ordinary refractive indices of the untwisted liquid crystal. In practice, only the first-order reflection ($m = 1$) appears at normal incidence [Sage, 2011]. At oblique incidence, higher-order reflections can also occur but are generally much weaker. To complicate things further, Bragg's law does not hold exactly when the helix axes are not uniformly aligned, with the incidence angle θ_1 and the observation angle θ_2 that can be different, leading to a more complex expression [Sage, 1992; Yang et al., 1997]. We should also note that the reflected light is circularly or elliptically polarized and has the same handedness as that of the helical structure of the cholesteric phase. A left-handed helix reflects left-handed light, whereas right-handed light passes unaffected (so 50% of the intensity of an unpolarized incident radiation is lost to reflection). Incidentally, this is the opposite of the behaviour of a mirror which reflects circular polarized light with opposite handedness.

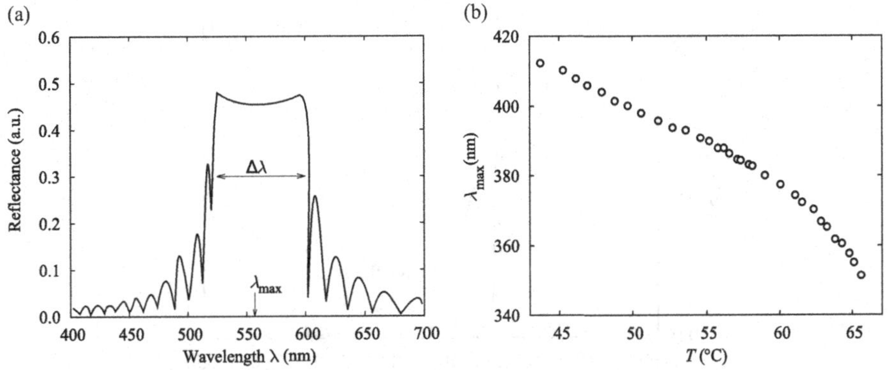

Figure 1.17 (a) Typical spectrum of the reflected light intensity from a cholesteric illuminated with linearly polarized light at normal incidence [Hong et al., 2003] showing the reflection band of width $\Delta\lambda$ centred at λ_{max}. (b) The temperature dependence of the reflected λ_{max}, proportional to the helix pitch, in cholesteryl chloride [Hanson et al., 1977].

A material capable of interacting with light in a well-defined way is called a *photonic material* and in this sense an oriented cholesteric is a one-dimensional (1D) photonic material showing, as described, a band gap around λ_{max}. A more complete treatment for arbitrary incidence can be found [e.g., Dreher et al., 1971; Yang and Wu, 2006]. Thin (5–10 μm thick) cholesteric films with planar alignment of the mesogens at the surface can be prepared (e.g. using a rubbed polyimide alignment layer) yielding a stable planar texture with helices as will be discussed in Section 1.4.2. On a black background, so that the light that is not selectively reflected and is more or less completely absorbed, we can have a reflection colour corresponding to a certain pitch. If we perturb the cholesteric, e.g. by changing temperature, the pitch will also change to a first approximation linearly with temperature and accordingly the reflected colour will change as well. This effect is commonly used to obtain a thermographic map of a surface underlying the cholesteric film. The possibility of easily locating hotspots in a surface has led to applications ranging, e.g. from the non-invasive detection of breast cancer to that of anomalous hot areas in electronic circuits [Sage, 2011]. In most cholesterol derivatives the helix winds itself when temperature increases, with a slow decrease in p on increasing temperature as we see in Fig. 1.17b, typically $\mathrm{d}p/\mathrm{d}T \approx -0.28\%/\,^{\circ}\mathrm{C}$ [Pindak and Ho, 1976], while non-steroid chiral nematics usually have $\mathrm{d}p/\mathrm{d}T \approx 0$ [Chilaya and Lisetski, 1986].

Another fascinating application of cholesterics is in the realization of cavity-less dye lasers. These employ a fluorescent dye dissolved in the cholesteric, e.g. the laser dye 4-(dicyanomethylene)-2-methyl-6-(4-di-methylamino-styryl)-4-H-pyran (DCM), that emits across the reflection band of the host (see Fig. 1.17a). After suitable optical pumping of the dye above a threshold intensity (typically $\approx 10^5$ W/cm^2), the fraction of the fluorescent light with emitted wavelength within the transmission bandgap is trapped inside the cholesteric that acts like a cavity resonator. A monochromatic laser light is instead emitted at the band edge (see Fig. 1.18). Note that for a helix with a certain handedness, the light with opposite

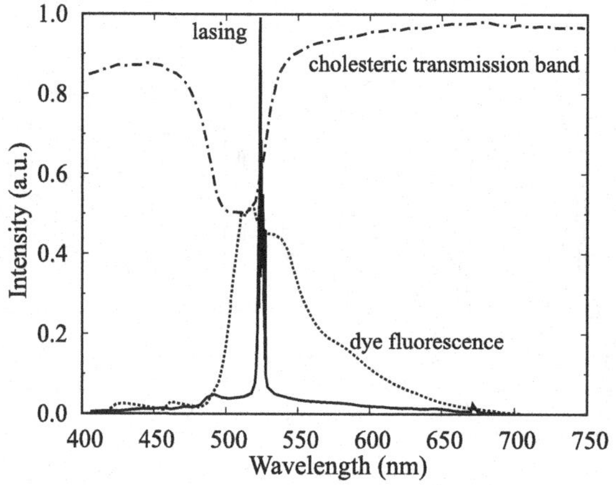

Figure 1.18 An example of cholesteric bandgap (dot-dashed), the emission spectrum of the dissolved dye (dashed) and the laser emission at the transmission band edge (continuous line) [De Santo, 2016].

handedness is unaffected by the structure while that with the same circular polarization shows the band gap. Thus, for instance, for a right-handed helix the emitted light with left circular polarization is transmitted, while the right-handed one is selectively reflected, as discussed before [Chandrasekhar, 1992] and shows lasing. A reason for particular interest in these lasers is that they can be tuned by varying the cholesteric pitch and a number of methods have been used: by the application of an external field to a change of concentration of a chiral agent, etc.

Systems with a cholesteric-like disposition of side mesogenic units have also been prepared [Blumstein, 1978]. It is particularly interesting and somehow surprising that the helical structure of chiral reactive mesogens can be consolidated by polymerization, leading to a well-defined cholesteric pitch, and reflection colour, that will not change with temperature any further.

1.4.2 Cholesteric Textures

Chiral nematics exhibit textures at the polarizing microscope that are quite different from those of nematics. The texture of a thin film of cholesterics will depend on the orientations of the mesogens (and then of the helices) with respect to the support surface [Demus et al., 1998]. In Fig. 1.19 we show the main texture types and a sketch of the molecular organizations originating them, well discussed in literature [de Gennes, 1974]. The planar organization selectively reflects incident light as discussed before, while the polydomain one strongly scatters.

A planar configuration realized with a cholesteric film sandwiched between two plates with proper surface treatment can be transformed in a homeotropic one with the helix unwound if the LC has a positive dielectric anisotropy ($\Delta\epsilon > 0$) and a field higher than the critical value $E_c = (\pi^2/p)\sqrt{(K_{22}/(\epsilon_0 \Delta\epsilon)}$ is applied [Yang et al., 1997]. Alternatively, applying an intermediate strength field the reflecting planar configuration can be transformed into a strongly scattering focal-conic one where helical domains still exist but are disordered in space. It is interesting to note that both textures, when created, can last without an applied field, providing the basis for a bistable, low power consumption, device. These changes of optical properties, that can be driven pixel by pixel are the basis for a class of cholesteric based reflective LCD [Yang et al., 1997].

1.4.3 Blue Phases

For certain strongly chiral materials, typically among those that have a short pitch, other molecular organizations called *Blue Phases* (BPs) are formed between the cholesteric and the isotropic phase [Crooker, 1983; Kitzerow and Bahr, 2001; Kikuchi, 2008; Rahman et al., 2015]. For instance, cholesteryl oleate has the following sequence of phases, including one of layered smectic phases discussed Section 1.7 [Voets et al., 1989]:

$$\text{K} \overset{41.37^\circ\text{C}}{\longleftrightarrow} \text{S}_\text{A} \overset{49.09^\circ\text{C}}{\longleftrightarrow} \text{Chol} \overset{53.58^\circ\text{C}}{\longleftrightarrow} \text{BPI} \overset{54.39^\circ\text{C}}{\longleftrightarrow} \text{BPII} \overset{54.87^\circ\text{C}}{\longleftrightarrow} \text{I},$$

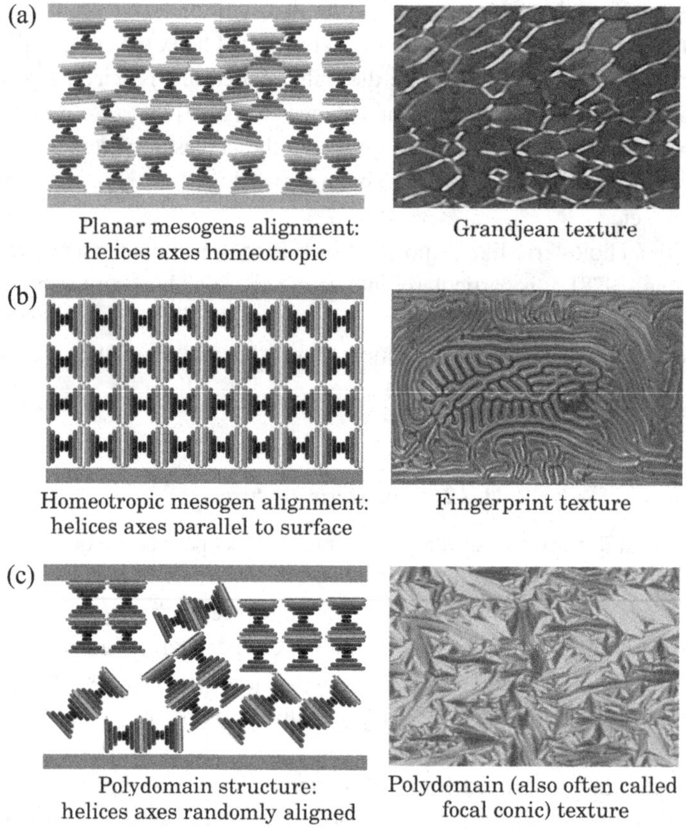

Figure 1.19 Sketches of the arrangement of cholesteric helices in films with different surface alignment (left) and corresponding observed textures between cross polarizers (right). (a) Planar structure with helix axes perpendicular to the support surface. The dark lines in the Grandjean texture (right) are called *oily streaks* [Dierking, 2003]. (b) Helices with local axis parallel to the surface (homeotropic alignment of mesogens) and fingerprint texture [Giordano et al., 1982]. (c) polydomain distribution of helices and real texture [Dierking, 2003].

and cholesteryl nonanoate has three BPs:

$$\text{Chol} \overset{90.29^\circ\text{C}}{\longleftrightarrow} \text{BPI} \overset{90.80^\circ\text{C}}{\longleftrightarrow} \text{BPII} \overset{90.88^\circ\text{C}}{\longleftrightarrow} \text{BPIII} \overset{91.07^\circ\text{C}}{\longleftrightarrow} \text{I}.$$

These BPs are, differently from cholesterics, optically isotropic: BPI and BPII are cubic, while BPIII has spherical symmetry as the isotropic phase [Crooker, 2001] even though they still have large optical activity. The simplest way of visualizing the structure of BPs is probably that of arbitrarily starting from one constituent molecule and a twist propagating from it, not only in a direction perpendicular to its axis, like in a cholesteric, but also in the other perpendicular direction. This structure is called a *double twist cylinder* (see Fig. 1.20). The various phases have complicated superstructures, formed by packing the double twist

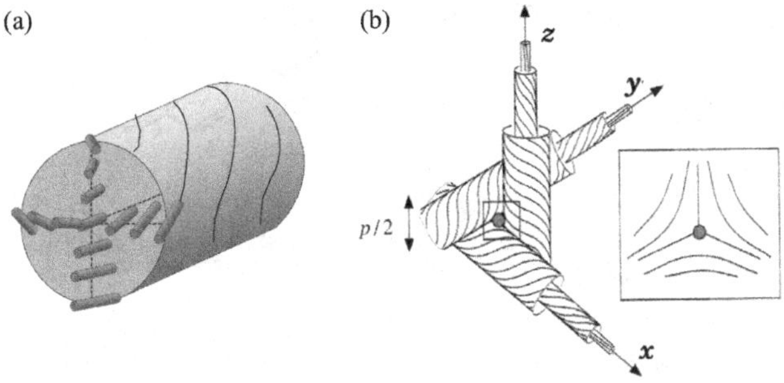

Figure 1.20 (a) Cylindrical structure with twist in two perpendicular directions away from the cylinder axis or *double twist*. The twist increases moving away from the centre of the cylinder and is around 45° at the tube boundary. (b) As three cylinders meet at a corner the directors form a defect [Cao et al., 2002; Kleman and Lavrentovich, 2003].

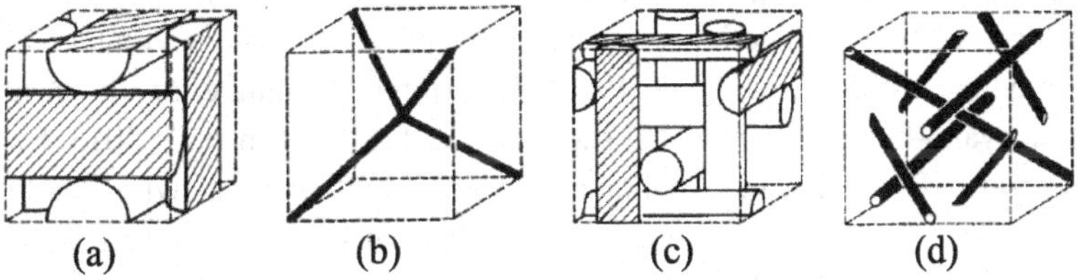

Figure 1.21 The simple cubic (sc) and body centred cubic (bcc) structure of blue phases (a) BPII and (c) BPI and the corresponding unit cells of defect lines (b, d) [Dubois-Violette and Pansu, 1988].

cylinders in various ways. In practice, the unit cells are body-centred cubic for BPI and simple cubic for BPII (Fig. 1.21).

The first BPs only existed in a narrow temperature range and thus were of limited interest for applications, but this has completely changed with the discovery of wide temperature range materials [Coles and Pivnenko, 2005] and of a procedure for stabilizing BPs, widening their temperature range of existence, by the inclusion of suitable polymers, e.g. obtained polymerizing the monomers directly in the BP [Kikuchi et al., 2002]. This stabilization has led to applications in displays, exploiting the very high susceptivity of the BP to an applied electric field: the Kerr effect already discussed (see Eq. 1.6). The field turns the optically isotropic BP into a birefringent medium and opens the possibility of realizing very fast response (<1 ms) displays based on BPs that do not require an aligning film. For instance, while no light would emerge between crossed polarizers at field off, light would be observed as the field is switched on (see Eq. 1.5), even though rather high voltages, of the order of 5 V/μm, some 10 times higher than those of a twisted nematic LCD seen before, are required [Endo et al., 2016].

The periodicity of BPs along different directions is similar to that of cholesterics and also gives the possibility of mirror-less dye lasing, but with emission in three directions rather than in just one as for cholesterics [Cao et al., 2002].

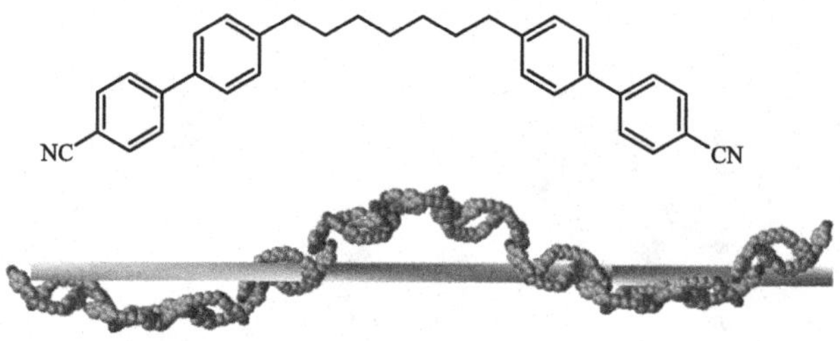

Figure 1.22 A possible model for the twist-bend phase of the cyano-biphenyl dimer CB7CB.

1.5 Twist-Bend Nematic Phase

We have seen how chiral mesogens originate spontaneously twisted nematic (cholesteric) structures with an overall chirality determined by the chirality of the mesogen. More intriguingly twisted nematics have also been predicted in systems of achiral molecules [Dozov, 2001] and observed first in materials containing flexible alkyl spacers with an odd number of CH_2 (methylene) units [Ungar et al., 1992]. In particular, a number of studies have been performed on cyano-biphenyl dimers [Cestari et al., 2011; Borshch et al., 2013; Chen et al., 2013; Zhu et al., 2016] like 4′,4′-(heptane-1,7-diyl)-bis([1′,1′-biphenyl]-4′-carbo-nitrile) (CB7CB), shown in Fig. 1.22, where the pitch of the structure is found to be around 8 nm using various experimental techniques and in particular soft X-ray resonant scattering at the absorption band edge of the carbon atoms of the molecules [Zhu et al., 2016]. It is amazing that in this phase formed from non-chiral mesogens the pitch is some 50 times shorter than that of typical cholesterics originated from chiral mesogens, as we have seen in Section 1.4. The original prediction was that the twist-bend phase should occur when $K_{33} < 0$ and $K_{11}/K_{22} > 2$ [Dozov, 2001]. Experimentally, these materials have a very low bend elastic constant, with $K_{33} < K_{11}$ [Adlem et al., 2013] but this situation does not seem general or clear [Parthasarathi et al., 2017]. The detailed structure of the twist-bend or heliconical phase N_{TB}, for example, its single or double helical structure, is still an object of much debate [Mandle et al., 2015; Tuchband et al., 2017; Mandle and Goodby, 2018]. In any case, the very short pitch of the structure promises to open the way to various novel applications.

1.6 Biaxial Nematics

The occurrence of nematic phases with biaxial symmetry (N_b), where a second director, corresponding to a preferred orientation of the transversal axis, perpendicular to the first, main one, exists, as shown in Fig. 1.23, has been theoretically predicted for a long time [Freiser, 1970] to occur for mesogens of biaxial, rather than uniaxial, symmetry. However, N_b systems were only relatively recently reported in compounds like *p*-dodecyloxy benzoate diester of 2,5-bis (*p*-hydroxyphenyl)-1,3,4-oxadiazole (ODBP-Ph-OC12) [Acharya et al., 2004; Luckhurst, 2004; Madsen et al., 2004; Merkel et al., 2004; Tschierske and Photinos, 2010] and organo siloxane tetrapodes [Merkel et al., 2004; Figueirinhas et al.,

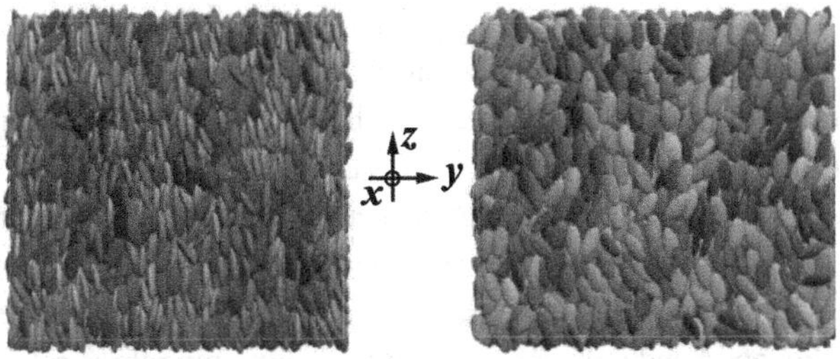

Figure 1.23 Two orthogonal transversal views of a biaxial nematic with the main director along the vertical (z-axis). The difference of the two images shows that molecules tend to stack while maintaining positional disorder [Berardi and Zannoni, 2000].

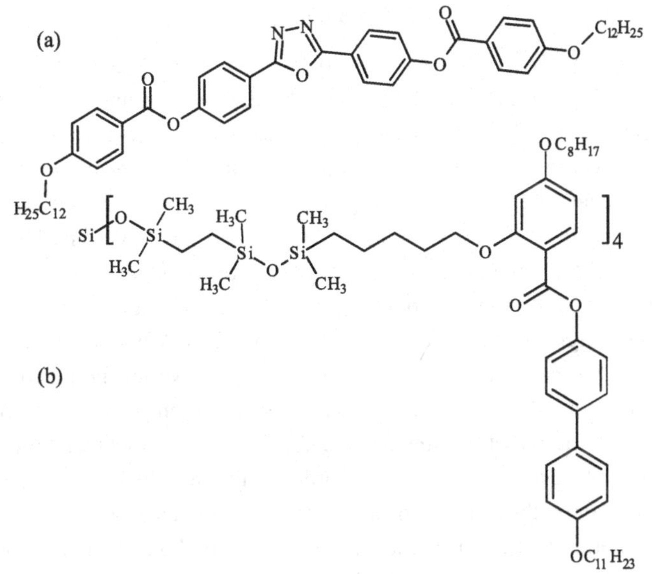

Figure 1.24 Biaxial nematic mesogens: (a) oxadiazole [Acharya et al., 2004; Madsen et al., 2004] and (b) organosiloxane tetrapode [Merkel et al., 2004; Figueirinhas et al., 2005].

2005] (see Fig. 1.24). This may seem strange, since hardly any mesogen is uniaxial, and just looking at molecular shapes they could be better approximated by particles with different thickness in the two directions transversal with respect to the long axis. However, it turns out that the factors favouring biaxial packing also tend to favour biaxial crystal and layered structures, rather than nematic organizations where molecules have to be able to slide one with respect to the other. The observation of a biaxial nematic is thus a difficult compromise to achieve. Biaxial nematics are a matter of active current study both theoretically and experimentally [Luckhurst and Sluckin, 2015]. One of the potential applications explored is the possibility of using an electric field to also control the orientation of the transverse director of a biaxial nematic LC, beyond the only one afforded by the standard uniaxial materials. In particular, the expectation is that switching this secondary director could be much faster, as also supported by computer simulations [Ricci et al., 2015].

The expression for the elastic free energy for biaxial nematics is much more complex than that for their uniaxial analogue (cf. Eqs. 1.8 and 1.9). The number of invariant terms in an expansion of the free energy to second order of the director derivatives increases to 12 bulk elastic constants and 3 additional surface-like coefficients [Govers and Vertogen, 1984; Stewart, 2015]. Three of the bulk elastic constants describe twist deformations, six of them refer to splay and bend deformations, and the remaining three refer to the coupling of bend and twist deformations.

1.7 Orthogonal Smectics

1.7.1 Smectic A

Smectics [Gray and Goodby, 1984; Oswald and Pieranski, 2006] have a layered structure, as shown in Fig. 1.25, with the mass density not constant as in a uniform fluid, but rather periodic in one dimension. They are thus more ordered than simple nematics and, although still fluids, are more viscous and generally more similar to crystalline phases than nematics and cholesterics. The layered structure corresponds to an ordering in the molecular positions reduced with respect to the three-dimensional one of crystals (cf. Table 1.1). There is a whole family of smectic structures [Goodby and Gray, 1979; Gray and Goodby, 1984; Sackmann, 1989; Goodby and Gray, 1999] corresponding to different molecular organizations within the layer and accordingly to different observable properties. We shall indicate a generic smectic phase with Sm and the various smectic types that we are going to introduce as $S_A, S_B, \ldots$ with further subscripts or accents as needed to identify subtypes. We describe first the simplest smectic structure, the smectic A (S_A), with molecules perpendicular to the layer and devoid of long-range positional order in the layer itself (Fig. 1.25). In these systems each layer can be thought of, in first approximation, as a two-dimensional liquid, even if it is worth noting that this is somewhat oversimplified. For instance, the molecular mobility along the director, i.e. across the layers, is in many cases actually higher than inside the layers, as found experimentally and with atomistic simulations (see Chapter 12) for 4-n-octyl-4′-cyano-biphenyl (8CB). Disrupting the layered structure is an energetically

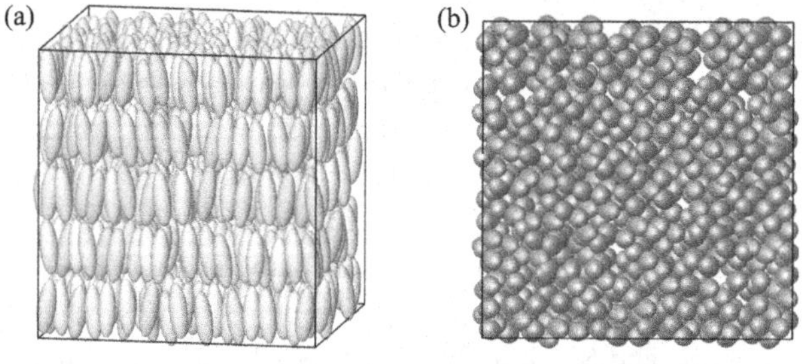

Figure 1.25 (a) The layered structure of a smectic A formed of cylindrically symmetric molecules and (b) a top view showing the fluid-like disorder inside each layer.

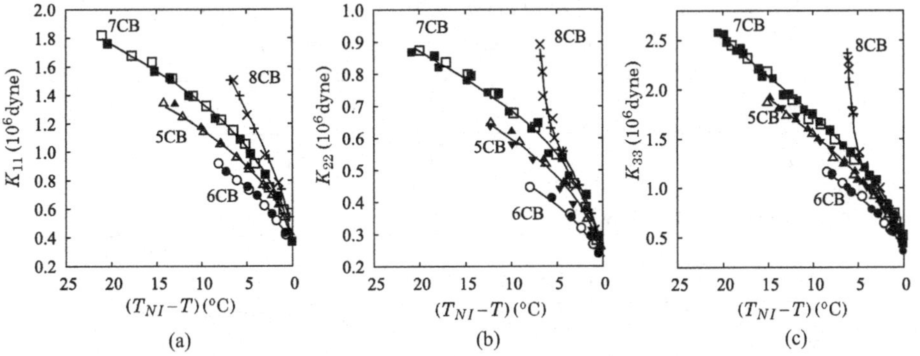

Figure 1.26 Elastic constants of nCBs ($n = 5, 6, 7, 8$) as a function of the relative temperature: (a) splay, K_{11}, (b) twist, K_{22} and (c) bend, K_{33}. Results of different experiments are marked separately [Stannarius, 1998b; Pasechnik et al., 2009].

costly process, so of the three Frank deformations seen earlier (Fig. 1.9), only splay (K_{11}) is allowed, while bend and twist are forbidden. Thus, the elastic constants K_{33} and K_{22} related to these deformations will increase and diverge at the transition from nematic to smectic, for a material that presents both phases, for instance 8CB, as shown in Fig. 1.26. In a smectic, the elastic free energy density $\mathcal{G}_{\text{el}}$ has to include a term which describes the layer elasticity [Kleman and Lavrentovich, 2003]. In a S_A, we can consider the deformations of a uniform smectic layer with director $\boldsymbol{d}$ along the layer normal, $\boldsymbol{d}||z$ and centred at position $z_0 = z_0(x, y)$ with layer spacing $\ell_{z0} = \ell_{z0}(x, y)$. When at rest the layer is flat, so that position z_0, layer spacing ℓ_{z0}, and director $\boldsymbol{d}$ are constant. As the smectic is slightly deformed so that the thickness can vary slightly (by a relative amount $\gamma \equiv (\ell_z - \ell_{z0})/\ell_{z0}$) but maintaining the layers parallel and the local director untwisted, we can write [Kleman and Lavrentovich, 2003; Oswald and Pieranski, 2006]

$$\mathcal{G}_{\text{el}} = \frac{K_{11}}{2} (\nabla \cdot \boldsymbol{d})^2 + \frac{B}{2} \gamma^2, \tag{1.16}$$

where B is a Young's modulus for layers dilation. The preferred organizations are thus those that can be formed without destroying the layers. As an example, we see in Fig. 1.27 a typical smectic A texture at the polarizing microscope, the *focal conic* [Gray and Goodby, 1984], whose origin is strictly linked to the layered organization. In fact, the smectic layers can easily roll and in general take on superstructures that do not involve breaking the layers themselves. In practice, a Swiss roll structure can close in a torus leaving a circular line of discontinuity (or an ellipsoidal one when deformed) as well as a straight line at the joining of the rolls. A particularly interesting way of studying the layered structure of smectics is in films freely suspended across the opening of a thin glass or metal plate, where they form a sort of membrane containing a controlled number of layers (from one or two to thousands). In Table 1.4 we give the transition temperatures for some smectogens. Note that the first one, HAB, is an analogue of the nematic PAA (see Table 1.2), from which it just differs for

Table 1.4. *A few smectic A liquid crystals. We report the melting, T_{KA}, the smectic A-nematic, T_{AN}, and the clearing temperature T_{NI} or T_{AI}*

Smectogen	T_{KA}(°C)	T_{AN} (°C)	T_{NI} (°C)
4,4′-heptyl azoxybenzene (HAB)	34	54	71
4-*n*-pentyl-4′-cyano-biphenyl (8CB)	21	32.5	40
Octyloxy-cyano-biphenyl (8OCB)	55	66.7	79.8
Smectogen	T_{KA}(°C)		T_{AI} (°C)
Diethyl 4,4′-azoxy benzenedicarboxylate	114		122.7

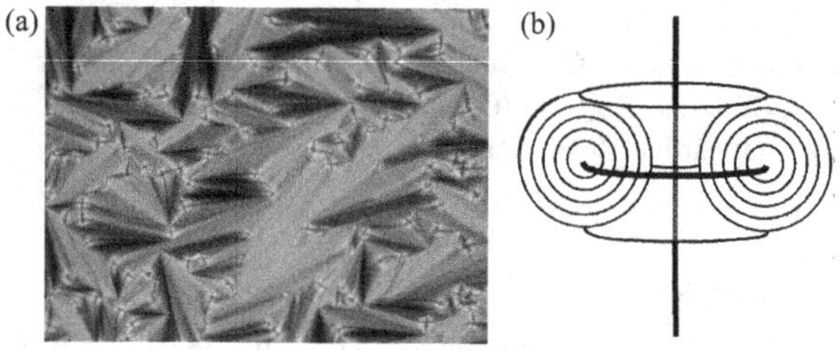

Figure 1.27 (a) The focal-conic fan texture of a smectic A phase observed under the microscope through crossed polarizers and (b) the director tends to lie in the plane of the substrate and the smectic layers are curved across the fans [Dierking, 2003].

a longer alkyl chain. Note also that we can have compounds that go directly from smectic to isotropic phase, as shown in the last entry in Table 1.4.

1.7.2 Strongly Polar Smectics

Only one type of smectic A exists for rod-like, head-tail symmetric, molecules. However, a number of subtypes have been found for smectics formed by asymmetric, polar, molecules, particularly those with strong dipolar end groups and long aromatic cores (see Fig. 1.28). Considering a highly ordered smectic, one has monolayer structures S_{A_1} with the polar molecules randomly oriented up and down in the layer and a periodicity similar to the molecular length ℓ_m. Especially with strong terminal dipoles (typically $-CN$ or $-NO_2$), bilayer phases S_{A_2} with (*antiferroelectric*) ordering of the molecules, up in one layer and down in the next and periodicity similar to $2\ell_m$, are formed. Partial bilayer structures S_{A_d} occur when there is some interdigitation, as we see in Fig. 1.28, with a resulting periodicity ℓ_z intermediate between ℓ_m and $2\ell_m$. For instance, in 8CB $\ell_z \approx 1.4\ell_m$ [Gray and Goodby, 1984]. In both these bilayer phases the dipoles of one layer are schematized as up (down) and their effect is cancelled by those of the adjacent layer. In the *antiphase* smectic A or $S_{\tilde{A}}$ there is instead a modulation of dipole orientation inside each layer, with regions of opposite dipoles separated by defect walls which can additionally form a regular stripe or array structure [Levelut et al., 1981; Berardi et al., 1996a]. A change from one type

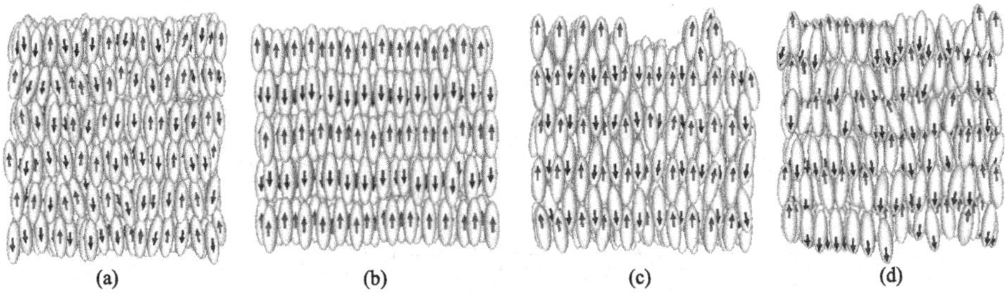

Figure 1.28 Schematic organization of mesogens with a strong axial dipole (black arrow) in smectic A phases showing (a) monolayers in S_{A_1}, (b) bilayers in S_{A_2}, (c) interdigitated structures in S_{A_d} and (d) alternating stripes of up and down dipoles in $S_{\tilde{A}}$ [Berardi et al., 2002].

of organization to another can occur with temperature. For instance, in *p*-heptylphenyl-*p*′-nitrobenzoyloxybenzoate (DB_7NO_2) we have the sequence

$$S_{A_2} \overset{68^\circ C}{\longleftrightarrow} S_{\tilde{A}} \overset{87^\circ C}{\longleftrightarrow} S_{A_1} \overset{94^\circ C}{\longleftrightarrow} N.$$

The behaviour is also observed in mixtures, e.g. of pentylphenyl cyanobenzoyloxy benzoate (DB_5CN) and cyanobenzoyloxypentyl stilbene (C5 stilbene) [Young et al., 1994]. It is interesting that in some of these strongly associated systems, after cooling from nematic to smectic, a new, *re-entrant*, nematic is obtained upon further cooling (see, for instance, Chapter 2). The different types of pairing lead to different effective repeat units and correspondingly to different X-ray patterns (cf. [Ostrovskii, 1993] and references therein). The possibility of positional order within the layers opens the way to a number of smectic phases [Collings, 1990]. Here we see only a few important ones.

1.7.3 Twist Grain Boundary Smectics

When a cholesteric also has a smectic A, e.g. for cholesteryl nonanoate and decanoate, the pitch diverges at the smectic transition temperature T_{AN}. A typical temperature dependence is [Pindak et al., 1974; Pindak and Ho, 1976]

$$p(T) = p_0 + a(T - T_{AN})^{-\nu}, \tag{1.17}$$

where p_0 is the pitch far away from the smectic transition and ν is a critical exponent (see Chapter 2). For cholesteryl nonanoate and decanoate $\nu \approx 0.66$, similar to theoretical predictions, but other values have been found in different materials [Chilaya and Lisetski, 1986]. In the smectic A phase formed of chiral molecules or S_A^*, Renn and Lubensky [1988] have predicted and Goodby et al. [1989] have observed a helical phase constituted of smectic slabs of a certain thickness ℓ_b ($\approx$ 20 nm) separated by surfaces containing regularly spaced screw dislocations or twist grain boundaries. This phase that has on the one hand

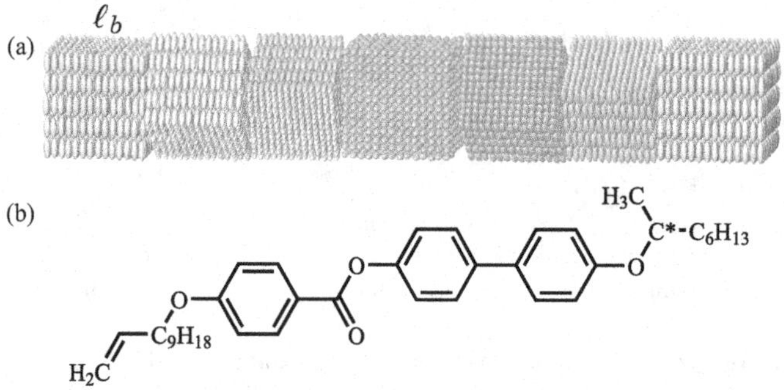

$$K \overset{72.2°C}{\longleftrightarrow} S_I^* \overset{76.9°C}{\longleftrightarrow} S_C^* \overset{91°C}{\longleftrightarrow} S_A \overset{102°C}{\longleftrightarrow} TGBA^* \overset{108°C}{\longleftrightarrow} N^* \overset{113.2°C}{\longleftrightarrow} BPI \overset{115.9°C}{\longleftrightarrow} I.$$

Figure 1.29 (a) A schematic representation of the TGB smectic phase, showing the twisting of S_A slabs of typical thickness ℓ_b [Zhang et al., 2006a]. (b) The chiral compound 4-[4′-(1-methyl heptyloxy)] biphenyl-4-(10-undecenyloxy) benzoate (11EB1M7) and its phase transitions. The S_I^* and S_C^* are chiral versions of the tilted smectics of Fig. 1.32.

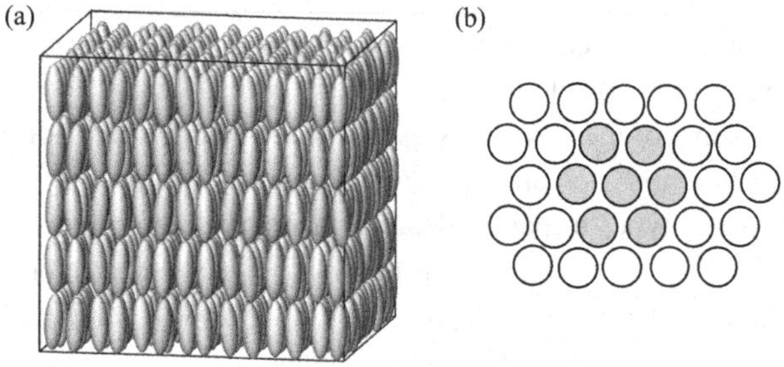

Figure 1.30 (a) The layered structure of a smectic B and (b) a sketch of its local hexagonal clustering.

smectic planes and on the other an average director twist is called *twist grain boundary* (TGB or TGBA*) (see Fig. 1.29).

1.7.4 Smectic B

Smectic B (S_B) systems have a layered structure with molecules perpendicular, on average, to the layer normal and are therefore superficially similar to smectic A (cf. Fig. 1.30a). However, they have a higher viscosity than smectic A and, differently from them, here we have some hexagonal clustering of the molecular centres, as schematically represented in Fig. 1.30b. This kind of ordering is called *bond order* (see Section 4.9.2) and the smectic B phase can be considered as a bond-orientationally ordered version of a smectic A [Brock et al., 1989]. The clusters of hexagonally ordered molecules can themselves be orientationally ordered or disordered with respect to one another. In this last case the

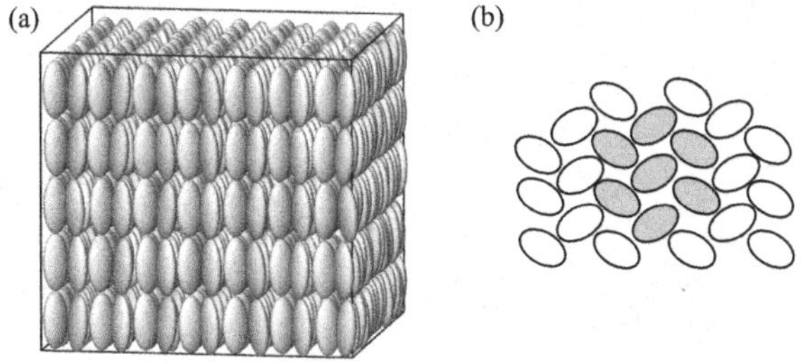

Figure 1.31 (a) The structure of a smectic E and (b) a sketch of the herringbone arrangement of the short molecular axes of its mesogens.

structure of what is called a crystal smectic B (S_{B_K}), to distinguish it from the locally hexatic (S_B) one, is really very similar to a true crystal. An example of a molecule giving smectic B phases is: 4′-*n*-butyloxy benzylidene-4-*n*-octylaniline (4O.8) [Birgeneau et al., 1981; Pershan et al., 1981; Gray and Goodby, 1984] which has the sequence [Juszynska et al., 2011]:

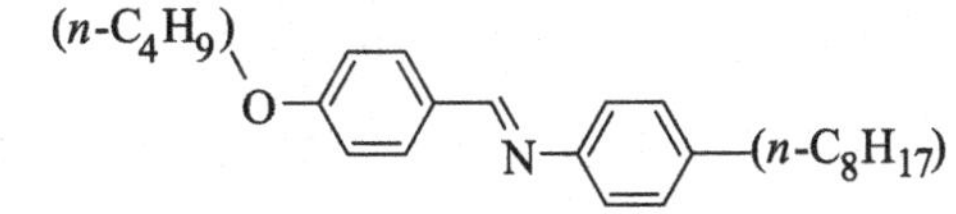

$$S_{B_K} \xleftrightarrow{37.35^\circ\mathrm{C}} S_B \xleftrightarrow{47.95^\circ\mathrm{C}} S_A \xleftrightarrow{62.15^\circ\mathrm{C}} N \xleftrightarrow{77.85^\circ\mathrm{C}} I.$$

1.7.5 Smectic E

In describing the smectic B phase, we have focussed on the layer structure, with molecules essentially orthogonal to the layer, and on the fact that the centres have a hexagonal order, more or less correlated. In reality we should consider also the fact that the molecules are not cylindrically symmetric and that the short axes can have a preferred orientation too. Indeed, while in the smectic B phase the short axes are disposed in a way that preserves a macroscopic uniaxial symmetry, in the smectic E phase, which is very similar to the smectic B, we have instead a herringbone arrangement of the short axes (cf. Fig. 1.31b) and the resulting mesophase is biaxial. As for the S_B, we can have a more crystal like variety, S_{E_K}. An example of a compound having a smectic B and a crystal E phase is *n*-hexyl-4′-*n*-pentyloxy biphenyl-4-carboxylate (65OBC). This has a two-dimensional hexatic structure and transitions [Van Roie et al., 2005]:

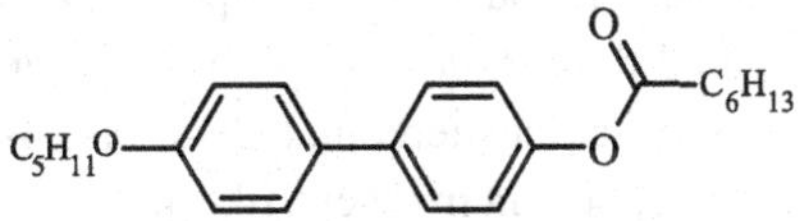

$$\text{solid} \xleftrightarrow{52.7^\circ\mathrm{C}} S_{E_K} \xleftrightarrow{60.3^\circ\mathrm{C}} S_B \xleftrightarrow{67.7^\circ\mathrm{C}} S_A \xleftrightarrow{83.7^\circ\mathrm{C}} I.$$

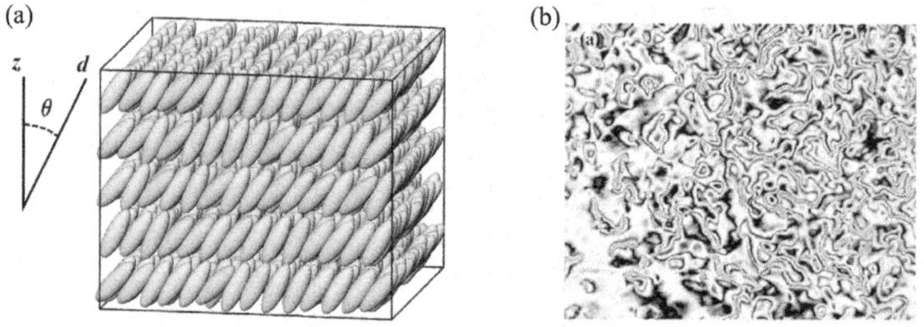

Figure 1.32 (a) Molecular organization in a monoclinic S_C phase and (b) schlieren texture of a smectic C [Sepelj et al., 2007].

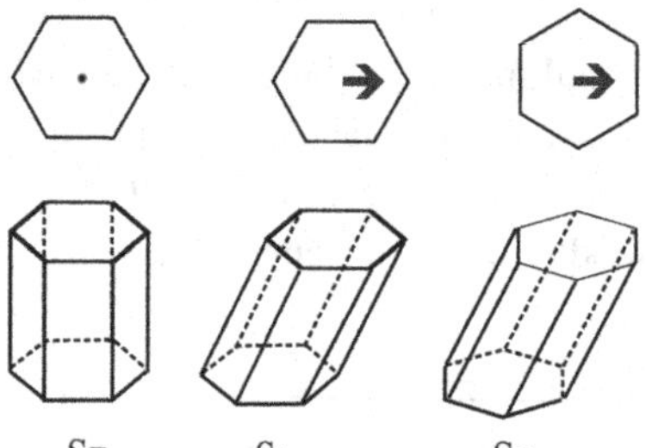

Figure 1.33 A scheme of the various smectic phases with local hexagonal order: the orthogonal smectic B and the tilted smectic I and F.

1.8 Tilted Smectics

1.8.1 Smectic C

The smectic C phase corresponds to a tilted version of a smectic A, as sketched in Fig. 1.32a, without long range positional ordering inside the layers. However, there exist tilted versions of the upright monolayer, S_{A_1}, and bilayer, S_{A_2}, phases discussed in Section 1.7.2, i.e. the S_{C_1} and S_{C_2}. In the same way there exist tilted versions of the smectic B phase. These can be further distinguished by the direction of the tilt with respect to the hexagonal neighbour structure, as can be seen in the summary Fig. 1.33. Thus, S_I and S_F are tilted S_B with inclination direction pointing towards a vertex of the hexagonal cluster of nearest neighbours or perpendicular to a hexagon side. As for the S_B, this positional hexagonal order is only local, if instead it has a long-range character, the S_J, S_G phases, the analogous versions of the crystal S_B are obtained. Yet another version, the smectic L (S_L), with tilt direction intermediate between that shown in Fig. 1.33 for S_I and S_F has been proposed [Chaikin and Lubensky, 1995] and found in some experiments [Chao et al., 1998]. The director $\boldsymbol{d}$ makes an angle, called the *tilt angle* θ, with the layer normal, which can vary with temperature in some materials, e.g. in terephtalydene-bis-(4-*n*-butylaniline) (for short TBBA) while it is temperature independent in others, like 4,4′-di-*n*-heptyloxyazoxybenzene (HAB). For materials that present both a S_A and a S_C phase, the layer spacing normally shrinks going from S_A to S_C and this corresponds to the simple idea of a rigid molecule

Table 1.5. *The melting, T_{KC}, smectic C-nematic, T_{CN}, and nematic-isotropic, T_{NI}, transition temperatures for a few smectic Cs*

Smectogen	T_{KC}(°C)	T_{CN}(°C)	T_{NI}(°C)
4,4′-heptyloxyazoxybenzene (HAB)	74	94	124
4,4′-octyloxyazoxybenzene (OAB)	80	108	126
4-octyloxy benzoic acid (OOBA)	101	108	145

tilting away from the normal of a tilt angle. However, this is not the case in the so-called DeVries smectics, that have recently attracted much interest. A possible explanation is that for these systems the molecules are distributed in a kind of conical (and thus uniaxial) way around the layer normal (see [Lagerwall and Giesselmann, 2006; Gorkunov et al., 2007]) even in the S_A phase. It is worth pointing out that some materials exhibit a rich LC polymorphism. For instance, TBBA shows the following cascade of phases [Doucet et al., 1973]:

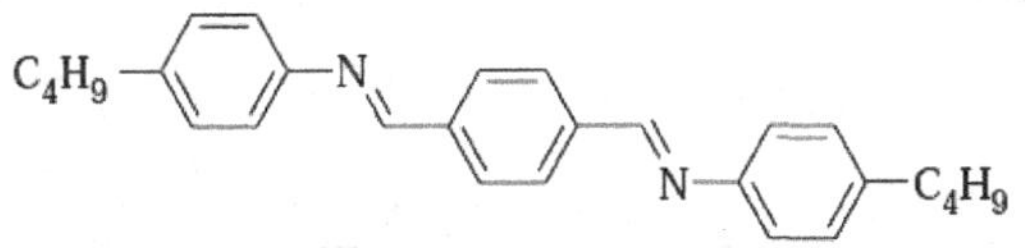

$$\text{solid} \overset{113.0°\text{C}}{\longleftrightarrow} S_B \overset{144.1°\text{C}}{\longleftrightarrow} S_C \overset{172.5°\text{C}}{\longleftrightarrow} S_A \overset{199.6°\text{C}}{\longleftrightarrow} N \overset{236.5°\text{C}}{\longleftrightarrow} I.$$

In Table 1.5 we give the temperature of transition for some common smectic C materials. An ordered smectic C has a C_{2h} symmetry, with the twofold axis perpendicular to the director and in the layer plane. The presence of the inversion centre is not compatible with a macroscopic polarization, which is a vector. The optical properties of a smectic C are similar to those of a biaxial crystal. A polarizing microscope texture of this phase is shown in Fig. 1.32b. Note the nematic-like aspect but with four streak-only centres. For polar smectogens monolayer, partial bilayer and bilayer smectic C phases can exist as for smectic A.

1.8.2 Chiral Smectic C

When the smectogen molecules are chiral or when a chiral solute is added to a smectic C, a phase with macroscopically helical symmetry called *chiral smectic C* or S_C^* can be generated. A centre of symmetry cannot exist any more and the symmetry of the phase is reduced from C_{2h} to C_2. If we visualize the director in a smectic layer as tilted, we can think that its tip rotates as we move through the layers (see Fig. 1.34), thus generating a helix of given pitch. If the molecules possess a transverse dipole the projection of the dipole in the layer plane is randomly distributed in the layer plane itself and thus there is no resulting net dipole. However, the dipoles can be easily oriented by applying an electric field along a direction parallel to the layer planes. The basic constraint of preserving the tilted structure can then be maintained with the macroscopic polarization along the C_2 axis.

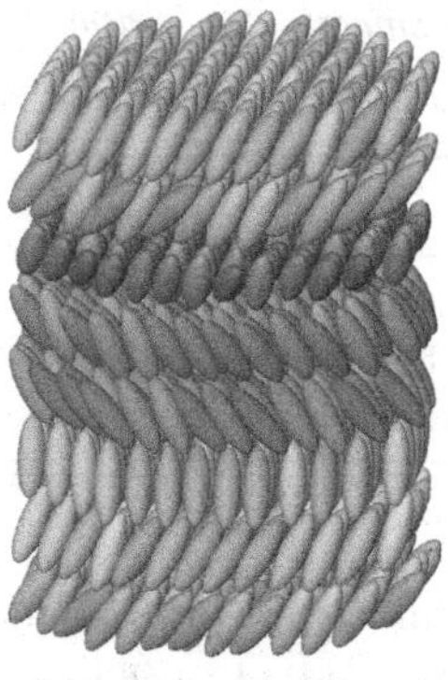

Figure 1.34 A sketch of the smectic C* molecular organization showing the tilted layers twisting along the layer normal.

The system is ferroelectric and is the basis for certain electro-optic displays [Gray, 1987]. An example of compound of this kind is (S)-4-*n*-decyloxy benzylidene amino-2′-methyl butyl cinnamate (DOBAMBC):

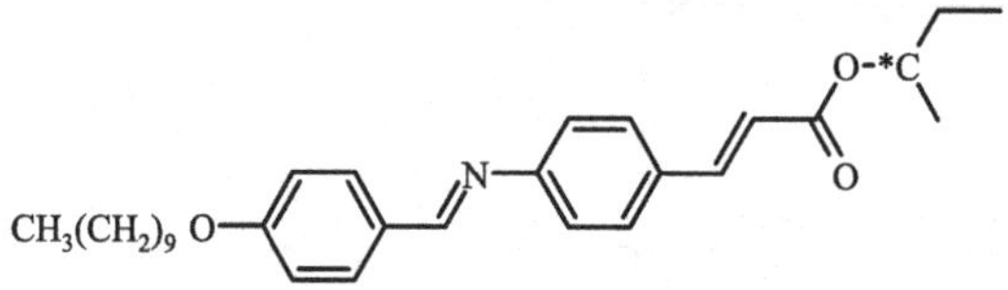

$$\text{solid} \overset{76^\circ C}{\longleftrightarrow} S_{C^*} \overset{95^\circ C}{\longleftrightarrow} S_A \overset{117^\circ C}{\longleftrightarrow} \text{I.}$$

Another compound is (S)-4′-(decyloxy)-4-[(1-methylheptyl)oxy]-2-nitro phenyl-[1,1′-biphenyl]-4-carboxylic acid ester (W314) [Jang et al., 2001]

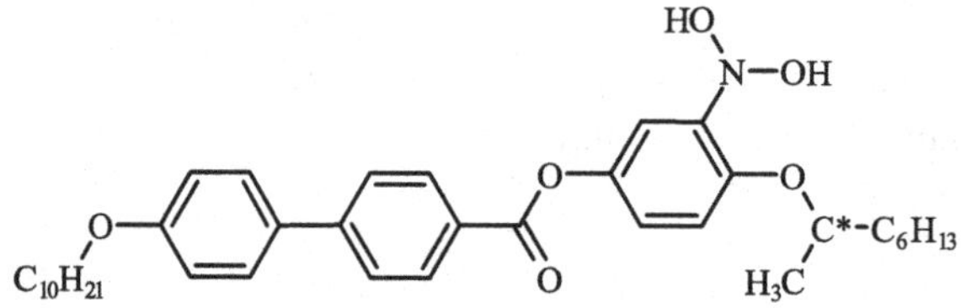

$$\text{solid} \overset{31.2^\circ C}{\longleftrightarrow} S_{C^*} \overset{89.9^\circ C}{\longleftrightarrow} S_{A^*} \overset{124^\circ C}{\longleftrightarrow} \text{I.}$$

1.8.3 Banana Phases

Ferroelectric S_C^* phases are formed by chiral mesogens, so that the discovery [Niori et al., 1996] that ferroelectric phases could be formed by non-chiral mesogens, with an unusual banana, rather than rod-like, shape (see Fig. 1.35) was particularly striking [Jákli et al., 2018]. Excluding a molecular chirality of some conformations, allowed by internal flexibility, a helical S_C^* structure can ensue from the supramolecular chirality of the smectic layers with tilted bow molecules endowed with a transversal dipole moment (see Fig.

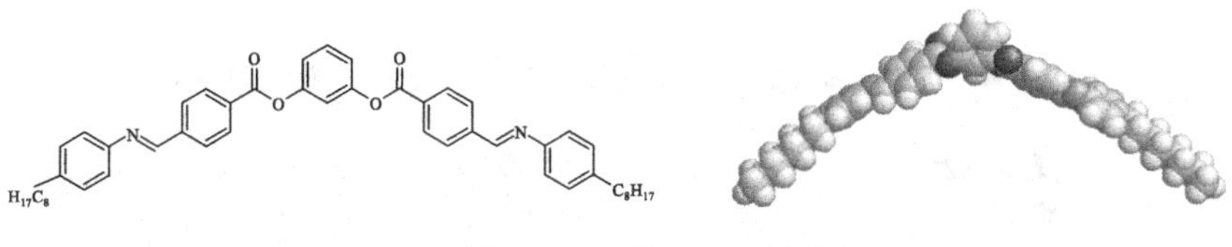

$$\mathrm{K} \underset{72.2^\circ\mathrm{C}}{\overset{97.7^\circ\mathrm{C}}{\rightleftarrows}} \mathrm{B}_3 \underset{139.9^\circ\mathrm{C}}{\overset{156.4^\circ\mathrm{C}}{\rightleftarrows}} \mathrm{B}_2 \underset{158.1^\circ\mathrm{C}}{\overset{161.4^\circ\mathrm{C}}{\rightleftarrows}} \mathrm{I}$$

Figure 1.35 Chemical structure and atomistic model of a bent shape mesogen exhibiting tilted smectic ('banana') mesophases (B_2, B_3) together with its transition temperatures [Niori et al., 1996; Pelzl et al., 1999].

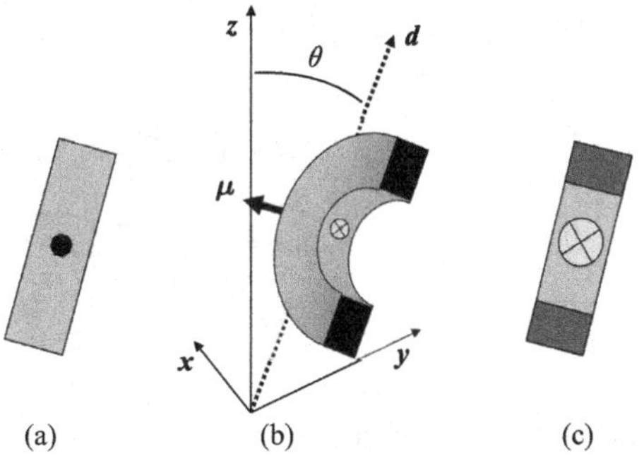

Figure 1.36 Chirality resulting from tilted bent shape mesogens belonging to a smectic layer. The dipole moment $\boldsymbol{\mu}$ is transversal to the bow. (b) A sketch of the banana molecule and of (a) a view from the front and (c) back sides.

1.36). The stack of layers can tilt in the same direction (*sinclinic*) or in opposite directions (*anticlinic*) with respect to the layer normal (see Fig. 1.37). Successive smectic layers can be either ferroelectric (with the same direction of polar order) or antiferroelectric (with opposite directions). It is worth noting that the packing of the banana molecules leads to smectic rather than nematic phases, even though nematic phases have been found, e.g. from 4-cyanoresorcinol with short terminal alkyl chains [Keith et al., 2010]. Indeed, a variety of smectic organizations has been found [Link et al., 1997; Pelzl et al., 1999; Coleman et al., 2003] and a specific nomenclature: $B_1, B_2, \ldots, B_7$ was suggested, with the subscript n corresponding to a conventional indexing of the different phases. This can be a bit confusing, since the phases are rather different and not smectic B after all, and a more recent approach to nomenclature has been based on the arrangement of layers and of layer polarity leading to ferroelectricity or antiferroelectricity (see Fig. 1.37). A detailed treatment of the classification is reported in the reviews by Reddy and Tschierske [2006] and Jákli et al. [2018]. As for the rather complex physical properties features of the various phases, these are definitely beyond the aims of this book, and we just refer to specific treatments, like those found in [Pelzl et al., 1999; Coleman et al., 2003; Reddy and Tschierske, 2006; Jákli et al., 2007; Takezoe and Eremin, 2017; Jákli et al., 2018].

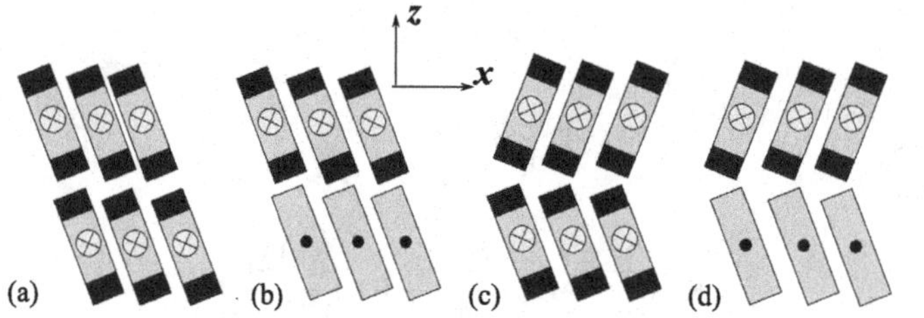

Figure 1.37 A sketch of the molecular organizations that can be obtained in two contiguous layers of banana molecules: (a) sinclinic ferroelectric (S_{CSPF}); (b) sinclinic antiferroelectric (S_{CSPA}); (c) anticlinic ferroelectric (S_{CAPF}); (d) anticlinic antiferroelectric (S_{CAPA}). The z-direction is along the smectic layer normal.

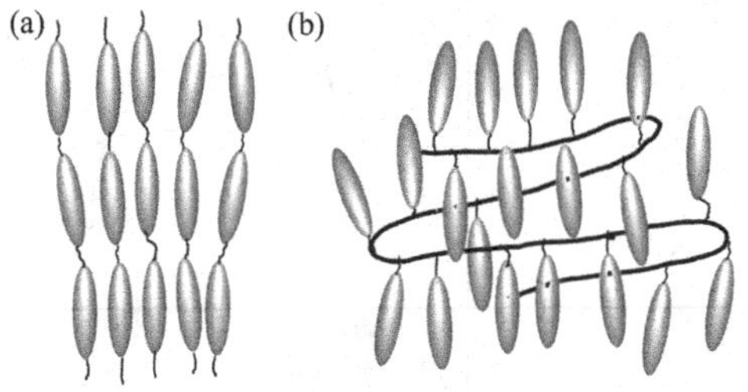

Figure 1.38 A sketch of (a) a main-chain and (b) a side-chain smectic LC polymer.

1.9 Smectic Liquid Crystal Polymers

Besides a nematic organization of the side chains in a linear polymer like that in Fig. 1.13, a smectic organization can also exist. Various other types of low-molar-mass liquid crystals presented earlier on have analogues in the polymer field. Here we see, in Fig. 1.38, a sketch of the structure of a smectic main-chain and side-chain polymer where strong lateral interactions of the monomers lead to layered organizations.

1.10 Discotic and Columnar Phases

Disc-like molecules can organize themselves, face parallel, and produce a variety of *discotic* phases [Chandrasekhar, 1982; Guillon, 1999; Bushby and Lozman, 2002; Kumar, 2002; Sergeyev et al., 2007; Bisoyi and Kumar, 2010; Wohrle et al., 2016; Gowda and Kumar, 2018]. The basic unit is typically a disc or board shape core with flexible chain substituents as, for example, in hexa-n-alkanoates of benzene and triphenylene or in a variety of phthalocyanines, both with and without central metal, or condensed aromatics, like hexabenzocoronene (see Table 1.6). Thus, we can have discotic nematics [Praefcke, 2001] formed by single mesogens, N_D, or by short columnar stacks N_C (see Fig. 1.39) or the chiral variety, nematic N_D^*, i.e. discotic cholesteric. Although nematic discotics N_D do exist, their occurrence is quite rare and more often the molecules organize in long columnar aggregates [Chandrasekhar, 1988] (cf. Fig. 1.40) and actually in the large majority of cases the transition is directly from the isotropic to a columnar phase. The discotic phases can

Table 1.6. *Some discotic molecules and their phase transitions (see text and Figs. 1.40 and 1.42)*

Molecule	Chemical structure
(a) HHTT $R = C_6H_{13}$ $K \xrightarrow{62^\circ C} D_h^h \xleftrightarrow{70^\circ C} D_h^d \xrightarrow{93^\circ C} I$	[Fontes et al., 1988] 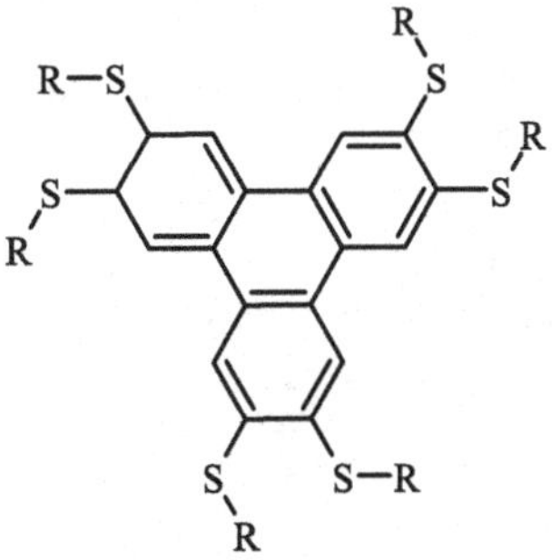
(b) Phthalocyanine $R = CH_2OC_{12}H_{25}$ $K \xrightarrow{79^\circ C} D \xrightarrow{185^\circ C} D_h \xrightarrow{260^\circ C} I$	[Sergeyev et al., 2007] 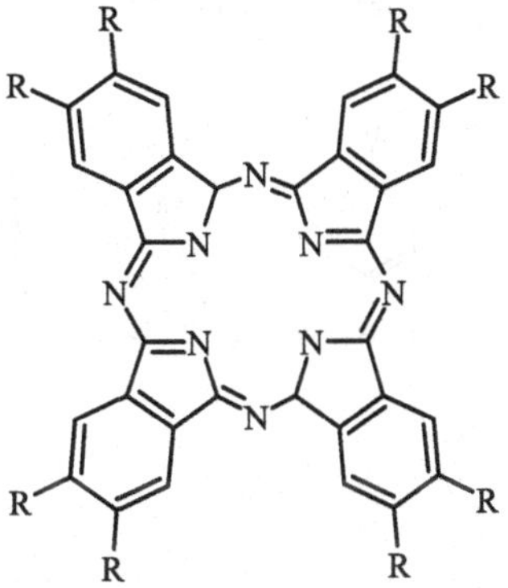
(c) Hexabenzocoronene: HBC-$(C_{14})_6$ $R = C_{14}H_{29}$ $K \xleftrightarrow{114^\circ C} D_h^o \xleftrightarrow{\approx 420^\circ C} I$	[Pisula et al., 2005] 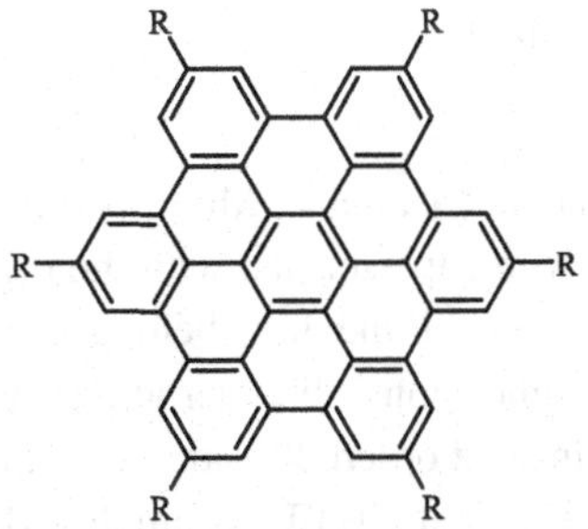

of course be polymerized, like we have seen for rod-like nematics, and in Fig. 1.41 we see an example of a discotic nematogen that can be made reactive by introducing acrylate end groups. There can be various kinds of columnar organizations, usually labelled generically D (or Col by other authors) phases [Luz et al., 1985] if additional details about the

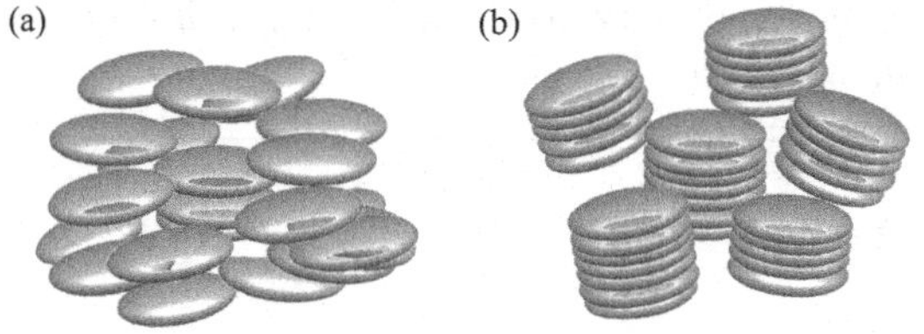

Figure 1.39 Schematic organization of (a) a nematic discotic, N_D and of (b) a columnar nematic, N_C.

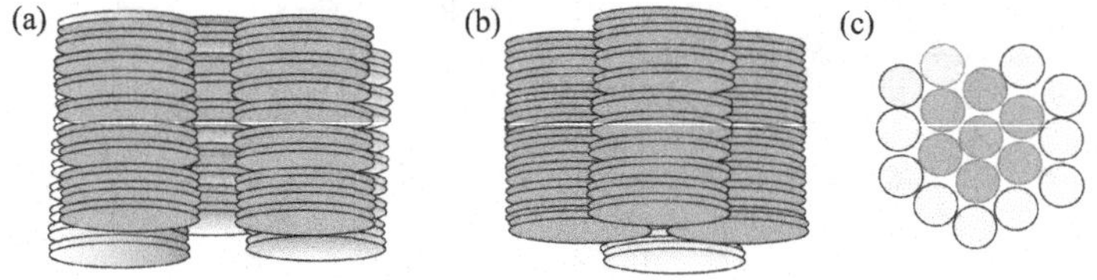

Figure 1.40 Uniaxial columnar liquid crystals with (a) molecular disorder (D_h^d) or (b) some positional order (D_h^o) inside the disc stacks and (c) hexagonal arrangement of the columns.

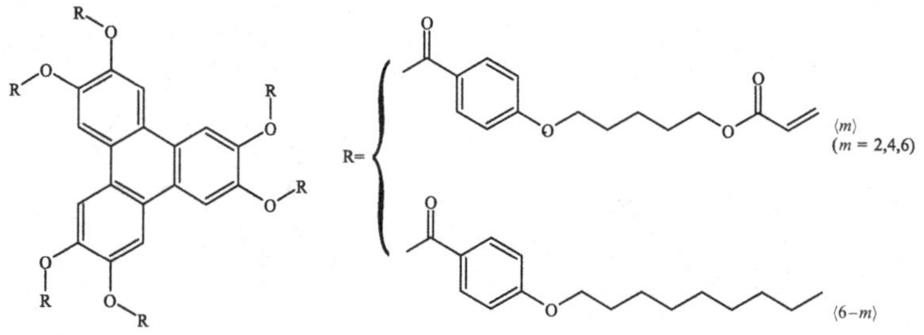

Figure 1.41 A reactive discotic nematogen consisting of a triphenylene core with a total of six alkoxybenzoate substituents (R), of which an average number $\langle m \rangle$, with $\langle m \rangle = 2, 4$ or 6 have reactive acrylate end groups (top right ones) and the remaining $\langle 6 - m \rangle$ are devoid of polymerizable acrylate and thus inactive [Kim et al., 2017].

structure are not available. More specific features of the arrangement inside the columns are indicated with superscripts, with the organization of the columns themselves indicated by subscripts. Thus, if the discs belonging to a column are not regularly positioned, so that we have one-dimensional disorder along the columns the further superscript d is added. If the discs are instead ordered inside the columns, a superscript o, or h if the order is helicoidal (like in the case of HHTT), is employed. A tilt of the mesogens with respect to the column axis is indicated by a superscript t.

The columns themselves can form a more or less regular two-dimensional (2D) array, corresponding to a 2D positional order of the columns, that is manifested with a subscript (h for hexagonal, r for rectangular, ...) as well as orientational order of the molecules. In Fig. 1.40 we see hexagonal, uniaxial phases with or without order inside the column: D_h^o

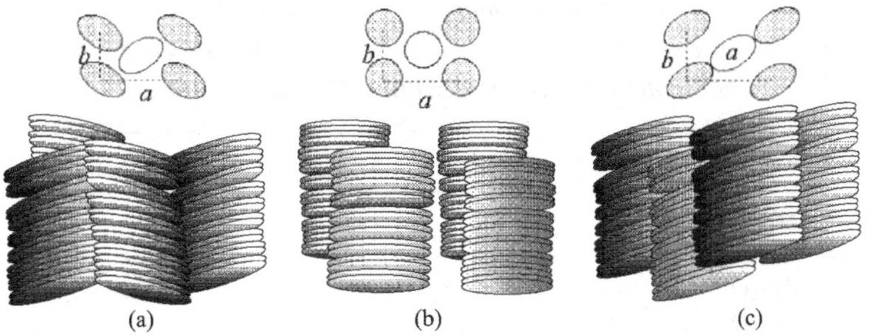

Figure 1.42 Liquid crystals with (a) a tilted arrangement of the columns and (b) a column arrangement which is on average upright, D_r^{dt}, D_r^d (see text) or (c) oblique disordered columns, D_{rt}^d.

and D_h^d. In the absence of further subscripts the columns are assumed to be upright with respect to a hypothetical support surface, while an oblique arrangement of the column axis is marked by a subscript t. Note that, even if chains attached to the periphery of the molecules are not explicitly shown, they play an important role in stabilization. Symmetric substituted discotics tend to give a hexagonal column arrangement and typically give a planar alignment of the discs, i.e. homeotropic alignment of the columns when deposited on a planar surface. Unsymmetrical discotic molecules tend to give tilted phases with rectangular columnar 2D arrangements and their surface alignment is typically non-homeotropic but side on, with the column axes parallel to the surface. Having said that, systems like some substituted phthalocyanines can change morphology with a change of temperature. We also see in Fig. 1.42 various (upright D_r^d, D_r^{dt} or oblique D_{rt}^d) biaxial phases with a rectangular 2D lattice of the columns. The organization shown for the D_r^{dt} phase corresponds to the model proposed by Goldfarb et al. [1983a] in their studies of truxene-hexalkanoate derivatives, with the molecules tilted with respect to the column axis in opposite (anticlinic) directions in alternating columns. The macroscopic symmetry of the D_r^{dt} phase is in any case orthorhombic, while that of the D_t phase is monoclinic. While columnar mesophases are easily found, N_D are available (see Table 1.7) but much more rare [Gasparoux, 1980; Praefcke, 2001; Bisoyi and Kumar, 2010]. The presence of substituents with large steric hindrance seems important in producing the discotic nematic as well as or instead of the stacked, columnar mesophases. Note that in a N_D phase the molecule symmetry axis is on average aligned with respect to the director, like in a normal calamitic nematic. However, the anisotropy in molecular properties with values larger in the plane of the rings (e.g. the polarizability for an aromatic core) tends to be negative rather than positive. Yet another possibility is that the columns themselves form a nematic phase, called a *columnar nematic* N_C. These thermotropic systems have been prepared from a mixture of strong electron donor disc molecules (superdiscs) with small electron acceptor molecules that very efficiently pack in columns [Praefcke et al., 1992]. Examples of superdiscs include pentakis (phenyl ethynyl) phenyl derivatives, while electron acceptor dopants include 2,4,7 trifluorenone. It

Table 1.7. *The crystal-columnar, T_{KD}, columnar-nematic, T_{DN_D}, and nematic-isotropic, T_{N_DI}, transition temperatures for a few hexa-R-benzoate triphenylene discotics [Gasparoux, 1980]*

R substituent	T_{KD}(°C)	T_{DN_D}(°C)	T_{N_DI}(°C)
$C_6H_{13}O$	186	193	274
$C_9H_{19}O$	154	183	227
$C_{10}H_{21}O$	142	191	212
$C_{11}H_{23}O$	145	179	185

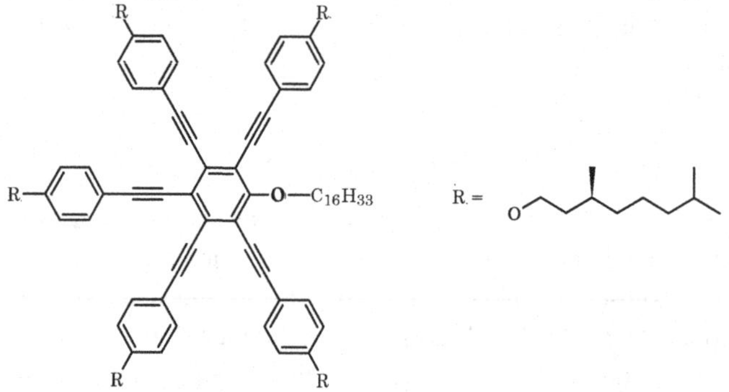

$$\text{solid} \overset{-36°\text{C}}{\longleftrightarrow} \text{N}_{\text{D}*} \overset{23.4°\text{C}}{\longleftrightarrow} \text{I}$$

Figure 1.43 The molecular structure and transitions of a chiral discotic nematic mesogen [Langner et al., 1995].

is interesting to realize that the liquid crystal is formed by the combination of molecules not necessarily mesogenic by themselves.

Chiral discotic nematic, N_D^*, are also fairly rare. However, an example [Langner et al., 1995] is shown in Fig. 1.43.

1.10.1 Columnar LC Properties

Columnar liquid crystals are particularly promising as semiconductors for organic electronic applications [Sergeyev et al., 2007; Bisoyi and Kumar, 2010; Kaafarani, 2011]. Liquid crystals are typically insulators in the isotropic phase, with charge mobilities of the order of 10^{-5} cm^2/(V s) like the majority of organic compounds. However, as the molecular organization changes from isotropic to columnar by lowering the temperature, the charge mobility increases by more than two orders of magnitude, as we see in Fig. 1.44. Note from the figure that the mobility is even higher in the crystalline phase, but the values refer to a nearly 'perfect crystal', while in practice it is difficult to avoid structural defects that considerably lower the performance in the crystal phase. These defects, or *grain boundaries* where different orientations of the crystal domain clash, can instead be avoided by the fluidity and self-healing capabilities typical of the liquid crystals, even in the rather stiff

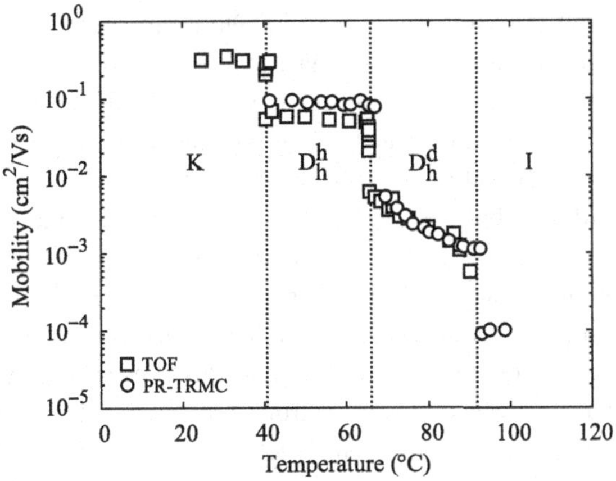

Figure 1.44 Charge mobility data of positive carriers in HHTT measured by time of flight (TOF) and pulse-radiolysis time-resolved microwave conductivity (PR-TRMC) [Warman and Van de Craats, 2003]. The dotted lines represent phase transitions.

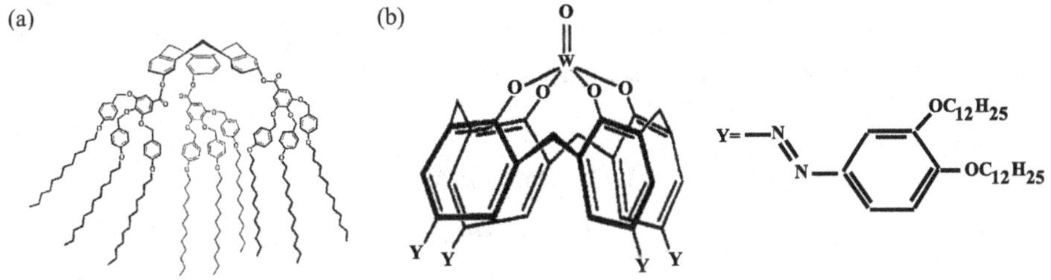

Figure 1.45 Two pyramidic molecules: (a) a fully organic one [Malthête et al., 1989] and (b) an organometallic one [Xu and Swager, 1993].

columnar phase. Another interesting feature is that the conductivity is strongly anisotropic, mainly directed along the columns, where the distance between the electron rich cores is of around 3.5–3.8 Å, the typical 'thickness' of a π−electron system, while the separation between the columns is, for HHTT, of around 20 Å. Even if this intercolumnar distance depends of course on the lengths of the chains and on the bulkiness of the substituents, it is so large anyway to reduce intercolumnar charge hopping, leading to strong conductivity anisotropy.

Other potential applications of columnar liquid crystals can be found in organic light-emitting diodes (OLED) [Benning et al., 2000] or in organic solar cells [Idé et al., 2014].

1.10.2 Bowlic Liquid Crystals

An interesting family of liquid crystals has been realized from bowlic or pyramidic shaped molecules, as, for example that in Fig. 1.45a, which is formed by a cone-shaped tribenzo-cyclononene (TBCN) core with three flat triangular substituents 3,4,5-tris-(p-n-dodecyloxy benzyloxy) benzoyloxyl (DOBOB) groups [Malthête et al., 1989]. This particular compound forms a single hexagonal columnar phase stable up to 145–150°C. Organometallic type molecules of this category have also been synthesized using a metal like tungsten or vanadium in a suitable hybridization state to orient its ligands in some appropriate directions [Xu and Swager, 1993; Atwood et al., 2001], like in Fig. 1.45b. Other bowlic mesogens have been put forward, e.g. based on tribenzocyclononenes [Zimmermann et al., 1985], cyclovtryveratrilene [Cometti et al., 1992], calix[4]arenes [Demus, 1989; Swager and Xu, 1994], and C_{60} fullerene molecules [Sawamura et al., 2002]. In the latter case, the C_{60} apex of each of these molecules fits perfectly into the cavity of a neighbouring one. These molecules can pile up in columns, similar to what discotics do, but with the important difference that the constituent particles are intrinsically asymmetric and that the stacking might preserve and enhance this asymmetry, yielding an overall polarity of the column. For a molecule with an axial dipole, like that of the mesogen in Fig. 1.45b, this offers in principle the possibility of creating simple ferroelectric fluid systems. Indeed, it can be shown, if the columns are arranged in a hexagonal structure [Guillon, 2000], that the dipole of overall polar individual columns cannot be balanced, for symmetry reasons, by that of an equal number of antiparallel columns with the result of obtaining a much sought-after ferroelectric columnar phase. Unfortunately, this objective, although not forbidden by some fundamental law, has not been realized, at least as yet.

1.11 Lyotropics

1.11.1 Micelles

We consider now phases, generically called *lyotropics*, formed by amphiphilic molecules A-B consisting of two parts A and B with different affinity to a certain host solvent C (see Fig. 1.46). A classic example is that of elongated molecules formed by lyophilic and lipophilic parts dissolved in a polar solvent (typically water) [Linden and Fox, 1975; Wennerström and Lindman, 1979; Tiddy, 1980; Charvolin and Hendrikx, 1985; Seddon, 1990; Figuereido Neto and Salinas, 2005; Fong et al., 2012]. When the solvent C is water, common amphiphilic molecules (also called *surfactants*) can be classified as:

- *Anionic* – with a negatively charged hydrophilic group carboxyl ($RCOO^- \, M^+$), sulphonate ($RSO_3^- \, M^+$), or sulphate ($RSO_4^- \, M^+$).
- *Cationic* – with the hydrophile bearing a positive charge, as for example, the quaternary ammonium halides ($R_4N^+ \, X^-$).
- *Non-ionic* – where the hydrophilic moiety has no charge but derives its water solubility from highly polar groups such as polyoxyethylene (POE) ($-OCH_2CH_2O-$), sugars or similar groups.
- *Amphoteric* (and *zwitterionic*) – in which the molecule is neutral overall, but has both a positive and a negative charge on its structure as, for instance, for the principal chain of the sulphobetaines (Fig. 1.46f), instead of needing a separate counterion.

A simple example of a system of the first or second kind is a soap-water mixture. The amphiphilic molecules tend to organize themselves so as to segregate the hydrophobic

(a)

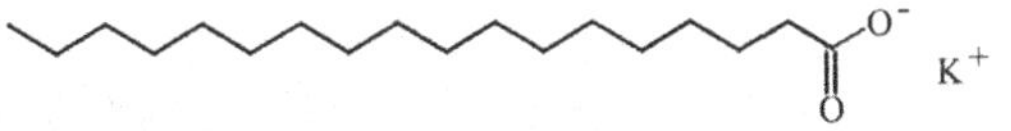

(b)

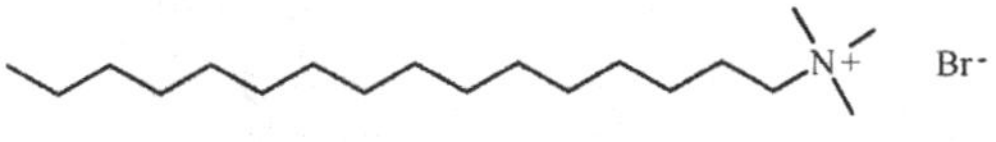

(c)

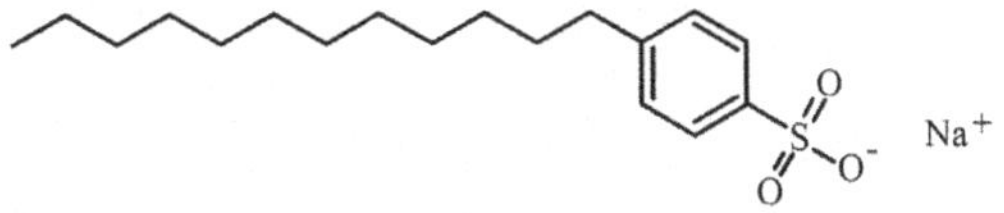

(d)

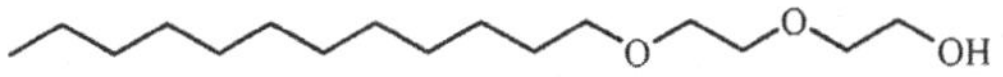

(e)

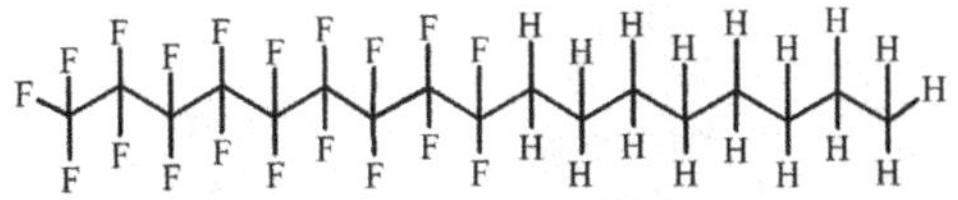

(f)

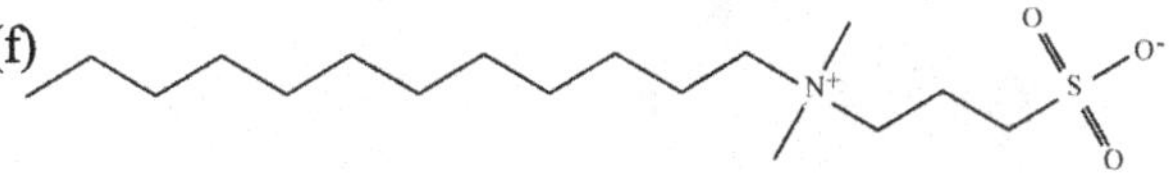

Figure 1.46 Some common water-soluble amphiphilic molecules: (a) potassium stearate (aliphatic, anionic); (b) cetyltrimethylammonium bromide (CTAB) (aliphatic, cationic); (c) sodium *p*-dodecyl benzene sulphonate (partly aromatic, anionic); (d) di (ethylene glycol) dodecyl ether (non-ionic); (e) a fluorinated-protonated amphiphile; (f) a zwitterionic molecule: lauryl sulphobetaine.

chains from the water. Different molecular arrangements occur, depending on the chemical structure of the amphiphile and on its concentration, as we see in Fig. 2.28 for potassium stearate, a typical example of these soaps [Luzzati et al., 1957].

When a certain critical micellar concentration (*CMC*) is reached, globular aggregates exposing the hydrophilic part of the molecule to the solvent, while having chains with a favourable reciprocal interaction on the inside (*micelles*), can be formed. The *CMC* of amphiphilic molecules with a net charge is normally very low (e.g. around 8 mM for SDS, sodium dodecyl sulphate [Wennerström and Lindman, 1979]), but strongly influenced by the ionic strength of the medium that acts, through a charge screening, towards dampening of the electrostatic repulsion between the ions. For instance, the *CMC* of the detergent sodium dodecyl sulphate is reduced tenfold when the NaCl concentration is raised from 0 M to 0.5 M. The shape of the micelles themselves can vary, as we sketch in Fig. 1.47, in order to optimize the local curvature, e.g. when we have a mixture of different amphiphiles where the size of the polar heads and of the chain lengths change [Israelachvili, 1992].

Spherical micelles. The simplest system is that of a suspension of isolated spherical micelles (Q phase). The polar heads are outside if the solvent is a polar one. The chains are directed toward the centre, although an arrangement that avoids excessive steric hindrance has to be found showing that the model in Fig. 1.47 can be too simple. Increasing the concentration of amphiphilic molecules in water means micelles come in contact and various structures can be formed. The diameter of these micelles is close to two molecular

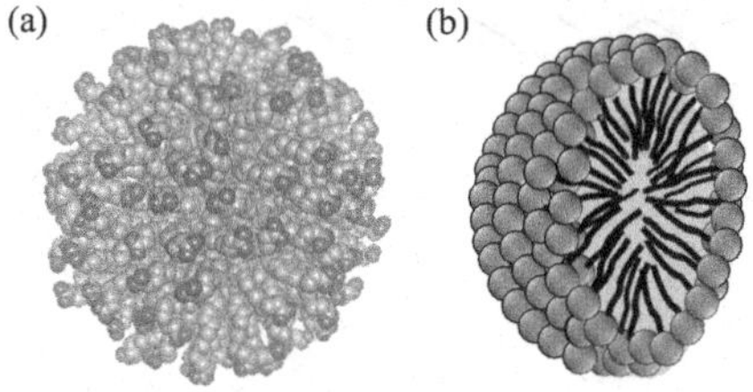

Figure 1.47 An atomistic representation of (a) a spherical micelle and (b) an idealized cartoon rendering of its section.

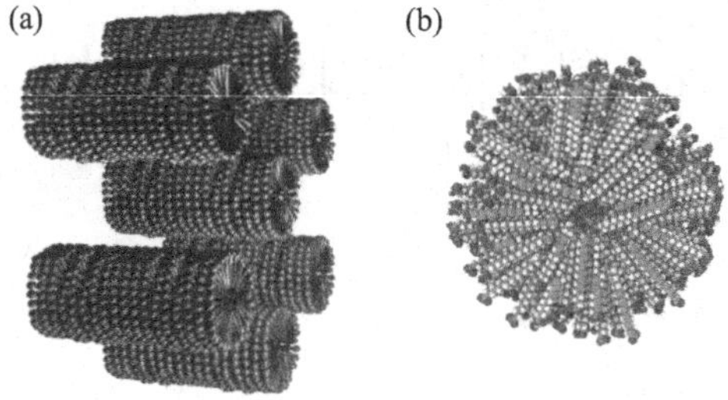

Figure 1.48 (a) Sketch of the hexagonal phase of cylindrical micelles and (b) of a section orthogonal to the micelle axis.

lengths (thus a few nanometres) and the connected high surface curvature is favoured by a cone-like molecular geometry typical of a polar head larger than the chain cross section.

Elongated micelles. Elongated micelles, e.g. cylindrical ones, can organize themselves so as to form, in turn, ordered structures. In Fig. 1.48 we see a sketch of the hexagonal, H, (also called *middle*) phase. All the phases we have seen have polar heads in contact with the solvent and inside chains, even though *reverse micelles* with chains pointing outside towards the solvent could be formed by suitably changing from water to a hydrophobic solvent [Tiddy, 1980]. More complex molecular architectures can be obtained increasing the concentration of amphiphiles so that micelles come into contact with each other [Luzzati et al., 1968].

Micellar architectures. Cubic arrays of spherical micelles can be obtained, e.g. with palmitoyl lysophosphatidyl choline (PLPC) [Landau et al., 1997]. In Fig. 1.49, we see instead an example of a so-called I (intermediate) or *cubic phase*, where the target of separating the immiscible water and lipid regions is met with a *fusion* process of the individual aggregates forming regular arrays of bilayer micellar units with a rather fascinating bicontinuous organization with water outside. In these systems a lipophilic solute molecule, e.g. a dye, should be able to migrate through all the sample without 'getting wet', so to speak. Phases of this type, e.g. 1-mono palmitoleoyl-*rac*-glycerol (MOG) have proved to be particularly useful in assisting the crystallization of proteins [Landau and Rosenbusch, 1996; Landau et al., 1997]. Note that, although the local surrounding of one of the mesogenic molecules is anisotropic, the overall macroscopic cubic symmetry gives optical isotropy to these phases

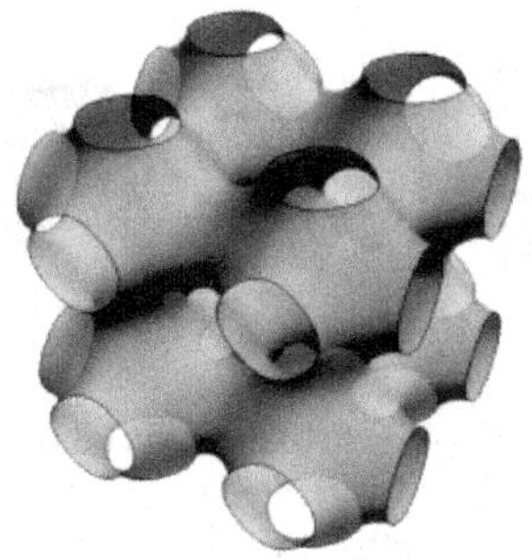

Figure 1.49 A bicontinuous cubic lyotropic phase of type I obtained by replicating the double layer micellar unit with water outside the structure as well as in the network channel inside.

that have the physical aspect of highly viscous, transparent materials. Other types of cubic phases, typically obtained from lipid bilayers, are based on a tetrahedral repeating unit and are denoted G (the *gyroid* type) and D (the diamond type) [Kulkarni, 2012].

Templating. In terms of applications, the ordered micellar arrangements are particularly important as templates in sol-gel synthesis of materials with a well-ordered and regular nanoporous structure [Raimondi and Seddon, 1999; Soler-Illia et al., 2002]. In practice, an organosilicate like tetraethyl orthosilicate (TEOS) that can hydrolyze and yield silica in the presence of water is added to the lyotropic system with the result of embedding the micelles in a silica matrix. A calcination is then performed, burning out the amphiphile forming the micelles, thus leaving empty nanopores. The extraordinary thing is that when performed with due care this rather invasive sequence of processes preserves the former micellar structure, leaving a solid nanoporous material with the original architecture [Attard et al., 1995]. Note also that the diameter of the spherical or cylindrical nanopores can be varied changing the chain length of the starting amphiphiles, providing a way of tuning the size of the cavities of these artificial zeolites and their effect on the molecules trapped inside, like significantly altering the melting point of water confined inside (see, e.g., [Findenegg et al., 2008]). The soft-templating methodology has been much extended and mesoporous (2–50 nm pore size) materials of many compositions beyond silicates, e.g. polymers, carbons, metals, metal oxides and of different dimensionality, have been prepared [Zhao et al., 2019]. Templating and fabrication applications are not the only available for lyotropics LCs and many others are used in drug delivery, as discussed, for example, by Chen et al. [2014].

1.11.2 Lamellar Phases, Bilayers and Membranes

Bilayer structures, like those shown in Fig. 1.50, with the chains separated from the water by the polar heads often form at higher amphiphile concentrations. *Lamellar phases* are similar to smectics, in having a one-dimensional periodic structure, but numerous organizations exist, corresponding to different arrangements of the chains. This variety has even caused some confusion in their nomenclature. Here we shall try to base our notation on the system

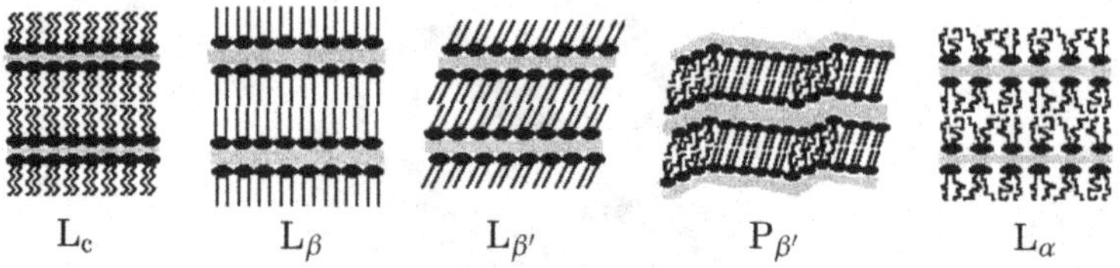

Figure 1.50 Cross-sectional sketch of some of the main bilayer phases. From the left: subgel (L_c), gel with untilted chains (L_β), gel with tilted chains ($L_{\beta'}$), rippled gel ($P_{\beta'}$) and liquid-crystalline fluid (L_α) [Koynova and Caffrey, 1998] .

initially proposed by Luzzati [Tiddy, 1980] where a code name is built using a capital letter for the phase type: L for (planar) lamellar, H for hexagonal and Q for spherical micelles, then roman numerals I, II to denote normal and reverse structures, if this needs to be specified [Seddon and Templer, 1993]. A subscript α is used to indicate, where appropriate, a fluid, liquid-like, while β is for a gel, solid-like, state of the molecular chains in the layers, and c for a solid-crystalline chain organization. In addition, the label γ indicates a phase with a sequence of α and β layers.

The tilt of the chains is indicated by adding a prime to the α, β or γ subscripts. For instance, the lamellar, hexagonal and reversed hexagonal illustrated earlier on could be called L_α, H_I and H_{II}. The L_α phase, also called liquid crystal, or *neat phase* in soap-water systems, has a fluid like organization of the chains above a certain temperature (*Kraft point*) but below it forms gels with networks of lamellar domain. *Coagels*, consisting of hydrated crystalline phases [Lo Nostro et al., 2003], also form. When the lamellar phase is rippled rather than just flat it is called P instead of L. Lamellar phases have a flat interface with water having a negligible curvature and they are more easily obtained from amphiphiles with nearly cylindrical overall shape. This is typically obtained with amphiphilic molecules with two chains attached to the polar head, like in the phospholipids that are the main constituents of biological membranes. By contrast, we may remark that the various spherical or cylindrical micelles that we have discussed until now are, typically, formed by a relatively large polar head and a single chain, that confer an overall cone-like, rather than cylinder-like shape to the amphiphiles [Israelachvili, 2011]. Upon shaking up a lamellar water suspension, the bilayers, particularly the phospholipid made ones, tend to close up upon themselves forming spherical vesicles like those shown in Fig. 1.51. These can be *monolamellar*, as in Fig. 1.51a, with diameter of 25–30 nm if the suspension is shaken at length with ultrasound (sonicated). Otherwise *multilamellar smectic* structures, with various bilayer shells, disposed onion-skin-like called *liposomes* [Deamer, 2010], are obtained. It is worth stressing that although liposomes can appear similar, at first sight, to spherical micelles, they are very different, as they have a water pool inside and their size is more than two orders of magnitude larger. Multilamellar liposomes typically range in diameter from 0.1 to 5 μm [Mabrey-Gaud, 1981] and are characterized by temperature dependent reversible order-disorder transitions occurring in a narrow temperature range. Multilamellar vesicles represent a more stable state with respect to the single layer ones and freshly prepared unilamellar vesicles tend

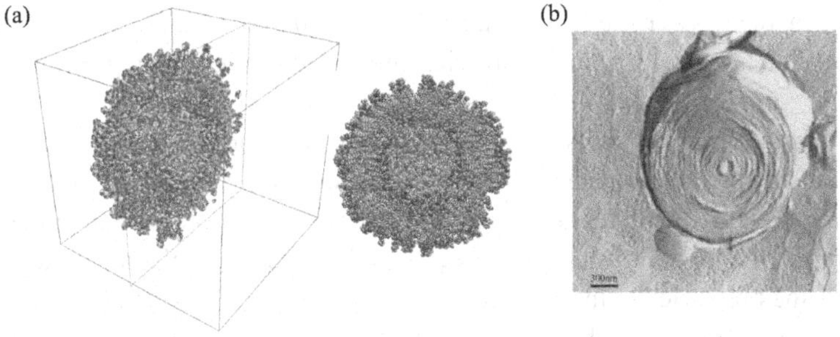

Figure 1.51 (a) Perspective and front view of an atomistic sketch of monolamellar vesicle (section). (b) A freeze fracture electron microscopy of 1,2-dioleoyl-sn-glycero-3-phosphocholine (DOPC) multilamellar liposomes [Francescangeli et al., 2003].

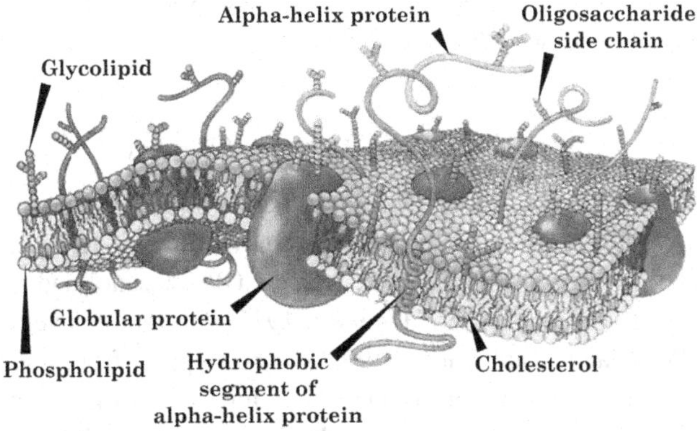

Figure 1.52 The fluid mosaic model of a cell membrane showing the liquid-crystalline organization of the lipids and the transmembrane proteins. The bush-like structures shown on the top are glyco-proteins [NIST Center for Neutron Research, 2016].

to fuse and give rise to multilamellar liposomes [Suurkuusk et al., 1976]. We also note that, although liposomes are macroscopically isotropic, they have local orientational order. Thus, they can show the same properties in every direction when a methodology probing at a macroscopic level, e.g. a rheological method to determine viscosity, is used and can appear ordered when a spectroscopic technique probing the structure at a local, molecular level is employed instead [Zannoni, 1981]. The bilayer structure common in phospholipid systems [Mabrey-Gaud, 1981] is particularly important because it models the organization found in the real membrane surrounding a cell. It is indeed similar to the 'enclosing envelope' of a eukaryote animal cell (see Fig. 1.52), i.e. a cell of an organism where a membrane encloses nucleus, mitochondria and other membrane bound organelles [Brown and Wolken, 1979; Lehninger et al., 2005]. The lipid cell membrane, called the *plasma membrane*, is of great biomedical importance because it is necessarily involved in the transport of chemicals and ions to and from the cell. The current, textbook, model for these lipid biomembranes is the *fluid mosaic* model [Singer and Nicolson, 1972] sketched in Fig. 1.52. In essence, it is based on a lipid double layer, with globular proteins embedded inside or going across the bilayer itself.

The amount of proteins, lipids and other components varies from system to system. In red blood cells membrane proteins constitute 60%–70% of the total dry weight, with lipids making up 20%−40%. Note that the microscopic organization of the membrane bilayer is essentially that of a liquid crystal. The influence of the state, e.g. the order and fluidity of the lipids on the properties of the membrane is a topic of great current interest. To complicate the picture, in a real membrane the lipids themselves are mixtures of many different kinds, so that we can have other phenomena like separation (segregation) of the various components in certain thermodynamic conditions. As already mentioned, phospholipids are typical components of the bilayer and artificial membranes can be prepared from them. In particular, model systems like dimyristoil-phosphatidyl choline (DMPC) and dipalmitoyl-phosphatidyl choline (DPPC) have been very well studied experimentally (see Section 2.14) and with computer simulations (Section 12.7.2).

1.11.3 Discoidal Micelles and Bicelles

Nematic lyotropics can be formed by discoidal micellar units. A nematic lyotropic system of this type, with a structure similar to that shown in Fig. 1.39a can be obtained, for example, with 30.35% potassium laurate (KL), 7.04% decanol and 62.61% water at a temperature of 17°C [Charvolin and Hendrikx, 1985]. Discotic bilayer micelles, sometimes called *bicelles*, [Sanders and Landis, 1995] can be prepared from suitable mixtures of phospholipids of different length. For instance, a mixture of dihexanol phosphatidylcoline (DHPC) and dimyristoyl phosphatidylcoline (DMPC) in aqueous solution at room temperature or just above goes from gel to a liquid crystal formed of bicelles of thickness ≈4 nm [Sanders and Schwonek, 1992] that has proved to be particularly useful as a solvent of weak and tunable anisotropy for Nuclear Magnetic resonance (NMR) studies of biomolecules, where they provide an environment similat to that of real biomembranes [Tjandra and Bax, 1997].

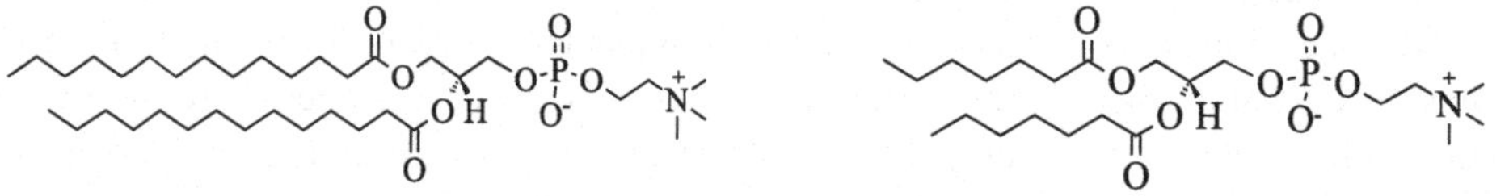

DMPC DHPC

It seems that in these systems a disc-like phospholipid bilayer formed by the longer chains (DMPC) is surrounded by a rim of the short chain lipids (DHPC). In a magnetic field these bicelles align with the disc axis, i.e. the bilayer normal, perpendicular to the field, although the addition of small amounts of lanthanides ions (e.g. Eu^{3+}) changes the alignment of the axis to being parallel to the field [Vold and Prosser, 1996]. The radius of the bicelles increases from roughly 1 nm to 100 nm as the ratio of DMPC to DHPC increases from 0.01 to 100.

While we have discussed up to now thermotropic systems that change molecular organization by varying temperature and lyotropics, where transformations are driven by modifications in composition of a solution, a number of systems, particularly natural ones, are amphitropic, in the sense that they show both types of behaviour. Phospholipids are an example [Chapman, 1975], phthalocyanines [Eichhorn et al., 1998] and many sugars are others [Blunk et al., 2009; Hashim et al., 2012].

1.12 Chromonics

Chromonics [Lydon, 1998, 2004] are a kind of lyotropic system but, differently from those seen until now, they are formed dissolving in water amphiphilic mesogens (ionic or non-ionic) that typically have an aromatic structure, with a fairly rigid and planar disc-like or plank-like core (Fig. 1.53), instead of the aliphatic, flexible, and elongated shape of common lyotropics. These mesogens have their polar, hydrophilic groups disposed around the periphery of the molecules rather than at one end. The molecules form one-dimensional stacks of various length in solution, without a critical concentration, rather than zero-dimensional (closed on themselves) micelles. The face-to-face aggregates can form a disordered nematic phase (Fig. 1.54a), but also more ordered arrays of columns, in particular, various hexagonal (H) phases (see, e.g., Fig. 1.54b). In the hexagonal phases the columns lie in a somewhat hexagonal array, differently from the more dilute nematic N phase where the columns are separated by a larger water amount. Some of these materials are in use as food dyes, and their suggestive common names (Fig. 1.53, caption) are inherited from that area. Chromonics are also used for optical devices, where a common orientation of the liquid-crystalline columns

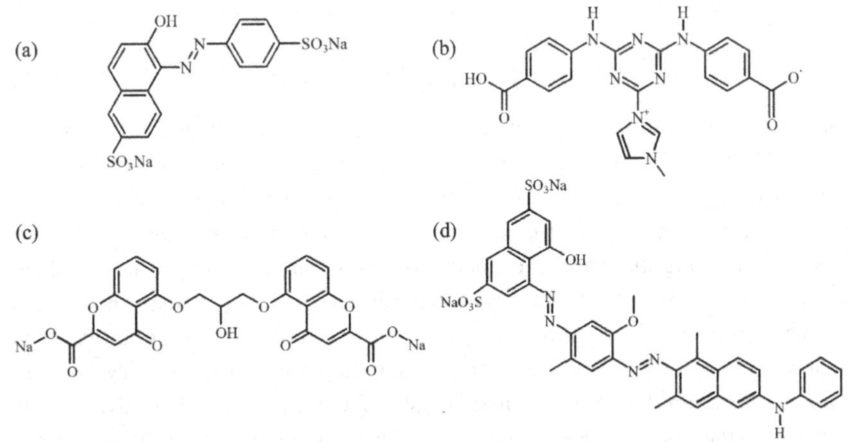

Figure 1.53 Some common chromonic mesogens: (a) 6-hydroxy-5-[(4-sulphophenyl) azo]-2-naphthalenesulphonic acid, also called Sunset Yellow (SSY) or Edicol [Edwards et al., 2008]; (b) 3-[6-(3-carboxyanilino)-4-(3-methyl-1H-imidazol-s-ium-1-yl)-1,3,5-triazin-1-ium-2-yl] aminobenzoate or just *n*-methyl imidazol (NMI) [Mohanty et al., 2006]; (c) disodium cromoglycate (DSCG), also called Cromolyn or Intal and (d) Direct Blue.

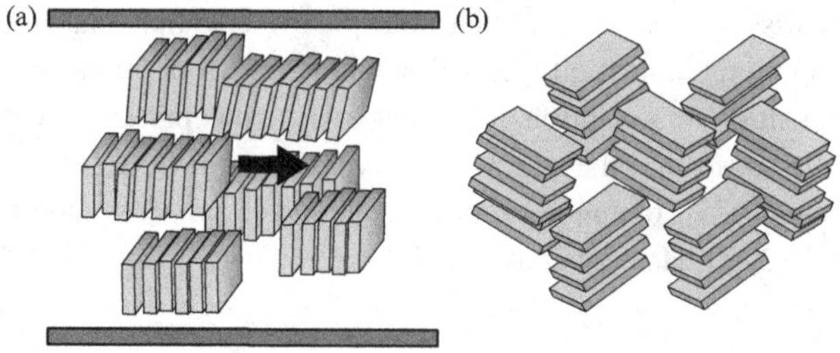

Figure 1.54 A sketch of a chromonic nematic (N) with (a) its columnar stacks aligned by flow in the direction of the arrow and (b) of a hexagonal (H) chromonic phase.

is achieved by a mechanical shearing force. Successive evaporation of the solvent can result in anisotropic solid films that can, e.g. linearly polarize incident light over a rather wide interval of wavelengths. If the columns are aligned parallel to the shearing direction (see Fig. 1.54a), the direction of the absorption transition moment (see Section 3.4.2 for the aromatic cores would typically be in the plane of the molecule, so that the component of an unpolarized light incident on the film with polarization along the stacks will be transmitted and the component with polarization perpendicular to it will be well absorbed.

It is interesting that the elastic constants of chromonics are rather different in relative magnitude from those of nematics. While in simple nematics K_{11}, K_{22} and K_{33} have values comparable to each other (see Table 1.3 and Fig. 1.26), in chromonics the twist constant K_{22} is nearly an order of magnitude smaller than the other two. For instance, for Sunset Yellow at a concentration of 29%, Zhou et al. [2012] found $K_{11} = 4.3 \pm 0.4$ pN, $K_{22} = 0.70 \pm 0.07$ pN and $K_{33} = 6.10 \pm 0.06$ pN.

1.13 Ionic Liquid Crystals

Thermotropic ionic liquid crystals [discussed in detail in Binnemans, 2005; Axenov and Laschat, 2011; Salikolimi et al., 2020] are a class of liquid-crystalline compounds typically obtained from anisotropic organic salts. Their defining property stems from being constituted of anions and cations, making them quite different from ordinary thermotropics, and it is not surprising that various of their properties, particularly their ion conductivity, differ significantly from those of conventional liquid crystals. Like ordinary, isotropic, ionic liquids, these compounds have a very low vapor pressure and thus low volatility. The ionic interactions tend to stabilize smectic-like lamellar mesophases. Most of the systems studied so far are imidazolium-derived ionic liquid crystals [Axenov and Laschat, 2011], as we see in Fig. 1.55. In the first type (Fig. 1.55a), the imidazolium group acts as a mesogenic core, which is substituted by one or multiple long aliphatic tails. In most of such cases, alkyl substituted imidazolium salts do not form nematics, but only smectic mesophases where the molecules are arranged in layers. This is due to the electrostatic interactions in the head group region and to the weaker van der Waals forces in the hydrophobic tails (see Chapter 5). The imidazolium group could also be connected via a flexible spacer to a conventional liquid

(a)

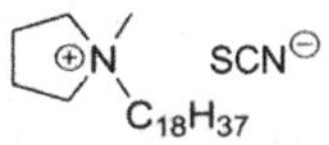

$$\mathrm{K} \overset{86^\circ\mathrm{C}}{\longleftrightarrow} \mathrm{S_A} \overset{164^\circ\mathrm{C}}{\longleftrightarrow} \mathrm{I}$$

(b)

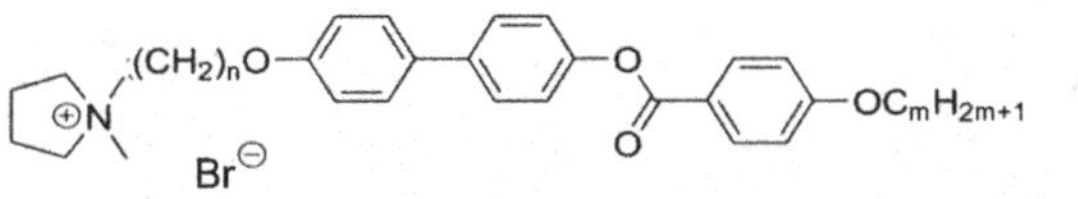

$(m=4, n=10)$

$$\mathrm{K} \overset{156^\circ\mathrm{C}}{\longleftrightarrow} \mathrm{S_A} \overset{215^\circ\mathrm{C}}{\longleftrightarrow} \mathrm{I}.$$

Figure 1.55 Some imidazolium-based ionic liquid crystal [Axenov and Laschat, 2011].

crystal mesogen on the tail ends (Fig. 1.55b). In these types of imidazolium-based materials, the liquid-crystalline properties originate from their strong amphiphilic character. The ionic interactions of the imidazolium groups stabilize both S_A and S_E phases. However, some imidazolium units with two pendant cyano-biphenyl groups show a monotropic nematic [Goossens et al., 2008]. Stable columnar phases have been obtained for the compounds, where imidazolium groups have been attached on the tail ends of discotic molecules.

A series of salts with two mesogenic cyano-biphenyl groups attached to a central pyridinium cationic unit via a flexible alkyl spacer was instead shown to exhibit only a nematic phase with a thermal range influenced by the various counterions employed and spacer length [Pana et al., 2016].

1.14 Colloidal Suspensions

An interesting and increasingly important [Mitov, 2012; Muševič, 2017] set of liquid crystals is obtained from suspensions of anisometric colloidal particles, where one or more of their dimensions is nanometric to micrometric in size. For particles up to this range of sizes the effects of gravity can, to a good approximation, be neglected and various isotropic and anisotropic phases [Manoharan, 2015] can be observed as a result of the balance between attractive forces leading to some sort of aggregation and repulsive forces tending to stabilize the suspension. These forces will be examined in Chapter 5, while for now we just wish to mention that the particles in liquid-crystalline suspensions can be of inorganic, mineral, polymeric or even biological origin. Amphiphilic particles (anisotropic Janus particles) can also be studied [Conradi et al., 2009]. For all these suspensions a very direct evidence of the formation of a liquid crystal is the observation of light transmission through crossed polarizers, due to the onset of anisotropy (see Fig. 1.56). Let us briefly examine the main types.

Mineral colloids. Classic examples are suspensions of particles of mineral origin [Davidson and Gabriel, 2005; Lekkerkerker and Vroege, 2013], e.g. rod-like bohemite (AlOOH) [Buining et al., 1994] or platelets of gibbsite (γAl(OH)$_3$) in toluene [van der Kooij et al., 2000].

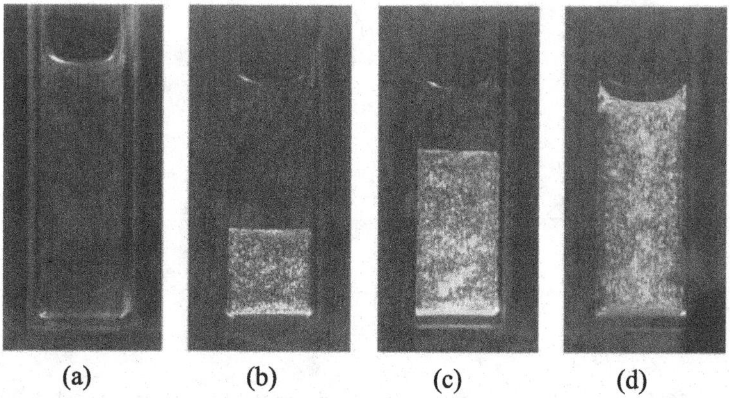

Figure 1.56 The formation of an anisotropic nematic in a suspension of platelets as their volume fraction increases from (a) to (d) monitored by the transmission through crossed polarizers due to the onset of birefringence [van der Kooij and Lekkerkerker, 1998].

In Fig. 1.56 we see such an example for a suspension of platelet shaped particles. Note that such a discotic nematic phase is obtained easily in colloidal suspensions, while it is extremely difficult to find it in thermotropic materials, where normally the transition is from isotropic to columnar upon cooling. It is also interesting that some of these suspensions convincingly appear to be biaxial nematics [Vroege, 2013], once again a phase very difficult to find, if at all, in thermotropic materials [Luckhurst and Sluckin, 2015].

Inorganic nanorods. Other liquid crystal assemblies (nematics and smectics) can be formed from suspensions of inorganic nanocrystals. In Fig. 1.57 we see an example of cadmium selenide nanorods from Alivisatos group [Li et al., 2002]. This is particularly interesting for semiconductor nanocrystals or *quantum dots* (QDs) in view of their optical absorption and fluorescence properties [Reed, 1993]. Indeed, when at least one of their sizes is in the range of a few nanometres, the quantum confinement is at the origin of fluorescence light emission at well-defined wavelengths depending on the particle size. Thus, a spherical or a rod-like QD can generate, in particular, a narrow red, green or blue emission band when illuminated with suitably UV light just choosing their diameter (for a sphere) or cross section (for a rod) to be of the appropriate size.

Note that the suspending host fluid does not have to be isotropic, but could be a liquid crystal. Indeed, novel hybrid systems have been obtained by Mundoor et al. [2018] using micron long, charged, inorganic colloidal nanorods of bare NaYF4:Yb/Er, with length-to-diameter ratios of 40–110. These nanorods tend to align perpendicularly to the 5CB director and form uniaxial and biaxial nematic phases of the composite mixture. They present very interesting optical properties due to the upconverting properties of the lanthanide-doped nanorods that can absorb multiple photons in the infrared and emit in the visible [Wang and Li, 2007].

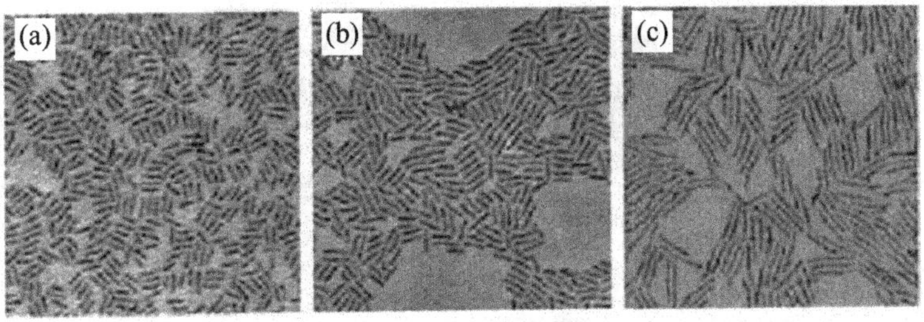

Figure 1.57 A liquid crystal suspension formed by CdSe nanorods of length (a) 11, (b) 20 and (c) 40 nm and width of ≈ 3.2 nm, respectively [Li et al., 2002].

Carbon nanotubes. Nematics can be formed by multiwall carbon nanotubes (CNT) in aqueous suspensions, e.g. CNT with average length ≈675 nm, average width ≈25 nm and concentration ≈4.8 vol% [Song et al., 2003a]. Single-wall nanotubes in strong acid suspensions [Rai et al., 2006] have also been studied.

Viruses. Viruses are of particular interest as model system of giant 'molecules' because, while being of colloidal size, each particle has the same length, diameter, mass and charge distribution, avoiding the polydispersity typical of other systems. This makes comparison with theory and computer simulations easier, eliminating the uncertainties and in general the effects due to an often unknown distribution of sizes and shapes. For instance, virus suspensions can form at a sufficiently high-volume fraction not only nematics, but also layered, smectic systems, that could be difficult to obtain from particles of different length. This was demonstrated in suspensions of tobacco mosaic virus (TMV) [Zasadzinski and Meyer, 1986; Dogic and Fraden, 2006] that, as shown in Fig. 1.58a and b, give nematic phases.

TMV is a plant virus, with rod-like shape [Caspar, 1964], 18 nm in diameter and 300 nm in length. TMV self-assembles from proteins and a single strand of RNA, ≈ 6000 nucleotides and ≈ 2130 copies of a single kind of polypeptide, the coat protein, each with 158 amino acids. The RNA forms a helical core, with a cylinder of protein subunits clustered around it. Another well-studied filamentous virus is *fd*. It has dimensions of about 880 nm in length and 6.6 nm in diameter. There are specialized proteins at the ends of the virus used for infection. The protein shell has a hollow core in the centre into which the DNA is contained and a helical coat, thus suspensions of *fd* virus are chiral nematics. Differently from TMV the virus is semiflexible, with a persistence length of 220–280 nm. *fd* as a virus is a bacteriophage that infects certain strains of *Escherichia coli* bacteria. Note that all the virus particles studied are strictly helical and thus chiral molecules. However, only some viruses form cholesterics (see Fig. 1.58c). Conventionally, when dealing with phases of nematic or smectic formed (as in Fig. 1.58d) the intrinsic chirality is neglected and they are considered effectively as achiral rods [Dogic and Fraden, 2006].

Figure 1.58 (a) Tobacco mosaic virus (TMV) water suspension, viewed between crossed polarizers, showing coexistence between the isotropic phase floating above the birefringent nematic one [Dogic and Fraden, 2006]. (b) Electron micrograph of a TMV nematic suspension [Gelbart and BenShaul, 1996]. (c) Image between crossed polarizers P_1 and P_2 of a *fd* virus cholesteric suspension phase. The pitch p corresponds to twice the repeat distance between the black stripes [Dogic and Fraden, 2000]. (d) Optical micrograph of a *fd* smectic. The *fd* particles lie in the plane of the photo with their long axis normal to the parallel lines (smectic layers). The layer spacing is 0.92 μm. [Dogic and Fraden, 1997, 2006].

Large differences in elastic constants can be expected in liquid crystals originated from suspensions of long rod particles, e.g. virus like tobacco mosaic virus (TMV) [Hurd et al., 1985], nanotubes [Song et al., 2003a] or nanocrystal suspensions [Li et al., 2002].

1.15 Lyotropic Liquid Crystal Polymers

Polymer liquid crystals have become quite important for industrial applications since they can yield materials with extremely interesting mechanical properties. To quantify this, we recall that in a tensile stress-strain test an increasing force per unit area, the *tensile stress* Σ, is applied to a specimen and the resulting changes in length (*strain* λ) are measured. For a small applied force the behaviour of the material is elastic, with the sample recovering its original shape when the stress is released and $\Sigma = E_Y \lambda$, the classical Hooke's law of springs, holds. This initial slope E_Y is called *Young's modulus* or tensile modulus. Upon pulling the specimen, a maximum load is reached before the sample develops a 'neck' and *tensile strength* is defined as the maximum load over original cross-section area. Let us consider, as an example, one of the most commonly used liquid crystal polymers, Kevlar (du Pont©), a para-substituted aromatic polyamid (aramid), whose structure is

Table 1.8. *Values of Young's modulus and tensile strength for some engineering materials [Anderson et al., 1990; Collyer, 1990]*

Material	Young's modulus, E_Y (GPa)	Tensile strength (MPa)
Polyamide 66	3	80
Polyamide 66–30 vol.% glass fibre	8	160
Vectra (thermotropic LCP)	10–40	140–240
Aluminium	71	80
Ultradrawn polyethylene	117	3670
Kevlar (LCP fibre)	139	2700
Mild steel	210	460

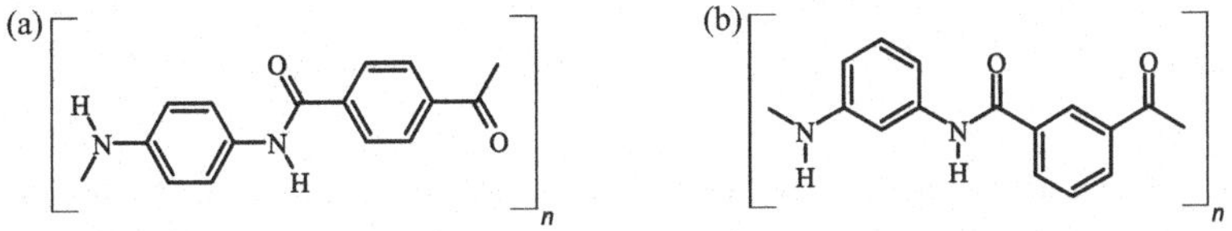

Figure 1.59 Chemical structure of two LCP. (a) Poly-para-phenylene terephthalamide or Kevlar (DuPont©). (b) Poly-meta-phenylene isophthalamide or Nomex (DuPont©).

shown in Fig. 1.59a, prepared polymerizing *p*-phenylene-diamine and terephtaloyl chloride. This main-chain polymer has high viscosity and is practically insoluble in organic solvents and only yields to very strong acids. The lyotropic liquid crystal obtained dissolving it in sulphuric acid can be spun to produce high-performance fibres. Chain alignment favours hydrogen-bonded sheet structures with the molecules forming stacks disposed radially with the chain backbone along the fibre axis. The Young's modulus of Kevlar, $E_Y = 139$ GPa and its tensile strength, 2.7 GPa, outperform those of aluminium [Anderson et al., 1990; Collyer, 1990] and compare favourably with other materials (see Table 1.8). Another similar LCP is Nomex (Fig. 1.59b).

1.16 Liquid Crystal Elastomers

An elastomer (rubber) is, in general, a molecular network formed by polymer chains with a certain, relatively low (e.g. 10%), percentage of cross links between the strands. Rubbers owe their characteristic elastic properties to this peculiar structure. In a liquid crystal elastomer (LCE), mesogenic units are embedded in the network [de Gennes, 1974; Anderson et al., 1990; Benning et al., 2000; Warner and Terentjev, 2003; de Jeu, 2012; Ohm et al., 2012] as shown in Fig. 1.60. As a result of this connected mesogenic structure a significant *thermal actuation* can arise, with changes in the form factor of an LCE sample that take place as the temperature crosses the isotropic-liquid crystal transition of the mesogenic

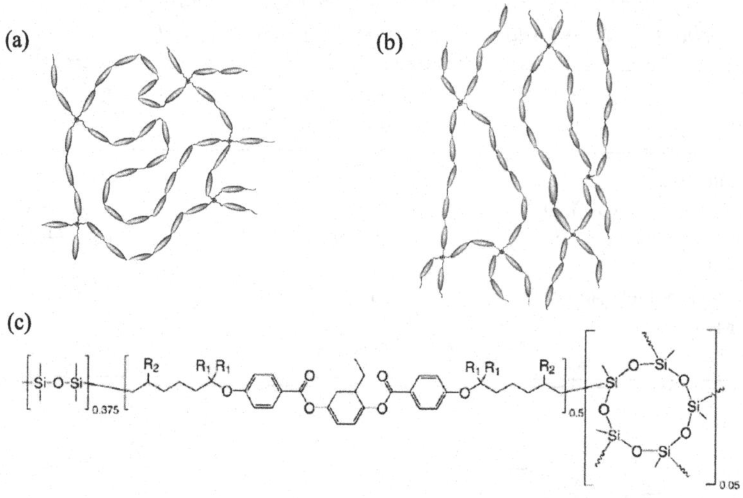

Figure 1.60 Sketch of a main-chain LCE in its (a) isotropic, disordered and (b) anisotropic, aligned nematic phase. (c) The chemical structure of a main-chain LCE [Brommel et al., 2011].

components. In Fig. 1.61a we report the actuation induced by temperature for four LCE materials from Finkemann group and we see that elongation can change by as much as a huge 400%. Various ways have been proposed to achieve the temperature change in a localized region (a pixel) by dispersing graphite or carbon nanotubes in the LCE and illuminating the specific area [Torras et al., 2013]. More generally, LCE unique mechanical properties are derived from the pronounced coupling between a macroscopic strain and the underlying mesogenic orientational order. As the latter can be controlled by a number of external stimuli [Ohm et al., 2012] beyond temperature, e.g. electric field, or ultraviolet light in the case of photo-responsive azobenzene-based elastomers, LCEs have great potential for application in various sensing devices, as well as for actuation. The potential applications of such actuators are fascinating, ranging from artificial muscles [de Gennes et al., 1997], heart valves, haptic displays [Torras et al., 2013] and other biomedical applications, to micro- and nano-electromechanical devices like soft robots, as well as electro-optical systems like adaptive lenses, where their high deformability and low weight make them particularly promising. Among the possible actuation stimuli, the external electric field is particularly appealing, even if quite difficult to implement since a rather strong field is required to induce deformations. Side-chain [Finkelmann and Rehage, 1984] and main-chain [Ortiz et al., 1998; Donnio et al., 2000] LCEs have been prepared. Side-chain polymeric materials are somehow easier to prepare and treat, even if main-chain LCEs show the bigger actuation. Fig. 1.61b shows the unusual and very interesting stress-strain curve of a main-chain LCE at a few different temperatures. An initial stress gives a linear, Hookean behaviour, with a relative steep slope (stiff modulus) but then a flat plateau, corresponding to a large deformation obtained with nearly zero stress (supersoft elasticity), while a further increase in pulling

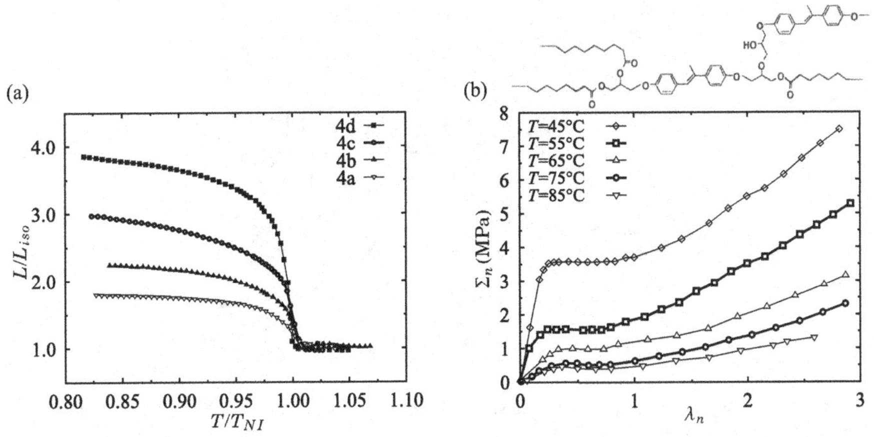

Figure 1.61 (a) The important elongation/contraction of monodomain LCE samples going through the NI transition, as shown by four different co-elastomer materials from Finkelmann group [Wermter and Finkelmann, 2001]. (b) The characteristic non-Hooke stress-strain trend, with a supersoft flat region of a main-chain LCE whose structure is shown above the stress-strain plot. Here Σ_n is the nominal or engineering stress (i.e. load divided by original cross-section area) and $\lambda_n = \Delta L/L$ is the applied nominal strain (i.e. deformation by unit length) [Ortiz et al., 1998].

gives again a normal Hooke behaviour. The phenomenon has been the object of intense theoretical study [Warner and Terentjev, 2003] and, recently, of computer simulations to be described later in Chapter 11 [Skačej and Zannoni, 2014].

1.17 Active Liquid Crystals

It is a common observation that many categories of living bodies show some sort of orientational if not positional order. On very different length scales, colonies of bacteria or fish shoals or flocks of birds or even herds of much larger animals are often spontaneously arranged in a manner far from disordered [Marchetti et al., 2013]. In particular, swarming and swimming represent instances of behaviour where motile bacteria migrate rapidly and collectively on surfaces. In many cases the constituent elements can be assimilated to rigid, self-propelled rods, but highly flexible, snake-like, active bodies, are also observed, like in the case of the strain of *Vibrio alginolyticus* shown in Fig. 1.62 [Böttcher et al., 2016].

It is worth noting that similar ordering effects take place not only in living systems endowed with some 'intelligence', but also in so-called active colloidal systems. A well-known case is that of Janus micron size colloidal particles [Ebbens et al., 2012; Jiang and Granick, 2012; Wang et al., 2015] with half body covered by a catalytic coating that reacts with the host solvent yielding a gas (in the classic example platinum coating and hydrogen peroxide solvent liberating oxygen).

Comparing all these, living or inanimate, systems with the anisotropic structures discussed until now for the various kinds of liquid crystals, it is apparent that some sort of LC

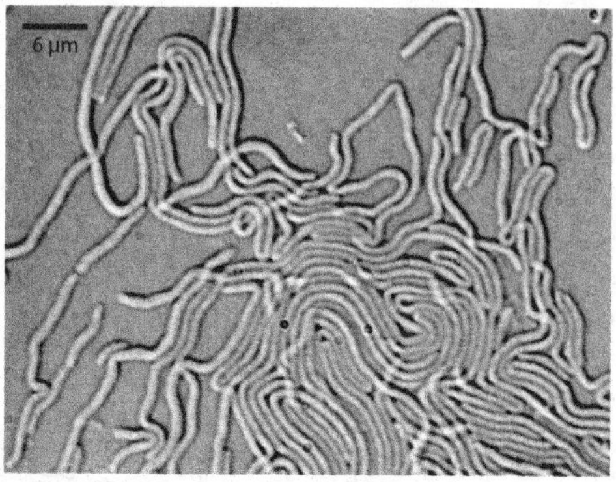

Figure 1.62 Swarming of a colony of snake-like *Vibrio alginolyticus* bacteria on a surface [Böttcher et al., 2016].

orientational order exists. There is, however, an important difference, since in all active systems the common characteristic is that the particles are propelled by some form of internal force and the resulting phases are thus not in equilibrium. The existence of an ordered state is dependent on the internal energy source (food, fuel or other) and can only persist until this is not exhausted. For reasons of space, we shall not treat in detail active systems in this book, even though the techniques discussed in this book will be of help in their microscopic description too.

2
Phase Transitions

Beside their famous technical applications in optics and electronic displays, liquid crystals can certainly be regarded as a paradise of the physics of phase transitions.

P. Barois, *Phase Transitions in Liquid Crystals, 1992*

2.1 Transitions between Phases

In Chapter 1 we have seen that matter does not just exist in three states of aggregation and that we can have, under fairly ordinary conditions, a variety of liquid crystalline phases with different molecular organizations and properties. We have seen that many of these materials have a cascade of phases between the most ordered and the fully disordered, and it is clear that the description and classification of the transitions from one microscopic organization to another is quite relevant for the study of liquid crystals. Here we approach the study of liquid crystal phase transitions from an elementary point of view, and, also to recall the terminology employed, we start with the most familiar thermotropic transitions: the solid-liquid and the liquid-gas. We focus, at least to start with, on one-component systems and on the thermodynamic classification of the different types of transition. We then present the powerful tool of Landau theory [Landau, 1965; de Gennes, 1974], which introduces a generic macroscopic order parameter, η, to be specified in detail case by case for the different phases involved, as a 'transformation coordinate' from one phase to another. We shall also consider models, such as the Ising, Heisenberg and XY ones originally employed for magnetic systems [Stanley, 1971; Ma, 1976; Uzunov, 2010]. This is important because phase transitions present characters of universality [Kadanoff, 1966; Spencer, 2000] that allow the classifying together of systems apparently extremely different and the language employed to describe the variety of liquid crystal transitions, particularly for those involving smectic phases, refers to these prototype models. Most of the phase transitions for the variety of liquid crystals suspensions, lyotropics and the other systems presented in Chapter 1 have not been studied with the same level of detail as thermotropics, and we shall just report the essential experimental facts currently available for each of the systems, particularly with a view to providing data to compare with the results of the computer simulations to be discussed in Chapters 10–12.

2.2 Phase Diagrams for One-Component Systems

Phase transformations are concerned, from the macroscopic point of view, with more or less sudden changes between thermodynamic states belonging to two different microscopic organizations or phases. We may start therefore by recalling that the equilibrium *state* of a system of N particles is defined by a set of thermodynamic variables such as pressure P, temperature T, and volume V. In a multicomponent system, additional variables, such as the concentration of the various components, need to be specified. Additional variables have also to be introduced when external fields, e.g. electric or magnetic, act upon the system. Here we start considering a one-component system not subjected to external fields. Not all of the three variables P, T and V are independent, and the equation linking them is called the *equation of state* for the system. A well-known and rather trivial example is the equation of state for a perfect gas, an idealized system assumed to be formed of non-interacting point size molecules with number density $\rho = N/V$. For such a system, elementary statistical thermodynamics gives the ideal gas law $P = Nk_BT/V$, where k_B is Boltzmann's constant, $k_B = 1.3806 \times 10^{-23}$ JK^{-1}. A diagram showing the relation between P, T and V is called a *phase diagram*. Quite often, constant volume or constant temperature sections of the (P, T, V) surface are considered. These are called respectively the (P, V) (or Andrews) and the (P, T) diagram. In the ideal gas case the (P, V) diagram consists of just a set of hyperbolas, each corresponding to a different temperature. Since the ideal gas model assumes no intermolecular interactions, it shows no condensation and thus no phase transitions. For real systems, even gaseous ones, the equation of state is not known in its analytic form and has to be determined experimentally or put forward in some approximate form coming from theory or computer simulations. In Fig. 2.1 we show, as an example, the experimental (P, V) diagram for benzene (data from [NIST, 2017]). At high temperatures the diagram somehow resembles the ideal gas one, but below a critical temperature T_C the system shows the familiar condensation (see, e.g., [Pryde, 1969; Atkins, 1978]) transition, characterized by a jump in volume from the gas (V_G) to the liquid (V_L), that becomes more pronounced as we lower the temperature. The point where the gas, isotropic liquid and solid phases coexist is called the *triple point*, indicated in this case with a subscript GIS or just T. The point (T_C, P_C) on the critical isotherm where the curvature changes sign is called the *critical point*. As an example, for benzene the critical point is at $T_C = 288.85$°C with $P_C = 48.26$ atm and $\rho_C = 3.9$ mol/l [NIST, 2017], while the triple point is at $T_T = 5.35$°C with $P_T = 0.0477$ atm. For CO_2, the critical point is at $T_C = 30.98$°C with $P_C = 72.81$ atm, while the triple point is at $T_T = -56.60$°C with $P_T = 5.11$ atm [NIST, 2017]. The liquid does not exist below T_T and since P_T is above 1 atm for CO_2, in this case we have a direct transformation from solid to gas (a *sublimation*) by increasing the temperature at standard atmospheric pressure.

At the critical point we have $V_L = V_G = V_C$, the so-called critical volume V_C. Thus, the liquid-gas transition along the critical isotherm is a continuous one and has a qualitatively different behaviour. Fluids with $T > T_C$ and $P > P_C$ are called *supercritical* and have interesting properties: densities much lower than those of normal liquids and viscosities and diffusion characteristics of gases. This makes some of them particularly useful as extraction

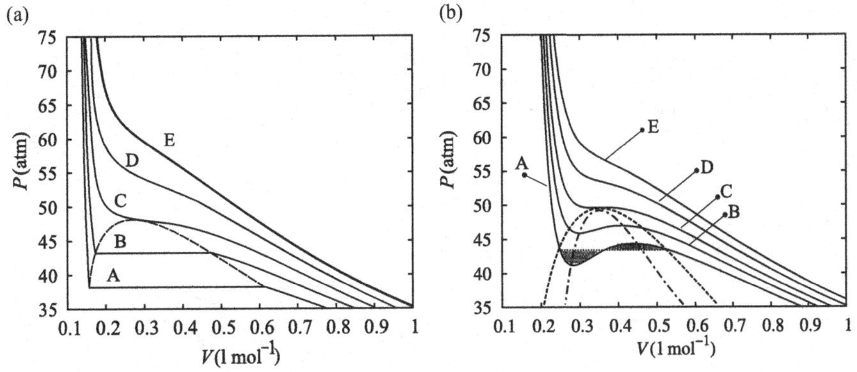

Figure 2.1 (a) The (P, V) diagram for benzene with a few isotherms: (A) 270°C, (B) 280°C, (C) 288.90°C, (D) 300°C, (E) 310°C. (b) The corresponding van der Waals diagram, showing the liquid-gas coexistence (- - -), the spinodal (- - - - -) curves and also, shaded, a Maxwell equal area construction.

solvents. For instance, supercritical CO_2 is used to extract caffeine from coffee beans. After filtration, and a return to normal conditions, the CO_2 evaporates away from the solid caffeine residue (and can be recompressed for another cycle). Fluids where only T or P are higher than the critical values are called *subcritical*: those with $T < T_C$ and $P > P_C$ that can be turned to liquid changing pressure could be called *subcritical liquids*, and those where $T > T_C$ and $P < P_C$ are *subcritical gases*. This classification is often more detailed than necessary if we are interested only in the general properties, and we shall often use the generic term fluid (F) for super- and subcritical fluids. A well-known approximate equation of state for fluids is the van der Waals (vdW) one [Kipnis et al., 1996]:

$$\left(P + a\frac{N^2}{V^2}\right)(V - Nb) = Nk_BT, \tag{2.1}$$

where a, b are positive constants that allow the equation to be parameterized for different gases and are related to the molecules' mutual attraction and to their excluded volume, respectively. The equation is plotted in Fig. 2.1b for a set of isotherms. Note that for a certain region of the van der Waals (P, V) diagram (Figure 2.1b) we have $(\partial P/\partial V) > 0$, a result not physically plausible and that implies thermal instability, since the volume should decrease by increasing the pressure. The region is limited by the metastability limit (*spinodal*) curve, dashed in Fig. 2.1b. The absence of a flat liquid-vapour line similar to experiment is mended by drawing a horizontal tie line, positioned so that the areas comprised above and below the line and the van der Waals curve are equal (the so-called *Maxwell construction* [Stanley, 1971; Swendsen, 2012]). For benzene the vdW parameters are $a = 18.062$ atm l^2 mol^{-2} and $b = 0.1193$ l mol^{-1}, while for CO_2 we have $a = 3.592$ atm l^2 mol^{-2} and $b = 0.04286$ l mol^{-1}. The critical variables T_C, P_C and V_C can be determined from the condition that the critical isotherm has a point of inflection, i.e.

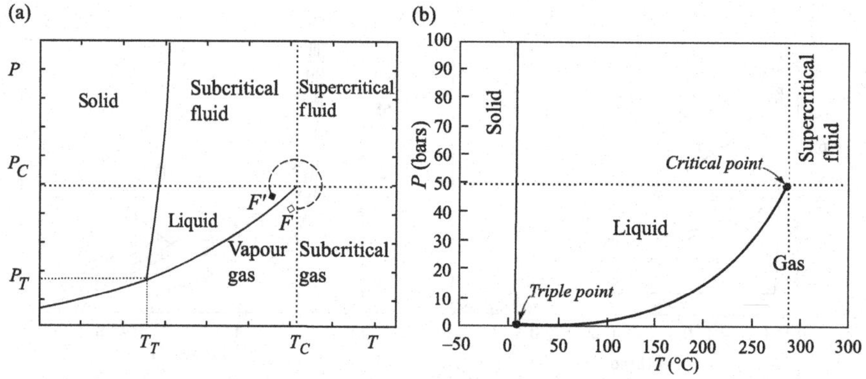

Figure 2.2 (a) Schematic pressure-temperature, (P, T), diagram for an ordinary fluid. The solid lines indicate phase transitions. The dotted lines are not transition lines, but just a guide to locate the triple point at T_T, P_T and the critical point at T_C, P_C. The $F \to F'$ path around the critical point shows a condensation to liquid without crossing a transition line. (b) The (P, T) diagram for benzene.

that the first and second derivatives equal 0: $(\partial P/\partial V)_{T_C} = 0$ and $\left(\partial^2 P/\partial V^2\right)_{T_C} = 0$. This gives $T_C = 8a/(27bk_B)$, $P_C = a/(27b^2)$, $V_C = 3Nb$ and $P_C V_C = 3T_C/8$. The van der Waals equation predicts that all fluids should obey a universal equation of state when using reduced units:

$$\left(P^* + 3/V^{*2}\right)(3V^* - 1) = 8T^*, \tag{2.2}$$

which is a cubic in V^*: $3P^*V^{*3} - P^*V^{*2} - 8T^*V^{*2} + 9V^* - 3 = 0$, where $T^* \equiv T/T_C$, $P^* \equiv P/P_C$ and $V^* \equiv V/V_C$. The (P, T) diagram of a system (see Fig. 2.2) offers another representation of the condensation transition and also allows us to easily include in the same diagram the melting transition. We see from the generic sketch in Fig. 2.2a and more specifically in Fig. 2.2b for benzene, that the various states of aggregation correspond to different regions of the phase diagram separated one from the other by *transition lines*, where the phase changes take place. It may be useful to summarize what could be the overall view of the (P, V, T) state diagram, and a sketch of the whole three-dimensional diagram, of which the (P, V) and (P, T) diagrams represent orthogonal sections, is shown in Fig. 2.3. Note that according to this definition there is no real distinction between a liquid and a gas. In fact, the vapour pressure line ends up in the critical point, making it quite possible to go from liquid to gas and vice versa by going around the critical point, e.g. with the path from F to F' in Fig. 2.2a thus avoiding the crossing of a phase transition line. In some (paradoxical) way, as Fisher [1972] commented: 'This happily disposes of the problem of liquids, they are just dense gases!'. The situation on the melting line is quite different, where there does not seem to be a critical point. In other words, it is not possible to go from a crystalline solid to liquid or vice versa without crossing a transition line and thus undergoing a phase transition. We may wonder if this is just because we have not gone high enough in pressure or if it should be like that in any case. The answer, substantiated by rigorous symmetry arguments, is that

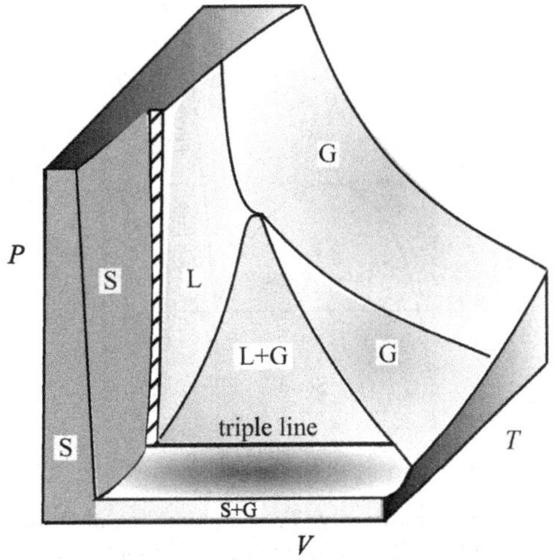

Figure 2.3 A sketch of a 3D (P, V, T) phase diagram. The gas (G), liquid (L) and solid (S) regions as well as the coexistence regions for the mixtures gas-liquid and gas-solid are marked explicitly, while the solid-liquid region is hatched. Note that since we focus on the structure of each phase we do not distinguish between a gas below the critical isotherm (often called vapour) and that above it (adapted from [Glasser, 2002]).

there is no critical point on the melting line [Landau, 1965; Fisher, 1972]. That is, there is no thermodynamic condition beyond which liquids and crystals become identical, because the symmetry of the two phases is different and the organizations of constituent molecules in the two phases are mutually exclusive, at least when the system is a true bulk one and when it is not subject to external symmetry breaking fields. Molecules are either orderly arranged, and we have a crystal, or disordered in which case we have a liquid (cf. Fig. 1.1). Symmetry, however, does not tell us anything on the fluidity or solidity of a system, and indeed liquids can become kinetically arrested at low temperature and turn into disordered solid *glasses*.

In Fig. 2.4 we show a sketch of yet another commonly used representation of the phase diagram, the temperature-density one. This (T, ρ) plot is quite useful in showing also the shape and extent of regions where two phases coexist in equilibrium, with an interface surface separating them. For benzene $T_T = 5.35\,^\circ\mathrm{C}$, $P_T = 0.0477$ atm and $\rho_T = 0.002074$ mol/l for the vapour and 11.4766 mol/l for the liquid. The boiling point at $P = 1$ atm is at $T_B = 80.15^\circ\mathrm{C}$, with $\rho_B = 0.035687$ mol/l for the gas in equilibrium with the liquid (*vapour*) and $\rho_B = 10.4075$ mol/l for the liquid. Note that, although this form of the diagram, exhibiting solid, liquid and gas, is the most common one, it does not have to be universal. In particular, obtaining a liquid just by cooling from the gas phase is related to the existence of attractive forces. For instance, a fluid made of hard repulsive spheres (see Chapter 5) does not have a liquid phase but only fluid and solid phases. This would ideally correspond to moving up the triple point so as to eliminate the liquid state pocket (hatched area in Fig. 2.4) when it goes above the critical point, as might be the case for C_{60} fullerene [Hagen et al., 1993]. All that has been said thus far applies to ordinary isotropic liquids. Let us now see what we expect the (P, T) phase diagram of a liquid crystal to be.

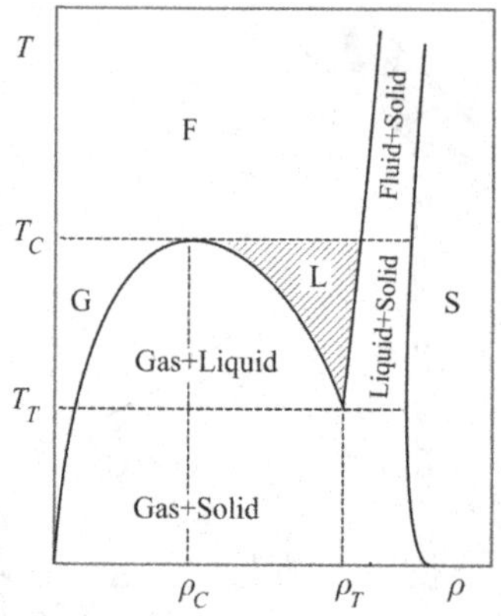

Figure 2.4 Schematic temperature-density (T, ρ) diagram for an ordinary fluid showing gas (G), liquid (L), supercritical fluid (F) and solid (S) phases and the coexistence regions. The subscripts C and T refer to the critical and the triple points, respectively. The liquid exists only in the hatched area between T_T and T_C.

In an idealized situation we could imagine, for a system showing a nematic phase, to have a new region of the phase diagram, located between the liquid and the solid, and for this reason often called a *mesophase*. In a liquid crystal we can have more than one triple point, e.g. solid-nematic-gas and nematic-liquid-gas. However, the situation is much more complicated as we have many different phases and topologies possible. In Fig. 2.5 we show two possible phase diagrams [de Miguel et al., 1996] where, by increasing temperature at a certain pressure, the nematic can only be brought in equilibrium with the supercritical fluid, without passing through an isotropic liquid (Fig.2.5a) or where the nematic can sublimate to vapour directly or go to vapour via the isotropic phase if the temperature is below T_C (Fig.2.5b). The diagrams do not show the crystal solid phase K, and considering its position would offer further possibilities of phase sequences and coexistence.

In Fig. 2.6 we see part of the experimental phase diagram obtained for 5CB and 8CB [Shashidhar and Venkatesh, 1979]. Of course, additional regions will be present if different solid phases or different liquid crystal phases exist. The full phase diagram of a system showing the variety of phases is difficult to obtain. However, some studies are available for fairly large portions of the (P, T) phase diagram. In Fig. 2.7 we see two examples of part of the experimental phase diagram (a) for the smectogen *p*-ethoxybenzoic acid [Chandrasekhar, 1992] and (b) for *n*-heptyloxyphenyl-4′-*n*-decyloxy benzoate (70PDOB) [Kalkura et al., 1982].

In general, we can say that a liquid crystal phase will correspond to an area of the phase diagram comprised between regions belonging to other thermodynamic phases, possibly including other liquid crystal phases, hence the name *mesophases* often used as a synonym for liquid crystals, as we have seen in Chapter 1. There are still many possible shapes of the phase diagram. For instance, for the two experimentally obtained portions of the (P, T) diagrams reported in Fig. 2.7, no nematic-gas direct transition line appears, differently from

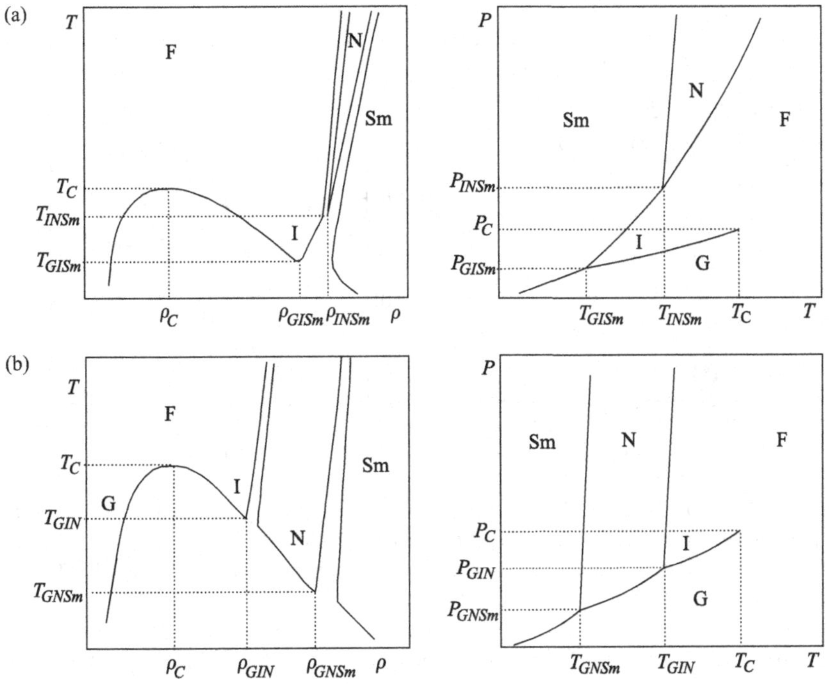

Figure 2.5 Two possible types of temperature-density (T, ρ) and pressure-temperature (P, T) phase diagrams for liquid crystals showing isotropic liquid (I), supercritical fluid (F), nematic (N) and smectic (Sm) phases. (a) The top plates correspond to a system where N is only stable at high pressures and temperatures. (b) The bottom plates refer to a case where nematic-gas phase coexistence occurs for a range of temperatures between the gas-nematic-smectic T_{GNSm} and the gas-liquid-nematic triple point T_{GIN} (adapted from [de Miguel et al., 1996]).

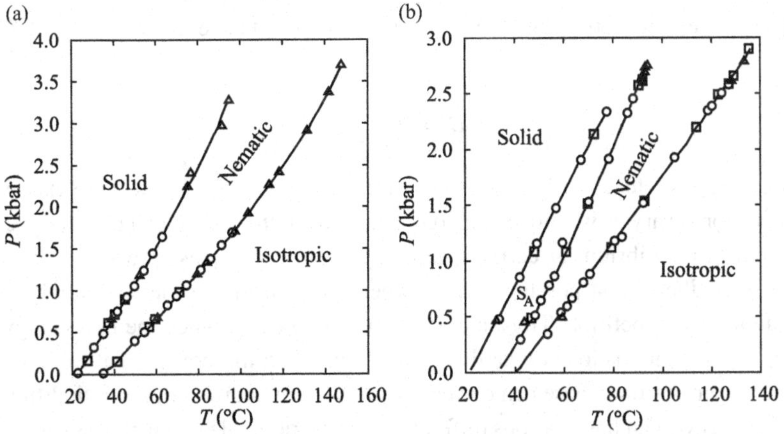

Figure 2.6 A portion of the experimental (P, T) phase diagram for the liquid crystals (a) 5CB and (b) 8CB. The symbols indicate independent sets of measurements. The lines are a guide for the eye [Shashidhar and Venkatesh, 1979].

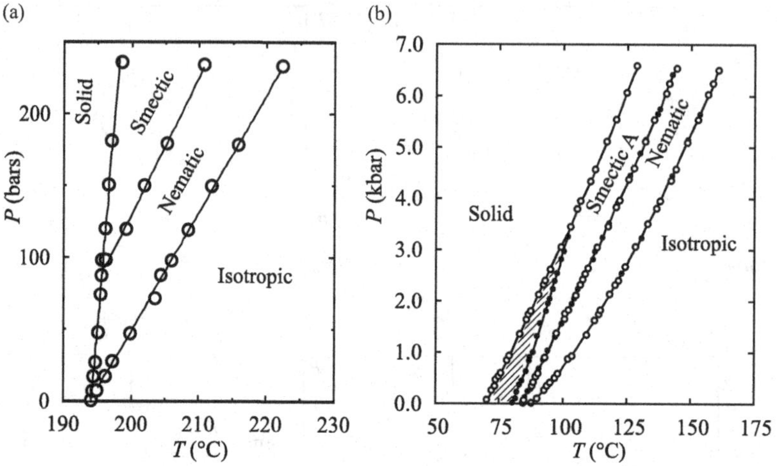

Figure 2.7 (a) (P, T) diagram for p-ethoxybenzoic acid showing the solid-smectic-nematic triple point [Chandrasekhar, 1992] and (b) that for 70PDOB showing also the S_C phase (hatched region) and the solid-smectic C-smectic A triple point [Kalkura et al., 1982].

the hypothetical cases in Fig. 2.5. It may be of course that experiments cannot cover a wide enough portion of the phase diagram and it is difficult to draw general conclusions from just a limited portion. Computer simulations can help a little in this respect, as we shall see in Chapters 10–12. To understand the various types of mesophases and phase transitions as well as to differentiate between them we shall have to resort to statistical thermodynamics, as discussed in Chapters 3, 4 and 7. Let us start however with some recollections of ordinary thermodynamics. We know that the behaviour of a one-component system with internal energy $\mathcal{U}$ and entropy $\mathcal{S}$ at a temperature T is regulated by its Helmholtz free energy $\mathcal{A}$ (see, e.g., [Pippard, 1966]), $\mathcal{A} = \mathcal{U} - T\mathcal{S}$, as long as we work at constant volume. If, instead, we work at a constant pressure P and temperature, T, we shall have to consider the Gibbs free energy:

$$\mathcal{G} = \mathcal{H} - T\mathcal{S}, \tag{2.3}$$

where we have introduced the enthalpy function $\mathcal{H} = \mathcal{U} + PV$. We wish to examine now how either $\mathcal{A}$ or $\mathcal{G}$ vary as we move from one phase to another. We recall that the free energy of a system in equilibrium at certain thermodynamic conditions (pressure, temperature, etc.) should be the lowest possible one between those of the various potential molecular organizations in competition. Thus, a crystal phase can occur when the free energy of this candidate organization is lower than that of the liquid, but, moreover, what we have simply indicated as a crystalline phase can correspond to various morphologies with different free energy and thus we can have various different and well-defined crystal phases (*polymorphs*) by varying external conditions. The phenomenon is very common for inorganic (sulphur and quartz being classic examples [Eggers et al., 1964]) and even more for molecular crystals from organic materials [Bernstein, 2007].

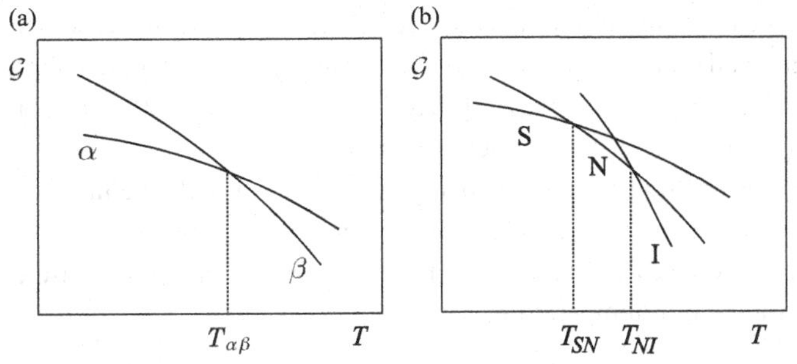

Figure 2.8 (a) Sketch of the free energy $\mathcal{G}$ for two virtual phases α and β as a function of temperature T. In $T_{\alpha\beta}$ the two curves cross and a transition between α and β occurs. (b) Similar sketch for three competing molecular organizations: solid (S), nematic (N) and isotropic (I).

2.3 Ehrenfest Classification of Phase Transitions

An astonishing feature of many transitions, also very familiar ones, like the crystallization of liquid water, is the fact that the molecular organization of a huge number of molecules can suddenly change from a disordered one to a beautifully ordered crystal structure at a very precise temperature. This is unusual, recalling the statement dating back to Lucretius '*natura non facit saltus*' (nature does not make jumps) normally obeyed at least for classic, non-quantum, systems. However, the discontinuous behaviour can be understood considering the free energy of each of the two phases that are interconverted at the transition, e.g. the isotropic liquid and crystal or isotropic liquid and nematic, or in general, α and β. Changing the thermodynamic driving variables, the temperature say, the free energy of each of the two molecular organizations candidate to exist will vary, indeed continuously and without jumps for each of the two potential phases, as we see in Fig. 2.8a. However, the two curves cross at a temperature $T_{\alpha\beta}$ where the free energy of phase β becomes lower than that of phase α. Since the phase that we observe at equilibrium is the one with the lowest free energy among all the candidate ones, for temperatures lower than $T_{\alpha\beta}$ the stable phase will be α, while for temperatures above $T_{\alpha\beta}$ the stable phase will be β. At the microscopic level we shall have a switch from the molecular organization corresponding to phase α to that of phase β, clearly more stable in the assigned conditions. At the transition point the free energy of the two phases is the same. Thus, at constant volume, $\mathcal{U}_\alpha - T_{\alpha\beta}\mathcal{S}_\alpha = \mathcal{U}_\beta - T_{\alpha\beta}\mathcal{S}_\beta$, and we have at $T_{\alpha\beta}$ a change in energy, the *latent heat*:

$$\mathcal{U}_\beta - \mathcal{U}_\alpha = T_{\alpha\beta}(\mathcal{S}_\beta - \mathcal{S}_\alpha). \tag{2.4}$$

Quite similarly we have, at constant pressure, $\mathcal{H}_\beta - \mathcal{H}_\alpha = T_{\alpha\beta}(\mathcal{S}_\beta - \mathcal{S}_\alpha)$. If more than one molecular organization with similar free energy is involved, as is often the case for liquid crystals, the argument can be repeated considering successive intersections of the respective curves and a succession of transitions takes place, as shown in Fig. 2.8b.

Let us now consider one transition point and examine the different ways that a transition can occur. In general, the two free energy curves in Fig. 2.8a may cross with different slopes in $T_{\alpha\beta}$ and in this case the resultant free energy for the stable phase will have an edge point, with the slope being different on the two sides of the curve. In this case, we may expect a sudden change of all quantities connected to free energy derivatives at a phase transition. In particular, the first partial derivative of the free energy $\mathcal{G}$ (or $\mathcal{A}$) with respect to temperature, i.e. the entropy $\mathcal{S}$, will undergo a sudden change at $T_{\alpha\beta}$.

We can express this mathematically using the definition of the delta function in terms of a formal derivative of the Heaviside step function $H(x-a)$ which is 1 for $x > a$ and 0 for $x < a$ (see Appendix D). Considering the free energy $\mathcal{G}$ as a piecewise continuous and differentiable function of temperature, we have:

$$\mathcal{G}(T) = \mathcal{G}_\alpha(T)\, H(T_{\alpha\beta} - T) + \mathcal{G}_\beta(T)\, H(T - T_{\alpha\beta}) \tag{2.5}$$

and from the partial derivative with respect to temperature

$$\mathcal{S}(T) = -\left(\frac{\partial \mathcal{G}}{\partial T}\right)_P = [\mathcal{S}_\alpha(T) - \mathcal{S}_\beta(T)]\, H(T_{\alpha\beta} - T), \tag{2.6}$$

where we have used the fact that $\mathcal{G}_\alpha(T) = \mathcal{G}_\beta(T)$ at $T_{\alpha\beta}$. An attempt at classifying phase transitions based on the behaviour of free energy derivatives is due to Ehrenfest [Pippard, 1966]. According to his scheme a transition is called of order n if the first $(n-1)$ derivatives of the free energy (either $\mathcal{G}$ or $\mathcal{A}$) are continuous across the transition, while the nth derivative shows a discontinuity. We sketch the free energy derivatives for such a *first-order* and *second-order* phase transition in Fig. 2.9. As we see, the free energy first derivative (entropy)

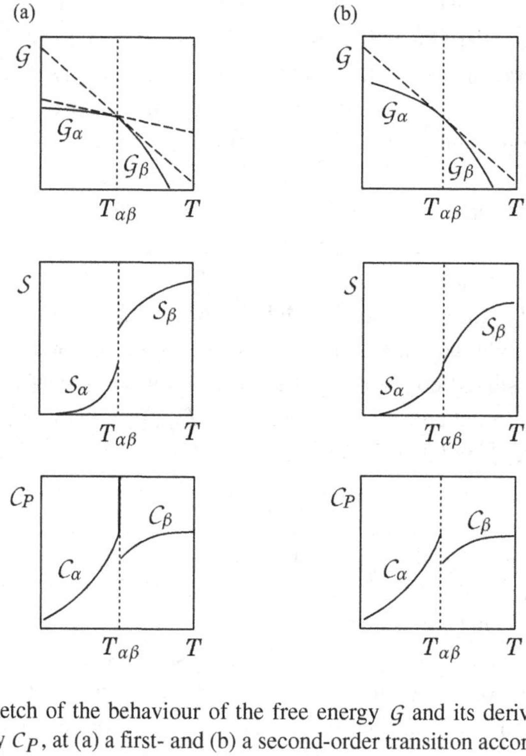

Figure 2.9 A sketch of the behaviour of the free energy $\mathcal{G}$ and its derivatives, entropy $\mathcal{S}$ and heat capacity C_P, at (a) a first- and (b) a second-order transition according to Ehrenfest classification.

makes a finite jump $\Delta S_{\alpha\beta}$ at a first-order phase transition, with the product $T_{\alpha\beta}\Delta S_{\alpha\beta}$ being the latent heat that we have already introduced. Examples are the liquid-gas transition for temperatures below T_C, the solid crystal-liquid, the solid-gas sublimation (see Fig. 2.3) or the nematic-isotropic phase change. Recalling that the heat capacity C_P can be written as the ratio of the infinitesimal amount of heat absorbed by the sample to produce a correspondingly small temperature increment, it can be written as:

$$C_P = \left(\frac{\partial \mathcal{H}}{\partial T}\right)_P = T\left(\frac{\partial \mathcal{S}}{\partial T}\right)_P = -T\left(\frac{\partial^2 \mathcal{G}}{\partial T^2}\right)_P. \tag{2.7}$$

The heat capacity at $T_{\alpha\beta}$, being the derivative of a step function, can be formally written as a delta function-like peak (see Appendix D), corresponding to the fact that temperature does not increase while we provide the energy needed to complete the passage from one phase to the other:

$$C_P(T) = [C_\alpha(T) - C_\beta(T)]\, H(T_{\alpha\beta} - T) + T(S_\beta - S_\alpha)\, \delta(T - T_{\alpha\beta}). \tag{2.8}$$

Very roughly we expect therefore the latent heat to be bigger when the transition is associated with a large change in the structure and organization of the system. We can forecast a transition corresponding to a profound disordering of the system, like the change from crystal to ordinary liquid, to produce a large change in entropy and thus to have a significant latent heat. Even if this is a very familiar situation, it does not imply that a discontinuity in the free energy slope is a characteristic of every transition. Indeed, for a second type of transition, the change is continuous with $S_\beta = S_\alpha$ at $T = T_{\alpha\beta}$ (see Fig. 2.9b). According to the Ehrenfest scheme, the jump we have observed in the first derivative should appear in the specific heat, i.e. in the second derivative of the free energy for a second-order transition. An analogous treatment can be done at constant volume. In that case the constant volume heat capacity C_V at a certain temperature can be written as

$$C_V = \left(\frac{\partial \mathcal{U}}{\partial T}\right)_V = T\left(\frac{\partial \mathcal{S}}{\partial T}\right)_V = -T\left(\frac{\partial^2 \mathcal{A}}{\partial T^2}\right)_V. \tag{2.9}$$

Incidentally, we have $C_P - C_V = -T\,(\partial V/\partial T)_P^2\,(\partial P/\partial V)_T$ and, since $(\partial P/\partial V)_T < 0$, the right-hand side (RHS) will be positive and $C_P > C_V$. The nematic-isotropic transition, as we see from Table 2.1, is a weak first-order one in the sense that the typical transition entropy is an order of magnitude smaller than that for melting. The volume change at the NI transition is also much smaller than that at melting. Typical volume changes are 3%–9% at the crystal-mesophase, 0.1%–0.4% at the nematic-isotropic, and 0.01%–0.2% at the smectic A-nematic transitions, respectively [Pestov and Vill, 2005]. For PAA we have $\Delta V_{KN}/V_K \approx 11.03\%$ and $\Delta V_{NI}/V_N \approx 0.36\%$ [McLaughlin et al., 1964].

In some circumstances the less stable phase can be continued beyond its range of stability and we talk in this case of *metastable* states. For example, a liquid can be supercooled below its solidification point.

While the Ehrenfest description is in accord with experiment for a first-order phase transition, this is not quite the case for continuous transitions, like the liquid-gas one at the critical point or the transition from ferromagnetic to paramagnetic behaviour, taking place at the Curie temperature, the order-disorder transition in alloys, some polymorphic

Table 2.1. *Entropy of transition* ($\Delta S_{KN}/R$) *for crystal-nematic and nematic - isotropic* ($\Delta S_{NI}/R$)*, with R the gas constant, for various nematogens (see Table 1.2). Data from [Pestov and Vill, 2005]*

	$\Delta S_{KN}/R$	T_{KN}(°C)	$\Delta S_{NI}/R$	T_{NI}(°C)
MBBA	6.15	22.0	0.43	48.0
5CB	6.98	24.0	0.4	35.3
6CB	10.18	14.3	0.16	30.1
7CB	15.0	10.8	0.36	30.0
PAA	9.09	118	0.17	135.3
PCH5	7.14	30.0	0.33	55.0

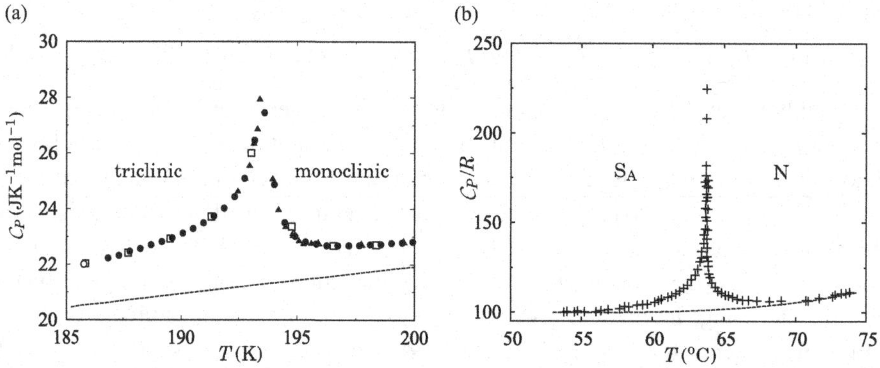

Figure 2.10 (a) Molar heat capacity vs temperature for the second-order triclinic-monoclinic transition of *p*-terphenyl occurring at 193.55 K [Chang, 1983] and (b) excess over ideal value heat capacity for the second-order NS_A transition of 4O.8 [Birgeneau et al., 1981]. The dashed lines represent the background variation and each symbol a set of measurements.

transformations in solids and the fluid-superfluid transitions in Helium and, as we shall see, some smectic-nematic transitions. There is no a priori reason why the free energy should behave according to the Ehrenfest scheme at a continuous transition and indeed normally it does not. In particular, for transitions not of the first order, the shape of the heat capacity peak is typically different from that predicted by Ehrenfest and shown in Fig. 2.9. Instead of a finite jump in heat capacity we have a divergence of the heat capacity with a certain functional dependence on temperature. In Fig. 2.10a, we see, as an example, the solid state transition between the triclinic and monoclinic phase of *p*-terphenyl that takes place over a few degrees [Chang, 1983]. The shape of this curve is reminiscent of a greek lambda and this kind of transition is referred to as a λ *transition*. Often the term *second order* is used for all transitions higher than the first, adopting the Ehrenfest nomenclature beyond its original meaning. A somewhat similar example of continuous transition for liquid crystals is shown in Fig. 2.10b for a smectic-nematic transition.

In general, first-order phase transitions correspond to situations with a sharp change of microscopic organization of some kind, while higher-order transitions occur by continuous modifications of the structure from one phase to the other. The possibility of continuous change in second-order transitions is normally associated with precursor or *pretransitional* phenomena that imply a manifestation of some of the properties of the new phase before this is reached. Indeed, one can consider the probability that a chance fluctuation in the disordered phase (β) will create a cluster of m ordered molecules (molar fraction x) with the local structure of the new phase (α). The probability that this takes place is proportional to the Boltzmann exponential weight factor and will be very low if the entropy of transition $\Delta \mathcal{S}_{\alpha\beta}$ is relatively large and negative, as at a first-order transition. However, if $\Delta \mathcal{S}_{\alpha\beta}$ is 0, or very small as at a second- or nearly second-order transition (cf. Table 2.1), the pretransitional effects will be non-negligible. A classic example is the liquid-gas transition at the critical point. This change of phase normally occurs with a change of volume and a latent heat, i.e. as a first-order transition. However, at a critical point this is not so, and fluctuations become so extensive to lead to inhomogeneous clusters of any size and even comparable to the wavelength of visible light, leading to a strong light scattering called *critical opalescence* [Stanley, 1971]. It is interesting to see that predicting the order of a phase transition from the molecular structure is extremely difficult, if even at all feasible. This is illustrated in Fig. 2.26 where the transitions for very similar n-alkyl-cyano-biphenyl mesogens only differing in the length of the alkyl (methylene) chain show rather different behaviour.

2.4 The Clausius–Clapeyron Equation

Going back to the (P,T) phase diagrams, we see (e.g. in Fig. 2.2) that they consist of transition lines, collections of points where the equilibrium between the two phases exist as P and T change. The slope of the (P,T) diagram transition lines can change from one type of material and of transition to another. If we consider two close equilibrium points (T_0, P_0) and (T,P) on a transition line (see Fig. 2.11) we have $\mathcal{G}_\alpha(T_0, P_0) = \mathcal{G}_\beta(T_0, P_0)$

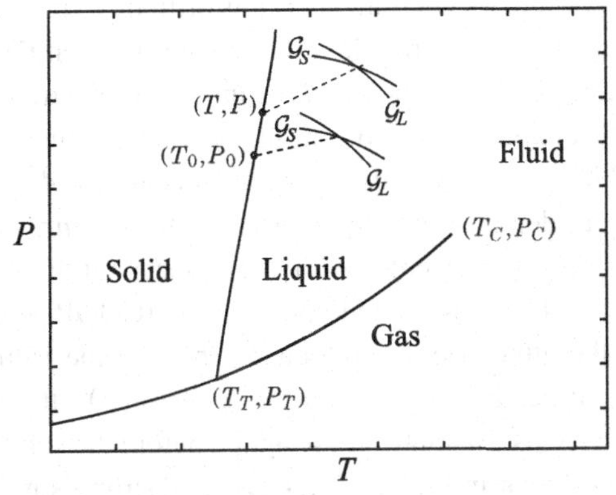

Figure 2.11 Two nearby equilibrium points (T_0, P_0) and (T,P) on a transition line used to derive the Clausius–Clapeyron equation for its slope. Each point corresponds to the intersection of the free energy curves for the solid and liquid phases (see Fig. 2.8).

but also $\mathcal{G}_\alpha(T, P) = \mathcal{G}_\beta(T, P)$. Since the two state points are close, we can write $\mathcal{G}_\alpha(T, P)$ and $\mathcal{G}_\beta(T, P)$ as the first terms in a Taylor expansion, starting from an equilibrium point at T_0, P_0:

$$\mathcal{G}_\alpha(T, P) = \mathcal{G}_\alpha(T_0, P_0) + \left(\frac{\partial \mathcal{G}_\alpha}{\partial P}\right)_{T_0} \mathrm{d}P + \left(\frac{\partial \mathcal{G}_\alpha}{\partial T}\right)_{P_0} \mathrm{d}T + \cdots, \tag{2.10a}$$

$$\mathcal{G}_\beta(T, P) = \mathcal{G}_\beta(T_0, P_0) + \left(\frac{\partial \mathcal{G}_\beta}{\partial P}\right)_{T_0} \mathrm{d}P + \left(\frac{\partial \mathcal{G}_\beta}{\partial T}\right)_{P_0} \mathrm{d}T + \cdots. \tag{2.10b}$$

Subtracting the two equations, and recalling that the partial derivative of the free energy with respect to pressure is the volume $V = (\partial \mathcal{G}/\partial P)_T$, while the partial derivative with respect to temperature is the entropy $\mathcal{S} = -(\partial \mathcal{G}/\partial T)_P$, we have then $V_\alpha \mathrm{d}P - \mathcal{S}_\alpha \mathrm{d}T = V_\beta \mathrm{d}P - \mathcal{S}_\beta \mathrm{d}T$. If $(V_\beta - V_\alpha) \neq 0$, this is more often written as the *Clausius–Clapeyron* equation for the slope of a transition line:

$$\frac{\mathrm{d}P}{\mathrm{d}T} = \frac{(\mathcal{S}_\beta - \mathcal{S}_\alpha)}{(V_\beta - V_\alpha)}, \tag{2.11}$$

which expresses a general result for first-order thermodynamic transitions. The slope of a transition line corresponding to a change of phase $\alpha \rightarrow \beta$ that reduces the entropy (e.g. a crystallization) will be positive if the volume per particle of the low entropy phase is smaller (e.g. along the liquid-solid transition line for CO_2 and most compounds), and negative otherwise. Thus, for materials that expand when crystallizing, like water (4%), gallium (1%), silicon (10%) and many silicates, the slope of the melting curve is negative. For these materials we also expect melting by just increasing the pressure.

Turning to liquid crystals, the nematic-isotropic (NI) transition is a *weak first-order* one, i.e. the entropy (and latent heat) and the volume (and density) variations for an NI transition are small and in any case much smaller than for a crystal-nematic, KN, or crystal-isotropic fluid, KI, (see, e.g., [Würflinger and Sandmann, 2001]). Since both $\Delta \mathcal{S}_{NI}$ and ΔV_{NI} decrease with respect to the melting ones, it is difficult to predict in general which of the KN or NI transitions will be steeper. It is worth stressing that the Clausius–Clapeyron relation is a local one, and that it does not imply a linearity of the transition lines. Let us see an example. The smectic phase normally occurs at lower temperatures than the nematic one, which is quite reasonable since, as we said earlier on, it is more ordered. However, this does not have to be the case, as demonstrated by the so called *re-entrant nematics*, that arise for instance in smectic formed by molecules with strong terminal dipoles (cf. Section 1.7.2). In these systems, e.g. the 4-cyano-4′-octyloxy biphenyl (COOB) compound, in a certain pressure range above the atmospheric, cooling down from the isotropic phase becomes nematic and then, by further cooling, a smectic A phase. On the other hand, a further decrease in temperature yields a nematic phase again, before the solid. The relevant portion of the phase diagram is shown in Fig. 2.12a. There is a strong similarity of this diagram with the universal one proposed a long time ago by Tamman showing melting by cooling (or inverse melting) in a region of the phase diagram (see Fig. 2.12b) discussed by Greer [2000] and experimentally observed in certain polymer systems [Rastogi et al., 1999].

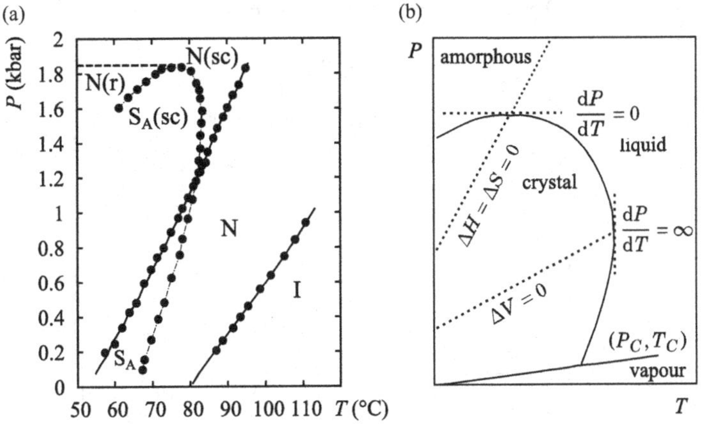

Figure 2.12 (a) (P,T) diagram of COOB showing isotropic (I), nematic (N), supercooled nematic (N(sc)) , supercooled smectic A (S_A(sc)), and re-entrant nematic (N(r)) [Cladis et al., 1977]. (b) Sketch of universal Tamman phase diagram [Greer, 2000].

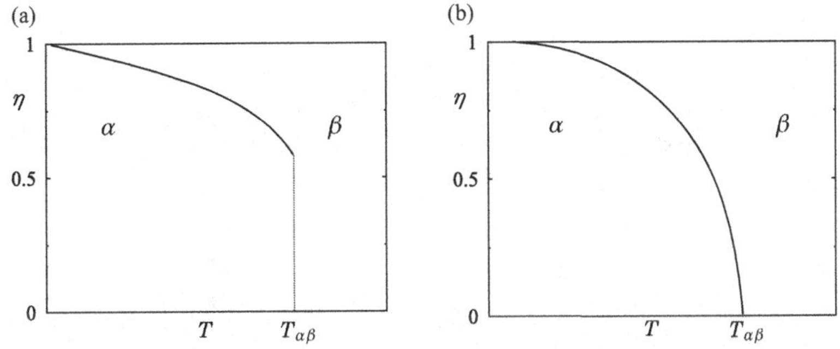

Figure 2.13 A sketch of the temperature variation of the order parameter η at (a) a first- and (b) second-order phase transition.

2.5 Empirical Order Parameters

What we have seen until now about phase transitions concerns macroscopic thermodynamic quantities. However, transitions involve changes in molecular organization, as we have very schematically seen in Table 1.1. A fundamental microscopic quantity, that we shall discuss at length in Chapter 3 is the order parameter, η say, that for the moment we intend simply as a suitable property that changes at the phase transition going to 0 in the more disordered phase. At a first-order phase transition the order parameter changes discontinuously to 0 (see in Fig. 2.13a) upon increasing temperature, while at a higher-order transition, the order parameter approaches 0 continuously, as we see in Fig. 2.13b. A common example of continuous transition, as a function of temperature, is the liquid-gas one, where an empirical order parameter η describing the approach to the critical temperature T_C can be taken proportional to the difference between the density of the liquid and the gas, $\rho_L - \rho_G$, at a

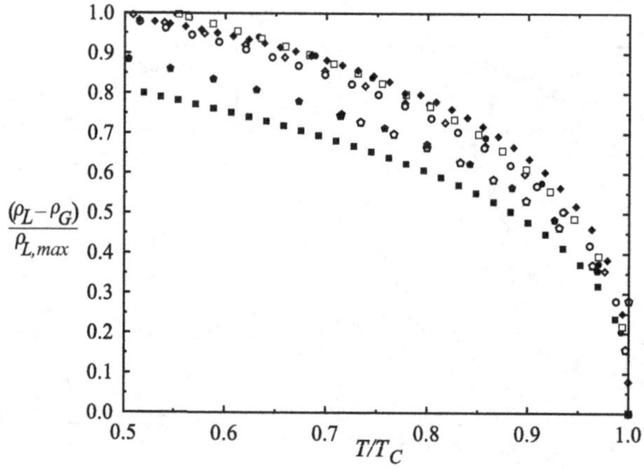

Figure 2.14 Variation with reduced temperature T/T_C of the liquid-gas transition order parameter $(\rho_L - \rho_G)/\rho_{L,max}$ in terms of the coexisting liquid, ρ_L, and gas, ρ_G, densities and scaled by the highest liquid density $\rho_{L,max}$. Data for Ne (●), Kr (□), N_2 (◇), CH_4 (○) [Guggenheim, 1945] and for H_2O (♦), benzene (■) [NIST, 2017].

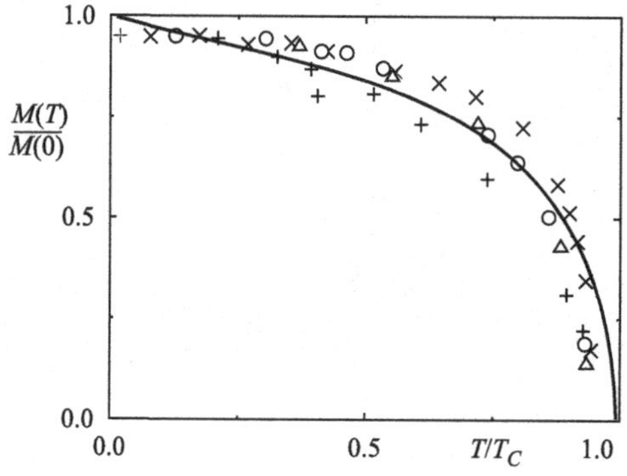

Figure 2.15 Continuous decay to 0 of order parameter η as reduced magnetization $M(T)/M(0)$ vs scaled temperature for a cobalt ($\triangle$), iron ($\times$), magnetite ($+$), nickel (○) ferromagnet on approaching the critical Curie point at T_C [Pathria, 1972]. The line is just a guide for the eye.

certain temperature. It is possible to see in Fig. 2.14 that the decay of this order parameter is actually fairly universal for simple fluids formed of particles of nearly spherical shape, also of different chemical nature, at the same reduced temperature T/T_C. This principle of *corresponding states* [Guggenheim, 1945] does not hold very well for more complex molecules, even for relatively simple ones, like benzene (Fig. 2.14). Let us consider, as a second example of continuous transition, ferromagnets, where nothing very apparent takes place in the (P, T) phase diagram, except that, by increasing the temperature, the macroscopic magnetization M is reduced and eventually destroyed at the *Curie temperature*, as we see in Fig. 2.15. For magnetic systems a suitable order parameter η is the absolute value M of the magnetization vector $\boldsymbol{M}$. Also, in this case, we see from Fig. 2.15 that magnets with different chemical nature have quite similar behaviour at the same reduced temperature T/T_C. For nematics, the most characteristic feature when approaching the clearing transition from below is the decrease in the anisotropy of tensor properties, e.g. the difference between the parallel and perpendicular components of the refractive index tensor $\mathbf{n}$, and its sudden disappearance at T_{NI}. A natural empirical order parameter could then be the anisotropy $\Delta n = n_{\parallel} - n_{\perp}$ and we could take $\eta(T) \propto \Delta n(T)$, scaled to have $0 \leq \eta \leq 1$, if the refractive index anisotropy is positive. The proportionality constant is not so easy to estimate without a molecular theory (see Chapter 3), but as we can see from Fig. 1.5b, the refractive index anisotropy, differently from the two cases just seen, jumps discontinuously to 0 at T_{NI}. Another way of seeing this discontinuity, is to realize that having order and disorder at the same time corresponds to a coexistence between the ordered and disordered phases. This behaviour is a characteristic feature of first-order transitions.

As long as we are interested in orientational order, the birefringence could be used also to characterize the anisotropy in colloidal suspensions, or lyotropics or even smectics.

2.6 Critical Exponents

The critical behaviour, i.e. the asymptotic features of physical properties approaching a continuous phase transition, is typical of a certain class of systems and has been studied in detail over the last few decades. The topic is relevant for liquid crystals, since various smectic-nematic transitions are of second-order type and even the nematic-isotropic transition, while being first order, is a very weak one, and has many features of a continuous phase change. We mentioned before that near a second-order transition fluctuations that involve clusters of all lengths, i.e. of any length scale, take place. This gives rise to scale invariance and suggests that the behaviour of certain properties may also be scale invariant, which in turn hints that their temperature dependence on approaching the transition may be scale free, like in particular a power law. Indeed, for a power law $f(x) = ax^{-\xi}$ scaling the argument by a constant c maintains the same functional form

$$f(x) = ax^{-\xi} \overset{x:=cx}{\Longrightarrow} a(cx)^{-\xi} = c^{-\xi} f(x) \propto f(x). \tag{2.12}$$

Here we just summarize the concept of critical exponent and show some classic examples, that we shall later use in Sections 2.7, 2.8 and 2.12 as reference cases to compare with liquid crystal transitions. The temperature dependence of the heat capacity can be represented by the asymptotic power law near the critical point

$$C_P = \begin{cases} (A/\alpha)\Delta T_r^{-\alpha} + B, & \text{for } T > T_C \\ (A'/\alpha')(-\Delta T_r)^{-\alpha'} + B', & \text{for } T < T_C, \end{cases} \tag{2.13}$$

where $\Delta T_r \equiv (T - T_C)/T_C$ and α, α' are the heat capacity *critical exponents*, characterizing the curves above and below T_C. The ratio $(A/\alpha)/(A'/\alpha')$, important in determining the shape of the heat capacity peak is called *amplitude ratio*. A related quantity obtained experimentally is the excess heat capacity, also fitted by the same exponent [Thoen, 1995], e.g. $\bar{C}_P = \alpha A \Delta T_r^{-\alpha}/(1-\alpha)$, for $T > T_C$. The critical exponents represent an important characteristic of a second-order phase transition. Clearly, classical Ehrenfest behaviour which corresponds to a finite jump, rather than a divergence, would give exponents $\alpha = \alpha' = 0$. The same arguments can obviously be applied also to C_V. Critical exponents can also be defined for changes in other quantities that play an analogous role in apparently very different transitions [Stanley, 1971]. This similarity proves very useful in transferring mathematical methods and ideas from one field to the other. The continuous decay to 0 of η (e.g. Figs. 2.13b–2.15) as T_C is approached from below can be represented by a power law with a critical exponent β:

$$\eta \approx (-\Delta T_r)^{\beta}, \; T < T_C. \tag{2.14}$$

Yet another type of important critical behaviour is that of susceptibilities relative to a (vanishingly small) applied external field. When considering a magnetic field $\boldsymbol{H}$, this material response quantity, $\boldsymbol{\chi}$, is the susceptibility for magnetic or non-magnetic, diamagnetic, systems like liquid crystals, respectively: $\boldsymbol{\chi} = (\mathrm{d}\boldsymbol{M}/\mathrm{d}\boldsymbol{H})_{H=0}$, where $\boldsymbol{M}$ is the magnetization. Approaching the Curie temperature, where the magnetization goes to 0, the susceptibility should vary as $\chi \approx (-\Delta T_r)^{-\gamma}$. The corresponding quantity for a simple fluid is the *isothermal compressibility*, with the pressure playing the role of the external field

$$\kappa_T = -\frac{1}{V}\left(\frac{\partial^2 \mathcal{G}}{\partial P^2}\right)_T = \frac{1}{V}\left(\frac{\partial V}{\partial P}\right)_T. \tag{2.15}$$

It is worth noting that these exponents are not all independent but, since they describe the behaviour of quantities linked by thermodynamic relations, they are also connected by important equations called *scaling laws* [Widom, 1996] that are either independent on space dimensionality n_d or that instead involve this dimensionality explicitly [Stanley, 1971]. According to the *universality* hypothesis for second-order phase transitions [Nelson, 1977; Spencer, 2000; Bellini et al., 2001] the critical exponents depend on the fundamental properties of the transition, e.g. the spatial dimensionality, the range of interaction and the symmetry of the system, rather than on the fine details of the potential between the particles. In Table 2.2 we give a brief summary of experimental values of exponents for some classical transitions: that for magnets at the Curie point and that for fluids at the critical point. The similarity of exponents, e.g. for decay to 0 of the order parameter, is nothing short of impressive. Phase transitions, even corresponding to extremely different materials, can thus be grouped in a limited number of *universality classes* with well-defined exponents. Moreover, phase transitions corresponding to seemingly extremely different systems can be understood in terms of simpler models that share the same universality class. In the next section we briefly look at some of these simple prototype models.

Note that, as we have listed them, the critical exponents do not appear to be related to the anisotropy of a material, except for the order parameter one. However, experiments sensitive to structural order and correlation in the positions and orientations between constituent particles in different directions, such as X-ray diffraction, will require anisotropic critical exponents in the data analysis [Barois, 1992].

Table 2.2. *A summary of experimental exponents for various continuous transitions: magnetization-demagnetization (Curie point) and critical point liquid-gas phase transitions. Here: F: Ferromagnet, AF: Antiferromagnet, LG: Liquid-Gas. References: (a) [Ma, 1976]; (b) [Widom, 1996]; (c) [Stanley, 1971]; (d) [Zemansky and Dittman, 1997].*

System	T_c(°C)	α	α'	β	γ	γ'	Ref.
F: Fe	770.9	−0.120	−0.120	0.34	1.33	1.33	(*a*)
F: Ni	358.48	−0.10	−0.10	0.33	1.32	1.32	(*a*)
F: $YFeO_3$	369.9			0.354	1.33		(*a*)
AF: FeF_2	−194.84	0.11	0.11				(*a*)
AF: $RbMnF_3$	−190.05	−0.139	−0.139	0.32	1.40	1.40	(*a*)
LG: CO_2	31.06	0.124	0.124	0.345	1.20	1.20	(*a*)
LG: Ar	−122.29	< 0.4	< 0.25	0.362	1.20	1.20	(*d*)
LG: Xe	16.64	0.08	0.08	0.34	1.203	1.203	(*a*)
LG: ^{4}He	−267.911	0.127	0.159	0.355	1.170	1.170	(*c*)
LG: vdW		0	0	1/2	1	1	(*b*)

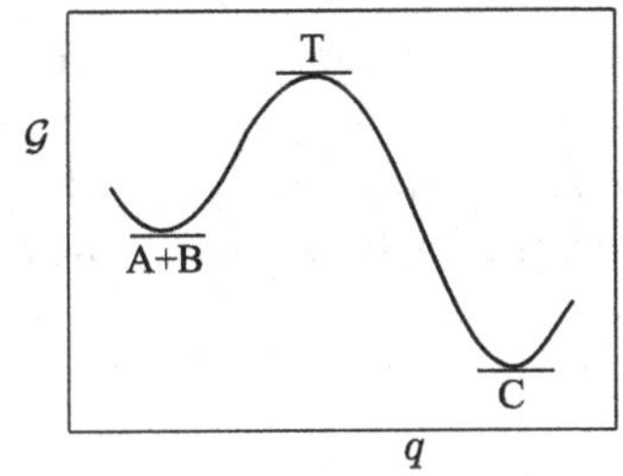

Figure 2.16 A sketch of the free energy $\mathcal{G}$ for a system of two reagents A and B going to the product C as a function of a reaction coordinate q.

2.7 Landau Theory

One of the most successful theories for phase transitions was put forward by Landau many years ago [Landau, 1965; Landau and Lifshitz, 1980]. In his approach the free energy of a system that undergoes a transition between two different phases is studied as a function of one or more order parameters differentiating the various types of microscopic organizations for the system (e.g. positional and orientational order parameters for smectics, or just orientational for nematics). The specific nature of η will change for different materials and reflect the symmetry of the phases, as we have seen in Section 2.5, but does not need to be detailed at this stage. Indeed, each η can be considered as a kind of generalized coordinate describing the passage from one phase to the other, a bit like the *reaction coordinate* used in chemistry (see Fig. 2.16) describes in a general (and inevitably generic) way the change from the free energy of a set of reagents to that of a set of products. The aim is then to describe a wide range of different physical systems and capture the general behaviour of free energy and related properties at second- and first-order-type transitions. The theory, originally applied with success to magnets and polymorphic transformations in crystals, has been extended by de Gennes [1974] and by a number of other authors [Stephen and Straley, 1974; Gramsbergen et al., 1986a; Vertogen and de Jeu, 1988] to nearly every liquid crystal transition and other systems [Binder, 1987; Tolédano and Tolédano, 1987]. In particular, the theory for various types of smectics has been discussed in detail by Pikin [1991]. Here we keep the treatment very simple, considering first the case of a system characterized by a scalar order parameter, η, that vanishes in the more disordered phase. The basis of Landau theory consists in assuming that the (non-equilibrium) free energy in the proximity of the phase transition can be written as a power series in terms of the order parameter η, supposed to be small enough for the expansion to converge. The Gibbs free energy density (or equivalently, the free energy, since we have a uniform system) is therefore written, at constant pressure, as

$$\mathcal{G} = \mathcal{G}_0 + \left(\frac{\partial \mathcal{G}}{\partial \eta}\right)_{\eta=0} \eta + \frac{1}{2!}\left(\frac{\partial^2 \mathcal{G}}{\partial \eta^2}\right)_{\eta=0} \eta^2 + \frac{1}{3!}\left(\frac{\partial^3 \mathcal{G}}{\partial \eta^3}\right)_{\eta=0} \eta^3 + \frac{1}{4!}\left(\frac{\partial^4 \mathcal{G}}{\partial \eta^4}\right)_{\eta=0} \eta^4 + \cdots, \tag{2.16}$$

where $\mathcal{G}_0$ is the free energy of the disordered, $\eta = 0$, state. Since we are interested in variations with respect to this reference state we can let $\mathcal{G}_0 = 0$ without loss of generality. At a given temperature and retaining a finite number of terms this is a polynomial in η, whose minima, identified by the conditions $(\partial \mathcal{G}/\partial \eta)_{\eta=0} = 0$ and $(\partial^2 \mathcal{G}/\partial \eta^2)_{\eta=0} > 0$, represent stable configurations. In particular, if the free energy has to be a minimum, its first derivative with respect to η at a fixed T (and/or P) will have to be 0. We can then write, more concisely

$$\mathcal{G}(\eta, T) = G_2(T)\eta^2 + G_3(T)\eta^3 + G_4(T)\eta^4 + G_5(T)\eta^5 + G_6(T)\eta^6 + \cdots, \tag{2.17}$$

where $G_2(T) \equiv \frac{1}{2}\left(\partial^2 \mathcal{G}/\partial \eta^2\right)_{\eta=0}$, $G_3(T) \equiv \frac{1}{6}\left(\partial^3 \mathcal{G}/\partial \eta^3\right)_{\eta=0}$, $G_4(T) \equiv \frac{1}{24}\left(\partial^4 \mathcal{G}/\partial \eta^4\right)_{\eta=0}$, $\ldots$, are unknown coefficients that can, in general, depend on temperature T. The free energy, Eq. 2.17, cannot be studied as such, but it is interesting to see its behaviour when making some (drastic) assumptions on the G_i coefficients. Landau suggested that they can be assumed to be constants, $G_3(T) = -g_3$, $G_4(T) = g_4$, $G_5(T) = g_5$, $G_6(T) = g_6$, with: $g_i \geq 0, \ldots$ (but notice the minus sign in front of g_3), while the variation of $G_2(T)$ with temperature is taken to be

$$G_2(T) = a(T - T^*), \tag{2.18}$$

where T^* is a certain characteristic temperature and $a > 0$. In this way, the coefficient $G_2(T)$ can invert its sign, changing the curvature of the free energy curve and transforming the system from stable to unstable at some temperature. With these assumptions, the free energy $\mathcal{G}$ for an isotherm at a given temperature T will just depend on η and, truncating for now at η^4, we have

$$\mathcal{G}(\eta, T) = a(T - T^*)\eta^2 - g_3\eta^3 + g_4\eta^4. \tag{2.19}$$

The value of the equilibrium order parameters η at that T will be obtained minimizing the free energy. We can now consider various scenarios resulting from different choices of coefficients.

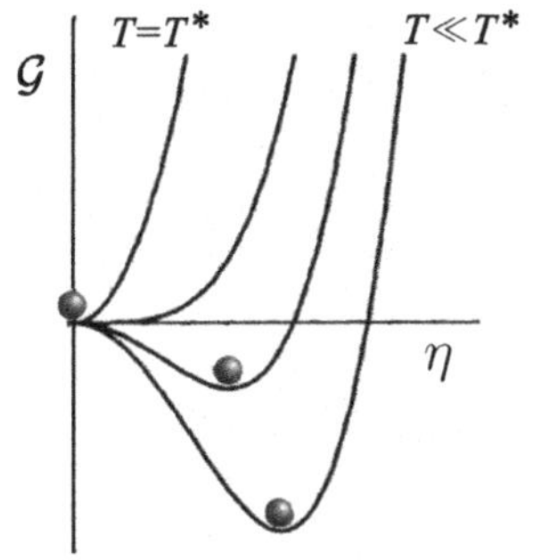

Figure 2.17 Sketch of the free energy at various temperatures below and up to the transition at a second-order phase change (here we only show the $\eta \geq 0$ part of the plot).

2.7.1 Second-Order Phase Transitions

Let us describe first what happens if, for a certain physical system, we have $g_3 = 0$, e.g. if, because of symmetry reasons, $\mathcal{G}(\eta, T) = \mathcal{G}(-\eta, T)$, as for the magnetization transition. Then the free energy will be just

$$\mathcal{G}(\eta, T) = a(T - T^*)\eta^2 + g_4\eta^4. \tag{2.20}$$

Minimization of $\mathcal{G}(\eta, T)$ gives $\mathcal{G}'(\eta, T) = 2a(T - T^*)\eta + 4g_4\eta^3 = 0$, showing that one, trivial, solution is the disordered one with $\eta = 0$. At $T > T^*$ only this disordered state will be stable. When $T < T^*$ the equilibrium state is instead an ordered one (see Fig. 2.17) with

$$\eta(T) = \pm\sqrt{a(T^* - T)/(2g_4)}. \tag{2.21}$$

The order decreases continuously approaching the transition at $T = T^*$, so the behaviour is that of a second-order transition, with $T_C = T^*$ as we see in Figs. 2.13b and 2.15. It is also clear that the critical exponent for the order parameter is in this case $\beta = 1/2$.

2.7.2 First-Order Phase Transitions

We can now consider what happens if the cubic contribution is non-vanishing ($g_3 \neq 0$) and verify that this can produce quite a dramatic effect. The first derivative of the free energy in Eq. 2.19 is

$$\mathcal{G}'(\eta, T) = 2a(T - T^*)\eta - 3g_3\eta^2 + 4g_4\eta^3 = 0. \tag{2.22}$$

Excluding again the trivial $\eta = 0$, isotropic solution, giving $\mathcal{G}' = 0$, Eq. 2.22 reduces, for $\eta \neq 0$, to the quadratic $2a(T - T^*) - 3g_3\eta + 4g_4\eta^2 = 0$, which has a solution for an order parameter

$$\eta_\pm(T) = \frac{3g_3}{8g_4}\left[1 \pm \sqrt{1 - \frac{32ag_4(T - T^*)}{9g_3^2}}\right], \tag{2.23}$$

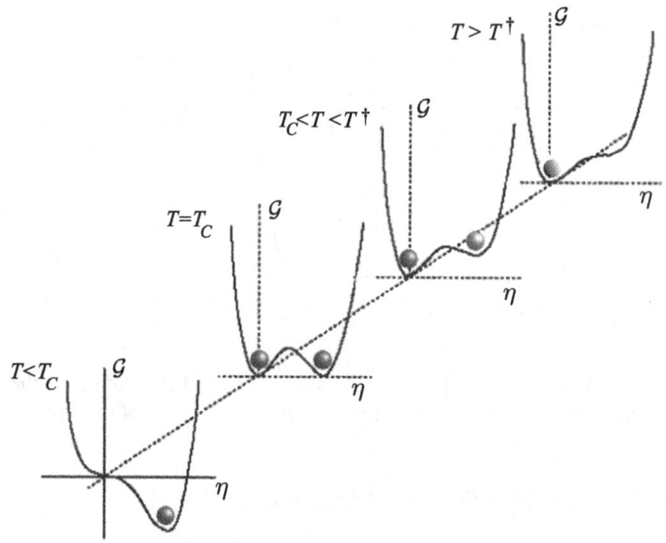

Figure 2.18 A sketch of the free energy $\mathcal{G}$ vs order parameter η at four temperatures near a first-order phase transition: $T < T_C$, $T = T_C$, $T_C < T < T^\dagger$, $T \gg T_C$, where T_C is the transition temperature ($T_C = T_{NI}$ for a nematic-isotropic transition).

where a local minimum exists when argument of the square root is non-negative, i.e as long as the temperature is lower than a temperature $T^\dagger = T^* + 9g_3^2/(32ag_4)$. The order parameter at the temperature $T^\dagger$, which is the limit of metastability of the ordered phase upon heating (*superheating* temperature), is $\eta(T^\dagger) = g_3/g_4$. Above this temperature, the only stable solution is the disordered one, $\eta = 0$. Substituting $\eta_\pm$ in $\mathcal{G}(\eta, T)$ and solving for $\mathcal{G}(\eta_\pm, T) = 0$, we see that we can have a double minimum in $\mathcal{G}$ with equal well depths (Fig. 2.18), corresponding to the coexistence of a disordered and of an ordered phase, characteristic of a first-order phase transition, at the transition temperature T_C:

$$T_C = T^* + \frac{g_3^2}{4ag_4}. \tag{2.24}$$

Replacing T_C in Eq. 2.23 gives

$$\eta_C = \frac{g_3}{2g_4} = \frac{1}{2}\eta(T^\dagger). \tag{2.25}$$

The entropy of transition can be obtained differentiating the free energy with respect to T, giving $\Delta S_C = (ag_3^2)/(18g_4^2)$. Thus, the transition has a latent heat, $ag_3^2 T_C/(18g_4^2)$, as long as $g_3 \neq 0$. Elimination of a, g_3 and g_4 in terms of T_C, T^* and η_C using the formulas just obtained, allows reformulating the temperature dependence of the order parameter in Eq. 2.23 as

$$\eta(T) = \frac{3}{4}\eta_C\left[1 + \left(1 - \frac{8(T - T^*)}{9(T_C - T^*)}\right)^{\frac{1}{2}}\right]. \tag{2.26}$$

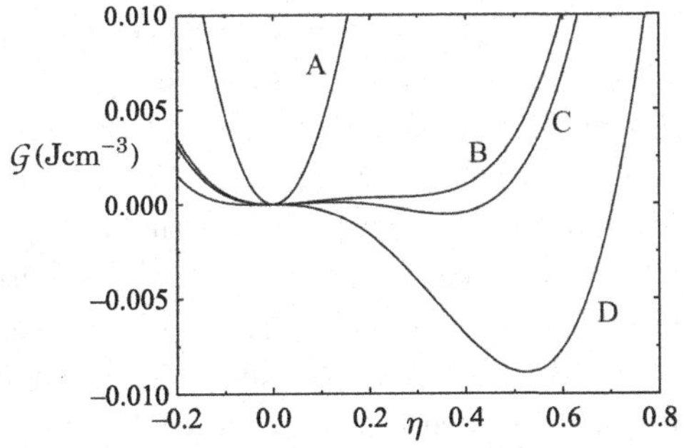

Figure 2.19 The free energy density $\mathcal{G}$ for the liquid crystal MBBA plotted as a function of the Landau order parameter η at a temperature of 55°(A), 46.1°(B), 45.9°(C) and 45°C (D). Data from Poggi et al. [1976].

We see that Landau theory predicts a (pseudo) critical exponent for the order parameter $\beta = \frac{1}{2}$. Clearly, η_C will tend to 0 if (g_3/g_4) does the same and in this case $T_C = T^*$ suggesting that T^* is the transition temperature of a normally virtual second-order transition.

It is apparent that the symmetry of the phases involved has a profound effect on the type of transition, as we have already alluded in various points before. We can summarize them recalling some general considerations put forward by Anderson [1981], who suggested a classification in three typologies:

(i) When the two phases have the same symmetry (e.g. the liquid-vapour transition), the transition may be first order. However, a switch from first to second order can occur at an isolated critical point, a *Landau point*, if the cubic term, g_3, in the free energy expansion depends on some external variable which can be used to vary it and bring it to 0. For the liquid-gas, the role of this external field can be played by the pressure. When the pressure reaches P_C, the order parameter, i.e. the density difference between the two fluids, $\rho_L - \rho_G$, vanishes, and the first-order character of the transition disappears.

(ii) When the two phases have different symmetry, but the symmetry group of the more ordered phase, G_1, is a subgroup of that of the other more symmetric phase, G_0. In this case, the symmetry is lowered at the transition, and we talk of *broken symmetry*. The transition may be first or second order depending on details, but there can never be a disappearance of the transition line, since symmetry cannot change continuously, according to what Anderson [1981] has called the *First Theorem of Condensed Matter Physics*.

(iii) If the symmetry of the two phases involved in the transition is intrinsically different, e.g. an ordinary liquid and a crystal, the cubic term g_3 will always be different from 0 and the transition will accordingly be first order. For instance, we do not have a critical point on the liquid-solid crystal transition line (see, e.g., Fig. 2.2) [Landau and Lifshitz, 1980].

For liquid crystals the most common transition is that between a nematic with a uniform order parameter η and an isotropic phase occurring at the transition temperature $T_{NI} = T_C$.

The properties of a uniform nematic are unchanged by rotation around the director, so the symmetry of the phase is uniaxial or cylindrical ($D_{\infty h}$), i.e. that of a subgroup of the special orthogonal group, SO(3), of rotations in space corresponding to the isotropic phase, like in case (ii) just mentioned. The NI transition is first order, even though a weak one, i.e. with a small latent heat. T^* is slightly lower than T_C, suggesting that the virtual second-order transition is not observed on cooling down from the isotropic phase because of the preliminary occurrence of the first-order one. At T^* the order parameter would be $\eta^* = \eta(T^*) = (3g_3)/(4g_4) = 3\eta_C/2$. We do not expect a critical point on the NI transition line because of the difference between the spherical symmetry of the isotropic and the uniaxial symmetry of the nematic, already mentioned in Section 1.2. However, a critical point may be recovered, at least in principle, if a liquid crystal with positive dielectric (or diamagnetic) anisotropy is subjected to an external uniaxial electric (or magnetic) field, so that both the nematic and the para-nematic phases share the same uniaxial symmetry [Nicastro and Keyes, 1984; Hornreich, 1985; Vause, 1986]. In practice the fields required appear to be extremely high and systems where the critical point can be determined have not been found to date.

The application of a magnetic or electric field $\boldsymbol{F}$ [Lelidis and Durand, 1993, 1994] induces in any case some order η above T_C and contributes a term $-\alpha\eta F$ to the free energy, that becomes

$$\mathcal{G}(\eta, T) = a(T - T^*)\eta^2 - g_3\eta^3 + g_4\eta^4 - \alpha\eta F. \tag{2.27}$$

Minimization with respect to η gives $2a(T - T^*)\eta - 3g_3\eta^2 + 4g_4\eta^3 - \alpha F = 0$. Since we are above T_C and η is very small we can neglect the term from η^2, finding the pretransitional order parameter

$$\eta = \frac{\alpha F}{2a(T - T^*)}. \tag{2.28}$$

The pretransitional susceptivity, $\chi = \alpha\eta/F = \alpha^2/[2a(T - T^*)]$, should increase with critical exponent $\gamma = 1$ on approaching the transition from above. In Fig. 2.19 we show the free energy versus order parameter curve calculated at four different temperatures with the parameters experimentally determined by Poggi et al. [1976] for the nematic MBBA, i.e. $a = 0.045\ \text{J cm}^{-3}\text{K}^{-1}$, $g_3 = 0.197\ \text{J cm}^{-3}$ and $g_4 = 0.307\ \text{J cm}^{-3}$. We show the free energy for a temperature in the isotropic phase (A), where $\eta = 0$ is the only stable solution, a temperature just above and just below the nematic-isotropic transition of 46.0°C (B and C) and for a temperature inside the nematic phase (D). In this case, the only stable solution is the ordered one. Note that only the minima of $\mathcal{G}(\eta, T)$ represent equilibrium points, while the probability of observing a fluctuation corresponding to a state with order parameter η at temperature T can be written [Landau and Lifshitz, 1980] as

$$P(\eta, T) \propto \exp[-\mathcal{G}/(k_B T)]. \tag{2.29}$$

As we shall see in Sections 10.2 and 12.4, computer simulations can provide histograms of the order parameters observed at a certain T, i.e. essentially of $P(\eta, T)$ and thus establish a link with Landau theory.

2.7.3 Tricritical Behaviour

Judging just from an algebraic point of view it is clear that other transitional behaviours could become available for different combinations of g_n coefficients in the Landau expansion, in particular when the non-negligible fifth, g_5, or sixth, g_6, order term appear in the free energy expansion Eq. 2.19. We have seen before that the absence of the cubic term leads to a second-order transition when the expansion of $\mathcal{G}$ is truncated to η^4. However, the inclusion of higher terms in the expansion can lead to obtaining a first-order transition even when the cubic term, or more generally, the odd terms are 0 by symmetry. In particular, the case $g_3 = 0, g_4 < 0, g_5 = 0, g_6 > 0$ of the expansion truncated at the sixth order,

$$\mathcal{G}(\eta, T) = a(T - T^*)\eta^2 + g_4\eta^4 + g_6\eta^6, \tag{2.30}$$

leads to a change from second order to first order when $g_4 = 0$, and to a *tricritical point* (TCP), where the second order becomes a first-order one, when $g_4 < 0$ [Gramsbergen et al., 1986a; Chaikin and Lubensky, 1995]. In this last case the prediction is of an order parameter exponent $\beta = \frac{1}{4}$ and of specific heat exponents below and above the transition $\alpha = 0, \alpha' = 0.5$. Thus, experimental determination of the exponents could shed some light on the type of transition [Mukherjee and Saha, 1997]. Unfortunately, a consensus does not seem to exist as yet on these experimental values and a variety of exponents have been found [Gramsbergen et al., 1986a]. From our point of view one disappointing aspect of this type of treatment is the lack of molecular interpretation for the coefficients $a, g_3, g_4, \ldots$ We shall see in Chapter 7 that this interpretation can be, in some instances, obtained by a comparison with molecular field theories.

2.8 Lattice Models

Much of our knowledge on the fundamentals of phase transitions is obtained from some remarkably simple and yet very rich models, such as the Ising, Heisenberg and XY models [Ma, 1985; Uzunov, 2010; Friedli and Velenik, 2017]. These were mainly developed for phase transitions in magnetic systems, but we shall briefly summarize them here since they are often used, also in the liquid crystal field, as prototypes of certain categories of transitions and of their critical exponents.

It is in many ways amazing that simple lattice models can also be useful for investigating the phase transitions of mesogens with the complex chemical structures seen in Chapter 1. This is due to the fact that continuous and nearly continuous transitions are of general, fairly universal, nature and can be classified into universality classes, each characterized by its space dimensionality, n_d, number of angular degrees of freedom, n_o, and critical exponents. In turn, different systems can be expected to belong to the same universality class, and share its critical exponents, if their particles position and orientation are respectively defined in real spaces $\mathbb{R}^{n_d}$ and $\mathrm{O}(n_o)$ with the same space and orientational dimensionality $n_d = 1, 2, 3$ and $n_o = 1, 2, 3$, respectively. Here we briefly introduce some of the most studied models and their critical exponents.

2.8.1 Ising Model

In its simplest form the model consists of a set of interaction centres μ_i, (often called *spins*) which can take the values ± 1 for spin up or down on a regular lattice. The pair potential is nearest neighbours, attractive, and the total energy, U_N, for N spins is

$$U_N = -\frac{1}{2} \sum_{\langle ij \rangle} \epsilon_{ij} \mu_i \mu_j. \tag{2.31}$$

The variable μ can take the values ± 1 for spin up or down (so that the number of orientational degrees of freedom for the spins is $n_o = 1$). The constant ϵ_{ij} is different from 0 and equal to ϵ only for nearest neighbour pairs, indicated as $\langle ij \rangle$, and can be chosen to be positive to describe *ferromagnetic* or negative to describe *antiferromagnetic* coupling. The determination of the phase transition of the model is simple when the spins are on a one-dimensional ($n_d = 1$) lattice (the only case solved by Ising). In this case, there is no phase transition. In two dimensions (2D), this $n_d = 2$, $n_o = 1$ model was solved analytically in landmark work by Onsager [1944]. The heat capacity for the square lattice has a logarithmic divergence at the transition temperature $k_B T_C/\epsilon = 2(\sinh^{-1} 1)^{-1} \approx 2.26918$. The average value of the variable μ represents the magnetization and thus the order parameter for the system and is [Ma, 1985]

$$M = \frac{\cosh^2[2\epsilon/(k_B T)]}{\sinh^4[2\epsilon/(k_B T)]} \left[\sinh^2 \left(\frac{2\epsilon}{k_B T} \right) - 1 \right]^{1/8}, \tag{2.32}$$

and near the critical point $M \propto (T_C - T)^{1/8}$. In three dimensions (3D), or $n_d = 3, n_o = 1$, the Ising model has not been solved analytically, as indeed no 3D model has. However, it has been the subject of an enormous number of approximate treatments and computer simulations. The transition temperature for the simple 3D cubic lattice is $k_B T_C/\epsilon \approx 4.5115$ [Hasenbusch et al., 1999]. One reason for particular interest in the Ising model is that it can be used to study fluids, and in particular the liquid-gas transition, as well as magnets. An equivalent *lattice gas* model [Stanley, 1971] for a fluid in a volume V can be introduced by dividing up the volume in N cells of volume $\mathcal{V}$ (of molecular dimension). A cell can be occupied by one (and only one) molecule or it can be empty. The interaction between two cells i and j is assumed to be

$$U_{ij} = \begin{cases} \infty, & \text{if } i \text{ and } j \text{ try to occupy the same cell} \\ -\epsilon, & \text{for } i \text{ and } j \text{ neighbouring occupied cells} \\ 0, & \text{otherwise.} \end{cases} \tag{2.33}$$

A correspondence between lattice gas and Ising models can be established defining a cell occupation variable $o_i = \frac{1}{2}(1 + \mu_i)$ from a fictitious 'spin' μ_i. In each configuration one has $\sum_{i=1}^{N} o_i = N$. The mapping of properties from the lattice gas to the Ising model is well described elsewhere (see, e.g., [Stanley, 1971]). Critical exponents for the 3D Ising model are given in Table 2.3. Comparing these exponents with the experimental ones for various transitions in Table 2.2, we can see the excellent agreement with those for the liquid-gas transition approaching the critical point. The two transitions are then said to belong to the

Table 2.3. *Critical exponents for some classical lattice models from (a) [Stanley, 1971], (b) [Uzunov, 2010], (c) [Lau and Dasgupta, 1989], (d) [Holm and Janke, 1993, 1997], (e) [Campostrini et al., 2001], (f) [Stephenson, 1971] and (g) Landau theory exponents for various combinations of the g_i expansion coefficients [Gramsbergen et al., 1986a]*

System	α	α'	β	γ	γ'	ν	Ref.
Ising 2D	0	0	1/8	7/4	7/4	1	(a, b)
Ising 3D	0.125	0.125	0.324	1.25	1.27		(f)
Heisenberg 2D			0.447				(a)
Heisenberg 3D	−0.118		0.364	1.390		0.704	(c, d)
XY 3D	−0.015	−0.02	0.348	1.318	1.32	0.671	(e)
Landau second order ($g_3 = 0, g_4 > 0, g_5 = 0, g_6 = 0$)			1/2				(g)
Landau first order ($g_3 > 0, g_4 > 0, g_5 = 0, g_6 = 0$)	0	0	1/2	1	1		(g)
Landau tricritical ($g_3 = 0, g_4 = 0, g_5 = 0, g_6 < 0$)	1/2	1/2	1/4	1	1		(g)

same universality class. The Ising model, with only two spin orientations allowed, is not particularly suited to describe orientational transitions, like the nematic-isotropic in liquid crystals, even if it can be applicable to some isotropic-isotropic ones (see Section 2.10) [Lubensky and Stark, 1996]. In any case it is worth mentioning that the exact solution of the 2D version of the Ising model provides an invaluable way of testing some of the computer simulation methodologies described in Chapters 8 and 10.

2.8.2 Potts Model

The Potts model [Straley, 1974; Wu, 1982; Binder, 1987] is a lattice system where the spins can assume a set of q quantized and equally spaced orientations and interact directly only with nearest neighbours. The lattice is typically assumed to be a 2D rectangular one, i.e. to have $n_d = 2$, but the model has been also generalized to other dimensions or other lattice types. In 2D its total energy is

$$U_N = -\frac{1}{2} \sum_{\langle ij \rangle} \epsilon_{ij} \cos(\phi_i - \phi_j), \tag{2.34}$$

with $\phi_j = 2\pi j/q$ and $j = 1, \ldots, q$, $\epsilon_{ij} = \epsilon$ for nearest neighbours $\langle ij \rangle$ and 0 otherwise. The 2D model with $\epsilon > 0$ has a first-order transition if $q \geq 5$ [Peczak and Landau, 1989]. When $q \leq 4$ a continuous transition is observed, as in the Ising model, to which it reduces when $q = 2$. The model has been recently generalized to a 'dynamic' version where the spins can hop from an occupied to a non-occupied lattice site. This Potts model has also been used for liquid crystal studies [Bailly-Reyre and Diep, 2015] and it is interesting as an example of weak first-order phase transition when $q = 5$.

2.8.3 Heisenberg Model

The classical version of the 3D Heisenberg model consists again of a regular lattice where each site is described by a unit vector $\boldsymbol{u}_i$ (spin) that can assume continuously varying orientations in space, with say polar angles (α_i, β_i) and with nearest neighbour interaction ϵ_{ij}. The energy of the system of spins is

$$U_N = -\frac{1}{2} \sum_{\langle ij \rangle} \epsilon_{ij} P_1(\cos \beta_{ij}), \tag{2.35}$$

where β_{ij} is the relative orientation between spins i and j, $P_1(\cos \beta_{ij}) = \boldsymbol{u}_i \cdot \boldsymbol{u}_j$ is a first-rank Legendre polynomial (see Eqs. A.48) and the constant $\epsilon_{ij} = \epsilon$ can be chosen to be positive to give a preference to parallel alignment of the spins, i.e. to describe *ferromagnetic* behaviour, or negative to describe *antiferromagnetic* coupling. The transition temperature has been estimated from computer simulations [Chen et al., 1993; Holm and Janke, 1993] as $k_B T_C/\epsilon \approx 1.4430$ for the simple cubic lattice, and there is a conjecture [Brown and Ciftan, 1996] that it might be $k_B T_C/\epsilon = 1/\ln 2$. The Heisenberg universality class corresponds to the critical behaviour of isotropic magnets, for instance the Curie transition in ferromagnets such as Fe, Ni and EuO [Campostrini et al., 2002], as we can see comparing the critical exponents in Table 2.3.

2.8.4 XY Model

In the XY or *plane rotator* model each site is characterized by a vector (or 'spin') lying in a plane (so the number of components of the vector is $n_o = 2$) and interacting to yield

$$U_N = -\frac{1}{2} \sum_{\langle ij \rangle} \epsilon_{ij} \cos \phi_{ij}, \tag{2.36}$$

where ϕ_{ij} is the angle between the spins at sites i and j, so that the system has a continuous symmetry. For dimensions $n_d \leq 2$ the model should strictly have no true transition, according to Mermin and Wegner's theorem [Mermin and Wagner, 1966; Hohenberg, 1967], that states that systems with continuous symmetries and sufficiently short-range interactions cannot spontaneously order, breaking this symmetry, at some finite temperature, if $n_d \leq 2$. However, Kosterlitz and Thouless [1973] (KT) have shown that the spins can give stable pairs of structures (*vortices*) that allow a sort of fairly long-range correlation in the system with power law decay below a transition temperature and that unbind with correlations rapidly decaying above. In the LC field, the smectic A-smectic C transition falls into the XY-universality class: its order parameter can be given as a two-component vector (d_x, d_y),

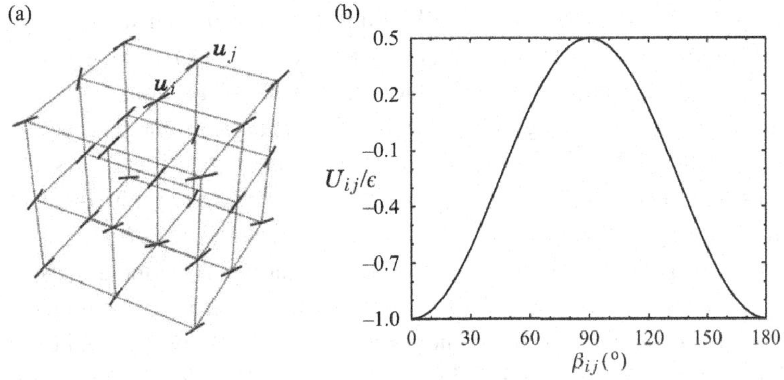

Figure 2.20 (a) The 3D LL cubic lattice model. The particles sitting at the lattice sites have direction $\boldsymbol{u}_i$ and interact with their nearest neighbours $\boldsymbol{u}_j$ with the potential U_{ij}/ϵ in Eq. 2.37 shown in (b) as a function of the relative orientation angle β_{ij}.

projection of the director onto the smectic layer plane. Disclinations, i.e. discontinuous changes of the direction of (n_x, n_y), correspond here to the defects of the KT model. Quasi 2D, freely suspended smectic films with thickness from a few nanometres to microns and their transitions have been studied and discussed in terms of their compliance with XY behaviour [Young et al., 1978; Bahr, 1994; de Jeu et al., 2003].

For a 3D cubic lattice (3D XY, $n_d = 3$, $n_o = 2$) Monte Carlo simulations [Gottlob and Hasenbusch, 1993] indicate that the model has a transition at $k_B T_C/\epsilon = 2.20167$. The heat capacity exponent is $\alpha = -0.002$. The model has been used to explain the so-called λ *transition* from normal to superfluid helium but also in classifying certain smectic transitions. The 3D XY model has been used for discussing various LC transitions (see Section 2.12.1) as, ideally, the nematic-smectic A transition should belong to the same universality class [Garland and Nounesis, 1994].

2.8.5 Lebwohl–Lasher Model

A natural generalization of the Heisenberg model to systems, like nematics, where the constituent particles are characterized by a molecular direction, rather than by a polar vector, is the Lebwohl and Lasher [1972] (LL) lattice model, where molecules (or small tight clusters of molecules) are represented by centres of interaction ('spins') placed at the sites of a simple cubic lattice and interacting with a nearest neighbour second-rank pair potential. The Hamiltonian is

$$U_N = -\frac{1}{2}\sum_{i,j} \epsilon_{ij} P_2(\cos\beta_{ij}) \equiv -\frac{\epsilon}{2}\sum_{\langle ij\rangle}\left(\frac{3}{2}(\boldsymbol{u}_i \cdot \boldsymbol{u}_j)^2 - \frac{1}{2}\right) \equiv -\frac{\epsilon}{2}\sum_{\langle ij\rangle} U_{ij}, \tag{2.37}$$

where ϵ_{ij} is a positive constant ϵ, for neighbouring sites i and j, indicated with $\langle ij \rangle$, and 0 otherwise. β_{ij} is the relative orientation between the i and j spins, i.e. $\cos \beta_{ij} = \boldsymbol{u}_i \cdot \boldsymbol{u}_j$ (see Fig. 2.20). No analytic solution exists for the orientational phase transition of the model, but we shall see that many analogies exists with the nematic-isotropic phase transition and that, more generally, the model is an extremely useful one in that it captures the essential orientational properties of nematics. In particular, the model has a weak first-order transition from an ordered state with $\langle P_2 \rangle > 0$ to a disordered, isotropic one at $k_B T_{NI}/\epsilon \approx 1.1224$. The temperature dependence of the order and its value at the transition are also similar to experiment for many real nematics. This system has been studied by a number of authors using a great variety of theoretical techniques in particular, computer simulations [Fabbri and Zannoni, 1986; Zhang et al., 1992; Shekhar et al., 2012], as we shall see in detail later in Chapter 10. A discrete version of the LL model, analogous to the Potts model but with a P_2 interaction, like in Eq. 2.37, was studied by Lasher [1972]. Even for a rather large set of allowed orientations (twelve) for the spins, the model presents marked differences with the continuous LL version, with a much too strong order parameter at the order-disorder transition.

After this general discussion on phase transitions, their classifications and main features, and a brief introduction to the foremost classical lattice models, we now turn to examining the main liquid crystal transitions.

2.9 The Nematic-Isotropic Transition

2.9.1 Main Features

The nematic-isotropic transition is arguably the most important of the transitions presented by liquid crystalline materials. Upon heating it takes place at the point where liquid crystals lose their orientational order and become ordinary isotropic fluids. In Table 2.1 we have reported the transition entropies for some popular nematics both for the transition from crystal to nematic and from nematic to isotropic. On the basis of the classification scheme that we have seen before, the nematic-isotropic transition is definitely first order, since it has a non-zero latent heat. However, its transition entropy is very low compared, for example, to the ordinary entropies of melting (see Table 2.1) and some characters of a second-order transition are present. In particular, important pretransitional effects appear in the isotropic phase as the nematic-isotropic transition is approached [Stinson and Litster, 1970]. As an example, the experimental heat capacity versus temperature curve for MBBA is shown in Fig. 2.21 [Anisimov, 1987]. The small variation in entropy suggests that the change in orientational order and the rather dramatic disappearance of anisotropy on going from liquid crystal to liquid is not associated with a big structural change [Luckhurst and Zannoni, 1977]. We shall see in Chapter 12 that this is confirmed by detailed atomistic simulations. The volume change at the nematic-isotropic transition is also much smaller than that at the crystal to nematic transition, as already mentioned in Section 2.3 [Würflinger and Sandmann, 2001], hinting that the molecular organization of a liquid crystal is much

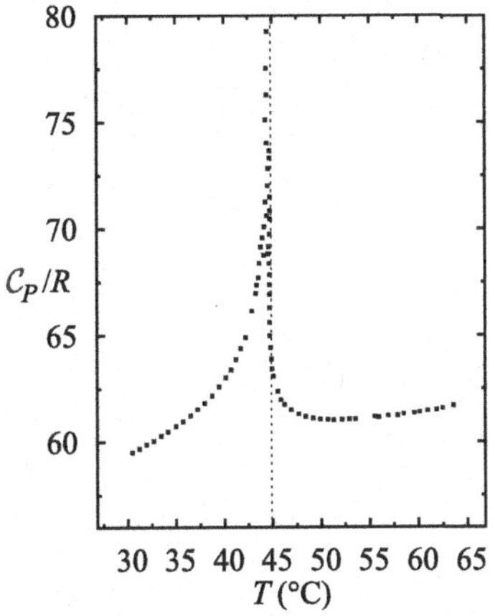

Figure 2.21 The temperature dependence of the heat capacity C_P near the nematic-isotropic transition for MBBA [Anisimov, 1987].

more similar to its liquid than to its solid phase. Note that, even if the entropy change is very small, the fact that the volume change is also small accounts (see Eq. 2.11) for the relatively large slope of the nematic-isotropic transition line (see Figs. 2.6 and 2.7). The pressure normally has a stabilizing effect on the ordered phase that remains stable at higher temperatures. Indeed, there are instances of materials that do not present a nematic phase at ordinary atmospheric conditions, while they do have one under a sufficiently high pressure.

2.9.2 The Effect of Molecular Structure

It is reasonable to assume that in thermotropic systems an increase in the nematic-isotropic transition temperature indicates a system more resilient to disordering (thermally robust), so that we can take a higher transition temperature as an indication of a higher stability of the nematic phase. A classic demonstration is the so called *odd-even* effect, which corresponds to a large alternation in properties and in mesophase transition temperatures, in particular for homologous series containing n methylene, CH_2, units as n varies from even to odd. Examples of this effect have been known in particular from the chemical synthesis work of Gray et al. [1973] and Gray [1979] and in Fig. 2.22 we show one such case for the ω-phenyl-n-alkyl 4-p-cyano benzylidene amino cinnamates series. Even though the lowest NI transitions are monotropic ones, observed only upon undercooling, the changes in clearing temperatures upon heating are impressive for such small, and apparently innocuous, chemical changes in the series. The large change in clearing temperature has been explained qualitatively considering the most stretched conformation of each molecule and observing that the terminal phenyl is in line with the long molecular axis only for the even terms (Fig. 2.22). While this has certainly an element of truth, it is clear that there are

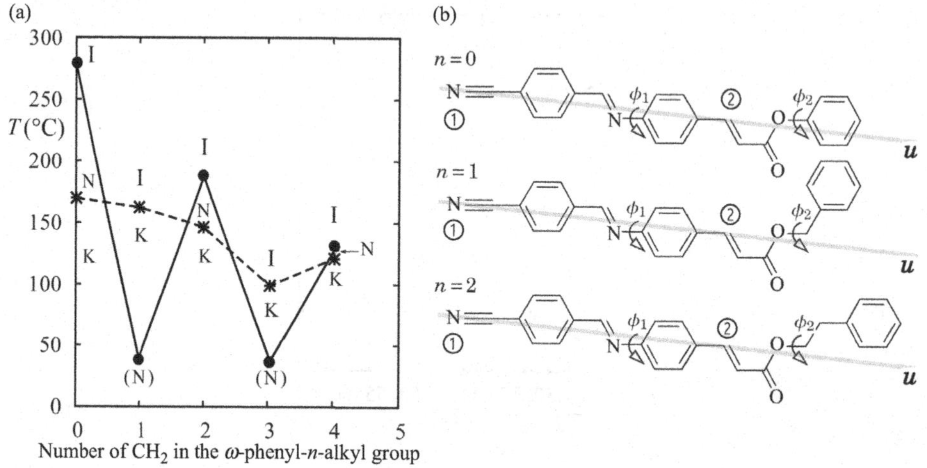

Figure 2.22 (a) The odd-even alternation of the NI transition temperature (●) in the ω-phenyl-n-alkyl 4-p-cyano benzylidene amino cinnamates series. (N) indicates a monotropic nematic transition. Asterisks ($*$) indicate a transition to crystal (K), while the lines are just a guide for the eye [Gray and Harrison, 1971]. (b) The chemical structure of three homologues ($n = 0, 1, 2$), showing the torsional angles ϕ_1 and ϕ_2 and the odd-even shape change in the fully stretched conformation. The thick grey lines through atoms ① and ② indicate the chosen reference molecular axes $\boldsymbol{u}$ [Berardi et al., 2004a].

a large number of conformations for each of these molecules and that the fully stretched one is unlikely to be the only (or perhaps even the most) populated one anyway, leaving the argument on shaky ground. We shall see later, in Section 12.3, how atomistic simulations [Berardi et al., 2004a] can clarify this issue.

2.10 Blue Phases

Blue phases have been well studied theoretically [Lubensky and Stark, 1996] by scanning calorimetric techniques, in particular by Thoen [1988] and Crooker [2001]. Note that the isotropic phase and BPIII have the same isotropic symmetry and that, according to the discussion in Section 2.7, when varying the chirality χ a liquid-gas-like critical point P_χ terminating a line of coexistence can then be expected in the temperature-chirality phase diagram sketched in Fig. 2.23a [Lubensky and Stark, 1996]. Such a critical point has been experimentally found [Garland, 2001]. In Fig. 2.23 a certain chirality is chosen, in practice by selecting a suitable compound: cholesteryl nonanoate and the sequences of BP phases observed by Thoen [1988] is reported. The strongest transition is by far the isotropic to BPIII, with a latent heat more than one order of magnitude larger than the other two.

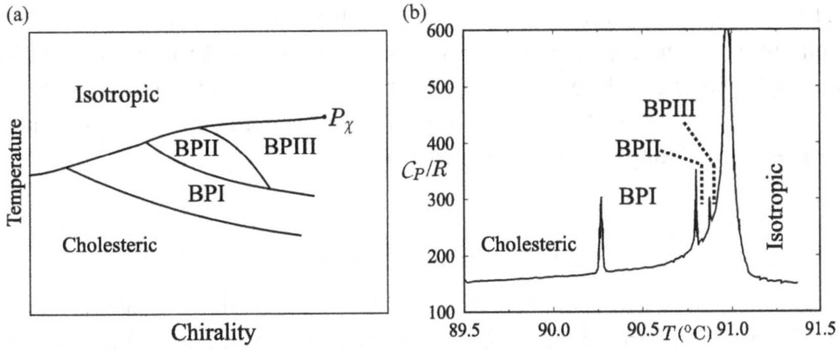

Figure 2.23 (a) Theoretically proposed qualitative BP phase diagram as chirality increases [Lubensky and Stark, 1996] up to critical point P_χ. (b) Adiabatic calorimetry of cholesteryl nonanoate, showing the transitions between the cholesteric, BPI, BPII, BPIII and isotropic phases. The BP latent heats (in J/mol) are: 18 (Cholesteric-BPI); 5.8 (BPI-BPII); 1.9 (BPII-BPIII); 170 (BPIII-Iso), much weaker than those typical of nematic-isotropic transitions [Thoen, 1988; Crooker, 2001].

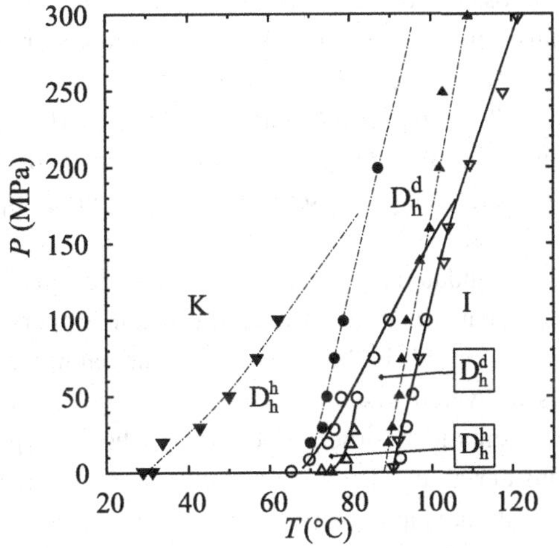

Figure 2.24 The phase diagram for the discotic HHTT obtained using wide angle X-ray diffraction (WAXD). The sequence crystal, hexagonal columnar, isotropic: K $\rightarrow$ D_h^d $\rightarrow$ I on heating and I $\rightarrow$ D_h^d $\rightarrow$ D_h^h $\rightarrow$ K transition sequences on cooling isotropic (I), columnar disordered hexagonal (D_h^d), helicoidal (D_h^h) and crystalline (K) are shown. Solid and broken lines refer to the phase sequences observed during heating and cooling scans, respectively [Maeda et al., 2003].

2.11 Columnar Liquid Crystals

The number of experimental phase diagrams available for discotics systems is even smaller than that of LCs formed by rod-like molecules seen until now. In Fig. 2.24 we see one of the few examples available for a rather large pressure range, that of the columnar phases of HHTT, introduced in Section 1.10 and whose charge transport properties have been shown in Fig. 1.44. The system exhibits two triple points: one at 40 MPa and 77°C for the $K-D_h^h-D_h^d$ coexistence [Maeda et al., 2003] and the other, extrapolated at 285 MPa, 118°C for the $K - D_h^d - I$ phases [Maeda et al., 2001]. Rather large differences are also shown in Fig. 2.24 between heating and cooling thermal scans.

2.12 Smectic Transitions

Given the variety of smectic liquid crystals, it is nearly impossible to examine all the different combinations of transitions they present. Thus, we shall concentrate on what are probably the most important and well studied to date: the smectic A-nematic and the smectic A-smectic C.

2.12.1 Smectic A-Nematic Transitions

In the variety of liquid crystal transitions taking place in materials exhibiting smectic phases some are foreseen to be possibly of second order. In particular, for materials that have a nematic and smectic transition, the smectic-nematic transition is predicted to change character from first to second order as the width of the nematic range increases, e.g. as signalled by the decrease in the ratio T_{AN}/T_{NI} [McMillan, 1971, 1972]. McMillan theory, discussed later in Chapter 7 assumes that the change in behaviour can be driven by systematically changing some molecular feature, e.g. in a series of homologous compounds like the nCBs, by a change in the chain length.

At a macroscopic level, Landau theory can be applied to the Sm-N transition, expanding the free energy around the nematic at the smectic transition temperature. A generalization needed with respect to Section 2.7 is that now two order parameters need to be considered, since a smectic is a layered system, endowed with positional, as well as orientational, order. As for the orientational order, we could use the same parameter η previously used for the nematic considering the deviation of the order from the value at the transition $\delta\eta \equiv \eta - \eta_{NS}$. For the positional parameter order we can use an empirical parameter τ that is 0 in the nematic phase and different from 0 in a periodic, structured phase. If we refer for simplicity to a smectic phase formed by virus particles, where the layers are directly visible, like in Fig. 1.58d we could consider as positional order parameter

$$\tau = \langle \cos(2\pi z/\ell_z) \rangle , \tag{2.38}$$

where z is the position of the particles and ℓ_z the spacing along the layer normal (see Section 3.2). Clearly, τ is 0 for uniform, nematic or isotropic arrangements of the particles and can be normalized to 1 for a perfectly periodic structure. The free energy expanded in terms of the two order parameters can be written as [Barois, 1999; Oswald and Pieranski, 2006]

$$\mathcal{G}(\tau, \eta, T) = a(T - T^*)\tau^2 + g_{40}\tau^4 - g_{21}\tau^2\delta\eta + g_{02}\delta\eta^2 + \cdots . \tag{2.39}$$

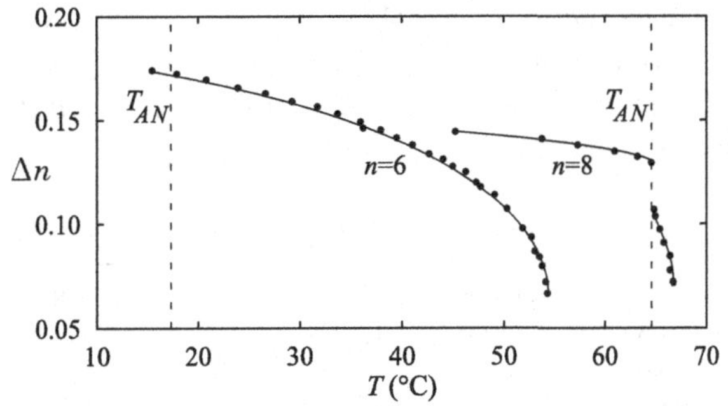

Figure 2.25 The refractive index anisotropy Δn (at a wavelength $\lambda = 632.8$ nm) of di-n-hexyl and di-n-octyl azoxybenzene plotted as a function of temperature near their S_A-nematic transition indicating a continuous second order transition for $n = 6$ and a discontinuous first-order one for $n = 8$ [de Jeu, 1973].

The first and second subscripts in the expansion coefficients g_{mn} refer to the power of the positional and orientational parameters, respectively. For $g_{21} > 0$ and $g_{02} > 0$, the energy has a minimum in $\delta\eta$ for $\delta\eta = g_{21}\tau^2/(2g_{40})$. Substituting, this gives

$$\mathcal{G}(\tau, T) = \frac{1}{2}a(T - T^*)\,\tau^2 + \frac{1}{4}b\tau^4 + \cdots, \tag{2.40}$$

where $b = g_{40} - g_{21}^2/g_{20}$. The transition can be of first order if b becomes negative and this could happen if $\delta\eta$ is large, as we could expect if the nematic range is small. This is to a good extent verified experimentally [Thoen, 1995]. The simplified description we have just seen gives only an idea of the Landau theory applications to the variety of liquid crystal transitions. In this generalized approach, after identification of a suitable order parameter (scalar, vector, tensor) the free energy is expanded in a generalized Taylor series in terms of invariant combinations of the order parameters. A detailed treatment, particularly for smectics, can be found in the classic book of de Gennes [1974] and a number of other works (e.g. [Chu and McMillan, 1977; Kovshev et al., 1977; Gramsbergen et al., 1986a; Tolédano and Tolédano, 1987; Pikin, 1991]).

From the microscopic theory standpoint, we shall see in Chapter 7 that the Mean Field Theory developed by McMillan [1971, 1972] predicts a first-order transition for T_{AN}/T_{NI} sufficiently close to 1 and a switch from first to second order for $T_{AN}/T_{NI} < 0.87$ at a Landau *tricritical point*. The TCP is predicted for $T_{AN}/T_{NI} = 0.87$. Thus, small nematic ranges would give a first-order transition and large nematic ranges a second-order one. This variation has been studied experimentally in certain homologous series. For instance, de Jeu [1973] has studied the series of di-n-alkyl azoxybenzenes (nAB). The change in behaviour is very apparent looking at the temperature dependence of the birefringence for two members of the series, as we do in Fig. 2.25. While the longer homologue shows a clear break at the smectic-nematic, there is hardly anything showing for the hexyl compound. The calorimetry data are given in Table 2.4. Thus, some liquid crystal transitions are experimentally found to have essentially second-order character in the sense that the entropy jump is 0 within

Table 2.4. *Latent heat* ΔH (kJ/mol) *for the phase transitions in some nCB and nAB (n-alkyl-4-azobenzene) mesogens: crystal-smectic A (KA), smectic A- nematic (AN), nematic-isotropic (NI), smectic A-isotropic (AI) (a) [Marynissen et al., 1983] and (b) [de Jeu, 1973]*

Mesogen	ΔH_{KA}	ΔH_{AN}	ΔH_{NI}	ΔH_{AI}	T_{AN}/T_{NI}	Ref.
8CB	25.7	$\leq$0.0004	0.612	–	0.978	(*a*)
9CB	34.5	$\leq$0.005	1.20	–	0.994	(*a*)
10CB	36.0	–	–	2.83	–	(*a*)
11CB	43.2	–	–	3.8	–	(*a*)
6AB	11.7	0.02	0.57	–	0.887	(*b*)
7AB	12.6	0.16	1.1	–	0.952	(*b*)
8AB	19.4	2.30	2.30	–	0.993	(*b*)

experimental error. For pure *n*CB only two cases, 8CB and 9CB, can be used to test the McMillan prediction and do not seem to quite verify it. A further test has been done on mixtures on 9CB and 10CB and the results indicate a stronger first-order transition as T_{AN}/T_{NI} gets closer to 1, i.e. for narrow nematic ranges [Thoen, 1995]. The type of smectic A polymorphism also seems to affect the observed heat capacity anomaly. For NA and NA_d-type transitions in materials with moderately large nematic ranges, the critical heat capacity contribution becomes very small (or even undetectable). However, quite large C_P anomalies are observed for NA_1 and NA_2 transitions to SA_1 and SA_2 phases for compounds with very large nematic ranges [Thoen, 1995]. As another example, the transition entropies for the cascade of phases of TBBA, given as $\Delta S/R$, i.e. in dimensionless form, are:

$$\text{Solid} \overset{5.661}{\longleftrightarrow} S_B \overset{1.084}{\longleftrightarrow} S_C \overset{\approx 0}{\longleftrightarrow} S_A \overset{0.074}{\longleftrightarrow} N \overset{0.177}{\longleftrightarrow} I.$$

We can see that the entropy change is quite different at the various transitions, e.g. the S_C to S_A transition is second order. We report in Table 2.4 the heats of transition for various cyano-biphenyls with smectic phases, measured by Marynissen et al. [1983].

In Fig. 2.26 we show a calorimetric scan for these four cyano-biphenyls. We see that the behaviour of the heat capacity is not consistent with the simplified Ehrenfest scheme, since rather than showing a jump, the heat capacity diverges as T_{AN} is approached. The heat capacity critical exponents can be determined by fitting the experimental C_P data to Eqs. 2.13, obtaining for 8CB ($\alpha = 0.31 \pm 0.03$) and for 9CB ($\alpha = 0.50 \pm 0.05$) [Marynissen et al., 1983]. The McMillan [1971] criterion applied to 8CB ($T_{AN}/T_{NI} = 0.978$) and 9CB ($T_{AN}/T_{NI} = 0.994$) would predict these transitions to be first order, in view of the short nematic range, but the transitions are second order. The N$-$$S_A$ transition in 9CB seems actually very close to a tricritical point, where the transition switches from second order to first order, since the expectations at this point are of $\alpha = \alpha' = 0.5$ (see Table 2.3). According to a conjecture of de Gennes [1972], the fact that the S_A order parameter has two components suggests that the NA transitions should be similar to that of superfluid helium and belong to the 3D XY universality class, that has critical exponent values ($\alpha = \alpha' = -0.02$), albeit

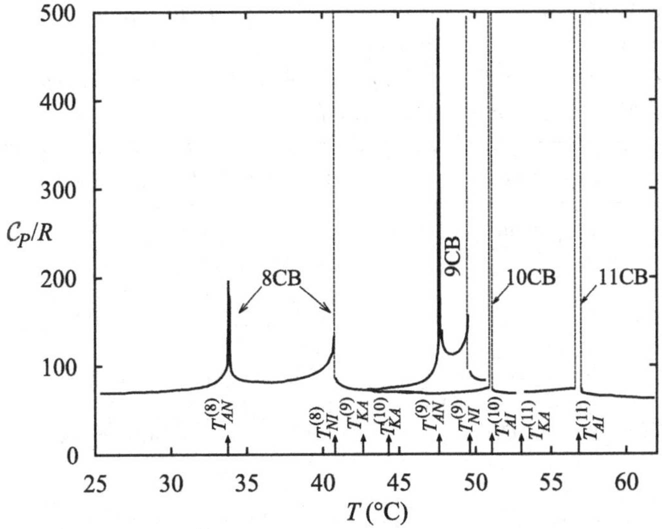

Figure 2.26 Molar heat capacity C_P for four cyano-biphenyls: 8CB, 9CB, 10CB, 11CB as a function of temperature. $T_{KA}^{(n)}$, $T_{AN}^{(n)}$, $T_{NI}^{(n)}$, $T_{AI}^{(n)}$ indicate the transition temperatures for the nth homologue. The vertical dashed lines indicate the first-order transitions and the separation between the lines at a given transition indicates the width of a two-phase region [Marynissen et al., 1983].

with an inverted C_P amplitude. The asymmetry of the heat capacity peak appears to be fairly consistent with this inverted XY behaviour at least for 8CB. In particular, $C_P(T)$ decays on the hot side of the transition to a value higher than that on the cold side. However, it now seems from a number of experimental observations (see also [Garland et al., 1983]) that this transition exhibits a non-universal critical behaviour and that it does not follow the 3D XY model. It is expected that XY behaviour will be observed when the McMillan ratio $T_{AN}/T_{NI} < 0.94$ [Marynissen et al., 1983]. In practice the situation seems even more complicated because of the presence of various types of smectic phases, as discussed in Chapter 1. For instance, crossover from 3D XY to tricritical behaviour has been reported [Nounesis et al., 1991] for N–S_{A_1} mixtures of octyloxyphenyl-nitrobenzoyloxy benzoate (DB_8ONO_2) and decyloxyphenyl-nitrobenzoyloxy benzoate [Nounesis et al., 1991]. The transition in dimer-like smectics S_{A_2} to nematic has been investigated by Wen et al. [1991]. Choosing 4′-n-heptyloxy-carbonylphenyl-4′-(4″-cyano-benzoyloxy) benzoate (7APCBB),

$$\text{Solid} \xrightarrow{121.50°\text{C}} S_{C_2} \overset{141.61°\text{C}}{\longleftrightarrow} S_{A_2} \xrightarrow{144.41°\text{C}} \text{N} \xrightarrow{209.41°\text{C}} \text{I},$$

a material with a wide nematic range and thus a rather small McMillan ratio, 0.86, so as to avoid as far as possible the possibility of first-order transition or tricritical behaviour they found a second-order transition well characterized in terms of 3D XY exponents ($\alpha = -0.07$) and non-inverted amplitude ratio. A brief summary of exponents for various LC transitions is given in Table 2.5.

Table 2.5. *A summary of experimental results for various liquid crystal phase transitions. Abbreviations: 1st, first order; 2nd, second order; CP, critical point; TCP, tricritical point. BP is any of the three Blue Phases (adapted from [Garland, 2001], where more details can be found)*

Transition	Experiment	Comments and additional Refs.
N–I	1st, pretrans.	[Lubensky and Priest, 1974]
BPIII–I	1st → CP	[Lubensky and Stark, 1996]
BP–BP	1st	See Fig. 2.23b, [Thoen, 1988]
N*–BP	1st	See Fig. 2.23b, [Thoen, 1988]
S_A–I	1st	Weak pretransition
N–S_{A_m}	2nd → TCP → 1st	Monolayer Sm from non-polar mesogens
N–S_{A_d}	2nd → TCP → 1st	See Fig. 1.28
N–S_{A_1}	XY → TCP	Monolayer Sm from polar mesogens
N–S_{A_2}	1st and 2nd seen	
N–S_C	1st	[Swift, 1976]
S_{A_1}–S_{A_2}	Ising-like	
S_{A_1}–$S_{\tilde{A}}$	1st	Unusual C_p wings
$S_{\tilde{A}}$–S_{A_2}	Broad 1st coexistence	
S_{A_d}–S_{A_2}	1st → CP (unusual)	
S_{A_d}–S_{A_1}	1st → N(r) region	Re-entrant
S_A–S_{B_H}	Unusual 2nd/weak 1st	
S_A–S_{B_K}	1st	Weak pretransition C_p wings

2.13 Liquid Crystal Polymers

The thermal behaviour of liquid crystal polymers (LCPs) is to some extent similar to that of low-molar-mass liquid crystals, but with the important difference that, like for most polymers, a completely crystalline polymer is very hard, if at all possible, to obtain, and on cooling down from the isotropic melt a glass transition may occur. The LCP materials that can show a crystalline phase are principally the main-chain ones and, in this case, the polymer above the melting and before the isotropic transition behaves similarly to low molecular weight liquid crystals, including flow, even though with a much higher viscosity. A different situation is observed for non-crystallizable polymers, most polymers with mesogenic side group. In this case, the LC state is limited below by the glass transition temperature T_g, shown in the calorimetry by the typical step, rather than a peak (see Fig. 2.27). The glass transition is not a true thermodynamic one as the ones described in previous sections, but rather corresponds to a situation where the increase of viscosity when cooling down a melt is so high that a number of degrees of freedom of the molecules are effectively frozen in, not allowing the system to explore all the positions, orientations and conformations eventually leading to a true crystalline state. The value of T_g then needs to be defined according to an experimental protocol, and in particular specifying a cooling rate (e.g. in Fig. 2.27 it is 10°C/min).

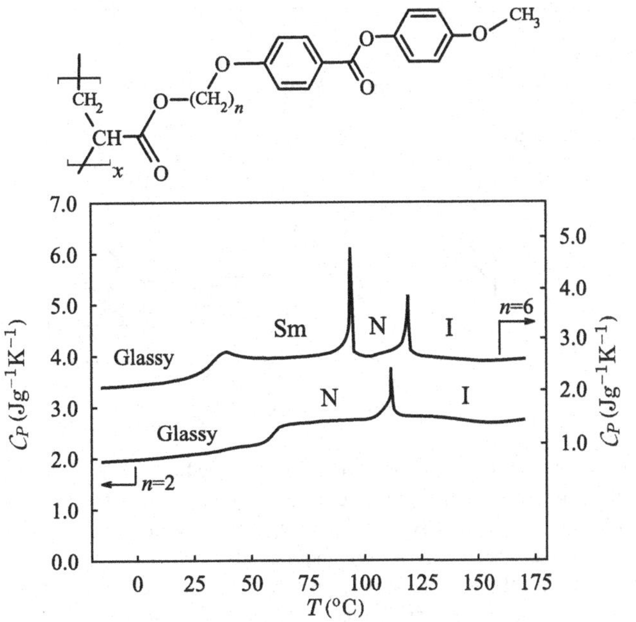

Figure 2.27 DSC curves, showing the dependence of C_P on temperature for two liquid crystalline side-chain polymers with a polyacrylate backbone (structure on top) and $n = 2$ and $n = 6$. The phases observed are: isotropic melt (I), nematic (N), smectic (Sm) and glassy liquid crystal [Portugall et al., 1982; Wassmer et al., 1985].

2.14 Lyotropics

The situation regarding the transitions from one molecular organization to another is even more complicated in the case of mixtures, instead of the pure materials we have treated until now. In particular, in the lyotropic systems obtained from amphiphilic molecules in a solvent, that we have qualitatively described in Section 1.11, phase changes are not only driven by temperature, but are also brought about by changing the composition, e.g. the amount of solvent with respect to the amphiphiles (see, e.g., [Tiddy, 1980]). Given these additional complications with respect to one-component thermotropics, it is not surprising that many details on the transitions involved, like the critical exponents or the type of transition, are normally not available. This difficulty is even more pronounced for computer simulation studies aimed at investigating the full phase diagrams of lyotropic systems, in view of the huge parameter space to be considered when changes in chemical compositions have to be taken into account. We, anyway, report a few examples from the better investigated systems.

2.14.1 Micellar Systems

A relatively simple example is that of a binary system, that of potassium stearate-water, whose phase diagram in Fig. 2.28 [Charvolin and Hendrikx, 1985] shows the variety of phases that can arise changing the concentration of the amphiphilic molecules and temper-

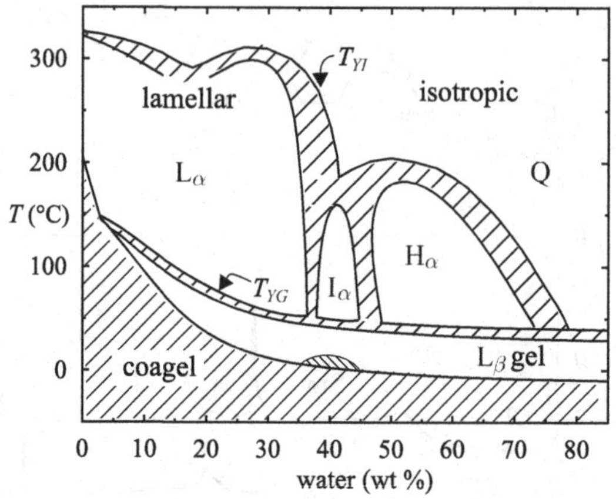

Figure 2.28 Phase diagram of the potassium stearate (see Fig. 1.46)-water system. Isotropic solutions are obtained above the lyotropic-isotropic temperatures T_{YI} and gel-coagel waxy non-equilibrium phases below T_{YG}. A phase, Q, of spherical micelles, occurring at high water content, gives rise to hexagonal and lamellar phases increasing the amphiphile concentration in the intermediate temperature region. An intermediate phase (I_α) is also observed. The hatched regions are biphasic [McBain and Sierichs, 1948; Charvolin and Hendrikx, 1985].

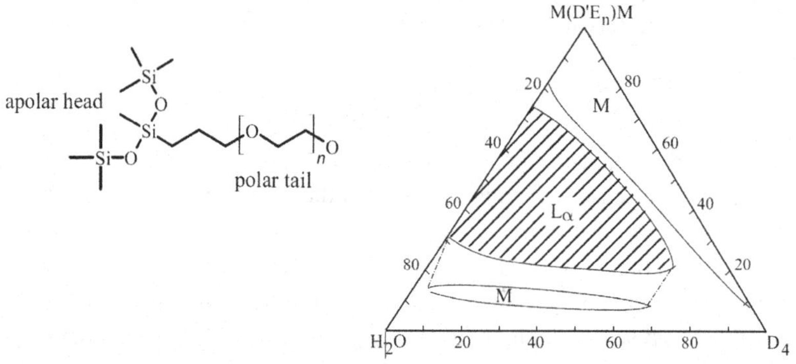

Figure 2.29 Ternary phase behaviour of a silicon oil-water-trisiloxane surfactant M(D′En)M with a polar head and polar chain. Here M is a microemulsion and L_α a lamellar phase [Li et al., 1996] .

ature, allowing to switch the molecular organization between the various micellar systems as summarized in Chapter 1.

Another example, this time of a three-component system is reported in Fig. 2.29. Note that, although the typical amphiphile with a polar head and lipid chain is most common, the opposite situation can also be found, as illustrated in this system, where the syloxane head is essentially non-polar, while the chains are polar because of the oxygens.

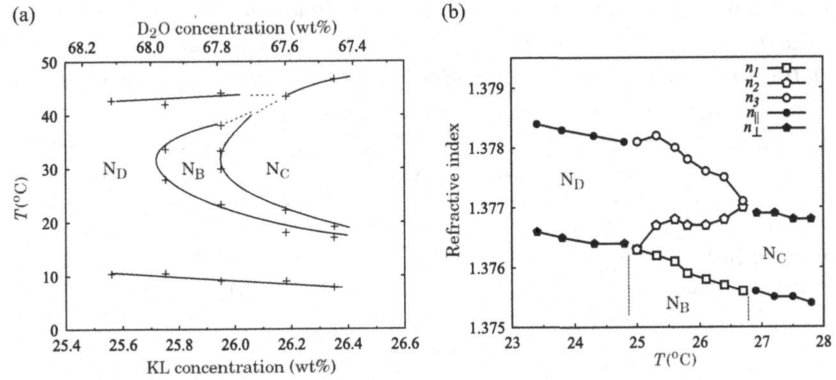

Figure 2.30 (a) Phase diagram for the ternary system KL, decanol and D_2O, with 6.24 wt% of 1-decanol [Yu and Saupe, 1980] and (b) refractive indices showing the occurrence of a biaxial nematic phase in this lyotropic system [Santoro et al., 2006].

2.14.2 Biaxial Micellar Phases

The first experimentally found of the much sought after biaxial nematic phases [Luckhurst and Sluckin, 2015] and probably the one which has received the most convinced consensus on its nature, is actually a lyotropic micellar one, discovered by Yu and Saupe [1980] in an appropriate range of concentration and temperature of a ternary system of potassium laurate (KL) with 1-decanol and D_2O. The N_B phase is bracketed between two uniaxial phases that appear to be formed by cylindrical micelles, N_C and by disc-like bilayer micelles N_D as seen in Fig. 2.30. The evidence for phase biaxiality was based in particular on optical observations. Fig. 2.30 shows the change from two (parallel and perpendicular to the director) to three refractive indices on moving from the uniaxial to biaxial phase.

2.14.3 Membrane Bilayers

As we have seen in Chapter 1, suspensions of phospholipids in water can give rise under suitable conditions to bilayer structures [Luzzati and Tardieu, 1974; Chapman, 1975; Mabrey-Gaud, 1981]. In large multilamellar liposomes of phosphatidyl-cholines (lecithins), such as dimyristoil-phosphatidyl choline (DMPC) and dipalmitoyl-phosphatidyl choline (DPPC, also called DPL) the calorimetric scan typically shows two peaks, as we see in Fig. 2.31: one, normally at lower temperature and weaker, called the *pretransition* and a sharper one at higher temperature, called the *main* transition. Transition enthalpies and entropies for some typical lipid systems are reported in Table 2.6. Qualitatively the interpretation is that below the main transition the lipids are in a *gel* state, with the chains typically in the stretched, all-trans, conformation. The chain positions themselves can have different organizations (Fig. 2.32). Thus, below the pretransition, the structure in DPPC has crystalline tilted chains, $L_{\beta'}$ and in the intermediate phase between the pretransition and T_M it becomes hexagonal $P_{\beta'}$. This intermediate phase is further complicated by the existence of bilayer

Table 2.6. *Transition temperature (°C), enthalpy* (kJ/mol) *and entropy (R units) jump for some liposomes. We give the 'pretransitional' (P) and 'main' (M) values (see text) for DMPC (a) [Mabrey-Gaud, 1981] and for DPPC. For DPPC we also show data for freshly prepared unilamellar vesicles (b) [Suurkuusk et al., 1976]*

Lipid	T_P	ΔH_P	ΔS_P	T_M	ΔH_M	ΔS_M	Ref.
DMPC	14.2	4.2	1.75	23.9	22.8	9.22	(a)
DPPC multilamellar	35.4	6.7	2.6	41.2	34.3	13.1	(b)
DPPC monolamellar	36.9	16.3	3.7	41.2	12.1	15.5	(b)

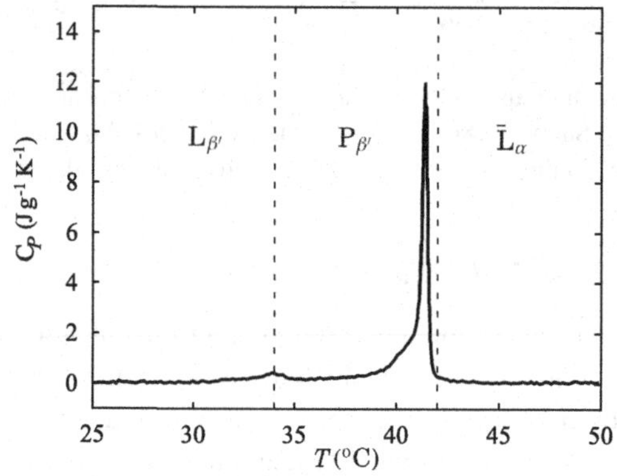

Figure 2.31 Calorimetric scan for DPPC multilamellar vesicles in a buffer at pH 7.4. The $L_{\beta'}$ – $P_{\beta'}$ (pretransitional) and the $P_{\beta'}$ –$\dot{L}_\alpha$ (main) phase changes (see Fig. 1.50) are indicated by the dashed lines [Losada-Perez et al., 2014] .

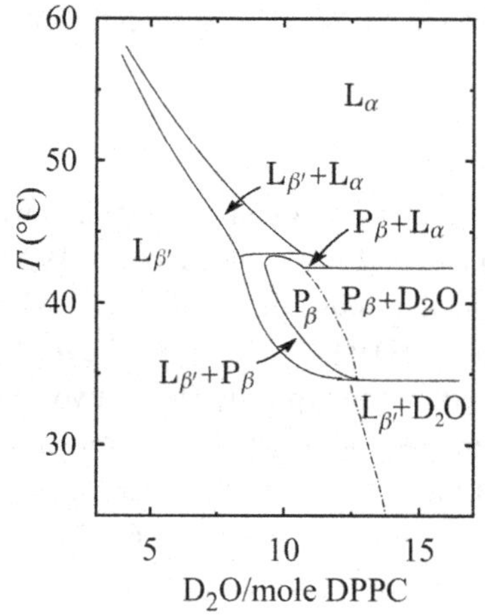

Figure 2.32 The DPPC-D_2O phase diagram. L denotes a one-dimensional lamellar structure; P denotes a two-dimensional monoclinic lattice. In the α and β phases, the hydrocarbon chains are liquid-like and solid-like, respectively. The prime indicates that the carbon chains are tilted with respect to the lamellar plane. The broken lines indicate that the phase boundary is uncertain [Ulmius et al., 1977; Jönsson et al., 1984].

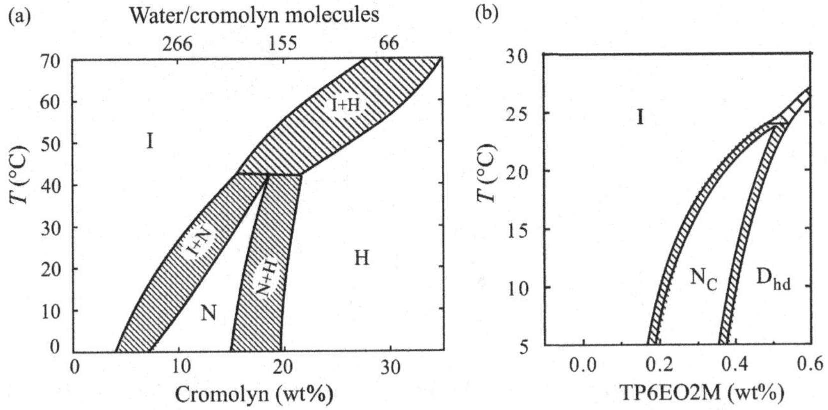

Figure 2.33 (a) Phase diagram for disodium cromoglycate (Cromolyn)-water showing the various phases formed [Cox et al., 1971]. (b) A portion of the phase diagram for the nonionic chromonic TP6EO2M when changing the amphiphile concentration [Boden et al., 1986]. The coexistence regions are hatched.

undulations (*ripples*) with wavelengths of about 200 Å. Above the *main transition* we have *chain melting* through *trans-gauche* isomerization and the bilayer can be roughly assimilated to a two-dimensional liquid crystal fluid phase called L_α. In real membranes we have mixtures of various lipids as well as other components and the combination of transition behaviour gives rise to a smooth continuous change [Martonosi, 1974].

2.14.4 Chromonics

As for ordinary lyotropics, various chromonic phases can be obtained by changing temperature and concentration, even if the change in molecular organization is continuous, rather than depending on some threshold molar fraction. Upon increasing the concentration of amphiphilic chromonic, hexagonal phases are often formed, as we see for instance in Fig. 2.33a for DSCG and in Fig. 2.33b for the non-ionic discotic 2,3,6,7,10,1 l-hexa-(1,4,7-trioxa-octyl)-triphenylene (TP6EO2M), obtained functionalizing a discotic triphenylene with hydrophilic groups on the periphery, dissolved in water [Boden et al., 1985, 1986; Boden, 1990].

2.15 Phase Diagrams for Colloidal Suspensions

Although the equations of state for gases are very far from those of liquid crystals, at least thermotropic ones, it is worth noting an important analogy with colloidal suspensions, where the building blocks are particles from nano up to micron size that can form isotropic and anisotropic suspensions, as we have seen in Section 1.14. Colloidal suspensions have phase diagrams similar, in some sense, to atomic or molecular systems and

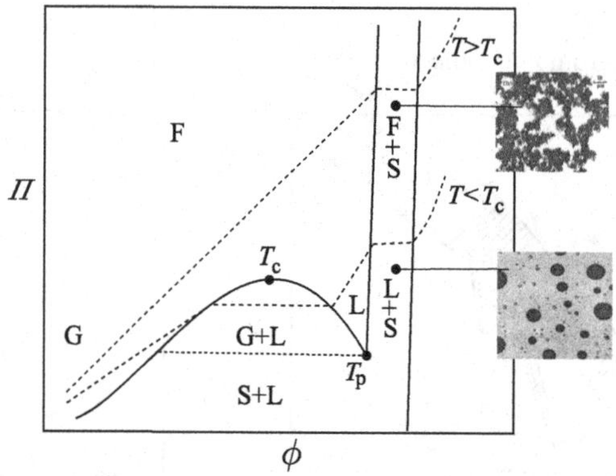

Figure 2.34 Schematic (Π, ϕ) theoretical diagram for a colloidal suspension that presents gas, solid and liquid phases: G = gas, L = liquid, S = solid, F = fluid; T_c, critical point; T_p, triple point. The dashed lines represent two isotherms. The insets represent experimental examples from an aqueous dispersion of γFe_2O_3 magnetic particles, surface coated with tri-sodium citrate molecules (pH $\approx$ 7) [Dubois et al., 2000].

phases equivalent to gas, liquid, fluid and solid can exist. In carrying on this very useful analogy the osmotic pressure Π replaces the pressure P and the *volume fraction* of particles ϕ is used instead of the density ρ employed for simple liquids. The diagram (Π, ϕ) thus replaces the standard diagram (P, ρ), as we see in Fig. 2.34. The colloidal gas phase corresponds to a solution of low volume fraction, with particles free to explore the whole sample volume, the liquid corresponds to a dense phase with positionally disordered particles in solution, while the crystal phase corresponds to a regularly organized solid phase. Liquid-crystalline phases can be obtained for anisotropic colloidal suspensions, as we have seen in Chapter 1, upon increasing the concentration of anisometric particles. Phase coexistence is possible, and easily detected by direct observation through crossed polarizers, as we have seen, for example, in Fig. 1.56. However, colloidal dispersions differ from simple molecular systems in some aspects. First, as the osmotic pressure is more difficult to measure than the pressure P, the (Π, ϕ) diagram is experimentally difficult to build directly and not many such representations are available in literature. Second, and more fundamentally, colloidal particles are usually polydisperse, differently from atomic or molecular systems, and present a certain spread in size around the average, a situation that has a deep influence on the phase diagrams, complicating the comparison with corresponding low molecular systems.

2.15.1 Carbon Nanotubes

The phase behaviour of single-wall carbon nanotubes (SWNTs) in acids is quite similar to that of rod-like macromolecule dispersions, and in particular they show the so-called Flory chimney [Zhang et al., 2006b], recalling a theory developed for suspension of

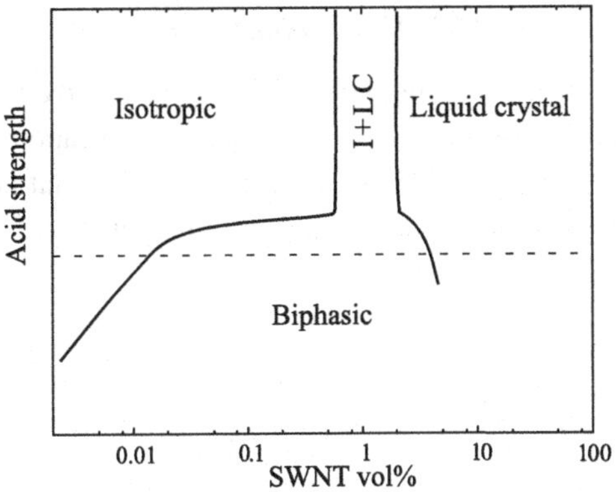

Figure 2.35 Postulated phase diagram of SWNTs in superacids. In 102% sulphuric acid (oleum), the biphasic region extends from a concentration of 100 ppm to 8%. These critical concentrations are expected to vary with the strength of the acid used [Rai et al., 2006].

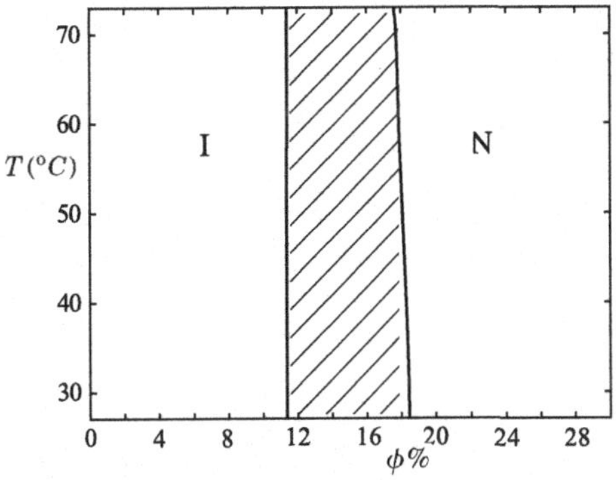

Figure 2.36 Phase diagram of CdSe nanorods suspensions showing isotropic, nematic and biphasic (hatched) regions [Li et al., 2004].

rod-like molecules such a TMV [Flory, 1956] and applied to other systems, e.g. benzyl alcohol solutions of the synthetic polypeptide poly(γ-benzyl-α, L-glutamate) or PBLG [Horton et al., 1990]. Fig. 2.35 shows a hypothetical phase diagram of SWNTs in 102% H_2SO_4 [Rai et al., 2006]. With increasing concentration, SWNT-superacid dispersions transition from an isotropic solution, where individual SWNTs are randomly oriented, to a biphasic system, where a birefringent ordered liquid phase is in equilibrium with the isotropic phase. These biphasic dispersions of rods should be separable into isotropic and ordered phases by application of external forces, such as ionic or centrifugal forces (or just gravity) [Rai et al., 2006]. Raising the concentration increases the proportion of ordered phase until the system becomes a fully liquid crystalline system. The critical concentrations at which this occurs is extremely low (a few parts per million!), which is reasonable considering the very high aspect ratio of the carbon nanotubes.

2.15.2 Mineral Suspensions

The phase diagram of suspensions colloidal particles will depend on their shape anisotropy. For CdSe nanorods with an aspect ratio (length/diameter) of about 15, the experimental phase diagram is shown in Fig. 2.36 [Li et al., 2004]. We see that at around 12% volume fraction the system enters a biphasic region, then becomes nematic, as shown by the pictures in Fig. 1.57.

3
Order Parameters

If I am to know an object, though I need not know its external properties, I must know all its internal properties.

L. Wittgenstein, *Tractatus Logico-Philosophicus, 1922*

… we can imagine liquid crystals as bodies in which the molecules, or more precisely their centres of mass, are distributed completely randomly, as in ordinary liquids. Anisotropy of the liquid crystal is caused by the equal orientation of its molecules; for instance, if the molecules have an elongated shape, then all of them can be arranged with their axes in one direction.

L. D. Landau, *Collected Papers of L. D. Landau, 1965*

3.1 Single Particle Distributions

In Chapter 2 we discussed how to describe phase transitions from a macroscopic point of view. We also saw how Landau's theory of phase transitions relies on the introduction of at least one phenomenological order parameter, η. Here we wish to tackle the problem of giving a molecular level description of a liquid crystal phase and of its order parameters. This is not so trivial, given not only the variety of liquid crystals that we have briefly described, but also that of the constituent, mesogenic, particles ranging from more or less complex molecules, as in thermotropic systems, to colloidal particles some 2–3 orders of magnitude larger. For a truly bulk system, but even for a very thin film containing perhaps an order of magnitude of 10^{12} molecules, the microscopic description we are looking for is necessarily a statistical one [Balescu, 1975; Landau and Lifshitz, 1980]. This does not have to mean that *configuration*s (i.e. sets of individual positions, orientations and possibly other degrees of freedom of all the molecules) are always not available. This is true for the bulk samples mentioned earlier, but not for systems of a few thousand to perhaps a few million molecules at equilibrium in certain thermodynamics conditions studied with computer simulations. These techniques, e.g. the Monte Carlo (MC) and Molecular Dynamics (MD) methods, that we will describe in some detail in Chapters 8 and 9, do actually generate molecular configurations and the problem then becomes that of reducing to the essential the huge amount of information available, rather than in its inaccessibility. We can also think of another example, when dealing with colloidal suspensions observed with some microscopy technique, where again a large set of positions and orientations of the particles can be extracted with some image analysis software. In general, we could adopt the definition of Fano [1957], that the *state* of a system is represented by the information required to calculate all the average properties of interest. To be more specific, let us consider a system of N identical molecules in a given state of aggregation, or phase (cf. Chapter 1) at certain thermodynamic conditions (temperature, pressure, ...). Since we focus on liquid crystals and they are formed by rather large molecules, as we have seen in Chapter 1, these can be considered as classical objects, and their quantum mechanical nature will only determine their physical properties (e.g. the bond lengths, angles, partial charges, etc.) and interactions. For a collection of such molecules where a complete, atomistic level description is available, we call a *configuration* the set of positions of all the N_a, atoms for each molecule. The complete information about a system of N such molecules at time, t, is represented by its configuration $\widetilde{\mathbf{X}}(t) = (\boldsymbol{r}_1(t), \boldsymbol{r}_2(t), \ldots, \boldsymbol{r}_{N_a \times N}(t))$, i.e. by the set of positions $\boldsymbol{r}_i(t)$ of all the atoms of all the molecules. This can be a bit overdetailed, if we want to avoid 'missing the forest for the trees' and in many generic (i.e. not chemically specific) models of liquid crystals, a molecular, rather than atomistic, level description is adopted, with the molecules assumed to be represented by classical, rigid particles with centre of mass at position $\boldsymbol{r}_i$ and orientation Ω_i. This simplified description applies also when we consider colloidal particles, where their inner structure is not relevant or unknown. The configuration of such a system of N particles at time t is then represented by $\widetilde{\mathbf{X}}(t) = (\boldsymbol{X}_1(t), \boldsymbol{X}_2(t), \ldots, \boldsymbol{X}_N(t))$, i.e. by the set of generalized coordinates $\boldsymbol{X}_i(t)$ of all the particles, with $\boldsymbol{X}_i(t) \equiv \boldsymbol{r}_i(t), \Omega_i(t)$ with $\boldsymbol{r}_i$ the coordinates of the centre of mass and $\Omega_i \equiv \Omega_{\mathrm{M}_i\mathrm{L}} \equiv \Omega_{i\mathrm{L}}$, the orientation of particle i frame, M_i, with respect to a chosen laboratory frame L and specified by a sufficient set of parameters (angles, or as we shall see later, quaternion components). In many cases, a further assumption of effective cylindrical symmetry of the particles is made, which means that $\Omega_i = (\alpha_i, \beta_i)$, with $0 \leq \alpha_i \leq 2\pi$, $0 \leq \beta_i \leq \pi$ as illustrated in Fig. 3.1a, is sufficient to specify the molecular orientation. When a particle, while being still rigid, is non-uniaxial, an additional angle, γ_i, $0 \leq \gamma_i \leq 2\pi$, is needed to fully determine the molecular orientation (Fig. 3.1b). Indeed, for any rigid particle the orientation can be given by three Euler angles (α, β, γ) [Rose, 1957].

It is worth pointing out that, even when retaining a coarse-grained, rather than atomistic, level description, real molecules are typically non-rigid, for they can have intramolecular rigid fragments that can move, e.g. flexible chains or rings that can rotate with respect to each other [Maruani and Toro-Labbe, 1983; Zannoni, 1985] and internal degrees of freedom ϕ_i specifying the conformations will need to be specified. We shall have therefore to expect that, beyond a certain level of sophistication, features like deviation from cylindrical symmetry and flexibility will have to be taken into account [Zannoni, 1985] and we shall discuss these complications in Sections 3.10–3.12. The enormous number of positional and orientational coordinates specifying the various configurations can be used in calculating average properties. Let us consider a single molecule property A depending on particle position and orientation. The value observed for this property as obtained, for instance, by

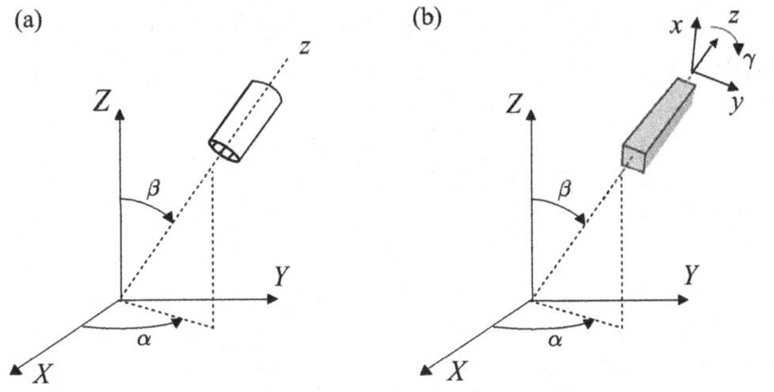

Figure 3.1 (a) The two angles α, β defining the orientation of a cylindrically symmetric particle and (b) the three Euler angles α, β, γ [Rose, 1957] required for a generic rigid particle. Here, to avoid confusion, we indicate laboratory frame axes with capital letters, X, Y, Z and molecular frame with lower case x, y, z.

some spectroscopic technique, can be determined as an average over all the molecules in the sample at a given time t, followed by an average over many independent samples (taken at the same time, or over many times, or both):

$$\langle A \rangle = \Big\langle \langle A\left(\boldsymbol{X}_i(t)\right) \rangle_S \Big\rangle_t, \tag{3.1}$$

where the angular brackets indicate an average and the additional subscripts, if any, specify some detail of the type of average. Here $\langle \ldots \rangle_S$ is a sample average, taken over all the N molecules in the sample at a certain time, and $\langle \ldots \rangle_t$ is a time average over a sufficiently large number N_t of snapshots of the system at different times:

$$\langle A \rangle = \frac{1}{N} \left\langle \sum_{i=1}^{N} A(\boldsymbol{X}_i(t)) \right\rangle_t = \frac{1}{N_t} \sum_{j=1}^{N_t} \left[\frac{1}{N} \sum_{i=1}^{N} A(\boldsymbol{X}_i(t_j)) \right]. \tag{3.2}$$

The procedure has to be repeated when we want to calculate any other single molecule average property, say $\langle B \rangle$, which is not very convenient. However, if we have a sufficiently large number of molecular configurations available, we can also divide the total range of the variable $\boldsymbol{X}$ into sufficiently small multi-dimensional intervals $\Delta \boldsymbol{X}_J$, $J = 1, \ldots, M$ and count the number of molecules that for all the available configurations fall in each such bin, creating a histogram of the number of molecules $n(\boldsymbol{X}_J(t))$ populating each bin, out of the total of N. For instance, for rigid molecules we can count the number of molecules $n(\boldsymbol{r}_J(t), \Omega_J(t))$ that have a position-orientation belonging to one of the M bins we have divided their full range into, at time t. In practice we could use a binning function, that we can write as

$$\Delta(\boldsymbol{X} - \boldsymbol{X}_J) = \begin{cases} 1, & \text{if } \boldsymbol{X} \text{ falls into the } J\text{th bin of width } \Delta \boldsymbol{X}_J \\ 0, & \text{otherwise.} \end{cases} \tag{3.3}$$

To avoid confusion, we use lower case subscripts to indicate molecules and capital letters to label the bins. We can then write

$$\langle A\rangle \approx \sum_{J=1}^{M} \frac{1}{N}\left\langle \sum_{i=1}^{N} \Delta\big(X_i - X_J\big)\right\rangle_t A(X_J), \tag{3.4a}$$

$$\approx \sum_{J=1}^{M} \frac{\big\langle n\big(X_J(t)\big)\big\rangle_t}{N} A(X_J), \tag{3.4b}$$

$$= \sum_{J=1}^{M} P(X_J)\,\Delta X_J\, A(X_J), \tag{3.4c}$$

where we have introduced the *probability density* $P(X_J)$, with dimensions inverse to those of X, for a molecule to be found at a certain generalized coordinate bin X_J. Taking the limit of a series of bins of decreasing width, the binning function becomes a delta function (see Appendix D) and we can obtain the classical statistical expression

$$\langle A\rangle \equiv \langle A(X)\rangle_X = \int \mathrm{d}X\, P(X)A(X). \tag{3.5}$$

It is very useful to write $P(X)$ using delta functions (Appendix D) as

$$P(X) = \big\langle \delta(X - X')\big\rangle_{X'}. \tag{3.6}$$

The formula can be easily verified using the definition of single particle average. Thus,

$$\langle A(X)\rangle_X = \left\langle \int \mathrm{d}X'\,\delta(X - X')\; A(X')\right\rangle_X = \int \mathrm{d}X' A(X')\big\langle \delta(X - X')\big\rangle_X, \tag{3.7a}$$

$$= \langle A(X')\rangle_{X'}. \tag{3.7b}$$

This is simply related to the probability of finding any one of the N molecules at coordinate X, i.e. $P^{(1)}(X)$, that we need in order to calculate global, rather than single, molecule properties that require considering all the molecules in the sample, e.g. the *number density* $P^{(1)}(X) = NP(X)$. Note that at the moment we have not given any prescription to obtain $P(X)$ from the molecular interactions, arguably the main task of statistical mechanics (see Chatpers 4 and 7) and of computer simulations, which will be discussed in Chapters 8 and 9. The multidimensional histogram obtained is averaged over a sufficiently large set of sample configurations, e.g. recorded at a set of different times as in the MD technique (Chapter 9) or, as we shall see in Chapter 8, generated with a suitable stochastic process, as in the MC method. To be more specific, in the simple case of rigid particles with position $\boldsymbol{r}$ and orientation Ω we have $P(X) = P(\boldsymbol{r}, \Omega)$ and $\mathrm{d}X = \mathrm{d}\boldsymbol{r}\mathrm{d}\Omega$ with $\mathrm{d}\boldsymbol{r} = \mathrm{d}x\mathrm{d}y\mathrm{d}z$ and $\mathrm{d}\Omega$, respectively, $\mathrm{d}\alpha \sin\beta \mathrm{d}\beta$ and $\mathrm{d}\alpha \sin\beta \mathrm{d}\beta \mathrm{d}\gamma$ for a rigid molecule of uniaxial or arbitrary symmetry. Thus, the integration over positions gives the volume V of the sample: $\int_V \mathrm{d}\boldsymbol{r} = \int \mathrm{d}x\mathrm{d}y\mathrm{d}z = V$ and the integration over orientations gives the total angular measure, V_Ω. Thus, $\int_{V_\Omega} \mathrm{d}\Omega = V_\Omega$, where, in 3D, $V_\Omega = 8\pi^2$ or just $V_\Omega = 4\pi$ for cylindrical symmetry where γ is not present (Fig. 3.1). In general, to simplify notation, we shall not write explicitly the integration range unless needed, intending the integrals to be extended to all relevant space.

The probability density $P(\boldsymbol{r}, \Omega)$ contains all the microscopic information needed to calculate one-particle properties [Zannoni, 1979c]. In turn, the structure and ordering of the system will be reflected by $P(\boldsymbol{r}, \Omega)$. We are interested in studying this single-particle (or *singlet*) distribution in various phases and in examining how it changes at the various phase transitions. Before getting involved in details concerning the calculation of distribution functions it is worth examining if we can make some general statements about them. The first is that $P(\boldsymbol{r}, \Omega)$ is non-negative and normalized, as:

$$\int_{V_X} \mathrm{d}X\, P(X) = \int_{V, V_\Omega} \mathrm{d}\boldsymbol{r}\mathrm{d}\Omega\, P(\boldsymbol{r}, \Omega) = 1, \tag{3.8}$$

since we must have our particle somewhere in space.

Let us now consider as a first example a *uniform* system, i.e. a system whose properties do not change if we translate the sample or, equivalently, translate the origin of the coordinate system. This could be the case of an ordinary isotropic liquid or of a nematic, but not of a smectic or a crystal, where molecular positions are regularly arranged in one or three dimensions. For such a uniform fluid the single particle probability density will be independent of the position of molecules with respect to the laboratory frame and:

$$P^{(1)}(\boldsymbol{r}, \Omega) = N P(\boldsymbol{r}, \Omega) = \rho P(\Omega), \tag{3.9}$$

where the proportionality factor, $\rho \equiv N/V$, is the number density and $P(\Omega)$ is a purely orientational distribution function normalized as $\int_{V_\Omega} \mathrm{d}\Omega P(\Omega) = 1$. For an ordinary isotropic fluid $P(\Omega)$ must be a constant, so that $P(\Omega) = 1/V_\Omega$. In this case, if we limit ourselves to one-particle properties, all that can change at the liquid-gas transition is just the density. We can then take as order parameter the difference between the density of the liquid and that of the gas at coexistence [Stanley, 1971], suitably normalized. We have already done this, purely on macroscopic grounds, in Section 2.5 (see, in particular, Fig. 2.14), where we have shown the order parameter versus temperature for a number of simple fluids. The situation is, however, quite different in anisotropic systems, as we shall see in detail in the next sections.

3.2 Positional Order

3.2.1 One-Dimensional Order: Smectics

We now start examining the description of the one-dimensional (1D) positional order present in smectics. We can visualize how such a system looks, considering for simplicity a smectic formed by nanorods (Fig. 3.2). To simplify the issue we only consider 1D positional order such as the molecular centres of mass ordering along the director $\boldsymbol{d}$ ($\boldsymbol{d} \| z$ say), in a smectic A. When this ordering is perfect, as sketched in Fig. 3.3a, the centres of mass lay exactly on regularly spaced parallel layers and the positional distribution $P(z)$ consists of a series of Dirac delta functions separated by the lattice spacing ℓ_z. If the order is not complete the peaks of the distribution will become broader (Fig. 3.3b). In the limit of no positional order (e.g. a nematic), the distribution becomes flat. In any case for

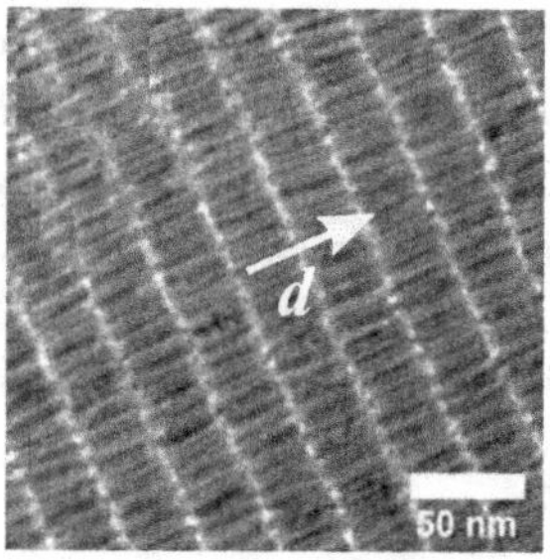

Figure 3.2 Transmission electron microscopy (TEM) image of a smectic formed by self-assembled CdSe/CdS nanorods on a polar liquid (dimethylacetamide). The inorganic nanorods are 28.4 nm in length and 5.8 nm in diameter and have a capping layer of octadecyl phosphinic acid [Diroll et al., 2015]. The arrow indicates the director ***d***.

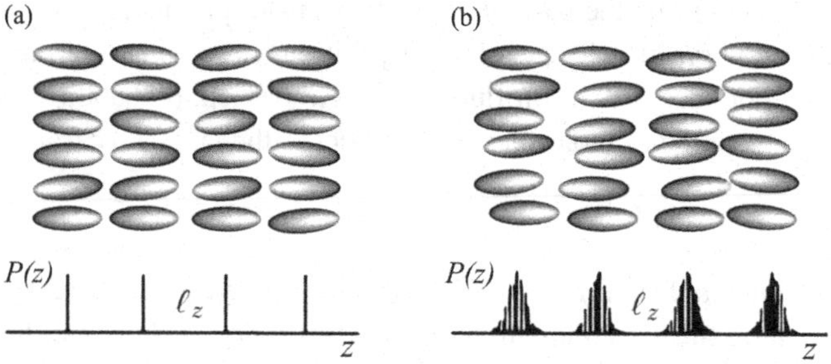

Figure 3.3 A 1D periodic system with (a) complete and (b) incomplete positional order in one dimension. On the bottom we have the probability density $P(z)$ of finding the particle centre at position z.

a smectic like system, such as the ones in Fig. 3.3, $P(z)$ remains a periodic function of position z.

$$P(z) = P(z + k\ell_z), \quad k = \pm 1, \pm 2, \dots . \tag{3.10}$$

This implies that we can limit ourselves to considering $P(z)$ with $0 \leq z \leq \ell_z$. Moreover, we can expand $P(z)$, like any other periodic function, in Fourier series [Arfken and Weber, 1995], i.e. write it as a suitable combination of sines and cosines or harmonics (cf. Appendix E). If the distribution is an even one, i.e. $P(z) = P(-z)$, it will suffice to consider a basis set of similarly even functions: the cosines: $\{\cos(2\pi nz/\ell_z)\}$. We have therefore

$$P(z) = \sum_{n=0}^{\infty} p_n \cos(2\pi nz/\ell_z) = p_0 + p_1 \cos(2\pi z/\ell_z) + p_2 \cos(4\pi z/\ell_z) + \cdots . \tag{3.11}$$

The coefficient p_m can be obtained multiplying both sides of Eq. 3.11 by the mth basis function, $\cos(2\pi mz/\ell_z)$, and integrating over z,

$$\int_0^{\ell_z} dz P(z) \cos\left(\frac{2\pi m z}{\ell_z}\right) = \sum_{n=0}^{\infty} p_n \int_0^{\ell_z} dz \cos\left(\frac{2\pi n z}{\ell_z}\right) \cos\left(\frac{2\pi m z}{\ell_z}\right),$$

$$= \sum_{n=0}^{\infty} \frac{p_m \ell_z}{2} (\delta_{m,0}\delta_{n,0} + \delta_{m,n}), \tag{3.12}$$

where we have used the orthogonality of the cosine functions (Eq. A.44). We find therefore that the coefficients

$$p_m = \frac{2}{\ell_z(\delta_{m,0}+1)} \int_0^{\ell_z} dz P(z) \cos\left(\frac{2\pi m z}{\ell_z}\right) = \frac{2}{\ell_z(\delta_{m,0}+1)} \left\langle \cos\left(\frac{2\pi m z}{\ell_z}\right)\right\rangle \tag{3.13}$$

are positional averages of the basis functions and that the averages

$$\tau_n = \left\langle \cos\left(\frac{2\pi n z}{\ell_z}\right)\right\rangle \tag{3.14}$$

represent our set of *positional order parameters*. They tend to 1 when the layer distribution is perfectly regular (Fig. 3.3a), and thanks to the orthogonality of the basis (Eq. A.44), to 0 when instead the distribution of positions is random, so that $P(z)$ is a constant. We see that τ_1 is just the positional order parameter τ that we introduced empirically when discussing Landau theory for smectics in Section 2.12. The order, e.g. $\tau_1 = \langle\cos(2\pi z/\ell_z)\rangle$, normally decreases with increasing temperature so that the observables connected to it will decrease as well. We can in turn write $P(z)$ as

$$P(z) = \frac{1}{\ell_z} + \frac{2}{\ell_z}\langle\cos(2\pi z/\ell_z)\rangle \cos(2\pi z/\ell_z) + \cdots, \tag{3.15a}$$

$$= \frac{1}{\ell_z} + \frac{2}{\ell_z}\sum_{m=1}^{\infty} \tau_m \cos(2\pi m z/\ell_z). \tag{3.15b}$$

The result also follows from the delta function expansion in Eq. D.28 ,

$$P(z) = \langle\delta(z-z')\rangle_{z'} = \frac{1}{\ell_z} + \frac{2}{\ell_z}\sum_{n=1}^{\infty}\langle\cos\left(n2\pi(z-z')/\ell_z\right)\rangle_{z'}, \tag{3.16a}$$

$$= \frac{1}{\ell_z} + \frac{2}{\ell_z}\sum_{n=1}^{\infty}\left\langle\cos\left(n2\pi z'/\ell_z\right)\right\rangle_{z'} \cos\left(n2\pi z/\ell_z\right), \tag{3.16b}$$

for $-\ell_z/2 \le z \le \ell_z/2$, using the standard trigonometric expression $\cos(\theta - \psi) = \cos(\theta)\cos(\psi) + \sin(\theta)\sin(\psi)$ and averaging, since $\langle\sin(n2\pi z'/\ell_z)\rangle = 0$. It might be worth mentioning that the treatment we just discussed means that if long-range positional order exists, it can be described in terms of order parameters without, however, proving the existence of such an order. Indeed, Peierls [1936] has shown that the fluctuations of layers positions in a 1D lattice diverge logarithmically with the linear size of the sample and should destroy the possibility of having such long-range order when increasing size to the asymptotic limit. However, in practice the effect should be noticeable only on huge sizes of the order of 1 km [Kleman and Lavrentovich, 2003], while experiments on smectics are typically for samples of 10^{-4}–10^{-3} m and quasi-long-range order, if not strictly true, long-range order is definitely observable.

3.2.2 Two-Dimensional Order

Two-dimensional positional order can be found, e.g. in certain discotic phases, where we have columns of mesogenic molecules which have no positional order inside the column, while the column themselves are arranged on a hexagonal ($\mathrm{D}_\mathrm{h}^\mathrm{d}$) or rectangular lattice ($\mathrm{D}_\mathrm{r}^\mathrm{d}$) (cf. Section 1.10). The concept of 2D positional order may also be useful to describe the structural order of molecules adsorbed on a surface. In both cases we have to consider the probability distribution of finding the particles at a certain position (x, y) of the laboratory plane. Now we assume that the molecular organization we are considering is periodic on average along x and y, with repeat distances ℓ_x, ℓ_y, respectively. Expanding in a product of harmonics relative to the x- and y-direction we have

$$P(x,y) = \frac{1}{\ell_x\ell_y} + \frac{4}{\ell_x\ell_y}\sum_{n_x,n_y}\left\langle \cos\left(\frac{2\pi n_x x}{\ell_x}\right)\cos\left(\frac{2\pi n_y y}{\ell_y}\right)\right\rangle \times \cos\left(\frac{2\pi n_x x}{\ell_x}\right)\cos\left(\frac{2\pi n_y y}{\ell_y}\right), \tag{3.17}$$

with $n_x, n_y > 0$. The averages $\left\langle \cos\left(\frac{2\pi n_x x}{\ell_x}\right)\cos\left(\frac{2\pi n_y y}{\ell_y}\right)\right\rangle$ are in this case the order parameters. Arguments are available against the existence of true long-range positional order also in 2D [Landau, 1965; Fisher, 1972], although a little less severe than those for 1D systems, but again the problem does not forbid systems of realistic, observable size [Hoover et al., 1974; Denham et al., 1980].

3.2.3 Three-Dimensional Order: Crystals

The treatment can be easily generalized to 3D positional order, as needed for crystals, expanding $P(x, y, z)$ in harmonics over x, y, z.

$$P(x,y,z) = \frac{1}{\ell_x\ell_y\ell_z} + \frac{8}{\ell_x\ell_y\ell_z} \times \sum_{n_x,n_y,n_z}\left\langle \cos\left(2\pi n_x x/\ell_x\right)\cos\left(2\pi n_y y/\ell_y\right)\cos\left(2\pi n_z z/\ell_z\right)\right\rangle \times \cos\left(2\pi n_x x/\ell_x\right)\cos\left(2\pi n_y y/\ell_y\right)\cos\left(2\pi n_z z/\ell_z\right). \tag{3.18}$$

3.3 Orientational Order for Uniaxial Molecules

We now turn to the description of long-range orientational order. It is a central issue for liquid crystals, since this is the only kind of order common to all the various mesophases, and the one allowing us to distinguish anisotropic from isotropic liquids. We start by considering the simplest case of a molecule that can be considered of effective uniaxial symmetry, be it a rod-like or disc-like shape, and that is embedded in a uniaxial phase. We do not distinguish by now if the molecule is a solute or one of the mesogens. Similarly to what we have done for the description of positional order, we start with the probability

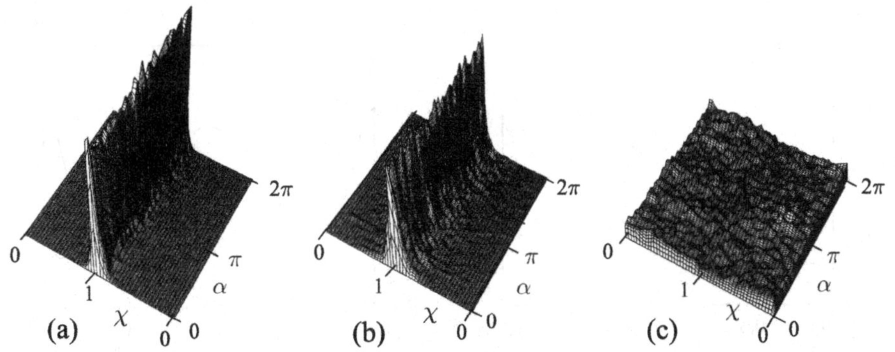

Figure 3.4 A histogram of $P(\alpha, \chi)$, $\chi \equiv | \cos\beta |$ for a model liquid crystal as obtained by MD simulation of the Lebwohl–Lasher lattice model at three dimensionless temperatures: (a) $T^* = 0.50$, (b) 0.88 in the nematic and (c) $T^* = 1.30$ in the isotropic phase [Zannoni and Guerra, 1981].

of finding the molecule at a certain orientation with respect to the axis of the mesophase, i.e. the director. As we have already mentioned these single-particle polar angles give the molecular orientation of the particle in question, i.e. $P(\Omega) \propto P(\alpha, \beta)$, if our molecules have cylindrical symmetry. The detailed form of $P(\alpha, \beta)$ is of course unknown, but some constraints imposed on it by symmetry can nevertheless be easily taken into account. If we take the laboratory z-axis parallel to the director, and if the mesophase is uniaxial around the director then rotating the sample about z should leave all observable properties unchanged. This means that the probability for a molecule having an orientation (α, β) should be the same whatever the angle α, i.e. $P(\alpha, \beta) \propto P(\beta)$. For example, we show in Fig. 3.4 a histogram of the full singlet orientational distribution obtained from a molecular dynamics computer simulation of the Lebwohl–Lasher model (see Eq. 2.37) [Zannoni and Guerra, 1981], discussed in Chapter 10. We see that the distributions in the ordered phase strongly depend on the angle β but not on α corresponding to the macroscopic uniaxiality around the director just stated. For such a system, we are thus justified in considering from now on only the dependence on the angle β. Another experimental finding for nematics and most smectics is that nothing changes on turning the aligned sample upside down. Thus, we should have

$$P(\beta) = P(\pi - \beta). \tag{3.19}$$

This is quite reasonable if we think of the molecules of interest as cylindrically symmetric objects in which head and tail are not distinguishable (cf. Fig. 3.1). However, most mesogen molecules are polar and, like for instance 5CB (see Fig. 1.10), have dipole moments. In practice, the symmetry expressed by Eq. 3.19, that is verified experimentally in nematics, means that the molecular arrangement is such as to have, on average, the same number of molecules pointing up and down, so that no overall polarization (no *ferroelectricity*) results. Note that no fundamental argument forbids uniaxial ferroelectric nematics and indeed these have been predicted by theory and simulations [Biscarini et al., 1991; Berardi et al., 2001],

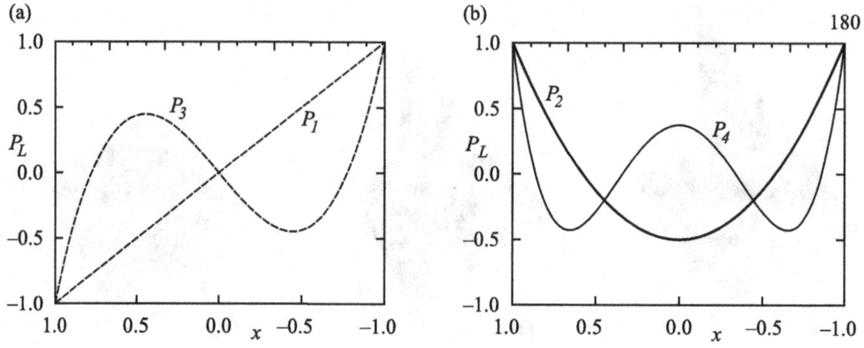

Figure 3.5 The first two (a) odd and (b) even Legendre polynomials $P_L(x)$ vs $x \equiv \cos\beta$.

although experimental confirmation is only very recent [Chen et al., 2020]. Here we just limit ourselves to the common case of non-polar nematics. It is convenient to normalize $P(\beta)$ so that

$$\int_0^\pi d\beta \sin\beta P(\beta) = \int_{-1}^1 dx\, P(x) = 1, \tag{3.20}$$

since there is the certainty of finding the molecule at some angle β. Note that the change of variable from β to $x = \cos\beta$ is particularly convenient, as it absorbs the volume element and here we shall normally use $P(x)$, which is flat for an isotropic system. In a real experiment it would be extremely difficult to get the kind of complete information on the orientational distribution pictured in the histogram. A useful approach is, however, that of trying to approximate $P(x)$ in terms of a set of quantities that we can obtain from experiment. Reasoning as we have done for the positional distribution, we can try to expand $P(x)$. For this we need a set of functions that are orthogonal when integrated over $dx \equiv d\beta \sin\beta$. Such a set of functions is that of Legendre polynomials $P_L(x)$ (see Appendix A and [Rose, 1957]), for which we have the orthogonality relation Eq. A.50. The explicit form of these polynomials is really very simple and the first few terms are given in Eqs. A.48. In Fig. 3.5 we show a graph of $P_1(x)$, $P_3(x)$ and $P_2(x)$, $P_4(x)$ versus x, showing the useful property that $P_L(x)$ is an even function of x if the rank L is even and an odd one if L is odd, i.e. $P_L(x) = (-1)^L P_L(-x)$. Since $\cos(\pi - \beta) = -\cos\beta$ this means that in writing our even orientational distribution in terms of $P_L(x)$ functions only even L terms need be retained. Clearly, the odd terms will be present if $P(x)$ is not even, as for ferroelectric liquid crystal phases. Limiting ourselves to the more common case of non-polar nematics (see Eq. 3.19) we can write

$$P(x) = \sum_{J=0}^{\infty} p_J P_J(x), \quad J \text{ even}. \tag{3.21}$$

The Lth expansion coefficient can be found multiplying both sides of Eq. 3.21 by $P_L(x)$ and integrating over dx:

$$\langle P_L \rangle = \int_{-1}^{1} \mathrm{d}x \, P(x) \, P_L(x) = \sum_{J=0}^{\infty} p_J \int_{-1}^{1} \mathrm{d}x \, P_J(x) P_L(x) = \frac{2}{(2L+1)} p_L, \tag{3.22}$$

with $p_L = (2L+1)\langle P_L \rangle/2$. The averages $\langle P_L \rangle$ represent our orientational order parameters. In particular, recalling the explicit expressions (Eqs. A.48d and A.48f),

$$\langle P_2 \rangle = \left\langle \frac{3}{2}x^2 - \frac{1}{2} \right\rangle, \tag{3.23}$$

$$\langle P_4 \rangle = \left\langle \frac{35}{8}x^4 - \frac{30}{8}x^2 + \frac{3}{8} \right\rangle. \tag{3.24}$$

The knowledge of the (infinite) set of $\langle P_L \rangle$ would completely define the distribution:

$$P(x) = \frac{1}{2} + \frac{5}{2}\langle P_2 \rangle P_2(x) + \frac{9}{2}\langle P_4 \rangle P_4(x) + \cdots + \frac{2L+1}{2}\langle P_L \rangle P_L(x) + \cdots. \tag{3.25}$$

It is worth mentioning that this expansion could be obtained more directly by writing the distribution as the average of a delta function (Eq. 3.6) and expanding the delta function in Legendre polynomials (Eq. D.29):

$$P(x) = \langle \delta \left(x - x' \right) \rangle_{x'} = \sum_{L=0}^{\infty} \frac{2L+1}{2} \langle P_L \rangle \, P_L\,(x)\,. \tag{3.26}$$

Again, we can limit L to even values, since all odd $\langle P_L \rangle$ will be 0 by symmetry. On going from an ordered to a disordered system, the order parameters jump discontinuously to 0 since the nematic-isotropic transition is of the first-order type, even if weakly so (cf. Section 2.9).

3.4 Experimental Determination of Orientational Order Parameters

The second-rank order parameter $\langle P_2 \rangle$ is proportional to the anisotropy in various experimentally measurable properties. We shall examine this relation for molecules of different symmetry in a rather general way in Section 3.10 and Appendix G. However, just to make contact with real life measurements, we briefly consider here some simple examples of experiments for the determination of the order parameters $\langle P_2 \rangle$ and $\langle P_4 \rangle$ for uniaxial molecules (or colloidal particles) in a uniaxial liquid crystal phase. To determine $\langle P_2 \rangle$ we start with a macroscopic method (*diamagnetic anisotropy*) useful for liquid crystal materials as such and with a spectroscopic method suitable for determining the order of solute dye molecules dissolved in liquid crystals (*linear dichroism*).

3.4.1 Diamagnetic Susceptivity

Placing a diamagnetic material in an external magnetic field $\boldsymbol{H}$, a magnetization $\boldsymbol{M}$ (per unit volume), is induced [Leenhouts et al., 1979] and $\boldsymbol{M} = \boldsymbol{\chi} \boldsymbol{H}$, where $\boldsymbol{\chi}$ is the magnetic susceptibility, a second-rank tensor. The work $\mathrm{d}w$ performed by a small change in the field, $\mathrm{d}\boldsymbol{H}$, is

$$\mathrm{d}w = \boldsymbol{M} \cdot \mathrm{d}\boldsymbol{H} = \sum_{a,b} \chi_{ab} H_b \, \mathrm{d}H_a; \; a,b = X,Y,Z. \tag{3.27}$$

If we have a field parallel to the laboratory Z-axis: $\boldsymbol{H}(Z) = (0,0,H(Z))$ which varies from 0 to H along Z with a certain constant gradient $\partial H/\partial Z$, then the sample is subject to a force f_Z,

$$f_Z = \chi_{ZZ} \frac{\partial H}{\partial Z} H, \tag{3.28}$$

that can be simply measured with a suitable balance [O'Connor, 1982]. Thus, if H and $\partial H/\partial Z$ are known, the diamagnetic susceptivity can be determined. In a uniaxial liquid crystal two components $\chi_{\parallel}^{\text{LAB}} \equiv \langle \chi_{ZZ}^{\text{LAB}} \rangle$ and $\chi_{\perp}^{\text{LAB}} = \langle \chi_{XX}^{\text{LAB}} \rangle = \langle \chi_{YY}^{\text{LAB}} \rangle$ corresponding to the director parallel or perpendicular to the magnetic field direction, can be obtained. The difference between the parallel and perpendicular components is related to the orientational order in the system. To see this, we recall first that intermolecular magnetic interactions are of negligible strength and thus the measured $\boldsymbol{\chi}^{\text{LAB}}$ is the average of independent single molecule contributions in the lab frame. $\boldsymbol{\chi}^{\text{LAB}}$ can then be related to the magnetic susceptibility $\boldsymbol{\chi}^{\text{MOL}}$ in the molecular frame where it is diagonal, through the Cartesian rotation matrix $\mathbf{R}$ (see Section B.1). Thus, assuming that $\boldsymbol{\chi}^{\text{MOL}}$ is uniaxial,

$$\chi_{\parallel}^{\text{LAB}} = \langle \chi_{ZZ}^{\text{LAB}} \rangle = \sum_{a,b} \langle R_{Za} \chi_{ab}^{\text{MOL}} R_{bZ}^{T} \rangle \delta_{a,b} = \langle R_{Zz}^2 \rangle \chi_{\parallel}^{\text{MOL}} + [\langle R_{Zx}^2 \rangle + \langle R_{Zy}^2 \rangle] \chi_{\perp}^{\text{MOL}} \tag{3.29a}$$

$$= \langle \cos^2 \beta \rangle \chi_{\parallel}^{\text{MOL}} + \langle \sin^2 \beta \rangle \chi_{\perp}^{\text{MOL}} = \bar{\chi}^{\text{MOL}} + \frac{2}{3} \Delta \chi^{\text{MOL}} \langle P_2 \rangle, \tag{3.29b}$$

since $\langle \cos^2 \beta \rangle = (1/3) + (2/3)\langle P_2 \rangle$ and $\langle \sin^2 \beta \rangle = (2/3) + (2/3)\langle P_2 \rangle$. Similarly,

$$\chi_{\perp}^{\text{LAB}} = [\langle \chi_{XX}^{\text{LAB}} \rangle + \langle \chi_{YY}^{\text{LAB}} \rangle]/2, \tag{3.30a}$$

$$= \frac{1}{2} \sum_{a,b,a',b'} \left(\left\langle R_{Xa} \chi_{ab}^{\text{MOL}} R_{bX}^{T} \right\rangle \delta_{a,b} + \left\langle R_{Ya'} \chi_{a'b'}^{\text{MOL}} R_{b'Y}^{T} \right\rangle \delta_{a',b'} \right), \tag{3.30b}$$

$$= \langle \sin^2 \beta \rangle \chi_{\parallel}^{\text{MOL}} + \frac{1}{2} \langle 1 + \cos^2 \beta \rangle \chi_{\perp}^{\text{MOL}} = \bar{\chi}^{\text{MOL}} - \frac{1}{3} \Delta \chi^{\text{MOL}} \langle P_2 \rangle, \tag{3.30c}$$

where $R_{ab} \equiv R_{ab}(\alpha, \beta, \gamma)$ are elements of the Cartesian rotation matrices connecting the laboratory to the molecule fixed system [Rose, 1957] (see Eq. B.8 for an explicit expression) and $\bar{\chi}^{\text{MOL}} = \frac{1}{3}(\chi_{\parallel}^{\text{MOL}} + 2\chi_{\perp}^{\text{MOL}})$ is the scalar susceptivity. We have then

$$\Delta \chi^{\text{LAB}} \equiv \chi_{\parallel}^{\text{LAB}} - \chi_{\perp}^{\text{LAB}} = \Delta \chi^{\text{MOL}} \langle P_2 \rangle. \tag{3.31}$$

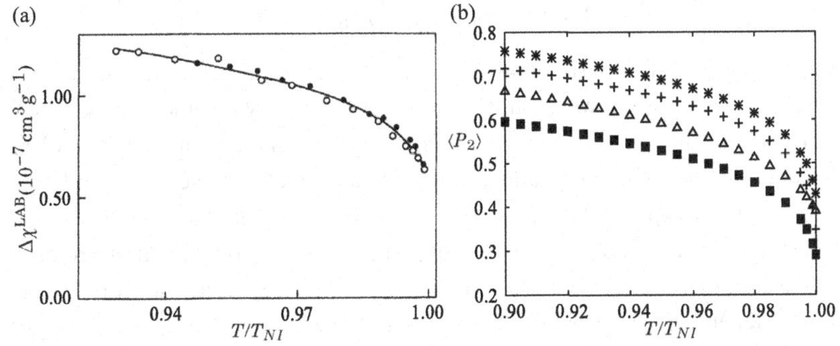

Figure 3.6 (a) OHMBBA diamagnetic anisotropy, $\Delta\chi^{\mathrm{LAB}}$, vs reduced temperature, T/T_{NI} for two sets of measurements (○) and (●) [Leenhouts et al., 1979]. (b) $\langle P_2 \rangle$ for MBBA (■), APAPA (△), MBCA (+) [Leenhouts et al., 1979] and 5CB (∗) [Pohl and Finkenzeller, 1990] as obtained from diamagnetic anisotropy as a function of T/T_{NI}.

Thus, determining $\Delta\chi^{\mathrm{LAB}}$ (see Fig. 3.6a) immediately gives $\langle P_2 \rangle$ if $\Delta\chi^{\mathrm{MOL}} \equiv \chi_{\parallel}^{\mathrm{MOL}} - \chi_{\perp}^{\mathrm{MOL}}$ is known. A small set of values for common nematics is reported in Table 1.3.

Temperature dependence of $\langle P_2 \rangle$. De Jeu and coworkers [Leenhouts et al., 1979] have measured the diamagnetic susceptivity anisotropy in a series of Schiff's base nematics which include the popular mesogens MBBA (cf. Table 1.2), anisylidene-*p*-aminophenyl acetate (APAPA), 4-methoxybenzylidene-4′-cyanoaniline (MBCA) and in Fig. 3.6b we report the dependence of the $\langle P_2 \rangle$ they obtained on reduced temperature $T_R \equiv T/T_{NI}$. In Fig. 3.6b we also show a plot of the orientational order of the nematic 5CB, as measured by Pohl and Finkenzeller [1990]. We see that the order decreases with increasing temperature and then suddenly jumps to 0, as expected for this first-order phase transition. The trend is similar for the different compounds, even though the detailed behaviour is not universal. The temperature dependence of the order parameter is empirically well represented by the so-called Haller [1975] equation

$$\Delta\chi(T_R) = \Delta\chi(0)\,(1 - T_R)^{\beta_H}, \tag{3.32}$$

where $\Delta\chi(0)$ and β_H are fitting parameters. The exponent β_H that also describes the temperature dependence of $\langle P_2 \rangle$ when approaching the transition has values $\beta_H = 0.17{-}0.22$ for most liquid crystals. For the diamagnetic results on Schiff bases in Fig. 3.6b, Leenhouts et al. [1979] found $\beta_H = 0.17$ (MBBA) and 0.185 (APAPA), while for MBCA the value was rather different: $\beta_H = 0.134$. A study of the order parameter variation with temperature has been performed with many techniques, some of which we are now going to explore. The results, even for the same material, are not as unequivocal as might be expected. The experimental values, even for the most studied materials are fairly scattered, e.g. for 5CB the reported exponents β_H range from 0.172 [Wu and Cox, 1988] to 0.19 [Horn, 1978]. In an analysis of refractive index anisotropy data, Chirtoc et al. [2004] have actually stated that all their data for 5, 6, 7, 8CB can be fitted with a unique exponent $\beta_H = 0.25$.

3.4.2 Linear Dichroism (LD)

As a second example of experimental determination of $\langle P_2 \rangle$, let us consider an optical absorption experiment, where we assume to have a cylindrically symmetric (rod-like or disc-like) dye dissolved at low concentration in a uniaxial liquid crystal, and that its molecules be aligned to some extent by the surrounding liquid crystal solvent. We send a linearly polarized beam of light on the sample and we measure the absorption parallel and perpendicular to the director. The probability of absorption of light plane polarized along the unit vector $\boldsymbol{e}$ can be written, if we assume the exciting light beam to be of relatively weak intensity, according to standard time dependent perturbation theory [Atkins, 1983; Michl and Thulstrup, 1986] as

$$P_{abs} \propto \langle |\boldsymbol{e} \cdot \boldsymbol{\mu}^{(a)}|^2 \rangle = \sum_{i,j} e_i e_j^* \langle \mu_i^{(a)} \mu_j^{(a)*} \rangle = \sum_{i,j} E_{ij} \langle A_{ij}^{(a)} \rangle, \quad i,j = x,y,z, \tag{3.33}$$

with $\boldsymbol{\mu}^{(a)}$ the absorption transition moment [Michl and Thulstrup, 1986]:

$$\boldsymbol{\mu}^{(a)} \equiv \langle \psi_{exc} | \hat{\boldsymbol{\mu}} | \psi_0 \rangle, \tag{3.34}$$

i.e. the matrix element of the electric dipole operator $\hat{\boldsymbol{\mu}}$ between the ground and excited states with wave functions ψ_0 and ψ_{exc}. The vector $\boldsymbol{\mu}^{(a)}$ is, to a good approximation, a molecular quantity for a certain excited state, i.e. for the state reached with the incident radiation frequency. We have also introduced the absorption transition tensor containing the relevant molecular information as the direct product $\mathbf{A}^{(a)} = \boldsymbol{\mu}^{(a)} \otimes \boldsymbol{\mu}^{(a)*}$. Similarly, we have also defined a light polarization tensor, containing all the information about the experimental disposition of the polarizer [Zannoni, 1979d] as the direct product $\mathbf{E} = \boldsymbol{e} \otimes \boldsymbol{e}^*$. Eq. 3.33 holds for an arbitrary orientation of the transition moment with respect to the molecular axis. Here, however, we assume for simplicity to have chosen a cylindrically symmetric dye molecule with $\boldsymbol{\mu}^{(a)}$ parallel to the molecule axis, so that $\boldsymbol{\mu}^{(a)} = \mu^{(a)}(0,0,1)$ and $[A^{(a)}]_{ij}^{\mathrm{MOL}} = [\mu^{(a)}]^2 \delta_{i,j}$. In the laboratory frame, that we take with the Z-axis parallel to the director, the transition moment components will be $\boldsymbol{\mu}^{(a)} = \mu^{(a)}(\sin\beta\cos\alpha, \sin\beta\sin\alpha, \cos\beta)$. Thus, the intensity of light, polarized parallel or perpendicular to the director, that is absorbed by the sample will be

$$\langle [A^{(a)}]_{\parallel}^{\mathrm{LAB}} \rangle = \langle [A^{(a)}]_{ZZ}^{\mathrm{LAB}} \rangle \propto [\mu^{(a)}]^2 \langle \cos^2\beta \rangle = [\mu^{(a)}]^2 \Big(\frac{1}{3} + \frac{2}{3} \langle P_2 \rangle \Big), \tag{3.35a}$$

$$\langle [A^{(a)}]_{\perp}^{\mathrm{LAB}} \rangle = \frac{1}{2} \langle [A^{(a)}]_{XX}^{\mathrm{LAB}} + [A^{(a)}]_{YY}^{\mathrm{LAB}} \rangle \propto \frac{1}{2} [\mu^{(a)}]^2 \langle \sin^2\beta \rangle = [\mu^{(a)}]^2 \Big(\frac{1}{3} - \frac{1}{3} \langle P_2 \rangle \Big). \tag{3.35b}$$

The anisotropy of the intensity of light of a certain wavelength absorbed, i.e. the *linear dichroism,* $\langle \Delta [A^{(a)}]^{\mathrm{LAB}} \rangle \equiv \langle [A^{(a)}]_{\parallel}^{\mathrm{LAB}} - [A^{(a)}]_{\perp}^{\mathrm{LAB}} \rangle$, will be proportional to the order parameter of the solute dye [Michl and Thulstrup, 1986]:

$$\big\langle \Delta [A^{(a)}]^{\mathrm{LAB}} \big\rangle \propto \frac{1}{3} [\mu^{(a)}]^2 \langle P_2 \rangle, \tag{3.36}$$

and should drop to 0 when the system becomes isotropic. This anisotropy can therefore be used to monitor orientational phase transitions and to determine $\langle P_2 \rangle$ of the dye. Although this is of course not the same as the host nematic, it can mimic it if the probe has been judiciously chosen to be as similar as possible to that of the host nematogens.

3.4.3 Fluorescence Depolarization and $\langle P_4 \rangle$

A difficulty in determining the fourth-rank order parameter $\langle P_4 \rangle$ is the lack of convenient fourth-rank tensorial quantities to measure. A way out is to employ a technique that can provide the square of a second-rank property. Continuing with optical methods, such a technique is Fluorescence Depolarization (FD) and Polarized Raman [Southern and Gleeson, 2007; Sanchez-Castillo et al., 2010], that we shall not discuss here, another one. In the FD technique, the orientational order and the rotational motion of fluorescent dyes (chromophores) is studied by first exciting them with short pulses of plane polarized light and then observing the polarization of their emitted fluorescence as a function of time on a nanosecond scale. Fluorescence is an extremely sensitive technique and a dye concentration of 10^{-3} w/w (or even much lower according to the type of dye) can be used, with the advantage of causing only a little perturbation of the liquid crystals solvent. In an idealized FD experiment, the system is probed with an extremely short light pulse, plane polarized in a certain direction $\boldsymbol{e}_{\mathrm{i}}$. What we mean by extremely short is that the pulse duration should be much shorter than the fluorescence and reorientation time scales in the experiment. This condition is of course only approximately met in real experiments and normally numerical deconvolution techniques [Arcioni et al., 1990] will have to be applied to correct for the finite duration of the pulse as well as for the instrument response time. The emitted fluorescence light is collected through a second polarizer (*analyzer*) set at a direction of polarization $\boldsymbol{e}_{\mathrm{o}}$ placed on a certain observation direction. In such an idealized experiment the fluorescence intensity can be recorded at a time t elapsed from the initial pulse. In more detail, if light of a suitable wavelength is employed, absorption takes place and an excited state is formed. This excited state may undergo some, usually rapid, internal conversion process, normally followed by emission from the lowest vibrational level of the excited singlet state (see Fig. 3.7). If we assume these processes to be independent from each other, we can write the fluorescence intensity emitted from a molecule at a time t after excitation as a product:

$$I^F(t) \propto \left\langle P_{abs}(0) P_{em}(t) F(t) \right\rangle, \tag{3.37}$$

where $P_{abs}(0)$ stands for the probability that the molecule is excited at time $t = 0$, $F(t)$ that it is still in the excited state at time t, and $P_{em}(t)$ the emission probability [Zannoni, 1979d]. The form of the intrinsic fluorescence decay $F(t)$ depends on the detailed photophysics of the dye molecule and in practice it is often approximated with an exponential or a sum of exponentials. In general, we can assume that $F(t)$ is characterized by an effective decay time τ_F. This characteristic time is normally in the picosecond to nanosecond range, so that using fluorescence, we can hope to study motional processes that take place on this time scale, since these are the ones that can effectively modulate the decay. The probability of absorption in Eq. 3.37 can be written down as before, in terms of the transition moment $\boldsymbol{\mu}^{(a)}$ between the ground and excited state (Eq. 3.34). Quite similarly the emission probability

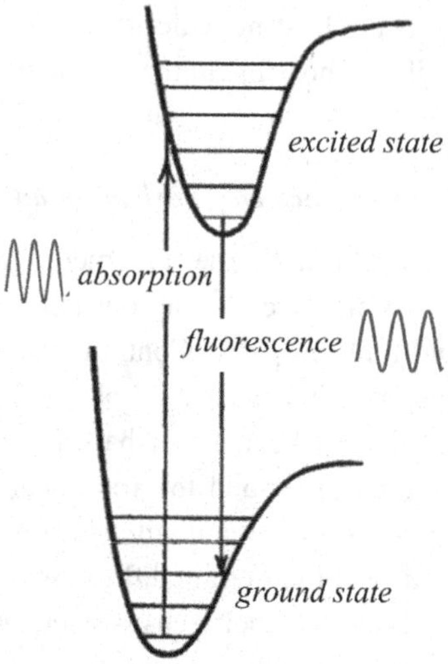

Figure 3.7 A schematic representation of the absorption and fluorescence processes. The photon absorbed promotes the molecule to a vibrational level (horizontal line) of the excited state at fixed nuclei position (Franck–Condon vertical transition). After a fast internal relaxation to the lowest vibrational level of the excited state, the molecule emits a lower energy (longer wavelength) fluorescence photon.

will involve an emission transition moment $\boldsymbol{\mu}^{(e)} = \langle \psi_0 \mid \hat{\boldsymbol{\mu}} \mid \psi'_{exc} \rangle$, where ψ'_{exc} is the wave function of the emitting state, that can typically be different from the initially excited state ψ_{exc}, because of very fast internal electronic relaxation [Michl and Thulstrup, 1986]. Thus,

$$I^F_{\mathrm{io}}(t) = \langle |\boldsymbol{e}_{\mathrm{i}} \cdot \boldsymbol{\mu}^{(a)}(0)|^2 \; |\boldsymbol{e}_{\mathrm{o}} \cdot \boldsymbol{\mu}^{(e)}(t)|^2 \rangle F(t), \tag{3.38}$$

where we have assumed an isotropic fluorescence decay $F(t)$ and we indicate with angular brackets an average over all the motions experienced by the probe molecule up to time t. When the chosen dye emits so fast that the molecule has not had time to reorient, $I^F_{\mathrm{io}}(t) \approx I^F_{\mathrm{io}}(0)$. In particular, referring to the experiment represented schematically in Fig. 3.8 we have, when $\boldsymbol{\mu}^{(a)} \| \boldsymbol{\mu}^{(e)} \| \boldsymbol{u}$, with $\boldsymbol{u}$ the effective molecular symmetry axis, the intensities depend explicitly on $\langle P_2 \rangle$ and $\langle P_4 \rangle$.

$$\frac{I^F_{ZZ}(0)}{F(0)} = [\mu^{(a)}]^2[\mu^{(e)}]^2 \langle \cos^4 \beta \rangle = [\mu^{(a)}]^2[\mu^{(e)}]^2 \left[\frac{8}{35}\langle P_4 \rangle + \frac{4}{7}\langle P_2 \rangle + \frac{1}{5} \right], \tag{3.39a}$$

$$\frac{I^F_{ZX}(0)}{F(0)} = [\mu^{(a)}]^2[\mu^{(e)}]^2 \langle \cos^2 \beta \, \sin^2 \beta \rangle = [\mu^{(a)}]^2[\mu^{(e)}]^2 \left[-\frac{8}{35}\langle P_4 \rangle + \frac{2}{21}\langle P_2 \rangle + \frac{2}{15} \right]. \tag{3.39b}$$

This is approximately the case for the probes 1,6, diphenyl-hexatriene (DPH) [Zannoni, 1979d; Zannoni et al., 1983] and p-dimethylamino-p-nitro-stilbene (DMANS) [Dozov and Penchev, 1980] shown in Fig. 3.9. It is convenient to introduce the depolarization ratio $r(t)$,

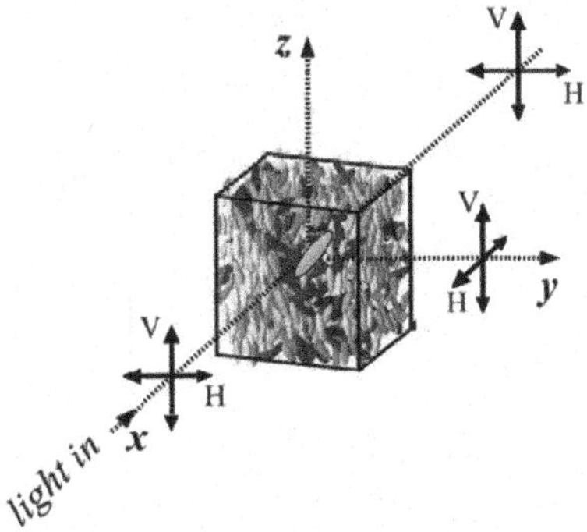

Figure 3.8 A schematic representation of a fluorescence polarization experiment. Light with polarization $\boldsymbol{e}_{\mathrm{i}}$, vertical (V) or horizontal (H), impinges from direction $\boldsymbol{x}$ on a monodomain liquid crystal sample ($\boldsymbol{d}||z$) containing fluorescent dyes. The emitted radiation intensity is observed either at a right angle, along $\boldsymbol{y}$ or (filtering out the residual incoming light) in the forward direction $\boldsymbol{x}$ through a polarizer $\boldsymbol{e}_{\mathrm{f}}$ set either vertically or horizontally.

$$r(t) = \frac{I^F_{ZZ}(t) - I^F_{ZX}(t)}{I^F_{ZZ}(t) + 2I^F_{ZX}(t)}, \tag{3.40}$$

that can be determined combining the results of a right angle and a forward geometry experiment (see Fig. 3.8). For 'long' times, i.e. for an emission fluorescence time so long that the molecule has had the time to fully reorient, the average of the product: $\langle |\boldsymbol{e}_{\mathrm{i}} \cdot \boldsymbol{\mu}^{(\mathrm{a})}(0)|^2 \, |\boldsymbol{e}_{\mathrm{o}} \cdot \boldsymbol{\mu}^{(\mathrm{e})}(t)|^2 \rangle$ in Eq. 3.38 becomes the product of the averages: $\langle |\boldsymbol{e}_{\mathrm{i}} \cdot \boldsymbol{\mu}^{(\mathrm{a})}(0)|^2 \rangle \, \langle |\boldsymbol{e}_{\mathrm{o}} \cdot \boldsymbol{\mu}^{(\mathrm{e})}(\infty)|^2 \rangle$ and the limiting value of $r(t)$ for long times becomes just the orientational order parameter of the probe:

$$r(\infty) = \langle P_2 \rangle. \tag{3.41}$$

However, at the other limiting case, for $t = 0$, the depolarization ratio is

$$r(0) = \left[\frac{2}{5} + \frac{11}{7}\langle P_2 \rangle + \frac{36}{35}\langle P_4 \rangle\right] \Big/ \left[1 + 2\langle P_2 \rangle\right]. \tag{3.42}$$

Thus, $r(t)$ starts from a value depending on $\langle P_2 \rangle$ and $\langle P_4 \rangle$ and goes to a plateau value equal to $\langle P_2 \rangle$. The order parameter $\langle P_4 \rangle$ can then be extracted from the initial value $r(0)$. As an example, $\langle P_2 \rangle$ and $\langle P_4 \rangle$ and their temperature dependence have been determined [Wolarz and Bauman, 2006] for the stilbene dye DMANS dissolved ($\approx$3%w/w) in the nematics 5CB, 7CB and PCH7 (Fig. 3.9). Note that $\langle P_2 \rangle$ obtained from LD (Section 3.4.2) and FD are in fairly good agreement. What can be obtained in practice from a certain experiment will depend on the relative time scales of the fluorescence decay and reorientation process, thus on the molecular feature of the probe and the medium fluidity. We shall discuss the information on the reorientation dynamics of a fluorescent probe obtainable from the full-time dependence of $r(t)$ in Chapter 6.

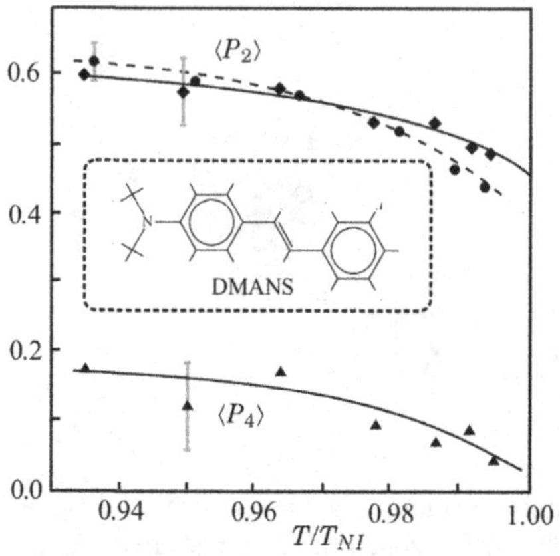

Figure 3.9 Order parameters of the fluorescent stilbene dye, DMANS, dissolved in the nematic, PCH7, as a function of reduced temperature T/T_{NI}, assuming transition moments parallel to the dye long axis. The lines are a guide for the eye, the bars give an error estimate. Shown: $\langle P_2 \rangle$ evaluated from absorption dichroic ratio (●) and $\langle P_2 \rangle$ (♦) and $\langle P_4 \rangle$ (▲) obtained from fluorescence measurements [Wolarz and Bauman, 2006].

3.5 Orientational Order from Computer Simulations

The calculation of orientational order parameters is clearly of particular importance in computer simulations of model liquid crystals. It also requires the development of some algorithms as compared to simulations of isotropic fluids, while the thermodynamic observables are calculated essentially with the standard techniques developed for ordinary liquids. In practice, computer simulations can provide, as discussed later in Chapters 8 and 9, a number of equilibrium configurations of a system formed by N particles and we can consider the determination of the second- and fourth-rank parameters in a way similar to setting up a virtual experiment.

3.5.1 Second Rank

The second-rank order parameter $\langle P_2 \rangle$ can be calculated in principle by averaging $P_2(x)$ over the normalized probability $P(x)$ of finding the molecule at an orientation $x = \cos\beta = \boldsymbol{u} \cdot \boldsymbol{d}$ of the molecular axis (represented by the unit vector $\boldsymbol{u}$) with respect to the director $\boldsymbol{d}$:

$$\langle P_2 \rangle = \int_{-1}^{1} \mathrm{d}x \; P(x) P_2(x). \tag{3.43}$$

The problem in using this definition is that in simulations we do not work in a director frame but in an arbitrary laboratory frame, and normally, in the absence of an external field that aligns the director along a desired direction. Thus, we do not know the orientation of $\boldsymbol{d}$ in each configuration and we have no guarantee that it will not change with time and remain the same. Because $P_2(x)$ is not a scalar, we cannot normally calculate $\langle P_2 \rangle$ as in Eq. 3.43 in each configuration and then average the result over many of these. To find a way out it is helpful to think that a computer simulation can be considered as an experimental technique where we can choose our observables at will. Thus, we can introduce a simple virtual single-molecule matrix property **A**, whose only non-vanishing component is along a suitable particle fixed direction [Fabbri and Zannoni, 1986]. For a uniaxial particle this could be a unit vector parallel to the molecular symmetry axis, $\boldsymbol{u} = (0, 0, 1)$, and we can define **A** as the direct square (see Appendix A),

$$\mathbf{A} = \boldsymbol{u} \otimes \boldsymbol{u}, \tag{3.44}$$

with Cartesian components, in the molecule fixed frame, $A_{ab}^{\mathrm{MOL}} = \delta_{a,z}\delta_{b,z}$, for $a, b = x, y, z$. This matrix, apart from a proportionality constant, is the same as the absorption tensor we introduced earlier in Eq. 3.34, if the transition moment $\boldsymbol{\mu}^{(\mathrm{a})}$ is parallel to molecular effective symmetry axis $\boldsymbol{u}$ [Michl and Thulstrup, 1986]. Of course, we do not need to push the analogy too far, but it may be helpful to think of simulations as an experiment that produces pseudo-experimental data. $\langle \mathbf{A}^{\mathrm{LAB}} \rangle$, the sample average of **A** in our arbitrary laboratory frame, is obtained writing the components of $\mathbf{A}^{\mathrm{LAB}}$ in terms of the molecule fixed components and summing over all the particles:

$$\langle A_{ab}^{\mathrm{LAB}} \rangle_S = \frac{1}{N} \sum_{i=1}^{N} \left\{ \sum_{a'b'} [R_i]_{aa'} [A_i]_{a'b'}^{\mathrm{MOL}} [R_i^T]_{b'b} \right\} = \langle R_{az} R_{bz} \rangle_S \equiv \frac{2}{3} Q_{ab} + \frac{1}{2} \delta_{a,b}, \tag{3.45}$$

where we have introduced the *ordering matrix* **Q** with elements

$$Q_{ab} = \frac{3}{2} \Big\langle [R_i]_{aZ} [R_i]_{bZ} - \frac{1}{2} \delta_{a,b} \Big\rangle_S, \tag{3.46}$$

where $\langle \ldots \rangle_S$ indicates an average over all the i molecules in the sample, and $[R_i]_{aa'}$ are the elements of the Cartesian rotation matrix (see Eq. B.8) transforming from the laboratory frame to the frame attached to molecule i, that we have already used in Section 3.4.1. The sample ordering will in general change with time, so that

$$\mathbf{Q}(t) = \frac{1}{2N} \sum_{i=1}^{N} \left[3\boldsymbol{u}_i(t) \otimes \boldsymbol{u}_i(t) - \mathbf{1} \right] = \frac{1}{2} \langle 3\boldsymbol{u}_i(t) \otimes \boldsymbol{u}_i(t) - \mathbf{1} \rangle_S. \tag{3.47}$$

We can write explicitly the instantaneous ordering matrix **Q** of the sample configuration as (time argument omitted)

$$\mathbf{Q} = \frac{3}{2N} \sum_{i=1}^{N} \begin{pmatrix} \sin^2\beta_i \cos^2\alpha_i - \frac{1}{3} & \sin^2\beta_i \cos\alpha_i \sin\alpha_i & \sin\beta_i \cos\beta_i \cos\alpha_i \\ \sin^2\beta_i \cos\alpha_i \sin\alpha_i & \sin^2\beta_i \sin^2\alpha_i - \frac{1}{3} & \sin\beta_i \cos\beta_i \sin\alpha_i \\ \sin\beta_i \cos\beta_i \cos\alpha_i & \sin\beta_i \cos\beta_i \sin\alpha_i & \cos^2\beta_i - \frac{1}{3} \end{pmatrix}. \tag{3.48}$$

Note that **Q** is symmetric and that the sum of its diagonal elements is 0, i.e. it is traceless: $\mathrm{Tr}\mathbf{Q} = 0$. Diagonalization of the sample averaged $\langle \mathbf{A}^{\mathrm{LAB}} \rangle_S$ with the unitary matrix **X** identifies the director frame where

$$\langle A_{ZZ}^{\mathrm{DIR}} \rangle_S = \sum X_{aZ} X_{bZ} \langle A_{ab}^{\mathrm{LAB}} \rangle_S = \langle \cos^2\beta \rangle_S = \frac{2}{3} \langle P_2 \rangle_S + \frac{1}{3}, \tag{3.49}$$

and **Q** is also diagonal

$$\mathbf{Q} = \begin{pmatrix} -\frac{1}{2}\langle P_2\rangle_S - \xi & 0 & 0 \\ 0 & -\frac{1}{2}\langle P_2\rangle_S + \xi & 0 \\ 0 & 0 & \langle P_2\rangle_S \end{pmatrix}. \tag{3.50}$$

The sample biaxiality parameter ξ makes the ordering with respect to the laboratory x- and y-directions different and will tend to 0 at large sample sizes if the mesophase has uniaxial symmetry. It is now obvious that the rotation diagonalizing $\langle \mathbf{A}^{\mathrm{LAB}}\rangle_S$ or equivalently **Q** defines the orientation of the director frame in our laboratory frame. The director itself is defined by the eigenvector corresponding to the largest eigenvalue, λ_{max}, of **Q**. If we take this to define the z-axis of the director frame, and we call $\boldsymbol{l}$ and $\boldsymbol{m}$ the unit vectors along the other two axes, we can also write

$$\mathbf{Q} = \langle P_2\rangle \left(\boldsymbol{d}\otimes\boldsymbol{d} - \frac{1}{3}\mathbf{1}\right) + \xi\left(\boldsymbol{m}\otimes\boldsymbol{m} - \boldsymbol{l}\otimes\boldsymbol{l}\right), \tag{3.51}$$

with $\boldsymbol{m} = \boldsymbol{d}\times\boldsymbol{l}$. The second-rank order parameter referred to the director in the sample, $\langle P_2\rangle_\lambda$, is obtained from this λ_{max} as $\langle P_2\rangle_{\lambda,S} = \lambda_{max}$. Thus, we can define a **Q** tensor for every configuration, say $\mathbf{Q}^{(j)}$ for the jth one. By diagonalizing $\mathbf{Q}^{(j)}$, we obtain an order parameter $P_2^{(j)}$ and a director $\boldsymbol{d}^{(j)}$. Even if the director can change from one configuration to another, since the $P_2^{(j)}$, being the eigenvalues of a matrix, are rotationally invariant (i.e. scalars), we can calculate

$$\langle P_2\rangle_\lambda = \frac{1}{M}\sum_{j=1}^{M}(\lambda_{max})^{(j)}, \tag{3.52}$$

where $(\lambda_{max})^{(j)}$ is the largest eigenvalue of the matrix $\mathbf{Q}^{(j)}$. Note that $(\lambda_{max})^{(j)} \geq 0$, thus $\langle P_2\rangle_\lambda$ is never going to be negative and even for an isotropic system it will approach 0 from above with a value of the order of $1/\sqrt{N}$. Eppenga and Frenkel [1984] have suggested that a better estimate of the order parameter in the isotropic phase can be obtained by averaging the intermediate, rather than the largest eigenvalue of the matrix. The resulting $\langle P_2\rangle_{\lambda_2}$ approaches 0 faster in the isotropic phase [Eppenga and Frenkel, 1984; Fabbri and Zannoni, 1986]. The calculation of the orientational distribution $P(x)$ with respect to the director strictly involves transforming the orientations, after each diagonalization, to the new director frame. An alternative procedure is to transform the order parameters to the director frame after every diagonalization. Once the rotation $(\Omega_{\mathrm{L'L}})$ carrying the old laboratory frame into the new one has been found, the order parameters calculated in the old frame can be transformed to the new one. Using the closure property of Wigner matrices (cf. Eq. F.14) we have in fact that $\langle \mathscr{D}^L_{m,n}(\Omega_{\mathrm{ML}})\rangle = \sum_{q=-L}^{L}\mathscr{D}^L_{q,m}(\Omega_{\mathrm{L'L}})\langle \mathscr{D}^L_{q,n}(\Omega_{\mathrm{ML'}})\rangle$. A simpler situation arises if the director orientation is known and fixed, e.g. when $\boldsymbol{d}$ is pinned along the z-direction by an external field or a surface. In this rather special case we can simply write $\langle P_2\rangle_{\mathrm{LAB}}$ as an average over M equilibrium configurations of the sample order parameter, $\langle P_2\rangle^{(j)}$. We have

$$\langle P_2\rangle_{\mathrm{LAB}} = \frac{1}{M}\sum_{j=1}^{M}\langle P_2\rangle^{(j)}, \tag{3.53}$$

where $\langle P_2 \rangle^{(j)} \equiv \frac{1}{N}\sum_{i=1}^{N} P_2(\boldsymbol{u}_i \cdot \boldsymbol{d})$ is the order parameter computed for the jth configuration. In this particular case it may be even simpler to calculate a histogram for the probability density $P(x)$ and subsequently determine the desired order parameters $\langle P_L \rangle_{\text{LAB}}$ by integration.

3.5.2 Fourth-Rank Order Parameter

We can generalize the frame independent procedure by defining a convenient fourth-rank virtual molecular property as the direct square $\mathbf{F} = \mathbf{A} \otimes \mathbf{A}$, of the matrix $\mathbf{A} = \boldsymbol{u} \otimes \boldsymbol{u}$ [Fabbri and Zannoni, 1986]. We note that in the molecule fixed frame the fourth-rank property defined in this way has only one non-vanishing component, F^{MOL}_{zzzz}, i.e. $F^{\text{MOL}}_{abcd} = \delta_{a,z}\delta_{b,z}\delta_{c,z}\delta_{d,z}$, where $a,b,c,d = x,y,z$. The sample average of F is

$$\langle \mathbf{F}^{\text{LAB}} \rangle_S = \langle \mathbf{A}^{\text{LAB}} \otimes \mathbf{A}^{\text{LAB}} \rangle_S = \langle (\mathbf{X}\mathbf{A}^{\text{DIR}}\mathbf{X}^T) \otimes (\mathbf{X}\mathbf{A}^{\text{DIR}}\mathbf{X}^T) \rangle_S, \tag{3.54a}$$

$$= (\mathbf{X} \otimes \mathbf{X})\langle \mathbf{A}^{\text{DIR}} \otimes \mathbf{A}^{\text{DIR}} \rangle_S (\mathbf{X}^T \otimes \mathbf{X}^T) = (\mathbf{X} \otimes \mathbf{X})\langle \mathbf{F}^{\text{DIR}} \rangle_S (\mathbf{X}^T \otimes \mathbf{X}^T). \tag{3.54b}$$

We then write the components in the director frame in terms of those in the molecular frame using the rotation matrix $\mathbf{R}$:

$$\left\langle \mathbf{F}^{\text{DIR}} \right\rangle_S = \left\langle (\mathbf{R}\mathbf{A}^{\text{MOL}}\mathbf{R}^T) \otimes (\mathbf{R}\mathbf{A}^{\text{MOL}}\mathbf{R}^T) \right\rangle_S, \tag{3.55a}$$

$$= \left\langle (\mathbf{R} \otimes \mathbf{R})\mathbf{A}^{\text{MOL}} \otimes \mathbf{A}^{\text{MOL}} (\mathbf{R}^T \otimes \mathbf{R}^T) \right\rangle_S = \left\langle (\mathbf{R} \otimes \mathbf{R})\mathbf{F}^{\text{MOL}} (\mathbf{R}^T \otimes \mathbf{R}^T) \right\rangle_S. \tag{3.55b}$$

In particular

$$\langle F^{\text{DIR}}_{ZZZZ} \rangle_S = \langle (R_{ZZ})^4 \rangle_S = \langle x^4 \rangle_S = \sum_{a,b,c,d} X_{aZ} X_{bZ} X_{cZ} X_{dZ} \langle F^{\text{LAB}}_{abcd} \rangle_S, \tag{3.56}$$

giving the fourth-rank order parameter $\langle P_4 \rangle$ as $\langle P_4 \rangle_S = \dfrac{35}{8}\langle x^4 \rangle_S - \dfrac{30}{8}\langle x^2 \rangle_S + \dfrac{3}{8}$. A further average of $\langle P_4 \rangle_S$ over equilibrium configurations yields $\langle P_4 \rangle_\lambda$. We shall see later, in Chapter 10, an application of this methodology to obtain $\langle P_4 \rangle$ for the Lebwohl–Lasher model (see Fig. 10.4).

3.6 Landau–deGennes Q-Tensor Approach

The $\mathbf{Q}$ matrix (tensor) order parameter can be employed to write a Landau expression for the anisotropic free energy avoiding the explicit use of the director field $\boldsymbol{d}$ which is singular at defect points. The free energy is independent of an arbitrary rotation of the laboratory system and thus can be expanded in powers of invariants of the tensor $\mathbf{Q}$. For a symmetric traceless tensor like $\mathbf{Q}$, these invariants are just (see Eq. B.13), Tr $\left(\mathbf{Q}^2\right)$ and $\frac{1}{3}\text{Tr}(\mathbf{Q}^3)$. The Landau–deGennes phenomenological expression for the free energy density in the vicinity of the nematic-isotropic transition and in the presence of an external field $\boldsymbol{H}$ can then be written as [de Gennes, 1971; Gramsbergen et al., 1986a; Mori et al., 1999; Mottram and Newton, 2014]

$$\mathcal{G}_{\text{tot}} = \mathcal{G}_{\text{L}} + \mathcal{G}_{\text{surf}} + \mathcal{G}_{\text{field}}, \tag{3.57}$$

where the field coupling term is, e.g. for a magnetic field $\boldsymbol{H}$,

$$\mathcal{G}_{\text{field}} = -\frac{1}{2}\Delta\chi H_\alpha Q_{\alpha\beta} H_\beta, \tag{3.58}$$

while the Landau free energy term $\mathcal{G}_{\text{L}}$ depends on the elements of $\mathbf{Q}$ and all derivatives of its elements,

$$\mathcal{G}_{\text{L}} = \mathcal{G}_0 + \frac{1}{2}A(T)Q_{\alpha\beta}Q_{\alpha\beta} - \frac{1}{3}B(T)Q_{\alpha\beta}Q_{\beta\gamma}Q_{\gamma\alpha} + \frac{1}{4}C(T)Q_{\alpha\beta}Q_{\alpha\beta}Q_{\gamma\delta}Q_{\gamma\delta} + \cdots, \tag{3.59}$$

where $\mathcal{G}_0$ is the free energy of the disordered state, and we use the convention of implicit summation over repeated Greek subscripts. We also assume, like in Section 2.7, that $A(T) = a(T - T^*)$ with $a > 0$, and temperature independent $B(T) = b$, $C(T) = c$. Beyond this Landau-type expansion, the Frank elastic deformation energy (see Section 1.2.3) that penalizes small spatial non-uniformities can also be written in a rotationally invariant form in terms of tensor components $Q_{\alpha\beta}(\boldsymbol{r})$ and their derivatives with respect to Cartesian coordinates, e_α, e_β, e_γ. This $\mathbf{Q}(\boldsymbol{r})$ expansion can be written as [Mori et al., 1999; Mottram and Newton, 2014]:

$$\begin{aligned}\mathcal{G}_{\text{el}} = &\left[\frac{L_1}{2}\left(\frac{\partial Q_{\alpha\beta}}{\partial e_\gamma}\right)^2 + \frac{L_2}{2}\frac{\partial Q_{\alpha\beta}}{\partial e_\beta}\frac{\partial Q_{\alpha\gamma}}{\partial e_\gamma} + \frac{L_3}{2}\frac{\partial Q_{\alpha\gamma}}{\partial e_\beta}\frac{\partial Q_{\alpha\beta}}{\partial e_\gamma}\right] \\ &+ \left[\frac{L_4}{2}\varepsilon_{\lambda\alpha\gamma}Q_{\lambda\beta}\frac{\partial Q_{\alpha\beta}}{\partial e_\gamma} + \frac{L_6}{2}Q_{\lambda\gamma}\frac{\partial Q_{\alpha\beta}}{\partial e_\lambda}\frac{\partial Q_{\alpha\beta}}{\partial e_\gamma}\right] + \cdots.\end{aligned} \tag{3.60}$$

Here L_i are elastic constants that can be obtained from a comparison with the Frank expression for the free energy. In the case of a uniaxial nematic

$$Q_{\alpha\beta}(\boldsymbol{r}) = \frac{3}{2}\langle P_2\rangle\left(d_\alpha(\boldsymbol{r})d_\beta(\boldsymbol{r}) - \frac{1}{3}\delta_{\alpha,\beta}\right), \tag{3.61}$$

where $d_\alpha(\boldsymbol{r})$, $d_\beta(\boldsymbol{r})$ are components of the director at position $\boldsymbol{r}$, and the explicit connection to the Frank elastic constants K_{ii} introduced in Section 1.2.3 is [Mori et al., 1999; Mottram and Newton, 2014]:

$$L_1 = (K_{33} - K_{11} + 3K_{22})/\left(6\langle P_2\rangle^2\right), \tag{3.62a}$$

$$L_2 = (K_{11} - K_{22} - K_{24})/\langle P_2\rangle^2, \tag{3.62b}$$

$$L_3 = K_{24}/\langle P_2\rangle^2, \tag{3.62c}$$

$$L_4 = 2q_0K_{22}/\langle P_2\rangle^2, \tag{3.62d}$$

$$L_6 = (K_{33} - K_{11})/\left(2\langle P_2\rangle^3\right), \tag{3.62e}$$

where q_0 is the wave vector of the phase (0 for a nematic and $\neq 0$ for a chiral nematic liquid crystal). Writing the free energy deformation density in this way, instead of the Frank expression in terms of the director, presents various numerical advantages, fully discussed elsewhere for many applications, e.g. pretransitional phase transitions in boundary layers [Sheng, 1982] and liquid crystal droplets [Kralj et al., 1991], wetting phenomena [Sluckin and Poniewierski, 1985; Nobili and Durand, 1992], surface-induced bulk alignment [Zhuang et al., 1994] and defects [Schopohl and Sluckin, 1987; Fukuda et al., 2002] so shall not be discussed in detail here. It is also worth mentioning that the theory has also been generalized to the treatment of smectic phases [Biscari et al., 2007; Mei and Zhang, 2015]. The surface free energy will also have to be introduced according to the real (or virtual) experimental conditions [Mottram and Newton, 2016]. In static equilibrium situations, minimization of the total free energy, obtained by integration over the sample, leads, following the classic procedures of calculus of variations, to a set of Euler–Lagrange differential equations in the bulk of the material and at the surface, for each of the dependent variables. The solution of the bulk equations subject to the surface boundary conditions provides the equilibrium configuration of the ordering tensor components $Q_{\alpha\beta}(\boldsymbol{r})$ across the sample [Mottram and Newton, 2016].

3.7 Physical Significance of Order Parameters

Quite similarly to what we have said for the second-rank order parameter, the higher-order parameters $\langle P_4 \rangle$, $\langle P_6 \rangle$, etc., are respectively 1 for complete order and 0 for an isotropic system. We may therefore wonder if there is an advantage in considering more than one order parameter. That this is the case becomes apparent if we refer to Fig. 3.10 where the Legendre polynomials $P_2(x)$ and $P_4(x)$ are shown as a function of $x = \cos\beta$. We have indicated with A, B, C, D regions which correspond to different combinations of the signs of the Legendre polynomials. As we can deduce from Fig. 3.10, if we measure $\langle P_2 \rangle$ and find that $\langle P_2 \rangle > 0$, (e.g. $\langle P_2 \rangle = 0.6$ in Fig. 3.10) this will mean that the majority of molecules have an orientation β between 0 and the so-called magic angle, $\beta = \cos^{-1}\sqrt{\frac{1}{3}} \approx 54.7°$,

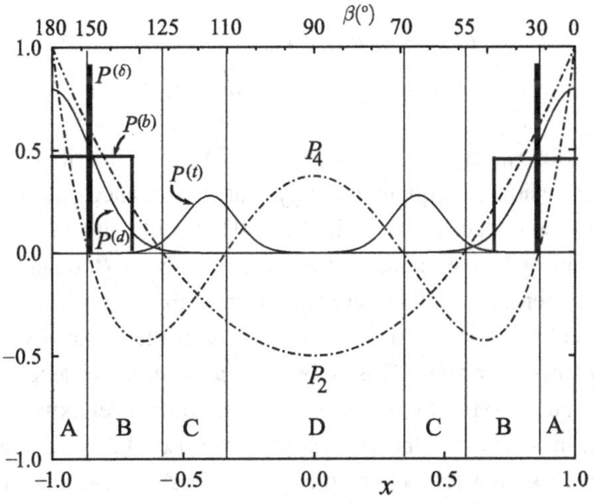

Figure 3.10 The angular dependence of the second- and fourth-rank Legendre polynomials $P_2(x)$, $P_4(x)$ (dash-dot lines) and of four (not normalized) distributions compatible with the same $\langle P_2 \rangle = 0.6$ but different $\langle P_4 \rangle$: $P^{(d)}(x)$ peaked along the director, a tilted one, $P^{(t)}(x)$ peaked at a tilt angle in region C, a delta-like at a cone-angle β_t, $P^{(\delta)}(x)$ and a box-like $P^{(b)}(x)$. The angular regions A, B, C, D correspond to combinations of the Legendre polynomials of different sign.

where $P_2(x) = 0$. This is what we normally expect for an elongated molecule and could be compatible with the distribution of molecular orientations $P^{(d)}$ shown as the continuous line in Fig. 3.10. If, on the other hand, $\langle P_2 \rangle < 0$, then we may infer that on average molecules will have an orientation giving a negative P_2 (e.g. in region D, with β between 54.7 and 90 degrees). This could be the case for a disc-like solute molecule in a liquid crystal. In many instances the orientation of the effective symmetry axis of the molecule of interest with respect to the director will not be at all obvious, and a determination of $\langle P_2 \rangle$ provides important information. For instance, if our molecule was dissolved in an oriented bilayer membrane (Figs. 1.50 and 1.52) we would learn from the sign of $\langle P_2 \rangle$ if the molecule is on average aligned parallel or perpendicular to the bilayer normal.

Let us now consider the fourth-rank Legendre polynomial $P_4(x)$ and its average $\langle P_4 \rangle$. The zeros of P_4 between 0° and 90° fall at $\beta \approx 30.56°$ and $\beta \approx 70.12°$. Suppose we have now measured $\langle P_2 \rangle$ and $\langle P_4 \rangle$ with some experimental technique. If $\langle P_2 \rangle > 0$ and $\langle P_4 \rangle > 0$, then the distribution of orientations will be such that the majority of molecules has a long axis orientation between 0° and 30.5° (region A). We might, however, find a positive $\langle P_2 \rangle$, as before, and a negative $\langle P_4 \rangle$. This would suggest a different type of orientational distribution, for example, a tilted one, $P^{(t)}$, with a peak between 30.5 and 54.7 degrees, as shown in Fig. 3.10. Other possible distributions can also be examined and tested. One is a delta-like distribution $P^{\delta}(x) = \delta(x - x_{\delta})$ where only one orientation: $x_{\delta} = \cos\beta_{\delta} = \sqrt{(4\langle P_2 \rangle + 3)/6}$. This implies also a well-defined value of $\langle P_4 \rangle$. In our example, $\langle P_2 \rangle = 0.6$ would correspond to $x_{\delta} = 0.9487$, i.e. $\beta_0 \approx 31°$ which in turn would give $\langle P_4 \rangle = -0.02$. Yet another possibility is that of box function $P(x) = const$ for $x_0 < x < 1$, or $0 \leq \beta \leq \beta_0$ which gives [Luckhurst and Yeates, 1976]

$$\langle P_2 \rangle = x_0\,(x_0 + 1)\,/2, \tag{3.63}$$

$$\langle P_4 \rangle = x_0\,(x_0 + 1)\,(7x_0^2 - 3)/8, \tag{3.64}$$

which can be solved for x_0. In our example, $x_0 = 0.704$. This distribution corresponds to a cone fully populated inside. The qualitative physical significance of other combinations of order parameters can be deduced in a similar way. This could also be extended if we knew $\langle P_6 \rangle$ etc. Every higher-order parameter restricts the bounds on $P(x)$ and thus increases our knowledge on the system. It is also interesting to note that the introduction of additional order parameters, and thus of additional coefficients in the expansion, does not lead to a revision of the lower-order ones. This stability is a general characteristic of orthogonal expansions as opposed to non-orthogonal ones, e.g. simple power expansions. What we are saying does not mean, of course, that the expansion in Eq. 3.21, is so rapidly convergent that we only need the first few terms to reconstruct $P(x)$. Actually, this will not be the case, at least in general, even though a knowledge of $\langle P_4 \rangle$ as well as that of $\langle P_2 \rangle$ can be very useful to discriminate between various models of molecular organization inside the liquid crystal. An example of a distribution function of this type will be reported later, together with other distributions, in Fig. 3.12. As an illustration of the kind of inconsistencies that one can obtain reconstructing the distribution from the first few terms, consider again an example where we have $\langle P_2 \rangle = 0.6$. Then $P(x)$, as given by the orthogonal expansion truncated at P_2, is shown in Fig. 3.12 as the dashed line. We see that $P(x)$ constructed in this way can even become negative, which is certainly unphysical when we think that $P(x)$ is the probability of finding a molecule at a certain orientation. Note that any property depending only on $\langle P_2 \rangle$ is calculated correctly using this $P(x)$. However, $\langle P_4 \rangle$ and the higher-order parameters calculated with the second-rank approximation are 0, because of the orthogonality of the Legendre polynomials. Thus, the orthogonal approximation is exact for terms that we have included but very bad if we want higher terms. Inclusion of more terms in the orthogonal expansions can of course improve the approximation. For instance, we show again in Fig. 3.12 that inclusion of the fourth-rank term (dot-dashed curve) avoids at least in this case the negative region of $P(x)$. We should also remember that all $\langle P_L \rangle$ tend to 1 for complete order, so that we expect the neglect of higher terms to be bad at low temperatures, where the order is higher.

3.8 Maximum Entropy

As we have seen, orthogonal expansions of the orientational distribution provide a formally exact way of writing down the order parameters in terms of quantities, the first few of which we might be able to determine. The representation of the distribution obtained when we know only the terms of rank 2, 4 or anyway of the first few symmetry allowed ranks, is exact up to that level but gives a poor approximation of the whole behaviour of the distribution. The method of Maximum Entropy (ME) [Jaynes, 1957b, 1957a; Levine and Tribus, 1978; Zannoni, 1988] approaches the problem of making use of the limited experimental information available on the distribution from a very different point of view. It tries to find the most probable or *least biased* distribution amongst the infinite number of them that agrees with the available data, typically a set of averages of some quantity. From this point of view, we cannot predetermine the form of the distribution in the absence of a certain amount of evidence, so that before actually knowing something about the distribution we should assume it is a flat, constant one with all the states equally probable. To be clearer, we can examine how to construct an ME distribution for a discrete system.

3.8.1 A System with Discrete States

Let us consider a system that can exist in a set of n discrete states (e.g. a coin or a die with n faces). We introduce a probability density vector $\boldsymbol{p} = (p_1, p_2, \ldots, p_n)$ containing the probabilities p_i of finding the system in the ith state. We require p_i to be dimensionless and non-negative, $p_i \geq 0$, and the distribution to be normalized, so that $\sum_{i=1}^{n} p_i = 1$. We can now try to obtain some information on the p_i by performing measurements on a set of M independent observable properties $\{A^{(k)}\} = (A^{(1)}, A^{(2)}, \ldots, A^{(M)})$. In practice, we try to infer the best estimate of the set p_i from the knowledge of the experimentally determined mean values of the set of properties. The average of the kth observable is

$$\langle A^{(k)} \rangle = \sum_{i=1}^{n} p_i A_i^{(k)}, \quad k = 1, \ldots, M. \tag{3.65}$$

In the trivial case of a two-state system (say a coin), $n = 2$, the average of one observable determines $\boldsymbol{p}$: $\langle A^{(k)} \rangle = p_1 A_1^{(k)} + (1 - p_1) A_2^{(k)}$. In most practical cases, when the number of states exceeds the number of constraints, i.e. of experimental observables, there is an infinity of distributions satisfying the given constraints. This is clearly the case of a continuous distribution of orientations Ω_i or positions $\boldsymbol{r}_i$. According to the ME principle [Jaynes, 1957b, 1957a] the most likely distribution is the one that maximizes the Boltzmann–Shannon entropy, that we can write as

$$\mathcal{S}(\boldsymbol{p}) = -\sum_{i=1}^{n} p_i \ln p_i, \tag{3.66}$$

subject to the given constraints. Introducing a set of $M + 1$ Lagrange multipliers λ_j we reduce the problem to that of searching the unconstrained maximum of the extended function $\mathscr{L}$ [Zannoni, 1988]

$$\mathscr{L} = -\sum_{i=1}^{n} p_i \ln p_i - \lambda_0 \left(\sum_{i=1}^{n} p_i - 1 \right) - \sum_{j=1}^{M} \lambda_j \left(\sum_{i=1}^{n} A_i^{(j)} p_i - \langle A^{(j)} \rangle \right). \tag{3.67}$$

Thus, at the extremum point

$$\frac{\partial \mathscr{L}}{\partial p_i} = -\ln p_i - 1 - \lambda_0 - \sum_{j=1}^{M} \lambda_j A_i^{(j)} = 0, \tag{3.68a}$$

$$\frac{\partial \mathscr{L}}{\partial \lambda_0} = 1 - \sum_{i=1}^{n} p_i = 0, \tag{3.68b}$$

$$\frac{\partial \mathscr{L}}{\partial \lambda_j} = \sum_{i=1}^{n} A_i^{(j)} p_i - \langle A^{(j)} \rangle = 0. \tag{3.68c}$$

The solution gives an approximation to the distribution that is of exponential nature, i.e.

$$p_i = \frac{1}{Z_0} \exp\left[\lambda_1 A_i^{(1)} + \lambda_2 A_i^{(2)} + \cdots + \lambda_M A_i^{(M)} \right], \tag{3.69}$$

with the normalization constant, $Z_0 = \exp(-\lambda_0)$, given by

$$Z_0 = \sum_{i=1}^{n} \exp\left[\lambda_1 A_i^{(1)} + \lambda_2 A_i^{(2)} + \cdots + \lambda_M A_i^{(M)} \right]. \tag{3.70}$$

The distribution obtained a posteriori from a set of known (experimentally determined) quantities $A^{(k)}$ can be tested by calculating (predicting) the value of new, different observables. If they are not satisfactorily obtained, they can be included in the procedure to give a refined version of the distribution and the procedure repeated.

3.8.2 Orientational Distributions

The problem of finding the best, in the sense of least-biased approximation to the whole continuous distribution $P(x)$, or in general $P(\Omega)$, starting from a knowledge of a set of order parameters $\langle P_L \rangle$, say up to rank L', can be approached using Information Theory [Jaynes, 1957b, 1957a] again. In practice, we proceed by generalizing the example just seen for a discrete variable to a set of continuous variables representing the orientations. Correspondingly, we shall have integrals over the angles rather than sums and we shall have to choose a set of independent observables $A^{(k)}(\Omega)$. For effectively uniaxial molecules in a uniaxial phase our input is a set of order parameters $\langle P_L \rangle$, that we assume to be known [Bower, 1981, 1982]. We then determine the distribution maximizing the entropy function

$$\mathcal{S} \propto -\int_{-1}^{1} \mathrm{d}x \, P(x) \ln P(x), \tag{3.71}$$

with $P(x)$, as x, dimensionless and subject to the constraints

$$\int_{-1}^{1} \mathrm{d}x \, P(x) = 1, \tag{3.72a}$$

$$\int_{-1}^{1} \mathrm{d}x \, P(x) \, P_L(x) = \langle P_L \rangle, \tag{3.72b}$$

where $x = \cos\beta$, for the known $\langle P_L \rangle$ (e.g. those from $L = 0, 2, \ldots$ to L_{max}). Using the Lagrange multipliers technique as for the discrete case, we find that the ME distribution has the form

$$P(x) = \exp\left\{ \sum_{\substack{L=2 \\ L \text{ even}}}^{L_{max}} a_L P_L(x) \right\} = \exp(a_0) \exp\left\{ \sum_{\substack{L=2 \\ L \text{ even}}}^{L_{max}} a_L P_L(x) \right\} \equiv \exp\{-\mathscr{U}(x)\}/Z_0, \tag{3.73}$$

where the function $\mathscr{U}(x) \equiv \mathscr{U}(\cos\beta)$ plays the role of a dimensionless angular dependent pseudo-potential. The coefficients a_L are obtained imposing the constraints that the $\langle P_L \rangle$, $L = 0, 2, \ldots, L_{max}$ calculated from $P(x)$ have the known (experimental) values, while the normalization condition corresponds of course to $\langle P_0 \rangle = 1$. The information theory approach is thus in any case an a posteriori one. It allows constructing an approximate full distribution from available information, but on the other hand, it can make no prediction on what the distribution will be at, say, a different temperature. The approach also does not say anything on the molecular origin of the distribution itself. It is just a way of translating the experimental information into the most probable distribution compatible with the data themselves. As more and more order parameters or in general observables become available the estimate of $P(x)$ can be refined. The method does not rely on a priori assumptions and as the number of terms increases the sequence of maximum entropy approximations converges to the true one. It is also important to stress that at any level of approximation the distribution obtained is positive and of exponential character. It may be worth discussing in some detail the differences between the orthogonal and the ME approximations.

We now consider briefly what inferences can be made about the molecular organization starting from a knowledge of a small number of order parameters and in particular of $\langle P_2 \rangle$ and $\langle P_4 \rangle$.

(i) **Knowing $\langle P_2 \rangle$ only.** To start with we suppose that only the second-rank order parameter, $\langle P_2 \rangle$, has been determined. The ME distribution associated with this $\langle P_2 \rangle$ will be

$$P(x) = \exp[a_2 P_2(x)] \,/\, Z_0 \tag{3.74}$$

with a_2 determined by the condition

$$\langle P_2 \rangle = \int_{-1}^{1} \mathrm{d}x\; P_2(x)\, \exp[a_2 P_2(x)] \,/Z_0 \tag{3.75}$$

and where the normalization constant is

$$Z_0 = \int_{-1}^{1} \mathrm{d}x\, \exp[a_2 P_2(x)] = 2M\Big(\frac{1}{2}, \frac{3}{2}, \frac{3}{2}a_2\Big), \tag{3.76}$$

in terms of $M(a, b, z)$, the Kummer confluent hypergeometric function [Erdélyi et al., 1953], $M\Big(n+\frac{1}{2}, n+\frac{3}{2}, z\Big) \equiv (2n+1)\int_0^1 \mathrm{d}x x^{2n}\, \mathrm{e}^{zx^2}$. It is easy to see that [Zannoni, 1975] $\langle P_2 \rangle = M\Big(\frac{3}{2}, \frac{5}{2}, \frac{3}{2}a_2\Big) / M\Big(\frac{1}{2}, \frac{3}{2}, \frac{3}{2}a_2\Big)$. Z_0 and $\langle P_2 \rangle$ can also be written in terms of another special function, the Dawson function [Abramowitz and Stegun, 1965]:

$$D(x) \equiv \mathrm{e}^{-x^2} \int_0^x \mathrm{d}t\, \mathrm{e}^{t^2}, \tag{3.77}$$

that gives $Z_0 = 2\exp(a_2) D\Big(\sqrt{\frac{3a_2}{2}}\Big) / \sqrt{3a_2/2}$. Expressing the result in terms of special functions immediately puts at our disposal the very large number of recurrence relations, asymptotic expansions, etc. available for them. We shall see later that an identical functional form for the average order parameter can be obtained by various approximate theoretical treatments, like the Mean Field theories of Maier and Saupe [1958] (Section 7.2) and Onsager [1949] (Section 7.6). In view of this wide applicability, it is worth discussing this distribution in some detail. Eq. 3.75 can be solved for a_2 in terms of $\langle P_2 \rangle$ and in Fig. 3.11

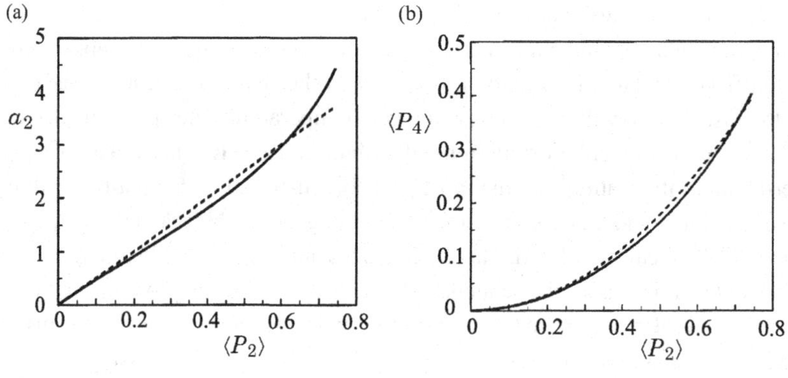

Figure 3.11 (a) The maximum entropy parameter a_2 plotted against $\langle P_2 \rangle$ obtained from Eq. 3.75 (—) and the analytic approximation $a_2 = 5\langle P_2 \rangle$ (- - -). (b) $\langle P_4 \rangle$ vs $\langle P_2 \rangle$ as obtained by integration over the distribution in Eq. 3.74 (—) and the approximate analytic expression $\langle P_4 \rangle = 5\langle P_2 \rangle^2/7$ (- - -) [Fabbri and Zannoni, 1986].

we show the resulting curve for positive $\langle P_2 \rangle$ as the full line. We see that for positive $\langle P_2 \rangle$ the distribution is peaked at $x = \pm 1$ (or $\beta = 0, \pi$), so that the majority of molecules will be parallel to the director. This is normally the case when we dissolve an elongated molecule in a nematic.

A useful estimate of a_2 from $\langle P_2 \rangle$ can be obtained by first expanding $\langle P_2 \rangle$ in Eq. 3.75 as a series of powers of a_2: $\langle P_2 \rangle = \frac{1}{5}a_2 + \frac{1}{35}a_2^2 - \frac{1}{175}a_2^3 - \cdots$. Reversion of this series [Abramowitz and Stegun, 1965] gives a_2 in a power series in $\langle P_2 \rangle$ [Fabbri and Zannoni, 1986]

$$a_2 = 5\langle P_2 \rangle - \frac{25}{7}\langle P_2 \rangle^2 + \frac{425}{49}\langle P_2 \rangle^3 - \cdots . \tag{3.78}$$

The series diverges at $\langle P_2 \rangle = 1$ but it can still be useful for realistic order parameters found in nematics. The very simple approximation

$$a_2 = 5\langle P_2 \rangle, \tag{3.79}$$

shown as the dashed line in Fig. 3.11a is useful to get a good idea of a_2 and thus, of the distribution at least up to $\langle P_2 \rangle = 0.6$. Having determined a_2 we can immediately plot the distribution $P(x)$. For example, if we assume $\langle P_2 \rangle = 0.6$, as in Section 3.7, we obtain the approximate ME distribution as the continuous line in Fig. 3.12. We note that a_2 becomes negative as $\langle P_2 \rangle$ changes sign and that the corresponding distribution becomes peaked at $x = 0, (\beta = \pi/2)$. Physically this will normally happen when we study a disc-like molecule dissolved in a nematic, since in this case its plane will tend to align with the host molecules and its molecular z-axis (the disc symmetry axis) will be preferentially aligned perpendicular to the director.

The distribution in Eq. 3.74 can be used to calculate the fourth-rank order parameter $\langle P_4 \rangle$ by integration (see Eq. 3.72b). The curve obtained is shown in Fig. 3.11b as the continuous line. A simple analytic approximation for the dependence of $\langle P_4 \rangle$ on $\langle P_2 \rangle$ can be obtained

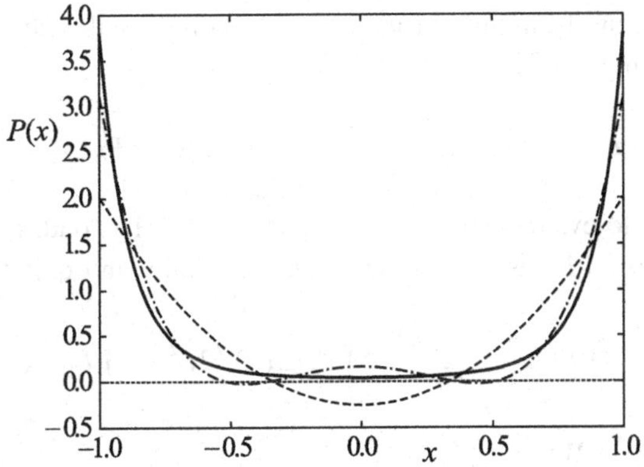

Figure 3.12 The distribution $P(x)$ in Eq. 3.74 corresponding to $\langle P_2 \rangle = 0.60$ and $\langle P_4 \rangle = 0.244$ (—) and its approximations using the orthogonal expansion in Eq. 3.25 truncated at P_2 (- - -) and at P_4 (-·-·-).

expanding $\langle P_4 \rangle$ in powers of a_2 and substituting in Eq. 3.78. This gives a series with large terms of alternating sign and poorly convergent unless terms are properly grouped together. However, the simplest approximation which retains just the first term, i.e.

$$\langle P_4 \rangle = \frac{5}{7} \langle P_2 \rangle^2, \tag{3.80}$$

is actually a good approximation up to $\langle P_2 \rangle \approx 0.6$ as we see from the dashed line in Fig. 3.11b. Note that $\langle P_4 \rangle$ is correctly predicted to be positive even for negative $\langle P_2 \rangle$. In a similar way we can obtain an approximate expression for $\langle P_6 \rangle$:

$$\langle P_6 \rangle = \frac{375}{1001} \langle P_2 \rangle^3, \tag{3.81}$$

showing that, at least for the simple distribution Eq. 3.74, the order parameter $\langle P_6 \rangle < \langle P_4 \rangle < \langle P_2 \rangle$. Actually, it is interesting to note that a recurrence relation can be obtained for the order parameters obtained from Eq. 3.74 [Zannoni, 1975]. Let us start from the even moments of the distribution:

$$\langle x^{2n} \rangle = \frac{\int_0^1 x^{2n} \exp\left(\xi x^2\right) \mathrm{d}x}{\int_0^1 \exp\left(\xi x^2\right) \mathrm{d}x} \equiv \frac{Z_{2n}(\xi)}{Z_0(\xi)} \langle x^{2n} \rangle = \frac{1}{2n+1} \frac{M\left(n+\frac{1}{2}, n+\frac{3}{2}, \xi\right)}{M\left(\frac{1}{2}, \frac{3}{2}, \xi\right)}, \tag{3.82}$$

where $\xi \equiv \frac{3}{2} a_2$ and we have used the derivatives of Kummer functions [Abramowitz and Stegun, 1965]

$$Z_{2n}(\xi) = \frac{\mathrm{d}^n}{\mathrm{d}\xi^n} Z_0 = \frac{\mathrm{d}^n}{\mathrm{d}\xi^n} M\left(\frac{1}{2}, \frac{3}{2}, \xi\right) = \frac{1}{2n+1} M\left(n + \frac{1}{2}, n + \frac{3}{2}, \xi\right). \tag{3.83}$$

Substitution of the moments in the general expression for the Legendre polynomials yields the order parameters of any rank in terms of ξ or of $\langle P_2 \rangle$. We can also derive a recurrence relation for $\langle P_L \rangle$ starting from that for the Legendre polynomials (Eq. A.51). Multiplying by $P(x)$ and integrating both sides gives

$$L \langle P_L \rangle = (2L-1) \langle x P_{L-1} \rangle - (L-1) \langle P_{L-2} \rangle. \tag{3.84}$$

The problem is then to evaluate $\langle x P_{L'} \rangle$, where $L' = L - 1$ is odd. Integrating by parts, using $P_{L'}(0) = 0$ for odd L', as well as the derivative relation in Eq. A.52,

$$\frac{\mathrm{d}}{\mathrm{d}x} P_{L'}(x) = \left(2L'-1\right) P_{L'-1} + \left(2L'-5\right) P_{L'-3} + \left(2L'-9\right) P_{L'-5} + \cdots + P_0, \tag{3.85}$$

we obtain [Zannoni, 1975]

$$\langle P_L \rangle = \frac{(2L-1)}{3La_2} \left\{ \frac{e^{\frac{3}{2}a_2}}{Z_0} - \sum_{n=0}^{(L-2)/2} \left[(2L-3-4n) + \frac{3a_2(L-1)}{(2L-1)} \delta_{0,n} \right] \langle P_{L-2-2n} \rangle \right\}. \tag{3.86}$$

The first few terms are

$$\langle P_2 \rangle = \frac{1}{2a_2}\left\{\frac{e^{\frac{3}{2}a_2}}{Z_0} - (1+a_2)\right\}, \tag{3.87a}$$

$$\langle P_4 \rangle = \frac{7}{12a_2}\left\{\frac{e^{\frac{3}{2}a_2}}{Z_0} - \left(5+\frac{9a_2}{7}\right)\langle P_2 \rangle - 1\right\}, \tag{3.87b}$$

$$\langle P_6 \rangle = \frac{44}{36a_2}\left\{\frac{e^{\frac{3}{2}a_2}}{Z_0} - \left(9+\frac{30a_2}{22}\right)\langle P_4 \rangle - 5\langle P_2 \rangle - 1\right\}. \tag{3.87c}$$

Asymptotic expressions for high values of a_2 (i.e. high order) can also be obtained from the corresponding formulae for the Kummer functions [Abramowitz and Stegun, 1965]. In particular, in this high-order limit we find

$$\langle P_2 \rangle = 1 - \frac{1}{a_2} - \frac{1}{3a_2^2} + \mathcal{O}\left(a_2^{-3}\right), \tag{3.88a}$$

$$\langle P_4 \rangle = 1 - \frac{10}{3a_2} + \frac{25}{9a_2^2} + \mathcal{O}\left(a_2^{-3}\right), \tag{3.88b}$$

where $\mathcal{O}\left(a_2^{-3}\right)$ indicates the order of magnitude of the neglected terms.

(ii) **Knowing $\langle P_2 \rangle$ and $\langle P_4 \rangle$**. We now turn to the case where both $\langle P_2 \rangle$ and $\langle P_4 \rangle$ have been experimentally determined. The first thing we might try is to test if the distribution derived using just the information on $\langle P_2 \rangle$, is consistent with the observed $\langle P_4 \rangle$. If this is not the case, and the experimental (or simulated) $\langle P_4 \rangle$ does not fall on the curve in Fig. 3.11b, we can improve our maximum entropy distribution using Eq. 3.73 with $L = 0, 2, 4$. To do this we have to find a_2 and a_4 from our given $\langle P_2 \rangle$ and $\langle P_4 \rangle$. The first thing to observe is that the domain of the functions $a_2(\langle P_2 \rangle, \langle P_4 \rangle)$, $a_4(\langle P_2 \rangle, \langle P_4 \rangle)$ consists of the set of allowed values of $\langle P_2 \rangle, \langle P_4 \rangle$. To find these bounds it is useful to invoke Schwarz inequality [Abramowitz and Stegun, 1965]

$$\left(\int_a^b \mathrm{d}x\, u(x)v(x)\right)^2 \leq \left(\int_a^b \mathrm{d}x\, u^2(x)\right)\left(\int_a^b \mathrm{d}x\, v^2(x)\right), \tag{3.89}$$

where the functions $u(x)$, $v(x)$ are arbitrary as long as the integrals exist. Taking a, b to be the extremes of the angular range together with $u(x) \equiv f(x)P(x)^{1/2}$ and $v(x) \equiv g(x)P(x)^{1/2}$ with $P(x)$ the normalized distribution, we have

$$\langle f(x)g(x)\rangle^2 \leq \langle f^2(x)\rangle\langle g^2(x)\rangle. \tag{3.90}$$

In particular, if we take $g(x) = 1$, we have $\langle f(x)\rangle^2 \leq \langle f^2(x)\rangle$. A useful pair of inequalities for our purposes is $\langle \cos^2\beta\rangle^2 \leq \langle \cos^4\beta\rangle \leq \langle \cos^2\beta\rangle$. The explicit form of $P_2(x)$ and $P_4(x)$, Eq. A.48, together with these inequalities yields a lower and upper bound on $\langle P_4 \rangle$:

$$\frac{35}{18}\langle P_2 \rangle^2 - \frac{5}{9}\langle P_2 \rangle - \frac{7}{18} \leq \langle P_4 \rangle \leq \frac{5}{12}\langle P_2 \rangle + \frac{7}{12}, \tag{3.91}$$

that define the region of space where possible values of $\langle P_2 \rangle$, $\langle P_4 \rangle$ consistent with their respective trigonometric form should lie. It goes without saying that it makes sense to check

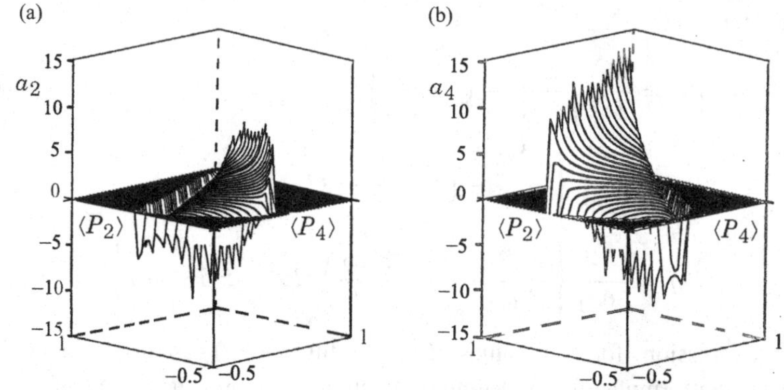

Figure 3.13 (a) The maximum entropy coefficients a_2 and (b) a_4 in the distribution $P(\beta) \propto \exp[a_2 P_2(\cos\beta) + a_4 P_4(\cos\beta)]$ shown as a function of $\langle P_2 \rangle$ and $\langle P_4 \rangle$ [Chiccoli et al., 1988c].

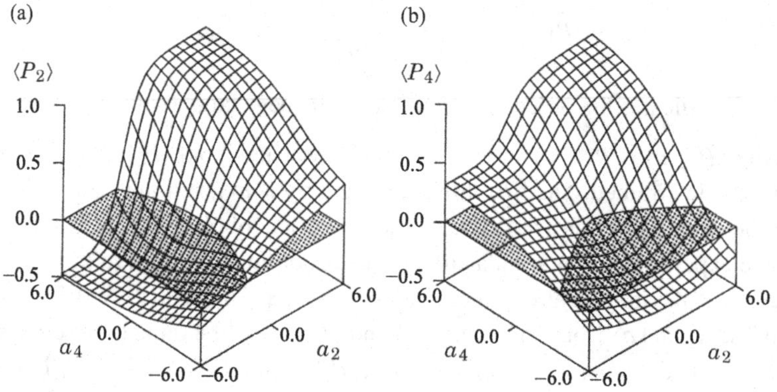

Figure 3.14 (a) The order parameters $\langle P_2 \rangle$ and (b) $\langle P_4 \rangle$ as a function of the maximum entropy coefficients a_2 and a_4 in the distribution $P(\beta) \propto \exp[a_2 P_2(\cos\beta) + a_4 P_4(\cos\beta)]$ [Chiccoli et al., 1988c].

that experimental values do fall within this area. The parameters a_2, a_4 can be explicitly determined from $\langle P_2 \rangle$, $\langle P_4 \rangle$ solving the non-linear system

$$\langle P_2 \rangle = \int_{-1}^{1} \mathrm{d}x\, P_2(x) \exp[a_2 P_2(x) + a_4 P_4(x)] \Big/ \int_{-1}^{1} \mathrm{d}x \exp[a_2 P_2(x) + a_4 P_4(x)], \quad (3.92a)$$

$$\langle P_4 \rangle = \int_{-1}^{1} \mathrm{d}x\, P_4(x) \exp[a_2 P_2(x) + a_4 P_4(x)] \Big/ \int_{-1}^{1} \mathrm{d}x \exp[a_2 P_2(x) + a_4 P_4(x)]. \quad (3.92b)$$

and the results are shown in Fig. 3.13. We also plot, in Fig. 3.14, $\langle P_2 \rangle$ and $\langle P_4 \rangle$ in terms of a_2 and a_4. We note that, although we normally expect $\langle P_2 \rangle$ to be greater than $\langle P_4 \rangle$ as it was the case in the P_2 distribution [Pottel et al., 1986] (see Fig. 3.11), a range of solutions exists also for $\langle P_4 \rangle$ greater than $\langle P_2 \rangle$. Indeed, an interesting case is that of $\langle P_4 \rangle > \langle P_2 \rangle$, with the values falling on a curve like the continuous one in Fig. 3.15a. This unusual behaviour has been found to be consistent with fluorescence depolarization data of the probe

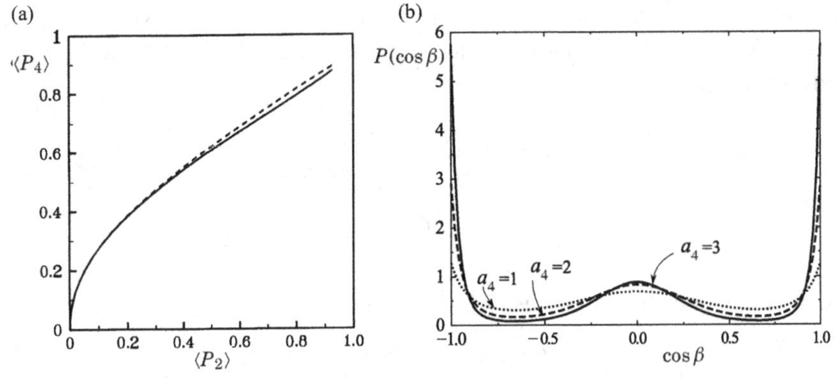

Figure 3.15 (a) the dependence of the fourth-rank order parameter $\langle P_4 \rangle$ on the second-rank one $\langle P_2 \rangle$ for a purely fourth-rank distribution, Eq. 3.93 (—). We also show the analytical approximation (see text) in Eq. 3.97 (- - -). (b) The angular variation of the normalized pure P_4 distribution $P(\cos\beta) \propto \exp[a_4 P_4(\cos\beta)]$ in Eq. 3.93 with $a_4 = 1, 2, 3$.

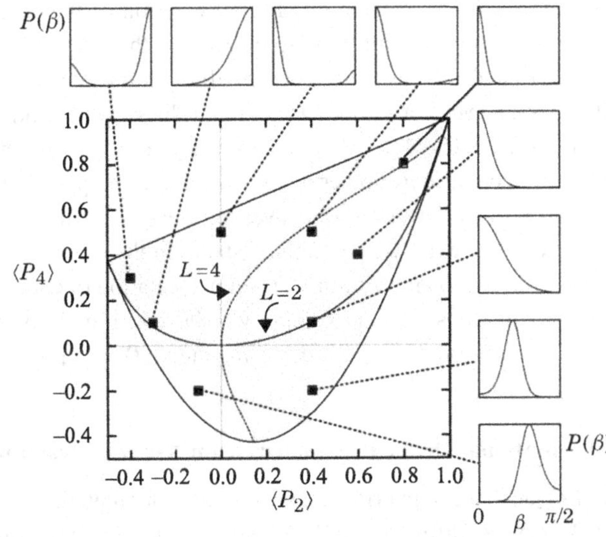

Figure 3.16 ME uniaxial orientational distributions $P(\beta)$ corresponding to various input $\langle P_2 \rangle$–$\langle P_4 \rangle$ pairs inside the region of allowed values (in grey) comprised within the bounds of Eq. 3.91. The curves marked $L = 2$, $L = 4$ correspond to a pure P_2 (Maier–Saupe like, Eq. 3.74) and pure P_4 (Eq. 3.93) distribution.

diphenyl-hexatriene obtained for the fluorescent probe DPH embedded in DMPC membrane vesicles [Pottel et al., 1986]. In turn, the behaviour agrees with that predicted by a pure P_4 effective potential model [Zannoni, 1979b], which corresponds to a probability density

$$P(x) = \exp\left[a_4 P_4(x)\right] \bigg/ \int_{-1}^{1} \mathrm{d}x \exp\left[a_4 P_4(x)\right]. \tag{3.93}$$

We can obtain, also in this case, a simple approximation to the $\langle P_4 \rangle$ versus $\langle P_2 \rangle$ curve [Zannoni, 1988]. A Taylor expansion of the expressions for $\langle P_2 \rangle$ and $\langle P_4 \rangle$ with respect to a_4 yields the first few terms as

$$\langle P_2 \rangle = \frac{10a_4^2}{693} + \frac{10a_4^3}{3003} + \frac{1010a_4^4}{26189163} + \cdots, \tag{3.94}$$

$$\langle P_4 \rangle = \frac{a_4}{9} + \frac{9a_4^2}{1001} - \frac{1367a_4^3}{1378377} + \frac{457a_4^4}{2909907} + \cdots, \tag{3.95}$$

Reversion of the series for $\langle P_4 \rangle$ gives a_4 in terms of $\langle P_4 \rangle$

$$a_4 = 9\langle P_4 \rangle - \frac{6561\langle P_4 \rangle^2}{1001} + \frac{273458673\langle P_4 \rangle^3}{17034017} + \cdots \tag{3.96}$$

and, with a little of computer algebra (see, e.g., [Harris, 2014])

$$\langle P_4 \rangle = \sqrt{\frac{77}{90}}\langle P_2 \rangle^{\frac{1}{2}} - \frac{69}{260}\langle P_2 \rangle + \frac{7794479}{1007760\sqrt{770}}\langle P_2 \rangle^{\frac{3}{2}} + \cdots. \tag{3.97}$$

This simple power series in $\langle P_2 \rangle^{\frac{1}{2}}$ gives a good representation of the curve for $\langle P_2 \rangle$ up to 0.9. In Fig. 3.15a we show the analytical approximation to the $\langle P_4 \rangle$ vs. $\langle P_2 \rangle$ curve from the truncation in Eq. 3.97 and the curve obtained by direct numerical integration. Using Eq. 3.97 it is quite easy to test if a set of $\langle P_2 \rangle$, $\langle P_4 \rangle$ values corresponds to a pure P_4 distribution as plotted in Fig. 3.15b. Note that this distribution shows a maximum not only for molecules parallel to $\boldsymbol{d}$, but also a smaller one for molecules perpendicular to it.

In essence, the ME approach offers a way of converting measured values of $\langle P_2 \rangle$ and $\langle P_4 \rangle$, whose significance is not immediately easy to understand, in the most likely distributions compatible with those values, allowing an immediate interpretation of experimental results in terms of preferred orientations. As an example we show in Fig. 3.16 the most likely distributions obtained from various pairs of order parameters $\langle P_2 \rangle$, $\langle P_4 \rangle$.

3.9 Orientational Order Parameters from X-ray Diffraction

An experimental technique that can in principle provide also orientational order parameters of rank higher than 4, is X-ray diffraction (XRD), whose basic principles are described in Appendix J. Let us start considering a system of N rigid molecules, each with M point scattering centres (atoms in real molecules or in a realistic model). Then the scattered intensity at scattering vector $\boldsymbol{q}$ is

$$I(\boldsymbol{q}) \propto \left\langle \left| \sum_{i=1}^{N} \sum_{a \in i}^{M} E_{a,i}(\boldsymbol{q}) \right|^2 \right\rangle, \tag{3.98a}$$

$$\propto \sum_{i=1}^{N} \sum_{a \in i}^{M} \sum_{j=1}^{N} \sum_{b \in j}^{M} b_{a,i}(\boldsymbol{q})\, b_{b,j}^{*}(\boldsymbol{q}) \left\langle e^{i\boldsymbol{q}\cdot(\boldsymbol{r}_{a,i}-\boldsymbol{r}_{b,j})} \right\rangle, \tag{3.98b}$$

where $E_{a,i}$ is the field scattered by centre a on molecule i and the sums in Eq. 3.98b run on the N molecules i and j, as well as on the M atoms a, b belonging to each of them and located at $\boldsymbol{r}_{a,i}$, $\boldsymbol{r}_{b,j}$, respectively. Eq. 3.98b can be used to obtain $I(\boldsymbol{q})$ if the coordinates of all atoms are available, e.g. from the atomistic simulations discussed in Chapter 12, possibly approximating $b_{a,i}$ with the number of electrons Z_a on that atom. Simpler, and perhaps more illuminating, formulas can be written if we can assume the same scattering factor, $b(\boldsymbol{q})$, for all centres. Then

$$I(\boldsymbol{q}) \propto |b(\boldsymbol{q})|^2 S(q), \tag{3.99}$$

where we have introduced the *structure factor* $S(q)$

$$S(q) = \frac{1}{N} \sum_{i,j=1}^{N} \sum_{\substack{a,b=1 \\ a\in i, b\in j}}^{M} \left\langle e^{i\boldsymbol{q}\cdot(\boldsymbol{r}_{a,i}-\boldsymbol{r}_{b,j})} \right\rangle, \tag{3.100a}$$

$$= \frac{1}{N} \sum_{i,j=1}^{N} \sum_{\substack{a,b=1 \\ a\in i, b\in j}}^{M} \left\{ \delta_{i,j}\delta_{a,b} + \delta_{i,j} \left\langle e^{i\boldsymbol{q}\cdot(\boldsymbol{r}_{a,i}-\boldsymbol{r}_{b,i})} \right\rangle + (1-\delta_{i,j}) \left\langle e^{i\boldsymbol{q}\cdot(\boldsymbol{r}_{a,i}-\boldsymbol{r}_{b,j})} \right\rangle \right\},$$

$$= 1 + S_s(q) + S_d(q). \tag{3.100b}$$

Here, apart from the first constant term which is irrelevant, we have regrouped terms in the 'self' contribution, from atoms in the same molecule:

$$S_s(q) = \frac{1}{N} \sum_{i}^{N} \sum_{a,b\in i}^{M} \left\langle e^{i\boldsymbol{q}\cdot(\boldsymbol{r}_{a,i}-\boldsymbol{r}_{b,i})} \right\rangle \tag{3.101}$$

and the 'distinct' one, where each of the nuclei involved belongs to two different molecules:

$$S_d(q) = \frac{1}{N} \sum_{\substack{i,j=1 \\ i\neq j}}^{N} \sum_{\substack{a,b=1 \\ a\in i, b\in j}}^{M} \left\langle e^{i\boldsymbol{q}\cdot(\boldsymbol{r}_{a,i}-\boldsymbol{r}_{b,j})} \right\rangle. \tag{3.102}$$

If nuclei a and b belong to molecules i and j then $\boldsymbol{r}_{a,i} - \boldsymbol{r}_{b,j} = \boldsymbol{r}_{ij} + \boldsymbol{h}_{a,i} - \boldsymbol{h}_{b,j}$, where $\boldsymbol{h}_{a,i}$, $\boldsymbol{h}_{b,j}$ locate the nuclei a, b in their respective molecular frames and $\boldsymbol{r}_{ij}$ is the separation vector, joining the centres of the two molecules. The single molecule and intermolecular term can, to some extent, be separated performing the XRD experiment in different conditions or looking at different q. We shall assume that this is the case and consider only the single molecule scattering term, while the intermolecular one will be treated in Chapter 4.

Writing the laboratory frame position of each atomic centre as $\boldsymbol{r}_{a,i} = \boldsymbol{r}_i + \boldsymbol{h}_{a,i}$, where $\boldsymbol{r}_i$ is the position of the molecular frame origin of the ith molecule and $\boldsymbol{h}_{a,i}$ the position of atom a in that frame, we have, leaving out the now redundant i label, that the single molecule contribution becomes

$$S_s(\boldsymbol{q}) = \sum_{a,b=1}^{M} \left\langle e^{i\boldsymbol{q}\cdot\boldsymbol{h}_{ab}} \right\rangle, \quad a \neq b, \tag{3.103}$$

where $\boldsymbol{h}_{ab} = \boldsymbol{h}_a - \boldsymbol{h}_b$ is the intramolecular separation vector between two scattering centres. The Bragg-like reflection condition will be obtained for $q = 2\pi / h_{ab}$. To obtain a relation with the order parameters we can use the Rayleigh plane wave expansion (see Eq. J.7) to write

$$S_s(\boldsymbol{q}) = \sum_L \sum_{a,b=1}^{M} i^L (2L+1) j_L(q h_{ab}) \langle \mathscr{D}^L_{0,0}(\hat{\boldsymbol{q}} \cdot \hat{\boldsymbol{h}}_{ab}) \rangle, \quad a \neq b, \tag{3.104}$$

with $j_L(x)$ a spherical Bessel function [Abramowitz and Stegun, 1965] and the cap over the vectors indicating their unit length. Now, applying the closure relation of Wigner rotation matrices, Eq. F.13 we can write the relative orientation between the interatomic vector $\hat{\boldsymbol{h}}_{ab}$ and $\hat{\boldsymbol{q}}$ in terms of rotations of Ω_{qd} from director $\boldsymbol{d}$ to $\hat{\boldsymbol{q}}$, of the rotation Ω_{Md} from director to molecule fixed frame M and of the further rotation $\Omega_{h_{ab}\mathrm{M}}$ that brings from M to $\hat{\boldsymbol{h}}_{ab}$ as follows:

$$\langle \mathscr{D}^L_{0,0}(\hat{\boldsymbol{q}} \cdot \hat{\boldsymbol{h}}_{ab}) \rangle = \sum_m \langle \mathscr{D}^{L*}_{0,m}(\Omega_{dh_{ab}}) \rangle \mathscr{D}^{L*}_{m,0}(\Omega_{qd}), \tag{3.105a}$$

$$= \sum_{m,p} \mathscr{D}^{L*}_{m,0}(\Omega_{qd}) \langle \mathscr{D}^L_{m,p}(\Omega_{\mathrm{M}d}) \rangle \mathscr{D}^L_{p,0}(\Omega_{h_{ab}\mathrm{M}}), \tag{3.105b}$$

$$= \sum_p \mathscr{D}^{L*}_{0,0}(\Omega_{qd}) \langle \mathscr{D}^L_{0,p}(\Omega_{\mathrm{M}d}) \rangle \mathscr{D}^L_{p,0}(\Omega_{h_{ab}\mathrm{M}}), \tag{3.105c}$$

where we have used the uniaxiality of the LC phase to write $\langle \mathscr{D}^L_{m,p}(\Omega_{\mathrm{M}d}) \rangle = \delta_{m,0} \langle \mathscr{D}^L_{0,p}(\Omega_{\mathrm{M}d}) \rangle$. Substituting Eq. 3.105c into Eq. 3.104 shows that the single molecule scattering depends on a series of order parameter of increasing rank L. If the scattering centres are along the molecular axis, with direction along the unit vector $\boldsymbol{u}$, then $\Omega_{h_{ab}\mathrm{M}} = (0,0,0)$ and $\mathscr{D}^L_{p,0}(0,0,0) = \delta_{p,0}$ for all a, b pairs. Thus,

$$S_s(\boldsymbol{q}) = \sum_{\substack{L \\ even}} \sum_{\substack{a,b=1 \\ a \neq b}}^{M} i^L (2L+1) j_L(q h_{ab}) \langle P_L(\hat{\boldsymbol{q}} \cdot \boldsymbol{d}) \rangle_d \langle P_L \rangle, \tag{3.106}$$

where $\langle P_L \rangle$ is an orientational order parameter of rank L with respect to the director $\boldsymbol{d}$: $\langle P_L \rangle = \langle \mathscr{D}^L_{0,0}(\Omega_{\mathrm{M}d}) \rangle = \langle P_L(\boldsymbol{u} \cdot \boldsymbol{d}) \rangle$. For a nematic or an orthogonal non-polar smectic (e.g. a $\mathrm{S_A}$), L has to be even. The geometry of the various experiments normally performed on liquid crystals is shown in Fig. J.5. The average $\langle \ldots \rangle_d$ is instead over the distribution of directors, assumed to be static during the experiment. Thus, if we have an isotropic polydomain in the area observed, $\langle P_L(\boldsymbol{u} \cdot \boldsymbol{d}) \rangle = \delta_{L,0}$. Then $S_s(q) = j_0(q\ell_z)$ and the information on the order parameters is lost, even if the system is locally oriented (e.g. a spherical bilayer membrane vesicle, as in Fig. 1.51). If instead the sample is a monodomain with $\hat{\boldsymbol{q}} \| \boldsymbol{d}$, then $\langle P_L(\hat{\boldsymbol{q}} \cdot \boldsymbol{d}) \rangle = (-1)^L$ and we have

$$S_s(\boldsymbol{q} \| \boldsymbol{d}) = \sum_{L\ even} i^L (2L+1) j_L(q\ell_z) \langle P_L \rangle, \tag{3.107a}$$

$$\approx j_0(q\ell_z) - 5 j_2(q\ell_z) \langle P_2 \rangle + 9 j_4(q\ell_z) \langle P_4 \rangle + \cdots. \tag{3.107b}$$

The first few spherical Bessel functions have a very simple form, e.g. $j_0(x) = \sin x/x$, $j_2(x) = (3 - x^2)\sin x/x^3 - 3\cos x/x^2$. At the Bragg reflection condition, when the argument $q\ell_z = 2\pi$ we have $j_0(2\pi) = 0$, $j_2(2\pi) = -(3/4\pi^2)$, $j_4(2\pi) = [5/(2\pi^2) - 105/(16\pi^4)]$. Thus, the experimental data contain in principle information on order parameters higher than four. However, if the order parameters of higher rank get smaller,

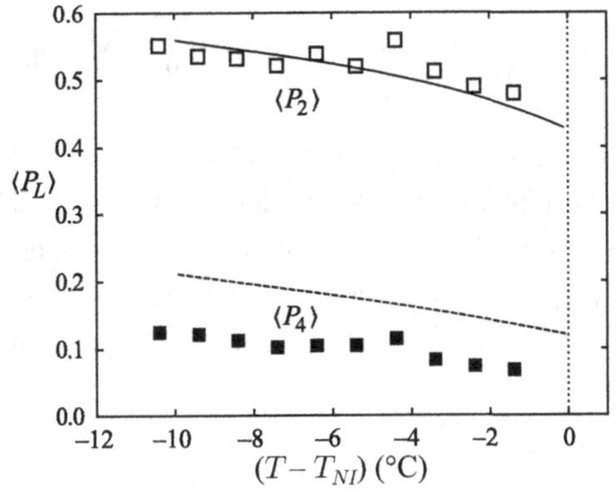

Figure 3.17 Experimental XRD $\langle P_2 \rangle$ (□) and $\langle P_4 \rangle$ (■) order parameters for 5CB close to $(T - T_{NI})$ (°C). The lines are from a pure P_2, Maier–Saupe-like distribution (see Chapter 7) [Sanchez-Castillo et al., 2010].

e.g. if$\langle P_6 \rangle < \langle P_4 \rangle < \langle P_2 \rangle$, higher reflections will attenuate and be difficult to obtain, also recalling the decrease of the atomic scattering factor $b(q)$ with increasing q [Leadbetter and Norris, 1979]. In general, order parameters higher than $L = 2$ remain hard to obtain with XRD [Jenz et al., 2016]. However, in Fig. 3.17 we see the order parameters $\langle P_2 \rangle$ and $\langle P_4 \rangle$ for 5CB obtained from XRD [Sanchez-Castillo et al., 2010].

3.10 Non-Cylindrical Molecules in Uniaxial Phases

In Sections 3.3–3.9 we have considered in some detail the treatment of cylindrically symmetric particles. This allows us to skip some explicit steps, since the logic here is the same, even though the algebra is somewhat more complicated. To start with, we note that when the rigid molecule of interest, which we still assume to be dissolved in a uniaxial phase, cannot be assimilated to a rod-like or a disc-like particle, we need an extra angle in defining its orientation (Fig. 3.1b). Thus, if β is the angle between the z-axis of the particle and the director (taken to be the laboratory z-axis), the extra angle γ is an angle of rotation around the molecular z-direction [Rose, 1957]. The probability of finding the molecule at a specific orientation, $P(\beta,\gamma)$, can be expanded like any other function of the two Euler angles (β,γ), in a complete basis set of spherical harmonics $\mathscr{D}^L_{0,n}(\beta,\gamma)$. These are often written as spherical harmonics, $Y_{L,m}(\beta,\alpha) = [(2L+1)/4\pi]^{1/2}\mathscr{D}^{L*}_{0,n}(\beta,\gamma)$ (cf. Eq. F.11) and the first few are given explicitly in Eqs. A.54a–A.54f. Sticking to our Wigner-type notation,

$$P(\beta,\gamma) = \sum_{L=0}^{\infty} \sum_{n=-L}^{L} p_{Ln} \mathscr{D}^L_{0,n}(\beta,\gamma). \tag{3.108}$$

Orthogonality of the basis set immediately permits identifying the coefficients p_{Ln} and obtaining

$$P(\beta,\gamma) = \frac{1}{4\pi}\sum_{L=0}^{\infty}\sum_{n=-L}^{L}(2L+1)\langle\mathscr{D}_{0,n}^{L*}\rangle\mathscr{D}_{0,n}^{L}(\beta,\gamma). \tag{3.109}$$

The averaged Wigner orientation matrices $\langle\mathscr{D}_{0,n}^{L}\rangle$ provide a complete characterization of $P(\beta,\gamma)$ and represent the orientational order parameters for the problem. Since $P(\beta,\gamma)$ is real, recalling the expression for the complex conjugate of a Wigner function (Eq. F.6), we have $\langle\mathscr{D}_{0,n}^{L*}\rangle = (-1)^n\langle\mathscr{D}_{0,-n}^{L}\rangle$, and the number of independent quantities is correspondingly reduced. At second-rank level, $L = 2$, there are at most five independent order parameters $\langle\mathscr{D}_{0,n}^{2}\rangle$, $n = -2, -1.0, 1, 2$.

3.10.1 Molecular Symmetry

Whatever the formalism used, the relevant order parameters for molecules belonging to a certain point group can be listed. A fairly detailed treatment of the effect of symmetry on the order parameters is given in Appendix G. Here, in Table 3.1, we just report the results for various molecular symmetries and a uniaxial LC phase.

3.10.2 Cartesian Description: Ordering Matrices

An alternative definition for order parameters in uniaxial phases can be obtained by expanding the singlet orientational distribution $P(\beta,\gamma)$ in terms of the components (or direction

Table 3.1. *The second- and fourth-rank independent orientational order parameters $\langle\mathscr{D}_{0,n}^{L}\rangle$ for molecules of various point group symmetry (we use the Schönflies notation, see, e.g., [Lax, 1974]) in a uniaxial phase; n_2, n_4 are the number of independent terms [Zannoni, 1979c]*

Point group	n_2	$\langle\mathscr{D}_{0,n}^{2}\rangle$	n_4	$\langle\mathscr{D}_{0,n}^{4}\rangle$
C_1, C_i	5	$\langle\mathscr{D}_{0,0}^{2}\rangle, \langle\mathscr{D}_{0,1}^{2}\rangle$ $\langle\mathscr{D}_{0,2}^{2}\rangle$	9	$\langle\mathscr{D}_{0,0}^{4}\rangle, \langle\mathscr{D}_{0,1}^{4}\rangle, \langle\mathscr{D}_{0,2}^{4}\rangle$ $\langle\mathscr{D}_{0,3}^{4}\rangle, \langle\mathscr{D}_{0,4}^{4}\rangle$
C_s, C_2, C_{2h}	3	$\langle\mathscr{D}_{0,0}^{2}\rangle, \langle\mathscr{D}_{0,2}^{2}\rangle$	5	$\langle\mathscr{D}_{0,0}^{4}\rangle, \langle\mathscr{D}_{0,2}^{4}\rangle, \langle\mathscr{D}_{0,4}^{4}\rangle$
C_{2v}, D_2, D_{2h}	2	$\langle\mathscr{D}_{0,0}^{2}\rangle$, Re$\langle\mathscr{D}_{0,2}^{2}\rangle$	3	$\langle\mathscr{D}_{0,0}^{4}\rangle$, Re$\langle\mathscr{D}_{0,2}^{4}\rangle$, Re$\langle\mathscr{D}_{0,4}^{4}\rangle$
C_3, S_6	1	$\langle\mathscr{D}_{0,0}^{2}\rangle$	3	$\langle\mathscr{D}_{0,0}^{4}\rangle, \langle\mathscr{D}_{0,3}^{4}\rangle$
C_4, C_{4h}, S_4	1	$\langle\mathscr{D}_{0,0}^{2}\rangle$	3	$\langle\mathscr{D}_{0,0}^{4}\rangle, \langle\mathscr{D}_{0,4}^{4}\rangle$
C_{3v}, D_3, D_{3d}	1	$\langle\mathscr{D}_{0,0}^{2}\rangle$	2	$\langle\mathscr{D}_{0,0}^{4}\rangle$, Re$\langle\mathscr{D}_{0,3}^{4}\rangle$
$C_{4v}, D_{2d}, D_{4h}, D_4$	1	$\langle\mathscr{D}_{0,0}^{2}\rangle$	2	$\langle\mathscr{D}_{0,0}^{4}\rangle$, Re$\langle\mathscr{D}_{0,4}^{4}\rangle$
C_5, C_{5h}, C_{5v}	1	$\langle\mathscr{D}_{0,0}^{2}\rangle$	1	$\langle\mathscr{D}_{0,0}^{4}\rangle$
$D_{4d}, D_5, D_{5h}, D_{5d}$	1	$\langle\mathscr{D}_{0,0}^{2}\rangle$	1	$\langle\mathscr{D}_{0,0}^{4}\rangle$
$C_{3h}, C_6, C_{6h}, C_{6v}$	1	$\langle\mathscr{D}_{0,0}^{2}\rangle$	1	$\langle\mathscr{D}_{0,0}^{4}\rangle$
$D_{3h}, D_6, D_{6h}, D_{6d}$	1	$\langle\mathscr{D}_{0,0}^{2}\rangle$	1	$\langle\mathscr{D}_{0,0}^{4}\rangle$
$C_\infty, C_{\infty v}, C_{\infty h}, D_{\infty h}$	1	$\langle\mathscr{D}_{0,0}^{2}\rangle$	1	$\langle\mathscr{D}_{0,0}^{4}\rangle$

cosines) $d_a \equiv d_{za}(\beta,\gamma)$, $(a = x, y, z)$ of the director ($\sum_a d_a^2 = 1$) with respect to a molecule fixed frame [Saupe, 1966; Buckingham, 1967a; Zannoni, 1979c; Turzi, 2011; Kwak and Kim, 2016]. Thus, for non-polar phases

$$P(\beta,\gamma) = \frac{1}{4\pi} + \frac{5}{4\pi}\sum_{a,b} S_{ab} d_a d_b + \frac{9}{4\pi}\sum_{a,b,c,d} S_{abcd} d_a d_b d_c d_d + \cdots. \tag{3.110}$$

For non-polar phases, like nematics, only the terms with an even number of indices (even rank tensors) are different from 0. In particular, the symmetric and traceless second-rank matrix with elements

$$S_{ab} = \frac{3}{2}\langle d_a d_b\rangle - \frac{1}{2}\delta_{a,b}, \tag{3.111}$$

is called the *Saupe ordering matrix* and

$$\begin{aligned} S_{abcd} = {} & \frac{35}{8}\langle d_a d_b d_c d_d\rangle - \frac{5}{8}\big(\langle d_a d_b\rangle \delta_{cd} + \langle d_a d_c\rangle \delta_{b,d} + \langle d_a d_d\rangle \delta_{b,c} + \langle d_b d_c\rangle \delta_{a,d} \\ & + \langle d_b d_d\rangle \delta_{a,c} + \langle d_c d_d\rangle \delta_{a,b}\big) + \frac{1}{8}\big(\delta_{a,b}\delta_{c,d} + \delta_{a,c}\delta_{b,d} + \delta_{a,d}\delta_{b,c}\big). \end{aligned} \tag{3.112}$$

The following bounds [Buckingham, 1967b] hold:

$$-\frac{1}{2} \leqslant S_{11}, S_{22}, S_{33} \leqslant 1; \quad -\frac{3}{4} \leqslant S_{31}, S_{32}, S_{12} \leqslant \frac{3}{4}; \quad -\frac{3}{7} \leqslant S_{3333} \leqslant 1. \tag{3.113}$$

As we see the Cartesian description rapidly becomes quite complicated and even at fourth-rank level is rather heavy. The description of ordering in terms of Wigner matrices is normally more convenient for theoretical manipulations in view of the simple transformation properties of these functions under rotation (cf. Appendix F). However, the ordering matrix formalism is very convenient and most frequently used in the analysis of experimental data, which normally involve only second-rank tensors anyway, as we see in the next subsections of Section 3.10.

3.10.3 Brick-like Biaxial Molecules

In many physical situations, the assumption is made that the molecules of interest can be considered as effective brick-like (D_{2h}) biaxial particles [Straley, 1974; Cuetos et al., 2017]. This case, which includes many compounds relevant in experimental studies, e.g. anthracene, perylene, pyrene, etc., will now be discussed in some detail. First, we choose our molecular frame axis along the three twofold symmetry, C_2, axes. Since we can turn our biaxial particle upside down without changing anything we only need to retain in Eq. 3.109 functions that are invariant for this transformation. Remembering that the spherical harmonics $\mathscr{D}^L_{0,n}(\beta,\gamma)$ are multiplied by $(-1)^L$ under the same operation (cf. Appendix F), we see that we only need to expand in Wigner matrices with even rank L.

Since the principal frame of the ordering matrix is determined by symmetry, at second-rank level there are two relevant order parameters, $\langle\mathscr{D}^2_{0,0}\rangle$, $\mathrm{Re}\langle\mathscr{D}^2_{0,2}\rangle$ or, e.g. S_{zz}, $S_{xx} - S_{yy}$. While $\langle\mathscr{D}^2_{0,0}\rangle$ measures the alignment of the z molecular axis with respect to the director, as we have seen for cylindrical molecules, $\mathrm{Re}\langle\mathscr{D}^2_{0,2}\rangle$ is a molecular biaxiality parameter.

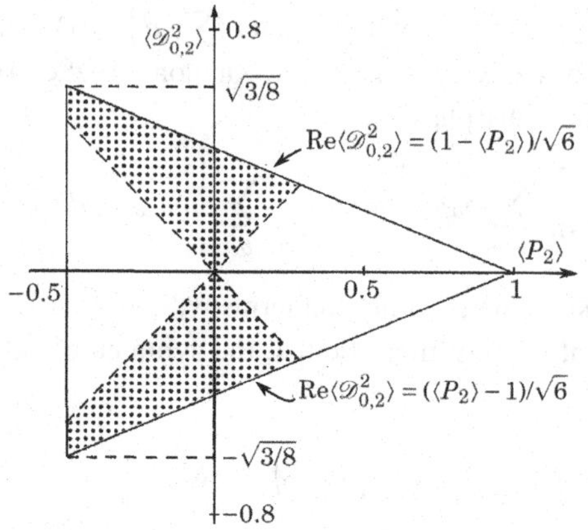

Figure 3.18 Allowed regions for the order parameters, $\langle\mathscr{D}^2_{0,2}\rangle$ and $\langle P_2\rangle$.

It provides the difference in ordering of the x and y molecular axes in that liquid crystal solvent and at the given thermodynamic conditions. We have, explicitly,

$$\mathrm{Re}\langle\mathscr{D}^2_{0,2}\rangle = \sqrt{(3/8)}\,\langle\sin^2\beta\cos 2\gamma\rangle = \langle[1 - P_2(\cos\beta)]\cos 2\gamma\rangle\,/\sqrt{6}. \tag{3.114}$$

The maximum of biaxiality is obtained when the single particle orientational distribution $P(\beta,\gamma)$, instead of being uniformly distributed in γ, is so peaked that only $\gamma = 0$, or respectively $\gamma = \pi/2$, are possible. In these two limits, $P(\beta,\gamma)$ reduces to $P(\beta)\delta(\gamma)$ and $P(\beta)\delta(\gamma - \pi/2)$, so in general the biaxiality order parameter is bounded as

$$(\langle P_2\rangle - 1)/\sqrt{6} \le \mathrm{Re}\langle\mathscr{D}^2_{0,2}\rangle \le (1 - \langle P_2\rangle)/\sqrt{6}. \tag{3.115}$$

In Fig. 3.18 we show the regions of allowed order parameters determined in this way. When $\langle P_2\rangle = 1$ the biaxiality reduces to 0. In general, $\mathrm{Re}\langle\mathscr{D}^2_{0,2}\rangle$ can take values inside the triangular region defined by Eq. 3.115 as $\langle P_2\rangle$ varies. An interpretation of the physical significance of the various combinations of $\langle P_2\rangle$ and $\mathrm{Re}\langle\mathscr{D}^2_{0,2}\rangle$, that is perhaps more immediate, can be obtained by constructing approximate molecular distributions consistent with a given set of these order parameters with the ME technique introduced earlier for cylindrical molecules.

3.10.4 Linear Dichroism

The ordering matrix elements are easily related to the measurement of the averaged components of a second-rank tensor. For example, let us consider a spectroscopic technique that measures the components of a certain tensor **A** in the laboratory frame. The technique could be again absorption spectroscopy and then the tensor would be the absorption tensor, but it could equally well be entirely different. For instance, in a magnetic resonance technique, **A** would be identified with a magnetic interaction tensor (see Appendix I). The average of **A** along the director in our laboratory frame with $\boldsymbol{d}||\boldsymbol{Z}$ is:

$$\langle A^{\mathrm{LAB}}_{ZZ}\rangle = \sum_{k,l}\langle R_{Zk}A^{\mathrm{MOL}}_{kl}R^T_{lZ}\rangle = \sum_{k,l}\langle d_k d_l\rangle A^{\mathrm{MOL}}_{kl}. \tag{3.116}$$

We can rewrite Eq. 3.116 adding and subtracting the scalar, isotropic value of the tensor, i.e. $a = \frac{1}{3}\text{Tr}\mathbf{A}^{\text{MOL}} = \frac{1}{3}\text{Tr}\langle\mathbf{A}^{\text{LAB}}\rangle = \frac{1}{3}\sum_{k,l} A_{kl}\delta_{k,l}$, where we assume the laboratory and molecular components to be connected by rotations. We find

$$\langle A_{\parallel}^{\text{LAB}}\rangle = a + \frac{2}{3}\sum_{k,l} S_{kl} A_{kl}^{\text{MOL}} \tag{3.117}$$

in terms of the Saupe ordering matrix **S** (Eq. 3.111). We also have the component perpendicular to the director

$$\langle A_{\perp}^{\text{LAB}}\rangle \equiv \langle A_{XX}^{\text{LAB}}\rangle = \langle A_{YY}^{\text{LAB}}\rangle = \frac{1}{2}(3a - \langle A_{\parallel}^{\text{LAB}}\rangle), \tag{3.118a}$$

$$= a - \frac{1}{3}\sum_{k,l} S_{kl} A_{kl}^{\text{MOL}} = a - \frac{1}{3}\text{Tr}(\mathbf{S}^T\mathbf{A}^{\text{MOL}}). \tag{3.118b}$$

The observed anisotropy is then

$$\Delta A^{\text{LAB}} \equiv \langle A_{\parallel}^{\text{LAB}}\rangle - \langle A_{\perp}^{\text{LAB}}\rangle = \sum_{k,l} S_{kl} A_{kl}^{\text{MOL}} = \text{Tr}(\mathbf{S}^T\mathbf{A}^{\text{MOL}}). \tag{3.119}$$

The Saupe ordering matrix **S** describes the average orientation of the chosen molecule frame in the laboratory. In the molecular frame diagonalizing **S**, which we call the *ordering matrix frame*, we have

$$\Delta A^{\text{LAB}} = A_{xx}^{\text{MOL}} S_{xx} + A_{yy}^{\text{MOL}} S_{yy} + A_{zz}^{\text{MOL}} S_{zz} \tag{3.120}$$

and

$$\langle A_{\parallel}^{\text{LAB}}\rangle = a + \frac{1}{3}(2A_{zz}^{\text{MOL}} - A_{xx}^{\text{MOL}} - A_{yy}^{\text{MOL}})S_{zz} + \frac{1}{3}(A_{xx}^{\text{MOL}} - A_{yy}^{\text{MOL}})(S_{xx} - S_{yy}). \tag{3.121}$$

The **S** matrix is, as we have already mentioned, traceless and symmetric. Its components are explicitly

$$\mathbf{S} = \begin{pmatrix} \langle\frac{3}{2}\sin^2\beta\cos^2\gamma - \frac{1}{2}\rangle & \langle\sin^2\beta\cos\gamma\sin\gamma\rangle & \langle\sin\beta\cos\beta\cos\gamma\rangle \\ \langle\sin^2\beta\cos\gamma\sin\gamma\rangle & \langle\frac{3}{2}\sin^2\beta\sin^2\gamma - \frac{1}{2}\rangle & \langle\sin\beta\cos\beta\sin\gamma\rangle \\ \langle\sin\beta\cos\beta\cos\gamma\rangle & \langle\sin\beta\cos\beta\sin\gamma\rangle & \langle\frac{3}{2}\cos^2\beta - \frac{1}{2}\rangle \end{pmatrix}. \tag{3.122}$$

The order parameters can be easily converted from the Saupe to the Wigner rotation matrix (spherical harmonics) form

$$S_{xx} - S_{yy} = \sqrt{6}\,\text{Re}\langle\mathscr{D}_{0,2}^2\rangle, \tag{3.123a}$$

$$S_{xy} = -\sqrt{\frac{3}{2}}\,\text{Im}\,\langle\mathscr{D}_{0,2}^2\rangle, \tag{3.123b}$$

$$S_{xz} = -\sqrt{\frac{3}{2}}\,\text{Re}\langle\mathscr{D}_{0,1}^2\rangle, \tag{3.123c}$$

$$S_{yz} = \sqrt{\frac{3}{2}}\,\text{Im}\,\langle\mathscr{D}_{0,1}^2\rangle, \tag{3.123d}$$

$$S_{zz} = \langle\mathscr{D}_{0,0}^2\rangle = \langle P_2\rangle. \tag{3.123e}$$

In particular, the observed anisotropy of a second-rank property **A** is

$$\Delta A^{\text{LAB}} = \langle P_2 \rangle \Big[A_{zz}^{\text{MOL}} - \frac{1}{2} A_{xx}^{\text{MOL}} - \frac{1}{2} A_{yy}^{\text{MOL}} \Big] + \sqrt{\frac{3}{2}} \text{Re} \langle \mathscr{D}_{0,2}^2 \rangle \Big[A_{xx}^{\text{MOL}} - A_{yy}^{\text{MOL}} \Big]. \quad (3.124)$$

It should be stressed that other equivalent formulations for describing orientational order have been given. For instance, a set of ordering constants particularly used in optical spectroscopy is that of *orientation factors* [Michl and Thulstrup, 1986] $K_{ab} = \langle (\boldsymbol{a} \cdot \mathbf{Z})(\boldsymbol{b} \cdot \mathbf{Z}) \rangle = \frac{2}{3} S_{ab} + \frac{1}{3} \delta_{a,b}$, where $\boldsymbol{a}$, $\boldsymbol{b}$ are unit vectors parallel to the $\boldsymbol{x}$, $\boldsymbol{y}$ or $\boldsymbol{z}$ molecular axes and $\boldsymbol{Z} \parallel \boldsymbol{d}$. The Cartesian formulation can be extended to higher ranks both for the **S** matrices and the orientation factors although it becomes progressively more complicated than the spherical one as the rank increases, as mentioned earlier. The same complications also arise when we want to establish relations between measured quantities of arbitrary rank and their molecular counterpart. An alternative and simpler possibility is to use, rather than the standard Cartesian components A_{ab} of a tensor quantity, certain suitable combinations $A^{L,m}$, called *spherical components* that transform between themselves in a convenient way under rotation (see Appendix B). In particular, the $(2L+1)$ spherical components of rank L transform under rotation as

$$A_{\text{LAB}}^{L,m} = \sum_{n=-L}^{L} \mathscr{D}_{m,n}^{L*}(\alpha, \beta, \gamma) \, A_{\text{MOL}}^{L,n}, \quad (3.125)$$

where the Wigner function $\mathscr{D}_{m,n}^{L*}(\alpha, \beta, \gamma) \equiv \mathscr{D}_{m,n}^{L*}(\Omega_{\text{ML}})$ of the three Euler angles define the rotation from laboratory to molecule frame.

Here we just wish to start showing the immediate applicability of this formalism to our linear dichroism problem. First, we find from Eq. B.22i the explicit expression for A_{ZZ}. This is, in the lab frame where measurements are made,

$$A_{ZZ}^{\text{LAB}} = -\frac{1}{\sqrt{3}} A_{\text{LAB}}^{0,0} + \sqrt{\frac{2}{3}} A_{\text{LAB}}^{2,0}. \quad (3.126)$$

The first term on the right is a scalar, $-A_{\text{LAB}}^{0,0}/\sqrt{3} = \text{Tr}\mathbf{A}\,/\,3 = a$ (Eq. B.21a) and does not change with rotations. For the second, we just use Eq. 3.125. The averaged measurable spherical components will be

$$\langle A_{\text{LAB}}^{L,m} \rangle = \sum_n \langle \mathscr{D}_{m,n}^{L*} \rangle A_{\text{MOL}}^{L,n}. \quad (3.127)$$

For a uniaxial phase the invariance for rotation around z requires:

$$\langle \mathscr{D}_{m,n}^{L} \rangle = \delta_{m,0} \langle \mathscr{D}_{0,n}^{L} \rangle. \quad (3.128)$$

The measured absorption parallel to the director can then be written as

$$\langle A_{\parallel}^{\text{LAB}} \rangle = \langle A_{ZZ}^{\text{LAB}} \rangle = a + \sqrt{\frac{2}{3}} \sum_n \langle \mathscr{D}_{0,n}^{2*} \rangle A_{\text{MOL}}^{2,n}. \quad (3.129)$$

Quite similarly the measured perpendicular component will be

$$\langle A_{\perp}^{\text{LAB}} \rangle = \frac{1}{2}\Big(\langle A_{XX}^{\text{LAB}} \rangle + \langle A_{YY}^{\text{LAB}} \rangle\Big) = a - \sqrt{\frac{1}{6}} \sum_n \langle \mathscr{D}_{0,n}^{2*} \rangle A_{\text{MOL}}^{2,n}. \tag{3.130}$$

For a biaxial molecule the observable anisotropy of $\langle \mathbf{A} \rangle$ is

$$\langle A_{\parallel}^{\text{LAB}} \rangle - \langle A_{\perp}^{\text{LAB}} \rangle = \sqrt{\frac{3}{2}} \Big[A_{\text{MOL}}^{2,0} \langle \mathscr{D}_{0,0}^{2} \rangle + 2\text{Re}\Big(A_{\text{MOL}}^{2,2} \langle \mathscr{D}_{0,2}^{2*} \rangle \Big) \Big]. \tag{3.131}$$

Thus, the measurement of at least two anisotropy values is required to determine both $\langle \mathscr{D}_{0,0}^{2} \rangle$ and $\langle \mathscr{D}_{0,2}^{2} \rangle$. Moreover, the parameter of deviation from cylindrical symmetry, $\langle \mathscr{D}_{0,2}^{2} \rangle$, only becomes measurable when the tensor $\mathbf{A}$ has an off axis component so that $A_{\text{MOL}}^{2,2} \neq 0$. If the molecule has effective cylindrical symmetry, in the sense that the order parameters

$$\langle \mathscr{D}_{0,n}^{2} \rangle = \langle \mathscr{D}_{0,0}^{2} \rangle \delta_{n,0} = \langle P_2 \rangle \delta_{n,0} \tag{3.132}$$

within experimental error, then we recover the uniaxial case:

$$\langle P_2 \rangle = \frac{\langle A_{\parallel}^{\text{LAB}} \rangle - \langle A_{\perp}^{\text{LAB}} \rangle}{(A^{\text{MOL}})_{\parallel} - (A^{\text{MOL}})_{\perp}}. \tag{3.133}$$

We should be well aware of the fact that the order parameter $\langle P_2 \rangle$ measured for a molecule dissolved in a liquid crystal is not the same as that of the pure liquid crystal, since solute-solvent terms in the anisotropic potential acting on the molecule are different from the solvent-solvent ones. This also means that except special cases where the solute is very similar to the solvent, probe techniques give information on the behaviour of solutes in anisotropic phases and thus, only indirectly report on the phase itself. While this has been initially perceived as a limitation of these class of measurements, there is instead a lot of scope for learning about the behaviour of interesting classes of molecules in liquid crystals [Burnell and de Lange, 2003].

3.10.5 Nuclear Magnetic Resonance Dipolar Couplings

The effective spin Hamiltonian for a pair of nuclear spins $\hat{\mathbf{I}}_i$ and $\hat{\mathbf{I}}_j$ (e.g. ^{1}H or ^{13}C) interacting through their magnetic dipoles can be written (see Appendix I) as $\hat{\mathscr{H}}_{\text{D}} = \hat{\mathbf{I}}_i\,[\mathbf{T}_{ij}]\hat{\mathbf{I}}_j$, where $[\mathbf{T}_{ij}]$ is the second-rank dipolar coupling tensor, whose explicit form is given in Eq. I.6. In the usual high-field approximation [Emsley and Lindon, 1975] only the part of the interaction $\hat{\mathbf{I}}_i[\mathbf{T}_{ij}]\hat{\mathbf{I}}_j$ commuting with $[\hat{\mathbf{I}}_i]_Z$ is retained. Here $[\hat{\mathbf{I}}_i]_Z$ is the nuclear spin projection operator, quantized along the magnetic field $\boldsymbol{H}$ of the spectrometer, which in turn is assumed to define the laboratory z-axis. Within the limits of the approximation this means that the energy levels separation and thus the observed splittings are proportional to the average couplings $\langle [T_{ij}^{\text{LAB}}]_{ZZ} \rangle$, where the average indicated by the angular brackets is over all the molecular reorientations taking place during the NMR observation time windows. As mentioned in Appendix I, this is normally of the order of microseconds, sufficient to cover equilibration of molecular motions for a 'small' molecule reorienting in low viscosity ordinary liquids or nematics. This in turn means that these are the quantities

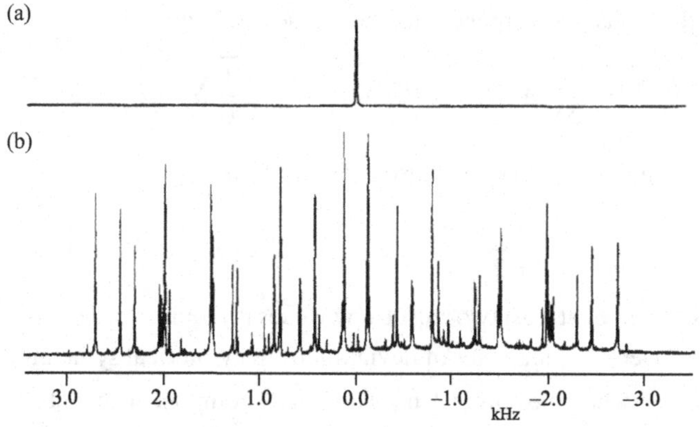

Figure 3.19 (a) The proton NMR spectrum of benzene in the isotropic phase and (b) in a liquid crystal solvent [Schmidt-Rohr et al., 1994].

that we can extract from the spectra, i.e. our observables. Since the trace of $[\mathbf{T}_{ij}]$ is 0, NMR dipolar couplings cannot be measured in isotropic fluid phases, while the averaged $\langle [T_{ij}]_{ZZ} \rangle$ components are non-vanishing and measurable in anisotropic phases. In Fig. 3.19 we see for example the proton NMR spectrum of benzene in the isotropic phase and as a solute dissolved in a liquid crystal. The measured dipolar couplings, normally called D_{ij}, are obtained fitting the spectra calculated from the spin Hamiltonian to the experimental one and we can write:

$$D_{ij} \equiv \langle [T_{ij}^{\text{LAB}}]_{ZZ} \rangle = \sqrt{(2/3)} \langle [T_{ij}]_{\text{LAB}}^{2,0} \rangle, \tag{3.134a}$$

$$= -\frac{h\,\gamma_i\,\gamma_j}{4\pi^2} \left\langle \frac{P_2(\hat{\boldsymbol{r}}_{ij} \cdot \hat{\boldsymbol{H}})}{r_{ij}^3} \right\rangle = -\frac{\mu_0}{8\pi^2} \gamma_i \gamma_j \hbar \left\langle \frac{3\cos^2\theta_{ij} - 1}{2r_{ij}^3} \right\rangle, \tag{3.134b}$$

where $\hat{r}_{ij}$ and $\hat{H}$ are unit vectors along the internuclear vector r_{ij} and the spectrometer magnetic field, respectively, with $\cos\theta_{ij} = \hat{r}_i j \cdot \hat{H}$, while $\mu_0 = 4\pi \times 10^{-7}\text{T}^2\text{J}^{-1}m^3$ is the magnetic permeability in vacuum, $\gamma_i = g_i \mu_N / \hbar$ is the nuclear gyromagnetic ratio of nucleus $i, \mu_N = 5.051 \times 10^{-27}\text{JT}^{-1}\text{m}^3$ is the so called nuclear magneton. In general, $\boldsymbol{r}_{ij}$ will depend on molecular orientation and conformation, thus dipolar couplings can in principle be useful for recovering geometrical distribution functions. To relate the observed $\langle T_{ij}^{2,0} \rangle$ to molecular constants and order parameters, we write down the coupling $T_{ij}^{2,0}$ in the laboratory frame and then transform it to a molecule frame. This requires introducing a few auxiliary coordinates systems:

(i) L (or LAB) frame: this is our laboratory coordinate system, with $z||\boldsymbol{H}$ the spectrometer magnetic field direction.
(ii) d frame, with $z||\boldsymbol{d}$ the average director.
(iii) d' frame, with $z||\boldsymbol{d}'$ the instantaneous director orientation.
(iv) M (or MOL) frame. This is a reference system fixed on the molecule, e.g. the frame that diagonalizes its ordering matrix.

$$\left\langle [T_{ij}]^{2,0}_{\mathrm{LAB}} \right\rangle = \sum_{m,m',n} \mathscr{D}^{2*}_{0,m}(\Omega_{d\mathrm{L}}) \left\langle \mathscr{D}^{2*}_{m,m'}(\Omega_{d'd}) \mathscr{D}^{2*}_{m',n}(\Omega_{\mathrm{M}d'}) [T_{ij}]^{2,0}_{\mathrm{MOL}} \right\rangle, \tag{3.135}$$

where the Wigner rotation matrix $\mathscr{D}^2_{p,q}(\Omega_{BA})$ connects frame A to frame B and the angular brackets indicate an average over all the relevant motions. In an ordinary NMR experiment with a nematic solvent the director aligns along the spectrometer magnetic field or perpendicular to it according to the positive or negative sign of the diamagnetic anisotropy $\Delta\chi$ of the liquid crystal host. It can usually be assumed that director fluctuations can be neglected, and if this is the case, $\mathscr{D}^2_{m,m'}(\Omega_{d'd}) = \delta_{m,m'}$. If, furthermore, the mesophase is uniaxial, i.e. invariant for an arbitrary rotation around the director, a $\delta_{m,0}$ restriction in Eq. 3.128 follows. The observed, average couplings $\left\langle [T_{ij}]^{2,0}_{\mathrm{LAB}} \right\rangle$ can thus be written as

$$\left\langle [T_{ij}]^{2,0}_{\mathrm{LAB}} \right\rangle = P_2(\boldsymbol{d} \cdot \boldsymbol{H}) \sum_n \left\langle \mathscr{D}^{2*}_{0,n} [T_{ij}]^{2,n}_{\mathrm{MOL}} \right\rangle. \tag{3.136}$$

We note that to be able to obtain a motionally averaged spectrum the dynamics of the molecule has to be fast on the NMR time scale. If this is not true, we have in fact that the NMR spectrum will be a superposition of spectra coming from the differently oriented domains, a so-called *polycrystalline* or powder or poly-liquid-crystalline-type spectrum. If we can assume our probe molecule to be a rigid one, the T_{ij} couplings will be constant in the chosen molecule frame, and

$$\left\langle [T_{ij}]^{2,0}_{\mathrm{LAB}} \right\rangle = P_2(\boldsymbol{d} \cdot \boldsymbol{H}) \sum_n \langle \mathscr{D}^{2*}_{0,n} \rangle [T_{ij}]^{2,n}_{\mathrm{MOL}}, \tag{3.137}$$

where $\langle \mathscr{D}^2_{0,n} \rangle$ are orientational order parameters. If the molecule is rigid so that the internuclear distances r_{ij} are essentially constant (neglecting vibration fluctuations), then the average in Eq. 3.134b can be approximated with $\langle P_2(\cos\theta_{ij}) \rangle / r^3_{ij}$.

3.10.6 Deuterium Nuclear Magnetic Resonance

One of the most convenient and most often used observables obtainable from NMR spectra measured in anisotropic solvents, is a set of nuclear quadrupolar splittings for nuclei with spin $I_k \geq 1$, notably deuterons, ^{2}H or D [Emsley and Lindon, 1975; Photinos et al., 1990; Ernst et al., 1991; Dong, 1997; Sugimura and Luckhurst, 2016]. For such a nuclear spin I_k in a given local chemical environment (e.g. the Deuterium of an aliphatic or aromatic C-D bond), the quadrupole coupling tensor $\mathbf{q}_k$, written in a suitable molecular frame is $[\mathbf{q}_k]_{\mathrm{M}_q} = eQ_k\mathbf{V}_k / [2\hbar I_k(2I_k - 1)]$, where Q_k and $\mathbf{V}_k$ are the quadrupole moment and the electric field gradient tensor at the site of nucleus k, measured with respect to the local molecular frame M_q which makes $\mathbf{q}_k$ diagonal and with quadrupolar biaxiality $\eta_k \equiv ([q_k]_{yy} - [q_k]_{xx})/[q_k]_{zz}$. For the case of a C-D bond, the principal system of the quadrupolar tensor, M_q has z-axis along the bond. The tensor $\mathbf{q}_k$ is traceless and thus its motionally averaged value can be measured only in anisotropic phases. In particular, for uniaxial liquid crystals, the only relevant component is $\langle [q_k]^{\mathrm{LAB}}_{ZZ} \rangle$, where Z is the laboratory magnetic field direction that we take to be parallel to the mesophase director. Using spherical components we can write, similarly to what we did in Sections 3.10.4 and 3.10.5,

$$\langle [q_k]^{2,0}_{\text{LAB}} \rangle = \sum_{n=-2}^{2} \langle \mathscr{D}^{2*}_{0,n}(\Omega_{\text{M}d}) \rangle [q_k]^{2,n}_{\text{MOL}}, \tag{3.138a}$$

$$= [q_k]^{2,0}_{\text{M}_q} \sum_{n=-2}^{2} \langle \mathscr{D}^{2*}_{0,n}(\Omega_{\text{M}d}) \rangle \left\{ \mathscr{D}^{2*}_{n,0}(\Omega_{q\text{M}}) - \frac{\eta_k}{\sqrt{6}} \left[\mathscr{D}^{2*}_{n,2}(\Omega_{q\text{M}}) + \mathscr{D}^{2*}_{n,-2}(\Omega_{q\text{M}}) \right] \right\}, \tag{3.138b}$$

where $\Omega_{q\text{M}}$ represents the orientation of the principal quadrupole frame M_q (i.e. of the C-D bond in essence), in the molecular ordering matrix (M $\equiv$ MOL) eigenframe and we have assumed the director to be along the magnetic field. If this is not the case, and the director makes an angle with $\boldsymbol{H}$, the RHS should be multiplied by $P_2(\boldsymbol{d} \cdot \boldsymbol{H})$. Since in Eq. 3.138b one has products of geometric factors and order parameters, the determination of absolute values of the latter requires knowing the geometry of the solute and quadrupolar tensor parameters. In practice, quadrupolar splitting constant and asymmetry parameter are relatively standard [Jacobsen and Pedersen, 1981]. For C-D bonds the value of $q_{zz} = \sqrt{2/3}[q_k]^{2,0}_{\text{M}_q}$ changes with the hybridization: for sp $\approx$ 200 kHz, sp^2 $\approx$ 185 kHz, sp^3 $\approx$ 170 kHz [Millet and Dailey, 1972]. The biaxiality parameter η is much smaller ($\approx$0.05 kHz) and can normally be neglected. For a given deuteron the Deuterium NMR (DNMR) spectrum contains a pair of lines (a *doublet*) and the average value of this quadrupolar component for nucleus k obtained from the splitting can be written, neglecting the small quadrupolar biaxiality η, as

$$\Delta\nu_k = \frac{3}{2} \langle [q_k]_{ZZ} \rangle \equiv \sqrt{\frac{3}{2}} \langle [q_k]^{2,0}_{\text{LAB}} \rangle, \tag{3.139}$$

where $\langle [q_k]^{2,0}_{\text{LAB}} \rangle \approx [q_k]^{2,0}_{\text{M}_q} \sum_{n=-2}^{2} \langle \mathscr{D}^{2*}_{0,n}(\Omega_{\text{M}d}) \rangle \mathscr{D}^{2*}_{n,0}(\Omega'_{q\text{M}})$. For a molecule with approximate brick-like shape only $n = 0, \pm 2$ terms appear. In Fig. 3.20 we show as an example the order parameter $S_{yy} - S_{xx} = \sqrt{6}\text{Re}\langle \mathscr{D}^2_{0,n}(\Omega_{\text{M}d}) \rangle$ versus $-S_{zz} = -\langle \mathscr{D}^2_{0,0}(\Omega_{\text{M}d}) \rangle$ for nitrobenzene in various nematics. In all cases a $q_{zz} = 185$ kHz was assumed [Catalano et al., 1983].

3.10.7 Maximum Entropy Distributions

Having determined a set of biaxial order parameters $\langle \mathscr{D}^L_{0,n} \rangle$ we can obtain the 'best' (flattest) distribution compatible with them using Information Theory and the ME method [Jaynes, 1957a; Zannoni, 1988] with a simple generalization of the uniaxial case discussed in Section 3.8.2 as

$$P(\beta,\gamma) = \exp\left\{ \sum_{L,n} a_{Ln} \mathscr{D}^L_{0,n}(\beta,\gamma) \right\} \equiv \frac{1}{Z_0} \exp\left\{ -\mathscr{U}(\beta,\gamma) \right\} \tag{3.140}$$

with

$$-\mathscr{U}(\beta,\gamma) \equiv \sum_{\substack{L=2 \\ L\ even}}^{L_{max}} \sum_{n} a_{Ln} \mathscr{D}^L_{0,n}(\beta,\gamma) \tag{3.141}$$

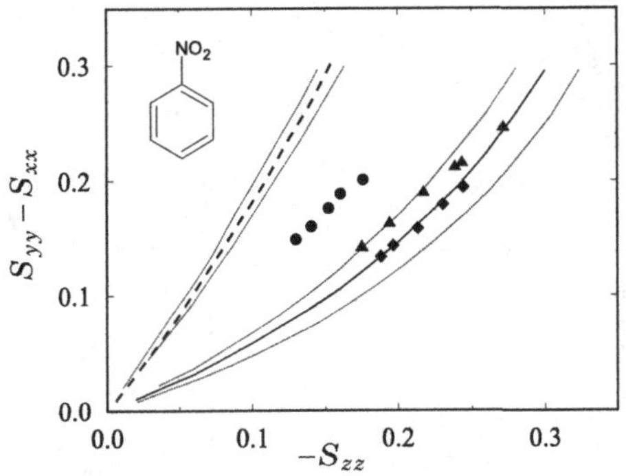

Figure 3.20 Ordering matrix components for fully deuterated nitrobenzene in EBBA(●), PCH7(■), ZLI-1167(▲) from DNMR. The lines correspond to biaxiality estimates for the same solute obtained assuming the attractive (dispersive) (——) or repulsive (- - -) interactions discussed in Chapters 5 and 7 with the dotted lines (·····) indicating their uncertainty limits [Catalano et al., 1983].

and where the coefficients $a_{L,n}$ are obtained solving the non-linear system of consistency constraints

$$\langle \mathscr{D}^L_{0,n} \rangle = \int_0^\pi \mathrm{d}\beta \sin\beta \int_0^{2\pi} d\gamma\, \mathscr{D}^L_{0,n}(\beta,\gamma) \exp\Big\{\sum_{L,n} a_{Ln} \mathscr{D}^L_{0,n}(\beta,\gamma)\Big\}, \tag{3.142}$$

with $a_{0,0}$ derived from the normalization constraint $\langle \mathscr{D}^0_{0,0} \rangle = 1$. For a biaxial solute where $\langle \mathscr{D}^2_{0,0} \rangle$ and $\mathrm{Re}\langle \mathscr{D}^2_{0,2} \rangle$ are determined, we have simply

$$P(\beta,\gamma) = \frac{\exp\left\{a[P_2(\cos\beta) + \xi \mathrm{Re}\mathscr{D}^2_{0,2}(\beta,\gamma)]\right\}}{\int_0^\pi \mathrm{d}\beta \sin\beta \int_0^{2\pi} d\gamma\, \exp\left\{a[P_2(\cos\beta) + \xi \mathrm{Re}\mathscr{D}^2_{0,2}(\beta,\gamma)]\right\}}, \tag{3.143}$$

with $a \equiv a_{2,0}$, $\xi \equiv a_{2,2}/a_{2,0}$. The parameter ξ is a measure of deviation from cylindrical symmetry, that reduces to 0 for uniaxial molecules. To illustrate the interplay between order parameters and distributions, we show in Fig. 3.21 a few examples of distributions corresponding to particles of different shapes: biaxial particles similar to a distorted disc (a), (c), uniaxial discs (b) and rods (d), elongated biaxial objects with $\langle P_2 \rangle \geq 0$ and $\mathrm{Re}\langle \mathscr{D}^2_{0,2} \rangle = \pm 0.05$ (e,f). In cases (a–c) the particle has a greater probability of having its plate-like core aligned parallel to the surrounding host molecules and thus the normal to that core (taken as z-axis) perpendicular to the director. The sign of the order parameter tells us which of the two axes in the plane is most aligned. It is interesting to note that biaxiality effects are somewhat magnified for oblate molecules. If we recall the explicit expression of $\mathrm{Re}\langle \mathscr{D}^2_{0,2} \rangle$ (Eq. 3.114), we see that, for a rod-like molecule, β tends to approach 0 as the order increases and hence $\sin^2\beta$ and ultimately $\mathrm{Re}\langle \mathscr{D}^2_{0,2} \rangle$ will vanish (see Fig. 3.22). This behaviour is different for an oblate-like molecule, where β in a similar situation approaches $\pi/2$ and $\sin^2\beta$ approaches 1, thus allowing the dependence on the angle γ to emerge. Notice that here we have no means of knowing if ξ is a single molecule property or not. The ME formalism just converts order parameters in distributions, without offering a molecular interpretation to what is observed.

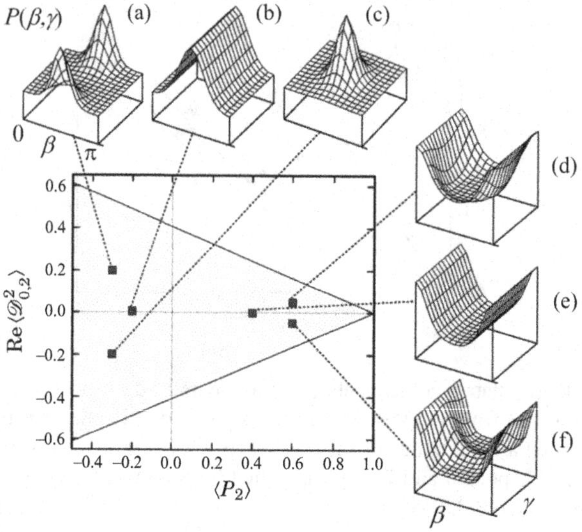

Figure 3.21 Maximum entropy orientational distributions $P(\beta, \gamma)$ for a biaxial solute molecule in a uniaxial host phase. The solute order parameters $(\langle P_2\rangle, \mathrm{Re}\langle\mathscr{D}^2_{0,2}\rangle)$ are: (a) $(-0.3, 0.2)$, (b) $(-0.2, 0.0)$, (c) $(-0.3, -0.2)$, (d) $(0.6, 0.05)$, (e) $(0.4, 0.0)$, (f) $(0.6, -0.05)$. The light grey area of allowed order parameters is limited by the inequalities in Eq. 3.115.

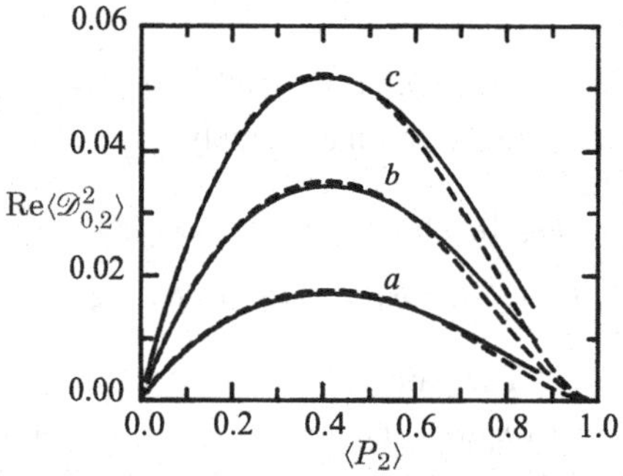

Figure 3.22 A plot of the order parameter $\mathrm{Re}\langle\mathscr{D}^2_{0,2}\rangle$ vs $\langle P_2\rangle$ for the biaxial distribution in Eq. 3.143 and for ξ = (a) 0.2, (b) 0.4, (c) 0.6 calculated either by numerical integration (continuous lines) or from the approximate analytic expansion Eq. 3.147 (dashed lines).

However, as we shall see in Chapter 7, Eq. 3.143 is formally identical to that obtained with Mean Field Theory [Luckhurst et al., 1975], e.g. starting from dispersive interactions. Curves of $\mathrm{Re}\langle\mathscr{D}^2_{0,2}\rangle$ versus $\langle\mathscr{D}^2_{0,0}\rangle$ or equivalently of $(S_{xx} - S_{yy})$ vs S_{zz} at constant ξ are often used when analyzing experimental data. In Fig. 3.22 we see such a family of curves for various values of the biaxiality parameter ξ.

We shall now try to find some simple approximations for the biaxial order parameters calculated for integration over the distribution in Eq. 3.143. To do this we consider a fixed

ξ and start with an expansion in terms of a. The first few terms are:

$$\langle P_2 \rangle = \frac{1}{5}a - \frac{(\xi^2 - 2)}{70}a^2 - \frac{(\xi^2 + 2)}{350}a^3 + \frac{3\xi^4 + 12\xi^2 - 20}{7700}a^4 + \cdots, \tag{3.144}$$

$$\mathrm{Re}\langle \mathscr{D}^2_{0,2} \rangle = \frac{\xi}{10}a - \frac{\xi}{35}a^2 - \frac{(\xi^3 + 2\xi)}{700}a^3 + \frac{3\xi^3 + 2\xi}{1925}a^4 + \cdots. \tag{3.145}$$

Reversion of Eq. 3.144 gives

$$a = 5\langle P_2 \rangle + \frac{(25\xi^2 - 50)\langle P_2 \rangle^2}{14} + \frac{(125\xi^4 - 325\xi^2 + 850)\langle P_2 \rangle^3}{98} + \cdots. \tag{3.146}$$

Substituting a in Eq. 3.145 and regrouping we find

$$\mathrm{Re}\langle \mathscr{D}^2_{0,2} \rangle = \langle P_2 \rangle \big(\langle P_2 \rangle - 1\big)^2 \left\{ \frac{\xi}{2} + \frac{5\xi^3 - 2\xi}{28} \langle P_2 \rangle + \cdots \right\}. \tag{3.147}$$

We see from Fig. 3.22 that the simple approximation in Eq. 3.147 (dashed line) is quite reasonable throughout the range and very good for $\langle P_2 \rangle$ up to 0.6–0.7. At low order parameters, e.g. when we study the order induced by a field in the isotropic phase (pretransitional phenomena), the biaxiality order parameter is linear in $\langle P_2 \rangle$ and the slope immediately gives a hint of ξ: $\mathrm{Re}\langle \mathscr{D}^2_{0,2} \rangle \approx \frac{1}{2}\xi \langle P_2 \rangle$.

An example: pyridine in nematics. The second-rank order parameters for pyridine in the commercial nematic 4-cyano-4′-alkyl bicyclohexane mixture ZLI-1167 and in 4-ethoxybenzylidene-4′-*n*-butylaniline (EBBA) at different temperatures, determined using proton NMR [Catalano et al., 1983], are shown in Fig. 3.23a. The molecular coordinate system assumed has the z-axis perpendicular to the pyridine plane and the y-axis going through the positions of the nitrogen and of the hydrogen in para position. We see that the behaviour in the two solvents is very different, showing that order parameters are in general solute-solvent rather than just solute properties. While on one hand this represents a source of complication, it also offers an interesting handle towards probing specific interactions in the fluid phase. The construction of distributions corresponding to these different situations, as shown in Fig. 3.21, can help in determining the most probable orientation. As a specific example we show in Fig. 3.23b the ME probability distributions for pyridine in the liquid crystal mixture, ZLI-1167, at the lowest temperature employed. A similar plot for pyridine in EBBA hardly shows a dependence on the angle γ because of the small biaxiality values (cf. Fig. 3.23a).

3.11 Orientational Order in Biaxial Phases

Biaxial phases have been found in smectics and more recently in nematics (see Section 1.6). It is easy to see that the stacking of biaxial particles could give rise to a macroscopic biaxially symmetric phase, as shown in Fig. 1.23. In this case the short axes are aligned in addition to the long ones. The symmetry of the most common liquid crystal phases is summarized

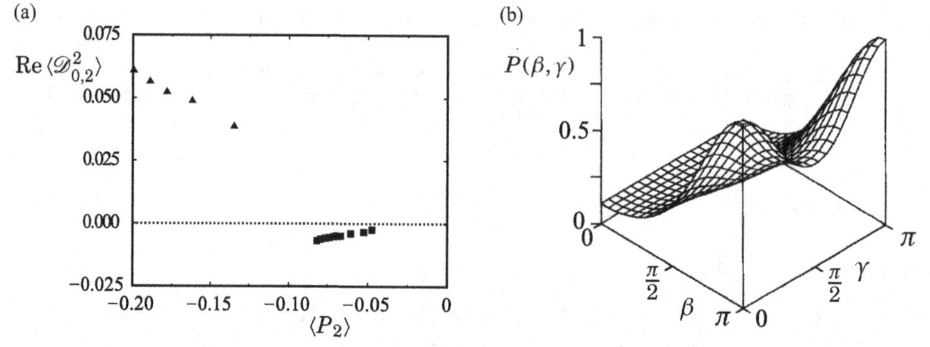

Figure 3.23 (a) The molecular biaxiality order $\text{Re}\langle \mathscr{D}^2_{0,2}\rangle$ vs $\langle P_2\rangle$ for pyridine dissolved in the nematics EBBA (squares) and ZLI-1167 (triangles) [Catalano et al., 1983]. (b) The ME probability distribution $P(\beta,\gamma)$ for pyridine in ZLI-1167 at $\langle P_2\rangle = -0.207$, $\text{Re}\langle \mathscr{D}^2_{0,2}\rangle = 0.0624$ [Zannoni, 1988].

in Table 3.2. In a non-uniaxial phase the purely orientational distribution will depend in general on the three Euler angles, α, β, γ (Fig. 3.1b). Thus,

$$P(\alpha,\beta,\gamma) = \sum_{Lmn} p_{Lmn}\mathscr{D}^L_{m,n}(\alpha,\beta,\gamma) = \frac{1}{8\pi^2}\sum_{L,m,n}\langle \mathscr{D}^{L*}_{m,n}\rangle \mathscr{D}^L_{m,n}(\alpha,\beta,\gamma), \tag{3.148}$$

where we have exploited the fact that Wigner matrices $\mathscr{D}^L_{m,n}$ [Rose, 1957; Zannoni, 1979c] are an orthogonal basis set over the molecular orientations (α,β,γ) (see Eq. F.10). The orientational order parameters will thus be averages of Wigner rotation matrices $\langle \mathscr{D}^L_{m,n}\rangle$. In the presence of molecular and/or phase symmetries (Appendix G), only suitably symmetrized combinations need to be considered. For brick–like, D_{2h}, biaxial molecules and phase this means that only 4 parameters, instead of the 25 possible $\langle \mathscr{D}^L_{m,n}\rangle$ are required for the rank $L = 2$. Using the shorthand $\mathscr{R}^J_{q,p}$ for the D_{2h} symmetrized basis functions [Biscarini et al., 1995],

$$\mathscr{R}^J_{q,p} = \frac{1}{2}\text{Re}\left[\mathscr{D}^J_{q,p} + \mathscr{D}^J_{q,-p}\right] = \frac{1}{4}\left[\mathscr{D}^J_{q,p} + \mathscr{D}^J_{-q,-p} + \mathscr{D}^J_{q,-p} + \mathscr{D}^J_{-q,p}\right], \tag{3.149a}$$

$$= \frac{1}{2}\Big[\cos(q\alpha)\cos(p\gamma)\big(d^J_{-qp}(\beta) + d^L_{qp}(\beta)\big) + \sin(q\alpha)\sin(p\gamma)\big(d^J_{-qp}(\beta) - d^J_{qp}(\beta)\big)\Big], \tag{3.149b}$$

where J, q, p are even. Explicitly [Berardi and Zannoni, 2015]

$$\langle \mathscr{R}^2_{0,0}\rangle = \frac{3}{2}\left\langle \cos^2\beta\right\rangle - \frac{1}{2} = \frac{3}{2}\left\langle (\boldsymbol{z}\cdot\boldsymbol{d})^2\right\rangle - \frac{1}{2}, \tag{3.150a}$$

$$\langle \mathscr{R}^2_{0,\pm 2}\rangle = \sqrt{\frac{3}{8}}\left\langle \sin^2\beta\cos 2\gamma\right\rangle = \sqrt{\frac{3}{8}}\left\langle (\boldsymbol{x}\cdot\boldsymbol{d})^2 - (\boldsymbol{y}\cdot\boldsymbol{d})^2\right\rangle, \tag{3.150b}$$

$$\langle \mathscr{R}^2_{\pm 2,0}\rangle = \sqrt{\frac{3}{8}}\left\langle \sin^2\beta\cos 2\alpha\right\rangle = \sqrt{\frac{3}{8}}\left\langle (\boldsymbol{z}\cdot\boldsymbol{l})^2 - (\boldsymbol{z}\cdot\boldsymbol{m})^2\right\rangle, \tag{3.150c}$$

Table 3.2. *Symmetry of the main liquid crystal phases [Singh, 2000]. Notation for point group symmetry (cf. Appendix G) as in [Cotton, 1990]. $T(n)$ indicates the translational symmetry group in n dimensions*

Liquid crystal phase	Point group and translational degrees of freedom
I. Achiral phases:	
Calamitic, micellar, nematic (N)	$D_{\infty h} \times T(3)$
Nematic discotic (N_D), columnar nematic (N_C)	$D_{\infty h} \times T(3)$
Biaxial nematic (N_b)	$D_{2h} \times T(3)$
Calamitic orthogonal smectic or lamellar (S_A)	$D_{\infty h} \times T(2)$
Tilted smectic phase (S_C)	$C_{2h} \times T(2)$
Orthogonal and lamellar hexatic phase (S_B)	$D_{6h} \times T(1)$ locally $D_{6h} \times T(2)$ globally
Tilted and lamellar hexatic phases (S_F and S_I)	$C_{2h} \times T(1)$ or $T(2)$
Discotic columnar: hexagonal order of columns, ordered or not within columns (D_h^o or D_h^d)	$D_{6h} \times T(1)$
Rectangular array of columns (D_r^o or D_r^d)	$D_{2h} \times T(1)$
Molecules tilted within columns (D^{to} or D^{td})	$C_{2h} \times T(1)$
II. Chiral phases:	
Cholesteric, twisted nematic (N*)	$D_{\infty} \times T(3)$
Tilted smectic C phase (S_C^*)	$C_2 \times T(2)$
Tilted and lamellar hexatic phases (S_F^* and S_I^*)	$C_2 \times T(1)$ or $T(2)$

$$\langle \mathscr{R}^2_{\pm 2,\pm 2} \rangle = \frac{1}{4} \langle (1+\cos^2\beta)\cos 2\alpha \cos 2\gamma \rangle - \frac{1}{2}\langle \cos\beta \sin 2\alpha \sin 2\gamma \rangle,$$
$$= \frac{1}{4}\left\langle (\boldsymbol{x}\cdot\boldsymbol{l})^2 - (\boldsymbol{x}\cdot\boldsymbol{m})^2 - (\boldsymbol{y}\cdot\boldsymbol{l})^2 + (\boldsymbol{y}\cdot\boldsymbol{m})^2 \right\rangle. \tag{3.150d}$$

Note that the notation includes the special case of uniaxial phases in Table 3.1 where $q = 0$. In the Cartesian expressions for the $\mathscr{R}^2_{q,p}$ we have scalar products of the three orthogonal unit vectors $\boldsymbol{l}, \boldsymbol{m}, \boldsymbol{d}$ defining the laboratory director frame and the corresponding axes $\boldsymbol{x}$, $\boldsymbol{y}$, and $\boldsymbol{z}$ of the molecular frame.

Unfortunately, a variety of notations, explicitly compared by Rosso [2007] have been used for order parameters. A few commonly used that we report for convenience are [Bisi et al., 2007; Rosso, 2007]:

$$S = \langle \mathscr{R}^2_{0,0} \rangle,\ T = \sqrt{2/3}\langle \mathscr{R}^2_{2,0} \rangle,\ S' = \sqrt{6}\langle \mathscr{R}^2_{0,2} \rangle,\ T' = 2\langle \mathscr{R}^2_{2,2} \rangle, \tag{3.151a}$$
$$S = \langle \mathscr{R}^2_{0,0} \rangle,\ P = \sqrt{2}\langle \mathscr{R}^2_{2,0} \rangle,\ U = \sqrt{2}\langle \mathscr{R}^2_{0,2} \rangle,\ F = 2\langle \mathscr{R}^2_{2,2} \rangle. \tag{3.151b}$$

3.11.1 Biaxial Order Parameters from Simulations

Let us consider the determination of order parameters from simulated configurations. We shall assume our sample here to be a monodomain orthorhombic (i.e. with D_{2h} symmetry) biaxial nematic, call it a 'block' for simplicity (Fig. 1.23) even though, as stressed by Photinos and colleagues, lower symmetry phases may occur [Vanakaras and Photinos, 2008; Karahaliou et al., 2009; Peroukidis et al., 2009]. Even if the sample has this simple symmetry, in a simulated sample, we have two problems: finding the orientation of the block (its principal or director frame) and the order of the molecules with respect to that frame. Let us start assuming that our block is aligned with the lab axes.

In an ideal experiment the director frame and the order parameters could be determined choosing one or more second-rank tensor properties $\mathbf{A}^{(a)}$ whose components $\mathbf{A}^{(a)}_{\text{MOL}}$ are assumed to be known and measuring all its components, i.e $\langle \mathbf{A}^{(a)}_{\text{LAB}} \rangle$ in the laboratory frame $\boldsymbol{XYZ}$. The director frame $\boldsymbol{l}, \boldsymbol{m}, \boldsymbol{d}$ is, reducing the matrix to its diagonal form, identified by the eigenvector matrix $\mathbf{X}$.

$$\mathbf{X}^T \langle \mathbf{A}^{(\alpha)}_{\text{LAB}} \rangle_S \mathbf{X} = \mathbf{a}^{(\alpha)} = \text{diag}\,(a_l^{(\alpha)}, a_m^{(\alpha)}, a_d^{(\alpha)}). \tag{3.152}$$

Note that we have omitted the property superscript label for the eigenvector matrix, since the director frame should be the same for different observable properties. In mathematical terms, this means that the matrices corresponding to different properties should commute.

The observed principal values of a second-rank tensor $\mathbf{A}^{(\alpha)}$are connected to the biaxial order parameters. For a biaxial mesogen (or a biaxial solute dissolved in a biaxial phase) [Biscarini et al., 1995] we have:

$$a_l^{(\alpha)} = -\frac{1}{\sqrt{3}}[A^{(\alpha)}]^{0,0} - \frac{1}{\sqrt{6}}\Big[\langle \mathscr{D}^2_{0,0}\rangle [A^{(\alpha)}]^{2,0}_{\text{MOL}} + 2\text{Re}\langle \mathscr{D}^{2*}_{0,2}\rangle A^{2,2}_{\text{MOL}}\Big]$$
$$+\text{Re}\langle \mathscr{D}^{2*}_{2,0}\rangle [A^{(\alpha)}]^{2,0}_{\text{MOL}} + \text{Re}\langle \mathscr{D}^{2*}_{2,2} + \mathscr{D}^{2*}_{2,-2}\rangle [A^{(\alpha)}]^{2,2}_{\text{MOL}}, \tag{3.153a}$$

$$a_m^{(\alpha)} = -\frac{1}{\sqrt{3}}[A^{(\alpha)}]^{0,0} - \frac{1}{\sqrt{6}}\Big[\langle \mathscr{D}^2_{0,0}\rangle [A^{(\alpha)}]^{2,0}_{\text{MOL}} + 2\text{Re}\langle \mathscr{D}^{2*}_{0,2}\rangle [A^{(\alpha)}]^{2,2}_{\text{MOL}}\Big]$$
$$-\Big[\text{Re}\langle \mathscr{D}^{2*}_{2,0}\rangle [A^{(\alpha)}]^{2,0}_{\text{MOL}} + \text{Re}\langle \mathscr{D}^{2*}_{2,2} + \mathscr{D}^{2*}_{2,-2}\rangle [A^{(\alpha)}]^{2,2}_{\text{MOL}}\Big], \tag{3.153b}$$

$$a_d^{(\alpha)} = -\frac{1}{\sqrt{3}}[A^{(\alpha)}]^{0,0} + \sqrt{\frac{2}{3}}\Big[\langle \mathscr{D}^2_{0,0}\rangle [A^{(\alpha)}]^{2,0}_{\text{MOL}} + 2\text{Re}\langle \mathscr{D}^{2*}_{0,2}\rangle [A^{(\alpha)}]^{2,2}_{\text{MOL}}\Big], \tag{3.153c}$$

where $[A^{(\alpha)}]^{2,n}_{\text{MOL}}$ are spherical components of $\mathbf{A}^{(\alpha)}$. In a real experiment the director frame could be defined by the procedure for obtaining a properly aligned monodomain, perhaps with the help of an electric and a magnetic external field so as to align the main director $\boldsymbol{d}$ and a secondary transversal one, say $\boldsymbol{l}$.

Computer simulations present some differences. On one hand we would normally like to avoid applying external fields since, given the limited number of molecules they would probably have to be rather strong to be effective and thus, possibly influence properties. This means that even if a biaxial monodomain is formed, it can be at an unknown and fluctuating orientation. On the other hand, however, we have the advantage that in simulations we can choose our observable properties at will. The simplest we can imagine, $[A^{(\alpha)}_{\text{MOL}}]_{i,j} = \delta_{\alpha,i}\delta_{\alpha,j}$, with $\alpha = x, y, z$, so that we have three matrices $\mathbf{A}^{(x)}$, $\mathbf{A}^{(y)}$, $\mathbf{A}^{(z)}$

with only the xx or yy or zz element, respectively, equal to 1 and all the others, 0. We can calculate first the sample average of each of the three matrices as

$$\langle \mathbf{A}_{\mathrm{LAB}}^{(\alpha)} \rangle_S = \langle \mathbf{R}^{(\alpha)T} \mathbf{A}_{\mathrm{MOL}}^{(\alpha)} \mathbf{R}^{(\alpha)} \rangle_S \tag{3.154}$$

and then diagonalize each one, finding its eigenvalues matrix $\mathbf{a}^{(\alpha)}$. For an ideal orthogonal (D_{2h}) biaxial phase the three matrices should commute, with $[\langle \mathbf{A}_{\mathrm{LAB}}^{(\alpha)} \rangle_S, \langle \mathbf{A}_{\mathrm{LAB}}^{(\beta)} \rangle_S] = 0, \alpha \neq \beta$ and should be simultaneously diagonalized by the same eigenvector rotation matrix, $\mathbf{X}$.

$$\mathbf{X}^T \langle \mathbf{A}_{\mathrm{LAB}}^{(\alpha)} \rangle_S \mathbf{X} = \mathbf{a}^{(\alpha)}. \tag{3.155}$$

However, the average matrices may not exactly commute because of numerical errors due to the finite size of the sample or because of fluctuations, or for the formation of local cybotactic clusters. If the matrices $\langle \mathbf{A}_{\mathrm{LAB}}^{(\alpha)} \rangle_S$ do not exactly commute, they could still have fairly similar eigenvectors. In this case it is still possible to define biaxial order parameters if a similarity transformation to the nearly diagonal form, with off diagonal elements non-zero but smaller than a given threshold, can be found. The problem can be tackled numerically as described by Berardi and Zannoni [2015], e.g. using the algorithms of Flury and Constantine [1985] and Flury and Gautschi [1986]. The transformation to nearly diagonal form identifies the eigenvalues $a_i^{(\alpha)}$, with $i = l, m$, and d, that can then be used to define the order parameters. The eigenvectors can be labelled to give a right-handed frame (i.e. $\boldsymbol{l} \times \boldsymbol{m} = \boldsymbol{d}$, $\boldsymbol{m} \times \boldsymbol{d} = \boldsymbol{l}$, and $\boldsymbol{d} \times \boldsymbol{l} = \boldsymbol{m}$), but in any case reflection of one eigenvector, say $\boldsymbol{l} \to -\boldsymbol{l}$, does not affect the order parameters. This is also true for permutations of the molecular labels since the axis of preferential alignment can be different in various phases. Conventionally, the z_i molecular axis is assigned to the direction of preferential alignment, e.g. for elongated mesogens this is usually the longer axis, while for disc-like mesogens the shortest one. However, particularly for biaxial mesogens, making a priori the proper assignment on the basis of molecular symmetry, i.e. the permutation which provides the most physically meaningful $\langle \mathscr{R}^2_{m,n} \rangle$ (see Eqs. 3.156a–3.156d), is not always simple. For instance, the (wrong) assignment of swapped $\boldsymbol{m}$, and $\boldsymbol{d}$ axes can result in a deceivingly high $\langle \mathscr{R}^2_{2,2} \rangle$ and rather small $\langle \mathscr{R}^2_{0,0} \rangle$. In practice, a convenient conservative criterion is that of selecting the permutations giving the highest values of $\langle \mathscr{R}^2_{0,0} \rangle$, and the smallest positive values of $\langle \mathscr{R}^2_{2,2} \rangle$. The first part of this prescription is consistent with the standard algorithm for calculating order parameters [Zannoni, 1979c], while the second part prevents overestimating the phase biaxiality. Once the physically meaningful axis labelling described previously has been performed, the second-rank order parameters can be computed from the eigenvalues of the rearranged ordering matrices for the three molecular axes as [Biscarini et al., 1995; Berardi and Zannoni, 2015]

$$\langle \mathscr{R}^2_{0,0} \rangle = \frac{3}{2} a_d^{(z)} - \frac{1}{2} = 1 - \frac{3}{2} \left(a_d^{(x)} + a_d^{(y)} \right), \tag{3.156a}$$

$$\langle \mathscr{R}^2_{2,0} \rangle = \sqrt{\frac{3}{8}} \left(a_l^{(z)} - a_m^{(z)} \right) = \sqrt{\frac{3}{8}} \left(-a_l^{(x)} + a_m^{(x)} - a_l^{(y)} + a_m^{(y)} \right), \tag{3.156b}$$

$$\langle \mathscr{R}^2_{0,2} \rangle = \sqrt{\frac{3}{8}} \left(a_d^{(x)} - a_d^{(y)} \right) = \sqrt{\frac{3}{8}} \left(-a_l^{(x)} - a_m^{(x)} + a_l^{(y)} + a_m^{(y)} \right), \tag{3.156c}$$

$$\langle \mathscr{R}^2_{2,2} \rangle = \frac{1}{4} \left(a_l^{(x)} - a_m^{(x)} - a_l^{(y)} + a_m^{(y)} \right). \tag{3.156d}$$

This procedure can also be generalized to biaxial mixtures [Berardi and Zannoni, 2015]. Average $\langle \mathscr{R}^2_{m,n} \rangle$ obtained with computer simulations will be shown in Chapter 10 (see Fig. 10.8) and in Chapter 11 (see Fig. 11.25). Considering only the case of rigid biaxial molecules in biaxial phases, an application of the symmetry arguments mentioned earlier shows that we can have four order parameters: $\langle \mathscr{D}^2_{0,0} \rangle$, $\text{Re}\langle \mathscr{D}^2_{0,2} \rangle$, $\text{Re}\langle \mathscr{D}^2_{2,0} \rangle$, $\text{Re}\langle \mathscr{D}^2_{2,2} + \mathscr{D}^2_{-2,2} \rangle$. Note that the observation of phase biaxiality does not require an off axis molecular tensor or even biaxiality in the molecular tensor (i.e. it can be obtained even when we have $A^{2,2}_{\text{MOL}} = 0$). Indeed, phase biaxiality can be examined using a uniaxial probe. For example, the DNMR spectrum of fully deuterated benzene C_6D_6 is often used. The phase biaxiality can split the 'perpendicular' quadrupole lines in the spectra in an x and a y pair, as verified experimentally in various systems [Allender and Doane, 1978; Goldfarb et al., 1983b; Doane, 1985a, 1985b]. In a real situation observation or not of phase biaxiality will of course depend on the relative magnitude of the terms in Eqs. 3.153b–3.153c and on the sensitivity of the experiment. The difference between a_l and a_{mm} will be, for a uniaxial tensor $\mathbf{A}$,

$$a_l - a_m = 2\text{Re}\langle \mathscr{D}^{2*}_{2,0} \rangle A^{2,0}_{\text{MOL}}. \tag{3.157}$$

3.11.2 A Tilted Biaxial Phase: Smectic C

We now turn to the case of tilted biaxial phases, such as the smectic C and some of the columnar liquid crystals described in Chapter 1. The purely orientational order parameters, which define the distribution, $P(\Omega)$, can be classified simply according to the point group for the system. Consider an ideal smectic C formed of uniaxial molecules as an example (see Fig. 1.32). According to this simple model we can choose a laboratory frame with axes $\boldsymbol{x}$, $\boldsymbol{y}$, $\boldsymbol{z}$ where the director $\boldsymbol{d}$ makes an angle θ, the tilt angle, with $\boldsymbol{z}$ parallel to the layer plane and $\boldsymbol{y}$ along the layer normal. As discussed in Appendix G the symmetry operations of the C_{2h} group give the restrictions

$$\langle \mathscr{D}^L_{m,n} \rangle = (-1)^m \langle \mathscr{D}^L_{m,n} \rangle = (-1)^{L+m} \langle \mathscr{D}^L_{m,n} \rangle. \tag{3.158}$$

Thus, L and m have to be even. The relevant second-rank order parameters should then be $\langle \mathscr{D}^2_{0,n} \rangle$and $\langle \mathscr{D}^2_{2,n} \rangle = (-1)^n \langle \mathscr{D}^{2*}_{-2,n} \rangle$. Note that, even if the particles constituting the phase have cylindrical symmetry, we still have two independent order parameters, corresponding to the biaxiality of the phase. This is perhaps more transparent in Cartesian coordinates. We can define a mesophase ordering matrix $\mathbf{Q}$

$$Q^{\text{LAB}}_{AB} = (3\langle d_A d_B \rangle - \delta_{A,B})/2; \quad A, B = x, y, z, \tag{3.159}$$

formally identical to the molecular (Saupe) ordering matrix. Here, however, it is the phase that is not cylindrically symmetric instead of the particles, and so we take $d_A \equiv d_{Az}$ to be the direction cosines for a molecule in the laboratory frame. Now, from the assumed symmetry of the smectic C, it is clear that $\mathbf{Q}^{\text{LAB}}$ can be diagonalized by a rotation of the angle θ, with $\tan 2\theta = 2Q^{\text{LAB}}_{xy}/(Q^{\text{LAB}}_{yy} - Q^{\text{LAB}}_{xx})$, around z. The tilt angle θ represents the orientation of the director in the (xy) plane. The eigenvalues of $\mathbf{Q}$ are

$$q_1 = -\frac{1}{2}Q_{zz}^{\mathrm{LAB}} + \rho, \quad q_2 = -\frac{1}{2}Q_{zz}^{\mathrm{LAB}} - \rho, \quad q_3 = Q_{zz}^{\mathrm{LAB}}, \tag{3.160}$$

where $\rho = \sqrt{\frac{1}{2}\left[(Q_{xx}^{\mathrm{LAB}} - Q_{yy}^{\mathrm{LAB}})^2 + 4(Q_{xy}^{\mathrm{LAB}})^2 \right]}$ is a phase biaxiality parameter. If, on the other hand, the molecules forming the biaxial phase are themselves biaxial there will be four order parameters: $\langle \mathscr{R}_{0,0}^2 \rangle$, $\langle \mathscr{R}_{2,0}^2 \rangle$, $\langle \mathscr{R}_{0,2}^2 \rangle$, $\langle \mathscr{R}_{2,2}^2 \rangle$ like in Eqs. 3.150a–3.150d. Other combinations of molecular and mesophase symmetries can be treated along the same lines as the need arises employing the results given in Section 3.11. The methods presented here also apply to the various types of discotic mesophases (see Chapter 1).

3.11.3 Cartesian Formulation

Let us consider again the measurement of an observable average property which behaves as a second-rank tensor $\langle \mathbf{A} \rangle$. We imagine for completeness that all nine components of the representative matrix can be measured in a given laboratory frame. The laboratory frame components will be connected to their molecular counterparts through a simple rotation:

$$\langle A_{ij}^{\mathrm{LAB}} \rangle = \sum_{k,l} \langle R_{ik} A_{kl}^{\mathrm{MOL}} R_{lj}^T \rangle, \tag{3.161}$$

R_{ik} is the rotation matrix connecting laboratory to molecule frame with components, e.g. $R_{ik} = d_{ik}$, the direction cosine of the molecular k-axis. If the molecule is rigid or if the tensor A_{kl}^{MOL} does not fluctuate in the molecular frame (e.g. we have a dipolar coupling between two nuclei belonging to the same rigid fragment) then

$$\langle A_{ij}^{\mathrm{LAB}} \rangle = \sum_{k,l} \langle R_{ik} R_{lj}^T \rangle A_{kl}^{\mathrm{MOL}}. \tag{3.162}$$

We rewrite this equation by adding and subtracting $a\delta_{i,j}$ on the right-hand side, with $a = \mathrm{Tr}\mathbf{A}^{\mathrm{LAB}}/3 = \mathrm{Tr}\mathbf{A}^{\mathrm{MOL}}/3$ as

$$\langle A_{ij}^{\mathrm{LAB}} \rangle = a\delta_{i,j} + \frac{2}{3}\sum_{i,j} C_{ij,kl} A_{kl}^{\mathrm{MOL}}, \tag{3.163}$$

where $\mathbf{C}$ is a fourth-rank tensor (see Appendix B), defined as

$$C_{ij,kl} \equiv \frac{3}{2}\langle R_{ik} R_{jl} \rangle - \frac{1}{2}\delta_{i,j}\delta_{k,l}, \tag{3.164}$$

and can be called a *superordering matrix*, reducing to 0 in an isotropic liquid. We define in general a laboratory principal frame through the transformation which makes the $\mathbf{C}$ matrix block diagonal in the first two indices. Using this formalism a suitable tensor $\mathbf{A}$ can be defined in the molecular frame and all its average components $\langle A_{ij}^{\mathrm{LAB}} \rangle$, determined in the lab frame. In the principal laboratory frame the average tensor should be diagonal with eigenvalues

$$\langle a_{ii}^{\mathrm{LAB}} \rangle = \langle A_{ij}^{\mathrm{LAB}} \rangle \delta_{i,j} = a + \frac{2}{3}\sum_{k,l} C_{iikl} A_{kl}^{\mathrm{MOL}}. \tag{3.165}$$

In a uniaxial phase there are only two independent quantities of the average tensor $\langle \mathbf{A}^{\mathrm{LAB}} \rangle$, e.g. the trace and the components parallel to the director, i.e. $\langle a_{zz}^{\mathrm{LAB}} \rangle$. Thus, in turn only one block of the superordering matrix needs to be specified, i.e. $S_{kl} = C_{zzkl}$. In other words, in a uniaxial phase the superordering matrix reduces to the usual Saupe matrix. Another case mentioned earlier is that of a biaxial phase formed of uniaxial particles. In this case we only need $Q_{ij} = C_{ijzz}$ and the superordering matrix can be reduced to the $\mathbf{Q}$ tensor discussed earlier.

3.11.4 Triclinic Biaxial Phases

When the positions are on average distributed on some regular lattice, defined by the primitive vectors $\boldsymbol{a}_1, \boldsymbol{a}_2, \boldsymbol{a}_3$ we can write

$$P(\boldsymbol{r}, \Omega) = \sum_{\boldsymbol{k}, L, m, n} \left[\frac{(2L+1)}{8 v_0 \pi^2} \right] \left\langle \mathrm{e}^{-i\boldsymbol{k}\cdot\boldsymbol{r}'} \mathscr{D}_{m,n}^{L*}(\Omega') \right\rangle \mathrm{e}^{i\boldsymbol{k}\cdot\boldsymbol{r}} \mathscr{D}_{m,n}^{L}(\Omega), \tag{3.166}$$

with $\boldsymbol{k}$ a point in the reciprocal lattice. As a particular case, when k=0 th system is uniform as a nematic but with a tilted set of directors, instead of an upright one like we have treated until now. Phases of this type and their order parameters have been discussed by Karahaliou et al. [2002, 2009] and by Kwak and Kim [2016]. A full treatment considering various molecular symmetries could be carried out as previously seen but, since not many experimental or simulated data have been determined, even for the much simpler biaxial systems already considered, we shall not dwell any more into these phases here.

3.12 Flexible Molecules

As our description of molecular order becomes more realistic, dealing with detailed chemical structures, rather than generic particles, it becomes important to develop ways of describing orientational order in flexible molecules. We now wish to briefly mention how the present treatment of order parameters can be generalized to molecules with internal degrees of freedom. This is an important problem because most molecules of practical interest, including molecules forming liquid crystals, possess some internal flexibility [Orville-Thomas, 1974; Maruani and Serre, 1983]. Here we shall only consider one mechanism for internal flexibility, that of internal rotation, since this often represents the most important mechanism able to provide large changes in molecular shape.

Typically, we might have multiring molecules with some degree of internal rotation or molecules with floppy chain substituents (see Fig. 3.24) [Zannoni, 1985; Berardi et al., 1996b, 1998b]. In addition to this, solutes with internal degrees of freedom dissolved in liquid crystals or membranes can be studied with various techniques. The assumption of rigidity (and quite often that of cylindrical symmetry) has usually been made in the past on the grounds of simplicity and of the inadequacy of experimental data in studying molecular structures at this level of sophistication. This is not necessarily true anymore. As a matter of fact, the quality and quantity of data becoming available, particularly from NMR techniques applied to isotopically substituted molecules, now often demands going beyond the rigid molecule approximation. The problem of a general formalism for describing orientational order in non-rigid particles is also relevant in the field of computer simulations, given

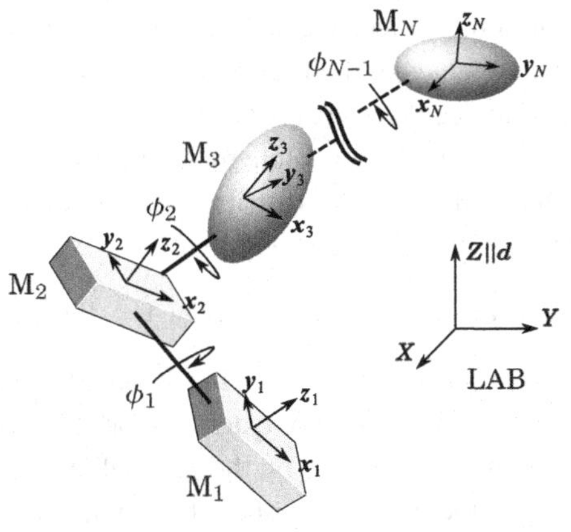

Figure 3.24 A sketch of a multirotor molecule consisting of N linearly connected anisotropic rigid fragments $\text{M}_1, \text{M}_2, \ldots, \text{M}_N$ with their local coordinate frame. Flexibility results from the internal rotations of dihedral angles $\phi_1, \phi_2, \ldots, \phi_{N-1}$ as shown.

the huge amount of information they provide. As an example, in an atomistic Molecular Dynamics simulation, all the positions and orientations of the submolecular fragments would be determined at every time step considered. Therefore, the need arises to condense this embarrassingly large amount of data into a set of quantities which, on one hand, contains the relevant information and, on the other, can hopefully be experimentally determined.

We consider the case of a flexible multirotor molecule, treated as a set of N rigid fragments linked by $N-1$ bonds [Zannoni, 1985; Berardi et al., 1996c] shown in Fig. 3.24. Each rotor has its local reference system M_k and the relative conformation of adjacent fragments M_k, M_{k+1} is defined by a dihedral angle ϕ_k. The overall conformational state is then specified by the set $\mathbf{\Phi} \equiv (\phi_1, \phi_2, \ldots, \phi_{N-1})$ and we assume as reference geometry that with all angles $\phi_k = 0$. In the case of *continuous* degrees of freedom, each angle ϕ_k can assume any real value in the range $[0, 2\pi]$. However, in a number of cases, a *discrete* treatment, exemplified by the so-called rotational isomeric state (RIS) model [Flory, 1969], in which the angles ϕ_k take only some discrete values $\phi_k^{(j)}$, or more simply j_k, is well established. For alkyl chains, these values are the so-called *trans*, *gauche*$^+$ and *gauche*$^-$ states with $\phi_k = 0, \pm 2\pi/3$ as typical values. As in previous sections, the Euler angles (α, β, γ) describe the rotation $\Omega \equiv \Omega_{\text{M}_1\text{L}}$ from the LAB system, with the Z-axis along the director, to the frame attached to the first (*'rigid'*) molecular fragment M_1 (Fig. 3.24) and define accordingly its molecular orientation.

3.12.1 One Internal Degree of Freedom

We start examining in detail the simplest case of a non-rigid molecule formed by two rigid distinguishable parts M_1 and M_2 connected by a bond and rotating one with respect to the other. This could be, for instance, an unsymmetrically substituted biphenyl or bithienyl. Let us take one of the fragments, M_1, say, as the 'rigid' part, where we place the molecular frame. The second fragment of the molecule can rotate through an angle ϕ around a certain bond directed along $\boldsymbol{r}$. We now imagine the molecule embedded in a uniform isotropic or anisotropic fluid phase. The molecule we consider could be a mesogen in its nematic or isotropic phase or it could equally well be a solute molecule dissolved in such a phase. The single molecule orientational-conformational distribution function for our internal rotor can be written for a uniform system like a nematic as

$$P(\boldsymbol{r}_1, \Omega_{\mathrm{M_1L}}, \boldsymbol{r}_2, \Omega_{\mathrm{M_2L}}) = P_{\boldsymbol{r}}(\Omega_{\mathrm{1L}}, \Omega_{\mathrm{2L}}) = P_{\boldsymbol{r}}(\Omega_{\mathrm{1L}}, \phi), \tag{3.167}$$

where $\boldsymbol{r}_1, \boldsymbol{r}_2$ give the position of the two fragments in the laboratory and the inter-fragment vector $\boldsymbol{r}$ is taken to be constant while internal or external reorientation takes place. We shall omit writing down explicitly this dependence on the internal axis orientation and distance parameter r from now on, when there is no risk of confusion. In the biphenyl type molecule mentioned above the internal rotation axis $\boldsymbol{r}$ is along the ring-ring 'long' axis. It should be noticed that by writing the distribution as a function of the two labelled orientations Ω_1 and Ω_2 we have made use of the physical distinguishability of the two groups, i.e. that it is possible to devise an experiment that can tell which part we are looking at. If, on the contrary, there is permutation symmetry $P_{1,2}$ for the two parts we should consider just one independent orientation Ω and write $P(\Omega, \phi)$, dropping unnecessary subscripts. We can now write the one particle distribution as

$$P(\Omega, \phi) = \langle \delta(\Omega - \Omega')\, \delta(\phi - \phi') \rangle = \sum \frac{2L+1}{16\pi^3} p_{Lmn|q} \mathscr{D}^{L*}_{m,n}(\Omega)\, \mathrm{e}^{-iq\phi}, \tag{3.168}$$

where we have used the representation of the two types of angular δ functions in an orthogonal basis of Wigner matrices (see Eq. F.10) and, for the internal rotation, in a Fourier basis, i.e.

$$\delta(\phi - \phi') = \frac{1}{2\pi} \sum_{q=-\infty}^{\infty} \mathrm{e}^{iq(\phi-\phi')}. \tag{3.169}$$

The coefficients $p_{Lmn|q}$ are, as in previous similar expansions, the order parameters. The vertical line in the coefficients $p_{Lmn|q}$ separates the indices L, m, n referring, as before, to rotations from the internal ones, here: $q = 0, \pm 1, \pm 2, \ldots$ and as usual the sum is extended to all the coefficients not appearing on the left-hand side. The angle ϕ, $0 \leq \phi \leq 2\pi$ is the dihedral rotation angle around $\boldsymbol{r}$. The orthogonality relation of the chosen basis immediately allows definition of the expansion coefficients. For nematics

$$p_{Lmn|q} = \langle \mathscr{D}^{L}_{m,n}(\Omega)\, \mathrm{e}^{iq\phi} \rangle, \tag{3.170}$$

where the angular brackets denote an orientational average over the distribution $P(\Omega, \phi)$. We have here three types of order parameters:

(i) **Purely orientational order parameters.**

$$\langle \mathscr{D}^{L}_{m,n}(\Omega) \rangle = p_{Lmn|0}. \tag{3.171}$$

This type of expansion coefficient is essentially an ordinary orientational order parameter for the molecular frame. It gives the average orientation of the chosen rigid part of the molecule with respect to the director frame. For a truly rigid molecule the ordering matrix is of course unique and the parameters determined give information on the total ordering matrix. Thus, in the rigid molecule limit, if we wish to employ symmetry to reduce the number of independent parameters, it is the overall point symmetry of the molecule that should be employed and not the local one. If, instead, the rigid fragment is connected to the rest of the molecule via a single bond (i.e. if it is an internal rotor), then local symmetry operations may become feasible and be applied. As an example in the two-ring-type molecules already mentioned if the fragment where the molecular frame has been located possesses D_2 symmetry, then the single particle distribution can be simplified using this local symmetry independently on the second ring orientation. This means that feasible local symmetry operations of this kind can be treated exactly as we have seen for rigid molecules. In the example given we shall have the restriction that L has to be even. The possibility of having independent fragments implies, however, that in a flexible molecule the orientation of one subunit does not automatically determine the orientation of the other fragments. We also note that the expansion separates completely the laboratory frame operations from the internal motion. The rotation $\mathscr{D}^L_{m,n}(\Omega) \equiv \mathscr{D}^L_{m,n}(\Omega_{\mathrm{M_1L}})$ is the one affected by mesophase symmetry operations. Thus, for example, we have for uniaxial mesophases, $p_{Lmn|q} = \delta_{m,0}\, p_{L0n|q}$, whatever the conformation. Similarly, if the phase has a plane of symmetry perpendicular to the director, then L has to be even. Other operations can be applied as already seen for rigid molecules.

(ii) **Internal order parameters**. We define purely internal order parameters the averages

$$\langle \mathrm{e}^{iq\phi} \rangle = p_{000|q}, \quad q = 0, \pm 1, \pm 2, \ldots. \tag{3.172}$$

These parameters describe the ordering of the second part of the molecule with respect to the first one, irrespective of the overall orientation. They can be different from 0 even in the isotropic phase if there is some preferential orientation of the second fragment around the internal axis. The internal order parameters for a rigid molecule with the second fragment at an angle ϕ_0 from the first one will just be $\exp(iq\phi_0)$. By suitable definition of axis system this becomes just 1. A flexible molecule can be defined as a molecule with internal order parameters deviating from the theoretical rigid value, whatever the frame. At the other extreme, we have $p_{000|q} = 0$ when the distribution $P(\Omega, \phi)$ does not depend on the internal angle ϕ. Apart from the trivial case of a cylindrical fragment this indicates a uniaxial distribution of the second subunit around the rotation axis. The coefficient $p_{000|q} = 0$ thus indicates complete internal disorder around the rotation axis. Internal order parameters are determined by the energy barrier to rotation. Thus, if the energy barrier in the medium is $U(\phi)$, then the internal order parameters are obtained from the Boltzmann expression

$$p_{000|q} = \int \mathrm{d}\phi\, \mathrm{e}^{iq\phi}\, \mathrm{e}^{-U(\phi)/(k_B T)} \Big/ \int \mathrm{d}\phi\, \mathrm{e}^{-U(\phi)/(k_B T)}. \tag{3.173}$$

If the medium effect on the barrier can be neglected, $U(\phi)$ is the same as the gas phase potential $U^{\mathrm{G}}(\phi)$ obtainable in principle from an *ab initio* or semi-empirical theoretical calculation [Leach, 2001]. In turn, the determination of the parameters $p_{000|q}$ can provide information on internal barriers, and medium effects can be obtained by comparison with separate quantum chemistry calculations in the gas phase.

(iii) **Mixed internal-external order parameters**. The third type of coefficients in Eq. 3.170 are

$$p_{Lmn|q} = \langle \mathscr{D}^L_{m,n}(\Omega)\, e^{iq\phi} \rangle. \tag{3.174}$$

These terms describe coupling between internal and external degrees of freedom. They can be used to recover purely orientational order parameters for the second subunit from those of the first one. In fact, writing down explicitly the transformations between frames, we have

$$\langle \mathscr{D}^L_{m,n}(\Omega_{M_2L}) \rangle = \sum_{m'} \langle \mathscr{D}^L_{m,m'}(\Omega_{M_1L}) \mathscr{D}^L_{m',n}(\Omega_{M_2M_1}) \rangle, \tag{3.175a}$$

$$= \sum_{m'} e^{-im'\alpha_{M_2M_1}} d^L_{m',n}(\beta_{M_2M_1}) \langle \mathscr{D}^L_{m,m'}(\Omega_{M_1L})\, e^{-in\phi} \rangle, \tag{3.175b}$$

$$= \sum_{m'} e^{-im'\alpha_{M_2M_1}} d^L_{m',n}(\beta_{M_2M_1}) p_{Lmm'|-n}, \tag{3.175c}$$

where $(\alpha_{M_2M_1}, \beta_{M_2M_1})$ indicate the orientation of the internal rotation axis. Note that here $-L \leq m', n \leq L$ so that only a subset of mixed order parameters is needed to obtain the ordering matrix for the second ring. This provides a sort of sum rule for the mixed order parameters. For a rigid molecule it reduces to

$$\langle \mathscr{D}^L_{m,n}(\Omega_{M_2L}) \rangle = \sum_{m'} \langle \mathscr{D}^L_{m,m'}(\Omega_{M_1L}) \rangle \mathscr{D}^L_{m',n}(\Omega_{M_1M_2}), \tag{3.176}$$

since the transformation linking the two frames is a time independent one. In this case the orientation $\Omega = (\Omega_{M_1L})$ is sufficient to completely define the state of the particle. Eq. 3.176 simply gives the relation between the orientational order parameter when expressed in the two frames. For a molecule where the internal rotation axis coincides with the molecular frame z-axis, we have $d^L_{p,n}(0,0) = \delta_{p,n}$ and

$$\langle \mathscr{D}^L_{m,n}(\Omega_{M_2L}) \rangle = \left\langle \mathscr{D}^L_{m,n}(\Omega_{M_1L})\, e^{-in\phi} \right\rangle = p_{Lmn|-n}, \tag{3.177}$$

so that in this case the sum rule reduces to just one term. We have already mentioned when considering the purely orientational parameters that in the present treatment the different frame transformations employed to specify the orientational and internal state are well separated, thus making it possible to apply symmetry simplifications referring to internal or external degrees of freedom. We have also said that mesophase symmetry does not put restrictions on the internal order parameters. Here, however, we have the possibility of implementing fragment symmetry. In simple cases this can be done by direct inspection. For example, the order parameters for a molecule with a para (i.e. along the z-axis) biaxial substituent are $p_{Lmn|q}, q = 0, \mp 2, \mp 4, \ldots$. Similarly, if we have a methyl substituent or, more generally, a substituent with C_3 symmetry around the internal rotation axis, then $q = 0, \mp 3, \ldots$. In more complex cases the methods developed within the group theory of non-rigid molecules can be employed [Altmann, 1977; Renkes, 1981; Maruani and Serre, 1983]. Our purpose here is not that of giving a systematic treatment for various rotor symmetries but it is useful to be aware that the formalism allows full exploitation of local symmetry if needed.

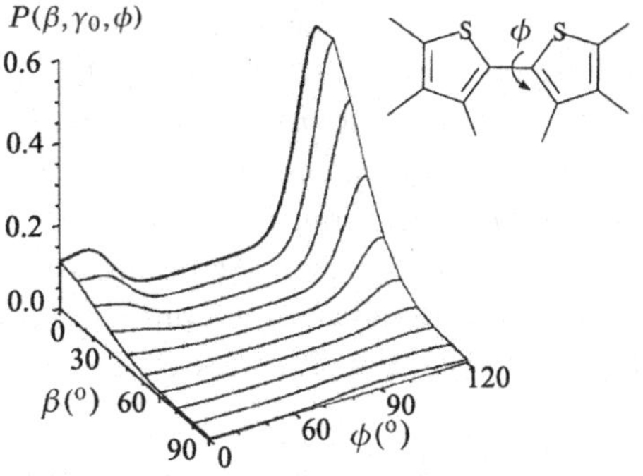

Figure 3.25 Structure of the cis-form of 2,2′-bithienyl (DTH) with the torsional angle ϕ describing a rotation around the axis z connecting the two rings. The orientational-conformational distribution: $P(\beta,\gamma_0,\phi)$ obtained by Berardi et al. [1994] analyzing data from Ter Beek et al. [1991] is shown for $\gamma_0 = 15°$.

An example of rotameric molecule: bithienyl. To give a specific example of a molecule with one degree of internal rotation made up of two rigid fragments, we consider the bithienyl shown in Fig. 3.25. The first thing to point out is that the set of three Euler angles Ω we have used until now is only sufficient to specify the state of a rigid fragment, e.g. it can describe the orientation of a suitable molecular frame. When the molecule has additional degrees of internal freedom more variables have to be introduced. For a two-ring molecule an angle ϕ giving the orientation of one ring with respect to the other could do. Thus, we can define an orientational-conformational state (Ω,ϕ) by choosing a molecular frame M_1 on one molecular fragment and giving its orientation $\Omega \equiv (\Omega_{M_1L})$ with respect to the laboratory frame and then giving the angle ϕ that the second ring makes with the first one. We write the probability of finding the molecule in a certain orientational-conformational state as the probability of finding the first fragment at orientation Ω with respect to the laboratory director frame and the second fragment at an angle ϕ from the first, i.e. $P(\Omega,\phi)$. This one particle distribution is then expanded in a composite Wigner–Fourier basis set [Zannoni, 1985]. We have for a molecule dissolved in a uniaxial phase, where $\Omega = (\beta,\gamma)$,

$$P(\beta,\gamma,\phi) = \frac{1}{8\pi^2}\sum_{\substack{L=\\0,2,\dots}}^{\infty}\sum_{n=-L}^{L}\sum_{q=-\infty}^{\infty}(2L+1)p_{L0n|q}\mathscr{D}^{L}_{0,n}(\beta,\gamma)\,\mathrm{e}^{iq\phi}, \tag{3.178}$$

where in general $q = 0,\pm1,\pm2,\dots$. The angle ϕ, with $0 \le \phi \le 2\pi$ is the dihedral rotation angle around the inter-fragment vector connecting the two thiophenes. A particular subset of these parameters allows the recovery of purely orientational order parameters for the second subunit. The ME method outlined earlier can be generalized to yield the best distribution compatible with a given set of order parameters. For instance, if an experiment determines a set of second-rank order parameters $p_{20n|q}$, this distribution will be of the form

$$P(\beta,\gamma,\phi) = \exp\left\{\sum_{n,q} a_{n|q}\mathscr{D}^{2*}_{0,n}(\beta,\gamma)\,\mathrm{e}^{iq\phi}\right\}, \tag{3.179}$$

where the coefficients $a_{n|q}$ are obtained by minimizing the squared difference between the measured quantities and those obtained by integrating Eq. 3.179. The ME approach has been applied to analyzing the experimental proton NMR dipolar couplings of a number of solutes [Catalano et al., 1991; Berardi et al., 1992]. In the case of bithienyl the purely orientational order parameters for the two rings as well as an approximate rotamer distribution can be obtained. In Fig. 3.25 we show the results for the average purely internal distribution $\bar{P}(\phi)$ and for the orientational-conformational distribution $P(\beta,\gamma,\phi)$ in 2,2′-bithienyl obtained analyzing data from Ter Beek et al. [1991] with the ME method described in Berardi et al. [1994].

3.12.2 Multirotor Molecules

We can now extend the treatment to the more general case of a flexible molecule in a uniform anisotropic solution, which has the orientational-conformational distribution $P(\Omega,\boldsymbol{\Phi})$ with the normalization condition $\int d\Omega d\boldsymbol{\Phi} P(\Omega,\boldsymbol{\Phi}) = 1$. The purely conformational distribution $P(\boldsymbol{\Phi})$ is obtained by integration over the orientational variables $P(\boldsymbol{\Phi}) = \int d\Omega P(\Omega,\boldsymbol{\Phi})$. The orientational-conformational distribution can be formally considered as an averaged product of Dirac delta functions that counts the particles in the various intervals

$$P(\Omega,\boldsymbol{\Phi}) = \left\langle \delta(\Omega-\Omega')\,\delta(\phi_1-\phi_1')\,\delta(\phi_2-\phi_2')\ldots\delta(\phi_{N-1}-\phi_{N-1}')\right\rangle_{\Omega',\boldsymbol{\Phi}'}, \tag{3.180}$$

where the symbol $\langle\ldots\rangle_{\Omega',\boldsymbol{\Phi}'}$ represents the average over primed variables. Eq. (3.180) can be expanded as

$$P(\Omega,\boldsymbol{\Phi}) = \sum_{\substack{L,m,n;\\ \boldsymbol{q}}} p_{Lmn|\boldsymbol{q}} \left[\frac{2L+1}{8\pi^2(2\pi)^{N-1}}\right] \mathcal{W}^{L}_{m,n;\boldsymbol{q}}(\Omega,\boldsymbol{\Phi}), \tag{3.181}$$

where, generalizing the previous notation for the expansion coefficients, the indices before the vertical bar sign | in $p_{Lmn|\boldsymbol{q}}$, i.e. L,m,n, refer to orientational order (we would add positional order indices after a semicolon as we did before if we had positional order too) and the set $\boldsymbol{q} \equiv (q_1,\ldots,q_{N-1})$ refers to the internal order for non-rigid molecules. The mixed Wigner–Fourier basis functions $\mathcal{W}$ are defined as $\mathcal{W}^{L}_{m,n;\boldsymbol{q}}(\Omega,\boldsymbol{\Phi}) \equiv \mathcal{D}^{L*}_{m,n}(\Omega)\, e^{iq_1\phi_1+\cdots+iq_{N-1}\phi_{N-1}}$. The expansion coefficients $p_{Lmn|\boldsymbol{q}} \equiv p_{Lmn|q_1,\ldots,q_{N-1}}$ are the *order parameters* for the orientational-conformational problem

$$p_{Lmn|\boldsymbol{q}} = \int d\Omega d\boldsymbol{\Phi}\, P(\Omega,\boldsymbol{\Phi}) \mathcal{W}^{L*}_{m,n;\boldsymbol{q}}(\Omega,\boldsymbol{\Phi}) \equiv \left\langle \mathcal{W}^{L*}_{m,n;\boldsymbol{q}}(\Omega,\boldsymbol{\Phi})\right\rangle_{\Omega,\boldsymbol{\Phi}}. \tag{3.182}$$

These order parameters form an infinite set that fully describes the molecular orientational-conformational order. In particular, we find as special cases the usual Saupe ordering matrix components $S_{zz} = \langle P_2\rangle = p_{200|0,\ldots,0}$, $S_{xx} - S_{yy} = \sqrt{6}\,\mathrm{Re}\,\langle \mathcal{D}^2_{0,2}\rangle = \sqrt{6}\,\mathrm{Re}\, p_{202|0,\ldots,0}$, $S_{xz} = \sqrt{3/2}\,\mathrm{Re}\,\langle \mathcal{D}^2_{0,1}\rangle = \sqrt{3/2}\,\mathrm{Re}\, p_{201|0,\ldots,0}$, where the S_{ij} were given in Eq. 3.111. From the expansion coefficients we can compute the orientational order parameters of the single fragments, k measuring the average orientation of reference frame M_k with respect to the LAB system. For clarity, we consider simply connected structures, as sketched in

Fig. 3.24 and we call $\Omega_2 \equiv \Omega_{M_2M_1}$ the Euler angles giving the orientation of frame M_2 with respect to M_1 in the reference conformation, $\Omega_3 \equiv \Omega_{M_3M_1}$ the orientation of frame M_3 with respect to M_1 and so on, until Ω_k. We also call α_k and β_k the polar angles, measured with respect to frame M_k, defining the local orientation of the chemical bond associated to the dihedral angle ϕ_k, as shown in Fig. 3.24. Using these definitions and measuring all orientations and angles with respect to the reference conformation we obtain the purely orientational order parameters of frame M_k as

$$\langle \mathscr{D}^L_{m,n}(\Omega_{M_kL}) \rangle = \sum_{\substack{1,\dots,k-1\\1,\dots,k-1}} p_{Lm1|1,\dots,k-1,n,0,\dots,0}\, G^L_{1,\dots,k-1,1,\dots,k-1,n}, \tag{3.183}$$

where we write $p_{Lm,1|1,\dots,k-1,0,\dots,0}$ to indicate $p_{Lm,1|1,\dots,k-1,k,\dots,N-1}$ with all subscripts, if any, from k to $N-1$ equal to 0, since the internal degrees of freedom for the fragments following the one of interest do not enter in the expression. The $G^L_{1,\dots,k-1,1,\dots,k-1,n}$ coefficients:

$$G^L_{1,\dots,k-1,1,\dots,k-1,n} = \sum_{r_1,\dots,r_{k-1}} \left[\prod_{s=1}^{k-1} e^{-ia_s\alpha} d^L_{a_s,b_s}(\beta_s)\, e^{+ir_s\alpha_s} d^L_{r_s,b_s}(\beta_s) \right] \times \left[\prod_{s=2}^{k-1} \mathscr{D}^L_{r_{s-1},s}(\Omega_s) \right] \mathscr{D}^L_{r_{k-1},n}(\Omega_k), \tag{3.184}$$

collect all the chemical bond geometric information known from the molecular architecture and supposedly unaffected by the surrounding, while $d^L_{m,n}(\beta)$ are small Wigner matrices (Appendix F). The order parameters, $\langle \mathscr{D}^L_{m,n}(\Omega_{M_kL}) \rangle$ are then a linear combination of coefficients $p_{Lmn|1,\dots,N-1}$, weighted by the constant geometrical G coefficients. The formalism can be useful since the purely orientational order parameters of one or all fragments with respect to the lab frame can be obtained by some experiments, e.g. proton or deuterium NMR, and these can help to obtain information of the internal, conformational order of the molecule. As an example we can now apply these general transformation rules to the case of alkyl chains ($R(CH_2)_nR'$), or just of an alkane if $R = R'$ is a CH_3 [Rosen et al., 1993; Berardi et al., 1996b, 1998b]. For simplicity, we fix the reference frame M_1 in one of the terminal molecular fragments R. The remaining systems M_k are collinear to M_1, with their z_k-axis parallel to the direction of full molecular elongation and the x_k-axis pointing along the symmetry axis of each $\widehat{HCH}$ group, on the same side of the H atoms. Assuming the $\widehat{CCC}$ angle θ to be equal for all fragments, we define $\psi \equiv (\pi - \theta)/2$ and using Eq. 3.183, we write the kth frame order parameter as

$$\langle \mathscr{D}^L_{m,n}(\Omega_{M_kL}) \rangle = \sum_{\substack{1,\dots,k-1\\1,\dots,k-1}} (-1)^{2+4+\dots+[k/2]}\, p_{Lm1|1,\dots,k-1,0,\dots,0} \times d^L_{1,1}(\psi) \dots d^L_{k-1,k-1}(\psi)\, d^L_{2,1}(\psi) \dots d^L_{n,k-1}(\psi), \tag{3.185}$$

where $[k/2]$ is the integer part of $k/2$. In the case of *discrete* conformations, like those implied in the RIS model mentioned before [Flory, 1969; Dill and Bromberg, 2011] we can introduce the distribution function $P(\Omega, j)$ representing the probability of finding a molecule with orientation within the range $[\Omega, \Omega + \mathrm{d}\Omega]$ and conformation $\boldsymbol{j} \equiv (j_1, \ldots, j_{N-1})$. In this case, the normalization condition becomes $\sum_{\boldsymbol{j}} \int \mathrm{d}\Omega \, P(\Omega, \boldsymbol{j}) = 1$. Again, the conformational distribution $P(\boldsymbol{j})$ is obtained integrating over the other degrees of freedom Ω

$$P(\boldsymbol{j}) = \int \mathrm{d}\Omega P(\Omega, \boldsymbol{j}). \tag{3.186}$$

The orientational-conformational order parameters are

$$p_{Lmn|j_1,\ldots,j_{N-1}} = \int \mathrm{d}\Omega \, P(\Omega, \boldsymbol{j}) \, \mathscr{D}^L_{m,n}(\Omega), \tag{3.187}$$

and they are the expansion coefficients of the orientational distribution function for the molecule in a given conformation $\boldsymbol{j}$. For alkyl chains described using the *discrete* RIS model we have (cf. Eq. 3.183)

$$\begin{aligned}\langle \mathscr{D}^L_{m,n}(\Omega_{\mathrm{M}_k\mathrm{L}}) \rangle = \sum_{j} p_{Lmn|j_1,\ldots,j_{N-1}} \sum_{\substack{1,\ldots,k-1 \\ 1,\ldots,k-1}} (-1)^{2+4+\cdots+[k/2]} \, \mathrm{e}^{-i[_1\phi_1^{(j_1)}+\cdots+_{k-1}\phi_{k-1}^{(j_{k-1})}\}} \\ \times d^L_{1,1}(\psi) \ldots d^L_{k-1,k-1}(\psi) \, d^L_{2,1}(\psi) \ldots d^L_{n,k-1}(\psi).\end{aligned} \tag{3.188}$$

The knowledge of these order parameters for fragment M_k is sufficient to calculate any single particle and bond observable of an alkyl chain. For instance, the order parameter for a CH bond in a methylene group, often measured from DNMR experiments (cf. Appendix I) after replacing the hydrogens with deuterons as S_{CD}, can be written as $S_{CD} = -\frac{1}{2} S_{zz} + \frac{1}{2}(S_{xx} - S_{yy}) \cos \chi$, where $S_{zz} = \langle \mathscr{D}^2_{0,0} \rangle_{\mathrm{M}_k}$, $S_{xx} - S_{yy} = \sqrt{6} \mathrm{Re} \langle \mathscr{D}^2_{0,2} \rangle_{\mathrm{M}_k}$ and χ is the $\widehat{\mathrm{HCH}}$ bond angle. It is important to realize that NMR is not the only technique that can be used to get information on the conformations of flexible molecules dissolved in LC and indeed this generalized order parameters formalism, coupled with the ME approach, can be applied to the analysis of experimental data from a variety of techniques, also combined together, as shown in Berardi et al. [1998b].

3.13 Order in Smectics

After having considered the description of purely positional and orientational order, we now briefly introduce ordering in smectic mesophases, where both types of order exist at the same time.

Cylindrically symmetric particles and smectic. We start considering a smectic A formed of cylindrically symmetric particles. We assume the phase to be uniformly aligned and to have uniaxial symmetry around the director (parallel to the z laboratory axis). We also assume the centres of mass of the molecules to be randomly distributed in the x, y plane while they may organize in layers along the z-direction. The single particle distribution function can thus be written as $P(z, x)$, $x = \cos \beta$ and expanded as

$$P(z,x) = \sum_{L=0}^{\infty} \sum_{n_z=0}^{\infty} p_{L;n_z} P_L(x) \cos(n_z 2\pi z/\ell_z), \quad L \text{ even}, \tag{3.189}$$

in a product basis set of Legendre polynomials for orientations and Fourier harmonics for position. The distribution is normalized as $\int_{-1}^{1} \mathrm{d}x \int_{0}^{\ell_z} \mathrm{d}z P(z,x) = 1$. Orthogonality of the basis gives at once the coefficients $p_{L;n_z}$

$$p_{L;n_z} = \frac{2L+1}{\ell_z} \int_{-1}^{1} \mathrm{d}x \int_{0}^{\ell_z} \mathrm{d}z P(z,x) P_L(x) \cos\left(\frac{n_z 2\pi z}{\ell_z}\right),$$
$$= \frac{2L+1}{\ell_z} \left\langle P_L(x) \cos\left(\frac{n_z 2\pi z}{\ell_z}\right)\right\rangle, \quad n_z \neq 0, \tag{3.190a}$$

where the angular brackets $\langle \ldots \rangle$ have been used to indicate a positional-orientational average over $P(z, \cos\beta)$. Explicitly, $p_{0;0} = 1/\ell_z$, $p_{2;0} = 5\langle P_2\rangle/2\ell_z$, $p_{0;1} = \langle \cos(2\pi z/\ell_z)\rangle/\ell_z \equiv \tau_1/\ell_z$, $p_{2;1} = 5\langle P_2(x)\cos(2\pi z/\ell_z)\rangle/\ell_z = 5\sigma/\ell_z, \ldots$. These first few terms show the three kinds of order parameters present in a smectic phase. Thus, $\langle P_2\rangle$ is the usual orientational order parameter familiar from work on nematics, while $\tau_1 \equiv \langle\cos(2\pi z/\ell_z)\rangle$ is a purely positional order parameter expressing how effectively molecules are arranged in layers. The last type of parameter, σ, is a mixed order parameter related to the extent of translational orientational coupling. We shall comment on their relation to XRD experimental results in Chapter 4.

Non-cylindrical molecules in smectics. We now consider a rigid non-cylindrically symmetric molecule in a system with a layer structure and positional disorder inside the layer, such as smectic A or smectic C. The probability of finding the molecule at a specific position-orientation, $P(z,\beta,\gamma)$, can be expanded as we have seen earlier on in a complete basis set of spherical harmonics. Thus we get, generalizing Eq. 3.108,

$$P(z,\beta,\gamma) = \sum_{L,n;n_z} p_{Ln;n_z} \cos(2\pi n_z z/\ell_z) \mathscr{D}^L_{0,n}(\beta,\gamma). \tag{3.191}$$

The orthogonality of the basis set immediately gives the coefficients $p_{L,n;n_z}$. We can thus write the distribution as

$$P(z,\beta,\gamma) = \frac{1}{4\pi\ell_z} + \frac{1}{2\pi} \sum_{n_z=1}^{\infty} \sum_{L=0}^{\infty} \sum_{n=-L}^{L} \frac{2L+1}{\ell_z} \left\langle \cos(2\pi n_z z/\ell_z) \mathscr{D}^{L*}_{0,n}\right\rangle$$
$$\times \cos(2\pi n_z z/\ell_z) \mathscr{D}^L_{0,n}(\beta,\gamma), \quad n_z \neq 0. \tag{3.192}$$

The set of positional-orientational order parameters $\langle\cos(2\pi n_z z/\ell_z)\mathscr{D}^{L*}_{0,n}\rangle$ yields a complete characterization of $P(z,\beta,\gamma)$.

Flexible molecules in smectics. The treatment introduced in Section 3.12 for flexible molecules can be generalized to a smectic. Since a general treatment would bring considerable complications, difficult to justify, we only consider, like in Section 3.12.1, the simple example of a single internal rotation characterizsed by an angle ϕ. The one particle distribution for the molecule formed by two connected rigid fragments, Eq. 3.167 embedded in a smectic A is now

$$P(\boldsymbol{r}_1, \Omega_{1L}, \boldsymbol{r}_2, \Omega_{2L}) = P_r(z, \Omega_1, \Omega_2) = P_r(z, \Omega_1, \phi), \tag{3.193}$$

that can be written,

$$P(z,\Omega,\phi) = \langle \delta(z - z')\,\delta(\Omega - \Omega')\,\delta(\phi - \phi')\rangle_{z',\Omega',\phi'}, \tag{3.194a}$$

$$= \sum \frac{2L+1}{2\ell_z 16\pi^3}(1 + \delta_{p,0})\, p_{Lmn;\,p|a} \mathscr{D}^{L*}_{m,n}(\Omega)\cos(q_p z)\, \mathrm{e}^{-ia\phi}, \tag{3.194b}$$

where $q_p \equiv (2p\pi z/\ell_z)$ and $a = 0, \pm 1, \pm 2, \ldots$. In this case we have positional-orientational-internal order parameters $\langle \mathscr{D}^{L*}_{m,n}(\Omega)\cos(q_p z)\, \mathrm{e}^{-ia\phi}\rangle$ describing not only the purely orientational, positional or internal order, but also, in the mixed ones, the coupling of the different types of degrees of freedom, such as the effect of layering on conformations, a field that is still quite unexplored.

In plane orientational order. It is convenient to consider the limiting case of a smectic where the long axis ($\boldsymbol{u}$, say) of the mesogens is nearly completely aligned while the transversal is only partially ordered. In a smectic B the transversal molecular axis ($\boldsymbol{v}$, say) is isotropically distributed, while in a smectic E, there is a herringbone order (see Section 1.7). The orientational order parameter for molecules lying flat on the surface and thus reorienting in two dimensions can be obtained with simple modifications of what we have seen until now. We define a 2×2 matrix property $\mathbf{B}$, with elements $B_{ab} = \delta_{a,y}\delta_{b,y}$. The sample average of $\mathbf{B}$ in our arbitrary laboratory frame is obtained relating the Cartesian components of $\mathbf{B}^{\mathrm{LAB}}$ to the molecule fixed components and summing over all particles:

$$\langle B^{\mathrm{LAB}}_{ab}\rangle_S = \frac{1}{N}\sum_{i=1}^{N}\left\{\sum_{a'b'}[R_i]_{aa'}[B_i]^{\mathrm{MOL}}_{a'b'}\delta_{a',y}\delta_{b',y}[R_i^T]_{b'b}\right\} = \langle R_{ay}R_{by}\rangle_S, \tag{3.195}$$

where we have used the 2D rotation matrix connecting lab to molecule frame $\mathbf{R}(\phi) = \begin{pmatrix} \cos\phi & \sin\phi \\ -\sin\phi & \cos\phi \end{pmatrix}$. We can introduce a 2×2 traceless ordering matrix $\mathbf{P}$, analogous to the 3D ordering matrix $\mathbf{Q}$ (Eq. 3.50), as

$$\mathbf{P} = 2\begin{pmatrix} \langle \sin^2\phi\rangle - \frac{1}{2} & \langle \sin\phi\cos\phi\rangle \\ -\langle \sin\phi\cos\phi\rangle & \langle \cos^2\phi\rangle - \frac{1}{2} \end{pmatrix} \tag{3.196}$$

and we can identify the 2D order parameter as its largest eigenvalue:

$$\langle T_2\rangle = \langle 2\cos^2\phi - 1\rangle = \langle \cos(2\phi)\rangle, \tag{3.197}$$

where $T_n = \cos(n\phi)$ is a Chebishev polynomial of order n [Abramowitz and Stegun, 1965]. $\langle T_n\rangle$ is nothing but the first term in the orthogonal expansion of the distribution $P(\phi)$,

$$P(\phi) = \sum_n c_n T_n(\phi). \tag{3.198}$$

For a smectic B (Section 1.7.4) we expect $\langle T_2\rangle = 0$, while for a smectic E (see Section 1.7.5), $\langle T_2\rangle \neq 0$ [Luckhurst et al., 1987; Baggioli et al., 2019].

3.14 Columnar Phases

The positional orientational distribution for a non-cylindrically symmetric molecule in a columnar phase with some positional order of the columns in the x, y plane, e.g. a $\mathrm{D_{rd}}$ rectangular discotic phase [Goldfarb et al., 1983b; Kats and Monastyrsky, 1984], with the column axes parallel to z, will be

$$P(x,y,\beta,\gamma) = \frac{1}{4\pi \ell_x \ell_y} + \frac{1}{2\pi} \sum_{n_x,n_y} \sum_{Ln} \frac{2L+1}{\ell_x \ell_y} \left\langle \cos\left(\frac{2\pi n_x}{\ell_x}x\right) \cos\left(\frac{2\pi n_y}{\ell_y}y\right) \mathscr{D}^{L*}_{0,n}(\beta,\gamma) \right\rangle$$
$$\times \cos\left(\frac{2\pi n_x}{\ell_x}x\right) \cos\left(\frac{2\pi n_y}{\ell_y}y\right) \mathscr{D}^{L}_{0,n}(\beta,\gamma), \tag{3.199}$$

with $n_x, n_y \neq 0$ and $\int_0^\pi \mathrm{d}\beta \sin\beta \int_0^{2\pi} \mathrm{d}\gamma \int_0^{\ell_x} \mathrm{d}x \int_0^{\ell_y} \mathrm{d}y \, P(x,y,\beta,\gamma) = 1$. The positional-orientational order parameters

$$p_{Ln;n_x n_y} \propto \left\langle \cos\left(\frac{2\pi n_x}{\ell_x}x\right) \cos\left(\frac{2\pi n_y}{\ell_y}y\right) \mathscr{D}^{L*}_{0,n}(\beta,\gamma) \right\rangle \tag{3.200}$$

express any regularity in the 2D arrangement of the columns and the coupling to molecular orientation. If the mesophase is formed by disc-like particles we can reduce Eq. 3.199 to

$$P(x,y,\beta) = \frac{1}{2\ell_x \ell_y} + 2 \sum_{n_x,n_y} \sum_{L} \frac{2L+1}{\ell_x \ell_y} \left\langle \cos\left(\frac{2\pi n_x}{\ell_x}x\right) \cos\left(\frac{2\pi n_y}{\ell_y}y\right) P_L(\cos\beta) \right\rangle$$
$$\times \cos\left(\frac{2\pi n_x}{\ell_x}x\right) \cos\left(\frac{2\pi n_y}{\ell_y}y\right) P_L(\cos\beta), \tag{3.201}$$

with $\int_0^\pi \mathrm{d}\beta \sin\beta \int_0^{\ell_x} \mathrm{d}x \int_0^{\ell_y} \mathrm{d}y P(x,y,\beta) = 1$.

4

Distributions

The statistical mechanics of liquids is difficult; the statistical mechanics of nematics is still worse! Even for the simplest physical models, no exact solution has been worked out.

P. G. de Gennes, *The Physics of Liquid Crystals, 1974*

4.1 Phase Space Distributions

As we have seen in Chapter 3, the microscopic configurations of a system of classical particles (atoms or molecules or colloids) correspond to their set of generalized coordinates $\widetilde{\boldsymbol{X}} \equiv (\boldsymbol{X}_1, \boldsymbol{X}_2, \ldots, \boldsymbol{X}_N)$. These in turn can be detailed according to the level of description adopted. In computer simulations of liquid crystals both atomistic and generic models, where mesogenic molecules are replaced by some suitably shaped rigid body, are important and we shall try to take into account both the 'atomistic' description in terms of spherical particles and the one for rigid anisotropic bodies.

For an *atomistic* description, the configuration is given by $\widetilde{\boldsymbol{X}} \equiv (\boldsymbol{r}_1, \boldsymbol{r}_2, \ldots, \boldsymbol{r}_N)$, where $\boldsymbol{r}_i$ is just the position of each of the N atoms and, as before, we use an upper tilde to indicate the whole set of coordinates. The same description is also appropriate, on a different length scale, to the description of suspensions of spherical nanoparticles or quantum dots.

For rigid, non-spherical, particles each $\boldsymbol{X}_i$ is instead given by the centre of mass positions, $\boldsymbol{r}_i$, and the orientations Ω_i, referred to some laboratory frame, i.e. $\Omega_i \equiv \Omega_{i\mathrm{L}}$ of the N rigid bodies. In this case $\widetilde{\boldsymbol{X}} \equiv (\widetilde{\boldsymbol{r}}, \widetilde{\Omega}) \equiv (\boldsymbol{r}_1, \Omega_1, \boldsymbol{r}_2, \Omega_2, \ldots, \boldsymbol{r}_N, \Omega_N)$. If we are also interested in the dynamics of the system (Chapter 6), e.g. so as to be able to generate molecular trajectories like in the molecular dynamics simulations technique described later (Chapter 9), we will need to complement the description with the velocities, $\dot{\boldsymbol{r}}_i$, where we use the upper dot to indicate a time derivative, or the moments, $\boldsymbol{p}_i = m\dot{\boldsymbol{r}}_i$, of all atoms. This complete description of a *dynamic configuration* corresponds to a point in the $6N$ dimensional *phase space*, that we will indicate with the notation $\widetilde{\mathcal{X}} \equiv (\widetilde{\boldsymbol{X}}, \dot{\widetilde{\boldsymbol{X}}})$. In particular $\widetilde{\mathcal{X}} \equiv (\widetilde{\boldsymbol{r}}, \widetilde{\boldsymbol{p}}) \equiv (\boldsymbol{r}_1, \boldsymbol{p}_1, \boldsymbol{r}_2, \boldsymbol{p}_2, \ldots, \boldsymbol{r}_N, \boldsymbol{p}_N)$ for a collection of atoms or of spherical particles or, in the case of a system of N rigid anisotropic particles, $\widetilde{\mathcal{X}} \equiv \left(\widetilde{\boldsymbol{r}}, \widetilde{\Omega}, \widetilde{\boldsymbol{p}}, \widetilde{\boldsymbol{J}}\right) \equiv \left(\boldsymbol{r}_1, \Omega_1, \boldsymbol{p}_1, \boldsymbol{J}_1, \ldots, \boldsymbol{r}_N, \Omega_N, \boldsymbol{p}_N, \boldsymbol{J}_N\right)$, where, together with the centre of mass positions $\boldsymbol{r}_i$ and momenta $\boldsymbol{p}_i$ we have also introduced orientations Ω_i and angular momenta $\boldsymbol{J}_i = \mathbf{I}\,\dot{\boldsymbol{\Omega}}_i$

with $\mathbf{I}$ the inertia tensor and $\dot{\mathbf{\Omega}}_i$ the angular velocity of the ith particle [Goldstein et al., 2001]. As the system of particles evolves in time this point in phase space will change and its trajectory provides a description of the time evolution of the system (see Chapter 6). Here, however, we are only dealing with the description of systems at equilibrium and static instead of dynamic (i.e. time or frequency dependent) properties. We recall that in Chapter 3 we introduced the probability density of finding the system in a certain configuration in an a posteriori way, basically supposing to have at our disposal a sufficiently large number of configurations and counting the frequency of occurrence of certain coordinates $\boldsymbol{X}_i$. We have also focussed on single molecule properties. Now we wish instead to examine how the probability of observing a certain configuration can be linked 'bottom up' to the interaction potential energy between molecules and to external thermodynamic variables like, for example, temperature and/or pressure. It is reasonable to expect that this link will be a function of the energy of a particular set of coordinates and of temperature, but for the explicit form, not at all obvious, we have to resort to statistical mechanics. To proceed, the first thing is to specify the 'experimental' conditions, i.e. how our sample interacts with the environment. Once these are established, and assuming our system is at equilibrium, we can either study our sample following its evolution in time or somehow consider an arbitrarily large number of samples (an *ensemble*). If the system can evolve in time covering all its phase space, i.e. it is *ergodic,* a necessary condition for the equivalence of a time average for a single system followed for a sufficiently long time and the average for an ensemble of systems at equilibrium is satisfied. However, from a rigorous point of view stricter conditions should apply, e.g. that the dynamics of the system is 'chaotic', with trajectories that diverge exponentially in time, no matter how close they are initially [Coveney and Wan, 2016]. A brief list of the main sets of conditions for real or computer simulated experiments is:

Canonical (NVT). The system of N particles, contained in a volume V, is in contact with a thermostat, and thus can exchange energy with the environment, to keep the temperature T constant. At equilibrium the total energy is an observable property and will not be exactly constant but will show very small fluctuations around its average value.

Isothermal-isobaric (NPT). For this ensemble the system of N particles is maintained at constant temperature and pressure, but the volume and then the density can vary.

Grand canonical (μVT). In this case the system can exchange energy and molecules with the surrounding environment. Both the total energy and number of particles will be subject to small fluctuations about the mean of both quantities.

Microcanonical (NVE). The system is completely isolated from the environment. Here the number of molecules N, and volume V are fixed for the system and energy E, linear momentum $\boldsymbol{p}_T$, angular momentum $\boldsymbol{J}$ will be constant in time. The temperature will be an observable and at equilibrium it will fluctuate in time around an average value.

The detailed treatment of the different ensembles is the task of statistical mechanics and beyond the scope of this book, particularly since there are many excellent books dealing with that in general (see, e.g., [Balescu, 1975; Mazenko, 2000; Tuckerman, 2010; Swendsen, 2012]) and more specifically for liquids [Gray and Gubbins, 1984; Santos, 2016]. Here we wish to summarize the main facts, without any attempt at rigorous demonstrations, with the main aim of summarizing the expressions needed to set up later on computer simulations (see Chapters 8 and 9) for the calculation of physical properties in the various ensembles for systems that can be anisotropic fluids. We start from the, perhaps most common, canonical conditions.

4.2 Canonical Conditions

The problem of determining the functional form of the probability distributions in terms of the internal energy of a certain system was solved by Boltzmann and then Gibbs [1902], who showed that for a homogeneous system of N particles in equilibrium in a certain volume V at temperature T, the equilibrium distribution is [Balescu, 1975]

$$\varrho(\widetilde{X},\dot{\widetilde{X}}) = \exp[-\mathcal{H}_N(\widetilde{X},\dot{\widetilde{X}})/(k_BT)] \Big/ \int d\widetilde{X}d\dot{\widetilde{X}} \exp[-\mathcal{H}_N(\widetilde{X},\dot{\widetilde{X}})/(k_BT)], \tag{4.1a}$$

$$= \frac{1}{\mathcal{Z}(N,V,T)} \exp[-\mathcal{H}_N(\widetilde{X},\dot{\widetilde{X}})/(k_BT)], \tag{4.1b}$$

where $\varrho(\widetilde{X},\dot{\widetilde{X}})$ is the probability density, such that $\varrho(\widetilde{X},\dot{\widetilde{X}})\, d\widetilde{X}d\dot{\widetilde{X}}$ is the probability of occurrence of a configuration with particle 1 in the phase space infinitesimal volume element $dX_1d\dot{X}_1$ centred at $X_1, \dot{X}_1$, particle 2 inside $dX_2d\dot{X}_2$ centred at $X_2, \dot{X}_2$, and so on. Thus,

$$\mathcal{Z}(N,V,T) = \int d\widetilde{X}d\dot{\widetilde{X}} \exp[-\mathcal{H}_N(\widetilde{X},\dot{\widetilde{X}})/(k_BT)], \tag{4.2}$$

provides the normalization and is called the *phase integral* [Gray and Gubbins, 1984]. The Hamiltonian $\mathcal{H}_N$,

$$\mathcal{H}_N(\widetilde{X},\dot{\widetilde{X}}) = U_N(\widetilde{X}) + K_N(\dot{\widetilde{X}}), \tag{4.3}$$

is the sum of the total interaction energy between the N particles $U_N(\widetilde{X})$ that will be discussed in detail later, in Chapter 5, and the kinetic energy contributions $K_N(\dot{\widetilde{X}})$. This canonical Boltzmann–Gibbs distribution $\varrho(\widetilde{X},\dot{\widetilde{X}})$ can be immediately obtained, applying the Jaynes *Maximum Entropy principle* introduced in Section 3.8.2, when the input observed quantity is the total average energy $\mathcal{E} = \langle U_N(\widetilde{X}) + K_N(\dot{\widetilde{X}})\rangle$ [Jaynes, 1957b, 1957a; Santos, 2016]. The problem becomes that of finding the distribution that maximizes the associated Gibbs entropy functional

$$\mathcal{S}[\varrho] = -k_B\langle \varrho(\widetilde{X},\dot{\widetilde{X}}) \ln[\nu_N \varrho(\widetilde{X},\dot{\widetilde{X}})]\rangle = -k_B \int d\widetilde{X}d\dot{\widetilde{X}} \varrho(\widetilde{X},\dot{\widetilde{X}}) \ln[\nu_N \varrho(\widetilde{X},\dot{\widetilde{X}})], \tag{4.4}$$

where ν_N is a constant making the argument of the log dimensionless. It is often taken as $\nu_N \equiv N!\,h^{dN}$, with h Planck's constant and d the space dimensionality, while $N!$ is inserted to allow for the indistinguishability of particles. The distribution should reproduce the observed energy $\mathcal{E}$, so that

$$\mathcal{E} = \int \mathrm{d}\widetilde{X}\mathrm{d}\dot{\widetilde{X}}\, \mathcal{H}_N(\widetilde{X}, \dot{\widetilde{X}})\, \varrho(\widetilde{X}, \dot{\widetilde{X}}), \tag{4.5}$$

as well as obey the normalization condition $\int \mathrm{d}\widetilde{X}\mathrm{d}\dot{\widetilde{X}}\, \varrho(\widetilde{X}, \dot{\widetilde{X}}) = 1$. To find the ϱ maximizing $\mathcal{S}[\varrho]$ subject to these conditions, we introduce as in Section 3.8 a Lagrange multiplier, λ_i, for each relation that has to be obeyed, reducing the constrained maximization problem of the extended functional $\mathcal{L}$,

$$\mathcal{L} = -k_B \int \mathrm{d}\widetilde{X}\mathrm{d}\dot{\widetilde{X}}\, \varrho(\widetilde{X}, \dot{\widetilde{X}})\, \ln[\nu_N \varrho(\widetilde{r}, \widetilde{p})] - \lambda_1 \left(\mathcal{E} - \int \mathrm{d}\widetilde{X}\mathrm{d}\dot{\widetilde{X}}\, \varrho(\widetilde{X}, \dot{\widetilde{X}}) \mathcal{H}_N(\widetilde{X}, \dot{\widetilde{X}})\right)$$
$$- \lambda_0 \left(1 - \int \mathrm{d}\widetilde{X}\mathrm{d}\dot{\widetilde{X}}\, \varrho(\widetilde{X}, \dot{\widetilde{X}})\right), \tag{4.6}$$

to the unconstrained one. Proceeding as in Section 3.8 gives the maximum entropy distribution as

$$\varrho(\widetilde{X}, \dot{\widetilde{X}}) = \frac{1}{\mathcal{Z}(N, V, T)} \exp\left[\frac{\lambda_1}{k_B} \mathcal{H}_N(\widetilde{X}, \dot{\widetilde{X}})\right], \tag{4.7}$$

with the normalization condition determining λ_0

$$\mathcal{Z}(N, V, T) = \int \mathrm{d}\widetilde{X}\mathrm{d}\dot{\widetilde{X}}\, \exp\left[\frac{\lambda_1}{k_B} \mathcal{H}_N(\widetilde{X}, \dot{\widetilde{X}})\right] = \nu_N \exp\left(1 - \frac{\lambda_0}{k_B}\right). \tag{4.8}$$

Differentiating $\mathcal{S}$ in Eq. 4.4 with respect to energy, after inserting $\varrho(\widetilde{X}, \dot{\widetilde{X}})$ from Eq. 4.7, gives

$$\frac{\mathrm{d}\mathcal{S}}{\mathrm{d}\mathcal{E}} = \frac{\mathrm{d}}{\mathrm{d}\mathcal{E}} \left[-k_B \ln \nu_N - \lambda_1 \mathcal{E} + \ln \mathcal{Z}(N, V, T)\right] = \frac{1}{T} = -\lambda_1. \tag{4.9}$$

Identifying λ_1 with $-1/T$ gives the *Gibbs–Boltzmann distribution*:

$$\varrho(\widetilde{X}, \dot{\widetilde{X}}) = \frac{1}{\mathcal{Z}(N, V, T)} \exp\left[-\mathcal{H}_N(\widetilde{X}, \dot{\widetilde{X}})/(k_B T)\right]. \tag{4.10}$$

We also have, as a link to thermodynamics $-k_B \ln \mathcal{Z}(N, V, T) = (\mathcal{E} - T\mathcal{S})/T = \mathcal{A}/T$, where $\mathcal{A}$ is the *Helmholtz free energy*, the thermodynamic potential of the canonical ensemble. We shall now consider in turn the special cases of spherical and anisotropic constituent particles.

4.2.1 Spherical Particles

For a system of N identical spherical particles of mass m, that we can call 'atoms' (A) for simplicity, we have the classic Hamiltonian

$$\mathcal{H}_N^{\mathrm{A}}(\widetilde{r}, \widetilde{p}) = U_N^{\mathrm{A}}(\widetilde{r}) + K_N^{\mathrm{T}}(\widetilde{p}), \tag{4.11}$$

with the total interaction energy between particles $U^{\mathrm{A}}(\widetilde{r})$ depending on radial separations and the translational kinetic energy $K_N^{\mathrm{T}}(\widetilde{p}) = \sum_{i=1}^{N} (\boldsymbol{p}_i \cdot \boldsymbol{p}_i)/(2m_i)$. The integral

$\mathcal{Z}^{\mathrm{A}}(N,V,T)$ in Eq. 4.10 is often written, dividing by the previously defined ν_N, as the *canonical partition function*

$$\mathcal{Q}^{\mathrm{A}}(N,V,T) = \frac{1}{\nu_N}\int \mathrm{d}\widetilde{\boldsymbol{p}} \exp[-K_N^{\mathrm{T}}(\widetilde{\boldsymbol{p}})/(k_B T)] \int \mathrm{d}\widetilde{\boldsymbol{r}} \exp[-U_N^{\mathrm{A}}(\widetilde{\boldsymbol{r}})/(k_B T)], \tag{4.12a}$$

$$= \frac{1}{\nu_N}\mathcal{Z}^{\mathrm{A}}(N,V,T) = \frac{1}{\nu_N}\mathcal{Q}_P^{\mathrm{T}}\, Z^{\mathrm{A}}(N,V,T), \tag{4.12b}$$

where $\mathcal{Q}^{\mathrm{A}}(N,V,T)$ is dimensionless and essentially the classical limit of its quantum analogue [Balescu, 1975; Hansen and McDonald, 2006]. Here, and in what follows, we shall assume $d = 3$ unless otherwise specified, but it is worth mentioning that for particles confined to a surface, $d = 2$. We also have

$$\mathcal{Q}_P^{\mathrm{T}} = \frac{1}{h^{3N}}\int \mathrm{d}\widetilde{\boldsymbol{p}} \exp[-\sum_a \frac{\widetilde{p}_a^2}{2mk_B T}] = \lambda^{-3N}, \tag{4.13}$$

where $\lambda = \sqrt{h^2/(2\pi m k_B T)}$ is called the *thermal* or *De Broglie wavelength.* As long as the Hamiltonian is separable in a sum of a potential and a kinetic term, as in Eq. 4.11, we can write: $\varrho^{\mathrm{A}}(\widetilde{\boldsymbol{r}}, \widetilde{\boldsymbol{p}}) = \frac{1}{\nu_N} P^{\mathrm{A}}(\widetilde{\boldsymbol{r}}) P^{\mathrm{T}}(\widetilde{\boldsymbol{p}})$, where $P^{\mathrm{A}}(\widetilde{\boldsymbol{r}})$ is

$$P^{\mathrm{A}}(\widetilde{\boldsymbol{r}}) = \exp[-U_N^{\mathrm{A}}(\widetilde{\boldsymbol{r}})/(k_B T)]\,/\,Z^{\mathrm{A}}(N,V,T), \tag{4.14}$$

while $P^{\mathrm{T}}(\widetilde{\boldsymbol{p}})$ is the Maxwell distribution of linear momenta,

$$P^{\mathrm{T}}(\widetilde{\boldsymbol{p}}) = \prod_{i=1}^{N} P^{\mathrm{A}}(\boldsymbol{p}_i) = \frac{1}{\lambda^{3N}} \prod_{i=1}^{N} \exp\Big[-\frac{\boldsymbol{p}_i \cdot \boldsymbol{p}_i}{2mk_B T}\Big]. \tag{4.15}$$

In the case of no interaction between particles, $U_N = 0$, (ideal gas limit), the partition can be evaluated explicitly

$$\mathcal{Q}^{\mathrm{A,id}}(N,V,T) = \frac{1}{N!}\Big[\frac{1}{\lambda^3}\int \mathrm{d}\boldsymbol{p} \exp\Big[-\frac{\boldsymbol{p}\cdot\boldsymbol{p}}{2mk_B T}\Big]\Big]^N \Big[\int_V \mathrm{d}\boldsymbol{r}\Big]^N = \frac{V^N}{N!\,\lambda^{3N}}. \tag{4.16}$$

The usage of classical mechanics in our treatment is valid as long as $\lambda \ll \bar{r}$, the typical interparticle separation distance. It is easy to see that this is well satisfied not only for the mesogenic molecules we have seen in Chapter 1, but also for most molecular fluids, except perhaps hydrogen and helium, not of central interest here [Hansen and McDonald, 2006]. The free energy of the ideal gas becomes, using the Stirling approximation for large N, i.e. $\ln N \approx N(\ln N - 1)$,

$$\mathcal{A}^{\mathrm{A,id}}(N,V,T) = -k_B T \ln \mathcal{Q}^{\mathrm{A,id}}(N,V,T) = k_B T N[\ln(\rho\lambda^3) - 1]. \tag{4.17}$$

Having written the partition function as a product of a configurational and a kinetic term, we obtain in general the Helmoltz free energy as a sum of an ideal term for non-interacting molecules and an excess one containing the particle-particle interactions:

$$\mathcal{A}^{\mathrm{A}}(N,V,T) = \mathcal{A}^{\mathrm{A,id}}(N,V,T) + \mathcal{A}^{\mathrm{A,ex}}(N,V,T), \tag{4.18a}$$

$$= k_B T N[\ln(\rho\lambda^3) - 1] - k_B T \ln\Big[Z^{\mathrm{A}}(N,V,T)/V^N\Big]. \tag{4.18b}$$

The configurational integral for a single-component system with pairwise radial interactions can be rewritten, using the so called Mayer expansion:

$$e^{-U_N^A(\tilde{r})/(k_BT)} = \prod_{i=1}^{N-1}\prod_{j=i+1}^{N} e^{-U_{ij}^A/(k_BT)} = 1 + \sum_{i>j}\Phi_{ij} + \sum \Phi_{ij}\Phi_{i'j'} + \cdots, \quad (4.19)$$

where $\Phi_{ij} = \exp[-U_{ij}^A/(k_BT) - 1]$. This gives the *virial series* [Santos, 2016], in powers of the number density $\rho = N/V$

$$\frac{\mathcal{A}^A(N,V,T)}{Nk_BT} = [\ln\rho - 1] + B_2\rho + \frac{1}{2}B_3\rho^2 + \cdots. \quad (4.20)$$

The *virial coefficients* B_2 and B_3 are proportional to β_1 and β_2, the first two irreducible *cluster integrals*, β_1, β_2,

$$\beta_1 = -2B_2 = \beta_1(j,j') = \frac{1}{V}\int \mathrm{d}\boldsymbol{r}_j \mathrm{d}\boldsymbol{r}_{j'}\Phi_{jj'}, \quad (4.21a)$$

$$\beta_2 = -\frac{3}{2}B_3 = \beta_2(j,j',j'') = \frac{1}{V}\int \mathrm{d}\boldsymbol{r}_j \mathrm{d}\boldsymbol{r}_{j'}\mathrm{d}\boldsymbol{r}_{j''}\Phi_{jj'}\Phi_{j'j''}\Phi_{j''j}. \quad (4.21b)$$

4.2.2 Rigid Anisotropic Particles

Turning now to a system of N identical rigid anisotropic particles or 'bodies' (B), which could be molecules or nanoparticles or viruses (particles) still in canonical conditions, the situation is a little more complicated. The classical Hamiltonian $\mathcal{H}_N(\tilde{\boldsymbol{X}}, \dot{\tilde{\boldsymbol{X}}})$ can be written as

$$\mathcal{H}_N^B(\tilde{\boldsymbol{r}}, \tilde{\Omega}, \tilde{\boldsymbol{p}}, \tilde{\boldsymbol{J}}) = U_N^B(\tilde{\boldsymbol{r}}, \tilde{\Omega}) + K_N^B(\tilde{\boldsymbol{p}}, \tilde{\boldsymbol{J}}). \quad (4.22)$$

The kinetic energy is in this case the sum over the N particles of the translational and rotational contributions, i.e.

$$K_N^B(\tilde{\boldsymbol{p}}, \tilde{\boldsymbol{J}}) = \sum_{i=1}^{N}\left[\frac{\boldsymbol{p}_i\cdot\boldsymbol{p}_i}{2m} + \frac{1}{2}\dot{\boldsymbol{\Omega}}_i\cdot\mathbf{I}_i\cdot\dot{\boldsymbol{\Omega}}_i\right] = \sum_{i=1}^{N}\sum_{\substack{\alpha=\\x,y,z}}\left[\frac{p_{i,\alpha}^2}{2m} + \frac{J_{i,\alpha}^2}{2I_{i,\alpha}}\right], \quad (4.23)$$

where m is the mass and $\mathbf{I}$ is the inertia tensor of the particles in their molecule fixed frame, with principal components I_x, I_y, I_z. Note that, even if all the molecules have the same inertia tensor, we need the particle subscript i in Eq. 4.23 since we work in a laboratory system common to all particles, where each of them has orientation $\Omega_{i\mathrm{L}}$ and will have a different inertia tensor, $\mathbf{I}_i^{\mathrm{LAB}} \equiv \mathbf{I}_i^{\mathrm{LAB}}(\Omega_{i\mathrm{L}})$ connected to $\mathbf{I} = \mathbf{I}_i^{\mathrm{MOL}}$ as $\mathbf{I}_i^{\mathrm{LAB}} = \mathbf{R}(\Omega_{i\mathrm{L}})\,\mathbf{I}\,\mathbf{R}^T(\Omega_{i\mathrm{L}})$. In the case of a rigid molecule formed by a set of n_b spherical particles of mass $m_{k,i}$ placed at position $\boldsymbol{h}_{k,i} = \boldsymbol{r}_{k,i} - \boldsymbol{r}_i$ with respect to the *centre of mass* of the molecule $\boldsymbol{r}_i$ at $\boldsymbol{r}_i = \frac{1}{m_i}\sum_{k=1}^{n_b} m_{k,i}\boldsymbol{r}_{k,i}$, where $m_i = \sum_{k=1}^{n_b} m_{k,i}$ is the ith particle mass, the Cartesian components of the inertia tensor of particle i can be written as

$$[I_i]_{ab} = \sum_{k=1}^{n_b} m_{k,i}\left[\delta_{a,b}h_{k,i}^2 - (h_{k,i})_a(h_{k,i})_b\right], \quad a,b = x,y,z, \quad (4.24)$$

with $h_{k,i}^2 = \sum_a (h_{k,i})_a^2$. If we have instead a rigid particle with a certain continuous distribution of mass, $\rho_m(\boldsymbol{h})$, the sums over the tensor of inertia components become integrals:

$$I_{ab} = \int_{\mathcal{V}_P} \mathrm{d}\boldsymbol{h}\, \rho_m(\boldsymbol{h})\, (\delta_{a,b} h^2 - h_a h_b), \tag{4.25}$$

where the integral is over the particle shape domain $\mathcal{V}_P$. In matrix form

$$\mathbf{I} = \int_{\mathcal{V}_P} \mathrm{d}h_x \mathrm{d}h_y \mathrm{d}h_z\, \rho(h_x, h_y, h_z) \begin{pmatrix} h_y^2 + h_z^2 & -h_x h_y & -h_x h_z \\ -h_x h_y & h_z^2 + h_x^2 & -h_y h_z \\ -h_x h_z & -h_y h_z & h_x^2 + h_y^2 \end{pmatrix}. \tag{4.26}$$

For a uniform system $\rho(h_x, h_y, h_z)$ reduces to a constant and can be taken outside the integral. It is convenient to choose the particle fixed coordinate system as that diagonalizing $\mathbf{I}$. For symmetric bodies this system might be trivial to locate by inspection. As a few relevant examples, we have, for a biaxial ellipsoid of mass m with semiaxes a, b, c, $I_a = \frac{1}{5}m\left(b^2 + c^2\right)$, $I_b = \frac{1}{5}m\left(a^2 + c^2\right)$ and $I_c = \frac{1}{5}m\left(a^2 + b^2\right)$ in the particle fixed coordinate frame defined by the ellipsoid symmetry axes. For a uniaxial ellipsoid, two of the eigenvalues are the same: $I_{\parallel}$, $I_{\perp}$ and the particle is elongated (*prolate*) if $I_{\parallel} < I_{\perp}$ and squashed or discotic (*oblate*) if $I_{\parallel} > I_{\perp}$. For the limiting case of a sphere of radius $r = a = b = c$, that separates the two previous cases, $I_a = I_b = I_c = I = (2/5)mr^2$. Similarly, for a solid cylinder of radius r, height h and mass m, $I_{\parallel} = \frac{1}{2}mr^2$, $I_{\perp} = I_x = I_y = \frac{1}{12}m\left(3r^2 + h^2\right)$. As a last example, for a solid orthogonal block of height h, width w, depth d, and mass m: $I_h = \frac{1}{12}m\left(w^2 + d^2\right)$, $I_w = \frac{1}{12}m\left(d^2 + h^2\right)$ and $I_d = \frac{1}{12}m\left(w^2 + h^2\right)$. For a more realistic molecular model (see, e.g., Fig. 1.10) symmetry would typically be lacking. However, a simple numerical diagonalization [Press et al., 1992]: $\mathbf{I} = \mathbf{U}\Lambda_{\mathbf{I}}\mathbf{U}^T$, with $\Lambda_{\mathbf{I}} = \mathrm{diag}(I_a, I_b, I_c)$ determines the principal values, while the columns of the eigenvector matrix $\mathbf{U}$ define the three axes of the body fixed inertia frame. Going back to the overall distribution and the partition function, they can be factorized as [Gray and Gubbins, 1984]: $P^{\mathrm{B}}(\widetilde{\boldsymbol{r}}, \widetilde{\Omega}, \widetilde{\boldsymbol{p}}, \widetilde{\boldsymbol{J}}) = P_P^{\mathrm{T}}(\widetilde{\boldsymbol{p}}) P_J^{\mathrm{B}}(\widetilde{\boldsymbol{J}}) P^{\mathrm{B}}(\widetilde{\boldsymbol{r}}, \widetilde{\Omega})$ and

$$\mathcal{Q}^{\mathrm{B}}(N, V, T) = Q_P^{\mathrm{T}} Q_J^{\mathrm{B}} Q^{\mathrm{B}}(N, V, T), \tag{4.27}$$

where we have used the labels (N, V, T) for configurational and P, J for linear and angular momentum. We can now consider each of the three terms in turn. The translational contribution (and thus Q_P^{T} and $P_P^{\mathrm{T}}(\widetilde{\boldsymbol{p}})$), is the same as before. The angular momentum partition function is

$$Q_J^{\mathrm{B}} = \frac{V_\Omega^N}{h^{(f-3)N}} \int \mathrm{d}\widetilde{\boldsymbol{J}}\ \mathrm{e}^{-\sum_{i,\alpha} J_{i\alpha}^2/(2k_B T I_\alpha)} = \Lambda_J^{-N}. \tag{4.28}$$

with f the number of degrees of freedom per particle (5 for linear, 6 for non linear ones). For non-linear molecules, V_Ω, the angular volume, is $8\pi^2$ and

$$\Lambda_J = \left(\frac{h^2}{8\pi^2 k_B T I_x}\right)^{\frac{1}{2}} \left(\frac{h^2}{8\pi^2 k_B T I_y}\right)^{\frac{1}{2}} \left(\frac{h^2}{8\pi^2 k_B T I_z}\right)^{\frac{1}{2}} \text{(non-linear)}, \tag{4.29}$$

while, for linear molecules, $V_\Omega = 4\pi$ and $\Lambda_J = \frac{h^2}{8\pi^2 I k_B T}$. The positional-orientational configurational probability density $P(\widetilde{\boldsymbol{r}}, \widetilde{\Omega})$ studied earlier in Chapter 3 is

$$P^{\mathrm{B}}(\widetilde{\boldsymbol{r}}, \widetilde{\Omega}) = \exp[-U_N^{\mathrm{B}}(\widetilde{\boldsymbol{r}}, \widetilde{\Omega})/(k_B T)]/[N!\, Q^{\mathrm{B}}(N, V, T)], \tag{4.30}$$

where

$$Q^{\mathrm{B}}(N, V, T) = \frac{1}{N!} \int \mathrm{d}\widetilde{\boldsymbol{r}} \mathrm{d}\widetilde{\Omega}\ \mathrm{e}^{-U_N^{\mathrm{B}}(\widetilde{\boldsymbol{r}}, \widetilde{\Omega})/k_B T} = \frac{Z(N, V, T)}{N!} \tag{4.31}$$

is the purely configurational partition. Having written the full partition function as a product of configurational and kinetic terms, we obtain the free energy as a sum of an ideal gas term for non-interacting molecules and a configurational or excess one containing the interactions:

$$\mathcal{A}^{\mathrm{B}} = \mathcal{A}^{\mathrm{id,B}} + \mathcal{A}^{\mathrm{ex,B}}, \tag{4.32}$$

where the kinetic (translational and rotational) ideal term,

$$\mathcal{A}^{\mathrm{id,B}} = -k_B T \ln\left[Q_P^{\mathrm{T}} Q_J^{\mathrm{B}}\right] = -k_B T \ln\left[\frac{V^N V_\Omega^N}{N!\, \lambda^{3N} \Lambda_J^N}\right] = N k_B T \ln\left(\frac{\rho \lambda^3 \Lambda_J}{\mathrm{e} V_\Omega}\right), \tag{4.33}$$

has the analytic form obtained from Eqs. 4.13 and 4.28 and the configurational free energy, normally the really relevant one, is

$$\mathcal{A}^{\mathrm{ex,B}} = -N k_B T \ln[Q(N, V, T)]. \tag{4.34}$$

4.3 Isobaric-Isothermal Ensemble

Let us now go back to the simpler case of a system formed of atoms or spherical particles (now omitting the superscript A since no confusion can arise). If necessary, the rigid body expressions just derived can be generalized following the same procedure. In this case pressure, rather than volume, as well as T and N, is a fixed parameter. These conditions correspond to the vast majority of ordinary experiments and standard tabulated values for physical properties are often reported at N, P, T fixed, e.g. at $P = 1$ atm and room temperature (25°C). The relevant free energy (thermodynamic potential) is the Gibbs free energy, $\mathcal{G} = \mathcal{E} + PV - T\mathcal{S}$. The distribution is

$$\varrho(\widetilde{\boldsymbol{r}}, \widetilde{\boldsymbol{p}}) = \frac{\exp[-(\mathcal{H}_N + PV)/(k_B T)]}{\int_0^\infty \mathrm{d}V \int \mathrm{d}\widetilde{\boldsymbol{r}} \mathrm{d}\widetilde{\boldsymbol{p}}\ \exp[-(\mathcal{H}_N + PV)/(k_B T)]} \tag{4.35}$$

and the isothermal-isobaric partition function

$$\Delta(N, P, T) = \frac{1}{V_0 \nu_N} \int_0^\infty \mathrm{d}V \int \mathrm{d}\widetilde{\boldsymbol{r}} \mathrm{d}\widetilde{\boldsymbol{p}}\ \exp[-(\mathcal{H}_N + PV)/(k_B T)], \tag{4.36a}$$

$$= \frac{1}{V_0} \int_0^\infty \mathrm{d}V \exp[-PV/(k_B T)]\, Q(N, P, T), \tag{4.36b}$$

where V_0 is a reference volume making the expression dimensionless.

4.4 Grand Canonical Ensemble

In this case, the number of particles can change and the input constraints when maximizing the entropy functional entail the number of particles, the energy, as well as the normalization:

$$\sum_{N=0}^{\infty}\int \mathrm{d}(\widetilde{\boldsymbol{r}},\widetilde{\boldsymbol{p}}:N)\,\varrho(\widetilde{\boldsymbol{r}},\widetilde{\boldsymbol{p}}:N)=1, \tag{4.37a}$$

$$\sum_{N=0}^{\infty}\int \mathrm{d}(\widetilde{\boldsymbol{r}},\widetilde{\boldsymbol{p}}:N)\,\varrho(\widetilde{\boldsymbol{r}},\widetilde{\boldsymbol{p}}:N)\mathscr{H}_N(\widetilde{\boldsymbol{r}},\widetilde{\boldsymbol{p}}:N)=\mathcal{E}, \tag{4.37b}$$

$$\sum_{N=0}^{\infty}N\int \mathrm{d}(\widetilde{\boldsymbol{r}},\widetilde{\boldsymbol{p}}:N)\,\varrho(\widetilde{\boldsymbol{r}},\widetilde{\boldsymbol{p}}:N)=\langle N\rangle, \tag{4.37c}$$

where we have introduced the notations $(\widetilde{\boldsymbol{r}},\widetilde{\boldsymbol{p}}:N)$ and $\mathrm{d}(\widetilde{\boldsymbol{r}},\widetilde{\boldsymbol{p}}:N)$ for the set of coordinates and momenta $(\widetilde{\boldsymbol{r}},\widetilde{\boldsymbol{p}})$ appropriate for a certain number N of particles and for the corresponding volume element. The entropy maximization gives

$$\varrho(\widetilde{\boldsymbol{r}},\widetilde{\boldsymbol{p}}:N)=\frac{\exp[-(\mathscr{H}_N-N\mu)/(k_BT)]}{\nu_N\,\Xi}, \tag{4.38}$$

where μ is the chemical potential and

$$\Xi=\sum_{N=0}^{\infty}\frac{\mathrm{e}^{\mu N/k_BT}}{\nu_N}\int \mathrm{d}\widetilde{\boldsymbol{r}}\mathrm{d}\widetilde{\boldsymbol{p}}\,\exp[-\mathscr{H}_N/(k_BT)], \tag{4.39}$$

is the *grand partition function*. The grand canonical average of a generic property A will be

$$\langle A\rangle=\sum_{N=0}^{\infty}\int \mathrm{d}(\widetilde{\boldsymbol{r}},\widetilde{\boldsymbol{p}}:N)\,\varrho(\widetilde{\boldsymbol{r}},\widetilde{\boldsymbol{p}}:N)\,A(\widetilde{\boldsymbol{r}},\widetilde{\boldsymbol{p}}:N) \tag{4.40}$$

and the *grand potential* [Santos, 2016]: $\Omega(T,V,\mu)=-PV=-k_BT\ln\Xi_{TV\mu}$.

4.5 Microcanonical Conditions

The microcanonical conditions correspond to a system of particles in isolation obeying Hamilton equations of motion, thus with constant total energy, momentum and angular momentum. The partition function is in this case [Tuckerman, 2010]

$$\Omega(N,V,E)=\frac{E_0}{N!\,h^{3N}}\int \mathrm{d}\widetilde{\boldsymbol{p}}\int \mathrm{d}\widetilde{\boldsymbol{r}}\,\delta(\mathscr{H}_N(\widetilde{\boldsymbol{r}},\widetilde{\boldsymbol{p}})-\mathcal{E}), \tag{4.41}$$

where E_0 makes the RHS dimensionless and $N!$ comes from the indistinguishability of the particles. An average observable will be

$$\langle A\rangle=\frac{\int \mathrm{d}\widetilde{\boldsymbol{r}}\mathrm{d}\widetilde{\boldsymbol{p}}A(\widetilde{\boldsymbol{r}},\widetilde{\boldsymbol{p}})\,\delta(\mathscr{H}_N(\widetilde{\boldsymbol{r}},\widetilde{\boldsymbol{p}})-\mathcal{E})}{\int \mathrm{d}\widetilde{\boldsymbol{r}}\mathrm{d}\widetilde{\boldsymbol{p}}\,\delta(\mathscr{H}_N(\widetilde{\boldsymbol{r}},\widetilde{\boldsymbol{p}})-\mathcal{E})}. \tag{4.42}$$

Microcanonical constant energy conditions are not easy to implement for real experiments but represent rather natural conditions for following the evolution of an isolated system of particles by solving their Newton equations of motion. This is the basis of the molecular dynamics simulation method discussed in Chapter 9 .

4.6 Structural Properties

We note that, as long as we are interested in static structural properties, $A(\widetilde{\boldsymbol{X}})$, depending only on positions (and orientations), the calculation of their average only requires the configurational distribution as the kinetic part at the numerator and denominator will factor out. Considering canonical conditions, here and in the following sections, unless stated otherwise, the probability of finding particle 1 at $\boldsymbol{X}_1$, particle 2 at $\boldsymbol{X}_2$ and so on up to particle N at $\boldsymbol{X}_N$ is the N particle distribution [Balescu, 1975; Landau and Lifshitz, 1980]:

$$P(\widetilde{\boldsymbol{X}}) = \exp\left[-U_N(\widetilde{\boldsymbol{X}})/(k_B T)\right] \,/\, Z(N,V,T), \tag{4.43}$$

where $Z(N,V,T) = \int \mathrm{d}\widetilde{\boldsymbol{X}}\ \exp\left[-U_N(\widetilde{\boldsymbol{X}})/(k_B T)\right]$, with $\mathrm{d}\widetilde{\boldsymbol{X}} \equiv \mathrm{d}\boldsymbol{X}_1,\ldots,\mathrm{d}\boldsymbol{X}_N$ is the configurational integral. The average of a collective property depending on the coordinates of all the N particles, $A(\widetilde{\boldsymbol{X}})$, will then be

$$\langle A\rangle = \int \mathrm{d}\widetilde{\boldsymbol{X}}\ A(\widetilde{\boldsymbol{X}}) \exp\left[-U_N(\widetilde{\boldsymbol{X}})/(k_B T)\right] / \ Z(N,V,T). \tag{4.44}$$

For instance, the average potential energy is

$$\mathcal{U} = \left\langle U_N(\widetilde{\boldsymbol{X}})\right\rangle = \int \mathrm{d}\widetilde{\boldsymbol{X}}\ \ U_N(\widetilde{\boldsymbol{X}}) \exp\left[-U_N(\widetilde{\boldsymbol{X}})/(k_B T)\right] / \ Z(N,V,T). \tag{4.45}$$

In particular, for rigid, non-spherical particles, the N particle distribution can be written in terms of a full set of molecular positions and orientations as

$$P(\widetilde{\boldsymbol{r}},\widetilde{\Omega}) = \exp\left[-U_N(\widetilde{\boldsymbol{r}},\widetilde{\Omega})/(k_B T)\right] / \ Z(N,V,T), \tag{4.46}$$

where the normalization factor $Z_N = Z(N,V,T) = \int \mathrm{d}\widetilde{\boldsymbol{r}}\,\mathrm{d}\widetilde{\Omega} \exp\left[-U_N(\widetilde{\boldsymbol{r}},\widetilde{\Omega})/(k_B T)\right]$ ensures that the integral of the probability density in Eq. 4.46 over all space is 1. If we start from this N particle distribution, the probability of finding particle 1 at $(\boldsymbol{r}_1,\Omega_1)$ and the others anywhere will be obtained from Eq. 4.46 by integrating over all coordinates except those of particle 1 (see, e.g., [Feynman et al., 1963]). The probability density of finding any one particle at $(\boldsymbol{r}_1,\Omega_1)$ and the others anywhere is N times this, i.e.

$$P^{(1)}(\boldsymbol{r}_1,\Omega_1) = \frac{N}{Z(N,V,T)} \int \mathrm{d}\boldsymbol{r}_2\mathrm{d}\Omega_2 \ldots \mathrm{d}\boldsymbol{r}_N\mathrm{d}\Omega_N\ \mathrm{e}^{-U_N(\widetilde{\boldsymbol{r}},\widetilde{\Omega})/(k_B T)}. \tag{4.47}$$

We have already discussed the single particle distribution $P^{(1)}(\boldsymbol{r},\Omega)$, or rather $P(\boldsymbol{r},\Omega) = P^{(1)}(\boldsymbol{r},\Omega)/N$ and its expansion in Chapter 3, but the importance of Eq. 4.47 is that it offers a way of constructing the distribution bottom up, from molecular information. Here we wish anyway to proceed by defining the 'generic' probability of finding any two particles at $(\boldsymbol{r}_1,\Omega_1,\boldsymbol{r}_2,\Omega_2)$ as the pair distribution

$$P^{(2)}(\boldsymbol{r}_1,\Omega_1,\boldsymbol{r}_2,\Omega_2) = \frac{N(N-1)}{Z(N,V,T)} \int \mathrm{d}\boldsymbol{r}_3\mathrm{d}\Omega_3 \ldots \mathrm{d}\boldsymbol{r}_N\mathrm{d}\Omega_N\ \mathrm{e}^{-U_N(\widetilde{\boldsymbol{r}},\widetilde{\Omega})/(k_B T)}. \tag{4.48}$$

The procedure can be extended [Zannoni, 1979c; Hansen and McDonald, 2006], to introduce the probability $P^{(n)}$ of finding any subset of n particles in an infinitesimal volume element at $(\boldsymbol{r}_1, \Omega_1, \boldsymbol{r}_2, \Omega_2, \ldots, \boldsymbol{r}_n, \Omega_n)$. The general n-particle distribution $P^{(n)}$ represents all the information needed to compute any equilibrium observable of the system. Normally we need to concern ourselves only with single-particle and two-particle properties. For these properties the information given by the single particle (see Chapter 3) and, respectively, the pair distribution is complete. In fact, they constitute all we need to know to be able to calculate the average of any single-particle property dependent on position and orientation, $A(\boldsymbol{r}, \Omega)$, as we saw in Chapter 3 and of any pair property $A(\boldsymbol{r}_1, \Omega_1, \boldsymbol{r}_2, \Omega_2)$. We can write this two-particle average as

$$\langle A(\boldsymbol{X}_1, \boldsymbol{X}_2)\rangle = \frac{\int \mathrm{d}\boldsymbol{X}_1 \mathrm{d}\boldsymbol{X}_2 P^{(2)}(\boldsymbol{X}_1, \boldsymbol{X}_2) A(\boldsymbol{X}_1, \boldsymbol{X}_2)}{\int \mathrm{d}\boldsymbol{X}_1 \mathrm{d}\boldsymbol{X}_2 P^{(2)}(\boldsymbol{X}_1, \boldsymbol{X}_2)}. \tag{4.49}$$

The pair distribution just defined is normalized as

$$\int \mathrm{d}\boldsymbol{X}_1 \mathrm{d}\boldsymbol{X}_2 P^{(2)}(\boldsymbol{X}_1, \boldsymbol{X}_2) = N(N-1). \tag{4.50}$$

The generic n-particle distribution

$$P^{(n)}(\boldsymbol{X}_1, \boldsymbol{X}_2, \ldots, \boldsymbol{X}_n) = \frac{N!}{(N-n)!\, Z(N,V,T)} \int \prod_{i=n+1}^{N} \mathrm{d}\boldsymbol{X}_i \, \exp[-U_N(\widetilde{\boldsymbol{X}})/(k_B T)] \tag{4.51}$$

is normalized to the number of n-tuples:

$$\int \prod_{i=1}^{n} \mathrm{d}\boldsymbol{X}_i \, P^{(n)}(\boldsymbol{X}_1, \boldsymbol{X}_2, \ldots, \boldsymbol{X}_n) = \frac{N!}{(N-n)!}. \tag{4.52}$$

As we have seen in Chapter 3, a useful way of writing the one-particle distribution is through the introduction of Dirac delta functions (see Appendix D). In a similar way we can define the pair distribution in terms of positional and orientational delta functions. In particular, for rigid particles, where $\boldsymbol{X}_i = (\boldsymbol{r}_i, \Omega_i)$

$$P^{(2)}(\boldsymbol{r}_1 \Omega_1, \boldsymbol{r}_2, \Omega_2) = N(N-1)\langle \delta(\boldsymbol{r}_1 - \boldsymbol{r}_1')\, \delta(\Omega_1 - \Omega_1')\, \delta(\boldsymbol{r}_2 - \boldsymbol{r}_2')\, \delta(\Omega_2 - \Omega_2')\rangle, \tag{4.53a}$$

$$\equiv N(N-1) P(\boldsymbol{r}_1 \Omega_1, \boldsymbol{r}_2, \Omega_2), \tag{4.53b}$$

with the normalization $\int \mathrm{d}\boldsymbol{r}_1 \mathrm{d}\Omega_1 \mathrm{d}\boldsymbol{r}_2 \mathrm{d}\Omega_2 \, P(\boldsymbol{r}_1, \Omega_1, \boldsymbol{r}_2, \Omega_2) = 1$. Even though an explicit functional form of the pair distribution is of course not available, a few general results can be provided. As the separation between the particles becomes very large the joint probability of finding molecule 1 at $\boldsymbol{X}_1$ and molecule 2 at $\boldsymbol{X}_2$ will just be the product of these two independent events and the pair distribution will tend to the product of two single particle ones.

$$\lim_{r\to\infty} P(\boldsymbol{r}_1\Omega_1,\boldsymbol{r}_2,\Omega_2) = \lim_{r\to\infty} \left\langle \delta(\boldsymbol{r}_1-\boldsymbol{r}_1')\,\delta(\Omega_1-\Omega_1')\,\delta(\boldsymbol{r}_2-\boldsymbol{r}_2')\,\delta(\Omega_2-\Omega_2')\right\rangle, \quad (4.54a)$$

$$= \left\langle \delta(\boldsymbol{r}_1-\boldsymbol{r}_1')\,\delta(\Omega_1-\Omega_1')\right\rangle \left\langle \delta(\boldsymbol{r}_2-\boldsymbol{r}_2')\,\delta(\Omega_2-\Omega_2')\right\rangle, \quad (4.54b)$$

$$= P(\boldsymbol{r}_1,\,\Omega_1)\,P(\boldsymbol{r}_2,\,\Omega_2). \quad (4.54c)$$

We can introduce, for uniform systems like nematics, the 'reduced' pair distributions $G(\boldsymbol{r},\Omega_1,\Omega_2)$ and $g(\boldsymbol{r},\Omega_1,\Omega_2)$ [Zannoni, 1979c]:

$$P^{(2)}(\boldsymbol{r}_1,\Omega_1,\boldsymbol{r}_2,\Omega_2) = N^2 P(\boldsymbol{r}_1,\Omega_1,\boldsymbol{r}_2,\Omega_2) = \rho^2 G(\boldsymbol{r}_1,\Omega_1,\boldsymbol{r}_2,\Omega_2), \quad (4.55a)$$

$$= \rho^2 V G(r,\Omega_1,\Omega_2,\Omega_r), \quad (4.55b)$$

$$= \rho^2 V\, P(\Omega_1)\, P(\Omega_2)\, g(r,\Omega_1,\Omega_2,\Omega_r), \quad (4.55c)$$

where $\boldsymbol{r} \equiv \boldsymbol{r}_2 - \boldsymbol{r}_1 = r\hat{\boldsymbol{r}}$, $\mathrm{d}\boldsymbol{r} = r^2\mathrm{d}r\mathrm{d}\Omega_r$. The *pair correlation* $g(r,\Omega_1,\Omega_2,\Omega_r)$ has the long-range behaviour $\lim_{r\to\infty} g(\boldsymbol{r},\Omega_1,\Omega_2) = 1$, while for the other reduced pair distribution function or spatial-orientational correlation function we have

$$\lim_{r\to\infty} G(r,\Omega_1,\Omega_2,\Omega_r) = P(\Omega_1)P(\Omega_2), \quad (4.56)$$

and the normalization $\int \mathrm{d}\boldsymbol{r}\mathrm{d}\Omega_1\mathrm{d}\Omega_2 G(r,\Omega_1,\Omega_2,\Omega_r) = V$, with $\mathrm{d}\boldsymbol{r} = \mathrm{d}r r^2$. Thus, at large separations, the only orientational correlation between particles will be that indirectly coming from the fact that both molecule 1 and 2 are separately parallel to the same director, if that exists. In particular, no long-range orientational correlation between particles exists in a normal isotropic fluid. The correlation $G(r,\Omega_1,\Omega_2,\Omega_r) = P(\Omega_1)\,P(\Omega_2)\,g(r,\Omega_1,\Omega_2,\Omega_r)$, although very informative, is difficult to visualize and in LCs we may also choose, as will be discussed later in Section 4.9.1 to introduce a further reduced *radial-angular distribution* representing the distribution in space of the vector connecting molecular positions:

$$g(r,\Omega_r) = \int \mathrm{d}\boldsymbol{r}_1\,\mathrm{d}\Omega_1\,\mathrm{d}\Omega_2\, G(\boldsymbol{r}_1,\Omega_1,\boldsymbol{r},\Omega_r,\Omega_2)\Big/\int \mathrm{d}\boldsymbol{r}_1\mathrm{d}\Omega_1\mathrm{d}\Omega_2, \quad (4.57a)$$

$$= \langle g(\boldsymbol{r}_1,\Omega_1,r,\Omega_r,\Omega_2)\rangle_{\boldsymbol{r}_1,\Omega_1,\Omega_2}, \quad (4.57b)$$

which tends to 1 for large separations and, for uniaxial molecules and uniform uniaxial phases, reduces to $g(r,\beta_r)$. Going back to the full distribution, $g(\boldsymbol{r},\Omega_1,\Omega_2)$, the limiting value is often subtracted to define the *total correlation function* $h(\boldsymbol{r},\Omega_1,\Omega_2)$ which tends to 0 at large separations

$$h(\boldsymbol{r},\Omega_1,\Omega_2) = g(\boldsymbol{r},\Omega_1,\Omega_2) - 1, \quad (4.58)$$

or more generally, $h(X_1,X_2) = g(X_1,X_2) - 1$. Similarly, we can define for anisotropic fluids

$$H(\boldsymbol{r},\Omega_1,\Omega_2) = G(\boldsymbol{r},\Omega_1,\Omega_2) - P(\Omega_1)P(\Omega_2), \quad (4.59)$$

removing again the appropriate asymptotic limit. The pair correlation function can be inverted *ex-post* in terms of an effective *potential of mean force and torque* $W(\boldsymbol{r},\Omega_1,\Omega_2)$:

$$W(\boldsymbol{r},\Omega_1,\Omega_2) = -k_B T \ln g(\boldsymbol{r},\Omega_1,\Omega_2), \quad (4.60)$$

which is of course changing with temperature or other state variables. For very dilute fluid systems, when the density is sufficiently low, configurations with three or more particles interacting simultaneously are extremely rare and, as long as they can be neglected, we can expect

$$g(\boldsymbol{r}, \Omega_1, \Omega_2) \approx \exp[-U(\boldsymbol{r}_1, \Omega_1, \boldsymbol{r}_2, \Omega_2)/(k_B T)] \tag{4.61}$$

and

$$G(\boldsymbol{r}, \Omega_1, \Omega_2) \approx P(\Omega_1)\, P(\Omega_2)\, \exp[-U(\boldsymbol{r}_1, \Omega_1, \boldsymbol{r}_2, \Omega_2)/(k_B T)]. \tag{4.62}$$

We can imagine this low-density approximation to be more applicable to dilute suspensions of anisotropic colloidal particles (see Section 1.14) rather than to dense condensed thermotropic systems.

Another limiting situation can be obtained for very short pair separations. If the molecules have a hard impenetrable core, there is vanishing probability of finding a second particle nearer than a contact distance $\sigma(\boldsymbol{r}, \Omega_1, \Omega_2)$ from the first one. In this case,

$$g(\boldsymbol{r}, \Omega_1, \Omega_2) = 0, \quad \text{when} \quad \sigma(\boldsymbol{r}, \Omega_1, \Omega_2) \to \infty. \tag{4.63}$$

The condition is expressed much more simply for spherical particles, even if they possibly have an anisotropic interaction (e.g. an embedded dipole or quadrupolar term). In this case,

$$g(\boldsymbol{r}, \Omega_1, \Omega_2) = g(r, \Omega_1, \Omega_2, \Omega_r) = 0, \quad \text{if } r < \sigma. \tag{4.64}$$

If the constituent particles are even simpler, say not only of spherical shape, but also endowed with isotropic interactions, the pair correlation $g(r, \Omega_1, \Omega_2, \Omega_r)$ will only depend on the interparticle distance r, so that

$$g(r, \Omega_1, \Omega_2, \Omega_r) = g(r), \tag{4.65}$$

where $g(r)$ is the important radial distribution that we shall now examine more closely. From a practical point of view the *radial distribution* function, $g(r)$, [Hansen and McDonald, 2006] is just the average number density of molecules at a distance r from one chosen as the origin (Fig. 4.1a), divided by the bulk density:

$$g(r) = \frac{\langle \rho(r) \rangle}{\rho} = \frac{1}{4\pi r^2 \rho} \langle \delta(r - r_{12}) \rangle_{r_{12}}. \tag{4.66}$$

Thus, $g(r)$ represents the probability, relative to the bulk one, of finding the centre of mass of a second particle at a distance r from one chosen as the origin. For the limiting case of no interactions at all (like in a perfect gas), we have $U(\widetilde{X}) = 0$ and $g(r) = 1$. Recalling the normalization of $P^{(2)}$ in Eq. 4.50, we obtain at once the normalization for $g(r)$ as

$$\int_0^{2\pi} \mathrm{d}\alpha \int_0^{\pi} \mathrm{d}\beta \sin\beta \int_V \mathrm{d}r\, r^2\, g(r) = \frac{V}{N} \int \mathrm{d}r\, \langle \delta(r - r_{12}) \rangle_{r_{12}} = V. \tag{4.67}$$

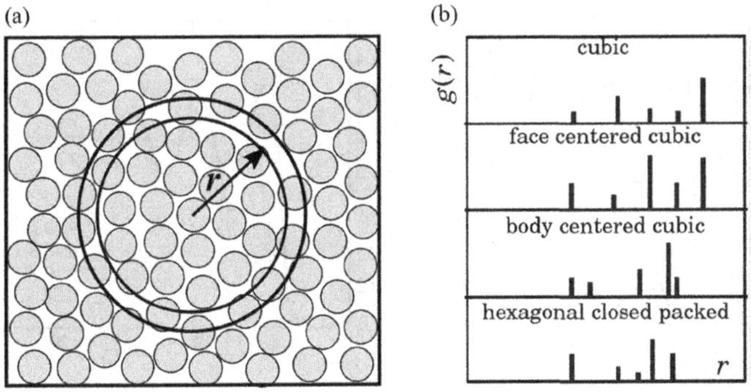

Figure 4.1 (a) Calculation of the radial distribution $g(r) = \langle \rho(r) \rangle / \langle \rho \rangle$, where $\rho(r) = n(r)/V(r)$ is the density of the shell at r. (b) $g(r)$ for some ideal 3D lattices.

In a crystalline solid made up of spherical particles, neighbours are located only at certain well-defined distances r_k typical of the lattice type and the radial distribution will consist of a sequence of spikes with intensity proportional to the number of neighbours z_k at each r_k:

$$g(r) = \frac{1}{4\pi r^2 \rho} \sum_k z_k \, \delta(r - r_k). \tag{4.68}$$

In Fig. 4.1b we show as an example the radial distribution for various simple lattices. In an ideal crystal the pattern of 'spikes' will be regular over distances as large as we wish. This is a manifestation of the positional long-range order in a crystal. The fact that only spikes appear depends on the certainty with which we can define the position of the particles building up our crystal. Even though in a real crystal we might have less than complete certainty due to thermal oscillations etc., the radial distribution will still consist of a sequence of very well-defined peaks.

4.6.1 Site-Site Radial Distribution Functions

We have seen that $g(r)$ is a complete representation of the pair distribution for spherical particles in isotropic liquids. For particles of arbitrary symmetry, it is possible and often convenient to have a radial distribution for the centres of mass. However, we can always define site-site radial distributions g^{AB} considering a point (or *site*) A on particle 1 and a point B on particle 2 and their distance vector $\boldsymbol{r}_{\mathrm{AB}}$. As an example in a simulation of water, it could be interesting to look at different oxygen-oxygen, oxygen-hydrogen and hydrogen-hydrogen site-site correlations $g^{\mathrm{OO}}(r)$, $g^{\mathrm{OH}}(r)$ and $g^{\mathrm{HH}}(r)$, as well as the one between centres of mass, $g(r)$. $g^{\mathrm{AB}}(r)$ can be obtained by counting the density of particles that have their site-site separation $r_{\alpha\beta}$ at distance r normalized by the average density ρ, i.e.

$$g_{\mathrm{AB}}(r) = \frac{1}{4\pi \rho r^2} \langle \delta(r - r_{\mathrm{AB}}) \rangle_{r_{\mathrm{AB}}} = \overline{g(r_{\mathrm{AB}}, \Omega_1, \Omega_2, \Omega_r)}(r), \tag{4.69}$$

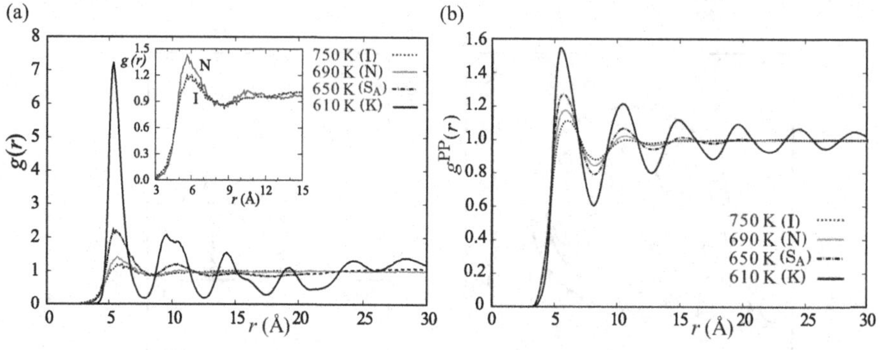

Figure 4.2 (a) Centre of mass radial distribution of p-quinquephenyl (P5). $g(r)$ at temperatures $T = 750$ K (isotropic), 690 K (nematic), 650 K (S_A) and 610 K (crystal) and inset of a close-up across the NI transition. (b) Radial distribution $g^{PP}(r)$ between the centre of a phenyl ring on a P5 molecule with that of a phenyl ring on another P5 from atomistic simulations of p-quinquephenyl [Olivier et al., 2014].

where we have used the overbar to indicate that $g_{AB}(r)$ can be formally obtained integrating $g(r_{AB}, \Omega_1, \Omega_2, \Omega_r)$ over orientations at separation $r_{AB} = r$, as indicated by the argument. Here, unless differently specified, we shall write simply $g(r)$ when the two sites correspond to the c.o.m. of the two particles, whatever their symmetry. In Figure 4.2a we show the radial distributions, $g(r)$, of p-quinquephenyl (P5) in the nematic phase, obtained using as reference centre the molecular centre of mass. Even though LCs are characterized by the long-range order described in Chapter 3, also some short-range order exists in liquid crystals, as shown in Fig. 4.2a for the $g(r)$ of ρ-quinquephenyl c.o.m. obtained from atomistic computer simulations [Olivier et al., 2014] (see Section 12.2). As expected, this short-range order, manifested by the relatively sharp first few peaks in $g(r)$, is more pronounced in the smectic than in the nematic and decreases as the temperature of the liquid is increased. In Fig. 4.2b we show an example of site-site radial distribution, where each site is the centre of one of the phenyl rings.

4.7 The Pair Distribution in Various Phases

We can now return to the pair distribution $G(\mathbf{r}, \Omega_1, \Omega_2) = G(r, \Omega_1, \Omega_2, \Omega_r)$ and examine its features for various types of phases and symmetry. Although this distribution allows us to calculate any pairwise static property, we should point out that it is very difficult to obtain and to visualize G in its full form. Indeed, even for rigid particles, G depends on up to $3+3+3+2 = 11$ variables (three positional, and the rest angular) implying that even if we have a large number M of configurations (snapshots) of the system, obtained from some computer simulation technique, it would be hardly possible to build a multidimensional histogram of G with sufficient resolution (in terms of the number of bins for every variable) and adequate population. In particular, we would need to fill each bin with a number of events n_H sufficiently high to make the errors, roughly $\approx 1/\sqrt{n_H}$, acceptable. The two options available are: (i) expand $G(r, \Omega_1, \Omega_2, \Omega_r)$ in a suitable orthogonal basis set and concentrate on calculating a small but significant number of expansion coefficients or (ii) calculate and represent histograms not of the full pair distribution, but of reduced functions obtained by integrating the full distribution over some of the variables. In Sections 4.8 and 4.9 we shall discuss both these approaches.

4.8 Invariant Expansion of the Pair Distribution

For a uniform fluid we can start from the distribution $G(r, \Omega_1, \Omega_2, \Omega_r)$, (Eq. 4.55c), where $\boldsymbol{r} = \boldsymbol{r}_2 - \boldsymbol{r}_1$ is the intermolecular vector of length r and orientation $\Omega_r = (\alpha_r, \beta_r)$. In the absence of an external field our system can be assumed to be macroscopically isotropic, in the sense that its properties have to be independent from an arbitrary rotation of the coordinate frame. This applies to an ordinary fluid, but also to a polydomain nematic obtained from cooling down from its isotropic phase without an external aligning field. In this case, nothing should change for an arbitrary rotation of the sample (or of the laboratory system). In computer simulations we deal necessarily with finite samples of a few hundred to a few hundred thousand particles, according to the different models adopted and thus at any instant of time t a sample director $\boldsymbol{d}(t)$ will always exist. If we recall that the turbidity of an unoriented liquid crystal indicates that the size of a typical oriented domain approaches micron size (see Fig. 1.4), thus containing hundreds of millions of molecules, it seems unavoidable that the simulated sample will correspond to a correlated domain and exhibit 'long-range' order. This sample director will, however, fluctuate in time in the absence of an external field so as to eventually re-establish the original spherical symmetry of the starting isotropic system. The pair distribution will thus depend on the Wigner rotation matrices $\mathscr{D}^{L_1}_{m_1,n_1}(\Omega_{1\mathrm{L}})$, $\mathscr{D}^{L_2}_{m_2,n_2}(\Omega_{2\mathrm{L}})$ and $\mathscr{D}^{L_3}_{m_3,0}(\Omega_{r\mathrm{L}})$, but only through their rotationally invariant combinations [Jepsen and Friedman, 1963; Steele, 1963; Blum and Torruella, 1972; Stone, 1978; Zannoni, 1979c]. In Appendix G we discuss how to build these scalar and orthogonal combinations $S^{k_1,k_2}_{L_1,L_2,L}(\Omega_{1\mathrm{L}}, \Omega_{2\mathrm{L}}, \Omega_{r\mathrm{L}})$ which, in the notation of Stone [1978] (cf. Appendix G, Eq. G.19) are

$$S^{k_1,k_2}_{L_1,L_2,L}(\Omega_{1\mathrm{L}}, \Omega_{2\mathrm{L}}, \Omega_{r\mathrm{L}}) = \frac{(i)^{L_1-L_2+L}}{\sqrt{2L+1}} \times \sum_{q_1,q_2} C(L_1, L_2, L; q_1, q_2)\, \mathscr{D}^{L_1 *}_{q_1,k_1}(\Omega_{1\mathrm{L}})\, \mathscr{D}^{L_2 *}_{q_2,k_2}(\Omega_{2\mathrm{L}})\, \mathscr{D}^{L}_{q_1+q_2,0}(\Omega_{r\mathrm{L}}). \tag{4.70}$$

In Appendix G we provide a tabulation of the first few of these Stone invariants in Cartesian coordinates. Note that with this formulation and the convention of Rose [1957], that we have adopted for the Euler angles, the indices k_1 and k_2 are referred to the particles and their symmetries, as we saw in Section 3.10. In particular, for ranks $L = 2, 4$, the values that k_1 and k_2 can assume for the various point groups are listed in Table 3.1. We can now expand the pair correlation, similarly to what we did in Chapter 3 for the single-particle distribution, as

$$G(r,\Omega_{1\mathrm{L}},\Omega_{2\mathrm{L}},\Omega_{r\mathrm{L}}) = \sum_{\substack{L_1,k_1,\\ L_2,k_2,L}} G^{k_1,k_2}_{L_1,L_2,L}(r)\, S^{k_1,k_2}_{L_1,L_2,L}(\Omega_{1\mathrm{L}},\Omega_{2\mathrm{L}},\Omega_{r\mathrm{L}}), \tag{4.71}$$

where the expansion coefficients can be written, thanks to the orthogonality of the S functions (Eq. G.23) , as

$$G^{k_1,k_2}_{L_1,L_2,L}(r) = \int \mathrm{d}\Omega_{1\mathrm{L}}\,\mathrm{d}\Omega_{2\mathrm{L}}\,\mathrm{d}\Omega_{r\mathrm{L}}\, G(r,\Omega_{1\mathrm{L}},\Omega_{2\mathrm{L}},\Omega_{r\mathrm{L}})\, S^{k_1,k_2*}_{L_1,L_2,L}(\Omega_{1\mathrm{L}},\Omega_{2\mathrm{L}},\Omega_{r\mathrm{L}}), \tag{4.72a}$$

$$\equiv c_{L_1,L_2,L}\overline{S^{k_1,k_2*}_{L_1,L_2,L}(r)}, \tag{4.72b}$$

where $c_{L_1,L_2,L} \equiv (2L_1+1)(2L_2+1)(2L+1)/(V_{\Omega_1}V_{\Omega_2}V_{\Omega_r})$ and from now on we indicate with a long overbar an average over all the variables not appearing on the left-hand side, performed while keeping the remaining ones, in inside brackets, fixed. In practice, this is an average over $(\Omega_{1\mathrm{L}},\Omega_{2\mathrm{L}},\Omega_{r\mathrm{L}})$ for particles at distance r:

$$\overline{S^{k_1,k_2}_{L_1,L_2,L}(r)} = \left\langle \delta\,(r-r_{12})\, S^{k_1,k_2}_{L_1,L_2,L}(\Omega_{1\mathrm{L}},\Omega_{2\mathrm{L}},\Omega_{r\mathrm{L}})\right\rangle, \tag{4.73}$$

where we have concisely written $\langle\ldots\rangle$ instead of $\langle\ldots\rangle_{r_{12},\Omega_{1\mathrm{L}},\Omega_{2\mathrm{L}},\Omega_{r\mathrm{L}}}$, as we shall do in general, unless confusion arises. The product $V_{\Omega_1}V_{\Omega_2}V_{\Omega_r}$ corresponds to the 'angular normalization constant' for the angles required to specify the orientations. If all three Euler angles are needed for Ω_1 and for Ω_2, then $V_{\Omega_1}V_{\Omega_2}V_{\Omega_r} = (8\pi^2)(8\pi^2)(4\pi) = 256\pi^5$. If the molecules have axial symmetry, we have $V_{\Omega_1}V_{\Omega_2}V_{\Omega_r} = (4\pi)(4\pi)(4\pi) = 64\pi^3$. In this way

$$\overline{S^{0,0}_{0,0,0}(r)} = \langle \delta\,(r-r_{12})\rangle = 4\pi r^2 \rho\, g(r). \tag{4.74}$$

The quantities $\overline{S^{k_1,k_2}_{L_1,L_2,L}(r)}$ are in a way two-particle analogues of the order parameters of Chapter 3. They provide important structural quantities for the description of a certain mesophase. While the expansion Eq. 4.71 is hardly usable to reconstruct the whole pair distribution, it can be very useful to identify in a systematic way the first few relevant pair correlation functions $\overline{S^{k_1,k_2}_{L_1,L_2,L}(r)}$, for assigning phase type and for structural characterization. To see this and to establish a relation with the single-particle order parameters, it is useful to consider that in the limit of large intermolecular distances, the average of the product of rotation matrices becomes the product of the averages and

$$\lim_{r\to\infty} \overline{S^{k_1,k_2}_{L_1,L_2,L}(r)}$$
$$= \frac{(i)^{L_1-L_2+L}}{\sqrt{2L+1}} \times \sum_{q_1,q_2} C(L_1,L_2,L;q_1,q_2)\,\langle \mathscr{D}^{L_1*}_{q_1,k_1}(\Omega_{1\mathrm{L}})\rangle\,\langle \mathscr{D}^{L_2*}_{q_2,k_2}(\Omega_{2\mathrm{L}})\rangle \langle \mathscr{D}^{L}_{q_1+q_2,0}(\Omega_{r\mathrm{L}})\rangle. \tag{4.75}$$

Here we have for simplicitly indicated, rather loosely, with $r\to\infty$ a scale of separations much larger than the typical nearest neighbours distances (say many nanometres), but not a truly macroscopic one.

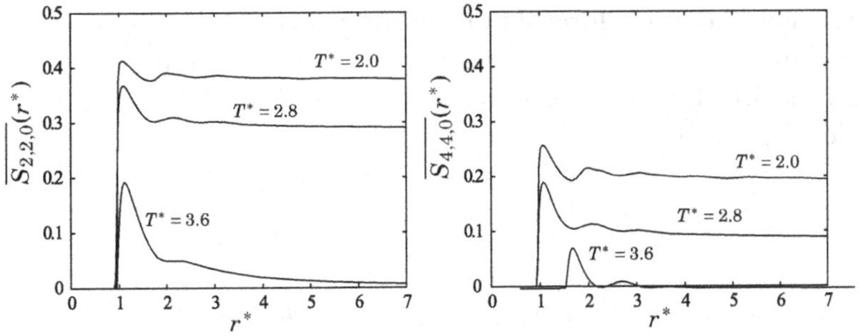

Figure 4.3 The rotationally invariant pair correlations $\overline{S_{2,2,0}(r^*)}$ and $\overline{S_{4,4,0}(r^*)}$ as a function of dimensionless distance r^* for a model LC (specifically the uniaxial Gay–Berne model [Berardi et al., 1993] discussed in Chapters 5 and 11) at three dimensionless temperatures in the isotropic ($T^* = 3.6$), nematic ($T^* = 2.8$) and smectic ($T^* = 2.0$) phase.

4.8.1 Macroscopic Disorder with Local Order

In a number of cases, e.g. spherical micelles or liposomes, the system can appear isotropic on a scale larger than the aggregates, while being definitely ordered on the local scale of constituent molecules or of solutes used as molecular probes for fluorescence polarization [Zannoni, 1981] or ESR spectroscopy [Zannoni et al., 1981] studies. For an isotropic distribution of the intermolecular vectors at large distances we have $\langle \mathscr{D}^L_{q_1+q_2,0}(\Omega_{r\mathrm{L}})\rangle = \delta_{L,0}\,\delta_{-q_1,q_2}$ and, recalling that according to Eq. F.27, $C(L_1, L_2, 0; q_1, -q_1) = (-1)^{L_1-q_1}\delta_{L_1,L_2}/\sqrt{2L_1+1}$, we find the limiting value to be a sum of products of order parameters

$$\lim_{r\to\infty} \overline{S^{k_1,k_2}_{L_1,L_2,0}(r)} = \frac{\delta_{L_1,L_2}}{\sqrt{2L_1+1}} \sum_{q_1} (-1)^{L_1-q_1} \langle \mathscr{D}^{L_1\,*}_{q_1,k_1}\rangle \langle \mathscr{D}^{L_1\,*}_{-q_1,k_2}\rangle. \tag{4.76}$$

In the large separation limit, i.e. $r \gg \sigma_c$, with σ_c the scale of molecular dimensions, but again not asymptotically large, the average invariants tail to a plateau corresponding to a sum of products of the order parameters of the same rank L_1 when $L_1 = L_2$, or 0 otherwise [Zannoni, 1979a]. Let us now briefly see the most common cases.

(i) **Locally uniaxial phases**. Phase uniaxiality around the local director taken as z-axis gives $\delta_{q_1,0}$ and

$$\lim_{r\to\infty} \overline{S^{k_1,k_2}_{L_1,L_2,0}(r)} = \frac{\delta_{L_1,L_2}}{\sqrt{2L_1+1}} (-1)^{L_1} \langle \mathscr{D}^{L_1\,*}_{0,k_1}\rangle \langle \mathscr{D}^{L_1\,*}_{0,k_2}\rangle, \tag{4.77}$$

where L_1 has to be even for a non-polar liquid crystal, e.g. an ordinary nematic, but can be odd for ferroelectric phases. For non-cylindrical particles,

$$\lim_{r\to\infty} \overline{S^{2,2}_{2,2,0}(r)} = \frac{1}{\sqrt{5}} \langle \mathscr{D}^{2\,*}_{0,2}\rangle \langle \mathscr{D}^{2\,*}_{0,2}\rangle. \tag{4.78}$$

As we have seen in Chapter 3 , the biaxial order parameter $\langle \mathscr{D}^{2\,*}_{0,2}\rangle$ is rather small and, given unavoidable numerical errors, it might be difficult to distinguish a real plateau from the reference null background.

(ii) **Locally uniaxial phases formed by uniaxial particles.** The pair correlation for mesogenic particles with effective uniaxial symmetry can be obtained letting $\delta_{k_1,0}$ and $\delta_{k_2,0}$ (see Table 3.1). Considering also the $\delta_{q_1,0}$ coming from phase uniaxiality, as in the previous case, yields

$$G(r, \Omega_{1L}, \Omega_{2L}, \Omega_{rL}) = \sum_{L_1, L_2, L} c_{L_1,L_2,L} \overline{S_{L_1,L_2,L}(r)}\, S_{L_1,L_2,L}(\Omega_{1L}, \Omega_{2L}, \Omega_{rL}), \tag{4.79}$$

using the notation $\overline{S_{L_1,L_2,L}(r)} \equiv \overline{S^{0,0}_{L_1,L_2,L}(r)}$. At large separations

$$\lim_{r\to\infty} \overline{S_{L_1,L_2,0}(r)} = \frac{(-1)^{L_1}\delta_{L_1,L_2}}{\sqrt{2L_1+1}} \langle P_{L_1}\rangle^2. \tag{4.80}$$

In particular,

$$\overline{S_{2,2,0}(r)} = \frac{1}{\sqrt{5}} \left\langle \delta\,(r - r_{12}) \left(\frac{3}{2} \left(\hat{z}_1 \cdot \hat{z}_2 \right)^2 - \frac{1}{2} \right) \right\rangle_{r_{12}}, \tag{4.81}$$

$$\overline{S_{4,4,0}(r)} = \frac{1}{3} \left\langle \delta\,(r - r_{12}) \left(+ \frac{35}{8} \left(\hat{z}_1 \cdot \hat{z}_2 \right)^4 - \frac{30}{8} \left(\hat{z}_1 \cdot \hat{z}_2 \right)^2 + \frac{3}{8} \right) \right\rangle_{r_{12}} \tag{4.82}$$

and, at large inter-particle distances (see Fig. 4.3),

$$\lim_{r\to\infty} \overline{S_{2,2,0}(r)} = \frac{1}{\sqrt{5}} \langle P_2 \rangle^2 \text{ and} \tag{4.83a}$$

$$\lim_{r\to\infty} \overline{S_{4,4,0}(r)} = \frac{1}{3} \langle P_4 \rangle^2. \tag{4.83b}$$

Thus, examining if these pair correlations decay to a plateau, rather than to 0, we can confirm that a phase is anisotropic if the plateau value is above the estimated background error. In a cooling down sequence of simulations starting from the isotropic phase, the onset of a plateau in these invariants indicates an isotropic-liquid crystal phase transition.

(iii) **Polar phases.** Nematics are not polar, i.e. the director is not a true vector endowed with a sense as well as an orientation, and indeed in Chapter 3 we have employed reflection symmetry (Eq. 3.19). However, the occurrence of polar nematics, and in particular of ferroelectric ones if the molecules are endowed with a suitable electric dipole, is not forbidden by some fundamental law. Thus, computer simulations on simple models can and have been employed to try to find some molecular design features of help in the search for ferroelectric nematics. The general invariant limit, Eq. 4.77, can be used to test the existence of phase polarity. In particular, a convenient pair invariant is $\overline{S_{1,1,0}(r)} = -(1/\sqrt{3})\langle \delta(r - r_{12})(\hat{z}_1 \cdot \hat{z}_2)\rangle$. For large particle-particle separations we have

$$\lim_{r\to\infty} \overline{S_{1,1,0}(r)} = -\frac{1}{\sqrt{3}} \langle P_1 \rangle^2, \tag{4.84}$$

that is different from 0 only for a polar, e.g. for a ferroelectric, phase. In Fig. 4.4 we show an example for a model system of tapered particles introduced in Berardi et al. [2001, 2004b] that exhibits a polar phase. We see that after a separation corresponding

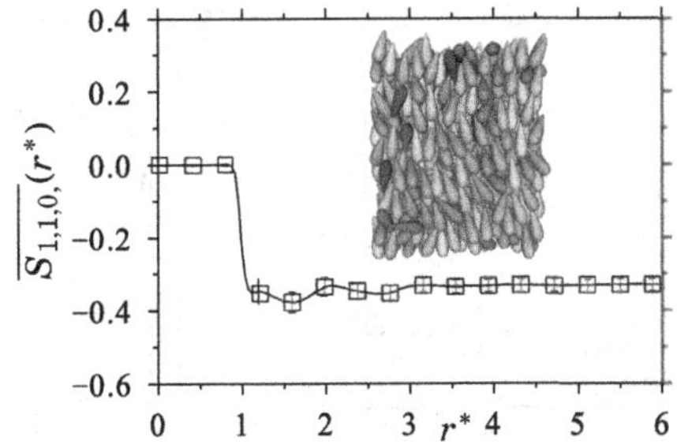

Figure 4.4 The pair correlation $\overline{S_{1,1,0}(r^*)}$ for a polar nematic system of tapered model particles (see inset) as a function of dimensionless distances, r^* [Berardi et al., 2001].

to particle size where a second particle cannot be present and thus $\overline{S_{1,1,0}(r)} = 0$, the tapered particles are on average parallel rather than antiparallel giving a negative plateau as from Eq. 4.84, thus confirming the overall phase polarity.

(iv) **Biaxial phase of uniaxial particles**. We could imagine such a phase as a system of rods adopting on average a cross-like configuration, as observed in the suspensions of inorganic nanorods in 5CB briefly mentioned in Section 1.6 [Mundoor et al., 2018] or in systems of discs organized in a 'house of cards' style. In this case,

$$\lim_{r\to\infty} \overline{S_{2,2,0}(r)} = \frac{1}{\sqrt{5}}\left(\langle \mathscr{R}^2_{0,0}\rangle^2 + 2\langle \mathscr{R}^2_{2,0}\rangle^2\right), \tag{4.85}$$

where $\mathscr{R}^L_{m,n}$ are the D_{2h} symmetry adapted Wigner functions [Biscarini et al., 1995] defined in Eq. 3.149 . In the limit of vanishing biaxiality this reduces to Eq. 4.83a .

(v) **Biaxial phases of biaxial particles.** For an orthogonal biaxial nematic with D_{2h} phase symmetry, we have, for $L_1 = L_2 = 2$, that $q_1 = 0, \pm 2$ and $k_1, k_2 = 0, \pm 2$. Some relevant expressions for second-rank invariants and their limiting values are

$$\overline{S^{2,0}_{2,2,0}(r)} = \frac{\sqrt{3}}{2\sqrt{10}}\left\langle \delta\left(r - r_{12}\right)\left(\left(\hat{\boldsymbol{x}}_1\cdot\hat{\boldsymbol{z}}_2\right)^2 - \left(\hat{\boldsymbol{y}}_1\cdot\hat{\boldsymbol{z}}_2\right)^2\right)\right\rangle, \tag{4.86}$$

$$\lim_{r\to\infty} \overline{S^{2,0}_{2,2,0}(r)} = \frac{1}{\sqrt{5}}\left(\langle \mathscr{R}^2_{0,0}\rangle\langle \mathscr{R}^2_{0,2}\rangle + 2\langle \mathscr{R}^2_{2,0}\rangle\langle \mathscr{R}^2_{2,2}\rangle\right) \tag{4.87}$$

and

$$\begin{aligned}\overline{S^{2,2}_{2,2,0}(r)} = \frac{1}{4\sqrt{5}}\Big\langle \delta\left(r - r_{12}\right)\big(\left(\hat{\boldsymbol{x}}_1\cdot\hat{\boldsymbol{x}}_2\right)^2 - \left(\hat{\boldsymbol{x}}_1\cdot\hat{\boldsymbol{y}}_2\right)^2 - \left(\hat{\boldsymbol{y}}_1\cdot\hat{\boldsymbol{x}}_2\right)^2 + \left(\hat{\boldsymbol{y}}_1\cdot\hat{\boldsymbol{y}}_2\right)^2 \\ - 2\left(\hat{\boldsymbol{x}}_1\cdot\hat{\boldsymbol{y}}_2\right)\left(\hat{\boldsymbol{y}}_1\cdot\hat{\boldsymbol{x}}_2\right) - 2\left(\hat{\boldsymbol{x}}_1\cdot\hat{\boldsymbol{x}}_2\right)\left(\hat{\boldsymbol{y}}_1\cdot\hat{\boldsymbol{y}}_2\right)\big)\Big\rangle,\end{aligned} \tag{4.88}$$

$$\lim_{r\to\infty} \overline{S^{2,2}_{2,2,0}(r)} = \frac{1}{\sqrt{5}}\left(\langle \mathscr{R}^2_{0,2}\rangle^2 + 2\langle \mathscr{R}^2_{2,2}\rangle^2\right). \tag{4.89}$$

This asymptotic limit provides a useful route to check the single-particle order parameters obtained as described in Chapter 3. However, only products or squares of order

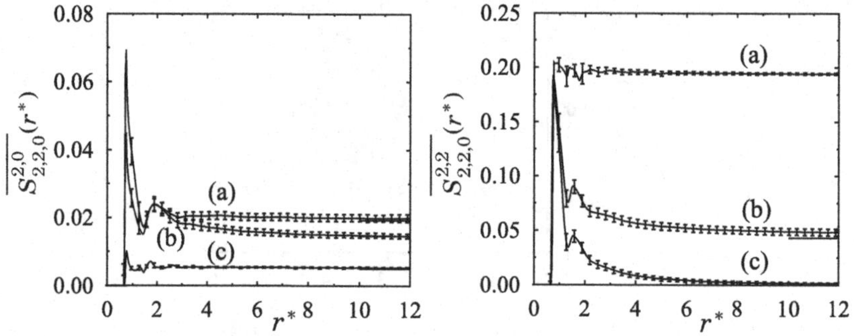

Figure 4.5 Scalar $\overline{S^{2,0}_{2,2,0}(r^*)}$ and $\overline{S^{2,2}_{2,2,0}(r^*)}$ vs dimensionless distance r^* for a system of biaxial particles showing biaxial and uniaxial phases: (a) $T^* = 2.6$ (biaxial smectic), (b) 2.8 (biaxial nematic), and (c) 3.0 (uniaxial nematic) [Berardi and Zannoni, 2000].

parameters are obtained in this way, thus missing their individual signs. From Fig. 4.5 we see clearly that these invariants and particularly their tail can detect the onset of a biaxial phase. Thus, $\overline{S^{2,0}_{2,2,0}(r)}$ and $\overline{S^{2,2}_{2,2,0}(r)}$ decay to 0 in isotropic and uniaxial phase but to a plateau if the phase is biaxial [Berardi and Zannoni, 2000].

(vi) **Cholesteric and other chiral phases**. Recalling that the effect of the inversion operator on the Wigner matrices composing the Stone invariant will produce a factor $(-1)^{L_1+L_2+L_3}$ (see Appendix G), it follows that the correlations $\overline{S^{k_1,k_2}_{L_1,L_2,L}(r)}$ with $(L_1 + L_2 + L_3)$ odd will be 0 in a non-chiral phase. To follow the onset of a chiral phase we can then monitor invariant correlations with $(L_1 + L_2 + L_3)$ odd. In particular, the following invariant

$$\overline{S_{2,2,1}(r)} = \Big\langle \delta(r - r_{12})\, S_{2,2,1}(\Omega_{1\mathrm{L}}, \Omega_{2\mathrm{L}}, \Omega_{r\mathrm{L}}) \Big\rangle,$$
$$= -\sqrt{\frac{3}{10}} \Big\langle \delta(r - r_{12}) \left(\hat{z}_1 \cdot \hat{z}_2 \times \hat{\boldsymbol{r}}\right) \left(\hat{z}_1 \cdot \hat{z}_2\right) \Big\rangle \tag{4.90}$$

has been employed for various chiral systems [Berardi et al., 1998c; Memmer, 1998; Berardi et al., 2003a; Chiccoli et al., 2013]. In Fig. 4.6 we see, as an example, the invariant calculated for discotic particle characterized by a chirality strength parameter χ [Memmer, 1998]. $\overline{S_{2,2,1}(r)}$ is 0 for the non-chiral discs and upon increasing χ gives a cholesteric and a Blue Phase.

4.8.2 Isotropic Phases

If the intermolecular vector has an isotropic distribution not only in the very large separation limit, but also in the range of a few particle dimensions, then we can consider also in this intermediate range, $\langle \mathscr{D}^L_{-q_1-q_2,0}(\Omega_{r\mathrm{L}})\rangle = \delta_{L,0}\delta_{q_1+q_2,0}$ which in turn gives: $C(L_1, L_2, 0; q_1, -q_1) = (-1)^{L_1-q_1}\delta_{L_1,L_2}/\sqrt{2L_1+1}$ (formally just as in the previous cases). However, the orientations of the two particles will be, in general, correlated each other, so that the average Stone invariants become

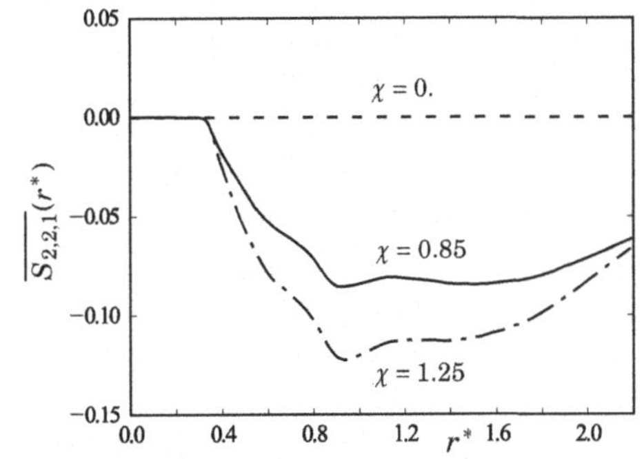

Figure 4.6 The radial pseudoscalar orientational pair correlation functions $\overline{S_{2,2,1}}(r^*)$ as a function of scaled particle separation r^* for different values of a chirality strength parameter χ: in the discotic nematic phase ($\chi = 0$, - - -), in the discotic cholesteric phase ($\chi = 0.85$, —) and in a discotic Blue Phase ($\chi = 1.25$, –·–·–) [Memmer, 1998].

$$\overline{S^{k_1,k_2}_{L_1,L_2,0}}(r) = \frac{(-1)^{L_1}\delta_{L_1,L_2}}{\sqrt{2L_1+1}} \sum_{q_1} (-1)^{-q_1} \langle \mathscr{D}^{L_1 *}_{q_1,k_1}(\Omega_{1L}) \mathscr{D}^{L_1 *}_{-q_1,k_2}(\Omega_{2L}) \rangle, \tag{4.91a}$$

$$= \frac{(-1)^{L_1+k_2}\delta_{L_1,L_2}}{\sqrt{2L_1+1}} \langle \mathscr{D}^{L_1}_{k_1,-k_2}(\Omega_{21}) \rangle, \tag{4.91b}$$

showing that for isotropic systems formed by rigid particles of arbitrary symmetry, only relative orientations of the two particles are relevant. Relabelling, for convenience, $C_{J,k_1} \equiv c_{J,J,0}(-1)^{J+k_1}/\sqrt{2J+1}$, we have, using the notation of Eq.4.57,

$$g(r,\Omega_{21}) = \sum_{J,k_1,k_2} C_{J,k_1} \overline{\mathscr{D}^J_{k_1,-k_2}(\Omega_{21})(r)}\, \mathscr{D}^J_{k_1,-k_2}(\Omega_{21}), \tag{4.92}$$

where $\mathscr{D}^J_{k_1,-k_2}(\Omega_{21}) = S^{k_1,k_2}_{J,J,0}(\Omega_{21})$. For uniaxial molecules with a principal axis $\boldsymbol{u}$, symmetry requires $\delta_{k_1,0}$ and $\delta_{k_2,0}$, while $c_{L_1,L_1,0} \equiv (2L_1+1)^2/(16\pi^2)$ and $g(r,\Omega_{21})$ reduces to $g(r,\beta_{21})$ with β_{12} the angle between the symmetry axes of the particles. In Fig. 4.7 we show this radial angular distribution for benzene obtained from neutron scattering experiments [Headen et al., 2010]. The peaks indicate that some structuring still persists in the liquid at short range. However, after just two closest approach distances the peaks have already disappeared. In this sense this is just *short-range order*. We note that this description, while it concerns isotropic liquids, also formally applies to simple lattice models where the interparticle vector has a sufficiently high symmetry. Thus, for example, it applies to the Heisenberg and to the Lebwohl–Lasher models with the spins at the site of a simple cubic lattice. We can expand $g(r,\beta_{21})$ in terms of Legendre polynomials of the relative orientation:

$$g(r,\beta_{21}) = \sum_J C_J G_J(r)\, P_J(\cos\beta_{21}), \tag{4.93}$$

with $C_J \equiv (-1)^J C_{J,0} = (-1)^J c_{JJ0}/\sqrt{2J+1}$ and the normalization

$$\int \mathrm{d}\beta_{12} \sin\beta_{12}\, g(r,\beta_{21}) = g(r). \tag{4.94}$$

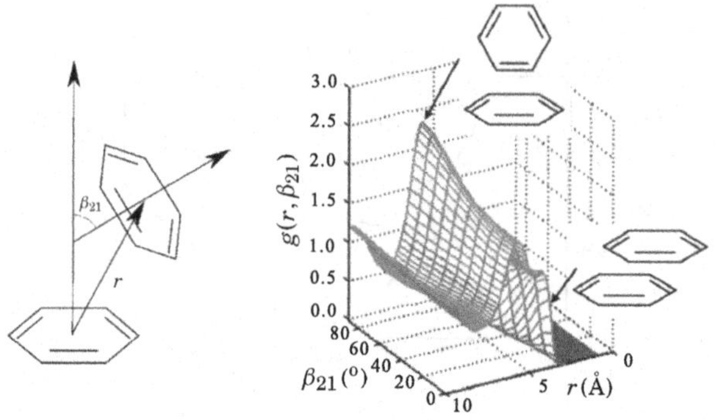

Figure 4.7 The radial-angular distribution $g(r, \beta_{21})$ of liquid benzene obtained from neutron scattering [Headen et al., 2010].

Here we have introduced the angular-radial correlation of rank J, $G_J(r)$

$$G_J(r) = \overline{P_J(\boldsymbol{u}_1 \cdot \boldsymbol{u}_2)}(r) = \overline{\mathscr{D}^J_{0,0}(\Omega_{21})}(r), \tag{4.95a}$$

$$= \sum_{q=-J}^{J} \left\langle \delta(r - r_{12}) \mathscr{D}^{J*}_{q,0}(\Omega_{1L}) \mathscr{D}^{J}_{q,0}(\Omega_{2L}) \right\rangle, \tag{4.95b}$$

with

$$G_0(r) = \overline{S_{0,0,0}}(r) = \langle \delta(r - r_{12}) \rangle = 4\pi r^2 \rho\, g(r). \tag{4.96}$$

In an ordinary liquid, long-range order is absent and $G_J(\infty) = \delta_{J,0}$. A particularly, important case is the second-rank space-orientational correlation

$$G_2(r) = \overline{P_2(\boldsymbol{u}_1 \cdot \boldsymbol{u}_2)}(r) = \frac{1}{2} \left\langle \delta(r - r_{12}) \left(3(\boldsymbol{u}_1 \cdot \boldsymbol{u}_2)^2 - 1 \right) \right\rangle_{r_{12}}, \tag{4.97}$$

shown in Fig. 4.8 which decays to a plateau $\langle P_2 \rangle^2$ [Zannoni, 1979a].

4.8.3 Thin Film Systems

In most applications, liquid crystals are not employed in bulk quantities but in very thin (nano- to micro-) thick films and then it is important to study and simulate and describe LCs in these confined environments. In the case of thin films confined between two plane parallel slabs, it is natural to study the various types of order across the film. Considering a laboratory system with the z-axis perpendicular to the surfaces, it may be convenient to adapt the scalar correlations we have just introduced for particles at various distances z_1, z_2 from one of the two surfaces. In particular, we can define, using $r_{12} \| z$ [Berardi et al., 1998a],

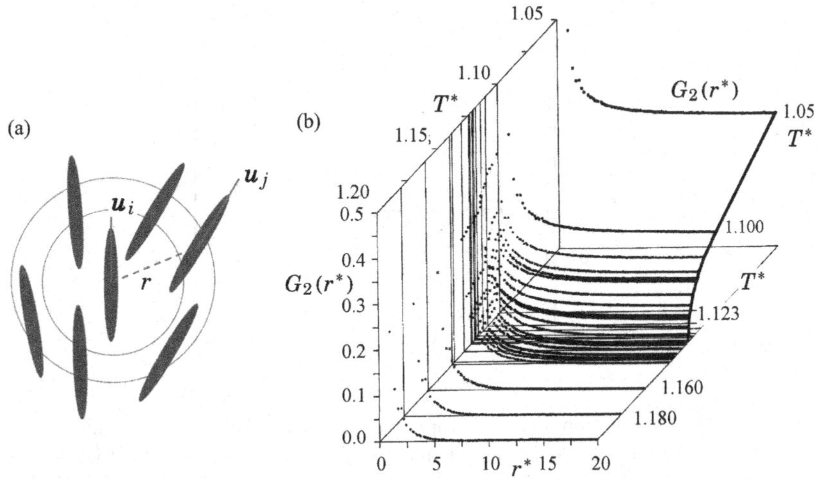

Figure 4.8 (a) Calculating the orientational pair correlation. (b) The angular-radial correlation $G_2(r^*)$ for the Lebwohl–Lasher lattice model at various temperatures T^* below and above the transition, as indicated on the graph. The distance r^* is measured in lattice units [Fabbri and Zannoni, 1986].

(i) **Surface-particle radial distribution.** A density profile across film normalized by the average density:

$$g_0(z) \equiv \overline{S_{0,0,0}}(z) = \frac{1}{L_x L_y \rho} \langle \delta(z_i - z) \rangle_{z_i}. \tag{4.98}$$

In the case of smectics and particularly if the layer planes are parallel to the surfaces, we would expect this to be some sort of sinusoid.

(ii) **Orientational correlations inside the film.**

$$\overline{S_{0,2,2}}(z) = \frac{1}{\sqrt{5}\, L_x L_y \rho} \left\langle \delta(z_i - z) \left(\frac{3}{2} u_{zi}^2 - \frac{1}{2} \right) \right\rangle_{z_i}. \tag{4.99}$$

This could allow us to see, e.g. the propagation of ordering of a certain type from some surface alignment layer. We can also look at the correlation between particles at different distances z_1 and z_2 from the support surface (at $z = 0$) inside the film:

$$\begin{aligned} \overline{S_{L_1,L_2,L_3}}(z_1, z_2) &= \frac{1}{(L_x L_y \rho)^2} \int \mathrm{d}x_1 \mathrm{d}y_1 \mathrm{d}\Omega_1 \mathrm{d}x_2 \mathrm{d}y_2 \mathrm{d}\Omega_2 \\ &\quad \times P^{(2)}(\boldsymbol{r}_1, \Omega_{1\mathrm{L}}, \boldsymbol{r}_2, \Omega_{2\mathrm{L}})\, S_{L_1 L_2 L_3}(\Omega_{1\mathrm{L}}, \Omega_{2\mathrm{L}}, \Omega_{r_z}), \\ &= \frac{1}{(L_x L_y \rho)^2} \left\langle \delta(z_i - z_1)\, \delta(z_j - z_2)\, S_{L_1 L_2 L_3}(\Omega_{1\mathrm{L}}, \Omega_{2\mathrm{L}}, \Omega_{r_z}) \right\rangle_{z_i, z_j}. \end{aligned} \tag{4.100}$$

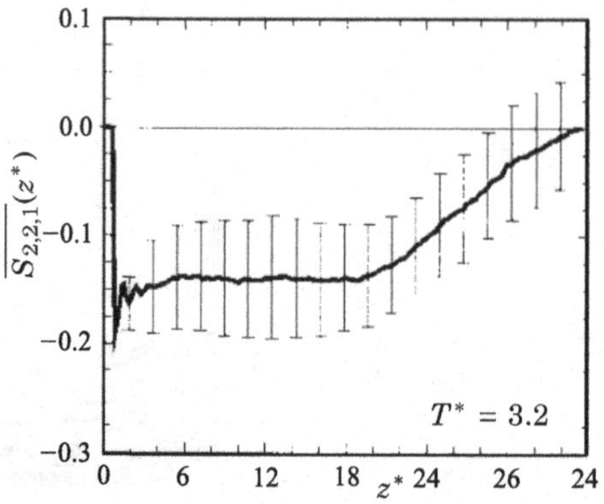

Figure 4.9 Chiral correlation $\overline{S_{2,2,1}}(z^*)$ for a nematic sandwiched between a chiral surface at $z^* = 0$ and $T^* = 3.2$ and a planar achiral one at $z^* = 35.6$ (in particle width units) [Berardi et al., 1998c]. We see chirality propagating from the surface inside the film and then vanishing as the achiral surface approaches.

For instance:

$$\overline{S_{2,2,0}}(z_1, z_2) = \frac{1}{2\sqrt{5}(L_x L_y \rho)^2} \left\langle \delta(z_i - z_1)\,\delta(z_j - z_2)\, \left(3(\boldsymbol{u}_i \cdot \boldsymbol{u}_j)^2 - 1\right)\right\rangle_{z_i, z_j}. \tag{4.101}$$

(iii) **Chirality across the film**. An example is that of a nematic film confined between two planar surfaces, one of which is chiral and the other a flat, planar achiral monodomain [Berardi et al., 1998c]. The propagation of chirality away from the surface can be monitored by the quantity

$$\overline{S_{2,2,1}}(z) = \frac{\sqrt{3}}{\sqrt{10}\, L_x L_y \rho} \langle \delta(z_i - z)\, u_{xi} u_{yi} \rangle_{z_i}, \tag{4.102}$$

as shown in Fig. 4.9. The twisting of the director between two different distances, z_1 and z_2, away from the chiral surface could instead be followed using

$$\overline{S_{2,2,1}}(z_1, z_2) = \frac{-\sqrt{3}}{\sqrt{10}\,(L_x L_y \rho)^2} \langle \delta(z_i - z_1)\,\delta(z_j - z_2)\,(\boldsymbol{u}_i \cdot \boldsymbol{u}_j \times \hat{\boldsymbol{r}}_z)\,(\boldsymbol{u}_i \cdot \boldsymbol{u}_j)\rangle_{z_i, z_j}. \tag{4.103}$$

4.8.4 Non-Uniform Fluids: Smectics

For smectics, and in general for non-uniform fluids, the general pair distribution $G(\boldsymbol{r}_1, \Omega_1, \boldsymbol{r}_2, \Omega_2) = G(r_1, \Omega_{r_1}, \Omega_1, r, \Omega_r, \Omega_2)$, with $\boldsymbol{r} = \boldsymbol{r}_2 - \boldsymbol{r}_1$. To carry out an invariant expansion like we did for uniform fluids we need to build and employ rotational invariant combinations of four, rather than three, rotation matrices. This can be done (see, e.g., [Longa et al., 2001]), however, it becomes very cumbersome and not easily usable. It is more convenient to introduce reduced distributions, as we shall now describe.

4.9 Reduced Distributions

As mentioned in Section 4.7, in view of the difficulties in examining the pair distribution in full, it is often useful to introduce reduced distributions, involving only a subset of positional and/or orientational coordinates. The reduction is typically done in two ways. One is integrating over some of the angular or positional coordinates. The other is freezing some coordinates at a certain fixed value.

4.9.1 Anisotropic Radial Distributions

As a first example, we go back to the angular-radial distribution $g(r, \Omega_r)$ in Eq. 4.57b, the distribution in space of the intermolecular vector $\boldsymbol{r}$ with length r, the distance of the centres of mass and orientation $\Omega_r = (\alpha_r, \beta_r)$ with respect to the laboratory frame. For systems that are uniaxial around the director we can first locate $\boldsymbol{d}$ by calculating and diagonalizing the ordering matrix $\mathbf{Q}$. Then we can consider orientations defined with respect to a laboratory system with $z \parallel \boldsymbol{d}$. As long as the system has uniaxial symmetry its properties are invariant for an arbitrary rotation around the director and we do not need to consider the angle, α_r, so that the intermolecular vector distribution reduces to $g(\boldsymbol{r}) = g(r, \Omega_r) = g(r, \cos\beta_r)/2\pi$. It is worth noting that this reduced distribution can be obtained both for uniform and non-uniform systems like smectics or crystals. As an example, in Fig. 4.10 we see a set of histograms of $g(r, \beta_r)$ for a model liquid crystal (a system of the so-called Gay–Berne ellipsoidal particles that we shall discuss in Chapters 5 and 11) at various temperatures and phases [Berardi et al., 1993]. The three-dimensional representation shows at once that the distribution changes quite significantly with temperature. Not surprisingly it is flat (and thus devoid of angular dependence) in the isotropic phase, but far from isotropic in the nematic phase. The very low temperature one ($T^* = 1.8$) shows that for molecules along the director ($\cos\beta_r = 1$) a second molecule is found slightly below the particle length. However, if we move transversally to the director ($\cos\beta_r = 0$), very sharp, well-defined peaks appear. At least six orders of peaks occur, indicating a high degree of structure in the layer. A more careful look at the spiltting of the second peak suggests the presence of hexagonal ordering , as expected in a smectic B or a crystalline layer structure. In particular we have first the sharp nearest neighbours peak, then a double peak with similar intensities around $r^* = 2$, another double peak with 2 : 1 intensity factor between $r^* = 2.6$ and 3.0. These would appear at $r^* = 1.73$ and 2 and at 2.646 and 3 for the perfect triangular lattice. At $T^* = 2.0$ the structure is quite similar, although less resolved. We still have order in the layer but the characteristic features of hexagonal ordering are not evident. Apart from

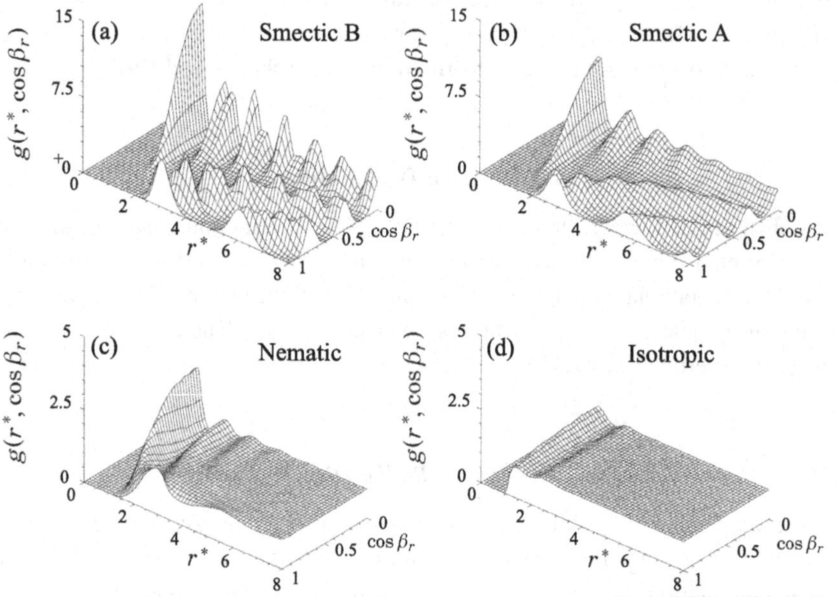

Figure 4.10 The anisotropic radial distribution $g(r^*_; \cos\beta_r)$ for a system of rod-like molecules in the highly ordered smectic B (a), smectic A (b), nematic (c) and isotropic (d) phases. The specific cases refer to the so-called Gay–Berne GB(3,5,1,3) model of liquid crystals, discussed in detail in Sections 5.6.3 and in 11.5.1, at $T^* = 1.8$ (a), 2.0 (b), 2.8 (c), 3.5 (d). Here r^* is the interparticle separation scaled by the particle width, β_r the angle between the interparticle vector and the director and T^* the dimensionless temperature [Berardi et al., 1993].

looking at $g(r, \cos\beta_r)$ as a whole, it is also convenient to expand it, as

$$g(r, \cos\beta_r) = g(r, \hat{\boldsymbol{r}} \cdot \boldsymbol{d}) = g(r) \sum_{L>0} (2L+1) g_L(r) P_L(\hat{\boldsymbol{r}} \cdot \boldsymbol{d}), \tag{4.104}$$

where $g(r) = \frac{1}{2} \int \mathrm{d}\cos\beta_r \, g(r, \cos\beta_r)$ is the standard radial distribution. The set of quantities $g_L(r)$ represent a sort of order parameters of the intermolecular vector with respect to the director:

$$g_L(r) = \frac{1}{2\, g(r)} \int \mathrm{d}\cos\beta_r \, g(r, \cos\beta_r) P_L(\cos\beta_r) = \langle P_L(\cos\beta_r) \rangle. \tag{4.105}$$

Note that, even if these coefficients appear to be similar to the $G_L(r)$ in Eq. 4.95, they are actually quite different since those express the angular-radial correlation between two

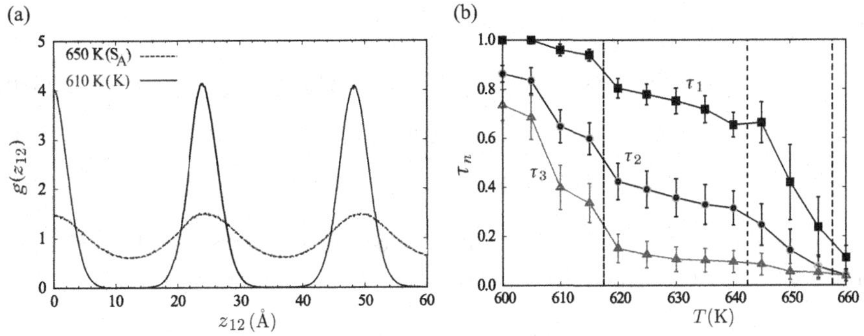

Figure 4.11 (a) The two-particle density correlation function $g(z_{12})$ in a crystal (—) and smectic A (- - -) phase and (b) the first few positional order parameters τ_n as a function of temperature for p-quinquephenyl as obtained from atomistic simulations [Olivier et al., 2014].

molecules, while the $g_L(r)$ refers to the orientational distribution in space of the molecular centres. We can also define a positional pair correlation function along the director

$$g(z_{12}) = \frac{\int \mathrm{d}r\; r^2 \mathrm{d}\beta_r \sin\beta_r \delta(r\cos\beta_r - z_{12})\, g(r, \cos\beta_r)}{\int \mathrm{d}r\; r^2 \mathrm{d}\beta_r \sin\beta_r\, \delta(r\cos\beta_r - z_{12})}. \tag{4.106}$$

This provides an important way of characterizing the onset of a smectic phase, where $g(z_{12})$ will have a sinusoidal profile. In Fig. 4.11a we see an example for p-quinquephenyl resulting from analyzing atomistic configurations generated with molecular dynamics computer simulations [Olivier et al., 2014] (see Chapters 9 and 12). It is convenient to expand $g(z_{12})$ in a Fourier series:

$$g(z_{12}) = \sum_{n=0}^{\infty} g_n \cos(q_n z_{12}). \tag{4.107}$$

Note that this expansion is very (and possibly confusingly) similar to that of the single-particle positional distribution Eq. 3.15b. However, the expansion coefficients are approximately proportional to squares of the positional order parameters, i.e. to τ_n^2 rather than to τ_n,

$$g(z_{12}) = \sum_{n=0}^{\infty} \left(\frac{2\langle \cos(q_n z_{12})\rangle}{\ell_z(\delta_{m,0}+1)} \right) \cos(q_n z_{12}) \approx \sum_{n=0}^{\infty} \left(\frac{2\tau_n^2}{\ell_z(\delta_{m,0}+1)} \right) \cos(q_n z_{12}), \tag{4.108a}$$

$$= \frac{1}{\ell_z} + \frac{2}{\ell_z} \sum_{n=1}^{\infty} \tau_n^2 \cos(q_n z_{12}), \tag{4.108b}$$

where $q_n \equiv 2\pi n/\ell_z$. The approximation comes from

$$g_n = \langle \cos[q_n z_2] \cos[q_n z_1] + \sin[q_n z_2] \sin[q_n z_1] \rangle, \tag{4.109a}$$

$$\approx \langle \cos[q_n z_2] \rangle \langle \cos[q_n z_1] \rangle + \langle \sin[q_n z_2] \rangle \langle \sin[q_n z_1] \rangle, \tag{4.109b}$$

$$\approx \langle \cos[q_n z_2] \rangle \langle \cos[q_n z_1] \rangle = \tau_n^2, \tag{4.109c}$$

where we have used the even parity of the single-particle distribution. In Fig. 4.11b we report the first few positional order parameters τ_n for *p*-quinquephenyl.

4.9.2 Hexatic Bond Order

The reduction scheme we have just described leads to various distributions of the intermolecular vector with respect to the director, taken as the z-axis of our laboratory system. We have also demonstrated that these distributions can be quite telling to distinguish isotropic, nematic and smectic systems, However, as we have seen in Chapters 1 and 2 there are a number of different smectic phases characterized by a different ordering inside the layers. The description of this structuring in smectic layers is not simple, when thinking of the variety of smectic phases characterized by some regular clustering of the particles in the smectic layer at short and/or long range. For instance, the local clustering of centres of mass in a smectic B has a hexagonal character, while in smectic A this is absent. Quite similarly, if we consider the ordering of molecular stacks in an oriented columnar system, we can also have hexagonal or rectangular arrangements in the planes transversal to the column axis. To quantify these situations, we can focus first on molecules belonging to the same layer and their centres of mass. Introducing an order parameter like the *hexatic order parameter* employed by Halperin and Nelson [1978] to quantify order in a two-dimensional crystal of discs, where the close-packing is hexagonal. To do this let us introduce again an intermolecular vector distribution but this time for the projection of the intermolecular vectors on the smectic planes. We can start by defining a right-handed coordinate system $(\boldsymbol{e}_1, \boldsymbol{e}_2, \boldsymbol{e}_3)$, with the z-axis along the layer normal (for an upright smectic this coincides with the director $\boldsymbol{d}$), $\boldsymbol{e}_1$ an arbitrary reference axis in the layer plane, and $\boldsymbol{e}_2 = \boldsymbol{e}_3 \times \boldsymbol{e}_1$. We then consider a particle k, with position $\boldsymbol{r}_k$ and identify the set $\mathscr{S}(\boldsymbol{d}, \boldsymbol{r}_k)$ of first neighbour particles, at position $\boldsymbol{r}_j$. To look at the structuring around the particle we can consider the unit vectors $\hat{s}_{kj}$, such that $\boldsymbol{e}_3(\boldsymbol{r}_k) \cdot s_{kj} = 0$, joining each neighbour j in the same layer to particle k chosen as the origin (these are normally called *bonds* even if of course no chemical bond is involved). In general, the cluster $\mathscr{S}(\boldsymbol{d}, \boldsymbol{r}_k)$ for the particle at position $\boldsymbol{r}_k$ can be defined by its Voronoi cell, Vor($\boldsymbol{r}_k$), i.e. the region of space no further from the point at $\boldsymbol{r}$ than to any other point $\boldsymbol{r}_j$ in the set, i.e.

$$\mathrm{Vor}(\boldsymbol{r}_k) = \left\{ \boldsymbol{r} : |\boldsymbol{r} - \boldsymbol{r}_k| \leqslant \left| \boldsymbol{r} - \boldsymbol{r}_j \right|, \text{for all particles } \boldsymbol{r}_j \right\}. \tag{4.110}$$

These cells (see Fig. 4.12) are convex non-overlapping polyhedra that share common faces, and their partition of the whole sample space is called the *Voronoi tessellation* [Torquato and Stillinger, 2010]. We can introduce a radial, in plane, distribution.

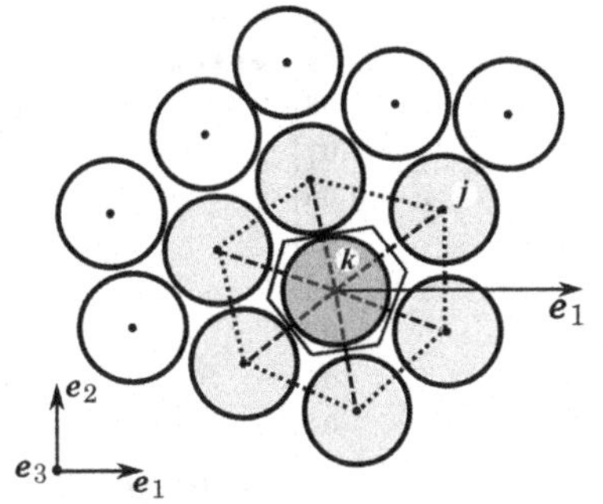

Figure 4.12 Hexagonal order in a smectic layer. The continuous and dotted lines are respectively the edges of the Voronoi and Delaunay cells around molecule k (dark grey) resulting from tessellation of space [Torquato and Stillinger, 2010]. The hexatic order parameter is calculated considering for every neighbour in the layer (light grey) the angle θ_{kj} from the arbitrary reference frame axes $\boldsymbol{e}_1$ to $\boldsymbol{s}_{kj}$ around the layer normal $\boldsymbol{e}_3$ at position $\boldsymbol{r}_k$.

$$G(\boldsymbol{r}_k, s_{kj}, \alpha_{kj}) \propto \sum_m \psi_m(s_{kj})\, e^{im\alpha_{kj}}, \tag{4.111}$$

where $s_{kj} = [(x_j - x_k)^2 + (y_j - y_k)^2]^{\frac{1}{2}}$ is the radial distance between the two particles. If the distribution of neighbours around particle k has local C_{n_i} symmetry around $\boldsymbol{e}_3$, e.g. tetradic, with nearest neighbour number $n_k = 4$ or hexatic $n_k = 6$, we can express the bond order with respect to the ideal tetradic or hexatic distribution where the bonds have an angular spacing $2\pi/n_i$ as

$$\psi_{n_k}(\boldsymbol{r}_k) = \frac{1}{n_k}\sum_{j=1}^{n_k} \mathrm{e}^{i[\theta_{kj}(\boldsymbol{r}_i) - j2\pi/n_k]} = \frac{1}{n_k}\sum_{j=1}^{n_k} \mathrm{e}^{in_k\theta_{kj}(\boldsymbol{r}_k)}. \tag{4.112}$$

The sum on j runs over the n_k neighbours of this particle and θ_{kj} is the angle between the particles k and j and the reference axis $\boldsymbol{e}_1$. If $n_k = 6$ and the system has hexagonal symmetry around $\boldsymbol{d}$, we have that the local hexatic bond angle order parameter [Steinhardt et al., 1983; Brock et al., 1986; Selinger and Nelson, 1988; Kamien, 1996; Torquato and Stillinger, 2010] is then, for particle k,

$$\psi_6(\boldsymbol{r}_k) = \frac{1}{6}\sum_{j=1}^{6} \mathrm{e}^{i6\theta_{kj}(\boldsymbol{r}_k)} \tag{4.113}$$

and the global hexatic order parameter is a sample average

$$\langle\psi_6\rangle = \left|\frac{1}{N}\sum_{k=1}^{N}\psi_{6,k}\right|. \tag{4.114}$$

The bond order is not necessarily short-ranged across the layer. It is thus useful to also introduce a space correlation function of the hexatic order:

$$G_6^H(\boldsymbol{r}) = \langle\psi_6(\boldsymbol{r}_0)\psi_6^*(\boldsymbol{r}_0 + \boldsymbol{r})\rangle \tag{4.115}$$

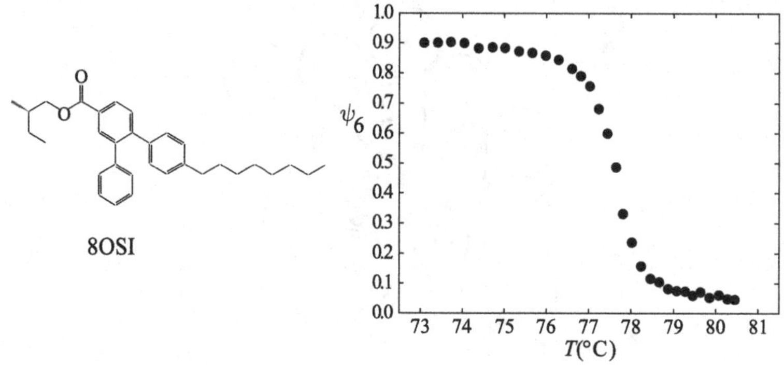

Figure 4.13 Hexatic bond order parameter ψ_6 vs temperature for the S_C-S_I transition in a monodomain of (racemic) 4-(2-methylbutyl) phenyl 4′-(octyloxy)-(1, 1′)-biphenyl-4-carboxylate (8OSI) with the following sequence of cooling down transitions [Brock et al., 1986]: $I \overset{174.5°C}{\longrightarrow} N \overset{170.0°C}{\longrightarrow} S_A \overset{133.4°C}{\longrightarrow} S_C \overset{79.9°C}{\longrightarrow} S_I \overset{75.1°C}{\longrightarrow} S_J \overset{61.7°C}{\longrightarrow} S_K$.

and its asymptotic limit for large separations between the two clusters:

$$\lim_{|\boldsymbol{r}|\to\infty} G_6^H(\boldsymbol{r}) = \langle \psi_6(\boldsymbol{r}_0)\rangle \langle \psi_6^* \boldsymbol{r})\rangle, \tag{4.116}$$

which will be 0 in a normal smectic B, where the local neighbour cluster is not just replicated across the sample but can be deformed and/or rotated (cf. Chapter 1). In a crystal smectic B, the local order hexatic bond order decays instead as a power law and is maintained at fairly long $\boldsymbol{r}$, with the axis of the hexagonal cluster remaining the same. Quite similarly we can describe the positional regularity of columns of flat mesogens aligned with their axes perpendicular to a surface, considering virtual transversal slices and the in plane (e.g. tetradic or hexatic) bond order of the discotic mesogens (see Fig. 4.12). It is worth noting that, notwithstanding their name, the bond-orientational order parameters just described are pairwise quantities, differently from the single-particle positional order parameters described in Chapter 3. It is also not forbidden by symmetry to have a phase of matter in which the positional order is short range as in a liquid or glass, but the bond order is long range as in a crystalline solid [Kosterlitz and Thouless, 1973; Halperin and Nelson, 1978; Brock et al., 1986]. In a way, this is the case of a transition between a S_A, S_B and S_{B_K} where the hexagonal order changes but the positional order and the layer spacings are very similar. Clearly, the concept of hexatic order applies also to tilted smectic phases (see Section 1.8) and in Fig. 4.13 we show an example of ψ_6 for a S_C-S_I transition [Brock et al., 1986].

4.10 Some Thermodynamic Properties

The distribution functions introduced earlier can be used to write down expressions for the various thermodynamic functions. Quite often these will be too complicated to be practically applicable as such, but they nevertheless constitute the basis for approximate formulations or for algorithms to be used in computer simulations. Some relevant formulas particularly useful for this purpose are given here.

Free energy. The canonical Helmoltz free energy (Chapter 2) is related to the configurational partition:

$$\mathcal{A} = -k_B T \ln Q(N, V, T), \tag{4.117}$$

with $Q(N, V, T) \equiv Z(N, V, T)/N!$ Note that the free energy as such is not a directly observable quantity, unlike its derivatives with respect to temperature and volume.

Entropy.

$$\mathcal{S} = -\left(\frac{\partial \mathcal{A}}{\partial T}\right)_V = -k_B \left\langle \ln P(\widetilde{X})\right\rangle = -k_B \int d\widetilde{X}\; P(\widetilde{X}) \ln P(\widetilde{X}). \tag{4.118}$$

Energy. We start from the classical thermodynamic formula and obtain a formula in terms of $Q(N, V, T)$

$$\mathcal{U} = \mathcal{A} + T\mathcal{S} = \mathcal{A} - T\left(\frac{\partial \mathcal{A}}{\partial T}\right)_V = -T^2\left(\frac{\partial}{\partial T}\left(\frac{\partial \mathcal{A}}{\partial T}\right)_V\right)_V = \left(\frac{\partial(\mathcal{A}/T)}{\partial(1/T)}\right)_V, \tag{4.119a}$$

$$= k_B T^2 \left(\frac{\partial \ln Q(N, V, T)}{\partial T}\right)_V. \tag{4.119b}$$

The total configuration energy of a system of N particles is most often assumed to be a sum of pairwise contributions, as we shall discuss in detail in Chapter 5, and in turn the observed average can be written in terms of the pair distribution. For pairwise interactions,

$$U_N = \frac{1}{2}\sum_i^N \sum_{j \neq i}^N U(\boldsymbol{X}_i, \boldsymbol{X}_j) \tag{4.120}$$

and the average internal energy is

$$\mathcal{U} = \langle U_N \rangle = \frac{1}{2N(N-1)} \int d\boldsymbol{X}_1 d\boldsymbol{X}_2 P^{(2)}(\boldsymbol{X}_1, \boldsymbol{X}_2)\, U(\boldsymbol{X}_1, \boldsymbol{X}_2). \tag{4.121}$$

For rigid anisotropic particles,

$$\mathcal{U} = \frac{1}{2}\rho^2 \int d\boldsymbol{r}_1 d\Omega_1 d\boldsymbol{r}_2 d\Omega_2\, G(\boldsymbol{r}_1 \Omega_1, \boldsymbol{r}_2, \Omega_2)\, U(\boldsymbol{r}_1, \Omega_1, \boldsymbol{r}_2, \Omega_2) \tag{4.122}$$

and for a uniform system, e.g. a nematic,

$$\mathcal{U} = \frac{V}{2}\rho^2 \int d\boldsymbol{r} d\Omega_1 d\Omega_2 P(\Omega_1) P(\Omega_2) g(\boldsymbol{r}, \Omega_1, \Omega_2)\, U(\boldsymbol{r}, \Omega_1, \Omega_2). \tag{4.123}$$

Heat capacity. The constant volume specific heat is given by Eqs. 2.7 and 2.9. Differentiating the microscopic expression for the energy (Eq. 4.45) with respect to temperature we can show, that the heat capacity is related at constant N, V, T to the mean square fluctuations in the energy:

$$C_V = \left(\frac{\partial \mathcal{U}}{\partial T}\right)_V = \left(\frac{\partial \langle U \rangle}{\partial T}\right)_V = \frac{1}{k_B T^2}\left(\langle U_N^2 \rangle - \langle U_N \rangle^2\right), \tag{4.124a}$$

$$= \frac{1}{4k_B T^2}\left\{ \sum_{\substack{i,j,k,l=1 \\ i\neq j\neq k\neq l}}^{N} \left[\langle U(\mathbf{X}_i,\mathbf{X}_j)U(\mathbf{X}_k,\mathbf{X}_l)\rangle - \langle U(\mathbf{X}_i,\mathbf{X}_j)\rangle\langle U(\mathbf{X}_k,\mathbf{X}_l)\rangle \right] \right.$$

$$\left. + \sum_{\substack{i,j,=1 \\ i\neq j}}^{N} \langle U(\mathbf{X}_i,\mathbf{X}_j)^2\rangle - \langle U(\mathbf{X}_i,\mathbf{X}_j)\rangle^2 \right\}, \tag{4.124b}$$

$$= \frac{1}{2}\rho^2 \int \mathrm{d}\boldsymbol{r}_1 \mathrm{d}\Omega_1 \mathrm{d}\boldsymbol{r}_2 \mathrm{d}\Omega_2\, U(\boldsymbol{r}_1,\Omega_1,\boldsymbol{r}_2,\Omega_2)\, \frac{\partial}{\partial T} G(\boldsymbol{r}_1\Omega_1,\boldsymbol{r}_2,\Omega_2), \tag{4.124c}$$

where the last equation applies to rigid molecules and temperature independent potentials. Eq. 4.124a shows that C_V is a non-negative quantity. This in turn implies from Eq. 2.9 that the free energy has to be a downward concave function of temperature (see Fig. 2.8), since its second derivative with respect to temperature has to be negative. Note that the heat capacity is not a pairwise quantity, even if the potential is a pairwise one. C_V does not depend only on the pair distribution at a given temperature but also on its derivative. If we try to perform the derivative we see that the microscopic expression depends on more than two particles simultaneously. Thus, the specific heat is really a collective property and it is reasonable that it can change significantly and diverge at a phase transition where the collective organization changes. Similar formulas hold for the constant pressure heat capacity in terms of the enthalpy $\mathcal{H}$ of the N particle system:

$$C_P = \left(\frac{\partial \mathcal{H}}{\partial T}\right)_P = \frac{1}{k_B T^2}\left(\langle \mathcal{H}^2 \rangle - \langle \mathcal{H} \rangle^2\right). \tag{4.125}$$

Pressure. We start from the thermodynamic definition of pressure for a system of N particles in equilibrium at canonical conditions:

$$P = -\left(\frac{\partial \mathcal{A}}{\partial V}\right)_T = k_B T \left(\frac{\partial \ln Q(N,V,T)}{\partial V}\right)_T = \frac{k_B T}{Q(N,V,T)}\left(\frac{\partial \ln Q(N,V,T)}{\partial V}\right)_T. \tag{4.126}$$

To derive a molecular expression for the pressure we can render the volume dependence of the configurational integral $Q(N,V,T)$ an explicit one by changing the positional variable $\boldsymbol{r}_i$ to dimensionless units $\boldsymbol{s}_i$. Thus, letting

$$\boldsymbol{r}_i = V^{\frac{1}{3}} \boldsymbol{s}_i, \tag{4.127}$$

for $i = 1, \ldots, N$ and, using as before a tilde to indicate collectively the set of N coordinates

$$Q(N,V,T) = \frac{V^N}{N!}\int \mathrm{d}\widetilde{s}\, \mathrm{d}\widetilde{\Omega}\; \mathrm{e}^{-U_N(\widetilde{s},\, \widetilde{\Omega})/(k_B T)}. \tag{4.128}$$

$$P = \frac{Nk_BT}{V} - \frac{V^N}{Z_N}\int \mathrm{d}\widetilde{s}\,\mathrm{d}\widetilde{\Omega}\,\frac{\partial U_N(\widetilde{s},\widetilde{\Omega})}{\partial V}\,\mathrm{e}^{-U_N(\widetilde{s},\widetilde{\Omega})/(k_BT)}, \tag{4.129a}$$

$$= \frac{Nk_BT}{V} - \left\langle \frac{\partial U_N(\widetilde{\boldsymbol{r}},\widetilde{\Omega})}{\partial V}\right\rangle, \tag{4.129b}$$

where the first term is the ideal gas contribution. For pairwise interactions, (Eq. 4.120) we can write

$$\frac{\partial U_N(\widetilde{\boldsymbol{r}},\widetilde{\Omega})}{\partial V} = \sum_{i<j}\frac{\partial U(\boldsymbol{r}_{ij},\Omega_i,\Omega_j)}{\partial \boldsymbol{r}_{ij}}\cdot\frac{\partial \boldsymbol{r}_{ij}}{\partial V}, \tag{4.130}$$

where, differentiating Eq. 4.127, $\partial \boldsymbol{r}_{ij}/\partial V = \boldsymbol{r}_{ij}/(3V)$. This gives the *virial* equation for pairwise interactions. For rigid anisotropic particles

$$P = \rho k_BT - \frac{1}{3V}\sum_i\sum_{j>i}\boldsymbol{r}_{ij}\cdot\left\langle\nabla_{\boldsymbol{r}_{ij}}U(\boldsymbol{r}_{ij},\Omega_i,\Omega_j)\right\rangle_{\Omega_i,\Omega_j}, \tag{4.131a}$$

$$= \rho k_BT - \frac{1}{3V}\sum_i\sum_{j>i}r_{ij}\left\langle\frac{\partial U\left(r_{ij},\Omega_i,\Omega_j\right)}{\partial r_{ij}}\right\rangle_{\Omega_i,\Omega_j}, \tag{4.131b}$$

i.e. P depends on derivatives of the centre of mass separation between particles i and j taken at constant orientation of the molecules [Allen and Tildesley, 2017]. For a system of spherical particles (or atoms) we have simply

$$P = \rho k_BT + \frac{1}{3V}\sum_i\sum_{j>i}\langle\boldsymbol{r}_i\cdot\boldsymbol{f}_{ij}\rangle, \tag{4.132}$$

recalling that the force acting on a particle i due to a particle j is the gradient on the potential energy felt by that particle: $\boldsymbol{f}_{ij} = -\nabla_i U_{ij} = -\boldsymbol{f}_{ji}$ and the virial equation of state

$$P = \rho k_BT - \frac{2}{3}\pi\rho^2\int \mathrm{d}r_{ij}\,g(r_{ij})\,r_{ij}^3\frac{\mathrm{d}U_{ij}(r)}{\mathrm{d}r_{ij}}. \tag{4.133}$$

Writing the temperature in terms of the kinetic energy K, and using the equipartition theorem which assigns $k_B/2$ to each of the n_f independent degrees of freedom for each particle:

$$T = \frac{2K}{n_fNk_\mathrm{B}} = \frac{1}{n_fNk_\mathrm{B}}\sum_{i=1}^{N}m_i\boldsymbol{v}_i\cdot\boldsymbol{v}_i, \tag{4.134}$$

we can get an expression suitable for microscopic calculations. More generally, one can define a *pressure tensor* $\boldsymbol{\Pi}$, with elements

$$\Pi_{ab} = \frac{1}{V}\sum_{i=1}^{N}\left(m_iv_{i,a}v_{i,b} + r_{i,a}f_{i,b}\right), \quad a,b = x,y,z. \tag{4.135}$$

For bulk fluids the pressure is just $P = \frac{1}{3}\mathrm{Tr}(\boldsymbol{\Pi})$ [Heinz et al., 2005]. This is obvious for an isotropic liquid, but it has been shown to hold also for nematic liquid crystals [Allen and Masters, 1993]. Note, however, that the thermodynamics expressions for the pressure just shown give a global, sample averaged value. It may be convenient, particularly for

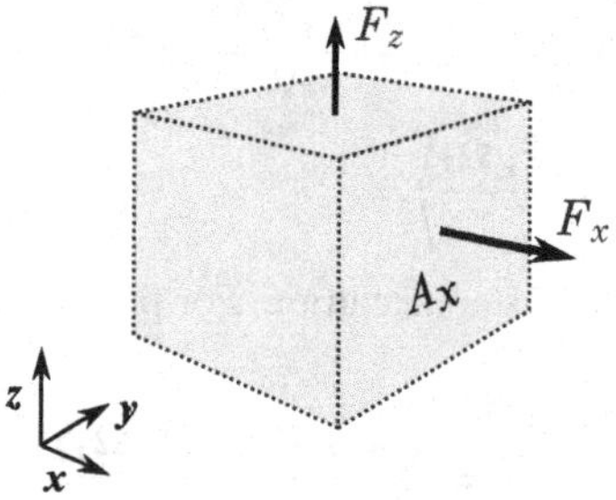

Figure 4.14 A small volume element $\mathcal{V}$ used to define local mechanical pressure or stress. We show a facet A_x perpendicular to $\boldsymbol{x}$ and the force components F_x and F_z.

inhomogeneous systems (e.g. when interfaces are present) to resort to the mechanical definition [Heinz et al., 2005; Berendsen, 2007] of pressure as a force acting over a certain local surface (Fig. 4.14). We can imagine dividing the sample into a 3D grid of very small virtual cubes (voxels) of volume $\mathcal{V}$ centred at a point $\boldsymbol{r}$ that contains the c.o.m. of $N(\boldsymbol{r})$ particles and limited by surfaces A_b perpendicular to the laboratory frame axes. If $F_a(\boldsymbol{r})$ is the component of the force in the material along axis α and acting on the surface A_β perpendicular to β, then the element $\Pi_{ab}(\boldsymbol{r})$ of the local pressure tensor $\boldsymbol{\Pi}(\boldsymbol{r})$ is:

$$\Pi_{ab}(\boldsymbol{r}) = \frac{F_a(\boldsymbol{r})}{A_b}. \tag{4.136}$$

Note that this is the negative of the stress tensor $\boldsymbol{\sigma}$ used to characterize hard and soft solids, polymers and elastomers, in particular liquid crystal elastomers (see Sections 1.15 and 1.16). The diagonal components $\sigma_{xx}, \sigma_{yy}, \sigma_{zz}$ are the normal (*tensile* or *compressive*) stresses. The off-diagonal components, $\sigma_{xy}, \sigma_{xz}, \sigma_{yz}, \ldots$, are instead *shear stresses*. The force component $F_a(\boldsymbol{r})$ across the surface A_b can be calculated summing over the pairwise forces $f_{ij,a}$ between two particles i, j whose separation vector $\boldsymbol{r}_{ij}$ intersects surface A_b [Irving and Kirkwood, 1950; Heinz et al., 2005]

$$\Pi_{ab}(\boldsymbol{r}) = \frac{1}{\mathcal{V}} \sum_{\substack{i=1 \\ i \in V(\boldsymbol{r})}}^{N(\boldsymbol{r})} m_i v_{i,a} v_{i,b} + \frac{1}{A_b} \sum_{\substack{i=1 \\ r_{ij} \cap A_\beta}}^{N(\boldsymbol{r})} f_{ij,a}. \tag{4.137}$$

The sign convention assumed is f_{ij} is the force on the particle with higher a coordinate. The averages involved in the previous equations can be formally written in term of the pair distribution, e.g.

$$P = \rho k_B T - \frac{\rho^2}{6V} \int \mathrm{d}\boldsymbol{X}_1 \mathrm{d}\boldsymbol{X}_2 \, G(\boldsymbol{X}_1, \boldsymbol{X}_2) \left(\boldsymbol{r} \cdot \nabla_1 U(\boldsymbol{X}_1, \boldsymbol{X}_2)\right). \tag{4.138}$$

Thermal expansion coefficient. We just recall the definition

$$\alpha_p \equiv \frac{1}{V}\left(\frac{\partial V}{\partial T}\right)_{P,N} = \frac{1}{V}\left(\frac{\partial^2 \mathcal{G}}{\partial T \partial P}\right)_N = -\frac{1}{V}\left(\frac{\partial \mathcal{S}}{\partial P}\right)_{T,N}. \tag{4.139}$$

Isothermal compressibility. The isothermal compressibility, also called inverse isothermal bulk modulus, is defined as (see Eq. 2.15)

$$\kappa_T = -\frac{1}{V}\left(\frac{\partial V}{\partial P}\right)_T = -\frac{1}{V}\left(\frac{\partial^2 G}{\partial P^2}\right)_T = \frac{1}{\rho}\left(\frac{\partial \rho}{\partial P}\right)_T. \tag{4.140}$$

The compressibility can be related to the mean square fluctuations in the volume or in the number of particles [Allen and Tildesley, 2017]

$$\kappa_T = \frac{V}{k_B T}\frac{\langle N^2\rangle - \langle N\rangle^2}{\langle N\rangle^2} = \frac{1}{\rho k_B T}\frac{\langle(\delta N)^2\rangle}{\langle N\rangle} = \frac{\langle(\delta V)^2\rangle_{NPT}}{k_B T\langle V\rangle_{NPT}}, \tag{4.141}$$

where $\delta N = N - \langle N\rangle$. Thus, $\kappa_T > 0$. The last expression is particularly convenient to obtain $\kappa_T > 0$ from simulations. We can rewrite κ_T in terms of distributions writing $\langle N^2\rangle - \langle N\rangle^2 = \langle N\rangle + \langle N(N-1)\rangle - \langle N\rangle^2$ and recalling the normalization of $P^{(n)}$ (Eq. 4.52), we have for large N,

$$\kappa_T = \frac{1}{\rho k_B T} + \frac{V}{k_B T}\int \mathrm{d}\boldsymbol{X}_1\mathrm{d}\boldsymbol{X}_2\big[P^{(2)}(\boldsymbol{X}_1,\boldsymbol{X}_2) - P^{(1)}(\boldsymbol{X}_1)P^{(1)}(\boldsymbol{X}_2)\big], \tag{4.142}$$

where $\rho = \langle N\rangle/V$. For a uniform fluid, e.g. a nematic formed of rigid mesogens, we can write $P^{(1)}(\boldsymbol{X}_1) = \rho P(\Omega_1)$ and $P^{(2)}(\boldsymbol{X}_1,\boldsymbol{X}_2) = \rho^2 V G(r,\Omega_1,\Omega_2,\Omega_r) = \rho^2 V P(\Omega_1)$ $P(\Omega_2)g(r,\Omega_1,\Omega_2,\Omega_r)$, so that we have, integrating over r_1,

$$\kappa_T = \frac{1}{\rho k_B T} + \frac{1}{k_B T}\int \mathrm{d}\boldsymbol{r}\mathrm{d}\Omega_1\mathrm{d}\Omega_2\big[G(r,\Omega_1,\Omega_2,\Omega_r) - P(\Omega_1)P(\Omega_2)\big]. \tag{4.143}$$

In simple isotropic fluids, $P(\Omega_1) = 1/V_{\Omega_1}$ and $g(r,\Omega_1,\Omega_2,\Omega_r) = g(r,\Omega_{12})$

$$\kappa_T = \frac{1}{\rho k_B T} + \frac{4\pi}{k_B T V_\Omega}\int \mathrm{d}r r^2 \mathrm{d}\Omega_{12}\big[g(r,\Omega_{12}) - 1\big]. \tag{4.144}$$

For spherical particles,

$$\kappa_T = \frac{1}{\rho k_B T} + \frac{4\pi}{k_B T}\int \mathrm{d}r r^2\,[g(r) - 1]\,. \tag{4.145}$$

4.10.1 Virial Expansion

The equation of state for a realistic gas is approximated as a power series in the density ρ, the so called *virial expansion*:

$$\frac{P}{k_B T} = z\rho = \rho + B_2(T)\rho^2 + B_3(T)\rho^3 + \cdots, \tag{4.146}$$

where z is the *compressibility factor* $z = P/(\rho k_B T)$ and $B_2, B_3, \ldots, B_n$ are the second, third and nth virial coefficients. If the potential energy can be written as a sum of pair-wise interactions terms $U(\boldsymbol{X}_i,\boldsymbol{X}_j)$ (see, later. Eq. 5.2b), they are related to the integrals of combinations of the Mayer functions Φ_{ij}

$$\Phi_{ij} \equiv \Phi(\boldsymbol{X}_i,\boldsymbol{X}_j) \equiv \mathrm{e}^{-U(\boldsymbol{X}_i,\boldsymbol{X}_j)/(k_B T)} - 1. \tag{4.147}$$

In particular, for rigid particles, we have in terms of the configurational integrals Z_n

$$B_2(T) = \frac{-1}{2VV_\Omega^2}\left(Z_2 - Z_1^2\right) = \frac{-1}{2VV_\Omega^2}\int \mathrm{d}\boldsymbol{r}_1\mathrm{d}\boldsymbol{\Omega}_1\mathrm{d}\boldsymbol{r}_2\mathrm{d}\boldsymbol{\Omega}_2\, \Phi_{12}(\boldsymbol{r}_{12},\Omega_1,\Omega_2). \tag{4.148}$$

and, for fluids of spherical particles:

$$B_2(T) = -\frac{2\pi}{V}\int \mathrm{d}r\, r^2\left\{\mathrm{e}^{-U(r)/(k_BT)} - 1\right\}. \tag{4.149}$$

B_2 and the other virial coefficients depend on temperature. Moreover, $B_2 > 0$ for purely repulsive interaction and, we expect it to be positive, if the repulsive part of the potential dominates, and viceversa if the attractive part prevails. We can get an idea of this also taking the second virial from the van der Waals equation of state, Eq. 2.1, $B_2^{\mathrm{vdW}} = b - (a/Nk_BT)$. Vliegenthart and Lekkerkerker [2000] noticed that for isotropic attractive repulsive potentials of various types $B_2(T_c) \approx -6\mathcal{V}$ at the critical point, with $\mathcal{V} = (\pi/6)\sigma^3$ the spherical particle volume. We shall see some explicit examples of second virial coefficients in Chapter 5. Going to the third virial coefficient [Boublik and Nezbeda, 1986]:

$$B_3(T) = -\frac{1}{3VV_\Omega^3}\left(Z_3 - 3\frac{Z_2^2}{Z_1} + 3Z_2Z_1 - Z_1^3\right), \tag{4.150a}$$

$$= -\frac{1}{3VV_\Omega^3}\int \mathrm{d}\boldsymbol{X}_1\mathrm{d}\boldsymbol{X}_2\mathrm{d}\boldsymbol{X}_3\, \Phi(\boldsymbol{X}_1,\boldsymbol{X}_2)\,\Phi(\boldsymbol{X}_1,\boldsymbol{X}_3)\,\Phi(\boldsymbol{X}_2,\boldsymbol{X}_3). \tag{4.150b}$$

A virial series for the free energy has been obtained, going through the grand canonical ensemble, using diagrammatic methods [Balescu, 1975; Santos, 2016] to represent all the interactions among the particles as linear graphs and properly simplifying and reducing them to a cluster expansion. The virial series can also be obtained directly in the canonical ensemble approximating the configurational integral, Q_N, and Eq. 4.126, following Van Kampen [1961] (but see [Pulvirenti and Tsagkarogiannis, 2012] for a mathematically rigorous derivation).

4.10.2 Surface Tension

The thermodynamic definition relates the surface tension γ at constant temperature and pressure to the variation in free energy $\mathcal{A}$ as the area $A = A_{MM'}$ of the interface between two materials M, M' is changed

$$\gamma_{MM'} = \left(\frac{\partial \mathcal{A}}{\partial A}\right)_{T,V} = -k_BT\left(\frac{\partial \ln Q(N,V,T)}{\partial A}\right)_{T,V}. \tag{4.151}$$

To derive a molecular expression for the surface tension we can proceed as we did for the pressure to get a virial type expression, by making the surface dependence of the configurational integral $Q(N,V,T)$ an explicit one by changing the positional variable to dimensionless units. In practice, we can use $r_a = A^{\frac{1}{2}}s_a$, and $\partial r_a/\partial A = a/(2A)$, with $a = x, y, z$.

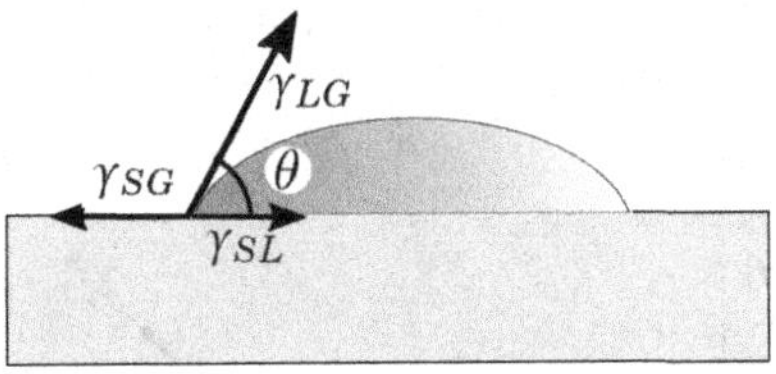

Figure 4.15 The contact angle θ for a liquid in contact with a solid surface. γ_{SG}, γ_{SL} and γ_{LG} are the surface tensions at the solid-gas, solid-liquid and liquid-gas interfaces.

The surface tension can be written in terms of the average of the potential energy derivative with respect to the area as [Harasima, 1958]

$$\gamma = \left\langle \frac{\partial U_N(\widetilde{\boldsymbol{r}}, \widetilde{\Omega})}{\partial A} \right\rangle = \frac{1}{2A} \sum_{i<j} \left\langle \boldsymbol{r}_{ij} \cdot \frac{\partial U(\boldsymbol{r}_{ij}, \Omega_i, \Omega_j)}{\partial \boldsymbol{r}_{ij}} - 3z_{ij} \frac{\partial U(\boldsymbol{r}_{ij}, \Omega_i, \Omega_j)}{\partial z_{ij}} \right\rangle, \tag{4.152a}$$

$$= \frac{1}{2V_z} \int_{-z'}^{z'} \mathrm{d}z \sum_{i<j} \left\langle \left(x_{ij} \frac{\partial U_{ij}}{\partial x_{ij}} + y_{ij} \frac{\partial U_{ij}}{\partial y_{ij}} \right) - 2z_{ij} \frac{\partial U_{ij}}{\partial z_{ij}} \right\rangle, \tag{4.152b}$$

where V_z is the volume of a thin element of thickness $2z'$ and area A equal to the sample cross section [Martin del Rio et al., 1995]. Surface tension values for some smectic materials have been obtained by literally blowing bubbles of a free-standing smectic film [Stannarius and Cramer, 1997] and using the simple relation between the pressure difference p and the local radius of curvature R, $p = 4\gamma/R$.

A balance of the forces at a contact point (see Fig. 4.15) gives, for an isotropic liquid and surface, the classical Young equation

$$\gamma_{SG} = \gamma_{SL} + \gamma_{LG} \cos\theta, \tag{4.153}$$

where θ is the *contact angle*. It tends to 0 if the liquid wets the surface and will be above 90° if, on the contary, the liquid is surface repellent. In the terminolgy developed for the common case of a water droplet, the surface would be called hydrophilic, and in the second case hydrophobic, or superhydrophobic if the contact angle tends to approach 180° with the droplet resembling a sphere just touching the surface. However, for an anisotropic fluid, like a liquid crystal, we can have [Vanzo et al., 2016]

$$\gamma_{SG} = \gamma_{SL}(\phi) + \gamma_{LG}(\phi) \cos\theta(\phi). \tag{4.154}$$

A few experimental results for the surface tension of nematics are: $\gamma = (26.8 \pm 0.3) \times 10^{-3}$ N/m at $T = 26.8°$ C for ZLI 4237-100 and $\gamma = (24.0 \pm 0.3) \times 10^{-3}$ N/m at $T = 26°$ for 8CB.

Note that the phase transition temperatures in a thin smectic film change from the bulk values. For instance, 8CB presents in bulk a S_A between 22 and 33°, while in free-standing films the smectic A has a wider range, between 12° and 33° [Dash and Wu, 1997].

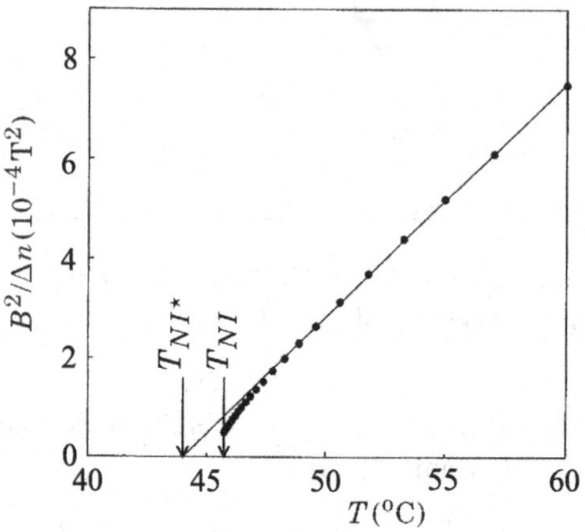

Figure 4.16 Inverse optical birefringence $B^2/\Delta n$ in the isotropic phase of MBBA as induced by an applied magnetic field with induction $B = 1.7\mathrm{T}$. The pretransitional optical birefringence Δn diverges at $T_{NI^\star}$, slightly below $T_{NI} = 45.7^\circ\mathrm{C}$ [Blachnik et al., 2000].

4.11 Pretransitional Behaviour

We shall now examine how the correlation coefficients are related to pretransitional behaviour in a nematic, considering the isotropic phase of a nematogen material above T_{NI}. As we have already seen the pair distribution for a macroscopically isotropic system depends in general on the distance and the relative orientation of the two molecules (see Eq. 4.92). As mentioned in Chapter 1, when a suitable external field is applied to an isotropic liquid a small long-range order is induced. The magnitude of this ordering is proportional to the intensity of the applied field while the proportionality constant is a material *susceptivity*. The induced anisotropy can be detected optically by monitoring the refractive index anisotropy Δn (see Fig. 1.5) or by using any other sensitive enough technique. These field induced effects are rather small for normal liquids, but very large ordering effects corresponding to electric and magnetic birefringence perhaps tens of times those of an ordinary liquid, take place in nematogen materials above their isotropic transition [Chandrasekhar, 1992]. A particularly neat demonstration of magnetic field (Cotton–Mouton) effect detected by NMR has been given [Attard et al., 1982; Luckhurst, 1988], while an NMR detected electric field (Kerr effect) has been reported by Hilbers and MacLean [1972] and Ruessink et al. [1988]. A typical feature in all these experiments is that the induced ordering increases as the temperature approaches the transition. However, these pretransitional effects do not diverge at the transition temperature but at a slightly lower one, $T_{NI^\star}$ [Stinson and Litster, 1970; Mukherjee, 1998], as we see, for example, in Fig. 4.16 [Blachnik et al., 2000] and we already mentioned in Sections 1.2.1, 2.7.2 and 2.9.1. The straight line in Fig. 4.16 is the linear fit corresponding to Landau theory expectations (see Eq. 2.28). Notice the deviation close to the transition where the simple theory is not adequate [Blachnik et al., 2000]. Similar deviations are also observed [Skačej and Zannoni, 2021] in computer simulations of the Lebwohl–Lasher lattice model (Chapter 10).

The conventional wisdom is that T_{NI*} is the temperature of transition of a hypothetical second-order phase transition that does not take place because it is intercepted by the occurrence of the first-order one. We can write down some theory of pretransitional effects and show that they depend on the pair orientational correlations existing between molecules even above the transition temperature. We start by considering the ordering induced by a uniform field which tends to align the liquid crystal molecules parallel to itself and which we could imagine to be a magnetic field, just to fix ideas. The average of a property A in the presence of a perturbation $\lambda U' = \lambda \mathscr{B} \phi(\widetilde{X})$ of intensity $\mathscr{B}$ dependent on the orientation of all molecules can be written as a series expansion in the small parameter λ as[1]

$$\langle A \rangle = \frac{\int \mathrm{d}\widetilde{X}\, A\, \mathrm{e}^{-\mathscr{U}_0}\, \mathrm{e}^{-\lambda \mathscr{U}'}}{\int \mathrm{d}\widetilde{X}\, \mathrm{e}^{-\mathscr{U}_0}\, \mathrm{e}^{-\lambda \mathscr{U}'}} = \frac{\int \mathrm{d}\widetilde{X} A\, \mathrm{e}^{-\mathscr{U}_0}[1 - \lambda \mathscr{U}' + \lambda^2 \mathscr{U}'^2 - \cdots]}{\int \mathrm{d}\widetilde{X}\, \mathrm{e}^{-\mathscr{U}_0}[1 - \lambda \mathscr{U}' + \lambda^2 \mathscr{U}'^2 - \cdots]}, \tag{4.155a}$$

$$= \langle A \rangle_0 - \lambda \langle A \mathscr{U}' \rangle_0 + \lambda \langle A \rangle_0 \langle \mathscr{U}' \rangle_0 + \mathscr{O}(\lambda^2), \tag{4.155b}$$

where $\mathscr{U}_0 \equiv U_0(\widetilde{X})/(k_B T)$ and $\mathscr{U}_0 \equiv U'(\widetilde{X})/(k_B T)$ are dimensionless and

$$\langle A \rangle_0 = \int \mathrm{d}\widetilde{X}\, A(\widetilde{X}) \exp[-\mathscr{U}_0(\widetilde{X})] \;\Big/ \int \mathrm{d}\widetilde{X} \exp[-\mathscr{U}_0(\widetilde{X})] \tag{4.156}$$

is the field-off average. If the phase is isotropic and A is anisotropic, then $\langle A \rangle_0 = 0$ and similarly, if the perturbation $\mathscr{U}'$ is anisotropic like in our case, then $\langle \mathscr{U}' \rangle_0 = 0$. Then

$$\langle A \rangle = -\lambda \langle A \mathscr{U}' \rangle_0 + \mathscr{O}(\lambda^2). \tag{4.157}$$

As a specific example, the anisotropic magnetic energy of a molecule with magnetic susceptibility anisotropy $\Delta\chi$ in a magnetic field $\boldsymbol{H}$ is

$$\mathscr{U}'_{\text{mag}} = -\frac{1}{3\mu_0} \frac{\Delta\chi H^2}{k_B T} \sum_{i=1}^{N} P_2(\cos\beta_i), \tag{4.158}$$

where β_i is the angle between $\boldsymbol{H}$ and the principal axis of the molecular frame that diagonalizes the diamagnetic susceptivity, so that in this case $\lambda = -\Delta\chi/(3\mu_0)$, $\mathscr{B} = H^2$. Another common example is that of an electric field $\boldsymbol{E}$, where for purely dielectric interactions with particles with dielectric anisotropy $\Delta\epsilon$, we would have

$$\mathscr{U}'_{\text{el}} = -\frac{\epsilon_0}{3} \frac{\Delta\epsilon E^2}{k_B T} \sum_{i=1}^{N} P_2(\cos\beta_i) \tag{4.159}$$

and $\lambda = -\epsilon_0 \Delta\epsilon/3$ and $\mathscr{B} = E^2$, while β_i is the angle between field direction and the principal axis of molecular susceptibility tensor, with its z-axis along its major value. The order parameter with respect to the applied field, 0 in the isotropic phase, in the presence of the field becomes (see Eq. 4.157)

1

$$\frac{a - \lambda b + \lambda^2 c}{1 - \lambda d + \lambda^2 e} = a + (ad - b)\lambda + \left(-ae + c + ad^2 - bd\right)\lambda^2 + O\left(\lambda^3\right).$$

$$\langle P_2 \rangle_{T,\mathscr{B}} = \xi \mathscr{B} = \frac{\lambda \mathscr{B}}{N k_B T} \left\langle \sum_{i=1}^{N} P_2(\cos \beta_i) \sum_{j=1}^{N} P_2(\cos \beta_j) \right\rangle_{T,\mathscr{B}} . \tag{4.160a}$$

The order parameter induced by the field is thus, in a linear regime, i.e. for not exceedingly strong fields, just proportional to the strength of the applied field, while the field susceptibility ξ is the derivative evaluated at 0 field

$$\xi = \lim_{\mathscr{B} \to 0} \frac{\partial}{\partial \mathscr{B}} \langle P_2 \rangle_{T,\mathscr{B}}. \tag{4.161}$$

Differentiating Eq. 4.160a with respect to $\mathscr{B}$, the susceptibility can thus be rewritten, recalling that that in the isotropic phase the unperturbed (zero field) order parameter is zero, as

$$\xi = (\lambda / N k_B T) \sum_i \sum_j \langle P_2(\cos \beta_i) P_2(\cos \beta_j) \rangle_{\mathscr{B}=0}, \tag{4.162a}$$

$$= (\lambda / k_B T) \sum_j \langle P_2(\cos \beta_1) P_2(\cos \beta_j) \rangle_{\mathscr{B}=0}. \tag{4.162b}$$

The two-particle average in Eq. 4.162b can be formally performed using the pair distribution for a macroscopically isotropic fluid, $g(r, \beta_{12})$. This in turn is rewritten in terms of the pairwise correlation $G_2(r)$ obtaining

$$\langle P_2(\cos \beta_1) P_2(\cos \beta_2) \rangle = \frac{1}{V} \int \mathrm{d}\boldsymbol{r} \, \mathrm{d}\cos \beta_1 \mathrm{d}\cos \beta_2 \; g(r, \beta_{12}) P_2(\cos \beta_1) P_2(\cos \beta_2),$$

$$= \frac{1}{5V} \int \mathrm{d}r r^2 \, g(r) \, G_2(r) = \frac{1}{5N} g_2, \tag{4.163}$$

where the coefficient g_2 is defined as the zero-field average

$$g_2 = \frac{5}{N} \sum_i \sum_j \langle P_2(\cos \beta_i) P_2(\cos \beta_j) \rangle_{\mathscr{B}=0} . \tag{4.164}$$

The field susceptibility ξ then becomes $\xi = (\lambda / k_B T) g_2$. The quantity g_2 contains the molecular information on the existing orientational pair correlations and is sometimes called a second-rank Kirkwood coefficient by analogy with the first-rank dipolar correlation coefficients introduced in the study of dielectric properties [Böttcher et al., 1973]. Having determined g_2, the divergence temperature $T_{NI^\star}$ can then be obtained by fitting T/g_2 versus T [Fabbri and Zannoni, 1986].

4.12 Pair Correlations and X-ray Scattering

Regularities, or in general non-uniformities, in a molecular organization can produce interferences with radiation of suitable wavelength and produce diffraction patterns that depend and somehow contain information on the organization itself. To show this, we consider again the X-ray experiment discussed in Appendix J and in Section 3.9. We have already examined the single molecule, self, contribution in Section 3.9 and we now focus on the intermolecular term to extract positional correlations and with some approximations, positional order parameters. Let us thus return to XRD experiments and consider the intermolecular contribution,

$$S_d(q) = \frac{1}{N} \sum_{\substack{i,j=1 \\ i \neq j}}^{N} \sum_{\substack{a,b=1 \\ a \in i, b \in j}}^{M} \left\langle e^{i[\boldsymbol{q} \cdot (\boldsymbol{r}_{a,i} - \boldsymbol{r}_{b,j})]} \right\rangle, \tag{4.165}$$

and see how it can provide information on pair correlations. We write $\boldsymbol{r}_{a,i} - \boldsymbol{r}_{b,j} = \boldsymbol{r}_{ij} + \boldsymbol{h}_{a,i} - \boldsymbol{h}_{b,j}$, where $\boldsymbol{h}_{a,i}$, $\boldsymbol{h}_{b,j}$ locate the nuclei a, b in their respective molecular frames and $\boldsymbol{r}_{ij}$ is the separation vector, joining the centres of molecules i and j. With these notations

$$S_d(q) = \frac{1}{N} \sum_{\substack{i,j=1 \\ i \neq j}}^{N} \sum_{\substack{a,b=1 \\ a \in i, b \in j}}^{M} \left\langle e^{i\boldsymbol{q} \cdot (\boldsymbol{h}_{a,i} - \boldsymbol{h}_{b,j})} e^{i\boldsymbol{q} \cdot \boldsymbol{r}_{ij}} \right\rangle. \tag{4.166}$$

We then use the Rayleigh expansion (see Eq. J.7) and find:

$$\begin{aligned} &\left\langle e^{i\boldsymbol{q} \cdot (\boldsymbol{h}_{a,i} - \boldsymbol{h}_{b,j})} e^{i\boldsymbol{q} \cdot \boldsymbol{r}_{ij}} \right\rangle \\ &\quad = \sum_{L,L',L''} i^{L+L'} (-i)^{L''} (L+1)(2L'+1)(2L''+1) \\ &\quad \times j_{L'}(q h_{a,i}) j_{L''}(q h_{b,j}) j_L(q r_{ij}) \left\langle \mathscr{D}^{L'}_{0,0}(\hat{\boldsymbol{q}} \cdot \hat{\boldsymbol{h}}_{a,i}) \mathscr{D}^{L''}_{0,0}(\hat{\boldsymbol{q}} \cdot \hat{\boldsymbol{h}}_{b,j}) \mathscr{D}^{L}_{0,0}(\hat{\boldsymbol{q}} \cdot \hat{\boldsymbol{r}}_{ij}) \right\rangle, \end{aligned} \tag{4.167a}$$

where $j_L(qr)$ is a spherical Bessel function [Abramowitz and Stegun, 1965]. Now, repeatedly applying the closure relation of Wigner matrices we have:

$$\mathscr{D}^{L'}_{0,0}(\hat{\boldsymbol{q}} \cdot \hat{\boldsymbol{h}}_{a,i}) = \sum_{m'} \mathscr{D}^{L'}_{0,m'}(\Omega_{dh_{a,i}}) \mathscr{D}^{L'}_{m',0}(\Omega_{qd}), \tag{4.168a}$$

$$= \sum_{m',n'} \mathscr{D}^{L'*}_{m',0}(\Omega_{qd}) \mathscr{D}^{L'}_{m',n'}(\Omega_{\mathrm{M}_i d}) \mathscr{D}^{L'}_{n',0}(\Omega_{h_{a,i}\mathrm{M}_i}) \tag{4.168b}$$

and

$$\mathscr{D}^{L''}_{0,0}(\hat{\boldsymbol{q}} \cdot \hat{\boldsymbol{h}}_{bj}) = \sum_{m'',n''} \mathscr{D}^{L''*}_{m'',0}(\Omega_{qd}) \mathscr{D}^{L''}_{m'',n''}(\Omega_{\mathrm{M}_j d}) \mathscr{D}^{L''}_{n'',0}(\Omega_{h_{bj}\mathrm{M}_j}), \tag{4.169}$$

$$\mathscr{D}^{L}_{00}(\hat{\boldsymbol{q}} \cdot \hat{\boldsymbol{r}}_{ij}) = \sum_{m} \mathscr{D}^{L*}_{m,0}(\Omega_{qd}) \mathscr{D}^{L}_{m,0}(\Omega_{r_{ij}d}), \tag{4.170}$$

where $(\Omega_{\mathrm{M}_i d})$ is the rotation from the lab (director) frame d to the ith molecule frame M_i, $(\Omega_{h_{a,i}\mathrm{M}_i})$ the rotation from the molecular frame to scattering centre a. Collecting terms, we find

$$\begin{aligned} &\left\langle \mathscr{D}^{L'}_{0,0}(\hat{\boldsymbol{q}} \cdot \hat{\boldsymbol{h}}_{a,i}) \mathscr{D}^{L''}_{0,0}(\hat{\boldsymbol{q}} \cdot \hat{\boldsymbol{h}}_{b,j}) \mathscr{D}^{L}_{0,0}(\hat{\boldsymbol{q}} \cdot \hat{\boldsymbol{r}}_{ij}) \right\rangle \\ &= \sum \left\langle \mathscr{D}^{L''*}_{m'',0}(\Omega_{qd}) \mathscr{D}^{L'*}_{m',0}(\Omega_{qd}) \mathscr{D}^{L*}_{m,0}(\Omega_{qd}) \right\rangle_d \mathscr{D}^{L'}_{n',0}(\Omega_{h_{a,i}\mathrm{M}_i}) \mathscr{D}^{L''}_{n'',0}(\Omega_{h_{bj}\mathrm{M}_j}) \\ &\quad \times \left\langle \mathscr{D}^{L'}_{m',n'}(\Omega_{\mathrm{M}_i d}) \mathscr{D}^{L''}_{m'',n''}(\Omega_{\mathrm{M}_j d}) \mathscr{D}^{L}_{m,0}(\Omega_{r_{ij}d}) \right\rangle, \end{aligned} \tag{4.171a}$$

where we have considered an average over the distribution of directors and a (separate) average over molecular reorientations. Director fluctuations are normally much slower than molecular reorientations and $\langle \ldots \rangle_d$ can be considered a static sum of contributions from the distribution of directors. We now consider two limiting cases, (i) for a locally ordered but macroscopically isotropic sample and (ii) for a monodomain.

(i) **Poly-liquid crystalline samples.** A first possibility is that of a non-aligned poly-liquid crystalline system, with an isotropic distribution of directors. For this, using the Gaunt integral of three Wigner rotations matrices (Eq. F.32), we get

$$\begin{aligned}
&\left\langle \mathscr{D}^{L'}_{0,0}(\hat{\boldsymbol{q}} \cdot \hat{\boldsymbol{h}}_{a,i}) \mathscr{D}^{L''}_{0,0}(\hat{\boldsymbol{q}} \cdot \hat{\boldsymbol{h}}_{b,j}) \mathscr{D}^{L}_{0,0}(\hat{\boldsymbol{q}} \cdot \hat{\boldsymbol{r}}_{ij}) \right\rangle \\
&\quad = \sum C(L', L'', L; 0, 0) \mathscr{D}^{L'}_{n',0}(\Omega_{h_{a,i}\mathrm{M}_i}) \mathscr{D}^{L''}_{n'',0}(\Omega_{h_{bj}\mathrm{M}_j}) \frac{(i)^{-L'+L''-L}}{\sqrt{2L+1}} \\
&\quad \times (-1)^{-n'-n''} \frac{(i)^{L'-L''+L}}{\sqrt{2L+1}} \sum_{m,m'} C(L', L'', L; -m', -m''), \\
&\left\langle \mathscr{D}^{L'*}_{-m',-n'}(\Omega_{\mathrm{M}_i d}) \mathscr{D}^{L''*}_{-m'',-n''}(\Omega_{\mathrm{M}_j d}) \mathscr{D}^{L}_{m,0}(\Omega_{r_{ij} d}) \right\rangle \qquad (4.172a) \\
&\quad = \sum C(L', L'', L; 0, 0) \mathscr{D}^{L'}_{n',0}(\Omega_{h_{a,i}\mathrm{M}_i}) \mathscr{D}^{L''}_{n'',0}(\Omega_{h_{bj}\mathrm{M}_j}) \frac{(i)^{-L'+L''-L}}{\sqrt{2L+1}} \\
&\quad \times (-1)^{-n'-n''} \left\langle S^{-n',-n''}_{L',L'',L}(\Omega_{1\mathrm{L}}, \Omega_{2\mathrm{L}}, \Omega_{r\mathrm{L}}) \right\rangle, \qquad (4.172b)
\end{aligned}$$

when recalling the definition of Stone invariants (Eq. G.19).

(ii) **Smectic A monodomain and positional-orientational order parameters.** We now derive, following Palermo et al. [2013], an explicit expression for the distinct molecules contribution to the scattered intensity $S_d(q)$ in the particular case of X-ray reflections from the smectic A monodomain planes with $\boldsymbol{q} = (0, 0, q_n)$, $q_n \equiv 2\pi n/d$, thus if the scattering vector is parallel to the laboratory, z axis, $\boldsymbol{q}||z$.

$$S_d(00n) = \frac{1}{N} \sum_{\substack{i,j \\ i \neq j}}^{N} \sum_{a \in i, b \in j}^{M} \left\langle \mathrm{e}^{i q_n z_{ij}} \, \mathrm{e}^{i q_n d \cdot (h_{a,i} - h_{b,j})} \right\rangle \qquad (4.173)$$

$$\begin{aligned}
&\left\langle \mathrm{e}^{i q_n z_{ij}} \, \mathrm{e}^{i \boldsymbol{q} \cdot \boldsymbol{h}_{a,i}} \, \mathrm{e}^{-i \boldsymbol{q} \cdot \boldsymbol{h}_{b,j}} \right\rangle \\
&\quad = \sum_{L,L'} e^{i\pi(L'-L'')/2} (2L'+1)(2L''+1) j_{L'}(q h_{a,i}) j_{L''}(q h_{b,j}) \sum_{\substack{m',n' \\ m'',n''}} \mathscr{D}^{L'*}_{m',0}(\Omega_{qd}) \mathscr{D}^{L''*}_{m'',0}(\Omega_{qd}) \\
&\quad \times \mathscr{D}^{L'}_{n',0}(\Omega_{h_{ai}\mathrm{M}_i}) \mathscr{D}^{L''}_{n'',0}(\Omega_{h_{bj}\mathrm{M}_j}) \left\langle \cos(q_n z_{ij}) \mathscr{D}^{L'}_{m',n'}(\Omega_{\mathrm{M}_i d}) \mathscr{D}^{L''}_{m'',n''}(\Omega_{\mathrm{M}_j d}) \right\rangle \qquad (4.174)
\end{aligned}$$

where $z_{ij} = z_i - z_j$. It is clear that the distinct term contains information relevant for smectic positional correlations, since it depends on molecule-molecule distances. Assuming a uniaxial smectic the director along z, as well as effective uniaxial molecular symmetry we have, in the particular case of reflections from the smectic A planes with $\boldsymbol{q} = (0, 0, q_n)$, $q_n \equiv 2\pi n/\ell_z$,

$$S_d(00n) = \frac{1}{N}\sum_{\substack{i,j\\ i\neq j}}^{N}\sum_{\substack{a\in i\\ b\in j}}^{M}\left\langle e^{iq_n z_{ij}}\, e^{iq_n \mathbf{z}\cdot \boldsymbol{h}_{a,i}}\, e^{-iq_n \mathbf{z}\cdot \boldsymbol{h}_{b,j}}\right\rangle, \tag{4.175}$$

where $z_{ij} = z_i - z_j$. Before dealing with Eq. 4.175 for a molecule containing a number of atoms located at a generic position $\boldsymbol{h}_a$ it is useful to see, with a simple special case, how information on translational order is contained in this expression. To do this, let us assume that the molecules contain only one relevant scattering centre (e.g. a metal atom) at the molecular origin, so that $\boldsymbol{h}_{a,i} = \boldsymbol{h}_{b,j} = 0$. In this case,

$$S_d(00n) = \frac{1}{N}\sum_{\substack{i,j\\ i>j}}^{N}\left\langle e^{iq_n z_{ij}}\right\rangle \approx \frac{1}{N}\sum_{\substack{i,j\\ i>j}}^{N}\left\langle e^{iq_n z_i}\right\rangle\left\langle e^{-iq_n z_j}\right\rangle \propto (\tau_n)^2. \tag{4.176}$$

So, as long as the approximation applied to Eq. 4.176 is valid, the dominant term is

$$S_d(001) \propto \tau_1^2 = \langle \cos(2\pi/\ell_z)\rangle^2. \tag{4.177}$$

We can now go back to Eq. 4.175 and apply the Rayleigh expansion, Eq. J.7, to the exponentials, e.g.

$$\exp[iq_n \mathbf{z}\cdot \boldsymbol{h}_{bj}] = \sum_L i^L (2L+1) j_L(q_n h_{bj}) \mathscr{D}^L_{0,0}(\hat{\mathbf{z}}\cdot \hat{\boldsymbol{h}}_{bj}). \tag{4.178}$$

Applying again the closure relation of Wigner rotation matrices (see Eq. F.13) we can introduce the rotation from (in this case parallel to the scattering vector) to molecular frame $\Omega_{M_j d}$ and finally that from molecular frame to atom b: $\Omega_{h_{bj}M_j}$

$$\mathscr{D}^L_{0,0}(\hat{\mathbf{z}}\cdot \hat{\boldsymbol{h}}_{bj}) = \mathscr{D}^L_{0,0}(\boldsymbol{d}\cdot \hat{\boldsymbol{h}}_{bj}) \tag{4.179a}$$

$$= \sum_p \mathscr{D}^{L*}_{0,p}(\Omega_{M_j d}) \mathscr{D}^{L*}_{p,0}(\Omega_{h_{bj}M_j}), \tag{4.179b}$$

In view of the assumed effective cylindrical symmetry of the molecules we have

$$\begin{aligned} S_d(00n) &= \frac{1}{N}\sum_{L,L'}\sum_{\substack{i,j\\ i\neq j}}^{N}\sum_{a\in i, b\in j}^{M} c_{nLL'}\left\langle \cos(q_n z_{ij}) P_L(\cos\beta_{ai}) P_{L'}(\cos\beta_{bj})\right\rangle_{ij}, \\ &= c_{n00}\left\langle \cos(q_n z_{ij})\right\rangle_{ij} + 2c_{n02}\left\langle \cos(q_n z_{ij}) P_2(\cos\beta_j)\right\rangle_{ij} \\ &\quad + c_{n22}\left\langle \cos(q_m n_{ij}) P_2(\cos\beta_i)\, P_2(\cos\beta_j)\right\rangle ij + \cdots . \end{aligned} \tag{4.180}$$

The first term is the only one retained in the classical formulation [Leadbetter, 1979; Seddon, 1998], which assumes, moreover,

$$\left\langle \cos(q_n z_{ij})\right\rangle_{ij} \approx \langle \cos(q_n z_i)\rangle^2. \tag{4.181}$$

Thus, for the first two reflections, the intensity is proportional to the first two translational order parameters: $S_d(001) \approx c_{100}\tau_1^2$ and $S_d(002) \approx c_{200}\tau_2^2$, and so on. The experimental determination of translational order parameters is, however, rather difficult and has been reported for only a very limited number of cases [Leadbetter and Norris, 1979; Kapernaum and Giesselmann, 2008]. Explicit results for positional and mixed translational-orientational order parameters have been obtained using atomistic simulations (see Chapter 12) by Palermo et al. [2013].

5

Particle–Particle Interactions

> It seems that in these liquid crystals, though the molecules are freely mobile, just as are those of water, they are yet subject to, or endowed with, a 'directive force', a force which confers upon them a definite configuration or 'polarity', the 'Gestaltungskraft' of Lehmann.
>
> *D'Arcy Thompson, On Growth and Form, 1917*

5.1 Intermolecular Interactions

Up to now we have seen what liquid crystals are and how their molecular organization can be described in terms of single and pair particle distributions. We have discussed how these can be expanded in terms of a linear combination of suitable orthogonal functions, but we have not yet started to examine which properties of nematogenic molecules or particles can give rise to a certain phase behaviour. The link between microscopic and macroscopic properties is based, as we have seen in Chapter 4, on the Gibbs–Boltzmann distributions for the various sets of experimental conditions. In particular, we have seen in Section 4.2 that the probability of occurring of a certain molecular configuration, at a certain temperature T and density ρ (canonical conditions) depends on the potential energy between molecules. Indeed, the starting point of any molecular modelling is the energy of interaction between molecules or more generally particles, and here we examine briefly some of the most important and most often used interactions. Starting for simplicity with spherical particles, the pair interaction potential $U_{ij}(r)$ will depend on their separation, and we expect it to contain a *steric* repulsive part at short distance (the space occupied by a particle being excluded to the others) as well as an attractive part at much larger separations. The two contributions will balance at some intermediate distance giving the classical profile sketched in Fig. 5.1. Both types of interactions are important and it is often believed that packing effects are mainly responsible for the molecular organization of solid and liquid phases [Chandler et al., 1983] and that attractive contributions mainly determine the density at given temperature and pressure and the vapour to liquid condensation. It is natural these days to think that the existence of liquid crystal phases results from a balance of various intermolecular interactions which are present in every compound, even though in different amounts. However, it is striking that, around one hundred years ago, the existence of fluid ordered phases was still so mysterious that a special driving force was invoked [Thompson, 1917].

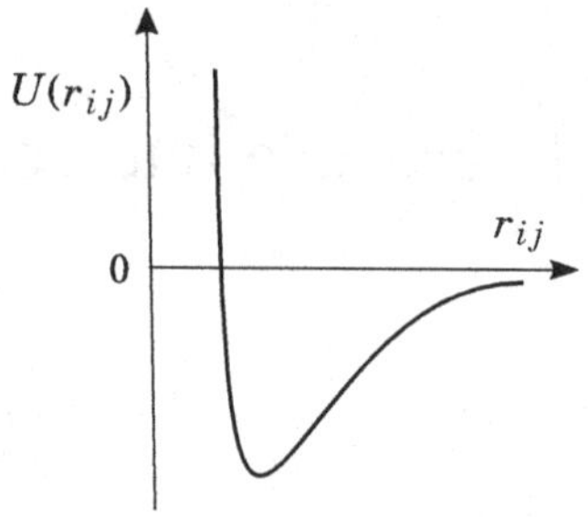

Figure 5.1 Schematic drawing of the pair potential $U(r_{ij})$ between two spherical particles i and j as a function of their separation r_{ij}.

Even if the total interaction energy of a system of N particles, U_N will depend on the coordinates of all of them, it would be unfeasible to write a functional form for this dependence. The potential energy of the system is then normally written as a sum of effective interaction terms involving more and more particles [Sinanoglu, 1967]:

$$U_N = U_N^{\text{total}} - \sum_i U_i^{\text{isolated}}.$$

$$= \frac{1}{2}\sum_{i \neq j} U_{i,j}^{\text{isolated}} + \frac{1}{3!}\sum_{i \neq j \neq k} U_{i,j,k}^{\text{isolated}} + \frac{1}{4!}\sum_{i \neq j \neq k \neq l} U_{i,j,k,l}^{\text{isolated}} + \cdots, \tag{5.1}$$

where the terms $U_{i,j}^{\text{isolated}}$, $U_{i,j,k}^{\text{isolated}}, \ldots$ represent the contribution from the interaction between two, three and increasingly larger sets of distinct constituents in the absence of the others. U_N tends to 0 when the particles are so far apart that they can be considered to be isolated, Fortunately, it is often a good approximation to assume that U_N can be written as a sum of 'effective' pairwise interactions U_{ij} so that:

$$U_N = U_N(\widetilde{\boldsymbol{X}}) \equiv U_N(\boldsymbol{X}_1, \boldsymbol{X}_2, \ldots, \boldsymbol{X}_N), \tag{5.2a}$$

$$= \frac{1}{2}\sum_{i \neq j} U_{i,j} = \sum_{i=1}^{N}\sum_{j>i}^{N} U(\boldsymbol{X}_i, \boldsymbol{X}_j), \tag{5.2b}$$

where $\boldsymbol{X}_i$ is, as in previous chapters, the set of coordinates specifying the state of molecule i and $\widetilde{\boldsymbol{X}}$ the collective set of $\boldsymbol{X}_i, i = 1, \ldots, N$. The interaction potential U_{ij} between two particles at a certain distance and orientation (or *pair potential*) corresponds to the difference between the energy of the two molecules in that configuration and, respectively, at infinite separation. Note that the word 'effective' for the pair potential underlines the fact that $U(\boldsymbol{X}_i, \boldsymbol{X}_j)$ is not the potential between two molecules in vacuum at 0 degrees Kelvin, as could be obtained with some Quantum Chemistry (QC) technique, even at a high level of sophistication (see, e.g., [Cramer, 2004]). Indeed, the 'effective' pair interaction normally contains a few parameters that, properly tuned, somehow take into account the effect of being in condensed medium, with the unavoidable presence of multiparticle interactions. Apart from this, many model potentials for the simulation of liquids or liquid crystals make other drastic assumptions, like that of considering the interacting particles as rigid bodies, neglecting internal degrees of freedom, or of assuming a particle symmetry higher than that of the real molecule. Looking at the many explicit chemical structures shown in Chapter 1 (see, e.g., Table 1.2) it is clear that this is a drastic approximation, given the nearly inevitable presence of flexible chains or of internal rotations between fragments that can alter the overall molecular shape as the conformations change. However, the approximation of rigidity is often used in so-called molecular resolution models where a generic particle rather than a detailed chemical structure is considered. If the aim is that of simulating systems of real molecules and predicting their properties, this can be viewed at most as a drastic form of preliminary coarse graining. In any case, leaving aside the quest of realism, it is of great interest to consider simplified models where only some type of interactions are retained (e.g. only repulsive or only attractive) to see what the minimal sufficient conditions are to obtain certain types of LC phases (e.g. nematics, or smectics). Moreover, as we have seen in Chapter 1, many liquid crystal phases are formed by suspensions of nano or colloidal particles or viruses, where dealing with chemical details is out of question. We shall thus consider both generic and atomistically detailed models, and start by dealing with rigid particles as constituents.

For rigid molecules the pair potential will in general depend on the positions $\boldsymbol{r}_i$, $\boldsymbol{r}_j$ and the orientations Ω_i, Ω_j of the two molecules with respect to some coordinate frame, i.e.

$$U_{ij} = U(\boldsymbol{r}_i, \Omega_i, \boldsymbol{r}_j, \Omega_j) = U(\boldsymbol{r}_{ij}, \Omega_i, \Omega_j), \tag{5.3}$$

where $\boldsymbol{r}_{ij} = \boldsymbol{r}_i - \boldsymbol{r}_j$ and the second equality holds since there is no privileged position in free space (translational invariance) as is the case in the absence of external inhomogeneous fields. Still talking in general terms, the potential will also be invariant for space inversion (changing the sign of all coordinates). In some cases, it can be useful to write the pair potential as a combination of terms with well-defined angular dependence. We can do this by expanding this unknown function of the orientations of the two particles and of the intermolecular vector $\boldsymbol{r}$ (see Fig. 5.2), i.e. of $\Omega_i \equiv (\alpha_i, \beta_i, \gamma_i)$, $\Omega_j \equiv (\alpha_j, \beta_j, \gamma_j)$, $\Omega_r \equiv (\alpha_r, \beta_r)$ in a basis set of products of Wigner matrices or, in the case of uniaxial molecules,

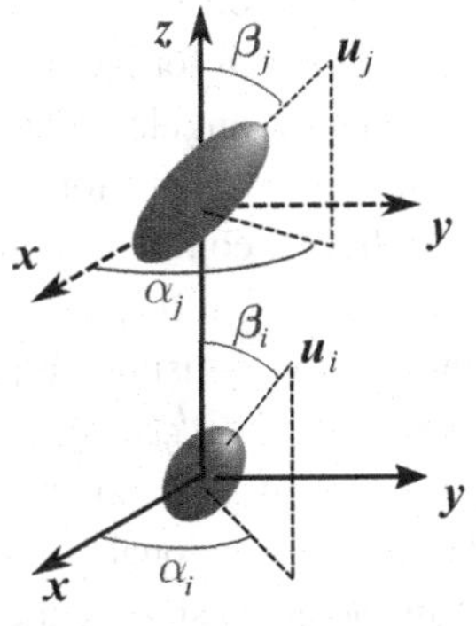

Figure 5.2 The intermolecular (IM) frame with its z-axis along the centre-centre vector $\boldsymbol{r}_{ij}$. Also shown are two uniaxial particles with their axes $\boldsymbol{u}_i$, $\boldsymbol{u}_j$ at angles (α_i, β_i), (α_j, β_j) with respect to the IM frame.

of spherical harmonics $\mathscr{D}^{L_i *}_{m_i,0}(\alpha_i, \beta_i)\mathscr{D}^{L_j}_{m_j,0}(\alpha_j, \beta_j)\ \mathscr{D}^{L}_{m,0}(\alpha_r, \beta_r)$, as we did in Chapters 3 and 4 for the single and pair orientational distributions (cf. also Appendix G). There are two coordinate frames that are, in different ways, convenient for this expansion: one is a laboratory (LAB) fixed one. The other, that we shall consider to begin with, is the intermolecular (IM) frame. In this system, that has the origin in one of the two molecules (typically in its centre of mass) and the z-axis pointing along the intermolecular vector $\boldsymbol{r}_{ij}$ joining the centres of the two molecules, we have [Steele, 1963; Sweet and Steele, 1967; Humphries et al., 1972]

$$U(r_{ij}, \Omega_{ir_{ij}}, \Omega_{jr_{ij}}) = \sum_{\substack{L_i, L_j,\\ m_i, m_j}} u_{L_i L_j m_i; n_i n_j}(r_{ij}) \delta_{m_i, m_j}\ \mathscr{D}^{L_i}_{m_i, n_i}(\Omega_{ir_{ij}}) \mathscr{D}^{L_j *}_{m_j, n_j}(\Omega_{jr_{ij}}), \qquad (5.4)$$

where the coefficients $u_{L_i L_j m_i; n_i n_j}(r_{ij})$ are related to the specific interactions between the two molecules. Note that we have taken $m_i = m_j$, since the potential has to be invariant for an arbitrary rotation, of say, an angle ϕ around the IM frame (z-axis here). Such a rotation would give, recalling the form of the Wigner matrices (Eq. F.4), a multiplication for $\exp[-i\phi(m_i - m_j)]$, so that invariance requires δ_{m_i, m_j}. For uniaxial molecules a similar argument for an arbitrary rotation around molecular axes z_i, z_j gives the restriction $\delta_{n_i,0}$, $\delta_{n_j,0}$, thus Eq. 5.4 simplifies, for cylindrically symmetric molecules to a spherical harmonics expansion:

$$U(r_{ij}, \Omega_{ir_{ij}}, \Omega_{jr_{ij}}) = \sum_{L_i, L_j, m} u_{L_i L_j m}(r_{ij}) \mathscr{D}^{L_i}_{m,0}(\Omega_{ir_{ij}}) \mathscr{D}^{L_j *}_{m,0}(\Omega_{jr_{ij}}), \qquad (5.5a)$$

$$= \sum_{L_i, L_j, m} y_{L_i L_j m}(r_{ij}) Y_{L_i, m}(\beta_{ir_{ij}}, \alpha_{ir_{ij}}) Y_{L_j, -m}(\beta_{jr_{ij}}, \alpha_{jr_{ij}}), \qquad (5.5b)$$

with $y_{L_i L_j m}(r_{ij}) \equiv 4\pi(-1)^m u_{L_i L_j m}(r_{ij})/\sqrt{(2L_i + 1)(2L_j + 1)}$ and the index m going from $-\min(L_i, L_j)$ to $\min(L_i, L_j)$. Since the potential has to be real, the coefficients have the property $u_{L_i L_j m} = u_{L_j L_i -m}$. If molecules i and j are identical, a permutation of the two particles should also leave the potential invariant. This implies that $u_{L_i L_j m} = u_{L_j L_i m}$ in Eq. 5.5a. The general expansion will of course be truncated retaining only a few terms and some empirical form will often be chosen for the remaining coefficients, but Eq. 5.5a shows, at every level of truncation, what the angular dependence will be.

Another form of the potential, more convenient for computer simulations, is obtained transforming Eq. 5.4 to a laboratory frame, common to all ij molecular pairs. To emphasize the axis systems involved it is convenient to rewrite the orientations referred to the IM vector $\boldsymbol{r}_{ij}$ as rotations from the laboratory using the closure relation of Wigner matrices (cf. Eq. F.13), i.e. $\mathscr{D}^{L}_{m,n}(\Omega_{ir_{ij}}) = \sum_{q=-L}^{L} \mathscr{D}^{L*}_{qm}(\Omega_{r_{ij}\mathrm{L}}) \mathscr{D}^{L}_{q,n}(\Omega_{i\mathrm{L}})$. The potential should also be invariant for an arbitrary rotation of the laboratory frame. It is then convenient to expand the pair potential in terms of the rotational invariants (see Appendix G) [Stone, 1978] that we have already used in the expansion of the pair correlation in Section 4.8. We then have

$$U(r_{ij}, \Omega_{i\mathrm{L}}, \Omega_{j\mathrm{L}}, \Omega_{r_{ij}\mathrm{L}}) = \sum u^{k_1 k_2}_{L_i L_j L}(r_{ij}) S^{k_i, k_j}_{L_i, L_j, L}(\Omega_{i\mathrm{L}}, \Omega_{j\mathrm{L}}, \Omega_{r_{ij}\mathrm{L}}), \qquad (5.6)$$

where the rotationally invariant functions $S^{k_i,k_j}_{L_i,L_j,L}(\Omega_i,\Omega_j,\Omega_{r_{ij}})$ are defined as in Eq. G.19. The coefficients of the laboratory and intermolecular expansion can be related to each other using the properties of rotation matrices (cf. Appendix G), obtaining

$$u^{n_i n_j}_{L_i L_j L}(r_{ij}) = \sum (-)^{m_i} C(L_i L_j L; -m_i m_i)\, u_{L_i L_j m_i; n_i n_j}(r_{ij}), \tag{5.7}$$

where $C(L_i L_j L; -m_i m_i)$ are Clebsch–Gordan coefficients (Section F.1.1). In practice, we shall find the expansion in the IM frame more convenient when we examine the explicit orientation dependence of the intermolecular potential, while the LAB frame expression in terms of invariants is more apt to formal theoretical treatments, e.g. the Mean Field Theory discussed in Chapter 7. It is, however, worth pointing out that the general expansions in Eqs. 5.4 and 5.6 that start from spherically symmetric potentials and introduce anisotropy by adding angular dependent contributions of progressively increasing rank L_i, L_j, although quite elegant do not necessarily converge. In particular, they do not converge well for molecules that significantly deviate from spherical symmetry, like the rod-like or disc-like particles that typically give rise to LCs. In practice, we will then have to resort to either empirical or atomistic potentials without systematically expand their angular dependence. Here we shall discuss both, starting with spherical particles, that can be particularly useful, when suitably assembled together, for building models of particles of complex shape.

5.2 Spherical Particles

This first term in the pair potential expansion contains every type of interaction depending on inter-particle separation alone: $U(r_{ij}) = u_{000}(r_{ij})$. This is of course the only term present in systems formed of truly spherical particles such as noble gases or, more appropriately for soft matter, many colloidal particles. A number of empirical potentials have been introduced and we briefly discuss a few of the most important and commonly used ones in what follows.

5.2.1 Hard Spheres (HS)

In this case we have no attraction between particles of diameter σ_h, but also no possibility for interpenetration. The pair potential is

$$U^{\text{HS}}(r_{ij}) = \begin{cases} 0, & \text{for } r_{ij} > \sigma_h \\ \infty, & \text{for } r_{ij} \leq \sigma_h \end{cases}. \tag{5.8}$$

Note that for this potential, like for any other hard particle interaction, the effect of temperature, normally arising through the Boltzmann factor is cancelled out. Thus, the partition function and all other thermodynamic quantities resulting from this type of potential will depend on density alone and the driving forces for going from one phase to another will be the entropy of the system at a certain density. A number of quantities that we have previously introduced can be calculated exactly for this potential. An example is the second virial coefficient (see Eq. 4.20) for the hard sphere model, trivially obtained by integration as

$$B_2^{\text{HS}} = 2\pi \int_0^{\sigma_h} dr_{ij} r_{ij}^2 = \frac{2\pi}{3}\sigma_h^3 = 4V_h, \tag{5.9}$$

where $\mathcal{V}_h$ is the particle volume. A number of higher virial coefficients have been obtained, analytically up to B_4^{HS} [Santos, 2016]:

$$B_3^{\mathrm{HS}} = \frac{5}{8}\left(B_2^{\mathrm{HS}}\right)^2, \quad B_4^{\mathrm{HS}} = \frac{2707}{4480} + \frac{219\sqrt{2}}{2240\pi} - \frac{4131\cos^{-1}(1/3)}{4480\pi}\left(B_2^{\mathrm{HS}}\right)^3 \tag{5.10}$$

and then, up to B_{11}^{HS} numerically [Schultz and Kofke, 2014]. The HS model has a first-order freezing transition, as shown since the early days of computer simulations by Alder and Wainwright [1957] and Wood and Jacobson [1957]. This is somehow surprising since the behaviour for a hard-particle system is entropy controlled and common sense would suggest the entropy of the crystal be lower than that of the disordered fluid. This turns out not to be the case as discussed by Chaikin et al. [2006]. To briefly summarize the argument, we can start from the entropy per particle of the ideal gas: $\mathcal{S}_i = k_B \ln V$. The van der Waals correction for the available volume gives

$$\mathcal{S} = k_B \ln(V - V_p) = k_B \ln[V(1-\phi)], \tag{5.11}$$

with $\phi = V_p/V$ the packing fraction i.e. the volume fraction V_p occupied by the particles. In a similar way, for a hard particle system, considering that not all the unoccupied volume is reachable by the particles and thus available, there is an actual maximum volume V_{max}, so that

$$\mathcal{S} = k_B \ln[V\{1 - (\phi/\phi_{max})\}]_{\phi\to\phi_{max}}, \tag{5.12}$$

where $\phi_{max} = V_{max}/V$ corresponds to the maximum packing ratio that can be reached for a certain particle organization (crystal, disordered, ...). Finding the densest packing for a given particle shape is a basic problem in geometry. According to the famous Kepler conjecture, that has only been proved rather recently [Lagarias, 2011], the densest arrangement of spheres in 3D has a packing fraction $\phi_{max} = \pi/\sqrt{18} \approx 0.7405$, and is crystalline, realized by stacking variants of the face-centered cubic (FCC) lattice packing [Donev et al., 2004a]. If instead we are trying to find the densest packing experimentally (e.g. by shaking arrays of spheres) or by computer simulations, we see that for a disordered (random) organization of particles, the systems gets stuck at a packing ϕ_J and that we are not able to add more spheres beyond that maximum volume fraction. Thus,

$$\mathcal{S} = \mathcal{S}_i + k_B \ln\{1 - (\phi/\phi_J)\}_{\phi\to\phi_J}. \tag{5.13}$$

To decide if the lowest free energy (or highest entropy, being the system devoid of internal energy) is for this random (glassy) state or for the crystal, we should consider that if the organization is a crystalline one, the maximum packing will be ϕ_{cryst}. When $\phi_{cryst} > \phi_J$ there will be a crystallization for entropic effect in the HS system. In practice, for hard spheres $\phi_{cryst} = 0.74$, while $\phi_J \approx 0.64$ [Torquato and Stillinger, 2010] and thus crystallization takes place! The crystal formed stacking the spheres can be FCC or hexagonal closed packed (HCP), both with the same closed packed volume. Woodcock [1997] has found by computer simulations a slight but significative free energy difference, with the FCC more stable by about 0.005 RT per mol.

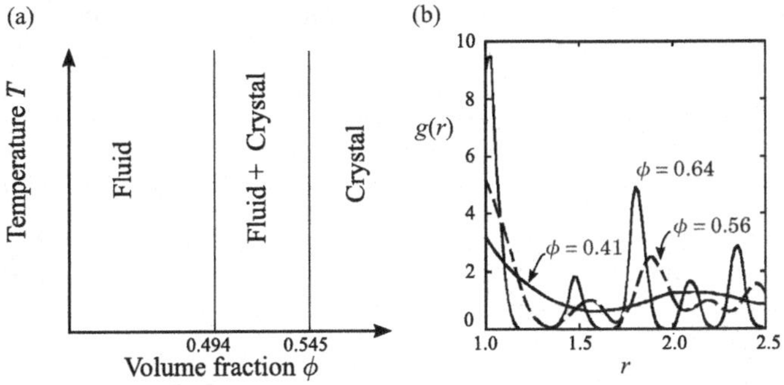

Figure 5.3 (a) The phase diagram for hard spheres [Lekkerkerker and Anderson, 2002] and (b) its pair distribution $g(r)$, as a function of scaled distance r in units of σ_h at volume fraction $\phi = 0.41$ (fluid), $\phi = 0.56$ (approaching fluid-solid coexistence) and $\phi = 0.64$ (crystal) [Rosenbluth and Rosenbluth, 1954].

The HS system has no liquid-vapour transition but is, however, quite important because it can produce at the right density a system that closely mimics a real simple fluid or a suspension of microspheres (see the phase diagram in Fig. 5.3a). Indeed, the success of these simple models with no attractive part has contributed to generating the widely shared view that only steric forces control the structure and dynamic of liquids [Chandler et al., 1983]. In Fig. 5.3b, the HS pair distribution $g(r)$ obtained with the Monte Carlo computer simulation methods discussed in Chapter 8 is shown. Clearly, $g(r)$ is 0 for $r \leq \sigma_h$. In terms of the packing (or volume) fraction ϕ, which for HS is just $\phi = N\mathcal{V}_h/V = \pi\rho\sigma_h^3/6$, the equation of state of a system of hard spheres can be very well approximated by the Carnahan and Starling [1969] heuristic equation, written as a Padé approximant [Baker and Gammel, 1970], i.e. a ratio of two polynomials in ϕ, for the compressibility factor:

$$z^{\mathrm{HS}} = \frac{P}{\rho k_B T} = \frac{1+\phi+\phi^2-\phi^3}{(1-\phi)^3}. \tag{5.14}$$

This equation provides an important reference for a variety of thermodynamic perturbation theories [Santos, 2016]. It has also been generalized to m component mixtures of hard spheres of possibly different size σ_{h_i} and mole fraction [Mansoori et al., 1971].

5.2.2 Square Well Potential

The square well potential, U^{SW}, introduces particle-particle attraction considering spheres with hard repulsive core of diameter σ_h surrounded by an outer shell of diameter $\sigma_a = \lambda\sigma_h$ with an attractive range of thickness $(\sigma_a - \sigma_h)$ and well depth ϵ:

$$U^{\mathrm{SW}}(r_{ij}) = \begin{cases} 0, & \text{for } r_{ij} > \sigma_a \\ -\epsilon, & \text{for } \sigma_h < r_{ij} < \sigma_a \\ \infty, & \text{for } r_{ij} \leq \sigma_h \end{cases}. \tag{5.15}$$

Table 5.1. *A few square well potential parameters [Croxton, 1975]*

System	σ_h(nm)	$(\sigma_a - \sigma_h)$(nm)	λ	ϵ/k_B(K)
Neon	0.238	0.187	1.786	19.5
Argon	0.316	0.185	1.585	69.4
Nitrogen	0.330	0.187	1.567	53.7
Carbon dioxide	0.392	0.183	1.467	119.0
Methane	0.290	0.127	1.438	692.0
Water	0.261	0.120	1.460	1290.0

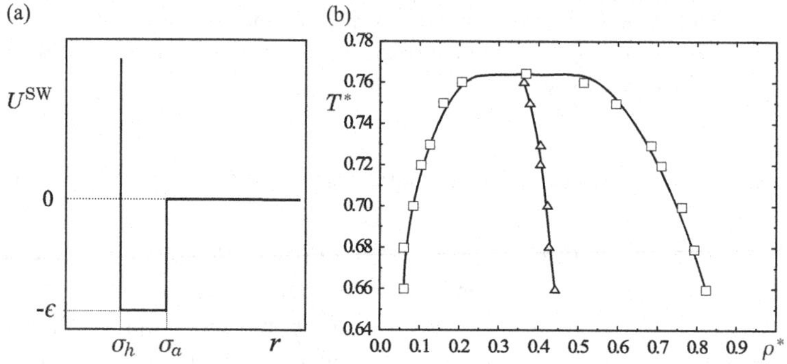

Figure 5.4 (a) The square well potential U^{SW} showing the hard repulsive core diameter σ_h and the attractive well of thickness $(\sigma_a - \sigma_h)$ as a function of particle separation $r = r_{ij}$. (b) The liquid-vapour coexistence for $\lambda \equiv \sigma_a/\sigma_h = 1.25$. Reduced density $\rho^* = \rho\sigma_h^3$ and temperature $T^* = k_B T/\epsilon$ are used [Vega et al., 1992].

In Fig. 5.4a we sketch the square well potential and in Table 5.1 we list the width and depth parameters for a few molecules. The square well potential has been extensively studied with a variety of theoretical techniques and with computer simulations (see, e.g., [Vega et al., 1992]). Differently from the HS model, U^{SW} can have a liquid and gas phase, instead of just a fluid one. In Fig. 5.4b we show the liquid-vapour coexistence for a square well fluid obtained [Vega et al., 1992] using Monte Carlo computer simulations. The shape of the coexistence curve changes with the range. The apparent critical exponent for the liquid-gas transition order parameter $\eta \propto \rho_L - \rho_G$, (see Eq. 2.14 and Fig. 2.14): $\beta = (\partial \ln(\rho_L - \rho_G))/(\partial \ln |\Delta T_r|)$, where $\Delta T_r = (T - T_c)/T_c$, has the near-universal value of $\beta \approx 0.325$, for $\lambda = 1.25, 1.375, 1.5$ and 1.75, while for the system with a larger attractive range: $\lambda = 2$, a near classical value of $\beta \approx 0.5$ is obtained [Vega et al., 1992]. The square well potential can be considered a sticky sphere version of the HS one, which it includes as a special case. The second virial coefficient for the SW potential, easily calculated by integration:

$$B_2^{SW}(T) = B_2^{HS}\left[1 - \left(\lambda^3 - 1\right)\left(e^{-\epsilon/(k_B T)} - 1\right)\right], \tag{5.16}$$

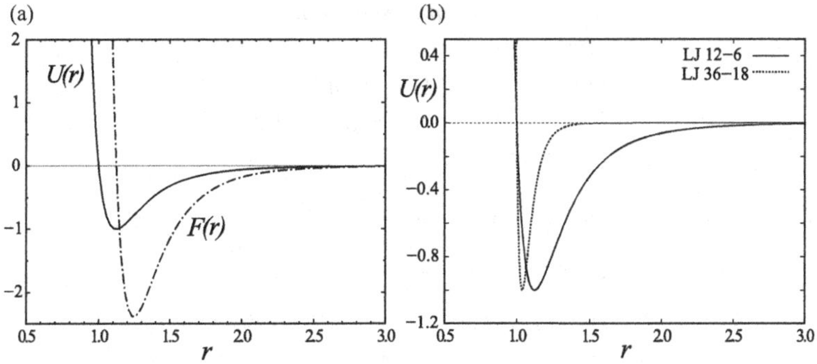

Figure 5.5 (a) The 12–6 LJ potential and the corresponding LJ force. (b) The (12–6) and short-range (36–18) LJ potentials. Distances in units of σ, energies in units of ϵ.

shows a change of sign with temperature as the balance between the positive (repulsive) contribution and the negative (attractive) one, where aggregation occurs. The temperature that corresponds to an ideal gas behaviour ($B_2 = 0$) is known as the *Boyle temperature* $T_{\rm B}$ [Gamez and Caro, 2015] and for this potential is: $T_{\rm B}^* = k_B T_{\rm B}/\epsilon = \left\{\ln\left(\lambda^3/(\lambda^3-1)\right)\right\}^{-1}$.

5.2.3 Lennard–Jones Potentials

Each pair potential of this family, first put forward by Mie in 1903 (see, e.g., Israelachvili [2011]) is the sum of a repulsive and an attractive term varying with separation as inverse power laws with exponents (m, n):

$$U^{\rm Mie}(r_{ij}) = 4\epsilon\left\{\left(\frac{\sigma}{r_{ij}}\right)^m - \left(\frac{\sigma}{r_{ij}}\right)^n\right\}, \tag{5.17}$$

where ϵ, σ are positive constants chosen to model the strength and the range of the interaction and m, n are positive integers. The most often used choice for two spherical particles moving in 3D space is the Lennard–Jones (LJ) 12–6 potential with $m = 12$, $n = 6$:

$$U^{\rm LJ}(r_{ij}) = 4\epsilon\left\{\left(\frac{\sigma}{r_{ij}}\right)^{12} - \left(\frac{\sigma}{r_{ij}}\right)^6\right\}. \tag{5.18}$$

The potential changes from attractive to repulsive for $r_{ij} = \sigma$ and has its minimum at $r_{ij} = 2^{1/6}\sigma$, with well depth ϵ. Thus, σ gives the sphere diameter and $\sigma/2$, called the van der Waals (vdW) radius, or the corresponding vdW volume $\pi\sigma^3/6$ are often used to give an idea of the size of the particle represented by an LJ. In Fig. 5.5a we show as an example the LJ potential in reduced units ($\sigma = 1, \epsilon = 1$), together with the force exerted on an LJ particle by another approaching one:

$$F^{\rm LJ}(r_{ij}) = -\frac{dU^{\rm LJ}(r_{ij})}{dr_{ij}} = -\frac{24\epsilon}{\sigma}\left\{2\left(\frac{\sigma}{r_{ij}}\right)^{13} - \left(\frac{\sigma}{r_{ij}}\right)^7\right\}. \tag{5.19}$$

Table 5.2. *A small selection of parameters for the Lennard–Jones (12–6)-potential between two identical particles [Atkins, 1978]*

Particle	ϵ/k_B (K)	σ(nm)
Ar	119.8	0.340
Xe	221.0	0.410
H_2	38.0	0.292
O_2	118.0	0.346
N_2	95.1	0.370
Cl_2	257.0	0.440
Br_2	520.0	0.427
CO_2	189.0	0.449
CH_4	148.0	0.382

In Fig. 5.5b we see that changing from 12–6 exponents to 36–18 has the effect of shortening the range of the potential. Lennard–Jones parameters are available for a variety of systems and are used in the construction of more complex potentials. In Table 5.2 we report a small compilation of LJ parameters. In computer simulation it is often convenient to neglect the interaction for particles at a distance larger than a cut off r_c and use the cut and shifted potential U^{LJS}:

$$U^{\text{LJS}}(r_{ij}) = \begin{cases} U^{\text{LJ}}(r_{ij}) - U^{\text{LJ}}(r_c), & \text{for } r_{ij} < r_c \\ 0, & \text{for } r_{ij} \geq r_c \end{cases}. \tag{5.20}$$

When the interaction is between two different particles the parameters are normally obtained with a linear average of the diameters and a geometric average of the well depths (*Lorentz–Berthelot* mixing rules), i.e.

$$\sigma_{AB} = \frac{\sigma_{AA} + \sigma_{BB}}{2}, \tag{5.21a}$$

$$\epsilon_{AB} = \sqrt{\epsilon_{AA}\,\epsilon_{BB}}. \tag{5.21b}$$

Other prescriptions also employed for mixed parameters are, the 'geometric': $\sigma_{AB} = \sqrt{\sigma_A \sigma_B}$, $\epsilon_{AB} = \sqrt{\epsilon_A \epsilon_B}$ and the 'sixth power' one: $\sigma_{ij} = [(\sigma_i^6 + \sigma_j^6)/2]^{1/6}$, $\epsilon_{AB} = 2\sqrt{\epsilon_A \epsilon_B}\sigma_A^3 \sigma_B^3/(\sigma_A^6 + \sigma_B^6)$ available, for instance, in the computer simulation package LAMMPS [Plimpton, 1995].

A system of LJ particles (*Lennard–Jonesium*), in itself one of the most widely studied models for condensed systems, gives rise to solid, liquid and vapour. In Fig. 5.6 we show the LJ radial distribution $g(r)$ in various phases. Although the (12–6) LJ potential is by far the most common one used to deal with simple fluids, other exponents can be chosen, e.g. to change the range of the attractive and repulsive forces, as might be appropriate, for instance, for large colloidal particles where the interaction range extends effectively for only

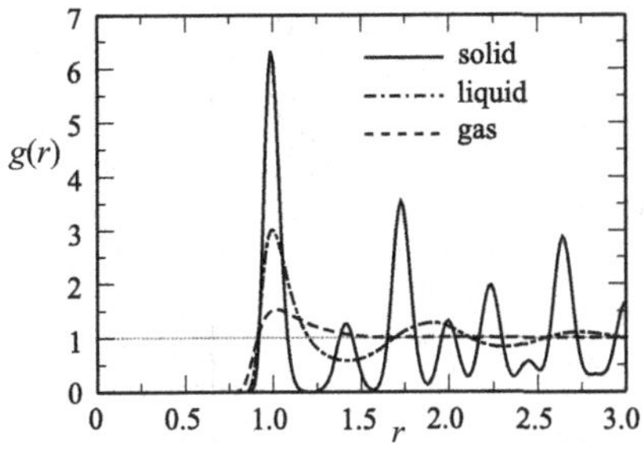

Figure 5.6 Radial distribution $g(r)$ for the (12–6) LJ model, as a function of interparticle distance r in units of σ at three different temperatures in the crystal, liquid and gas phase. The solid corresponds to a molecular dynamics simulation of Argon at 50 K [Rowley, 2015].

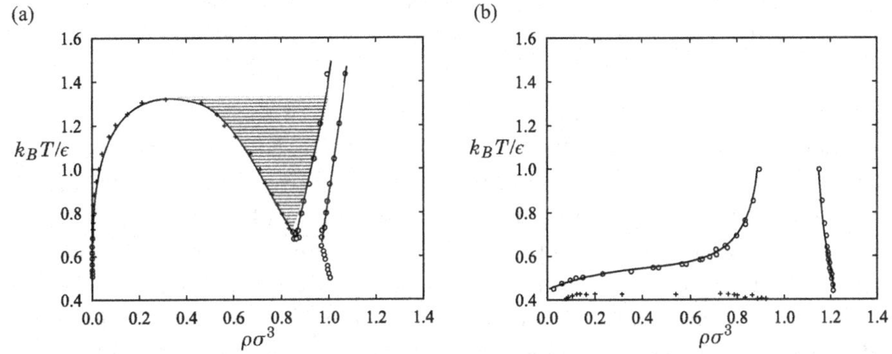

Figure 5.7 (a) Temperature-density ($k_B T/\epsilon$ vs. $\rho\sigma^3$) phase diagram for (12–6) and (b) (36–18) LJ potentials showing that only the longer ranged (12–6) one forms a liquid (hatched region) [Vliegenthart and Lekkerkerker, 2000].

a fraction of the particle diameter. In Fig. 5.7 we show the effect of changing the exponents on the potential and the resulting T, ρ phase diagram [Vliegenthart and Lekkerkerker, 2000]. Notice the suppression of the liquid phase for the short-range variant.

The second virial coefficient B_2 can be written in closed form for a r_{ij}^{-n} interaction in terms of gamma function $\Gamma(x)$ [Abramowitz and Stegun, 1965], giving [Balescu, 1975]

$$B_2^{\text{LJ}}(T) = \frac{2\pi}{3}\sigma^3 \left[\left(\frac{4\epsilon}{k_B T}\right)^{1/4} \Gamma\left(\frac{3}{4}\right) - \left(\frac{4\epsilon}{k_B T}\right)^{1/2} \Gamma\left(\frac{1}{2}\right) \right], \tag{5.22}$$

where $\Gamma(3/4) = 1.22541$, $\Gamma(1/2) = \sqrt{\pi}$. Higher virial coefficients have been calculated numerically and B_2 to B_5 are reported as a function of temperature by Barker and Henderson [1976].

Since the LJ potential has both an attractive and a repulsive part, it is interesting to examine their relative importance. In the so-called van der Waals picture it is only the repulsive forces that determine the structure of a (non-associated) liquid, while the attractive forces may be neglected or introduced at a perturbative level [Andersen et al., 1976; Chandler et al., 1983]. This is the case, e.g. for an LJ fluid where the radial distribution of the repulsive core

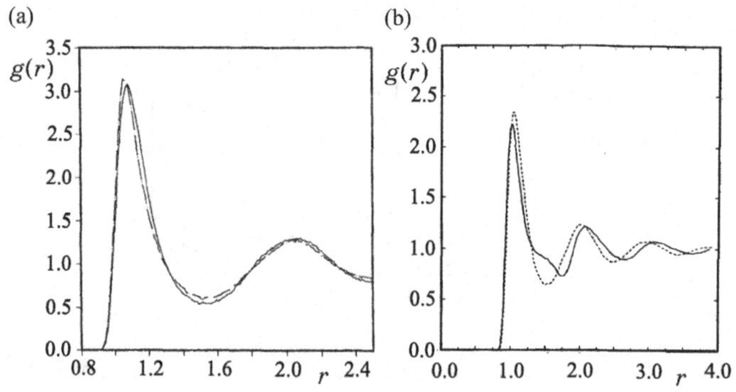

Figure 5.8 (a) The pair distribution $g(r)$ vs r for the LJ fluid (——) compared with that of the repulsive part of the potential (- - -) at $T^* = 0.75$ and $\rho^* = 0.87$ [Weeks and Broughton, 1983]. (b) $g(r)$ for the Hexon liquid (——) and for its core liquid (----) at $T^* = 0.60$ and $\rho^* = 0.8895$. Distances in units of σ [LaViolette and Stillinger, 1985].

and the full potential give essentially the same result (cf. Fig. 5.8b). However, it should be remembered that this approach has limitations, since attractive interactions are necessary to justify the condensation of gases and the formation of liquids. In the words of Fisher [1972] '... if one wants some sort of phase transitions - if one wants a gas to condense - there must be attractive forces, otherwise the atoms will not particularly like being next to one another!'. Properties that depend on the presence of attractive forces include the liquid-vapour interface and the ability of fluids to resist to a certain extent of stress, that is to a negative pressure [Zheng et al., 1991]. In many ways the latent heat of vaporization can also be associated with attractive forces and indeed attractive dispersive interactions (cf. Section 5.8) have been used to estimate it with some success [Homer and Mohammadi, 1987]. It is worth noting that, while Fig. 5.8a shows that the LJ radial distribution at short range is determined by the repulsive forces, this may not be the general case. In an interesting exercise LaViolette and Stillinger [1985] introduced a simple pair potential whose liquid structure cannot be satisfactorily predicted from the repulsions alone. Its form is

$$U^{\text{Hexon}}(r_{ij}) = \begin{cases} A(r_{ij}^{-m} - r_{ij}^{-n})\exp[1/(r_{ij} - a)], & \text{for } r_{ij} < a \\ 0, & \text{for } r_{ij} \geq a, \end{cases} \tag{5.23}$$

with $m = 12$, $n = -3$, $a = 2$, $A = 1.914166098$. The potential, fairly similar to the LJ one for this parametrization, has been called *Hexon* to recall both noble gases and the hexagonal symmetry of its crystal. We see that the pair distribution for Hexon (Fig. 5.8b) shows non-negligible differences from the purely repulsive structure.

5.3 Buckingham Potential

Another popular potential, that we can consider as a variant of the LJ potential where the repulsive part is taken as an exponential, is the Buckingham potential $U^{\text{B}}(r_{ij})$ [Buckingham, 1938]

$$U_{ij}^{\mathrm{B}} = U_{ij}^{\mathrm{B}}(r_{ij}) = A\exp(-Br_{ij}) - \frac{C}{r_{ij}^{6}}, \tag{5.24}$$

where A, B and C are constants. Even if its application to LC has been limited, the potential is often used for solid molecular crystals [Gavezzotti, 1997] and is included in various simulation packages, like LAMMPS [Plimpton, 1995]. A potential problem is that the repulsive term is finite as $U_{ij}^{\mathrm{B}} \rightarrow 0$, while the attractive one tends to $-\infty$. Modifications of the potential that mend this are available [Stone, 1996].

5.4 Atomistic Force Fields

In a sense the ultimate model based on multi-spherical beads is the class of atomistic potentials, where the spherical beads correspond to atoms placed at the appropriate geometry so as to represent a specific chemical structure. This category of potentials is the closest to a realistic modelling of chemical structures, allowing for proper chemical composition and structure as well as complete conformational freedom. The total potential energy of a molecular system is calculated considering the molecules and their constituent atoms as classical objects, whose properties (geometrical arrangements, sizes, charges, etc.) are determined using quantum mechanical methods or empirically estimated. The potential energy of such a *molecular mechanics* system can be written as a sum of effective multiparticle interaction terms somewhat similar to Eq. 5.1:

$$U_N = \sum_{A}\Big[\sum_{i<j} U_{i,j}^{(A)} + \sum_{i<j<k} U_{i,j,k}^{(A)} + \sum_{i<j<k<l} U_{i,j,k,l}^{(A)} + \cdots\Big] + \sum_{A}\sum_{B>A}\sum_{i\in A,\, j\in B} U_{i,j}^{(A,B)}, \tag{5.25}$$

where in Eq. 5.25 the terms are grouped in terms of molecules, considering contributions where the atoms belong to the same molecule (A) or, respectively to two different molecules (A, B) with, say, $i \in A$, $j \in B$. In practice, the intramolecular sums usually involve, as we shall see, not more than four-body terms. The contributions are usually parametrized to reproduce some key experimental property and in this way the contributions are really effective potential terms including some collective contributions. A set of atomistic potentials, also called *Force Field* (FF) allows us to model molecules with a certain chemical composition and structure and appropriate internal degrees of freedom.

Several formulae and parametrization schemes have been proposed, which can be organized in various broad classes of FFs [Leach, 2001; Roscioni and Zannoni, 2016] of different level of sophistication. The so-called 'class I' FFs are based on a division of the total energy of the system into bonded (bonds, angles and torsions) and non-bonded (dispersion and electrostatic) interactions. This category, that includes AMBER, GROMOS, CHARMM and OPLS (see Table 5.3) is the one we shall describe in more detail and employ in LC simulations, with the aim of reproducing molecular morphologies and thermochemical properties such as density, transition temperatures and phase organizations.

Other classes of FFs also aim at reproducing more specific details or spectral features, e.g. vibrational spectra. In the class I molecular mechanics FFs, a predefined connectivity

Table 5.3. *Some popular FFs relevant for the simulation of liquids, liquid crystals and functional organic materials*

AMBER	Assisted Model Building and Energy Refinement, widely used also for proteins and DNA [Weiner et al., 1984; Cornell et al., 1995]
GAFF	General AMBER FF [Wang et al., 2004; Yang et al., 2006]
CHARMM	Chemistry at Harvard Molecular Mechanics, used both for small and macromolecules [Brooks et al., 1983, 2009; Zhu et al., 2012; Yu and Klauda, 2020]
CVFF	Consistent Valence Force Field, used for small building block fragments [Braun et al., 1999], macromolecules, liquid crystals [Ennari et al., 1997] and cholesterol derivatives [Liang et al., 1995]
COSMOS-NMR	Hybrid QM/MM FF with experimental order parameter constraints. Includes semi-empirical calculation of atomic charges and NMR properties [Sternberg et al., 2007]
DREIDING	A generic FF for organic, biological and main-group inorganic molecules [Mayo et al., 1990]
GROMOS	Groningen Molecular Simulation FF and package, a general-purpose package for the study of biomolecular systems [Scott et al., 1999]
MM1$^{(a)}$, MM2$^{(b)}$ MM3$^{(c)}$, MM4$^{(d)}$	Molecular Mechanics code, with successive versions $^{(a)}$[Allinger et al., 1971], $^{(b)}$[Allinger, 1977], $^{(c)}$[Allinger et al., 1989] $^{(d)}$[Allinger et al., 1996]
OPLS	Optimized Potential for Liquid Simulations [Jorgensen et al., 1996; Sambasivarao and Acevedo, 2009]
UFF	A Universal FF with parameters for the full periodic table [Rappé et al., 1992]

between atoms is assumed, precluding simulations that involve bonds to break and form, as in chemical reactions. Methods to overcome this limitation allowing for bond breaking, like the ReaxFF one, that employ a bond strength order formalism, have been introduced [van Duin et al., 2001; Senftle et al., 2016], but will not be discussed here. Going back to non-reactive FFs, several expressions and parametrization schemes have been proposed (see [Muccioli et al., 2014] for a brief summary) and some of the FFs typically used for LC simulations are reported in Table 5.3. Note that the optimization of these FFs and their adaptation to modern, e.g. graphic processing units (GPUs), based computer architectures continues [Jasz et al., 2020]. Classical FFs, e.g. MM3/MM4 [Allinger et al., 1989], are also often included as part of quantum mechanics/molecular mechanics (QM/MM) mixed schemes, where a small region of interest is studied with an *ab initio* method, while the extended molecular environment is modelled classically. They have found wide use for chemical problems, but not much for applications to liquid crystals. Some exceptions are for systems, like azobenzene based LCs, where the molecular geometry may change due to photoexcitation [Mukherjee et al., 2012].

It is worth pointing out that the quality of FFs, i.e. their ability to reproduce and predict properties, depends heavily of their parameters being adjusted for a certain 'basket' of compounds. Thus, excellent FFs are available for biological systems (lipid membranes, proteins, ...) where extensive tuning has been performed over the years (see Section 12.7.2). Much less exists for functional organic materials and LCs, although efforts have been made [Cook and Wilson, 2001a; Tiberio et al., 2009; Boyd and Wilson, 2015, 2018]. We shall not discuss in detail the various atomistic FFs here, but we wish to illustrate the general terms that contribute in somewhat different forms to all of them. To do this we write the interaction energy as a sum of contributions coming from atoms bonded or not-bonded. Bonded atoms clearly belong to the same molecule, while non-bonded interactions can be from atom pairs belonging to the same molecule or to different ones. A typical expression is

$$\begin{aligned} U_N &= U^{\text{bonded}} + U^{\text{non-bonded}}, \\ &= \left(U^{\text{stretch}} + U^{\text{bend}} + U^{\text{DHT}} + U^{\text{DHI}}\right) + \left(U^{\text{LJ}} + U^{\text{e}}\right), \end{aligned} \tag{5.26}$$

determining the force field as a sum of contributions from chemically bonded atoms including stretching: U^{stretch}, bending: U^{bend}, proper and improper torsional terms: U^{DHT}, U^{DHI} as well as of non-bonded Lennard–Jones (LJ), U^{LJ}, and electrostatic, U^{e}, terms. We shall now describe these various contributions.

5.4.1 Intramolecular Bonded Interactions

In a molecular mechanics model each atom is a spherical bead of a certain type t_i. The atom types correspond to the chemical nature and to a certain hybridization in the molecular structure. For instance, a carbon connected with a single bond like that in a methylene chain (sp^3 hybridization), one connected to a double bond like in ethylene (sp^2 hybridization) and one connected to a triple bond like in acetylene (sp hybridization) will correspond to different types. This is quite reasonable since the bond corresponds to different strength and geometries. Although the detailed form of the energy for a bond stretching or bending is generally unknown, in a stable chemical compound we are normally dealing with small deviations from the equilibrium values of the respective bond length $r_{eq}^{t_it_j}$ and bond angle $\theta_{eq}^{t_it_jt_k}$, for atoms i, j, k of type t_i, t_j, t_k, values that correspond to a minimum of the potential energy. Additional atom types can be introduced to allow for differences in the local chemical environment even if atom type and hybridization are the same. For instance, in the MM2, MM3 and MM4 force fields of Allinger and collaborators (see, e.g., Allinger et al. [1996]) in addition to sp^3, sp^2, and sp, the following types of carbon atom are distinguished: carbonyl, cyclopropane, radical, cyclopropene and carbonium ion [Leach, 2001]. To reduce the number of particles and consequently computer demands, it is also fairly common to employ for large molecules a United Atom (UA) coarse graining, where CH, CH_2 and CH_3 groups are considered as suitably parameterized spherical interaction sites (pseudo-atom types).

As long as the potential function is continuous and differentiable with respect to the relevant coordinate we can Taylor expand it around its minimum (corresponding to the equilibrium value) obtaining a polynomial. Normally only the quadratic terms are retained,

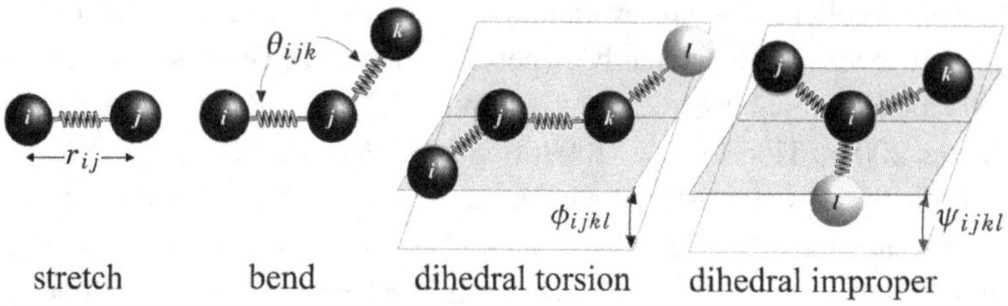

Figure 5.9 A sketch of the main intramolecular potential contributions (see text): stretch, bend, dihedral torsion (DHT) and improper dihedral (DHI) with the respective variables: $r_{ij}, \theta_{ijk}, \phi_{ijkl}, \psi_{ijkl}$.

obtaining harmonic potentials describing oscillations around the equilibrium values. More explicitly we have bond stretching:

$$U^{\text{stretch}} = \sum_{\text{stretchings}} \frac{1}{2} K_r^{t_i t_j} \left(r_{ij} - r_{eq}^{t_i t_j} \right)^2, \tag{5.27}$$

where r_{ij} is the interatomic separation, $r_{ij} = |\boldsymbol{r}_{ij}| = |\boldsymbol{r}_j - \boldsymbol{r}_i|$ and

$$U^{\text{bend}} = \frac{1}{2} \sum_{\text{bendings}} K_\theta^{t_i t_j t_k} \left(\theta_{ijk} - \theta_{eq}^{t_i t_j t_k} \right)^2 \tag{5.28}$$

describes the bending fluctuations of the angle, $\theta_{ijk} = \cos^{-1}\left[-(\boldsymbol{r}_{ij} \cdot \boldsymbol{r}_{jk})/(r_{ij} r_{jk})\right]$, as in Fig. 5.9. The stretching and bending force constants $K_s^{t_i t_j}$ and $K_b^{t_i t_j t_k}$ are typically obtained by fitting spectroscopic observables or from QC calculations, but these parameters are fairly transferable and good libraries exist (Table 5.3), in particular the Molecular Mechanics series MM1 [Allinger et al., 1971], MM2 [Allinger, 1977], MM3 [Allinger and Lii, 1987], MM4 [Allinger et al., 1996]; AMBER [Weiner and Kollman, 1981; Cornell et al., 1995].

A small selection of force constants, helping to give an idea of the order of magnitude of the terms involved, is reported in Tables 5.4, 5.5 and 5.6. Table 5.5 shows an example of the same stretching parameters in different FFs, showing that even for these 'simple' parameters differences up to 30% in the elastic constant K_r can occur. Clearly, if the United Atoms approximation mentioned before is employed, the parameters have to be modified with respect to those of the full or 'All Atom' (AA) description (see, e.g., [Yang et al., 2006; Tiberio et al., 2009] and references therein).

Contributions involving four atoms i, j, k, l are typically introduced to describe the relative orientation of a molecular fragment with respect to another connected to it by the i-j bond, e.g. the torsion of one ring with respect to another one in a polyphenyl or polythienyl or a methylene in an alkyl chain. The dihedral contribution is typically calculated using QC at some level of theory, e.g. density functional theory (DFT) or one of the many other methods available [Cramer, 2004] for determining the optimized energy for a number of dihedral angles ϕ_{ijkl} (see Fig. 5.9),

Table 5.4. *Bond lengths* $r_{eq}^{t_i t_j}$ *and stretching constant* $K_r^{t_i t_j}$ *for carbon bonds of various types (hybridization). Bond lengths* r_{eq} *in* Å *and force constants in* kcal mol^{-1}Å^{-2} *[Allinger, 1977]*

Bond type	$r_{eq}^{t_i t_j}$	$K_r^{t_i t_j}$
$Csp^3 - Csp^3$	1.523	634
$Csp^3 - Csp^2$	1.497	634
$Csp^2 = Csp^2$	1.337	1380
$Csp^3 - O$	1.208	222
$Csp^3 - Nsp^3$	1.438	734
C-Namide	1.345	1438

Table 5.5. *Bond stretching parameters for three FFs. Bond lengths* r_{eq} *in* Å *and force constants* K_r *in* kcal mol^{-1}Å^{-2} *[Rappé and Casewit, 1997]*

	AMBER		CHARMM		DREIDING	
Atom pair	r_{eq}	K_r	r_{eq}	K_r	r_{eq}	K_r
$N - C_{amide}$	1.334	980	1.330	942	1.34	700
$C_{amide} - O$	1.229	1140	1.215	1190	1.25	1400
$C_{amide} - Csp^3$	1.522	670	1.524	374	1.46	700
$N - Csp^3$	1.526	620	1.530	471	1.53	700
$C\,sp^3 - H$	1.449	710	1.450	844	1.41	700
$N - H$	1.090	622	1.100	660	1.09	700

Table 5.6. *Typical bond angles* θ (deg) *and bending force constants* K_θ (kcal mol^{-1}deg^{-1}) *for various common chemical bonds [Allinger, 1977]*

Angle type	$\theta_{eq}^{t_i t_j t_k}$	$K_\theta^{t_i t_j t_k}$
$Csp^3 - Csp^3 - Csp^3$	109.47	0.0198
$Csp^3 - Csp^3 - H$	109.47	0.0158
$H - Csp^3 - H$	109.47	0.0140
$Csp^3 - Csp^2 - Csp^3$	117.2	0.0198
$Csp^3 - Csp^2 = Csp^2$	121.4	0.0242
$Csp^3 - Csp^2O$	122.5	0.0202

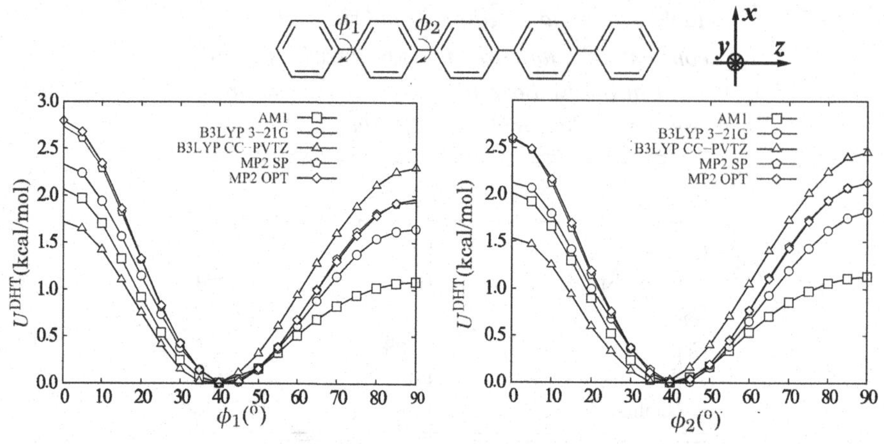

Figure 5.10 Quantum chemistry torsional potential U^{DHT}(kcal/mol) for the angles ϕ_1 and ϕ_2 of P5 (top chemical sketch) and different levels of theory as indicated (see text). [Olivier et al., 2014].

$$\phi_{ijkl} = \cos^{-1}\left[\frac{(\boldsymbol{r}_{ij} \times \boldsymbol{r}_{jk})\cdot(\boldsymbol{r}_{jk} \times \boldsymbol{r}_{kl})}{|\boldsymbol{r}_{ij} \times \boldsymbol{r}_{jk}|\,|\boldsymbol{r}_{jk} \times \boldsymbol{r}_{kl}|}\right] \tag{5.29}$$

and fitting it to a Fourier series:

$$U^{\mathrm{DHT}} = \sum_{\substack{\text{torsions}\\ ijkl}} \sum_{n} V_n^{ijkl}\left[1 + \cos\left(n\phi_{ijkl} - \gamma_{ijkl}\right)\right], \tag{5.30}$$

where the sum is over all torsions and γ_{ijkl} is a phase factor. Some FFs (e.g. AMBER) retain only one term in the sum, but in general as many terms as needed for a good fit can be used. In Fig. 5.10 we see, as an example, the torsional potential energy as a function of the outer and inner $\phi_1(^{\circ})$, $\phi_2(^{\circ})$ internal angles for *p*-quinquephenyl (P5) calculated with the Gaussian09 [Frisch et al., 2009] code as described in [Olivier et al., 2014]. The rather obscure codewords in Fig. 5.10 are needed to identify the various level of QC theory employed. The ones referred to here are: AM1 (Austin Model 1, a QC code for solving the electronic Schrödinger equation at semiempirical level [Dewar et al., 1985]) and MP2 (Møller–Plesset second-order perturbation theory [Cramer, 2004; Lewars, 2016; Irikura, 2019]). The acronym MP2 OPT corresponds to optimizing geometry with MP2 CC-pVDZ calculations, while MP2 SP refers to single point B3LYP/cc-pVTZ/MP2/cc-pVDZ.

An additional type of contribution can come from four atoms belonging to a branched, rather than linearly connected, set of atoms. These ‘improper dihedral’ bending terms are often used to enforce a specific out of plane angle or, more commonly, planarity or to maintain chirality at certain positions using in some cases a harmonic term or, also,

$$U^{\mathrm{DHT}} = \sum_{\substack{\text{improper}\\ \text{torsions } ijkl}} \sum_{m=0}^{m_{max}} V_m^{ijkl} \cos(m\psi_{ijkl} - \eta_{ijkl}), \tag{5.31}$$

Table 5.7. *A small set of LJ parameters for the OPLS AAFF [Jorgensen et al., 1996]*

Atom type	σ (nm)	ϵ (kcal mol^{-1})
O	0.296	0.210
N	0.325	0.170
C in C=O	0.375	0.105
Other C	0.350	0.066
H on C	0.250	0.030

where η_{ijkl} is a phase angle and the so-called Wilson off-plane angle ψ_{ijkl} (see Fig. 5.9), defined by $\psi_{ijkl} = \sin^{-1}\left[\frac{-\boldsymbol{r}_{il}\cdot(\boldsymbol{r}_{ij}\times\boldsymbol{r}_{ik})}{r_{il}\,|\boldsymbol{r}_{ij}\times\boldsymbol{r}_{ik}|}\right]$, measures the angle between $\boldsymbol{r}_{il}$ and the plane defined by $\boldsymbol{r}_{ij}$ and $\boldsymbol{r}_{ik}$ for three-coordinated atoms (see Fig. 5.9). An example could be the use of an improper torsion to constrain a benzene ring to be planar or restrict movements of the cyclobutanone oxygen so as to keep the molecule planar [Leach, 2001]. Other cases of chemical moieties, where these geometry constraining terms are useful, are given by Schlick [2002].

5.4.2 Non-Bonded Lennard–Jones Interactions

The non-bonded interactions include intramolecular and intermolecular pairwise terms modelling steric repulsive and attractive van der Waals interactions. In practice, in Eq. 5.26 these are represented by (12–6) Lennard–Jones terms, U^{LJ} analogous to Eq. 5.18, that we rewrite for convenience as

$$U_{ij}^{\text{LJ}} = 4\sum_{\text{LJ pairs}} f_{ij}^{\text{LJ}}\,\epsilon_{t_i t_j}\left[\left(\frac{\sigma_{t_i t_j}}{r_{ij}}\right)^{12} - \left(\frac{\sigma_{t_i t_j}}{r_{ij}}\right)^{6}\right]. \tag{5.32}$$

Here $\sigma_{t_i t_j}$ is the pseudo-contact distance, where the interaction changes from attractive to repulsive, $\epsilon_{t_i t_j}$ is the strength of interaction (potential well depth) and f_{ij}^{LJ} is a scaling parameter for the i, j pair introduced to avoid including non-bonded contributions for atoms that are already chemically bonded (labelled as 1–2) and for atoms sharing a common bonded atom (1–3), while for atoms which are 1–4 connected they are scaled by a factor. Following the recommendations suggested for using OPLS parameters [Jorgensen et al., 1996] within the AMBER force field: if atoms i and j belong to the same molecule and are separated by less than three consecutive bonds, $f_{ij}^{\text{LJ}} = 0$; if i and j are separated by three consecutive bonds $f_{ij}^{\text{LJ}} = 1/8$, while for all other cases $f_{ij}^{\text{LJ}} = 1$. Other choices are used in different FFs [Schlick, 2002]. Tabulations of LJ parameters are available (see Table 5.7) for various atom types while LJ parameters for mixed interactions are usually obtained using Lorentz–Berthelot rules (Eq. 5.21b), i.e. linear additivity for atomic diameters and geometric mean for the potential well depths [Allen and Tildesley, 2017]. Note that this is an empirical rule and not a unique choice. Another fairly common choice is to use geometric means for both σ and ϵ [Abraham et al., 2014]. LJ interactions are normally calculated within a certain

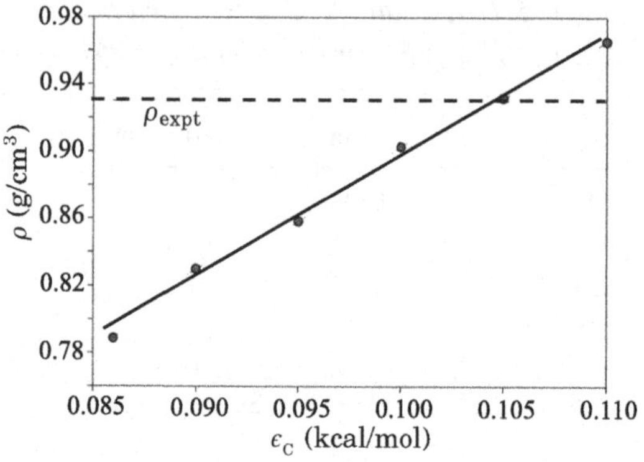

Figure 5.11 Density ρ as a function of the LJ well depth ϵ_C (kcal/mol) for an orthorhombic sample of *p*-quinquephenyl at 660 K using the AMBER values for the remaining LJ parameters ($\sigma_C = 3.816$ Å; $\epsilon_H = 0.015$ kcal/mol, $\sigma_H = 2.918$ Å. The dashed horizontal line corresponds to the experimental density ρ_{expt} [Olivier et al., 2014].

cut-off distance of a few σ. However, it is important to stress that LJ parameters, particularly ϵ, capture in an effective pairwise interaction some of the many body effects of the system studied. Thus, rather than using QC values, even of high quality, but obtained in the gas phase, it is important to optimize the specific LJ parameters comparing the simulated results for some relevant observable with available experimental ones. As an example, we show in Fig. 5.11 the effect of varying the carbon atom LJ well depth, ϵ_C, on the mass density ρ of *p*-quinquephenyl (P5) obtained from computer simulations [Olivier et al., 2014], comparing it with the experimental one. Standard AMBER parameters ($\epsilon_C = 0.086$ kcal/mol, $\sigma_C = 3.816$ Å; $\epsilon_H = 0.015$ kcal/mol, $\sigma_H = 2.918$ Å) [Jorgensen et al., 1996] give a density significantly lower than experiment (0.78 g/cm^3 at 660 K). Parameters from another popular FF, CHARMM ($\epsilon_C = 0.07$ kcal/mol, $\sigma_C = 3.9848$ Å; $\epsilon_H = 0.022$ kcal/mol, $\sigma_H = 2.64$ Å) give no stable condensed phase at this temperature. A parameterization based on high level QC calculations for benzene dimers ($\epsilon_C = 0.115$ kcal/mol, $\sigma_C = 3.844$ Å; $\epsilon_H = 0.011$ kcal/mol, $\sigma_H = 2.46$ Å) [Sherrill et al., 2009] favours instead solid over liquid phases and too high densities ranging from 1.07 to 1.21 g/cm^3 in the 600–720 K temperature range. However, tuning the single well depth ϵ_C at a certain T while leaving the other parameters unchanged gives a value ($\epsilon_C = 0.105$ kcal/mol) that provides satisfactory results for the nematic-isotropic transition and the cascade of P5 morphologies [Olivier et al., 2014]. A toolkit for fitting non-bonded FF parameters is described in Hedin et al. [2016].

It is worth pointing out that even though transferability is an important desirable feature of an, FF, in practice, we can only expect a specifically tuned FF to be applicable to similar molecules, but not universally.

5.4.3 Computing Partial Charges

Even in an overall neutral molecule, every atom is expected to have a different electron affinity and thus a different partial charge, e_k. Quantum chemistry offers a number of ways for assigning a net atomic charge within a molecule, which unfortunately means that there is no unique way of partitioning the overall electron density, as there is no partial charge observable. The simplest method is the Mulliken population analysis [Szabo and Ostlund, 1996; Cramer, 2004], which starts considering a closed shell molecule with N electrons paired in $N/2$ occupied molecular orbitals (MO) ψ_i, written as a linear combination of normalized atomic orbital (AO) basis functions φ_r, i.e. $\psi_i(\boldsymbol{r}) = \sum_{r=1}^{\text{AO}} c_{ir}\varphi_r(\boldsymbol{r})$, where each c_{ir} is the coefficient of the basis function r in the ith MO. The overall electron population is

$$N = 2\sum_i^{\text{MO}} \int \mathrm{d}\boldsymbol{r}\, \psi_i(\boldsymbol{r})\, \psi_i^*(\boldsymbol{r}) = 2\sum_{r,s}^{\text{AO}} \left(\sum_{i=1}^{\text{MO}} c_{ir}c_{is}^*\right)\left(\int \mathrm{d}\boldsymbol{r}\varphi_r(\boldsymbol{r})\, \varphi_s^*(\boldsymbol{r})\right), \tag{5.33a}$$

$$= \sum_r^{\text{AO}} P_{rr} + \sum_{\substack{r,s \\ r\neq s}}^{\text{AO}} P_{rs}S_{rs} = \sum_{r,s}^{\text{AO}} P_{rs}S_{rs} = \mathrm{Tr}(\mathbf{PS}), \tag{5.33b}$$

where $P_{rs} = 2\sum_i^{\text{MO}} c_{ir}c_{is}^*$ is the *charge density matrix* and $\mathbf{S}$ is an AO overlap matrix (cf. Section A.2.1). The first term in Eq. 5.33b refers to electrons associated with only single basis functions that thus belong entirely to the atom on which that basis function is centred. The second term in the same equation, involves sharing of electrons between atoms and assigning charges which has long been an argument of debate. In the original form proposed by Mulliken, in order to compute the atomic population N_k, the electrons are equally shared between the two bonded atoms. Considering that the partial charge on atom k of atomic number Z_k is $e_k = Z_k - N_k$, the difference between the positive charge of the protons and the electron population obtained summing on all basis functions centred at k:

$$e_k = Z_k - \sum_{r\in k}^{\text{AO}} (\mathbf{PS})_{rr}. \tag{5.34}$$

In practice, the resulting partial charges can be very sensitive to the basis set orthogonality and size, even though robustness can be improved performing a Löwdin orthogonalization (Section A.2.1) over the atomic orbitals, so that the second term in Eq. 5.33b vanishes. The Mulliken method has the advantage of not adding extra costs to a QC calculation, but there are many other approaches to estimate partial charges, in particular, at least for relatively small molecules, those that assign the effective charge on every atom by fitting $V^{\text{QC}}(\boldsymbol{r})$, the molecular electrostatic potential (MESP) obtained from a QC calculation,

$$V^{\text{QC}}(\boldsymbol{r}) = \sum_{k=1}^{M} \frac{Z_k}{|\boldsymbol{r} - \boldsymbol{R}_k|} - \int \frac{\mathrm{d}\boldsymbol{r}'\rho(\boldsymbol{r})}{|\boldsymbol{r}' - \boldsymbol{r}|}, \tag{5.35}$$

where $\boldsymbol{R}_k$ is the position of nucleus k of atomic number Z_k and $\rho(\boldsymbol{r})$ the electron density. This is calculated on a grid of points around the molecule and partial charges are determined by a direct fit to this MESP [Singh and Kollman, 1984; Chirlian and Francl, 1987; Besler et al., 1990; Bayly et al., 1993; Berardi et al., 2004d]. There are a variety of methods depending on the type of grid around the molecules, e.g. based on spherical shells at a certain distance apart, centred on each atom with points symmetrically distributed on the surface [Chirlian and Francl, 1987] or based on a cubic grid in CHELPG [Breneman and Wiberg, 1990]. A good survey of these methods is reported by Leach [2001]. Another possibility that has been used in the literature is more empirical, with atomic charges adjusted to fit the computer simulated properties of crystals or liquids. In particular, the OPLS non-bonded parameters are derived by fitting the enthalpy of vaporization and density of liquids determined by Monte Carlo calculations [Jorgensen and Tirado-Rives, 1988].

5.4.4 Computing Electrostatic Energy: Ewald Sums

Once partial charges are known, electrostatic interactions are described, in principle very simply, by the classical Coulomb law, which in vacuum, with dielectric constant ε_0 is

$$U^{\mathrm{e}} = \frac{1}{2}\sum_{k,m}^{N}{}' \frac{e_k\, e_m}{4\pi\,\varepsilon_0 r_{km}}, \tag{5.36}$$

where the prime indicates that the sum excludes pairs of atoms belonging to an intramolecular set $\mathcal{M}$ with $k = m$ or k and m either (1–2), (1–3) or (1–4) connected (see definitions in Section 5.4.2). In some schemes connected atoms 1–4 also contribute, but with a scaling factor (1/2) [Jorgensen et al., 1996]. Although very simple to write, Eq. 5.36 has the fundamental problem of being only conditionally convergent. In practice, if we try to limit the contribution of the interacting charges to those within a certain cut-off distance r_c, i.e. limit the sum to $r_{km} \leq r_c$, very different results can be obtained when the sum is truncated at different cut-offs. A way to transform the sum in a convergent one is the classic Ewald method [de Leeuw et al., 1980a, 1980b; Frenkel and Smit, 2002; Arnold and Holm, 2005; LeSar, 2013], which applies to an infinite periodic lattice. This is particularly appealing since this is always the case when the actual sample is replicated using the so-called periodic boundary conditions, as normally employed in computer simulations at least when simulating a bulk system (see Chapters 8 and 9). In practice, the electrostatic energy is computed assuming a periodic lattice consisting of the overall neutral sample box of edges $(\boldsymbol{a}_1, \boldsymbol{a}_2, \boldsymbol{a}_3)$ surrounded by infinitely many replicas at $\boldsymbol{R}_{\boldsymbol{n}} = n_i\, \boldsymbol{a}_1 + n_j\, \boldsymbol{a}_2 + n_k\, \boldsymbol{a}_3$, with $\boldsymbol{n} = (n_i, n_j, n_k)$ and n_i, n_j, n_k integers. In the Ewald method the Coulomb energy is rewritten as a sum of two absolutely convergent series: one, $U^{\mathrm{e}}_{\mathrm{dir}}$, in direct (real) space and the other, $U^{\mathrm{e}}_{\mathrm{rec}}$ in reciprocal Fourier space [de Leeuw et al., 1980a; Essmann et al., 1995]:

$$4\pi\,\varepsilon_0 U^{\mathrm{e}} = U^{\mathrm{e}}_{\mathrm{dir}} + U^{\mathrm{e}}_{\mathrm{rec}} = \frac{1}{2}\sum_{\boldsymbol{R}_{\boldsymbol{n}}}\sum_{k,m=1}^{N}{}'' \frac{e_k\, e_m}{|\boldsymbol{r}_k - \boldsymbol{r}_m + \boldsymbol{R}_{\boldsymbol{n}}|}, \tag{5.37}$$

where, $\boldsymbol{r}_k$ is the position of the charge e_k and the double prime indicates that pairs of the set $\mathcal{M}$ introduced before are excluded in the primary box, i.e. when $\boldsymbol{R}_n = 0$,

$$U^{\mathrm{e}}_{\mathrm{dir}} = \frac{1}{2}\sum_{\boldsymbol{R}_n}\sum_{k,m}^{N}{}'' e_k e_m \frac{\mathrm{erfc}\,(\alpha\,|\boldsymbol{r}_{km}+\boldsymbol{R}_{\mathbf{n}}\,|)}{|\boldsymbol{r}_{km}+\boldsymbol{R}_{\mathbf{n}}|} - \frac{1}{2}\sum_{k,m\in\mathcal{M}}^{N} e_k e_m \frac{\mathrm{erf}\,(\alpha\,|\boldsymbol{r}_{km}|)}{|\boldsymbol{r}_{km}|}, \tag{5.38a}$$

$$U^{\mathrm{e}}_{\mathrm{rec}} = \frac{1}{2\pi V}\sum_{\boldsymbol{q}\neq 0}^{\infty} \frac{\exp\left(-\pi^2|\boldsymbol{q}|^2/\alpha^2\right)}{|\boldsymbol{q}|^2}\,|S(\boldsymbol{q})|^2 - \frac{\alpha}{\sqrt{\pi}}\sum_{k=1}^{N} e_k^2. \tag{5.38b}$$

Here $V = \boldsymbol{a}_1 \cdot \boldsymbol{a}_2 \times \boldsymbol{a}_3$ is the sample volume, $\boldsymbol{q} = h_1\boldsymbol{b}_1 + h_2\boldsymbol{b}_2 + h_3\boldsymbol{b}_3$ is a point in the reciprocal lattice, with h_i integers. The reciprocal lattice vectors $\boldsymbol{b}_i$ are $\boldsymbol{b}_i = \boldsymbol{a}_j \times \boldsymbol{a}_k/V$, and $\boldsymbol{a}_i \cdot \boldsymbol{b}_j = \delta_{i,j}$, with i, j, k an even permutation of 1, 2, 3, while $\mathrm{erfc}(x) = 1 - \mathrm{erf}(x)$ is the complementary error function [Abramowitz and Stegun, 1965]. The separation parameter α determines which part of the Coulomb sum is performed in real or in reciprocal space and is typically chosen so that $\mathrm{erfc}(\alpha r_c)$ is small at the real space cut-off distance r_c. Finally, $S(\boldsymbol{q}) = \sum_{m=1}^{N} e_j \exp(i2\pi\boldsymbol{q}\cdot\boldsymbol{r}_m)$ is the electrostatic structure factor. We should point out that in the Ewald potential given above, the so-called *tinfoil* boundary conditions are implicitly assumed: the sample and its retained images are immersed in a perfectly conducting medium and hence the dipole term on the surface of the Ewald sphere is 0 [de Leeuw et al., 1980a]. It is worth noting that this choice can significantly affect long-range properties, like the overall ferroelectric order of an LC polar sample, which may appear or not if conducting boundaries or a vacuum surrounding the sample are used, as shown in [Wei and Patey, 1992; Ayton and Patey, 1996].

The standard Ewald summation is very demanding in terms of resources as it scales with the number of particles, N, as $N^{\frac{3}{2}}$ [Perram et al., 1988]. However, optimized implementations such as the particle mesh Ewald (PME) [Darden et al., 1993] and smooth particle mesh Ewald (SPME) [Essmann et al., 1995] based on interpolation methods [Hockney, 1989] that use the very efficient fast Fourier transform (FFT) for the reciprocal lattice contribution allow this to reach $N \log N$ scaling. Further modifications and developments of the technique have been proposed, e.g. the staggered mesh Ewald (StME) algorithm [Cerutti et al., 2009] further improves the accuracy of computed forces by an order of magnitude.

5.4.5 Tuning the Atomistic Force Fields

As we have already noted, for most practical cases it is necessary to adjust at least in part the Force Field. In doing that it is important to avoid double counting as all intramolecular electrostatic and dispersive interactions between the atoms are implicitly included in the QC calculation. For instance, when obtaining the torsional barrier relative to the variation of an angle ϕ, one possibility is to subtract from the calculated QC energy $U^{\mathrm{QC}}(\phi)$ the molecular mechanics one, calculated from the original FF with the torsional contribution set to 0, $U_c^{\mathrm{MM}}(\phi)$. In practice, to evaluate this, one can compute a Boltzmann distribution $P_c^{\mathrm{MM}}(\phi; T)$ by performing a short (a few nanoseconds) MD simulation of an isolated molecule at a temperature so high that the conformations corresponding to all angles are well populated (even something like $T = 1000$ K or more), and that a thorough exploration of the potential energy surface (PES) can be achieved. $U_c^{\mathrm{MM}}(\phi)$ can then be obtained through a Boltzmann inversion of the distribution [Berardi et al., 2005; Pizzirusso et al., 2011]:

$$U_c^{\mathrm{MM}}(\phi) = -k_B T \ln P_c^{\mathrm{MM}}(\phi; T) - U_0, \tag{5.39}$$

where k_B is the Boltzmann constant and U_0 shifts to 0 the minimum of the energy. The corrected force field contribution is therefore:

$$U^{\mathrm{DHT}}(\phi) = U^{\mathrm{QC}}(\phi) - U_c^{\mathrm{MM}}(\phi) \tag{5.40}$$

and can be expanded in Fourier series (see Eq. 5.30) for an easy on the fly evaluation during the simulation.

5.5 Hard Anisotropic Particles (Shape Matters!)

This very simple class of purely repulsive potentials is based on the idea that the structure of a crystal or of a liquid is just determined by optimal packing. The idea is probably one of the oldest, and indeed, structures of crystals, based on the wood particle models of Professor Daniell (Kings' College, London), kept at the Science Museum in London, date back to around 1830. In line with this idea, a number of models with only anisotropically shaped hard particles, without attractions, have been among the first to be studied, particularly by Frenkel and collaborators [Allen et al., 1989, 1993]. They are still actively investigated and have been recently reviewed [Mederos et al., 2014]. Hard particle (HP) models represent a natural generalization of the hard-core potentials between spherical particles:

$$U^{\mathrm{HP}}(\boldsymbol{r}_{ij}, \Omega_i, \Omega_j) = \begin{cases} \infty, & \text{for } r_{ij} \le \sigma_h(\hat{\boldsymbol{r}}_{ij}, \Omega_i, \Omega_j) \\ 0, & \text{for } r_{ij} > \sigma_h(\hat{\boldsymbol{r}}_{ij}, \Omega_i, \Omega_j) \end{cases} \tag{5.41a}$$

$$= \begin{cases} \infty, & \text{if } \boldsymbol{r}_{ij} \in V_{\mathrm{ex}}\left(\Omega_i, \Omega_j\right) \\ 0, & \text{otherwise,} \end{cases} \tag{5.41b}$$

where the contact function $\sigma_h(\hat{\boldsymbol{r}}_{ij}, \Omega_i, \Omega_j)$ gives the minimum distance between two non-overlapping particles at fixed orientations and fixed relative interparticle vector. The overlap condition defines a region of space $V_{ex}\left(\Omega_i, \Omega_j\right)$ excluded to particle j because of the presence of particle i (and vice versa) [Allen et al., 1993]. Note that the Boltzmann factor for hard particles does not depend on temperature and thus their molecular organization is controlled by the entropy, which in turn depends on the volume accessible to the particles. In a system of particles devoid of attractive energy a phase change from disordered (e.g. liquid) to ordered (e.g. crystal or nematic) necessarily requires an increase in entropy (see Sec.5.2.1). The only relevant parameter in the canonical ensemble is then the density ρ or, equivalently, the volume fraction ϕ which for a monodisperse system, is $\phi = \rho \mathcal{V}$, where $\mathcal{V}$ is the volume of a particle. The simplest particles to study are rigid *convex bodies*, [Kihara, 1963; Allen et al., 1993] characterized by the geometrical requirement that a line segment connecting any two points on their surface lies entirely within it and that a tangential plane (the local supporting plane) can touch their surface at one and only one contact point. Considering an origin inside a convex body, the surface is described, in polar coordinates, by a vector

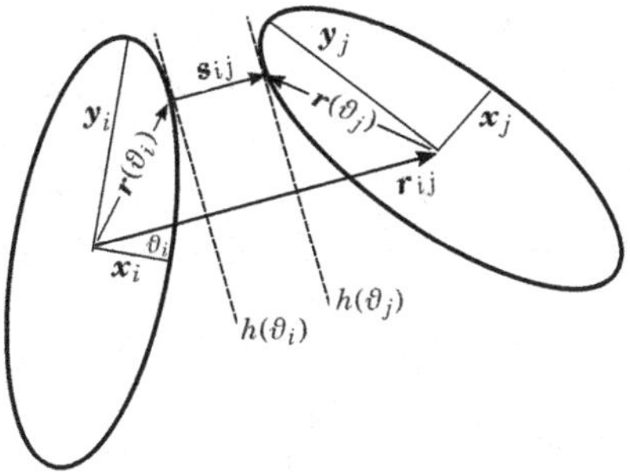

Figure 5.12 A sketch of two convex particles i, j in 2D and of their support functions $h(\vartheta_i)$, $h(\vartheta_j)$ with close contact points $\boldsymbol{r}(\vartheta_i)$, $\boldsymbol{r}(\vartheta_j)$ and surface-surface distance s_{ij}.

function $\boldsymbol{r}(\vartheta,\varphi)$. Another description is in terms of the *supporting function* $h(\theta,\varphi)$ the perpendicular distance from the origin to the supporting plane (see Fig. 5.12). We can write $h(\theta,\varphi) = \boldsymbol{r}(\vartheta,\varphi)\cdot\hat{\boldsymbol{n}}$, where $\hat{\boldsymbol{n}}$ is the unit vector in the direction of the normal to the tangent plane. There is a variety of differently shaped particles relevant as model mesogens that fall into the definition, e.g. spheres, prolate and oblate spherocylinders and ellipsoids, also if tapered at one end. It is interesting that these different shapes, like any convex body, can be characterized by three metrics: volume, $\mathcal{V}$ surface area, $\mathcal{S}$, and mean curvature integral, $\mathcal{R}$, normalized by 4π:

$$\mathcal{R} = \frac{1}{4\pi}\iint d\vartheta d\varphi\, \boldsymbol{r}\left(\frac{\partial\hat{\boldsymbol{n}}}{\partial\vartheta}\times\frac{\partial\hat{\boldsymbol{n}}}{\partial\varphi}\right) = \frac{1}{4\pi}\iint d\vartheta d\varphi\, h(\theta,\varphi)\sin\vartheta, \tag{5.42a}$$

$$\mathcal{S} = \iint d\vartheta d\varphi\, \hat{\boldsymbol{n}}\left(\frac{\partial\boldsymbol{r}}{\partial\vartheta}\times\frac{\partial\boldsymbol{r}}{\partial\varphi}\right), \tag{5.42b}$$

$$\mathcal{V} = \frac{1}{3}\iint d\vartheta d\varphi\, \boldsymbol{r}\left(\frac{\partial\boldsymbol{r}}{\partial\vartheta}\times\frac{\partial\boldsymbol{r}}{\partial\varphi}\right). \tag{5.42c}$$

The scalar (isotropically averaged) second virial coefficient for any pair of hard convex bodies is given by the exact Isihara–Hadwiger relation [Kihara, 1967; Boublik and Nezbeda, 1986] as

$$B_2^{\mathrm{HP}} = \mathcal{R}\mathcal{S} + \mathcal{V} = (1+3\xi)\,\mathcal{V} \tag{5.43}$$

in terms of $\mathcal{R},\mathcal{S},\mathcal{V}$, while ξ is a size independent asphericity parameter, $\xi = \mathcal{R}\mathcal{S}/3\mathcal{V}$. For instance, given a sphere of radius $\sigma_h/2$, $\mathcal{V} = \pi\sigma_h^3/6$, $\mathcal{S} = \pi\sigma_h^2$, $\mathcal{R} = \sigma_h/2$, we find $B_2^{\mathrm{HS}} = 2\pi\sigma_h^3/3 = 4\mathcal{V}_h$ as it should (Eq. 5.9). The concept of *support plane* is useful when we have to determine the surface-surface distance between two convex particles i and j at separation $\boldsymbol{r}_{ij}$ (see Fig. 5.12), which amounts to the distance s_{ij} between the two support planes. The computational aspects are discussed in detail in Allen et al. [1993].

5.5.1 Hard Ellipsoids (HE)

Ellipsoids with semi-axes a, b and c can be obtained by an affine deformation of a sphere. A linear pulling or squeezing along one axis gives a uniaxial prolate or, respectively, oblate

Table 5.8. *Mean radius of curvature $\mathcal{R}^{\mathrm{HE}}$, surface, $\mathcal{S}^{\mathrm{HE}}$ and volume $\mathcal{V}^{\mathrm{HE}}$ of prolate hard ellipsoids ($a = b < c$) where the ellipticity $\eta = \sqrt{(a^2 - c^2)}/a$, and the same for oblate ($a < b = c$) ones, with ellipticity $\epsilon = \sqrt{(b^2 - a^2)}/b$ [Singh and Kumar, 2001]*

metric	prolate	oblate
$\mathcal{R}^{\mathrm{HE}}$	$\frac{c}{2}\left[1 + \frac{1-\eta^2}{2\eta}\ln\left(\frac{1+\eta}{1-\eta}\right)\right]$	$\frac{a}{2}\left(1 + \frac{\sin^{-1}\epsilon}{\epsilon\left(1-\epsilon^2\right)^{1/2}}\right)$
$\mathcal{S}^{\mathrm{HE}}$	$2\pi a^2\left[1 + \frac{\sin^{-1}\eta}{\eta\sqrt{1-\eta^2}}\right]$	$2\pi b^2\left(1 + \frac{1-\epsilon^2}{2\epsilon}\ln\frac{1+\epsilon}{1-\epsilon}\right)$
$\mathcal{V}^{\mathrm{HE}}$	$\frac{4\pi}{3}a^2 c$	$\frac{4\pi}{3}b^2 a$

ellipsoid, while scaling along two axes gives a biaxial ellipsoid. The function $S(\boldsymbol{h})$ describing the surface of a hard ellipsoidal particle is given by $S^{\mathrm{HE}} = (x^2/a^2) + (y^2/b^2) + (z^2/c^2) = 1$, where a, b, and c are three semi-axes of the ellipsoid and (x, y, z) are the components of the centre-to-surface vector $\boldsymbol{h} = \hat{\boldsymbol{a}}x + \hat{\boldsymbol{b}}y + \hat{\boldsymbol{c}}z$, with $(\hat{\boldsymbol{a}}, \hat{\boldsymbol{b}}, \hat{\boldsymbol{c}})$ the basis vectors along the principal axes. The ratio of the largest to the smallest semi-axes, is the maximal *aspect ratio* of the ellipsoid. An expression for the second virial coefficient of identical ellipsoidal particles can be obtained using the Isihara–Hadwiger formula, Eq. 5.43. For the general case of biaxial ellipsoids, expressions for $\mathcal{R}^{\mathrm{HE}}$ and $\mathcal{S}^{\mathrm{HE}}$ have been derived by Singh and Kumar [1996, 2001]. Explicitly, the mean radius of curvature and the surface are given by

$$\mathcal{R}^{\mathrm{HE}} = \frac{a}{2}\left[\sqrt{\frac{1+\epsilon_b}{1+\epsilon_c}} + \sqrt{\epsilon_c}\left\{\frac{1}{\epsilon_c}F(\varphi, k_1) + E(\varphi, k_1)\right\}\right], \tag{5.44a}$$

$$\mathcal{S}^{\mathrm{HE}} = 2\pi a^2\left[1 + \sqrt{\epsilon_c(1+\epsilon_b)}\left\{\frac{1}{\epsilon_c}F(\varphi, k_2) + E(\varphi, k_2)\right\}\right], \tag{5.44b}$$

$$\mathcal{V}^{\mathrm{HE}} = \frac{4\pi}{3}abc, \tag{5.44c}$$

where $F(\varphi, k)$ is an elliptic integral of the first kind and $E(\varphi, k)$ is an elliptic integral of the second kind [Abramowitz and Stegun, 1965], with the amplitude $\varphi = \tan^{-1}\left(\sqrt{\epsilon_c}\right)$ and $k_1 = \sqrt{(\epsilon_c - \epsilon_b)/\epsilon_c}$, $k_2 = \sqrt{[(\epsilon_b(1+\epsilon_c)]/[\epsilon_c(1+\epsilon_b)]}$, while the ellipsoid anisotropy is $\epsilon_b = [(b/a)^2 - 1]$, $\epsilon_c = [(c/a)^2 - 1]$. These, together with Eq. 5.43, can give numerical values of the second virial coefficient for identical hard biaxial ellipsoids [McBride and Lomba, 2007]. The expressions become much more manageable for the special case of uniaxial ellipsoids, shown in Table 5.8.

5.5.2 Hard Spherocylinders (HSC)

A *spherocylinder*, also called a *capsule*, is a cylinder of length l and diameter $d = 2w$ with hemispherical caps of the same diameter d at both ends, thus with length $l + d$. The potential between two hard spherocylinders is

$$U^{\mathrm{HSC}}(\boldsymbol{r}_{ij}, \boldsymbol{u}_i, \boldsymbol{u}_j) = \begin{cases} 0, & \text{if there is no overlap} \\ \infty, & \text{otherwise.} \end{cases} \tag{5.45}$$

The potential, like that for hard ellipsoids, is particularly suitable to model elongated objects and it reduces to that of hard spheres when $l = 0$. For a prolate hard spherocylinder the fundamental measures are:

$$\mathcal{R}^{\mathrm{HSC}} = \frac{1}{4} l + \frac{1}{2} d, \tag{5.46a}$$

$$\mathcal{S}^{\mathrm{HSC}} = \pi d(d + l), \tag{5.46b}$$

$$\mathcal{V}^{\mathrm{HSC}} = \pi\, d^2 \left(\frac{1}{6} d + \frac{1}{4} l \right). \tag{5.46c}$$

The Isihara–Hadwiger relation, Eq. 5.43, allows calculating the coefficient B_2 needed for the imperfect gas virial expansion. When $l = 0$ this reduces to $8\mathcal{V}^{\mathrm{HS}}$ for the two-molecule system; if this volume is distributed among the two particles, it gives the conventional result of four times the volume of a hard sphere, an important quantity in the van der Waals equation of state. The coefficient B_2, as considered until now, is an isotropic average of the anisotropic Mayer function (Eq. 4.147). We shall see in Section 7.6 that the coefficient $B_2^{\mathrm{HSC}}(\beta_{12})$, representing the excluded volume of two spherocylinders, i.e. the volume denied to the second spherocylinder by the presence of first when their relative orientation is fixed, is key in setting up Onsager theory for anisotropic phases.

5.5.3 Hard Polyhedra Particles (HPP)

Suspensions of sufficiently anisotropic hard rod-like and disc-like particles give liquid crystal phases upon increasing concentration (Section 1.14). However, it is interesting to see if rigid polyhedra of other shapes can, just on the basis of entropy optimization, yield LC phases or other molecular organizations, possibly with novel material properties. The problem is of increasing importance in view of the variety of nanoparticle shapes that have and are being synthesized [Glotzer et al., 2004; Vogel et al., 2015; Boselli et al., 2020] as we also see from the few examples in Fig. 5.13 comprising both convex (a, b, c, f) and non-convex particles (d, e). At high density the thermodynamically stable phase for a system of hard particles is typically the one with the highest packing density [Chaikin et al., 2006]. Extensive numerical simulations of polyhedra packing have been performed, particularly in the groups of F. A. Escobedo [John et al., 2008; Agarwal and Escobedo, 2011; Escobedo, 2014] and of S. Glotzer [Damasceno et al., 2012] and the formation of columnar, nematic and smectic phases has been observed. Interestingly, none of the polyhedra yielding LCs

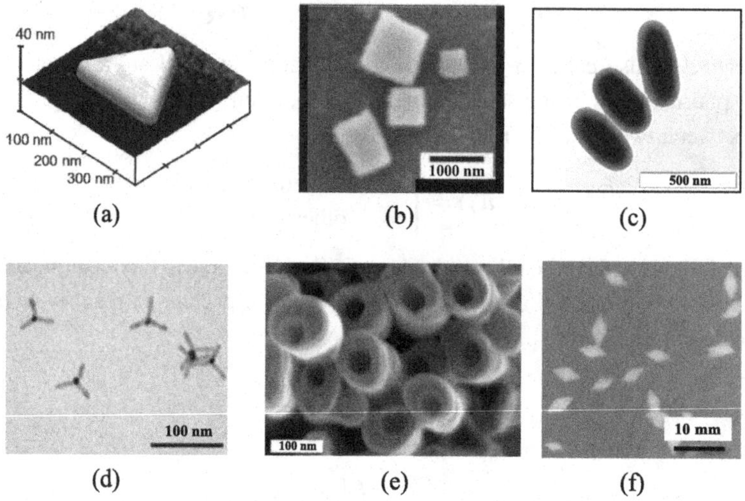

Figure 5.13 Nanoparticles and microparticles of different shapes experimentally prepared: (a) A triangular Au prism [Malikova et al., 2002], (b) Pt nanocubes [Chan et al., 2012], (c) silica ellipsoids covered by haematite [Sacanna et al., 2006], (d) CdSe/CdTe tetrapods [Fiore et al., 2009], (e) Au nanorings [Ozel et al., 2015] and (f) rhombic $GeSe_2$ flakes on a mica substrate [Zhou et al., 2017].

corresponds to simple Platonic (regular and identical faces) or Archimedean solids (regular faces with identical vertices [Damasceno et al., 2012]. The thermodynamic behaviour of hard particles can be understood through entropy maximization. Packing efficiency plays an increasingly important role towards higher density and induces a preferential alignment of flat facets. Because packing efficiency increases with contact area, the alignment can be interpreted as the result of an effective, many-body directional entropic force arising from the increased number of configurations available to the entire system, causing suitably faceted polyhedra to order. Virial coefficients for a number of convex polyhedral particles have been discussed [Irrgang et al., 2017] and in particular second virial coefficients have been obtained using Eq. 5.43, after showing that the (normalized) integrated mean curvature for a polyhedron is given by $\mathcal{R}^{\mathrm{HP}} = \sum_i l_i(\pi - \theta_i)/(8\pi)$, where the summation runs over all edges with edge length l_i and dihedral angle θ_i between adjacent faces. Volume, $\mathcal{V}^{\mathrm{HP}}$, and surface, $\mathcal{S}^{\mathrm{HP}}$, of the HPP are easily calculated by summing over suitably carved-up components.

5.6 Attractive-Repulsive Rigid Particles

Given that real molecules have both attractive and repulsive interactions, it is desirable to find a simple generalization of the LJ potential for anisotropic particles. Clearly, the contact distance (size) σ, and the interaction energy strength (attractive well depth), ϵ, parameters in an LJ-like potential, Eq. 5.17, will need to depend on positions and orientations of the two molecules, becoming $\sigma(\boldsymbol{r}_{ij}, \Omega_i, \Omega_j) = \sigma(r_{ij}, \Omega_i, \Omega_j, \Omega_{r_{ij}})$ and $\epsilon(\boldsymbol{r}_{ij}, \Omega_i, \Omega_j) = \epsilon(r_{ij}, \Omega_i, \Omega_j, \Omega_{r_{ij}})$. In the next few sections we introduce a way of obtaining these anisotropic expressions starting from the Gaussian model for ellipsoidal molecules.

5.6.1 Gaussian Overlap Models

In order to develop a soft attractive-repulsive model to handle anisotropic particles, a very useful suggestion was that of Berne and Pechukas [1972], that is to assume that the interaction energy of two identical ellipsoidal particles can be taken to be proportional to their overlap. To calculate explicitly this overlap a molecule can be represented by a 3D Gaussian distribution, so as to take advantage of the fact that the overlap between two Gaussian functions is another Gaussian. Thus, a particle i of a certain type A with centre at $\boldsymbol{r}_i$ and orientation, $\Omega_{i\mathrm{L}}$ with respect to the laboratory frame, corresponds to a Gaussian $G(\boldsymbol{r}_i, \Omega_{i\mathrm{L}}, \boldsymbol{r})$ that, except for a normalization factor, can be written as [Berardi et al., 1995]

$$G_i = G(\boldsymbol{r}_i, \Omega_{i\mathrm{L}}, \boldsymbol{r}) = \exp\left[-\frac{1}{2}(\boldsymbol{r} - \boldsymbol{r}_i)^T \mathbf{M}_i^T [\mathbf{S}]^{-2} \mathbf{M}_i (\boldsymbol{r} - \boldsymbol{r}_i)\right], \tag{5.47}$$

where $\mathbf{S}$ is a diagonal *shape* matrix whose elements, $\sigma_x, \sigma_y, \sigma_z$, are the length of the axes of the ellipsoid (in units of a certain convenient σ_0), while $\mathbf{M}_i \equiv \mathbf{R}(\Omega_{i\mathrm{L}})$ is the rotation matrix transforming from laboratory to the ith molecule frame. A particle shape biaxiality can be defined as

$$\lambda_S = \sqrt{\frac{3}{2}} \frac{\sigma_x - \sigma_y}{2\sigma_z - \sigma_x - \sigma_y}. \tag{5.48}$$

To make the definition unambiguous, we conventionally assign the axes x, y and z so as to obtain the lowest biaxiality. In practice, for a prolate object we choose $\sigma_z > \sigma_x \geqslant \sigma_y$ while for an oblate particle we adopt $\sigma_y \geqslant \sigma_x > \sigma_z$. Using this prescription λ_S can vary from 0, corresponding to a uniaxial object as considered in the original treatment by Gay and Berne [1981], to $1/\sqrt{6}$ for an object of spherical symmetry, i.e. when all three σ_i tend to σ with the ordering above. When the first molecule is centred at the origin, $\boldsymbol{r}_i = 0$, and the second at a separation vector, $\boldsymbol{r}_{ij}$ from it, we have the overlap integral

$$Q_{ij}(\boldsymbol{r}_{ij}, \Omega_i, \Omega_j)$$
$$= \int_0^\infty \mathrm{d}\boldsymbol{r}\, G(\boldsymbol{r}_i, \Omega_i, \boldsymbol{r})\, G(\boldsymbol{r}_j, \Omega_j, \boldsymbol{r}), \tag{5.49a}$$
$$= \int_0^\infty \mathrm{d}\boldsymbol{r}\, \exp\left\{-\frac{1}{2}\left[\boldsymbol{r}^T \mathbf{M}_i^T [\mathbf{S}]^{-2} \mathbf{M}_i \boldsymbol{r} - (\boldsymbol{r}_{ij} - \boldsymbol{r})^T \mathbf{M}_j^T [\mathbf{S}]^{-2} \mathbf{M}_j (\boldsymbol{r}_{ij} - \boldsymbol{r})\right]\right\}. \tag{5.49b}$$

The great advantage of using Gaussians is that the product of two of them is another Gaussian, so their overlap integral can be calculated analytically. A convenient way is to write the convolution integral Eq. 5.49b as the inverse Fourier transform of the product of Fourier transforms of G_i and G_j (see Section E.2.1 and [Bracewell, 2000]). We find in this way

$$Q_{ij}(\boldsymbol{r}_{ij},\Omega_i,\Omega_j) = \det([\mathbf{S}]^2)\left(\frac{\pi^3}{8\det([\mathbf{A}_{ij}])}\right)^{1/2}\exp\left[-\frac{1}{2}\boldsymbol{r}_{ij}^T[\mathbf{A}_{ij}]^{-1}\boldsymbol{r}_{ij}\right], \tag{5.50}$$

where $\det([\mathbf{S}]^2) = (\sigma_x\sigma_y\sigma_z)^2$ and $[\mathbf{A}_{ij}]$ is a symmetric matrix defined as

$$[\mathbf{A}_{ij}] \equiv [\mathbf{A}(\Omega_i,\Omega_j)] = \mathbf{M}_i^T[\mathbf{S}]^2\mathbf{M}_i + \mathbf{M}_j^T[\mathbf{S}]^2\mathbf{M}_j. \tag{5.51}$$

Armed with these results for the overlap, one can proceed towards developing an (empirical) expression for the pair potential [Berne and Pechukas, 1972].

5.6.2 Berne–Pechukas Potential

For uniaxial molecules (with $\sigma_x = \sigma_y \neq \sigma_z$) both $\det([\mathbf{A}_{ij}])$ and $[\mathbf{A}_{ij}]^{-1}$ can be calculated analytically [Berne and Pechukas, 1972] and one can assume, for two identical uniaxial molecules, an interaction potential

$$U(r_{ij},\boldsymbol{u}_i,\boldsymbol{u}_j,\hat{\boldsymbol{r}}_{ij}) = \epsilon_0\,\epsilon(\boldsymbol{u}_i,\boldsymbol{u}_j)\exp\left\{-\frac{r_{ij}^2}{\sigma^2(\boldsymbol{u}_i,\boldsymbol{u}_j,\hat{\boldsymbol{r}}_{ij})}\right\}, \tag{5.52}$$

where $\boldsymbol{u}_i$, $\boldsymbol{u}_j$, $\hat{\boldsymbol{r}}_{ij}$ are unit vectors giving the orientations $\Omega_{i\mathrm{L}}, \Omega_j, \Omega_{r_{ij}\mathrm{L}}$ of the two particles and of the intermolecular vector respectively and the pre-exponential coefficient $\epsilon(\boldsymbol{u}_i,\boldsymbol{u}_j)$ is: $\left\{1-\chi^2(\boldsymbol{u}_i\cdot\boldsymbol{u}_j)^2\right\}^{-\frac{1}{2}}$, where χ is a shape anisotropy parameter,

$$\chi = \frac{(\sigma_\mathrm{e})^2-(\sigma_\mathrm{s})^2}{(\sigma_\mathrm{s})^2+(\sigma_\mathrm{e})^2} = \frac{\kappa^2-1}{\kappa^2+1}, \tag{5.53}$$

with $\kappa = \sigma_\mathrm{e}/\sigma_\mathrm{s}$ the aspect ratio, and a contact distance $\sigma(\boldsymbol{u}_i,\boldsymbol{u}_j,\hat{\boldsymbol{r}}_{ij})$ related to the molecular shape:

$$\sigma(\boldsymbol{u}_i,\boldsymbol{u}_j,\hat{\boldsymbol{r}}_{ij}) = \sigma_0\left\{1-\frac{\chi}{2}\left[\frac{(\hat{\boldsymbol{r}}_{ij}\cdot\boldsymbol{u}_i+\hat{\boldsymbol{r}}_{ij}\cdot\boldsymbol{u}_j)^2}{1+\chi(\boldsymbol{u}_i\cdot\boldsymbol{u}_j)}+\frac{(\hat{\boldsymbol{r}}_{ij}\cdot\boldsymbol{u}_i-\hat{\boldsymbol{r}}_{ij}\cdot\boldsymbol{u}_j)^2}{1-\chi(\boldsymbol{u}_i\cdot\boldsymbol{u}_j)}\right]\right\}^{-\frac{1}{2}}, \tag{5.54}$$

where ϵ_0, σ_0 are scaling parameters. It should be noted that this 'soft' contact distance is similar but not identical to the 'hard' geometrical one between impenetrable ellipsoids (see [Everaers and Ejtehadi, 2003; Zheng and Palffy-Muhoray, 2007] for a discussion). An expression for ϵ and σ in terms of the rotational invariants $S_{L',L'',L} \equiv S_{L',L'',L}(\boldsymbol{u}_i,\boldsymbol{u}_j,\hat{\boldsymbol{r}}_{ij})$ for uniaxial ellipsoids (Eq. G.25) was given by Stone [1979] as:

$$\epsilon(\boldsymbol{u}_i,\boldsymbol{u}_j) = \sqrt{3}\epsilon_0\left(3-\chi^2-\sqrt{20}\chi^2 S_{2,2,0}\right)^{-\frac{1}{2}}, \tag{5.55}$$

$$\sigma(\boldsymbol{u}_i,\boldsymbol{u}_j,\hat{\boldsymbol{r}}_{ij}) = 3\sigma_0\left(\epsilon_0/\epsilon(\boldsymbol{u}_i,\boldsymbol{u}_j)\right)\Big\{9-6\chi-\chi^2+\sqrt{280}\chi^2 S_{2,2,2} - \sqrt{20}\left[\chi^2 S_{2,2,0}+(3\chi-2\chi^2)(S_{2,0,2}+S_{0,2,2})\right]\Big\}^{-\frac{1}{2}}. \tag{5.56}$$

The interaction parameters $\epsilon(\boldsymbol{u}_i,\boldsymbol{u}_j)$ and $\sigma(\boldsymbol{u}_i,\boldsymbol{u}_j,\hat{\boldsymbol{r}}_{ij})$ obtained in this way can be, somewhat arbitrarily, plugged in the classic Lennard–Jones potential to produce an anisotropic (12–6) pair potential. This model has been used in various early off-lattice simulations

[Kushick and Berne, 1973b; Tsykalo and Bagmet, 1978; Kabadi and Steele, 1985; Tsykalo, 1991] and in statistical mechanics studies [Monson and Gubbins, 1983] but presents various limitations. One is that its well depth only depends on the relative orientation of the two particles. Thus, for instance, the potential well depth for two parallel molecules is the same if the molecules lie side by side or head-to-head, which is not realistic (cf. [Stone, 1979]). Moreover, the well width varies with the molecular orientation with respect to the intermolecular vector.

5.6.3 Uniaxial Gay–Berne Potentials

An improved version of the previous potential is the Gay–Berne one [Gay and Berne, 1981; Adams et al., 1987], which has a shifted, rather than scaled, Lennard–Jones-like form. Omitting for conciseness the particle type label A, since all particles are of the same type,

$$\begin{aligned} U^{\mathrm{GB}}(\boldsymbol{r}_{ij},\boldsymbol{u}_i,\boldsymbol{u}_j) &= 4\epsilon_0\, \epsilon^{(\mu,\nu)}(\boldsymbol{u}_i,\boldsymbol{u}_j,\hat{\boldsymbol{r}}_{ij}) \\ &\times \left[\left(\frac{\sigma_c}{r_{ij}-\sigma(\boldsymbol{u}_i,\boldsymbol{u}_j,\hat{\boldsymbol{r}}_{ij})+\sigma_c}\right)^{12} - \left(\frac{\sigma_c}{r_{ij}-\sigma(\boldsymbol{u}_i,\boldsymbol{u}_j,\hat{\boldsymbol{r}}_{ij})+\sigma_c}\right)^{6}\right], \end{aligned} \tag{5.57}$$

where ϵ_0 is an energy scale, while the length scale σ_0 is contained in $\sigma(\boldsymbol{u}_i,\boldsymbol{u}_j,\hat{\boldsymbol{r}}_{ij})$, as defined in Eq. 5.54. The other parameter with dimensions of length, σ_c, is a range of interactions that typically, but not necessarily equals the unit distance σ_0. For instance, $\sigma_c < \sigma_0$ can be used to describe interactions in systems of interacting colloidal particles, where we can expect the effective interaction range to be shorter than the particle size. Apart from this rather non-standard case, σ_c depends on the ellipsoid being prolate or oblate, as will be discussed in Section 11.5.1. The potential is 0 at the contact distance $r_{ij} = \sigma(\boldsymbol{u}_i,\boldsymbol{u}_j,\hat{\boldsymbol{r}}_{ij})$. The energy term is written as a product:

$$\epsilon^{(\mu,\nu)}(\boldsymbol{u}_i,\boldsymbol{u}_j,\hat{\boldsymbol{r}}_{ij}) = \left[\epsilon_1(\boldsymbol{u}_i,\boldsymbol{u}_j,\hat{\boldsymbol{r}}_{ij})\right]^{\mu} \left[\epsilon_2(\boldsymbol{u}_i,\boldsymbol{u}_j)\right]^{\nu}, \tag{5.58}$$

where

$$\epsilon_1(\boldsymbol{u}_i,\boldsymbol{u}_j,\hat{\boldsymbol{r}}_{ij}) = 1 - \frac{\chi'}{2}\left[\frac{(\boldsymbol{u}_i\cdot\hat{\boldsymbol{r}}_{ij}+\boldsymbol{u}_j\cdot\hat{\boldsymbol{r}}_{ij})^2}{1+\chi'(\boldsymbol{u}_i\cdot\boldsymbol{u}_j)} + \frac{(\boldsymbol{u}_i\cdot\hat{\boldsymbol{r}}_{ij}-\boldsymbol{u}_j\cdot\hat{\boldsymbol{r}}_{ij})^2}{1-\chi'(\boldsymbol{u}_i\cdot\boldsymbol{u}_j)}\right], \tag{5.59a}$$

$$\epsilon_2(\boldsymbol{u}_i,\boldsymbol{u}_j) = \left[1-\chi^2(\boldsymbol{u}_i,\boldsymbol{u}_j)^2\right]^{-1/2}. \tag{5.59b}$$

The attractive anisotropy parameter χ' is defined as

$$\chi' = \frac{\epsilon_s^{1/\mu} - \epsilon_e^{1/\mu}}{\epsilon_s^{1/\mu} + \epsilon_e^{1/\mu}} = \frac{\kappa'^{1/\mu}-1}{\kappa'^{1/\mu}+1}, \tag{5.60}$$

with $\kappa = \sigma_e/\sigma_s = \sigma_\parallel/\sigma_\perp$ and $\kappa' = \epsilon_s/\epsilon_e = \epsilon_\parallel/\epsilon_\perp$, where e stands for a configuration of particles i and j corresponding to $\boldsymbol{u}_i \| \boldsymbol{u}_j = 1$, $\boldsymbol{u}_i \| \hat{\boldsymbol{r}}_{ij} = 1$, $\boldsymbol{u}_j \| \hat{\boldsymbol{r}}_{ij} = 1$ and s to that with $\boldsymbol{u}_i \| \boldsymbol{u}_j = 1$, $\boldsymbol{u}_i \perp \hat{\boldsymbol{r}}_{ij} = 0$, $\boldsymbol{u}_j \perp \hat{\boldsymbol{r}}_{ij} = 0$. For a rod-like particle the subscripts e and s thus refer, respectively, to an end-to-end and side-by-side configuration, the length to breadth ratio is $\kappa > 1$, while the ratio of the well depths is $\kappa' > 1$. In this case, the

Table 5.9. *The GB strength of interaction ϵ for end-to-end* (ee)*, side-by-side* (ss)*, cross* ($\times$)*, and tee* (T) configurations and for $\kappa \equiv \sigma_e/\sigma_s = 3$, $\kappa' \equiv \epsilon_s/\epsilon_e = 5$, $\epsilon_0 = 1$, with exponents ($\mu = 2$, $\nu = 1$) or ($\mu = 1$, $\nu = 3$) [Berardi et al., 1993]

Config.	$\epsilon(\boldsymbol{u}_i, \boldsymbol{u}_j, \hat{\boldsymbol{r}}_{ij})/\epsilon_0$	GB(3, 5, 2, 1)	GB(3, 5, 1, 3)
(ee)	$(1/\kappa')(1-\chi^2)^{-\nu/2}$	$\frac{1}{3}$	$\frac{25}{27}$
(ss)	$(1-\chi^2)^{-\nu/2}$	$\frac{1}{15}$	$\frac{5}{27}$
($\times$)	1	1	1
(T)	$[2/(\kappa'^{1/\mu}+1)]^{\mu}$	$4/(5^{1/2}+1)^2$	$8/(5^{1/3}+1)^3$

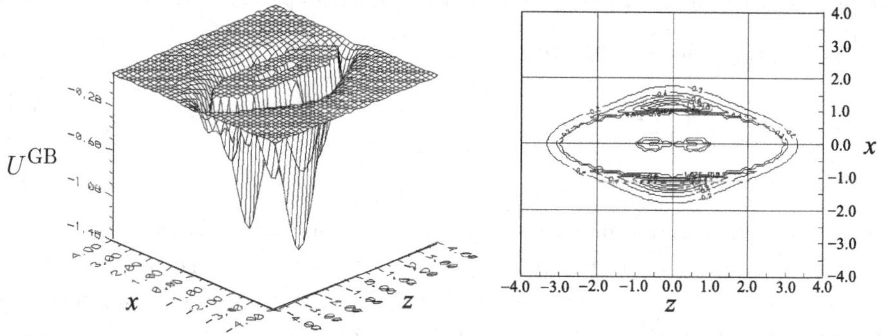

Figure 5.14 (a) The Gay–Berne GB(3,5,2,1) potential for two particles parallel to the z-axis as the position of the centre of the second molecule moves in the zx plane around the first one placed at the origin. The two deep minima correspond to side-side positions. (b) A contour plot of the same potential map. Notice the spurious unphysical well very close to the origin. All distances are expressed in σ_0 units and energies in ϵ_0 units.

length parameter one takes $\sigma_c = \sigma_0 = \sigma_s$. For a discotic the subscript e corresponds to a face-to-face approach, so that we have $\kappa < 1$, $\kappa' < 1$ and the usual choice made [Bates and Luckhurst, 1996] is $\sigma_c = \kappa\sigma_0 = \sigma_f$. The potential also depends on two parameters, μ and ν that can help to tune the shape of the potential well. In Table 5.9 we show the effect of changing μ and ν at certain special orientations. In Fig. 5.14 we see a representation of the GB potential surface as in the original formulation, i.e. for $\mu = 2, \nu = 1$ [Gay and Berne, 1981]. In Fig. 5.15a we consider some sections for the most significant orientations for two approaching molecules: side-by-side (ss), end-to-end (ee) and transversal both for $\mu = 2, \nu = 1$ and for $\mu = 1, \nu = 3$, as proposed by Berardi et al. [1993]. We see that, even though the profile is qualitatively similar in both cases, the well depth and profile change significantly with the different parameterizations. In particular, for $\mu = 1, \nu = 3$ the ss interaction is stronger, favouring parallel alignment and leading, as we shall see in Chapter 11, to a wider nematic range. A convenient notation to summarize the parameters determining a GB potential has been introduced by Bates and Luckhurst [1999a] as GB($\kappa, \kappa', \mu, \nu$). The choice GB($\kappa$,$\kappa'$,0,0) has been used by Andrienko et al. [2001], while GB(1,1,0,0) reduces

Table 5.10. *Dimensionless units for the Gay–Berne model. Here σ_0, m_0, ϵ_0, τ_0 are the units of length, mass, energy and time. For rod-like molecules $\sigma_0 = \sigma_s$, while for discotics $\sigma_0 = \sigma_f$ (see text). Conversion to real units is effected choosing a specific mesogen. For example, for 8CB: $\sigma_0 = 0.5$ nm, $m_0 = 4.8 \times 10^{-22}$ kg, $\epsilon_0 = 1.4 \times 10^{-21}$ J, giving $\tau_0 = 0.3$ ns and $P_0 = 112$ bar. [Vanzo et al., 2012]*

Distance	$r^* \equiv r/\sigma_0$
Mass	$m^* \equiv m/m_0$
Volume	$V^* \equiv V/\sigma_0^3$
Number density	$\rho^* \equiv \rho\sigma_0^3, \quad \rho = N/V$
Temperature	$T^* \equiv T/T_0, \quad T_0 = \epsilon_0/k_B$
Energy	$U^* \equiv U/\epsilon_0$
Pressure	$P^* \equiv P/P_0, \quad P_0 = \epsilon_0/\sigma_0^3$
Time	$t^* \equiv t/\tau_0, \quad \tau_0 = \sqrt{m\sigma_0^2/\epsilon_0}$
Force	$f^* \equiv \sigma_0 f/\epsilon_0$
Inertia moment	$I^* \equiv I/I_0, \quad I_0 = m/\sigma_0^2$
Charge	$q^* \equiv q/q_0, \quad q_0 = \sqrt{4\pi\varepsilon_0\epsilon_0\sigma_0}$
Viscosity	$\eta^* \equiv \sigma_0^2\eta/\sqrt{m\epsilon}$
Elastic constants	$K_\alpha^* = K_\alpha\sigma_0/(k_B T)$

to the ordinary spherical LJ potential. It is convenient to employ scaled, dimensionless variables for all quantities, that we report for convenience in Table 5.10 and that, unless otherwise specified, we indicate with an asterisk. For an elongated molecule it is assumed that $\sigma_0 = \sigma_s$. The GB potential can also be used to model discotic mesogens employing oblate ellipsoids with thickness σ_f and diameter σ_s. In Fig. 5.15b we see an example with a parametrization [Emerson et al., 1994] based on the dimensions of a triphenylene core, namely, shape anisotropy $\kappa = \sigma_f/\sigma_s = 0.345$ and interaction anisotropy $\kappa' = \epsilon_s/\epsilon_f = 0.2$, but using instead energy parameters $\mu = 1$ and $\nu = 3$ as in Bacchiocchi and Zannoni [1998]. Note that the unit length σ_0 for discotic, although assumed to be $\sigma_0 = \sigma_s$ in early simulations [Emerson et al., 1994] is more reasonably assumed to be $\sigma_0 = \sigma_f$ [Bates and Luckhurst, 1996]. Various applications of uniaxial GB potentials for elongated or discotic particles to liquid crystals will be discussed in Section 11.5.2.

5.6.4 Biaxial Gay–Berne Potentials

While modelling rod-like or disc-like molecules with elongated or squashed ellipsoids is quite reasonable as a first option, a simple observation of many of the mesogens presented

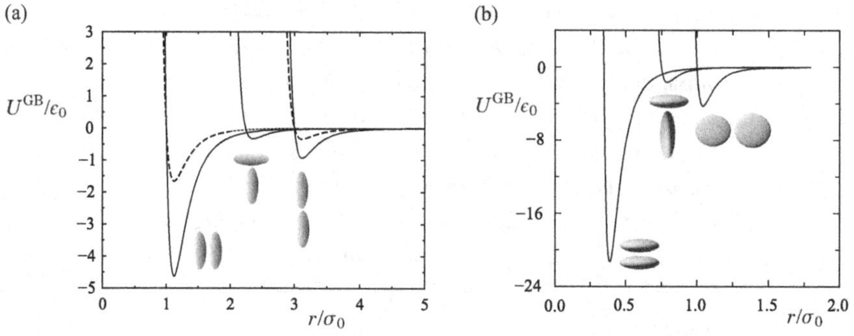

Figure 5.15 (a) A plot of the GB potential U^{GB}/ϵ_0 as a function of intermolecular separation: r/σ_0 for the GB(3,5,2,1) (- - -) and GB(3,5,1,3) (——) prolate models and side-by-side, T, and end-to-end configurations [Berardi et al., 1993]. (b) The same for two discotic mesogens GB(0.345,0.2,1,3), i.e. with thickness to diameter ratio $\sigma_{ff}/\sigma_{ss} = 0.345$, well depth ratio $\epsilon_{ss}/\epsilon_{ff} = 0.2$ and parameters $\mu = 1, \upsilon = 3$. Here $\sigma_0 = \sigma_s$ in both cases [Bacchiocchi and Zannoni, 1998; Zannoni, 2001b].

in Chapter 1 (e.g. in Table 1.2), shows that they are better represented as biaxial, rather than uniaxial, particles with different length, breadth and width. It is possible to generalize the GB potential to a biaxial version following the procedure we have just seen but starting with the overlap between two Gaussian ellipsoids particles with axes lengths $\sigma_x^{(P)}$, $\sigma_y^{(P)}$, $\sigma_z^{(P)}$ and attractive interaction strengths $\epsilon_x^{(P)}$, $\epsilon_y^{(P)}$ and $\epsilon_z^{(P)}$, for particles of type P = A or B [Berardi et al., 1998a]. Assuming the two particles to be not necessarily identical is essential to studying mixtures of two different nematogens [Berardi and Zannoni, 2015; Querciagrossa et al., 2017], but also to treat solutes, not necessarily mesogenic, in an LC host solvent, as is typically the case for spectroscopic studies. The coefficients $\epsilon_a^{(P)}$ are related to the well depths for the homogeneous side-by-side, face-to-face, and end-to-end interactions. Note that a particle P can have both a shape biaxiality

$$\lambda_\sigma^{(P)} = \sqrt{\frac{3}{2}} \frac{\sigma_x^{(P)} - \sigma_y^{(P)}}{2\sigma_z^{(P)} - \sigma_x^{(P)} - \sigma_y^{(P)}} \tag{5.61}$$

and an attractive interaction biaxiality

$$\lambda_\epsilon^{(P)} = \sqrt{\frac{3}{2}} \frac{\left(\epsilon_x^{(P)}\right)^{-1/\mu} - \left(\epsilon_y^{(P)}\right)^{-1/\mu}}{2\left(\epsilon_z^{(P)}\right)^{-1/\mu} - \left(\epsilon_x^{(P)}\right)^{-1/\mu} - \left(\epsilon_y^{(P)}\right)^{-1/\mu}}. \tag{5.62}$$

Both parameters can now be used to model particles and it is important to realize that $\lambda_\epsilon^{(P)}$ and $\lambda_\sigma^{(P)}$ can be quite distinct. For instance, we can imagine that molecules of essentially the same shape can be endowed with some lateral substituents that, thanks to some specific attractive interaction, favour side-by-side attraction enhancing the tendency to transversal, rather than a face-to-face arrangement of the molecules. We shall see some examples of that in Chapter 11. The resulting generalized anisotropic GB potential for two, possibly dissimilar particles, A, B, is [Berardi et al., 1998a],

$$U_{ij}^{\rm GBX} \equiv U_{\rm AB}^{\rm GBX}(r_{ij},\Omega_i,\Omega_j,\hat{\boldsymbol{r}}_{ij}) = 4\epsilon_0\,\epsilon_{\rm AB}^{(\mu\nu)}(\Omega_i,\Omega_j,\hat{\boldsymbol{r}}_{ij}) \times \left\{\left[\frac{\sigma_{\rm c}}{r_{ij}-\sigma_{\rm AB}(\Omega_i,\Omega_j,\hat{\boldsymbol{r}}_{ij})+\sigma_{\rm c}}\right]^{12} - \left[\frac{\sigma_{\rm c}}{r\ ij} - \sigma_{\rm AB}(\Omega_i,\Omega_j,\hat{\boldsymbol{r}}_{ij})+\sigma_{\rm c}\right]^{6}\right\}, \tag{5.63}$$

where the 'minimum contact distance' σ_c determines the width of the potential wells and lies in the range $0 < \sigma_{\rm c} \leqslant \left(\min[\sigma_a^{(\rm A)}] + \min[\sigma_a^{(\rm B)}]\right)/2$. In practice, we typically assume $\sigma_{\rm c} = \frac{1}{2}(\sigma_c^{(\rm A)} + \sigma_c^{(\rm B)})$, the average of the parameters $\sigma_c^{(\rm P)}$. The orientations Ω_i, Ω_j of the two particles with respect to a common laboratory frame can be expressed as Euler angles, $(\alpha_i,\beta_i,\gamma_i)$ and $(\alpha_j,\beta_j,\gamma_j)$, or more conveniently for numerical calculations, as the two unitary quaternions $\mathbf{u}_i$ and $\mathbf{u}_j$ defining the rotations linking laboratory and molecular frames (see Appendix H). The distance and energy units are σ_0 and ϵ_0. The pseudo-contact distance is

$$\sigma_{\rm AB}(\Omega_i,\Omega_j,\hat{\boldsymbol{r}}_{ij}) = (2\,\hat{\boldsymbol{r}}_{ij}^T\,[\mathbf{A}_{\rm AB}(\Omega_i,\Omega_j)]^{-1}\,\hat{\boldsymbol{r}}_{ij})^{-1/2} \tag{5.64}$$

and, similarly to the uniaxial case [Gay and Berne, 1981], the anisotropic interaction term is written as the product

$$\epsilon_{\rm AB}^{(\mu\nu)}(\Omega_i,\Omega_j,\hat{\boldsymbol{r}}_{ij}) = \epsilon_{\rm AB}^{\nu}(\Omega_i,\Omega_j)\,\epsilon_{\rm AB}^{\prime\mu}(\Omega_i,\Omega_j,\hat{\boldsymbol{r}}_{ij}), \tag{5.65}$$

where μ and ν are empirical exponents. The strength coefficient is

$$\epsilon_{\rm AB}(\Omega_i,\Omega_j) = \left[\frac{2\sigma_{\rm c}^{(\rm A)}\sigma_{\rm c}^{(\rm B)}}{\det\left[\mathbf{A}_{\rm AB}\left(\Omega_i,\Omega_j\right)\right]}\right]^{1/2}, \tag{5.66}$$

where the scaling constants $\sigma_c^{(\rm P)}$ are $\sigma_c^{(\rm P)} = \left(\sigma_x^{(\rm P)}\sigma_y^{(\rm P)} + \sigma_z^{(\rm P)}\sigma_z^{(\rm P)}\right)\left(\sigma_x^{(\rm P)}\sigma_y^{(\rm P)}\right)^{1/2}$, for particles $\rm P = A, B$. The overlap matrix $[\mathbf{A}_{\rm AB}]$ (Eq. 5.51) for a pair of particles A, B with orientation Ω_i, Ω_j is

$$[\mathbf{A}_{\rm AB}] = \mathbf{A}_{\rm AB}(\Omega_i,\Omega_j) = \mathbf{M}_i^T\,[\mathbf{S}^{(\rm A)}]^2\mathbf{M}_i + \mathbf{M}_j^T\,[\mathbf{S}^{(\rm B)}]^2\mathbf{M}_j. \tag{5.67}$$

The other dimensionless interaction parameter ϵ' is written as

$$\epsilon'_{\rm AB}(\Omega_i,\Omega_j,\hat{\boldsymbol{r}}_{ij}) = 2\,\hat{\boldsymbol{r}}_{ij}^T[\mathbf{B}_{\rm AB}]^{-1}\hat{\boldsymbol{r}}_{ij}, \tag{5.68}$$

where the matrix $[\mathbf{B}_{\rm AB}]$ is defined as $[\mathbf{B}_{\rm AB}] = \mathbf{M}_i^T\,[\mathbf{E}^{(\rm A)}]\mathbf{M}_i + \mathbf{M}_j^T\,[\mathbf{E}^{(\rm B)}]\mathbf{M}_j$ in terms of the diagonal interaction matrix $[\mathbf{E}^{(\rm P)}]$ with elements $[E^{(\rm P)}]_{ab} = \delta_{a,b}(\epsilon_0/\epsilon_a^{(\rm P)})^{1/\mu}$. The generalized biaxial potential $U_{ij}^{\rm GBX}$ reduces to the standard $U_{ij}^{\rm GB}$ one [Gay and Berne, 1981] when the molecules become identical and uniaxial, except for the fact that a tunable σ_c is used instead of $\sigma_\perp$. For a given particle-particle configuration $(\boldsymbol{r}_{ij},\Omega_i,\Omega_j)$, the potential has two minima at $r_{ij}^{\pm} = \sigma_{\rm AB}(\Omega_i,\Omega_j,\hat{\boldsymbol{r}}_{ij}) - \sigma_c(1 \pm 2^{1/6})$ and is 0 at $r_{ij} = \sigma_{\rm AB}(\Omega_i,\Omega_j,\hat{\boldsymbol{r}}_{ij})$ and $r_{ij} = \sigma_{\rm AB}(\Omega_i,\Omega_j,\hat{\boldsymbol{r}}_{ij}) - 2\sigma_c$, thus σ_c is related to the width of the potential well. The minimum at r_{ij}^{+} is not physically meaningful, but if $\sigma_{\rm AB}(\Omega_i,\Omega_j,\hat{\boldsymbol{r}}_{ij}) > 2\sigma_c$ part of its branch of the curve could be found at positive (and thus potentially occurring) values of r_{ij}. This unphysical region should be avoided in Monte Carlo simulations (see Chapter 8),

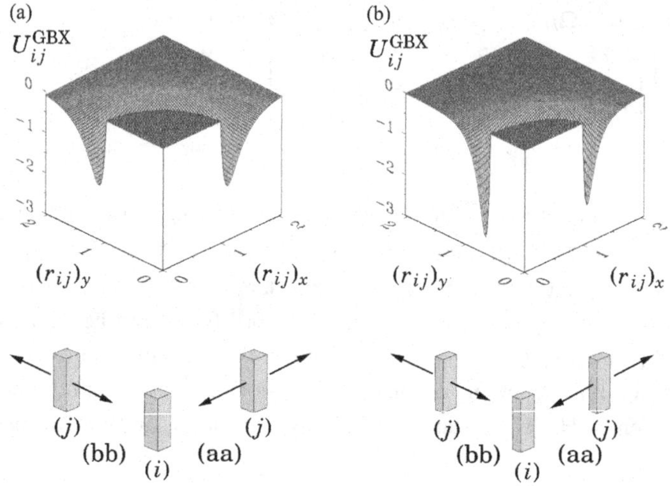

Figure 5.16 (a) Uniaxial and (b) biaxial interaction energy U_{ij}^{GBX} for two identical GB particles ($\mu = 2, \nu = 1$) with fixed orientations $\Omega_i = \Omega_j = (0, 0, 0)$ moving on the xy plane ($\hat{\boldsymbol{r}}_{ij} \perp \hat{\boldsymbol{z}}$). Approach along the x, y axes corresponds to the *aa*, *bb* configurations described in Table 5.11. In (a) we consider uniaxial molecules with $\sigma_x = \sigma_y = \sigma_c = 1, \sigma_z = 3$ ($\lambda_\sigma = 0$), and $\epsilon_x = \epsilon_y = 1, \epsilon_z = 0.2$. In (b) we show a biaxial case with $\sigma_y = \sigma_c = 0.6$ ($\lambda_\sigma = 0.111$), and $\epsilon_y = 1.25$. All distances are expressed in σ_0 units and energies in ϵ_0 units [Berardi et al., 1995].

where the molecules are allowed to move with not necessarily small coordinate jumps, since this would lead to unphysical and possibly unlockable configurations. It is also worth realizing that when $\sigma_c \gg \sigma_{\text{AB}}(\Omega_i, \Omega_j, \hat{\boldsymbol{r}}_{ij})$, as could be the case if σ_c is taken close to the in plane dimensions for two oblate molecules approaching face to face, $U_{ij}^{\text{GBX}}(0)$ can be finite and even small rather than tending to ∞ as desired. The introduction of the parameter σ_c, with $0 < \sigma_c \leq \min(\sigma_x, \sigma_y, \sigma_z)$, allows the extension of the potential to variously shaped objects. Having provided these general expressions for two possibly different molecules, it is convenient to focus on the important special case of two identical molecules. It is also quite useful to examine specific well-defined configurations and for the purpose of representation we consider the biaxial ellipsoids previously introduced as parallelepipeds with the faces orthogonal to $\boldsymbol{x}$, $\boldsymbol{y}$ and $\boldsymbol{z}$ labelled as a, b and c. In other words, face a is parallel to the yz plane, b to the xz plane and c to the xy one. Given two identical molecules with a fixed mutual orientation we then consider their interaction energy as a function of separation r. In particular, we choose the twelve configurations described in Table 5.11, where every axis of the second frame is parallel (antiparallel) or perpendicular to those of the first one. All the mutual *orthogonal* orientations can be generated starting with the two molecules aligned, i.e. $\boldsymbol{x}_i \| \boldsymbol{x}_j$, $\boldsymbol{y}_i \| \boldsymbol{y}_j$, $\boldsymbol{z}_i \| \boldsymbol{z}_j$, and performing a rotation of the second molecule of Euler angles $\Omega = (\alpha, \beta, \gamma)$. Every rotation generates three approaching configurations, each of them along one axis of the first molecule. Each of these is identified with a two-letter code formed by the names of the faces coming in contact so that axes perpendicular to both faces

Table 5.11. *Sketch of selected configurations for two identical biaxial GB particles, with* $\hat{\boldsymbol{r}}_{ij}$ *direction and* (α, β, γ) *angles of the second particle with respect to the first. We give* $\sigma(\Omega_i, \Omega_j, \hat{\boldsymbol{r}}_{ij})$, $\epsilon(\Omega_i, \Omega_j)$ *and* $\epsilon'(\Omega_i, \Omega_j, \hat{\boldsymbol{r}}_{ij})$ *values, with* $\tau \equiv \sqrt{2/[(\sigma_x^2+\sigma_z^2)(\sigma_x^2+\sigma_y^2)(\sigma_y^2+\sigma_z^2)]}$ *[Berardi et al., 1995; Ricci et al., 2019]*

Sketch	angles	$\sigma(\Omega_i, \Omega_j, \hat{\boldsymbol{r}}_{ij})$	$\epsilon(\Omega_i, \Omega_j)$	$\epsilon'(\Omega_i, \Omega_j, \hat{\boldsymbol{r}}_{ij})$
(aa) 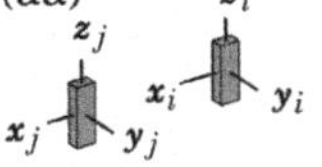	$\hat{\boldsymbol{r}}_{ij} \parallel \hat{\boldsymbol{x}}_i$ (0, 0, 0)	σ_x	$\dfrac{\sigma_x\sigma_y+\sigma_z^2}{2\sigma_z\sqrt{\sigma_x\sigma_y}}$	$\left(\dfrac{\epsilon_x}{\epsilon_0}\right)^{1/\mu}$
(bb) 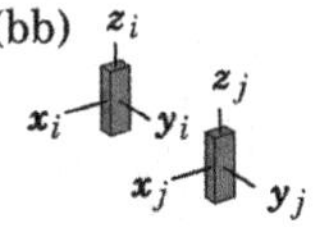	$\hat{\boldsymbol{r}}_{ij} \parallel \hat{\boldsymbol{y}}_i$ (0, 0, 0)	σ_y	$\dfrac{\sigma_x\sigma_y+\sigma_z^2}{2\sigma_z\sqrt{\sigma_x\sigma_y}}$	$\left(\dfrac{\epsilon_y}{\epsilon_0}\right)^{1/\mu}$
(cc) 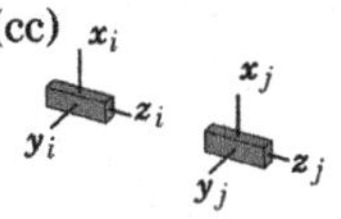	$\hat{\boldsymbol{r}}_{ij} \parallel \hat{\boldsymbol{z}}_i$ (0, 0, 0)	σ_z	$\dfrac{\sigma_x\sigma_y+\sigma_z^2}{2\sigma_z\sqrt{\sigma_x\sigma_y}}$	$\left(\dfrac{\epsilon_z}{\epsilon_0}\right)^{1/\mu}$
(ab') 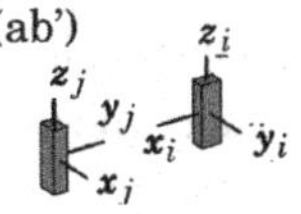	$\hat{\boldsymbol{r}}_{ij} \parallel \hat{\boldsymbol{x}}_i$ $(0, 0, \pi/2)$	$\sqrt{\dfrac{\sigma_x^2+\sigma_y^2}{2}}$	$\dfrac{(\sigma_x\sigma_y+\sigma_z^2)\sqrt{\sigma_x\sigma_y}}{(\sigma_x^2+\sigma_y^2)\sigma_z}$	$2/\left[\left(\dfrac{\epsilon_0}{\epsilon_x}\right)^{1/\mu}+\left(\dfrac{\epsilon_0}{\epsilon_y}\right)^{1/\mu}\right]$
(cc') 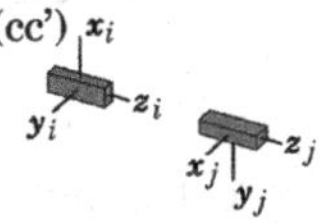	$\hat{\boldsymbol{r}}_{ij} \parallel \hat{\boldsymbol{z}}_i$ $(0, 0, \pi/2)$	σ_z	$\dfrac{(\sigma_x\sigma_y+\sigma_z^2)\sqrt{\sigma_x\sigma_y}}{(\sigma_x^2+\sigma_y^2)\sigma_z}$	$\left(\dfrac{\epsilon_z}{\epsilon_0}\right)^{1/\mu}$
(ac') 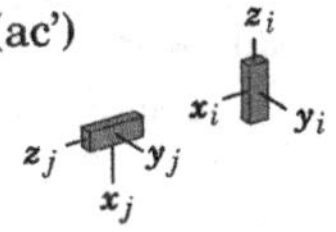	$\hat{\boldsymbol{r}}_{ij} \parallel \hat{\boldsymbol{x}}_i$ $(0, \pi/2, 0)$	$\sqrt{\dfrac{\sigma_x^2+\sigma_z^2}{2}}$	$\dfrac{(\sigma_x\sigma_y+\sigma_z^2)\sqrt{\sigma_x\sigma_y}}{(\sigma_x^2+\sigma_z^2)\sigma_y}$	$2/\left[\left(\dfrac{\epsilon_0}{\epsilon_x}\right)^{1/\mu}+\left(\dfrac{\epsilon_0}{\epsilon_z}\right)^{1/\mu}\right]$
(bb') 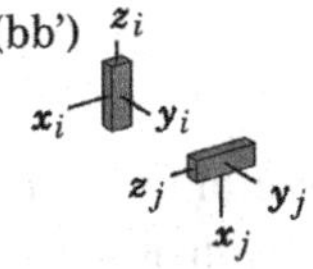	$\hat{\boldsymbol{r}}_{ij} \parallel \hat{\boldsymbol{y}}_i$ $(0, \pi/2, 0)$	σ_y	$\dfrac{(\sigma_x\sigma_y+\sigma_z^2)\sqrt{\sigma_x\sigma_y}}{(\sigma_x^2+\sigma_z^2)\sigma_y}$	$\left(\dfrac{\epsilon_y}{\epsilon_0}\right)^{1/\mu}$
(ac) 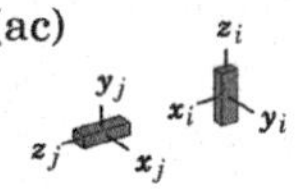	$\hat{\boldsymbol{r}}_{ij} \parallel \hat{\boldsymbol{x}}_i$ $(0, \pi/2, \pi/2)$	$\sqrt{\dfrac{\sigma_x^2+\sigma_z^2}{2}}$	$\sqrt{\sigma_x\sigma_y}(\sigma_x\sigma_y+\sigma_z^2)\tau$	$2/\left[\left(\dfrac{\epsilon_0}{\epsilon_x}\right)^{1/\mu}+\left(\dfrac{\epsilon_0}{\epsilon_z}\right)^{1/\mu}\right]$
(ba) 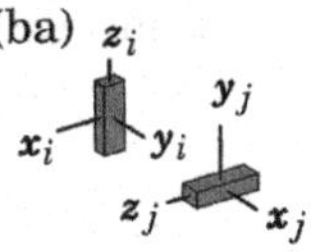	$\hat{\boldsymbol{r}}_{ij} \parallel \hat{\boldsymbol{y}}_i$ $(0, \pi/2, \pi/2)$	$\sqrt{\dfrac{\sigma_x^2+\sigma_y^2}{2}}$	$\sqrt{\sigma_x\sigma_y}(\sigma_x\sigma_y+\sigma_z^2)\tau$	$2/\left[\left(\frac{\epsilon_0}{\epsilon_x}\right)^{1/\mu}+\left(\frac{\epsilon_0}{\epsilon_y}\right)^{1/\mu}\right]$

Table 5.11. *Continued*

Sketch	angles	$\sigma(\Omega_i,\Omega_j,\hat{\boldsymbol{r}}_{ij})$	$\epsilon(\Omega_i,\Omega_j)$	$\epsilon'(\Omega_i,\Omega_j,\hat{\boldsymbol{r}}_{ij})$
(cb) 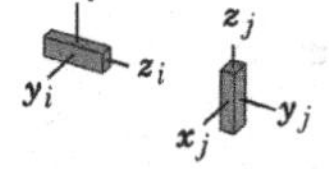	$\hat{\boldsymbol{r}}_{ij}\|\hat{\boldsymbol{z}}_i$ $(0,\pi/2,\pi/2)$	$\sqrt{\frac{\sigma_y^2+\sigma_z^2}{2}}$	$\sqrt{\sigma_x\sigma_y}(\sigma_x\sigma_y+\sigma_z^2)\tau$	$2/\left[\left(\frac{\epsilon_0}{\epsilon_y}\right)^{1/\mu}+\left(\frac{\epsilon_0}{\epsilon_z}\right)^{1/\mu}\right]$
(aa') 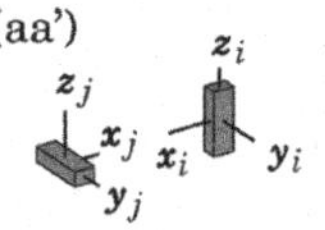	$\hat{\boldsymbol{r}}_{ij}\|\hat{\boldsymbol{x}}_i$ $(\pi/2,\pi/2,\pi/2)$	σ_x	$\frac{(\sigma_x\sigma_y+\sigma_z^2)\sqrt{\sigma_x\sigma_y}}{(\sigma_y^2+\sigma_z^2)\sigma_x}$	$\left(\frac{\epsilon_x}{\epsilon_0}\right)^{1/\mu}$
(bc') 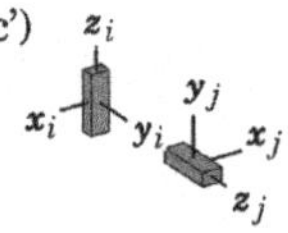	$\hat{\boldsymbol{r}}_{ij}\|\hat{\boldsymbol{y}}_i$ $(\pi/2,\pi/2,\pi/2)$	$\sqrt{\frac{\sigma_y^2+\sigma_z^2}{2}}$	$\frac{(\sigma_x\sigma_y+\sigma_z^2)\sqrt{\sigma_x\sigma_y}}{(\sigma_y^2+\sigma_z^2)\sigma_x}$	$2/\left[\left(\frac{\epsilon_0}{\epsilon_y}\right)^{1/\mu}+\left(\frac{\epsilon_0}{\epsilon_z}\right)^{1/\mu}\right]$

define the intermolecular vector. In Fig. 5.16 we see a representation of the potential surface for two molecules approaching side by side for uniaxial and biaxial particles ($\lambda_\sigma = 0$ and $\lambda_\sigma = 0.111$). Switching on the biaxiality allows different approach distances and interaction strengths for the two shorter axes of the molecules. In Table 5.11 we list the independent configurations (12 out of 15) obtained rotating β and γ equal to 0 or $\pi/2$ and with their codes and we give the respective analytic expressions for σ, ϵ and ϵ'. For a uniaxial object there are only four unique configurations of this type: aa $\equiv$ ab$'$ $\equiv$ bb (*side-by-side*), cc $\equiv$ cc$'$ (*end-to-end*), ac $\equiv$ ac$'$ $\equiv$ cb$'$ $\equiv$ bc$'$ $\equiv$ cb $\equiv$ (*tee*), aa$'$ $\equiv$ ba $\equiv$ bb$'$ (*cross*). Specific numerical examples of dimensions and interaction coefficients for the representation of oligophenyls, from biphenyl to *p*-quinquephenyl, approximated with rigid biaxial GB ellipsoids are given in [Berardi et al., 1998a]. A few other versions of biaxial pair potentials inspired by Gaussian overlap have been put forward by Cleaver et al. [1996], Perram et al. [1996], and Everaers and Ejtehadi [2003].

5.6.5 Soft Core Gay–Berne Potential

Although the GB potential in its uniaxial and biaxial variants is extremely useful in simulating LC phases, as we shall see in some detail in Chapter 11, in a number of situations it can be convenient to consider a softer version of the potential where molecules can to some extent overlap. Berardi et al. [2009, 2011] have introduced such a soft-core variant of the standard GB potential where the attractive part, corresponding to $U_{ij}^{\text{GBS}} < 0$, is still given by the GB potential U_{ij}^{GB}, while the repulsive ($U_{ij}^{\text{GBS}} \geq 0$) branch is replaced by a function growing more slowly than the standard one $\propto r_{ij}^{-12}$. It is also convenient to assume that the soft-core term does not diverge to infinity for vanishing values of distance but takes a finite (possibly large compared to $k_B T$) value. We thus have

$$U_{ij}^{\mathrm{GBS}} = \left[1 - f(r_{ij}, \Omega_i, \Omega_j, \hat{\boldsymbol{r}}_{ij})\right] U^{\mathrm{GB}}(r_{ij}, \Omega_i, \Omega_j, \hat{\boldsymbol{r}}_{ij}) + f(r_{ij}, \Omega_i, \Omega_j, \hat{\boldsymbol{r}}_{ij}) U^{\mathrm{SC}}(r_{ij}, \Omega_i, \Omega_j, \hat{\boldsymbol{r}}_{ij}), \tag{5.69}$$

where U^{GB} (or U^{GBX} for biaxial particles) is the standard Gay–Berne potential, given explicitly in Eq. 5.57 (or Eq. 5.63 for biaxial particles), and we assume identical molecules. The soft-core repulsive potential is assumed to have a linear dependence on the particle-particle separation:

$$U^{\mathrm{SC}}(r_{ij}, \Omega_i, \Omega_j, \hat{\boldsymbol{r}}_{ij}) = m[r_{ij} - \sigma(\Omega_i, \Omega_j, \hat{\boldsymbol{r}}_{ij})], \tag{5.70}$$

where $\sigma(\Omega_i, \Omega_j, \hat{\boldsymbol{r}}_{ij})$ is the anisotropic contact term of the GB potential. The sigmoidal seaming function is

$$f(r_{ij}, \Omega_i, \Omega_j, \hat{\boldsymbol{r}}_{ij}) = \frac{\exp\left[k\left(r_{ij} - \sigma(\Omega_i, \Omega_j, \hat{\boldsymbol{r}}_{ij})\right)\right]}{1 + \exp\left[k\left(r_{ij} - \sigma(\Omega_i, \Omega_j, \hat{\boldsymbol{r}}_{ij})\right)\right]}, \tag{5.71}$$

with the parameter k defining the steepness of the logistic function at its inflection point. The GB and SC surfaces are smoothly joined at $r = \sigma(\Omega_i, \Omega_j, \hat{\boldsymbol{r}}_{ij})$, where the GB pair potential goes to 0. As we can see from Fig. 5.17 the softness can be varied by changing the slope parameter m. This could be useful to model, e.g. polymer particles or colloidal particles covered with a stabilizing layer of flexible chains in an attempt to allow for the possibility of two such particles to come closer than a formally impenetrable geometric barrier would allow. We shall see in Chapter 11 that another field of application is in the modelling of LC

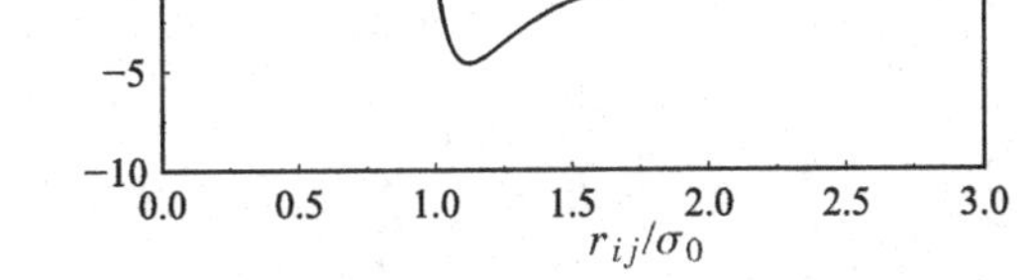

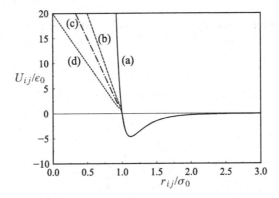

Figure 5.17 Side-by-side interaction energy U_{ij}/ϵ_0 as a function of separation r_{ij}/σ_0 of standard (a) and soft-core (b, c, d) uniaxial GB ellipsoids with $\sigma_\perp = \sigma_c = \sigma_0$, $\sigma_\| = 3\sigma_0$ and $\epsilon_\perp = \epsilon_0$, $\epsilon_\| = 0.2\epsilon_0$, $\mu = 1$ $\nu = 3$, and slopes $m/\left(\epsilon_0\sigma_0^{-1}\right) = ,\ -40$ (b), -30 (c), and -20 (d) [Berardi et al., 2009].

elastomers [Skačej and Zannoni, 2011, 2012, 2014] where avoiding the harsh repulsion of the standard GB potential is essential for the simulation to avoid being stuck.

5.7 Electrostatic Multipoles

We have discussed up to now various empirical potentials that try to take into account repulsions and attractions. An important missing contribution in these generic potentials is, however, the electrostatic one. As we have seen in Section 5.4, in atomistic or chemically detailed models the electrostatic contributions are typically calculated by summation of the charge-charge Coulomb interactions, in particular using the Ewald method (Section 5.4.4). In molecular scale, empirical models, where the information on the individual atoms and their position is not available, it is often interesting to study the effect of some simple charge distributions and introduce molecular properties linked to these charge distributions to be able to see their effect on physical properties and phase transitions with computer simulation techniques (see Chapter 11).

In this section we wish to introduce the multipolar approximation to a charge distribution, and in particular define important quantities like dipoles and quadrupoles, as these are very useful to 'decorate' a simple generic empirical potential like the spherocylindrical and ellipsoidal systems introduced so far.

We start by considering a certain set of charges e_k whose position $\boldsymbol{h}_k$ in the particle fixed frame of particle i is assumed to be known. The electrostatic potential as sensed by a unit test charge at a point $\boldsymbol{r}$ away from the origin (cf. Fig. 5.18), $V^{\mathrm{e}}(\boldsymbol{r})$ can be considered as a sum of contributions from the various charges given by Coulomb's law:

$$V^{\mathrm{e}}(\boldsymbol{r}) = \frac{1}{4\pi\varepsilon_0}\sum_k e_k \frac{1}{|\boldsymbol{r}-\boldsymbol{h}_k|}. \tag{5.72}$$

If we examine the charge distribution from a distance much larger than the distance of the point charges from the origin and between themselves (as sketched in Fig. 5.18), we can expand the potential in a Taylor series of the separation $|\boldsymbol{r}-\boldsymbol{h}_k|$ in powers of $\boldsymbol{h}_k$ starting from the point $\boldsymbol{r}$. Using the expression for the Taylor expansion of a scalar function of a vector $\boldsymbol{r}-\boldsymbol{h}_k$ around $\boldsymbol{r}$ (cf. Appendix C), we have, at second order,

$$4\pi\varepsilon_0\; V^{\mathrm{e}}(\boldsymbol{r}) = \sum_k e_k \frac{1}{|\boldsymbol{r}-\boldsymbol{h}_k|}, \tag{5.73a}$$

$$= \sum_k e_k \frac{1}{r} - \sum_k e_k \nabla\frac{1}{r}\cdot\boldsymbol{h}_k + \frac{1}{2!}\sum_k e_k \nabla\nabla\frac{1}{r}:\boldsymbol{h}_k\boldsymbol{h}_k + \cdots, \tag{5.73b}$$

$$= \frac{q^{\mathrm{e}}}{r} - \nabla\frac{1}{r}\cdot\Big(\sum_k e_k\boldsymbol{h}_k\Big) + \frac{1}{2}\nabla\nabla\frac{1}{r}:\Big(\sum_k e_k\boldsymbol{h}_k\boldsymbol{h}_k\Big) + \cdots, \tag{5.73c}$$

$$= \frac{q^{\mathrm{e}}}{r} - \nabla\frac{1}{r}\cdot\boldsymbol{\mu} + \frac{1}{3}\nabla\nabla\frac{1}{r}:\mathbf{Q} + \cdots, \tag{5.73d}$$

$$= \frac{q^{\mathrm{e}}}{r} - \mathbf{T}^{(1)}(\boldsymbol{r})\cdot\boldsymbol{\mu} + \frac{1}{3}\mathbf{T}^{(2)}(\boldsymbol{r}):\mathbf{Q} + \cdots, \tag{5.73e}$$

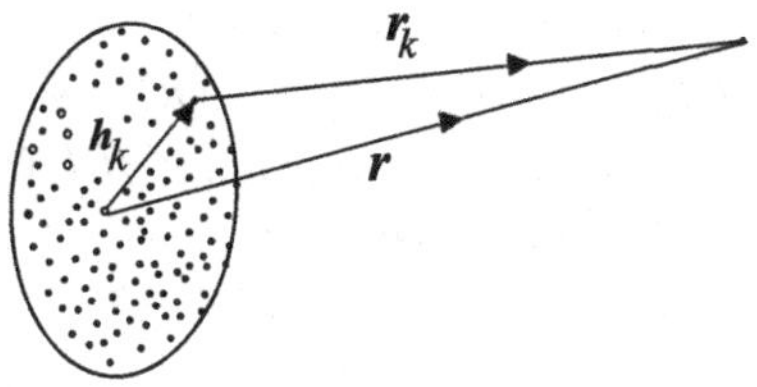

Figure 5.18 The distribution of set of charges k positioned at $\boldsymbol{h}_k$ in their molecular frame as viewed from a distant point $\boldsymbol{r}$.

where we have introduced the first two multipolar tensors $\mathbf{T}^{(n)}(\boldsymbol{r})$ (given explicitly in Eqs. C.9b–C.9d), i.e. the dipolar and quadrupolar ones. In particular, $\mathbf{T}^{(1)}(\boldsymbol{r})$ is a vector, $\mathbf{T}^{(2)}(\boldsymbol{r})$ is a second-rank tensor (Eq. C.10) and so on. In Eq. 5.73e we also use the contraction symbols for two tensors, implying a sum over all products of the two corresponding tensorial components (see Appendix B.5). The first few terms in the expansion correspond to:

(i) **Total charge** of the molecule, q^{e}, which is simply the algebraic sum of the various partial charges e_i distributed on the molecule itself, irrespective of their location:

$$q^{\mathrm{e}} = \sum_k e_k. \tag{5.74}$$

We deal mainly with thermotropics and uncharged molecules, although in lyotropics or in ionic liquid crystals this is not the case, and there the total charge will be extremely important. The second term depends on:

(ii) **Electric dipole** vector $\boldsymbol{\mu}$, which is determined by the separation between the centres of the positive and negative charges on the particle:

$$\boldsymbol{\mu} = \sum_k e_k \, \boldsymbol{h}_k. \tag{5.75}$$

The overall dipole is 0 for a symmetric molecule, like benzene or *p*-quinquephenyl. To a certain extent the total dipole of a molecule can be estimated by the vector addition of local dipoles assigned to chemical bonds between two atoms or between two moieties, and tables of group dipoles are available [Ouellette and Rawn, 2015]. A list of dipoles for some polar molecules and mesogens is given in Table 5.12. Tabulations of experimental dipoles are available [Le Fèvre, 1964], but very few indeed for LC forming molecules. Molecular dipoles can be obtained using a variety of QC codes using *ab initio* wave function or electronic density functional theory (DFT) methods. However, it is worth noting that the calculation of molecular dipoles using the widely used DFT QC codes that are optimized from the point of view of energy minimization, rather than electron distribution, may be troublesome and lead even to 'absurd' results as discussed in Hait and Head-Gordon [2018] and easily verified even for simple molecules like toluene.[1] An additional source of complication is that in the case of

[1] Calculated using ADF [te Velde et al., 2001] with DFT with a CGA-BLYP_D3BJ functional form and a triple zeta basis set with polarization (see ADF manual for details [Baerends et al., 2021].

Table 5.12. *Dipole moments (in Debye) of a few molecules and source. Refs. (a) experimental [Le Fèvre, 1964]; (b) QC, plane waves [Adam et al., 1997]; (c) experimental [Minkin et al., 1970]; (d) QC, DFT [Zannoni, 2020]; (e) [Pestov and Vill, 2005]*

Molecule	μ, $\boldsymbol{\mu}$ (D)	Ref.
Toluene	0.37	(a)
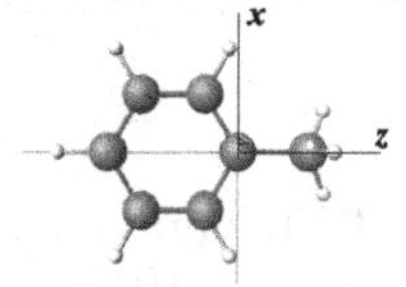	0.39, (0.00, 0.00, 0.39)	(d)
Fluorobenzene	1.57	(a)
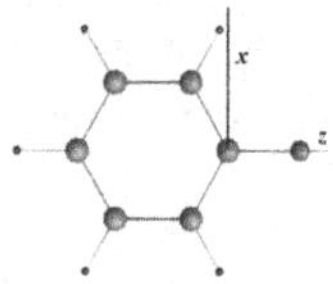	1.39, (0.00, 0.00, −1.39)	(d)
Nitrobenzene	4.24	(a)
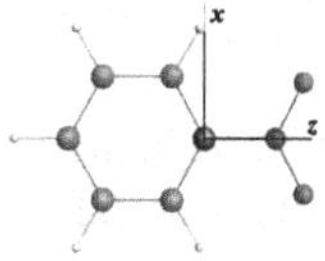	4.20, (0.00, 0.00, −4.20)	(c)
	5.1, (0.00, 0.00, −5.10)	(d)
PAA	2.3	(e)
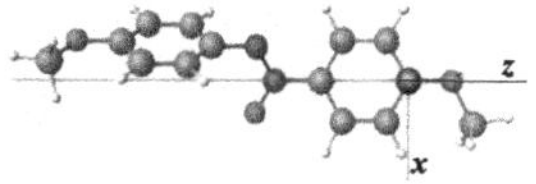	2.03, (−0.72, 1.54, −1.10)	(d)
MBBA	3.2	(e)
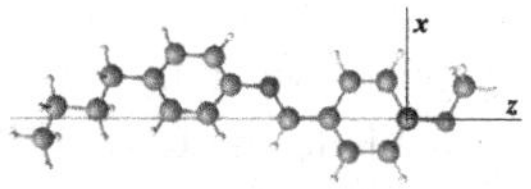	2.05, (−0.21, 0.00, 2.04)	(d)
5CB	6.50	(b)
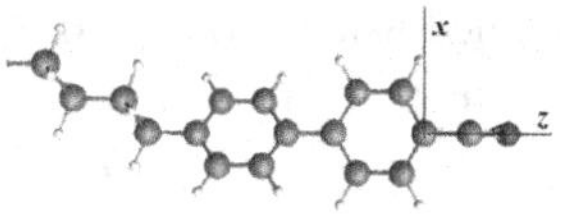	5.08 (in C_6H_{14} at 25°C)	(e)
	6.330, (−0.117. −0.194, −6.329)	(d)

non-rigid molecules, like practically all the LC mesogens (see Chapter 1), the molecular dipole, like other higher multipoles, is dependent on conformation, making the comparison between results of different provenance difficult. Here we have calculated the values in Table 5.12 using for this purpose the conformer obtained from the PubChem data base [Kim et al., 2016] reported to the molecular frame shown.

Most mesogens are actually endowed with electric dipoles, and one of the first theories about the molecular origin of liquid crystallinity by Max Born (no less!), assumed that dipoles were needed for the formation of LC phases [Sluckin et al., 2004]. This was discounted by the fact that symmetric, apolar, molecules like *p*-quinquephenyl can form LC phases. Indeed, simple theoretical models for mesogens, like the hard ellipsoids and spherocylinders or the Gay–Berne model yield LC phases even if apolar. However, the presence of one or more dipoles is important for establishing the properties of liquid crystals, as we have seen in Chapter 1, particularly for smectics (see Section 1.7.2). We shall examine this more in detail in Chapter 11 and see how important not only the existence of a dipole, but also its position and orientation in the molecule is in order to modify, sometimes radically, the type of LC phase obtained. Going back for now to the multipole expansion, the third term is:

(iii) **Electric quadrupole.** It contains a quadratic dependence on the distances of the charges from the centre and can be different from 0 also for molecules that have no net dipole moment or charge. For example, it can be different from 0 if we have four equal charges with alternating signs placed at the vertices of a square. Because of this, the contribution is called *quadrupolar*. Another, more common case at least for molecular systems (e.g. CO_2, *p*-difluorobenzene, ...), is that of an axial quadrupole with single charges of a given sign at the two ends of a rod-like particle and a double charge of opposite sign at the centre. It is worth noting that instead of using the charge distribution second moment tensor $\mathbf{Q}$, its traceless version, the electric quadrupole tensor $\mathbf{\Theta}$ [Buckingham, 1967b] is often used. Its Cartesian components are:

$$\Theta_{ab} = \frac{3}{2} Q_{ab} - \frac{1}{2} Q_{aa}\delta_{a,b} = \frac{1}{2}\sum_k e_k(3h_{k,a}h_{k,b} - h_{k,a}^2\delta_{a,b}). \tag{5.76}$$

The advantage of defining traceless electrostatic multipoles is that these vanish for a spherically symmetric distribution of charges. For a uniaxial molecule with symmetry axis along $\boldsymbol{u}$, $\mathbf{\Theta}$ reduces to $\mathbf{\Theta} = \frac{1}{2}\Theta\,(3\boldsymbol{u}\otimes\boldsymbol{u} - \mathbf{1})$ and depends on just one physical parameter $\Theta \equiv \Theta_{\|} \equiv \Theta_{zz} = \sqrt{\frac{2}{3}}\Theta^{2,0}$ in the spherical notation of Appendix B, since $\Theta_{zz} = -2\Theta_{xx} = -2\Theta_{yy}$. We give in Tables 5.13 and 5.14 a list of molecular quadrupole moments for some small molecules and mesogens. For tetrahedral symmetry the quadrupole vanishes (and so does the dipole of course) and it may be necessary to go to the next multipole in the expansion, the octupole.

In a multipolar expansion the focus is normally on the first non-zero component and, for molecules forming LCs, anisometric by nature, a quadrupole is typically the highest multipole needed so we do not discuss octupoles and higher multipoles [Kielich, 1972].

It is important to note that only the first, non-vanishing multipole is independent of the choice of origin [Buckingham, 1967b]. Thus, for an ion, the dipole and quadrupole moments vary with the origin. Similarly, in an uncharged but polar molecule, like fluorobenzene or 5CB, $\boldsymbol{\mu}$ is independent of the origin, i.e. the dipole $\boldsymbol{\mu}'$ relative to a new origin O' at the point $\boldsymbol{r}' = (r'_a, r'_b, r'_c)$ from the old origin O is equal to $\boldsymbol{\mu}$, while the quadrupole $\mathbf{\Theta}$ depends on the shift in origin and

$$\mathbf{\Theta}'_{ab} = \mathbf{\Theta}_{ab} - \left(\frac{3}{2}r'_a\mu_b + \frac{3}{2}r'_b\mu_a - \boldsymbol{r}'\cdot\boldsymbol{\mu}\,\delta_{a,b}\right). \tag{5.77}$$

Table 5.13. *Quadrupole tensor principal values for a few molecules. The z-axis is chosen along the highest symmetry axis. For naphthalene and anthracene, **z** is parallel to the long axis and **y** out of the molecule plane. Units are* Buckingham *(*1B=1DÅ*) (a) [Gray and Gubbins, 1984]; (b) [Zannoni, 2020]; (c) [Graham et al., 1998]; (d) [Johnson, 2019]; (e) Clark et al. [1997]*

Molecule	Θ_{xx}	Θ_{yy}	Θ_{zz}	Ref.
Nitrogen	0.7	0.7	−1.4	(a)
Benzene	4.35	4.35	−8.7	(a)
	5.05	5.05	−10.1	(b)
Carbon dioxide	2.139	2.139	4.728	(c)
Hexafluorobenzene	−4.75	−4.75	9.5	(a)
Naphthalene	8.71	−13.5	6.89	(a)
	6.52	−12.82	6.30	(d)
Anthracene	10.4	−18.30	7.94	(a)
	8.12	−16.15	8.03	(b)
Fluorobenzene	7.34	−5.82	−1.52	(a)
	8.13	−7.12	−1.02	(b)

Table 5.14. *QC calculated quadrupole tensor components (in* Buckingham*) for a few mesogens already shown in Table 5.12 (same coordinate system). For p-quinquephenyl, **z** is parallel to the long axis and **x** is across the terminal ring plane. Ref. (a) [Zannoni, 2020]*

Molecule	Θ_{xx}	Θ_{xy}	Θ_{xz}	Θ_{yy}	Θ_{yz}	Θ_{zz}	Ref.
PAA	−15.53	-9.95	−24.04	−9.47	33.71	25.00	(a)
MBBA	−6.18	-0.07	15.40	−4.43	−0.07	10.61	(a)
5CB	22.47	-4.66	1.32	4.83	1.91	−27.30	(a)
Quinquephenyl (P5)	−33.90			13.81		20.08	(a)

This makes it very difficult to compare quadrupoles from different calculations, particularly if the molecules are flexible. In Tables 5.13 and 5.14 we report quadrupoles both experimentally and from QC calculations and we see that they compare well for apolar molecules. Θ for 5CB has also been obtained by using a first-principles electronic structure plane wave QC code typically used for solid state applications [Clark et al., 1997; Clark, 2001], but the coordinate system is however different from the one used here which in the absence of full coordinate information, as mentioned, makes the comparison not meaningful.

The electrostatic field generated by a set of charges e_k at positions $\boldsymbol{h}_k$ with respect to the particle centre, as sensed by a test charge at a position $\boldsymbol{r}$ can be written as the gradient of the electrostatic potential:

$$4\pi\varepsilon_0\, \boldsymbol{E}^{\mathrm{e}}(\boldsymbol{r}) = \sum_k e_k \frac{\boldsymbol{r}-\boldsymbol{h}_k}{|\boldsymbol{r}-\boldsymbol{h}_k|^3} = -\nabla V^{\mathrm{e}}(\boldsymbol{r}), \tag{5.78}$$

where $\varepsilon_0 = 8.854 \times 10^{-12}$ C^2/Nm^2. The vector $\boldsymbol{E}^{\mathrm{e}}$ at a 'distant' position from the molecule can be viewed as the superposition of the fields generated by the charge q^e, the dipole $\boldsymbol{\mu}$, the quadrupole $\boldsymbol{\Theta}$, etc. of the molecule itself

$$4\pi\varepsilon_0 \boldsymbol{E}^{\mathrm{e}}(\boldsymbol{r}) = -q^{\mathrm{e}}[\mathbf{T}^{(1)}(\boldsymbol{r}) + \mathbf{T}^{(2)}(\boldsymbol{r})\cdot\boldsymbol{\mu} - \frac{1}{3}\mathbf{T}^{(3)}(\boldsymbol{r})\colon\boldsymbol{\Theta} + \cdots]. \tag{5.79}$$

The field generated by a dipole is $\boldsymbol{E}^{\mathrm{D}}(\boldsymbol{r}) = \frac{1}{4\pi\varepsilon_0 r^5}[3(\boldsymbol{\mu}\cdot\boldsymbol{r})\boldsymbol{r} - r^2\boldsymbol{\mu}]$. We recall that the direction of the field at a certain position is represented by the tangent to the field line at that point. The intensity of the field is given instead by the number of lines passing through a unit normal area (i.e. the flux). The field around a quadrupole is [Kielich, 1972]

$$\boldsymbol{E}^{\mathrm{Q}}(\boldsymbol{r}) = \frac{1}{4\pi\varepsilon_0 r^7}[5(\boldsymbol{r}\cdot\boldsymbol{\Theta}\cdot\boldsymbol{r})\boldsymbol{r} - 2r^2\boldsymbol{\Theta}\cdot\boldsymbol{r}]. \tag{5.80}$$

For uniaxial molecules this is $\boldsymbol{E}^{\mathrm{Q}}(\boldsymbol{r}) = \dfrac{3\Theta}{8\pi\varepsilon_0 r^5}[(5\cos^2\beta - 1)\boldsymbol{r} - 2r\boldsymbol{u}\cos\beta]$, where $\boldsymbol{u}$ is again the molecule axis.

5.7.1 Pairwise Electrostatic Interactions

The electrostatic interaction energy between two particles i and j with a certain point charge distribution (in vacuum) is just

$$4\pi\varepsilon_0 U^{\mathrm{e}}_{ij} = \sum_{k\in i}\sum_{m\in j} \frac{e_{k,i}e_{m,j}}{r_{km}}. \tag{5.81}$$

Assuming that the two molecules are far apart, the charge-charge separation distance r_{km} (Fig. 5.19), i.e. $r_{km} = |\boldsymbol{r}_k - \boldsymbol{r}_m| = |\boldsymbol{r}_{ij} + \boldsymbol{h}_{k,i} - \boldsymbol{h}_{m,j}|$, depends mainly on the separation between the two molecule centres $r_{ij} \equiv |\boldsymbol{r}_j - \boldsymbol{r}_i|$, much larger than the distance of the charges from the respective centres, given by vectors $\boldsymbol{h}_{k,i}$, $\boldsymbol{h}_{m,j}$. Expanding in a double Taylor series of these small displacement vectors we can rewrite the interaction energy in terms of the multipoles q^{e}, $\boldsymbol{\mu}$, $\boldsymbol{\Theta}$, of the two molecules and after some algebra we find, neglecting octupoles and higher terms [Buckingham, 1967b; Kielich, 1972]

$$\begin{aligned}4\pi\,\varepsilon_0 U^{\mathrm{e}}_{ij} = {} & \frac{q_i^{\mathrm{e}} q_j^{\mathrm{e}}}{r_{ij}} + \left(q_i^{\mathrm{e}}\boldsymbol{\mu}_j - \boldsymbol{\mu}_i q_j^{\mathrm{e}}\right)\cdot\mathbf{T}^{(1)}(\boldsymbol{r}_{ij}) \\ & + \left(\frac{1}{3}q_i^{\mathrm{e}}[\boldsymbol{\Theta}_j] + \frac{1}{3}[\boldsymbol{\Theta}_i]q_j^{\mathrm{e}} - \boldsymbol{\mu}_i\boldsymbol{\mu}_j\right)\colon\mathbf{T}^{(2)}(\boldsymbol{r}_{ij}) \\ & + \left(-\frac{1}{3}\boldsymbol{\mu}_i[\boldsymbol{\Theta}_j] + \frac{1}{3}[\boldsymbol{\Theta}_i]\boldsymbol{\mu}_j\right)\vdots\mathbf{T}^{(3)}(\boldsymbol{r}_{ij}) + \cdots .\end{aligned} \tag{5.82}$$

Note that the two molecules are not necessarily identical, as indeed could be the case of a solute with a certain set of electric multipoles in a isotropic or liquid crystal solvent. This

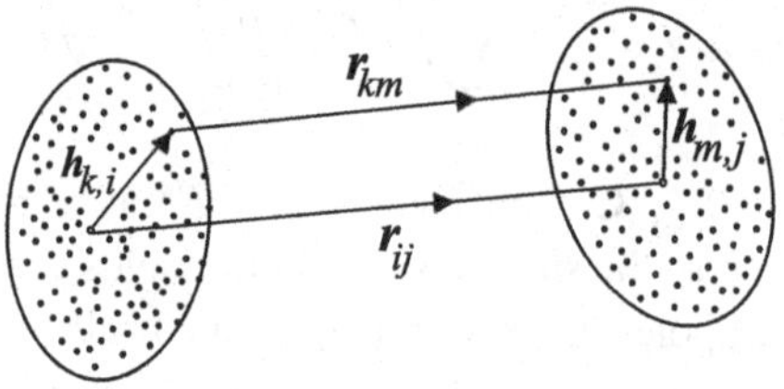

Figure 5.19 Two interacting charge distributions with centre-centre separation $\boldsymbol{r}_{ij}$ larger than those between the individual charges at positions $\boldsymbol{h}_{k,i}$, $\boldsymbol{h}_{m,j}$ and the origin of their particle coordinate frames.

case has been much studied both with computer simulations and experimentally, particularly using NMR [Burnell and de Lange, 2003; Pizzirusso et al., 2012b]. We now examine briefly some of the main contributions to the electrostatic pair potential:

(i) **Ion-ion.** Coulomb interaction between the two total charges $q_i^{\mathrm{e}}, q_j^{\mathrm{e}}$

$$U_{ij}^{\mathrm{II}} = \frac{q_i^{\mathrm{e}} q_j^{\mathrm{e}}}{4\pi\varepsilon_0 r_{ij}}. \tag{5.83}$$

(ii) **Dipole-ion.**

$$U_{ij}^{\mathrm{DI}} = -q_j^{\mathrm{e}} \left(\boldsymbol{\mu}_i \cdot \boldsymbol{r}_{ij}\right) / r_{ij}^3 = -q_j^{\mathrm{e}} \mu_i \cos\beta_i / r_{ij}^2. \tag{5.84}$$

Here the force is along the centre-centre direction, with the dipole orienting with its end of a certain sign approaching the opposite sign of the ion. A classical case showing the importance of this interaction is the easy dissolution of an ionic salt like NaCl in water.

(iii) **Dipole-dipole.** The explicit frame independent form of the dipole-dipole interaction picked from Eq. 5.82 is

$$U_{ij}^{\mathrm{DD}} = -\frac{1}{4\pi\varepsilon_0}\boldsymbol{\mu}_i \cdot \mathbf{T}^{(2)}(\boldsymbol{r}_{ij}) \cdot \boldsymbol{\mu}_j = -\frac{1}{4\pi\varepsilon_0 r_{ij}^5}[3(\boldsymbol{\mu}_i \cdot \boldsymbol{r}_{ij})(\boldsymbol{\mu}_j \cdot \boldsymbol{r}_{ij}) - r_{ij}^2(\boldsymbol{\mu}_i \cdot \boldsymbol{\mu}_j)]. \tag{5.85}$$

It is worth stressing that the interaction energy, although independent on a overall rotation of the coordinate frame (it only depends on scalar products), does depend on the orientation of the intermolecular vector. For instance, the energy of two dipoles parallel to one another but side by side (intermolecular vector parallel to $\boldsymbol{x}$, say) or head to tail (intermolecular vector parallel to z) is clearly different. In this last case, i.e. in the intermolecular frame, the explicit trigonometric dependence is:

$$U_{ij}^{\mathrm{DD}} = -\frac{\mu_i \mu_j}{4\pi\varepsilon_0\, r_{ij}^3}[2\cos\beta_i \cos\beta_j - \sin\beta_i \sin\beta_j \cos(\alpha_j - \alpha_i)] \tag{5.86}$$

and is plotted in Fig. 5.20a. Note that dipole-dipole interactions disfavour parallel (↑↑) and favour antiparallel (↑↓) side-by-side alignment, while they stabilize parallel (→→) and destabilize antiparallel (→←) end-to-end configurations. In units of

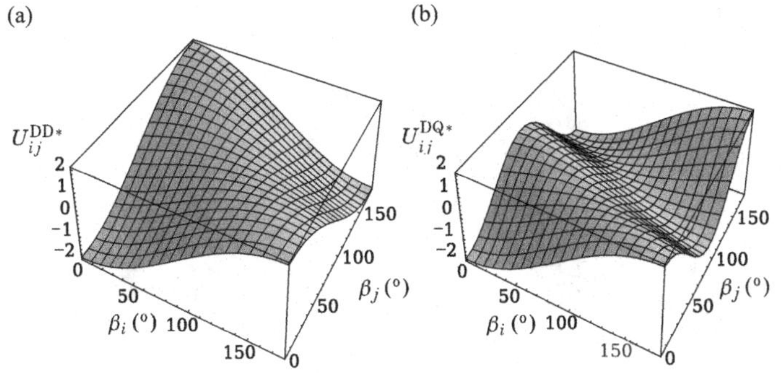

Figure 5.20 (a) Dipole-dipole U_{ij}^{DD*} and (b) dipole-quadrupole U_{ij}^{DQ*} interaction energy (dimensionless units) as a function of the angles β_i, β_j relative to the intermolecular frame z-axis.

$(\mu_i\mu_j)/(4\pi\varepsilon_0 r_{ij}^3)$ $U_{\rightarrow\rightarrow}^{DD} = -2, U_{\uparrow\downarrow}^{DD} = -1$, $U_{\uparrow\uparrow}^{DD} = +1$, $U_{\rightarrow\leftarrow}^{DD} = +2$. Notice also that if the two dipoles are both at the so-called *magic angle* of $\arccos(1/\sqrt{3}) \approx 54.74°$ from $\boldsymbol{r}_{ij}$, then $U_{\nearrow\nearrow}^{DD} = 0$.

(iv) **Quadrupole-ion.** In an arbitrary frame

$$U_{ij}^{QI} = \frac{1}{4\pi\varepsilon_0}\frac{q_j^e}{r_{ij}^5}(\boldsymbol{r}_{ij}\cdot[\boldsymbol{\Theta}_i]\cdot\boldsymbol{r}_{ij}) \tag{5.87}$$

and for axially symmetric molecules $U_{ij}^{QI} = \dfrac{1}{4\pi\varepsilon_0}\dfrac{\Theta_i q_j^e}{r_{ij}^3}P_2(\cos\beta_i)$.

(v) **Dipole-quadrupole.**

$$U_{ij}^{DQ} = -\frac{1}{4\pi\varepsilon_0 r_{ij}^7}[5(\boldsymbol{\mu}_i\cdot\boldsymbol{r}_{ij})(\boldsymbol{r}_{ij}\cdot[\boldsymbol{\Theta}_j]\cdot\boldsymbol{r}_{ij}) - 2r_{ij}^2(\boldsymbol{\mu}_i\cdot[\boldsymbol{\Theta}_j]\cdot\boldsymbol{r}_{ij})]. \tag{5.88}$$

For axially symmetric molecules

$$U_{ij}^{DQ} = -\frac{3\mu_i\Theta_j}{8\pi\varepsilon_0 r_{ij}^4}[5\cos\beta_i\cos^2\beta_j - 2\cos\beta_{ij}\cos\beta_j - \cos\beta_i]. \tag{5.89}$$

In Fig. 5.20b we show a plot of the dipole-quadrupole interaction in the intermolecular frame.

(vi) **Quadrupole-quadrupole.**

$$U_{ij}^{QQ} = \frac{1}{12\pi\varepsilon_0 r_{ij}^9}\Big\{35(\boldsymbol{r}_{ij}\cdot[\boldsymbol{\Theta}_i]\cdot\boldsymbol{r}_{ij})(\boldsymbol{r}_{ij}\cdot[\boldsymbol{\Theta}_j]\cdot\boldsymbol{r}_{ij}) - 20r_{ij}^2(\boldsymbol{r}_{ij}\cdot[\boldsymbol{\Theta}_i]\cdot[\boldsymbol{\Theta}_j]\cdot\boldsymbol{r}_{ij}) + 2r_{ij}^4[\boldsymbol{\Theta}_i]:[\boldsymbol{\Theta}_j]\Big\}. \tag{5.90}$$

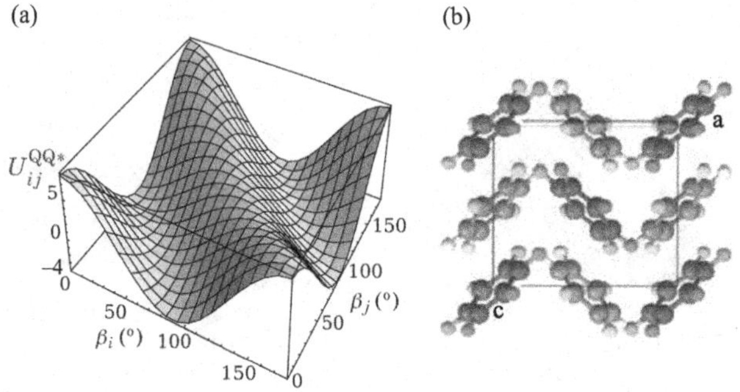

Figure 5.21 (a) A plot of the quadrupole-quadrupole interaction energy U_{ij}^{QQ*} (dimensionless units) in the intermolecular frame. Here $\alpha_i = \alpha_j$. (b) Crystal structure of benzene, viewed down the c-axis of the unit cell [Schweizer and Dunitz, 2006].

The trigonometric form of the potential for uniaxial molecules with axial quadrupole is (cf. Fig. 5.21a)

$$U_{ij}^{QQ}(\alpha_{ij}, \beta_i, \beta_j) = \frac{3}{16\pi\varepsilon_0}\frac{\Theta_i\Theta_j}{r_{ij}^5}\Big\{1 - 5\cos^2\beta_i - 5\cos^2\beta_j - 15\cos^2\beta_i\cos^2\beta_j + 2[4\cos\beta_i\cos\beta_j - \sin\beta_i\sin\beta_j\cos(\alpha_j - \alpha_i)]^2\Big\}. \quad (5.91)$$

Some special cases are $U^{QQ}(0,0,90) = -4$, $U^{QQ}(0,45,45) = -13/4$, $U^{QQ}(0,0,0) = +8$, $U^{QQ}(0,90,90) = +3$, $U^{QQ}(90,90,90) = +1$, $U^{QQ}(30,30,30) = +3/16$, all in units of $3\Theta_i\Theta_j/(16\pi\varepsilon_0 r_{ij}^5)$. Thus, the potential between two axial quadrupoles with the same sign (as, in particular, for two identical molecules) disfavours parallel side-by-side alignment unless the second molecule slides along the first bringing the positive and negative partial charges of the two molecules closer. Accordingly, when the two molecules are kept parallel a staggered configuration with an intermolecular vector tilted with respect to the two molecular axes is favoured. The potential also has a minimum at 90 degrees thus somewhat favouring perpendicular alignment of two molecules and collective herringbone-type molecular organizations. This is consistent with the crystal structure of CO_2, of acetylene as well as of benzene (see Fig. 5.21b) and flat condensed aromatics like coronene [Bannister et al., 2019] which form the inner core of discotic mesogens like HBC (see Table 1.6). More relevantly for liquid crystals, quadrupolar interactions have been invoked (see, e.g., [Goossens, 1987]) as a possible source of the tilt in smectic C phases. The importance of the sign of quadrupolar interactions in apolar molecules is also outlined by the T-shaped structure of binary complexes of benzene in the gas phase [Steed et al., 1979]. It is interesting that mixtures of molecules like benzene and hexafluorobenzene that have very similar shape but quadrupoles of opposite signs give rise to strong attractive interactions for parallel configurations. The complex resulting from an equimolar mixture of benzene and hexafluorobenzene melts at 23.7°C and thus is more stable than benzene that melts at 5.5°C or of hexafluorobenzene that melts at 3.9°C [Hird, 2007]. This type of attractive interaction has been invoked to

Table 5.15. *Expansion coefficients $y_{L_i L_j m}$ for the main types of electrostatic interactions between two uniaxial molecules (Eq. 5.5b)*

L_i	L_j	m	$4\pi\varepsilon_0 y_{L_i L_j m}(r_{ij})$	Interaction type
0	0	0	$q_i^e q_j^e / r_{ij}$	Ion-ion
1	0	0	$\mu_i q_j^e / \sqrt{3} r_{ij}^2$	Dipole-ion
1	1	0	$2\mu_i \mu_j / 3r_{ij}^3$	Dipole-dipole
1	1	1	$\mu_i \mu_j / 3r_{ij}^3$	Dipole-dipole
2	0	0	$\Theta_i q_j^e / \sqrt{5} r_{ij}^3$	Quadrupole-ion
2	1	0	$\sqrt{3}\Theta_i \mu_j / \sqrt{5} r_{ij}^4$	Quadrupole-dipole
2	1	1	$\Theta_i \mu_j / \sqrt{5} r_{ij}^4$	Quadrupole-dipole
2	2	0	$6\Theta_i \Theta_j / 5r_{ij}^5$	Quadrupole-quadrupole
2	2	1	$4\Theta_i \Theta_j / 5r_{ij}^5$	Quadrupole-quadrupole
2	2	2	$\Theta_i \Theta_j / 5r_{ij}^5$	Quadrupole-quadrupole

explain the stability of certain columnar arrangements obtained from a mixture of discotic mesogens [Bates and Luckhurst, 1998; Hughes et al., 2003]. However, even if the qualitative argument holds, it should be mentioned that on its basis alone the structures of benzene and hexafluorobenzene could be expected to be very similar, while they are not [Schweizer and Dunitz, 2006].

Other families of systems where we might expect quadrupolar interactions to be important are suspensions of crystalline flakes of clays like kaolinite, with negatively charged faces and positive edges, which can lead to a *house of cards*-type of self-aggregate arrangements.

Notice that these various interactions have different distance dependence as well as angular dependence. In general, the range of electrostatic interactions is $r_{ij}^{-L_i - L_j - 1}$ if L_i, L_j are the order of the multipoles on the two molecules. This is also shown in Table 5.15, where we give a summary of the expansion coefficients in the spherical harmonics expansion, Eq. 5.5b.

A limitation of the standard treatment with the charge distribution represented by a series of central multipoles is the need for a large separation between molecules. This requirement is clearly not satisfied for anisotropic molecules in contact in a condensed phase. In this case the distance between partial charges on different molecules can be much closer than the centre-centre distance rendering the multipolar series non-convergent. A generalization of the treatment due to Stone and colleagues considers instead distributed multipoles and polarizabilities [Stone, 1981, 1985; Price and Stone, 1983].

5.7.2 Electrostatic Interactions between Particles in Suspension

The importance of electrostatic interactions is of course very different for simple thermotropic materials and amphiphilic, chromonics, or DNA suspensions, or in general with ionic systems. This is not surprising since ionic interactions are dominant when they exist. When modelling mesogenic molecules, e.g. lyotropics or DNA or proteins, in solutions one possibility is simply to treat both solutes and solvent (typically water) explicitly at full atomistic level. On the other hand, this increases enormously the size of the calculation with the inclusion of solvent degrees of freedom that may not be fully relevant, e.g. to study phase morphologies. One possibility often adopted is that of modelling the solvent with a continuous medium with a certain dielectric constant ε. However, in evaluating the importance of interactions between charged nanoparticles suspended in some solvent, it is important to realize that the effective Coulomb interaction energy between two charges e_i, e_j embedded in a medium with dielectric constant ε is [Dill and Bromberg, 2011; Israelachvili, 2011]

$$W_{ij}^{\mathrm{e\,es}} = \frac{e_i e_j}{4\pi\varepsilon_0\varepsilon r_{ij}}, \tag{5.92}$$

which can be very different from that in vacuum ($\varepsilon = 1$). $W_{ij}^{\mathrm{e\,es}}$, albeit being very similar to the standard Coulomb expression, is actually quite different, since it is actually a free energy, and like other effective potentials obtained a posteriori by taking the log of a Boltzmann distribution (inverting it), it depends on temperature. From a practical point of view the presence of the material dielectric constant ε has the effect of reducing (screening) the attractive and repulsive electrostatic interactions. In particular, for the most common case of water suspensions, recalling that for water at room temperature $\varepsilon = 78$, the effects can clearly be big. Moreover, the screening effect can vary with temperature. In these systems changing the ionic strength, e.g. by adding salts to the solutions, can vary the importance of electrostatic interactions and even change the phase transition sequences. One way to account for this is to use expressions for screened Coulombic interactions, similar to Debye–Hückel theory, e.g. [Ninham and Lo Nostro, 2010]

$$W^{\mathrm{ee}} = \sum_{i<j}^{N} \frac{e_i e_j}{4\pi\varepsilon_0\varepsilon\, r_{ij}} \mathrm{e}^{-r_{ij}/\lambda_D}, \tag{5.93}$$

where λ_D is the Debye screening length. This is expected to be valid for the low-salt concentrations, like the physiological concentrations that are generally encountered in biological systems. For water solutions, the dielectric constant, ε, takes the value of water. The Debye length in a system with M different charged species, with the jth species of charge e_j and concentration c_j in a medium with permittivity ε is:

$$\lambda_{\mathrm{D}} = \kappa_D^{-1} = \left(\frac{\varepsilon_0\varepsilon k_{\mathrm{B}}T}{\sum_{j=1}^{M} c_j e_j^2}\right)^{1/2}. \tag{5.94}$$

Thus, changes in the salt concentrations can vary λ_D. For example, a system with $[\mathrm{Na}^+] = 50$ mM (similar to physiological conditions), gives $\lambda_D = 13.603$ Å [Knotts IV et al., 2007].

5.8 Inductive and Dispersive Interactions

The classical treatment we have just described is sufficient to treat the interaction between the permanent distribution of charges on two molecules. However, another important type of interaction deals with instantaneous deformations of the charge distribution of a molecule caused by the charge distribution of another one. This type of interactions has to be treated using quantum mechanics and indeed it consists of intermolecular electronic correlation effects. The different orbitals on each molecule have different size and shape and the deformation induced by the other molecule amounts to fluctuations in the electron density around an atom or molecule, which consists of promotions from the initial ground state to higher states and back. In practice, we consider the set of interaction terms that we have obtained (Eq. 5.81) as an operator $\hat{U}^{\mathrm{e}}$:

$$\hat{U}^{\mathrm{e}} = \frac{1}{4\pi\varepsilon_0}\sum_{k\in i}\sum_{m\in j}\frac{e_{k,i}e_{m,j}}{r_{km}}. \tag{5.95}$$

and apply QM perturbation theory [Buckingham, 1967b; Gray and Gubbins, 1984; Stone, 1996]

If we assume that molecules i, j have unperturbed Hamiltonians $\hat{\mathscr{H}}_i$, $\hat{\mathscr{H}}_j$ obeying their respective Schrodinger equations $\hat{\mathscr{H}}_i|n_i\rangle = \mathscr{E}^{(0)}_{n_i}|n_i\rangle$ and $\hat{\mathscr{H}}_j|n_j\rangle = \mathscr{E}^{(0)}_{n_j}|n_j\rangle$, where we indicate with $|n_i\rangle$ an eigenstate of $\hat{\mathscr{H}}_i$ with eigenvalue $\mathscr{E}^{(0)}_{n_i}$ (in practice a vibronic state), then for two molecules in the ground state at infinite separation

$$[\hat{\mathscr{H}}_i + \hat{\mathscr{H}}_j]|0_i0_j\rangle = [\mathscr{E}^{(0)}_{0_i} + \mathscr{E}^{(0)}_{0_j}]|0_i0_j\rangle, \tag{5.96}$$

with $|n_in_j\rangle = |n_i\rangle|n_j\rangle$. When the molecules get closer, even if still so far away that the overlap between their electron clouds is negligible, we can apply standard quantum mechanical perturbation theory [Landau and Lifshitz, 1958] and get the first-order energy correction to the ground state as

$$\mathscr{E}^{(1)} = \langle 0_i0_j|\hat{U}^{\mathrm{e}}|0_i0_j\rangle. \tag{5.97}$$

This gives back the standard electrostatic interaction energy, corresponding to the expectation value of the perturbation over the ground state $|0_i0_j\rangle$, that we have considered until now. The second-order correction is

$$\mathscr{E}^{(2)} = -\sum_{m_a,n_j}{}' \frac{\langle 0_i0_j|\hat{U}^{\mathrm{e}}|m_in_j\rangle\langle m_in_j|\hat{U}^{\mathrm{e}}|0_i0_j\rangle}{\mathscr{E}^{(0)}_{m_i} + \mathscr{E}^{(0)}_{n_j} - \mathscr{E}^{(0)}_{0_i} - \mathscr{E}^{(0)}_{0_j}}, \tag{5.98}$$

where the prime in the sum indicates that at least one of the indices has to be greater than 0, so that the energy denominator is non-vanishing. We see that at this stage different type of terms, corresponding to excitations of higher states on one or both of the molecules, and a return to the ground state, caused by the perturbation, can appear. The involvement of higher states with different wave functions corresponds to the intuitive classical notion of the electron distribution of one molecule being deformed and fluctuating as the second molecule approaches. The first type of terms, e.g.

$$U^{\text{ind}} = -\sum_{m_a>0} \frac{\langle 0_i 0_j | U^{\text{e}} | m_i 0_j \rangle \langle m_i 0_j | U^{\text{e}} | 0_i 0_j \rangle}{\mathscr{E}^{(0)}_{m_i} - \mathscr{E}^{(0)}_{0_i}}, \tag{5.99}$$

are called *inductive* and represent the distortion of the charge distribution on molecule i caused by molecule j (or vice versa). The second type of terms, called *dispersive*, describe the interaction between mutually distorted charge distributions, e.g.

$$U^{\text{disp}} = -\sum_{\substack{m_a>0\\ n_j>0}}{}' \frac{\langle 0_i 0_j | \hat{U}^{\text{e}} | m_i n_j \rangle \langle m_i n_j | \hat{U}^{\text{e}} | 0_i 0_j \rangle}{\mathscr{E}^{(0)}_{m_i} + \mathscr{E}^{(0)}_{n_j} - \mathscr{E}^{(0)}_{0_i} - \mathscr{E}^{(0)}_{0_j}}, \tag{5.100}$$

with a similar term for the fluctuations of the electron cloud of the first molecule under the effect of the second. Different physical contributions can be obtained inserting in these expressions the previously obtained multipolar expression for the interaction energy, Eq. 5.82, now treated as a perturbation operator.

5.8.1 Inductive Interactions and Polarizability

The permanent multipoles of a molecule interact with the multipoles they induce in neighbouring molecules by electronic polarization giving a multipole-induced -multipole-type interaction. Here we shall limit ourselves to induced dipoles $\boldsymbol{\mu}^{\text{ind}}$, that we can formally write expanding in a Taylor series of the field $\boldsymbol{E}$ and its gradient $\nabla \boldsymbol{E}$ acting on the molecule, obtaining its components as [Buckingham, 1967b]

$$\mu_a^{\text{ind}} = \sum_b \alpha_{ab} E_b + \frac{1}{2} \sum_{b,c} \beta_{abc} E_b E_c + \frac{1}{6} \sum_{b,c,d} \gamma_{abcd} E_b E_c E_d + \frac{1}{3} \sum_{b,c} A_{abc} \nabla_b E_c + \cdots, \tag{5.101}$$

where the linear susceptibility second-rank tensor $\boldsymbol{\alpha}$ is the molecular polarizability and the third rank $\boldsymbol{\beta}$ and the fourth rank $\boldsymbol{\gamma}$ are *hyperpolarizabilities* corresponding to non-linear higher-rank contributions, that have to be included if the acting field is particularly strong. For axially symmetric molecules we can write $\alpha_{ab} = \bar{\alpha}\delta_{a,b} + \frac{1}{3}\Delta\alpha\left(3u_a u_b - \delta_{a,b}\right)$, where $\boldsymbol{u}$ is a unit vector along the molecule axis and $\bar{\alpha} \equiv \frac{1}{3}\text{Tr}\boldsymbol{\alpha} = \frac{1}{3}\left(\alpha_\parallel + 2\alpha_\perp\right)$ the scalar polarizability and $\Delta\alpha \equiv \alpha_\parallel - \alpha_\perp$, its anisotropy. Similarly

$$\beta_{abc} = \beta_\perp \left(u_a \delta_{b,c} + u_b \delta_{c,a} + u_c \delta_{a,b}\right) + \left(\beta_\parallel - 3\beta_\perp\right) u_a u_b u_c. \tag{5.102}$$

The non-linear terms are normally not important for intermolecular interactions and will be neglected here. The third-rank tensor A_{abc} is to be taken into account in a non-uniform field, like that generated by the charge distribution of a neighbouring molecule. $\mathbf{A}$ is the dipole-quadrupole polarizability, describing the dipole induced by a field gradient. For an axially symmetric $C_{\infty v}$ molecule

$$A_{abc} = \frac{1}{2} A_\parallel u_a \left(3u_b u_c - \delta_{b,c}\right) + A_\perp \left(u_b \delta_{a,c} + u_c \delta_{a,b} - 2u_a u_b u_c\right). \tag{5.103}$$

Notice that only the first non-vanishing moment is independent of the choice of origin. Thus, for an ion, the dipole and quadrupole moments depend on the origin and an origin can be chosen as a charge centre so that $\mu = 0$, while for an uncharged molecule, μ is origin independent, while the quadrupole is not.

To give the polarizability $\boldsymbol{\alpha}$ a molecular interpretation, we can apply QM perturbation theory to calculate the change in energy of a molecule in the ground state ($n = 0$) subjected to an external electric field $\boldsymbol{E}$. We start from the perturbation energy $\mathscr{H}' = -\boldsymbol{E} \cdot \boldsymbol{\mu}^{\text{ind}}$, where $\boldsymbol{\mu}^{\text{ind}}$ is the molecular dipole in the presence of the field, that we rewrite, using Eq. 5.101 as

$$\sum_a E_a \mu_a^{\text{ind}} = \sum_{a,b} E_a \alpha_{ab} E_b + \frac{1}{2} \sum_{a,b,c} E_a \beta_{abc} E_b E_c + \frac{1}{6} \sum_{a,b,c,d} E_a \gamma_{abcd} E_b E_c E_d$$
$$+ \frac{1}{3} \sum_{a,b,c} E_a A_{abc} \nabla_b E_c + \cdots . \tag{5.104}$$

From QM perturbation theory we have instead that the equivalent perturbation energy for a molecule is, to second order,

$$\mathscr{E} = \mathscr{E}^1 + \mathscr{E}^2 = -\sum_a \langle 0|\hat{\mu}_a|0\rangle E_a - \sum_{k\neq 0} E_a \frac{\langle 0|\hat{\mu}_a|k\rangle\langle k|\hat{\mu}_b|0\rangle}{\mathscr{E}_k^{(0)} - \mathscr{E}_0^{(0)}} E_b, \tag{5.105}$$

with $a, b = x, y, z$. Comparing the two energy expressions Eqs. 5.104 and 5.105, we see that the polarizability tensor is

$$\boldsymbol{\alpha} = \sum_{k\neq 0} \frac{\langle 0|\hat{\boldsymbol{\mu}}|k\rangle\langle k|\hat{\boldsymbol{\mu}}|0\rangle}{\mathscr{E}_k^{(0)} - \mathscr{E}_0^{(0)}}. \tag{5.106}$$

The polarizability represents the deformability of the molecule electron cloud in the external field, obtained with virtual transitions from the ground to the excited levels and back. Given the different shape and extension of the wave functions of the different states, it corresponds to a quantum mechanical way of representing what would be classical fluctuations in shape. To see the physical significance of the polarizability it is useful to make the mean energy approximation [Unsöld, 1927] to replace the denominator:

$$\boldsymbol{\alpha} \approx \frac{1}{\mathscr{E}} \sum_{k\neq 0} \langle 0|\hat{\boldsymbol{\mu}}|k\rangle\langle k|\hat{\boldsymbol{\mu}}|0\rangle = \frac{1}{\mathscr{E}} \langle 0|\hat{\boldsymbol{\mu}} \otimes \hat{\boldsymbol{\mu}}|0\rangle - \frac{1}{\mathscr{E}} \langle 0|\hat{\boldsymbol{\mu}}|0\rangle \otimes \langle 0|\hat{\boldsymbol{\mu}}|0\rangle, \tag{5.107}$$

which shows that the polarizability is related to the mean square fluctuations of the dipole. Notice that the polarizability exists also for a non-polar molecule. We give in Table 5.16 a small list of molecular polarizabilities for some simple molecules. In Table 5.17 we give instead the full set of components of $\boldsymbol{\alpha}$ for some mesogens. We notice that the polarizability is origin independent [Stone, 1996], so we can compare the results with others in literature, e.g. those of Clark [2001] for 5CB. Going back to inductive interactions, the induced dipole will depend on the field at molecule i and on the polarizability $\boldsymbol{\alpha}_i$ of the molecule. In particular $\boldsymbol{\mu}_i^{\text{ind}} = \boldsymbol{\alpha}_i \cdot \boldsymbol{E}^{(ij)}$, where $\boldsymbol{E}^{(ij)}$ is the field produced by molecule j and felt by molecule i. The induced interaction energy is

$$U_{ij}^{\text{ind}} = -\frac{1}{2}[\boldsymbol{E}^{(ij)} \cdot \boldsymbol{\mu}_i^{\text{ind}} + \boldsymbol{E}^{(ji)} \cdot \boldsymbol{\mu}_j^{\text{ind}}], \tag{5.108}$$

Table 5.16. *Principal polarizabilities (Å^3) of a few molecules. We choose the z-axis along the highest symmetry axis and* **y** *out of the molecule plane for planar molecules. For naphthalene and anthracene z is parallel to the long axis and* **x** *in the molecule plane. Refs. (a) Zannoni [2020]; (b) [Le Fèvre, 1965]; (c) [Le Fèvre et al., 1967]; (d) [Gray and Gubbins, 1984]; (e) [Le Fèvre and Murthy, 1969]; (f) [Le Fèvre and Radom, 1967]*

Molecule	α_{xx}	α_{yy}	α_{zz}	Ref.
Cyclohexane	10.04	10.04	9.05	(a)
Benzene	11.2	11.2	7.3	(b)
	10.8	10.8	3.7	(a)
Fluorobenzene	11.1	7.73	11.3	(d)
Naphthalene	16.6	11.3	21.8	(f)
Anthracene	24.6	15.9	35.9	(f)
	24.56	11.38	44.06	(a)
Triphenylene	39.0	39.0	15.5	(e)

Table 5.17. *QC calculated polarizability tensor components (in* Buckingham*) for a few mesogens in the molecular frames of Table 5.12. Refs. (a) QC DFT [Zannoni, 2020]; (b) QC plane waves [Clark, 2001]. For p-quinquephenyl, z is along the long axis and* **x** *in the terminal ring plane*

Molecule	α_{xx}	α_{xy}	α_{xz}	α_{yy}	α_{yz}	α_{zz}	Ref.
PAA	20.12	−6.53	−7.17	25.77	5.41	67.32	(a)
MBBA	28.30	0.02	2.03	25.05	0.06	58.66	(a)
5CB	27.95	−1.53	1.20	21.14	2.39	59.11	(a)
	27.64	−2.07	−0.90	26.67	−2.65	66.15	(b)
Quinquephenyl	−33.90			13.81		20.08	(a)

where $\boldsymbol{E}^{(ij)}$ is the field generated on molecule i by molecule j. In particular, we have, using the expressions for the field already seen, the first few interaction contributions [Kielich, 1972]:

Ion-induced dipole. An ion with charge e_i induces a dipole in molecule j. In the following we write only the first of the two terms in Eq. 5.108. For two charged polarizable molecules another term with i,j exchanged has of course to be added

$$U_{ij}^{\mathrm{ID}'} = -\frac{e_i^2}{8\pi\varepsilon_0 r_{ij}^6}(\boldsymbol{r}_{ij}\cdot[\boldsymbol{\alpha}_j]\cdot\boldsymbol{r}_{ij}), \tag{5.109}$$

where we use D$'$ to indicate an induced dipole. If the polarizable molecule is axially symmetric, we have $U_{ij}^{\mathrm{ID}'} = -\mathrm{e}_i^2\bar{\alpha}_j\left[\frac{\Delta\alpha_j}{3\bar{\alpha}_j}P_2(\cos\beta_j)+\frac{1}{2}\right]/(4\pi\varepsilon_0 r_{ij}^4)$.

Permanent dipole-induced dipole. The permanent dipoles of polar molecules interact with the dipoles they induce in neighbouring molecules by electronic polarization giving a dipole-induced dipole interaction:

$$U_{ij}^{\mathrm{DD}'} = -\frac{1}{8\pi\varepsilon_0 r_{ij}^{10}}\big[9(\boldsymbol{r}_{ij}\cdot[\boldsymbol{\alpha}_j]\cdot\boldsymbol{r}_{ij})(\boldsymbol{\mu}_i\cdot\boldsymbol{r}_{ij})^2,$$
$$-6r_{ij}^2(\boldsymbol{r}_{ij}\cdot[\boldsymbol{\alpha}_j]\cdot\boldsymbol{\mu}_i)(\boldsymbol{\mu}_i\cdot\boldsymbol{r}_{ij})(\boldsymbol{\mu}_i\cdot\boldsymbol{r}_{ij}) + r_{ij}^4(\boldsymbol{\mu}_i\cdot[\boldsymbol{\alpha}_j]\cdot\boldsymbol{\mu}_i)\big]. \tag{5.110}$$

For axial symmetry this reduces to

$$U_{ij}^{\mathrm{DD}'} = -\frac{\mu_i^2}{8\pi\varepsilon_0\bar{\alpha}_j r_{ij}^6}\big[(\alpha_\perp)_j(3\cos^2\beta_j+1) + \Delta\alpha_j(3\cos\beta_i\cos\beta_j - \cos\beta_{ij})^2\big]. \tag{5.111}$$

Permanent quadrupole-induced dipole. This term, like the previous one, could be of interest for studies of solute alignment in LCs:

$$U_{ij}^{\mathrm{QD}'} = \frac{-1}{8\pi\varepsilon_0 r_{ij}^{14}}\Big\{25(\boldsymbol{r}_{ij}\cdot[\boldsymbol{\alpha}_j]\cdot\boldsymbol{r}_{ij})(\boldsymbol{r}_{ij}\cdot[\boldsymbol{\Theta}_i]\cdot\boldsymbol{r}_{ij})^2 + 4r_{ij}^4(\boldsymbol{r}_{ij}\cdot[\boldsymbol{\Theta}_i]\cdot[\boldsymbol{\alpha}_j]\cdot[\boldsymbol{\Theta}_i]\cdot\boldsymbol{r}_{ij})$$
$$-20r_{ij}^2(\boldsymbol{r}_{ij}\cdot[\boldsymbol{\alpha}_j]\cdot[\boldsymbol{\Theta}_i]\cdot\boldsymbol{r}_{ij})(\boldsymbol{r}_{ij}\cdot[\boldsymbol{\Theta}_i]\cdot\boldsymbol{r}_{ij})\Big\} \tag{5.112}$$

and for axially symmetric molecules

$$U_{ij}^{\mathrm{QD}'} = -\frac{9\Theta_i^2}{32\pi\varepsilon_0 r_{ij}^8}\Big\{(\alpha_\perp)_j(1-2\cos^2\beta_i+5\cos^4\beta_i) + \Delta\alpha_j(5\cos\beta_j\cos^2\beta_i$$
$$-2\cos\beta_{ij}\cos\beta_i - \cos\beta_j)^2\Big\}. \tag{5.113}$$

5.8.2 Dispersive Interactions

As we have seen, inductive effects need at least one of the two molecules to have permanent multipoles. However, even non-polar molecules can have an instantaneous dipole moment due to charge fluctuations and this instantaneous dipole will polarize another molecule and induce a new instantaneous dipole moment. The attractive interaction energy between the two induced dipoles is called a London, van der Waals or dispersive interaction [Buckingham, 1967b]. The dispersive interaction between two molecules i and j can be obtained, in the static polarizability approximation, by inserting the electrostatic energy operator in Eq. 5.100. For two uncharged molecules this operator is

$$4\pi\varepsilon_0\hat{U}_{ij}^{\mathrm{e}} = -\hat{\boldsymbol{\mu}}_i\hat{\boldsymbol{\mu}}_j{:}\mathbf{T}^{(2)}(\boldsymbol{r}_{ij}) + \left(-\frac{1}{3}\hat{\boldsymbol{\mu}}_i\hat{\boldsymbol{\Theta}}_j + \frac{1}{3}\hat{\boldsymbol{\Theta}}_i\hat{\boldsymbol{\mu}}_j\right)\vdots\mathbf{T}^{(3)}(\boldsymbol{r}_{ij}) + \cdots, \tag{5.114a}$$

$$= \hat{U}_{ij}^{\mu\mu} + \hat{\overline{U}}_{ij}^{\mu\Theta}, \tag{5.114b}$$

where $\hat{\overline{U}}_{ij}^{\mu\Theta} \equiv \hat{U}_{ij}^{\mu\Theta} + \hat{U}_{ij}^{\Theta\mu}$. In particular, retaining only the first and most important term, i.e. the dipole-dipole interaction, Eq. 5.85 in the second-order perturbation expression Eq. 5.100, we have

$$U_{ij}^{\mathrm{D'D'}} = -\sum_{\substack{m_i>0\\ n_j>0}}{}' \frac{\langle 0_i 0_j|\hat{U}_{ij}^{\mu\mu}|m_i n_j\rangle\langle m_i n_j|\hat{U}_{ij}^{\mu\mu}|0_i 0_j\rangle}{\mathcal{E}_{m_i}^{(0)}+\mathcal{E}_{n_j}^{(0)}-\mathcal{E}_{0_i}^{(0)}-\mathcal{E}_{0_j}^{(0)}}, \tag{5.115a}$$

$$= -\frac{1}{(4\pi\varepsilon_0)^2}\sum_{\substack{m_i>0\\ n_j>0}}{}' \frac{\langle 0_i 0_j|\hat{\boldsymbol{\mu}}_i\cdot\mathbf{T}_{ij}^{(2)}\cdot\hat{\boldsymbol{\mu}}_j|m_i n_j\rangle\langle m_i n_j|\hat{\boldsymbol{\mu}}_i\cdot\mathbf{T}_{ij}^{(2)}\cdot\hat{\boldsymbol{\mu}}_j|0_i 0_j\rangle}{\mathcal{E}_{m_i}^{(0)}+\mathcal{E}_{n_j}^{(0)}-\mathcal{E}_{0_i}^{(0)}-\mathcal{E}_{0_j}^{(0)}},$$

$$= -\frac{1}{(4\pi\varepsilon_0)^2}\sum_{\substack{m_i>0\\ n_j>0}}{}' \frac{\langle 0_i|\hat{\boldsymbol{\mu}}_i|m_i\rangle\langle m_i|\hat{\boldsymbol{\mu}}_i|0_i\rangle\cdot\mathbf{T}_{ij}^{(2)}:\langle 0_j|\hat{\boldsymbol{\mu}}_j|n_j\rangle\langle n_j|\hat{\boldsymbol{\mu}}_j|0_j\rangle\cdot\mathbf{T}_{ji}^{(2)}}{\mathcal{E}_{m_i}^{(0)}+\mathcal{E}_{n_j}^{(0)}-\mathcal{E}_{0_i}^{(0)}-\mathcal{E}_{0_j}^{(0)}}, \tag{5.115b}$$

where $\mathbf{T}_{ij}^{(2)} = \nabla\nabla(1/r_{ij})$ is the *dipolar tensor* (see Eq. C.10). Approximating the denominator as a product of denominators [Unsöld, 1927],

$$\frac{1}{\mathcal{E}_{m_i}^{(0)}+\mathcal{E}_{n_j}^{(0)}-\mathcal{E}_{0_i}^{(0)}-\mathcal{E}_{0_j}^{(0)}} \approx \mathcal{W}_{ij}\frac{1}{\mathcal{E}_{m_i}^{(0)}-\mathcal{E}_{0_i}^{(0)}}\frac{1}{\mathcal{E}_{n_j}^{(0)}-\mathcal{E}_{0_j}^{(0)}}, \tag{5.116}$$

where $\mathcal{W}_{ij}$ is a proportionality constant, an expression involving the polarizabilities of the two molecules can be obtained:

$$U_{ij}^{\mathrm{D'D'}} = -\frac{\mathcal{W}_{ij}}{(4\pi\varepsilon_0)^2}[\boldsymbol{\alpha}_i]\cdot[\mathbf{T}_{ij}^{(2)}]:[\boldsymbol{\alpha}_j]\cdot\mathbf{T}_{ij}^{(2)}]. \tag{5.117}$$

Rewriting this scalar tensorial contraction in a spherical tensor form (cf.Section B.4) gives an expression particularly convenient for theoretical manipulations [Luckhurst et al., 1975]. Using Eq. B.30 we find

$$U_{ij}^{\mathrm{D'D'}} = -\frac{\mathcal{W}_{ij}}{(4\pi\varepsilon_0)^2}\sum_{L,m}[\boldsymbol{\alpha}_i\cdot\mathbf{T}_{ij}^{(2)}]^{L,m}[\boldsymbol{\alpha}_j\cdot\mathbf{T}_{ij}^{(2)}]^{L,m^*}, \tag{5.118}$$

which can be rewritten decoupling the products of dipolar tensor and individual polarizabilities in terms of spherical components of the individual polarizability, $\alpha_i^{L,m}$, $\alpha_j^{L,m}$ and of the dipolar tensor $T_{ij}^{L_i,q}$, e.g.

$$[\boldsymbol{\alpha}_i\cdot\mathbf{T}_{ij}^{(2)}]^{L,m} = (-1)^{L+1}\sum_{L_i,L_j,q}\sqrt{(2L_i+1)(2L_j+1)}\;W(L_j,1,L_i,1;1,L)$$
$$\times\, C(L_j,L_i,L;m-q,q)\,T_{ij}^{L_i,q\,*}\alpha_j^{L_j,m-q}, \tag{5.119}$$

where $C(a,b,c;d,e)$ are Clebsch–Gordan (Eq. F.18b) and $W(a,b,c,d;e,f)$ are Racah coefficients (Eq. F.23). The previous expression is a scalar, thus frame independent. If we specialize in the intermolecular frame (Fig. 5.2), the dispersive interaction between two particles at a distance r_{ij} can be written as [Luckhurst et al., 1975]

$$U_{ij}^{\mathrm{D'D'}} = -\frac{\mathcal{W}_{ij}}{(4\pi\varepsilon_0)^2}\sum_{L_i,L_j,m} w_{L_iL_jm}^{\mathrm{D'D'}}(r_{ij})\,\alpha_i^{L_i,m\,*}\alpha_j^{L_j,m}, \tag{5.120}$$

with the potential coefficients

$$w^{\mathrm{D'D'}}_{L_iL_jm}(r_{ij}) = \frac{30}{r_{ij}^6}\sqrt{(2L_i+1)(2L_j+1)}\sum_L W(L_j,1,2,1;1,L)$$
$$\times W(2,1,L_i,1;1,L)\,C(L_j,2,L;m,0)C(2,L_i,L;0,m). \tag{5.121}$$

Transforming the polarizabilities to their principal molecular frame, the potential becomes

$$U^{\mathrm{D'D'}}_{ij} = \sum\left[\frac{-\mathcal{W}_{ij}}{(4\pi\varepsilon_0)^2} w^{\mathrm{D'D'}}_{L_iL_jm}(r_{ij})\,\alpha_i^{L_i,n_i\,*}\,\alpha_j^{L_j,n_j}\right]\mathscr{D}^{L_i}_{m,n_i}(\Omega_{ir_{ij}})\,\mathscr{D}^{L_j\,*}_{m,n_j}(\Omega_{jr_{ij}}), \tag{5.122}$$

with $L_p = 0,2$, $n_p = 0, \pm 2$, $p = i,j$. This is now in the form of Eq. 5.4, and

$$u^{\mathrm{D'D'}}_{L_iL_jm;n_in_j} = \left[\frac{-\mathcal{W}_{ij}}{(4\pi\varepsilon_0)^2}\, w^{\mathrm{D'D'}}_{L_iL_jm}(r_{ij})\,\alpha_i^{L_i,n_i\,*}\,\alpha_j^{L_j,n_j}\right]. \tag{5.123}$$

The rather fierce-looking equations above become rather simple for some important special cases. For two cylindrically symmetric molecules with polarizability, $\alpha_i^{2,n} = \alpha_i^{2,0}\delta_{n,0}$ or in Cartesian coordinates: $\alpha_\parallel$ and $\alpha_\perp$ components, the interaction becomes

$$U^{\mathrm{D'D'}}_{ij} = \sum\left[\frac{-\mathcal{W}_{ij}w^{\mathrm{D'D'}}_{L_iL_jm}(r_{ij})}{4\pi\varepsilon_0^2\sqrt{2L_i+1}\sqrt{2L_j+1}}\alpha_i^{L_i,0\,*}\,\alpha_j^{L_j,0}\right]Y^*_{L_i,m}(\Omega_{ir_{ij}})Y_{L_j,-m}(\Omega_{jr_{ij}}), \tag{5.124a}$$

$$= \sum_{L_i,L_j,m} y^{\mathrm{D'D'}}_{L_iL_jm}(r_{ij})Y_{L_i,m}(\Omega_{ir_{ij}})Y_{L_j,-m}(\Omega_{jr_{ij}}), \tag{5.124b}$$

where $\bar{\alpha} = (\alpha_\parallel + 2\alpha_\perp)/3$, $\Delta\alpha = (\alpha_\parallel - \alpha_\perp) = \sqrt{3/2}\alpha^{2,0}$. The explicit form is

$$U^{\mathrm{D'D'}}_{ij} = -\frac{3W_{ij}}{4(4\pi\varepsilon_0)^2r_{ij}^6}\left\{2\bar{\alpha}_i\bar{\alpha}_j - \frac{\bar{\alpha}_j\Delta\alpha_i}{3} - \frac{\bar{\alpha}_i\Delta\alpha_j}{3} + \Delta\alpha_i\alpha_{j,\perp}\cos^2\beta_i\right.$$
$$\left. + \Delta\alpha_j\alpha_{i,\perp}\cos^2\beta_j + \frac{\Delta\alpha_i\Delta\alpha_j}{3}(3\cos\beta_i\cos\beta_j - \cos\beta_{ij})^2\right\}. \tag{5.125}$$

and for two identical molecules:

$$U^{\mathrm{D'D'}}_{ij} = -\frac{W_{ij}}{4(4\pi\varepsilon_0)^2r_{ij}^6}\left\{6\bar{\alpha}^2 + \bar{\alpha}\Delta\alpha[3\cos^2\beta_i + 3\cos^2\beta_j - 2]\right.$$
$$\left. + (\Delta\alpha)^2(3\cos\beta_i\cos\beta_j - \cos\beta_{ij})^2\right\}. \tag{5.126}$$

The coefficients are:

$$y^{\mathrm{D'D'}}_{000}(r_{ij}) = -\frac{3\mathcal{W}_{ij}}{4\pi\varepsilon_0^2r_{ij}^6}\bar{\alpha}^2, \tag{5.127a}$$

$$y^{\mathrm{D'D'}}_{200}(r_{ij}) = y^{\mathrm{D'D'}}_{020}(r_{ij}) = \frac{\Delta\alpha}{\sqrt{5}\bar{\alpha}}y^{\mathrm{D'D'}}_{000}(r_{ij}), \tag{5.127b}$$

$$y^{\mathrm{D'D'}}_{220}(r_{ij}) = 3y^{\mathrm{D'D'}}_{222}(r_{ij}) = \frac{(\Delta\alpha)^2}{3\bar{\alpha}^2}y^{\mathrm{D'D'}}_{000}(r_{ij}). \tag{5.127c}$$

For atoms or spherical particles $\Delta\alpha = 0$, and Eq. 5.125 reduces essentialy to the classical London–van der Waals formula with the well-known r_{ij}^{-6} dependence on distance:

$$U_{ij}^{\mathrm{D'D'}} = -\frac{3W_{ij}}{2(4\pi\varepsilon_0)^2 r_{ij}^6}\bar{\alpha}_i\bar{\alpha}_j. \tag{5.128}$$

The importance of the attractive dispersive interaction increases as α increases and when this happens, we expect the stability of a condensed phase to increase. Since at the boiling point the phase changes from a condensed to a gas phase where molecules are far apart, we may take the boiling point temperature T_B as a rough indicator of the importance of dispersive forces in a set of particles otherwise as similar as possible. As we have seen before (Eq. 5.115a) α is linked to the mean square fluctuations of the induced dipole. If we recall that any dipole is the product of a charge on the negative and positive pole times their distance d, i.e. $\mu = ed$, it is easy to see that the polarizability is related to the space available for charge delocalization and to the number of electrons. For example, in the series of noble gas atoms, since α increases over 25 times from He to Rd, the boiling points T_B of the noble gases increase: helium −269°C, neon −246°C, argon −186°C, krypton −152°C, xenon −108°C, radon −62°C. More generally, and even more approximately, if we consider molecules with the same chemical composition (brute formula) and different shape, we expect polarizability to be larger for the larger one. For instance, the linear n-pentane has $T_B = 36.1$°C while the more spherical neopentane has $T_B = 9.5$°C.

Notice that dispersive interactions are always attractive and tend to bring two anisotropic molecules parallel to each other. Indeed, these interactions have been originally invoked [Maier and Saupe, 1958] as a possible source of the collective long-range alignment in nematics. However, these cannot be sufficient to explain the phenomenon [Kaplan and Drauglis, 1971] and other mechanisms, such as steric repulsions, should be considered, even if dispersion forces should certainly play an important role in molecular alignment. The scalar part also plays a role in ordering, in association with anisotropic repulsive forces, because of the incomplete space averaging following from the steric repulsion [Gelbart and Gelbart, 1977].

5.8.3 Interaction between Colloidal Particles

In closing this short section on dispersive interactions, it is worth pointing out that what we have discussed concerns only atomic or molecular scale interactions. When dealing with the interactions between colloidal particles with dimension one to two orders of magnitude larger than simple molecular systems, the colloidal particle can be considered as an assembly of interaction centres and the total interaction is obtained summing (integrating) over all the volume occupied by the two colloidal particles. In this way the centre-centre distance dependence will inevitably change with the shape of the particles [Israelachvili, 2011]. For instance, if we imagine each colloidal particle to be formed by a set of small molecular size volume elements (elementary voxels) with an r_{ij}^{-6} distance dependence and if $\rho_1(r_i)$, $\rho_2(r_j)$ is the density of these centres belonging to the two particles P_1 and P_2: $i \in P_1$ and $j \in P_2$, made of materials M_1, M_2 respectively, then the effective interaction energy between two rigid particles of arbitrary shape with centre-centre distance vector $\boldsymbol{R}_{12}$ and orientation Ω_1, Ω_2 will be, assuming additivity,

$$
\begin{aligned}
W^{\mathrm{D'D'}}_{M_1 M_2}(\boldsymbol{R}_{12}, \Omega_1, \Omega_2) &= -\int_{V_1} \mathrm{d}V_i \int_{V_2} \mathrm{d}V_j \frac{w_{12}\rho_1(r_i)\rho_2(r_j)}{r_{ij}^6} \\
&= A_{M_1 M_2} \int \mathrm{d}\Omega_1 \int \mathrm{d}\Omega_2 \mathscr{S}(\boldsymbol{R}_{12}, \Omega_1, \Omega_2),
\end{aligned} \tag{5.129}
$$

where the Hamaker constant,[2]

$$
A_{M_1 M_2} = \pi^2 w_{12} \rho_1 \rho_2, \tag{5.130}
$$

contains all the material dependent properties and we have assumed that particles 1, 2 have a uniform composition and density. Here $w_{ij} = w_{12}, \forall i \in P_1, \forall j \in P_2$ depends on the polarizabilities: $w_{12} \propto \alpha_1 \alpha_2$ of the voxels of particles P_1 and P_2 and a number of correction factors, for non-additivity, etc. The Hamaker constants can be determined empirically and extensive tabulations for various materials exist [Visser, 1972]. Their order of magnitude is typically of the order of 10^{-19} J [Israelachvili, 2011].

The original derivation of Hamaker was for spherical particles [Hamaker, 1937]. However, for colloidal particles giving LC suspensions (Section 1.14), anisometry is essential. The factor $\mathscr{S} \equiv \mathscr{S}(\boldsymbol{R}_{12}, \Omega_1, \Omega_2)$:

$$
\mathscr{S}(\boldsymbol{R}_{12}, \Omega_1, \Omega_2) = \int_{V_1(\Omega_1)} \mathrm{d}V_i \int_{V_2(\Omega_2)} \mathrm{d}V_j \frac{1}{\pi^2 r_{ij}^6} \tag{5.131}
$$

is the result of the integration over the volume of the two particles and will necessarily depend on the shape of the two particles and their distance and orientation. For simple geometries the integration can be performed analytically and is reported (e.g. in [Mahanty and Ninham, 1976; Israelachvili, 2011]). A few important cases are shown in Table 5.18. A very interesting example is that of Everaers and Ejtehadi [2003] for two possibly different hard biaxial ellipsoids with semiaxes $\sigma_e^{(i)}$ and where the term $\gamma_{12}\eta_{12}$ in Table 5.18 is

$$
\eta_{12}\chi_{12} = \frac{2}{\sigma}\left\{\left(\frac{1}{R_1} - \frac{1}{R_1'}\right)\left(\frac{1}{R_2} - \frac{1}{R_2'}\right)\sin^2\theta + \left(\frac{1}{R_1} + \frac{1}{R_2}\right)\left(\frac{1}{R_1'} + \frac{1}{R_2'}\right)\right\}^{-\frac{1}{2}}, \tag{5.132}
$$

while $\mathbf{M}_1$, $\mathbf{M}_2$ are the rotation matrices specifying the orientation of the two ellipsoids (see Section 5.6.1). Order of magnitude calculations show that for colloidal particles the total attractive energy is of the same order of magnitude as thermal energies when their mean size ($\approx \mathcal{V}^{1/3}$, where $\mathcal{V}$ is the volume) is of the order of magnitude of the particle separation, regardless of whether the particles are spherical, rod-like, or plate-shaped. At smaller separations the order of attractive energies is plates > rectangular rods > cylinders > spheres [Vold, 1957]. At separations such that $W \approx 10 k_B T$ the attraction between spheres varies nearly as R^{-1}, but for rods and cylinders it varies approximately as R^{-2} and for platelets as R^{-3}. Likewise, we have that, at small separations, a parallel orientation is

[2] The factor π^2 is conventionally included.

Table 5.18. *Geometrical factor $\mathscr{S} = \mathscr{S}(R_i, \Omega_A, \Omega_B)$ for (a) two spheres of radius a_1 and a_2 [Hamaker, 1937]; (b) two facing blocks [Russel et al., 1989]; (c) two parallel and (d) two crossed cylinders [Mahanty and Ninham, 1976]; (e) two hard biaxial ellipsoids [Everaers and Ejtehadi, 2003]. These should be multiplied by the Hamaker constant H_{12} to get the attraction energy*

Disposition	Expression	Ref.
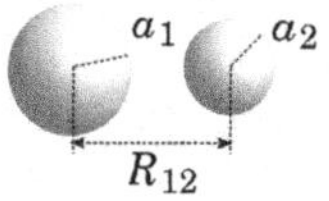	$-\frac{1}{6}\left(\frac{2a_1a_2}{R_{12}^2-(a_1+a_2)^2}+\frac{2a_1a_2}{R_{12}^2-(a_1-a_2)^2}+\ln\frac{R_{12}^2-(a_1+a_2)^2}{R_{12}^2-(a_1-a_2)^2}\right)$ $-\left(\frac{16}{9}\right)\frac{a_1^3a_2^3}{R_{12}^6}$, for $R_{12} \gg a_1+a_2$	(a)
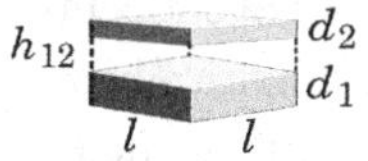	$-\frac{l^2}{12\pi}\left(\frac{1}{h_{12}^2}+\frac{1}{(h_{12}+d_1+d_2)^2}-\frac{1}{(h_{12}+d_1)^2}-\frac{1}{(h_{12}+d_2)^2}\right)$	(b)
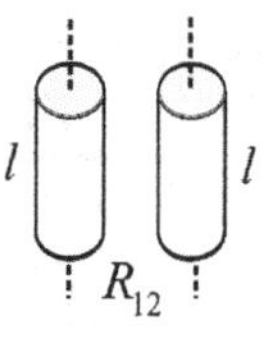	$-\frac{1}{24}\frac{l}{a}\left(\frac{a}{R_{12}-2a}\right)^{3/2}\left(1-\frac{R_{12}-2a}{a}+\frac{1}{\sqrt{2\pi}}\ln\frac{R_{12}-2a}{a}+\cdots\right)$, for $a \gg R_{12}-2a$ $-\frac{3\pi}{8}\frac{la^4}{R_{12}^5}\left(1+\frac{25}{4}\frac{a^2}{R_{12}^2}+31.9\frac{a^4}{R_{12}^4}+150.7\frac{a^6}{R_{12}^6}+\cdots\right)$, for $l \geqslant R_{12} \geqslant a$ $-\frac{a^4l^2}{R_{12}^6}\left(1-\frac{l^2}{2R_{12}^2}+\cdots\right)$, for $l \ll R_{12}$	(c)
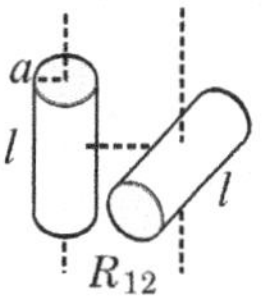	$-\frac{1}{6}\frac{a}{R_{12}-2a}\left(1-\frac{3}{2}\frac{R_{12}-2a}{a}+\cdots\right)$, for $a \gg R_{12}-2a$ $-\frac{\pi}{2}\left(\frac{a}{R_{12}}\right)^4\left(1+\frac{5a^2}{R_{12}^2}+21.87\frac{a^4}{R_{12}^4}+\cdots\right)$, for $a \geqslant R_{12}$	(d)
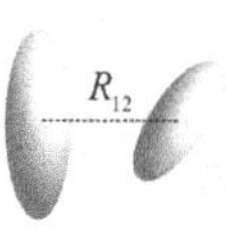	$U_A^{\text{RE-squared}}(\mathbf{M}_1,\mathbf{M}_2,\boldsymbol{R}_{12}) = -\frac{1}{36}\left(1+3\eta_{12}\chi_{12}\frac{\sigma_c}{h_{12}}\right)$ $\times\prod_{i=1}^{2}\prod_{e=x,y,z}\left(\frac{\sigma_e^{(i)}}{\sigma_e^{(i)}+h_{12}/2}\right)$	(e)

greatly favoured over a perpendicular one for rods while, for rectangular rods, orientation with the largest faces opposite each other is preferred. These differences diminish as the particle separation increases but remain important so long as the van der Waals' attraction itself is of the order of k_BT or larger.

When considering the interaction between the two particles, we have implicitly assumed until now that they are in vacuum. However, in normal applications, like the suspensions

Table 5.19. *A small selection of Hamaker constants [Parsegian, 2006]*

Material (M)	A_{MM}(zJ) in water	A(zJ) in vacuum
Water	–	55.1
Polystyrene	13	79
Hydrocarbon (tetradecane)	3.8	47
Polymethyl methacrylate	1.47	58.4
Diamond (IIa)	138	296
Mica (monoclinic)	13.4	98.6
Mica (muscovite)	2.9	69.6
Quartz (silicon dioxide)	1.6	66
Aluminium oxide	27.5	145
Rutile (titanium dioxide)	60	181

of viruses or nanoparticles giving rise to LCs we have a liquid and we have to consider the effect of an intervening medium. The theory of the effect is clearly outside the scope of this book and well treated elsewhere [Parsegian, 2006; Israelachvili, 2011].

At an approximate level we can account for the intervening medium with a modified Hamaker constant $A_{M_1M_3M_2}$ for two particles acting across the medium, replacing $A_{M_1M_2}$. On the basis of a theory due to Lifshitz, $A_{M_1M_3M_2}$ is always positive for two identical particles, hence their interaction always attractive. Approximate values of $A_{M_1M_2M_1}$ can be obtained as

$$A_{M_1M_2M_1} \approx A_{M_2M_1M_2} \approx \left(A_{M_1M_1}^{1/2} - A_{M_2M_2}^{1/2}\right)^2, \tag{5.133}$$

where $A_{M_1M_1}$ and $A_{M_2M_2}$ are the Hamaker constants of the individual media, respectively. In the case of surfaces with adsorbed layers, for instance particles coated with a polymer or with a stabilizing layer, Eq. 5.133 can be written in terms of the Hamaker constants of the individual media: core particle, layer and solvent medium (see Table 5.19). Other approximate relations are $A_{M_1M_3M_2} \approx \pm\left(A_{M_1M_3M_1}\, A_{M_2M_3M_2}\right)^{1/2}$ or alternatively $A_{M_1M_3M_2} \approx \left(\sqrt{A_{M_1M_1}} - \sqrt{A_{M_3M_3}}\right)\left(\sqrt{A_{M_2M_2}} - \sqrt{A_{M_3M_3}}\right)$. Note that for two dissimilar materials the sign can be negative, hence the apparent paradoxical result of vdW interaction leading to repulsion even if originated from always attractive dispersions forces. Israelachvili [2011] also reports a more theoretically sound approximate relation for the non-retarded Hamaker constant for two macroscopic phases 1 and 2 interacting across a medium 3:

$$\begin{aligned} A_{M_1M_3M_2} \approx\; & \frac{3}{4} k_B T \left(\frac{\varepsilon_1 - \varepsilon_3}{\varepsilon_1 + \varepsilon_3}\right)\left(\frac{\varepsilon_2 - \varepsilon_3}{\varepsilon_2 + \varepsilon_3}\right) \\ & + \frac{3h\nu_e}{8\sqrt{2}} \frac{\left(n_1^2 - n_3^2\right)\left(n_2^2 - n_3^2\right)}{\left(n_1^2 + n_3^2\right)^{1/2}\left(n_2^2 + n_3^2\right)^{1/2}\left[\left(n_1^2 + n_3^2\right)^{1/2} + \left(n_2^2 + n_3^2\right)^{1/2}\right]}, \end{aligned} \tag{5.134}$$

where ν_e is the main electronic absorption frequency in the UV and n is refractive index of the media in the visible (where $n^2 = \varepsilon_{vis}(\nu)$). We shall not discuss this any further, except for noting that matching of refractive indices can reduce or cancel the vdW effects.

5.9 Distributed Effective Charges

The various permanent or induced and dispersive potential terms described in the last few sections are important, when added to a generic pair potential like the purely repulsive or repulsive-attractive models introduced in this chapter to test the variety of phases that these simple molecular features can generate. We shall see some examples of this approach, particularly for dipoles and quadrupoles, in Chapter 11. The multipolar expansion, although elegant, has, however, the major disadvantage of not being convergent at short distances. Indeed, the single-centre multipolar expansion is not justified for densely packed anisometric molecules. It is easy to see that this is the case for long rods or large discs, where the distance between some point charges on two adjacent molecules can be easily shorter than the respective charge-centre distance on each molecule. This problem is particularly severe in the simulations of LC phases, where the details of the charge distribution can generate a variety of molecular organizations, e.g. different smectic phases with bilayer, interdigitated or striped molecular arrangements [Berardi et al., 1996a]. Some of these structures and of the even more complex phases based on polyphilic mesogens [Tschierske, 2001] would probably be impossible to obtain without considering a realistic distribution of the charges.

A reasonable alternative between the fully atomistic and the central multipole approaches, comes from suitably placing a reduced set of effective charges that produces an electrostatic field equivalent to the one coming from the full charge atomic distribution [Berardi et al., 2004d]. Since a generic model does not have the atoms by definition, it is necessary to identify how many charges to decorate the models with, where to locate them and what charges to assign in order to reasonable reproduce the subtleties of a specific charge distribution. To illustrate the method, we can consider, as an example, the case of two discotic molecules, hexa-thio triphenylene (HT-T) and the very similar hexa-thio-aza-triphenylene (HT-AT) [Orlandi et al., 2007] represented in Fig. 5.22. The hexa-alkyl substituted moieties show LC phases [Maeda et al., 2001; Kumar, 2004, 2006] while the corresponding HT-ATs, even if similar in shape, diameter, volume and absence of permanent dipole, do not form mesophases [Roussel et al., 2003], if not modified by the inclusion of hydrogen bonding side groups [Gearba et al., 2003; Palma et al., 2006]. In order to understand the cause of the different behaviour, we can focus on the central core, shown in Fig. 5.22, modelled with a GB uniaxial oblate ellipsoid with parameters [Orlandi et al., 2007]: $\mu = 1$, $\nu = 0$, diameter $\sigma_e = 19.25$ Å, face-to-face thickness $\sigma_f = 3.75$ Å, interaction energies $\varepsilon_e = 9\,\text{kcal/mol}$ and $\varepsilon_f = 60\,\text{kcal/mol}$. In order to endow the simple GB model with a charge distribution appropriate to the molecule of interest, the first step is to start from an atomistic level model and obtain the atomic charges by one of the available methods outlined in Section 5.4.3. For our example case of HT-T and HT-AT these were calculated at their equilibrium geometry by a direct fit of the molecular electrostatic potential (ESP) [Besler et al., 1990]. These partial charges could be used to

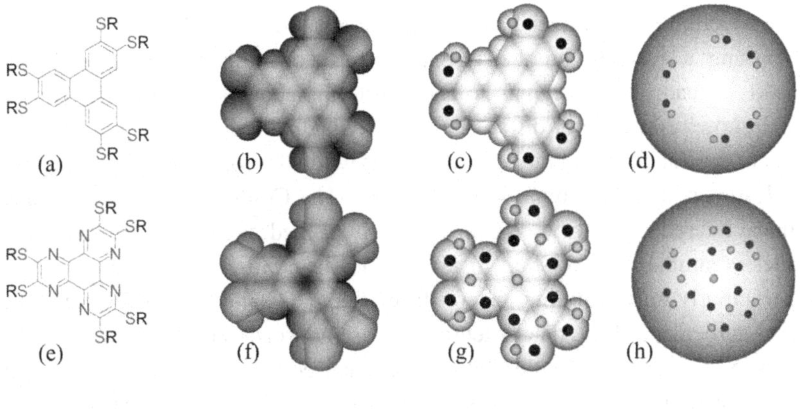

Figure 5.22 Chemical structure of HT-T (a) and HT-AT (e) with SR an n-alkylthio chain together with their atomistic space-filling model grayscaled according to their atomistic (b), (f) and effective (c), (g) charge distributions (negative charges in black, positive grey), On the right oblate GB models for HT-T and HT-AT decorated with their effective charges (d),(h) [Orlandi et al., 2007]. For octyl-HT-T (K $\overset{55^\circ\mathrm{C}}{\longleftrightarrow}$ $\mathrm{D}^{\mathrm{d}}_{\mathrm{h}}$ $\overset{87^\circ\mathrm{C}}{\longleftrightarrow}$ I) [Gramsbergen E. and B., 1986b]. For octyl-HT-AT (K_1 $\overset{49^\circ\mathrm{C}}{\longleftrightarrow}$ K_2 $\overset{93^\circ\mathrm{C}}{\longleftrightarrow}$ I) [Roussel et al., 2003].

directly decorate the GB discs previously parametrized. However, exploring the whole phase diagram with such a large number of charges is not very practical because of the computationally overwhelming number of electrostatic interacting pairs to be evaluated. We can then try to get a sufficiently good approximation to the ESP using a smaller number n_e of effective charges, determining their optimal positions and values by fitting their electrostatic potential to the atomistic one. The global minimization can be performed in different ways, but one convenient way is to employ a genetic algorithm (GA) [Goldberg, 1989] as in [Berardi et al., 2004d] to explore the space of charge positions, $\boldsymbol{g}^{(i)} = (\boldsymbol{r}_1^{(i)}, \ldots, \boldsymbol{r}_{n_e}^{(i)})$. The actual charge values can instead be determined with a least square fit to the QC reference ESP, U^{e}, calculated at M grid points $\boldsymbol{p}_j$ outside the molecular surface and stored as a M-dimensional vector $\boldsymbol{u}^{\mathrm{e}}$. For each individual attempt $\boldsymbol{g}^{(i)}$, the corresponding n_e dimensional vector of effective charges, $\boldsymbol{e}^{(i)}$, is computed as the least square solution to the overdetermined linear equation system $[\mathbf{D}^{(i)}]\,\boldsymbol{e}^{(i)} = \boldsymbol{v}^{e}$, which has on the left a $(M+1)\times n_e$ matrix $\mathbf{D}^{(i)} = \mathbf{D}^{(i)}\left(\boldsymbol{g}^{(i)}\right)$ containing in the first M rows the electrostatic potential generated by the trial charges and a last row with equal elements w (e.g. $w = M/50$) added to constrain the total trial charge equal within an error threshold to the real one q^{e}. Thus,

$$[\mathbf{D}^{(i)}]\boldsymbol{e}^{(i)} = \begin{bmatrix} \frac{1}{|\boldsymbol{r}_1^{(i)} - \boldsymbol{p}_1|} & \cdots & \frac{1}{|\boldsymbol{r}_{nQ}^{(i)} - \boldsymbol{p}_1|} \\ \cdots & \cdots & \cdots \\ \frac{1}{|\boldsymbol{r}_1^{(i)} - \boldsymbol{p}_M|} & \cdots & \frac{1}{|\boldsymbol{r}_{nQ}^{(i)} - \boldsymbol{p}_M|} \\ w & \cdots & w \end{bmatrix} \begin{pmatrix} e_1^i \\ \cdots \\ e_{n_e}^i \end{pmatrix} = \boldsymbol{v} = \begin{pmatrix} U_1^{\mathrm{e}} \\ \cdots \\ U_M^{\mathrm{e}} \\ wq^{\mathrm{e}} \end{pmatrix} \tag{5.135}$$

and, formally $e^{(i)} = \left([\mathbf{D}^{(i)}]^T[\mathbf{D}^{(i)}]\right)^{-1}[\mathbf{D}^{(i)}]^T \boldsymbol{v}$. Other rows with additional constraints (e.g. reproduction of the molecular dipole) can also be added. For the specific cases we are discussing a sufficient number of charges turns out to be $n_e = 12$ for HT-T and 22 for HT-AT instead of the original 36 and 30 atomic charges (see Fig. 5.22 (d) and (h)). Fig. 5.22 shows that a major difference of the two charge distributions is the presence of an effective central charge in the case of HT-AT, while the core region is essentially neutral for HT-T, thus making columnar stacking easier. The Coulombic potential obtained from the full and the reduced sets of charges, for fairly packed, liquid like configurations of two molecules is similar even at short distances and for configurations that can be expected to be particularly important, like face-to-face for the discotic HT-T [Orlandi et al., 2007]. The good agreement of the pair potential at close approach and for a variety of approach directions is important for the simulation of condensed phases, where molecules are likely to be in close contact. The electrostatic energy for a model HT-T system of N=1000 GB decorated with 12 charges, deviates from that of the full set of 36 charges for less than 5% both in the isotropic and in the columnar phases. The method has proved to be a useful tool in modelling complex mesogens, e.g. for the coarse-grain simulation of the phases of banana molecules [Francescangeli et al., 2009] or of triphenylenes [Lamarra et al., 2012] and even for complex organic molecules used in organic electronics [Ricci et al., 2019]. While the potential described are of quite general use, they have not included two wide classes of LCs described in Chapter 1: chiral phases and lyotropics, and we shall now briefly deal with these.

5.10 Chiral Interactions

The introduction of a chiral term in the pair potential is essential to describe the various chiral phases, in particular cholesterics. Chiral discriminating terms are normally quite small in absolute terms, but they are not averaged out by molecular reorientations as is the case, e.g. for permanent dipole-dipole interactions, and they can have a large effect on phase organization. An example is the change from nematic to a long-pitch cholesteric for the simple substitution of a hydrogen with a deuterium which makes a certain organic mesogen chiral [Coates and Gray, 1973]. Chiral terms can arise from:

Hard particle chiral models. These should be particularly important to model suspensions of chiral viruses and other biological systems [Barry et al., 2006]. The simplest model is probably obtained by arranging a set of hard spheres on a helix [Frezza et al., 2013] and we shall see some examples of the phases obtained in Chapter 11. Another possibility is that of having twisted hard particles, e.g. hard chiral polyhedra [Belli et al., 2014; Dussi et al., 2015].

Gay–Berne with a twist. The GB potential provides an effective model for non-chiral mesogens with attractive as well as soft repulsive interactions, so it seems worth modifying it by adding a chiral contribution in an attempt to simulate chiral phases. Cholesterics, in particular, are just chiral nematics with essentially the same local structure of normal nematics, but with a twisting director. In building an empirical model potential, we recall that the pair potential has to be rotationally invariant and comply with head-to-tail molecular symmetry, thus looking at the table of chiral invariant in Appendix G it is reasonable to add to the GB potential a term that has an angular form like the pseudoscalar in Eq. G.64, i.e. $S_{2,2,1}(\boldsymbol{u}_i,\boldsymbol{u}_j,\hat{\boldsymbol{r}}_{ij}) = -\sqrt{3/10}[\boldsymbol{u}_i \cdot \boldsymbol{u}_j \times \hat{\boldsymbol{r}}_{ij}](\boldsymbol{u}_i \cdot \boldsymbol{u}_j)$, where $\boldsymbol{u}_i$, $\boldsymbol{u}_j$, $\hat{\boldsymbol{r}}_{ij}$ are unit vectors along the two molecular axes and the centre-centre vector, respectively. However, a distance dependence has still to be (arbitrarily) chosen, e.g. by analogy with pseudoscalar interactions of dispersive nature. Going back to the second-order dispersion energy expression, Eq. 5.100, and retaining the first two terms in the electrostatic energy operator in Eq. 5.114a, i.e. $U_{ij}^{\mu\mu} + \overline{U}_{ij}^{\mu\Theta}$, perturbation theory will now give a sum of three types of terms. The distance dependence comes from the multipolar tensors $\mathbf{T}^{(n)}$ and the first term, proportional to the square of the matrix elements of $U_{ij}^{\mu\mu}$, will contain $\mathbf{T}^{(2)}\mathbf{T}^{(2)}$ and have, as we have already seen, a r_{ij}^{-6} distance dependence. The second, mixed, term $U_{ij}^{\mu\mu}\overline{U}_{ij}^{\mu\Theta}$, will contain $\mathbf{T}^{(2)}\mathbf{T}^{(3)}$ and then go as r_{ij}^{-7} and the last term $\overline{U}_{ij}^{\mu\Theta}\overline{U}_{ij}^{\mu\Theta}$ containing $\mathbf{T}^{(3)}\mathbf{T}^{(3)}$ will depend on r_{ij}^{-8}. Thus, the mixed chiral pseudoscalar term depends on r_{ij}^{-7} [Goossens, 1971; Vandermeer and Vertogen, 1979]. Consistently with this, Memmer et al. [1993] assumed the empirical twist term as

$$\begin{aligned} U_{ij}^{\chi} &= U^{\chi}(\boldsymbol{r}_{ij},\boldsymbol{u}_i,\boldsymbol{u}_j) \\ &= 4\epsilon\left(\boldsymbol{u}_i,\boldsymbol{u}_j,\hat{\boldsymbol{r}}_{ij}\right)\left(\frac{\sigma_0}{r_{ij}-\sigma\left(\boldsymbol{u}_i,\boldsymbol{u}_j,\hat{\boldsymbol{r}}_{ij}\right)+\sigma_0}\right)^7 \left[\left(\boldsymbol{u}_i \times \boldsymbol{u}_j\right)\cdot\hat{\boldsymbol{r}}_{ij}\right]\left(\boldsymbol{u}_i\cdot\boldsymbol{u}_j\right), \end{aligned} \tag{5.136}$$

where all quantities are the same as in the standard GB potential in Eq. 5.57. This chiral potential has been used to study bulk systems [Memmer et al., 1993, 1996; Memmer, 1998] and the induction of chirality in a nematic from a chiral surface in a non-chiral nematic film [Berardi et al., 1998c]. It is worth mentioning that chiral attractive-repulsive models can also be built combining two or more non-chiral GB particles so as to form a chiral object. An example is that of chiral discotics formed by two different biaxial GB particles, used to study chiral nematic [Memmer et al., 1996] or columnar [Berardi et al., 2003a] phases.

5.11 Hydrogen Bonds

The 'hydrogen bond' or just 'H-bond' [Pimentel and McClellan, 1960; Gilli and Gilli, 2009] is more difficult to define precisely than other chemical bonds. Citing from the classical definition [Pauling, 1960]: '... under certain conditions an atom of hydrogen is attracted by rather strong forces to two atoms, instead of only one, so that it may be considered to be acting as a bond between them ...'. This implies that there are three atoms forming the hydrogen bond: the hydrogen donor $\mathcal{D}$, the acceptor $\mathcal{A}$ and the hydrogen atom between these two atoms. According to Pauling [1960], 'the hydrogen bond is largely ionic in character, and is formed only between the most electronegative atoms.' A more modern and general definition from IUPAC [Arunan et al., 2011] is that 'The hydrogen bond is an attractive interaction between a hydrogen atom from a molecule or a molecular fragment X – H in which X is more electronegative than H, and an atom or a group of atoms in the same or a different molecule, in which there is evidence of bond formation'. More practically, hydrogen bonds are specific, short-range, and directional partially bonded interactions. They occur between hydrogen atoms, bound covalently to an electronegative atom (usually F, N, or O), and thus acquiring a partial positive charge, and an additional electronegative atom that can be the same as X or different. Distances of 2.5–3.2 Å between hydrogen-bond donor X and Y and X–H... Y angles of 130–180° are typically found, but the strength is higher when the three atoms involved are aligned with the hydrogen donor pointing directly at the acceptor electron pair. As a result of its at least partial electrostatic nature, the strength of a hydrogen bond depends also on its microscopic environment and on the local dielectric constant ε of the surrounding medium. The free energy for hydrogen bonding can vary between that of a physical van der Waals interaction $\approx 1-5$ kJ/mol to one approaching a chemical covalent bond ≈ -150 kJ/mol [Perrin and Nielson, 1997], but usually is in the range of ≈ -20 kJ/mol. Although their strength is weaker than ionic or covalent bonds, they provide in general the dominant contribution to specific molecular recognitions in biological systems. Most important is their help in forming and maintaining the double helical structure of DNA, through the hydrogen bonding between the base pairs linking one complementary strand to the other and enabling replication. Hydrogen bonds are also essential to maintaining the structural integrity of α-helix and β-sheet conformations of peptides and proteins and in folding them into the specific shape appropriate to their biochemical functions [Wermuth et al., 2015]. Turning to LCs proper, a key molecular ingredient for the formation of mesophases is the presence of a sufficiently pronounced anisotropy of shape or of some other suitable interaction. Even if this rules out very many candidate molecules, some of these would-be mesogens endowed with appropriate chemical structure could attach one to the other and dimerize through terminal hydrogen bonds, yielding newly formed dimers that are sufficiently elongated or 'squashed' to be able to form a liquid crystal phase. The existence of such pre-mesogenic species that can generate nematic, smectic or columnar liquid crystalline phases via H-bond-induced self-assembly is widely reported in the literature [Gray, 1962; Kato, 1998]. In particular, a classic case is that of carboxylic acids like benzoic acids that can dimerize maintaining the rod-like shape while increasing the length of the resulting aggregate (see Fig. 5.23). In a model proposed by Berardi et al. [1993] having the benzoic acids in mind, an anisotropic H-bond pair potential is added to the usual GB interaction (Section 5.6.3), representing the molecule as an ellipsoid with axes $\sigma_{\parallel}$ and $\sigma_{\perp}$. The interaction of a donor-acceptor (DA) site in molecule i with an acceptor-donor site in molecule j leading to the formation of a double H-bond is

$$U_{ij}^{XX'} = U_{ij}^{DA} + U_{ji}^{DA}, \tag{5.137}$$

where

$$U_{ij}^{DA} = 15\epsilon_0\, \epsilon_{DA}\left(\hat{\boldsymbol{r}}_{DA}, \hat{\boldsymbol{u}}_D, \hat{\boldsymbol{u}}_A\right)\left\{\left(\frac{\sigma_{DA}}{r_{DA}}\right)^{12} - \left(\frac{\sigma_{DA}}{r_{DA}}\right)^{10}\right\} \tag{5.138}$$

with

$$\epsilon_{DA}\left(\hat{\boldsymbol{r}}_{DA}, \hat{\boldsymbol{u}}_D, \hat{\boldsymbol{u}}_A\right) = \frac{\epsilon_{DA}^{+}}{\epsilon_0}\left\{\left(\frac{1 + \hat{\boldsymbol{u}}_D \cdot \hat{\boldsymbol{r}}_{DA}}{2}\right)\left(\frac{1 - \hat{\boldsymbol{u}}_A \cdot \hat{\boldsymbol{r}}_{DA}}{2}\right)\right\}^{\xi}, \tag{5.139}$$

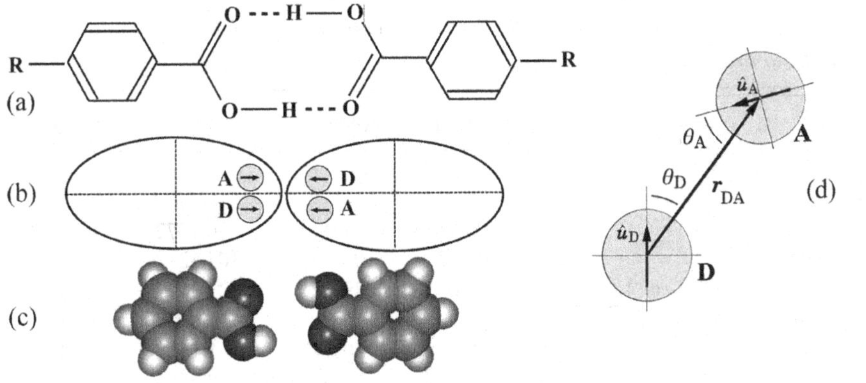

Figure 5.23 (a) Dimerization of two benzoic acids (R is a substituent, e.g. H or OC_nH_{2n+1}) shown with chemical structures, (b) Gay–Berne models decorated with H-bond donor (**D**) and acceptor (**A**) sites (b) and (c) space filling atomistic models with R = H. (d) Schematic of interaction of donor and acceptor with orientation $\hat{\boldsymbol{u}}_A$, $\hat{\boldsymbol{u}}_D$ and site-site separation $\boldsymbol{r}_{DA} \equiv \boldsymbol{r}_D - \boldsymbol{r}_A = r_{DA}\,\hat{\boldsymbol{r}}_{DA}$ [Berardi et al., 1993].

where ϵ^+_{DA} is the well depth for the DA interaction, while the exponent ξ gives an angular sensitivity which determines the amplitude of the DA well. The configuration that maximizes the interaction is that with sites D and A aligned parallel and antiparallel to the inter-site vector (see Fig. 5.23d). We assume the donor and acceptor sites to be placed at a certain fixed distance r_{DA} between them. For the benzoic acids we take D and A to be so close that $r_{DA} = 0$ and that they are positioned in near terminal position along the monomer axis at a distance d from the centre. This potential has been used to simulate nematics [Berardi et al., 1999] showing that in a system of short ($\sigma_\parallel/\sigma_\perp = 2$) dimerizing rods, the nematic-isotropic transition temperature is increased by as much as 25% by the preliminary formation of a significant fraction of elongated dimers in the isotropic phase.

5.11.1 Water Pair Potential

Water is of course one of the foremost cases of a system where the H-bond is important [Gallo et al., 2016]. A water potential is essential for simulation of most biological systems and of lyotropic LCs at atomistic resolution and water has been investigated with computer simulations for a few decades, since the pioneering works of Barker and Watts [1969] and Rahman and Stillinger [1971], while it still remains an extremely active area of research [Gallo et al., 2016]. Some general facts for water are:

(i) Every water molecule shares on average three to four hydrogen bonds with neighbouring water molecules.
(ii) The strength of a H-bond in the liquid phase is $\approx$ 12kJ/mol weaker than a covalent bond but stronger than dispersion forces

Table 5.20. *Parameters for common water pair potentials (see text and Fig. 5.24). References: (a) [Jorgensen et al., 1983], (b) [Berendsen et al., 1987], (c) [Abascal and Vega, 2005]*

Model	SPC	TIP3P	SPC/E	TIP4P/2005	TIP4P
Ref.	(a)	(a)	(b)	(c)	(a)
r_{OH}, Å	1.0	0.9572	1.0	0.9572	0.9572
θ_{HOH}, deg	109.47	104.52	109.47	104.52	104.52
$A_{OO} \times 10^{-3}$, kcal Å^{12}/mol	629.4	582.0	629.4	731.3	600.0
C_{OO}, kcal Å^6/mol	625.5	595.0	625.5	736.0	610.0
e_O	−0.82	−0.834	−0.8476	−1.1128	0.0
e_H	0.41	0.417	0.4238	0.5564	0.52
e_M	0.0	0.0	−0.98	−1.07	−1.04
r_{OM}, Å	0.0	0.0	0.15	0.15	0.15

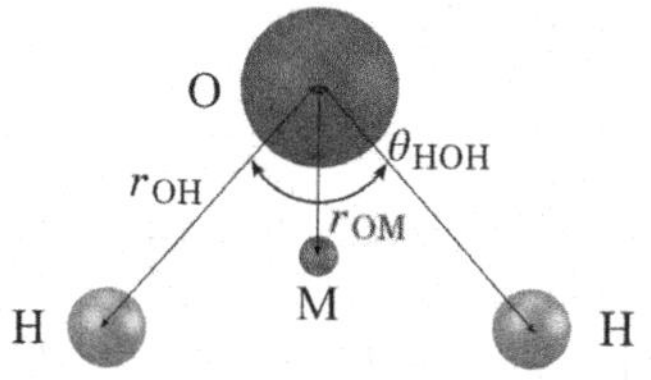

Figure 5.24 The water TIP4P model potential (see text). M is the location of the negative charge, that in the SPC and TIP3P models resides on the oxygen.

(ii) The hydrogen bond network is a space-filling random network with a high local tetrahedral structure, but a large number of defects to the fifth neighbour, that is interacting weakly, favouring the mobility of water molecules.

The most popular simple rigid geometry models for water: the simple point charge (SPC) [Berendsen et al., 1981, 1987] and the 'transferable intermolecular potential with 3 points' (TIP3P) consists of a non-polarizable set of one LJ and three charges: two positive ones, e_H on the hydrogens and a negative one, $e_O = -2e_H$, on the oxygen [Jorgensen and Jenson, 1998] in which the total interaction energy is:

$$U_{ij}^{ww} = \frac{A_{OO}}{(r_{OO})^{12}} - \frac{C_{OO}}{(r_{OO})^6} + \frac{1}{4\pi\varepsilon_0}\sum_{\substack{a=1\\a\in i}}^{3}\sum_{\substack{b=1\\b\in j}}^{3}\frac{e_a e_b e^2}{r_{ab}}, \qquad (5.140)$$

where the parameters A_{OO}, C_{OO} are determined by fitting to bulk properties (TIP3P) or to MD data (SPC). The extended SPC (SPC/E) model adds an average polarization correction to the potential energy function of around 1.25 kcal/mol (5.22 kJ/mol). The SPC/E model results in a better density and diffusion constant than the SPC model.

In the TIP4P model, apart from changes in the parameters, the site of negative charge is moved away from the oxygen to a point M located 0.15 Å along the bisector of the HOH angle (see Fig. 5.24). The modified version TIP4P/2005 [Abascal and Vega, 2005] is parametrized to provide a qualitatively correct description of the phase diagram of water and a good description of the vapour-liquid equilibria, surface tension, global phase diagram, ice properties, maximum in its density, structure, equation of state at high pressures, diffusion coefficient and viscosity. The parametrization of these 3- and 4-points models is given in Table 5.20.

6

Dynamics and Dynamical Properties

Flowing crystals! Is that not a contradiction in terms?

O. Lehmann, Über fliessende krystalle, 1889

6.1 Introduction

We have examined up to now time-independent, static, structural properties. Here we wish instead to discuss time, or equivalently frequency dependent observable properties, like the dielectric constant or the viscosity of an isotropic or anisotropic fluid. To do this we consider the dynamic evolution of a molecular system, starting with a formal classical mechanics approach [Berne and Pecora, 2000]. We then introduce the concept of *correlation functions*, key to the interpretation of many spectroscopic properties and describe their general properties. For liquid crystals, the distinctive feature is, as we have seen, the bias in the distribution of molecular orientations due to the existence of a preferred orientation (director) and of non-vanishing order parameters. This affects many properties dependent on molecular orientation, introducing macroscopically observable anisotropies, as discussed at least for some typical experimental techniques and single molecule properties in Chapter 3. Here we shall similarly introduce the description of single-molecule time-dependent properties, first in general terms and then with application to some experimental techniques.

6.2 Dynamic Evolution of Molecular Properties

To start with, we consider a system of N particles at equilibrium in a volume V at temperature T (canonical conditions). The equilibrium distribution of the system in phase space, $\varrho_0(\widetilde{\boldsymbol{X}}, \dot{\widetilde{\boldsymbol{X}}})$, was introduced in the Chapter 4 (see Eq. 4.1) and we employ the notation for variables introduced there (adding for clarity a subscript 0 here for the equilibrium distribution). In particular, for a classical system described by a Hamiltonian $\mathscr{H} \equiv \mathscr{H}(\widetilde{\boldsymbol{r}}, \widetilde{\boldsymbol{p}})$ the equations of motion for coordinates[1] $\boldsymbol{r}_i$ and momenta $\boldsymbol{p}_i$ are

$$\frac{\partial \mathscr{H}}{\partial \boldsymbol{p}_i} = \dot{\boldsymbol{r}}_i, \quad \frac{\partial \mathscr{H}}{\partial \boldsymbol{r}_i} = -\dot{\boldsymbol{p}}_i. \tag{6.1}$$

[1] *Note that more generally X can consist of positions but also of orientational variables.*

The time derivative of a property A at the state point $\widetilde{\mathcal{X}}(t)$, will be

$$\dot{A}(t) \equiv \frac{\mathrm{d}A(t)}{\mathrm{d}t} = \frac{\partial A}{\partial t} + \sum_i \left(\frac{\partial A}{\partial \boldsymbol{p}_i} \dot{\boldsymbol{p}}_i + \frac{\partial A}{\partial \boldsymbol{r}_i} \dot{\boldsymbol{r}}_i \right) = \frac{\partial A}{\partial t} + \mathcal{H}^{\times} A, \tag{6.2}$$

where the differential *evolution operator* $\mathcal{H}^{\times}$ [2], with dimensions t^{-1}, is defined [Landau and Lifshitz, 1993] as

$$\mathcal{H}^{\times} \equiv -\sum_i \left(\frac{\partial \mathcal{H}}{\partial \boldsymbol{r}_i} \frac{\partial}{\partial \boldsymbol{p}_i} - \frac{\partial \mathcal{H}}{\partial \boldsymbol{p}_i} \frac{\partial}{\partial \boldsymbol{r}_i} \right) = \sum_i \left(\boldsymbol{f}_i \cdot \frac{\partial}{\partial \boldsymbol{p}_i} + \frac{1}{m_i} \boldsymbol{p}_i \cdot \frac{\partial}{\partial \boldsymbol{r}_i} \right), \tag{6.3}$$

where $\boldsymbol{f}_i$ is the force acting on particle i. The evolution operator is conveniently written also as $\mathcal{H}^{\times} = \{f, \mathcal{H}\}$ using the classical Poisson brackets definition, that for any two dynamical variables f, g is:

$$\{f, g\} = \sum_{i=1}^{N} \left[\frac{\partial f}{\partial \boldsymbol{r}_i} \frac{\partial g}{\partial \boldsymbol{p}_i} - \frac{\partial f}{\partial \boldsymbol{p}_i} \frac{\partial g}{\partial \boldsymbol{r}_i} \right]. \tag{6.4}$$

Thus, we can also write $\frac{\mathrm{d}A}{\mathrm{d}t} = \{A, \mathcal{H}\} + \frac{\partial A}{\partial t}$. This also has the advantage that, by replacing the Poisson brackets with its corresponding term in quantum mechanics, the *commutator*: $[\hat{f}, \hat{g}] = \hat{f}\hat{g} - \hat{g}\hat{f}]$ we can write an evolution equation for an operator $\hat{A}$ as

$$\frac{\mathrm{d}\hat{A}}{\mathrm{d}t} = -\frac{i}{\hbar}[\hat{A}, \hat{\mathcal{H}}] + \frac{\partial \hat{A}}{\partial t} = -\frac{i}{\hbar} \hat{\mathcal{H}}^{\times} A + \frac{\partial \hat{A}}{\partial t}, \tag{6.5}$$

where $\hat{\mathcal{H}}^{\times}$ is called the evolution superoperator. We shall use it only much later, when discussing the evolution of spin levels in magnetic resonance (Section 10.8.2), so by now we just stick to our classical mechanics treatment. If the property does not directly depend on time, $\partial A/\partial t = 0$, and the evolution of $A(t)$ will be given by the *Heisenberg evolution equation* [Friedman, 1985]

$$A(t) = \mathrm{e}^{t\mathcal{H}^{\times}} A(0). \tag{6.6}$$

Note that this evolution is purely mechanical (temperature does not appear). In particular, the time evolution of the equilibrium probability distribution $\varrho(\widetilde{\mathcal{X}})$ will be

$$\frac{\mathrm{d}\varrho}{\mathrm{d}t} = \frac{\partial \varrho}{\partial t} + \mathcal{H}^{\times} \varrho. \tag{6.7}$$

However, since the total density must be conserved in time, $\dot{\varrho}(\widetilde{\mathcal{X}}) = 0$, and evolution takes place through the *Liouville equation*

$$\frac{\partial \varrho}{\partial t} = -\mathcal{H}^{\times} \varrho. \tag{6.8}$$

In the absence of external interventions, the Hamiltonian, that we can call the unperturbed Hamiltonian, $\mathcal{H}_0$, does not modify the equilibrium distribution ϱ_0

$$\mathcal{H}_0^{\times} \varrho_0 = \mathcal{H}_0^{\times} \frac{1}{\mathcal{Z}_0} \mathrm{e}^{-\mathcal{H}_0/k_B T} = -\frac{1}{k_B T \mathcal{Z}_0} \mathrm{e}^{-\mathcal{H}_0/k_B T} \mathcal{H}_0^{\times} \mathcal{H}_0 = 0, \tag{6.9}$$

[2] The *Liouville operator* $\mathcal{L} \equiv -i\mathcal{H}^{\times}$ *is also often used to write dynamic evolution equations.*

with $\mathcal{Z}_0$ the canonical phase integral, Eq. 4.2. In other words, ϱ_0 is an eigenfunction of the operator $\mathcal{H}_0^\times$ corresponding to a 0 eigenvalue. At the same time, for such a stationary system in the absence of external time-dependent fields, $\mathcal{H}_0^\times \langle A \rangle = 0$, i.e. an average equilibrium observable is constant in time and, in turn, it does not provide information on the microscopic dynamics. A more useful way to study time-dependent effects is to consider the correlation between the value of the property A at an arbitrary initial time t_0 and at a successive time $t_0 + t$, as given by the equilibrium *autocorrelation function* $C_{AA}(t)$

$$C_{AA}(t) = \langle A(t_0) A^*(t_0 + t) \rangle. \tag{6.10}$$

More generally, we can consider the *cross-correlation* between two properties A and B, $C_{AB}(t)$, that can be written as

$$C_{AB}(t) = \langle A(t_0) B^*(t_0 + t) \rangle = \langle A(0) B^*(t) \rangle = \langle A(0)\, \mathrm{e}^{t\mathcal{H}_0^\times} B^*(0) \rangle, \tag{6.11a}$$

$$= \int \mathrm{d}\widetilde{X} A(\widetilde{X})\, \mathrm{e}^{t\mathcal{H}_0^\times} B^*(\widetilde{X})\, \varrho_0(\widetilde{X}). \tag{6.11b}$$

A time correlation function is thus an average of the product of a property A, at a certain time t_0 with the complex conjugate of a property B (possibly the same as A) taken at time $t_0 + t$. If our system is at equilibrium, its properties will be independent on the initial time for observations, i.e. there should be no dependence on t_0 and the system is *stationary*. The evolution operator is a combination of derivatives and thus it follows the standard rules of derivatives. For instance, $\mathcal{H}^\times(AB) = (\mathcal{H}^\times A)B + A(\mathcal{H}^\times B)$. In particular, since the average $\langle AB \rangle$ is constant in time, $\mathcal{H}_0^\times \langle AB \rangle = 0$, and given two properties A, B we have the antisymmetric relation,

$$\langle A \mathcal{H}_0^\times B \rangle = \int \mathrm{d}\widetilde{X} \varrho_0 A \mathcal{H}_0{}^\times B = -\int \mathrm{d}\widetilde{X} \varrho_0 \mathcal{H}_0^\times A B = -\langle (\mathcal{H}_0^\times A)\, B \rangle, \tag{6.12}$$

and, more generally,

$$\langle A\, \mathrm{e}^{t\mathcal{H}_0^\times} B \rangle = \langle (\mathrm{e}^{-t\mathcal{H}_0^\times} A) B \rangle. \tag{6.13}$$

In practice, the dependence of properties on microscopic motion is most often observed determining their dependence on the frequency of a certain measuring field. Instead of correlation functions, experimental results are in this case related to their Fourier–Laplace transforms, the so-called *spectral densities* at a certain frequency ω :

$$j_{AB}(\omega) = \int_0^\infty \mathrm{d}t\, \mathrm{e}^{i\omega t} C_{AB}(t) = \left\langle A(0) \frac{1}{\mathcal{H}^\times - i\omega} B^*(0) \right\rangle. \tag{6.14}$$

6.3 Single Particle Dynamics

In most experimental techniques for studying condensed phases dynamics the focus is on following the positions and orientations of the molecules in time (or correspondingly looking at the frequency dependence of dynamic observables). The single particle dynamics in liquids or liquid crystals cannot be described by the one particle distribution $P(X)$ that we have discussed in Chapter 3 just introducing time as an extra variable. Indeed, if we are at equilibrium and we have a stationary system, $P(X)$ is a purely static quantity that can be calculated just from configurational Boltzmann averages, ignoring the dynamics altogether. A more suitable quantity describing dynamics is instead the *joint distribution* $P(X_1 t_1; X_2 t_2)$ giving the probability that a molecule has position-orientation X_1 at time t_1 and X_2 at time t_2 (see Appendix K). We shall not consider, for the time being, linear and angular velocities of the particles although these will be discussed later in this chapter.

At equilibrium the origin of time is not relevant and the joint distribution can only depend on the time difference, $t = t_2 - t_1$. Thus, we simply consider the joint distribution as $P(X_0; X, t)$. This is a real, non-negative quantity normalized as

$$\int \mathrm{d}X_0 \mathrm{d}X P(X_0; X, t) = 1. \tag{6.15}$$

It is clear that the probability of finding the molecule at X_0 at time 0 and at X at time t will become for a very long time t, a product of two independent quantities: the probability of finding the molecule at X_0 and that of finding the molecule at X. Thus,

$$\lim_{t \to \infty} P(X_0; X, t) = P(X_0) P(X), \tag{6.16}$$

where $P(X_0)$, $P(X)$ are the single-particle distributions studied in previous chapters. At time 0, the molecule has not yet moved and the angles X_0, X will be the same: $P(X_0; X, 0) = P(X_0)\delta(X - X_0)$. It is also clear that the probability of finding the molecule at X will be the sum over all the possible X_0 of the joint distribution, i.e

$$\int \mathrm{d}X_0 P(X_0; X, t) = P(X), \tag{6.17}$$

and of course $\int \mathrm{d}X P(X_0; X, t) = P(X_0)$. In principle, it should be possible to compute $P(X_0; X, t)$ as a multidimensional histogram. In practice, this is not quite feasible, as it is easy to see from a quick consideration of the amount of storage required.

6.4 Orientational Correlation Functions

In general, the properties of interest can involve positions, orientations, linear and angular velocities, etc. [Berne, 1971]. Here, however, we shall mainly concentrate on properties that depend on orientations, since these are possibly the most characteristic ones of LC systems and, as we shall see, they are the ones needed to interpret a number of experimental observables. The correlation function between two single-particle properties $A(\Omega)$, $B(\Omega)$, varying with the molecular orientation $\Omega \equiv (\alpha, \beta, \gamma)$, can be written as

$$C_{AB}(t) = \langle A(0) B(t)^* \rangle = \int \mathrm{d}\Omega_0 A(\Omega_0) \int \mathrm{d}\Omega P(\Omega_0; \Omega, t)\, B(\Omega)^*, \tag{6.18}$$

where $P(\Omega_0; \Omega, t)$ is the joint probability distribution for orientations. In writing Eq. 6.18 we have already used the fact that at equilibrium any experiment should not depend on the time origin. The properties A, B are generally tensors and the relation to the Wigner matrices correlation functions is made more transparent when we consider their spherical components (see Appendix B). Thus (see Eq. B.18),

$$\left\langle A_{\mathrm{LAB}}^{L,m}(0) B_{\mathrm{LAB}}^{L',m'*}(t) \right\rangle = \sum_{n,n'} \phi_{mn;m'n'}^{LL'*}(t)\, A_{\mathrm{MOL}}^{L,n}\, B_{\mathrm{MOL}}^{L',n'*}, \tag{6.19}$$

where we have introduced the Wigner rotation matrix orientational correlation functions

$$\phi_{mn;m'n'}^{LL'}(t) = \left\langle \mathscr{D}_{m,n}^{L}(\Omega_0) \mathscr{D}_{m',n'}^{L'*}(\Omega_t) \right\rangle. \tag{6.20}$$

A most important case is that of autocorrelation functions, where $A = B$. Typical observables behave under rotation as tensor properties of first and second rank L, with spherical components $A^{L,m}$ (cf. Appendix B). We have in mind Infrared Absorption [Gordon, 1968; Dozov et al., 1984] and Dielectric Relaxation [Nordio et al., 1973; Williams, 1994], where rank 1 properties are studied and Raman [Gordon, 1968; Jen et al., 1973; Southern and Gleeson, 2007; Sanchez-Castillo et al., 2010] and NMR [Emsley, 1985; Dong, 1997] or Electron Spin Resonance (ESR) [Nordio, 1976; Freed et al., 1994] where second-rank properties are typically studied. Another useful technique is Fluorescence Depolarization (FD) [Zannoni, 1979d; Zannoni et al., 1983], where the cross-correlation between the second-rank absorption and emission tensors of a chromophore molecule are investigated. In all these experiments and in all the others that probe properties of well-defined tensorial rank, the information in the correlation functions of the corresponding rank is sufficient to interpret or predict the outcome of the experiment. Since, as we have mentioned, first- and second-rank properties are particularly prominent, it follows that first- and second-rank correlation functions provide a great deal of the information we are likely to need, even though they represent only a very little part of the information contained in $P(\Omega_0; \Omega t)$.

6.5 Orientational Joint Distributions

In Chapter 3 we saw the advantages of expanding the single-particle distribution in an orthogonal basis set. A similar expansion can be performed for times $t \neq 0$ in a product basis set of Wigner rotation functions at time 0 and time t. Considering only rigid particles and orientational variables the joint distribution can be written as

$$P(\Omega_0; \Omega, t) = \sum P_{mn,m'n'}^{LL'}(t)\, \mathscr{D}_{m,n}^{L*}(\Omega_0)\, \mathscr{D}_{m',n'}^{L'}(\Omega), \tag{6.21}$$

where the expansion coefficients can be identified at once, using the orthogonality of the basis set and the normalization condition Eq. 6.15, as the averages

$$P_{mn,m'n'}^{LL'}(t) = \frac{(2L+1)(2L'+1)}{64\pi^4} \left\langle \mathscr{D}_{m,n}^{L}(\Omega_0) \mathscr{D}_{m',n'}^{L'*}(\Omega) \right\rangle. \tag{6.22}$$

Thus, the expansion coefficients are essentially the Wigner rotation matrix reorientational correlation functions $\phi_{mn;m'n'}^{LL'}(t)$. As we see, the orientational correlation functions play for dynamics the same role that order parameters play for the statics. They represent a systematic way of approaching the information contained in the full distribution. Moreover, orientational correlation functions corresponding to certain ranks L, L' represent all the information we need to calculate correlation functions and thus indirectly experimental observables corresponding to those ranks and it is thus very important to extract them from computer simulations generating time dependent configurations (Chapter 9). The number of correlation functions up to a given rank is, however, very high, just judging from all the possible combinations of subscripts m, n, m', n' we can make. So, even if all the correlation functions can be calculated it is essential to use all the general arguments we can produce to identify the independent ones. We list the main ones here.

Reality. The reality of $P(\Omega_0; \Omega, t)$ and the general complex conjugation property of the Wigner rotation matrices (Eq. F.6) give

$$\phi^{LL'*}_{mn;m'n'}(t) = (-1)^{m-n+m'-n'} \phi^{LL'}_{-m-n;-m'-n'}(t). \tag{6.23}$$

Thus, correlation functions of the type $\phi^{LL'*}_{00;00}(t)$ should be real whatever the symmetry of the molecule and of the mesophase.

Time reversal symmetry. Under the operation of time reversal, the coordinates and the angular momentum are invariant, while the linear momentum changes sign: $t \to -t$; $\boldsymbol{r} \to \boldsymbol{r}$; $\boldsymbol{p} \to -\boldsymbol{p}$; $\boldsymbol{J} \to \boldsymbol{J}$. The Wigner matrices $\mathscr{D}^L_{mn}$ are invariant being only a function of coordinates, while their correlation functions follow the usual [Berne and Pecora, 2000] symmetry rule for a system following classical mechanics, i.e. that the real part of the correlation functions is invariant under time reversal: $\mathrm{Re}\phi^{LL'}_{mn;m'n'}(t) = \mathrm{Re}\phi^{LL'}_{mn;m'n'}(-t)$, while the imaginary part changes sign: $\mathrm{Im}\,\phi^{LL'}_{mn;m'n'}(t) = -\mathrm{Im}\,\phi^{LL'}_{mn;m'n'}(-t)$.

6.5.1 Mesophase Symmetry

The symmetry operations of the mesophase and of the molecule will limit the number of independent correlation functions, as we can demonstrate, using standard Group Theory methods,

(cf. Appendix G) in various essentially equivalent ways [Blum and Torruella, 1972; Lax, 1974; Versmold, 1977; Zannoni, 1979c; Briels, 1980; Lynden-Bell, 1980; Steele, 1980, 1983; Pick and Yvinec, 1983]. We shall only give a few examples here.

Isotropic phases. The requirement of invariance for an arbitrary rotation of the laboratory frame implies that only relative orientation correlation functions should be present. Thus only correlation functions: $\langle \mathscr{D}^L_{n,n'}(\Omega_{t0}) \rangle = \sum_{q=-L}^{L} \phi^{LL}_{qn;qn'}(t)$ can be different from 0. Notice in particular that in isotropic fluids we have no coupling between properties of different rank.

Uniaxial mesophases. In a uniaxial phase an arbitrary rotation about the director, taken as the z laboratory axis, should leave the system unchanged. Thus,

$$\phi^{LL'}_{mn;m'n'}(t) = \langle \mathscr{D}^L_{m,n}(\Omega_0) \mathscr{D}^{L'*}_{m',n'}(\Omega_t) \rangle \,\delta_{m,m'} \equiv \phi^{LL'}_{mn;n'}(t)\,\delta_{m,m'}, \tag{6.24}$$

where we have removed an unnecessary subscript. Notice that in LCs the number of independent correlation functions is much higher than in an isotropic fluid. Moreover, certain couplings between properties of different rank can be admissible, as we shall briefly see in Section 6.8.

6.5.2 Particle Symmetry

Here we consider transformations of the molecule system that leave the joint distribution unchanged. Strictly, one should consider the point symmetry of the molecule. However, this could be quite misleading, since most molecules of interest in spectroscopy of LC phases would have essentially no symmetry, while nematogen molecules can be quite successfully treated as effectively uniaxial or biaxial. This means that their reorientation can often be considered as that of an effective molecule of higher symmetry. This, however, leaves the possibility of lower symmetry observables being studied. For example, nematogen molecules are normally treated as uniaxial $D_{\infty h}$ molecules and yet they are often dipolar. The rationale behind this seemingly contradictory situation is that both the orientational and reorientational behaviour are dominated by interactions that are normally different from those actually used to probe the system and thus observed. For instance, most nematogens possess dipoles and yet the nematic phases that they form are not ferroelectric. Thus, a dipole can be carried around by a molecule without influencing its orientational properties in a determinant and perhaps even in a significant way. If a molecule with no symmetry interacts with its neighbours only through an effective second-rank tensor, then its molecular dynamics trajectories will look as those of molecules of true ellipsoidal shape. In the limiting case that the reorientation of a molecule is diffusional (see Section 6.14.1), then its lowest effective symmetry can be that of its diffusion tensor, i.e. biaxial. Another example of the need to carefully consider what symmetry we should invoke for a molecule is that of submolecular properties. A quadrupolar interaction of a C-D bond, say, can be observed by DNMR [Vold, 1985; Dong, 1997], as seen in Section 3.10.6, but is not determining the reorientation of the molecule it belongs to in any significant way. Thus, it would be wrong to eliminate it on the basis that it is not compatible with uniaxial symmetry. The non-realization of the difference between the effective symmetry group and the point group of the molecule including its observables has led to some erroneous or misleading statements in the literature. The discussion of symmetry selection rules is also often confused in the literature. For instance, it is claimed [Fisz, 1987] that correlation functions $\phi^{12}_{11;11}(t)$ ought to be 0 by symmetry in uniaxial systems. We shall see in Section 6.8 that this is not the case. Here we do not intend to tackle the topic of a systematic symmetry classification of orientational correlation functions. Rather we shall just list as examples the symmetry restrictions that result for the practically important cases of uniaxial and biaxial particles reorienting in a uniaxial mesophase.

Uniaxial particles. If the particles are uniaxial, an arbitrary rotation around the molecular frame should leave everything unchanged. Thus, eliminating an unnecessary subscript on the right-hand side, $\phi^{LL'}_{mn;n'}(t) = \phi^{LL'}_{mn}(t)\,\delta_{n,n'}$.

Biaxial particles. The following selection rules hold for D_{2h} symmetry [Dozov et al., 1987]:

$$\phi^{LL'}_{mn;n'}(t) = (-1)^{L+L'}\phi^{LL'}_{m-n;-n'}(t) = (-1)^{L+L'-n-n'}\phi^{LL'}_{-mn;n'}(t) = (-1)^{n+n'}\phi^{LL'}_{mn;n'}(t). \tag{6.25}$$

Table 6.1. *Initial ($t = 0$) values for the reorientational correlation functions of first rank $\phi^{11}_{mn;n'}(0) = \langle \mathscr{D}^1_{m,n}\mathscr{D}^{1*}_{m,n'}\rangle$ for biaxial molecules reorienting in uniaxial phases (left column). In the right column the $\phi^{11}_{mn}(0) = \phi^{11}_{mn;n'}(0)$ symmetry allowed for uniaxial molecules ($n = n'$)*

Biaxial molecule	Initial value	Uniaxial molecule
$\phi^{11}_{10;0}(0) = \phi^{11}_{01;1}(0) = \phi^{11}_{0-1;-1}(0)$	$-\frac{1}{3}\langle P_2\rangle + \frac{1}{3}$	$\phi^{11}_{10}(0) = \phi^{11}_{01}(0) = \phi^{11}_{0-1}(0)$
$\phi^{11}_{01;-1}(0) = \phi^{11}_{0-1;1}(0)$	$-\sqrt{\frac{2}{3}}\text{Re}\langle \mathscr{D}^2_{0,2}\rangle$	–
$\phi^{11}_{00;0}(0)$	$\frac{2}{3}\langle P_2\rangle + \frac{1}{3}$	$\phi^{11}_{00}(0)$
$\phi^{11}_{11;1}(0) = \phi^{11}_{1-1;-1}(0)$	$\frac{1}{6}\langle P_2\rangle + \frac{1}{3}$	$\phi^{11}_{11}(0) = \phi^{11}_{1-1}(0)$
$\phi^{11}_{11;-1}(0) = \phi^{11}_{1-1;1}(0)$	$\frac{1}{\sqrt{6}}\text{Re}\langle \mathscr{D}^2_{0,2}\rangle$	–

6.6 Correlations at Short and Long Times

Although the time dependence of correlation functions is of course unknown a priori and indeed an important target for studies of the dynamics of molecules, some general statements on their limiting values at short and long times and about their range of variations can be made. Here we briefly review some important properties.

Initial values. The initial values of the reorientational autocorrelation functions will just be averages of products of Wigner rotation functions. More specifically, for auto- or cross-correlations,

$$\phi^{LL'}_{mn;n'}(0) = \left\langle \mathscr{D}^L_{m,n}\mathscr{D}^{L'*}_{m,n'}\right\rangle = (-1)^{n-n'}\phi^{L'L}_{-m-n';-n}(0) = \phi^{LL'}_{-m-n;-n'}(0),$$

$$= (-1)^{m-n'}\sum_{J=|L-L'|}^{L+L'} C(L,L',J;m,-m)C(L,L',J;n,-n')\left\langle \mathscr{D}^J_{0,n-n'}\right\rangle, \quad (6.26a)$$

where $C(A,B,C;d,e)$ are the Clebsch–Gordan coefficients discussed in Appendix F and tabulated elsewhere [Pasini and Zannoni, 1984b]. In the next few tables we give explicit results for first (Table 6.1), mixed (Table 6.2) and second (Table 6.3) rank correlation functions for biaxial or uniaxial particles reorienting in a uniaxial phase. We see that these initial values can contain information on order parameters of rank higher than L, L'. The expressions for probes with uniaxial symmetry can be obtained from Eqs. 6.26a by letting the various biaxial order parameters, e.g. $\langle \mathscr{D}^2_{0,2}\rangle$, $\langle \mathscr{D}^4_{0,2}\rangle$, $\langle \mathscr{D}^4_{0,4}\rangle$ go to 0.

$$\phi^{LL'}_{mn}(0) = \left\langle \mathscr{D}^L_{m,n}\mathscr{D}^{L'*}_{m,n'}\right\rangle\delta_{n,n'} = \phi^{L'L}_{-m-n}(0) = \phi^{LL'}_{-m-n}(0), \quad (6.27a)$$

$$= (-1)^{m-n}\sum_{J=|L-L'|}^{L+L'} C(L,L',J;m,-m)C(L,L',J;n,-n)\langle P_J\rangle. \quad (6.27b)$$

Table 6.2. *Same as Table 6.1 but for mixed first and second rank* $\phi^{12}_{mn;n'}(0) = \langle \mathscr{D}^1_{m,n} \mathscr{D}^{2*}_{m,n'} \rangle$.

Biaxial molecule	Initial value	Uniaxial molecule
$\phi^{12}_{1-1;-1}(0) = -\phi^{12}_{11;1}(0)$	$-\frac{1}{2}\langle P_2 \rangle$	$\phi^{12}_{1-1}(0) = -\phi^{12}_{11}(0)$
$\phi^{12}_{1-1;1}(0) = -\phi^{12}_{11;-1}(0)$	$-\frac{1}{\sqrt{6}}\mathrm{Re}\langle \mathscr{D}^2_{0,2} \rangle$	0
$\phi^{12}_{10;-2}(0) = -\phi^{12}_{10;2}(0)$	$\frac{1}{\sqrt{3}}\mathrm{Re}\langle \mathscr{D}^2_{0,2} \rangle$	0
$\phi^{12}_{10;0}(0)$	0	0

Correlation functions bounds. In general, the initial value $C_{AB}(0) = \langle AB^* \rangle$. This has to be greater or equal to 0 if $B = A$, i.e. for autocorrelations. Applying the Schwarz inequality (see Eq. 3.89) yields $0 \leq \left|\langle A(0)B(t)^* \rangle\right| \leq \sqrt{\langle AA^* \rangle \langle BB^* \rangle}$. Moreover, $\langle [A(0) \pm B(t)][A(0) \pm B(t)]^* \rangle = \langle AA^* \rangle + \langle BB^* \rangle \pm 2\mathrm{Re}\langle A(0)B^*(t) \rangle$, and thus the further bound $\left|\mathrm{Re}C_{AB}(t)\right| \leq \frac{1}{2}\left[\langle AA^* \rangle + \langle BB^* \rangle\right]$ is obtained. For Wigner rotation matrices time correlation functions $\left|\mathrm{Re}\phi^{LL'}_{mn;m'n'}(t)\right| \leq \frac{1}{2}\left[\phi^{LL}_{mn;mn}(0) + \phi^{L'L'}_{m'n';m'n'}(0)\right]$. As a special case, if we have $A = B$, we see that the autocorrelation cannot take values greater than the initial one:

$$\left|\mathrm{Re}C_{AA}(t)\right| \leq \langle AA^* \rangle, \tag{6.28}$$

and in particular $\left|\mathrm{Re}\phi^{LL}_{mn}(t)\right| \leq \phi^{LL}_{mn}(0)$. Note that even the autocorrelation functions can be negative, at least this is not forbidden by the present inequalities, as long as Eq. 6.28 is satisfied.

Asymptotic long-time values. In view of Eq. 6.16 we have

$$\phi^{LL'}_{mn;m'n'}(\infty) = \langle \mathscr{D}^L_{m,n} \rangle \langle \mathscr{D}^{L'*}_{m',n'} \rangle, \tag{6.29}$$

so that the long-time limit is a product of order parameters. For a uniaxial phase we have $\delta_{m,m'}$ and thus

$$\phi^{LL'}_{mn;n'}(\infty) \equiv \phi^{LL'}_{mn;mn'}(\infty) = \langle \mathscr{D}^L_{0,n} \rangle \langle \mathscr{D}^{L'*}_{0,n'} \rangle \, \delta_{m,0}. \tag{6.30}$$

For a uniaxial molecule, which requires $\delta_{n,n'}$ as well, we can write

$$\phi^{LL'}_{mn}(\infty) \equiv \phi^{LL'}_{mn;mn}(\infty) = \langle P_L \rangle \langle P_{L'} \rangle \delta_{m,0} \delta_{n,0}, \tag{6.31}$$

where we have eliminated unnecessary subscripts. For example, the first-rank correlation function $\phi^{11}_{00;0}(t)$ will tend to 0, while the second-rank one $\phi^{22}_{00;0}(t)$ will tend to the square of the order parameter, $\langle P_2 \rangle^2$.

Correlation functions and exponential model. It is often convenient to define normalized autocorrelation functions

$$\mathscr{C}_{AA}(t) \equiv \frac{C_{AA}(t) - C_{AA}(0)}{C_{AA}(0) - C_{AA}(\infty)}, \tag{6.32}$$

Table 6.3. *Same as Table 6.1 but for second-rank correlation functions* $\phi^{22}_{mn;n'}(0) = \langle \mathscr{D}^2_{m,n} \mathscr{D}^{2*}_{m,n'} \rangle$

Biaxial molecules	Time 0 value	Uniaxial molecules
$\phi^{22}_{0-2;0}(0) = \phi^{22}_{02;0}(0)$ $= \phi^{22}_{00;-2}(0) = \phi^{22}_{00;2}(0)$	$\frac{3}{70}\sqrt{60}\mathrm{Re}\langle \mathscr{D}^4_{0,2}\rangle - \frac{2}{7}\mathrm{Re}\langle \mathscr{D}^2_{0,2}\rangle$	-
$\phi^{22}_{0-2;2}(0) = \phi^{22}_{02;-2}(0)$	$\frac{3}{35}\sqrt{70}\mathrm{Re}\langle \mathscr{D}^4_{0,4}\rangle$	-
$\phi^{22}_{01;1}(0) = \phi^{22}_{0-1;-1}(0)$ $= \phi^{22}_{10;0}(0)$	$-\frac{12}{35}\langle P_4\rangle + \frac{1}{7}\langle P_2\rangle + \frac{1}{5}$	$\phi^{22}_{01}(0) = \phi^{22}_{0-1;}(0)$ $= \phi^{22}_{10}(0)$
$\phi^{22}_{0-1;1}(0) = \phi^{22}_{01;-1}(0)$	$-\frac{6}{35}\sqrt{10}\mathrm{Re}\langle \mathscr{D}^4_{0,2}\rangle - \frac{1}{7}\sqrt{6}\mathrm{Re}\langle \mathscr{D}^2_{0,2}\rangle$	-
$\phi^{22}_{00;0}(0)$	$\frac{18}{35}\langle P_4\rangle + \frac{2}{7}\langle P_2\rangle + \frac{1}{5}$	$\phi^{22}_{00}(0)$
$\phi^{22}_{0-2;-2}(0) = \phi^{22}_{02;2}(0)$ $= \phi^{22}_{20;0}(0)$	$\frac{3}{35}\langle P_4\rangle - \frac{2}{7}\langle P_2\rangle + \frac{1}{5}$	$\phi^{22}_{0-2}(0) = \phi^{22}_{02}(0)$ $= \phi^{22}_{20}(0)$
$\phi^{22}_{1-2;-2}(0) = \phi^{22}_{12;2}(0)$ $= \phi^{22}_{2-1;-1}(0) = \phi^{22}_{21;1}(0)$	$-\frac{2}{35}\langle P_4\rangle - \frac{1}{7}\langle P_2\rangle + \frac{1}{5}$	$\phi^{22}_{1-2}(0) = \phi^{22}_{12}(0)$ $= \phi^{22}_{2-1}(0) = \phi^{22}_{21}(0)$
$\phi^{22}_{1-2;2}(0) = \phi^{22}_{12;-2}(0)$	$-\frac{2}{35}\sqrt{70}\mathrm{Re}\langle \mathscr{D}^4_{0,4}\rangle$	-
$\phi^{22}_{11;1}(0) = \phi^{22}_{1-1;-1}(0)$	$\frac{8}{35}\langle P_4\rangle + \frac{1}{14}\langle P_2\rangle + \frac{1}{5}$	$\phi^{22}_{11}(0) = \phi^{22}_{1-1}(0)$
$\phi^{22}_{11;-1}(0) = \phi^{22}_{1-1;1}(0)$	$\frac{4}{35}\sqrt{10}\mathrm{Re}\langle \mathscr{D}^4_{0,2}\rangle - \frac{1}{14}\sqrt{6}\mathrm{Re}\langle \mathscr{D}^2_{0,2}\rangle$	-
$\phi^{22}_{10;2}(0) = \phi^{22}_{12;0}(0) = \phi^{22}_{10;-2}(0) = \phi^{22}_{1;-2,0}(0)$	$-\frac{1}{35}\sqrt{60}\mathrm{Re}\langle \mathscr{D}^4_{0,2}\rangle - \frac{1}{7}\mathrm{Re}\langle \mathscr{D}^2_{0,2}\rangle$	-
$\phi^{22}_{2-2;-2}(0) = \phi^{22}_{22;2}(0)$	$\frac{1}{70}\langle P_4\rangle + \frac{2}{7}\langle P_2\rangle + \frac{1}{5}$	$\phi^{22}_{2-2}(0) = \phi^{22}_{22}(0)$
$\phi^{22}_{2-2;0}(0) = \phi^{22}_{22;0}(0) = \phi^{22}_{20;2}(0) = \phi^{22}_{20;-2}(0)$	$\frac{1}{140}\sqrt{60}\mathrm{Re}\langle \mathscr{D}^4_{0,2}\rangle + \frac{2}{7}\mathrm{Re}\langle \mathscr{D}^2_{0,2}\rangle$	-
$\phi^{22}_{2-2;2}(0) = \phi^{22}_{22;-2}(0)$	$-\frac{1}{70}\sqrt{70}\mathrm{Re}\langle \mathscr{D}^4_{04}\rangle$	-
$\phi^{22}_{2-1;1}(0) = \phi^{22}_{21;-1}(0)$	$-\frac{1}{35}\sqrt{10}\mathrm{Re}\langle \mathscr{D}^4_{0,2}\rangle + \frac{1}{7}\sqrt{6}\mathrm{Re}\langle \mathscr{D}^2_{0,2}\rangle$	-
$\phi^{22}_{22;-1}(0)$	$-\frac{1}{12}\langle P_2\rangle + \frac{1}{12}$	-

that vary between 1 and 0, in view of the bound Eq. 6.28. It is convenient to define also a characteristic *relaxation time*

$$\tau_A \equiv \int_0^\infty \mathrm{d}t\, \mathscr{C}_{AA}(t). \tag{6.33}$$

For each orientational correlation function, we would have a characteristic time for decaying to equilibrium

$$\tau_{Lmn} = \int_0^\infty \mathrm{d}t\, \frac{\phi^{LL}_{mn}(t) - \phi^{LL}_{mn}(\infty)}{\phi^{LL}_{mn}(0) - \phi^{LL}_{mn}(\infty)} = J_{Lmn}(0), \tag{6.34}$$

where $J_{Lmn}(\omega)$ is the *spectral density* associated with the correlation function, i.e. its Fourier–Laplace transform at frequency ω. Similar characteristic times $\tau_{Lmn,L'm'n'}$ can be associated to cross-correlations. Note that often (e.g. [Nordio and Segre, 1979; Arcioni et al., 1988]) the normalization factor in the denominator is not included and one uses characteristic times which are just areas under $\phi_{mn}^{LL}(t) - \phi_{mn}^{LL}(\infty)$. A very simple interpolating time dependence for the correlation function $C_{AA}(t)$ that has the correct time dependence at time 0 and at time infinity is one that assumes that the normalized autocorrelation function decays exponentially. For orientational autocorrelation functions this assumption gives the simple form

$$\phi_{mn}^{LL}(t) = \left[\phi_{mn}^{LL}(0) - \phi_{mn}^{LL}(\infty)\right] \mathrm{e}^{-t/\tau_{Lmn}} + \phi_{mn}^{LL}(\infty), \tag{6.35}$$

where τ_{Lmn} is the characteristic decay or relaxation time for that particular correlation function. With the further assumption that orientational autocorrelation functions of all ranks L relax with the same characteristic time, this becomes equivalent to the so-called *Strong Collision* model [Nordio and Segre, 1979]. The name comes from the fact that a dynamic consisting of short collisions, so strong as to cause the molecule to forget completely its starting orientation, would have such a type of decay. In anisotropic systems the weaker assumption that τ_{Lmn} reduces to τ_n, i.e. that each irreducible component in an equation like Eq. 6.35 relaxes with its characteristic time, has often been made [Luckhurst and Sanson, 1972].

Short-time expansion. The time dependence of the orientation correlation functions is of course unknown and we have no general way of predicting even the functional form of this time behaviour, even if often one or more exponential decays are assumed. If we perform molecular dynamics simulations, correlation functions can be calculated from the time trajectories. In this case, our problem may be an overabundance of data, since the number of orientational correlation functions, even though somewhat limited by symmetry, is still huge. Thus, it might be tempting to fit a set of correlation functions to some semiempirical time dependence (e.g. continued fraction or sum of exponentials or Gaussians). In this case, it is essential to restrict the fit so that as many general constraints as possible are implemented. We have already seen some of these, e.g. the initial and the asymptotic values of $\phi_{mn;m'n'}^{LL'}(t)$. Indeed, all the initial values and asymptotic values can be written in terms of a limited number of orientational order parameters appropriate to the system. Other such constraints can be obtained from a short-time Taylor expansion of $\phi_{mn;n'}^{LL'}(t)$:

$$\varphi_{mn;n'}^{LL'}(t) \equiv \mathrm{Re}\phi_{mn;n'}^{LL'}(t) = \sum_p \left[\frac{\mathrm{d}^{2p}}{\mathrm{d}t^{2p}} \varphi_{mn;n'}^{LL'}(t)\right]_{t=0} \frac{t^{2p}}{(2p)!}, \tag{6.36a}$$

$$= \varphi_{mn;n'}^{LL'}(0) - \frac{1}{2!}\ddot{\varphi}_{mn;n'}^{LL'}(0)t^2 + \frac{1}{4!}\ddddot{\varphi}_{mn;n'}^{LL'}(0)t^4 + \cdots . \tag{6.36b}$$

Note that odd time derivatives do not contribute, since here we are considering only the real part of the orientational correlation functions. The first few coefficients of the short-time expansion in isotropic fluids have been obtained for linear molecules [Gordon, 1968] and for symmetric tops [St. Pierre and Steele, 1981]. Coefficients up to the fourth for a symmetric top in a uniaxial environment have been obtained by Pasini and Zannoni [1984a].

This work has been extended to biaxial molecules reorienting in a uniaxial LC [Pasini et al., 1991] using Eq. 6.36b, and

$$\frac{\mathrm{d}^{2p}}{\mathrm{d}t^{2p}}\varphi_{mn}^{LL'}(t) = \mathrm{Re}\left\langle \frac{\mathrm{d}^p}{\mathrm{d}t^p}\mathscr{D}_{m,n}^{L}(\Omega_t)\frac{\mathrm{d}^p}{\mathrm{d}t^p}\mathscr{D}_{m,n}^{L'}(\Omega_t)^*\right\rangle, \tag{6.37}$$

that is proved using the stationarity relation. To evaluate the derivative

$$\dot{\mathscr{D}}_{mn}^{L}(\alpha\beta\gamma) = -i(m\dot{\alpha} + n\dot{\gamma})\mathscr{D}_{mn}^{L}(\alpha\beta\gamma) + e^{-i(m\alpha+n\gamma)}\dot{d}_{mn}^{L}(\beta), \tag{6.38}$$

we need the time derivatives of the Euler angles in terms of angular velocities. To do this, we recall that the angular velocity $\dot{\mathbf{\Omega}}$ is the time derivative of the rotation vector $\mathbf{\Omega}$ [Landau and Lifshitz, 1993] directed along the rotation axis. Following the convention of Rose [1957] for the Euler angles (see Fig. 3.1b) we have $\mathbf{\Omega} = \alpha\boldsymbol{Z} + \beta\boldsymbol{y}' + \gamma\boldsymbol{z}$, where $\alpha\mathbf{Z}$ is a positive, right-handed, rotation about the laboratory Z-axis, $\beta\mathbf{y}'$ a rotation of β about the new $\mathbf{y}'$-axis and, finally, $\gamma\mathbf{z}$ a rotation of γ about the body fixed $\mathbf{z}$. We can rewrite $\mathbf{\Omega}$ in terms of rotations $\Omega_X, \Omega_Y, \Omega_Z$, about the three laboratory axes. Using capitals for clarity, $\boldsymbol{X}, \boldsymbol{Y}, \boldsymbol{Z}$ to refer to laboratory axes and lower case, $\boldsymbol{x}, \boldsymbol{y}, \boldsymbol{z}$ to refer to the molecular frame, it is not difficult to prove that

$$\mathbf{\Omega} = \Omega_X\boldsymbol{X} + \Omega_Y\boldsymbol{Y} + \Omega_Z\boldsymbol{Z}. \tag{6.39}$$

We can now derive expressions for the angular velocity $\dot{\mathbf{\Omega}}$ [Landau and Lifshitz, 1993] starting with the rotation vector, $\delta\mathbf{\Omega} = \delta\alpha\boldsymbol{Z} + \delta\beta\boldsymbol{y}' + \delta\gamma\boldsymbol{z}$, where $\boldsymbol{x}' = -\boldsymbol{X}\sin\alpha + \boldsymbol{Y}\cos\alpha$, $\boldsymbol{y}' = \boldsymbol{X}\cos\alpha + \boldsymbol{Y}\sin\alpha$ and $\boldsymbol{z} = \boldsymbol{Z}\cos\beta + \boldsymbol{x}'\sin\beta = \boldsymbol{Z}\cos\beta + (\boldsymbol{X}\cos\beta + \boldsymbol{Y}\sin\alpha)\sin\beta$. Collecting terms, we have

$$\begin{aligned}\delta\mathbf{\Omega} &= (-\delta\beta\sin\alpha + \delta\gamma\cos\alpha\sin\beta)\boldsymbol{X} + (\delta\beta\cos\beta + \delta\gamma\sin\alpha\sin\beta)\boldsymbol{Y} + (\delta\alpha + \delta\gamma\cos\beta)\boldsymbol{Z}\\ &= \delta\Omega_X\boldsymbol{X} + \delta\Omega_Y\boldsymbol{Y} + \delta\Omega_Z\boldsymbol{Z}.\end{aligned} \tag{6.40}$$

The angular velocity components referred to the lab frame are therefore

$$\dot{\Omega}_X = -\sin\alpha\dot{\beta} + \sin\beta\cos\alpha\dot{\gamma}, \tag{6.41a}$$

$$\dot{\Omega}_Y = \cos\alpha\dot{\beta} + \sin\beta\sin\alpha\dot{\gamma}, \tag{6.41b}$$

$$\dot{\Omega}_Z = \cos\beta\dot{\gamma} + \dot{\alpha}. \tag{6.41c}$$

In a similar way we find the molecule fixed angular velocity components:

$$\dot{\Omega}_x = -\sin\beta\cos\gamma\dot{\alpha} + \sin\gamma\dot{\beta}, \tag{6.42a}$$

$$\dot{\Omega}_y = \sin\beta\sin\gamma\dot{\alpha} + \cos\gamma\dot{\beta}, \tag{6.42b}$$

$$\dot{\Omega}_z = \cos\beta\dot{\alpha} + \dot{\gamma}. \tag{6.42c}$$

From these, the time derivatives of α, β, γ can be written as

$$\dot{\alpha} = [-\dot{\Omega}_x\cos\gamma + \dot{\Omega}_y\sin\gamma]/\sin\beta, \tag{6.43a}$$

$$\dot{\beta} = \dot{\Omega}_x\sin\gamma + \dot{\Omega}_y\cos\gamma, \tag{6.43b}$$

$$\dot{\gamma} = [\dot{\Omega}_x\cos\gamma\cos\beta - \dot{\Omega}_y\sin\gamma\cos\beta + \dot{\Omega}_z\sin\beta]/\sin\beta. \tag{6.43c}$$

Second derivatives can be formally written recalling the angular equivalent of Newton's law, i.e. Euler equation [Landau and Lifshitz, 1993],

$$\mathbf{I}\ddot{\boldsymbol{\Omega}} + \dot{\boldsymbol{\Omega}} \times (\mathbf{I} \cdot \dot{\boldsymbol{\Omega}}) = \boldsymbol{N}, \tag{6.44}$$

where $\mathbf{I}$ is the inertia tensor (see Eq. 4.24) and $\boldsymbol{N}$ the *torque*, i.e. the moment of the force acting on the molecule. We assume to have a set of molecules with uniaxial inertia tensors, interacting through certain pairwise interactions, U_{ij}. If we just call U the potential acting on the molecule at $\boldsymbol{r}$ under consideration, $\boldsymbol{N} = -\boldsymbol{r} \times \nabla U$. The second derivatives of the Euler angles are thus expressed in terms of body frame angular velocities and moments of inertia $I_\parallel, I_\perp$ as well as of the torque components $N_\alpha = -\partial U/\partial\alpha, N_\beta = -\partial U/\partial\beta, N_\gamma = -\partial U/\partial\gamma$. After substitution we find [Zannoni, 1979a]

$$\ddot{\alpha} = \left[(I_\parallel - 2I_\perp)\dot{\alpha}\dot{\beta}\cos\beta\sin\beta + I_\parallel\dot{\beta}\dot{\gamma}\sin\beta + N_\alpha - N_\gamma\cos\beta\right]/\left(I_\perp\sin^2\beta\right), \tag{6.45a}$$

$$\ddot{\beta} = \left[-I_\parallel\dot{\alpha}\dot{\gamma}\sin\beta - (I_\parallel - I_\perp)(\dot{\alpha})^2\cos\beta\sin\beta + N_\beta\right]/I_\perp, \tag{6.45b}$$

$$\begin{aligned}\ddot{\gamma} = \Big\{&-I_\parallel^2\dot{\gamma}\dot{\beta}\cos\beta\sin\beta + [I_\perp + (I_\perp - I_\parallel)\cos^2\beta]I_\parallel\dot{\beta}\dot{\alpha}\sin\beta,\\ &+ [I_\perp + (I_\parallel - I_\perp)\cos^2\beta]N_\gamma - I_\parallel N_\alpha\cos\beta\Big\}/\left(I_\perp I_\parallel\sin^2\beta\right).\end{aligned} \tag{6.45c}$$

In particular, for a linear rotor, $I_\parallel = 0$ and Eqs. 6.45 reduce to

$$\ddot{\alpha} = -2\dot{\alpha}\dot{\beta}\cot\beta + N_\alpha/I_\perp\sin^2\beta, \tag{6.46a}$$

$$\ddot{\beta} = (\dot{\alpha})^2\sin\beta\cos\beta + N_\beta/I_\perp. \tag{6.46b}$$

Using the Maxwell distribution: $P(\dot{\boldsymbol{\Omega}}) \propto \exp\left[-\dot{\boldsymbol{\Omega}} \cdot \mathbf{I} \cdot \dot{\boldsymbol{\Omega}}/(2k_BT)\right]$ (Section 4.2.2) to average over angular velocities gives $\langle\dot{\Omega}_a\dot{\Omega}_b\rangle = 0$ for $a \neq b$ as well as $\langle\dot{\Omega}_a^{2n}\rangle = \dfrac{(2n)!}{n!}\left(\dfrac{k_BT}{2I_a}\right)^n$. We can then obtain, after some algebraic manipulation, the coefficients in Eq. 6.36b. Up to second rank these are [Pasini et al., 1991]

$$\varphi_{00;0}^{11}(0) = \frac{2}{3}\langle P_2\rangle + \frac{1}{3}, \tag{6.47a}$$

$$\ddot{\varphi}_{00;0}^{11}(0) = \frac{k_BT}{3}\left[\left(\frac{1}{I_{xx}} + \frac{1}{I_{yy}}\right)\left(1 - \langle P_2\rangle\right) + \sqrt{6}\left(\frac{1}{I_{yy}} - \frac{1}{I_{xx}}\right)\mathrm{Re}\langle\mathscr{D}_{0,2}^2\rangle\right], \tag{6.47b}$$

$$\varphi_{00;0}^{12}(0) = 0, \tag{6.47c}$$

$$\varphi_{11;1}^{12}(0) = \frac{1}{2}\langle P_2\rangle, \tag{6.47d}$$

$$\varphi_{00;0}^{22}(0) = \frac{18}{35}\langle P_4\rangle + \frac{2}{7}\langle P_2\rangle + \frac{1}{5}, \tag{6.47e}$$

$$\begin{aligned}\ddot{\varphi}_{00;0}^{22}(0) = &\frac{3k_BT}{35}\left(\frac{1}{I_{yy}} - \frac{1}{I_{xx}}\right)\left[6\sqrt{10}\mathrm{Re}\langle\mathscr{D}_{0,2}^4\rangle + 5\sqrt{6}\mathrm{Re}\langle\mathscr{D}_{0,2}^2\rangle\right]\\ &+ \frac{3k_BT}{35}\left(\frac{1}{I_{xx}} + \frac{1}{I_{yy}}\right)\left(-12\langle P_4\rangle + 5\langle P_2\rangle + 7\right).\end{aligned} \tag{6.47f}$$

We see that the correlation functions initial values depend just on order parameters, the second derivatives (curvature) at time 0 depend also on inertia moments. It is only at fourth order, as shown in Pasini et al. [1991], that a direct dependence of the orientational correlation functions on intermolecular torques comes in. This is true in general and provides an important set of constraints for the short-time behaviour of orientational autocorrelation functions. The formulas for the higher moments become very complicated, at least for a human (they are just a few lines of a high-level language, so coding them is no problem), but can be found in Pasini et al. [1991]. In the isotropic limit, where the order parameters reduce to 0, the individual m,n angular momentum components correlation functions become degenerate and one recovers, multiplying by the number of components $(2L+1)$, the known formulas for normal liquids [Wegdam et al., 1977; St. Pierre and Steele, 1981]

$$\varphi^{11}(0) = 3\varphi^{11}_{00;0}(0) = 1 \tag{6.48a}$$

$$\ddot{\varphi}^{11}(0) = 3\ddot{\varphi}^{11}_{00;0}(0) = k_B T\left(\frac{1}{I_{xx}} + \frac{1}{I_{yy}}\right), \tag{6.48b}$$

$$\varphi^{22}(0) = 5\varphi^{11}_{00;0}(0) = 1, \tag{6.48c}$$

$$\ddot{\varphi}^{22}(0) = 5\ddot{\varphi}^{22}_{00;0}(0) = 3k_B T\left(\frac{1}{I_{xx}} + \frac{1}{I_{yy}}\right). \tag{6.48d}$$

Note that the expressions obtained for second-order derivatives allow the definition of a Gaussian approximation for the short-time behaviour of the orientational correlation functions, e.g.

$$\phi^{LL'}_{mn}(t) = \phi^{LL'}_{mn}(0) - \ddot{\phi}^{LL'}_{mn}(0)t^2/2 + \cdots \cong \phi^{LL'}_{mn}(0)\exp\left(\frac{-\ddot{\phi}^{LL'}_{mn}(0)\,t^2}{2\phi^{LL'}_{mn}(0)}\right). \tag{6.49}$$

6.7 Translational Diffusion

Let us now consider translational motion and in particular the time dependence of the mean square displacement (MSD) of a particle from an arbitrary initial position $\boldsymbol{r}(t)$:

$$\langle \Delta r^2(t) \rangle = \frac{1}{N}\sum_{i=1}^{N} \langle \Delta r_i^2(t) \rangle. \tag{6.50}$$

The MSD can be evaluated from the particle trajectories obtained from real observations, e.g. on colloidal particles or obtained from computer simulations, like the molecular dynamics ones to be discussed in Chapter 9. Einstein has shown that in the case of Brownian motion in 3D,

$$\text{MSD} = \langle \Delta r^2(t) \rangle = 6D_{\text{T}}t, \tag{6.51}$$

where D_{T} is a translational diffusion coefficient and the average is over all initial conditions. To calculate the MSD on the LHS we start by writing $\boldsymbol{r}(t)$ in terms of successive displacements. Using

$$\Delta \boldsymbol{r}(t) = \boldsymbol{r}(t) - \boldsymbol{r}(0) = \int_0^t \mathrm{d}t' \frac{\mathrm{d}\boldsymbol{r}}{\mathrm{d}t'} = \int_0^t \mathrm{d}t'\, \boldsymbol{v}(t'), \tag{6.52}$$

we have, writing each step in detail,

$$\langle \Delta r^2(t)\rangle = \left\langle \left[\int_0^t \mathrm{d}t_1 \boldsymbol{v}(t_1)\right] \cdot \left[\int_0^t \mathrm{d}t_2 \boldsymbol{v}(t_2)\right]\right\rangle, \tag{6.53a}$$

$$= \int_0^t \mathrm{d}t_1 \int_0^t \mathrm{d}t_2 \langle \boldsymbol{v}(t_1)\cdot \boldsymbol{v}(t_2)\rangle = 2\int_0^t \mathrm{d}t_1 \int_0^{t_1} \mathrm{d}t_2 \langle \boldsymbol{v}(t_1)\cdot \boldsymbol{v}(t_2)\rangle, \tag{6.53b}$$

$$= 2\int_0^t \mathrm{d}t_1 \int_0^{t_1} \mathrm{d}t_2 \langle \boldsymbol{v}(0)\cdot \boldsymbol{v}(t_2 - t_1)\rangle = 2\int_0^t \mathrm{d}t_1 \int_0^{t_1} \mathrm{d}\tau\, \langle \boldsymbol{v}(0)\cdot \boldsymbol{v}(\tau)\rangle, \tag{6.53c}$$

$$= 2\int_0^t \mathrm{d}t_1 D_{\mathrm{T}}(t_1) \approx 2tn_d D_{\mathrm{T}}, \tag{6.53d}$$

where we used first the stationarity condition, i.e. the independence of the starting observation time and then a change of variable $\tau = t_1 - t_2$. The last equation follows from the definition of the translational diffusion coefficient (Eq. 6.51)

$$D_{\mathrm{T}} = \frac{1}{n_d}\int_0^\infty \mathrm{d}\tau \langle \boldsymbol{v}(0)\cdot \boldsymbol{v}(\tau)\rangle, \tag{6.54}$$

where $n_d = 1, 2, 3$ is the dimensionality of the space involved in the particle movement. The equation relating diffusion coefficient to linear velocity correlation function belongs to a family of similar relations linking transport coefficients to time correlation functions called Green–Kubo relations [Balescu, 1975; Kubo, 1986]. Note that the velocity correlation function is even in time, so that at short time the MSD has to be quadratic and not linear in time and it is the behaviour at longer times that becomes linear, as sketched in Fig. 6.1 for a diffusive process or in certain, more unusual conditions, sub-diffusive or super-diffusive (Fig. 6.1b). In aligned liquid crystals the diffusion can be different in different directions with respect to the director, and is described by a tensor $\mathbf{D}_{\mathrm{T}} = \langle \Delta \boldsymbol{r} \otimes \Delta \boldsymbol{r}\rangle$, with elements

$$(D_{\mathrm{T}})_{ab} = \frac{1}{n_d}\int_0^\infty \mathrm{d}\tau \langle v_a(0) v_b(\tau)\rangle, \quad a, b = x, y, z. \tag{6.55}$$

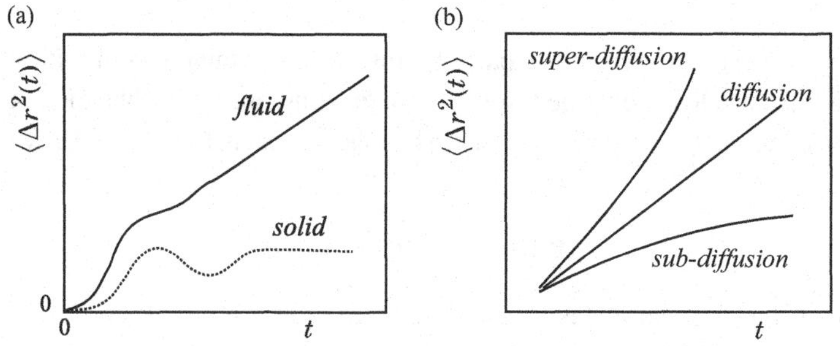

Figure 6.1 (a) A sketch of the MSD in a fluid, showing diffusive, linear, behaviour at long times, with the slope giving the diffusion constant and in solids (glasses or crystals) with only oscillations around an essentially constant position (b) Behaviour of the MSD $\propto t^\alpha$ for $\alpha = 1$ (diffusive), $\alpha < 1$ (sub-diffusive) and $\alpha > 1$ (super-diffusive) cases.

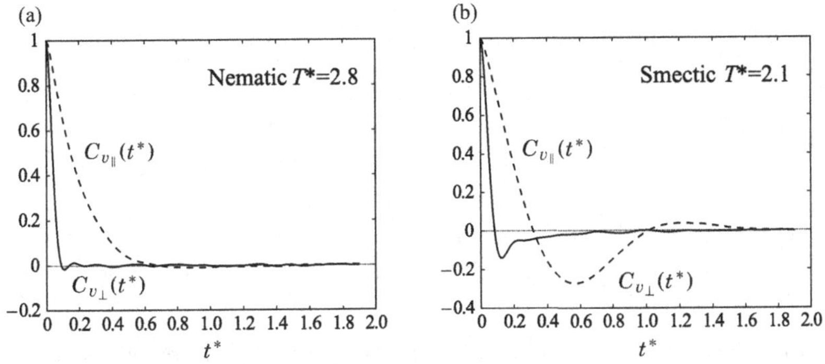

Figure 6.2 Velocity autocorrelation functions in the directions parallel, $C_{v_\parallel}(t^*) = \langle \boldsymbol{v}_\parallel(0)\, \boldsymbol{v}_\parallel(t^*)\rangle$ (- - -), and perpendicular, $C_{v_\perp}(t^*) = \langle \boldsymbol{v}_\perp(0)\, \boldsymbol{v}_\perp(t^*)\rangle$ (—), to the director in a (a) nematic and (b) smectic phase. Results are from MD simulations of a system of Gay–Berne GB(3,5,1,3) elongated ellipsoids (Section 5.6.3). Dimensionless temperature T^* and time t^* were defined in Table 5.10.

In an isotropic phase we only have $D_\mathrm{T} = \mathrm{Tr}(\mathbf{D}_\mathrm{T})$, corresponding to Eq. 6.54. In a nematic or a smectic the principal values are

$$(D_\mathrm{T})_\parallel = \frac{1}{n_d}\int_0^\infty \mathrm{d}\tau \langle \boldsymbol{v}_\parallel(0)\, \boldsymbol{v}_\parallel(\tau)\rangle, \tag{6.56}$$

with $\boldsymbol{v}_\parallel = (\boldsymbol{v}\cdot\boldsymbol{d})\,\boldsymbol{d} = \boldsymbol{v}_z$, for movements along the director $\boldsymbol{d}$ (with $\boldsymbol{d}||z$) and

$$(D_\mathrm{T})_\perp = \frac{1}{n_d}\int_0^\infty \mathrm{d}\tau \langle \boldsymbol{v}_\perp(0)\, \boldsymbol{v}_\perp(\tau)\rangle, \tag{6.57}$$

where $\boldsymbol{v}_\perp = \frac{1}{2}(\boldsymbol{v}_x + \boldsymbol{v}_y) = \boldsymbol{v} - (\boldsymbol{v}\cdot\boldsymbol{d})\boldsymbol{d}$. In Fig. 6.2 we report an example for a system of attractive-repulsive Gay–Berne particles (cf. Section 5.6.3). The diffusion coefficients at a certain temperature (and thus order parameter $\langle P_2\rangle$) can be related to those of the completely ordered system $D_\parallel^{(0)}$, $D_\perp^{(0)}$ as [Blinc et al., 1974]

$$(D_\mathrm{T})_\perp = \frac{1}{3}\langle \mathrm{Tr}\mathbf{D}_\mathrm{T}\rangle(1 - \langle P_2\rangle) + \langle P_2\rangle (D_\mathrm{T})_\perp^{(0)}, \tag{6.58a}$$

$$(D_\mathrm{T})_\parallel = \frac{1}{3}\langle \mathrm{Tr}\mathbf{D}_\mathrm{T}\rangle(1 - \langle P_2\rangle) + \langle P_2\rangle (D_\mathrm{T})_\parallel^{(0)}. \tag{6.58b}$$

In Fig. 6.3 we show, as an example, the MSD parallel and perpendicular to the director obtained from the simulation of a system of Gay–Berne model mesogens (described in detail in Sections 5.6.3 and 11.5). We see that, both in the nematic and smectic phase, the MSD is essentially the same in the two directions perpendicular to the director, consistently with the uniaxial symmetry of the LC. We also see that motion is an order of magnitude faster in the nematic than in the smectic phase, but also that in both cases the MSD is larger along the director than transversally to it. While this is to be expected in the nematic, it is hard to reconcile with the simple textbook model of a smectic as a stack of liquid like layers. Indeed, if that was the case, we would expect easier motion inside the layers than across

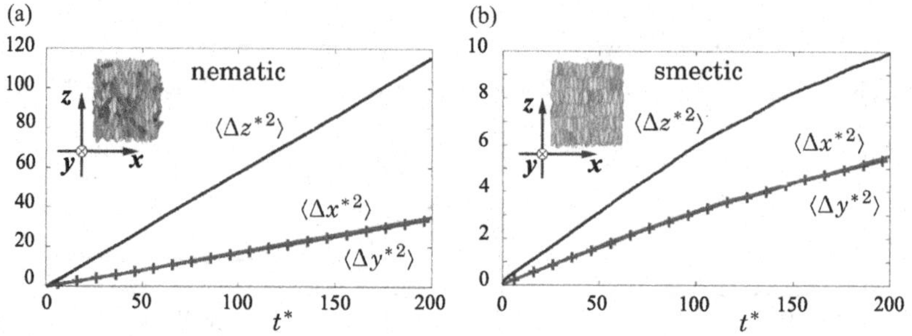

Figure 6.3 MSD $\langle(\Delta z^*)^2\rangle$, $\langle(\Delta x^*)^2\rangle$, $\langle(\Delta y^*)^2\rangle$ along $\boldsymbol{d}||z$ and transversal to it (x, y) of a Gay–Berne model LC (see Sections 5.6.3 and 11.5) in the (a) nematic and (b) smectic phase. Time t^* and lengths in dimensionless units (Table 5.10).

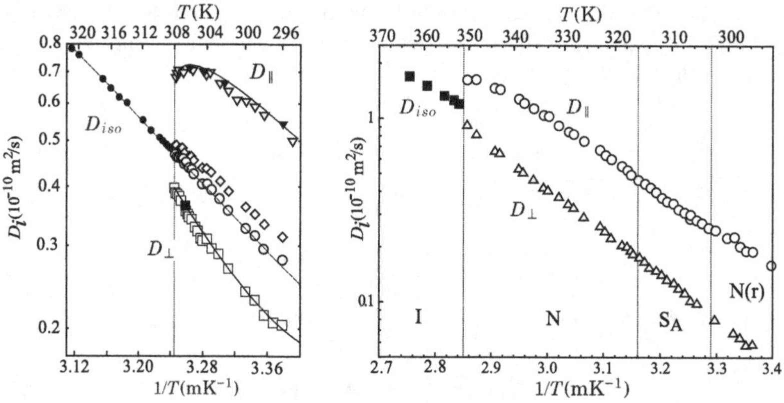

Figure 6.4 Translational diffusion coefficients parallel ($D_{\|}$), perpendicular ($D_{\perp}$) to the director in the nematic and isotropic (D_{iso}) phase of (a) 5CB from NMR measurements [Dvinskikh and Furó, 2001]. (b) The diffusion coefficients in the isotropic, nematic, smectic A and re-entrant nematic phase of a 6OCB-8OCB (27 wt% 6OCB) mixture [Dvinskikh and Furó, 2012].

them. This seems actually to be happening for another system studied in literature [Vaidya et al., 1994] where, for a perfectly aligned fluid of hard spherocylinders (Sections 5.5.2 and 11.4.2), molecular dynamics simulations find, in dimensionless units, $(D_{\mathrm{T}})^*_{\|}/(D_{\mathrm{T}})^*_{\perp} = 9.05$ (nematic) and $(D_{\mathrm{T}})^*_{\|}/(D_{\mathrm{T}})^*_{\perp} = 0.15$ (smectic). Experimentally the translational diffusion of molecules in nematic and smectic phases has been studied with NMR and in Fig. 6.4 we report some results on the isotropic and nematic phase of (a) 5CB [Dvinskikh and Furó, 2001] and (b) on an LC system of 6OCB-8OCB showing nematic, smectic and re-entrant nematic phases [Dvinskikh and Furó, 2012]. We see that in this case motion along the director is faster and $D_{\|} > D_{\perp}$ even in the smectic phase.

Considering now orientational variables, we can, quite similarly to the previous case, consider angular velocities $\dot{\boldsymbol{\Omega}}$ and introduce a rotational diffusion tensor with components

$$D_{ab} = \frac{1}{n_o}\int_0^{\infty} \mathrm{d}\tau \left\langle \dot{\boldsymbol{\Omega}}_a(0)\cdot\dot{\boldsymbol{\Omega}}_b(\tau)\right\rangle, \quad a,b = x,y,z. \tag{6.59}$$

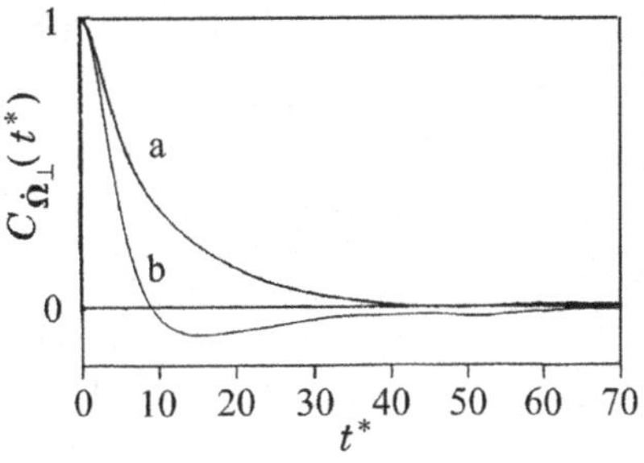

Figure 6.5 The correlation function of the transversal component of the angular velocity $C_{\dot{\Omega}_\perp}(t^*) = \langle \dot{\boldsymbol{\Omega}}_\perp(0) \cdot \dot{\boldsymbol{\Omega}}_\perp(t^*)\rangle / \langle \dot{\boldsymbol{\Omega}}_\perp(0) \cdot \dot{\boldsymbol{\Omega}}_\perp(0)\rangle$ as a function of scaled time t^* for the Gay–Berne GB(3,5,2,1) model for two-state points at scaled temperature and density (units as in Table 5.10) $T^* = 0.60$, $\rho^* = 0.19$ (a, isotropic), $T^* = 0.94$, $\rho^* = 0.33$ (b, nematic) [de Miguel et al., 1992].

The angular velocity correlation functions on the RHS can be obtained directly from molecular dynamics simulations. Some examples of interest for liquid crystals are by Kushick and Berne [1973a], Tsykalo and Bagmet [1978] and Tsykalo [1991]. For a symmetric top particle, the correlation function of $\dot{\Omega}_\perp$, the component of the angular velocity perpendicular to the orientation of the top axis is a particularly interesting observable. We expect $C_{\dot{\Omega}_\perp}(t)$ to reflect how hindered the reorientation of the long axis is and how this is affected by the onset of orientational ordering in a liquid crystal. As an example we show, in Fig. 6.5, $C_{\dot{\Omega}_\perp}(t^*)$ obtained from MD of the Gay–Berne model of liquid crystals introduced earlier (Section 5.6.3). We see that in the ordered phase the correlation function of the angular velocity perpendicular component exhibits a negative region corresponding to a change in the sense of rotation of the particle long axis which in turn indicates some sort of angular oscillatory (librational) motion in the LC phase, while the decay is similar to an exponential in the isotropic phase [de Miguel et al., 1992].

It is worth pointing out that extracting full angular velocity time correlations from experiments is hardly possible, and even obtaining their time integral (the rotational diffusion coefficients) normally requires assuming a certain stochastic model (e.g. a diffusional one) for molecular reorientations as we shall discuss later in Section 6.14.

6.8 Time Correlation Functions from Trajectories

As already mentioned, time correlation functions can be obtained from observed trajectories. For example, considering two properties $A(t)$, $B(t)$,

$$C_{AB}(t) = \langle A(0)B^*(t)\rangle = \lim_{\tau\to\infty} \frac{1}{\tau}\int_0^\tau \mathrm{d}t_0\, A(t_0)\, B^*(t_0+t) \tag{6.60}$$

can be evaluated from the sequence of values for the two properties obtained from the recorded values. In practice, trajectories consist of a sequence of configurations at successive times separated by a small time increment Δt. For a single molecule property A

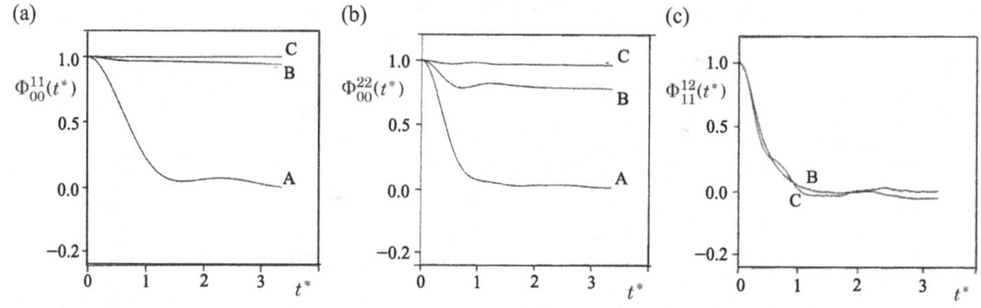

Figure 6.6 (a) The first-rank: $\Phi_{00}^{11}(t^*)$, (b) second-rank: $\Phi_{00}^{22}(t^*)$ and (c) mixed-rank: $\Phi_{11}^{12}(t^*)$, time correlation functions vs. reduced time t^* for the LL model [Zannoni and Guerra, 1981]. The curves correspond to dimensionless temperatures $T^* = 1.30\,(A)$, $0.95\,(B)$, $0.68\,(C)$.

that assumes the value A_i for molecule i, the time correlation at time $t = n\Delta t$ can be approximated with

$$C_{AB}(n\Delta t) \approx \frac{1}{N_m}\sum_{i=1}^{N_m}\frac{1}{M-n}\sum_{k=1}^{M-n} A_i(k\Delta t)\, B_i^*([k+n]\Delta t), \tag{6.61}$$

where N_m is the number of particles used as time origins (a number large enough to achieve sufficiently good statistics). The autocorrelation $C_{AA}(t)$ is of course just a special case of Eq. 6.61. An estimate of the errors $\sigma[\mathscr{C}_{AA}(t)]$ involved in calculating the normalized autocorrelation function $\mathscr{C}_{AA}(t)$ with the finite sum in Eq. 6.61 has been given by Zwanzig and Ailawadi [1969].

6.8.1 Orientational Correlation Functions: An Example

As we have discussed in the previous sections the Wigner rotation correlation functions $\phi_{mn}^{LL'}(t)$ provide a systematic approximation to the information contained in the joint distributions. We have calculated all the $\phi_{mn}^{LL'}(t)$ up to rank 2 for the Lebwohl–Lasher model [Zannoni and Guerra, 1981] and here we present a few illustrative examples. In Fig. 6.6a we see the normalized first-rank correlation function $\Phi_{00}^{11}(t) = \langle \mathscr{D}_{00}^{1*}(\Omega_0)\mathscr{D}_{00}^{1}(\Omega_t)\rangle / \langle \mathscr{D}_{00}^{1*}\mathscr{D}_{00}^{1}\rangle \equiv \phi_{00}^{11}(t)/\phi_{00}^{11}(0)$, that in a real situation could correspond to the reorientation of an axial dipole. We notice a rapid decay to 0 at the isotropic temperature (A), but a very slow decay indeed in the ordered system. The decay becomes slower as the temperature decreases (C). Note that the long-time limit of the first-rank correlation function is 0, so we have a very large increase in the dipolar correlation time, as is also found in real systems. In Fig. 6.6b we show instead the second-rank correlation $\Phi_{00}^{22}(t) \equiv \langle \mathscr{D}_{00}^{2*}(\Omega_0)\mathscr{D}_{00}^{2}(\Omega_t)\rangle / \langle \mathscr{D}_{00}^{2*}\mathscr{D}_{00}^{2}\rangle \equiv \phi_{00}^{22}(t)/\phi_{00}^{22}(0)$, still relative to long axis tumbling motion. At first glance the behaviour may seem similar, but this is not the case, because now the long-time asymptotic limit for the second-rank correlation function is non-zero in view of the long-range order limit (cf. Eq. 6.31). Finally, we show in Fig. 6.6c the mixed-rank correlation $\Phi_{11}^{12}(t) \equiv \langle \mathscr{D}_{11}^{1*}(\Omega_0)\mathscr{D}_{11}^{2}(\Omega_t)\rangle / \langle \mathscr{D}_{11}^{1*}\mathscr{D}_{11}^{2}\rangle \equiv \phi_{11}^{12}(t^*)/\phi_{11}^{12}(0)$. Only two curves, belonging to the LC phase, are shown, since mixed-rank correlations are 0 in the isotropic phase. We also notice that at very short times all the curves are rounded, consistent with the analytic results in Section 6.6. For the mixed-rank function (Fig. 6.6c), the initial value and the second moments are: $\phi_{11}^{12}(0) = \varphi_{11}^{12}(0) = \frac{1}{2}\langle P_2\rangle$ and $\ddot{\phi}_{11}^{12}(0) = \ddot{\varphi}_{11}^{12}(0) = \frac{1}{2}k_BT\langle P_2\rangle/I_{\parallel}$. We see that only when the order goes to 0 the initial value and the second moment vanish. Thus, this correlation function is not forbidden in an LC and it is only a matter of determining it using a suitable experimental technique (an example is discussed by Nordio and Segre [1977]). MD is an ideal technique from the point of view of determining unusual correlation functions, since having determined the trajectory, any dynamic quantity can, a posteriori, be calculated.

6.9 Contact with Experiment: Linear Response Theory

Let us consider the measurement of a property A of a molecular system through the application of a weak measuring field $F(t)$ [Berne, 1971; Friedman, 1985; Zannoni, 2000]. We assume canonical (N, V, T) conditions and a system with an unperturbed Hamiltonian $\mathcal{H}_0 \equiv \mathcal{H}_0(\widetilde{\boldsymbol{r}}, \widetilde{\boldsymbol{p}})$ and that a weak field $F(t)$ (the field used to perform a measurement of a certain property) is switched on at time $t = 0$ and that it interacts with the system through the perturbation Hamiltonian

$$\mathcal{H}_1 = -B(\widetilde{\boldsymbol{r}}, \widetilde{\boldsymbol{p}})F(t). \tag{6.62}$$

The property coupling to the field B depends in the most general case on the coordinates and the momenta of all molecules. We assume that the observed value of property A changes from its static equilibrium value in the absence of the field, i.e. $\langle A\rangle_0$. Since the applied field is weak the observed non-equilibrium value in the presence of the field at time t, $\langle \delta A(t)\rangle_F$, should be linear in the field strength. Considering that the system may not react instantaneously to the field, what we observe at time t is a sum of the contributions from all possible time lags τ between application and observation

$$\langle \delta A(t)\rangle_F = \sum_{\tau_i} K_{AB}(\tau_i)F(t-\tau_i) = \int_0^{\infty} \mathrm{d}\tau\, K_{AB}(\tau)F(t-\tau). \tag{6.63}$$

The observed value is a *convolution* of the field F with a 'kernel' K_{AB} whose functional form depends on the type of applied field and the observable property. The dynamics of molecular phenomena is most often explored measuring a certain observable property as a function of frequency rather than a direct time-dependence from a given starting event (there are exceptions of course, e.g. time-domain fluorescence depolarization experiments). We can write the Fourier–Laplace transform of the time-dependent response as

$$\langle \delta \tilde{A}(\omega)\rangle_F \equiv \int_0^{\infty} \mathrm{d}t\ \mathrm{e}^{i\omega t}\langle \delta A(t)\rangle_F = \int_0^{\infty} \mathrm{d}t\ \mathrm{e}^{i\omega t}\int_{-\infty}^{t} \mathrm{d}\tau\, K_{AB}(\tau)F(t-\tau), \tag{6.64a}$$

$$= \int_0^{\infty} \mathrm{d}u F(u)\int_0^{\infty} \mathrm{d}t\ \mathrm{e}^{i\omega t} K_{AB}(t-u), \tag{6.64b}$$

$$= \int_0^{\infty} \mathrm{d}u\, \mathrm{e}^{i\omega u} F(u)\int_0^{\infty} \mathrm{d}\tau\, K_{AB}(t-u)\,\mathrm{e}^{i\omega(t-u)}, \tag{6.64c}$$

$$= \chi_{AB}(\omega)\, \tilde{F}(\omega), \tag{6.64d}$$

where the Fourier–Laplace transform $\chi_{AB}(\omega) = \int_0^\infty d\tau K_{AB}(\tau) \exp(i\omega\tau)$ is called a *susceptibility* and we have used the fact that the Fourier transform of the convolution integral of two functions is the product of the Fourier transform of each function by the *convolution theorem* (Section E.2.1). Eq. 6.64d refers to macroscopic quantities, but it is clear that if we could obtain a microscopic expression for the susceptibility we would be able to calculate the response to a measuring field. The great importance of Linear Response Theory is that it gives a molecular interpretation to the susceptibility in terms of fluctuations of the unperturbed system. This is particularly important for computer simulations, since it allows us to perform the calculation of the observables for a variety of different techniques, each measuring the response to some different external field from the same simulation performed in the absence of any field, rather than repeating the simulation in the presence of each different field and experimental set up.

Let us now walk through the derivation of the desired micro-macro relation. We start adding to the unperturbed Hamiltonian the time-dependent perturbation $\mathcal{H}_1(t) = -B(\widetilde{\boldsymbol{r}}, \widetilde{\boldsymbol{p}})F(t)$, so that $\mathcal{H}(t) = \mathcal{H}_0 + \mathcal{H}_1(t)$. The perturbation produces an evolution of ϱ_0:

$$\mathcal{H}_1^\times(t)\varrho_0 = \mathcal{H}_1^\times(t)\frac{1}{\mathcal{Z}} e^{-\mathcal{H}_0/(k_BT)} = -\frac{1}{k_BT\mathcal{Z}} e^{-\mathcal{H}_0/(k_BT)} \mathcal{H}_1^\times(t)\mathcal{H}_0, \tag{6.65a}$$

$$= \frac{1}{k_BT}\varrho_0 \dot{B}(\widetilde{\boldsymbol{r}}, \widetilde{\boldsymbol{p}})F(t), \tag{6.65b}$$

since, using the definition of the evolution operator $\mathcal{H}^\times$ (Eq. 6.3),

$$\mathcal{H}_1^\times(t)\mathcal{H}_0 = -\mathcal{H}_0^\times \mathcal{H}_1(t) = \mathcal{H}_0^\times B(\widetilde{\boldsymbol{r}}, \widetilde{\boldsymbol{p}})F(t), \tag{6.66a}$$

$$= F(t)\mathcal{H}_0^\times B(\widetilde{\boldsymbol{r}}, \widetilde{\boldsymbol{p}}) = F(t)\dot{B}(\widetilde{\boldsymbol{r}}, \widetilde{\boldsymbol{p}}). \tag{6.66b}$$

In the presence of $\mathcal{H}_1(t)$ the distribution becomes, at first order, $\varrho(t) = \varrho_0 + \delta\varrho(t)$ and the non-equilibrium average change in the observable A is

$$\langle \delta A(t) \rangle_F = \int d\widetilde{\boldsymbol{r}} d\widetilde{\boldsymbol{p}}\, A(\widetilde{\boldsymbol{r}}, \widetilde{\boldsymbol{p}})\, \delta\varrho(t). \tag{6.67}$$

From the Liouville equation we have, keeping only linear terms,

$$\delta\dot{\varrho}(t) = -\mathcal{H}_0^\times \delta P - \mathcal{H}_1^\times(t)\varrho_0 = -\mathcal{H}_0^\times \delta\varrho - \frac{\varrho_0}{k_BT}\dot{\mathcal{H}}_1(t). \tag{6.68}$$

This is a simple first-order linear equation with general solution is[3]

$$\delta\varrho(t) = -\frac{1}{k_BT}\int_{-\infty}^t dt'\, e^{-(t'-t)\mathcal{H}_0^\times/(k_BT)} \varrho_0 \dot{\mathcal{H}}_1(t'). \tag{6.69}$$

If we substitute the perturbation, Eq. 6.62, we have

$$\langle \delta A(t) \rangle_F = \frac{1}{k_BT}\int_{-\infty}^t dt' F(t') \int d\widetilde{\boldsymbol{r}} d\widetilde{\boldsymbol{p}} A(\widetilde{\boldsymbol{r}}, \widetilde{\boldsymbol{p}}) e^{-(t'-t)\mathcal{H}_0^\times/(k_BT)} \dot{B}(\widetilde{\boldsymbol{r}}, \widetilde{\boldsymbol{p}})\varrho_0(\widetilde{\boldsymbol{r}}, \widetilde{\boldsymbol{p}}),$$

$$= \frac{1}{k_BT}\int_{-\infty}^0 d\tau F(t+\tau)\langle A(0)\dot{B}(\tau)\rangle = \frac{1}{k_BT}\int_{-\infty}^0 d\tau F(\tau+t) C_{A\dot{B}}(\tau), \tag{6.70a}$$

[3] The known differential equation is $\dot{y}(t) + Py = Q(t)$ if $P = \mathcal{H}_0^\times$, $Q(t) = -\beta\rho_0\dot{\mathcal{H}}_1$ that with $y = y_0$ when $t = t_0$ has the solution $e^{Pt}y - e^{Pt_0}y_0 = \int_{t_0}^t e^{Pt'}Q(t')dt'$.

where we have shifted the time origin letting $\tau = t' - t, t = t' - \tau$. We can do one further manipulation, noticing that

$$C_{A\dot{B}}(t) = \frac{\mathrm{d}}{\mathrm{d}t} C_{AB}(t) = -C_{A\dot{B}}(-t), \tag{6.71}$$

since $C_{AB}(t)$ is invariant for time reversal: $C_{AB}(t) = C^*_{AB}(-t)$. We can then write [Gordon, 1968; Hansen, 1977; Böttcher and Bordewijk, 1978]

$$\langle \delta A(t) \rangle_F = \frac{1}{k_B T} \int_0^\infty \mathrm{d}\tau \, C_{A\dot{B}}(\tau) F(\tau + t). \tag{6.72}$$

This beautiful equation shows that the observed response in a property A to the perturbation $-BF(t)$ can be obtained from the equilibrium time correlation functions $\langle A(0)\dot{B}^*(t)\rangle$ that we can calculate in the absence of the perturbation, for instance from molecular dynamics simulations of the unperturbed system. The *AB* susceptivity can be expressed in a particularly useful way for a monochromatic field: $F(t) = F_0 \exp(i\omega t)$ and we have, comparing with Eq. 6.63,

$$\chi_{AB}(\omega) = \frac{1}{k_B T} \int_0^\infty \mathrm{d}t \, C_{A\dot{B}}(t) \, \mathrm{e}^{i\omega t} \tag{6.73}$$

or, integrating by parts,

$$\chi_{AB}(\omega) = \frac{C_{AB}(0)}{k_B T} - \frac{i\omega}{k_B T} \int_0^\infty \mathrm{d}t \, C_{AB}(t) \, \mathrm{e}^{i\omega t}, \tag{6.74}$$

where $C_{AB}(t)$ is the *time correlation function* for the measured and field coupling properties A and B. We now proceed to show some examples of applications to experimental techniques, paying attention to ordered phases, identifying the relevant correlation functions.

6.10 Dielectric Properties

When a material is placed in an electric field $\boldsymbol{E}$, a *polarization* $\boldsymbol{P}_E$, i.e. a total electrical dipole moment $\boldsymbol{P}$ per unit volume, $\boldsymbol{P}_E = \boldsymbol{P}/V$ is observed and is proportional to the probing field $\boldsymbol{E}$ when this is weak,

$$\boldsymbol{P}_E = \boldsymbol{\chi}_{EE} \boldsymbol{E} = \frac{(\boldsymbol{\varepsilon} - \mathbf{1})}{4\pi} \boldsymbol{E}, \tag{6.75}$$

where $\boldsymbol{\chi}_{EE} \equiv \boldsymbol{\chi}^{(e)}$ is called the *dielectric susceptivity* and, like the *dielectric permittivity*, $\boldsymbol{\varepsilon}$, is a property of the material.[4] For a liquid crystal, $\boldsymbol{\chi}_{EE}$ and $\boldsymbol{\varepsilon}$ will be second-rank tensors. The permittivity is in general a complex quantity: its real part can in principle be obtained by measuring the ratio of the capacity of a flat cell (a parallel plate condenser) filled with the LC sample, C_S, and the capacity of the empty cell C_0. In the absence of aligning fields or surface effects, measurements will give $\varepsilon = C_S/C_0$, the scalar *dielectric constant* ($\varepsilon = \mathrm{Tr}\,\boldsymbol{\varepsilon}/3$). In practice, it is not convenient to use DC fields because of ionic impurities transport and other practical problems, so a low frequency AC field is used instead. Again, for practical

[4] In the SI system ε_0 appears instead of $1/4\pi$.

reasons, C_0 is determined using a reference material with a known dielectric constant (e.g. cyclohexane, that has $\varepsilon = 2.019$ at $T = 22°C$). The current through the capacitor in the frequency domain has two components: one out of phase, giving the real component ε': $I_C(\omega) = -i\omega\varepsilon' C_0 E(\omega)$ and one in phase, the imaginary part ε'': $I_L(\omega) = \omega\varepsilon'' C_0 E(\omega)$, related to the energy dissipation of the same cell. Their ratio, the so-called *loss tangent*: $\tan\delta = \varepsilon''/\varepsilon'$ is often employed. If the liquid crystal can be aligned, e.g. by a bias electric field, or using a magnetic field if the liquid crystal is a nematic, or some suitable surface treatment in a thin flat cell, specific components of the dielectric permittivity tensor $\boldsymbol{\varepsilon}$, can be measured. We can now try to connect $\boldsymbol{\varepsilon}$ to microscopic quantities. It should be said that the field felt by the molecules inside the sample is not necessarily the external applied one, and a correction depolarization factor should be applied to find the actual field acting on, say, a small virtual cavity containing a molecule. This correction factor has been developed for isotropic polar liquids [Glarum, 1960; Fatuzzo and Mason, 1967] and for anisotropic, liquid crystal fluids [Luckhurst and Zannoni, 1975]. Since we are focussing on molecular aspects, here we leave these complications out, since if needed they can be considered as separate corrections.

6.10.1 Theory of Dielectric Response

We start by considering a system of N molecules with permanent electric dipole moments $\boldsymbol{\mu}_i$ (Eq. 5.75) to which a uniform external electric field $\boldsymbol{E}(t) = \boldsymbol{E}_0 \exp(i\omega t)$ is applied. The system has an instantaneous total dipolc (polarization)

$$\boldsymbol{P} = \frac{1}{N}\sum_{i=1}^{N} \boldsymbol{\mu}_i, \tag{6.76}$$

while the general perturbation $\mathscr{H}_1 = -B(\widetilde{X})F(t)$ in Eq. 6.62 becomes

$$\mathscr{H}_1 = -\boldsymbol{P}(\tilde{\Omega}) \cdot \boldsymbol{E}(t). \tag{6.77}$$

This depends on all molecular orientations, but not on molecular positions, since the field is assumed to be uniform, i.e. essentially the same at every position. The Green-Kubo expression for the dielectric susceptivity follows from general Linear Response Theory (Eqs. 6.73 and 6.74) as:

$$\boldsymbol{\chi}_{\boldsymbol{PP}}(\omega) = \frac{1}{k_B T}\int_0^\infty dt\; e^{i\omega t}\, C_{\boldsymbol{P}\dot{\boldsymbol{P}}}(t) = \frac{1}{k_B T}\left(1 - i\omega\int_0^\infty dt\; e^{i\omega t}\, C_{\boldsymbol{PP}}(t)\right), \tag{6.78}$$

where $C_{\boldsymbol{PP}}(t)$ is the overall dipole correlation function matrix with, after removing the static long-time components (which will be 0 anyway for non-ferroelectric systems),

$$[C_{\boldsymbol{PP}}]_{ab}(t) = \langle P_a(0)P_b(t)\rangle - \langle P_a(0)\rangle\langle P_b(\infty)\rangle, \quad a,b = x,y,z. \tag{6.79}$$

Another equivalent expression is:

$$\boldsymbol{\chi}_{\boldsymbol{PP}}(\omega) = \frac{\langle \boldsymbol{P} \otimes \boldsymbol{P}\rangle}{k_B T}\left(1 - i\omega\int_0^\infty d\tau\; e^{i\omega t}\, \hat{C}_{\boldsymbol{PP}}(t)\right), \tag{6.80}$$

in terms of the scaled correlation $\hat{C}_{PP}(t) = C_{PP}(t)/C_{PP}(0)$. The dielectric susceptivity is recovered in the limit of very low frequency, $\omega \to 0$ as

$$\chi_{PP}(0) = \langle \boldsymbol{P} \otimes \boldsymbol{P} \rangle / (k_B T). \tag{6.81}$$

Alternatively, using the Fourier–Laplace transform $\mathscr{L}$ (defined in Appendix E)

$$\chi_{PP}(\omega) = \frac{\langle \boldsymbol{P} \otimes \boldsymbol{P} \rangle}{k_B T} \mathscr{L} \left\{ \frac{\mathrm{d}}{\mathrm{d}t} \hat{C}_{PP}(t) \right\}. \tag{6.82}$$

$C_{PP}(t)$ is the total dipole moment correlation function of a macroscopic volume containing N dipoles. To connect it to molecular correlation functions, we can write, using Eq. 6.76,

$$\langle \boldsymbol{P}(0) \otimes \boldsymbol{P}(t) \rangle = \sum_{i=1}^{N} \sum_{j=1}^{N} \langle \boldsymbol{\mu}_i(0) \otimes \boldsymbol{\mu}_j(t) \rangle, \tag{6.83a}$$

$$= \sum_{i=1}^{N} \langle \boldsymbol{\mu}_i(0) \otimes \boldsymbol{\mu}_i(t) \rangle + 2 \sum_{i=2}^{N} \sum_{j=1}^{i-1} \langle \boldsymbol{\mu}_i(0) \otimes \boldsymbol{\mu}_j(t) \rangle, \tag{6.83b}$$

for a system of identical molecules. The dipole correlation is therefore a sum of a single molecule correlation term:

$$\boldsymbol{\phi}_{\mu\mu}(t) = \langle \boldsymbol{\mu}(0) \otimes \boldsymbol{\mu}(t) \rangle, \tag{6.84}$$

depending on the motion of the dipole of a molecule and of an intermolecular dipole correlation term between the dipole of a molecule at time 0 and that of another molecule at a later time. This term, very difficult to evaluate, is most often neglected, and we will do the same here.

Empirical expressions. Before turning to LCs, it is convenient to consider some empirical correlation function expressions, frequently used when dealing with isotropic liquids, where the correlation matrix $\boldsymbol{\phi}_{\mu\mu}(t)$ reduces to the scalar $\phi_{\mu\mu}(t) = \mathrm{Tr}\boldsymbol{\phi}_{\mu\mu}(t)$. In particular, in analyzing experimental data, empirical relaxation models are often used. The simplest, called the Debye model, is a single exponential decay with a time τ_μ:

$$\phi_{\mu\mu}(t) = \exp(-t/\tau_\mu), \tag{6.85}$$

which, upon Fourier–Laplace transform (Eq. E.22), yields the complex Lorentzian expression,

$$\varepsilon(\omega) = \varepsilon_\infty + \frac{\varepsilon_0 - \varepsilon_\infty}{1 + i\omega\tau_\mu}, \tag{6.86}$$

with real and imaginary parts,

$$\varepsilon'(\omega) = \mathrm{Re}[\varepsilon(\omega)] = \varepsilon_\infty + \frac{\varepsilon_0 - \varepsilon_\infty}{1 + \omega^2\tau_\mu^2}, \tag{6.87a}$$

$$\varepsilon''(\omega) = \mathrm{Im}[\varepsilon(\omega)] = \frac{(\varepsilon_0 - \varepsilon_\infty)\omega\tau_\mu}{1 + \omega^2\tau_\mu^2}. \tag{6.87b}$$

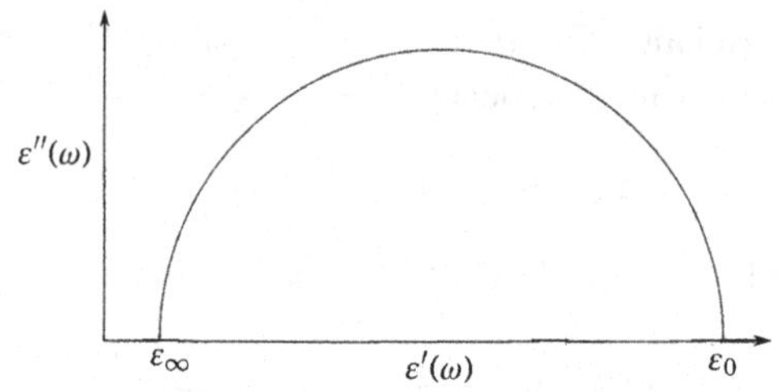

Figure 6.7 Cole–Cole plot for a single exponential time decay of the dipole correlation corresponding to a semicircle in frequency space.

This can be immediately generalized to a sum of similar profiles if the decay is a sum of exponentials with different relaxation times and pre-exponential factors. Experimental data are often presented as a Cole–Cole plot, showing $\varepsilon''(\omega)$ versus $\varepsilon'(\omega)$. For a single relaxation (Eq. 6.85), this is a semicircle (Fig. 6.7):

$$\varepsilon''(\omega)^2 = (\varepsilon_0 - \varepsilon'(\omega))(\varepsilon'(\omega) - \varepsilon_\infty). \tag{6.88}$$

Another common empirical expression is the stretched exponential or Kolrausch–Williams–Watts (KWW) decay [Kohlrausch, 1854; Williams and Watts, 1970]:

$$\phi(t) = \exp[-(t/\tau_\mu)^\beta], \quad \text{with } 0 < \beta \leq 1. \tag{6.89}$$

It is a very slow function of time giving broad dispersion curves. For $\beta = 1/2$ the complex permittivity has the simple analytic form

$$\frac{\varepsilon(\omega) - \varepsilon_\infty}{\varepsilon_0 - \varepsilon_\infty} = \sqrt{\frac{\pi}{i4\omega\tau_\mu}} \exp\left(\frac{1}{i4\omega\tau_\mu}\right) \operatorname{erfc}\left(\frac{1}{\sqrt{i4\omega\tau_\mu}}\right) \tag{6.90}$$

in terms of the complementary error function erfc(x).[5] This empirical curve is very widely used to characterize non-exponential relaxation data particularly in glass-forming systems, where the stretching exponent β can give a one-parameter characterization of the material. Another approach often used in the analysis of experimental data is to employ, instead of some other form for the time decay of the dipole correlation function, an empirical form for the frequency dependence of its Laplace transform, like the Havriliak–Negami equation [Böttcher and Bordewijk, 1978; Delafuente et al., 1994],

$$\varepsilon(\omega) = \varepsilon_\infty + \frac{\varepsilon_0 - \varepsilon_\infty}{\left[1 + (i\omega\tau)^{1-\alpha}\right]^\beta}. \tag{6.91}$$

The special case with $\beta = 1$ is called Cole–Cole equation [Cole and Cole, 1941; Ganzke et al., 2004].

[5] The complementary error function id defined as [Abramowitz and Stegun, 1965]

$$\operatorname{erfc} z = \frac{2}{\sqrt{\pi}} \int_z^\infty \mathrm{d}t\, e^{-t^2}.$$

Frequency-dependent permittivities. The orientational correlations $\phi^{11}_{mn}(\mathrm{t})$ needed to calculate the permittivities can be obtained from simulations, as we have seen in Section 6.8.1 for a very simple case. We shall also discuss, in Section 6.14.1, a theory of molecular rotational diffusion in LCs, that allows calculation of all the needed orientational correlation functions. We can now relate the single-particle dipole moment time correlation function, Eq. 6.84, to orientational correlations for a molecule reorienting in a liquid crystal. The relevant components of the correlation matrix measured in the laboratory director frame will be $\langle \mu_Z(0)\mu_Z(t)\rangle$ and $\langle \mu_X(0)\mu_X(t)\rangle$. Using spherical components (Appendix B): $\mu_Z^{\mathrm{LAB}} = \mu_{\mathrm{LAB}}^{1,0}$ and, $\mu_X^{\mathrm{LAB}} = \frac{1}{\sqrt{2}}\left(\mu_{\mathrm{LAB}}^{1,-1} - \mu_{\mathrm{LAB}}^{1,1}\right)$, we find

$$\langle \mu_{\mathrm{LAB}}^{1,m}(0)\mu_{\mathrm{LAB}}^{1,m'*}(t)\rangle = \sum_{q,p,q',p'} \left\langle \mathscr{D}^{1*}_{m,p}(\Omega_{0d})\mathscr{D}^{1}_{m',p'}(\Omega_{td})\right\rangle \mu_{\mathrm{MOL}}^{1,p}\,\mu_{\mathrm{MOL}}^{1,p'*}, \tag{6.92}$$

where the $\mu_{\mathrm{MOL}}^{1,p}$ are spherical components of the molecular dipole moment. For a uniaxial probe and mesophase the Wigner rotation matrix correlation functions are

$$\left\langle \mathscr{D}^{1*}_{m,p}(\Omega_{0d})\mathscr{D}^{1}_{m',p'}(\Omega_{td})\right\rangle = \left\langle \mathscr{D}^{1*}_{m,p}(\Omega_{0d})\mathscr{D}^{1}_{m',p'}(\Omega_{td})\right\rangle \delta_{m,m'}\delta_{p,p'} \equiv \phi^{11*}_{mp}(t). \tag{6.93}$$

Writing down the terms needed for the parallel and perpendicular dipole correlation function we have [Nordio et al., 1973]:

$$\langle \mu_Z(0)\mu_Z(t)\rangle = \mu_z^2\,\phi^{11*}_{00}(t) + (\mu_x^2+\mu_y^2)\,\phi^{11*}_{01}(t), \tag{6.94a}$$

$$\langle \mu_X(0)\mu_X(t)\rangle = \mu_z^2\,\phi^{11*}_{10}(t) + (\mu_x^2+\mu_y^2)\,\phi^{11*}_{11}(t). \tag{6.94b}$$

Thus, in the nematic phase we have up to 4 correlation functions if the molecule has an off axis dipole with components all different from 0:

$$\frac{\varepsilon_{\parallel}(\omega)-1}{4\pi} = \chi_{ZZ}(\omega) = \frac{1}{k_BT}\left(\langle \mu_Z\mu_Z\rangle - i\omega\int_0^{\infty} \mathrm{d}\tau\; \mathrm{e}^{i\omega t}\langle \mu_Z(0)\mu_Z(t)\rangle\right), \tag{6.95}$$

$$\frac{\varepsilon_{\perp}(\omega)-1}{4\pi} = \chi_{XX}(\omega) = \frac{1}{k_BT}\left(\langle \mu_X\mu_X\rangle - i\omega\int_0^{\infty} \mathrm{d}\tau\; \mathrm{e}^{i\omega t}\langle \mu_X(0)\mu_X(t)\rangle\right). \tag{6.96}$$

In general, the long axis correlation function $\phi^{11}_{00}(t)$ will decay in the nematic with a much longer time, giving rise to a characteristic peak in $\varepsilon''_{\parallel}(\omega)$ shifted down in frequency (in the MHz region) with respect to the others that are related to the spinning of the molecule around its long axis and that often have similar characteristic frequency, corresponding to decay times of the correlation functions $\phi^{11*}_{01}(t)$, $\phi^{11*}_{10}(t)$, $\phi^{11*}_{11}(t)$ of the same order of magnitude (in the 100 MHz–GHz region) [Ganzke et al., 2004].

In Fig. 6.8 we see that this is consistent with the case of a mesogen, PAA, whose dipole moment components were given earlier in Table 5.12. Considering $\varepsilon''_{\parallel}(\omega)$ we see indeed that one absorption peak, that corresponds to the long axis tumbling, is occurring at much lower frequency. Given the difficulty in observing molecular dynamics trajectories for times as

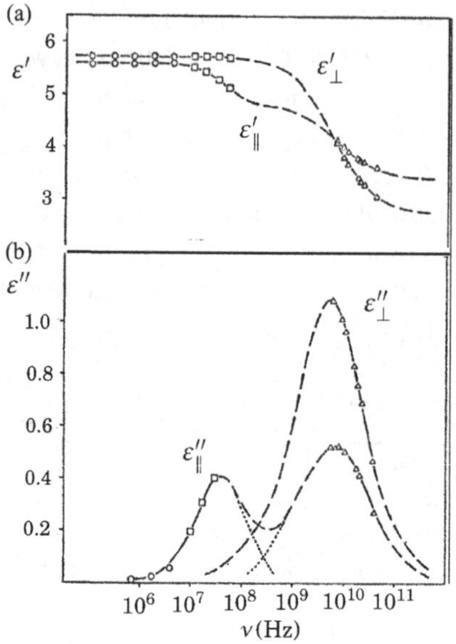

Figure 6.8 (a) Dielectric permittivities $\varepsilon'_{\parallel}, \varepsilon'_{\perp}$ and (b) dielectric losses $\varepsilon''_{\parallel}, \varepsilon''_{\perp}$ vs. frequency $\nu = \omega/2\pi$ for PAA at $T = 125°$C. The symbols are experimental data [Martin et al., 1971].

long as microseconds, comparison with experiments could be simpler for molecules with a transversal dipole rather than just one along the long axis (like nCB).

Dielectric constant. We have, considering the case of a zero-frequency experiment by letting $\omega \to 0$

$$\frac{(\varepsilon_{\parallel} - 1)k_B T}{4\pi} = \langle \mu_Z \mu_Z \rangle = \mu_z^2 \left(\frac{1}{3} + \frac{2}{3} \langle P_2 \rangle \right) + (\mu_x^2 + \mu_y^2) \left(\frac{1}{3} - \frac{1}{3} \langle P_2 \rangle \right), \tag{6.97a}$$

$$\frac{(\varepsilon_{\perp} - 1)k_B T}{4\pi} = \langle \mu_X \mu_X \rangle = \mu_z^2 \left(\frac{1}{3} - \frac{1}{3} \langle P_2 \rangle \right) + (\mu_x^2 + \mu_y^2) \left(\frac{1}{3} + \frac{1}{6} \langle P_2 \rangle \right). \tag{6.97b}$$

Thus, the anisotropy of the dielectric permittivity is related to

$$\Delta\varepsilon = \frac{4\pi}{k_B T} \left[\mu_z^2 \langle P_2 \rangle - \frac{(\mu_x^2 + \mu_y^2)}{2} \langle P_2 \rangle \right] = \frac{4\pi \mu^2}{k_B T} \langle P_2 \rangle P_2(\cos\theta), \tag{6.98}$$

where θ is the angle that the dipole makes with the molecular z-axis. We see that, as far as the permanent dipole moment (the only we consider in our simplified treatment) is concerned, the dielectric permittivity anisotropy can be positive or negative if the angle θ is below or above the magic angle ($\approx 55°$), where $P_2(\cos\theta)$ changes sign.

6.11 Ionic Conductivity

Let us consider a fluid system containing N mobile particles (ions,...) with charge e_k subjected to a perturbation consisting of a periodic electric field. The perturbation Hamiltonian is

$$\mathcal{H}' = -B(\widetilde{\boldsymbol{r}}, \widetilde{\boldsymbol{p}})\, \boldsymbol{E}\, \mathrm{e}^{-i\omega t} = -\left(\sum_k e_k \boldsymbol{r}_k\right) \boldsymbol{E}\, \mathrm{e}^{-i\omega t}. \tag{6.99}$$

We now calculate the current induced by the field, i.e. the flux of charges: $\boldsymbol{j}(t) = \sum_{k=1}^{N} e_k \langle \dot{\boldsymbol{r}}_k \rangle$, by applying again Linear Response Theory. Referring to the general treatment in Section 6.9, we see that in this case the measured property A is the time derivative of the field coupling property B, i.e. $A = \dot{B}$. In the frequency domain $\boldsymbol{j}(\omega) = \boldsymbol{\sigma}(\omega)\boldsymbol{E}$. The susceptibility, i.e. the frequency dependent conductivity $\sigma(\omega)$ can be obtained from the general result in Eq. 6.73 as

$$\boldsymbol{\sigma}(\omega) = \frac{1}{k_B T} \int_0^{\infty} \mathrm{d}t\, C_{jj}(t)\, \mathrm{e}^{i\omega t}, \tag{6.100}$$

where

$$C_{jj}(t) = \sum_{k,k'} e_k e_{k'} \langle \boldsymbol{v}_k(0) \boldsymbol{v}_{k'}(t) \rangle. \tag{6.101}$$

In the anisotropic case, e.g. for ions in LCs, the conductivity should be different in the different directions, yielding a tensor correlation:

$$[C_{jj}]_{ab}(t) = \sum_{k,k'} e_k e_{k'} \left\langle v_{k,a}(0) v_{k',b}(t) \right\rangle. \tag{6.102}$$

In particular, for an oriented, monodomain, liquid crystal we have in the director frame

$$\sigma_{aa}(\omega) = \frac{1}{k_B T} \sum_{k,k'} e_k e_{k'} \int_0^{\infty} \mathrm{d}t\, \langle v_{k,a}(0) v_{k',a}(t) \rangle\, \mathrm{e}^{i\omega t}, \tag{6.103a}$$

$$\approx \frac{1}{k_B T} \sum_{k} e_k^2 \int_0^{\infty} \mathrm{d}t\, \langle v_{k,a}(0) v_{k,a}(t) \rangle\, \mathrm{e}^{i\omega t}. \tag{6.103b}$$

As an example of application, the zero-frequency conductivity of an ionic solution with species A, B will be

$$\boldsymbol{\sigma}(0) = \frac{1}{k_B T} \int_0^{\infty} \mathrm{d}t \sum_{A,B} \left[\delta_{A,B}\, \rho_A \langle \boldsymbol{j}_A(0) \boldsymbol{j}_A(t) \rangle + V \rho_A \rho_B \langle \boldsymbol{j}_A(0) \boldsymbol{j}_B(t) \rangle\right]. \tag{6.104}$$

The ionic conductivity in liquid crystals is expected (and found) to be anisotropic. There is a limited number of examples available. In Fig. 6.9 we see dielectric and conductivity results for 5CB (nematic) and 8CB (nematic and smectic) [Jadzyn and Kedziora, 2006].

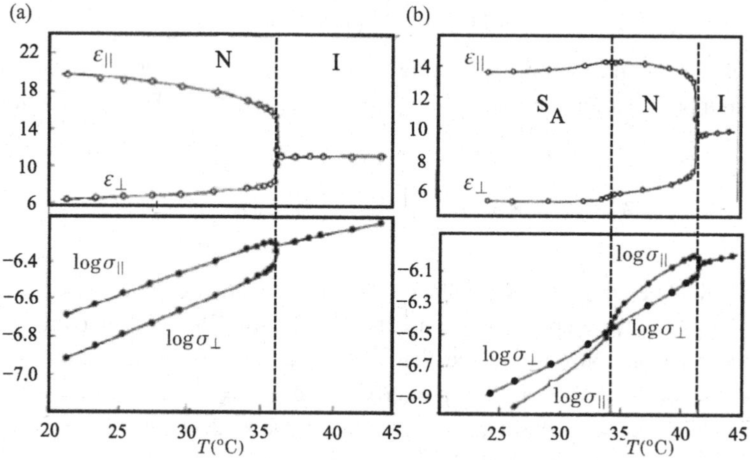

Figure 6.9 The dielectric constant and conductivity components ($\varepsilon_{\|}, \varepsilon_{\perp}$) and $\log \sigma_{\|}$, $\log \sigma_{\perp}$ for (a) 5CB in the isotropic and nematic phase and for (b) 8CB as a function of T(°C) in the isotropic, nematic and smectic phases [Jadzyn and Kedziora, 2006].

6.12 Thermal Conductivity

We can formulate thermal conductivity in the framework of Linear Response Theory [Sarman, 1994], by considering a heat field, $\mathbf{F}_{\rm e}$, with the property that the energy dissipated is proportional to $\boldsymbol{J}^{\rm Q}\mathbf{F}_{\rm e}$ and with the constraint that, with the thermostat off, the stationary condition for ϱ, the N particle distribution function, is satisfied ($\mathrm{d}\varrho/\mathrm{d}t = 0$). Given the expression for the thermal conductivity $\boldsymbol{\lambda}$ as heat current for small values of the heat field,

$$\lim_{\mathbf{F}_{\rm e}\to 0} \boldsymbol{J}^{\rm Q} = \boldsymbol{\lambda}\mathbf{F}_{\rm e}, \tag{6.105}$$

where the heat field $\mathbf{F}_{\mathbf{e}}$ is identified with the negative of the logarithmic temperature gradient $\nabla \ln T$, $\boldsymbol{J}_Q$ is the heat current, T is the absolute temperature. In an aligned anisotropic system, the heat conductivity is a tensor, with three independent components, λ_{aa}, $a = x, y,$ z in the director frame, where the components refer to the temperature gradient and the heat current in different directions.

In a nematic liquid crystal subjected to a temperature gradient ∇T a heat flow is induced and follows the generalized Fourier's law [Evans and Murad, 1989; Sarman and Laaksonen, 2011] which, in a uniaxial nematic liquid crystal (and assuming that the system remains in the same phase during the experiment), gives

$$\langle \boldsymbol{J}_Q \rangle = -\left[\lambda_{\|}\boldsymbol{dd} + \lambda_{\perp}(\mathbf{1} - \boldsymbol{dd})\right] \cdot \nabla T, \tag{6.106}$$

where $\boldsymbol{d}$ is the nematic director and $\lambda_{\|}$ and $\lambda_{\perp}$ are the parallel and perpendicular components of the heat conductivity tensor. The heat conductivity components can be obtained by MD simulations [Sarman and Laaksonen, 2011] in terms of an equilibrium time correlation function, using the Green–Kubo relation:

$$\lambda_{aa} = \frac{V}{k_{\rm B}T^2} \int_0^{\infty} \mathrm{d}t \left\langle J_a^{\rm Q}(t) J_a^{\rm Q}(0) \right\rangle, \tag{6.107}$$

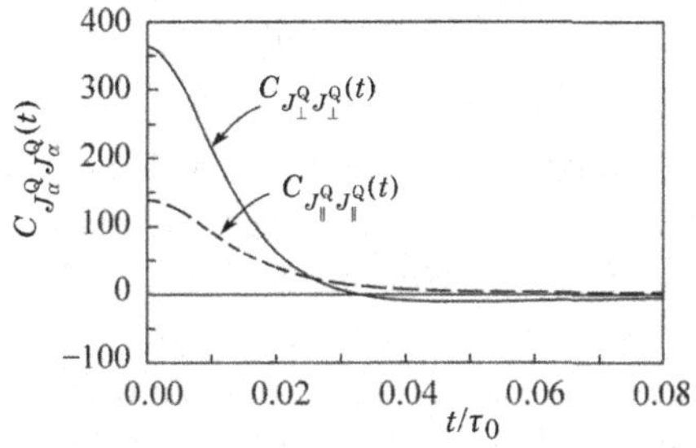

Figure 6.10 The time (scaled units) evolution of heat current correlation functions parallel $C_{J^Q_\parallel J^Q_\parallel}(t) = \frac{V}{k_BT}\left\langle J^Q_\parallel(t)J^Q_\parallel(0)\right\rangle$ (- - -) and perpendicular $C_{J^Q_\perp J^Q_\perp}(t) = \frac{V}{k_BT}\left\langle J^Q_\perp(t)J^Q_\perp(0)\right\rangle$ (——) to the cholesteric axis in a system of prolate chiral GB ellipsoids (see text) [Sarman and Laaksonen, 2013].

where J^Q_a and λ_{aa} are the heat flux vector and the heat conductivity in the a-direction, $a = x, y, z$, while V is the volume of the system. The heat flux vector for rigid molecules is given by [Ravi et al., 1992]

$$J^Q V = \frac{1}{2}\sum_{i=1}^{N}\frac{\boldsymbol{p}_i}{m}\left(\frac{p_i^2}{m} + \dot{\boldsymbol{\Omega}}_{\mathrm{M},i}\cdot\mathbf{I}_\mathrm{M}\cdot\dot{\boldsymbol{\Omega}}_{\mathrm{M},i} + \sum_{j=1}^{N}U_{ij}\right)$$
$$+\frac{1}{2}\sum_{i,j=1}^{N}\boldsymbol{r}_{ij}\left(\frac{\boldsymbol{p}_i\cdot\boldsymbol{f}_{ij}}{m} + \dot{\boldsymbol{\Omega}}_{\mathrm{M},i}\cdot\boldsymbol{N}_{\mathrm{M,ij}}\right), \tag{6.108}$$

where $\boldsymbol{p}_i$ and $\dot{\boldsymbol{\Omega}}_i$ are the linear momentum and angular velocity of particle i, U_{ij} is the pair potential between molecules i and j, and $\boldsymbol{f}_{ij}$ and $\boldsymbol{N}_{\mathrm{M,ij}}$ are the force and torque exerted on molecule i by molecule j, $\boldsymbol{r}_{ij} = \boldsymbol{r}_i - \boldsymbol{r}_j$ where $\boldsymbol{r}_j$ and $\boldsymbol{r}_i$ are the positions of particles i and j, while m is the molecular mass and $\mathbf{I}_\mathrm{M}$ is the inertia tensor in the molecule principal frame. Numerical results for a system of rigid molecules can be obtained from the heat flux vector and forming the heat current correlation function in an equilibrium molecular dynamics simulation (Chapter 9). Sarman has calculated the thermal conductivity of nematic [Sarman, 1994] and cholesteric LCs [Sarman and Laaksonen, 2013] consisting of various prolate and oblate soft ellipsoid fluids, respectively. In the nematic fluid, the thermal conductivity parallel to the director $\lambda_\parallel$, is greater than that perpendicular to the director $\lambda_\perp$. In the fluid of discotic particles, the reverse is true and $\lambda_\perp > \lambda_\parallel$. The torque exerted by the temperature gradient on the molecules favours twisting of the prolate ellipsoids towards the orientation perpendicular to the temperature gradient. The oblate ellipsoids are instead twisted towards the parallel orientation.

As for cholesterics, in Fig. 6.10 we see an example for a system of chiral mesogens [Sarman and Laaksonen, 2013], where the pair potential chosen is that originally proposed by Memmer et al. [1993]: $U_{ij}^{GB\chi} = U_{ij}^{GB} + \lambda_\chi U_{ij}^{\chi}$, where U_{ij}^{GB} is a Gay–Berne with the parameterization GB(4.4, 20, 1, 1) [Bates and Luckhurst, 1999a] (cf. Section 5.6.3) and U_{ij}^{χ} the pseudoscalar term introduced in Eq. 5.136 with the chiral parameter of $\lambda_\chi = 0.45$, which gives an equilibrium pitch of $45\sigma_0$. In this case, the thermal transport is easier transversally to the helix axis.

6.13 Viscosities

We consider a liquid (or liquid crystal) film of thickness h along y and we assume that a certain shear force (i.e. tangential force) $\boldsymbol{f}_x$ is applied along x (see Fig. 1.3). Introducing some terminology, we define the *shear strain*, i.e. the tangential displacement over thickness as $\varepsilon = \delta x/h$, and correspondingly the *shear rate* $\dot{\varepsilon} = \partial v_x/\partial y$. The *shear stress* or tangential force per area is $\sigma = f_x/A$. For an ideal viscous liquid $\sigma(t) = \eta\dot{\varepsilon}$, where η is called the *shear viscosity* or *dynamic viscosity*. To obtain a microscopic expression for the shear viscosity we consider an idealized experiment where a certain shear rate $\dot{\varepsilon}$ is applied and a certain shear stress σ_{xy} (force/area) observed. Thus, each layer of fluid is subject to a certain external velocity $\boldsymbol{v}(y)$ depending on y, and we have a velocity gradient. The elements of the *stress tensor* (the negative of the pressure tensor) are defined by:

$$\sigma_{ab} = \frac{1}{V}\Big[\frac{1}{m}\sum_i p_{ia}p_{ib} + \sum_i\sum_{j>i}(r_{ij})_a(f_{ij})_b\Big], \quad a, b = x, y, z, \tag{6.109}$$

where V is the volume of the system, $\boldsymbol{p}_i$ is the linear momentum of molecule i, and $\boldsymbol{r}_{ij}$ and $\boldsymbol{f}_{ij}$ are, respectively, the distance vector and force between molecules i and j. In our case we have

$$\sigma_{xy} = \frac{1}{V}\Big[\frac{1}{m}\sum_i p_{ix}p_{iy} + \sum_i\sum_{j>i}(r_{ij})_x(f_{ij})_y\Big]. \tag{6.110}$$

In order to apply Linear Response Theory we have to identify a perturbation in the form of Eq. 6.62. A body moving with a fixed velocity $\boldsymbol{V}_B$ has an extra term added to the Hamiltonian [Landau and Lifshitz, 1993] $\mathscr{H}_1 = \boldsymbol{p}_T \cdot \boldsymbol{V}_B$, where $\boldsymbol{p}_T$ is the total linear momentum of the molecular system: $\boldsymbol{p}_T = \sum_i m_i \boldsymbol{v}_i$. In our case each layer has its own momentum so that the perturbation is

$$\mathscr{H}_1 = \sum_i m_i \int_0^{y_i} \mathrm{d}y\,(v_{i_x})\,\frac{\partial V_{i_x}}{\partial y} \tag{6.111}$$

and the coupling property B employed in the Linear Response Theory (cf. 6.62) is thus formally $B = \sum_i m_i \int_0^{y_i} \mathrm{d}y(v_{ix})$ and[6]

$$\begin{aligned}\dot{B} &= \sum_i m_i \int_0^{y_i} \mathrm{d}y\dot{v}_{ix} + \Big(\sum_i m_i v_{ix}\Big)\dot{y}_i, \\ &= \sum_i m_i y_i a_{ix} + \Big(\sum_i m_i v_{ix}\Big)v_{iy} = \sum_i f_{ix}y_i + \sum_i m_i v_{ix}v_{iy} \equiv \sigma_{xy},\end{aligned} \tag{6.112}$$

where $\boldsymbol{f}_i$ is the force acting on molecule i. We see that the derivative of the coupling term is the stress tensor σ_{xy}. The viscosity is then determined by a shear stress autocorrelation If we imagine the applied perturbation to be sinusoidal with frequency ω we have

$$\eta(\omega) = \frac{1}{Vk_BT}\int_0^{\infty} \mathrm{d}t\, \mathrm{e}^{i\omega t}\langle\sigma_{xy}(0)\sigma_{xy}(t)\rangle \tag{6.113}$$

[6] We recall that $\frac{\mathrm{d}}{\mathrm{d}t}\int_{a(t)}^{b(t)} f(x,t)\mathrm{d}x = \int_{a(t)}^{b(t)} \frac{\partial f(x,t)}{\partial t}\mathrm{d}x + f(b,t)\frac{\mathrm{d}b}{\mathrm{d}t} - f(a,t)\frac{\mathrm{d}a}{\mathrm{d}t}$.

and the usual viscosity η when $\omega = 0$. This is the so-called Green–Kubo formula for the viscosity and allows $\eta(\omega)$ to be calculated from equilibrium molecular dynamics simulations.

Liquid crystal viscosities. Considering, more specifically, liquid crystals, the dynamical description of a compressible nematic requires six coefficients [Smondyrev et al., 1995]: three shear viscosities, ν_1, ν_2 and ν_3; two bulk viscosities, $\nu_4 - \nu_2$ and ν_5; and a director rotational viscosity, γ_1. To calculate these viscosities from correlation functions of the stress tensor and the director, we consider a coordinate system where the z-axis is parallel to the average orientation of the director, and the x and y axes are perpendicular to the director. Starting from the elements of the stress tensor defined by Eq. 6.109, the five viscosities $\nu_1, \nu_2, \nu_3, \nu_4$ and ν_5 associated with shear and compression are given in terms of Green–Kubo-like formulas [Forster, 1974]:

$$\nu_1 = \frac{V}{2k_BT}\int_0^\infty \mathrm{d}t\ \left\{\langle[\sigma_{zz}(t) - \sigma(t)][\sigma_{zz}(0) - \sigma(0)]\rangle - \langle\sigma_{xy}(t)\sigma_{xy}(0)\rangle\right\}, \tag{6.114a}$$

$$\nu_2 = \frac{V}{k_BT}\int_0^\infty \mathrm{d}t\ \langle\sigma_{xy}(t)\sigma_{xy}(0)\rangle, \tag{6.114b}$$

$$\nu_3 = \frac{V}{k_BT}\int_0^\infty \mathrm{d}t\ \langle\sigma_{xz}(t)\sigma_{xz}(0)\rangle, \tag{6.114c}$$

$$\nu_4 = \frac{V}{k_BT}\int_0^\infty \mathrm{d}t\ \langle\sigma(t)\sigma(0)\rangle, \tag{6.114d}$$

$$\nu_5 = \frac{V}{k_BT}\int_0^\infty \mathrm{d}t\ \langle\sigma_{zz}(t)\sigma(0)\rangle, \tag{6.114e}$$

where $\sigma(t) = \frac{1}{2}\left(\sigma_{xx}(t) + \sigma_{yy}(t)\right)$. The ν_i viscosity coefficients are related to the experimentally measurable Miesowicz viscosities η_i briefly described in Chapter 1 (see Fig. 1.3)

$$\eta_1 = \nu_3 + \frac{1}{4}\gamma_1(1-\lambda)^2 + \lambda\gamma_1, \tag{6.115a}$$

$$\eta_2 = \nu_3 + \frac{1}{4}\gamma_1(1-\lambda)^2, \tag{6.115b}$$

$$\eta_3 = \nu_2. \tag{6.115c}$$

The parameter λ is a reactive coefficient which determines the response of the director to shear flow and for PAA: $\lambda = 1.15 \pm 0.10$ [Forster, 1974]. Typical experimental results for the three Miesowicz shear viscosities are shown in Fig. 1.3. As the nematic order parameter increases, η_2 decreases at first as the temperature is lowered and then rises, as observed experimentally in PAA and MBBA [Langevin, 1972], and in the cyano-biphenyl homologues. The viscosity η_2 is associated with shear flow parallel to the director, so its value drops when the nematic order becomes appreciable.

The calculation of the director rotational viscosity, γ_1, is important for applications as it determines, e.g. the on and off switching times: $\tau_{\mathrm{on}} = \left(\frac{\gamma_1 h^2}{V^2 - V_0^2}\right)/(\varepsilon_0\Delta\varepsilon)$ and $\tau_{\mathrm{off}} = (\gamma_1 h^2)/(K\pi^2)$, of a twisted nematic display [Hirschmann and Reiffenrath, 1998] (see Fig. 1.7), where γ_1 is the twist viscosity, h is the cell thickness, K is a linear combination

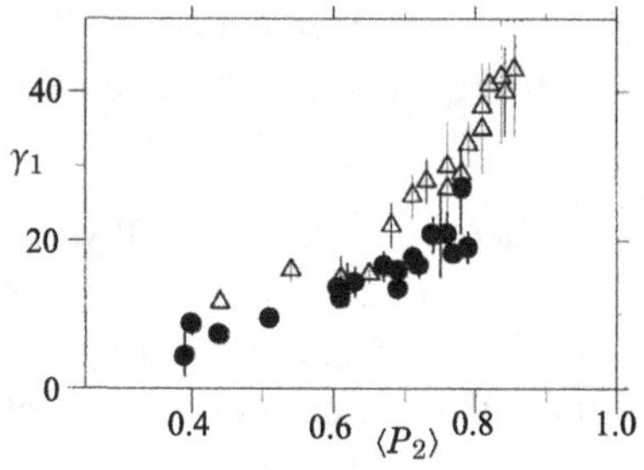

Figure 6.11 Rotational viscosity γ_1 vs. $\langle P_2 \rangle$ for two GB systems (Section 5.6.3): GB(3,5,2,1) at density $\rho^* = 0.345$ (●) and GB(3,1,2,1) at $\rho^* = 0.345$ (Δ) [Cuetos et al., 2002].

of elastic constants of the material [Tarumi et al., 1992], V is the voltage applied across the cell, V_0 is the threshold voltage, ε_0 is the permittivity of vacuum and $\Delta\varepsilon$ is the dielectric anisotropy. In terms of microscopic correlations [Sarman and Evans, 1993]

$$\gamma_1 = (k_B T)/[V \int_0^\infty \mathrm{d}t \langle \Omega_2^{(d)}(t)\, \Omega_2^{(d)}(0) \rangle]. \tag{6.116}$$

In Fig. 6.11 we show the results of N, V, T molecular dynamics simulations calculations [Cuetos et al., 2002] that proceeds by determining first the director, $\boldsymbol{d}$ and its first derivative, $\dot{\boldsymbol{d}}$, from which the director angular velocity, $\dot{\boldsymbol{\Omega}}_d = \boldsymbol{d} \times \dot{\boldsymbol{d}}$ is calculated. These are then transformed to a coordinate frame with the director along the z-axis, such that the director angular velocity, $\dot{\boldsymbol{\Omega}}_d^{(d)}$ becomes $\left(\dot{\Omega}_{d,x}^{(d)}, \dot{\Omega}_{d,y}^{(d)}, 0\right)$. The latter is important because the director can diffuse on the timescale of the simulations, while Eq. 6.116 was derived under the assumption that the director orientation is constant.

6.14 Molecular Reorientation as a Stochastic Process

In the previous sections we have discussed the generalities of dynamic processes in condensed media and liquid crystals and their description in terms of correlation functions. We have also established a connection between these correlations and measurable properties. However, many experimental techniques do not provide access to the full time (or frequency) dependence of correlations functions, or in other words, the relative raw experimental results cannot be fully mapped into them. The only practical way to analyze results is often to fit experimental data to some model of single-molecule motion, assuming, for instance, that reorientation takes place in a *Brownian* fashion through small incremental angular steps and then use experimental data to extract, by some sort of fitting, the parameters of the model, e.g. diffusion coefficient (integrals of correlation functions, like in Eq. 6.59). Comparison with computer simulation data is then also more likely to take place comparing these reduced information parameters with experiment. The problem is particularly severe (and interesting) in liquid crystals, where anisotropy multiplies the number of observables, and it is important to identify the quantities that experiments can in principle determine. Here we shall then consider stochastic models and particularly the most important one: rotational diffusion.

6.14.1 Rotational Diffusion

The diffusional model is based on the hypothesis that molecular reorientation can be considered a stochastic Markov process (Appendix K) evolving through small angular steps under the influence of collisions with the surrounding molecules and of any orienting torque provided by the long-range orientational order present in the medium. Angular momenta relaxation is assumed to be so fast with respect to orientational relaxation that it can be neglected when treating reorientations. The conditions under which this is possible have been discussed in detail in the literature (see, e.g., [Steele, 1976]). We only notice here that a rotational diffusion mechanism is particularly plausible when the reorienting molecule that is observed (e.g. a solute) is bulkier than the surrounding solvent ones. In practice, the diffusion model has been very successful in describing reorientation of molecules not only much larger but also comparable in size with those of the solvent. Indeed, the rotational diffusion model has been applied to the interpretation of a variety of experiments in liquids and in liquid crystals where a time decay or a linewidth is fitted by a dynamic model [Tarroni and Zannoni, 1991].

In an experiment probing molecular reorientation in liquids what is normally important are the correlation functions between spherical components of rank L, L' of the tensor properties $\mathbf{A}, \mathbf{B}$ under consideration or their Fourier transform at frequency ω, i.e. the spectral densities $J_{AB}^{LL'}(\omega)$ [Gordon, 1968], and a most important case is that of autocorrelation functions, where the two properties are the same. For instance, in magnetic resonance, $\mathbf{A}, \mathbf{B}$ could be some second-rank, $L = 2$, magnetic tensor relevant to the experiment being performed and in particular dipolar and quadrupolar interactions have been studied [Dong, 2016], Raman techniques [Kirov et al., 1985; Fontana et al., 1986; Wang et al., 1988]. In Infrared Dichroic [Dozov et al., 1984; Simova et al., 1988] and dielectric relaxation measurements [Nordio et al., 1973; Williams, 1994; Sebastian et al., 2017], tensors of rank $L = 1$ are studied. In a Fluorescence Depolarization experiment $A^{L,m}$, $B^{L,m}$ would be second-rank absorption and emission tensors, respectively (as we have seen in Sec.3.4.3) [Zannoni, 1979d; Arcioni et al., 1987; Bauman et al., 2008]. When we deal with rigid molecules and when intermolecular contributions to the observed quantities can be neglected, the various experiments mentioned can be interpreted in a unified way in terms of orientational correlation functions $\phi_{mn;m'n'}^{LL'}(t)$. Writing the laboratory fixed components in the chosen molecular frame we have in fact

$$\left\langle A_{\mathrm{LAB}}^{L,m}(0) B_{\mathrm{LAB}}^{L',m'*}(t) \right\rangle = \sum_{n,n'} \left\langle \mathcal{D}_{m,n}^{L}(0) \mathcal{D}_{m',n'}^{L'*}(t) \right\rangle A_{\mathrm{MOL}}^{L,n} B_{\mathrm{MOL}}^{L',n'*}, \tag{6.117a}$$

$$= \sum_{n,n'} \phi_{mn;m'n'}^{LL'*}(t) A_{\mathrm{MOL}}^{L,n} B_{\mathrm{MOL}}^{L',n'*}, \tag{6.117b}$$

with $\mathcal{D}_{m,n}^{L*}$ a Wigner rotation matrix (Appendix F) connecting the two frames and n, n' range from $-\min(L, L')$ to $+\min(L, L')$. If the relaxation of the angular variables is assumed to be a stochastic Markov process (Appendix K), the orientational correlation functions can be written as

$$\phi^{LL'}_{mn,m'n'}(t) = \int\int \mathrm{d}\Omega_0 \mathrm{d}\Omega\, P(\Omega_0)\mathcal{D}^{L}_{m,n}(\Omega_0)P(\Omega_0|\Omega t)\mathcal{D}^{L'*}_{m',n'}(\Omega), \tag{6.118}$$

where $\Omega \equiv (\alpha, \beta, \gamma)$ is the set of three Euler angles defining molecular orientation. $P(\Omega_0|\Omega t)$ is the so-called conditional probability, giving the probability of finding a molecule at orientation Ω at time t, if the orientation of the molecule was Ω_0 at time 0. The orientational distribution function $P(\Omega)$, represents the equilibrium probability of finding a molecule at orientation Ω that we discussed in Chapter 3, i.e.

$$P(\Omega) = \exp[-\mathcal{U}(\Omega)]/\int \mathrm{d}\Omega\, \exp[-\mathcal{U}(\Omega)], \tag{6.119}$$

with $\mathcal{U}(\Omega) \equiv U(\Omega)/k_B T$. The effective anisotropic potential acting on the reorienting molecule as a result of all the others could then be obtained from an orientational distribution obtained from Mean Field Theory (see, Chapter 7) or from some computer simulation (see, e.g., Fig. 3.4) as $\mathcal{U}(\Omega) = -\ln P(\Omega) + \mathcal{U}_0$ with $\mathcal{U}_0$ a constant. As we have seen in Chapter 3, this can also be obtained from some measured order parameters using the Maximum Entropy approach for uniaxial (Eq. 3.73) or biaxial (Eq. 3.141) molecules. Note that the distribution $P(\Omega)$ and the potential $\mathcal{U}(\Omega)$ obey the same symmetry as the mesophase. Here we assume the probe to be biaxial (or uniaxial as a special case), but the host solvent to be uniaxial, so that $P(\Omega) = P(\beta, \gamma)$ and the orientational order parameters are the average Wigner rotation matrix $\langle \mathcal{D}^{L}_{m,n}(\Omega)\rangle$ (see Section 3.10). The case of effectively uniaxial molecules reorienting in a biaxial phase has been treated in Berggren et al. [1993], that of biaxial molecules dissolved in a biaxial phase in Berggren and Zannoni [1995]. For a molecule undergoing rotational diffusion in an anisotropic potential $\mathcal{U}(\Omega)$, the conditional probability $P(\Omega_0|\Omega t)$ evolves in time according to the differential evolution equation [Nordio and Segre, 1979]:

$$\frac{\partial P(\Omega_0|\Omega t)}{\partial t} = -\hat{\mathbf{J}}\mathbf{D}\left[\hat{\mathbf{J}} + \hat{\mathbf{J}}\mathcal{U}(\Omega)\right] P(\Omega_0|\Omega t), \tag{6.120a}$$

$$= -\sum_{\substack{i=\\x,y,z}} D^{\mathrm{MOL}}_{ii}\left[\hat{\mathbf{J}}^2_i + \hat{\mathbf{J}}_i[\hat{\mathbf{J}}_i\mathcal{U}(\Omega)]\right] P(\Omega_0|\Omega t), \tag{6.120b}$$

where $\hat{\mathbf{J}} = (\hat{\mathbf{J}}_x, \hat{\mathbf{J}}_y, \hat{\mathbf{J}}_z)$ is the dimensionless angular momentum operator (Eqs. F.34 and F.39c) and $\mathbf{D}$ is the rotational diffusional tensor (Eq. 6.59). Note that only derivatives of the potential $\mathcal{U}(\Omega)$ enter, so an additive constant $\mathcal{U}_0$ is irrelevant. We assume that the molecule starts its evolution from a certain orientation Ω_0, i.e. that the initial condition is

$$P(\Omega_0|\Omega 0) = \delta(\Omega - \Omega_0). \tag{6.121}$$

In Eq. 6.120b we have implicitly chosen the molecular frame where $\mathbf{D}$ is diagonal. The analysis of experimental dynamic data offers the possibility of getting rotational diffusion coefficients for the molecule acting as spectroscopic probe, which may be different from the solvent ones or, at least in some spectroscopic techniques, identical to those of the solvent. When we assume rotation to be diffusional, all our knowledge of the molecule and of its symmetry is contained in its diffusion tensor. Thus, as long as we can identify the diffusion frame, the most general or the least symmetric case possible is that of a biaxial diffusion tensor, where $D_{xx}^{\text{MOL}} \neq D_{yy}^{\text{MOL}} \neq D_{zz}^{\text{MOL}}$. This case has been treated for isotropic liquids ($\mathscr{U}(\Omega) = 0$), by various authors and analytic solutions are available for symmetric and asymmetric rotors [Edwardes, 1892; Perrin, 1934; Freed, 1964; Huntress, 1970; Chuang and Eisenthal, 1972]. In this limit the Wigner rotation matrices are eigenfunctions of the diffusion operator, $\hat{\mathbf{J}}\mathbf{D}\hat{\mathbf{J}}$. Thus, even for asymmetric rotors, there are no mixed-rank correlation functions in isotropic fluids. An orientational autocorrelation function of rank L decays at most as a sum of $(2L+1)$ exponentials. In anisotropic solvents the problem is made more complicated by the absence of analytic solutions but at the same time more interesting by the increase in the number of observable correlation functions [Zannoni, 2000] and with the possibility to study anisotropic interactions. To determine $P(\Omega_0|\Omega t)$ the diffusion operator can be given a matrix representation in a basis of Wigner rotation matrices. The resulting matrix is not diagonal and a sufficiently large basis set of Wigner functions (up to a certain rank J_{max} is needed). In LCs a mixing of certain contributions of different rank can take place and orientational correlation functions are generally given by a sum of an infinite number of exponentials. After Nordio et al's. classical papers on uniaxial molecules in uniaxial orienting potentials [Nordio and Busolin, 1971; Nordio and Segre, 1979], the diffusion model has also been solved for molecules behaving like symmetric rotors ($D_{xx}^{\text{MOL}} = D_{yy}^{\text{MOL}}$) while having a non-negligible biaxial order [Polnaszek et al., 1973; Nordio and Segre, 1975; Bernassau et al., 1982; Dozov et al., 1987; Arcioni et al., 1988]. This is clearly important since the ordering matrix for a number of probe and liquid crystal molecules has been determined experimentally and found to deviate from cylindrical symmetry (see, e.g., papers in [Emsley, 1985]). Estimating the effective orienting potential from the measured biaxial order parameters $\langle \mathscr{D}_{0n}^{2} \rangle$, is possible, e.g. with the maximum entropy technique (seen in Chapter 3). Numerical, non-perturbative solutions were presented for the analysis of ESR spectra by Polnaszek et al. [1973]. Both in this case and in the simpler uniaxial case [Nordio and Busolin, 1971] it was found that terms of rank as high as $J_{max} \approx 20$ were needed in the matrix representation to obtain sufficiently accurate results over the entire $\langle P_2 \rangle$ range, even though a few terms are sufficient for low-order parameters or for some of the correlation functions. It should be stressed that a feature present in all the mentioned works was the assumption of a diffusional tensor with cylindrical symmetry, $D_{xx}^{\text{MOL}} = D_{yy}^{\text{MOL}}$, which somewhat contrasts with the lower symmetry of the anisotropic potential. The complications increase further when a fully asymmetric rotor $D_{xx}^{\text{MOL}} \neq D_{yy}^{\text{MOL}} \neq D_{zz}^{\text{MOL}}$ in a biaxial potential $U(\beta, \gamma)$ is considered. Here we describe the rotational diffusion of asymmetric rotors subject to biaxial potentials in a rather general way, following the treatment by Tarroni and Zannoni [1991]. Having a theory connecting observables to order parameters and the diffusion tensor components is essential for their determination. The theory has been applied to the analysis

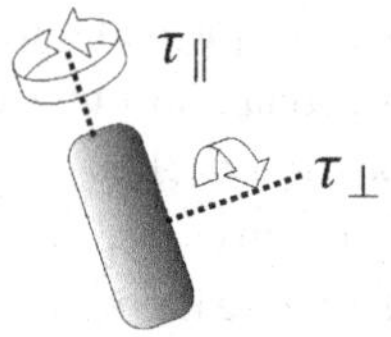

Figure 6.12 The spinning and tumbling reorientation processes for an elongated molecule with their respective characteristic times $\tau_{\parallel}$ (spinning) and $\tau_{\perp}$ (tumbling).

of NMR data in LCs [Dong and Shen, 1996; Dong, 1997; Domenici et al., 2005]. To start with we rewrite our biaxial rotational diffusion tensor in a more convenient form as:

$$\mathbf{D}^{\mathrm{MOL}} = D_\rho \begin{pmatrix} 1+\epsilon_{\mathrm{M}} & 0 & 0 \\ 0 & 1-\epsilon_{\mathrm{M}} & 0 \\ 0 & 0 & \eta_{\mathrm{M}} \end{pmatrix}, \tag{6.122}$$

where $D_\rho \equiv \frac{D_{xx}^{\mathrm{MOL}}+D_{yy}^{\mathrm{MOL}}}{2}$, $\epsilon_{\mathrm{M}} \equiv \frac{D_{xx}^{\mathrm{MOL}}-D_{yy}^{\mathrm{MOL}}}{D_{xx}^{\mathrm{MOL}}+D_{yy}^{\mathrm{MOL}}}$, $\eta_{\mathrm{M}} \equiv \frac{2D_{zz}^{\mathrm{MOL}}}{D_{xx}^{\mathrm{MOL}}+D_{yy}^{\mathrm{MOL}}}$. Here ϵ_{M} gives the diffusion tensor biaxiality and η_{M} the ratio between the *spinning* diffusion coefficient D_{zz}^{MOL}, which refers to rotations around the z-axis and that of the z-axis itself or *tumbling* (see Fig. 6.12). In the cylindrical symmetry limit ϵ_{M} reduces to 0, while D_ρ and η_{M} become $D_{\parallel}$ and $D_{\perp}$, respectively. With these definitions, we can rewrite Eq. 6.120b as:

$$\begin{aligned}
\frac{1}{D_\rho}\frac{\partial P(\Omega_0|\Omega t)}{\partial t} &= \Big\{-(1+\epsilon_{\mathrm{M}})[\hat{\mathbf{J}}_x^2 + \hat{\mathbf{J}}_x(\hat{\mathbf{J}}_x\mathcal{U})] - (1-\epsilon_{\mathrm{M}})[\hat{\mathbf{J}}_y^2 + \hat{\mathbf{J}}_y(\hat{\mathbf{J}}_y\mathcal{U})] \\
&\quad - \eta_{\mathrm{M}}[\hat{\mathbf{J}}_z^2 + \hat{\mathbf{J}}_z(\hat{\mathbf{J}}_z\mathcal{U})]\Big\} P(\Omega_0|\Omega t), \qquad (6.123a) \\
&= \Big\{-[\hat{\mathbf{J}}_x^2 + \hat{\mathbf{J}}_x(\hat{\mathbf{J}}_x\mathcal{U})] - [\hat{\mathbf{J}}_y^2 + \hat{\mathbf{J}}_y(\hat{\mathbf{J}}_y\mathcal{U})] - \eta_{\mathrm{M}}[\hat{\mathbf{J}}_z^2 + \hat{\mathbf{J}}_z(\hat{\mathbf{J}}_z\mathcal{U})] \\
&\quad - \epsilon_{\mathrm{M}}[\hat{\mathbf{J}}_x^2 + \hat{\mathbf{J}}_x(\hat{\mathbf{J}}_x\mathcal{U})] + [\hat{\mathbf{J}}_y^2 + \hat{\mathbf{J}}_y(\hat{\mathbf{J}}_y\mathcal{U})]\Big\} P(\Omega_0|\Omega t), \qquad (6.123b) \\
&\equiv \mathbf{\Gamma} P(\Omega_0|\Omega t). \qquad (6.123c)
\end{aligned}$$

The diffusion operator $\mathbf{\Gamma}$ or *propagator,* as written, is not self-adjoint, but detailed balance ensures that it can be symmetrized by the transformation $\bar{\mathbf{\Gamma}} = P^{-\frac{1}{2}}(\Omega)\mathbf{\Gamma}P^{\frac{1}{2}}(\Omega) = \exp\left[\frac{1}{2}\mathcal{U}(\Omega)\right]\mathbf{\Gamma}\exp\left[-\frac{1}{2}\mathcal{U}(\Omega)\right]$. We add an overbar (here and in what follows) to indicate a symmetrized operator or quantity. Applying the symmetrizing transformation yields

$$\exp\left[\frac{\mathcal{U}(\Omega)}{2}\right][\hat{\mathbf{J}}_i^2 + \hat{\mathbf{J}}_i(\hat{\mathbf{J}}_i\mathcal{U})]\exp\left[\frac{-\mathcal{U}(\Omega)}{2}\right] = \hat{\mathbf{J}}_i^2 - \frac{1}{4}(\hat{\mathbf{J}}_i\mathcal{U})^2 + \frac{1}{2}(\hat{\mathbf{J}}_i^2\mathcal{U});\ i = x, y, z. \tag{6.124}$$

Taking advantage of this property the symmetrized propagator can be rewritten as:

$$\begin{aligned}
\bar{\mathbf{\Gamma}} &= -\left(\nabla^2 + \frac{1}{2}(\nabla^2\mathcal{U}) - \frac{1}{4}(\hat{\mathbf{J}}_+\mathcal{U})(\hat{\mathbf{J}}_-\mathcal{U}) - \frac{1}{4}\eta_{\mathrm{M}}(\hat{\mathbf{J}}_z\mathcal{U})^2\right) \\
&\quad - \epsilon_{\mathrm{M}}\left(\frac{1}{2}(\hat{\mathbf{J}}_+^2 + \hat{\mathbf{J}}_-^2) + \frac{1}{4}[(\hat{\mathbf{J}}_+^2 + \hat{\mathbf{J}}_-^2)\mathcal{U}] - \frac{1}{8}[(\hat{\mathbf{J}}_+\mathcal{U})^2 + (\hat{\mathbf{J}}_-\mathcal{U})^2]\right) = \bar{\mathbf{\Gamma}}^a + \bar{\mathbf{\Gamma}}^b,
\end{aligned} \tag{6.125}$$

where $\nabla^2 = \hat{\mathbf{J}}_x^2 + \hat{\mathbf{J}}_y^2 + \eta\hat{\mathbf{J}}_z^2$ is the Laplacian operator in terms of angular momentum and $\hat{\mathbf{J}}_\pm = \hat{\mathbf{J}}_x \pm i\hat{\mathbf{J}}_y$ are step-up and step-down (or *ladder*) operators. The $\bar{\mathbf{\Gamma}}^b$ term follows from the non-cylindrical symmetry of the diffusion tensor. The symmetrized form of the diffusion equation is:

$$\frac{1}{D_\rho}\frac{\partial \bar{P}(\Omega_0|\Omega t)}{\partial t} = \bar{\mathbf{\Gamma}}\bar{P}(\Omega_0|\Omega t), \tag{6.126}$$

where $\bar{P}(\Omega_0|\Omega t)$ is the symmetrized conditional probability related to the unsymmetrized one by the transformation:

$$\bar{P}(\Omega_0|\Omega t) = \exp\left[\frac{\mathscr{U}(\Omega)}{2}\right]P(\Omega_0|\Omega t)\,\exp\left[\frac{-\mathscr{U}(\Omega_0)}{2}\right] = P^{-\frac{1}{2}}(\Omega)P(\Omega_0|\Omega t)P^{\frac{1}{2}}(\Omega_0). \tag{6.127}$$

The expression for the reorientational correlation functions becomes

$$\phi^{LL'}_{mn;m'n'}(t) = \int\int d\Omega_0 d\Omega \mathscr{D}^L_{m,n}(\Omega_0)\mathscr{D}^{L'*}_{m',n'}(\Omega)P^{\frac{1}{2}}(\Omega_0)P^{\frac{1}{2}}(\Omega)\bar{P}(\Omega_0|\Omega t). \tag{6.128}$$

The symmetrized diffusional equation Eq. 6.126 can be given a convenient matrix representation by expanding the anisotropic potential $\mathscr{U}(\Omega)$ as well as the symmetrized conditional probability $\bar{P}(\Omega_0|\Omega t)$ in a basis of Wigner matrices (Appendix F). For a uniaxial liquid crystal host phase

$$\mathscr{U}(\Omega) = \sum_{J,q} a_{Jq}\mathscr{D}^J_{0,q}(\Omega), \tag{6.129}$$

and

$$\bar{P}(\Omega_0|\Omega t) = \sum_{Lmn}\sqrt{\frac{2L+1}{8\pi^2}}C_{Lmn}(\Omega_0,\Omega t)\mathscr{D}^L_{m,n}(\Omega). \tag{6.130}$$

The expansion coefficients $C_{Lmn}(\Omega_0,\Omega t)$ at time 0 can be evaluated using the initial condition Eq. 6.121 and the representation of $\delta(\Omega-\Omega_0)$ in the Wigner matrices basis, Eq. F.10, so that the time 0 initial condition becomes

$$C_{Lmn}(\Omega_0,\Omega 0) = \sqrt{\frac{2L+1}{8\pi^2}}\mathscr{D}^{L*}_{m,n}(\Omega_0). \tag{6.131}$$

Substituting Eq. 6.130 into the diffusion equation Eq. 6.126, multiplying both sides on the left for $\mathscr{D}^{L'*}_{m',n'}(\Omega)$ and integrating over Ω we obtain the system of linear differential equations

$$\frac{1}{D_\rho}\frac{\partial}{\partial t}C_{L'm'n'}(\Omega_0,t) = \sum_{Lmn}\bar{R}_{L'm'n',Lmn}C_{Lmn}(\Omega_0,t), \tag{6.132}$$

where $\bar{R}_{L'm'n',Lmn}$ are matrix elements of the $\bar{\mathbf{\Gamma}}$ operator.

$$\bar{R}_{L'm'n',Lmn} = \frac{\sqrt{(2L'+1)(2L+1)}}{8\pi^2}\langle\mathscr{D}^{L'}_{m',n'}|\,\bar{\mathbf{\Gamma}}\,|\mathscr{D}^L_{m,n}\rangle, \tag{6.133a}$$

$$\equiv \frac{\sqrt{(2L'+1)(2L+1)}}{8\pi^2}\int d\Omega\mathscr{D}^{L'*}_{m',n'}(\Omega)\,\bar{\mathbf{\Gamma}}\mathscr{D}^L_{m,n}(\Omega). \tag{6.133b}$$

This matrix representation is real and symmetric [Nordio and Segre, 1979; Tarroni and Zannoni, 1991]. In the presence of an anisotropic potential $\mathscr{U}(\Omega)$ the matrix will not be diagonal, but instead the matrix will have a band form, with bandwidth determined by the maximum rank of the contributions to $\mathscr{U}(\Omega)$ in Eq. 6.129. If the medium is uniaxial, as we assume here, then there will be no coupling between terms with different m, so the diffusion matrix $\mathbf{R}$ will be block diagonal and we can employ m to label these blocks: $(\bar{R}^{(m)})_{Ln,L'n'} \equiv \bar{R}_{Lmn,L'm'n'}\,\delta_{m,m'}$. The explicit expressions for the matrix elements $(\bar{R}^{(m)})_{Ln,L'n'}$ are a bit cumbersome but are reported in Tarroni and Zannoni [1991], where the general formulas are adapted to the important specific case of a potential containing only second-rank interactions. Eq. 6.132 can be rewritten as the matrix equation for the coefficients:

$$\dot{\mathbf{C}}^{(m)}(t) = D_\rho \bar{\mathbf{R}}^{(m)}\, \mathbf{C}^{(m)}(t). \tag{6.134}$$

Then, if $\bar{\mathbf{X}}^{(m)}$ is the eigenvector matrix diagonalizing $\bar{\mathbf{R}}^{(m)}$ to $\bar{\mathbf{r}}^{(m)}$, i.e. $\bar{\mathbf{R}}^{(m)}\bar{\mathbf{X}}^{(m)} = \bar{\mathbf{X}}^{(m)}\,\bar{\mathbf{r}}^{(m)}$, the solution to Eq. 6.134 becomes

$$\mathbf{C}^{(m)}(t) = \bar{\mathbf{X}}^{(m)}\, \mathrm{e}^{[tD_\rho\, \bar{\mathbf{r}}^{(m)}]}\, (\bar{\mathbf{X}}^{(m)})^T\, \mathbf{C}^{(m)}(0), \tag{6.135}$$

where we have used the fact that $\bar{\mathbf{X}}$ is a unitary matrix, being the eigenvector matrix of a self-adjoint operator, and thus that its inverse is just its transpose. Inserting the initial time coefficient (Eq. 6.131) gives

$$C_{Jp}^{(m)}(t) = \sum (\bar{X}^{(m)})_{Jp,K}\, \mathrm{e}^{[tD_\rho \bar{r}_K^{(m)}]} (\bar{X}^{(m)})_{J'p',K} \sqrt{\frac{2J'+1}{8\pi^2}}\, \mathscr{D}_{mp'}^{J'*}(\Omega_0), \tag{6.136}$$

where we have used the single index K to label the eigenvalues of $\bar{\mathbf{R}}^{(m)}$. For $t \to \infty$ all the exponentials decay to 0 except the one corresponding to the eigenvalue $\bar{r}_0^{(0)}$, corresponding to the equilibrium distribution, recovered from the limiting equilibrium condition

$$\lim_{t\to\infty} P(\Omega_0|\Omega t) = P(\Omega). \tag{6.137}$$

The final expression for the LL' correlation function of a probe of arbitrary symmetry reorienting in a uniaxial medium, like a nematic or a smectic A, can be written as a series of exponentials

$$\phi_{mn;n'}^{LL'}(t) \equiv \phi_{mn;m'n'}^{LL'}(t)\delta_{m,m'} = \sum_K \left(b_{LL'}^{m,nn'}\right)_K \exp\,[tD_\rho \bar{r}_K^{(m)}], \tag{6.138}$$

where the explicit expression for the coefficients $\left(b_{LL'}^{m,nn'}\right)_K$ is given in Tarroni and Zannoni [1991]. We now consider the important case of second-rank correlation functions: $L = L' = 2$, that arise in most experiments mentioned until now (in particular NMR [Dong, 1997]), and we take an effectively uniaxial probe molecule. The two correlation functions more often accessed by experiments, and that are thus worth calculating from computer simulations will be:

$$\phi_{00}^{22}(t) = \langle P_2(\cos\beta(0))\ P_2(\cos\beta(t))\rangle\,, \tag{6.139}$$

$$\phi_{02}^{22}(t) = \sqrt{\frac{3}{8}}\left\langle \sin^2\beta(0)\sin^2\beta(t)\ e^{i2(\gamma(0)-\gamma(t))}\right\rangle. \tag{6.140}$$

Even in this relatively simple case the orientational correlation functions are a sum of an infinite number of exponentials

$$\phi_{mn}^{22}(t) = \sum \left(b^{mn}\right)_K \exp\left(t/\tau_{mn}^K\right), \tag{6.141}$$

where the reciprocal eigenvalues of the diffusion matrix, i.e.

$$\tau_{mn}^K = 1/\left\{D_\perp r_K^{mn} t - \left(D_\parallel - D_\perp\right) n^2\right\}, \tag{6.142}$$

play the role of decay times for the various exponentials and

$$(b_{22}^{mn})_K = \frac{1}{5}\sum_{J'}(2J'+1)(X^{(mn)})_{2,K}(X^{(mn)})_{K,J'}^{-1}\langle\mathscr{D}_{mn}^2\mathscr{D}_{mn}^{J'*}\rangle. \tag{6.143}$$

In Fig. 6.13 we show, as an example, the first three pre-exponential coefficients and the corresponding decay times for the long-axis reorientational correlation in Eq. 6.141. Note that this correlation function does not depend on $D_\parallel$ and then does not report on the speed of reorientation around the molecular symetry axis. As Fig. 6.13 shows, in this case the expansion can be quite safely truncated to the first term over a wide range of parameters $\langle P_2\rangle$ and thus of temperatures. On the other hand, such a quick convergence is not always guaranteed, especially at relatively high order. Note that, even if the correlation functions are a sum of many exponentials, all of these can be calculated from a knowledge of the relevant diffusion tensor components and the order parameters characterizing the potential. In analyzing real or computer imulated data, these few parameters are the only to be determined by fitting.

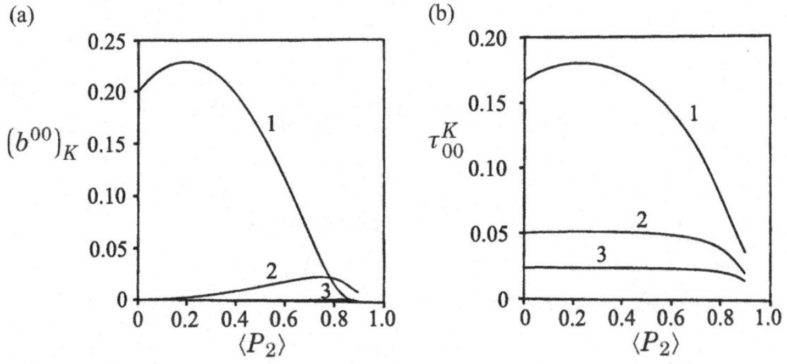

Figure 6.13 (a) The first three pre-exponentials $\left(b^{00}\right)_K$ and (b) the relative relaxation times τ_{00}^K in units of $D_\perp^{-1}$ (see Eq. 6.141) as a function of the order parameter $\langle P_2\rangle$ for the second rank correlation function $\phi_{00}^{22}(t)$ of a uniaxial probe reorienting in a P_2 potential [Zannoni et al., 1983].

6.14.2 Fluorescence Depolarization

Going back to the short description of a Fluorescence Depolarization (FD) experiment in Section 3.4.3 and the sketch in Fig. 3.8, that we invoked to describe the obtainment of fourth-rank order parameter $\langle P_4 \rangle$ in a special case of a fluorescence emission time τ_F much shorter than the typical fluorophore reorientation time τ_R, we now wish to consider the more general (and common) case of the two times being of the same order of magnitude, so that a time-dependent FD has to be considered. To do this we go back to Eq. 3.38, rewritten as:

$$I_{\rm io}^F(t) = F(t) \langle | \, \boldsymbol{e}_{\rm i} \cdot \boldsymbol{\mu}^{({\rm a})}(0) \, |^2 \; | \boldsymbol{e}_{\rm o} \cdot \boldsymbol{\mu}^{({\rm e})}(t) \, |^2 \rangle, \tag{6.144a}$$

$$= F(t) \langle | \mathbf{E}_{\rm i} : \mathbf{A}_{\rm LAB}^{({\rm a})}(\Omega_0, 0) | | \mathbf{E}_{\rm o} : \mathbf{A}_{\rm LAB}^{({\rm e})}(\Omega, t) | \rangle, \tag{6.144b}$$

$$= F(t) \sum_{L, L'} I_{\rm io}^{L, L'}(t) \, ; \quad L, L' = 0, 2, \tag{6.144c}$$

where $I_{\rm io}^{L,L'}(t) = \sum_{m,m'} E_{\rm i}^{L,m*} E_{\rm o}^{L',m'*} \langle [A_{\rm LAB}^{({\rm a})}(\Omega_0)]^{L,m} [A_{\rm LAB}^{({\rm e})}(\Omega_t)]^{L',m'} \rangle$ and $\mathbf{E}_{\rm i} = \boldsymbol{e}_{\rm i} \otimes \boldsymbol{e}_{\rm i}^*$, $\mathbf{E}_{\rm o} = \boldsymbol{e}_{\rm o} \otimes \boldsymbol{e}_{\rm o}^*$ are the input and output polarization tensors and $\mathbf{A}^{({\rm a})} = \boldsymbol{\mu}^{({\rm a})} \otimes \boldsymbol{\mu}^{({\rm a})*}$, $\mathbf{A}^{({\rm e})} = \boldsymbol{\mu}^{({\rm e})} \otimes \boldsymbol{\mu}^{({\rm e})*}$ the absorption and emission tensors. We have written the Cartesian contractions in Eq. 6.144b in the more convenient spherical tensor form, using the relations in Appendix B. Rotating the polarizer $\boldsymbol{e}_{\rm i}$ or the analyzer $\boldsymbol{e}_{\rm o}$ at a certain angle from the vertical, by a rotation around the incoming or outgoing axis, the various fluorescence intensities can be obtained. The expressions can be easily written in terms of the Wigner rotation matrices correlation functions by writing the lab frame absorption and emission tensor spherical components in terms of their molecular fixed counterparts (see Eq. 6.19). To be specific, let us assume that the input polarizer is set vertical, i.e. $\boldsymbol{e}_{\rm i} \parallel \mathbf{Z}$, then the required spherical components of the polarization tensors are $E_{\rm i}^{2,m} = \sqrt{\frac{2}{3}} \, \delta_{m,0}$. Moreover, if we set the analyzer at an angle ϵ to the vertical, the output polarizer tensor components will be $E_{\rm o}^{2,m'} = \sqrt{\frac{2}{3}} \, D_{m',0}^{2*}(0, \epsilon, 0)$. Considering a uniformly aligned sample with the director parallel to the laboratory Z-axis (Fig. 3.8), we find for molecules of arbitrary symmetry

$$I_{Z,\epsilon}^{0,0} = \frac{1}{9}, \tag{6.145a}$$

$$I_{Z,\epsilon}^{0,2} = \frac{1}{3}\sqrt{\frac{2}{3}} D_{0,0}^2(0,\epsilon,0) \sum_n \langle \, D_{0,n}^2 \, \rangle \big[A_{\rm MOL}^{({\rm e})} \big]^{2,n*}, \tag{6.145b}$$

$$I_{Z,\epsilon}^{2,0} = \frac{1}{3}\sqrt{\frac{2}{3}} D_{0,0}^2(0,\epsilon,0) \sum_n \langle \, D_{0,n}^{2*} \, \rangle \big[A_{\rm MOL}^{({\rm a})} \big]^{2,n}, \tag{6.145c}$$

$$I_{Z,\epsilon}^{2,2} = \frac{2}{3} D_{0,0}^2(0,\epsilon,0) \, \Phi_0(t), \tag{6.145d}$$

for vertical excitation and observation through a polarizer set at an angle ϵ from the vertical. Note that the dynamic information is all contained in the quantities $\Phi_q(t)$,

$$\Phi_q(t) \equiv \sum_{n,n'} \big[A_{\rm MOL}^{({\rm a})} \big]^{2,n} \big[A_{\rm MOL}^{({\rm e})} \big]^{2,n*} \phi_{qn;n'}^{22}(t). \tag{6.146}$$

Placing the analyzer at the magic angle $\epsilon_m \approx 54.7^\circ$ cancels the dynamic molecular reorientation contribution to the intensity as well as the first static contribution. Thus, this magic angle experimental setting can serve to obtain the fluorescence decay time. We also find in particular for the analyzer set at $\epsilon = 0$ and $\frac{\pi}{2}$ that

$$\frac{I_{ZZ}}{F(t)} = \frac{1}{9} + \frac{1}{3}\sqrt{\frac{2}{3}} \sum_n \langle D^{2*}_{0,n} \rangle \left(\left[A^{(e)}_{\text{MOL}}\right]^{2,n} + \left[A^{(a)}_{\text{MOL}}\right]^{2,n} \right) + \frac{2}{3}\Phi_0(t) \tag{6.147}$$

and

$$\frac{I_{ZX}}{F(t)} = \frac{1}{9} + \frac{1}{3\sqrt{6}} \sum_n \langle D^{2*}_{0,n} \rangle \left(2\left[A^{(a)}_{\text{MOL}}\right]^{2,n} - \left[A^{(e)}_{\text{MOL}}\right]^{2,n} \right) - \frac{1}{3}\Phi_0(t). \tag{6.148}$$

The polarization anisotropy ratio $r_{ZZ,ZX}(t)$ that can be determined from both a right angle and a parallel geometry experiment (see Fig. 3.8), becomes

$$r_{ZZ,ZX}(t) = \frac{I_{ZZ}(t) - I_{ZX}(t)}{I_{ZZ}(t) + 2I_{ZX}(t)} = \frac{\sqrt{\frac{1}{6}} \sum_n \left[A^{(e)}_{\text{MOL}}\right]^{2,n} \langle D^{2*}_{0,n} \rangle + \Phi_0(t)}{\frac{1}{3} + \sqrt{\frac{2}{3}} \sum_n \left[A^{(a)}_{\text{MOL}}\right]^{2,n} \langle D^{2*}_{0,n} \rangle}, \tag{6.149}$$

for a probe with arbitrary symmetry in a uniaxial mesophase. Note that in this idealized case the fluorescence decay $F(t)$ has factored out and $r_{ZZ,ZX}(t)$ depends only on ordering and reorientational dynamics. Limiting expressions for the fluorescence intensities and the polarization ratio $r_{ZZ,ZX}(t)$ for short times can be derived at once using the previously seen initial values of the orientational autocorrelation functions. The long-time limit of $r_{ZZ,ZX}(t)$ is obtained using Eq. 6.31. This gives a particularly simple result for $r_{ZZ,ZX}(\infty)$, i.e.

$$r_{ZZ,ZX}(\infty) = \sqrt{\frac{3}{2}} \sum_n \langle D^{2*}_{0,n} \rangle \left[A^{(e)}_{\text{MOL}}\right]^{2,n}. \tag{6.150}$$

Similar expressions can be obtained for other combinations of the input and output polarizers. In particular,

$$r_{XX,XZ}(t) = \frac{I_{XX}(t) - I_{XZ}(t)}{I_{XX}(t) + 2I_{XZ}(t)}, \tag{6.151a}$$

$$= \frac{-\sqrt{\frac{1}{6}} \sum_n \left[A^{(e)}_{\text{MOL}}\right]^{2,n} \langle D^2_{0n} \rangle + \frac{1}{2}\Phi_0(t) + \frac{1}{2}\Phi_2(t)}{\frac{1}{3} + \sqrt{\frac{1}{6}} \sum_n \langle D^2_{0n} \rangle \left[\left[A^{(e)}_{\text{MOL}}\right]^{2,n} - \left[A^{(a)}_{\text{MOL}}\right]^{2,n} \right] - \frac{1}{2}\Phi_0(t) + \frac{1}{2}\Phi_2(t)} \tag{6.151b}$$

and

$$r_{XX,XY}(t) = \frac{I_{XX}(t) - I_{XY}(t)}{I_{XX}(t) + 2I_{XY}(t)}, \tag{6.152a}$$

$$= \frac{\Phi_2(t)}{\frac{1}{3} - \sqrt{\frac{1}{6}} \sum_n \langle D^2_{0n} \rangle \left[\left[A^{(e)}_{\text{MOL}}\right]^{2,n} + \left[A^{(a)}_{\text{MOL}}\right]^{2,n} \right] + \frac{1}{2}\Phi_0(t) - \frac{1}{2}\Phi_2(t)}. \tag{6.152b}$$

We see that at least in principle the FD experiment may depend on a number of order parameters. In particular, the initial values of the intensity contain order parameters of rank up to 4 while the long-time limit second-rank ones. In practice, even though symmetry may limit the number of independent order parameters, these unknowns are normally too numerous. For example, if the molecule is biaxial, we have seen in Sec.3.10 that the independent order parameters up to rank 4 are $\langle D^2_{0,0}\rangle$, $\langle D^2_{0,2}\rangle$, $\langle D^4_{0,0}\rangle$, $\langle D^4_{0,2}\rangle$, $\langle D^4_{0,4}\rangle$, which are probably a bit too many to determine.

Uniaxial fluorescent probes. Considering the common case of a molecule approximately uniaxial, either rod-like like DPH (see Fig. 6.14a) or discotic, like perylene (Fig. 6.14b), we have $I_{AB} = F(t)\langle[\mu_A^{(a)}]^2(0)\,[\mu_B^{(e)}](t)\rangle$ with $A, B = X, Y, Z$. Thus,

$$\frac{I_{ZZ}}{F(t)} = \frac{1}{8} + \sqrt{\frac{2}{27}}\langle P_2\rangle\left([A^{(e)}_{\rm MOL}]^{2,0} + [A^{(a)}_{\rm MOL}]^{2,0}\right) + \frac{2}{3}\Phi_0(t), \tag{6.153a}$$

$$\frac{I_{ZX}}{F(t)} = \frac{1}{9} + \frac{1}{\sqrt{54}}\langle P_2\rangle\left(2[A^{(a)}_{\rm MOL}]^{2,0} - [A^{(e)}_{\rm MOL}]^{2,0}\right) - \frac{1}{3}\Phi_0(t), \tag{6.153b}$$

$$\frac{I_{YZ}}{F(t)} = \frac{1}{9} + \frac{1}{\sqrt{54}}\langle P_2\rangle\left(2[A^{(e)}_{\rm MOL}]^{2,0} - [A^{(a)}_{\rm MOL}]^{2,0}\right) - \frac{1}{3}\Phi_0(t), \tag{6.153c}$$

$$\frac{I_{YX}}{F(t)} = \frac{1}{3} - \frac{1}{\sqrt{54}}\langle P_2\rangle\left([A^{(e)}_{\rm MOL}]^{2,0} + [A^{(a)}_{\rm MOL}]^{2,0}\right) + \frac{1}{6}\Phi_0(t) - \frac{1}{2}\Phi_2(t), \tag{6.153d}$$

$$\frac{I_{XX}}{F(t)} = \frac{1}{9} - \frac{1}{\sqrt{54}}\langle P_2\rangle\left([A^{(a)}_{\rm MOL}]^{2,0} + [A^{(e)}_{\rm MOL}]^{2,0}\right) + \frac{1}{6}\Phi_0(t) + \frac{1}{2}\Phi_2(t), \tag{6.153e}$$

with $\Phi_L(t)$ as in Eq. 6.146 but with a $\delta_{n,n'}$. We can consider two common fluorescent probes. The first is a rod-like one, DPH, that we have already encountered in Section 3.4.3, where we mentioned that it has both absorption and emission transition moments parallel to the long axis. The second probe: perylene, is instead a flat aromatic molecule (Fig. 6.14b-top) with a shape roughly approximating that of a disc with a diameter of ≈ 8 Å. It is convenient to choose the molecular z-axis for perylene perpendicular to the ring so as to accentuate its near cylindrical symmetry about the short D_{2h} axis. Since perylene will presumably align with the ring parallel to the director when dissolved in an ordered phase, we expect the order parameter $\langle P_2\rangle$ of its highest symmetry axis, which is perpendicular to the molecule plane, to be negative. The UV-VIS absorption spectrum of perylene presents two main, well-structured, transitions [Berlman, 1971] corresponding to two transition moments lying in the ring plane and perpendicular to one another as shown in Fig. 6.14b (top).

6.14.3 Connection between Deterministic and Stochastic Approaches

We have introduced the diffusion equation as a stochastic model for the reorientation of molecules in liquid crystals and isotropic liquids. However, it is worth mentioning that such a microscopic, single-molecule equation is related to the overall molecular dynamics of a system of molecules, from which it can be formally derived by systematically projecting out from the collective time evolution all the unnecessary degrees of freedom. In this way

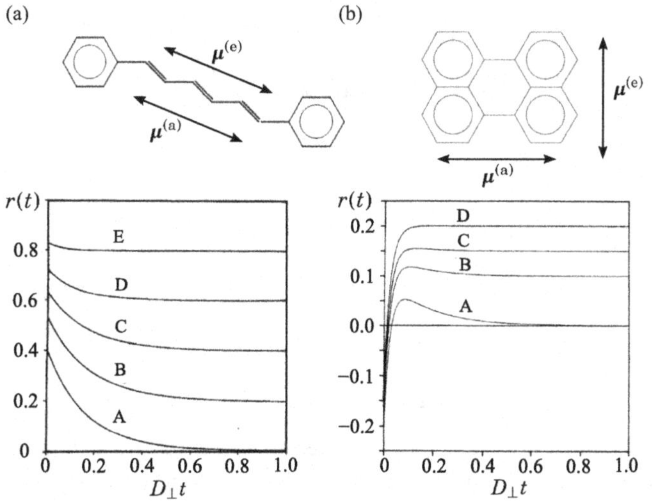

Figure 6.14 The time dependence of the FD ratio $r(t)$ for two common, nearly uniaxial, fluorescent probes dissolved in an oriented monodomain nematic. (a) DPH, a probe with absorption, $\boldsymbol{\mu}^{(a)}$, and emission, $\boldsymbol{\mu}^{(e)}$, dipole moments parallel to the long molecular axis. The results are for $\langle P_2 \rangle = 0\,(A), 0.2\,(B), 0.4\,(C), 0.6\,(D), 0.8\,(E)$, while for $\langle P_2 \rangle = 1$, $r(t) = 1$ [Zannoni, 1979d]. (b) A discotic probe, perylene, with transition moments in absorption at $\lambda_a \approx 252$ nm and emission at $\lambda_e \approx 410$ nm, that are perpendicular to one another and to the disk axis. The perylene order parameters $\langle P_2 \rangle$ are 0.0 (A), −0.2 (B), −0.3 (C), −0.4 (D) [Zannoni et al., 1983]. Rotational diffusion and $D_{\parallel}/D_{\perp} = 10$ have been assumed for both probes.

irreversibility in time is inevitably introduced as a consequence of the reduction in information that does not allow us to reverse time trajectories. The details of the various procedures, to achieve the projection from N particles to one particle dynamics, are clearly outside the scope of this book, but are treated in detail in various works, for example by Grabert [1982], and Zwanzig [2001].

7
Molecular Theories

7.1 Molecular Fields

Once we have assumed a certain form of intermolecular potential, however simple, for a system of particles of interest, it is worth realizing that no exact analytical solution is going to be available for its thermodynamics, or more generally, to predict the phases it can generate. Even if computer simulations will eventually be performed, it is often important to obtain an approximate preliminary description, e.g. to have a tentative location of the phase transition or, for liquid crystals, an estimate of the order parameters, employing some theoretical approach. Historically a first approach of this type is the *Mean Field* or *Molecular Field* Theory (MFT) one where, rather than describing in detail the effects of particle-particle interactions and their variation with separation, orientation, etc., the drastic assumption is made to consider just a single particle interacting with an effective field created by all the others in the system. Since each molecule can itself be contributing to the generation of this field for the others, a self-consistency condition is also introduced. The approach has been used, assuming different names, in most fields of physics: Weiss theory for magnetism [Stanley, 1971], Hartree-Fock for quantum mechanics [Schatz and Ratner, 1993], Debye-Hückel for ionic solutions [Chaikin and Lubensky, 1995], etc. The MFT of nematics put forward by Maier and Saupe [1958] was one of the first successful attempts on the theoretical front to understand liquid crystallinity. In their pioneering work they considered the mesogenic molecules to behave like uniaxial rods and London dispersion (and thus attractive) interactions (cf. Section 5.8.2) to be the dominant intermolecular force leading to anisotropic alignment. They then proceeded to evaluate the average anisotropic potential acting on a molecule by effect of all the others in the system. If we could monitor this mean field in a real situation, we would see it continuously fluctuating with time as a consequence of the surrounding molecules moving about. In MFT the effective potential acting on a molecule with a certain position-orientation is instead assumed to be constant in time, with a strength depending only on the order of the surrounding environment. The main advantage of the treatment lies in its simplicity and this in turn makes it applicable to a variety of situations. Thus, for example, MFTs have put forward not only for nematics, but also for smectic A [McMillan, 1971] and smectic C [McMillan, 1973; Selinger and Nelson, 1989] phases as well as, in a unified way for all of these together [Pajak and Osipov, 2013]. Molecular field descriptions for twisted, cholesteric systems have also been proposed [Van der Meer et al., 1976]. Importantly the predictions of MFT have proved to be, at least for most properties of nematics, very (perhaps even surprisingly) successful at a semi-quantitative level. It is interesting to understand why this is the case and especially if it is a fortunate coincidence or not, since this can serve as a guide for applying MFT to more complex and realistic cases. Since we have computer simulation results at hand the natural thing to do to try and assess MFT performance will be to compare those with the simulation results on the same model system. The plan of this chapter is thus the following. First, we illustrate the MFT procedure when applied to a simple LL or Maier–Saupe like model. We shall work through this example in detail since it is a good prototype for all mean field calculations even on more complex potentials. After this detailed and somewhat tutorial example, we shall develop a more general MFT for rigid molecules of arbitrary symmetry. In particular, we shall discuss the theories by Humphries et al. [1972] and Luckhurst et al. [1975] for uniaxial and biaxial particles.

7.2 Maier–Saupe Theory: A Simple Introduction

As in any, however simple, molecular theory the starting point is an intermolecular potential, and the mean field potential has to be obtained from this by some averaging procedure. In this example we choose a system of cylindrically symmetric particles and a pair of particles, say 1, 2, with axes $\boldsymbol{u}_1$, $\boldsymbol{u}_2$ interacting via the purely anisotropic pair potential [Zannoni, 1979b]

$$U_{12}^{(L)} = U^{(L)}(\boldsymbol{r}_{12}, \beta_{12}) = -u_L(\boldsymbol{r}_{12}) P_L(\cos\beta_{12}), \tag{7.1}$$

where P_L is a Legendre polynomial, $\cos\beta_{12} = \boldsymbol{u}_1 \cdot \boldsymbol{u}_2$ and $u_L(\boldsymbol{r}_{12}) \geq 0$. This contains as special cases the 3D ferromagnetic Heisenberg model (Section 2.8.3) when $L = 1$ (in this case the particles are polar, in the sense that head and tail can be differentiated), the dispersive [Saupe, 1974] and Lebwohl and Lasher [1972] potentials (Section 2.8.5), important for nematics, when $L = 2$ (non-polar particles) and also the case of $L > 2$ that could model molecules preferring to align along more than one relative orientation [Zannoni, 1979b]. The $u_L(\boldsymbol{r}_{12})$ in Eq. 7.1 gives the strength of the interaction as a function of the separation vector $\boldsymbol{r}_{12}$, however, to obtain a potential for a particle in a uniform mean field, we can formally average over $\boldsymbol{r}_{12}$ obtaining a purely orientational potential

$$\overline{U}_{12}^{(L)} = -\overline{u}_L P_L(\cos\beta_{12}), \tag{7.2}$$

where the overbar in $\overline{u}_L$ indicates an average over the distribution of intermolecular vectors, i.e. $\langle \ldots \rangle_{\boldsymbol{r}}$. Note that no dependence on the intermolecular vector direction or distance is present at this point, so that the effective potential is uniform in space, thus not suitable for a smectic, but still compatible with a nematic phase. We now need to average over the orientations of one of the two molecules, say molecule 2. We start by separating the relative orientations β_{12} in Eq. 7.1 recalling the spherical harmonics addition theorem (see Eq. F.15)

$$\overline{U}_{12}^{(L)} = -\overline{u}_L \; P_L(\cos\beta_{12}) = -\overline{u}_L \sum_{q=-L}^{L} \mathscr{D}_{q,0}^{L*}(\Omega_{1\mathrm{L}}) \mathscr{D}_{q,0}^{L}(\Omega_{2\mathrm{L}}), \tag{7.3a}$$

$$= -\overline{u}_L \sum_{q=-L}^{L} \mathrm{e}^{iq\alpha_{1\mathrm{L}}} d_{-q,0}^{L}(\beta_{1\mathrm{L}}) d_{q,0}^{L}(\beta_{2\mathrm{L}}) \, \mathrm{e}^{-iq\alpha_{2\mathrm{L}}}, \tag{7.3b}$$

where we have employed the explicit form of the Wigner rotation matrices $\mathscr{D}^L_{q,0}(\Omega_{2L})$, Eq. F.4. We can now average over the orientations of molecule 2, multiplying by the distribution $P(\Omega_{2L})$ and integrating over $\mathrm{d}\Omega_{2L} = \mathrm{d}\alpha_{2L}\mathrm{d}\beta_{2L} \sin\beta_{2L}$. Although $P(\Omega_{2L})$ is unknown, it should possess the symmetry of the phase we are trying to study. Thus, if we search for a uniaxial nematic phase with a director $\boldsymbol{d}$ along z, $P(\Omega_{2L}) = P(\cos\beta_2)$ and integration over $\mathrm{d}\alpha_{2L}$ just yields $\delta_{q,o}$ (cf. Eq. A.46) so that

$$\langle \overline{U}_{12}^{(L)} \rangle_{\alpha_2} = -\overline{u}_L \; P_L(\cos\beta_1) P_L(\cos\beta_2), \tag{7.4}$$

where we have used the fact that $d^L_{00}(\beta_i) = P_L(\cos\beta_i)$ and simplified the notation β_{iL} to β_i. Averaging on β_2 gives an effective field potential dependent on an order parameter $\langle P_L \rangle$

$$U_1^{(L)}(x) \equiv \langle U^{(L)} \rangle_{\alpha_2,\beta_2}(x) = -\overline{u}_L \; \langle P_L \rangle P_L(x), \tag{7.5}$$

where $x = \cos\beta_1$ and we have eliminated the subscript since only a molecule is now present. Clearly, if we use this potential to calculate the order parameter appearing in $U_1^{(L)}$, a self-consistency condition is obtained:

$$\langle P_L \rangle = \int_{-1}^{1} \mathrm{d}x \, P_L(x) \, \mathrm{e}^{\overline{u}_L \langle P_L \rangle P_L(x)/(k_B T)} / Z_1^{(L)}, \tag{7.6}$$

where the orientational configuration integral is

$$Z_1^{(L)} = \int_{-1}^{1} \mathrm{d}x \; \mathrm{e}^{\overline{u}_L \langle P_L \rangle P_L(x)/(k_B T)}. \tag{7.7}$$

On a qualitative level we remark that the pseudopotential Eq. 7.8 is quite reasonable since the anisotropic molecular field trying to align a molecule along the field is proportional to the order parameter and it reduces to 0 in the isotropic phase. In particular, for $L = 1$, we have a model for a polar anisotropic phase (a ferromagnetic or ferroelectric system). This is a system very well studied with MFT [Stanley, 1971] and with computer simulations [Chen et al., 1993; Holm and Janke, 1993]. For $L = 2$ we have the Maier–Saupe (MS) theory for nematics and using this as a superscript

$$U^{\mathrm{MS}}(x) \equiv U_1^{(2)}(x) = -\overline{u}_2 \langle P_2 \rangle P_2(x). \tag{7.8}$$

The pseudopotential can now be used to calculate any average single-particle property using the Boltzmann expression

$$P(x) = (1/Z_1^{\mathrm{MS}}) \, \mathrm{e}^{a_2 P_2(x)}, \tag{7.9}$$

with $a_2 \equiv \overline{u}_2 \langle P_2 \rangle / (k_B T)$ expressing the strength of the anisotropic molecular field and $Z_1^{\mathrm{MS}} = Z_1^{(2)}$ in Eq. 7.7. The order parameter $\langle P_2 \rangle$ is then

$$\langle P_2 \rangle = \left(1/Z_1^{\mathrm{MS}}\right) \int_{-1}^{1} \mathrm{d}x \; P_2(x) \, \mathrm{e}^{a_2 P_2(x)}. \tag{7.10}$$

We have already encountered this integral in Eq. 3.75 and expressed it in terms of special functions. The self-consistency requirement, of $\langle P_2 \rangle$ being the same on the two sides, constitutes a necessary condition, typical of MFT. A simple intuitive approach to solving

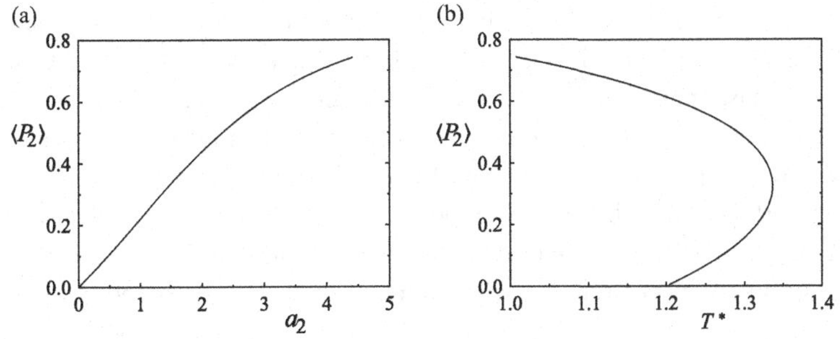

Figure 7.1 (a) The order parameter $\langle P_2 \rangle$ vs the molecular field strength parameter a_2. (b) The self-consistent values of $\langle P_2 \rangle$ vs scaled temperature $T^* = k_B T/\epsilon$, with $\epsilon = \bar{u}_2/z$, and $z = 6$ the nearest neighbour number for a cubic lattice to ease comparison with simulations.

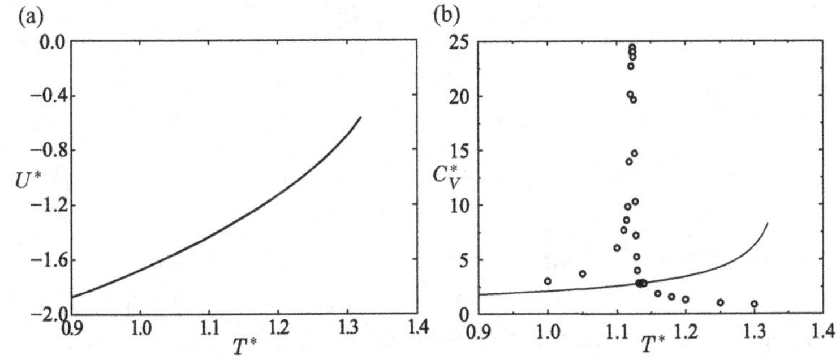

Figure 7.2 Lebwohl-Lasher model results. (a) MFT energy $-U^*$ vs temperature T^*. (b) The heat capacity $C_V^* = C_V^*/N$ from MFT (–) and MC simulations (∘). Dimensionless units [Fabbri and Zannoni, 1986].

the implicit Eq. 7.10 is a graphical one, locating the intersection between the $\langle P_2 \rangle$ versus a_2 curve, plotted in Fig. 7.1a and the straight line $\langle P_2 \rangle = a_2 k_B T/\overline{u}_2$. From this plot (or from a table of $\langle P_2 \rangle$ versus the strength of molecular field parameter a_2) and from Eq. 7.10 we can obtain the temperature corresponding to a certain $\langle P_2 \rangle$. The self-consistent $\langle P_2 \rangle$ is plotted versus scaled temperature in Fig. 7.1b. We now examine the various thermodynamic observables.

Mean field energy. The mean field energy is just the average of the effective potential. When calculating the total energy $\mathcal{U}$ of the whole system of N particles we have to multiply for a factor of (1/2) to avoid counting a particle twice (one as reference and the other as 'environment'). Thus, the total energy is $\mathcal{U} = -N\overline{u}_2 \langle P_2 \rangle^2/2$. In the special case of the LL model, discussed in detail in Chapter 10, where $\epsilon = \overline{u}_2/z$, with $\epsilon > 0$ ($z = 6$ for a cubic lattice used in simulations) the single-particle energy in dimensionless units is just $U^* = \mathcal{U}/(N\epsilon) = -3\langle P_2 \rangle^2$ and is plotted in Fig. 7.2a.

Heat capacity. As we have mentioned before, the heat capacity per particle $C_V^* = C_V^*/N$ can be obtained by differentiating the energy with respect to T (Eq. 4.124a) and in Fig. 7.2b we show the MFT result for C_V^* together with the Monte Carlo simulation results for the LL model [Fabbri and Zannoni, 1986]. From Fig. 7.2b we get a clear indication of the location of the phase transition. We see that MFT significantly overestimates the transition temperature (17% for the Lebwohl–Lasher model). Thus, it is much easier for molecules to form an ordered mesophase according to MFT than it is, for the same starting pair potentials, if we do not perform the rather drastic approximations we have performed. This is actually a rather general feature of MFT, also for more complex potentails. However, here we have the possibility of calculating entropy and free energy and of proceeding quantitatively from these.

Entropy. The entropy can be obtained from standard statistical thermodynamics (Chapter 4) as

$$\mathcal{S} = -Nk_B\langle \ln P(x)\rangle = -\frac{N}{T}\overline{u}_2\langle P_2\rangle^2 + Nk_B \ln Z_1^{\mathrm{MS}}. \tag{7.11}$$

Free energy. We have

$$\mathcal{A} = \mathcal{U} - T\mathcal{S} = -\frac{N}{2}U - Nk_BT \ln Z_1^{\mathrm{MS}} = +\frac{N\overline{u}_2}{2}\langle P_2\rangle^2 - Nk_BT \ln Z_1^{\mathrm{MS}}. \tag{7.12}$$

Note that this simple version of MFT is not fully thermodynamically consistent since the correct expression for $\mathcal{A}$ can be obtained in this way but not from the statistical thermodynamic expression from the average of ln P. In other words, Z_1^{MS} is not the complete partition function. The problem we are left with is that of determining if there is an orientational phase transition or not. To do this we should see if there is a temperature region where the free energy calculated from Eq. 7.12 is lower than that of the reference isotropic phase, taken as 0. We see that indeed at $T^* = kT/\epsilon \geq T_{NI} = 1.321$ a nematic phase has a lower free energy and thus is more stable than the disordered phase. The free energy curve meets the abscissa with a finite slope. Thus, the free energy of the most stable phase will have an edge point at the transition. Remembering the thermodynamic relations in Chapter 2 we see that MFT predicts the NI transition to be a first-order one. The values of the most important transition quantities are reported later in Table 7.1. It is worth noting that the absolute value of the transition temperature cannot really be predicted by MFT applied to induced dipole-induced dipole dispersive interactions (Section 5.8.2), as imagined in the original derivation by Maier and Saupe [1958, 1959]. Kaplan and Drauglis [1971] have shown that, using reasonable estimates for the polarizability anisotropy and other molecular parameters to calculate the effective field strength parameter $\overline{u}_2$, the NI transition obtained is completely wrong (not surprising since important terms like short-range steric interactions are completely missing). A rationalization proposed [Luckhurst and Zannoni, 1977] is that both short- and long-range forces are important in determining the molecular organization in a nematic phase but that they operate at quite different levels. Thus, short-range forces could be responsible for the formation of highly ordered groups or clusters of molecules, less anisometric than its constituent particles due to the local packing. As a consequence, the anisotropic short-range forces between clusters would become of minor importance.

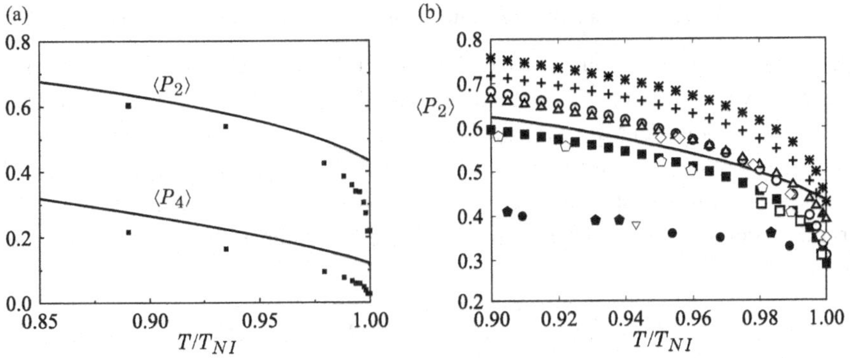

Figure 7.3 (a) The second-rank and fourth-rank order parameters $\langle P_2 \rangle$ and $\langle P_4 \rangle$ obtained from MC simulation of the LL model (■) [Fabbri and Zannoni, 1986] and from MFT (—) plotted against reduced temperature T/T_{NI}. (b) $\langle P_2 \rangle$ from MFT (—) and experiments on the nematics MBBA (■), OH-MBBA (⊙), MBCA (×), APAPA (△) [Leenhouts et al., 1979], 5CB (∗), HAB (□) [Picken, 2001], anisaldazine (◇) [Blinc et al., 1968], PAA (⬠) [Pines and Chang, 1974]; and three 4,4′-disubstituted benzoic acid-phenylesters: MBA5PE (⬟), HBA5PE (●), MBBAHPE (▽) [Ibrahim and Haase, 1976]. Full chemical names corresponding to the abbreviations are given in Appendix N.

The anisotropy in the polarizability of the cluster will be approximately proportional to the number of particles in the cluster and therefore the dispersive interaction energy between clusters would become dominant. The clusters are expected to be essentially the same above and below T_{NI}, in agreement with the small entropy of transition observed experimentally.

Order parameters. In Fig. 7.3a we compare the MFT $\langle P_2 \rangle$ and $\langle P_4 \rangle$ with the Monte Carlo results for the LL model (see Section 10.2.1 for details on the simulation). To compare MFT results for the order parameters with experiment, it is convenient to employ reduced temperatures: T/T_{NI}, since the NI transition temperature changes very significantly with chemical composition (see, e.g., Table 1.2). In Fig. 7.3b we compare MFT results for $\langle P_2 \rangle$ with experimental data for various nematics (some data were already shown in Fig. 3.6). The Maier–Saupe theory based on second-rank interactions predicts that all liquid crystals should have a universal behaviour for the order parameter $\langle P_2 \rangle$ versus reduced temperature curve, with $\langle P_2 \rangle_{NI} = 0.429$ and $\langle P_4 \rangle_{NI} = 0.12$ (see Table 7.1). Even if no universality can be claimed, and data for some compounds appear quite different at low temperature, the similarity of the curves is apparent. The agreement is, however, not perfect and shifts up and down with respect to the MS curve of the experimental results, as well as changes of slope, can be observed. The measured $\langle P_2 \rangle$ at the transition are, however, reasonably similar to the MFT ones. For instance, NMR results were: $\langle P_2 \rangle_{NI} = 0.338 \pm 0.001$ [Emsley et al., 1981] for 5CB and $\langle P_2 \rangle_{NI} = 0.404 \pm 0.005$ for 4′-n-propyl-4-n-cyano bicyclohexane (CCH3) [Dong et al., 1989]. It is worth mentioning that, differently from the results just mentioned, the order parameters experimentally measured are most often for probe molecules dissolved in liquid crystals and not for pure nematogens (see Section 7.3.3).

7.3 Generalized Mean Field Theory for Uniaxial Nematics

We saw in Chapter 3 how the order of biaxial molecules can be described in terms of Wigner matrix averages, or of distributions [Berardi et al., 2002; Tschierske, 2001]. Here we wish to write down an MFT for a uniform system formed of rigid molecules of arbitrary symmetry. In particular, we discuss the so called LZNS theory for non-cylindrical particles as developed by Luckhurst et al. [1975]. We start from the general pair potential between two rigid particles of arbitrary symmetry (see Eq. 5.6) written as an expansion in rotational invariants as

$$-U(\boldsymbol{r}_{12},\Omega_{1L},\Omega_{2L}) = \sum u^{k_1k_2}_{L_1L_2J}(r_{12}) S^{k_1,k_2}_{L_1,L_2,J}(\Omega_{1L},\Omega_{2L},\Omega_{rL}), \tag{7.13}$$

where the scalar functions $S^{k_i,k_2}_{L_1,L_2,J}(\Omega_{1L},\Omega_{2L},\Omega_{rL})$ [Stone, 1978] are the same already used in Chapter 4 and discussed in Appendix G. We now take the three averages described before. The first is performed over the orientations of the intermolecular vector Ω_{rL} assuming that they are distributed isotropically, which gives $\langle \mathscr{D}^J_{-m_1-m_2,0}(\Omega_r)\rangle_{\Omega_r} = \delta_{J,0}\delta_{m_1+m_2,0}$,

$$-U_{12} = U(r_{12},\Omega_1,\Omega_2) = \sum u^{k_1k_2}_{LL0}(r_{12}) S^{k_1,k_2}_{L,L,0}(\Omega_{1L},\Omega_{2L}), \tag{7.14a}$$

$$= \sum_L \frac{(-1)^L}{\sqrt{2L+1}} \sum_{k_1.k_2} (-1)^{k_2} u^{k_1k_2}_{LL0}(r_{12}) \mathscr{D}^L_{k_1,-k_2}(\Omega_{21}), \tag{7.14b}$$

after recalling the expressions for the Stone invariants (Eq. G.19) and the relevant Clebsch–Gordan coefficients (Eq. F.27), together with the Wigner matrices closure, Eq. F.14. We see that the assumption of isotropic distribution of intermolecular vectors has caused L_1 to equal L_2 (just called L from now on) and has made the pair potential dependent only on relative orientations. It is useful to note that in most theoretical treatments a specific interaction rank is chosen, typically $L = 2$. In this case, the expression Eq. 7.14b holds also for lattice models with sufficient symmetry, e.g. on a cubic lattice where the isotropic average is replaced by a sum over nearest neighbours taken over the six inter-particle vectors at the lattice distance $r_{12} = a$. Assuming $L = 2$, we have

$$-U_{12} = \frac{1}{\sqrt{5}} \sum_{k_1.k_2} (-1)^{k_2} u^{k_1k_2}_{220}(r_{12}) \mathscr{D}^2_{k_1,-k_2}(\Omega_{21}). \tag{7.15}$$

For biaxial (D_{2h}) particles only k_1, k_2 even (i.e. 0, ± 2) are allowed and

$$-U_{12} = U(r_{12},\Omega_{21}) = \frac{1}{\sqrt{5}} \sum_{k_1.k_2} u^{k_1k_2}_{220}(r_{12}) \mathscr{R}^2_{k_1,-k_2}(\Omega_{21}), \tag{7.16}$$

where $\mathscr{R}^2_{m,n}$ are the real symmetrized Wigner functions defined in Eq. 3.149b. For two identical molecules

$$-U'_{12} = \frac{u^{00}_{220}}{\sqrt{5}} \left\{ \mathscr{R}^2_{0,0}(\Omega_{21}) + 2\lambda_{20}[\mathscr{R}^2_{2,0}(\Omega_{21}) + \mathscr{R}^2_{0,2}(\Omega_{21})] + 4\lambda_{22}\mathscr{R}^2_{2,2}(\Omega_{21}) \right\}, \tag{7.17}$$

with $\lambda_{20} = u^{20}_{220}/u^{00}_{220} = u^{02}_{220}/u^{00}_{220}$, $\lambda_{22} = u^{22}_{220}/u^{00}_{220}$. This potential has been put forward by Straley [1974] and more recently by Sonnet et al. [2003] and its phase diagram has been studied with MC simulations by Preeti et al. [2011]. For a dispersive potential, where $u_{220}^{qn} = k_{12}\alpha_1^{2,q}\alpha_2^{2,n}$, or other separable potentials where the pair interaction coefficients can be factorized as a product of single molecule properties, one obtains the second-rank biaxial potential studied by Luckhurst and Romano [1980a] and Biscarini et al. [1991]

$$-U_{12}' = \frac{u_{220}^{00}}{\sqrt{5}}\left\{\mathscr{R}_{0,0}^{2}(\Omega_{21}) + 2\lambda[\mathscr{R}_{2,0}^{2}(\Omega_{21}) + \mathscr{R}_{0,2}^{2}(\Omega_{21})] + 4\lambda^2\mathscr{R}_{2,2}^{2}(\Omega_{21})\right\}, \quad (7.18)$$

with $\lambda = \alpha^{2,2}/\alpha^{2,0}$. Referring to Eq. 7.16, it is worth considering the case of two possibly different molecules and symmetries, as, for example, for a mixture of mesogens or for a solute (not necessarily mesogenic) in a nematic solvent. If the molecules are not identical we have $u_{220}^{qn} \neq u_{220}^{nq}$. For the case of a dispersive potential $u_{220}^{qn} = k_{12}\alpha_1^{2,q}\alpha_2^{2,n}$, so that the biaxiality parameters are $\lambda_2^{(1)} = \alpha_1^{2,2}/\alpha_1^{2,0}$, $\lambda_2^{(2)} = \alpha_2^{2,2}/\alpha_2^{2,0}$, $\lambda_{22} = \lambda_2^{(1)}\lambda_2^{(2)}$. We shall consider solutes in Section 7.3 and now return to the general expression in Eq. 7.14b for the pair potential. We can perform yet another average, albeit a formal one, on the molecular separations and obtain the purely orientational effective pair potential

$$-U_{12} = \sum \overline{u}_{Lk_1k_2}\, S_{L,L,0}^{k_1,k_2}(\Omega_1, \Omega_2), \quad (7.19)$$

where $-\overline{u}_{Lk_1k_2} \equiv \frac{1}{V}\int dr_{12} r_{12}^2 u_{LL0}^{k_1k_2}(r_{12}) g(r_{12})$, and $g(r_{12})$ is the radial distribution defined in Chapter 4. Note that in this way the mean field interaction between particles does not depend on their separation any more! The final average we have to take is over the orientations of the second molecule, keeping into account the symmetry of the candidate-oriented phase we are considering. Recalling the explicit expression for the Stone invariants Eq. G.19 we can write the mean field potential acting on one molecule in a uniform phase as

$$-U(\Omega_1) = \sum \overline{u}_{Lk_1k_2}\langle S_{L,L,0}^{k_i,k_2}(\Omega_1, \Omega_2)\rangle_{\Omega_2} = \sum \frac{(-1)^{L+q}\overline{u}_{L,k_1k_2}}{\sqrt{2L+1}}\langle \mathscr{D}_{q,-k_2}^{L}\rangle \mathscr{D}_{q,k_1}^{L}(\Omega_1). \quad (7.20)$$

If we wish to calculate the properties of a candidate phase with a certain symmetry (e.g. uniaxial or biaxial) we can first use this symmetry to find the order parameters $\langle \mathscr{D}_{q,-k_2}^{L}\rangle$ appropriate for such a phase using the techniques presented in Chapter 3 and Appendix G. This does not imply that the phase exists or that it is more stable than the isotropic or of some other competing molecular organization. We just calculate what would happen for such a mesophase. Then, after that, we shall compute the free energy of this candidate phase and compare it with that of the others (particularly the isotropic one) and decide which has the lowest free energy and thus is going to exist at certain thermodynamic conditions. We now proceed by specializing in various cases.

7.3.1 Uniaxial Phases of Uniaxial Mesogens

For a uniaxial phase, with the director defining the z laboratory axis, a rotation of an arbitrary angle α around the director should leave everything the same. This gives $\langle \mathscr{D}_{q,-k}^{L}\rangle = \langle \mathscr{D}_{0,-k}^{L}\rangle \delta_{0,q}$. Similarly, for a uniaxial molecule, the single-particle distribution will not depend on rotations of an arbitrary angle γ around the molecular z-axis, so

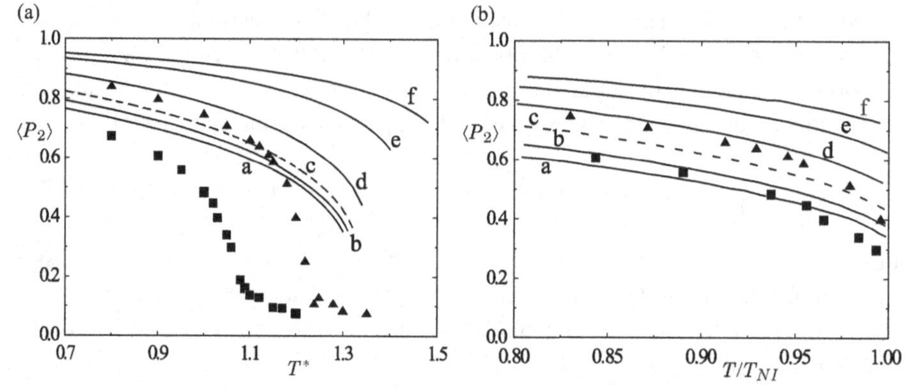

Figure 7.4 (a) $\langle P_2\rangle$ as a function of temperature T^* for the HJL potential with $\lambda_4 = -0.4$ (a), -0.2 (b), 0.0 (c), 0.2 (d), 0.4 (e), 0.6 (f). We also show the results of two Monte Carlo simulations for a $10 \times 10 \times 10$ lattice with $\lambda_4 = -0.2$ (■) and 0.2 (▲) [Fuller et al., 1985]. (b) The same plot vs reduced temperature T/T_{NI}. Curve (c) corresponding to the Maier–Saupe limit $\lambda_4 = 0$ is dashed.

that $\langle \mathscr{D}^L_{0,k_2}\rangle = \langle P_L\rangle \delta_{k_2,0}$. With the shorthand $\overline{u}_L \equiv (-1)^L/\sqrt{2L+1}\overline{u}_{LL0}$, the effective potential reduces to

$$-U(x) = \sum_L \overline{u}_L \langle P_L\rangle P_L(x), \tag{7.21}$$

with $x \equiv \cos\beta$. Only even L terms are allowed in a non-polar nematic, to comply with the $D_{\infty h}$ symmetry of the mesophase, which implies invariance for $\beta \longrightarrow \pi-\beta$ (cf. Chapter 3). The simplest of these potentials with the even symmetry required for a nematic, is that with $L=2$, corresponding to the Maier–Saupe case that we have treated in detail earlier in this chapter. The effect of fourth-rank terms on phases with $D_{\infty h}$ symmetry has been studied by Humphries–James–Luckhurst (HJL) truncating the series in Eq. 7.21 to the first two anisotropic contributions $L = 2, 4$ [Humphries et al., 1972].

$$-U^{\mathrm{HJL}}(x) = \overline{u}_2[\langle P_2\rangle P_2(x) + \lambda_4\langle P_4\rangle P_4(x)], \tag{7.22}$$

with $\lambda_4 \equiv \overline{u}_4/\overline{u}_2$. The orientational order parameters obtained from this potential will obey the consistency equations

$$\langle P_L\rangle = \left(1/Z_1^{\mathrm{HJL}}\right)\int_{-1}^{1} \mathrm{d}x\, P_L(x) \exp\left\{\frac{\overline{u}_2}{k_B T}[\langle P_2\rangle P_2(x) + \lambda_4\langle P_4\rangle P_4(x)]\right\}, \tag{7.23}$$

where $L = 2, 4$ and the single-particle orientational pseudo-partition function is

$$Z_1^{\mathrm{HJL}} = \int_{-1}^{1} \mathrm{d}x \exp\left\{\frac{\overline{u}_2}{k_B T}[\langle P_2\rangle P_2(x) + \lambda_4\langle P_4\rangle P_4(x)]\right\}. \tag{7.24}$$

The energy per particle is $U^{\mathrm{HJL}} = -\frac{1}{2}\overline{u}_2\left[\langle P_2\rangle^2 + \lambda_4\langle P_4\rangle^2\right]$, and the Helmholtz free energy per particle is $A^{\mathrm{HJL}} = \frac{1}{N}\mathscr{A} = \frac{1}{2}\overline{u}_2[\,\langle P_2\rangle^2 + \lambda_4\langle P_4\rangle^2] - k_B T \ln Z_1^{\mathrm{HJL}}$. The transition temperature for a given λ_4 can be determined by evaluating the order parameters $\langle P_2\rangle$ and $\langle P_4\rangle$ from

Table 7.1. *Transition temperatures and properties for the HJL potential. The $\lambda_4 = \infty$ results are for a pure P_4 potential*

λ_4	$k_B T_{NI}/\epsilon$	$\langle P_2 \rangle_{NI}$	$\langle P_4 \rangle_{NI}$	$\langle P_6 \rangle_{NI}$	$[C_V^*]_{NI}$	$[S_{net}]_{NI}$
−1.0	1.2787	0.282	0.033	−0.001	5.69	−0.184
−0.8	1.2826	0.296	0.039	0.000	5.92	−0.202
−0.6	1.2878	0.314	0.047	0.002	6.24	−0.226
−0.4	1.2948	0.338	0.060	0.005	6.68	−0.262
−0.2	1.3050	0.374	0.082	0.011	7.34	−0.319
0.0	1.3212	0.429	0.120	0.024	8.35	−0.418
0.2	1.3491	0.516	0.194	0.058	9.64	−0.609
0.4	1.3997	0.630	0.320	0.134	9.78	−0.938
0.6	1.4793	0.723	0.448	0.234	8.30	−1.304
0.8	1.5810	0.778	0.538	0.318	7.01	−1.587
1.0	1.6961	0.808	0.597	0.379	6.23	−1.787
∞	0.7508	0.430	0.561	0.300	19.58	−1.259

the consistency condition in Eq. 7.23 and using these to evaluate the Helmholtz free energy. The results are given in Table 7.1 for a few λ_4 values and including the pure P_2 and P_4 cases [Zannoni, 1979b]. As before we have considered $\overline{u}_2 = 6\epsilon$ to facilitate comparison with the cubic lattice version of the HJL model and its simulation results [Fuller et al., 1985], shown in Fig. 7.4. We note the increase in T_{NI} obtained for $\lambda_4 > 0$, indicating stabilization of the nematic phase and much smaller effect in the opposite direction for $\lambda_4 < 0$. In Fig. 7.4b we have plotted the second-rank order parameter versus reduced temperature T/T_{NI} for a number of λ_4 values. We note that at the same T/T_{NI} the order is shifted upwards or downwards with respect to the Maier–Saupe case ($\lambda_4 = 0$) as λ_4 increases or decreases and the order parameter at the transition changes correspondingly. The comparison of curves (c) and (e) with the two MC simulations corresponding to $\lambda_4 = \pm 0.2$ [Fuller et al., 1985] just shows, as for the LL case, the important overestimation of T_{NI} caused by the molecular field approximation. However, the curves are fairly parallel to one another and the slope of $\langle P_2 \rangle$ near transition temperature is quite similar. The effect on the $\langle P_4 \rangle$ versus $\langle P_4 \rangle$ curve is not too great. In particular, $\langle P_4 \rangle$ remains always positive.

7.3.2 Uniaxial Phases of Biaxial Mesogens

Even if a nematic phase is uniaxial, its constituent mesogens are not and their shape could often be better approximated with lath-like biaxial objects (see the formulas in Table 1.2). Assuming the mesogens to be all identical, as well as rigid and biaxial, we have, specializing Eq. 7.20 for the assumed uniaxial phase symmetry, that $\langle \mathscr{D}^{L_1}_{q_1,k_2} \rangle = \langle \mathscr{D}^{L_1}_{0,k_2} \rangle \delta_{q_1,0}$ and

$$-U(\beta,\gamma) = \sum \frac{(-1)^L}{\sqrt{2L+1}} \overline{u}_{Lk_1k_2} \langle \mathscr{D}^L_{0,k_2} \rangle \mathscr{D}^L_{0,k_1}(\beta,\gamma). \tag{7.25}$$

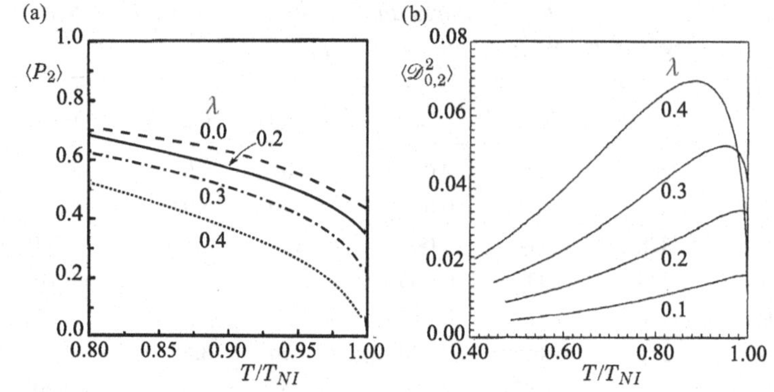

Figure 7.5 (a) The second rank uniaxial $\langle P_2 \rangle$ and (b) biaxial order parameter for uniaxial phases formed of biaxial particles with a certain biaxiality λ plotted as a function of reduced temperature [Zannoni, 1975].

This should hold both for polar and non-polar mesophases and it gives the effective average potential acting on a molecule by the combined action of all the other molecules in the system. This has to be taken together with the set of self-consistent equations

$$\langle \mathscr{D}^L_{0,k_2} \rangle = \left(1/Z_1^{\mathrm{UX}}\right) \int \mathrm{d}\beta \sin\beta \mathrm{d}\gamma\, \mathscr{D}^L_{0,k_2}(\beta,\gamma)$$
$$\exp\Big[\sum \frac{(-1)^L \overline{u}_{Lk_1k_2}}{k_B T\sqrt{2L+1}} \langle \mathscr{D}^L_{0,k_2}\rangle \mathscr{D}^L_{0,k_1}(\beta,\gamma)\Big], \tag{7.26}$$

with the normalization integral

$$Z_1^{\mathrm{UX}} = \int \mathrm{d}\beta \sin\beta \mathrm{d}\gamma \exp\Big[\sum \frac{(-1)^L \overline{u}_{Lk_1k_2}}{k_B T\sqrt{2L+1}} \langle \mathscr{D}^L_{0,k_2}\rangle \mathscr{D}^L_{0,k_1}(\beta,\gamma)\Big]. \tag{7.27}$$

The strength of the MFT potential is proportional to the orientational order parameters $\langle \mathscr{D}^L_{0,k_2} \rangle$. The effective potential for molecules of a certain symmetry can be obtained as special cases of Eq. 7.25 using the procedure in Appendix G. Here we consider only interactions of second rank ($L = 2$). Moreover, we assume that the molecules have biaxial D_{2h} symmetry, so that [Luckhurst et al., 1975]

$$-U(\beta,\gamma) = \frac{1}{\sqrt{5}} \sum_{k_1,k_2} \overline{u}_{2k_1k_2} \langle \mathscr{D}^2_{0,k_2}\rangle \mathscr{D}^2_{0,k_1}(\beta,\gamma), \tag{7.28}$$

with $k_1, k_2 = 0, \pm 2$. We consider, in particular, the case of two molecules interacting via dispersion interactions (Chapter 5), although the theory can really be applied to any second-rank interaction potential. This LZNS theory was developed by Luckhurst et al. [1975] for biaxial solutes in a uniaxial solvent. Theories for slightly different models of biaxial particles were developed by Freiser [1970] and Straley [1974] who looked particularly at the formation of biaxial nematic phases, rather than the effects of molecular biaxiality on uniaxial phases (see Section 7.3.3). In Fig. 7.5 we see the effect of going from uniaxial to biaxial mesogens on the order parameters $\langle P_2 \rangle$ and $\langle \mathscr{D}^2_{02} \rangle$. The dependence of $\langle \mathscr{D}^2_{02} \rangle$ on

$\langle P_2 \rangle$ for various biaxialities λ was already shown in Fig. 3.22 in the context of maximum entropy analysis of experimental data.

7.3.3 Biaxial Solutes in Uniaxial Phases

In many spectroscopic techniques a nematic is studied by adding a solute (probe) molecule whose spectrum is recorded and analyzed to obtain indirect information on the LC system. The probe concentration is kept as low as possible to reduce perturbations of the system and the amount needed depends on the sensitivity of the technique and can significantly vary. For instance, concentrations of a few percent w/w are typically needed for NMR, DNMR studies, of 10^{-3} w/w for ESR, and even lower for Fluorescence Depolarization experiments. In all these cases it is reasonable to assume that probe-probe interactions are negligible at these dilutions and that the surrounding of a probe is essentially an unperturbed nematic. If the solute is similar enough to the mesogens the probe technique can also be assumed to mimic their behaviour. If we assume second-rank interactions the effective potential acting on a probe molecule (P), $U^{\mathrm{XS}}(\beta, \gamma)$, in a uniaxial LC solvent (S) formed by biaxial mesogens and at infinite dilution is given by [Catalano et al., 1983]

$$-U^{\mathrm{P}}(\beta,\gamma)/(k_B T) = \eta_a P_2(\cos\beta) + \eta_b \mathscr{D}^2_{0,2}(\beta,\gamma), \tag{7.29}$$

where the effective field coefficients are $\eta_a = -\frac{1}{\sqrt{5}}[\overline{u}_{200}\langle P_2\rangle^{\mathrm{S}} + 2\overline{u}_{220}\langle \mathscr{D}^2_{0,2}\rangle^{\mathrm{S}}]/(k_B T)$ and $\eta_b = -\frac{2}{\sqrt{5}}[\overline{u}_{202}\langle P_2\rangle^{\mathrm{S}} + 2\overline{u}_{222}\langle \mathscr{D}^2_{0,2}\rangle^{\mathrm{S}}]/(k_B T)$. The coefficients $\overline{u}_{Lmn}$ are average solute-solvent interaction terms. The solvent orientational ordering is described by the order parameters $\langle \mathscr{D}^2_{0,2}\rangle^{\mathrm{S}}$. The solute order parameters are related to $U^{\mathrm{P}}(\beta,\gamma)$ by the Boltzmann averages:

$$\langle P_2\rangle^{\mathrm{P}} = \frac{1}{Z_1^{\mathrm{P}}} \int \mathrm{d}\beta \sin\beta \mathrm{d}\gamma \; P_2(\cos\beta) \; \exp[-U^{\mathrm{P}}(\beta,\gamma)/(k_B T)], \tag{7.30a}$$

$$\langle \mathscr{D}^2_{0,2}\rangle^{\mathrm{P}} = \frac{\sqrt{3}}{\sqrt{8} Z_1^{\mathrm{P}}} \int \mathrm{d}\beta \sin\beta \mathrm{d}\gamma \; \sin^2\beta \cos 2\gamma \; \exp[-U^{\mathrm{P}}(\beta,\gamma)/(k_B T)], \tag{7.30b}$$

where the normalization integral is $Z_1^{\mathrm{P}} = \int \mathrm{d}\beta \sin\beta \mathrm{d}\gamma \exp[-U^{\mathrm{P}}(\beta,\gamma)/k_B T]$. Since the MFT biaxial potential is formally identical to the maximum entropy expression discussed in Section 3.10.7, we can also recall the approximate analytic expression Eq. 3.147 obtained using computer algebra [Zannoni, 1988], i.e. with $\xi = 2\lambda$, $\mathrm{Re}\langle \mathscr{D}^2_{02}\rangle \approx \lambda\langle P_2\rangle\big(\langle P_2\rangle - 1\big)^2$. The apparent solute biaxiality is in this model

$$\lambda^{\mathrm{P}} = \frac{\eta_b}{2\eta_a} = \frac{\overline{u}_{202}\langle \mathscr{D}^2_{0,0}\rangle^{\mathrm{S}} + 2\overline{u}_{222}\langle \mathscr{D}^2_{0,2}\rangle^{\mathrm{S}}}{\overline{u}_{200}\langle \mathscr{D}^2_{0,0}\rangle^{\mathrm{S}} + 2\overline{u}_{220}\langle \mathscr{D}^2_{0,2}\rangle^{\mathrm{S}}}. \tag{7.31}$$

In general, λ^{P} will depend on the solute-solvent coefficients $\overline{u}_{2mn}$ and on the solvent order parameters and be in general temperature dependent. However, two important special cases should be considered. First, if the solvent biaxial order parameter, $\langle \mathscr{D}^2_{0,2}\rangle^{\mathrm{S}}$ is negligible, λ^{P} is a constant for that solute-solvent combination. Indeed, putting $\langle \mathscr{D}^2_{0,2}\rangle^{\mathrm{S}} = 0$ and combining equations gives $\lambda^{\mathrm{P}} = \overline{u}_{202}/\overline{u}_{200}$. Second, there will not only be temperature independence but also no solvent dependence in $\lambda^{\rm P}$ if the solute-solvent interactions can be decomposed into a product of averaged single-particle second-rank properties with elements $X_{\rm P}^{2,n}$, $X_{\rm solv}^{2,m}$ for the solute and solvent, respectively, so that, e.g.

$$\overline{u}_{2mn} \propto X_{\rm S}^{2,m} X_{\rm P}^{2,n}. \tag{7.32}$$

Thus, the solute biaxiality is

$$\lambda^{\rm P} = \frac{X_{\rm P}^{2,2}}{X_{\rm P}^{2,0}} = \sqrt{\frac{3}{2}} \frac{[X_{\rm P}]_{xx} - [X_{\rm P}]_{yy}}{2[X_{\rm P}]_{zz} - [X_{\rm P}]_{xx} - [X_{\rm P}]_{yy}}. \tag{7.33}$$

Note that even though we have assumed separability of the solute-solvent interaction coefficients (Eq. 7.32) we have not specified as yet the physical nature of the solute (or solvent) molecular biaxiality. Two natural, but by no means unique, choices could be traced to the biaxiality in the polarizability if dispersive forces are considered to be the dominant interaction or, in the molecular shape anisotropy, if packing and steric repulsion are thought to be the main contribution to ordering. We now consider some models proposed for evaluating λ.

Dispersion forces model. As we have already seen, when the orienting potential is determined by anisotropic dispersive interactions [Luckhurst et al., 1975] the solute-solvent coefficients actually separate in a product of molecular polarizability tensors $\boldsymbol{\alpha}_{\rm P}$, $\boldsymbol{\alpha}_{\rm S}$ for the probe solute and the solvent, thus verifying in this case Eq. 7.32 and we have simply $\overline{u}_{2qn} = k_{12}\alpha_{\rm S}^{2,q}\alpha_{\rm P}^{2,n}$. The probe biaxiality is $\lambda_\alpha^{\rm P} = \alpha^{2,2}/\alpha^{2,0} = \sqrt{\frac{3}{2}}\,(\alpha_{xx} - \alpha_{yy})/(2\alpha_{zz} - \alpha_{xx} - \alpha_{yy})$.

Straley box model. The effect of steric forces is not easy to evaluate for a biaxial object. Straley [1974] proposed to consider the solute molecule enclosed in the smallest possible orthogonal box of length L, breadth B, width W. Within this approach the coefficients for two equal molecules say of type (A) are [Luckhurst et al., 1975]

$$\overline{u}_{200}^{(\rm AA)} = \left\{-2B\left(W^2 + L^2\right) - 2W\left(L^2 + B^2\right) + L\left(W^2 + B^2\right) + 8WBL\right\}/3, \tag{7.34a}$$

$$\overline{u}_{220}^{(\rm AA)} = \left(L^2 - BW\right)(B - W)/\sqrt{6}, \tag{7.34b}$$

$$\overline{u}_{222}^{(\rm AA)} = -L(W - B)^2/2. \tag{7.34c}$$

In general, when considering steric models, we have to point out that separability is assumed, but not demonstrated. In particular, if we (arbitrarily) assume that a geometric mean relation holds for the coefficient obtained from $\overline{u}_{2qp}^{(\rm AB)} = \sqrt{\overline{u}_{2qp}^{(\rm AA)}\,\overline{u}_{2qp}^{(\rm BB)}}$.

Shape tensor. We can also get an estimate of the shape anisotropy considering a molecule as set of m steric interaction centres (spheres in lieu of the atoms or groups of atoms) at position $\boldsymbol{r}^i$ and defining a shape tensor $\mathbf{F}$ [Catalano et al., 1983] depending on the distribution of these centres in the molecule. We take this tensor to have the same form of an inertia tensor but with sizes replacing the masses: $F_{ab} = \sum_{i=1}^m d_i \left\{[\boldsymbol{r}_i \cdot \boldsymbol{r}_i - (r_{i,a})^2]\delta_{a,b} - r_{i,a}r_{i,b}\right\}$, $a, b = x, y, z$ where i runs over the particles forming the molecules and d_i is a typical size of the spherical centres. In practice, we can take d_i to be the cubic root of the appropriate van der Waals volume $\mathcal{V}^{\rm vdW}$ [Bondi, 1964].

Van der Est–Burnell continuum elastic model. In their analysis of solute order in nematics studied with NMR, Van der Est et al. [1987] assumed that the ordering potential acting on a molecule could be phenomenologically interpreted as an elastic potential acting on the solute. Thus, the liquid crystal host is viewed as a deformable continuum and the deformation is assumed to be

$$U(\beta,\gamma) = k_{\epsilon}[\mathscr{C}(\beta,\gamma)]^2, \tag{7.35}$$

where $\mathscr{C}(\beta,\gamma)$ is the contour length of the projection of the molecule on the plane perpendicular to the director. This in turn is evaluated modelling the molecule as a collection of van der Waals spheres and taking the contour as that of a minimal stretchable tube containing the molecule.

Zimmermann–Burnell. In an extended version of the model, Zimmerman and Burnell [1990] assumed

$$U(\beta,\gamma) = k_{\epsilon}[\mathscr{C}(\beta,\gamma) - k_{\xi}z(\beta,\gamma)]\mathscr{C}(\beta,\gamma), \tag{7.36}$$

where $z(\beta,\gamma)$ is the length of the projection of the solute molecule along the director:

$$z(\beta,\gamma) = Z_{\max}(\beta,\gamma) - Z_{\min}(\beta,\gamma) \tag{7.37}$$

with $Z_{\max}(\beta,\gamma)$, $Z_{\min}(\beta,\gamma)$ projections of the molecule onto the director $\boldsymbol{d}||Z$ and k_{ϵ}, k_{ξ} parameters. The model has been further extended and tested [Burnell and de Lange, 1998].

Ferrarini–Moro–Nordio–Luckhurst (FMNL) model. In the model of Ferrarini et al. [1992] the potential of mean torque acting on a rigid molecule is assumed to be proportional to the surface exposed to the surrounding environment in a certain direction and is written as

$$U(\beta,\gamma)/(k_BT) = \epsilon \int_S \mathrm{d}S\, P_2(\hat{\boldsymbol{n}}_S \cdot \boldsymbol{d}), \tag{7.38}$$

where the integral is over the surface, S, of the particle, $\hat{\boldsymbol{n}}_S$ is a unit vector normal to the surface in a certain point and $\boldsymbol{d}$ is the director. The parameter ϵ, expressing the strength of interaction, has dimensions of a length to the minus 2 to make the right-hand side dimensionless. This expression can be brought to a form similar to previous ones introducing a *surface tensor* with spherical components

$$S^{2,m} = -\int_S \mathrm{d}S\, \mathscr{D}^2_{0,m}(\theta,\phi), \tag{7.39}$$

where θ,ϕ give the orientation of the surface vector $\boldsymbol{n}_S$

$$U(\beta,\gamma)/(k_BT) = -\epsilon \sum_m S^{2,m*}\mathscr{D}^2_{0,m}(\beta,\gamma), \tag{7.40}$$

For a rectangular block the surface of each face can be easily evaluated giving $S^{2,0} = -[2A_{zz} - (A_{xx} - A_{yy})]/\sqrt{6}$ and $S^{2,\pm 2} = -(A_{xx} - A_{yy})/2$, where A_{aa} is the area of the face perpendicular to axis $a = x,y,z$. Note that a set of tangent spheres however disposed has a surface tensor and biaxiality zero. Overlap will reduce the problem, but the biaxiality is then somewhat dependent on the specific prescription employed. Thus, for a molecule having a lath shape with length L, width W and breadth B, the molecular biaxiality is according to the model [Ferrarini et al., 1992, 1994; Luckhurst, 2001],

$$u_{200} = [(W + B)L - 2BW]^2, \tag{7.41a}$$

$$u_{220} = \sqrt{3/2}\, L(W - B)\,[(W + B)L - 2BW], \tag{7.41b}$$

$$u_{222} = (3/2)L^2(W - B)^2, \tag{7.41c}$$

with $\lambda = \sqrt{\frac{3}{2}}[L(B - W)]/[L(B + W) - 2BW]$.

7.4 Biaxial Nematic Phases

We now depart from the assumption that the phase will be necessarily uniaxial even for biaxial mesogens and derive an MFT for the dispersive interactions (Chapter 5) between identical biaxial molecules [Straley, 1974; Remler and Haymet, 1986]. Starting again from Eq. 7.20 and considering rank $L = 2$ interactions,

$$-U_{12} = \sum (-1)^q \overline{u}_{Ln_1n_2}(r_{12}) \langle \mathscr{D}^L_{q,n_1}(\Omega_{1L}) \rangle \mathscr{D}^L_{-q,n_2}(\Omega_{2L}), \tag{7.42}$$

where $\overline{u}_{Ln_1n_2} = \sum_m (-1)^m \overline{u}_{Lmn_1n_2}/(2L+1)$. Now, we consider the possibility of candidate mesophases with orthorhombic D_{2h} symmetry. In this case, $\mathscr{D}^L_{q,n_1} = \mathscr{D}^L_{q,-n_2}$ and for identical particles the permutation symmetry of the pairwise intermolecular potential ensures that $\overline{u}_{Lqp} = \overline{u}_{Lpq}$. The reality of the pair potential requires that $\overline{u}_{Lqp} = (-1)^{q+p}\overline{u}_{L-q-p}$. Thus, using the set of symmetrized Wigner functions $\mathscr{R}^2_{m,n}$ (Eq. 3.149), one obtains the anisotropic single-particle mean-field potential as:

$$\begin{aligned} -U^{\text{BX}}(\Omega) = {} & [\overline{u}_{200}\langle \mathscr{R}^2_{0,0}\rangle + 2\overline{u}_{220}\langle \mathscr{R}^2_{0,2}\rangle]\mathscr{R}^2_{0,0}(\Omega) + [\overline{u}_{220}\langle \mathscr{R}^2_{0,0}\rangle + 2\overline{u}_{222}\langle \mathscr{R}^2_{0,2}\rangle]\mathscr{R}^2_{0,2}(\Omega) \\ & + [\overline{u}_{200}\langle \mathscr{R}^2_{2,0}\rangle + 2\overline{u}_{220}\langle \mathscr{R}^2_{2,2}\rangle]\mathscr{R}^2_{2,0}(\Omega) + [\overline{u}_{220}\langle \mathscr{R}^2_{2,0}\rangle + 2\overline{u}_{222}\langle \mathscr{R}^2_{2,2}\rangle]\mathscr{R}^2_{2,2}(\Omega). \end{aligned} \tag{7.43}$$

If the potential is separable, e.g. for the interaction coefficients for dispersive interactions [Luckhurst et al., 1975], $\overline{u}_{2qp} = X_{2q}X_{2p}$ and

$$-U^{\text{BX}}(\Omega) = \overline{u}_{200}\Big\{\eta_U[\mathscr{R}^2_{0,0}(\Omega) + \lambda\mathscr{R}^2_{0,2}(\Omega)] + \eta_B[\mathscr{R}^2_{2,0}(\Omega) + \lambda\mathscr{R}^2_{2,2}(\Omega)]\Big\}, \tag{7.44}$$

with $\eta_U = \langle \mathscr{R}^2_{0,0}\rangle + 2\lambda\langle \mathscr{R}^2_{0,2}\rangle$ and $\eta_B = \langle \mathscr{R}^2_{2,0}\rangle + 2\lambda\langle \mathscr{R}^2_{2,2}\rangle$ and where the angular brackets indicate the average on the single-particle orientational distribution. Thus, as a result of the separable-type potential, we have only two independent parameters: $\overline{u}_{200}$ and $\lambda = \alpha^{2,2}/\alpha^{2,0}$ and two independent effective mean field components η_U and η_B, acting on a single molecule. The MFT internal energy is given by:

$$\frac{-U^{\text{BX}}}{k_BT} = \frac{\overline{u}_{200}}{2k_BT}\Big\{\big[\langle \mathscr{R}^2_{0,0}\rangle + 2\lambda\langle \mathscr{R}^2_{0,2}\rangle\big]^2 + 2\big[\langle \mathscr{R}^2_{2,0}\rangle + 2\lambda\langle \mathscr{R}^2_{2,2}\rangle\big]^2\Big\}, \tag{7.45}$$

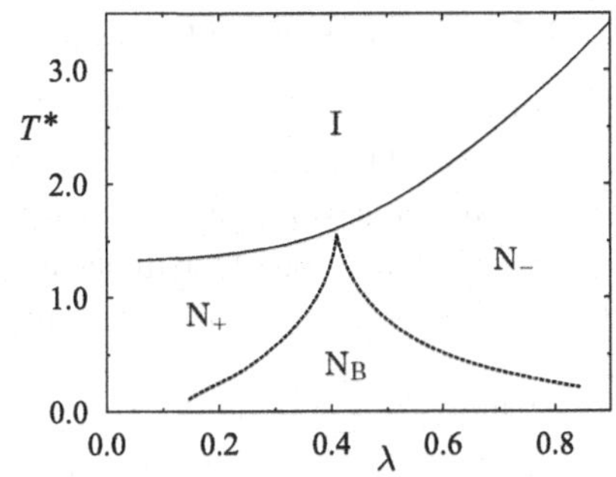

Figure 7.6 MFT phase diagram for biaxial and uniaxial phases. The scaled transition temperatures T^*_{NI} (—) and $T^*_{N_BN}$ (····) with $T^* = k_BT/\epsilon$, $\overline{u}_{200} = \varepsilon z$, are plotted as a function of the molecular biaxiality λ [Biscarini et al., 1995].

while the entropy, referred to one particle is:

$$\frac{S^{\mathrm{BX}}}{k_B} = -\frac{\overline{u}_{200}}{k_BT}\Big\{\big[\langle \mathscr{R}^2_{0,0}\rangle + 2\lambda\langle \mathscr{R}^2_{0,2}\rangle\big]^2 + 2\big[\langle \mathscr{R}^2_{2,0}\rangle + 2\lambda\langle \mathscr{R}^2_{2,2}\rangle\big]^2\Big\} + \ln Z_1^{\mathrm{BX}}. \quad (7.46)$$

The mean field free energy is obtained from its standard expression as:

$$\frac{A^{\mathrm{BX}}}{k_BT} = \frac{\overline{u}_{200}}{2k_BT}\left(\eta_U^2 + 2\eta_B^2\right) - \ln Z_1^{\mathrm{BX}}, \quad (7.47)$$

where the single-particle pseudo-partition function Z_1^{BX} is :

$$Z_1^{\mathrm{BX}} = \int_0^{2\pi} \mathrm{d}\alpha \int_0^{\pi} \mathrm{d}\beta \sin\beta \int_0^{2\pi} \mathrm{d}\gamma \exp\left[-U^{\mathrm{BX}}(\alpha,\beta,\gamma)/(k_BT)\right]. \quad (7.48)$$

The complete set of order parameter up to rank 4 describing the system is: $\langle P_2\rangle$, $\langle \mathscr{R}^2_{0,2}\rangle$, $\langle \mathscr{R}^2_{2,0}\rangle$, $\langle \mathscr{R}^2_{2,2}\rangle$, $\langle P_4\rangle$, $\langle \mathscr{R}^4_{0,2}\rangle$, $\langle \mathscr{R}^4_{0,4}\rangle$, $\langle \mathscr{R}^4_{2,0}\rangle$, $\langle \mathscr{R}^4_{4,0}\rangle$, $\langle \mathscr{R}^4_{2,2}\rangle$, $\langle \mathscr{R}^4_{2,4}\rangle$, $\langle \mathscr{R}^4_{4,2}\rangle$ and $\langle \mathscr{R}^4_{4,4}\rangle$. All the order parameters are non-vanishing if the system has a stable biaxial phase. In a uniaxial phase, the order parameters $\langle \mathscr{R}^L_{m,n}\rangle$ with non-zero m vanish. In the isotropic phase, all the order parameters are of course 0. Comparing the free energies for the isotropic, uniaxial and biaxial phases gives the phase diagram in Fig. 7.6. The curves for the second-rank order parameters: $\langle P_2\rangle$, $\langle \mathscr{R}^2_{0,2}\rangle$, $\langle \mathscr{R}^2_{2,0}\rangle$ and $\langle \mathscr{R}^2_{2,2}\rangle$ are also reported, together with the MC computer simulation results for the same potential in Fig. 10.8 [Biscarini et al., 1995]. To ease comparison with that nearest neighbour lattice representation of the model (see Section 10.3.1), the temperature in Fig. 7.6 is expressed as $T^* = k_BT/\epsilon$, with $\epsilon = \overline{u}_{200}/z$ and $z = 6$ nearest neighbours number for a cubic lattice. Increasing the biaxiality we see a transition from calamitic uniaxial N_+ to biaxial nematic N_B. The N_B phase becomes more stable and exists up to higher temperatures, until at $\lambda = 1/\sqrt{6}$ the deformed rod (prolate) shape becomes cubic and then further squashing brings the shape to be more flat-like (oblate). For $\lambda > 1/\sqrt{6}$ the nematic phase becomes a discotic one N_-, but thermodynamic results for a prolate particle at (λ, T^*) in the phase diagram can be mapped onto the point $(\lambda', T^{*\prime}) = \Big([(3-\lambda\sqrt{6})/(\sqrt{6}+6\lambda)], [(24T^*)/[(6\lambda+\sqrt{6})^2]\Big)$ for the dual oblate particle.

7.5 Uniaxial Smectic Phases

A simple and successful MFT of the smectic A phase, originally developed by McMillan [1971, 1972], and more recently revisited [Marguta et al., 2006] is based on the assumption that smectic ordering is promoted by the anisotropic attractive interactions, extending the Maier–Saupe model for nematics, with inclusion of the molecular separation, as well as the relative orientation in the starting pair potential, which is taken as, for uniaxial molecules:

$$-U(\Omega_1, \boldsymbol{r}_1) = \sum \left\langle u_{L00}(r_{12}) S^{0,0}_{L,L,0}(\Omega_1, \Omega_2) \right\rangle_{r_2, \Omega_2}, \tag{7.49a}$$

$$= \sum \frac{(-1)^{L+q}}{\sqrt{2L+1}} \left\langle u_L(r_{12}) \mathscr{D}^L_{q,0}(\Omega_2) \right\rangle_{r_2, \Omega_2} \mathscr{D}^L_{q,0}(\Omega_1), \tag{7.49b}$$

where the simplified notation $u_L(r_{12}) = u_{L,00}(r_{12})$ is used. The potential is an empirical one and, like in the Maier–Saupe one, steric repulsions are not taken into account. McMillan made the ansatz that the positional-orientational pair potential is separable and

$$-U_{12}\,(r_{12}, \cos\beta_{12}) = u_0(r_{12}) + u_2(r_{12}) P_2(\cos\beta_{12}) = -w\, \mathrm{e}^{-(r_{12}/\sigma)^2} [\delta_2 + P_2(\cos\beta_{12})], \tag{7.50}$$

with $u_0(r_{12}) = \delta_2 u_2(r_{12}) = -\delta_2 w\, \mathrm{e}^{-(r_{12}/\sigma)^2}$, where σ is the range of the interaction, w is the interaction strength of the anisotropic term and $\delta_2 = u_0/u_2$. Using the Fourier series of a Gaussian (Eq. E.11)

$$g(x) = \frac{\sqrt{\pi}\sigma}{d} \sum_{n=-\infty}^{\infty} \alpha_n \cos\left(\frac{n 2\pi x}{d}\right), \tag{7.51}$$

with $\alpha_n \equiv \mathrm{e}^{-[n\pi\sigma/d]^2}$, we have $u_0(r_{12}) \approx \dfrac{-\delta_2 w \sqrt{\pi}\sigma}{d} \left[1 + \alpha_1 \cos\left(\dfrac{2\pi r_{12}}{d}\right)\right]$ and

$$-U_{12} \approx \frac{\sqrt{\pi}\sigma w}{d} \left[1 + \alpha_1 \cos\left(\frac{2\pi r_{12}}{d}\right)\right] [\delta_2 + P_2(\cos\beta_{12})], \tag{7.52a}$$

$$\approx \frac{\sqrt{\pi}\sigma w}{d} \left[\delta_2 + P_2(\cos\beta_{12}) + \alpha_1 \delta_2 \cos\left(\frac{2\pi r_{12}}{d}\right) + \alpha_1 \cos\left(\frac{2\pi r_{12}}{d}\right) P_2(\cos\beta_{12})\right]. \tag{7.52b}$$

Averaging over the orientations and positions of one of the two molecules and assuming the candidate mesophase of which we are testing the stability against nematic and isotropic to be of smectic A type, i.e uniaxial around the director, say the z-axis of the lab, and with positions periodically distributed along z, we have, as seen in Eq. 3.189

$$P(z, x) = \sum_{L=0}^{\infty} \sum_{n_z=0}^{\infty} p_{L;n_z} \cos(n_z q z) P_L(x), \quad L \text{ even}, \tag{7.53}$$

with $x = \cos\beta_1$, $q = 2\pi/\ell_z$ and ℓ_z the layer spacing, while the expansion coefficients $p_{L;n}$ are order parameters given by

$$p_{L;n} = \langle \cos(nqz) P_L(x) \rangle\,. \tag{7.54}$$

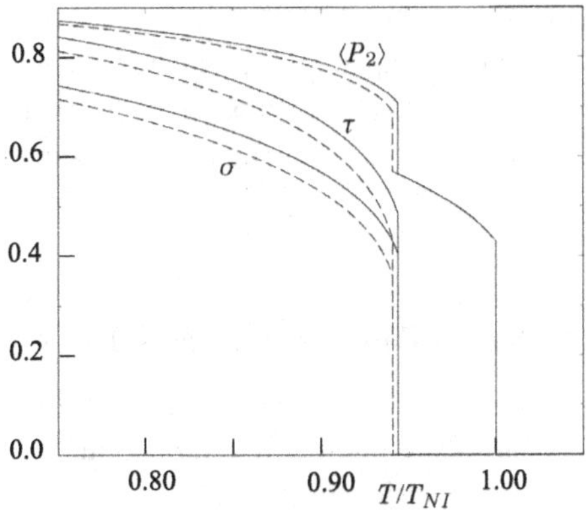

Figure 7.7 Positional (τ), mixed (σ), and orientational ($\langle P_2 \rangle$) and order parameters vs reduced temperature T/T_{NI} according to McMillan theory for smectic A with parameters $\alpha = 0.85$, and $\delta_2 = 0$. The solid lines correspond to the exact solution; the dashed lines correspond to McMillan's approximation [Marguta et al., 2006].

McMillan makes some further approximations, but the numerical calculation can be performed in full, as shown by Marguta et al. [2006] (see Fig. 7.7). The effective potential, neglecting an irrelevant scalar term, is

$$-U^{\text{SM}}(z,x) = u_2\left[\langle P_2 \rangle P_2(x) + c_{0:1}\tau \cos\left(\frac{2\pi z}{\ell_z}\right) + c_{2;1}\sigma \cos\left(\frac{2\pi z}{\ell_z}\right) P_2(x)\right], \quad (7.55)$$

with $c_{0:1} = \alpha_1\delta_2/u_2$, $c_{2;0} = 1$, $c_{2;1} = \alpha_1/u_2$. The order parameters have to be determined self-consistently as usual. The expressions can be integrated analtically on z, giving [Kventsel et al., 1985]

$$\langle P_2 \rangle = (\ell_z/Z_1^{\text{SM}}) \int_{-1}^{1} \mathrm{d}x\, P_2(x) \exp\{(u_2/k_B T)\langle P_2 \rangle P_2(x)\}\, I_0(\zeta), \quad (7.56a)$$

$$\sigma = (\ell_z/Z_1^{\text{SM}}) \int_{-1}^{1} \mathrm{d}x\, P_2(x) \exp\{(u_2/k_B T)\langle P_2 \rangle P_2(x)\}\, I_1(\zeta), \quad (7.56b)$$

$$\tau = (\ell_z/Z_1^{\text{SM}}) \int_{-1}^{1} \mathrm{d}x\, \exp\{(u_2/k_B T)\langle P_2 \rangle P_2(\cos\beta)\}\, I_1(\zeta), \quad (7.56c)$$

where $I_1(\zeta)$, $I_0(\zeta)$ are modified Bessel functions of order n [Abramowitz and Stegun, 1965]: $I_n(z) = \frac{1}{\pi}\int_0^{\pi} \mathrm{d}\theta\, e^{z\cos\theta}\cos(n\theta)$, with $\zeta = u_2\left(c_{2;1}\sigma P_2(x) + c_{0:1}\tau\right)/(k_B T)$. The pseudo-partition is

$$Z_1^{\text{SM}} = \ell_z \int_{-1}^{1} \mathrm{d}x\, \exp\{(u_2/kT)\langle P_2 \rangle P_2(\cos\beta)\}\, I_0(\zeta). \quad (7.57)$$

In Fig. 7.7 the order parameters are plotted as a function of the reduced temperature T/T_{NI}. Beyond the pioneering McMillan study, various extensions have been proposed. An MFT for biaxial molecules has been put forward by Teixeira et al. [2006]. The MFT has also been extended to smectic C phases [McMillan, 1973]. A fairly general density functional theory including smectics has also been developed by Lipkin and Oxtoby [1983].

The connection between mean field and Landau–de Gennes theories has been studied by Katriel et al. [1986] and more recently in rigorous mathematical terms by Ball and Majumdar [2010].

7.6 Density Functional Onsager Theory

In general, a statistical mechanics Density Functional Theory (DFT) considers the free energy as a suitable functional of the one-particle probability density $P(\boldsymbol{X})$. In particular, this can consist of a virial expansion of the pressure (or of the osmotic pressure as appropriate, e.g. if we are describing a colloidal suspension) or equivalently of the free energy in terms of the number density ρ (or of the volume fraction ϕ). This approach is probably more appropriate for lyotropic liquid crystals or colloidal suspensions where composition (volume fraction) is the relevant thermodynamic variable driving the disorder-order transition. As we have seen in Section 1.14, suspensions of rod-like or plate-like particles undergo a isotropic-nematic phase transition as a function of rod concentration. In this case, anisotropic shape and aspect ratio are likely the key ingredients to liquid crystalline behaviour, as assumed in hard-core interaction models (Section 5.5). While at low rod concentrations a suspension is isotropic, the particles tend to align to avoid the particle-particle contact with their high cost in energy. At the same time the greater translational freedom along the director can actually bring a counter-intuitive increase of entropy in the nematic phase. Onsager [1949] considered nematic ordering in a system of hard, rigid rods devoid of attractive interactions and with aspect ratio $L/D \gg 1$, where L is the length and D is the diameter of the rod. At low densities it is possible to express the free energy of such a system in the form of the virial expansion, as obtained with the 'trick' of considering a particle at a certain orientation Ω_i as the ith different component of a mixture. In this way the free energy can be built from the mixing entropy of a multicomponent mixture. Given its importance, we reproduce here the essential steps of the Onsager [1949] derivation. We start by considering a system of N rigid particles in a volume V at temperature T (canonical conditions) and write the Helmoltz free energy $\mathcal{A}$, starting from the configurational partition $Q(N,V,T)$ given in Eq. 7.58

$$Q^{\mathrm{B}}(N,V,T) = \frac{1}{N!}\int \mathrm{d}\widetilde{\boldsymbol{r}}\mathrm{d}\widetilde{\Omega}\, \exp\left[-U_N^{\mathrm{B}}\left(\widetilde{\boldsymbol{r}},\widetilde{\Omega}\right)/(k_BT)\right] = Z(N,V,T)/N! \tag{7.58}$$

and

$$\mathcal{A}(N,V,T) = \mathcal{A}^{\mathrm{id}}(N,V,T) - Nk_BT\ln[Q^{\mathrm{ex}}(N,V,T)], \tag{7.59}$$

where the ideal term for non-interacting particles given in Eq. 4.33 is

$$\mathcal{A}^{\mathrm{id}}(N,V,T) = Nk_BT\ln[\lambda^3\Lambda_J/V_\Omega] + Nk_BT[\ln\rho - 1]. \tag{7.60}$$

Following Onsager [1949] we can define a configuration integral for a suspension of N_p spherical particles in the volume V, as

$$Q^{\text{ex}}(N_p, V, T) = \int \mathrm{d}\tilde{r}\, \exp\left[-U/(k_B T)\right]/N_p! \tag{7.61}$$

and we have from the virial expansion Eq. 4.20 that

$$\ln Q^{\text{ex}}(N_p, V, T) = N_p \left\{1 - \ln\left(\frac{N_p}{V}\right) + \frac{1}{2}\beta_1\left(\frac{N_p}{V}\right) + \frac{1}{3}\beta_2\left(\frac{N_p}{V}\right)^2 + \cdots\right\}, \tag{7.62}$$

where β_i are the so-called irreducible integrals defined in Eqs. 4.21a and 4.21b, and in particular $\beta_1 = \beta(j, j')$ and $\beta_2 = \beta(j, j', j'')$ involve interactions of two and three particles (j, j') and (j, j', j'') . Quite similarly, for a mixture of $N_1, \ldots, N_s, \ldots$ particles of different type in suspension, we have

$$\begin{aligned}\ln Q^{\text{ex}}(N_p, V, T) = &\sum_j N_j[1 - \ln(N_j/V)] + \frac{1}{2V}\sum_{j,j'} \beta_1\left(j, j'\right) N_j N_{j'} \\ &+ \frac{1}{3V^2}\sum_{j,j',j''} \beta_2\left(j, j', j''\right) N_j N_{j'} N_{j''} + \cdots .\end{aligned} \tag{7.63}$$

In a system of anisotropic particles, we can formally consider that the particle types correspond to their orientation. Dividing the angular space in n_s solid angle bins $\Delta\Omega_1, \ldots, \Delta\Omega_i, \ldots, \Delta\Omega_s$ centred at the respective orientations $\Omega_1, \ldots, \Omega_i, \ldots, \Omega_{n_s}$, these bins will have relative populations of particles $\Delta N_i/N_p = P(\Omega_i)\,\Delta\Omega_i$, with $i = 1, 2, \ldots, n_s$ and where $P(\Omega_i)$ is the orientational distribution density introduced in Chapter 3. The irreducible integrals $\beta_1(j, j')$, $\beta_2(j, j', j'')$ correspond formally to $\beta_1(\Omega_j, \Omega_{j'})$, $\beta_2(\Omega_j, \Omega_{j'}, \Omega_{j''})$ for fixed orientations Ω_j, $\Omega_{j'}$, $\Omega_{j''}$ of the particles involved. Taking a continuum limit and replacing the sums by integrals we have

$$\begin{aligned}\ln Q^{\text{ex}}(N_p, V, T) = N_p \Bigg\{ &1 - \ln\left(\frac{N_p}{V}\right) - \int \mathrm{d}\Omega_1 P(\Omega_1) \ln\left[V_\Omega P(\Omega_1)\right] \\ &+ \left(\frac{N_p}{2V}\right) \int \mathrm{d}\Omega_1 \mathrm{d}\Omega_2 \beta_1\left(\Omega_1, \Omega_2\right) P(\Omega_1) P(\Omega_2) \\ &+ \left(\frac{N_p^2}{3V^2}\right) \int \mathrm{d}\Omega_1 \mathrm{d}\Omega_2 \mathrm{d}\Omega_3 \beta_2\left(\Omega_1, \Omega_2 \Omega_3\right) P(\Omega_1) P\left(\Omega_2\right) P\left(\Omega_3\right) + \cdots \Bigg\}\end{aligned} \tag{7.64}$$

and, simplifying the notation, we have the free energy as

$$\frac{\Delta\mathcal{A}}{N k_B T} = [\ln\rho - 1] + \int \mathrm{d}\Omega_1 P(\Omega_1) \ln[V_\Omega P(\Omega_1)] + B_2\rho + B_3\rho^2 + \cdots \tag{7.65}$$

in terms of the virial coefficients appearing in the density expansion of the (osmotic) pressure, where

$$B_2 = -\frac{1}{2V}\int d\Omega_1 \int d\Omega_2 \beta_1(\Omega_1, \Omega_2)\, P(\Omega_1) P(\Omega_2), \tag{7.66a}$$

$$B_3 = -\frac{1}{3V}\int d\Omega_1 \int d\Omega_2 \int d\Omega_3 \beta_2(\Omega_1, \Omega_2, \Omega_3)\, P(\Omega_1) P(\Omega_2)\, P(\Omega_3). \tag{7.66b}$$

The second term in Eq. 7.65 is the orientational entropy and the third one is the packing entropy that is related to B_2, the second virial coefficient for two rigid particles. The equilibrium orientational distribution function is obtained by minimizing the reduced free energy with respect to $P(\Omega)$, subject to the normalization constraint $\int d\Omega_1 P(\Omega_1) = 1$. Introducing a Lagrange multiplier λ, as we did in Sections 3.8 and 4.2, and minimizing yields

$$P(\Omega_1) \propto \exp\Big\{ -2\rho \int d\Omega_2 B_2(\Omega_1, \Omega_2)\, P(\Omega_2) - \frac{3}{2}\rho^2 \int d\Omega_2 d\Omega_3 B_3(\Omega_1, \Omega_2, \Omega_3) P(\Omega_2) P(\Omega_3) + \cdots \Big\}, \tag{7.67}$$

where the proportionality constant is determined by the normalization of $P(\Omega)$. In Onsager theory only terms in B_2 are retained, while the third and higher terms are neglected. This gives the single-particle distribution

$$P(\Omega_1) = \frac{\exp\Big[-2\rho \int d\Omega_2 B_2(\Omega_1, \Omega_2)\, P(\Omega_2)\Big]}{\int d\Omega_1 \exp\Big[-2\rho \int d\Omega_2 B_2(\Omega_1, \Omega_2)\, P(\Omega_2)\Big]}. \tag{7.68}$$

To proceed with a numerical calculation, a model for the particles has to be assumed and an expression for $B_2(\Omega_1, \Omega_2)$ obtained. Onsager theory is particularly convenient for hard particles, where B_2 is easier to evaluate, even though it is not limited to these potentials. More specifically the Mayer function (Eq. 4.147) for hard-core potentials is just:

$$\Phi(\boldsymbol{r}_{12}, \Omega_1, \Omega_2) = e^{[-U_{12}^{\rm HP}/(k_B T)]} - 1 = \begin{cases} -1, & \text{if particles overlap} \\ 0, & \text{if particles do not overlap} \end{cases} \tag{7.69}$$

and $B_2(\Omega_1, \Omega_2)$ is

$$B_2(\Omega_1, \Omega_2) = \int dr_{12}\, \Phi(\boldsymbol{r}_{12}, \Omega_1, \Omega_2) = \frac{1}{2} V_{\rm exc}^{\rm HP}(\Omega_1, \Omega_2), \tag{7.70}$$

with $V_{\rm exc}^{\rm HP}(\Omega_1, \Omega_2)$ the *excluded volume* or *covolume* [Vroege and Lekkerkerker, 1992], i.e. the volume of space where a particle cannot enter because of the presence of another one. The orientationally averaged value entering the free energy expression Eq. 7.65 will be

$$B_2 = \frac{1}{2}\langle V_{\rm exc}\rangle = \frac{1}{2}\int d\Omega_1 d\Omega_2 P(\Omega_1) P(\Omega_2) V_{\rm exc}(\Omega_1, \Omega_2). \tag{7.71}$$

For non-attracting particles all terms will be purely entropic in nature and the system will be athermal. The excluded volume has been studied in detail for some simple particle shapes.

Hard cylinders. In the original treatment Onsager dealt with hard cylinders of diameter d_1, d_2 and length l_1, l_2 and cylinders capped with hemispheres of the same diameter (the spherocylinders introduced in Section 5.5.2). The excluded volume of two cylinders of diameter d_1 and d_2 and length l_1 and l_2 is [Onsager, 1949]:

$$\begin{aligned} V_{\text{exc}}^{\text{CYL}} &(\Omega_1, \Omega_2) \\ &= -\beta_1^{\text{CYL}}(l_1, d_1; l_2, d_2; \beta_{12}) = 2B_2^{\text{CYL}}(\Omega_1, \Omega_2), \\ &= (\pi/4)d_1 d_2 (d_1 + d_2) \sin\beta_{12} + l_2 \left\{(\pi/4)d_2^2 + d_1 d_2 E(\sin\beta_{12}) + (\pi/4)d_1^2 |\cos\beta_{12}|\right\} \\ &\quad + l_1 \left\{(\pi/4)d_1^2 + d_1 d_2 E(\sin\beta_{12}) + (\pi/4)d_2^2 |\cos\beta_{12}|\right\} + l_1 l_2 (d_1 + d_2) \sin\beta_{12}, \end{aligned} \tag{7.72}$$

where β_{12} is the angle between the axes of the two rods and

$$E(\sin\beta_{12}) = \frac{1}{4}\int_0^{2\pi} d\psi \left(1 - \sin^2\beta_{12}\sin^2\psi\right)^{1/2} \tag{7.73}$$

is a complete elliptic integral of the second kind [Abramowitz and Stegun, 1965].

Hard spherocylinders. For two spherocylinders the excluded volume $V_{\text{exc}}^{\text{HSC}}$ is instead

$$V_{\text{exc}}^{\text{HSC}} = \frac{\pi}{4}(d_1 + d_2)^2 (l_1 + l_2) + \frac{\pi}{6}(d_1 + d_2)^3 + l_1 l_2 (d_1 + d_2) \sin\beta_{12}. \tag{7.74}$$

For identical HSCs this is [Onsager, 1949; Gelbart and Gelbart, 1977]

$$V_{\text{exc}}^{\text{HSC}} = 8\mathcal{V}^{\text{HSC}} + 2l^2 d\,|\sin\beta_{12}| = 2l^2 d\,|\sin\beta_{12}| + 2\pi l d^2 + \frac{4}{3}\pi d^3, \tag{7.75}$$

where $\mathcal{V}^{\text{HSC}} = \frac{\pi}{4} l d^2 + \frac{\pi}{6} d^3$ is the volume of a HSC (Eq. 5.46c). For long rods ($l \gg d$) the term containing l^2 is dominant and $B_2(\Omega_1, \Omega_2) = \frac{1}{2} V_{\text{exc}}^{\text{HSC}} \approx dl^2 |\sin\beta_{12}|$, while for hard spheres the excluded volume is $V_{\text{exc}}^{\text{HS}} = 2B_2^{\text{HS}} = \frac{4}{3}\pi d^3 = 8\mathcal{V}^{\text{HS}}$. In this limit the equation of state for a HSC suspension is given by

$$\frac{P}{k_B T} = \rho + \frac{1}{2}\rho^2 \left(8\mathcal{V}^{\text{HSC}} + 2l^2 d\langle|\sin\beta_{12}|\rangle\right). \tag{7.76}$$

Hard ellipsoids. We have introduced this popular model in Section 5.5.1. An analytic (albeit too complex to report here) expression for the second virial has been given by Isihara [1951], Straley [1973] and Camp et al. [1996]. Very often, however, the hard Gaussian overlap (HGO) model [Bhethanabotla and Steele, 1987; Rigby, 1989; Schmid and Phuong, 2002] is used, where the potential energy U_{ij}^{HGO} between molecules is assumed to be purely repulsive, and given by

$$U_{ij}^{\text{HGO}} = \begin{cases} \infty & \text{if} \quad r \leqslant \sigma\left(\hat{\boldsymbol{r}}, \boldsymbol{u}_i, \boldsymbol{u}_j\right) \\ 0 & \text{if} \quad r > \sigma\left(\hat{\boldsymbol{r}}, \boldsymbol{u}_i, \boldsymbol{u}_j\right) \end{cases}, \tag{7.77}$$

with $\sigma\left(\hat{\boldsymbol{r}}, \boldsymbol{u}_i, \boldsymbol{u}_j\right)$ given by Eq. 5.54 interpreted as the distance of closest approach between a pair of molecules of the ellipsoid aspect ratio $\kappa = \sigma_{\parallel}/\sigma_{\perp}$. This function gives the exact HE contact distance if the particles i and j are approaching colinearly ('head-on'), while it overestimates it by a factor of at most $\sqrt{2}/\sqrt{1 + 2\kappa/(\kappa^2 + 1)} < \sqrt{2}$ otherwise:

$$B_2^{\text{HGO}}(\cos\beta_{12}) = \frac{2\pi\sigma_0^3}{3(1-\chi)}\sqrt{1 - \chi^2(\cos\beta_{12})^2}. \tag{7.78}$$

In general, one way used for the numerical calculation consists of expanding the excluded volume in a series of Legendre polynomials P_L:

$$B_2(\beta_{12}) = B_{2,0} + \sum_{L=1}^{\infty} B_{2L} P_L(\cos\beta_{12}), \tag{7.79}$$

where the expansion coefficients $B_{2,L}$ are given in Tjipto-Margo and Evans [1990]. However, the expansion is slowly convergent, and another approach is that of Camp et al. [1996]. Returning to Eq. 7.68 we see that it provides a self-consistent expression for $P(\Omega)$ that can be solved numerically. In his original derivation Onsager [1949] made the functional ansatz

$$P(\cos\beta) = a\cosh(a\cos\beta)\,/\,[4\pi\,\sinh(a)], \tag{7.80}$$

for the single-particle orientational distribution. In the isotropic phase, $\alpha = 0$, while in the N phase $\alpha > 0$. Note that, in the hard-needle limit, the solution only depends on the scaled density $\rho^* = \rho_0 l^2 d$.

The theory predicts a first-order phase transition between the I and N phases for the following values of packing fraction $\phi = (\pi d/4l)\rho^*$: $\phi_I = 3.340\frac{d}{l}$, and $\phi_N = 4.486(d/l)$. We see this graphically in Fig. 7.8. The density gap becomes smaller as the particles go towards the hard-needle limit. As the ratio l/d is reduced, end-particle effects in the excluded volume become important, and there is no universal scaled density. An alternative, more systematic, procedure for determining $P(\cos\beta)$ [Isihara, 1951; Lasher, 1970] is to expand ln $P(\cos\beta)$ in Legendre polynomials. i.e. to write

$$P(\cos\beta) \propto \exp\left\{\sum_{L=0}^{J} a_L P_L(\cos\beta)\right\}, \tag{7.81}$$

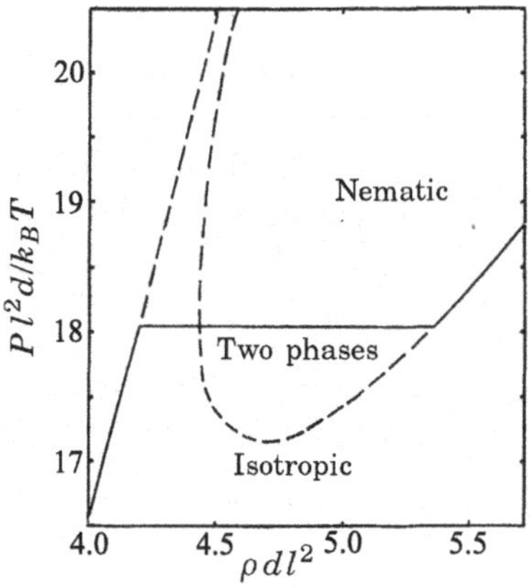

Figure 7.8 The equation of state for the Onsager model. At the liquid crystal transition, an anisotropic phase of density $\rho = 5.3l^{-2}d^{-1}$ is in equilibrium with an isotropic one of density 4.2. The order parameter in the dense phase at the transition is $\langle P_2\rangle = 0.78$ [Straley, 1973].

truncating at a certain J and minimizing in terms of α_L. After having obtained $P(\Omega)$, one can find the pressure by differentiation of the Helmholtz free energy with respect to the volume.

Onsager theory predicts a ratio of nematic to isotropic phase densities at the NI transition $\rho_N^*/\rho_I^* \approx 1.34$, a density change $\Delta\rho^*/\rho^* \approx 26\%$, much larger than the real one of $\Delta\rho^*/\rho^* \leq 0.5\%$ (cf. Section 2.3). The values of the entropy jump $\Delta \mathcal{S}_{\mathrm{NI}} \approx 8\mathrm{R}$ and of the order parameter $\langle P_2 \rangle_{\mathrm{NI}} \approx 0.85$ at the transition are also much larger than the typical nematic-isotropic experimental ones.

The theory works well for long rods and has been successfully applied in predicting properties of lyotropic nematic phases made of stiff rod-like molecules such as viruses [Straley, 1973; Fraden et al., 1993], long DNA double strands [Nakata et al., 2007; Bellini et al., 2012], peptide nanotubes [Bucak et al., 2009], etc. but does not predict nematic phases for aspect ratios l/d of a few units [Straley, 1973]. Indeed, for short rods the critical packing fraction probably becomes larger than the maximum close-packing fraction and, below a minimum molecular shape anisotropy the system might freeze before the critical density for the isotropic-nematic transition is reached [Lee, 1987].

The theory was originally developed for rods, but can equally well be applied to disc-like particles, defining the anisotropy ratio x as a ratio of the smallest to the largest dimension for either rods or plates (i.e. for rods $x = d/l$, plates have $x = l/d$, while isotropic particles will have $x = 1$, and completely anisotropic particles $x = 0$). This is quite relevant since, as we mentioned in Section 1.14 (cf. Fig. 1.56) many LC suspensions are obtained from flat-like particles of clays or other minerals. In Fig. 7.9 the transition concentrations are plotted for both rods and plates as a function of the anisotropy ratio x [Forsyth et al., 1978]. It is immediately apparent that the behaviour of plates differs markedly from that of rods. When $x = 0$, the transition concentration c_t of rods is infinite, while for thin plates, c_t is finite. As x

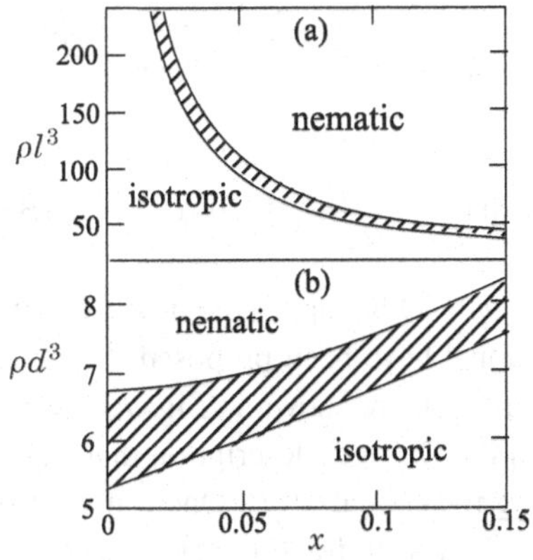

Figure 7.9 Density of (a) hard-rod or (b) hard-plate dispersions at the Onsager phase transition as a function of the anisotropy ratio x, the ratio of the smallest to the largest dimension for either rods or plates (i.e. for rods $x = d/l$, plates have $x = l/d$). The coexistence regions are hatched [Forsyth et al., 1978].

increases, c_t for rods decreases, while c_t for plates increases. Note that the virial expansion is actually performed in powers of the packing fraction $\phi = \rho \mathcal{V}^{\text{HSC}} \approx \pi \rho d^2 l \leqslant 1$ if $l/d \geqslant 1$, where $\mathcal{V}^{\text{HSC}}$ is the volume of a spherocylinder. At a very low volume fraction of rods the higher-order terms in the expansion can be neglected.

It is worth mentioning that the theory has also been extended to the smectic phase [Mederos et al., 2014].

7.7 Generalized Onsager Theories

The most successful theories of nematic ordering in systems of hard convex bodies make use of the Parsons–Lee correction to the excess free energy term. The theories of Parsons [1979] and Lee [1987] essentially amount to interpolating between the accurate limiting equations of state of Carnahan–Starling for hard spheres and of Onsager for infinitely long, hard spherocylinders. The general idea is to map the thermodynamic properties of the hard convex-body system into those of a hard-sphere system or, if possible, of some other reference system, in particular the isotropic phase of the corresponding hard-body system. Within the Onsager formulation coupled with the Parsons–Lee (PL) correction, the free energy is expressed in terms of the second virial contribution and the excess free energy of the system of interest is assumed to be proportional to that of a hard-sphere system (see Section 5.2.1). Using the Carnahan and Starling [1969] equation of state, which is an excellent approximation for HS, along with the relationships $B_2^{\text{HS}} = 4\mathcal{V}_{\text{HS}}$ and $\phi = N\mathcal{V}_{\text{HS}}/V$, yields

$$\frac{\mathcal{A}_{\text{ex}}}{Nk_BT} = \frac{\langle B_2(\Omega_1, \Omega_2)\rangle}{B_2^{\text{HS}}} \frac{\mathcal{A}_{\text{ex}}^{\text{HS}}(\phi_{\text{HS}})}{Nk_BT} = \frac{\langle B_2(\Omega_1, \Omega_2)\rangle}{4\mathcal{V}_{\text{HS}}} \frac{\mathcal{A}_{\text{ex}}^{\text{HS}}(\phi_{\text{HS}})}{Nk_BT}, \tag{7.82a}$$

$$= \frac{\langle B_2(\Omega_1, \Omega_2)\rangle}{4} \frac{(4 - 3\phi)}{(1-\phi)^2}, \tag{7.82b}$$

where $\mathcal{V}_{HS}$ is the volume of a hard sphere and ϕ_{HS} is the packing fraction of the hard-sphere system $(\phi_{\text{HS}} = \rho \mathcal{V}_{\text{HS}})$ and the angular average is like in Eq. 7.71. Keeping into account the correction

$$P(\cos\beta) = \frac{1}{Z} \exp\Big[-\frac{\rho\, G(\phi)}{2\pi} \int_{-1}^{1} \mathrm{d}\cos\beta' \, V_{\text{exc}}(\beta, \beta') P(\cos\beta')\Big], \tag{7.83}$$

with $G(\phi) = (4 - 3\phi)/(2 - 2\phi)^2$. Other approximations, e.g. the extended Rosenfeld (ER) which uses a different functional have been proposed [Cinacchi and Schmid, 2002]. The results of the PL and ER correction are shown in Fig. 7.10 for HSC (7.10a) or HE (7.10b) and significantly improve on the Onsager description that can make quantitatively correct predictions only for very long rods (small d/l). One can see from Fig. 7.10 [Cinacchi and Schmid, 2002] that the critical packing fraction approaches the close-packing fraction for length/breadth ≈ 5 so that the system may be expected to crystallize instead of becoming an LC. Another interesting result of the computations is that the discontinuity in the packing fraction at the isotropic-nematic phase transition becomes smaller as the molecular shape anisotropy decreases. This PL approach, which manages to incorporate some many-body

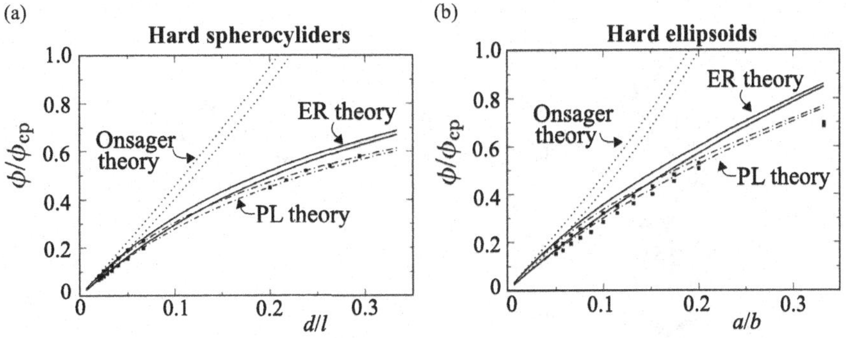

Figure 7.10 (a) The relative packing fractions referred to close-packing: $\phi/\phi_{cp} = \rho/\rho_{cp}$ as a function of d/l, the diameter to cylinder length ratio, for HSC and (b) the same as a function of inverse aspect ratio a/b for hard ellipsoids. The points are from computer simulations [Camp et al., 1996], while the lines are from Parsons–Lee (PL), extended Rosenfeld (ER) and Onsager theory as indicated [Cinacchi and Schmid, 2002].

effects in a very simple way, has been shown to be surprisingly successful at predicting the IN-transition parameters, even for short rods. Calculations and MC results for HE [Camp et al., 1996] are reported in Fig. 7.10b.

7.7.1 Non-convex Particle: Hard Helices

As an example of application of generalized Onsager theory to more elaborate mesogenic shapes we show some results from Frezza et al. [2013] for a system of hard helices formed by n_s fused spheres of diameter σ and with the same contour length $L_{\mathscr{C}} = 10\sigma$, but different pitches and radii. The spheres are positioned at $x_i = r\cos(2\pi t_i)$, $y_i = r\sin(2\pi t_i)$ and $z_i = pt_i$, $1 \le i \le n_s$, where r is the radius and p the pitch of the helix. The long axis of the helix $\hat{\boldsymbol{h}}$, passes through the centre of the helix. As shown in Fig. 7.11, the centres of the beads lie on an inner cylinder of radius r. The diameter of the outer cylinder $W = (2r+\sigma)$ is the width of the helix r_{max}, while $\Lambda = z_{n_s} - z_1$ is the Euclidean length, so that the end-end distance is $\Lambda + \sigma$ and the aspect ratio $(\Lambda+\sigma)/W$. As the spheres are arranged in different helical fashions L and $L_{\mathscr{C}}$ will differ. Given the values of r, p and $L_{\mathscr{C}}$, the increment $\Delta t = t_{i+1} - t_i$ is determined by the equation $L_{\mathscr{C}}/(n_s - 1) = 2\pi\Delta t\sqrt{r^2 + (p/(2\pi))^2}$. Simulation results can be compared with theoretical results using Parsons [1979] and Lee [1987, 1988] (PL) based on the assumption that the excess free energy is proportional to that of a system of hard spheres (HS) at the same packing fraction (ϕ). Since PL theory was developed for convex bodies, a modified PL (MPL) theory for non-convex particles like the present ones was introduced [Varga and Szalai, 2000], employing an effective volume, $\mathscr{V}_{\text{ef}}$, defined as the volume of the non-convex particle that is inaccessible to other particles, instead of the hard spheres volume. This effective volume, larger than the geometrical volume, has been evaluated for linear chains of hard spheres by Abascal and Lago [1985].

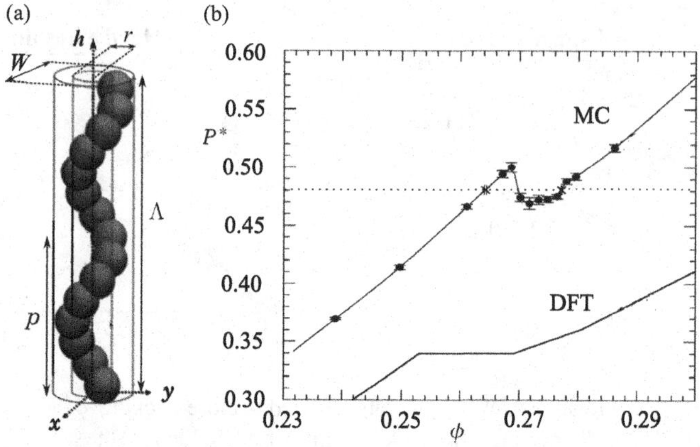

Figure 7.11 (a) Model rigid helix made up of $n_s = 15$ partially fused hard spheres of diameter σ and centres arranged on a helical string of fixed contour length $L = 10\sigma$, diameter r and pitch p. (b) Pressure $P^* = P/(k_B T D^3)$ vs volume fraction $\phi = V_{helices}/V$ equation of state from MC simulations with interpolating splines and equal area construction (line and symbols) and Onsager theory results with modified Parsons–Lee correction (MPL–DFT) for helices with $p = 4$ and $r = 0.2$ [Kolli et al., 2014b].

Results from the MPL density functional theory are compared with MC simulations in Fig. 7.11 [Kolli et al., 2014a]. Interestingly, this system forms a *screw-nematic* phase, where the helix twofold symmetry axes spiral around the main phase director, a phase qualitatively similar to that observed in experiments on systems of colloidal suspensions of helical flagella isolated from prokaryotic bacteria [Barry et al., 2006].

8

Monte Carlo Methods

8.1 Introduction

The problem of calculating the thermodynamic observables of a fluid starting from a given intermolecular potential is so complex that, apart from a few exceptional cases, there are only two possibilities open: one is to use approximate theories (see Chapter 7), the other to resort to computer simulations, consisting of numerical solutions to the problem of many interacting particles. In this chapter we describe one of these techniques, the Monte Carlo (MC) method, that aims to calculate averages of relevant observables by suitably generating equilibrium configurations in a certain statistical ensemble. We shall discuss computer simulations mainly with applications to liquid crystals in mind, but we begin by describing the general theory, aiming at a relatively concise summary since the topic is well covered in specific books [Newman and Barkema, 1999; Landau and Binder, 2000; Allen and Tildesley, 2017]. Monte Carlo methods have been defined in general as 'that branch of experimental mathematics which is concerned with experiments on random numbers' [Hammersley and Handscomb, 1965]. In condensed matter physics, however, the term is now universally reserved for the technique devised by Metropolis et al. [1953] to evaluate statistical averages. Consider a system of N particles where the intermolecular potential is known, for example, to be one of the pairwise interactions $U(\boldsymbol{X}_i, \boldsymbol{X}_j)$ discussed in Chapter 5. For a *configuration* defined by the set of positional-orientational coordinates $\widetilde{\boldsymbol{X}} = (\boldsymbol{X}_1, \mathbf{X}_2, \dots, \boldsymbol{X}_N)$ the potential energy is that seen in Eq. 5.2b. In the canonical ensemble (constant N,V,T) any time independent property of interest, depending on positions and orientations, can be written as the equilibrium average

$$\langle A \rangle = \int \mathrm{d}\widetilde{X}\; A(\widetilde{X}) \exp[-U_N(\widetilde{X})/(k_B T)] \;\Big/ \int \mathrm{d}\widetilde{X}\; \exp[-U_N(\widetilde{X})/(k_B T)], \tag{8.1}$$

where the denominator is the configurational integral $Z(N, V, T)$ (see Sec. 4.6). Note that we have not included velocities here. Indeed, since they are not contained in the potential and we only consider configurational properties $A(\widetilde{X})$, depending on coordinates but not on velocities, the integration over these kinetic contributions can be factored out and disappears from the expression for the average. Since the intermolecular potential is assumed to be known, the problem of calculating $\langle A \rangle$ is 'reduced' to performing a $\approx 6N$ dimensional integration. Let us now imagine a real system and some idealized way of taking 'snapshots' in configurational space. From the jth configuration, i.e. from the set of positions and orientations specifying the state of the system in this jth photograph, we could calculate the value of property A, call this $A^{(j)}$. We could repeat the process, M times say. It is clear that the average Eq. 8.1 would now be simply the arithmetic mean

$$\langle A \rangle = \frac{1}{M} \sum_{j=1}^{M} A^{(j)}, \tag{8.2}$$

where M should be large enough to reduce the statistical uncertainty in $\langle A \rangle$ to acceptable values. In this virtual experiment the substitution of the extremely complicated integral Eq. 8.1 with the simple sum in Eq. 8.2 is possible since in a system at equilibrium a certain configuration, say the jth, occurs with a frequency given by its Boltzmann factor $\exp[-U_N^{(j)}/(k_B T)]$. The order according to which configurations are produced just changes the order of the terms in Eq. 8.2 and is thus irrelevant, just the frequency they appear with is significant. The idea behind the Metropolis MC method is to choose an arbitrary convenient way of updating configurations consistent with the prescription above. It is also desirable that the necessarily arbitrary or otherwise user-chosen starting state is 'forgotten' after a sufficiently high number of configuration updates, so that what we obtain is independent on that initial configuration. Another way of saying this is that the process should have the desired *asymptotic* properties. In a list of wishes we could also add that the new configuration should be produced by the smallest possible number of previous configurations, since we should otherwise store all these in memory in order to produce the new state. It turns out that evolution processes with these prerequisites exist in the realm of processes which evolve in time with probabilistic laws, or as they are normally called *stochastic* processes. The Metropolis technique consists of introducing a stochastic Markov process in which asymptotically (i.e. for indefinitely long chains of updates) each configuration recurs with a frequency proportional to its Boltzmann factor. We have briefly summarized the characteristics of Markov processes in Appendix K.

8.2 Metropolis Method

The probability w_j that a system, whose evolution is taken as Markovian, is to be found in a state j, e.g. in a configuration $\widetilde{X}^{(j)}$ with energy $U^{(j)}$ at time step t, depends only on its state at time $(t-1)$. More precisely, a discrete Markov process with m states is completely described in terms of an $m \times m$ transition probability matrix whose elements $\Pi_{j,k} \equiv \Pi(\widetilde{X}^{(j)} \to \widetilde{X}^{(k)})$, which give the probability of going from configuration j to k, have the general properties (non-negativity and normalization) reported in Appendix K. We consider discrete changes from one configuration to the next, but of course a continuous process will be recovered for arbitrarily small steps. In our case we wish to model a condensed phase where every state of the Markov process represents a configuration and we require w_j to obey, for a system of N particles in a volume V at a certain temperature T (Section 4.2), a Boltzmann distribution

$$w_j \propto \exp[-U_N^{(j)}/(k_B T)]. \tag{8.3}$$

The problem becomes that of selecting a transition matrix with elements Π_{jk} that will yield in the asymptotic limit (Eq. K.5) the statistical weights in Eq. 8.3. One possible way (not the only one) of achieving this is to impose the condition of microscopic reversibility or *detailed balance*, that is

$$w_j \Pi_{jk} = w_k \Pi_{kj}, \tag{8.4}$$

whatever j and k. This gives the ratio of transition probabilities

$$\Pi_{jk}/\Pi_{kj} = \exp\left[-\left(U_N^{(k)} - U_N^{(j)}\right)/(k_B T)\right]. \tag{8.5}$$

Thus, if we are in configuration j with energy $U_N^{(j)}$ and we wish to jump to configuration k with energy $U_N^{(k)}$ the ratio of the transition probabilities in the two directions is related to the energy difference and the temperature by a Boltzmann factor. An increase in temperature will tend to make the transition between the two configurations equally likely even if the energy factor is unfavourable. We shall try to concoct a transition matrix that satisfies Eq. 8.5 even though we should note from the outstart that this constraint still will not specify a unique Markov process. That in turn implies that different transition matrices which generate configurations with the correct frequency factor can be introduced and, why not, that some may be more efficient than others in exploring phase space. The prescription proposed by Metropolis et al. [1953] is to choose the transition probability for going from j to k as

$$\Pi_{jk} = \begin{cases} a_{jk}, & \text{if} \quad U_N^{(k)} \leq U_N^{(j)}, \\ a_{jk} \exp[-(U_N^{(k)} - U_N^{(j)})/(k_B T)], & \text{if} \quad U_N^{(k)} > U_N^{(j)}, \end{cases} \tag{8.6}$$

with a_{jk} constants and $a_{jk} = a_{kj}$ and the probability of remaining in the original configuration $\Pi_{jj} = 1 - \sum_{k \neq j} \Pi_{jk}$. It is easily verified that this transition probability obeys Eq. 8.5. In practice, the process is realized rather simply by moving one particle at a time in the following way. A starting configuration is chosen, typically with the N particles in a box with a certain surrounding environment (boundary conditions) as described later (Section 8.2.1) and the energy of this configuration, $U_N^{(j)}$, is calculated. One particle is then chosen, either sequentially or at random, and a new configuration is generated by giving this particle a random displacement $\delta\mathbf{X}$. The energy, $U_N^{(k)}$, of this trial configuration is calculated, and the move is accepted with a relative probability $\exp\left[-(U_N^{(k)} - U_N^{(j)})/(k_B T)\right]$. Thus, if $\left(U_N^{(k)} - U_N^{(j)}\right) < 0$ the new configuration is accepted. Indeed, in this case the Boltzmann factor in Eq. 8.5 is greater than 1 so the probability of jumping to the new configuration amounts to a certainty. If $\left(U_N^{(k)} - U_N^{(j)}\right) > 0$, the move is instead accepted with a relative probability $\exp\left[-(U_N^{(k)} - U_N^{(j)})/(k_B T)\right]$ using a simple rejection technique. In practice, if we have to generate values of a variable x in the range $a \leq x \leq b$ distributed according to a given non-negative distribution $f(x)$ with a maximum value f_{max} we could do the following. First extract a random value x' in the range and calculate the value of the function at this point $f(x')$. Now, to decide if the value x' should be accepted or not, extract another random number ξ in the range $0 \leq \xi \leq f_{max}$. If $\xi < f(x')$, then x' is accepted, otherwise it is rejected. A sufficiently large set of x' values generated in this way has the correct distribution $f(x)$. In the present case the random variable to be tested is the displacement $\delta\mathbf{X}$ and thus a uniformly distributed random number between 0 and 1 is generated. If this is less than $\exp[-(U_N^{(k)} - U_N^{(j)})/(k_B T)]$ the move is accepted; if not, the original configuration is restored and counted again. This procedure is then repeated, typically for a few hundred thousand to a few million times in order to equilibrate the system. When equilibrium has eventually been reached the new configurations generated can be used to calculate averages according to Eq. 8.2. A flow diagram for the algorithm is shown in Fig. 8.1. It is worth

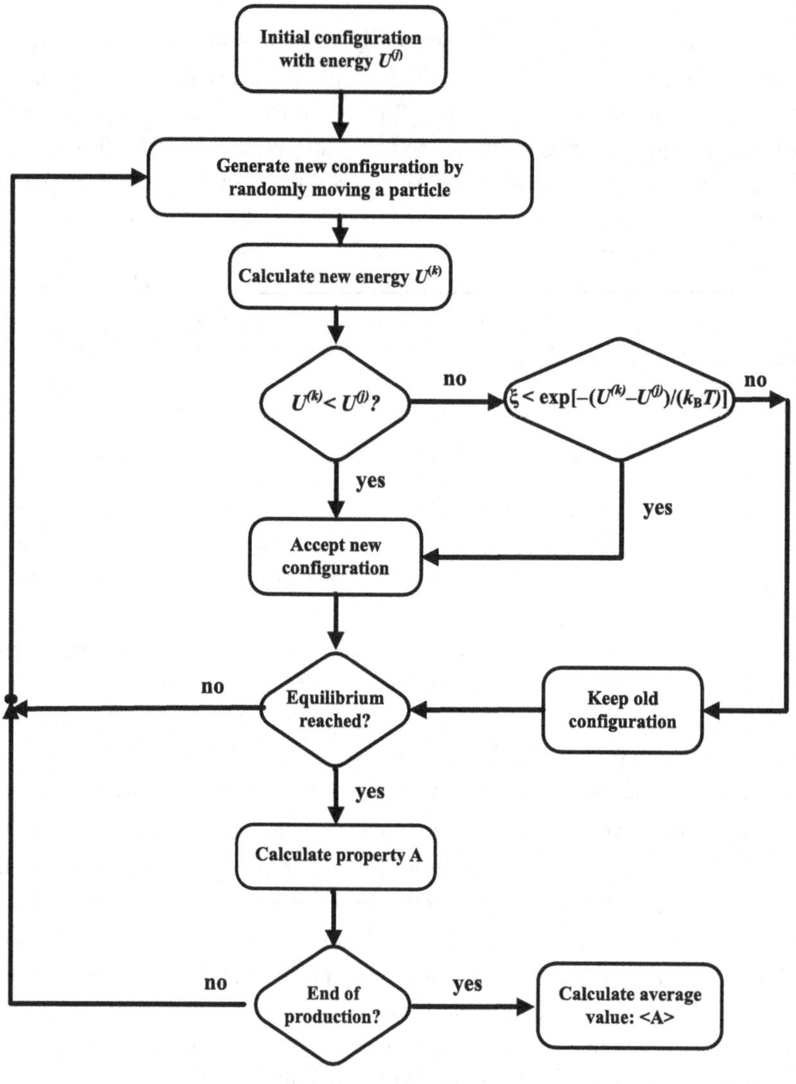

Figure 8.1 A flow diagram of the Metropolis MC method.

stressing that the time involved in proceeding along the MC Markov chain has in principle nothing to do with real time. Indeed, in the MC method the true trajectory of the system in phase space is replaced by the artificial Markovian trajectory chosen by us. In the common realization of the algorithm, with one particle at a time moving, subsequent configurations will be strongly correlated to the previous ones and some sort of dynamics is implied. If the evolution process is a plausible physical process, e.g. a small step Brownian motion, we may approach a somewhat realistic dynamic evolution, with at least some significance attached to evolution time. Indeed, a timescale can be worked out by comparison with some known dynamic information. On the other hand, one may also take advantage of the lack of strict specifications on the Markov process to explore phase space in a much more efficient, albeit non-dynamically realistic way. For instance, we may flip molecules in very ordered systems, allowing accelerations of many orders of magnitude with respect to small step tumbling. This is of great advantage in many cases where reaching equilibrium requires going through some bottleneck. A particularly significant example is near a phase transition where a local algorithm, with the update of one particle at a time, suffers from an abnormally large time needed to reach equilibrium or *critical slowing down*, just as for real systems. Indeed, a large group of molecules with the same orientation can be difficult to unlock and its 'melting' will essentially proceed at the boundary and propagate over distance proportionally to the square of the distance. This will in turn require a power law increase in equilibration rate with sample size near the transition [Swendsen, 1991]. In any case, it is useful to point out the formal analogy of the MC evolution with an (artificial) dynamics for the system, where the natural time unit is a *cycle* or *sweep* (an attempted move for every particle). For example, we shall be able to use the same methods for calculating thermodynamic quantities, order parameters, etc. both for the MC method and the Molecular Dynamics (MD) technique to be described in Chapter 9. The difference with MD is that there the true dynamics is followed, while in MC all we can guarantee is that, by construction, the process will, after a sufficiently large number of steps, lead to equilibrium, in the sense that configurations will occur with a frequency proportional to their Boltzmann factor. In practice, it is hardly possible to gauge a priori how large this number of steps will be. We expect the convergence to depend in some way on how efficiently we sample the configurational space and so it is clear that we want to reach some balance between the number of configurations accepted and rejected. This acceptance ratio can in turn be affected by the magnitude of the random displacement, $\delta \mathbf{X}$, assigned to the particle. Since the maximum jump length $\delta \mathbf{X}$ is not dictated by the method, it can be adjusted to speed up convergence. As a rule of thumb, the maximum displacement is typically chosen so that approximately half of the configurations are accepted and half rejected. It is worth checking that this constraint, especially when built in some automatic feedback loop, does not require in turn a displacement so low that the molecules hardly move. In that case, a lower acceptance ratio should be accepted to get the evolution going.

We shall now comment on some general points such as boundary conditions, the choice of the initial configuration and the calculation of thermodynamic observables. We shall then discuss the calculation of order parameters and orientational pair correlation functions.

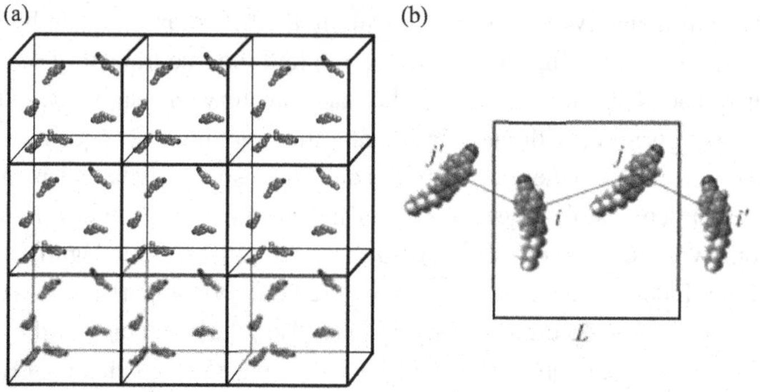

Figure 8.2 (a) A schematic representation of periodic boundary conditions (PBC) in 3D. The sample box at the centre is surrounded by replicas on every facet. For a cubic box of side L a particle at position $\boldsymbol{r} = (x, y, z)$ will have replicas with the same orientation, velocity, etc. at $(x \pm L, y, z), (x, y \pm L, z), (x, y, z \pm L)$, and so on. If a particle goes out of the box on one side, one of its replicas will enter the box from the opposite side. (b) Sketch of minimum image convention (MIC). The nearest distance between molecules i and j is between i and j', not j, in the cubic sample box of side length L with PBC.

8.2.1 Boundary Conditions

Even though we are always forced to study a finite and relatively small system of interacting particles, we have to consider clearly what type of system we are interested in studying. In many cases we are not interested in the properties of the small sample itself but rather in predicting those of the corresponding macroscopic system. In doing this we are hampered not only by having to deal with a limited number of particles, but also by surface effects. In order to reduce the latter, we do not normally use *free boundary conditions*, which correspond to an isolated system of N particles in *vacuum*, but instead employ the artifact of periodic boundaries where the sample box is surrounded by exact replicas of itself, as shown in Fig. 8.2. A particle at position (x, y, z) in a cubic box of side length L will have 26 nearest ghost images at $(x \pm L, y, z)$, $(x, y \pm L, z)$, $(x, y, z \pm L)$, etc. These are in turn surrounded by similar images, ad infinitum. The space-filling system obtained in this way does not have free surfaces at all so we have remedied that part of the problem. Note, however, that an additional spurious periodic correlation between particles has been introduced, as a particle will be perfectly correlated with its images. It is worth mentioning that we do not actually need to store copies of the sample box. The periodicity condition is equivalent to say that the position of particle i in the sample box is defined modulo the box length in every direction. Thus, $x_i := x_i (\text{mod}\, L)$ which means that the position can be considered as the remainder of the division of the original position by the side length (same as the hours in a day, that in a 24 hour clock we give mod 24). When using periodic conditions, the distance $\boldsymbol{r}_{ij}$ between two different particles, i and j, is usually taken according to the *minimum image convention* (MIC), i.e. as the distance between the c.o.m. of i and the nearest image of j. Thus, every particle is the centre of an identical box and accordingly a given molecule i

interacts only with the image of another molecule j which is the nearest. If the range of the molecular interactions is less than $L/2$ this comprises all interactions. The MIC ensures that every particle interacts with every other particle at most once. When applying the minimum image convention to multisite molecules these are considered to be inseparable entities and the minimum image convention is applied to the separation vector between the centres of mass of two multisite molecules i and j only. The total potential energy is the sum over all site-site interactions of the different particles. If the particles are rigid bodies, only interactions between the possibly anisotropic (e.g. ellipsoidal) sites belonging to distinct molecules ($i \neq j$) are taken into account:

$$U = \sum_{i,j>i} \sum_{\substack{a\in i \\ b\in j}} U(\boldsymbol{r}_{ab}, \Omega_a, \Omega_b). \tag{8.7}$$

The pair potential energy depends on the separation vector $\boldsymbol{r}_{ab}$ between the sites a and b and, if the sites are anisotropic, on the orientations Ω_{ai} and Ω_{bj} of the individual sites with respect to the common lab frame. If $\boldsymbol{h}_{a,i}$ is the vector from $\boldsymbol{r}_i$, the centre of mass of molecule i to the position of the ath site of molecule i, $\boldsymbol{r}_a = \boldsymbol{r}_i + \boldsymbol{h}_{a,i}$, the separation vector between two distinct sites $a \in i$ and $j \in b$ is $\boldsymbol{r}_{ab} = \boldsymbol{r}_{ij}^{\mathrm{MIC}} + \boldsymbol{h}_{a,i} - \boldsymbol{h}_{b,j}$. The MIC implies some care, e.g. for pressure calculations (cf. Section 4.10). The virial is generally derived as $W = (1/3V) \sum_i \boldsymbol{r}_i \cdot \boldsymbol{F}_i$, where $\boldsymbol{F}_i$ is the total force acting on the centre of mass of molecule i (V being the volume). When introducing the separation vector $\boldsymbol{r}_{ij}^{\mathrm{MIC}}$ between two molecules i and j we can express the virial by pairwise forces $\boldsymbol{F}_{ij}$ between the centres of mass of molecules i and j;

$$W = \frac{1}{3V} \sum_{i,j>i} \boldsymbol{r}_{ij}^{\mathrm{MIC}} \cdot \boldsymbol{F}_{ij} = -\frac{1}{3V} \sum_{i,j>i} \boldsymbol{r}_{ij}^{\mathrm{MIC}} \cdot \nabla_{\boldsymbol{r}_{ij}} U. \tag{8.8}$$

Now we express the pair forces between the centres of mass $\boldsymbol{F}_{ij} \equiv -\nabla_{\boldsymbol{r}_{ij}} U$ by pair forces $\boldsymbol{F}_{a,b} \equiv -\nabla_{\boldsymbol{r}_{a,b}} U$ that act between the centres of two distinct sites $a \in i$ and $b \in j$. Due to $\nabla_{\boldsymbol{r}_{kl}} \boldsymbol{r}_{ai,bj} = \delta_{ik}\,\delta_{jl}\,\mathbf{1}$ we can apply the chain rule for differentiation. This immediately yields

$$W = \frac{1}{3V} \sum_{i,j>i} \sum_{\substack{a\in i \\ b\in j}} \boldsymbol{r}_{ij}^{\mathrm{MIC}} \cdot \boldsymbol{F}_{ab} = -\frac{1}{3V} \sum_{i,j>i} \sum_{\substack{a\in i \\ b\in j}} \boldsymbol{r}_{ij}^{\mathrm{MIC}} \cdot \nabla_{\boldsymbol{r}_{ab}} U_{a,b}. \tag{8.9}$$

If the intermolecular potential is very long range, e.g. for Coulomb interactions between charged or dipolar systems, the minimum image convention for calculating the interaction energy is abandoned. Instead, the complete, space-filling system of the box and all its periodic images is considered and special techniques, like the Ewald summation [Allen and Tildesley, 2017] introduced in Section 5.4.4, are employed. Yet another method in use for long-range interactions is to simply truncate the intermolecular potential at a given cut-off distance and to take into account the long-range tail of the interaction using some form of perturbation theory.

PBCs are introduced when trying to predict bulk properties while doing calculations on small samples. There are, however, a number of cases where the system of interest is

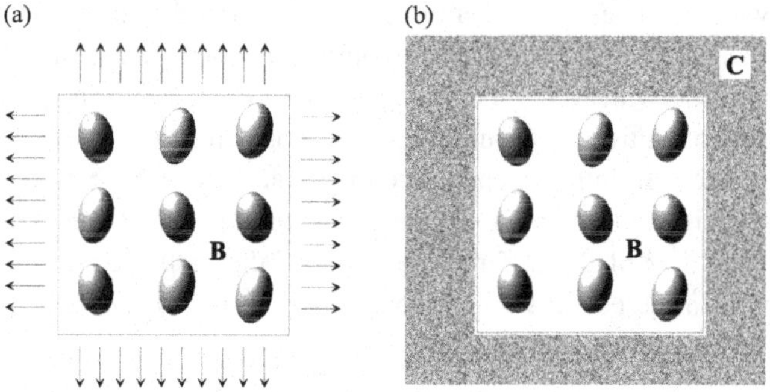

Figure 8.3 Non-periodic boundary conditions, sketched in 2D: (a) free (FBC), with empty space around the sample box **B** and (b) confined (CBC) with a certain 'frozen in' confining material **C** surrounding the sample.

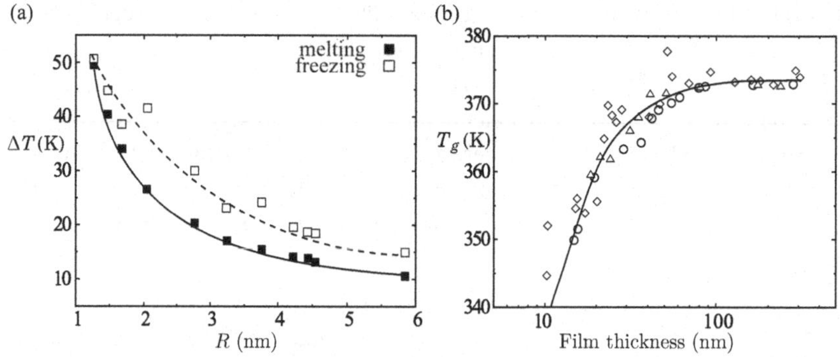

Figure 8.4 (a) Shift of the water melting (■) and freezing (□) temperature ΔT inside MCM41 and SBA15 artificial zeolite nanopores as a function of their pore radius R [Schreiber et al., 2001]. (b) The glass transition temperature T_g as a function of film thickness for polystyrene of three molecular weights (MW):120,000 (△); 500,800 (○); 2,900,000 (◇) [Keddie et al., 1994]. The lines are a guide for the eye.

actually small. One is the case of nanodroplets, with air (vacuum) outside the sample (see Fig. 8.3a). It is important to realize that in this case the properties can be very significantly different from those of the corresponding bulk, e.g. the melting temperature decrease (of a few hundred degrees!) with respect to the bulk in CdS nanocrystals of a few nanometres [Goldstein et al., 1992] or the decrease in ferroelectric phase transition temperature of some 50°C for $PbTiO_3$ nanocrystals of $\approx$25 nm [Ishikawa et al., 1988]. The other case is that of a material confined, e.g. to a nanopore (see Fig. 8.3b). We see in Fig. 8.4a the shift of the water melting and freezing temperatures in nanopores of the artificial zeolites MCM41 and SBA15 (we mentioned their fabrication by templating from micelles in Section 1.11.1) as a function of their pore radius. The glass transition in nano-thick polymer films also changes significantly, as we can see in Fig. 8.4b [Keddie et al., 1994].

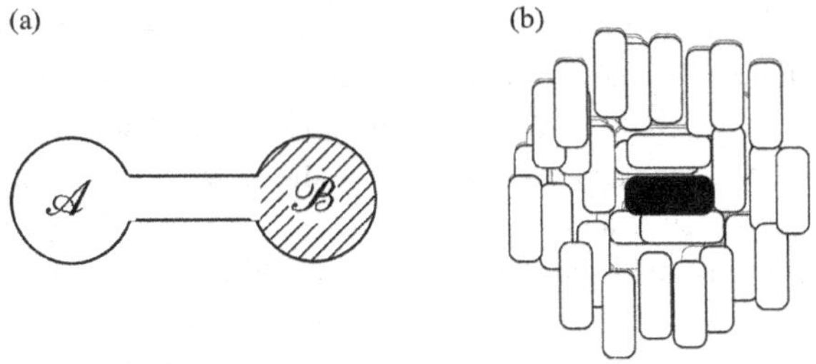

Figure 8.5 (a) An illustration of a possible factor hindering convergence: two regions of phase space $\mathscr{A}$ and $\mathscr{B}$ joined by a bottleneck. (b) A group of molecules that may be difficult to unlock by single particle moves [Zannoni, 1979a].

8.2.2 *Influence of the Initial Configuration*

We have shown that, whatever the initial configuration, the MC procedure should, after an asymptotically large number of steps, converge to equilibrium. Thus, in principle any configuration, be it completely ordered, completely disordered or whatever, could be chosen as a starting point. However, since we always perform a finite number of steps it is unwise to overlook the choice of starting configuration. For instance, in simulations of simple fluids it is normally very hard to start from an isotropic fluid and generate a crystal structure, even if this would be the thermodynamically stable state at the temperature and density chosen. If we want our system to reach a given region of phase space in a reasonable number of steps, and hence computer time, then the starting point and the evolution strategy are not unimportant. To visualize this, let us consider the situation sketched in Fig. 8.5a and imagine that the system finds itself in region $\mathscr{A}$ of the configuration space, while we would like it to reach region $\mathscr{B}$. This is clearly not impossible since there exists an open path from $\mathscr{A}$ to $\mathscr{B}$, but it is easy to convince ourselves that the process may well require a very long time due to the bottleneck shape of the pathway. A non-equilibrium state like $\mathscr{A}$ having a relatively long lifetime is sometimes called a *metastable state*. The problem can be particularly important in simulations of liquid crystals formed of elongated hard particles. In this case, configurations like the one shown in Fig. 8.5b may prove very difficult to unlock by moving one molecule at a time, as discussed. If at all feasible, obtaining the same results for energy, order parameters or other properties from simulations started from two different thermodynamic conditions, e.g. from previously generated configurations at lower and higher temperatures or concentrations as appropriate for the system studied, can be the touchstone for assessing if equilibrium instead of a long-lived metastable state has been reached.

8.2.3 *Evolution*

The generation of new configurations normally proceeds selecting one molecule at a time and giving it a random displacement. The particle is chosen sequentially or at random between those that have not yet been selected. In any case, a set of N attempted moves (a *cycle*) represents an update for the system. For example, to generate a random displacement from the position $\boldsymbol{r}_i = (x_i, y_i, z_i)$ of the ith particle we have to generate a new position $\boldsymbol{r}'_i = (x'_i, y'_i, z'_i)$. We may decide first on a maximum allowed displacement for the coordinates: $\Delta x, \Delta y, \Delta z$. Then for each coordinate, say the x, we extract a random number ξ uniformly sampled between 0 and 1 and $x'_i = x_i + \xi \Delta x$. The maximum displacement is chosen to ensure a reasonable proportion of accepted to rejected moves. According to common wisdom, the ratio of these two (the *acceptance ratio*) should be, as previously mentioned, about 0.5. This is not mandatory, but in practice it is important to ensure that the displacements required to get this rejection ratio are not too small, so that the system actually samples configuration space. It is also worth remembering that MC is an integration technique and that the variables (and the displacements) should be sampled with a weight consistent with the relevant measure (volume element). For orientations α, β, γ the volume element is $\mathrm{d}\alpha \sin\beta \mathrm{d}\beta \mathrm{d}\gamma$ so for α and γ we can proceed exactly as for positions. For the angle β we could generate random displacements in $x \equiv \cos\beta$, but generating the new $\cos\beta$ in this way does not give a command on the extent of the angular jump. Thus, the following procedure of Barker and Watts [1969] is often used. First, a random number is used to choose the rotation axis: $\boldsymbol{x}$, $\boldsymbol{y}$, $\boldsymbol{z}$ then an angle θ is generated as another random number between 0 and θ_{max}. Thus, if ξ is a random number uniformly distributed between 0 and 1 an angle $\theta = \xi\theta_{max}$ will be generated. Then a rotation from the previous orientation to the new one is performed. For example, if we have a uniaxial molecule whose axis is defined by the vector $\boldsymbol{u}$ and if a rotation around the y-axis has been randomly selected, we shall have

$$\begin{pmatrix} u_x^{\text{new}} \\ u_y^{\text{new}} \\ u_z^{\text{new}} \end{pmatrix} = \begin{pmatrix} \cos\theta & 0 & \sin\theta \\ 0 & 1 & 0 \\ -\sin\theta & 0 & \cos\theta \end{pmatrix} \begin{pmatrix} u_x^{\text{old}} \\ u_y^{\text{old}} \\ u_z^{\text{old}} \end{pmatrix}. \tag{8.10}$$

It is clear that generating good, unbiased random numbers is the key to the success of the method. This is far from easy [Knuth, 1998] as exemplified by plotting the would-be

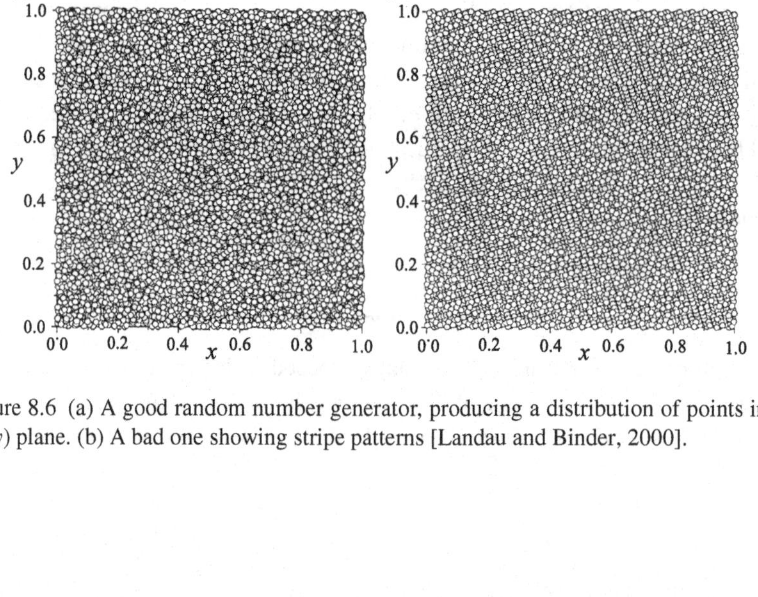

Figure 8.6 (a) A good random number generator, producing a distribution of points in the (x, y) plane. (b) A bad one showing stripe patterns [Landau and Binder, 2000].

random numbers obtained from two different generators as shown in Fig. 8.6. While the one employed in Fig. 8.6a produces a (visually at least) uniform covering of (x, y) plane, the one in Fig. 8.6b gives origin to geometrical patterns (stripes) that indicate an unwanted correlation in the generated pairs of random numbers.

It is worth mentioning that in MC simulations of hard particles, trial moves are selected using a random number generator, accepted if they do not lead to particle overlap and rejected if they do.

8.3 Simulations in Non-Canonical Ensembles

8.3.1 Constant Pressure

In the isobaric isothermal ensemble (constant N, P, T) introduced in Section 4.3, any time-independent configurational property of interest can be written as the equilibrium average

$$\langle A\rangle_{NPT} = \frac{\int \mathrm{d}\widetilde{\mathbf{X}}\; A(\widetilde{\mathbf{X}}) \exp[-\mathcal{H}_N(\widetilde{\mathbf{X}})/(k_B T)]}{\int \mathrm{d}\widetilde{\mathbf{X}} \exp[-\mathcal{H}_N(\widetilde{\mathbf{X}})/(k_B T)]}, \tag{8.11}$$

where $\mathcal{H}_N = \mathcal{U}_N + PV$ is the enthalpy function. Clearly, in this constant pressure condition the volume of the system will adjust and change towards an equilibrium value and represents a new degree of freedom we should integrate upon, even if this is not apparent from Eq. 8.11. To be able to apply the MC method we should, however, find a way of making the shape and volume dependence explicit and of bringing it into the exponent of Eq. 8.11 as an effective pseudopotential. This can be done writing explicitly $\boldsymbol{X} = (\boldsymbol{r}, \Omega)$ and transforming the particle positions to dimensionless variables as in Section 4.10 with the geometric transformation $\boldsymbol{r}_i = (\mathbf{H}\,\boldsymbol{s})_i$ where the matrix $\mathbf{H}$ contains the components of the sample cell axis $\boldsymbol{a}, \boldsymbol{b}, \boldsymbol{c}$ that are now allowed to vary:

$$\mathbf{H} \equiv \begin{pmatrix} a_1 & b_1 & c_1 \\ a_2 & b_2 & c_2 \\ a_3 & b_3 & c_3 \end{pmatrix}. \tag{8.12}$$

The Jacobian determinant of the transformation is the volume of the sample box (see Appendix A), and as already mentioned, we have to integrate over the volume as a new degree of freedom

$$\langle A\rangle_{NPT} = \frac{\int \mathrm{d}V V^N \int \mathrm{d}\tilde{s}\mathrm{d}\tilde{\Omega} A(\tilde{s}, \tilde{\Omega}) \exp[-(U_N(\tilde{s}, \tilde{\Omega}) + PV)/(k_B T)]}{\int \mathrm{d}V V^N \int \mathrm{d}\tilde{s}\mathrm{d}\tilde{\Omega} \exp[-(U_N(\tilde{s}, \tilde{\Omega}) + PV)/(k_B T)]}. \tag{8.13}$$

It is convenient to write $V^N = \exp(N \ln V)$, so that

$$\langle A\rangle_{NPT} = \frac{\int \mathrm{d}V \int \mathrm{d}\tilde{s}\mathrm{d}\tilde{\Omega} A(\tilde{s}, \tilde{\Omega}) \exp[-(U_N(\tilde{s}, \tilde{\Omega}) + PV)/k_B T + N \ln V]}{\int \mathrm{d}V \int \mathrm{d}\tilde{s}\mathrm{d}\tilde{\Omega} \exp[-(U_N(\tilde{s}, \tilde{\Omega}) + PV)/k_B T + N \ln V]}. \tag{8.14}$$

This is in the form of the standard canonical MC integral that we have discussed before, except that volume moves $V^{\text{old}} \to V^{\text{new}}$ are now allowed according to the Boltzmann factor

$$\Pi_{old \to new} = \exp\left[-(P\Delta V + \Delta U_N)/(k_B T) + N \ln \frac{V^{\text{new}}}{V^{\text{old}}}\right], \tag{8.15}$$

where ΔV $(-\Delta V_{max} < \Delta V < \Delta V_{max})$. In practice, the volume move is usually attempted with probability $1/N$ (i.e. every cycle) and the maximum allowed volume jump ΔV_{max} is adjusted, as the maximum jump lengths of the other degrees of freedom, to give a satisfactory acceptance ratio.

This procedure is most convenient if the box is cubic and after the move is accepted the change of box side is recalculated isotropically. However, as long as the pressure is isotropic we could use the same method by changing the box sides lengths L_a, L_b, L_c. This is important for ordered anisotropic systems, e.g. smectic phases, formed of elongated molecules where changing the shape of the box is essential to avoid perturbing the sample too much.

8.3.2 Constant Stress

An important way of studying the mechanical properties of polymer systems, including LC polymers and LC elastomers is that of stress-strain experiments, that we briefly mentioned in Section 1.15 when discussing materials like Kevlar and Nomex. Such mechanical experiments (see Fig.1.61) can be simulated with *isostress simulations* [Skačej and Zannoni, 2011], a variant of the standard MC method where an external pulling force (or stress) is applied along a certain direction, e.g. the z-axis. In this case, the applied stress induces a deformation, $\Delta\lambda_z = \Delta L_z/L_z$, of the sample box length L_z along the z-axis. Given an applied engineering stress Σ_{zz} for stretching/compression along the z-axis, the probability of acceptance for the sample strain move is obtained adding the deformation work to the internal energy, giving

$$\Pi_{old \to new} = \min\{1, \exp\left[-(\Delta U - \Sigma_{zz} V \Delta\lambda_z)/(k_B T)\right]\}, \tag{8.16}$$

where V denotes sample volume and T temperature. The rest of the MC evolution procedure involving single-molecule translation and rotation moves, leading to the contribution ΔU in Eq. 8.16, is the same as for the standard MC and shall not be discussed again in this book. We shall see some examples of applications of the isostress procedure to determine the elastic modulus of LC elastomers [Skačej and Zannoni, 2011, 2012, 2014] in Chapter 11.

8.4 Calculation of Thermodynamic Observables

The various thermodynamic properties can be calculated in principle from their expression as statistical averages over, say, M configurations starting from an already equilibrated initial one J_{in}

$$\langle A \rangle = \lim_{M\to\infty} \frac{1}{M} \sum_{J=J_{in}}^{J_{in}+M} A^{(J)}. \tag{8.17}$$

Let us now examine the error involved in the calculation of an observable. Since we use a finite chain of M steps instead of an infinite one, we expect that our estimate will deviate from the true expectation value $\langle A \rangle_{true}$. It is known from the theory of Markov chains that this deviation from $\langle A \rangle_{true}$ has an asymptotically Gaussian distribution [Binder, 1976]. Now, to estimate the standard deviation in a computer experiment we divide the chain of states into a number of subchains. We then calculate averages of the quantity A in every subchain, together with the usual average $\langle A \rangle$ over the complete chain. The statistical error on $\langle A \rangle$ is then estimated as the standard deviation from the average:

$$\sigma_A = \frac{1}{M_\alpha(M_\alpha - 1)} \left[\sum_{\alpha=1}^{M_\alpha} \left(\langle A \rangle_\alpha - \langle A \rangle \right)^2 \right]^{\frac{1}{2}}, \tag{8.18}$$

where M_α is the number of sub-averages. For this estimate to be reliable we need the various terms to be independent samples. It is not obvious that this is the case since in the Metropolis method configurations are produced by a sequence of small single-particle moves. Thus, we assume subchains to be long enough or distant enough so that they are statistically uncorrelated. To comply with this restriction, one obviously requires the number of sub-averages, M_α, to be as large as possible. In general, the MC average, $\langle A \rangle$, tends to the true value as the number of configurations, M, increases according to $\langle A \rangle = \langle A \rangle^{\text{true}} + \mathcal{O}(M^{-\frac{1}{2}})$, where $\mathcal{O}(x)$ indicates the order of magnitude of its argument x. It is useful to examine the behaviour of the subchain average to decide if convergence to equilibrium has been achieved. If this is the case the sub-average should simply oscillate about the average and not exhibit any systematic drift.

Any property of interest, A, is evaluated at every MC cycle and, after a certain number of cycles m_J (typically between 1000 and 2000), the values A_i are averaged effectively coarse graining the trajectory. A further overall average is then computed as the weighted average over M of such supposedly uncorrelated segments

$$\langle A \rangle = \frac{1}{M_{\text{C}}} \sum_J^M m_J A^J = \frac{1}{M} \sum_J^M \frac{1}{m_J} \sum_i^{m_J} A_i, \tag{8.19}$$

where $M_{\text{C}} = \Sigma_J^M m_J$ is the total number of production cycles. The attendant weighted standard deviation from the average σ_{A} is also calculated and gives the error estimates. Error estimates for correlated series of data have been discussed by Flyvbjerg and Petersen [1989].

The energy is the simplest property to calculate since its evaluation is already part of the MC prescription. The pressure can be obtained from the average virial for differentiable potentials or from the pair distribution at contact point for hard cores. We now wish to comment briefly on some observables: energy, heat capacity, free energy and order parameters.

Energy. The calculation of energy presents no particular problem and is besides already part of the MC evolution procedure. The average energy can be obtained by keeping track of the ΔU_N at every move or, more simply, by computing the energy $U_N^{(j)}$ at the needed configurations, for instance every cycle or every few cycles and calculating the average value $\langle U_N \rangle$ as in Eq. 8.17. Normally the results are presented referring to a single molecule $\langle U \rangle = \langle U_1 \rangle = \langle U_N \rangle / N$.

Heat capacity. The constant volume specific heat, C_V, is defined as the temperature derivative of the average internal energy (see Eq. 4.124a). Thus, we can calculate the internal energy for a series of temperatures, and perform a numerical differentiation of $\langle U \rangle$ versus T to find $C_V(T)$. However, the differentiation of experimental or, in general, noisy data is a well-known, ill-posed problem and every small error in the data may cause huge errors in the numerical results. A first way to tackle this difficulty is through a smoothing interpolation, for example, using suitable spline functions, i.e. a piecewise interpolating curve constructed joining together polynomials of degree n at a certain set of points (knots) before numerical differentiation [Fabbri and Zannoni, 1986]. An alternative is to use an inversion method [Tikhonov and Arsenin, 1977], which consists of solving the integral equation [Chiccoli et al., 1987]

$$U(T) = U(T_0) + \int_{T_0}^{T} \mathrm{d}T' C_V(T'). \tag{8.20}$$

In our case energies are known at a set of temperatures, $U(T_i)$, which can be used to build an M-component vector of energy differences $\boldsymbol{U}$. We can thus write

$$\int_{(i)} \mathrm{d}T' C_V(T') = \mathcal{U}_i, \tag{8.21}$$

where the integral is extended to the ith energy interval. Choosing to calculate a vector $\boldsymbol{C}$ containing C_V at a grid of M temperatures and employing a suitable numerical integration formula we reduce the integral equation to the matrix equation $\mathbf{WC} = \mathbf{U}$, where $\mathbf{W}$ is a matrix containing the weigths for the chosen type of numerical integration (e.g. trapezoidal). $\mathbf{W}$ will normally not be square, but rather rectangular, and the problem of finding $\mathbf{C}$ is solved in terms of the generalized inverse matrix $\mathbf{W}^{-1}$ that can be obtained using a numerical method (e.g. that by Rust et al. [1966]).

The estimate of errors in heat capacity calculations is rather complicated because of the numerical schemes employed. A useful simulation procedure [Chiccoli et al., 1987] is as follows. First, a rather large number (e.g. 100 or more) of plausible energy versus temperature curves is generated by sampling energy values at each temperature from a Gaussian distribution of width corresponding to the (known) standard deviation from the mean energy at that point. Repeating the heat capacity calculation for every curve yields a set of C_V values whose average and standard deviation are our final results and errors.

A third procedure to obtain the heat capacity consists in determining average energy fluctuations. We have (see Eq. 4.124a)

$$C_V^*/k_B = \left[\frac{1}{M} \sum_j^M \left(U^{(j)}\right)^2 - \left(\frac{1}{M} \sum_j^M U^{(j)} \right)^2 \right] / (k_B T)^2, \tag{8.22}$$

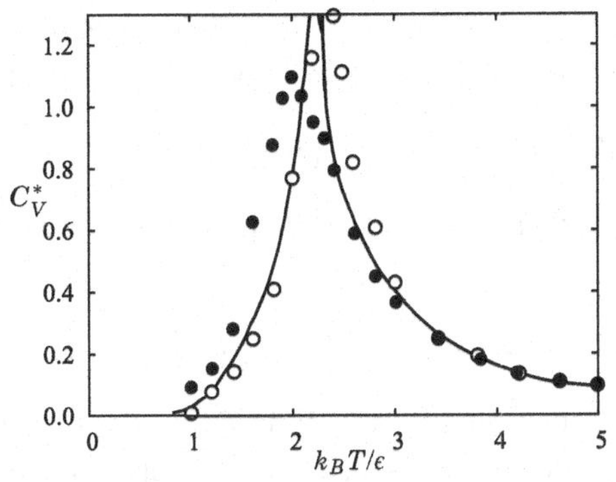

Figure 8.7 The heat capacity per particle C_V^* as a function of temperature $k_B T/\epsilon$ (dimensionless units) for a 12×12 2D Ising model obtained [Yang, 1961] with periodic boundary conditions (○) and free boundary conditions (●) and the true analytical curve (—) determined by Onsager [1944].

where M is the number of snapshots considered. The dimensionless quantity on the right-hand side of Eq. 8.22 or the analogous for the heat capacity at constant pressure C_P with enthalpy $\mathcal{H}$ instead of energy U, can be obtained directly form the simulations. However, since it is a fluctuation determined quantity, it is often affected by large errors.

The heat capacity is a key quantity in systems showing phase transitions, since it is often by an examination of its temperature dependence that a transition is located (see Chapter 2). It is important to appreciate the effect of various boundary conditions on a phase transition by looking at the changes induced on the heat capacity. We would intuitively expect the space correlations between particles to be greatest for periodic boundary conditions (PBCs) and a minimum for a system with free surfaces. Morerover, we would expect the 'true' bulk result to lie somewhere in between these two extremes. Consequently, periodic and free boundaries should give, respectively, an upper and lower bound for the transition temperature

T_K: $T_{\text{FBC}} < T_{\text{C}} < T_{\text{PBC}}$. In the same way we would expect order parameters obtained with periodic boundaries to be greater than those calculated for a system with free surfaces at the same temperature. The predicted trend is actually borne out by the classical MC simulations of Yang [1961] on the 2D Ising model, shown in Fig. 8.7, by comparison with the exact Onsager solution. Qualitatively similar behaviour is obtained for the 3D Ising model [Landau, 1976] where, however, analytic solutions are not available. Increasing the number of particles has the effect of restricting the upper and lower bounds; thus, for instance, the peak in the specific heat shifts to higher temperatures for the isolated system and to lower temperatures for periodic boundaries.

Free Energy. In the N, V, T ensemble the relevant free energy is the Helmholtz function $\mathcal{A} = \mathcal{U} - T\mathcal{S}$. This is a rather difficult quantity to evaluate because of the presence of the entropy term. The fact is that quantities easily calculated with MC methods are average quantities and we do not have such an expression for A. Nonetheless, we can actually make some progress in this direction. Given an arbitrary quantity A we can write its average as Eq. 8.1. Thus, if we formally choose $A = \exp[U_N(\widetilde{X}\})/k_BT]$, we have

$$Z(N, V, T) = V V_\Omega \,/\langle \exp[+U_N(\widetilde{X})/k_BT]\rangle. \tag{8.23}$$

The partition function can therefore be calculated, in principle, as an average and the free energy obtained from

$$\mathcal{A} = -k_B\big(\ln Z(N, V, T) \; - \ln N!\big). \tag{8.24}$$

Unfortunately only very poor estimates of $Z(N, V, T)$ can be obtained by this method when the system is somewhat ordered. In this case, in fact, important contributions to the summation come from configurations of high energy, with a small Boltzmann factor, which are very poorly sampled by the MC procedure. This results in very slow convergence except when the system is relatively disordered. A common method of evaluating the free energy exploits an interpolating relation obtained from the Gibbs–Helmholtz equation

$$\mathcal{U}(\beta) = \left(\frac{\partial \beta_T \mathcal{A}}{\partial \beta_T}\right)_V, \tag{8.25}$$

where $\beta_T \equiv 1/(k_BT)$, which gives upon integration

$$\int_{\beta_{T'}}^{\beta_{T''}} \mathrm{d}\beta_T \; \mathcal{U}(\beta_T) = \beta_{T''}\mathcal{A}(\beta_{T''}) - \beta_{T'}\mathcal{A}(\beta_{T'}), \tag{8.26}$$

and then the free energy at $\beta_{T''}$ as

$$\mathcal{A}(\beta_{T''}) = (\beta_{T'}/\beta_{T''})\mathcal{A}(\beta_{T'}) + \frac{1}{\beta_{T''}} \int_{\beta_{T'}}^{\beta_{T''}} \mathrm{d}\beta_T \, \mathcal{U}(\beta_T). \tag{8.27}$$

Thus, if the free energy at a temperature $T' = 1/(k_B\beta_{T'})$ is known together with the temperature dependence of the internal energy in the interval $T' \leq T \leq T''$, the free energy at the new temperature T'' can be calculated from Eq. 8.25. This method was used, e.g. by Lebwohl and Lasher [1972] in their simulations of a simple lattice model of liquid crystals that we shall discuss in detail in Chapter 10. Similarly, one gets for the free energy as a function of density [Frenkel, 1986]

$$\mathcal{A}(\rho'') = \mathcal{A}(\rho') + N \int_{\rho'}^{\rho''} \mathrm{d}\rho\rho^{-2} P(\rho). \tag{8.28}$$

The free energy is calculated first at a very low density from, say, a virial expansion. Thermodynamic integration of the density dependence of the pressure can then be used to extrapolate the free energy to the high densities of interest. Other methods for calculating free energy differences between the systems of interest, characterized by a configurational potential energy U_N, and a reference system with potential U_N^0 have been proposed by Valleau and Torrie [1977]; that is

$$\beta_T(\mathcal{A} - \mathcal{A}_0) = -\ln\left\langle \exp[-\beta_T(U_N - U_N^0)]\right\rangle_0, \tag{8.29}$$

where the angular brackets indicate the canonical ensemble average over the reference system. The calculation of free energies in computer simulations was advanced and discussed in detail by Frenkel [1986].

8.5 Pair Correlation Coefficients

The reduced pair distribution $G(\boldsymbol{r}_{12}, \Omega_1, \Omega_2)$ is not easy to evaluate in full from a computer simulation. One possibility is to produce it as a multidimensional histogram, dividing the range of separations and orientations into a number of intervals, then assigning the orientations and separations $(\boldsymbol{r}_{12}, \Omega_1, \Omega_2)$ of every pair of molecules produced by the simulations to the appropriate volume elements increasing the relative counters. The histogram of the pair distribution is obtained by repeating the sorting process for a sufficient number of equilibrium configurations and then normalizing. The procedure is limited on one hand by the storage required which poses a limit to the resolution that can be obtained in the histogram and, more seriously, by the need to have a number of counts in every bucket of the histogram sufficiently high to have good statistics. An alternative procedure is to evaluate the expansion coefficients of $G(\boldsymbol{r}_{12}, \Omega_1, \Omega_2)$ in a product basis of Wigner rotation matrices. For example, consider the case of a reduced pair distribution independent of the orientation of the intermolecular vector, as for simulation on a cubic lattice or in overall isotropic fluid, the most general, rotationally invariant form of $G(r_{12}, \Omega_1, \Omega_2)$ is (cf. Chapter 4) $G(r_{12}, \Omega_1, \Omega_2) = g(r_{12}, \Omega_{12})$, a function of relative orientations Ω_{12}. For cylindrically symmetric particles we have just $g(r_{12}, \cos\beta_{12})$ which can be expanded in Legendre polynomials $P_L(\cos\beta_{12})$ as in Eq. 4.93, with

$$g(r_{12}, \cos\beta_{12}) = g(r_{12}) \sum_L \frac{2L+1}{2} G_L(r_{12}) P_L(\cos\beta_{12}) \tag{8.30}$$

and expansion coefficients $G_L(r_{12})$ that represents an Lth rank angular correlation as discussed in Chapter 4:

$$G_L(r_{12}) = [1/g(r_{12})] \int \mathrm{d}\cos\beta_{12} g(r_{12}, \cos\beta_{12}) P_L(\cos\beta_{12}) = \langle P_L(\cos\beta_{12}) \rangle_{r_{12}}, \tag{8.31}$$

where $g(r_{12})$ is the radial distribution. The averages $\langle P_L(\cos\beta_{12}) \rangle_{r_{12}}$ can be calculated during the course of a simulation using the following procedure. First, the range of interparticle separations is divided into a number of intervals (buckets) of width r each labelled by an integer. Thus, to every separation r_{12} we can assign an integer number labelling one of the buckets. Then, for a given configuration, a particle, i, is chosen as an origin and the quantity of interest, $P_L(\cos\beta_{ij})$, is computed and added for every pair i, j into the bucket corresponding to the separation r_{ij}. The process is then repeated choosing another particle as an origin and so on. Normalization is achieved by dividing the content of every bucket by the number of pairs it holds. The sorting process is then repeated, as usual, for sufficiently many equilibrium configurations to obtain the MC estimate of $G_L(r_{12})$. Also, in this case, the size of the buckets should be chosen large enough to have a sufficiently high number of counts n_c, remembering that the error in estimating the function will be inversely proportional to $\sqrt{n_c}$ and this in turns limits the resolution that can be achieved.

8.6 The Cluster Monte Carlo Method

We have already mentioned that, in general, the effect of periodic boundary conditions is that of enhancing inter-particle correlations and thus increasing the transition temperature with respect to the true one. On the other extreme, free-space boundary conditions underestimate the interaction of our sample with the outside world and generally underestimate the transition temperature. To improve on this limitation and reduce finite size effects, another type of boundary condition easily applicable, at least for lattice systems, to anisotropic systems can be employed [Zannoni, 1986]. This method creates an environment outside the sample box which has on average the same properties as the inside by using self-consistency and maximum entropy (Section 3.8) [Levine and Tribus, 1978] principles. We briefly discuss the method for a generic lattice system, then show an application to the LL model. We begin by considering our sample box B of N particles as part of a global system G of N_G identical particles. The molecules are characterized by their orientation Ω_i and interact through a pair potential of a certain finite range. This length defines in turn a natural boundary area between the sample of N molecules inside the virtual box B and the world W of N_W particles outside (see Fig. 8.8a for a nearest neighbour model). The energy U_G of the global system of $N_G = N + N_W$ particles is

$$U_G = U_B + U_{BW} + U_W, \tag{8.32}$$

where $U_B = \sum_{i=1}^{N} \sum_{j=1}^{N} U_{ij}; i < j$ is the contribution from particles which are all inside the sample box, while $U_{BW} = \sum_{i=1}^{N} \sum_{j=N+1}^{N_G} U_{ij}$ comes from the interaction between molecules inside and outside. Finally, U_W is a purely external energy, i.e. $U_W = \sum_{i=N+1}^{N_G} \sum_{j=N+1}^{N_G} U_{ij}\; i < j$. The global average of a quantity $A(\widetilde{\Omega}) = A(\Omega_1, \Omega_2, \ldots, \Omega_N)$ dependent only on the orientations of the particles inside our virtual box will be

$$\langle A \rangle_G = \frac{1}{Z_G} \int \prod_{i=1}^{N} \mathrm{d}\Omega_i\, A(\widetilde{\Omega})\, \mathrm{e}^{-U_B/(k_B T)} \int \prod_{i=N+1}^{N_G} \mathrm{d}\Omega_i\, \mathrm{e}^{-(U_{BW}+U_W)/(k_B T)}, \tag{8.33}$$

where $Z_G = \int \prod_{i=1}^{N_G} \mathrm{d}\Omega_i \exp[-U_G/(k_B T)]$. We are typically interested in one- and two-particle observables, e.g. order parameters and pair correlation functions, and our aim is to rewrite $\langle A \rangle_G$ in a form amenable to some kind of MC calculation. We note first that for a certain configuration W of the outside world the average of A is

$$\langle A \rangle_{[W]} = \frac{1}{Z_{[BW]}} \int \prod_{i=1}^{N} \mathrm{d}\Omega_i\, \mathrm{e}^{-(U_B+U_{BW})/(k_B T)}\, A(\widetilde{\Omega}), \tag{8.34}$$

where we have defined the configuration integral for the molecules inside the box when surrounded by a fixed configuration $[W]$ of the outside world as $Z_{[BW]} = \int \prod_{i=1}^{N} \mathrm{d}\Omega_i\, \mathrm{e}^{-(U_B+U_{BW})/(k_B T)}$. We can now rewrite $\langle A \rangle_G$ as an average over the outside configurations $\langle \ldots \rangle_W$ of $\langle A \rangle_{[W]}$ since

$$\langle A \rangle_G = \frac{1}{Z_G} \int \prod_{i=N+1}^{N_G} \mathrm{d}\Omega_i\, \mathrm{e}^{-U_W/(k_B T)} \langle A \rangle_{[W]} Z_{[BW]} = \langle\, \langle A \rangle_{[W]}\, \rangle_W. \tag{8.35}$$

The average over the outside world can be performed through importance sampling, considering a finite number of configurations M_W for the molecules outside the box, i.e.

$$\langle A\rangle_G \approx \frac{1}{M_W}\sum_{[W]}\langle A\rangle_{[W]}, \tag{8.36}$$

where each average $\langle A\rangle_{[W]}$ can be calculated with ordinary MC when the outside configuration is known. The problem of approximating $\langle A\rangle_G$ based on just an N particle simulation then becomes that of generating suitably sampled outside configurations with a distribution P_{out}. In an ideal simulation over all N_G molecules our N-particles MC sample is just a virtual subsystem of a very large one without surface effects. In this case, all the (m) particles distributions inside and outside the virtual box (as well as across the interface) have to be the same: $P_{in}^{(m)}(\Omega_1, \Omega_2, \ldots, \Omega_m) = P_{out}^{(m)}(\Omega_{1'}, \Omega_{2'}, \ldots, \Omega_{m'})$; $m \leq N$, giving a self-consistency condition. In particular, the single-particle distribution outside the sample should equal the one inside, i.e. $P_{in}^{(1)}(\Omega) = P_{out}^{(1)}(\Omega)$, and the order parameters inside and outside should be the same. Thus, for a uniaxial phase made of uniaxial particles, $\langle P_L\rangle_{in} = \langle P_L\rangle_{out}$. We can estimate the single-particle distribution for the molecules inside, either by direct construction of a histogram if the sample is big enough, or using the order parameters inside the box and maximum entropy. Assuming that there exists a symmetry-breaking field director $\boldsymbol{d}$, say along the z laboratory axis we can calculate the order parameters $\langle P_2\rangle$, $\langle P_4\rangle$, etc. with respect to this direction. Armed with these observables we can construct the best maximum entropy inference for the molecular distribution as the exponential approximation (Eq. 3.73) with the coefficients α_L determined from the constraint that the available $\langle P_L\rangle$ can be reobtained by averaging $P_L(x)$ over the distribution in Eq. 3.73. Having done this, the Ω_i necessary to replace the missing interactions can be generated sampling from this distribution. The simplest approximation is that obtained from a knowledge of just $\langle P_2\rangle_{in}$, which gives the least biased inference for the single-particle distribution for the ghost molecules as $P(x) = (1/Z)\exp[a_2 P_2(x)]$, with Z a normalization coefficient and a_2 determined from the constraint that $\langle P_2\rangle_{out} = \langle P_2\rangle_{in}$. Ghost molecule orientations are sampled from this distribution thus creating in a self-consistent way a privileged laboratory direction. In a sense we are simulating interactions with the outside with an inhomogeneously fluctuating field whose average strength is proportional to the order parameter inside. The ghost surrounding the sample is refreshed to sample a sufficient number of configurations $[W]$. The director pinning effect will thus be larger at lower temperatures while it will essentially vanish in the isotropic phase.

An application to the Lebwohl–Lasher model. The results of CMC simulations [Zannoni, 1986] for the dimensionless energy $U^* = \langle U\rangle/N\epsilon$ as a function of temperature show clearly that a sharp change of slope occurs suggesting the onset of a first-order transition. The calculation of $\langle P_2\rangle$ and higher-rank order parameters presents no particular problem in the CMC method because of the existence of a symmetry breaking direction. The $\langle P_2\rangle$ calculated at different temperatures with respect to this direction are plotted in Fig. 8.8b for the $10 \times 10 \times 10$ system. The method has been successfully applied to the LL [Zannoni, 1986] and Heisenberg [Chiccoli et al., 1993] models.

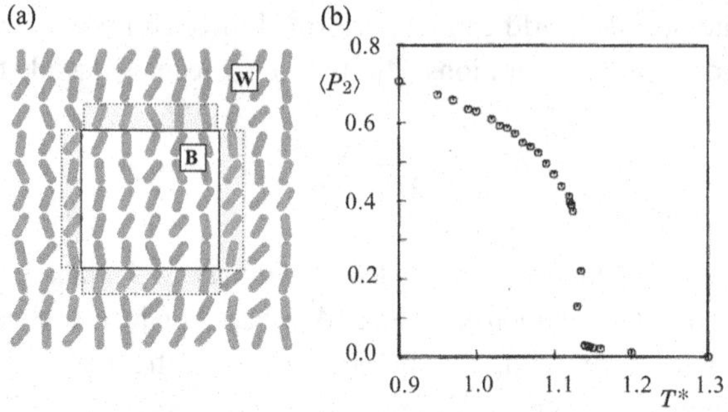

Figure 8.8 (a) partition of a lattice system in a sample box B, a surrounding external world W and a boundary region (grey shaded). (b) The second-rank order parameter calculated with respect to the laboratory symmetry breaking direction, $\langle P_2 \rangle$, as a function of reduced temperature T^* for the $10 \times 10 \times 10$ LL lattice [Zannoni, 1986].

In summary, the cluster MC procedure just described builds a ghost environment outside a 'small but significative' MC sample that mimics the true one and its fluctuations at least approximately. Note that in an ordinary PBC MC calculation the orientations Ω_i of the particles of the sample replicas surrounding the sample are identical to those inside, while in CMC only the distributions (or the order parameters) inside and outside, are constrained to be the same, reducing spurious space correlations. In essence, if MFT (Chapter 7) corresponds to a molecule in the effective field of all the others, CMC corresponds to N particles in the effective field of all the others.

8.7 Reweighting Techniques

Ferrenberg and Swendsen [1988] proposed a clever method for extending the range of temperatures studied starting from a relatively small number of cases. In practice, simulations are performed for a set of temperatures T and used to build histograms $H(U,T)$. The normalized probability distribution function $P(U,T)$ of the energy is then given by

$$P(U,T) = \frac{H(U,T)}{\sum\limits_{U} H(U,T)}. \tag{8.37}$$

Given this distribution function at T, the Ferrenberg–Swendsen method allows the calculation of thermodynamic quantities at a different temperature T' in the neighbourhood of T. Specifically, thermodynamic quantities at T can be calculated using the distribution function $P(U,T')$ where

$$P\left(U,T'\right) = \frac{H(U,T)\,\mathrm{e}^{-\left[U/(k_B T') - U/(k_B T)\right]}}{\sum\limits_{U} H(U,T)\,\mathrm{e}^{-\left[U/(k_B T') - U/(k_B T)\right]}}. \tag{8.38}$$

This and various other "smart" MC methods are described in detail in [Allen and Tildesley, 2017].

9

The Molecular Dynamics Method

> I wish we could derive the rest of the phenomena of nature by the same kind of reasoning from mechanical principles; for I am induced by many reasons to suspect that they may all depend upon certain forces by which the particles of bodies, by some causes hitherto unknown, are either mutually impelled towards each other, and cohere in regular figures, or are repelled and recede from each other; which forces being unknown, philosophers have hitherto attempted the search of nature in vain; but I hope the principles here laid down will afford some light either to that or some truer method of philosophy.
>
> *I. Newton, preface to Principia, 1686.*

9.1 Introduction

The calculation of dynamical quantities by computer simulations has to be performed by a method that allows following of the time evolution of the system explicitly. The Molecular Dynamics technique [Rahman and Stillinger, 1971; Barker and Henderson, 1976; Erpenbeck and Wood, 1977; Zannoni, 1979a; Frenkel and Smit, 2002; Rapaport, 2004; Hansen and McDonald, 2006; Allen and Tildesley, 2017] provides a practical way of achieving this objective. It consists of setting up and solving time step after time step the Newton–Euler equations of motion (in their generalized form) [Landau and Lifshitz, 1993] for a system of N molecules. As a result of the calculation, we obtain fairly complete information on the system. In Chapter 4 we considered two types of descriptions to be particularly relevant for liquid crystals, called for convenience: atomistic and generic particle models. In the first case the system is assumed to be formed of atoms, or more generally, of spherical particles linked in some way so as to form molecules or some complex nanoparticle. For a system described at atomistic level, we need the coordinates and velocities for each atom: $\{\boldsymbol{r}_i, \dot{\boldsymbol{r}}_i\}$, where as usual we use upper dots to indicate time derivatives, while for a system of anisotropic rigid particles we need three coordinates to define the position of the centre of mass, $\boldsymbol{r}_i$, and three Euler angles to define Ω_i, three linear velocity components to define $\dot{\boldsymbol{r}}_i$ and three angular velocity components to define $\dot{\Omega}_i$ in some laboratory frame. In phase space every configuration of the system i.e. the set of coordinates and velocities, e.g. in the last case $\{\boldsymbol{r}_i, \Omega_i, \dot{\boldsymbol{r}}_i, \dot{\Omega}_i\}$, is represented by a point. As the system evolves in time this point will move describing a trajectory in phase space. In both cases what the Molecular Dynamics method does by solving the appropriate equations of motion is to produce the information at a sequence of successive times $t_1, t_2, \ldots t_n$, discretizing the time trajectory. We have already mentioned (Chapter 4) that the approximation of classical behaviour holds if particle separations are much larger than the de Broglie wavelength. In the case of timescales, quantum effects can be neglected for times much longer than $\hbar/(k_B T)$ which, at room temperature, means for times longer than $\approx 10^{-14}$ s.

In this chapter (as in previous ones) we indicate, when possible, with $\boldsymbol{X}_i$ a generic set of variables for the ith particle, with $\widetilde{\boldsymbol{X}}(t)$ the collection of the same variables for all the particles and with $\dot{\widetilde{\boldsymbol{X}}}(t)$ the relative velocities. In Chapter 5 we saw a variety of interaction potentials. Here we shall confine ourselves to introducing the method for continuous differentiable potentials, since hard-core systems require some relatively ad hoc techniques [Erpenbeck and Wood, 1977] and develop the equations necessary to treat orientational properties.

Variants of the method allow simulation to take place at constant temperature or in other ensembles [Andersen, 1980; Abraham, 1986]. Here, however, we shall start from the traditional microcanonical conditions.

Periodic boundary conditions (Section 8.2.1) are implied, as is commonly done in computer simulations, to minimize surface effects. We recall that this amounts to considering the sample box surrounded by identical replicas of itself. Thus, for a cubic box of side L a particle at position $\boldsymbol{r} = (x, y, z)$ will have replicas with the same orientation, velocity, etc. at $(x \pm L, y, z)$, $(x, y \pm L, z)$, $(x, y, z \pm L)$, and so on. If a particle goes out the box on one side, then one of its replicas will enter the box from the opposite side. In this way the density of particles remains constant. The introduction of periodic boundary conditions leads inevitably to some spurious spatial correlation between particles with a separation of the order of the sample box. As discussed in Chapter 8, this effect can be particularly important when systems in the vicinity of a phase transition are studied, as is often the case for simulation of mesophases. In Molecular Dynamics, periodicity also places a limit on the longest time that can be meaningfully studied. This will be the time taken for propagating a perturbation across the box, since for longer times we would have it coming back from the other wall [Kushick and Berne, 1977]. In any case, this limit is normally of no practical concern, since trajectories are followed for much shorter periods due to computational restrictions.

9.2 Equations of Motion for Atomistic Systems

In the atomistic simulations considered here, the individual particles are spherical and correspond to real atoms (in the so called 'all-atom' MD) or to *united atoms* where H atoms are not described as such but are incorporated in the heavier atom they are attached to. This slightly coarse-grained description is typically adequate to describe molecular organizations and even to obtain phase transition temperatures [Tiberio et al., 2009], but should be avoided if observables involving the hydrogen atoms themselves (e.g. proton NMR dipolar couplings) are of interest [Pizzirusso et al., 2012b]. Other systems that we could consider in this section are colloidal suspensions of spherical particles ('colloidal atoms') or their aggregates. In any case, atoms are treated as classical particles of mass m_i having, in a chosen laboratory frame, positions $\boldsymbol{r}_i$ and momenta $\boldsymbol{p}_i = m_i\dot{\boldsymbol{r}}_i = m_i(\dot{r}_{i,x}, \dot{r}_{i,y}, \dot{r}_{i,z})$ related to the components of the particle velocity $\boldsymbol{v}_i$. Thus, $\widetilde{\mathcal{X}} \equiv (\widetilde{\boldsymbol{r}}, \widetilde{\boldsymbol{p}}) \equiv (\boldsymbol{r}_1, \boldsymbol{p}_1, \boldsymbol{r}_2, \boldsymbol{p}_2, \ldots, \boldsymbol{r}_N, \boldsymbol{p}_N)$ is the set of coordinates and momenta needed to specify a point in phase space. The system has an N-particle Hamiltonian $\mathscr{H}$, which is the sum of the potential $U(\widetilde{\boldsymbol{r}})$ and kinetic $K(\widetilde{\boldsymbol{p}})$ energy contributions:

$$\mathscr{H}(\widetilde{\boldsymbol{r}}, \widetilde{\boldsymbol{p}}) = U(\widetilde{\boldsymbol{r}}) + K(\widetilde{\boldsymbol{p}}). \tag{9.1}$$

We have already discussed the dynamic evolution in general terms in Chapter 6 for a system described by a Hamiltonian $\mathscr{H}$ and the equations of motion for coordinates and momenta. In particular, omitting for simplicity the arguments, we have that [Goldstein, 1980; Landau and Lifshitz, 1993]

$$\frac{\partial \mathscr{H}}{\partial \boldsymbol{p}_i} = \dot{\boldsymbol{r}}_i \tag{9.2}$$

and

$$\frac{\partial \mathscr{H}}{\partial \boldsymbol{r}_i} = \frac{\partial U}{\partial \boldsymbol{r}_i} = -\dot{\boldsymbol{p}}_i = -m_i\ddot{\boldsymbol{r}}_i. \tag{9.3}$$

In the absence of dissipative contributions, the force acting on particle i (e.g. an atom) depends only on coordinates and corresponds to the gradient of the potential U_i acting on that particle by effect of all the others:

$$\boldsymbol{f}_i = -\frac{\partial U_i}{\partial \boldsymbol{r}_i} = -\nabla_i U_i = -\nabla_i \sum_{j \neq i} U_{ij}. \tag{9.4}$$

Thus, Eq. 9.3 takes the familiar Newton equations of motion form

$$\boldsymbol{f}_i = m_i \boldsymbol{a}_i, \tag{9.5}$$

where $\boldsymbol{a} \equiv \ddot{\boldsymbol{r}}$ is the acceleration. The first item to examine is thus the potential energy between particles, as we have already discussed in Chapter 5, so we can assume the forces to be available when needed.

9.3 Integration of the Atomistic Equations of Motion

The set of coupled Newton equations (Eq. 9.3) can be integrated to compute positional and angular coordinates and velocities as a function of time, i.e. to generate the trajectory of the system in phase space. The numerical integration can be performed using one of the several algorithms available. Thus, suppose we have positions $\boldsymbol{r}_i(t)$ and velocities $\dot{\boldsymbol{r}}_i(t)$ for every particle at a time t and that we want to find their value at $(t + \Delta t)$, with Δt a small time step (typically of the order of 1 fs). A first-order Taylor expansion gives a simple *finite difference* approximation for the derivatives $\dot{\boldsymbol{r}}$ and $\ddot{\boldsymbol{r}}$ and yields the new velocity as

$$\dot{\boldsymbol{r}}_i(t + \Delta t) = \dot{\boldsymbol{r}}_i(t) + \frac{1}{m_i}\boldsymbol{f}_i(t)\Delta t + \mathcal{O}(\Delta t^2), \tag{9.6}$$

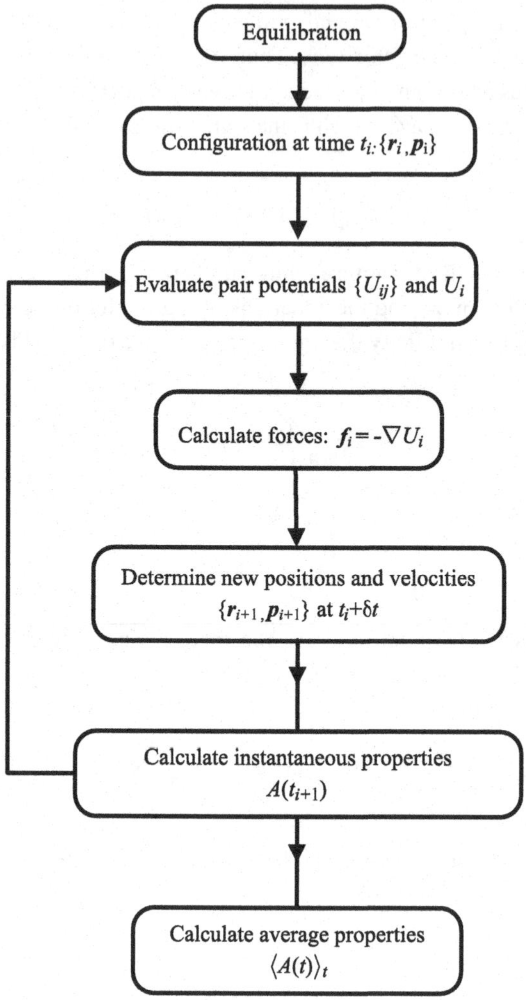

Figure 9.1 A sketchy flow diagram of the Molecular Dynamics method.

where $\boldsymbol{f}_i(t) \equiv \boldsymbol{f}_i(\tilde{\boldsymbol{r}}(t))$ is the force acting on the ith particle at time t, and we indicate with $\mathcal{O}(\Delta t^2)$ the order of magnitude of the terms neglected (the 'error'). The new position would be

$$\boldsymbol{r}_i(t + \Delta t) = \boldsymbol{r}_i(t) + \dot{\boldsymbol{r}}_i(t)\,\Delta t + \mathcal{O}(\Delta t^2). \tag{9.7}$$

This method, sometimes called Euler method, is very easy and simple to understand but performs very poorly. In general, the accuracy of these finite difference methods with respect to the true solution is affected by the inevitable truncation errors intrinsic to the algorithm. Even though they may seem negligible, given the mentioned small Δt, this is not really so, as many millions of time steps have to be performed in a typical simulation. More sophisticated algorithms used in solving other differential equations can obviously be used.

Note, however, that the improved accuracy in estimating the new coordinates and velocities is counterbalanced by the need to calculate and store additional quantities. In particular, while second derivatives of positions (forces) are easily available, higher derivatives would be progressively much more expensive to calculate. An essential sketch of the MD workflow is shown in Fig. 9.1 and in the next sections we list just a few of the most common and useful algorithms employed to perform the various steps of the procedure.

9.3.1 Verlet Algorithm

A useful compromise between speed and accuracy is given by Verlet's algorithm [Verlet, 1967; Allen and Tildesley, 2017]. If we carry the Taylor expansion Eq. 9.7 to third-order for positive and negative time increments:

$$\boldsymbol{r}_i(t+\Delta t) = \boldsymbol{r}_i(t) + \dot{\boldsymbol{r}}_i(t)\Delta t + \frac{1}{2}\ddot{\boldsymbol{r}}_i(t)\Delta t^2 + \frac{1}{6}\dddot{\boldsymbol{r}}_i(t)\Delta t^3 + \mathcal{O}(\Delta t^4), \tag{9.8a}$$

$$\boldsymbol{r}_i(t-\Delta t) = \boldsymbol{r}_i(t) - \dot{\boldsymbol{r}}_i(t)\Delta t + \frac{1}{2}\ddot{\boldsymbol{r}}_i(t)\Delta t^2 - \frac{1}{6}\dddot{\boldsymbol{r}}_i(t)\Delta t^3 + \mathcal{O}(\Delta t^4), \tag{9.8b}$$

and add them together, we find a third order predictor for the new position

$$\boldsymbol{r}_i(t+\Delta t) = 2\boldsymbol{r}_i(t) - \boldsymbol{r}_i(t-\Delta t) + \frac{\Delta t^2}{m_i}\boldsymbol{f}_i(t) + \mathcal{O}(\Delta t^4). \tag{9.9}$$

Thus, the truncation error is now in principle $\mathcal{O}(\Delta t^4)$ instead of $\mathcal{O}(\Delta t^2)$, even if with the finite digits of a computer adding large and small terms can lead to errors. Note that the velocity does not appear in this formula. It can be obtained afterwards, subtracting the two Eqs. 9.8, as the central difference

$$\dot{\boldsymbol{r}}_i(t) = \frac{1}{2\Delta t}[\boldsymbol{r}_i(t+\Delta t) - \boldsymbol{r}_i(t-\Delta t)] + \mathcal{O}(\Delta t^2), \tag{9.10}$$

which has, however, errors at $\mathcal{O}(\Delta t^2)$. In certain applications, e.g. calculation of kinetic energy and particularly where accelerations depend on velocities as well as on positions as in the modified evolution equations used for constant temperature methods, this can be a disadvantage.

A 'velocity' version of Verlet's algorithm can, however, be introduced as

$$\dot{\boldsymbol{r}}_i(t+\Delta t/2) = \dot{\boldsymbol{r}}_i(t) + \frac{1}{2m_i}\boldsymbol{f}_i(t)\Delta t, \tag{9.11a}$$

$$\boldsymbol{r}_i(t+\Delta t) = \boldsymbol{r}_i(t) + \dot{\boldsymbol{r}}_i(t+\Delta t/2)\,\Delta t, \tag{9.11b}$$

$$\dot{\boldsymbol{r}}_i(t+\Delta t) = \dot{\boldsymbol{r}}_i(t) + \left[\boldsymbol{f}_i(t) + \boldsymbol{f}_i(t+\Delta t)\right]\frac{\Delta t}{2m_i} + \mathcal{O}(\Delta t^3). \tag{9.11c}$$

The velocity Verlet algorithm explicitly provides the velocity at each step, is time reversible and is 'self-starting' from the positions and velocities at the initial time.

9.3.2 Leapfrog Integrator

A very popular method is the half-step or leapfrog integrator [Allen and Tildesley, 2017] that takes its name from the fact that velocities and positions are not calculated simultaneously, but alternatively at each half step. Velocities (Eq. 9.13) are obtained by combining two Taylor expansions of the velocity at time step $\pm\Delta t/2$:

$$\dot{\boldsymbol{r}}_i(t+\Delta t/2) = \dot{\boldsymbol{r}}_i(t) + \frac{\boldsymbol{f}_i(t)\Delta t}{2m_i} + \mathcal{O}(\Delta t^2), \tag{9.12a}$$

$$\dot{\boldsymbol{r}}_i(t-\Delta t/2) = \dot{\boldsymbol{r}}_i(t) - \frac{\boldsymbol{f}_i(t)\Delta t}{2m_i} + \mathcal{O}(\Delta t^2), \tag{9.12b}$$

to give an update of the velocity at $t+\Delta t/2$

$$\dot{\boldsymbol{r}}_i(t+\Delta t/2) = \dot{\boldsymbol{r}}_i(t-\Delta t/2) + \frac{\boldsymbol{f}_i(t)\Delta t}{m_i} + \mathcal{O}(\Delta t^2), \tag{9.13}$$

while the position is obtained using a half time step and shifting the time of another half time step:

$$\boldsymbol{r}_i(t+\Delta t) = \boldsymbol{r}_i(t) + \dot{\boldsymbol{r}}_i(t+\Delta t/2)\Delta t + \mathcal{O}(\Delta t^3). \tag{9.14}$$

The instantaneous velocities at time t can be computed as the average of the velocities at time $t-\Delta t/2$ and $t+\Delta t/2$. The method is time reversible and symplectic (i.e. it conserves phase space volume).

9.3.3 Gear Predictor-Corrector

Rahman and Stillinger [1971] have used the following [Gear, 1971] *predictor-corrector* method

$$\boldsymbol{r}'_i(t+\Delta t) = \boldsymbol{r}_i(t-\Delta t) + 2\Delta t\,\dot{\boldsymbol{r}}_i(t) + \mathcal{O}(\Delta t^3), \tag{9.15a}$$

$$\dot{\boldsymbol{r}}'_i(t+\Delta t) = \dot{\boldsymbol{r}}_i(t-\Delta t) + \frac{\Delta t}{m_i}\boldsymbol{f}_i(t) + \mathcal{O}(\Delta t^3), \tag{9.15b}$$

giving

$$\boldsymbol{r}_i(t+\Delta t) = \boldsymbol{r}_i(t) + \frac{1}{2}\Delta t\left[\dot{\boldsymbol{r}}_i(t) + \dot{\boldsymbol{r}}'_i(t+\Delta t)\right] + \mathcal{O}(\Delta t^3), \tag{9.16a}$$

$$\dot{\boldsymbol{r}}_i(t+\Delta t) = \dot{\boldsymbol{r}}_i(t) + \frac{\Delta t}{2m_i}\left[\boldsymbol{f}_i(t) + \boldsymbol{f}'_i(t+\Delta t)\right] + \mathcal{O}(\Delta t^3). \tag{9.16b}$$

Note, however, that the improved accuracy in estimating the new coordinates and velocities is counterbalanced by the need to calculate and store additional quantities. There are of course many other methods that try to increase Δt and thus the observation time span while maintaining accuracy [Leimkuhler and Matthews, 2015], but this typically requires paying the price of some complicated algorithm and additional calculations. This leaves out, in particular, methods requiring more than one evaluation of forces, normally the most computationally expensive ones to calculate.

9.3.4 Timescales

Having obtained the new configuration at time $(t + \Delta t)$ the procedure is repeated. Thus, the continuous trajectory of the system in phase space is replaced by a series of points at time intervals Δt. A necessary condition for a good integration is that energy should be conserved as long as the system is isolated, but this is not necessarily sufficient [Toxvaerd, 1983]. A true trajectory should also be time reversible and certain algorithms intrinsically satisfy this, e.g. the simple Verlet's one, Eq. 9.9, but reversibility can be obtained also for time intervals so large as to give an otherwise incorrect evolution [Toxvaerd, 1982]. It is worth stressing that whatever integration algorithm is used, it has to be stable in the sense that the solution obtained should not diverge exponentially with time from the true one. This requirement also determines the magnitude of the increment Δt.

An important point to consider when simulating realistic systems of molecules is that motions corresponding to different atoms or molecular fragments can have rather different timescales (e.g. bond vibrations will typically be significantly faster than molecular translations). This determines the need for a very short integration time step for a proper integration and can be very demanding as forces between particles have to be calculated at every time step, implying a waste of resources for properly dealing with bond stretching and bending that are of somewhat secondary importance. Because each particle interacts with all particles within the interaction range of the potential, the longer the range of the potential the larger the number of force contributions that must be calculated at each time. One rather drastic way of tackling the problem is that of fixing bond lengths, or more generally, the distances between mass centres by implementation of a set of geometric (*holonomic*) constraints. One such algorithm, called SHAKE [Ryckaert et al., 1977; Ciccotti and Ryckaert, 1986] is widely used particularly for large molecules [McCammon and Harvey, 1987] and implemented in some of the available MD engine packages like GROMOS [van Gunsteren and Berendsen, 1987]. Among the various other algorithms that implement geometrical constraints we mention RATTLE [Andersen, 1983], SHAPE [Tao et al., 2012] and LINCS [Hess et al., 1997].

Instead of introducing constraints, more general approaches to the integration of the equations of motion that allow instead multiple timescales (MTSs) can be formulated starting from the time evolution of the mechanical system of atoms with the help of the Liouville operator [Friedman, 1985; Tuckerman, 2010]

$$\mathcal{L}(\ldots) = -i\{\mathcal{H}, \ldots\} = -i\mathcal{H}^{\times}(\ldots), \tag{9.17}$$

where the classical Poisson brackets $\{\mathcal{H}, \ldots\}$ [Goldstein, 1980; Friedman, 1985; Tuckerman, 2010] were defined in Eq. 6.4. The formal solution of the equation of motion $\mathrm{d}A(t)/\mathrm{d}t = i\mathcal{L}A(t)$ for a dynamical variable $A(t) \equiv A(\{\boldsymbol{r}(t), \boldsymbol{p}(t)\})$ leads, for a time independent Hamiltonian, to the formal solution

$$A(t + \Delta t) = \mathrm{e}^{\Delta t \mathcal{H}^{\times}} A(t), \tag{9.18}$$

where the propagator $e^{\Delta t \mathcal{H}^\times}$ is a unitary operator so that its inverse, the propagator for going back in time, is its adjunct. If we just take $A(t) = (\boldsymbol{r}_i(t), \boldsymbol{v}_i(t))$ we can reobtain, for a very small Δt,

$$\boldsymbol{r}_i(t + \Delta t) = \boldsymbol{r}_i(t) + \boldsymbol{v}_i \Delta t + \cdots, \tag{9.19a}$$

$$\boldsymbol{v}_i(t + \Delta t) = \boldsymbol{v}_i(t) + \frac{1}{m_i} \boldsymbol{f}_i \Delta t + \cdots. \tag{9.19b}$$

The propagator formulation helps to ensure that the desired properties of MD integrators, as time reversibility, symplecticity, and good conservation of total energy, are implemented. If the Hamiltonian (and thus the Liouville operator) can be split into a sum of different terms, each leading to different timescales, the propagator can be rewritten using some known theorems [Tuckerman et al., 1991a, 1991b]. In particular, for any two linear operators $\mathcal{A}$, $\hat{B}$ the so-called symmetric Sprang–Trotter formula holds:

$$e^{(\mathcal{A}+\mathcal{B})t} = \lim_{n\to\infty} \left(e^{\mathcal{B}t/2n} e^{\mathcal{A}t/n} e^{\mathcal{B}t/2n} \right)^n. \tag{9.20}$$

Applying this to an evolution operator consisting of a 'slow' (S) and 'fast' (F) part, i.e.

$$\mathcal{H}^\times = \mathcal{H}_S^\times + \mathcal{H}_F^\times, \tag{9.21}$$

we have, considering a relatively long time step Δt, adequate for the integration of the slow contribution, that

$$e^{\Delta t \mathcal{H}^\times} = e^{\Delta t(\mathcal{H}_S^\times + \mathcal{H}_F^\times)} = e^{\frac{\Delta t}{2} \mathcal{H}_S^\times} (e^{\Delta t \mathcal{H}_F^\times}) e^{\frac{\Delta t}{2} \mathcal{H}_S^\times} + \mathcal{O}(\Delta t^3), \tag{9.22a}$$

$$= e^{\frac{\Delta t}{2} \mathcal{H}_S^\times} (e^{\frac{\Delta t}{n} \mathcal{H}_F^\times})^n e^{\frac{\Delta t}{2} \mathcal{H}_S^\times} + \mathcal{O}(\Delta t^3), \tag{9.22b}$$

$$= e^{\frac{\Delta t}{2} \mathcal{H}_S^\times} (e^{\Delta t_F \mathcal{H}_F^\times})^n e^{\frac{\Delta t}{2} \mathcal{H}_S^\times} + \mathcal{O}(\Delta t^3), \tag{9.22c}$$

where $\Delta t_F = \Delta t/n$ is now a 'short' time step suitable for the fast motion. Thus, the evolution can be performed doing a slow evolution step with $\mathcal{H}_S^\times$ followed by a sequence of evolution steps performed only on the fast contribution with its appropriate short time step and so on. The procedure can be repeated if more than two separable timescales can be individuated leading to MTS algorithms [Procacci and Marchi, 2000; Tuckerman, 2010]. In practice, there are a number of ways in which the separation can take place. In atomistic simulations, bending and stretching are good candidates for the 'fast' timescale, while expensive to calculate long-range interactions could be assigned to the 'slow' part, but also long-range electrostatic contributions can be further partitioned between direct and reciprocal space, carefully choosing the separation parameter α in Eqs. 5.38a and 5.38b. Time reversibility of the trajectory is a necessary requirement. The propagators developed as described are clearly unitary, therefore time reversible, automatically leading to more accurate approximations of the true discrete time propagator [De Raedt and De Raedt, 1983; Yoshida, 1990].

The Liouville approach coupled with a suitable Sprang- Trotter factorization of the total propagator allows developing efficient multiple time step integrators, e.g. the *reversible reference system propagator algorithm* (r-RESPA) [Tuckerman et al., 1992]. The approach is described, for example, in Tuckerman et al. [1991b], Procacci and Marchi [2000] and Frenkel and Smit [2002], and is implemented in various MD packages, for instance,

LAMMPS [Plimpton, 1995], ORAC [Procacci et al., 1997] and NAMD [Phillips et al., 2005].

Similar factorizations allow the integration of the equations of rotational motion when non-spherical particles are studied [Kamberaj et al., 2005; Allen and Tildesley, 2017] even though care must be taken to avoid singularities and this is usually accomplished by using quaternions [Altmann, 2005; Berardi et al., 2008a]. Multiple time-steps algorithms improve efficiency allowing each timescale an appropriately chosen time step. However, such approaches are hampered by possible resonance phenomena that enhance errors and limit the largest time step to around 5–6 fs. The development of integration methods to increase Δt while maintaining accuracy continues and the results [Leimkuhler et al., 2013] seem to indicate the possibility of an increase by more than an order of magnitude. More sophisticated algorithms can obviously be used. Methods that do not require the storing of previous information have the advantage of being self-starting. A class of methods of this kind, that we shall not describe in detail, but that is well known in the solution of ordinary differential equations are the Runge–Kutta methods in its various versions (see, e.g., [Moin, 2010]).

9.4 Equations of Motion for Rigid Anisotropic Particles

While atomistic simulations are needed to provide predictions on properties and phase transitions for realistic, chemically detailed, models of mesogens, the majority of simulations of liquid crystals has been and is still performed on the anisotropic molecular models introduced in Chapter 5. Here we discuss the equations of motion for a system of anisotropic rigid particles that obey classical mechanics.

Each particle will be characterized by position $\boldsymbol{r}$, velocity $\dot{\boldsymbol{r}}$, orientation with respect to the fixed laboratory frame rotation vector $\boldsymbol{\Omega}$ and angular velocity $\dot{\boldsymbol{\Omega}}$. For these systems, the evolution of the centres of mass is exactly similar to what we have seen up to now in terms of equations of motion (Newton's ones) and their integration.

The potential acting on the ith rotor because of its interaction with the other particles is, assuming effective pairwise additivity,

$$U_i = \sum_{j \neq i} U(\boldsymbol{r}_i, \Omega_i, \boldsymbol{r}_j, \Omega_j). \tag{9.23}$$

The summation in Eq. 9.23 is extended in principle to all the particles interacting with the particle considered, even though in practice only those within a given cut-off range are often taken into account. We do not enter here into the nature of the effective two body potentials that we have discussed in Chapter 5, but we shall assume it to be a continuous and differentiable function of positional and orientational coordinates as well as time independent. Hard particles can be treated too with specific methods [Allen et al., 1993]. The kinetic energy K of a system of rotors of individual mass m_i and inertia tensor $\mathbf{I}_i$ is the sum of the translational and rotational contributions K^{tr} and K^{rot}:

$$K^{\text{tr}} = \frac{1}{2} \sum_{i=1}^{N} m_i \dot{\boldsymbol{r}}_i \cdot \dot{\boldsymbol{r}}_i, \tag{9.24a}$$

$$K^{\text{rot}} = \frac{1}{2} \sum_{i=1}^{N} \dot{\boldsymbol{\Omega}}_i \cdot \mathbf{I}_i \cdot \dot{\boldsymbol{\Omega}}_i. \tag{9.24b}$$

The inertia tensor components in the laboratory frame depend on the orientation of particle i, while in the molecule fixed system its Cartesian components have been defined in Eq. 4.24 and, for continuous bodies, in Eq. 4.25. The dynamic evolution of the rotor can be described by Lagrange's equations [Landau and Lifshitz, 1993]

$$\frac{\mathrm{d}}{\mathrm{d}t}\frac{\partial \mathcal{L}}{\partial \dot{q}_\alpha} - \frac{\partial \mathcal{L}}{\partial q_\alpha} = 0, \alpha = 1, 2, \ldots, n_f, \tag{9.25}$$

where $\mathcal{L} = K - U$ is the Lagrangian function for the rotor under consideration, q_α a generalized coordinate component and n_f the number of degrees of freedom. For translational degrees of freedom, $q_a = (r_x, r_y, r_z)$, we obtain again Newton's equation, like in Eq. 9.5. The linear acceleration $\ddot{\boldsymbol{r}}$ is obtained from the force, as in Eq. 9.3, acting on the particle with centre of mass at $\boldsymbol{r}_i$, that is calculated as the gradient of the potential on the rotor. Similarly, we can deduce the rotational equation of motion in the laboratory frame

$$\boldsymbol{N}_i^{\mathrm{LAB}} = \frac{\mathrm{d}}{\mathrm{d}t}\left(\mathbf{I}_i^{\mathrm{LAB}} \cdot \dot{\boldsymbol{\Omega}}_i^{\mathrm{LAB}}\right), \tag{9.26}$$

where $\boldsymbol{N}_i$ is the torque produced on the rotor under consideration by the effect of all the others:

$$\boldsymbol{N}_i = \sum_k (\boldsymbol{h}_k - \boldsymbol{r}_i) \times \boldsymbol{f}_k. \tag{9.27}$$

Eq. 9.26 can be rewritten in a frame moving with the molecule by using the vector formula for the time derivative of a vector $\boldsymbol{a}$ attached to a frame moving with angular velocity $\dot{\boldsymbol{\Omega}}$ with respect to a reference frame:

$$\dot{\boldsymbol{a}}^{\mathrm{LAB}} = \dot{\boldsymbol{a}}^{\mathrm{MOL}} + \dot{\boldsymbol{\Omega}} \times \boldsymbol{a}. \tag{9.28}$$

Thus, we find from Eq. 9.28 the Euler equation describing the reorientation of the particle [Landau and Lifshitz, 1993]

$$\boldsymbol{N}^{\mathrm{LAB}} = \mathbf{I}^{\mathrm{MOL}}\ddot{\boldsymbol{\Omega}}^{\mathrm{MOL}} + \dot{\boldsymbol{\Omega}}^{\mathrm{MOL}} \times (\mathbf{I}^{\mathrm{MOL}}\dot{\boldsymbol{\Omega}}^{\mathrm{MOL}}). \tag{9.29}$$

If our molecular fixed-coordinate system is the principal frame of the inertia tensor we have explicitly for the jth particle

$$[N_j]_a = [I_j]_{aa}[\ddot{\Omega}_j]_a + \varepsilon_{\alpha\beta\gamma}[\dot{\Omega}_j]_\beta[I_j]_\gamma[\dot{\Omega}_j]_\gamma, \tag{9.30}$$

where $\varepsilon_{\alpha\beta\gamma}$ is the Levi-Civita symbol (Eq. A.9) and to avoid clutter we have omitted the superscripts indicating the coordinate frame. Summation on repeated Greek subscripts is implied as usual when using $\varepsilon_{\alpha\beta\gamma}$. In Eq. 9.30, I_{aa}, $a = x, y, z$, are the eigenvalues of the inertia tensor of the particle and $\dot{\Omega}_a$, N_a are, respectively, the components of the angular velocity and of the torque around the principal axes. As an example, again for the jth rotor,

$$\boldsymbol{N}_j = -\boldsymbol{r}_j \times \nabla U_j = i\boldsymbol{r}_j\hat{\mathbf{J}}U_j, \tag{9.31}$$

where we have introduced the dimensionless quantum mechanics angular momentum operator $\hat{\mathbf{J}}$ (Eq. F.34) [Rose, 1957]. This is quite useful to write down at once the torque components when the potential is written in terms of Euler angles (α, β, γ). Using the well-known expressions for the angular momentum operator in terms of Euler angles, we get the molecule fixed torque components

$$N_x = \cos\gamma \cot\beta N_\gamma + \sin\gamma N_\beta - \frac{\cos\gamma}{\sin\beta} N_\alpha, \tag{9.32a}$$

$$N_y = -\sin\gamma \cot\beta N_\gamma + \cos\gamma N_\beta - \frac{\sin\gamma}{\sin\beta} N_\alpha, \tag{9.32b}$$

$$N_z = N_\gamma, \tag{9.32c}$$

where $N_\alpha \equiv \frac{\partial U}{\partial \alpha}$, $N_\beta \equiv \frac{\partial U}{\partial \beta}$, $N_\gamma \equiv \frac{\partial U}{\partial \gamma}$, while the space fixed components of the torque are

$$N_X = \cos\alpha \cot\beta N_\alpha + \sin\alpha N_\beta - \frac{\cos\alpha}{\sin\beta} N_\gamma, \tag{9.33a}$$

$$N_Y = \sin\alpha \cot\beta N_\alpha - \cos\alpha N_\beta - \frac{\sin\alpha}{\sin\beta} N_\gamma, \tag{9.33b}$$

$$N_Z = -N_\alpha. \tag{9.33c}$$

From the point of view of numerical integration of the equations of motion, these Euler equations have the very inconvenient feature of diverging every time the molecule has occasionally $\beta = 0, \pi$. Historically, a first attempt to remedy this problem was that of adopting two coordinate frames and switching when the risk of singularity appeared [Barojas et al., 1973]. This is clearly difficult to handle and error prone. The best approach is to avoid the use of Euler angles altogether with use of quaternions.

9.4.1 Equations of Motion in Terms of Quaternions

For a linear molecule the Euler angles β, α become simply the polar angles (often called θ and ϕ) of the particle, but the problem of the unnecessary singularity due to $\sin\beta$ at the denominator in the rotational equations of motion remains. A simple and elegant way to overcome the problem for the linear rotors is that of replacing the representation of orientations in terms of two angles (α, β) with one in terms of three *direction cosines* (the Cartesian components of a unit vector directed along the axis of the rotor) and a normalization condition. For rigid molecules of arbitrary symmetry which require the three Euler angles, a similar approach involves representing the orientations in terms of four-component quaternions [Evans, 1977; Zannoni and Guerra, 1981; Fincham and Heyes, 1985] (see Appendix H). Thus, instead of using three Euler angles $\Omega_{\text{ML}} = (\alpha, \beta, \gamma)$, a unitary quaternion $\mathbf{u}_{\text{ML}} = (u_0, u_1, u_2, u_3)$ with $u_0^2 + u_1^2 + u_2^2 + u_3^2 = 1$ is introduced to describe the orientation of each molecule. As we show in Appendix H the two representations are connected by simple equations: Eqs. H.22a–H.22f. The time derivative of quaternions can be written in terms of angular velocities in the following way. First, we can use Eq. 9.28 to write the time derivative of a constant length vector $(\partial a/\partial t = 0)$

$$\dot{\boldsymbol{a}}^{\text{LAB}} = \dot{\boldsymbol{\Omega}} \times \boldsymbol{a} = \Xi\, \boldsymbol{a}, \tag{9.34}$$

where Ξ is the antisymmetric angular velocities matrix

$$\Xi = \begin{pmatrix} 0 & -\dot{\Omega}_z & +\dot{\Omega}_y \\ +\dot{\Omega}_z & 0 & -\dot{\Omega}_x \\ -\dot{\Omega}_y & +\dot{\Omega}_x & 0 \end{pmatrix} \tag{9.35}$$

and $\dot{r}_\alpha^{\mathrm{LAB}} = (\dot{\Omega} \times \boldsymbol{r})_\alpha = \varepsilon_{\alpha\beta\gamma}\dot{\Omega}_\beta r_\gamma$. The vector $\boldsymbol{r}$ in the lab fixed frame can be related to the molecular one:

$$\boldsymbol{r}^{\mathrm{LAB}} = \mathbf{R}^T \boldsymbol{r}^{\mathrm{MOL}}, \tag{9.36}$$

with the time derivative

$$\dot{\boldsymbol{r}}^{\mathrm{LAB}} = \dot{\mathbf{R}}^T \boldsymbol{r}^{\mathrm{MOL}} = \dot{\mathbf{R}}^T \mathbf{R}\, \boldsymbol{r}^{\mathrm{LAB}} = \Xi\, \boldsymbol{r}, \tag{9.37}$$

where the Cartesian rotation matrix $\mathbf{R} = \mathbf{R}(\mathbf{u}_{\mathrm{ML}})$, function of the orientation written as a quaternion $\mathbf{u}_{\mathrm{ML}}$, i.e., connects lab and body fixed frames and is given explicitly in terms of quaternions in Eq. H.24. Thus, $\dot{\mathbf{R}}^T\mathbf{R} = \Xi$ or, explicitly,

$$\Xi_{\alpha\gamma} = \varepsilon_{\alpha\beta\gamma}\dot{\Omega}_\beta = \varepsilon_{\gamma\alpha\beta}\dot{\Omega}_\beta = R^T_{\alpha\beta}\, \dot{R}_{\beta\gamma} = R_{\beta\alpha}\dot{R}_{\beta\gamma}. \tag{9.38}$$

Multiplying both sides for $\varepsilon_{\gamma\alpha\eta}$ and using Eq. A.13b, $\varepsilon_{\gamma\alpha\eta}\varepsilon_{\gamma\alpha\beta} = 2\,\delta_{\eta,\beta}$, we have

$$\varepsilon_{\gamma\alpha\eta}\varepsilon_{\gamma\alpha\beta}\dot{\Omega}_\beta = 2\delta_{\eta,\beta}\dot{\Omega}_\beta = \varepsilon_{\gamma\alpha\eta}\, R_{\beta\alpha}\, \dot{R}_{\beta\gamma}, \tag{9.39}$$

and

$$\dot{\Omega}_\eta = \frac{1}{2}\varepsilon_{\gamma\alpha\eta} R_{\beta\alpha}\dot{R}_{\beta\gamma}. \tag{9.40}$$

Recalling the explicit definition of the Cartesian rotation matrix $\mathbf{R}$ in terms of quaternion components, Eq. H.24, this gives the angular velocities in terms of quaternions and their derivatives:

$$\begin{pmatrix} 0 \\ \dot{\Omega}_x \\ \dot{\Omega}_y \\ \dot{\Omega}_z \end{pmatrix} = 2 \begin{pmatrix} +u_0 & +u_1 & +u_2 & +u_3 \\ -u_1 & +u_0 & +u_3 & -u_2 \\ -u_2 & -u_3 & +u_0 & +u_1 \\ -u_3 & +u_2 & -u_1 & +u_0 \end{pmatrix} \begin{pmatrix} \dot{u}_0 \\ \dot{u}_1 \\ \dot{u}_2 \\ \dot{u}_3 \end{pmatrix} = 2\mathbf{W}^T\dot{\mathbf{u}}. \tag{9.41}$$

The 4×4 $\mathbf{W}$ matrix has been obtained adding equation: $+u_0\dot{u}_0 + +u_1\dot{u}_1 + u_2\dot{u}_2 + u_3\dot{u}_3 = 0$ resulting by differentiation of $\sum_{i=0}^{3} u_i^2 = 1$. The quaternion derivatives can be obtained immediately, since the matrix $\mathbf{W}$ is orthogonal, so that its inverse is just its transpose:

$$\begin{pmatrix} \dot{u}_0 \\ \dot{u}_1 \\ \dot{u}_2 \\ \dot{u}_3 \end{pmatrix} = \frac{1}{2} \begin{pmatrix} +u_0 & -u_1 & -u_2 & -u_3 \\ +u_1 & +u_0 & -u_3 & +u_2 \\ +u_2 & +u_3 & +u_0 & -u_1 \\ +u_3 & -u_2 & +u_1 & +u_0 \end{pmatrix} \begin{pmatrix} 0 \\ \dot{\Omega}_x \\ \dot{\Omega}_y \\ \dot{\Omega}_z \end{pmatrix} = \frac{1}{2}\mathbf{W}\,\dot{\boldsymbol{\Omega}}. \tag{9.42}$$

If the angular velocities are obtained from Euler equation and the torques

$$\boldsymbol{N} = -\frac{1}{2}\mathbf{W}^T\frac{\partial U}{\partial \mathbf{u}}, \tag{9.43}$$

the integration of Eq. 9.42 can be carried out with some of the standard methods examined before, e.g. leapfrog. However, it should be recalled that the sum of squares of the quaternion components should be normalized, applying a correction factor at every time step, or at least, frequently enough [Allen, 1984]. A Molecular Dynamics simulation of rigid bodies using quaternions [Evans, 1977; Evans and Murad, 1977; Zannoni and Guerra, 1981; Fincham and Heyes, 1985; Berardi et al., 2008a] is most convenient, since the equations of rotational motion are free of singularities and avoid using trigonometric functions. The use of quaternions is now standard and is included in packages such as LAMMPS (Large-scale Atomic/Molecular Massively Parallel Simulator) from Sandia Labs [Plimpton, 1995].

The orientational propagator can also be embedded in a simple multiple time step scheme [Tuckerman et al., 1991a] with a standard velocity-Verlet propagator for the translational motion to achieve an algorithm for the complete finite differences integration of the full equations of motion for a rigid body.

9.4.2 Equilibration

The Molecular Dynamics process is started from a configuration which is rarely an equilibrium one for the desired temperature and density. The initial positions and orientations can be chosen using existing information if available, e.g. an existing Monte Carlo configuration could be used. For a model liquid crystal simulation, a convenient possibility could also be to sample orientations from a molecular field distribution with a certain order parameter

$$P(\cos\beta) = \exp[a\cos^2\beta]/Z, \tag{9.44}$$

where Z is a normalization coefficient and the approximation $\langle P_2 \rangle \approx a/5$ (see Eq. 3.79) is adequate for the purpose. This initialization also ensures that some basic symmetry requirements are built in, e.g. that on average heads and tails of particles are equally represented.

Velocities can be sampled from a suitable Maxwell distribution (Eq. 4.15) to speed up the approach to equilibrium or they can just be set in some rather arbitrary way (for example, all with constant magnitude but random directions).

If a system is isolated, its total energy, linear and angular momentum $\boldsymbol{p}_T$ and $\boldsymbol{J}$ will be conserved (constant in time) while potential and kinetic energy, and therefore the temperature, of the system will fluctuate. The $\boldsymbol{p}_T$ and $\boldsymbol{J}$ conservation laws follow from general symmetries: the translational and rotational invariance of physical laws. It should be noted that in a simulation of a sample in an orthogonal box with periodic boundary conditions energy and linear momentum are conserved but not angular momentum, since the system is strictly not unchanged by rotation [Allen and Tildesley, 2017].

For an arbitrary choice there will be in general a net linear and angular momentum $\boldsymbol{p}_T$ and $\boldsymbol{J}$, respectively. These should be subtracted out, e.g.

$$\boldsymbol{p}_i := \boldsymbol{p}_i - \boldsymbol{p}_T/N, \tag{9.45}$$

where we use the symbol :=' to indicate that the variable on the LHS is replaced by that on the RHS, so that the box does not, so to speak, move around during the computation. We note that the MD methodology as described up to now has been purely mechanical and deterministic based, as we have seen, on setting up and solving the equations of motion for a system of N particles in a box isolated from the outside (microcanonical conditions, Section 4.5). We could say that, given adequate computers, Newton could have done it! However, the temperature of the system was not set or even defined until now, and this is the point at which statistical mechanics has to enter. Indeed, temperature can be obtained from the translational kinetic energy K_{tr}, using the equipartition theorem of statistical mechanics (see, e.g., [Pathria and Beale, 2011]) which assigns to every degree of freedom a $k_B T/2$ contribution. Thus,

$$K_{\text{tr}} = \frac{1}{2}\sum_i m_i \left\langle \dot{\boldsymbol{r}}_i \cdot \dot{\boldsymbol{r}}_i \right\rangle = \frac{1}{2} N k_B T_{\text{tr}} n_{\text{tr}}, \tag{9.46}$$

where n_{tr} is the number of translational degrees of freedom, from which we can obtain a posteriori a 'translational temperature', T_{tr}. Quite similarly, if our particles are anisotropic rigid bodies, we have for the rotational kinetic energy K_{rot}

$$K_{\text{rot}} = \frac{1}{2}\left\langle \sum_i \dot{\boldsymbol{\Omega}}_i \cdot \mathbf{I}_i \cdot \dot{\boldsymbol{\Omega}}_i \right\rangle = \frac{1}{2} N k_B T_{\text{rot}} n_{\text{rot}}, \tag{9.47}$$

where n_{rot} is the number of rotational degrees of freedom. If $N \gg 1$, we can neglect the number of conserved quantities that should be subtracted. Obviously for a system at equilibrium, we should have $T_{\text{rot}} = T_{\text{tr}} = T$.

In the first part of the simulation (the equilibration stage), the temperature needs, in general, to be adjusted. For instance, we may expect that if, in the arbitrarily chosen initial configuration, some particles are too close to each other, their high repulsive potential energy will be transformed into kinetic energy as the particles are repelling each other, causing them to rapidly burst away. If this is the case, the kinetic energy and temperature will jump to values very high and unrealistic. Controlling the temperature to a desired one, T_0, can be done by multiplying the velocity of each particle by a scaling factor $\lambda = \sqrt{(T_0/T)}$, where T is the actual, current, temperature [Rapaport, 2004]. The method seems very drastic (it is in a way) but amounts to a realization of the often invoked concept of a *thermal bath* in thermodynamics, i.e. of a reservoir at temperature T_0, which instantaneously adjusts the temperature of a system with which it is put in thermal contact without modifying its own temperature (the classic example is a bathtub full of water at T_0 while the system to be thermostated could be constituted by a small test tube immersed in it). Normally a series of these adjustments will be necessary since, of course, the temperature is not a constant of motion. After this preparation stage the system is left to equilibrate for an adequate number of time steps (a few thousand, say). As a check of the reliability of the integration, the constants of motion such as energy and angular momentum are monitored. They should not drift systematically but be conserved. The temperature will not be constant at the desired value but should only oscillate about its time average without a systematic drift. The mean square amplitude of the equilibrium temperature fluctuations is an indication of the heat capacity of the system. This has been shown, for spherical particles, by Lebowitz et al. [1967], who obtained some general equations for relating averages obtained in the microcanonical N, V, E ensemble to the canonical ones. These are (correct to order $\mathcal{O}(1/N)$) for a generic property A,

$$\langle A \rangle_{NVE} = \langle A \rangle_{NVT} - \frac{1}{2} k_{\text{B}} T^2 \frac{\partial}{\partial T}\left((N C_V)^{-1} \frac{\partial A}{\partial T} \right) \tag{9.48}$$

and for fluctuations

$$\langle \Delta A^2 \rangle_{NVE} = \langle \Delta A^2 \rangle_{NVT} - \frac{k_B T^2}{N C_V} \left(\frac{\partial A}{\partial T} \right)^2. \tag{9.49}$$

Thus, for mean square fluctuations in temperature under the microcanonical conditions characteristic of an MD simulation:

$$\frac{\langle (\Delta T)^2 \rangle}{T^2} = \frac{2}{3N} \left(1 - \frac{3k_B}{2C_V} \right). \tag{9.50}$$

The requirement of constant total energy in the MD ensemble implies also that the fluctuations in temperature, and thus in kinetic energy, have to be equal to the potential energy fluctuations. For LC systems this may imply [Zannoni and Guerra, 1981] that large fluctuations in temperature also cause large fluctuations in the order parameters since these determine the internal energy.

9.5 Constant Temperature Molecular Dynamics

The simplest method for keeping a constant temperature is that proposed by Woodcock [1971] and Rahman and Stillinger [1971], which amounts to scaling velocities, as we have seen in the previous section, not only during equilibration but also at every time step during production. This very simple *velocity scaling* method, although generating a constant kinetic energy (rather than canonical) ensemble, is often used and even recommended [Fincham and Heyes, 1985] because it is inexpensive and numerically stable [Abraham, 1986]. Much more rigorous methods for canonical ensemble thermostats have been proposed but there is no evidence to suggest that they produce different results if one accounts for the noise inherent in any MD experiment [Fincham and Heyes, 1985]. In any case, the method, which does not require additional user-defined parameters, as in other methods, is a very useful tool that can hardly give wrong structural results.

9.5.1 Berendsen Thermostat

A less drastic approach proposed by Berendsen et al. [1984] is the weak-coupling thermostat using a factor that depends on the deviation of the instantaneous kinetic energy K from the average value K_0, corresponding to desired temperature T_0. At each time step velocities are rescaled by the factor λ_B:

$$\lambda_B = \sqrt{1 + \frac{\Delta t}{\tau_T} \left(\frac{K}{K_0} - 1 \right)}, \tag{9.51}$$

where Δt is the MD time step and τ_T is a parameter with the dimension of a time defining the coupling with the thermostat. The method does not reproduce a canonical ensemble, as the condition of constant average kinetic energy does not correspond to the condition of constant temperature, i.e. the fluctuations of the temperature and kinetic energy follow different laws. Thus, this method leads to trajectories whose average values correspond to the ones of the canonical ensemble, but whose fluctuations do not [Fincham and Heyes, 1985; Frenkel and Smit, 2002].

9.5.2 Langevin Thermostat

In this approach the modification of the particle velocities to achieve the desired constant temperature is achieved assuming a stochastic process, according to the Langevin equation for each particle:

$$m_i \frac{\mathrm{d}\dot{\boldsymbol{r}}_i(t)}{\mathrm{d}t} = \boldsymbol{f}_i(t) - m_i \gamma_i \dot{\boldsymbol{r}}_i(t) + \boldsymbol{f}_i^{\mathrm{st}}(t), \tag{9.52}$$

where γ_i is a friction coefficient determining the strength of the coupling of atom i to the thermal bath, while $\boldsymbol{f}_i^{\mathrm{st}}(t)$ is a stochastic force, a Gaussian distributed stochastic variable with 0 mean, $\langle \boldsymbol{f}_i^{\mathrm{st}}(t) \rangle = 0$, and a Dirac delta time correlation

$$\left\langle \boldsymbol{f}_i^{st}(t) \cdot \boldsymbol{f}_i^{st}(t+\tau) \right\rangle = 6 m_i \gamma_i k_B T_0 \, \delta(\tau) \, \delta_{i,j}. \tag{9.53}$$

T_0 is the temperature of the virtual thermal bath and has no correlation with prior velocities. The friction coefficient should be small enough to avoid moving from the Newtonian deterministic regime to an overdamped, Brownian diffusion one.

9.5.3 Nosè–Hoover Thermostat

This algorithm generates a canonical ensemble making use of the extended Lagrangian technique: the coupling with an external degree of freedom is performed by adding additional coordinates to the classical Lagrangian. The method proposed by Nosé [1984] and Hoover [1985] introduces an additional degree of freedom η, describing the external bath, and its time derivative $\dot{\eta}$ corresponding to a pseudo friction term used to scale particle velocities. The following equations of motion are obtained:

$$\dot{\boldsymbol{r}}_i = \frac{\boldsymbol{p}_i}{m_i} \tag{9.54a}$$

$$\dot{\boldsymbol{p}}_i = \boldsymbol{f}_i - \dot{\eta} \boldsymbol{p}_i \tag{9.54b}$$

$$\ddot{\eta} = \frac{1}{M_f} \sum_{i=1}^{N} \frac{p_i^2}{2m_i} - \frac{3}{2} N k_B T_0. \tag{9.54c}$$

where M_f is a fictitious mass parameter determining the change of the friction η when the kinetic energy and thermal equipartition value at the target temperature T_0 in Eq.9.54c deviate from each other.

The whole system, that contains all "real" degrees of freedom plus η, is conservative and obeys Liouville equation. It can be shown by direct substitution that the canonical distribution $\varrho \propto \exp[-(K+U)/(k_B T)]$ is its stationary, time independent solution. Therefore, configurations sampled by this algorithm represent canonical ensemble, even though the method can be difficult to tune. Differently from the former thermostats, this is an integral thermostat, with the instantaneous values of η and $\dot{\eta}$ depending on all previous states of the system.

9.6 Constant Pressure Molecular Dynamics

Molecular dynamics can be performed under external conditions corresponding to the different ensembles introduced in Chapter 4, in particular the important (N, P, T) one. To do this, the first step is to obtain the system pressure tensor $\mathbf{\Pi}$ using the virial expression introduced in Chapter 4 written as sum of the kinetic energy contribution (ideal gas contribution, always positive) plus the interparticle energy contribution. The pressure P is then calculated from the trace of the pressure tensor $\mathbf{\Pi}$ (see Eq. 4.135). If a cut-off scheme is used, the virial is calculated from the pairwise forces instead of being calculated from the total force acting on each particle (see, e.g., [Paci and Marchi, 1996]).

The procedures for pressure control generally mimic the ones derived for thermostats: in particular, one of the most used barostats is the weak-coupling barostat as well as the more specific one by Parrinello and Rahman [1981]. The weak-coupling scheme can be applied to couple the system to a 'pressure bath' [Berendsen et al., 1984]. Once fixed the desired external pressure P_{ext}, the task can be accomplished by periodically rescaling all centre of mass coordinates and box size, either isotropically or anisotropically, following a first-order relaxation law. In the isotropic case, this is just

$$\frac{\mathrm{d}P}{\mathrm{d}t} = \frac{P_{\text{ext}} - P}{\tau_P}, \tag{9.55}$$

$$P(t + \Delta t) = P(t) + \big(P_{\text{ext}} - P(t)\big)\frac{\Delta t}{\tau_P}, \tag{9.56}$$

where τ_P is the pressure coupling time constant. The coordinate scaling factor μ is given by:

$$\mu = \left[1 + \frac{\Delta t}{\tau_P}\kappa_T(P - P_{\text{ext}})\right]^{\frac{1}{3}}, \tag{9.57}$$

and that for the volume by μ^3, with κ_T, ideally, the experimental isothermal compressibility of the system. When the latter is not known, it is common practice to use another reference compressibility (e.g. that of water), since κ_T influences only the pressure fluctuations frequency and not the pressure itself, and many liquids have similar values.

To obtain an anisotropic coupling and eventually run a simulation with a non-cubic or even non-orthogonal shaped box, which could be important for LC fluids made of strongly anisometric molecules, one should deal [Parrinello and Rahman, 1981] with the 3×3 $\mathbf{H}$ matrix, already introduced in Chapter 8 (see Eq. 8.12), whose elements are the components of the vectors defining the simulation cell and whose determinant is the cell volume. An approximate cell relaxation matrix $\mathbf{M}$ can be obtained from the pressure tensor $\mathbf{\Pi}$:

$$\mathbf{M} = \left[\frac{\kappa_T}{\tau_P}(\Pi - P_{\text{ext}}\mathbf{I})\right]. \tag{9.58}$$

The new **H** matrix is given by: $\mathbf{H}(t+\Delta t) = \mathbf{H}(t) + \mathbf{M}\,\mathbf{H}(t)\Delta t$. The coordinates scaling is then accomplished as follows:

$$r(t+\Delta t) \;=\; \mathbf{H}(t+\Delta t)\,\mathbf{H}^{-1}(t)\,r. \tag{9.59}$$

A treatment of the generalized MD equations of motion and their integration is clearly outside the scope of this simple introduction, but a detailed discussion is given, for instance, by Tuckerman [2010]

From a technical point of view, using the Parrinello–Rahman technique, the computational box can undergo a spurious, physically insignificant overall rotation [Nosé and Klein, 1983]. The problem can be overcome by fixing the absolute orientation of the box in the laboratory system used [Ferrario and Ryckaert, 1985].

In the extended Lagrangian formulation of Andersen [1980], each element of the matrix **H** is treated as an extra degree of freedom, and a barostat mass is introduced, that has the same effect of the τ_P of the weak-coupling scheme. The method works with a fully anisotropic box; to simulate more symmetric boxes it is necessary to impose some constraints on the evolution of **H**. This type of barostat can be used together with the Nosé–Hoover thermostat, i.e. using a single extended Lagrangian, leading to the so-called Nosé–Parrinello–Rahman scheme; the best choice of the associated masses is the one that allows coupling between the relative vibrational modes of the system.

As in the case of the thermostats, only the extended Lagrangian barostat is able to reproduce correctly the observables fluctuations proper of the (N, P, H), or the (N, P, T), ensemble; on the other hand, the correct average values can also be obtained with the simpler weak-coupling barostat [Paci and Marchi, 1996].

9.7 Calculation of Static and Dynamic Properties

The mean value of a quantity A can be calculated as time averages from the configurations obtained after the system has evolved to equilibrium

$$\langle A\rangle \;=\; \lim_{\tau\to\infty}\frac{1}{\tau}\int_0^{\tau} \mathrm{d}t\,A(t) \approx \frac{1}{M_J}\sum_{J=1}^{M_J} A(t_J). \tag{9.60}$$

Thus, in Molecular Dynamics we do not calculate ensemble averages, as in the Monte Carlo method (Chapter 5), but we assume that in its evolution the system explores, given enough time, all phase space or at least the part of it relevant for our study (e.g. all the translational and rotational motion needed for reaching equilibrium). From a practical point of view, the calculation of static properties (energy, order parameters, pair distribution, etc.) is obtained as an average over a significant number of M_J configurations (Eq. 9.60). Since in MD the evolution process is deterministic, immediately successive states are unavoidably correlated, so that it is normally worth averaging over states J that are sufficiently spaced in time to be considered independent.

9.8 Hybrid Molecular Dynamics-Monte Carlo Methods

We have up to now considered Monte Carlo and Molecular Dynamics as two completely independent methods. However, combining them in a hybrid methodology can be quite powerful. One way to do this is to consider that in MC we describe updating configurations as a Markov process, where a new candidate configuration is generated by moving one particle at a time. While this is fine in many cases, it can also be problematic in cases of particle clusters difficult to unlock (see Fig. 8.5b). Moreover, while it is practically very convenient to move one particle at a time, there is no prescription in the Metropolis MC procedure that this is mandatory. We could thus consider MD as a way of producing collective N particles updates and select a configuration from a MD trajectory as a candidate configuration subject to the usual MC acceptance test.

9.8.1 Hamiltonian Exchange Replica Method

In the hybrid replica methods several independent MD or MC simulations are run in parallel, each at a different temperature [Sugita and Okamoto, 1999; Lin and van Gunsteren, 2015] or using a different Hamiltonian $\mathscr{H}(\widetilde{\boldsymbol{X}}, \dot{\widetilde{\boldsymbol{X}}})$ [Berardi et al., 2009], as we shall describe here. After a chosen number of MC or MD time steps or exchange time period τ_{ex}, pairs of configurations (replicas) with different $\mathscr{H}$ are exchanged with a transition probability governed by the detailed balance condition.

The acceptance probability for the Hamiltonian replica exchange can be realized [Fukunishi et al., 2002; Liu et al., 2005] by considering two replicas with (slightly) different potentials $U_n(\widetilde{\boldsymbol{X}}_n)$ and $U_m(\widetilde{\boldsymbol{X}}_m)$, where $\widetilde{\boldsymbol{X}}_n$ and $\widetilde{\boldsymbol{X}}_m$ represent the configurational coordinates for the replicas n and m, respectively. Such an example is that of soft-core GB potentials with different softness [Berardi et al., 2009]. The equilibrium probability (Boltzmann distribution) for the nth replica can be written as

$$P_n(\widetilde{\boldsymbol{X}}_n) = \exp\left[-U_n(\widetilde{\boldsymbol{X}}_n)/(k_B T)\right]/Z_n. \tag{9.61}$$

Now considering the transition probability, $\Pi(\widetilde{\boldsymbol{X}}_n, U_n; \widetilde{\boldsymbol{X}}_m, U_m)$ that the configuration $\widetilde{\boldsymbol{X}}_n$ in the nth replica exchanges with the configuration $\widetilde{\boldsymbol{X}}_m$ in the mth replica, the detailed balance condition (see Section 8.2) can be written as [Frenkel and Smit, 2002]

$$P_n(\widetilde{\boldsymbol{X}}_n)P_m(\widetilde{\boldsymbol{X}}_m)\,\Pi(\widetilde{\boldsymbol{X}}_n, U_n; \widetilde{\boldsymbol{X}}_m, U_m) = P_n(\widetilde{\boldsymbol{X}}_m)P_m(\widetilde{\boldsymbol{X}}_n)\Pi(\widetilde{\boldsymbol{X}}_m, U_n; \widetilde{\boldsymbol{X}}_n, U_m). \tag{9.62}$$

Substituting Eq. 9.61 into Eq. 9.62 the ratio of the transition probabilities can be obtained as

$$\frac{\Pi(\widetilde{\boldsymbol{X}}_n, U_n; \widetilde{\boldsymbol{X}}_m, U_m)}{\Pi(\widetilde{\boldsymbol{X}}_m, U_n; \widetilde{\boldsymbol{X}}_n, U_m)} = \exp(-\Delta_{nm}), \tag{9.63}$$

where $\Delta_{nm} = \left\{\left[U_n(\widetilde{\boldsymbol{X}}_m) + U_m(\widetilde{\boldsymbol{X}}_n)\right] - \left[U_n(\widetilde{\boldsymbol{X}}_n) + U_m(\widetilde{\boldsymbol{X}}_m)\right]\right\}/(k_B T)$. This in turn yields a Metropolis-type acceptance criteria for the transition

$$\Pi(\widetilde{\boldsymbol{X}}_n, U_n; \widetilde{\boldsymbol{X}}_m, U_m) = \begin{cases} 1 & \text{if} \quad \Delta_{nm} \leq 0, \\ \exp(-\Delta_{nm}) & \text{if} \quad \Delta_{nm} > 0. \end{cases} \tag{9.64}$$

Note that we have considered only the potential part of the Hamiltonian, since the kinetic part is the same in this case. An example of application of the method to LC problems involving a modified soft-core Gay–Berne potential has been shown [Berardi et al., 2009, 2011] to be a convenient way for accelerating simulations of anisotropic phases.

9.9 Simulation Packages

In most practical cases Molecular Dynamics simulations are performed employing some more or less standard package as a basic MD 'engine' to generate trajectories, possibly with the proviso that FFs have to be adjusted and specific observables have to be calculated. Thus, we provide a small summary of the more common computer simulation packages suitable for molecular simulation studies of organic materials [Muccioli et al., 2014] that often come with an extensive support including manuals, tutorials, online forums. Moreover, the more popular codes are continuously maintained and updated, an important aspect to take into account as machines change architectures very often. It is worth noting that while the standard packages provide MD trajectories (as sequences of instantaneous configurations), the calculation of many observables of interest have to be specifically added by the end user and suitable algorithms have often to be devised. The available MD packages differentiate according to many factors:

(i) Features and capabilities, e.g. multiple timescale integration algorithms [Tuckerman et al., 1992; Frenkel and Smit, 2002; Leimkuhler and Matthews, 2015], representation of simulated objects (coarse grained or all-atoms), constrained dynamics (e.g. fixing bond lengths with SHAKE [Ryckaert et al., 1977]) with many codes allowing for multiple options.
(ii) Licence and cost, i.e. free academic, open source, commercial.
(iii) Portability, i.e. how easily the code can be compiled and run on different platforms, from very common desktop computers and workstations to computer clusters or even specifically designed hardware.
(iv) Performance and parallelization: the speed of a single processor has not increased much in recent years, thus the actual performance of the different computer codes is often due to the possibility of efficiently splitting work among multiple processors. Nowadays the support of GPU boards is very important and allows to deploy graphics cards alongside traditional CPUs.
(v) Extensibility of the code: the possibility of adding features which were not available or even foreseen in the original code in order to tackle specific problems or to implement, for instance, new Force Fields for different materials, improved time evolution algorithms, or methodologies for the evaluation of different observables.

The choice of a particular MD package depends on the first instance on the system being studied, particularly on the model used to describe the interactions between the particles and on its scalability (i.e. the computational efficiency as the number of processing units increases), which can be a limiting factor when the system size exceeds tens or even hundreds of thousand of particles. A necessarily partial list of computer programs to carry out

Table 9.1. *A small list of important simulation packages. Here C: Commercial, FA: Free Academic, F/C: Free and Commercial versions. Refs: (a) [Harvey et al., 2009], (b) [Cornell et al., 1995], (c) [Brooks et al., 2009], (d) [Bowers et al., 2006], (e) [Smith, 2002], (f) [Limbach et al., 2006], (g) [Berendsen et al., 1995], (h) [Morozov et al., 2011] (i) [Plimpton, 1995], (j) [Phillips et al., 2005], (k) [Doi, 2003], (l) [Eastman et al., 2013], (m) [Matthey et al., 2004], (n) [Rackers et al., 2018], (o) [Lagardere et al., 2018], (p) [Rühle et al., 2009]*

Name	Website	Licence	Ref.
ACEMD	**http://multiscalelab.org/acemd**	C	a
AMBER	**http://ambermd.org/**	F/C	b
CHARMM	**www.charmm.org/**	C	c
Desmond	**www.deshawresearch.com/**	FA	d
DL_POLY	**www.ccp5.ac.uk/DL_POLY/**	FA	e
ESPResSo	**http://espressomd.org/**	F	f
GROMACS	**www.gromacs.org/**	F	g
HOOMD	**http://codeblue.umich.edu/hoomd-blue/**	F	h
LAMMPS	**http://lammps.sandia.gov**	F	i
NAMD	**www.ks.uiuc.edu/Research/namd/**	F	j
OCTA	**http://octa.jp/**	F	k
OpenMM	**https://simtk.org/home/openmm**	F	l
ProtoMol	**http://protomol.sourceforge.net/**	F	m
Tinker	**https://dasher.wustl.edu/tinker/**	F	n
Tinker-HP	**http://tinker-hp.ip2ct.upmc.fr/**	F	o
VOTCA	**www.votca.org/**	F	p

MD simulations is reported in Table 9.1. The ones most extensively used for liquid crystals and organic functional materials are probably:

NAMD [Phillips et al., 2005] is a simulation package which works with AMBER and CHARMM potential functions, parameters and file formats and that is specifically designed for high-performance simulation of large systems. Recent versions are is able to run on heterogeneous architectures made up of multiple CPUs and GPUs.

LAMMPS [Plimpton, 1995] is a classical MD code implementing potentials for soft materials (biomolecules, polymers, liquid crystals), solid-state materials (metals, semiconductors), allowing simulations at atomistic and coarse-grained particle level. The code is designed to be easy to modify or extend with new functionalities. Most of its model potentials have been parallelized and run on systems with multiple CPUs and GPUs, granting very good speed-ups, especially for the most complicated anisotropic pair potential styles, like the Gay–Berne and other CG model potentials.

GROMACS [Hess et al., 2008; Abraham et al., 2014] is conceived to carry out simulations with millions of particles, primarily for biochemical molecules like proteins, lipids and nucleic acids, but also for polymers and organic functional materials. The package is particularly fast at calculating the nonbonded interactions. GROMACS is freely available and continuously extended by a large user community. It also has the advantage of providing a large selection of tools for trajectory analysis. Recent versions provide support for single and multiple GPUs and hybrid Quantum-Classical (QM/MM) simulations.

The output of an MD simulation typically includes a trajectory file containing the position of every particle, saved with a given time increment. Trajectory files can be visualized with specific codes such as VMD [Humphrey et al., 1996; Eargle et al., 2006], which offers also basic tools for data analysis like Jmol [Jmol, n.d.; Scalfani et al., 2016], a powerful and highly portable program written in Java, Mercury [Mercury, n.d.] particularly well suited to visualize and manipulate crystal structures, and Avogadro [Avogadro, n.d.; Hanwell et al., 2012], which offers advanced molecular modelling tools. Among the other visualizers available, we mention here OVITO [OVITO, n.d.; Stukowski, 2010], PyMOL [Schrödinger, 2010], Molekel [Varetto, n.d.], RasMol [Sayle and Milner-White, 1995; Sayle, 2000], FOX [Favre-Nicolin and Cerny, 2004, 2007], QMGA [QMGA, n.d.; Gabriel et al., 2008], UCSF Chimera [UCSF, n.d.; Pettersen et al., 2004] and BALLView [BALLView, n.d.; Moll et al., 2006b, 2006a].

The vast choice of available packages might be a bit intimidating at first, so we offer a few practical comments on some of the key issues. Given the high quality of many open-source packages, they probably are to be preferred over commercial ones with closed sources, both for their abundance of features (as everybody can contribute to further development) and because the sources can be directly inspected to check how algorithms are implemented and to modify them to suit particular needs. As many algorithms work well only under specific conditions and many compromises are usually made, this is in some cases the most reliable way of checking the validity of the simulation results.

As long as small samples and/or short timescales are needed and the package provides the requested features, it really does not matter which code is used, and the one that is easier to run is to be preferred. If instead the problem at hand requires pushing to the limits of what can be accomplished currently, a well optimized code becomes the only choice. Given the typical speed-ups of GPU systems, codes able to run on heterogeneous architectures (mixed CPU/GPU) environments are currently amongst the best performing ones.

10
Lattice Models

10.1 Introduction

As we saw in Chapter 2, lattice models play an important role in the theory of phase transitions. Here we shall discuss in some detail these models with their applications to liquid crystals and, in particular, in the next section, some results for the LL model for nematics [Lebwohl and Lasher, 1972] introduced in Section 2.8.5. It may seem paradoxical, at first, to choose a model of liquid crystals where the molecules are just represented with unit vectors or quaternions (conventionally called 'spins') constrained on lattice sites, since a characteristic of nematics is that of being ordered fluids in which orientational order coexists with translational freedom. The resolution of the paradox is that when choosing a lattice model, the aim is not to try and reproduce all the properties of a real liquid crystal, but only its orientational properties. In any case, the main justification for studying lattice models lies clearly in the hope that suitably chosen models will correctly contain, despite their simplicity, the essential features of symmetry, dimensionality and form of interaction needed to capture and reproduce the main features of nematics. If this proves to be the case, lattice models have the advantage that they can be studied in far greater detail, e.g. using larger number of particles, larger number of cycles, finer temperature grids, than atomistic or even off-lattice molecular models. Lattice systems of 10^4–10^5 spins are now standard to computer simulate and millions or many more can be employed if needed. In this perspective the problem is not if a lattice model is a reasonable representation of liquid crystals but rather if it can be satisfactory in simulating observed orientational properties such as ordering and topological defects as well as the nematic-isotropic phase transition. Numerical experiments on the LL and similar 3D systems have then a role somewhat similar to that of exactly soluble Ising type models in 2D since, even in the absence of analytic solutions, they can provide accurate results suitable for testing approximate theories. Possibly the conclusions reached for the model system will then be applicable to real systems, at least if the essential physics has been incorporated into the model. There is some hope that this can be the case for nematics. In particular, as we saw in Chapter 2, the behaviour of the order parameters, their critical exponents, etc. appears, near a second-order phase transition, to be if not universal, fairly independent on details such as the exact nature of intermolecular interactions and lattice structures. It is worth recalling, as an example, that an extremely simple representation of the gas-liquid transition is obtained with the lattice gas model [Stanley, 1971], where each lattice site can be empty or contain a particle and each particle interacts with its nearest neighbours only (Eq. 2.33). The agreement between the lattice gas coexistence curve and experiments for simple rare gases, nitrogen, etc., is remarkable, bearing in mind the simplicity of the model. We also note that the nematic isotropic transition, even though first order, is a very weak one, and many of its features are shared by the vast majority of nematics (see Section 2.9), even if the behaviour is not really universal.

10.2 Lebwohl–Lasher Model

Here we wish to consider in some detail the application of the Monte Carlo method to the Lebwohl and Lasher [1972] lattice model introduced in Section 2.8.5. In the LL model molecules (or rather tightly packed clusters of neighbouring molecules whose short-range order is assumed to be maintained through the temperature range examined [Pasini and Zannoni, 2000]) are represented by unit vectors $\boldsymbol{u}_i$ 'spins' placed at the sites of a simple cubic lattice and interacting (see Eq. 2.37) with a pair potential $U_{i,j} = -\epsilon_{ij} P_2(\cos \beta_{ij}) \equiv -\epsilon_{ij} P_2(\boldsymbol{u}_i \cdot \boldsymbol{u}_j)$, where ϵ_{ij} is a positive constant ϵ, for neighbouring sites i and j and 0 otherwise (see Fig. 2.20). Going back to the requirement that the basic physics and symmetries are kept into account, the LL potential in Eq. 2.37 has the appropriate symmetry for a microscopic model of nematogens. In particular, it is invariant for head-tail exchange (a coordinate inversion with respect to the centre of a particle) and for an arbitrary rotation. The last requirement is important and lattice models where orientations are quantized (e.g. Ising or Potts models) abandoning the continuous rotational symmetry and the possibility of changing molecular orientations of arbitrary angles produce significant differences in the order-disorder transition, as shown by simulations of Lasher [1972]. The model offers an opportunity of comparison with theoretical treatments such as the Mean Field Theory described in Chapter 7 [Maier and Saupe, 1960; Luckhurst, 1985], two-site cluster [Sheng and Wojtowicz, 1976] and three- and four-site cluster theory [Van der Haegen et al., 1980]. The model was originally studied by Lebwohl and Lasher [1972], using the Metropolis MC method discussed in Chapter 8, with periodic boundary conditions (PBCs) applied to the sample box to minimize surface effects. Over the years samples of various sizes have been studied. Lebwohl and Lasher [1972] and Jansen et al. [1977] have studied lattices from $10 \times 10 \times 10$ to $20 \times 20 \times 20$. Fabbri and Zannoni [1986], Cleaver and Allen [1991], Zhang et al. [1992, 1993] and Shekhar et al. [2012] have examined $30 \times 30 \times 30$ lattice systems also using other techniques. A series of larger lattice sizes, up to $30 \times 30 \times 70$ has been studied by Priezjev and Pelcovits [2001] with MC and histogram reweighting to examine the scaling behaviour of T_{NI} with size. A much larger system, $N = 200 \times 200 \times 200$, has also been studied Skačej and Zannoni [2021]. All these authors found that the model exhibits a weak first-order orientational phase transition. For a soft potential, lacking repulsive forces like the LL one, it is possible to reach virtually any ordered or disordered state from an arbitrary starting configuration (the system is *ergodic*). Indeed, a series of runs in a heating up or cooling down sequence lead to superimposable results except in the transition region, where some hysteresis, as typical of first-order transitions, appears. To increases the rate of approach to equilibrium, it remains of course convenient to start a new calculation from an equilibrium configuration at a nearby temperature. Given the simplicity and widespread use of the model, we shall use it as a case study to illustrate methods useful also for other LC systems.

10.2.1 Locating the Phase Transition

Energy. As we saw in Chapter 2 the most common experimental way of locating a thermal phase transition is by monitoring the heat capacity, i.e. at constant volume, the energy derivative (Eq. 2.9) with respect to temperature and this applies also to computer simulations (Section 8.4). The average energy for the LL lattice system with PBC is

$$\mathcal{U}_N = -\frac{1}{2}\sum_{\substack{i,j=1\\ j\neq i}}^{N} \epsilon_{ij}\left\langle P_2(\cos\beta_{ij})\right\rangle = -\frac{1}{2}Nz\epsilon G_2(a), \tag{10.1}$$

where the coupling constant $\epsilon_{ij} = \epsilon$, $\epsilon > 0$ if i, j are nearest neighbours and 0 otherwise, $z = 6$ is the number of nearest neighbours for a cubic lattice, a is the lattice spacing and $G_2(a) = \sigma_2 = \langle P_2(\cos\beta_{12})\rangle$ is the nearest-neighbours value of the second-rank pair correlation (cf. Eq. 4.97). The zero-temperature limit for the dimensionless single-particle energy $\mathcal{U}^* = \mathcal{U}_N/(N\epsilon)$ is just $z/2$ or 3 for the cubic lattice. Lebwohl and Lasher [1972] have calculated the high temperature expansion as $\mathcal{U}^* = -3/(5T^*) - 3/(35T^{*2}) + \cdots$ and the low temperature expansion as $\mathcal{U}^* = -3 + T^* + \cdots$. The results of an MC simulation for $\mathcal{U}^*$ against reduced temperature $T^* = k_B T/\epsilon$ [Fabbri and Zannoni, 1986] are shown in Fig. 10.1a. We see clearly that a sharp change of slope occurs around $T^* = 1.11$, suggesting the onset of a phase transition.

Heat capacity. The heat capacity C_V^* can be evaluated differentiating $\mathcal{U}^*$ with respect to T^* as discussed in Section 8.4. The LL results are shown in Fig. 10.1b for various lattice sizes, while the errors in C_V^* have been estimated by sampling the energy from inside its error bar and repeating the differentiation procedure a number of times. In Fig. 10.1b we see, as an example, the size dependence of the heat capacity curve obtained for a system of $N = 5^3$, 10^3 and 30^3 particles. Judging from the width of the C_V^* peak and scaling the temperature to that of a real room temperature nematic we see that for the 10^3 system the uncertainty is ±15K, which seems huge considering the simplicity of the model. Note that in studies of liquid crystals the transition region is very important and that it has to be investigated. Fortunately the effect, so significant for lattice models, may be somewhat less dramatic for systems where particles have more degrees of freedom, like off-lattice molecular [Berardi et al., 1993] and atomistic [Tiberio et al., 2009] systems (see Chapters 11 and 12). The order of the transition is rather difficult to evaluate, and the task is made more complicated by the fact that a true first-order transition cannot take place in a finite size system. It is, however, very useful to examine not just the average energy values but also the histograms of the frequency of occurrence of a certain energy value, $P(\mathcal{U}^*)$, during the simulation, i.e. looking at the distribution of instantaneous observable values that concur to yield the average. We would expect the histogram to be Gaussian away from the transition, while at the first-order transition two peaks, corresponding to the two coexisting states, should be observed

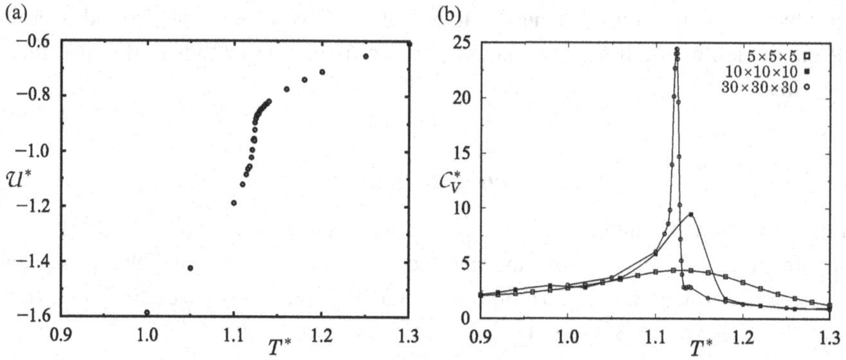

Figure 10.1 (a) The average dimensionless energy per particle $\mathcal{U}^* \equiv \langle U \rangle / N\epsilon$ vs T^* [Fabbri and Zannoni, 1986] and (b) the size dependent heat capacity $C_V^* = \mathrm{d}\mathcal{U}^*/\mathrm{d}T^*$ for the $5 \times 5 \times 5$, $10 \times 10 \times 10$, $30 \times 30 \times 30$ lattices. The lines are a guide for the eye.

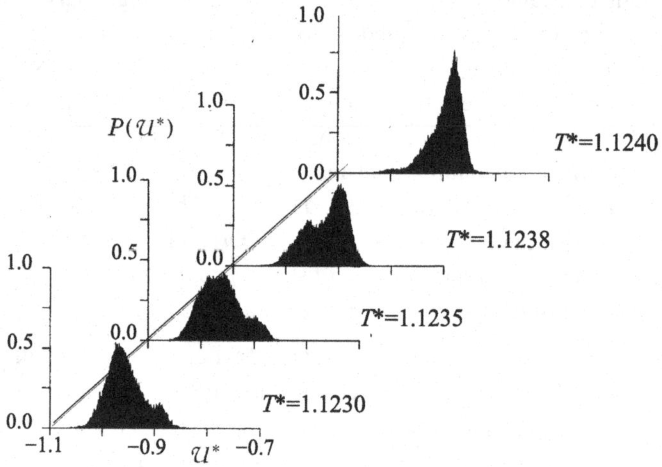

Figure 10.2 Histograms of the distribution of energy values $P(\mathcal{U}^*)$ obtained for four temperatures T^* in the transition region of the LL model [Fabbri and Zannoni, 1986].

according to Landau theory (Section 2.7.2). In general, the distributions consist of extremely narrow symmetric peaks, but the behaviour changes and becomes particularly interesting near the transition. In Fig. 10.2 we show four of these histograms in the temperature region $1.121 < T^* < 1.124$ that clearly present two peaks whose relative intensity changes with temperature, indicating that the transition is situated in this relatively narrow range. It is of course desirable to have some metrics associated with the histograms' shape and it is convenient to resort to quantities used in statistics for this purpose, i.e. moments and cumulants [Abramowitz and Stegun, 1965]. The nth *central moment* of the distribution $P(A)$ of a property A, $m_n^{(A)}$, is

$$m_n^{(A)} = \sum_j \left(A_j - \langle A \rangle\right)^n P(A_j), \tag{10.2}$$

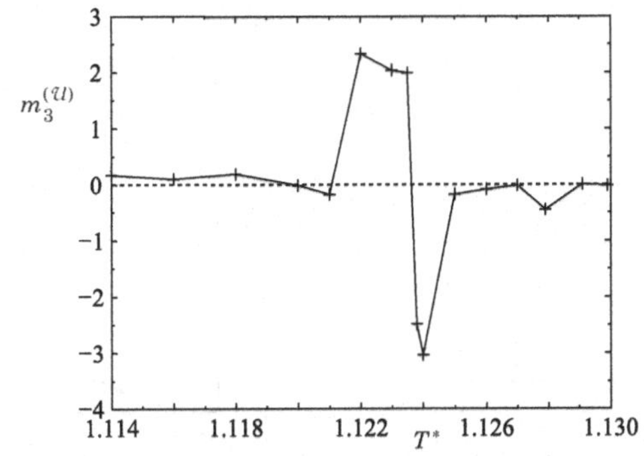

Figure 10.3 A close up of the third moment of the energy distribution, $m_3^{(\mathcal{U})}$ as a function of temperature T^*. The line through the points is just a guide for the eye [Fabbri and Zannoni, 1986].

where $P(A_j)$ is the population of the jth histogram bucket. It is convenient to also introduce the cumulants k_n of the same distribution defined in terms of the moments. In particular, for the first four cumulants of $P(A)$ we have the definitions $k_1^{(A)} = \langle A \rangle$, $k_2^{(A)} = m_2^{(A)} = \sigma_A^2$, $k_3^{(A)} = m_3^{(A)} = s_A\,(\sigma_A)^3$ and $k_4^{(A)} = m_4^{(A)} - 3\big(m_3^{(A)}\big)^2$. The first cumulant gives the centre and the second the variance, square of the standard deviation σ_A, while the third quantifies the asymmetry about the centre (it is proportional to the so-called *skewness* s_A). For a Gaussian distribution cumulants above the second are 0. In particular, $k_4^{(A)}$, called *kurtosis*, can be considered a measure of the non-Gaussianicity: it is positive for distributions with heavy tails and a peak at 0, and negative for flatter densities with lighter tails. An examination of the first four cumulants k_3, k_4 in comparison to k_2 [Fabbri and Zannoni, 1986] shows that the peaks are essentially Gaussian except near the transition, where the width of the peaks becomes about twenty times larger than that of the peaks in the isotropic phase. In the same region the third cumulant (Fig. 10.3) significantly differs from 0 and changes sign across the transition helping us to determine its precise location. The orientational transition temperature of the $30 \times 30 \times 30$ LL model was estimated to be $T_{NI}^* = 1.1232 \pm 0.0006$.[1] We also note that the heat capacity peak is asymmetric and much steeper on the hot (isotropic) side of the transition. We have seen in Chapter 2 that at a secoind-order transition the temperature dependence of various properties as they approach the transition can be characterized by critical exponents. The NI transition is a weak first-order one which appears to intercept a very close second-order one. Pseudo-critical exponents for C_V^* versus T^* can be estimated by non-linear least square fitting to the expression (cf. Chapter 2)

$$C_V^* = A\left[\frac{T^*}{T_{NI}^*} - 1\right]^{-\alpha} + B. \tag{10.3}$$

[1] For the $200 \times 200 \times 200$ LL model the estimate is $T_{NI}^* = 1.1224^{+0.00007}_{-0.00003}$ [Skačej and Zannoni, 2021]

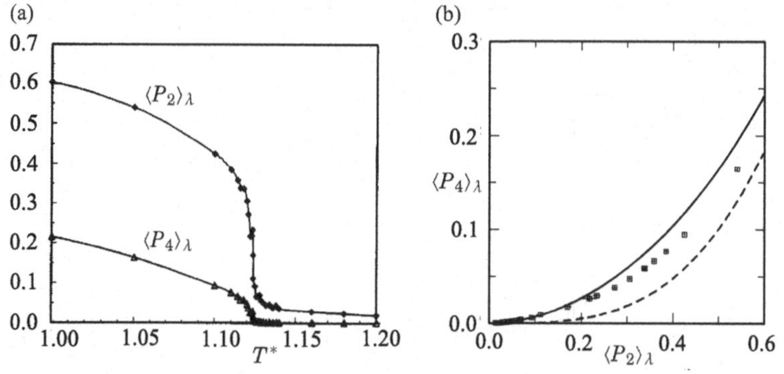

Figure 10.4 (a) The second- and fourth-rank order parameters $\langle P_2\rangle_\lambda$ and $\langle P_4\rangle_\lambda$ vs temperature T^* with the lines as a guide for the eye and (b) plotted one against the other. Also in part (b), the dashed curve is $\langle P_2\rangle^{(10/3)}$ as from the continuum theory of Faber [1980], while the solid line is obtained from the Maier–Saupe-like distribution in Eq. 3.74 [Fabbri and Zannoni, 1986].

Assuming the T^*_{NI} previously determined, the data can be fitted using $A = 0.363$, $B = -1.44$, $\alpha = 0.596$. We recall that experimentally values of α very close to 0.5 have been found (cf. Chapter 2).

Second- and fourth-rank order parameters. The order parameters for the LL model, calculated using the techniques described earlier on (Section 3.5) are shown in Fig. 10.4. In practice, the **Q** matrix, Eq. 3.47, is calculated and diagonalized every few cycles and its largest eigenvalue is averaged to give $\langle P_2\rangle_\lambda$, the order parameter for the system with respect to the instantaneous director (cf. Eq. 3.52) shown in Fig. 10.4a. The curve shows a smooth decrease with T^* becoming very steep in the transition region. We note that $\langle P_2\rangle_\lambda$ does not go to 0 above the transition, but rather to a value of the order of $1/\sqrt{N}$. This is of course to be expected since $\langle P_2\rangle_\lambda$, corresponding to the largest eigenvalue of a traceless matrix, is always non-negative. A better estimate of the order parameter in the isotropic phase is given by the intermediate eigenvalue of **Q**, $\langle P_2\rangle_{\lambda_2}$, that can fluctuate around 0 [Eppenga and Frenkel, 1984]. However, the problem is conceptual more than numerical and depends on the absence of a bona fide director above the orientational transition.

The fourth-rank order parameter $\langle P_4\rangle$, obtained as described in Section 3.5, is also shown as a function of $\langle P_2\rangle$ in Fig. 10.4b with a simple approximation, i.e. $\langle P_4\rangle = \langle P_2\rangle^{\frac{10}{3}}$, obtained from Faber's [1980] continuum theory of disordering by fluctuations. Although this represents well $\langle P_4\rangle$ versus $\langle P_2\rangle$ at high order, it largely underestimates $\langle P_4\rangle$ when $\langle P_2\rangle < 0.6$. A better approximation comes from $\langle P_4\rangle = \frac{5}{7}\langle P_2\rangle^2$, obtained from a $P(x) \propto \exp(a P_2(x))$ distribution, Eq. 3.74, and consistent also with that obtained by the Maier–Saupe MFT (Chapter 7).

Pair properties. In practice, we use the following algorithm to evaluate the orientational correlations defined in Chapter 4. Choose a total of M particles as origins, then pick an origin on particle i with orientation $\boldsymbol{u}_i$ and consider spherical shells of a certain resolution Δr.

Count the number of particles n_{r_i} within each shell (for a lattice this needs to be done just once beforehand). Calculate, for each particle j falling within the shell, the Legendre polynomial $P_L(\boldsymbol{u}_i \cdot \boldsymbol{u}_j)$. Repeating, for all the particles in the sample and for all M particles that we take as origins, gives the total average at r:

$$G_L(r) = \langle P_L(\cos\beta_{ij})\,\delta(r_{ij} - r)\rangle = \frac{1}{M}\sum_{i=1}^{M}\frac{1}{n_{r_i}}\sum_{j\neq i}^{N} P_L(\boldsymbol{u}_i \cdot \boldsymbol{u}_j)\Delta(r_{ij} - r), \qquad (10.4)$$

where the fuzzy delta function $\Delta(r_{ij} - r)$ is 1 when r_{ij} equal r within the chosen resolution and 0 otherwise. When using PBC, every distance dependent property is determined modulo the box length, i.e. as the remainder of the division between the distance and the box length. The correlations would tend to spuriously grow for distances exceeding half the box length. In particular, we have shown in Fig. 4.8 the angular pair correlation coefficient $G_2(r)$ at a number of temperatures from $T^* = 1.05$ to 1.20 for the $30 \times 30 \times 30$ LL model [Fabbri and Zannoni, 1986]. The build-up of the orientational correlation when approaching the transition from above is quite visible. If we concentrate on the right-hand side of the figure, we see that the correlation decays to 0 in the isotropic phase and to a plateau in the liquid crystal. The correlation functions $G_L(r)$, representing the expansion coefficients of the pair distribution, can be employed in investigating pretransitional behaviour.

Pretransitional behaviour. We saw in Section 4.11 that the susceptibility ξ of an isotropic fluid to an applied magnetic or electric field F can be written as

$$\xi = \frac{\lambda}{kT}g_2 = \frac{5}{N}\lim_{F\to 0}\sum_i\sum_j \langle P_2(\cos\beta_i)P_2(\cos\beta_j)\rangle_F, \qquad (10.5)$$

where g_2 is the second-rank Kirkwood coefficient [Ben-Reuven and Gershon, 1969; Luckhurst, 1988]. For a lattice

$$g_2 = \sum_k z_k G_2(r_k). \qquad (10.6)$$

Evaluation of g_2 with Eq. 10.6 has the advantage, with respect to a direct summation over the whole sample (Eq. 10.5), that spurious contributions to g_2 due to the build-up of the correlation at distances comparable to the box length are controllable by suitably truncating the sum. The divergence temperature of g_2 can then be determined by fitting T^*/g_2 versus T^*. However, the number of neighbours increases quadratically with distance and great care has to be taken to ensure that the results are not dependent on the arbitrary cut-off distance imposed. It is useful [Fabbri and Zannoni, 1986] to estimate the neglected tail contributions to g_2 by fitting a suitable analytical expression for the distance decay of $G_2(r)$ and use the analytic expression in conjunction with Eq. 10.7. For the LL model the Ornstein–Zernike form

$$G_2(r) = \frac{A}{r}\exp(-r/\xi_2), \qquad (10.7)$$

where ξ_2 is a correlation length, yields an excellent overall fit at the various temperatures above the transition with essentially the same parameters A and $k_c = 1/\xi_2$ obtained by fitting $G_2(r)$ truncated at different cut-offs (10, 15 and 20 lattice units). It is interesting to note that an Ornstein–Zernike form is expected theoretically [Chakrabarti and Bagchi, 2009] and that Landau theory (cf. Section 2.7) predicts that the correlation length squared should be linear in temperature. The MC results show that this is indeed the case. The temperature of divergence of the correlation length is found in this way to be $T^* = 1.1222 \pm 0.009$. To estimate a truncation correction, we can now substitute the Ornstein–Zernike form for $G_2(r)$ in Eq. 10.5 and extend the sum up to larger distances until convergence is reached. A linear regression of the corrected values for T^*/g_2 in temperature gives an estimate for the divergence temperature as $T_{NI^\star} = 1.1201 \pm 0.0006$, bringing it nearer to T_{NI}. Translating to the temperature scale for a room temperature nematic, $T_{NI^\star}$ would be about 1 degree below the transition, in agreement with what is found experimentally for MBBA [Stinson and Litster, 1970]. This result has been confirmed with a histogram technique by Zhang et al. [1992, 1993].

10.2.2 Molecular Dynamics Simulations

The Lebwohl–Lasher model can be slightly generalized to a set of symmetric top particles placed at the sites of a cubic lattice and interacting with the same pair potential. Molecular dynamic simulations on this dynamic LL models have been performed by Zannoni and Guerra [1981] and more recently by Chakrabarty et al. [2006]. The reduced units appropriate to the model and used later are $t^* = t\sqrt{\epsilon/I_\perp}$ and $T^* = k_B T/\epsilon$ for time and temperature, respectively. The integration of the equations of motion, written in quaternion form, was performed with a Runge–Kutta–Gill fourth-order procedure [Romanelli, 1960; Gear, 1971]. The MD method presents some advantages over the MC method because it explores the real-time trajectory. Thus, it can exploit the fact that the timescales for molecular and director reorientation are normally well separated; in a real system separations are usually larger than 3–4 decades. We can thus prepare the system with a certain director, e.g. choosing appropriately as already described the starting configuration, and let it reach orientational equilibrium, while preserving the average orientation. As long as this is true, we can calculate static molecular properties considering the director as essentially fixed in space. We can then obtain the full singlet orientational distribution discussed as a histogram, as shown in Fig. 3.4 for three temperatures: one in the isotropic and two in the orientationally ordered phase. We see that the distributions in the anisotropic phase strongly depend on the angle β but not on α which corresponds to rotations around the director, confirming the macroscopic uniaxiality around the director. The order parameters of rank L can then be calculated by direct integration as in Eq. 3.43. In this rather special case, it is not even necessary to calculate the full histogram if all we want is to calculate an order parameter. Thus, $\langle P_L \rangle_{\text{LAB}}$ can be simply calculated from an average over a sufficiently large number of M equilibrium configurations of the sample order parameter $\langle P_L \rangle_S$:

$$\langle P_L \rangle_{\text{LAB}} = \langle \langle P_L \rangle_S \rangle \equiv \frac{1}{M} \sum_{J=1}^{M} \langle P_L \rangle_S^{(J)} = \frac{1}{M} \sum_{J=1}^{M} \left[\frac{1}{N} \sum_{i=1}^{N} P_L(\cos \beta_i^{(J)}) \right]. \quad (10.8)$$

The method should be more accurate than that of integrating the histogram for the singlet orientational probability $P(\beta)$ over the Legendre polynomials, because of the limited

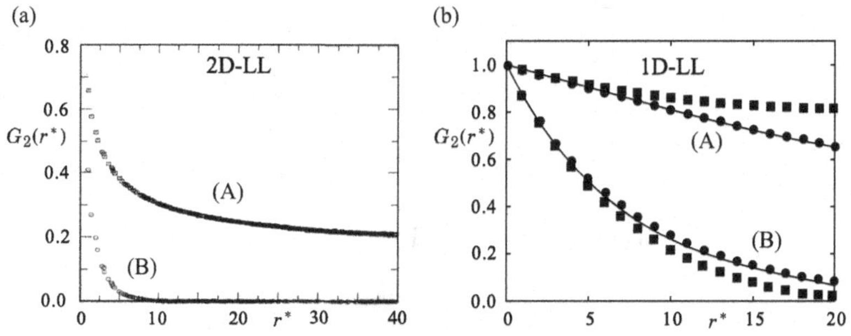

Figure 10.5 (a) The second-rank pair correlation coefficient $G_2(r^*)$ plotted vs distance r^* in lattice units for the planar LL model at temperatures $T^* = 0.54$ (A), 0.72 (B) from MC simulations of a $L \times L = 80 \times 80$ lattice [Chiccoli et al., 1988b]. (b) $G_2(r^*)$ vs r^* for the linear LL model with $L = 40$ (■) and $L = 1000$ (●) at temperatures $T^\star = 0.02$ (A) and $T^\star = 0.12$ (B) [Chiccoli et al., 1988a]. The continuous curves are the analytic results of Vuillermot and Romerio [1973a,1973b]

resolution in the angular grid that one needs to use in the histogram. The dynamic of the model was studied in detail [Zannoni and Guerra, 1981] and in Chapter 6 we saw its orientational time correlations functions of first, second and mixed rank (Fig. 6.6).

10.2.3 Space Dimensionality Effects: 2D and 1D Lattices

Systems of low dimensionality, e.g. monomolecular films forming a two dimensional layer, can be quite important, for instance as models of Langmuir-Blodgett films or of liquid crystalline patches on surfaces. The simplest interaction potential for these systems is still the Lebwohl–Lasher one, but with molecular centres confined to a planar square lattice. Since the molecules (or 'spins') can reorient in 3D we can call this a $n_d = 2, n_o = 3$ model, referring to lattice and spin dimensionality, while the ordinary 3D model we have studied up to now is a $n_d = 3, n_o = 3$. A 1D system ($n_d = 1, n_o = 3$) was also studied and solved analytically by Vuillermot and Romerio [1973b] who showed that it has no long-range order and the absence of a phase transition. It is interesting to see what an MC simulation [Chiccoli et al., 1987] would tell us in this case and if a false positive result is obtained. We start with the planar system, and we show in Fig. 10.5a the pair correlation coefficient $G_2(r^*)$, where $r^* = r/a$ is a dimensionless distance. The system should not have true long-range order. A non-linear least square fitting up to a certain cut-off length L_f of the second-rank correlation coefficients at various temperatures [Chiccoli et al., 1988b] can be performed to the following two expressions:

(i) exponential decay to a plateau A_e:

$$G_2(r^*) = (1 - A_e)\exp(-k_e r^*) + A_e, \tag{10.9}$$

(ii) power law decay to 0

$$G_2(r^*) = A_p/(r^*)^{k_p}. \tag{10.10}$$

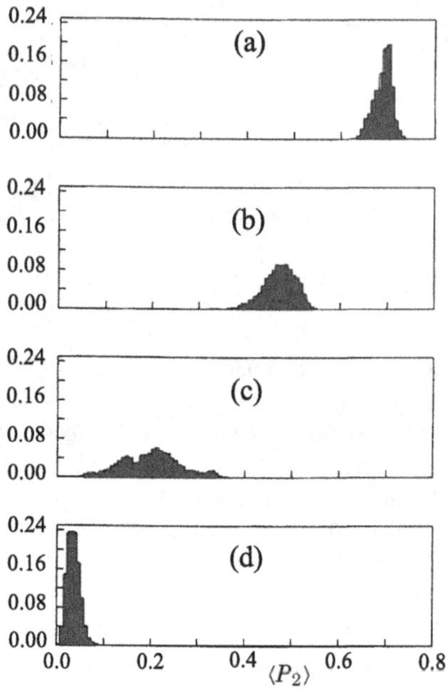

Figure 10.6 Histograms of frequency of occurrence of the second-rank order parameter $\langle P_2 \rangle_\lambda$ during simulations on the 80 × 80 LL lattice at temperatures $T^* = 0.40$ (a), 0.54 (b), 0.58 (c), 0.72 (d) [Chiccoli et al., 1988b].

We note that, different to the 3D model, when $T < T_{NI}$, the curve does not tail off to a plateau, at any temperature. The exponential decay law gives a better fit for the planar model results above T_C^* (with $A_e = 0$), while the algebraic decay law gives a better representation of the data below T_C^*, except near the order-disorder anomaly where the power law fit is rather poor. At temperatures below T_C^* the decay parameter $(1/k_p)$ increases regularly as the temperature decreases, tending to infinity at $T^* = 0$. Thus, the system presents some sort of 'long short-range order', even though not a truly long-range one. The histograms of the order parameter for the 2D LL model (Fig. 10.6) are quite different from those for the 3D model (Fig. 10.2): they do not show a double peak but just a single peak shifting to lower values as temperature increases, as expected for a second-order transition (Section 2.7.1). It is interesting to see if MC can give evidence of the absence of order in the 1D ($d = 1$, $n = 3$) linear system. The two-particle angular correlations $G_2(r)$ for a $N = 40$ and the $N = 1000$ linear system shows a very fast decay above the heat capacity anomaly, which becomes somewhat slower as the temperature decreases. In Fig. 10.5b we show $G_2(r)$ for two temperatures well below the heat capacity anomaly and the analytic results of Vuillermot and Romerio [1973b], i.e.

$$G_2(r) = \left[\frac{1}{2}\sqrt{\frac{3T^*}{2}} D^{-1} \left(\sqrt{\frac{3}{2T^*}} \right) - \frac{T^*}{2} - \frac{1}{2} \right]^r . \tag{10.11}$$

The size dependence of the heat capacity also shows [Chiccoli et al., 1988a] that the MC results, upon increasing size L, rapidly converge to the analytic result of Vuillermot and Romerio [1973a]:

$$\frac{C_v}{Nk_B} = \frac{1}{2} - \frac{3}{4T^*}\sqrt{\frac{3}{2T^*}}\left(\frac{1}{3}T^* - 1\right)D^{-1}\left(\sqrt{\frac{3}{2T^*}}\right) - \frac{3}{8T^*}D^{-2}\left(\sqrt{\frac{3}{2T^*}}\right), \quad (10.12)$$

where $D(x)$ is the Dawson function (Eq. 3.77). This indicates the absence of transition and is very different from the marked size dependence of the $n_d = 3, n_o = 3$ model (Fig. 10.1b). Thus, a careful analysis of the pair correlation function and of the heat capacity and of their size dependence, can give precise indications on the presence or not of a phase transition and of its character.

10.3 Biaxial Lattice Models

Biaxial nematics (N_B), fluids that possess two orthogonal preferred directions (directors) rather than the single one of standard nematics, were briefly introduced in Section 1.6. They have attracted considerable theoretical and experimental attention [Luckhurst and Sluckin, 2015] since their existence was predicted, decades ago, by Freiser [1970] for a fluid of biaxial particles interacting with a quadrupolar like potential. As most mesogenic molecules are biaxial, it would seem natural to find this N_B phase at least as frequently as the standard uniaxial one. On the contrary, the phase has proved to be extremely elusive and has long defied experimental attempts at preparing it (see [Luckhurst, 2001] for a review). The reasons for interest in N_B are both of fundamental and technological nature and start with the challenge to understand why a phase that was predicted by MFT (see Section 7.4) is so difficult to realize in practice. In particular, given that the approximations in MFT are known to enhance order, it is interesting to see if a minimal lattice model with biaxial symmetry can show biaxial phases when treated 'exactly' with computer simulations [Biscarini et al., 1995; Berardi et al., 2008b; Preeti et al., 2011]. Here we start from the most general purely orientational pair potential between two identical rigid particles with biaxial D_{2h} symmetry put forward by Straley [1974] and Luckhurst et al. [1975]. Assuming a 3D cubic lattice with biaxial particles, or 'spins', fixed at its sites and an isotropic average over the nearest-neighbours intermolecular vector directions $\hat{\boldsymbol{r}}_{ij}$, a biaxial model can be obtained as:

$$U_{ij} - \epsilon_{ij}\mathscr{R}^2_{0,0}(\Omega_{ij}) + 2u_{220}[\mathscr{R}^2_{0,2}(\Omega_{ij}) + \mathscr{R}^2_{2,0}(\Omega_{ij})] + 4u_{222}\mathscr{R}^2_{2,2}(\Omega_{ij}), \quad (10.13)$$

where the coupling parameter $\epsilon_{ij} \equiv [u_{200}]_{ij}$ is taken to be a constant $\epsilon > 0$ when i and j are nearest neighbours and 0 otherwise. $\mathscr{R}^L_{m,n}$ are combinations of the Wigner functions $\mathscr{D}^L_{mn}$ symmetry-adapted for the D_{2h} group of the two particles, as defined in Eqs. 3.150a–3.150d, which here are functions of the relative rotation angles from i to j, i.e. $\Omega_{ij} \equiv (\alpha_{ij}, \beta_{ij}, \gamma_{ij})$ [Biscarini et al., 1995]. The model reduces to the uniaxial LL potential [Fabbri and Zannoni, 1986] when $u_{220} = u_{222} = 0$. Using ϵ to renormalize temperature to $T^* \equiv k_B T/\epsilon$ the potential still contains two independent parameters, that make up a very large parameter space.

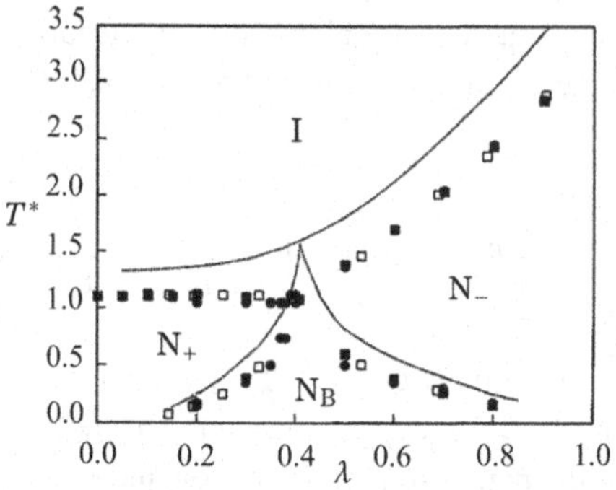

Figure 10.7 Dispersive biaxial model phase diagram showing the transition temperature T^* between the uniaxial (N_+, N_-) and the biaxial (N_B) nematic as well as isotropic (I) phases vs biaxiality λ. MC results for $10 \times 10 \times 10$ (■) and $8 \times 8 \times 8$ (●) as well as $10 \times 10 \times 10$ points mapped from (λ, T^*) onto $(\lambda', T^{*\prime})$ (□). The lines are the MFT results (see Fig. 7.6) [Biscarini et al., 1995].

10.3.1 Dispersive Biaxial Model

Assuming that this second-rank attractive pair potential is based on dispersion forces (Section 5.8.2), the coefficients in Eq. 10.13 can be written as $u_{2mn} = k_{ij}\alpha_i^{2,m}\alpha_j^{2,n}$, where $\alpha_i^{2,p}$, $\alpha_j^{2,p}$ are the spherical components of the polarizability tensor for molecules i, j and k_{ij} is a constant, so that $u_{220}/u_{200} \equiv \lambda$ and $u_{222}/u_{200} = \lambda^2$ [Luckhurst et al., 1975; Luckhurst and Romano, 1980a; Biscarini et al., 1991]. This gives (see Eq. 7.18),

$$U_{ij} = -\epsilon_{ij}\Big[P_2(\cos\beta_{ij}) + 2\lambda[\mathscr{R}^2_{0,2}(\Omega_{ij}) + \mathscr{R}^2_{2,0}(\Omega_{ij})] + 4\lambda^2\mathscr{R}^2_{2,2}(\Omega_{ij})\Big]. \quad (10.14)$$

The biaxiality parameter λ accounts for the deviation from cylindrical molecular symmetry: when λ is 0, the potential reduces to the LL P_2 potential, while for λ different from 0 the particles tend to align not only their major axis but also the other two. The value $\lambda = 1/\sqrt{6}$ marks the boundary between a system of prolate ($\lambda < 1/\sqrt{6}$) and oblate molecules ($\lambda > 1/\sqrt{6}$). This dispersive model has been simulated with the MC method by Luckhurst and Romano [1980a] and more extensively by Biscarini et al. [1995] who obtained the phase diagram shown in Fig. 10.7. The various orientational order parameters introduced in Chapter 3: $\langle P_2\rangle$, $\langle\mathscr{R}^2_{0,2}\rangle$, $\langle\mathscr{R}^2_{2,0}\rangle$ and $\langle\mathscr{R}^2_{2,2}\rangle$, can be obtained from computer simulations in a way similar to what we could do with real experiments introducing suitable virtual molecular observables and determining their average in the laboratory system [Biscarini et al., 1995]. Experimentally one would try to select a sufficient number of molecular properties $A^{(\alpha)}_{\text{MOL}}$ and measure their average values $\langle A^{(\alpha)}_{\text{LAB}}\rangle$. Then, through a diagonalization of the average tensors $\langle \mathbf{A}^{(\alpha)}_{\text{LAB}}\rangle$ one could determine their eigenvalues and from them the order parameters, as discussed in Section 3.11.1 and more in detail in Biscarini et al. [1995]. Here the procedure, applied to the three matrices $\mathbf{G} \equiv A^{(x)}_{\text{MOL}} = \boldsymbol{x} \otimes \boldsymbol{x}$, $\mathbf{H} \equiv A^{(y)}_{\text{MOL}} = \boldsymbol{y} \otimes \boldsymbol{y}$ and $\mathbf{F} \equiv A^{(z)}_{\text{MOL}} = \boldsymbol{z} \otimes \boldsymbol{z}$, gives the results shown in Fig. 10.8. These indicate that, even if MFT (Section 7.4) overestimates the transition temperatures, the topology of the phase

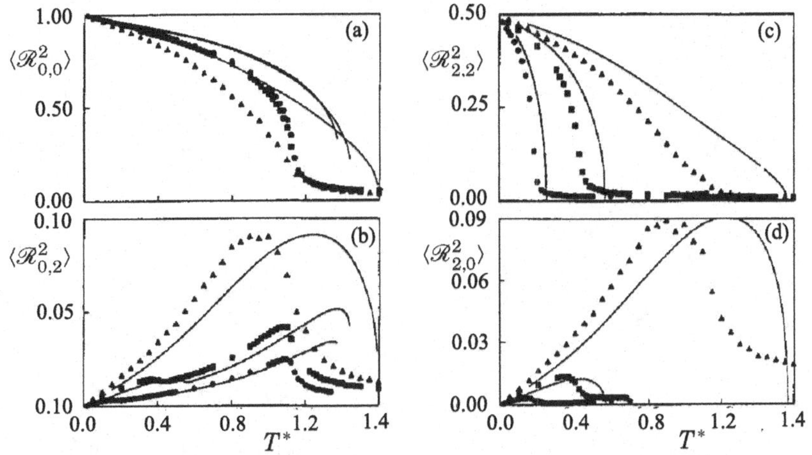

Figure 10.8 The order parameters $\langle \mathscr{R}^2_{0,0} \rangle \equiv \langle P_2 \rangle$ (a), $\langle \mathscr{R}^2_{0,2} \rangle$ (b), $\langle \mathscr{R}^2_{2,2} \rangle$ (c), $\langle \mathscr{R}^2_{2,0} \rangle$ (d) vs temperature obtained from MC: $\lambda = 0.2$ (○), $\lambda = 0.3$ (□), $\lambda = 0.40825$ (△) and from MFT (lines) [Biscarini et al., 1995].

diagrams obtained from MC simulations and MFT models is the same, with a Landau point (Section 2.7.2) where a direct transition from N_B to isotropic (I) phase takes place.

10.3.2 Straley Biaxial Model

The potential in Eq. 10.13 has also been studied beyond the dispersive interactions limit. In this general potential the coefficients u_{2mn} do not have a simple physical identification in terms of single-molecule properties, as would be desirable. Using $u_{200} = -\epsilon$ and the notation $\Gamma = \sqrt{(8/3)}(u_{220}/u_{200})$ and $\Lambda = -(2u_{222}/3u_{200})$, as employed by Romano [2004b], De Matteis et al. [2005] and Preeti et al. [2011] to define a reduced temperature $T^* \equiv -k_B T/\epsilon$, the potential in Eq. 10.13 still depends on both (Γ, Λ), while for dispersive interactions we had $\Lambda = \Gamma^2$. Simple models that give an explicit expression for u_{2mn} or Γ, Λ from the length, breadth and width L, B and W of brick-like molecules have been put forward. The model of Straley [1974], based on steric repulsions, gives the expressions already seen in Eq. 7.34a–7.34c. For that of Ferrarini et al. [1994] which assumes the interactions being proportional to the exposed surface in the three different directions, the coefficients u_{Lmn} were given in Eqs. 7.41a–7.41c, with $\lambda = \sqrt{3/2}L(W - B)/[(W + B)L - 2BW]$. The maximum biaxiality ($\lambda = 1/\sqrt{6}$) is obtained for $W^{-1} + L^{-1} = 2B^{-1}$. A phase diagram was also obtained for the specific choice of $\Gamma = 0$ [Sonnet et al., 2003]. It was shown that in this case a coexistence line between N_B and I phases is obtained, and these findings were confirmed by MC at selected state points [Romano, 2004a, 2004b; De Matteis et al., 2005]. However, for this case, a biaxial phase should be the first one found on cooling down from the isotropic, so it should be relatively easy to find, contrary to experimental evidence.

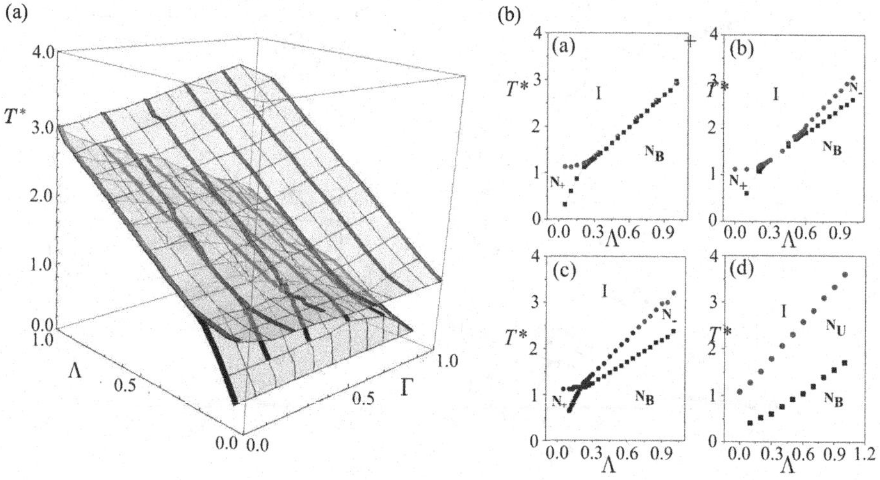

Figure 10.9 (a) Phase diagram for the Straley biaxial potential (Eq. 10.13) from MC simulations on a 20 × 20 × 20 lattice as the biaxiality parameters $\Gamma = \sqrt{(8/3)}(u_{220}/u_{200})$ and $\Lambda = -(2u_{222}/3u_{200})$ are varied. The two surfaces denote the $\mathrm{I}-\mathrm{N_u}$ and $\mathrm{N_u}-\mathrm{N_b}$ transitions. (b) (T^*, Λ) sections of the phase diagram for $\Gamma = 0$ (a), 0.2 (b), $0.4 \approx 1/\sqrt{6}$ (c), and 0.6 (d) [Preeti et al., 2011].

De Matteis et al. [2005] and Bisi et al. [2008] have analyzed the full Hamiltonian in Eq. 10.13, showing that it yields a stable minimum for calamitic states (parallel side-by-side blocks) only within a certain fan-shaped region of parameter space defined by $\Lambda - |\Gamma| + 1 > 0$ [Straley, 1974; Sonnet et al., 2003; De Matteis and Romano, 2009]. A detailed MC investigation of the phase diagram of the biaxial potential in Eq. 10.13 when varying (Γ, Λ) has given the phase diagram in Fig. 10.9 [Preeti et al., 2011]. The phase diagram shows that over a wide range of parameters the molecular organization obtained on cooling from the isotropic is a uniaxial nematic, while the biaxial phase is confined to low temperatures where realistic systems probably would become smectic or crystals. This is a rather more pessimistic view than that provided by the extended isotropic-biaxial nematic transition line found in Romano [2004a; 2004b] and De Matteis et al. [2005], but one which seems consistent with the persistent difficulty in finding biaxial nematics.

10.4 Confined Nematics: Films, Droplets, ...

Confined nematics are a class of composite materials of wide interest both from the technological and basic research point of view [Crawford and Žumer, 1996]. The first aspect is obvious since LCs are rarely used in bulk, but in displays and other electro-optical devices they are employed in thin films, polymer dispersed droplets, channels where they are confined in suitable geometries and boundaries. The academic interest is related to the effects that confinement induces on the phase transitions and on the molecular organization inside these systems. This in turn stems from a competition between the effects of surface boundary conditions, the nematic ordering inside the system and the disordering caused by temperature. Many experiments and theories have been put forward to understand these phenomena, but MC simulations seem to be a particularly useful method in studying relatively small lattices of confined nematics, particularly in the presence of complex geometries or boundary conditions not amenable to analytic solutions. The need to understand and predict experiments where orientational, rather than positional, ordering plays the key role, makes the simple spin models introduced in Chapter 2 a convenient and flexible tool to simulate fairly realistic experimental conditions. In particular, MC of lattice models has proved useful in investigating droplets with fixed surface anchoring [Pasini et al., 2000], mimicking polymer dispersed liquid crystals [Doane, 1990; Bunning et al., 2000; Bacchiocchi et al., 2009], nematic displays [Berggren et al., 1995; Chiccoli et al., 1998] and hybrid aligned nematic cells [Lavrentovich et al., 2001; Chiccoli et al., 2002]. Here we briefly present some illustrative examples of applications.

Since polarized optical microscopy (POM) textures are an important experimental technique to study LC devices [Ondris-Crawford et al., 1991], we start by discussing how these can be obtained from simulations.

10.5 Polarized Optical Microscopy Textures

The intensity of light transmitted through a liquid crystal device can be calculated starting from the simulated configurations [Berggren et al., 1994a] exploiting a matrix approach described in detail in Appendix L [Ondris-Crawford et al., 1991; Xu et al., 1992; Kilian, 1993; Berggren et al., 1994b]. The basic idea of the matrix approach is that geometric ray optics can be used and that the sample (droplet, film, ...) can be divided into small volume elements (voxels), or just the sites in a lattice model, whose effect on the propagating light beam is described by a Müller matrix (see Eq. L.3) [Schellman, 1998]. Then the light crossing the sample is retarded by the matrix resulting from the product of the Müller matrices $\mathbf{M}^{(j)}$ corresponding to each site along the light path. Each matrix involves the angles α_j, β_j, describing the director orientation of the jth voxel, taken from the simulation data, and the phase difference which depends on the thickness of the transversed layer, h, the wavelength of the incoming light, λ, and the LC refractive indices, n_0 and n_e. In the examples shown here we have chosen visible light, $\lambda = 545$ nm, and refractive indices $n_0 = 1.5$ and $n_e = 1.7$, similar to those of the nematic liquid crystal 5CB [Ondris-Crawford et al., 1991]. To model a real POM experiment we consider crossed polarizers, $\mathbf{P}^{in}$ and $\mathbf{P}^{out}$, placed at the front and back of the sample cell and obtain the Stokes vector of the outcoming polarized and retarded light beam as (see Eq. L.4)

$$s^{out} = \mathbf{P}^{out} \Big\langle \prod_j \mathbf{M}^{(j)} \Big\rangle \mathbf{P}^{in} s^{in}, \tag{10.15}$$

where s^{in} corresponds to the Stokes vector of the incoming unpolarized light. The intensity is proportional to the first element of the output Stokes vector, i.e. $[s^{out}]_0$, and can be grey coded with a normalized scale going from black (lowest) to white (highest) light intensity, with a certain number of different grey levels. To improve the quality of the optical image we further average over a suitable number of equilibrated configurations as indicated by the angular brackets in Eq. 10.15.

10.6 Topological Defects

Defects in a nematic liquid crystal (see Section 1.2.2) correspond to regions where the orientational order vanishes and thus a molecular director cannot be properly defined. Even though defects are treated quite exhaustively in many publications, particularly in terms of director distributions and continuum theory or of their mathematical properties [Mermin, 1979; Kleman, 1982; Kleman and Lavrentovich, 2006], computer simulations provide an important means of investigating defects in a number of situations where continuum theory might be difficult to apply. It is worth starting by briefly reviewing the main types of defects, considering that the curvature of the director around a given defect costs elastic energy. In a 2D thin film the theory of Frank elasticity, in the one constant approximation, i.e. assuming $K_{11} = K_{22} = K_{33} = K$, gives a simple expression for the director field at a point $\boldsymbol{p}(x, y)$ from a defect taken as the origin. If ϕ is the angle made by $\boldsymbol{d}$ at that point with the x-axis, the solutions take the simple form: $\boldsymbol{d} = (\cos\phi, \sin\phi, 0)$, $\phi = s\tan^{-1}(y/x) + c$, where s is the strength and c the initial orientation angle. In this 2D limit there are infinitely many different types of disclinations, each type being characterized by a half-integer or integer topological charge s [Mermin, 1979; Afghah et al., 2018]. Topology tells us the types of possible defects, but the ones actually observed should be those with lower elastic energy, since defects of different strength have a different cost in elastic energy. Assuming that a 'core' region extends from its origin to r_c, the energy per unit length of an isolated disclination line is given by [Nehring and Saupe, 1972; Hobdell and Windle, 1995]

$$\mathcal{G}_{\text{el}} = \mathcal{G}_{\text{c}} + \pi K s^2 \ln(R/r_c), \qquad (10.16)$$

with $\mathcal{G}_{\text{c}}$ the core elastic energy. Due to the s^2 dependence all defects of strength $|s| > \frac{1}{2}$ should be unstable with respect to defects of strength $\pm\frac{1}{2}$ only. From observed POM images the strength is obtained as the number of brushes divided by four, so essentially two brushes defects should appear. However, considering the classical schlieren texture in a thin film with tangential boundary conditions, quite stable four brushes defects of strength ± 1 are very frequently observed. One intuitive interpretation could be that each point represents a disclination line running vertically from one surface to the other, producing an essentially 2D structure, with all molecules parallel to the sample surfaces. However, Meyer [1973] and Cladis and Kleman [1972] showed that the elastic energy for a configuration with $s = +1$ without singularity (coreless) with the director escaping to the third dimension is independent of R and given by $\mathcal{G}_{\text{el}} = 3\pi K$, thus lower than that in Eq. 10.16. The picture emerging for a sample size of large R on a molecular scale is thus that of $s = \pm 1/2$ singularities at the surface connected by a line disclination. We shall examine this with MC simulations in Section 10.7.1. By contrast, if a nematic director field is defined in 3D, then there is only one type of disclination. It can be mathematicaly proven that all half-integer disclinations are topologically equivalent to each other, while integer disclinations are topologically equivalent to a defect-free configuration and hence are not defects at all. Thus, while in 2D the nematic director field can have infinitely many types of disclinations that are not equivalent to each other, in 3D the nematic director field can have only one type of disclination of strength $1/2$ (sign not assigned). The switch from one situation to the other, e.g. when the director is forced to a plane (so that the z-component of $\boldsymbol{d}$ tends to 0) by the action of an external electic field in a nematic with negative dielectric anisotropy has been studied with continuum theory based numerical simulations by Afghah et al. [2018].

10.6.1 Defect Detection

A non-trivial practical problem in computer simulations is that of examining sample snapshots to locate the presence of defects and assigning their topological charge. In real experiments, thanks to their birefringence they can be observed experimentally by optical microscopy, in particular between polarizers, the so-called Polarized Optical Microscopy (POM) technique. Quite naturally one possibility for simulation studies is to produce optical textures, as we describe in detail in Appendix L, and treat them as observed POM images (see examples later in this chapter). Another possibility is to define a suitable quantitative measure, a metric, related to the ordering at each lattice site i or, more generally, small volume (voxel) inside the sample. A convenient procedure is that of Westin et al. [2002] originally developed to visualize properties characterized by a 3×3 matrix with non-negative eigenvalues (a semi-positive second-rank tensor), in particular water diffusion tensors obtained from Magnetic Resonance Imaging (MRI) for medical purposes [Schultz et al., 2017]. In the more specific case of liquid crystals [Callan-Jones et al., 2006] the procedure starts by dividing the sample in a grid of voxels and defining for each voxel $\mathcal{V}^{(i)}$ a certain direct product matrix $\mathbf{U}^{(i)}$ obtained from the molecular unit vector $\boldsymbol{u}_j$ (in a lattice model the 'spin') for the jth particle:

$$\mathbf{U}^{(i)} \equiv \langle \boldsymbol{u}_j^{(i)} \otimes \boldsymbol{u}_j^{(i)} \rangle_j = \frac{1}{N_i} \sum_{j=1}^{N_i} \boldsymbol{u}_j \otimes \boldsymbol{u}_j, \text{ with } j \in \mathcal{V}^{(i)}. \tag{10.17}$$

where for a lattice $Ni = 1$ so that $i = j$, Note that this $\mathbf{U}$, with $\text{Tr}\mathbf{U} = 1$, is simply related to the $\mathbf{Q}$ matrix of the voxel as defined in Eq. 3.47: $\mathbf{U}^{(i)} = (2/3)\mathbf{Q}^{(i)} + (1/3)$. Diagonalizing $\mathbf{U}^{(i)}$ as $\mathbf{U}^{(i)}\mathbf{X}^{(i)} = \mathbf{X}^{(i)}\boldsymbol{\Lambda}^{(i)}$ and ordering the eigenvalues of $\mathbf{U}^{(i)}$ so that $\lambda_1^{(i)} \geq \lambda_2^{(i)} \geq \lambda_3^{(i)}$, the local director is parallel to the eigenvector corresponding to largest eigenvalue, λ_1^i. The Westin metric can be introduced as the eigenvalue combinations $c_l^{(i)} = \lambda_1^{(i)} - \lambda_2^{(i)}$, $c_p^{(i)} = 2(\lambda_2^{(i)} - \lambda_3^{(i)})$ and $c_s^{(i)} = 3\lambda_3^{(i)}$, which measure the *linearity*, *planarity* and *sphericity* of a second-rank tensor, respectively. The condition of unit trace allows a barycentric coordinate representation of the local tensor space, and a concise graphical (or glyph [Bi et al., 2019]) representation of the tensor as an ellipsoid as shown in Fig. 10.10. Regions of well-ordered uniaxial nematics correspond to $\lambda_1 \gg \lambda_2 \approx \lambda_3$, i.e. $c_l \approx 1$; planar ordering corresponds to $\lambda_1 \approx \lambda_2 \gg \lambda_3-$ i.e. $c_p \approx 1$; and the absence order (isotropy) corresponds to $\lambda_1 \approx \lambda_2 \approx \lambda_3-$ i.e. $c_s \approx 1$. In practice, a small threshold for c_l is arbitrarily chosen, similar to the highest values that could be obtained, given numerical errors, for a truly isotropic region of

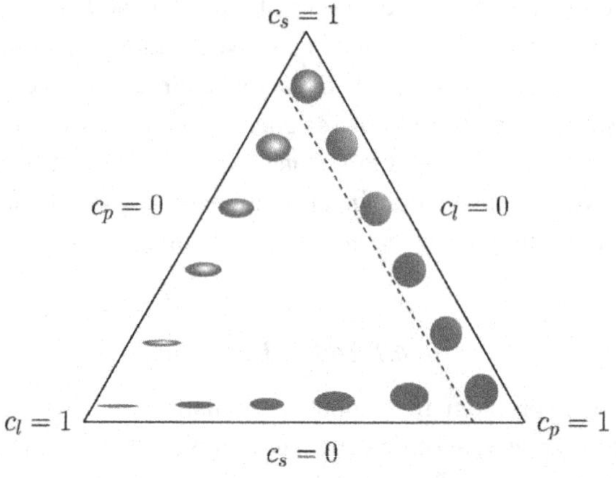

Figure 10.10 Barycentric coordinate system for Westin metrics. Note that $c_l + c_p + c_s = 1$ and $0 \le c_l \le 1, 0 \le c_p \le 1$. The dashed line is the threshold of c_l (here $c_l = 0.12$) which separates ordered uniaxial regions on the left of the line from the disordered defects core ones on the right. The ellipsoids have orientation and semi-axes given by eigenvectors and corresponding eigenvalues of the matrix **U** [Callan-Jones et al., 2006].

similar size, and a defect is assigned to regions with a value below that threshold. We shall show a few examples of both techniques. It is worth noting that the Westin metric approach is not limited to lattice models. For an atomistic or off-lattice model we could choose a voxel sufficiently large to contain enough molecules to offer a sufficient statistical quality to the results (the threshold mentioned before has to be meaningful). For a lattice model we could do the same assuming, e.g. that each voxel is (i) a small cube $N_i \times N_i \times N_i$. In this way we could analyze a particular 3D configuration looking for its instantaneous defects. However, we could also shrink the voxel down to one containing perhaps the lattice site and a neighbour shell and obtain a time average over a sufficiently high number of configurations: $\langle \mathbf{U}^{(i)} \rangle_t = \langle \, \langle \boldsymbol{u}_j^{(i)}(t) \otimes \boldsymbol{u}_j^{(i)}(t) \rangle_j \, \rangle_t$. If there exist stable defects, fixed at a certain position from some geometric or topological constraint, they should appear in the average $\langle \mathbf{U}^{(i)} \rangle_t$ (see Section 10.9).

10.7 Nematic Films

In thin nematic films, sandwiched between two flat surfaces with a null topological charge ($s = 0$), the occurrence and nature of defects can be controlled by imposing external fields or certain boundary conditions with the help of appropriate surface anchoring agents. Thus, defects should be absent in a flat cell with a uniform monodomain alignment either parallel to a certain direction of the surface, $\boldsymbol{r}$, or tilted with respect to it. More generally, when the film is so thin that the director can be considered to effectively live in 2D, homotopy theory shows that there can be in principle infinitely many types of disclinations in director orientations classified by their semi-integer or integer topological charge s [Mermin, 1979; Ohzono et al., 2017; Afghah et al., 2018]. These various defects have of course different free energy and in practice only the ones with smallest s are observed. The situation is

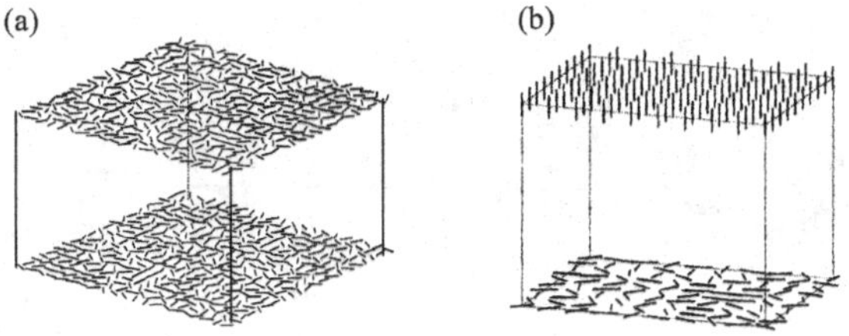

Figure 10.11 (a) A sketch of two cells with (a) random tangential boundary conditions and (b) hybrid ones with homeotropic and planar boundaries at the top and bottom surfaces (HAN cell).

different in a 3D bulk, where only half-integer defects are allowed, while integer ones can be eliminated, e.g. by the escape in the third-dimension mechanism [Cladis and Kleman, 1972].

10.7.1 Tangential Boundary Conditions: Schlieren Textures

As a first example of generating optical POM textures it is natural to try to reproduce the classic schlieren textures of nematics (Fig. 1.8). We can simulate these with nematic films confined between two parallel plates with tangential, in-plane disordered, boundary conditions (DTBC), as sketched in Fig. 10.11a, starting from the lattice model Hamiltonian,

$$U_N = \sum_{\substack{i,j \in \mathscr{F} \\ i<j}} U_{ij} + J_{\rm T} \sum_{\substack{i \in \mathscr{F} \\ j \in \mathscr{S}_{\rm T}}} V_{ij}^{(\rm T)} + J_{\rm B} \sum_{\substack{i \in \mathscr{F} \\ j \in \mathscr{S}_{\rm B}}} V_{ij}^{(\rm B)}, \tag{10.18}$$

where $\mathscr{F}$, $\mathscr{S}_{\rm T}$,$\mathscr{S}_{\rm B}$ are the set of particles (spins) inside the sample and at the top and bottom surfaces of the film, respectively, and U_{ij}, $V_{ij}^{(\rm T)}$, $V_{ij}^{(\rm B)}$ are the pair potentials for nematogen-nematogen and for nematogen-surface particles, depending on the relative orientation Ω_{ij} of the two spins. In the examples described here we assume $U_{ij} = V_{ij}^{(\rm T)} = V_{ij}^{(\rm B)} = -\epsilon_{ij} P_2(\boldsymbol{u}_i \cdot \boldsymbol{u}_j)$, i.e. the LL potential, even if of course this does not have to be the case and other interactions can be chosen, e.g. biaxial systems with dispersive interactions (Eq. 10.14) have also been studied [Chiccoli et al., 2002]. When modelling a film, the particle positions are fixed at the sites of a simple cubic lattice of dimensions $L \times L \times h$ and the strength of interaction ϵ_{ij} is $\epsilon \geqslant 0$, when i and j are nearest neighbours and 0 otherwise. The anchoring interactions at the interfaces couple the spins inside the sample cell to the outer layer of spins with a fixed orientation consistent with the boundary condition chosen and have strength J_T, J_B. Tangential boundary conditions are set by fixing random (x, y) in-plane orientations of the spins at the top and bottom surfaces (see Fig. 10.11a). Empty space is left at the four lateral faces of the lattice which is updated using the MC method (Chapter 8). Lattices of various sizes at some selected temperatures $T^* = k_B T/\epsilon$ and different couplings with the surface spins, in particular, $J_T = J_B = 0.5$, which corresponds to fairly weak anchoring have been studied. Simulations are normally started with spins perpendicular to the confining surfaces, so as to be as different as possible from

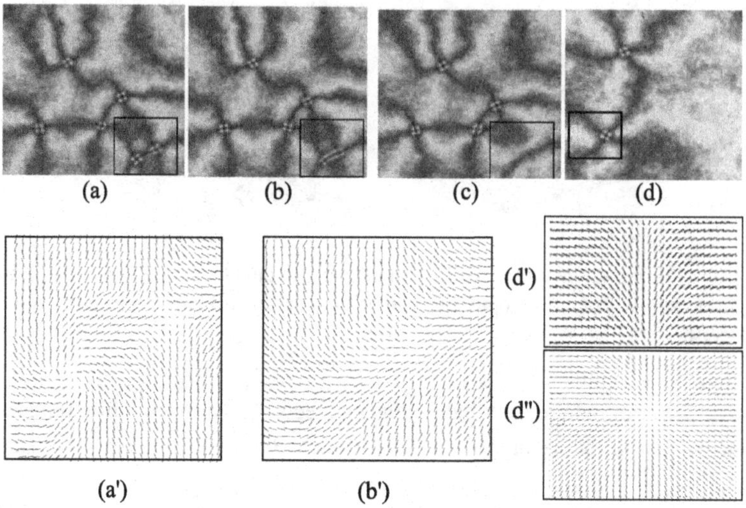

Figure 10.12 Evolution of a 120 × 120 × 16 LL film (Eq. 10.18) at $T^* = 0.4$ with random planar anchoring of strength $J = 0.5$. POM textures after $n = 6$ (a), 7 (b), 8 (c), 60 (d) MC kcycles, showing annihilation of a pair of $+1$ and -1 defects and formation of a schlieren texture. On the right the vertical (d′) and horizontal (d″) cross sections, five layers down the surface, through an $s = 1$ defect. Refractive indices $n_x = n_y = 1.5$ and $n_z = 1.7$, film thickness $d = 5.3$ μm and light wavelength $\lambda = 545$ nm [Chiccoli et al., 2002].

the expected final result. In Fig. 10.12 we show the formation of four brushes defects with strength $s = \pm 1$. The defects disappear by annihilation of a $+1$ and -1 pair as those enclosed by a square in Fig. 10.12. The defects $s = \pm 1$ are not singular, as we see from the section along the vertical axis, Fig. 10.12d′. This essentially agrees with the mechanism proposed by Meyer [1973] and Cladis and Kleman [1972] of an 'escape into third dimension' with a pair of point defect-boojums terminating the line at the confining surfaces even if a strong enough anchoring at the bounding plates can hinder, to some extent, the spins reorientation away from the surface normal.

10.7.2 Hybrid Aligned Nematic Films

Another type of confined system we consider here is a model of a hybrid cell, with random planar orientation at the bottom surface and homeotropic, normal orientation at the top (Fig. 10.11b). These conditions have been experimentally realized [Lavrentovich and Nastishin, 1990] placing an LC film of 5CB on top of an isotropic liquid substrate such as polyethylene glycol or glycerine and leaving a free air/LC surface, where these mesogens align essentially perpendicular to the interface. The cells can also be fabricated with photoalignment [Chiccoli et al., 2019] techniques. Lavrentovich and Nastishin [1990] have demonstrated that these hybrid aligned nematic (HAN) films produce very interesting POM textures, due to the presence of the two competing boundary conditions. MC simulations prove quite useful in studying these systems. The HAN cell [Lavrentovich

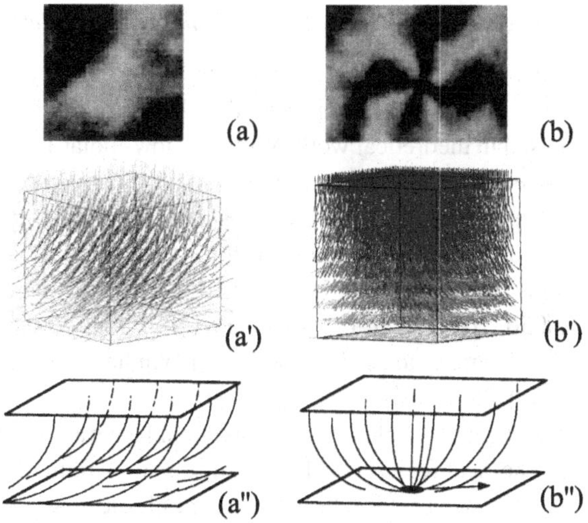

Figure 10.13 POM images of (a) a $30\times30\times10$ and (b) a $25\times25\times10$ portion of a $50\times50\times10$ HAN film from MC simulations together with snapshots of the MC molecular organization (a′), (b′) [Chiccoli et al., 1997]. On the bottom the predicted director distribution from continuum theory [Lavrentovich and Pergamenshchik, 1995] in a uniform HAN film (a″) and in a deformed state with a strength $|s| = 1$ point surface defect (b″). The Gaussian curvature κ_G of the director field $\boldsymbol{d}(x, y)$, i.e. the product of the principal curvatures, κ_x and κ_y, is 0 for the uniform HAN film and negative for the defect state (b″).

and Pergamenshchik, 1995] is mimicked [Chiccoli et al., 1997] assuming a $L \times L \times h$ lattice where the spins of the bottom layer, $z = 0$, have random fixed orientations in the horizontal (x, y) plane, while those of the top layer, $z = h$, are fixed along the surface normal. Open, i.e. empty space, boundary conditions are assumed on the four lateral faces of the cell [Chiccoli et al., 1997]. PBCs are avoided since the artificial periodicity might be incompatible with a ground state that contains topological defects. The system shows a $T_{NI} \approx 5\%–10\%$ lower than that of the bulk LL model [Chiccoli et al., 1999a], that is essentially unchanged upon increasing the horizontal dimensions of the cell. The molecular organizations obtained (Fig. 10.13) vary across the sample going from the disordered configuration of the bottom (first) layer to the aligned one of the top one. For films with small L/h, the director is just bent in a vertical plane and the POM appears black. The situation dramatically changes for large L/h (e.g. for $L \geq 50$ with $h = 10$) where the texture shows topological defects with four brushes emerging from the core [Kleman, 1982]. The defects are of strength $s = \pm 1$, and $|s| = 1$ is the lowest possible topological charge of a defect in a HAN film. Simulations show that the core of the defect is located near the lower surface; the distortions vanish as one moves towards the upper plate where the spins orient along the z-axis. A striking result of the MC simulations is that the model based exclusively on pure nearest-neighbours molecular interactions mimics the long-range deformations with topologically stable defects in agreement with continuum theory predictions (see Fig. 10.13).

10.7.3 Gruhn–Hess–Romano–Luckhurst Model

The LL model we have used up to now has an important limitation in that it yields the three elastic constants K_{ii} as equal, even if, actually, this corresponds to the one-constant approximation often used in theoretical work. While for low-molar-mass liquid crystals the difference between the elastic constants is normally small [Stannarius, 1998b], the need to account for the anisotropy in the elastic constants is particularly important for various systems where their values can differ considerably. In particular, as mentioned in Section 1.2.3, large differences in elastic constants can be expected in polymer liquid crystals [Kleman, 1991; Song et al., 2003b, 2005] or in LCs originated from long viruses like TMV [Hurd et al., 1985], nanotubes [Song et al., 2003a; Song and Windle, 2005] or other nanocrystal suspensions [Li et al., 2002].

The values and anisotropy of the elastic constants should have an effect on POM textures. Indeed, measuring geometrical features such as the disclination radii of the $s = 1/2$ defects has been suggested as a means to determine at least some elastic constants in polymeric LCs from transmission electron microscopy (TEM) images. In particular, Hudson and Thomas [1989] found in this way the elastic anisotropy $\varepsilon = (K_{11} - K_{33})/(K_{11} + K_{33})$ to vary from negative ($-0,15$) to positive ($+0.20$) in two very similar hydroquinone based polymers, TQT10H and TQT10M, differing only by a replacement of a hydrogen with a methyl. Unfortunately, this hints that it is extremely difficult to make predictions on elastic constants based on the similarity of molecular structures, differently from simple expectations such as that of de Jeu [1981], that $K_{11} : K_{22} : K_{33} = 1 : 1 : (l/w)^2$ for a molecule of length l and width w. Given the importance of relating elastic constants to texture, a number of numerical treatments based on minimization of the continuum free energy have been put forward, e.g. by Kilian and Hess [1989], Sonnet et al. [1995], and for polymer LCs, by Hobdell and Windle [1995]. Going back to microscopic simulations, the introduction of biaxial contributions [Chiccoli et al., 2002] or other more complex spatially anisotropic interactions in the pair potential can lift the degeneracy of elastic constants of the simple LL model. However, elastic constants are not directly predictable from the microscopic model and they would have to be obtained as observable properties, a task of considerable difficulty in itself even for lattice models [Cleaver and Allen, 1991]. If the aim is a study of the effect of elastic constants of the topological defects in nematic films this would imply modifying the pair potential in a more or less arbitrary way and calculating on one hand defects and on the other elastic constants and other related observables such as order parameters. Thus, since both elastic constants and defects are, in this fully microscopic approach, results of the simulation establishing a relation between them is a rather cumbersome process. A much more direct approach was provided by Gruhn and Hess [1996], Romano [1998] and Luckhurst and Romano [1999] by introducing an effective potential that directly depends on classic splay, twist and bend elastic constants K_{11}, K_{22} and K_{33}. The particles interact through the nearest-neighbour attractive pair pseudopotential

$$U_{ij} = -\varepsilon_{ij} \Big\{ \lambda \big[P_2 (\boldsymbol{u}_i \cdot \boldsymbol{s}_{ij}) + P_2 (\boldsymbol{u}_j \cdot \boldsymbol{s}_{ij}) \big] + \mu \big[(\boldsymbol{u}_i \cdot \boldsymbol{s}_{ij}) (\boldsymbol{u}_j \cdot \boldsymbol{s}_{ij}) (\boldsymbol{u}_i \cdot \boldsymbol{u}_j) - \frac{1}{9} \big]$$
$$+ P_2 (\boldsymbol{u}_i \cdot \boldsymbol{u}_j) \Big[\nu + \rho \big[P_2 (\boldsymbol{u}_i \cdot \boldsymbol{s}_{ij}) + P_2 (\boldsymbol{u}_j \cdot \boldsymbol{s}_{ij}) \big] \Big] \Big\}, \tag{10.19}$$

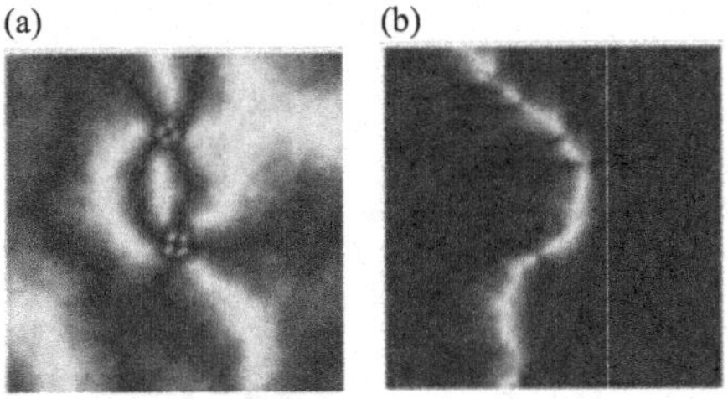

Figure 10.14 Effect of elastic constant anisotropy on simulated POM textures for a 100 × 100 × 12 GHRL nematic film in a schlieren geometry with surface anchoring couplings $J_T = J_B = J = 0.5$, (a) $K_{11} = 6.4$ pN, $K_{22} = 3.6$ pN, $K_{33} = 8.2$ pN (similar to MBBA). (b) $K_{11} = 1.0$ pN, $K_{22} = 0.22$ pN, $K_{33} = 8.8$pN (similar to TMV) [Chiccoli et al., 2010].

where $\varepsilon_{ij} = \varepsilon, \varepsilon > 0$ for i, j nearest neighbours and 0 otherwise, $\boldsymbol{s}_{ij} = \boldsymbol{r}_{ij}/|\boldsymbol{r}_{ij}|$, $\boldsymbol{r}_{ij} = \boldsymbol{r}_i - \boldsymbol{r}_j$, with $\boldsymbol{r}_i, \boldsymbol{r}_j$ the position vectors of the ith and jth lattice points; $\boldsymbol{u}_i, \boldsymbol{u}_j$ are unit vectors along the axis of the two particles (or 'spins') and P_2 is a second-rank Legendre polynomial. The parameters λ, μ, ν, ρ are defined in terms of the splay, twist and bend elastic constants as $\lambda = \frac{1}{3}\Lambda\,(2K_{11} - 3K_{22} + K_{33})$, $\mu = 3\Lambda\,(K_{22} - K_{11})$, $\nu = \frac{1}{3}\Lambda\,(K_{11} - 3K_{22} - K_{33})$ and $\rho = \frac{1}{3}\Lambda\,(K_{11} - K_{33})$ with Λ a factor with dimensions of length. We note that the pseudopotential in Eq. 10.19 is a mesoscopic, rather than a truly molecular one, as indicated by the fact that it contains elastic constants, and thus it is temperature dependent. The one-constant approximation, $K_{ii} = K$, yields $\lambda = \mu = \rho = 0$ and $\nu = -\Lambda K$, formally reducing Eq. 10.19 to the LL potential. Here we wish to show that a simple application of the Gruhn–Hess–Romano–Luckhurst (GHRL) pseudopotential coupled with MC simulations can be a very effective way to observe the onset of textures and their evolution in a variety of situations. We concentrate in particular on thin uniaxial films with random planar (schlieren) conditions and consider two cases with the U_{ij} in Eq. 10.19 with different choices of elastic constants. In Fig. 10.14 we see simulated POM textures for a nematic film in a schlieren geometry as obtained from an MC simulation of a GHRL potential for the three elastic constants taken from different materials. The values for Fig. 10.14a refer to a typical low-molar-mass nematic: MBBA [Dunmur, 2001], where the elastic constants are rather similar and, in Fig. 10.14b, to a suspension of TMV virus (see Section 1.14) with a large bending stiffness [Hurd et al., 1985]. The images are taken after 10^4 MC sample sweeps with the film placed between crossed polarizers. We see rather different textures consistently with the prevalence of two brush disclinations for long rigid particles like carbon nanotubes and main chain polymers [Kleman, 1991].

10.7.4 Hybrid Aligned Nematic Film with Partial Disorder

As we have seen in Section 10.7.2 the molecular organization inside thin HAN films can vary from a continuous bend type structure of the director connecting the planar degenerate and homeotropic boundaries to a defective structure for very thin films, with thickness $h \ll L$, the lateral size. It has also been reported that disclination lines can be formed in a

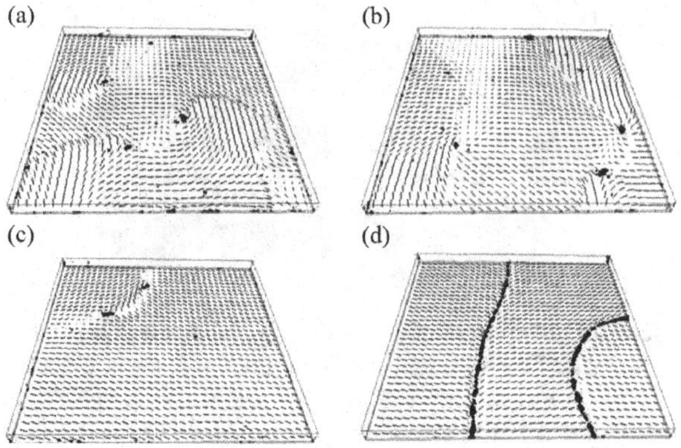

Figure 10.15 Plots of the local director at lattice sites and of the c_l Westin metric (Section 10.6.1) isosurfaces corresponding to a defect (black) for different values of ordering at the bottom surface ($z = 0$). The frames refer to the first overlayer for a substrate with $\langle T_2 \rangle = 0.00$, i.e. planar random (a), 2% (b), 20% (c), and 100%, i.e. perfect alignment along x (d) [Chiccoli et al., 2019].

hybrid nematic cell where the two facing flat surfaces enclosing the nematic are respectively treated to induce an orientation perpendicular to the surface at the top ($z = h$), as in the previous case, and a homogeneous anchoring, with full alignment along a specific in-plane direction at $z = 0$ [Buscaglia et al., 2010]. However, it is now possible to fabricate cells with controlled order at the surface by photopolymerization of reactive mesogens illuminated with suitably polarized light on a pixel-by-pixel basis [Kobashi et al., 2016; Chiccoli et al., 2019]. The alignment at the surface can be quantified with a 2D order parameter $\langle T_2 \rangle$ [Denham et al., 1980] $\langle T_2 \rangle = \langle \cos(2\alpha) \rangle$, where α is the orientation of a molecule belonging to the surface plane with respect to the preferred direction of the surface. The two limiting situations just described correspond to $\langle T_2 \rangle = 0$ (planar degenerate) or 1 (perfect alignment along one direction in the plane). The effect on the molecular organization inside a nematic film in an intermediate situation with only partial surface order is not obvious and computer simulations can be of use to see if there is some critical surface order that enforces the transition from points to lines or walls, everything else being the same, or actually which type of defects occurs. MC simulations and the Westin metric allow an inspection of the transition from one type of disclination to the other and in Fig. 10.15 we see an example. A comparison of MC simulations and real experiments [Chiccoli et al., 2019] indicates that fully random, degenerate, anchoring favours point defects, while homogeneous alignment favors the appearance of filament like defects. When the in-plane order of the bottom surface assumes values intermediate between the two extremes, the transition seems mainly due to the surface alignment rather than on the details of the nematic elastic constants. Moreover, it appears from the simulated configurations that the transition from planar to vertical alignment is a sudden one, starting from the bottom overlayer, rather than from an intermediate one. The combined use of MC simulations and POM measurements can be particularly useful. On one hand, the 2D polarized microscopy textures obtained from experiments can validate the MC simulations. On the other, the simulations, once validated, allow a detailed computational 3D visualization of the molecular organization across a film, complementing the 2D projections produced by POM experiments and offering an important way of discriminating between models.

10.8 Polymer-Dispersed Liquid Crystal Droplets

Polymer-dispersed liquid crystals (PDLCs) [Doane, 1990; Kitzerow, 1994; Drzaic, 1995; Crawford and Žumer, 1996] are composite materials that consist of microscopic nematic droplets, with typical radii from a few hundred Angström to more than a micron, embedded in a transparent polymer matrix. PDLC films are interesting for technical applications like smart windows changing between opaque and transparent by the application of an external field across the film [Drzaic, 1995; Drzaic and Drzaic, 2011] but PDLCs also represent practical realizations of systems of fundamental interest for the study of topological defects [Mermin, 1979; Kleman and Lavrentovich, 2003]. A number of experimental works have considered different boundary conditions at the droplet surface, for example, radial [Golemme et al., 1988b], axial [Ondris-Crawford et al., 1991], toroidal [Drzaic, 1988] and bipolar [Golemme et al., 1988b; Aloe et al., 1991; Ondris-Crawford et al., 1991] (see Fig. 10.16). These surface anchorings can be obtained by suitably choosing the polymer matrix and the preparation methods. Additional effects of interest come from the application of external, electric or magnetic fields [Golemme et al., 1988b; Kilian, 1993; Berggren et al., 1994b]. The PDLC model used in lattice simulations concentrates on a single droplet and consists of an approximately spherical sample $\mathscr{S}$ carved from a cubic lattice with spins interacting with the LL or GHRL potentials described before, while the surface effects are modelled with an external layer of 'ghost' spins, $\mathscr{G}$, with fixed orientations chosen to mimic the desired boundary conditions. The droplet Hamiltonian is

$$U_N^{\mathrm{D}} = \sum_{\substack{i,j\in\mathscr{S}\\ i<j}} U_{ij} + J \sum_{\substack{i\in\mathscr{S}\\ j\in\mathscr{G}}} V_{ij}, \tag{10.20}$$

with the surface interaction terms $V_{ij} = U_{ij}$ an LL potential, with the surface coupling parameter J determining the anchoring direction for the spins inside the droplet with respect to the ghost spins: parallel for $J > 0$ or perpendicular for $J < 0$, with $|J|$ the strength. The case $J = 0$ would be appropriate for a droplet in vacuum. Systems corresponding to

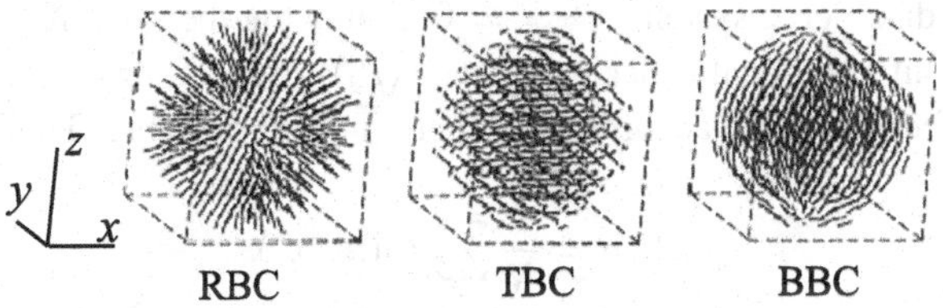

Figure 10.16 A sketch of droplets with radial (RBC), toroidal (TBC) and bipolar (BBC) boundary conditions [Pasini et al., 2000].

the following three different boundary conditions at the nematic-polymer interface (see Fig. 10.16) have been found to be experimentally relevant and have been studied by MC simulations [Pasini et al., 2000].

(i) **Radial boundary conditions** (RBCs) with the spins of the host matrix surrounding the droplet oriented normally to the local surface, i.e. towards the centre of the droplet.
(ii) **Toroidal boundary conditions** (TBCs) obtained when the spins at the polymer interface lie in planes perpendicular to the z-axis and are oriented tangentially to the droplet surface.
(iii) **Bipolar boundary conditions** (BBCs) with the ghost spins oriented tangentially to the droplet surface and belonging to planes parallel to the z-axis. MC simulations have been used to study the effects of different boundary conditions [Chiccoli et al., 1990, 1992; Berggren et al., 1994b], and of an external applied field on PDLCs [Berggren et al., 1992, 1994b; Chiccoli et al., 2000]. Particular attention has been devoted to simulating quantities observable in real experiments. For instance, methodologies to calculate powder deuterium NMR lineshapes and textures observable in polarized light experiments corresponding to the microscopic configurations obtained from computer simulations have been developed [Chiccoli et al., 1990; Berggren et al., 1994a; Chiccoli et al., 1995, 2000]. Here we briefly summarize some examples that can also be of general interest for other model droplets.

10.8.1 Ordering and Defects Inside Droplets

To examine the ordering inside the microdroplets, various second-rank orientational order parameters can be introduced, since the usual one, $\langle P_2 \rangle_\lambda$, obtained from the eigenvalues of the ordering matrix (see Section 3.5), is not always very informative for a spherical droplet, as it quantifies the nematic order with respect to a global director which may not exist as such for this geometry. For instance, in the case of RBCs, it is not possible to distinguish between a perfectly ordered radial configuration and a disordered one just from the value of $\langle P_2 \rangle_\lambda$ which would ideally vanish (as $\approx 1/\sqrt{N}$) in both cases. In this case, it is useful to define a radial order parameter, $\langle P_2 \rangle_R$ [Chiccoli et al., 1995, 2005]:

$$\langle P_2 \rangle_R = \frac{1}{N} \sum_{i=1}^{N} P_2(\boldsymbol{u}_i \cdot \boldsymbol{r}_i), \tag{10.21}$$

where $\boldsymbol{r}_i$ is the radial vector of the ith spin. For a perfect hedgehog configuration $\langle P_2 \rangle_R = 1$, while for a truly disordered system $\langle P_2 \rangle_R = 0$. Following the same reasoning it is possible to define a configurational order parameter, $\langle P_2 \rangle_C$, which tends to 1 for a molecular organization perfectly ordered according to a certain ideal structure. Thus,

$$\langle P_2 \rangle_C = \frac{1}{N} \sum_{i=1}^{N} P_2(\boldsymbol{u}_i \cdot \boldsymbol{c}_i), \tag{10.22}$$

where $\boldsymbol{c}_i$ is the direction corresponding to the local surface induced alignment. For example, in the bipolar case (Fig. 10.16b), $\boldsymbol{c}_i$ is a local meridian that belongs to the plane defined by

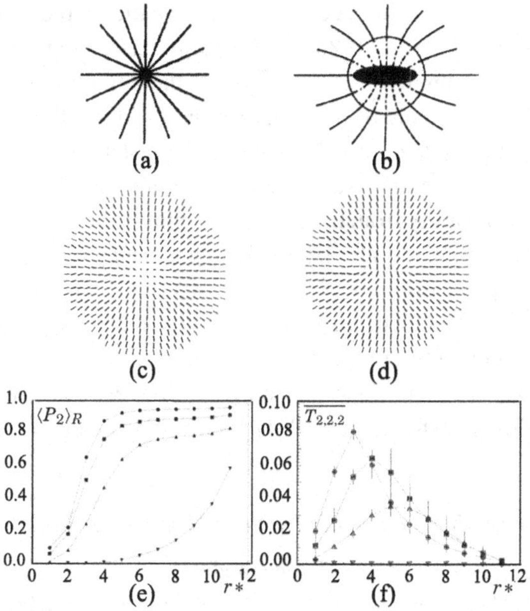

Figure 10.17 Sketch of (a) a radial and (b) a ring-core hedgehog in a droplet with RBCs and snapshots from MC simulations showing (c) an equatorial and (d) a vertical section of the nematic droplet with the aligned core in the middle. At the bottom (e) radial order parameter, $\langle P_2 \rangle_R$ and (f) rotational invariant $\overline{T_{2,2,2}} = \sqrt{70}\,\overline{S_{2,2,2}}/2$ vs distance $r*$ in lattice units from the centre of the droplet at reduced temperatures $T^* = 0.4(\bullet), 0.7(\blacksquare), 1.0(\blacktriangle), 1.3(\blacktriangledown)$ [Chiccoli et al., 1995].

the droplet axis (z-axis) and the radial vector $\boldsymbol{r}_i$, while being perpendicular to $\boldsymbol{r}_i$ itself. The configurational order across a droplet can be calculated by dividing the sphere in virtual concentric shells and calculating the relevant $\langle P_2 \rangle_C$ in each region. This in turn is important to test theories of defect type and molecular organization for various BCs. MC simulations for the three BCs in Fig. 10.16 have been performed [Berggren et al., 1992; 1994b; 1994a]. As an example, we report in Fig. 10.17c,d,e the molecular organization and the corresponding $\langle P_2 \rangle_R$ across an RBC droplet. To examine the question of the defect core being a hedgehog or a microscopic ring defect (Fig. 10.17a,b) it is useful to also calculate some of the average Stone invariants introduced in Section 4.8.1. In particular, the average $\overline{S_{2,2,2}}$, where

$$\overline{S_{2,2,2}}(r) = \Big\langle \delta\,(r - r_{12})\,\frac{1}{\sqrt{70}}\big[2 - 3\,(\hat{z}_1 \cdot \hat{z}_2)^2 - 3\,(\hat{z}_1 \cdot \hat{r})^2 - 3\,(\hat{z}_2 \cdot \hat{r})^2 + 9\,(\hat{z}_1 \cdot \hat{z}_2)\,(\hat{z}_1 \cdot \hat{r})\,(\hat{z}_2 \cdot \hat{r})\big]\Big\rangle \tag{10.23}$$

can discriminate between radial and hyperbolic hedgehogs, since for a radial hedgehog $\overline{S_{2,2,2}}(r) = 0$, while for the hyperbolic core it is expected to have a maximum at a certain distance from the centre. The results in Fig. 10.17d,f [Chiccoli et al., 1995] indicate the presence of a small ordered core at the centre of the nematic droplet, consistent with a ring disclination [Mori and Nakanishi, 1988], and a radius increasing with temperature. It is interesting to note that in this RBC case the size of the aligned core does not depend on the droplet size and has a radius of a few lattice units [Chiccoli et al., 1995] for all the system sizes studied, hinting that the core size is a true material property rather than being dependent on the droplet size and that for a large enough droplet the system appears to have a hedgehog structure away from the core. Interestingly the MC simulation for an RBC Heisenberg model (Section 2.8.3) droplet, modelling a magnetic droplet, shows a different behaviour, with a true point-like core [Chiccoli et al., 1995]. The similar the behaviour of properties calculated for different droplet sizes reinforces the argument that each of the spins could really be considered to represent a microdomain of some tens of particles, and that the results also are applicable to micron size PDLC droplets, that have been investigated experimentally [Ondris-Crawford et al., 1991] by optical techniques.

10.8.2 Calculating Deuterium Nuclear Magnetic Resonance Spectra

Simulated droplets configurations from lattice models can be used to calculate other observables used to study PDLCs [Chiccoli et al., 1995, 1999b]. In particular, Deuterium Nuclear Magnetic Resonance (DNMR) [Abragam, 1961; Emsley, 1985] is a powerful experimental technique (see Appendix I) frequently applied to investigate not only solutes ordering, as we have seen in Section 3.10.6, but also pure mesogens as long as they are, at least partially, deuterated, i.e. with some isotope ^{2}H (nuclear spin $I = 1$) replacing ordinary protons [Golemme et al., 1988b; Ambrozic et al., 1997]. The technique is very convenient for the study of these complex heterogeneous systems as it can selectively provide information on the deuterons inside the PDLC droplets, embedded in the non-deuterated polymer matrix ideally invisible to the experiment. An additional advantage is that DNMR is applicable also to submicron droplets, where optical methods fail to yield useful information. DNMR spectra provide information on the orientational ordering, director configurations and dynamic processes such as diffusion inside the droplets. Here we briefly describe how simulated configurations can be used to calculate DNMR spectra [Chiccoli et al., 2000]. As we have already seen in Section 3.10.6 the contribution of a ^{2}H nucleus k to the DNMR spectrum in an aligned nematic, for effectively uniaxial molecules, consists of two lines at an average frequency $\langle \nu_k \rangle$,

$$\langle \nu_k \rangle = \nu_Z \pm \langle \nu_Q \rangle = \nu_Z \pm \delta V_Q P_2(\boldsymbol{d} \cdot \hat{\boldsymbol{H}}) \langle P_2(\boldsymbol{d} \cdot \boldsymbol{u}_k) \rangle, \tag{10.24}$$

where $\boldsymbol{H}$ is the spectrometer magnetic field, $\langle P_2(\boldsymbol{d} \cdot \boldsymbol{u}_k) \rangle = \langle P_2 \rangle$ is the orientational order, $\delta_{VQ} = 3q_{zz} P_2(\cos\theta_k)/4$, with q_{zz} the magnetic quadrupole constant and $\theta_k = \arccos(\boldsymbol{u}_{D,k} \cdot \boldsymbol{u}_k)$ the angle between the unit vector $\boldsymbol{u}_{D,k}$ along the k-th C-D bond and the molecule axis $\boldsymbol{u}_k$. For instance, for the 4′-methoxy-4-cyano-biphenyl-D$_3$ (1OCB) molecule employed by Golemme et al. [1988b], $q_{zz} = 175$kHz and an angle $\theta_k = 59.45°$ for the C-D of the CD$_3$. In a monodomain sample the angle between $\boldsymbol{d}$ and $\hat{\boldsymbol{H}}$ is just determined by the sign of the diamagnetic anisotropy $\Delta\chi^{(m)}$. The splitting reduces to 0 in the isotropic phase, where a single line at the Zeeman frequency ν_Z (Appendix I) is observed.

Let us now turn to lattice simulations, where the local director can be approximated by the individual vector $\boldsymbol{u}_k$ that, we recall, essentially represents a short-range molecular cluster. The MC procedure provides directly each orientation $\boldsymbol{u}_k$ with respect to the lab frame with, for a nematic with positive magnetic susceptibility anisotropy $\Delta\chi^{(m)} > 0$, $\boldsymbol{z}||\boldsymbol{H}||\boldsymbol{d}$. If the dynamics can be considered neglible on the measurement timescale, the spectrum will be of the polydomain type, i.e. just a sum of contributions from all the N spins and the spectrum $\mathcal{I}(\nu)$, i.e. the absorption intensity as magnetic field frequency ν is scanned, will be

$$\mathscr{I}(\nu) = \propto \Big\langle \sum_{k=1}^{N} \delta(\nu - \nu_k) \Big\rangle = \Big\langle \sum_{k=1}^{N} \delta\big(\nu - \nu_z \mp \delta_Q P_2(\boldsymbol{d} \cdot \boldsymbol{u}_k)\big) \Big\rangle \tag{10.25}$$

where the angular brackets indicate an average over simulated configurations. The spectrum as written is constituted of delta functions (the so-called *stick spectrum*). In practice, however, each resonance peak will have a certain finite width, e.g. a Lorentzian shape with the intrinsic line width w (e.g. a typical w could be $\approx$200 Hz). In this case, the kth ^{2}H provides, considering the frequency origin at the Zeeman resonance, the double peak line shape

$$\mathcal{L}\left[\nu, \nu_k, w\right] = \frac{w}{[\nu - \nu_k]^2 + w^2} + \frac{w}{[\nu + \nu_k]^2 + w^2}, \tag{10.26}$$

so that the full spectrum will be $\mathcal{I}(\nu) \propto \big\langle \sum_{k=1}^{N} \mathcal{L}\left[\nu, \nu_k, w\right] \big\rangle$.

This static approach can be extended to also include dynamic effects, such as fluctuations of molecular long axes. This might seem strange because, even if MC simulations provide molecular orientations at a set of times, this is not a real-time sequence, but, as we have seen in Chapter 8, the result of an arbitrary stochastic Markov process. However, if the chosen MC sequence consists of physically plausible moves, e.g. of small rotations for each particle, we could use the simulation results also to examine the effect of molecular reorientations on the spectra, while leaving positions fixed, thus neglecting by now the effect of the slower translational diffusion. A further technical problem in considering the effect of molecular reorientations on $\mathcal{I}(\nu)$ is that the nuclear spin Hamiltonian terms containing $\hat{\mathbf{I}}_{\pm}$ operators do not commute with $\hat{\mathbf{I}}_Z$ and can induce magnetic transitions when changing orientations. However, at this level of generic, rather than atomistic, approach the so-called *adiabatic approximation* where these terms are neglected can be adopted, leading to a major simplification. In particular, in this limit we can use the simple form of the Linear Response (LR) Theory introduced in Section 6.9 to obtain an expression for the susceptivity to an applied probing field (here the radiofrequency magnetic field at frequency ν_k) in terms of correlation functions. To apply the general LR expression, Eq. 6.73, we identify the observable property A with the transversal nuclear magnetization component μ_X, $\mu_X = |\langle M|\hat{\mathbf{I}}_X|M'\rangle|$, the magnetic dipole transition moment (see Eq. I.12b), which in this case is also the physical property coupling to the probing magnetic field (Appendix I). Thus, using the general LR expression, Eq. 6.73,

$$\chi_{\mu_X\mu_X}(\omega) = \frac{\langle \mu_X \mu_X \rangle}{k_B T} - \frac{i\omega}{k_B T} \int_0^{\infty} \mathrm{d}t \, \langle \mu_X(0)\mu_X(t) \rangle \, \mathrm{e}^{i\omega t}, \tag{10.27a}$$

$$= \chi'(\omega) - i\chi''(\omega), \tag{10.27b}$$

where $\omega = 2\pi\nu$, $\chi'(\omega) \equiv \text{Re}\chi_{\mu_X\mu_X}(\omega)$, $\chi''(\omega) \equiv \text{Im}\,\chi_{\mu_X\mu_X}(\omega)$ and $\langle\mu_X(0)\mu_X(t)\rangle$ is the correlation function of μ_X [Anderson, 1954; Gordon and Messenger, 1972], also called the relaxation function $G(t)$. The absorption spectrum $\mathcal{I}(\omega)$ is obtained as the imaginary part of the susceptivity (similarly to the case of dielectric spectroscopy seen in Section 6.10) and the DNMR line shape $\mathcal{I}(\nu)$ is calculated as the Fourier–Laplace transform of $G(t)$ [Abragam, 1961; Golemme et al., 1988a]

$$\mathcal{I}(\nu) \propto \int \mathrm{d}t\, \mathrm{e}^{i2\pi\nu t} \langle\mu_X(0)\mu_X(t)\rangle = \int \mathrm{d}t\, \mathrm{e}^{i2\pi\nu t} G(t). \tag{10.28}$$

In our case, having adopted the adiabatic approximation, where spin flip terms are neglected, the spin Hamiltonians at different times commute and remain diagonal in the laboratory frame with $z \parallel \boldsymbol{H}_0$ even when molecular orientations and positions change, so that a new orientation just corresponds to a new resonance frequency. Using Heisenberg's equation of motion and the notation of Appendix I, we have

$$\frac{\mathrm{d}\mu_X(t)}{\mathrm{d}t} = \frac{\mathrm{d}}{\mathrm{d}t}\langle M|\hat{\mathbf{I}}_X(t)|M'\rangle = -\frac{it}{\hbar}\langle M|\hat{\mathscr{H}}_S^{\times}\hat{\mathbf{I}}_X|M'\rangle, \tag{10.29a}$$

$$= -\frac{it}{\hbar}[\langle M|\hat{\mathscr{H}}_S\hat{\mathbf{I}}_X|M'\rangle - \langle M|\hat{\mathbf{I}}_X\hat{\mathscr{H}}_S|M'\rangle], \tag{10.29b}$$

$$= -\frac{it}{\hbar}(\mathscr{E}_M - \mathscr{E}_{M'})\langle M|\hat{\mathbf{I}}_X|M'\rangle = -it2\pi\nu_{MM'}\,\mu_X = -it2\pi\nu_k\,\mu_X. \tag{10.29c}$$

Integrating over time we have $\mu_X(t) = \exp\left(i\int_0^t 2\pi\nu_k(t')dt'\right)\mu_X(0)$ and the relaxation function $G(t)$ is generated as

$$G(t) = \left\langle \exp\left[i2\pi\nu_Z t + i2\pi\int_0^t \mathrm{d}t'\nu_k\left(\boldsymbol{r}(t'), \boldsymbol{u}(t')\right)\right]\right\rangle, \tag{10.30a}$$

$$\approx \left\langle \exp\left[i2\pi\nu_Z t + i2\pi\sum_j \Delta t_j \nu_k(\boldsymbol{r}(t_j), \boldsymbol{u}(t_j))\right]\right\rangle. \tag{10.30b}$$

Here ν_Z is the Zeeman frequency, $\nu_k(\boldsymbol{r}(t_k), \boldsymbol{u}(t_k))$ is the instantaneous value of the quadrupolar frequency shift corresponding to the position $\boldsymbol{r}\left(t'\right)$ (here a lattice site) and $\langle\ldots\rangle$ stands for the average for the nuclei k. In this approach the time is divided into short intervals of length Δt_j so that the frequency ν_k can be represented by a single value at that 'instant'. This treatment does not include the effect of lifetime broadening, which can be introduced by convoluting $\mathcal{I}(\nu)$ with a Lorentzian or Gaussian curve of width w. The calculation of the $G(t)$ defined by Eq. 10.30b requires equilibrium molecular position, $\boldsymbol{r}(t)$ and orientation $\boldsymbol{u}(t)$ and to perform averages over all possible initial states and over all possible molecular 'paths'.

Translational diffusion can be further simulated by a simple random-walk process in which molecules jump between neighbouring lattice sites, always adjusting their orientation to the local, instantaneous $\boldsymbol{u}_i$. Both types of motional effects thus affect the instantaneous resonance frequency and, consequently, the line shape $\mathcal{I}(\nu)$. As an example, the DNMR spectrum can be calculated for an experiment where an electric field is applied to the PDLC droplets. If the effect of the NMR spectrometer magnetic field on the configuration is negligible, as is the case at least for submicrometre droplets

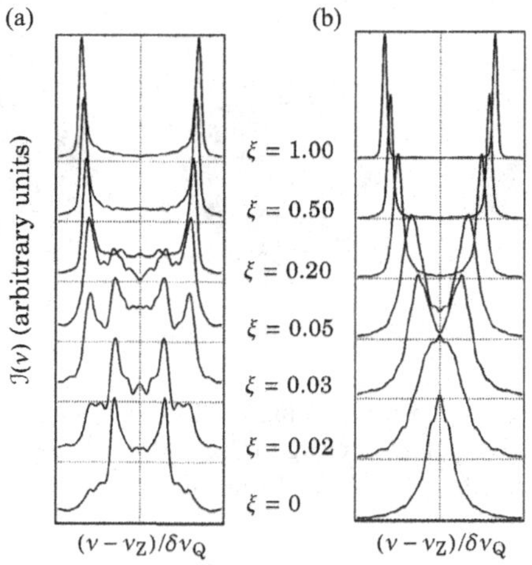

Figure 10.18 DNMR spectra $\mathcal{I}(\nu)$ as a function of frequency $(\nu - \nu_Z)/\delta\nu_Q$ for an LL droplet with radial BC in the nematic phase ($T^* = 0.8$) for different values of effective field strength ξ. (a) Rotational diffusion only. (b) Fast translational diffusion limit. A hedgehog-to-aligned structural transition occurs with increasing ξ [Chiccoli et al., 2000].

[Crawford et al., 1993], then the methodology can be used to study the effect of an external field $\boldsymbol{E}$ of different strengths on the molecular organization via the DNMR spectrum. The field can be introduced in the simulation by adding to the unperturbed droplet Hamiltonian U_N^{D} a field coupling term U_N^{F}:

$$U_N^{\mathrm{F}} = -\epsilon\xi \sum_{i=1}^{N} P_2(\boldsymbol{u}_i \cdot \hat{\boldsymbol{E}}), \tag{10.31}$$

where $\hat{\boldsymbol{E}}$ is a unit vector along the electric field direction and the effective strength parameter ξ depends on the anisotropy of the dielectric susceptivity anisotropy $\Delta\varepsilon$ and the field intensity $\propto E^2$. In Fig. 10.18 we see some simulated spectra where translational diffusion is either absent (Fig. 10.18a) or fast (Fig. 10.18b). It is apparent that the effect of increasing field strength is that of transforming the organization of the system from a hedgehog type, as we have seen before, to an aligned uniaxial one. In this case too we can define an appropriate order parameter, $\langle P_2 \rangle_E$, which now expresses the molecular alignment with respect to the field,

$$\langle P_2 \rangle_E = \frac{1}{N} \sum_{i=1}^{N} P_2(\boldsymbol{u}_i \cdot \hat{\boldsymbol{E}}). \tag{10.32}$$

10.9 Liquid Crystal Shells

A thin film of nematic that wets, with tangential anchoring, a spherical colloidal particle, has to exhibit defects with a total topological charge of 2, and actually four $s = 1/2$ defects

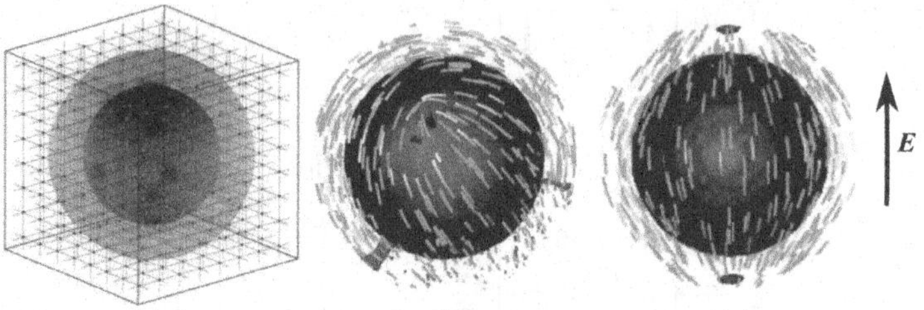

Figure 10.19 (a) A nematic thin film coating a sphere with tangential anchoring and (b) the four surface defects of topological charge 1/2. (c) Modification of the defects under the action of a uniform electric field [Skačej and Zannoni, 2008].

placed in a tetrahedral arrangement (Fig. 10.19a) appear, making the particle a kind of tetravalent 'colloidal atom' [Nelson, 2002]. Indeed, these defects represent spots of high-free energy, particularly suitable for a chemical attack, e.g. by nucleic acid strands, so as to build complex colloidal architectures using the now well-developed DNA technology [Song and Deng, 2017; Seeman, 2020]. Mathematical techniques have been used to locate the defects, through setting up and minimizing the Frank elastic energy like in Lubensky and Prost [1992]. However, it is important to find ways for varying the valence of these colloidal atoms, e.g. by application of some external field or by changing the particle shape, and computer simulations can be useful in this respect or also when additional features, e.g. particle biaxiality, are included.

MC simulations of LC ordering in spherical nematic shells can be based on the LL lattice model considering the shell comprised between two concentric spheres carved from the cubic lattice [Skačej and Zannoni, 2008]. Choosing inner and outer sphere radii of $30a$ and $40a$, yields a total of $N = 148,968$ shell particles. Tangential (planar degenerate) BCs at both surfaces can be imposed by an additional layer of 'ghost' particles, with $\epsilon_{ij} < 0$ for the ghost-nematic interaction (this corresponds to strong planaring anchoring). It is possible to visualize the nematic director field at its defects with the already described (Section 10.6.1) method of Callan-Jones et al. [2006], calculating the average components of the local ordering matrix $\mathrm{U}_i = \langle \mathbf{u}_i \otimes \mathbf{u}_i \rangle$ for each lattice site. The results in the absence of external fields (Fig. 10.19b) reproduce the analytical tetrahedral defect structure. We can now study the effect of applying a uniform electric field by adding to the LL Hamiltonian a coupling term, exactly like in Eq. 10.31. For a sufficiently strong field the four $s = 1/2$ defect lines partially coalesce to form two pairs of $s = 1$ surface point defects at the sphere poles (Fig. 10.19c) with the axis connecting the two poles parallel to the field direction. Note that the surface topological defect charge remains constant and equal to 2. The valence can be further changed by the application of suitable multipolar fields as shown in Skačej and Zannoni [2008].

The defect location, if not the valence, can also be changed, e.g. varying the shape of the colloidal core from spherical to ellipsoidal uniaxial and biaxial colloidal particles

[Bates et al., 2010]. Typically, four mutually repulsive half-strength defect lines penetrating the shell are observed, as for the spherical particles. For shells of constant thickness, the defect lines tend to accumulate in the high curvature regions or, if the thickness of the nematic coating varies across the surface, the defect lines tend to be located in the thinnest regions.

11

Molecular Simulations

11.1 Empirical Anisotropic Models

There are a number of convincing reasons to try to go beyond the lattice models of Chapter 10, even before requiring all the details of an atomistic approach. A very important one is the need to abandon the fixed positions inherent in lattice models, an essential step to simulate phases with partial positional order like the various smectic ones. Another reason is to start introducing some precious features of molecular structure, like shape, charge distribution, etc. and examine their relation to phase behaviour. Yet another reason is to find the minimal or essential molecular features required to obtain some desired phase. For instance: how elongated should a particle be to yield a nematic phase? How can we obtain a certain smectic phase, or avoid one if we only want a nematic, e.g. a biaxial, one? Empirical anisotropic models typically represent a molecule with a single particle or a few connected ones [Zannoni, 2001b; Care and Cleaver, 2005; Allen, 2019]. In many ways these are definitely more realistic than lattice models, as they allow molecular mobility and a cascade of positionally ordered or disordered phases like smectics and, respectively, nematics and isotropic (liquid and vapour) fluids. We introduced the main generic particle models of this type, as well as their interactions, in Chapter 5. Here we wish to focus on their applications to LCs. Due to their relative simplicity molecular models have allowed the investigation of molecular order, dynamics and bulk material properties in quite large-scale simulations (10^4-10^6 molecules) of a range of liquid crystalline phases [Allen, 2019]. Moreover, molecular potentials have been employed for the study of bulk properties of phases not yet experimentally found and for which molecular structures are not available, proving valuable in contributing design hints to synthetic chemists. Some examples include biaxial [Berardi and Zannoni, 2000] and ferroelectric nematics [Berardi et al., 2001] still actively pursued by synthetic chemists.

In terms of computer resources, molecular coarse-grained potentials, such as the Gay–Berne (GB) model (Section 5.6.3), provide three key advantages over atomistic potentials. Most obvious is the speed-up due to a reduction in the number of interacting sites with respect to an atomistic model, albeit handling anisotropic, rather than spherical, particles is unavoidably more demanding. However, almost as important, is the increase in time-step that can be achieved in a molecular dynamics simulation, with the associated longer time windows that can be explored. Moreover, the reduction in the number of degrees of freedom with respect to an atomistic model, can significantly speed up the phase space exploration [Hughes et al., 2008; Lintuvuori and Wilson, 2008]. For liquid crystalline systems, a major simulation cost is associated with taking a simulation through a phase transition from a disordered to a more ordered phase. For instance, while gradually cooling down an initially isotropic sample, the spontaneous onset of an ordered domain, large enough to seed the formation of a liquid crystalline phase, requires long simulation times, particularly if a first-order transition is involved and it is highly desirable to speed-up these processes. While it would be attractive to have a general formalism for molecular models including a variety of the interactions discussed in Chapter 5 and simultaneously present in real molecules, it is often useful to introduce empirical forms for the potential, which only include some basic features, one at a time avoiding unnecessary complications. This is particularly important in computer simulations, where a huge number of evaluations of the pair potential has to be performed. Here we report a few of the most relevant and commonly used of these model systems and their applications to liquid crystals.

11.2 Anisotropic Spherical Particles

Soft or hard spherical particles with an embedded anisotropic interaction are possibly the simplest type of off-lattice intermolecular potential that can be used to model LCs. Even if these systems inevitably lack the important shape anisotropy characteristic of real mesogens, they provide interesting models to test theories and here we briefly review some of the most important ones.

11.2.1 Maier–Saupe Spheres

The spherical particles models where the anisotropic contribution is a simple second-rank, pseudo-dispersive, P_2-type interaction essentially corresponds to the simple model used in Section 7.2 to introduce Maier–Saupe Molecular Field Theory, and thus they are sometimes referred to as Maier–Saupe spheres. A specific example is that of soft Lennard–Jones spheres (Section 5.2.3):

$$U^{\mathrm{MS}}(\boldsymbol{r}_{ij},\boldsymbol{u}_i,\boldsymbol{u}_j) = U^{\mathrm{LJ}}(r_{ij}) + u_2(r_{ij})P_2(\boldsymbol{u}_i \cdot \boldsymbol{u}_j). \tag{11.1}$$

The potential corresponds to LJ spheres (Eq. 5.18), each decorated at the centre with an anisotropic interaction site with orientation $\boldsymbol{u}_i$. The model reduces to the Lebwohl–Lasher one studied in Chapter 10, when molecule centres are fixed at the sites of a simple cubic lattice and the range of $u_2(r_{ij})$ is restricted to nearest neighbours. Two forms for $u_2(r_{ij})$ have been studied by Luckhurst and Romano [1980b]: (A) $u_2^{(\mathrm{A})}(r_{12}) = -4\lambda\epsilon\,(\sigma/r_{12})^6$ or

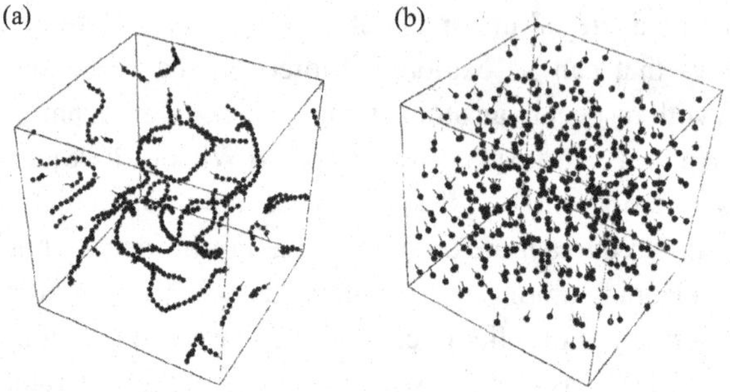

Figure 11.1 Instantaneous configuration of 500 dipolar hard spheres at $T^* = (1/\mu^*)^2 = 0.0816$ and reduced density (a) $\rho^* = 0.01$ and (b) $\rho^* = 0.8$, in a box of side length $L = 36.84\sigma$. Hard spheres belonging to chains of length larger than 20 beads are in black, the others in grey, with a thin line representing the direction of the dipole moment. The state in part (b) is ferroelectric with $\langle P_1 \rangle = 0.84$ [Levesque and Weis, 1994].

(B) $u_2^{(B)}(r_{12}) = -4\lambda\epsilon \left\{(\sigma/r_{12})^{12} + (\sigma/r_{12})^6\right\}$. It is worth noting that the repulsive part of the potential is isotropic and does not depend on the orientation of the intermolecular vector. MC simulations for the two forms of $u_2(r_{12})$ were run for $N = 256$ and 864 particles with the strength parameter $\lambda = 0.15$. It is found that both models yield a nematic phase with an orientational order parameter $\langle P_2 \rangle$ at the isotropic-nematic transition that appear slightly below that obtained with a mean field treatment, i.e. 0.429, as we see from Table 7.1 for the $\lambda_4 = 0$ case.

11.2.2 Dipolar Hard Spheres

A number of studies have discussed hard spherical particles with a permanent dipole, $\boldsymbol{\mu}$, embedded. The pair potential is:

$$U_{ij}^{\mathrm{DHS}} = U^{\mathrm{HS}}\left(r_{ij}\right) + U_{ij}^{\mathrm{DD}}, \tag{11.2}$$

where U^{HS} was given in Eq. 5.2.1, and the dipole-dipole interaction U_{ij}^{DD} is, recalling Eq. 5.86,

$$U_{ij}^{\mathrm{DD}} = -\frac{1}{r_{ij}^5}[3(\boldsymbol{\mu}_i \cdot \boldsymbol{r}_{ij})(\boldsymbol{\mu}_j \cdot \boldsymbol{r}_{ij}) - r_{ij}^2(\boldsymbol{\mu}_i \cdot \boldsymbol{\mu}_j)] \tag{11.3}$$

in units of $4\pi\varepsilon_0$. Patey and Valleau [1974] studied the model with MC simulations comparing the pressure calculated with the virial expression (Eq. 4.133) with various approximate statistical theories within the isotropic phase, but no LC phase was discussed. Levesque and Weis [1994] performed MC simulations for a system with reduced dipole moment $\mu^* = (\mu^2/\sigma^3 k_B T)^{1/2} = 3.5$ along a low temperature isotherm, finding that the dipolar hard spheres (DHS) system structure transforms from chain-like associations (Fig. 11.1a), at low reduced densities ($\rho^* \equiv \rho\sigma^3 \leq 0.2$), to a ferroelectric type ordering for densities $\rho^* > 0.6$. The aspect of these chains, kept together by the favourable head-to-tail dipolar interactions, is very similar to that of partially flexible polymers. de Gennes and Pincus [1970] predicted, using qualitative arguments, that this should occur in ferromagnetic colloids, for which the DHS model provides a simple model. Indeed, from a formal point of view, the dipole in Eq. 11.2 could be either an electric or a magnetic one and the respective polar phases would then be ferroelectric or ferromagnetic. Thus, for $\rho^* > 0.6$, DHS can exhibit, at sufficiently low temperatures, spontaneous orientational ordering of the dipole moments which is, different from that of ordinary nematics, ferroelectric. This can be put to test, since for normal, non-ferroelectric, nematics $\langle P_2 \rangle \neq 0$ and $\langle P_1 \rangle = 0$, while for ferroelectric nematics both $\langle P_2 \rangle$ and $\langle P_1 \rangle$ must be non-zero. A snapshot of a configuration at $\rho^* = 0.8$ showing clear polar ordering is reproduced in Fig. 11.1b. Its global dipole (polarization) $P = |\langle P_1 \rangle|$ is computed as the average value

$$P = \frac{1}{N} \Big| \sum_{i=1}^{N} \hat{\boldsymbol{\mu}}_i \cdot \boldsymbol{d} \Big|, \tag{11.4}$$

where $\hat{\boldsymbol{\mu}}_i$ is a unit vector along the dipole moment of particle i and $\boldsymbol{d}$ is the instantaneous director corresponding to the eigenvector associated with the largest eigenvalue of the ordering matrix $\mathbf{Q}$ (Eq. 3.47). The DHS systems, although apparently simple, turn out to be difficult to equilibrate. At the lower density, $\rho^* = 0.6$, a polarization $P = 0.83$ was obtained when starting from an initial perfectly polarized configuration. However, when the system was started from an unpolarized state with random distribution of the dipole moments, the polarization was extremely slow to build up and did not converge within the length of the run, yielding, after 110,000 trial moves per particle, only $P \simeq 0.30$. The isotropic-ferroelectric (IF) transition for DHS systems of different sizes at $\rho^* = 0.80$ and 0.88 was studied by Weis [2005] in a cooling-down temperature sequence (since $T^* = (1/\mu^*)^2$ this is equivalent to increasing the dipole strength), who found a transition at $T^*_{\text{IF}} \approx 0.212$ for $\rho^* = 0.80$ and at 0.263 for $\rho^* = 0.88$. The dipolar energy of the periodic system (N, V, T) was evaluated by means of an Ewald sum (Section 5.4.4) with conducting boundary conditions. The parameter α controlling the relative contributions to the Ewald sum of the direct and reciprocal space terms, was $\alpha L = 5.76$, allowing the real-space term to restrict contributions to the pair interaction truncated at half the box size of the simulation cell. The term in reciprocal space included all lattice vectors $\boldsymbol{k} = 2\pi \boldsymbol{n}/L$, $\boldsymbol{n} = (n_x, n_y, n_z)$ with $|\boldsymbol{n}|^2 \leqslant n^2_{max} = 64$. Weis and Levesque [2006] have studied DHS systems with the same technique and found a change from the isotropic-ferroelectric to the columnar ferroelectric phase in the density range $\rho^* = 0.92\text{–}0.95$. Solid phases of dipolar hard spheres with various crystal symmetries were also studied and, in particular, MC simulations at low temperatures [Levesque, 2017] point to the stability of a polarized solid phase with the unusual number of 11 nearest neighbours, the so-called primitive tetragonal packing or tetragonal close packing.

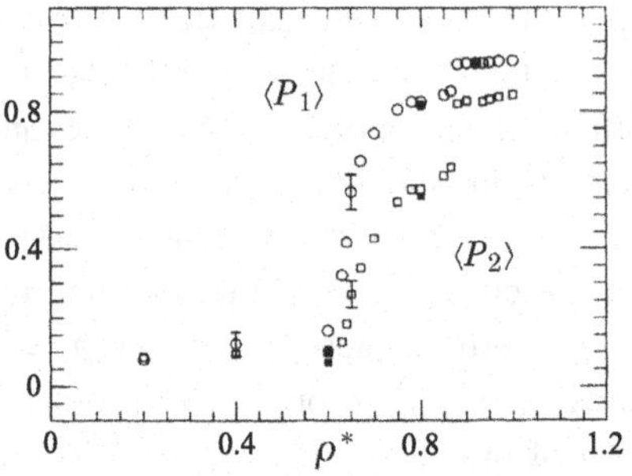

Figure 11.2 The orientational order parameters $\langle P_1 \rangle$ and $\langle P_2 \rangle$ as a function of reduced density ρ^* for a system of soft dipolar spheres with $N = 256$ (○), (□) and $N = 864$ (●), (■) particles [Wei and Patey, 1992].

11.2.3 Dipolar Soft Spheres

The pair potential of two LJ (cf. Eq. 5.18) spheres with a dipole embedded in the centre (also called the Stockmayer [1941] potential) is

$$U_{ij}^{\mathrm{DLJ}} = U_{ij}^{\mathrm{LJ}} + U_{ij}^{\mathrm{DD}}. \tag{11.5}$$

Clusters of Stockmayer spheres aggregate in chains with head-tail alignment of the dipoles that persist, for reduced dipoles $\mu^* = 1$, ($\mu^* = \mu/\sqrt{\epsilon\sigma^3}$) through the solid-liquid coexistence temperature. Wei and Patey [1992] considered a dipolar soft-sphere model defined by the pair potential

$$U_{ij}^{\mathrm{DSS}} = U_{ij}^{\mathrm{SS}} + U_{ij}^{\mathrm{DD}}, \tag{11.6}$$

where the soft-sphere (DSS) potential is the repulsive part of the LJ one: $U_{ij}^{\mathrm{SS}} = 4\epsilon(\sigma/r_{ij})^{12}$ and the dipole-dipole interaction was given in Eq. 11.3. Here ϵ and σ are the LJ parameters characterizing the soft-sphere potential, $\boldsymbol{\mu}_i$ is the dipole moment of particle i, $\boldsymbol{r}_{ij} = \boldsymbol{r}_j - \boldsymbol{r}_i$, and r_{ij} is the modulus of $\boldsymbol{r}_{ij}$. The possible existence of a nematic and/or ferroelectric nematic phase was monitored using molecular-dynamics simulations and calculating the first- and second-rank orientational order parameters, $\langle P_1 \rangle$ and $\langle P_2 \rangle$, respectively. For sufficiently high dipole moments, the system presents an orientationally ordered phase having ferroelectric order (Fig. 11.2). In summary, simulations have shown that the dipolar interactions alone are sufficient to induce an orientationally ordered liquid phase for spherical core systems both for the DSS potential and for the DHS potentials.

11.3 Anisotropic Aggregates of Spherical Particles

Even if the MS, DHS and DSS just seen yield some ordered phase, it is clear that in order to treat liquid crystals it is essential to allow for the anisotropy in the shape of particles. One simple possibility is to build such a potential by suitably joining together spherical particles to form an anisotropic object, e.g. a linear chain. We shall consider briefly the cases of hard and LJ beads.

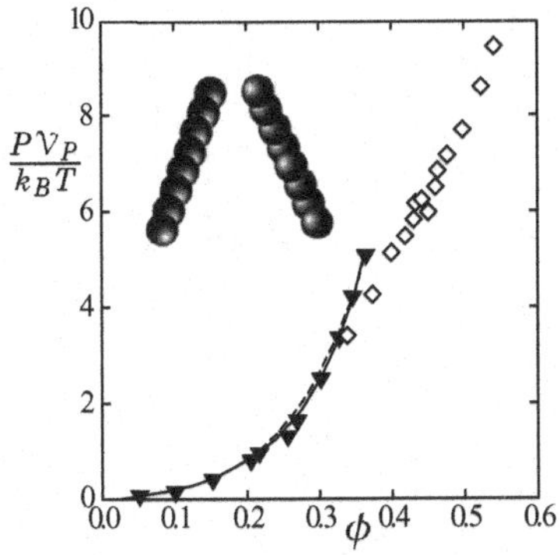

Figure 11.3 Equation of state for a system of rods made of eight fused spheres, with $L^* = 4.2$ showing the isotropic (▼) and the nematic-smectic branch (◇). The lines in the isotropic branch correspond to the Boublik [1975] expression, Eq. 11.7 (—) and to the best-fit y expansion mentioned in the text (- - -) [Whittle and Masters, 1991].

11.3.1 Linear Chains of Hard Spheres

Various models based on rigid chains of tangent or partially overlapping fused spheres have been considered. In particular, systems of linearly fused hard spheres have been found [Whittle and Masters, 1991] to exhibit nematic ordering for an overall length (L_{tot}) to breadth ratio $L^*_{tot} = L_{tot}/\sigma = 5.2$ (obtained with eight equally spaced fused spheres of diameter σ) as we show in Fig. 11.3. As the number of spheres per molecule increases, and their separation shrinks to 0, this system becomes equivalent to a hard spherocylinder fluid (see Section 11.4.2). In this case, the hard spheres linear chains, non-convex, can be approximated with spherocylinders (convex bodies). The isotropic branch is well represented by Boublik's [1975] equation of state for hard convex particles:

$$\frac{PV}{Nk_BT} = \frac{1 + (3\xi - 2)\phi + \left(3\xi^2 - 3\xi + 1\right)\phi^2 - \xi^2\phi^3}{(1-\phi)^3}, \tag{11.7}$$

where $\phi = \rho\mathcal{V}_P = N\mathcal{V}_P/V$ (with $\mathcal{V}_P$ the volume of the particle) is the packing fraction and ξ is the asphericity factor for convex bodies already employed in Section 5.5: $\xi = \mathcal{R}_P\mathcal{S}_P/3\mathcal{V}_P$, with $\mathcal{R}_P$ the mean curvature and $\mathcal{S}_P$ the surface area. Another useful expression for the equation of state is the so-called y expansion [Barboy and Gelbart, 1979; Hansen and McDonald, 2006]: $(P(\phi)\mathcal{V}_P)/(k_BT) = \sum_n C_n y^n$, where $y = \phi/(1-\phi)$ is the volume fraction. An interesting feature of modelling elongated particles with a connected set of spherical beads is the possibility of easily introducing flexibility, by allowing, e.g. spring-like, instead of perfectly rigid, links connecting some or all the spheres. A semi-flexible system of connected hard spheres was studied by Wilson and Allen [1993] and more recently tangent and fused hard spheres with a rigid and a flexible segment have been investigated [Cinacchi et al., 2005; Movahed et al., 2006; Oyarzun et al., 2013, 2015; van Westen et al., 2013].

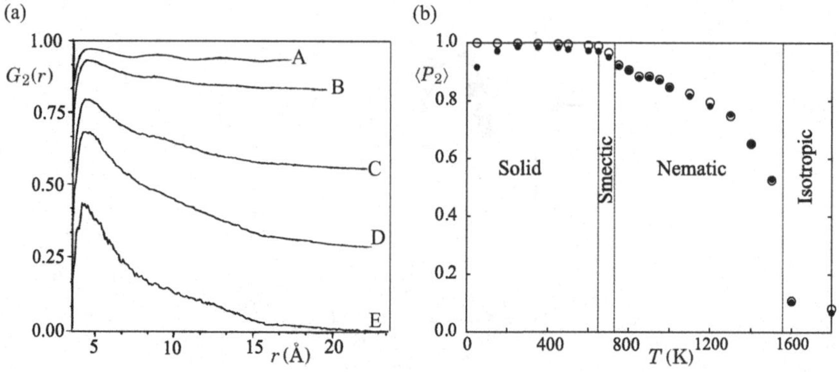

Figure 11.4 MD results for a system of 11 soft repulsive beads linear chains. (a) Orientational correlation $G_2(r)$ for temperatures (A) 700 K, (B) 800 K, (C)1300 K, (D) 1500 K, (E) 1800 K. (b) Order parameter $\langle P_2 \rangle$ from diagonalization of the ordering matrix (●) and from the large separation tail of $G_2(r) \approx \langle P_2 \rangle^2$ (Eq. 4.97) (○) [Paolini et al., 1993].

11.3.2 Linear Chains of Soft Repulsive Spheres

Rod-like anisotropic particles formed by a rigid linear chain of n_s centres of force that interact with other particles via a site-site potential corresponding to the repulsive part of the LJ potential

$$U(r_{ij}) = 4\epsilon \, (\sigma / r_{ij})^{12}, \tag{11.8}$$

where i and j are two sites belonging to different particles (with $n_s = 11$) were studied by Paolini et al. [1993]. The potential parameters used are $\sigma = 3.9$ Å and $\epsilon = 6.0 \times 10^{-22}$ J. Of the 11 sites of each molecule, only the 2 at the extremes are true beads, each of mass $m = 1.993 \times 10^{-23}$ g, while the others are just massless centres of force. NPT MD simulations of $N = 600$ molecules have shown liquid crystal phases, and in Fig. 11.4 we report the orientational order parameter $\langle P_2 \rangle$ and the orientational-radial correlation $G_2(r)$ (Eq. 4.97).

11.3.3 Lennard–Jones Linear Chains

Extensive simulations of linear chains of $n_s = 8, 10, 12$ LJ beads have been performed [Rivera et al., 2016] and also in this case partially flexible models formed by a rigid core and a flexible tail have been proposed. The freely jointed part is not subjected to any bond-bending or torsional potential; thus it is free to adopt any possible molecular configuration subject to the constraints of a rigid bond length and the pair interaction between segments. Intermolecular pair interactions are evaluated for beads of different molecules and for segments of the same molecule that are separated by two or more bonds. For longer LJ oligomeric chains, the isotropic-nematic equilibrium is shifted towards lower densities while the density difference between both coexisting phases is increased. An example of the phase diagram for a rigid chain system built from 10 LJ spheres [Rivera et al., 2016],

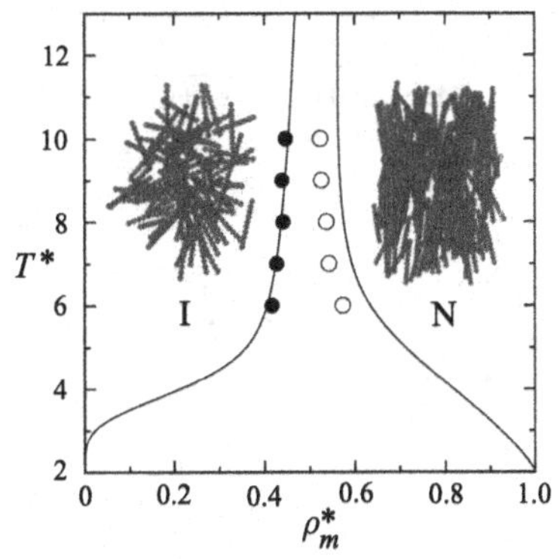

Figure 11.5 Reduced temperature T^* vs reduced monomer density ρ_m^* phase diagram for a system of 200 linear rigid chains of 10 LJ beads showing isotropic (I)-nematic (N) phase equilibria. Solid lines are from the analytical equation of state of van Westen et al. [2013], filled (●) and open (○) symbols are simulation results for the isotropic and nematic phase, respectively [Rivera et al., 2016].

shown in Fig. 11.5, compares, rather successfully, with the results of a generalized Onsager theory (Section 7.6) for non-convex molecules [van Westen et al., 2013].

11.4 Anisotropic Hard Particles

Various authors, in particular Frenkel [1987], Allen et al. [1993] and Bolhuis and Frenkel [1997] have shown that LC phases can be formed starting from hard ellipsoids and spherocylinders since, contrary to naive expectations, the formation of an anisotropic phase can lead to an increase in entropy, e.g. for sufficiently anisotropic particles or high densities (see Section 5.5). Here we focus on the LC phases obtained for the two most studied cases: hard ellipsoids and hard spherocylinders, whose pair potential were introduced in Sections 5.5.1 and 5.5.2, respectively.

11.4.1 Hard Ellipsoids

The first off-lattice simulation of elongated objects yielding an anisotropic liquid was probably that of a 2D system of hard ellipses performed by Vieillard-Baron [1972] who found, for an aspect ratio $\kappa = 6$, that the system showed isotropic, nematic and solid phases. The 2D system of hard ellipses has been recently revisited [Bautista-Carbajal and Odriozola, 2014] determining, with extensive MC runs, an area fraction-aspect ratio phase diagram for $1 \leq \kappa \leq 5$ and identifying four regions: isotropic, nematic, plastic and solid.

In 3D a system of hard ellipsoids (Section 5.5.1) is quite interesting since it gives rise for different elongations (see Fig. 11.6) to plastic crystals (when the ellipsoids are nearly spherical), isotropic liquids, nematics and crystals [Frenkel and Mulder, 1985; De Michele et al., 2007; Odriozola, 2012]. In particular, hard ellipsoids have been shown to form a

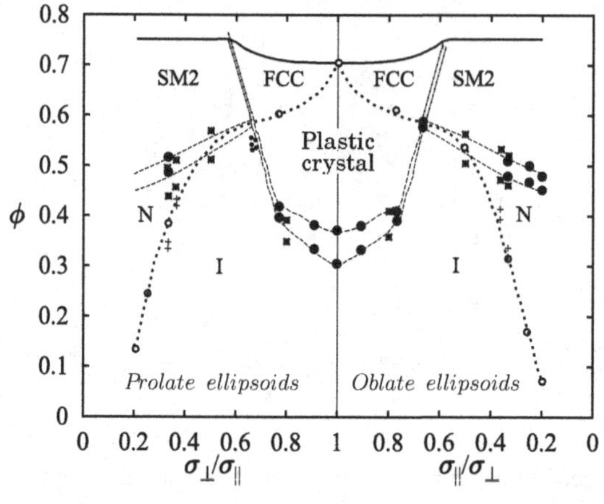

Figure 11.6 Phase diagram of hard uniaxial ellipsoids of length $\sigma_{||}$ and breadth $\sigma_\perp$ showing fluid-solid (■) and isotropic (I)-nematic (N) (○) transitions. Plastic crystal, face-centred cubic (FCC) and SM2 solid phases are observed. The solid line is the maximum achievable density [Donev et al., 2004a]. The dashed, dotted and dot-dashed lines are guides for the eye indicating the fluid-solid, the I-N and the FCC-SM2 [Radu et al., 2009] transitions, respectively. Some isotropic-nematic (+) and nematic-solid (∗) points are from Frenkel and Mulder [1985]. Adapted from [Odriozola, 2012].

nematic phase for length-to-breadth ratios $\kappa > 2.75$ [Frenkel and Mulder, 1985] without the need to invoke attractive forces. As for any hard repulsive system, the transition from one phase to the other will be dictated by the gain in total entropy and it is worth noting that going from a disordered isotropic fluid to a nematic, while the system loses orientational entropy, it gains translational entropy, e.g. by the increased possibilities of displacement along the molecular axis allowed by the orientational order and fluidity of the nematic phase. On the other hand, as it is apparent from the phase diagram in Fig. 11.6, the system does not give a smectic phase for elongated ellipsoids or a columnar phase for discotic ones and theoretical arguments have been put forward to justify this [Lebowitz and Perram, 1983; Frenkel, 1987]. In particular, Lebowitz and Perram [1983] argued that a system of hard parallel ellipsoids could be mapped into a system of spheres by a linear, volume preserving, affine transformation. Thus, a hypothetical fully ordered HE smectic would be mapped into a crystal of spheres instead of a stack of 2D liquid layers of spheres.

It is interesting to see how effectively ellipsoidal particles can pack. Given the packing fraction $\phi = N\pi\sigma_{||}\sigma_\perp^2/(6V)$, the densest arrangement of hard ellipsoids with axes σ_i for certain aspect ratios κ has been numerically found to be $\phi \approx 0.7707$ (see Fig. 11.7) [Donev et al., 2004a,b; Chaikin et al., 2006]. As the jamming density approaches, the particles touch to form the contact network of the packing, exerting compressive forces on each other but not being able to move despite thermal agitation. We can see that spheres ($\kappa = 1$) give the worst packing with a jamming density $\phi_J \approx 0.64$ (see Section 5.2.1) and that the density ϕ_J increases with κ, reaching densities as high as $\phi_J \approx 0.74$ for the self-dual ellipsoids ($\sigma_x = \sqrt{\sigma_z\sigma_y}$), which corresponds to $\beta = 1/2$, while for higher aspect ratios

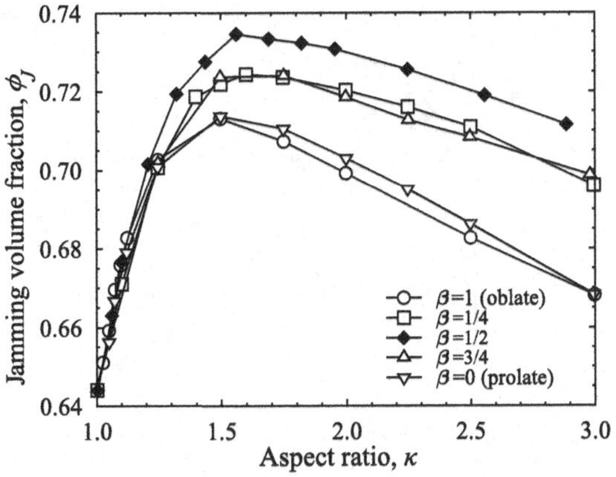

Figure 11.7 Jamming volume fraction ϕ_J vs aspect ratio κ obtained numerically for 10^4 hard biaxial ellipsoids with axes ratios $\sigma_x : \sigma_y : \sigma_z = 1 : \kappa^\beta : \kappa$, where $\kappa > 1$ is the aspect ratio and β, $(0 \leq \beta \leq 1)$ is the 'oblateness', with $\beta = 0$ corresponding to a prolate, $\beta = 1$ to an oblate uniaxial ellipsoid and $\beta = 1/2$ to a self-dual one [Chaikin et al., 2006].

there is a decrease in ϕ_J. Numerical simulations and experiments on hard spheroids [Donev et al., 2004b] have confirmed that ellipsoids can randomly pack more densely than spheres, up to $\phi = 0.68$ to 0.71 for spheroids with an aspect ratio close to that of M&M's© sweets and even approach $\phi \approx 0.74$ for ellipsoids with axes ratios near 1.25 : 1 : 0.8. Thus, the density of jammed random packings of ellipsoids can approach that of the crystal packing, raising the possibility of a thermodynamically stable glass [Man et al., 2005]. Careful free energy calculations have shown that the most stable crystal phase of hard ellipsoids, the SM2, has a simple monoclinic unit cell containing two ellipsoids with different orientation [Radu et al., 2009].

11.4.2 Hard Spherocylinders

This simple potential is extremely important because it has been shown by computer simulations [Frenkel, 1987] to give not only isotropic, nematic and crystal phases but also smectic ones. We already introduced the pair potential and some geometrical metrics for hard spherocylinders (HSCs) in Section 5.5.2 and in Section 7.6 we saw how HSCs have been studied with classical Onsager theory and with the extended version of Parsons and Lee starting from their second virial coefficient. Systems of HSCs with full orientational freedom have been extensively investigated with computer simulations by Bolhuis and Frenkel [1997], and the results show that the system presents a plastic crystal (rotator phase) at low aspect ratios $x \equiv (l + d)/d$ and, as x increases, an isotropic liquid; then for $x > 4.7$, nematic and smectic A phases and the high-density, orientationally ordered solid. Between $x = 1.35$ and $x = 4.1$, only two phases occur: the low-density isotropic phase and the high-density, orientationally ordered crystal phase. The smectic phase first becomes stable at the I-S_A-S triple point which is located at $x \approx 4.1$. The N-Sm transition occurs at density

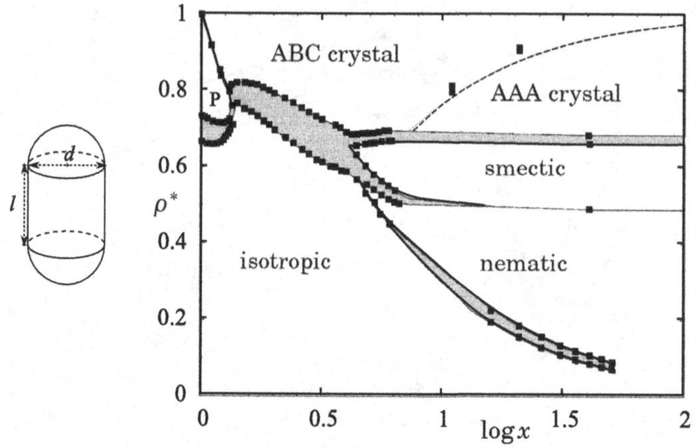

Figure 11.8 Phase diagram of hard spherocylinders showing reduced density ρ^* vs log of the aspect ratio, $x = (l + d)/d$, with $1 < x < 100$. Here $\rho^* = \rho/\rho_{\mathrm{cp}}$, with $\rho_{\mathrm{cp}} = 2/[\sqrt{2} + (l/d)\sqrt{3}]$ the density for regular close packing of spherocylinders. The phases shown are isotropic, nematic, plastic (P), smectic and solids (AAA, ABC). The dotted line is an estimate for the first-order AAA-ABC transition [Bolhuis and Frenkel, 1997].

$\rho^* \approx 0.5$, where it is clearly first order, but the density jump at the N-Sm transition is reduced upon increasing aspect ratio. The smectic to solid transition is located at $\rho^* = 0.66–0.68$ and is also first order [Bolhuis and Frenkel, 1997]. Here the reduced density $\rho^* = \rho/\rho_{cp}$, where $\rho_{cp} = 2/(\sqrt{2} + (l/d)\sqrt{3})$ is the density relative to that of regular close packing of spherocylinders.

The hard spherocylinder system is also important for the study of colloids formed by rod-like inorganic [van Bruggen et al., 1996] or virus particles [Fraden et al., 1993; Dogic and Fraden, 2006]. In this case, high anisometric ratios can be appropriate and the phase diagram for $0 < l/d < 100$ is shown in Fig. 11.8 [Bolhuis and Frenkel, 1997]. Two types of solids are found. In the AAA solid the stacking is such that every particle is exactly above a particle in the layer below, while in the ABC stacking the layers are shifted.

11.4.3 Dipolar Hard Spherocylinders

Weis et al. [1992] studied the orientational order in systems of dipolar hard spherocylinders of aspect ratio $x = (l + d)/d = 6$ (see Fig. 11.8) with a longitudinal dipole moment located at a distance $2.5d$ from the molecular centre. Starting from a close-packed structure with hexagonal symmetry and all dipole moments aligned in the same direction parallel to the z-axis, (N, P, T) MC calculations were performed allowing not only translations and rotations, but also head-tail flip moves to speed up dipole equilibration. At density $\rho^* = \rho d^3 \approx 0.13$ and pressure $P^* = P d^3/k_B T = 3.8$, the reduced dipole $\mu^* = \left(\mu^2/d^3 k_B T\right)^{1/2}$ was varied between 0 and $\sqrt{6}$. The simulations showed formation of a monolayer smectic A phase (cf. Fig. 1.28) with unpolarized layers. Different to the case of dipolar hard spheres, a ferroelectric phase was not found. Further extensions of this work [Levesque et al., 1993] showed that, irrespective of the characteristics of the dipole moment, such as position (from centre to the end of the molecule), strength (μ^* varying from 0 to $\sqrt{6}$) and direction (from parallel to perpendicular to the molecular axis), the stable smectic A phase was found to have a monolayer S_{A_1} structure and that in the case of an axial dipole moment the layers were unpolarized. Thus, the rich variety of strongly polar smectic A phases seen in Section 1.7.2 and in particular the rather common bilayer S_{A_2} or interdigitated partial bilayer S_{A_d} (like that found, e.g. in 8CB) was not observed.

11.5 Gay–Berne Models

The Gay–Berne (GB) pair potential [Gay and Berne, 1981], introduced in Section 5.6.3, is probably the one most extensively applied to liquid crystals and studied, both theoretically and with computer simulations. The uniaxial GB model (Eq. 5.57) has proved to be quite rich as it yields liquid, nematic, smectics and crystal condensed phases for elongated molecules as well as nematic, columnar and crystals ones for discotic mesogens [Bates and Luckhurst, 1999b; Zannoni, 2001a; Allen, 2019]. Other more complex structures can be obtained by modifying the ellipsoidal shape of the particles, e.g. to tapered [Berardi and Zannoni, 2000] or bowlic [Ricci et al., 2008] shapes or by combining various GB units with flexible spacers to model polymeric and elastomeric LCs [Berardi et al., 2004c; Skačej and Zannoni, 2014] as we shall see in the next few sections. An advantage of the GB potential is its analytic formulation and its differentiability which has eased simulations with both MC and MD. Initially simulation codes have typically been developed by individual groups, making maintenance over time and general availability difficult. However, the GB potential for uniaxial or biaxial particles has now been implemented in the open-source MD code LAMMPS (see Section 9.9), developed and maintained at Sandia Labs [Plimpton, 1995], making applications much easier. In what follows we briefly examine the main LC phases obtained by the GB mesogens, with no intention of a full coverage of the vast literature, but rather focussing on the modelling aspects more specific to LCs and showing representative examples. We start by briefly summarizing the results that are obtained from elongated and discotic uniaxial particles. We shall normally use scaled, dimensionless units, which were reported in Table 5.10. For uniaxial mesogens it is convenient to express the parameters defining the GB model in Eq. 5.57 by the concise notation of Bates and Luckhurst [1999a]: GB($\kappa, \kappa', \mu, \nu$), where $\kappa = \sigma_e/\sigma_s$ is the aspect ratio and $\kappa' = \epsilon_s/\epsilon_e$ is the ratio of lateral to axial attractive well depths.

11.5.1 Rod-like Gay–Berne Mesogens

Fig. 11.9a shows the phase diagram obtained by Chalam et al. [1991] and extended by Brown et al. [1998a], using MD and MC simulations, for the original version of the GB potential, the GB(3,5,2,1) whose parameters were originally chosen to model a line of four Lennard–Jones centres [Gay and Berne, 1981]. A close-up of the region inside the dashed rectangle including the fluid-nematic, highly ordered smectic B (or solid) phase as obtained by de Miguel [2002] using (N, V, T) and (N, P, T) MC simulations is shown in

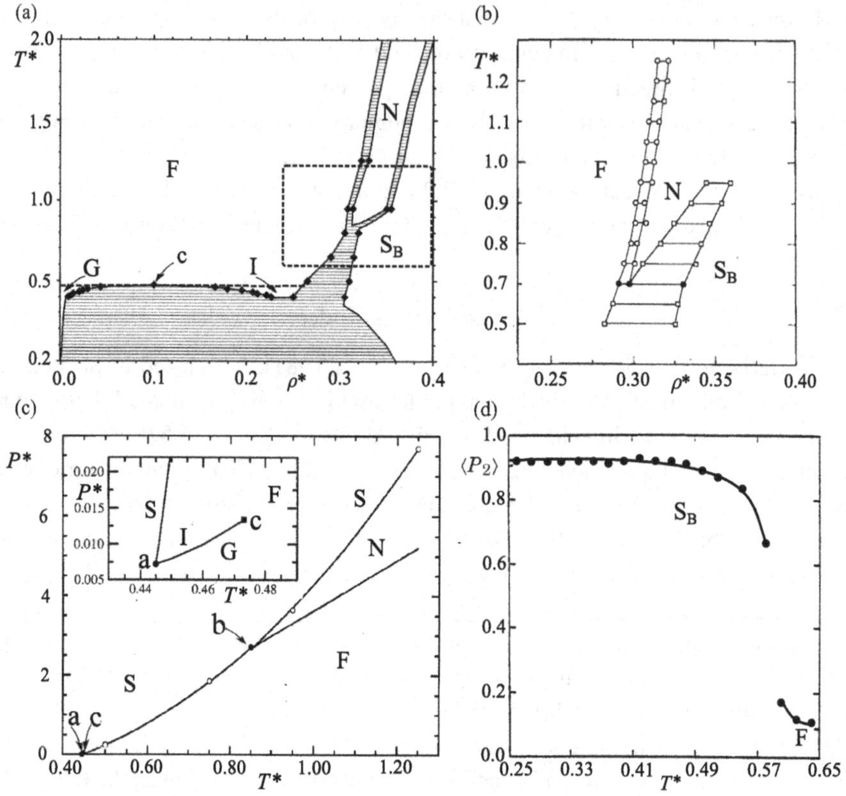

Figure 11.9 (a) The (T, ρ) phase diagram for the GB(3,5,2,1) model from MD simulations (♦) showing isotropic liquid (I), vapour (G), isotropic fluid (F) and smectic S_B (or solid S) regions and the critical point, c, [Brown et al., 1998a]. (b) Detail of the region enclosed by the dashed rectangle in part (a) [de Miguel, 2002]. (c) (P, T) phase diagram of the same model [de Miguel and Vega, 2002]. We see the ISG, a, and SNF, b, triple points as well as the critical point, c, (see inset). In all plates phase boundaries are drawn as a guide for the eye, with coexistence regions shaded. (d) MD results for the order parameter $\langle P_2 \rangle$ for the same GB system at $\rho^* = 0.27$ (on cooling) [Chalam et al., 1991]. Reduced units: T^*, P^*, ρ^* used.

Fig. 11.9b. Fig. 11.9c shows the global phase diagram of the model that has been gathered by de Miguel and Vega [2002] with careful free energy calculations and locating the vapour-isotropic liquid-solid triple point (GIS), a, at a temperature $T^*_{GIS} = 0.445$ and the isotropic fluid-nematic-solid (FNS) triple point, b, at $T^*_{FNS} = 0.85$. Thus, the liquid exists only in a very narrow range between $T^*_{GIS} = 0.445$ and the critical temperature $T^*_C = 0.473$ (see the sketch in Fig. 2.5a). This indicates that most LC-isotropic transitions discussed in the literature for the GB(3,5,2,1) model really refer to a high-density supercritical fluid rather than a liquid, as it would be intuitive to assume without a knowledge of the phase diagram. de Miguel and Vega [2002] also suggested that the high-density phase classified as smectic B could actually be a molecular solid (S) and not a smectic liquid crystal. However,

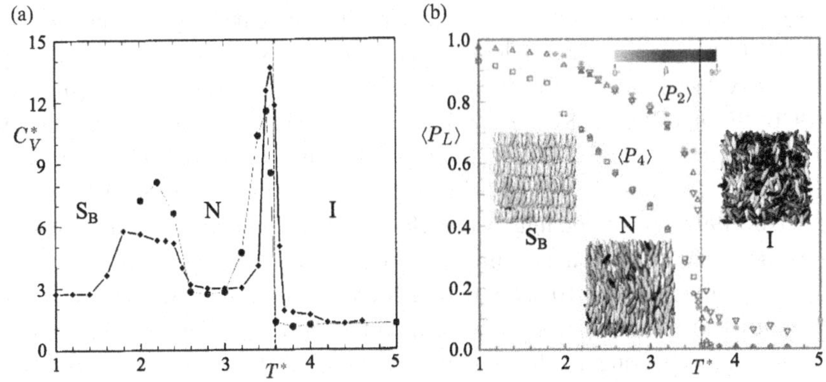

Figure 11.10 MC (N, V, T) results for (a) heat capacity C_V^* and (b) second- and fourth-rank order parameters for the GB(3,5,1,3) model as a function of temperature T^*at $\rho^* = 0.30$ (dimensionless units). C_V^* is shown for $N = 512$ (●) and $N = 1000$ (♦) systems. $\langle P_2 \rangle$, $\langle P_4 \rangle$ of the $N = 1000$ system are from both cooling (▽, ◇) and heating (△, □) sequences as well as results for $N = 512$ (●). The vertical dashed line in both plates indicates the NI transition [Berardi et al., 1993].

we already saw in Section 1.7.4 that highly ordered S_B are very similar to solids, so in the plates of Fig. 11.9 we leave the original assignment of the respective authors. In Fig. 11.9d we show the order parameter $\langle P_2 \rangle$ as a function of temperature exhibiting a clear jump at the isotropic fluid to high order smectic (solid). The nematic range of the GB(3,5,2,1) system is rather narrow, but keeping the same aspect ratio and potential well anisotropy, while choosing energy parameters $\mu = 1, \nu = 2$, i.e. GB(3,5,1,2) [Luckhurst et al., 1990], or $\mu = 1, \nu = 3$, i.e. GB(3,5,1,3) [Berardi et al., 1993], makes the side-side interaction of two molecules stronger (see Fig. 5.15). This in turn generates nematics with a wider temperature range. In Fig. 11.10 we show some results for (N, V, T) MC simulations with $N = 512$ and 1000 GB particles interacting with the GB(3,5,1,3) potential at a density $\rho^* = 0.30$ [Berardi et al., 1993]. Fig. 11.10a shows the heat capacity as a function of temperature and Fig. 11.10b the order parameters $\langle P_2 \rangle$ and $\langle P_4 \rangle$, calculated from diagonalization of the ordering matrix and subsequent averaging of the largest eigenvalue over configurations as described in Section 3.5. Two transitions are clearly visible at temperatures $T^* \approx 3.50$ and $T^* \approx 2.40$. We note that proper equilibration is confirmed by the superposition of the results for $\langle P_2 \rangle$, $\langle P_4 \rangle$ that were started from the isotropic phase and cooling (▽, ◇) or, respectively, that were started from the solid phase and heating (△, □). Taking the high temperature transition to be the isotropic to nematic, $T_{NI}^* \approx 3.50$ and considering a typical room temperature low-molar-mass rod-like mesogen, similar to 8CB say, we can estimate the GB units of length and mass for the GB(3,5,1,3) model as given in Table 5.10. The temperature dependence of $\langle P_2 \rangle$ in Fig. 11.10b is well represented by the Haller-type expression [Haller, 1975] often used to fit experimental data, as we have seen for the diamagnetic susceptivity in Eq. 3.32:

$$\langle P_2 \rangle = (1 - T/T_{NI})^{\beta_H} - \langle P_2 \rangle_{iso}, \tag{11.9}$$

where $\langle P_2 \rangle_{iso}$ is a small residual order parameter $\approx 1/\sqrt{N}$ appearing in simulation results due to the finite sample size. The Haller exponent turns out to be $\beta_H = 0.17$, in good agreement with the values found for Schiff bases [Leenhouts et al., 1979] ($0.17 \leq \beta_H \leq 0.22$), and for 5CB [Wu and Cox, 1988] ($\beta_H = 0.172$), discussed in Section 3.4.1. Order parameter data for the original GB(3,5,2,1) model can also be fitted by Eq. 11.9 but with a larger exponent, e.g. at densities $\rho^* = 0.30, 0.32, 0.35$, the exponents are $\beta_H = 0.37, 0.43, 0.45$ [Emsley et al., 1992], corresponding to a much steeper variation of $\langle P_2 \rangle$ with temperature.

The nature of the molecular organization in the different phases is not apparent from either of the plots in Fig. 11.10, even if formation of a nematic and a layered structure can be glimpsed from the representative snapshots reported in Fig. 11.10b. Much more informative is the radial-angular distribution $g(r, \cos\beta_r) \equiv g(r, \hat{\boldsymbol{r}} \cdot \boldsymbol{d})$ giving the density of particles at a distance r from one taken as the centre when the intermolecular vector $\boldsymbol{r}$ makes an angle β_r with respect to the director, so that $g(r, 1)$ refers to the distribution of particles along the director and $g(r, 0)$ that perpendicular to it. This distribution was introduced in Section 4.9 and was already shown there (Fig. 4.10), at four temperatures, as an example. From those pictures the formation of spatially homogeneous nematic and of layered smectic structures is apparent. At the lowest temperature a well-ordered S_B (or solid) with local hexagonal order is apparent from the peaks showing spacial correlations in the plane perpendicular to the director ($\beta_r = \pi/2$). An MD study of a much larger sample of the GB(3,5,1,3) model with $N = 8000$ particles [Allen and Warren, 1997], which investigated the pretransitional behaviour, located the clearing transition at $3.45 < T^*_{NI} < 3.50$. The long-range decay of the orientational correlation in the isotropic phase was found to fit an Ornstein–Zernike functional form, Eq. 10.7, giving a correlation length ξ_2 which, at the lowest isotropic temperature studied was $\xi_2 \approx 16.1\sigma_s$, comparable with half of the simulation box length. Also, for the GB(3,5,1,3), as for the LL lattice model (Section 10.2) the data fit $\xi_2 \propto (T - T^*)^{-1/2}$ quite well, as expected from Landau theory (Section 2.7), with a divergence temperature $T^*_{NI^*} = 3.47 \pm 0.02$, thus very close to T^*_{NI}, even if a precise location would require much larger samples.

The effect of changing the length to breadth ratio κ on the phase behaviour and on the dynamics of the GB(κ,5,2,1) model has been studied by Brown et al. [1998a], while the effect of the potential well anisotropy κ' for fixed aspect ratio $\kappa = 3$ and $\mu = 2, \nu = 1$ has been investigated by de Miguel et al. [1996] (see Table 11.1). It was found that smectic order is favoured at lower densities as κ' increases and that for $\kappa' \geqslant 5$ and at $T^* = 0.70$ the system goes directly from isotropic fluid to S_B with the transition density shifting down slightly as κ' increases. This is reasonable since increasing κ' enhances side-side aligned pair interactions, which in turn favours layer formation. When κ' is instead lowered, the nematic becomes more stable than the smectic phase, also at lower temperatures.

Nematic-vapour interface. A GB liquid-vapour coexistence has been found for different values of κ' using MC simulations. de Miguel et al. [1990] and de Miguel et al. [1996] have shown that the GB(3,κ',2,1) system with $\kappa' = 1$ and 1.25 exhibits a triple point, where the G, I and N phases coexist, at $T^*_{GIN} \approx 0.63$ for $\kappa' = 1$ and $T^*_{GIN} \approx 0.54$ for $\kappa' = 1.25$. From the theoretical point of view, the repulsive interactions should favour

Table 11.1. *A list of simulations performed for elongated uniaxial Gay–Berne models GB (κ,κ',μ,ν). Refs.: (a) [Adams et al., 1987], (b) [Luckhurst et al., 1990], (c) [Berardi et al., 1993], (d) [Allen and Warren, 1997], (e) [Brown et al., 1998a], (f) [de Miguel et al., 1996], (g) [Bates and Luckhurst, 1999a], (h) [Luckhurst and Satoh, 2003], (i) [Chalam et al., 1991], (j) [de Miguel et al., 1991], (k) [de Miguel, 2002], (l) [Allen et al., 1996b], (m) [Joshi et al., 2014], (n) [Caneda-Guzman et al., 2014], (o) [Emerson et al., 1997], (p) [Yildirim et al., 2011] , (q) [Mills et al., 1998], (r) [Huang et al., 2014], (s) Huang et al. [2015], (t) [Martin del Rio and de Miguel, 1997], (u) [Karjalainen et al., 2013]*

GB parameters (κ,κ',μ,ν)	Phases	Method and comments	Ref.
(2,5,1,2)	I, N, S_B	MD, (N, P, T)	(n), (q)
(3,5,2,1)	I, N, S_B	MD, MC	(a), (i), (j), (k)
		MD, $N = 8000$	(l), (m)
(3,5,1,2)	I, N, Sm	MC, $N = 256$	(b)
	I, N, S_A, S_B	MD, (N, P, T)	(n)
(3,5,1,3)	I, N, S_B	MC, (N, V, T), $N = 1000$	(c)
		MD, $N = 8000$	(d), (l)
		MD, (N, P, T), $N = 1024$, thermal conductivity	(p)[a]
(3,1.25,2,1)	I, N, G	MC, (N, V, T), $N = 2000$, N – V	(f), (o), (t)
$(\kappa, 4, 2, 1)$	I, N, S_A, S_B	Elongations: $\kappa = 3.0, 3.2, 3.6, 3.8$	(e)
$(3, \kappa', 2, 1)$	I, N, S_A, S_B	$\kappa' = 1, 1.25, 2.5, 5, 10, 25$, MD, (N, V, T)	(f)
(4.4, 20.0, 1, 1)	I, S_A, S_B	$P^* = 1$, MC, (N, P, T), $N = 2000$	(g)
	I, N, S_A, S_B	$P^* = 2$, MD, (N, P, T), $N = 16{,}000$	
		MC, (N, P, T), twist constant K_2	(h)
		MC, (N, P, T), field induced S_A	
		INS_A triple point	(r)
		S_A confined to nanoslits	(s)
		Confined to cylindrical nanopores	(u)

[a] In the paper the GB parameters κ,κ' are reported as χ, χ'.

perpendicular orientation at a free nematic interface, while the attractive components should enhance the tendency to be parallel to the free surface [Tjipto-Margo and Sullivan, 1988; de Miguel and Martin Del Rio, 1999]. For the GB(3,5,2,1) model, as we see from the phase diagram in Fig. 11.9a, the lowest temperature at which the nematic phase occurs is above the critical temperature, hence there is no nematic-vapour coexistence (unlike most real liquid crystals). However, de Miguel et al. [1996] studied the effect of varying the attraction anisotropy parameter κ' for the models GB(3,κ',2,1). For $\kappa' = 1$, 2.5, 5 they found the critical temperatures to be $T_c^* \approx 0.84$, 0.6, 0.47, respectively. Lowering κ' shifts down the nematic and the GB(3,1.25,2,1) and GB(3,1,2,1) models have gaseous-isotropic, liquid-nematic triple points at $T_{GIN}^* \approx 0.63$ and 0.54 for $\kappa' = 1$ and 1.25, respectively.

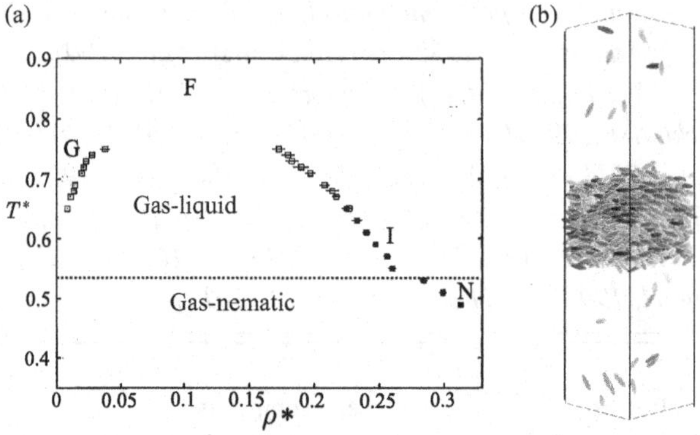

Figure 11.11 (a) Liquid-vapour ρ^* coexistence curves for the GB(3,1.25,2,1) fluid with aspect ratio $\kappa = 3$ and attractive anisotropy $\kappa' = 1.25$. The vapour-liquid-nematic triple point $T^*_{GIN} \approx 0.54$, corresponds to the horizontal dotted line. Open and filled symbols correspond to the MC methodologies employed as detailed by the authors in [de Miguel et al., 1996]. (b) Snapshot of the nematic-vapour interface for the GB(3,1.25,2,1) model showing planar alignment of the mesogens [Emerson et al., 1997].

Thus, for temperatures below this triple point, the G phase coexists with the nematic, as we see in Fig. 11.11a. The molecular organization at the nematic-vapour interface of the GB(3,1.25,2,1) model has been investigated [Martin del Rio and de Miguel, 1997; Emerson et al., 1997], finding the molecules to be preferentially aligned parallel to the free surface (Fig. 11.11b). Experimentally various types of alignment are found at a free interface e.g. planar for PAA and a few other LCs [Cognard, 1984] and tilted, with a dependence on temperature, for MBBA and EBBA [Chiarelli et al., 1983]. Alignment perpendicular to the free interface has been observed both in experiments and in atomistic simulations [Tiberio et al., 2009; Palermo et al., 2015] for cyano-biphenyls (see Section 12.4). A perpendicular alignment was also observed for a different GB system with shorter particles, the GB(2,5,1,2) [Mills et al., 1998]. Another extensive study for prolate and oblate GB particles was performed by Rull and Romero-Enrique [2017]. For all the prolate GB studied, i.e. GB$(4, \kappa', 2, 1)$, $\kappa' = 1, 0.5.0.25, 0.15$ and GB$(6, \kappa', 2, 1)$, $\kappa' = 0.5, 0.25$, the alignment at the N-G interface was always planar. For discotic particles they found instead both planar (P) and homeotropic (H) alignment for the cases: GB$(0.3, \kappa', 2, 1)$, $\kappa' = 0.4$(H), 0.3(H), 0.2(P) and GB$(0.5, \kappa', 2, 1)$, with $\kappa' = 1$(H), 0.7(H),0.6(H), 0.5(H), 0.4(P), 0.3(P), 0.2(P). Freely suspended nematic nanodroplets of GB(3,1.25,2,1) have been studied [Rull et al., 2012; Vanzo et al., 2012]. In these systems with tangential alignment the shape of the droplet can be of tactoid (spindle shape) depending on size. Moreover, a spontaneous chirality (of random sense) emerges in a range of dimensions due to the interplay of tangential anchoring and elastic energy [Vanzo et al., 2012].

Nematic-isotropic interface. Since the NI transition is of first order (Section 2.3), a coexistence of the two phases at T_{NI} is expected. In a real nematic, suitably prepared, a flat interface can exist between the two phases [Bechhoefer et al., 1989]. However, this is

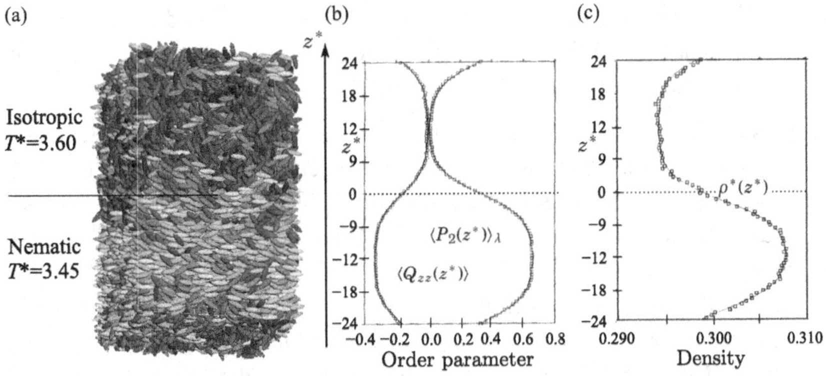

Figure 11.12 (a) Molecular organization at the NI interface for the GB(3,5,1,3) model. (N, V, T) MD results for $N = 12{,}960$ (periodic boundary conditions) with upper-half box isotropic at $T^* = 3.60$ and lower-half box nematic at $T^* = 3.45$. (b) The second-rank order parameters $\langle P_2(z^*)\rangle_\lambda$ (□) and $\langle Q^*_{zz}(z^*)\rangle$ (○) across the sample (see text). (c) Density profile $\rho^*(z^*)$ and its hyperbolic tangent fit (—), Eq. 11.10 [Bates and Zannoni, 1997].

hard to observe for simulated samples that typically fluctuate in time between ordered and disordered, making it impossible to answer by standard simulations to simple questions like: how do the molecules in the nematic align at the interface with the isotropic phase? The molecular organization at the NI coexistence for a GB(3,5,1,3) model was studied [Bates and Zannoni, 1997] with a slightly modified MD method, where the interface was prepared and maintained by separately thermostating the two halves of the sample cell containing a sufficiently large number of molecules ($N = 12{,}960$) at temperatures slightly above and below the transition. The system is thus kept slightly off equilibrium, but the computational device is similar to a simplified version of the temperature gradient hot stage, with the two ends of a sample heated at two different temperatures used to characterize polymer materials (see, e.g., [Kestenbach et al., 1999]). The molecules are free to cross the interface, but when they do so their velocity is scaled to adjust the temperature (see Section 9.5). The two thermostats ensure that the position of the interface does not move during the simulations. Starting from an isotropic sample with PBC, a planar interface is created and molecules in the nematic align parallel to the interface, as we can see not only glimpsing at the single snapshot in Fig. 11.12a, but also from the negative value of $\langle Q_{zz}(z^*)\rangle$ (see Eq. 3.48), the order parameter calculated with respect to z, the normal to the interface. The density across the sample is well fitted by a hyperbolic function

$$\rho^*\left(z^*\right) = \frac{1}{2}\left(\rho^*_N + \rho^*_I\right) + \frac{1}{2}\left(\rho^*_N - \rho^*_I\right)\tanh\left(\frac{z^* - Z^*}{2\delta^*}\right), \tag{11.10}$$

where ρ^*_N and ρ^*_I are the bulk densities of the coexisting nematic and isotropic phases, Z^* is the position of the dividing surface and δ^* is the interfacial thickness that a least squares fit to the simulation data gives as $\delta^* = 2.37$. The same MD procedure also gave a planar alignment for the GB(4.4,20.0,1,1) at the smectic-isotropic and smectic-nematic interfaces [Bates, 1998].

Experimentally a planar alignment has been reported for some liquid crystals like MBBA [Langevin and Bouchiat, 1973], although different types of tilted alignment are found for cyano-biphenyls and other nematics [Jérôme, 1991].

Elastic constants. Frank elastic constants are not easy to calculate, but various methods have been proposed. The most simple one to understand is probably that by Allen et al. [1996b], based on the inverse proportionality between quadratic fluctuations of the director and elastic constants [Group d'Etude des Cristaux Liquides, 1969; Forster, 1974]. Starting with a nematic aligned along the director $\boldsymbol{d}^0$, its ordering matrix (Section 3.5) will be uniform through the sample, with $\langle Q_{ab}(\boldsymbol{r})\rangle = \langle Q_{ab}\rangle = \langle P_2\rangle\left(d_a^0 d_b^0 - \frac{1}{3}\delta_{a,b}\right)$ and $a, b = x, y, z$ labelling the components in the lab coordinate frame. Upon application of magnetic field $\boldsymbol{H}(\boldsymbol{r})$, the director will be distorted to $\boldsymbol{d} = \boldsymbol{d}(\boldsymbol{r})$. In terms of continuum theory, the total elastic free energy $\mathcal{G}_{\rm el}^{tot}$ is obtained integrating over the sample volume V the local free energy density $\mathcal{G}_{\rm el}$ (see Eq. 1.8) and the field coupling term

$$\begin{aligned}\mathcal{G}_{\rm el}^{tot} &= \frac{1}{2}\int \mathrm{d}\boldsymbol{r}\Big[K_{11}\,(\nabla\cdot\boldsymbol{d})^2 + K_{22}[\boldsymbol{d}\cdot(\nabla\times\boldsymbol{d})]^2 + K_{33}|\boldsymbol{d}\times(\nabla\times\boldsymbol{d})|^2\Big]\\ &\quad - \frac{1}{2}\int \mathrm{d}\boldsymbol{r}\,\Delta\chi(\boldsymbol{H}\cdot\boldsymbol{d})^2, \qquad (11.11)\end{aligned}$$

where $\Delta\chi = \chi_\parallel - \chi_\perp$ is the diamagnetic susceptivity anisotropy (Section 3.4.1) and the splay, twist, bend elastic constants: K_{11}, K_{22}, K_{33} [Frank, 1958] express the resistance of the material to the three essential modes of deformation of the director: $(\nabla\cdot\boldsymbol{d})$, $[\boldsymbol{d}\cdot(\nabla\times\boldsymbol{d})]$, $|\boldsymbol{d}\times(\nabla\times\boldsymbol{d})|$, as sketched in Fig. 1.9. If the field induces only a small distortion of the director: $\boldsymbol{d}(\boldsymbol{r}) = \boldsymbol{d}^0 + \delta\boldsymbol{d}(\boldsymbol{r})$, then $\delta\boldsymbol{d}(\boldsymbol{r})$ has to be orthogonal to $\boldsymbol{d}^0$ in order to preserve the norm: $\boldsymbol{d}\cdot\boldsymbol{d} = \boldsymbol{d}^0\cdot\boldsymbol{d}^0 + 2\boldsymbol{d}^0\cdot\delta\boldsymbol{d}(\boldsymbol{r}) = 1$. Thus, $\delta\boldsymbol{d}$ has only two components in a frame with $z||\boldsymbol{d}^0$: $(\delta d_x, \delta d_y, 0)$. The free energy of distortion is, to second order in $\delta d(r)$,

$$\delta\mathcal{G}_{\rm el} = \frac{1}{2}\sum_{a,b}\int \mathrm{d}\boldsymbol{r}\int \mathrm{d}\boldsymbol{r}'\,\delta d_a(\boldsymbol{r})K_{ab}\left(\boldsymbol{r}-\boldsymbol{r}'\right)\delta d_b\left(\boldsymbol{r}'\right), \quad a,b = x,y, \qquad (11.12)$$

where $K_{ab}\left(\boldsymbol{r}-\boldsymbol{r}'\right)$ is the inverse of the susceptibility matrix $\chi_{ab}\left(\boldsymbol{r}-\boldsymbol{r}'\right)$. In continuum theory a local variable $A(\boldsymbol{r})$ is considered to be a function of the positions $\boldsymbol{r}_i$ and momenta $\boldsymbol{p}_i$ of only those particles in some small neighbourhood of the point $\boldsymbol{r}$ [Forster, 1974]. However, in computer simulations this 'neighbourhood' has to be defined by some prescription and it has to contain enough particles to ensure that a meaningful statistical error can be achieved. Assuming a sufficiently large sample box, the ordering matrix (Eq. 3.47) at a point $\boldsymbol{r}$ can be written as

$$\mathbf{Q}_S(\boldsymbol{r}) = \langle Q(\boldsymbol{r})\rangle_S = \frac{1}{2N}\sum_{j=1}^{N}[3\boldsymbol{u}_j\otimes\boldsymbol{u}_j - \mathbf{1}]\delta(\boldsymbol{r}-\boldsymbol{r}_j), \qquad (11.13a)$$

$$= \frac{1}{2N}\sum_{j=1}^{N}\int_{-\infty}^{\infty}\mathrm{d}\boldsymbol{k}\left[3\boldsymbol{u}_j\otimes\boldsymbol{u}_j - \mathbf{1}\right]\mathrm{e}^{i\boldsymbol{k}\cdot(\boldsymbol{r}-\boldsymbol{r}_j)} \equiv \frac{1}{V}\sum_{\boldsymbol{k}}\tilde{\mathbf{Q}}_S(\boldsymbol{k})\mathrm{e}^{i\boldsymbol{k}\cdot\boldsymbol{r}}, \qquad (11.13b)$$

using the Fourier representation of the 3D delta function (Eq. D.23). We assume an orthogonal box of size L_x, L_y, L_z with volume V and use units scaled by these for the wave vector $\boldsymbol{k}$ with components in units of $k_a = 2\pi n_a/L_a$. The sample ordering tensor in reciprocal space is

$$\tilde{\mathbf{Q}}_S(\boldsymbol{k}) = \frac{V}{N}\sum_{j=1}^{N}\frac{1}{2}\left(3\boldsymbol{u}_j\otimes\boldsymbol{u}_j - \mathbf{1}\right)\mathrm{e}^{-i\boldsymbol{k}\cdot\boldsymbol{r}_j}. \tag{11.14}$$

The order parameter $\langle P_2\rangle_S$ is the highest eigenvalue of $\langle\mathbf{Q}\rangle_S$ and the director $\boldsymbol{d}$ the corresponding eigenvector (Section 3.5). In the director frame with $\boldsymbol{d} = (0,0,1)$, $\mathbf{Q} \equiv \langle\mathbf{Q}\rangle_S$ is then diagonal and $Q_{xx} = Q_{yy} = -\frac{1}{2}\langle P_2\rangle_S$, $Q_{zz} = \langle P_2\rangle_S$ provided the phase is genuinely uniaxial (see Eq. 3.50). As ready mentioned, in this frame small fluctuations of the director may be expressed as $(\delta d_x, \delta d_y, 0)$ and considering $\delta(\boldsymbol{d}\otimes\boldsymbol{d}) = 2\delta\boldsymbol{d}\otimes\boldsymbol{d}$ we see that these are proportional to the elements Q_{xz} and Q_{yz}. Forster [1975] showed that the elastic constants can be extrapolated from the Fourier-transformed order tensor fluctuations. Static orientational fluctuations are described in terms of $\tilde{\boldsymbol{Q}}$, expressed in the director coordinate system. Taking the wave vectors to lie in the xz-plane, i.e. $\boldsymbol{k} = (k_x, 0, k_z)$,

$$\left\langle\tilde{Q}_{xz}(\boldsymbol{k})\tilde{Q}_{xz}(-\boldsymbol{k})\right\rangle \equiv \left\langle\left|\tilde{Q}_{xz}(\boldsymbol{k})\right|^2\right\rangle = \frac{9\langle P_2\rangle^2 V k_B T}{4[K_{11}k_x^2 + K_{33}k_z^2]}, \tag{11.15a}$$

$$\left\langle\tilde{Q}_{yz}(\boldsymbol{k})\tilde{Q}_{yz}(-\boldsymbol{k})\right\rangle \equiv \left\langle\left|\tilde{Q}_{yz}(\boldsymbol{k})\right|^2\right\rangle = \frac{9\langle P_2\rangle^2 V k_B T}{4[K_{22}k_x^2 + K_{33}k_z^2]}. \tag{11.15b}$$

The elastic constants are defined for long wavelength director fluctuations, and thus the above equations are valid only in the limit of small k. To extract the elastic constants from these expressions, one may employ the long-wavelength behaviour:

$$W_{xz}\left(k_x^2, k_z^2\right) \equiv \frac{9\langle P_2\rangle^2 V k_\mathrm{B} T}{4\left\langle\left|\tilde{Q}_{xz}(\boldsymbol{k})\right|^2\right\rangle} \to K_{11}k_x^2 + K_{33}k_z^2 \text{ as } k \to 0, \tag{11.16a}$$

$$W_{yz}\left(k_y^2, k_3^2\right) \equiv \frac{9\langle P_2\rangle^2 V k_\mathrm{B} T}{4\left\langle\left|\tilde{Q}_{yz}(\boldsymbol{k})\right|^2\right\rangle} \to K_{22}k_x^2 + K_{33}k_z^2 \text{ as } k \to 0. \tag{11.16b}$$

In practice, the K_{ii} are obtained by fitting Eqs. 11.16 for various k_x^2 and k_z^2 and extrapolating to $k_x = k_z = 0$. The dimensionless Frank elastic constants for the GB(3,5,2,1) and GB(3,5,1,3) models estimated in this way using MD by Allen et al. [1996b] are reported in Table 11.2 together with other available results. In particular, Humpert and Allen [2015a,b] have simulated very large systems with $N = 512{,}000$ particles, corresponding to sample dimensions $L > 100\sigma_0$, with results that we can probably consider as a reference. In this spirit, we note that the results of Stelzer et al. [1995] obtained with a combination of approximate statistical mechanics and simulations are at significant variance. Interestingly, Joshi et al. [2014] performed calculations with a different, free energy perturbation method, using MD with LAMMPS and the free energy plugin module PLUMED [Bonomi et al., 2009] on very few ($N = 338$) particles and found the results in Table 11.2 in good agreement with those of Humpert and Allen [2015a]. A similar calculation by Sidky et al. [2018] has obtained elastic constants as a function of temperature for various GB models. It is

Table 11.2. *Elastic constants for GB(3,5,2,1) and GB(3,5,1,3) models at a certain state point as obtained by various authors*

GB(3,5,2,1), $\rho^* = 0.33$, $T^* = 1.0$					
$\langle P_2 \rangle$	N	K_{11}^*	K_{22}^*	K_{33}^*	Reference
≈0.69	405	2.7	2.5	3.1	[Stelzer et al., 1995]
0.71	≤8000	0.70	0.72	2.27−2.59	[Allen et al., 1996b]
0.66	512,000	0.91	1.01	2.62−2.74	[Humpert and Allen, 2015a]
0.67	8000	0.70	0.78	2.31−2.41	[Humpert and Allen, 2015a]
0.66	512,000	1.04	1.02	2.65	[Humpert and Allen, 2015b]
≈0.78	338	0.96	0.91	2.44	[Joshi et al., 2014]
GB(3,5,1,3), $\rho^* = 0.30$, $T^* = 3.4$					
0.55	8000	2.17	1.71	3.95−3.97	[Allen et al., 1996b]
0.63	8000	2.92	2.60	5.32−5.43	[Humpert and Allen, 2015a]
0.614	512,000	3.17	2.80	5.85−5.89	[Humpert and Allen, 2015a]
0.68	338	2.83	2.82	4.2	[Sidky and Whitmer, 2016]

a little puzzling that these methodologies dealing with only $N = 338$ mesogens appear to be capable of obtaining very good elastic constants with apparently limited impact of finite size scaling [Joshi et al., 2014; Sidky and Whitmer, 2016], even though the order parameters and phase transition location are inevitably affected by the size.

11.5.2 Discotic Gay–Berne Systems

Discotic mesogens [Chandrasekhar, 1993; Kumar, 2004] can be modelled as squashed GB ellipsoids with thickness (face-face c.o.m separation) σ_{f} and diameter (lateral c.o.m. separation) σ_{s} (see Section 5.6.3). Emerson et al. [1994] first showed that a GB(0.345, 0.2,1,2) model based on the dimensions of a triphenylene core (cf. Table 1.6) gives an isotropic, nematic and columnar phase with rectangular packing of the columns at a scaled density $\rho^* = 3.0$, while at lower density $\rho^* = 2.5$, a hexagonal array is observed. The difference in structure of the two columnar phases is shown to be due to the increased interdigitation between the columns at the higher density by Emerson et al. [1994]. However, it is worth noting [Bates and Luckhurst, 1996] that for discotics the scaling distance σ_0 should be σ_f, instead of σ_s as used by Emerson et al. [1994] with large attractive well widths leading to interdigitation [Bates and Luckhurst, 1996]. The formation of very ordered columnar structures makes the constant volume simulations troublesome, as indicated by the development of cavities inside the sample. It is convenient to use constant pressure simulations (Section 8.3.1) where the sample volume and aspect ratio can change so as to adjust the system to the equilibrium state achieving its natural density avoiding the formation of spurious cavities in the sample [Bates and Luckhurst, 1996]. For a triphenylene mesogen (C_7OHBT with $T_{NDI} = 530$K at 1atm) Bates and Luckhurst [1996] took $\sigma_0 = \sigma_{\mathrm{s}} \approx 2.6$ nm and, assuming their simulated value of $T^*_{NDI} \approx 2.7$ at $P^* = 25$,

Table 11.3. *A summary of the phases observed for various discotic uniaxial Gay–Berne models GB (κ,κ',μ,ν) by: (a) [Emerson et al., 1994], (b) [Bacchiocchi and Zannoni, 1998], (c) [Bates and Luckhurst, 1996],(d) [Stelzer et al., 1997] (e) [Sidky and Whitmer, 2016], (f) [Caprion et al., 2003], (g) [Busselez et al., 2014], (h) [Orlandi et al., 2007], (i) [Lamarra et al., 2012]*

GB parameters (κ,κ',μ,ν)	Phases	Method and comments	Ref.
GB(0.345,0.2,1,2)	I, N, D_r	MD, (N, V, E), $\rho^* = 3(D_r)$, $\rho^* = 2.5(D_h)$	(a)
	I, N, D_r	MD, K_{ii}	(d)
	I, N, D_h^d	DOS MC, $(N, V, T), N = 338$, K_{ii}	(e)
	I, N, D_h	MC, $(N, P, T), N = 512$, $P^* = 25, 50, 75, 100$ $N = 2000$, $P^* = 25$, pair correlations	(c)
	I, N, D_h, K	MC, (N, P, T), $P^* = 50$, crystal phase	(f)
GB(0.2,κ',1,2)	I, D_h, K	MC, (N, P, T), $\kappa' = 0.1–0.8$, crystal phase	(f)
GB(0.2,0.1,1,2)[a]	I, D_h, K	MC, (N, P, T), bulk and confined	(g)
GB(0.345,0.2,1,3)	I, D_h.K	MC, (N, V, T), energy transfer	(b)
GB(0.195,0.15,1,0)	I, N, D_h, K	MC, (N, P, T), charge transport	(g), (h), (i)

[a] Due to a misprint the paper reports $\mu = 2$, $\nu = 1$.

estimated $\epsilon_o = k_B T_{NDI}/T^*_{NDI} \approx 2.6 \times 10^{-21}$ J. With these values $P^* = 25$, corresponds to ≈ 55 atm. At the same time, they also found a hexagonal columnar to nematic transition at $T^*_{D_h N_D} \approx 2.5$.

A list of some significant investigations of discotic GB systems is reported in Table 11.3. In particular, Caprion et al. [2003] studied the GB(0.345,0.2,1,2) model at $P^* = 50$, finding $T^*_{I-N} = 3.30 \pm 0.10$ and $T^*_{N-D_h} = 2.90 \pm 0.10$ and, at low temperatures, an orthorhombic crystalline phase. The slope of the transition line $\mathrm{d}P^*/\mathrm{d}T^*$ estimated according to Clapeyron's law (see Eq. 2.11) $dP^*/dT^* = \Delta H^*/(T^*\Delta V^*)$ is 35.0±4.5 and 60.5±7.6 for the IN phase boundary and the nematic-columnar phase boundary, respectively, in good agreement with Bates and Luckhurst's [1996] values: 35.5 ± 2.0 and 52.2 ± 4.0, respectively.

The energy anisotropy κ' is important in determining the phases diagram for a given shape anisotropy. In Fig. 11.13 we show the PT diagrams for discotics GB(0.2,κ',1,2) with thickness to diameter $\kappa = 0.2$, We note that at high κ', low anisotropy, the GB system is not able to build columns while at low κ', the system exhibits both orthorhombic crystal as well as hexagonal columnar phase over a wide range of pressures and temperatures. The domain of stability of the N phase is found to shift towards higher pressures as κ' decreases. The system GB(0.2,0.1,1,2) has also been studied in the bulk and confined to a cylindrical nanopore by Busselez et al. [2014]. The hexatic order parameter ψ_6 introduced in Section 4.9.2 for smectics is used to characterize the arrangement of columnar phases and calculated, with a small modification of Eq. 4.113 as

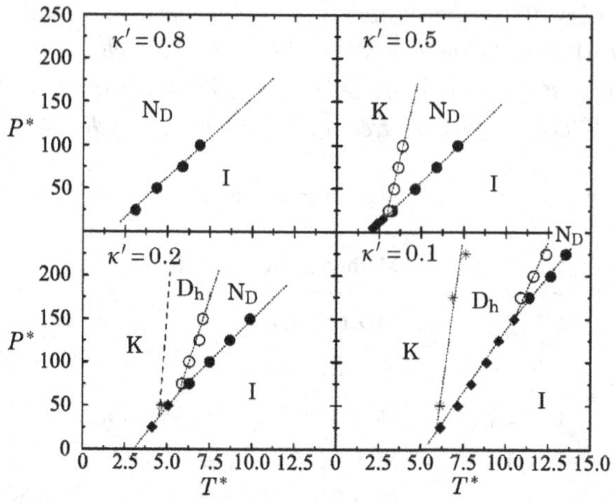

Figure 11.13 Phase diagrams for the model discotics GB(0.2,κ',1,2) with $\kappa' = 0.8$, $\kappa' = 0.5$, $\kappa' = 0.2$ $\kappa' = 0.1$ showing isotropic (I), discotic nematic (N$_D$), hexagonal columnar (D$_h$) and crystal (K) phases. The lines are guides for the eye [Caprion et al., 2003].

$$\psi_6 = \left| \frac{1}{N} \sum_j n_b^j \left(\sum_{\langle kl \rangle} w_{kl} e^{i6\theta_{kl}} \right) \right|, \tag{11.17}$$

where n_b^j is the number of pairs of nearest neighbours of the jth particle, the sum over kl goes over all possible pairs of neighbours k and l and θ_{kl} is the angle between the unit vectors along the projections of the intermolecular vectors between particles j and neighbours k and l onto a plane perpendicular to the orientation vector of particle j. The pre-exponential factor w_{kl} is assumed to be unity if the separation vectors $\boldsymbol{r}_{jk}$ and $\boldsymbol{r}_{jl}$ lie within a cylinder of radius $1.5\sigma_s$ and thickness $1.5\sigma_f$ centred on the jth particle and 0 otherwise. The usual order parameter $\langle P_2 \rangle$ and the hexatic order parameter ψ_6 for this discotic system are shown in Fig. 11.14. While order parameters, radial distributions and thermodynamic quantities have been calculated for GB discotics, very few other observables have been determined up to now, although the elastic constants have first been obtained, combining simulations and a statistical mechanics approach via a direct pair correlation functions route [Stelzer et al., 1997]. An ordering $K_{33}^* < K_{11}^* < K_{22}^*$, in agreement with experiment, has been found. The statistical mechanical route is however rather complex and, as we have seen in Table 11.2, at least for rod-like GB mesogens, it gives results at variance with other more direct methods. A more recent calculation [Sidky and Whitmer, 2016] by a different approach finds, at $\rho^* = 2.360$, that $K_{33} < K_{11} \approx K_{22}$. The same parameterization, but with a change to $\mu = 1$, $\nu = 3$ (Fig. 5.15b) has the effect of lowering the well depths of the face-to-face and side-by-side configurations, gives a hexagonal columnar structure and has been employed to estimate the anisotropy of energy transfer in the columnar phase with respect to the isotropic [Bacchiocchi and Zannoni, 1998].

One of the most promising fields of application of columnar materials is in organic electronics [Sergeyev et al., 2007] exploiting their strongly anisotropic, nearly 1D conduc-

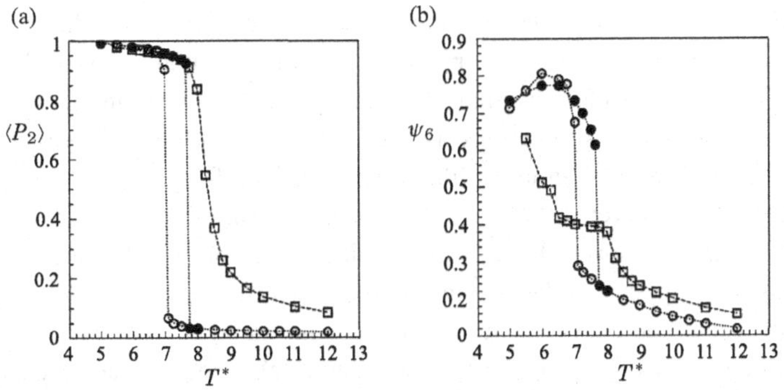

Figure 11.14 (a) The orientational order parameter $\langle P_2 \rangle$ and (b) the hexatic columns order, ψ_6 as a function of dimensionless temperature T^* for the discotic Gay–Berne GB(0.2,2,1,2) at $P^* = 100$. Bulk samples obtained heating from crystal (•) or cooling from isotropic phase (○). (N, P, T) MD ($N = 4096$) used. Also shown are results for the same system confined to a nanopore (□), with $N = 2529$ [Busselez et al., 2014].

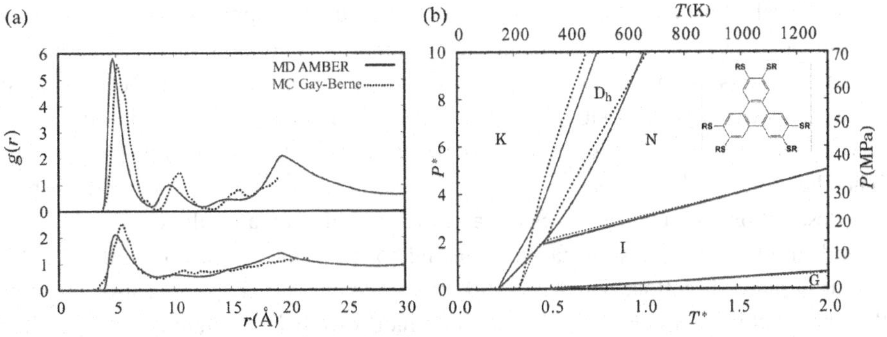

Figure 11.15 (a) Comparison of the centre of mass radial distribution $g(r)$ vs r(Å) of hexa-octyl-thio-triphenylene (8HTT) from atomistic MD at $T = 400$ K (isotropic, I) and $T = 300$ K (columnar, D_h), at $P = 3$ atm (—) and of Gay–Berne GB(0.1948,0.15,1,0) from MC simulations (····) at $T^* = 0.4$ (I) and $T^* = 0.33\,(D_h)$ at $P^* = 1.0$. (b) Pressure-temperature phase diagram of the same GB model for heating (—) and cooling (····) sequences of MC simulations showing pressure and temperature both in reduced: P^*, T^* and absolute: P (MPa), T (K) units. The phases observed are: crystal, K; columnar, D_h; nematic, N; isotropic, I; and gas, G. The lines are guides for the eye suggesting the phase boundaries [Orlandi et al., 2007].

tivity along the columns. Determining at least approximately the phase diagram of relevant discotic mesogens is thus important and can be the basis for charge transport calculations [Lamarra et al., 2012; Thompson et al., 2018]. Building such a phase diagram requires simulating a large number of state points, a nearly hopeless task at the atomistic resolutions level that we shall examine in Chapter 12. However, a small number of such simulations can be performed in order to parameterize a GB model tailored around a molecule or class of molecules of interest. In Fig. 11.15 we see, as an example, the modelling of alkythio-triphenylenes: a family of discotic mesogens with significant conductivity (see Fig. 1.44). In these aromatic systems the typical thickness is essentially that of the π electron cloud ($\approx$ 3.5 Å) while the effective diameter can be estimated from the radial distribution obtained from the atomistic simulations (Fig. 11.15a). The sequence of cooling runs (····) in Fig. 11.15b shows a phase diagram similar to that obtained from the heating runs (—) but with the K-I and K-D_h curves shifted to higher T^* values in the $P^* < 2.5$ region and to lower ones for $P^* > 2.5$. Moreover, the pressure stability range of the I-D_h coexistence region in the cooling scan appears to be strongly reduced with respect to the heating sequence and at low pressures the N $\rightarrow$ D_h transformation shows significant hysteresis. Relevant hysteresis effects and even monotropic behavior have also been observed experimentally (see Fig. 2.24) in this type of system [Maeda et al., 2001].

11.5.3 Modelling Devices

The most important application of LC materials is still certainly in LCDs [Semenza, 2007] and a reasonable question is if we can use simulations to model a simple LCD starting at molecular level. One possibility would be to use molecular computer simulations to obtain the parameters, e.g. elastic constants, needed as input in continuum theory equations. As we have seen, this is far from easy. The other possibility is to try to model directly the display and its working, without making use of continuum theory. This possibility has been explored [Ricci et al., 2010] to the most popular twisted nematic (TN) display (see Fig. 1.7) and here we briefly summarize methodology and results. The essential model is based on the description, already seen in Section 1.2.1, of a thin nematic film confined between two rigid parallel slabs whose surface is inducing homogeneous alignment of the LC along a certain in-plane direction, but twisted 90° from one another. The conventional wisdom is that the combined effect of the two surfaces induces a helical alignment which allows light transmission through input and output polarizers parallel to the alignment directions, and thus crossed. Correspondingly, the pixel has maximum transmittance in this OFF state. A suitable field applied between the two surfaces aligns the nematic and should destroy the helix, switching off transmission in the ON state. The process is in principle reversible and upon switching off the field, the nematic should spontaneously recover the helical state. A molecular simulation allows us to test if this is true or only compatible with experiment and to clarify the functioning mechanism at microscopic level. To model the display, a GB model is assumed for the nematic and also for the solid surfaces with the provision that the molecules forming the slabs are frozen in and do not move during the simulation, obviously differently from the fully mobile nematogens filling the cell. In Fig. 11.16 we see a sketch of the display (in section) and a plot of the transmittance upon switching off and on the field. In more detail, a geometrical scheme of the sample setup for the TN cell simulation is given in Fig. 11.16a. The internal dimensions of the TN cell were set to $L_x = L_y = 167.7906\sigma_0$ and $L_z = 93.8604\sigma_0$, which roughly correspond to a $0.100 \times 0.100 \times 0.067\ (\mu\text{m})^3$ volume, available to the $N_{\text{LC}} = 787320$ GB(3, 5, 1, 3) particles modelling the nematic fluid. The $\pi/2$ TN cell has planar aligning surfaces with periodic boundary conditions in the X and Y directions and confining surfaces along Z consisting

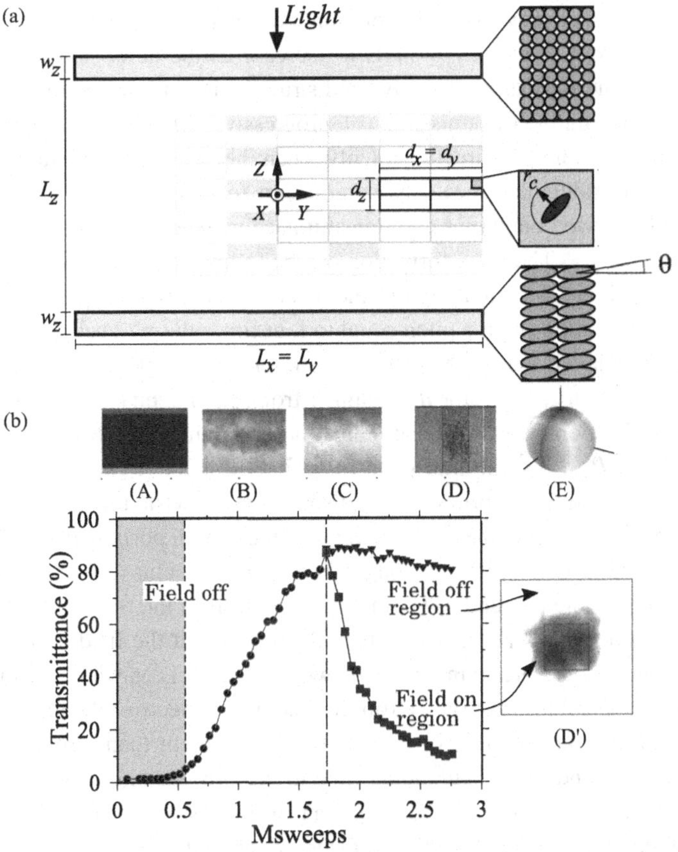

Figure 11.16 (a)Vertical (XY) section of the simulated TN cell of dimensions L_X, L_Y and thickness L_Z with a pre-tilt angle $\theta = 5°$. The entire cell (LC fluid + surfaces) is divided in 4 × 4 × 8 voxels with sides $d_x = d_y = 41.948\sigma_0$ and $d_z = 13.983\sigma_0$. The GB potential cut-off radius $r_c = 4\sigma_0$ is also shown. (b) Evolution of the integrated transmittance of linearly polarized light across the simulated TN cell (including crossed polarizers) for the relaxation from uniformly aligned (●), and for the central pixel switching (■) (see snapshot (D) and view from above (D′)). The transmittance of the external pixels during the field-on experiment is also given (▾). The vertical short-dashed line corresponds to a 5% transmittance, while the long-dashed one at 1.76×10^6 MC sweeps marks the end of the relaxation and the beginning of the switching experiment [Ricci et al., 2010].

of two slabs of thickness $w_z = 9\sigma_0$ cut from a lattice of parallel GB particles with a small pre-tilt angle of $\theta = 5°$ to remove energetical degeneracy. The frozen surface particles are also GB(3,5,1,3), implying the surface-fluid interactions to be of the same entity of the fluid-fluid ones, but this could of course be changed at will. The optics of the display (including crossed polarizers), is modelled using 4 × 4 Stokes matrices as described in Appendix L and the transmittance is plotted as a function of MC evolution cycles (sweeps) in Fig. 11.16b. On top of the transmittance plot, we also show small snapshots of the molecular organization, at various stages of the evolution from when the field is switched off and the transmittance from essentially 0 starts to increase as the helix starts to develop. We see, however, that the formation of the twisted structure is not immediate. The first stage, corresponding to negligible transmission, is a progressive alignment propagating from the two surfaces in perpendicular directions. Only when the two expanding domains come in contact, or at close enough distance, a twisting occurs to simultaneously accommodate the requirements of the two boundaries and transmittance increases up to its maximum with a certain 'MC time' τ_{OFF}. The helix formation can be followed quantitatively by computing order parameters giving the orientation of the local directors with respect to that of a perfectly helical organization (with a pitch equal to four times the $\pi/2$ cell thickness L_z). We consider N_s virtual layers across the cell, with a reference orientation $\hat{\boldsymbol{t}}_i$ corresponding to the ideal helix, and a local director $\boldsymbol{d}_i$ obtained from the diagonalization of the ith layer ordering matrix (Section 3.5). The local helical order parameters $\langle P_2\rangle_{\text{h}}$ can be defined as $\langle P_2\rangle_{\text{h}} = \frac{1}{N_s}\sum_{i=1}^{N_s} P_2(\boldsymbol{d}_i \cdot \hat{\boldsymbol{t}}_i)$ so that for a fully formed helix $\langle P_2\rangle_{\text{h}} \approx 1$. The evolution of $\langle P_2\rangle_{\text{h}}$ from the starting uniform director distribution to twisted nematic starts with the alignment of the LC closest to the surfaces, leaving the central portion of the $\langle P_2\rangle_{\text{h}}$ profiles essentially unaffected. The alignment along the two perpendicular surfaces then propagates towards the centre of the cell and $\langle P_2\rangle_{\text{h}}$ increases, as well as the transmittance, as seen in Fig. 11.16b, even if the helix is never perfect. Application of the field to a central region decreases the transmittance again in the region where the field is on, while the other remains transparent. It is apparent from Fig. 11.16b that the time to recover the dark state, τ_{ON}, is shorter. It should be said that this is more a proof of concept than a routine application because of high demand of computational time to perform the simulation, but this could soon be mended by the continuous development of high performance computers, giving full control of, e.g. surface details of the surfaces, which in this example we assumed to be perfectly flat, but could easily include roughness or other imperfections.

11.5.4 Tapered Gay–Berne Models

The development of ferroelectric nematic phases, i.e. of fluid phases with an overall polar order, is a goal of great fundamental and practical importance [Blinov, 1998; Guillon, 2000]. Currently known low-molar-mass ferroelectric LC phases, with possibly some recent exceptions [Chen et al., 2020], are in fact relatively complex tilted smectic phases from chiral or from bow-shaped molecules [Vita et al., 2018], and these layered phases clearly lack the high fluidity and self-healing characteristics that render nematics so useful in electro-optical devices. From a theoretical point of view, the existence of simple uniaxial polar nematics is not forbidden [Camacho-Lopez et al., 2004] but still very little is known on the shape and the features that a molecule should be endowed with in order to be a good candidate for exhibiting such a phase [Berardi et al., 2001]. A promising possibility is that of having polyphilic molecules that, thanks to the presence of suitable functional groups, favour side-side parallel rather than antiparallel ordering [Tournilhac et al., 1992; Guillon, 2000]. To investigate molecular models that can yield a polar nematic, N_p, phase we clearly have to resort to non-centrosymmetric objects, and this goes once more beyond simple GB systems.

Complex molecular structures could be simulated by a suitable combination of various ellipsoidal and spherical particles, e.g. a simple combination of an LJ sphere and a GB ellipsoid has been used to model pear-shaped molecules and to study the electrical polarization induced by a strain gradient (flexoelectric effect) [Stelzer et al., 1999]. However, in order to examine the role of molecular shape and of attractive forces in favouring or disfavouring the formation of polar nematics it is interesting to develop a simple candidate structure that is non-centrosymmetric both in its shape and attractive interactions using a one-site model. Berardi et al. [2001] have demonstrated the possibility of forming a polar phase with such a model developing an attractive-repulsive pair potential for tapered molecules (Fig. 11.17). In practice, the uniaxial GB type potential, Eq. 5.57 is generalized by writing $\sigma(\boldsymbol{u}_i,\boldsymbol{u}_j,\hat{\boldsymbol{r}}_{ij})$, and $\epsilon(\boldsymbol{u}_i,\boldsymbol{u}_j,\hat{\boldsymbol{r}}_{ij})$ in terms of rotational invariants, $S^*_{L_i,L_j,L}(\boldsymbol{u}_i,\boldsymbol{u}_j,\hat{\boldsymbol{r}}_{ij}) \equiv S^*_{L_i,L_j,L}(\Omega_i,\Omega_j,\Omega_r)$, [Stone, 1978; Zewdie, 1998], defined in Eq. 4.70 and tabulated in Appendix G, to obtain GB-like one-site interactions for different shapes. To start, the *soft-contact distance*, $\sigma(\boldsymbol{u}_i,\boldsymbol{u}_j,\hat{\boldsymbol{r}}_{ij})$, where $U=0$ and the potential changes from attractive to repulsive, is instead taken to be the *hard contact* distance $\ell(\boldsymbol{u}_i,\boldsymbol{u}_j,\hat{\boldsymbol{r}}_{ij})$ between the two particles considered as rigid uniaxial objects. Thus, we use a slightly generalized GB (GGB) pair potential, i.e.

$$U^{\text{GGB}}(\boldsymbol{r}_{ij},\boldsymbol{u}_i,\boldsymbol{u}_j)= 4\epsilon_0\,\epsilon^{(\mu,\nu)}(\boldsymbol{u}_i,\boldsymbol{u}_j,\hat{\boldsymbol{r}}_{ij}) \times\left[\left(\frac{\sigma_c}{r_{ij}-\ell(\boldsymbol{u}_i,\boldsymbol{u}_j,\hat{\boldsymbol{r}}_{ij})+\sigma_c}\right)^{12}-\left(\frac{\sigma_c}{r_{ij}-\ell(\boldsymbol{u}_i,\boldsymbol{u}_j,\hat{\boldsymbol{r}}_{ij})+\sigma_c}\right)^{6}\right], \tag{11.18}$$

where the symbols are the same as in Eq. 5.57, except for the replacement of $\sigma(\boldsymbol{u}_i,\boldsymbol{u}_j,\hat{\boldsymbol{r}}_{ij})$ with $\ell(\boldsymbol{u}_i,\boldsymbol{u}_j,\hat{\boldsymbol{r}}_{ij})$. For particles of arbitrary shape, an analytical expression for $\ell(\boldsymbol{u}_i,\boldsymbol{u}_j,\hat{\boldsymbol{r}}_{ij})$ is not available, but ℓ can be expanded, similarly to what was suggested by Zewdie [1998] for cylindrical particles, as

$$\ell(\boldsymbol{u}_i,\boldsymbol{u}_j,\hat{\boldsymbol{r}}_{ij}) \approx \sigma(\boldsymbol{u}_i,\boldsymbol{u}_j,\hat{\boldsymbol{r}}_{ij}) = \sum_{L_iL_jL} \sigma_{L_iL_jL} S^*_{L_iL_jL}(\boldsymbol{u}_i,\boldsymbol{u}_j,\hat{\boldsymbol{r}}_{ij}). \tag{11.19}$$

The Stone invariants $S^*_{L_iL_jL}(\boldsymbol{u}_i,\boldsymbol{u}_j,\hat{\boldsymbol{r}}_{ij})$ form an orthogonal basis set in the space of the orientations of the three unit vectors $(\boldsymbol{u}_i,\boldsymbol{u}_j,\hat{\boldsymbol{r}}_{ij})$ (Eq. G.23), thus the expansion coefficients $\sigma_{L_iL_jL}$ are just the normalized scalar products (Eq. A.35) between $\ell(\boldsymbol{u}_i,\boldsymbol{u}_j,\hat{\boldsymbol{r}}_{ij})$ and $S_{L_iL_jL}(\boldsymbol{u}_i,\boldsymbol{u}_j,\hat{\boldsymbol{r}}_{ij})$. This approach is fairly general and can be used for particles of various shapes, when the hard contact distance $\ell(\boldsymbol{u}_i,\boldsymbol{u}_j,\hat{\boldsymbol{r}}_{ij})$ can be determined numerically. Here we describe in some detail the procedure for tapered particles (see Fig. 11.17 [Berardi et al., 2001]) that can be easily adapted to other solids of rotation. The first step is to define geometrically the particle, and this is done by modelling a tapered 2D profile by joining Bézier curves, commonly used in computer graphics [Foley and Van Dam, 1982] (see Fig. 11.17a). This 2D contour is then rotated around its axis to obtain the 3D solid particle shape. The contact distance $\ell(\boldsymbol{u}_i,\boldsymbol{u}_j,\hat{\boldsymbol{r}}_{ij})$ for a pair of these particles M_i, M_j is obtained with the following algorithm (Fig. 11.17b): (i) Choose the centre of M_i as the origin and place M_j at $\boldsymbol{r}_{ij}$, with $r_{ij}=2R$, and R the radius of a sphere externally tangent to M_i (here $2R=3\sigma_0$). (ii) Define the surfaces of M_i and M_j at $\boldsymbol{u}_i$ and $\boldsymbol{u}_j$ using a set of

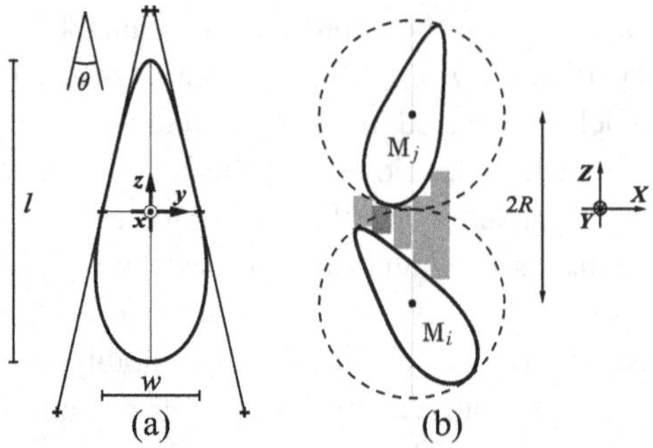

Figure 11.17 (a) Schematic section of a tapered particle with length to width ratio $l/w = 3$ and tip conic angle $\theta = 26°$. The tapered shape is drawn in 2D with Bézier curve [Foley and Van Dam, 1982] with control points (+). (b) Sketch of the contact distance ℓ (shown here as the darker bar) determination described in the text. The molecular (xyz) and laboratory (XYZ) frames are also shown [Berardi et al., 2001].

N_S evenly distributed sampling points. (iii) Rotate the two sets of points so that $\boldsymbol{r}_{ij} \parallel \hat{Z}$. (iv) Divide the relevant area of the XY-plane in N_B square bins (XY) and assign the Z-component of each surface point $P_b^{(P)} = \left(X_b^{(P)}, Y_b^{(P)}, Z_b^{(P)}\right)$ to a bin: $Z_b^{(P)} \rightarrow Z_{XY}^{(P)}$, where P = M$_i$, M$_j$ and $b = 1, 2, \ldots, N_S$ (for this example $N_S = 1452$ and $N_B = 1296$). (v) For all XY bins, subtract the distances $Z_{XY}^{(j)} - Z_{XY}^{(i)}$ between the two surfaces from $2R$. The contact distance $\ell(\boldsymbol{u}_i, \boldsymbol{u}_j, \hat{\boldsymbol{r}}_{ij})$ is the smallest of such differences $\ell = \min_{XY}\left[2\text{R} - \left(Z_{XY}^{(j)} - Z_{XY}^{(i)}\right)\right]$. (vi) Repeat steps (i)–(v) for all orientations of $\boldsymbol{u}_i$, $\boldsymbol{u}_j$ and $\hat{\boldsymbol{r}}_{ij}$ required to compute the expansion coefficients of the contact distance with the required accuracy. The attractive part of the potential can also be adapted to model the gradient of attraction that we expect from a polyphilic molecule. This is useful to compensate the antiparallel packing tendency stemming from the tapered shape and can be achieved replacing the standard $\epsilon(\boldsymbol{u}_i, \boldsymbol{u}_j, \hat{\boldsymbol{r}}_{ij})$ of a Gay–Berne model with a suitable combination of Stone invariants so as to favour: (i) parallel rather than antiparallel *side-by-side* molecular orientations and (ii) the interaction of particles approaching *bottom-on* rather than *tip-on*, i.e. with the *bottom* part closest to the centre of the other particle. In practice, this can be achieved using an ad hoc combination of Stone invariants (see [Berardi et al., 2001]) giving a difference in the parallel and antiparallel well depths. As an example, we can see the effect of combining shape and interaction polarity by simulating two model systems with the same tapered particles and with centrosymmetric (A) or non-centrosymmetric (B) models for the attractive terms. The pair potential for these two models is shown in Fig. 11.18 for some relevant configurations. (N, P, T) MC simulations of $N = 1024$ of these tapered particles [Berardi et al., 2001] show that nematic, polar nematic and smectic phases can be obtained for model (B), where attractive forces, in addition to shape are head-tail discriminating but not for model (A) with centrosymmetric attractions. In Fig. 11.18b we show a plot of the second-rank order parameter $\langle P_2 \rangle$ and of the polar one $\langle P_1 \rangle = \langle \boldsymbol{u} \cdot \boldsymbol{d} \rangle$. We see that cooling from the isotropic phase, the system presents a normal, apolar nematic phase N followed by a polar nematic N_P and a polar smectic S_P. The polar nematic has proved stable even after the introduction of a

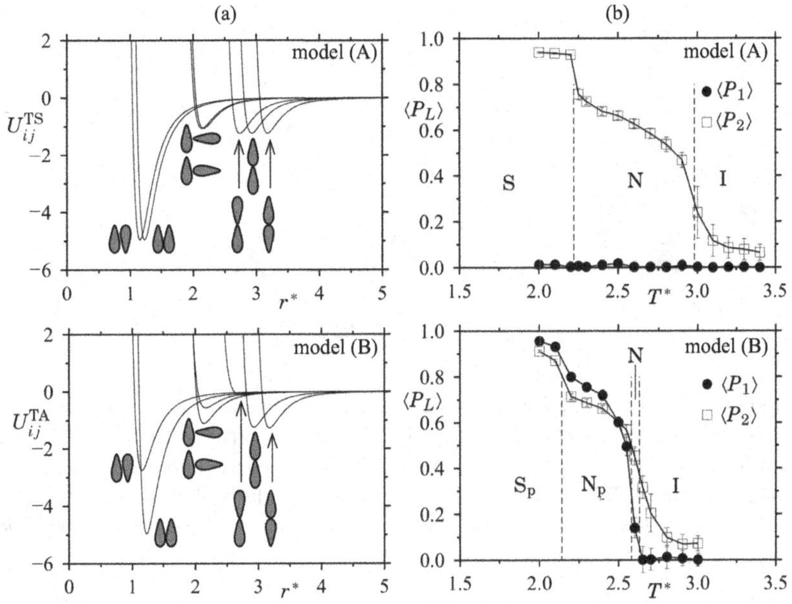

Figure 11.18 (a) The pair potential for the tapered shape model (A) with centrosymmetric (U_{ij}^{TS}) and (B) for non-centrosymmetric (U_{ij}^{TA}) attractions (in ϵ_0 units) vs scaled separation $r^* = r_{ij}/\sigma_0$. (b) The polar, $\langle P_1 \rangle$, and non-polar, $\langle P_2 \rangle$, order parameters $\langle P_L \rangle$ for models (A) and (B) obtained from tapered molecules vs scaled temperature T^*. Here I, N, N_p, S, S_p indicate isotropic, apolar nematic, polar nematic and polar smectic phases [Berardi et al., 2001].

small axial dipole, thus yielding a ferroelectric nematic [Berardi et al., 2001]. This indicates a mechanism where a non-centrosymmetric shape (a *steric dipole*) is not sufficient to create a ferroelectric fluid and a weak dipole is not sufficient to destroy it, but rather has the important consequence of causing the polar phase to become ferroelectric. A combination of non-centrosymmetric shape and attractive forces could point the way to a successful candidate for a ferroelectric nematic phase. This means that, at least in this model, the objective of obtaining a ferroelectric phase is not achieved by increasing the molecular dipole, but rather by suitably combining a steric dipole (tapered shape) and distributing different attractive chemical moieties from the head to the tail of the molecular backbone.

11.5.5 Bowlic Gay–Berne Models

As already mentioned in Sections 1.10 and 11.5.2 flat shape (*discotic*) mesogens have been employed to build columnar stacks for applications in organic electronics, exploiting their strongly anisotropic transport properties [O'Neill and Kelly, 2003; Sergeyev et al., 2007]. A very interesting feature of these systems is that the molecular columns are formed spontaneously, just by changing temperature or by solvent evaporation, like in spin-coating procedures. Unfortunately, columnar systems also present irregularities, in particular their

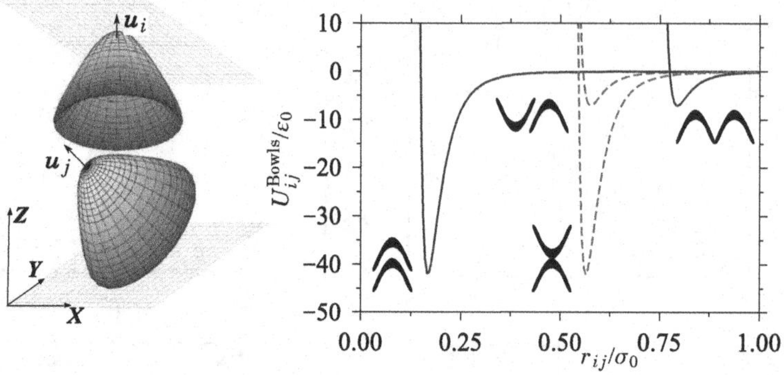

Figure 11.19 (a) Modelling of two bowl-shaped particles with diameter 0.375 σ_0, and height 0.3375 σ_0 at orientation vectors $\boldsymbol{u}_i$, $\boldsymbol{u}_j$. (b) The dimensionless pair potential $U_{ij}^{\text{Bowls}}/\epsilon_0$ vs separation r_{ij}/σ_0 for two bowl-shaped molecules with various parallel (—) and antiparallel (- - -) configurations. The well depths for parallel or antiparallel configurations are the same, and the polarity in the potential stems only from the geometrical shape [Ricci et al., 2008].

stacking columns can be interrupted by defects, e.g. disc slipping from the columns or swallow tail doubling of the columns. When designing mesogens that can pile up, an important problem is thus to control the size of the regular columnar domain. Using concave, bowlic rather than flat-shaped mesogens is intuitively expected to improve stacking and this in turn can be useful when trying to build achiral ferroelectric liquid crystal materials [Levelut et al., 1986; Xu and Swager, 1993; Blinov, 1998; Atwood et al., 2001]. The expectation is that if each bowlic molecule has an axial dipole, they should pile up head-to-tail yielding a macroscopic column dipole. The formation of a hexagonal columnar phase, as often found in columnar discotics, would ensure an overall ferroelectric order [Guillon, 2000; Tschierske, 2002] since up and down oriented columns cannot pair in equal number in such a structure. It is interesting to see if this qualitative argument is borne out by computer simulations and a simple molecular model has been developed for bowl-shaped molecules and their phase behaviour simulated with MC [Berardi et al., 2001]. To this effect, a hollow conical particle (see Fig. 11.19) is represented as the solid obtained by axial rotation of a 2D cross-section profile realized by joining four parametric Bézier curves [Foley and Van Dam, 1982]. As before, we replace the GB soft contact distance with the corresponding hard one (Eq. 11.18) and we now take $\sigma_c = 0.158\sigma_0$, with σ_0 the unit of distance. Since we only focus on the effects of concave shape, we retain a centrosymmetric attraction term $\epsilon(\boldsymbol{u}_i, \boldsymbol{u}_j, \hat{\boldsymbol{r}}_{ij})$ that does not add any energetic preference for *tip-to-tip*, *tip-to-bottom* or *bottom-to-bottom* interactions. In practice, $\epsilon(\boldsymbol{u}_i, \boldsymbol{u}_j, \hat{\boldsymbol{r}}_{ij})$ in Eq. 11.18, is modelled by using the centrosymmetric expression reported in Berardi et al. [2001], with coefficients $\epsilon_{000} = 81.25$, $\epsilon_{220} = -1.25$, $\epsilon_{202} = 3.75$, $\epsilon_{022} = 3.75$, $\lambda'_{220} = 0$, $\lambda_{220} = 1.5$ and $\lambda_{110} = 0$, $\epsilon_{101} = \epsilon_{011} = 0$, all given in units of energy ϵ_0. (N, P, T) MC PBC simulations of systems of $N = 1024, 8192, 32000$ bowl-shaped mesogens at scaled pressure $P^* = \sigma_0^3 P/\epsilon_0 = 8$, have been performed in both cooling-down and heating-up sequences showing isotropic (I), nematic (N) and columnar (D) phases. A triclinic sample box [Yashonath and Rao, 1985] has

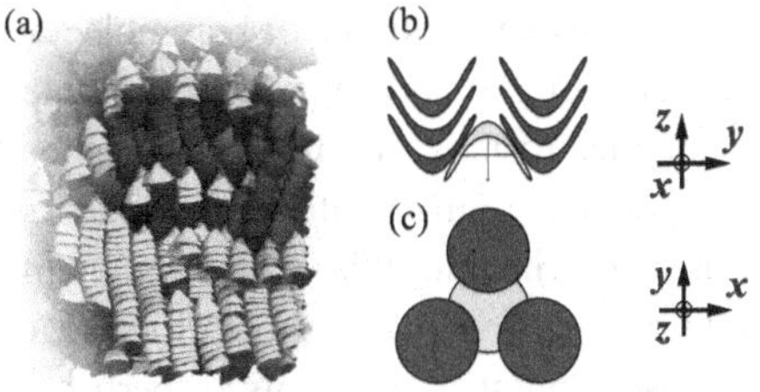

Figure 11.20 (a) Snapshot of a portion of an $N = 8192$ bowlic sample at $T^* = 1.37$ showing two polar domains formed by columns pointing upwards (grey), and downwards (black). Sketches of (b) lateral and (c) top views of the typical boundary region separating polar columnar domains with a particle of orientation antiparallel to that of three neighbouring columns [Ricci et al., 2008].

been used, allowing both sides and shape to evolve independently to avoid the formation of artifacts (e.g. cavities) in the low temperature samples. It is worth remarking that systems formed by bowlic molecules are experimentally known to give rise to glassy-like materials [Xu and Swager, 1993], and this behaviour considerably slows down equilibration. Fortunately 'large jump' moves can be used, as long they do not introduce spurious bias [Frenkel and Smit, 2002], in addition to the usual random translational and orientational MC updates, to enhance MC sampling efficiency and reduce the chance of being trapped in metastable states. Two types of extended moves are then introduced. The first move attempts to exchange particle tip with bottom by performing a 180° rotation around the molecular y-axis. Such flip moves are randomly attempted with $\approx 20\%$ frequency with respect to the standard rotation-translation MC ones. The second consists of attempts of collective roto-translations and tip-bottom flip of particle clusters [Frenkel and Smit, 2002; Berardi et al., 1999] (i.e. entire columns or stacks), tuned to attain an average acceptance ratio of $\approx 10\%$. The simulations show formation of polar columnar stacks that however do not extend across the whole sample (Fig. 11.20) and a net overall polarization is not observed. The interruption of the coherent stacks is surprising, but can be explained by the convenient hosting effect of a bowl of opposite polarity by a set of column ends, providing an effective way of terminating columns. This defect stabilization in turn indicates that changing from discotic to bowlic shapes may be insufficient to ensure the very long, coherent molecular stacks needed to optimize charge transport in organic electronic devices.

11.6 Adding Electrostatic Contributions to Gay–Berne Models

Specific electrostatic interactions are obviously present and important in many real systems and so a combination of empirical and dipole or higher electrostatic multipoles is certainly relevant. Indeed, we have already briefly examined the effect of adding dipoles to hard (Section 11.2.2) and soft (Section 11.2.3) spheres and hard spherocylinders (Section 11.4.3). However, these models, although interesting on their own, miss the attractive interactions needed to model real thermotropic mesogenic molecules, while GB models have at least the basic attractive as well as repulsive backbone of real molecules. If this is the case, one of the simplest possibilities for trying to add some relevant molecular details to a generic GB model is to include some electrostatic multipoles, corresponding to what a synthetic chemist could do introducing chemical groups that endow the molecules with a charge or one or more dipole moments. In principle, this is an ideal element of molecular design, since position, strength and orientation of one or more dipoles can be 'easily' inserted in a mesogen to change an LC phase to another or tune its properties. An understanding of their effects on phase organization, while essential, is however definitely not obvious, given the complex interplay between the different intermolecular contributions. Molecular models and simulations are particularly apt to study these effects, compared to atomistic ones that would require to input a specific chemical structure, while here the type of question could be like: where should a charge or dipole be placed to get certain LC phases or properties? The difficulty can be made clear by the examples that follow.

11.6.1 Ionic Liquid Crystals

One of simplest ways of modelling thermotropic ionic LCs (Section 1.13) is probably realized by adding a charge to a GB mesogen, obviously in a mixture with opposite charge particles, to ensure electroneutrality [Saielli et al., 2017; Margola et al., 2018]. Common examples of counterions are halides, tetrafluoroborate and hexafluorophosphate that can be reasonably represented with spherical LJs endowed with a charge opposite to the mesogen. The chains typically found in real ionic LCs (cf. Fig. 1.55) lead to micro-segregation from the ionic parts normally leading to the formation of layered smectic phases, while the nematics that would be desirable for application, are remarkably rare. Simulations could help in deciding how to choose systems more likely to produce the nematic. In the models of Saielli and Satoh [2019], GB(3,5,1,3) particles [Berardi et al., 1993] with a central positive charge were mixed with anionic LJ, in a certain stoichiometric ratio: $[\mathrm{GB}]_n[\,\mathrm{LJ}]_{\mathrm{m}}$ and MD simulations were run using LAMMPS [Plimpton, 1995]. The ionic nematic phase was observed and found to have a range of stability depending on the stoichiometric composition and favoured by increasing the ratio of anisotropic GB particles. Highly anisotropic GB(4.4, 20.0, 1, 1) systems were also studied [Margola et al., 2018] and found to produce ionic nematic and smectic phases in addition to the isotropic mixed phase and crystalline phases with honeycomb structures.

11.6.2 Dipolar Gay–Berne Systems

We consider uniaxial GB particles of length σ_e and breadth σ_s with an embedded electric dipole. The pair potential for these dipolar Gay–Berne molecules is thus $U_{ij}^{\mathrm{DGB}} = U_{ij}^{GB} + U_{ij}^{\mathrm{DD}}$, with the dipole-dipole interaction U_{ij}^{DD} as in Eq. 5.86. Our intention here is not to provide a review of the abundant literature [Berardi et al., 1996a; Satoh et al., 1996b,a; Houssa et al., 1998b,a, 1999; Berardi et al., 2002; Varga et al., 2002; Satoh, 2008; Jozefowicz and Longa, 2007; Bose and Saha, 2012], that has recently been surveyed [Allen, 2019]. We want to give instead an idea of the difficulty of using common sense in predicting the phase

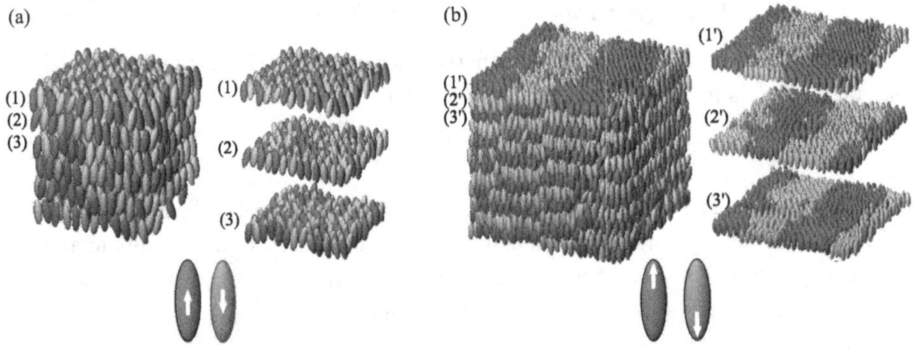

Figure 11.21 The molecular organization in the smectic phase of a system of dipolar GB (3,5,1,3) particles endowed with an axial point dipole $\boldsymbol{\mu}||\boldsymbol{u}$ of strength $\mu^* = 2$ at (a) the centre or (b) shifted of $d = \sigma_0$ towards the tip. Dark and light grey molecules represent dipoles pointing up and down, respectively. The shifted dipoles system forms a striped $S_{\tilde{A}}$ phase. Results from (N, V, T)MC, PBC simulations of $N = 1000$ (a) or $N = 8000$ (b) particles at $T^* = 2.0$, $\rho^* = 0.3$ [Berardi et al., 1996a].

structure choosing a GB with an axial dipole either at the centre or shifted towards the end of the molecule [Berardi et al., 1996a]. We recall that various types of structures have been experimentally found in smectics formed by elongated polar molecules (see Section 1.7.2 and Fig. 1.28). The dipoles can be randomly distributed up and down inside each layer giving a monolayer structure (S_{A_1}) or paired antiferroelectrically forming a uniform bilayer (S_{A_2}) or a local bilayer with modulation of domains having up and down dipoles in the *antiphase* $(S_{\tilde{A}})$ structures [Levelut et al., 1981]. The position of the dipole has a profound influence on the structure of the smectic phases obtained. In particular, according to MC simulations, a central axial dipole gives a smectic phase with dipoles randomly distributed up or down inside each layer (Fig. 11.21a) which is quite reasonable, thinking of the favourable attractive energy of two antiparallel side-side dipoles (Section 5.7.1). However, just shifting the dipole towards the end of the molecule, can yield, as we see in Fig. 11.21b, modulated antiferroelectric bilayer stripe domains [Berardi et al., 1996a] similar to the experimentally observed $(S_{\tilde{A}})$ structures [Levelut et al., 1981]. This major metamorphoses in the molecular organization following an apparently innocuous change in a molecular feature shows the essential role of computer simulations in providing, if not detailed predictions, at least non-trivial indications to the synthetic chemists making the model molecules real. It is worth noting that not only the position but also the strength of the dipole [Berardi et al., 2002; Houssa et al., 2009] and its orientations are important in determining the molecular organization, particularly of smectics, as discussed in Section 1.7.2. Estimates of molecular dipole can be obtained from quantum chemical calculations as seen in Section 5.7 and a few examples were tabulated in Table 5.12.

11.6.3 Quadrupolar Particles and Tilted Smectics

The next simple addition to a GB particle, particularly worth considering when modelling apolar molecules, is that of a quadrupole leading to a pair potential

$$U_{ij}^{\mathrm{QGB}} = U_{ij}^{GB} + U_{ij}^{\mathrm{QQ}}, \tag{11.20}$$

where U_{ij}^{QQ} has been given in Eq. 5.90 for the case of point quadrupoles, discussed in Section 5.7. The electric field generated by a quadrupole is rather different from a dipole and has been invoked both as a source of tilted smectic organizations and, together with molecular shape, of solute alignment in LC solvents [Burnell and de Lange, 1998].

Tilted smectics. The introduction of a quadrupole has been studied as a possible origin of tilt in smectic C or other canted phases, and is interesting also because it gives an opportunity to introduce the procedure for locating layers and their tilt from simulated configurations. The possibility of inducing tilt is suggested by the form of the U_{ij}^{QQ} pair potential. For instance, for an axial quadrupole the side-side aligned configuration of two particles, occurring in an orthogonal (S_A, S_B) smectic, is clearly unfavourable because of the repulsion between close partial charges of the same sign of neighbouring molecules (see Fig. 5.21a). This suggests that two neighbouring molecules should find it convenient to shift one with respect to the other, giving in the case of smectics a tilted layered structure. Neal and Parker [1998] studied systems of GB(4,5,2,1) particles with axial quadrupoles via (N, P, T) MD, finding that the quadrupole destabilizes the formation of the smectic phase, until it actually disappears altogether with high-magnitude quadrupoles. However, for weak quadrupoles, S_A, S_B and S_C with a tilt up to 17.28° are observed. It was also found [Neal and Parker, 1999] that, for the same GB model, a transverse quadrupole raises the smectic transition temperature, and that a large quadrupole stabilizes the S_A phase with respect to the GB reference fluid.

A quadrupolar term can also be modelled with two antiparallel dipoles (recall CO_2 in Section 5.7). In particular, a model based on a pair of dipoles pointing in opposite directions, inclined with respect to the long axis of a mesogenic molecule, was suggested in a pioneering work by McMillan [1973] as the origin of tilt in S_C and studied with MFT (Chapter 7). It is instructive to investigate a model of this type by simulations and Berardi et al. [2003b] have studied a GB(3,5,1,3) model with two embedded antiparallel point dipoles $\boldsymbol{\mu}_1 = \boldsymbol{\mu}_2$ located at position $(0,0,d^*)$ and $(0,0,-d^*)$, with $d^* = d/\sigma_s = 1.0$ and oriented at $\phi = 0, 60, 75$ and 90° from the long molecular axis (Fig. 11.22a). A dimensionless moment $\mu^* \equiv \left(\mu^2/\epsilon_s\sigma_s^3\right)^{1/2} = 1$, corresponding to ≈ 1.3D if we take $\sigma_s = 5$ Å and an energy scale $\epsilon_s/k_B = 100$ K, was used. A contour map of the pair potential energy obtained with a molecule placed at the origin oriented along z and a second one parallel to the first exploring the xz-plane, is shown in Fig. 11.22b for the case $\phi = 60°$ These systems and an additional case, also with $\phi = 60°$ but at position $d^* = 1.2$, were simulated with (N, P, T)MC. In Fig. 11.22c,d we report the uniaxial, $\langle P_2 \rangle$, and biaxial, $\langle R_{22}^2 \rangle$, order parameters showing the transitions occurring and the emerging biaxiality caused only by the introduction of the two dipoles (since the shape is uniaxial). The small insets also indicate layered tilted phases. However, to study the formation of tilted smectic phases, and the average tilt angle $\langle \theta \rangle$, that is the angle between the phase director $\boldsymbol{d}$ and the

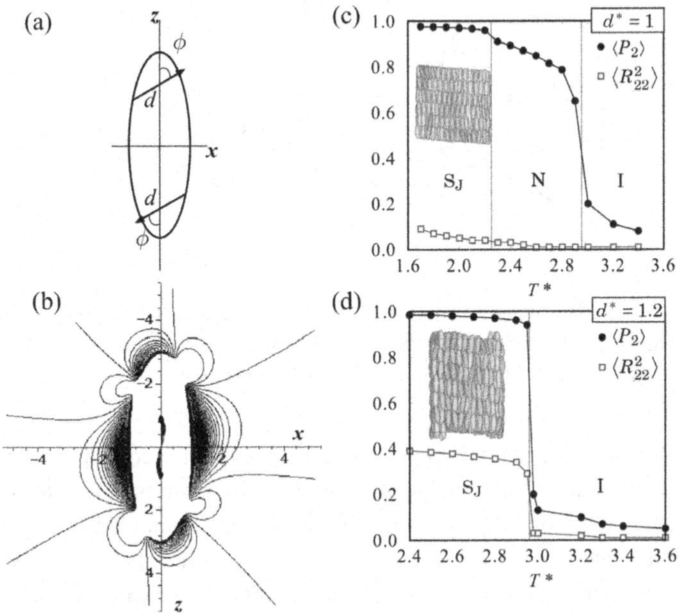

Figure 11.22 (a) A sketch of the molecular model showing position and orientation of the two dipoles within a GB(3,5,1,3) ellipsoidal particle. (b) Contour map of $U_{ij}^{\mathrm{QGB}*}$ for two molecules with $d^* = 1$, $\phi = 60°$ and long axes $\boldsymbol{u}_1 || \boldsymbol{u}_2 || z$ one at the origin and the other exploring the xz plane with molecular dipoles lying in the xz-plane. (c) Uniaxial $\langle P_2 \rangle$ and biaxial $\langle R_{22}^2 \rangle$ order parameters as a function of temperature T^*, for a system with $\phi = 60°$ and $d^* = 1$ and (d) the same when $d^* = 1.2$. The insets in (c) and (d) show snapshots of tilted smectic $\mathrm{S_J}$ phases. Results from (N, P, T)MC with $N = 1000$. Dimensionless units used [Berardi et al., 2003b].

normal to the layers $\boldsymbol{n}$ in these and other (also atomistic) systems, it is first necessary to identify the layers. An algorithm to this effect [Berardi et al., 2003b] consists in defining first a molecular layer as the set of N_k particles for which the first-neighbours distance is $r \leqslant f\sigma_s$ (e.g. a typical factor f could be $f = 1.3$). The kth layer plane can be expressed by the equation $a_k x + b_k y + c_k z = \boldsymbol{n}_k \cdot \boldsymbol{r} = d_k$, where the vector $\boldsymbol{n}_k = (a_k, b_k, c_k)$ is orthogonal to the kth layer plane.[1] All the N_l parallel planes of the sample in a certain configuration J should share the same normal vector, apart from the modulus expressing its length. We could then take two of the direction cosines to be shared by all N_l planes and determine the 'optimal' direction cosines a_J, b_J and c_J of the normal for a certain configuration J by minimizing $\chi_J^2 = \sum_{k=1}^{N_l} \sum_{i=1}^{N_k} \left(a_J x_i + b_J y_i + c_J z_i - d_k\right)^2$. The vector $\hat{\boldsymbol{n}}_J = (a_J, b_J, c_J)/\sqrt{(a_J^2 + b_J^2 + c_J^2)}$ is the best-fit unit vector normal to the layers for the single MC configuration. The tilt angle is calculated as an average over all configurations,

[1] If $\boldsymbol{P}$ and $\boldsymbol{Q}$ are vectors identifying points in the plane with equation $\boldsymbol{\ell} \cdot (x, y, z) = d$ and with $\boldsymbol{\ell} = (a, b, c)$, then $\boldsymbol{\ell} \cdot \boldsymbol{P} = d$ and $\boldsymbol{\ell} \cdot \boldsymbol{Q} = d$. Then $\boldsymbol{\ell} \cdot (\boldsymbol{Q} - \boldsymbol{P}) = d - d = 0$ and the coefficient vector $\boldsymbol{\ell}$ is orthogonal to any vector PQ between the arbitrary points P and Q of the plane, i.e. $\boldsymbol{\ell}$ is orthogonal to the plane.

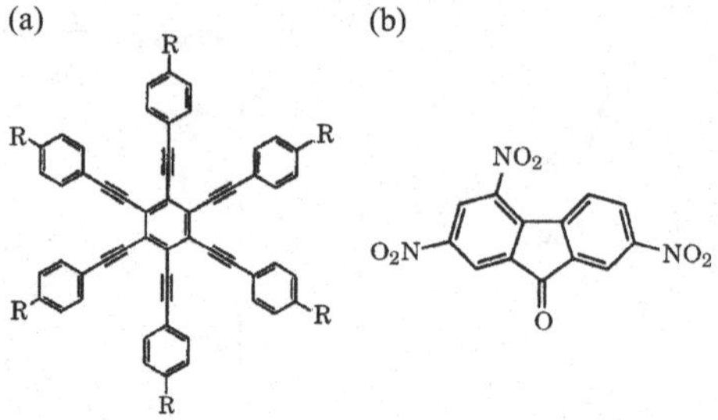

Figure 11.23 (a) Multiynes discotics ($R = OC_5H_{11}, OC_9H_9, CH_3, CN$) with an electron rich core and (b) 2,4,7-trinitrofluorenone, TNF with electron depleted core. Their mixture form a chemically induced liquid crystal phase [Praefcke et al., 1991].

with the director $\boldsymbol{d}_J$ obtained (Section 3.5) as the eigenvector corresponding to the largest eigenvalue of the ordering matrix $\mathbf{Q}_J$. The average tilt angle is $\langle\theta\rangle \equiv \left\langle\cos^{-1}(\boldsymbol{d}_J \cdot \hat{\boldsymbol{n}}_J)\right\rangle_J$. For the present quadrupolar GB system, and $\phi = 60°$, the tilt angle is $\approx 8°$, independent of temperature. It is interesting to note that the tilted phase obtained is not a S_C with liquid like positional disorder in the layer. Instead, it is characterized by hexatic bond order $\langle\psi_6\rangle \neq 0$ inside the layers, e.g. $\langle\psi_6\rangle = 0.8$ for $d^* = 1.2$ at $T^* = 2.7$. This hexatic order is long range as shown by the space correlation inside a layer $G_6^H(r)$ (see Eq.4.115) not decaying to 0. The phase is thus a tilted - crystal version of the S_B obtained for the dipole-less GB(3,5,1,3) [Berardi et al., 1993]. In particular, configurations show that the tilt is in the direction of a vertex of the hexagon formed by the nearest neighbours. On this basis these phases can be classified as a smectic-J.

Quadrupolar discotics. Chemically induced LC phases are formed in systems of non-mesogenic disc-like molecules that yield columnar phases when mixed. An example is that of mixtures of multiynes and 2,4,7-trinitrofluorenone (Fig. 11.23) [Praefcke et al., 1991]. Neither of these components is mesogenic, but their binary mixtures exhibit nematic and columnar phases. Various mechanisms have been invoked to explain this behaviour, but looking at the chemical formulas in Fig. 11.23 the two molecules have electron density close to the centre (the multyine) and pulled towards the periphery by the strongly electronegative NO_2 groups (TNF) and correspondingly have (nearly axial) quadrupoles of opposite sign. We recall that the expression for the interaction energy between two axial quadrupoles Θ_i and Θ_j (Eq. 5.90) has a multiplying factor $\Theta_i\Theta_j$ that, for discotics, should weaken the face-to-face attraction for like particles while strengthening it for particles with quadrupoles of opposite sign, that are thus expected to easily form columnar phases. This was verified to some extent by Bates and Luckhurst [1996] from (N,P,T)MC simulations of GB(0.345, 0.2, 1, 2) discotic particles endowed with an axial quadrupole Θ, which in reduced units is $\Theta^* = \Theta/\left(4\pi\epsilon_0\varepsilon_0\sigma_0^5\right)^{1/2} = 0.05$ and $\Theta^* = 0.10$ ($\approx 2.2 \times 10^{-38}\text{cm}^2$) [Bates and Luckhurst, 1998]. In the absence of quadrupoles, this system yields I, N and D_h columnar phases. The same authors studied the phase behaviour of the pure quadrupolar systems and their binary mixtures with molar fraction $x = 0.5, 0.75$. Columnar phases are obtained directly from the isotropic phase at a temperature higher than the nematic-isotropic

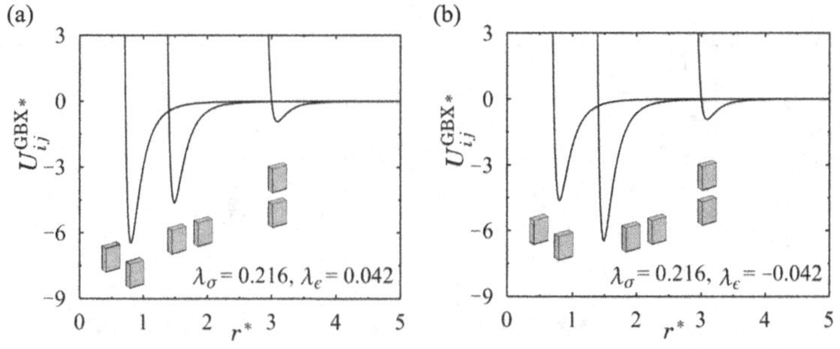

Figure 11.24 The biaxial Gay–Berne potential $U_{ij}^{\mathrm{GBX}*} = U_{ij}^{\mathrm{GBX}}/\epsilon_0$ as a function of separation $r^* = r/\sigma_0$ for *face-to-face*, *side-by-side* and *end-to-end* relative configurations using biaxialities $\lambda_\sigma = 0.216$ and (a) $\lambda_\epsilon = 0.042$ or (b) $\lambda_\epsilon = -0.042$ [Berardi and Zannoni, 2000].

transition temperature of the pure system. In addition, there is a strong tendency for unlike particles to be nearest neighbours, giving a structure in which the nature of the particles alternates on average along the column. The 75:25 mixture has a different phase behaviour in that it forms a nematic phase before undergoing a relatively weak transition to a columnar one.

11.7 Biaxial Nematics

As introduced in Section 1.6 biaxial nematics are anisotropic fluids defined as having two preferred directors, rather than a single one, as in uniaxial nematics [Luckhurst and Sluckin, 2015]. The existence of such LCs was predicted [Freiser, 1970], decades before the actual synthesis of some thermotropic materials with biaxial nematic (N_B) features took place and is still without a universal consensus [Luckhurst, 2004; Luckhurst and Sluckin, 2015]. The perspective of the availability of nematics that can be controlled in two directions rather than just one, opens extremely interesting possibilities of a new generation of faster LC displays [Ricci et al., 2015]. Unfortunately, current biaxial nematics are not really suitable for actual devices because of, e.g. the unfavourable temperature range and high viscosity. The synthesis of other more 'application friendly' materials has not been particularly successful up to now, possibly because the very existence of a biaxial fluid is in competition with the formation of smectics or crystalline biaxial phases missing the needed fluidity and possibility of easy alignment. One possible strategy to facilitate the observation of biaxial nematics can thus be that of disfavouring the formation of smectic or crystalline phases. If different molecular interactions contribute significantly to the pair potential, then each of them can have a different biaxiality parameter. For the attractive-repulsive biaxial GB we can have: (i) shape biaxiality λ_σ (see Eq. 5.61) and (ii) attractive interaction biaxiality λ_ϵ (Eq. 5.62). It is important to note that these parameters are relatively independent and can even have opposite signs. If we consider the interaction potential as two molecules approach with a certain orientation (say with two axes parallel) the position of the energy minimum is clearly linked to the closest approach distance and then on the dimensions σ_i

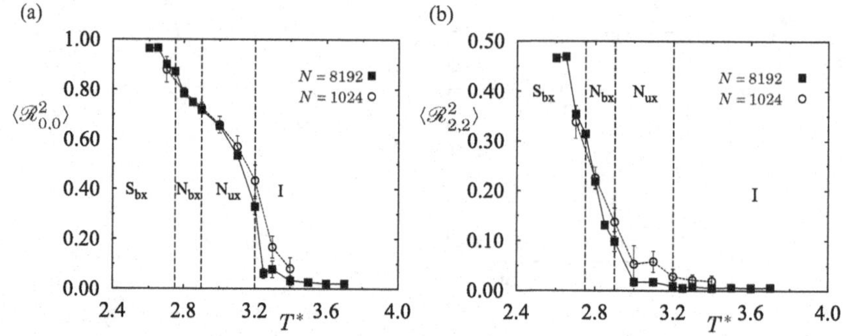

Figure 11.25 (a) Uniaxial and (b) biaxial order parameters for model mesogen particles biaxial GB parameters $\langle \mathscr{R}^2_{0,0} \rangle$ and $\langle \mathscr{R}^2_{2,2} \rangle$ vs scaled temperature $T^* = T/\epsilon_0$ with shape $\sigma_x = 1.4, \sigma_y = 0.714, \sigma_z = 3$ (in σ_0 units) and $\epsilon_x = 1.7, \epsilon_y = 1.2$ and $\epsilon_z = 0.2$ for the interaction strengths (in ϵ_0 units). Thus, we have shape biaxiality $\lambda_\sigma = 0.216$, interaction biaxiality $\lambda_\epsilon = -0.06$, model exponents $\mu = 1, \nu = 3$ and minimum contact distance $\sigma_c = \sigma_y = 0.714\sigma_0$ [Berardi and Zannoni, 2000].

and the chosen shape biaxiality. On the other hand, the interaction biaxiality λ_ϵ defines the relative importance of the strongest potential wells. In Fig. 11.24 we see an example where we have chosen $\lambda_\sigma = 0.216$ and $\lambda_\epsilon = \pm 0.042$, respectively. We see that having both anisotropy of the same positive sign, reinforces the face-face stacking configuration that in turn favours smectic organization, while in the case of $\lambda_\epsilon < 0$, this is compensated by the enhanced lateral edge-edge attraction. In this last case, a biaxial nematic phase is more easily obtained, as we can see from the uniaxial and biaxial order parameters $\langle \mathscr{R}^2_{0,0} \rangle$ and $\langle \mathscr{R}^2_{2,2} \rangle$ (Fig. 11.25) and by the averages of the biaxial Stone invariants $\overline{S^{2,0}_{2,2,0}}(r^*)$ and $\overline{S^{2,2}_{2,2,0}}(r^*)$ which we already showed in Fig. 4.5, decaying to a non-zero plateau in the biaxial phases at large separations.

It is worth noting that the addition of a dipole at various positions and orientations offers another possibility for extending the chance of observing biaxial phases and central and off-centre dipoles have been explored by Querciagrossa et al. [2013, 2018].

11.8 Rigid Multisite Gay–Berne

A simple way of representing a variety of rigid particles, even of rather complex shape, is that of modelling them as a set of assembled GB particles ('beads') arranged with the appropriate geometric structure. The pair potential between two such molecules i and j, respectively made up of n_i and of n_j GB ellipsoidal beads, will be

$$U^{\text{BGB}}_{ij} \equiv U^{\text{BGB}}\left(\boldsymbol{r}_{ij}, \Omega_i, \Omega_j\right) = \sum_{a=1, a\in i}^{n_i} \sum_{a'=1, a'\in j}^{n_j} U^{\text{GB}}(\boldsymbol{r}_{aa'}, \boldsymbol{u}_a, \boldsymbol{u}_{a'}). \tag{11.21}$$

Here $\boldsymbol{r}_{aa'} \equiv \boldsymbol{r}^{(i,j)}_{aa'} = \boldsymbol{r}^{(j)}_{a'} - \boldsymbol{r}^{(i)}_a$ stands for the vector joining the centre of bead a of molecule i and the centre of bead a' of molecule j and the sums are over all distinct pairs of beads.

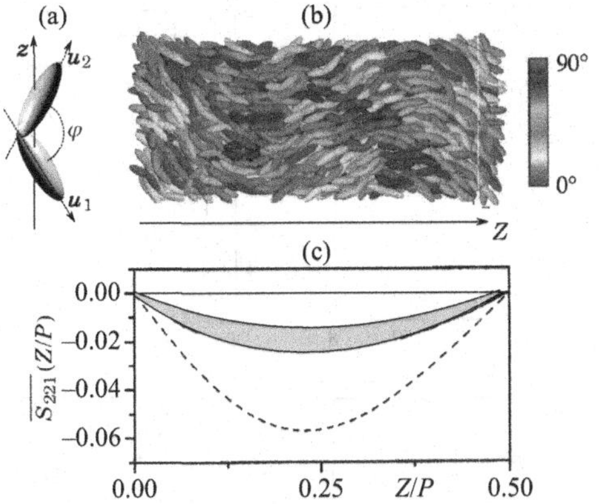

Figure 11.26 (a) Banana mesogen with two GB beads at a bent angle $\varphi = 140°$. (b) Snapshot of the molecular organization in the twisted nematic phase, at $T^* = 1.75$, visualized with a horizontal director $\boldsymbol{d}$ and grey-coded according to the azimuthal angle of the molecular long axis z. (c) A plot of the pseudoscalar invariant defined in Eq. 4.90 at the same T^*, calculated along the helical axis Z from MC (shaded area) and of that for a perfect helical phase: $\langle P_2 \rangle_\text{h} = 1$ with a right-handed spiralling local director and tilt angle $\theta = 20°$ (– – –) [Memmer, 2002].

GB beads belonging to the same molecules do not interact with each other. Electrostatic charges or multipoles can of course be added as for the case of single-bead particles, if needed. The interaction potential $U^{\text{GB}}(\boldsymbol{r}_{aa'}, \boldsymbol{u}_a, \boldsymbol{u}_{a'})$, between the beads of molecules i and j is the uniaxial or biaxial GB potential. Note also that some (or all) of the beads can be spherical, just a special case of GB, i.e. GB (1,1,0,0). We shall briefly discuss two rather different examples of this type of modelling, for bent-shape molecules and nanoparticles.

11.8.1 Banana Gay–Berne Mesogens

We saw in Section 1.8.3, that bent-shaped molecules [Niori et al., 1996] can form a variety of mesophases. Memmer [2002] modelled banana-shaped molecules by linking two identical GB(3,5,1,2) particles at a bending angle φ between the unit vectors $\boldsymbol{u}_1$ and $\boldsymbol{u}_2$ along the GB ellipsoids axes, to form an achiral biaxial particle, as shown in Fig. 11.26a. The pair potential between two such molecules i and j with orientations Ω_i and Ω_j, and separation vector $\boldsymbol{r}_{ij} = [(\boldsymbol{r}_{a'} + \boldsymbol{r}_{b'}) - (\boldsymbol{r}_a + \boldsymbol{r}_b)]/2$, is described by the sum of pair potentials between the GB particles of the two molecules, as in Eq. 11.21, with $n_i = n_j = 2$. The system forms isotropic, non-layered and layered phases. The non-layered nematic phase appears to form right-handed and left-handed domains with twist-bend structure [Dozov, 2001] (see Fig. 11.26b). In such a domain, the local director spirals around the helical axis with constant tilt angle θ and the helical repeat distance P. The helicity of the superstructure is confirmed by the pseudoscalar of odd total rank radial orientational correlation function $\overline{S_{221}}(r^*)$ defined in Eq. 4.90 and shown in Fig. 11.26c [Memmer, 2002]. Another banana

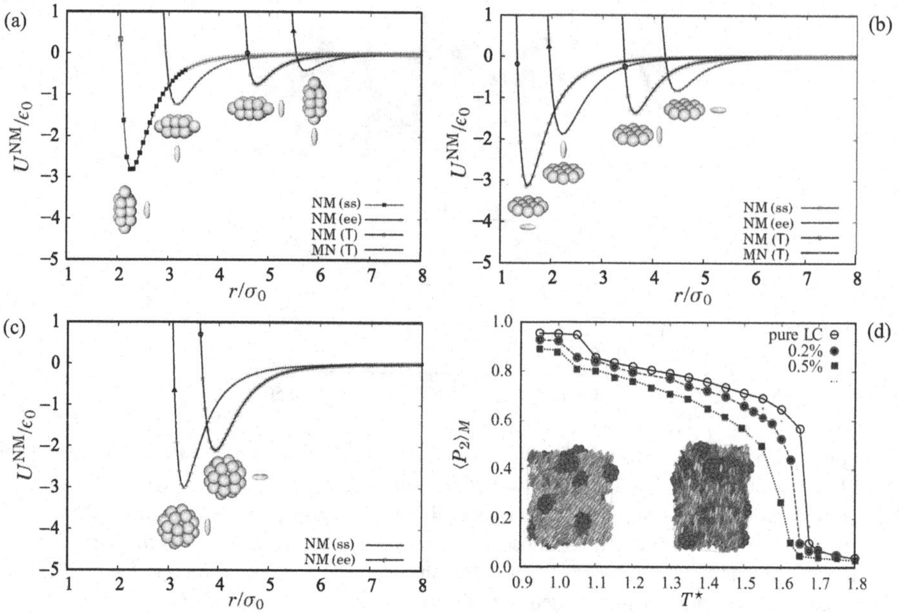

Figure 11.27 Pair potential $U^{\mathrm{NM}}/\epsilon_0$ as a function of distance between a nanoparticle N made of a cluster of LJ beads of (a) rod-like, (b) disc-like and (c) spherical shape, and a GB mesogen M. (d) shows the order parameter $\langle P_2 \rangle_M$ vs dimensionless temperature $T^\star$ for the pure mesogen system and for two spherical-NP/LC dispersions, at mole fraction $x_N = 0.2, 0.5\%$, with snapshots of the $x_N = 0.5\%$ one in the solid ($T^\star = 1.0$) and nematic ($T^\star = 1.4$) phase [Orlandi et al., 2016].

system with $n_i = n_j = 3$ GB beads with a central transversal dipole was studied by Orlandi et al. [2006], finding nematic and smectic phases but no evidence of a chiral phase.

11.8.2 Nanoparticles in Liquid Crystals

Nanoparticles (NPs) are often used as additives to LCs to modify their properties [Hegmann et al., 2007; Qi and Hegmann, 2008; Urbanski and Lagerwall, 2017; Singh, 2019] and, as seen in Section 5.5.3, a variety of NP shapes exist and new ones are continuously synthesized. Approaching their modelling could follow two routes. The first is to approximate each NP with a single shape, e.g. a polyhedral one [Damasceno et al., 2012]. The other is to model nanoparticles of different (even non-polyhedral) shapes by assembling GB or LJ particles [Orlandi et al., 2016]. We have seen in Section 5.8.3 that the overall dispersive interaction between colloidal size assembly of spherical beads can be analytically calculated for some simple colloidal particle geometries leading to the Hamaker expressions in Table 5.18. However, NPs are much smaller than colloidal particles and it is natural to calculate NP-NP interactions by numerical summation. Moreover, on many occasions, NPs are added to LCs in small concentrations as property modifiers and it is thus important to have models for NP-mesogen interactions in order to be able to study mixtures. Recalling that the Lennard–Jones potential is just a special case (GB (1,1,0,0)), applicable to spherical objects, of the GB potential between ellipsoids, the interaction between each pair of like or unlike sites, i and j, can be written in any case in terms of the heterogenous GB potential Eq. 5.63. In Fig. 11.27a, b, c we see, as an example, the modelling of nanoparticles of approximate rod-like, plate-like and spherical shape, and the effect on the LC host order parameter $\langle P_2 \rangle_M$ of the addition of various concentrations of spherical NPs, modelled with a spherical cluster Fig. 11.27d. In this case, the effect is that of lowering the order and the transition temperature of the pure LC, but the result can be different for other shapes [Orlandi et al., 2016]. The GB mesogenic hosts have the GB(3,5,2,1) parameterization of Chalam et al. [1991] for rod-like LC molecules, but clearly the method is general enough to allow for discotics or other shaped mesogens.

11.9 Flexible Multisite Gay–Berne and Liquid Crystal Polymers

Although the multi-beads models examined in the Section 11.8 correspond to rigid particles with a fixed geometry, they can be modified to represent the polymers and liquid crystal polymers (LCPs) introduced in Section 1.3. Coarse-grained polymer models are often based on chains of spherical beads connected with spring type bonds (see, e.g., [Binder, 1995]), although a small number of systems based on ellipsoidal beads have also been proposed [Lyulin et al., 1998; Hahn et al., 2001; Berardi et al., 2004c; Micheletti et al., 2005]. An ellipsoidal beads-based approach seems more reasonable, if the individual monomers are themselves mesogenic reactive monomers (RM) able to form LC phases. Here we describe LCP modelling based on the approach of Berardi et al. [2004c] representing polymerizable monomers with GB uniaxial particles decorated with reactive sites.

Polymerization process modelling. In studying polymeric and elastomeric LCs it is important to note that, different to low-molar-mass systems, many of their properties depend on the preparation history, e.g. if the polymerization is started in the isotropic or in the LC phase, and if an alignment stage (e.g. by flow, or spinning or by mechanical stress) takes place before completing polymerization. It is thus important to simulate the polymerization process itself, rather than just the final product. To do this we require at least two different molecular species: monomers and radical initiators (see Fig. 1.12), that can both be modelled with modified GB particles. For simplicity here they are both obtained from the same GB ellipsoids, decorating monomers with two or more reactive sites and initiators with only one. In more detail, the active sites embedded in each GB monomer can exist in three different states: non-bonded, reactive (propagating) and bonded. If the aim is that of simulating the formation of main chain polymers, as we shall assume in the examples shown here, the sites are placed in terminal position, like in Fig. 11.28, but they could be positioned otherwise to represent other types of LCPs. For instance, one or more additional sites could be placed transversally to model cross-linking in the case of elastomer formation [Skačej and Zannoni, 2011, 2012, 2014] (Section 1.16). As monomers link together and polymerization takes place, the total energy of the system, U^{LCP}, will consist of both non-bonded GB terms, U^{GB}, and bonded contributions, U^{bond}, and we can write

$$U^{\mathrm{LCP}} = U^{\mathrm{GB}} + U^{\mathrm{bond}}. \tag{11.22}$$

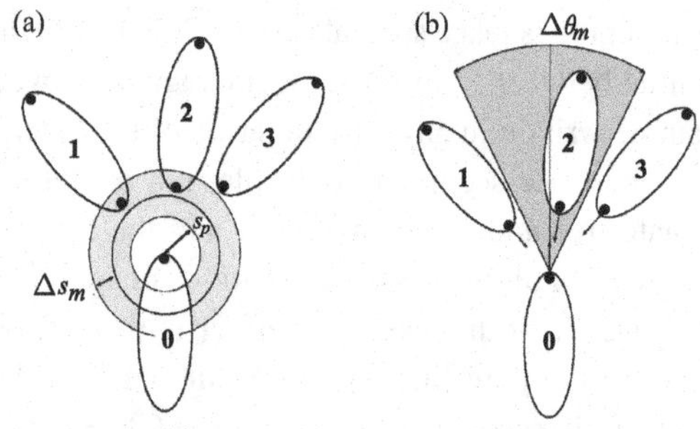

Figure 11.28 Cartoon of the (a) positional and (b) orientational criteria used to select candidate bi-functional monomers 1, 2, 3 for the MC chain propagation reaction with the initiator 0. Only monomer 2 satisfies both conditions, with the bond distance $s_p - \Delta s_m \leq s_{02} \leq s_p + \Delta s_m$ and the bond angle $\theta_p - \Delta\theta_m \leq \theta_{02} \leq \theta_p + \Delta\theta_m$ that are within the useful ranges for a bonding reaction [Berardi et al., 2004c].

The non-bonded term U^{GB} is calculated over all the non-directly linked pairs of monomers, as

$$U^{\text{GB}} = \sum_{i<j} (1 - w_{ij})\, U^{\text{GB}}(\boldsymbol{u}_i, \boldsymbol{u}_j, \boldsymbol{r}_{ij}), \tag{11.23}$$

with w_{ij} a switch function: $w_{ij} = 1$ if i and j are bonded and 0 otherwise and where $\boldsymbol{r}_{ij}$ is the vector joining the centres of mass of the i and j GB monomers with orientations $\boldsymbol{u}_i$ and $\boldsymbol{u}_j$. The pair potential between the two GB particles is that in Eq. 5.57, and already discussed in other applications in the current chapter. Upon bonding, the GB interaction between adjacent pairs of linked monomers is replaced by a sum of stretching and bending finitely extendable non-linear elastic (FENE) [Bird et al., 1971; Khare et al., 1996] spring-like contributions, so that the 'bond' energy is

$$U^{\text{bond}} = \sum_{i<j} w_{ij} \left[U^{\text{FENE}}(s_{ij}) + U^{\text{FENE}}(\theta_{ij}) + U_0 \right], \tag{11.24}$$

where s_{ij} and θ_{ij} are bond lengths and bond angles. The FENE stretching, $U^{\text{FENE}}(s_{ij})$, and bending, $U^{\text{FENE}}(\theta_{ij})$, energies between two reaction sites i, j can be written in general terms for a coordinate ξ as

$$U^{\text{FENE}}(\xi) = -\frac{1}{2} K_\xi (\delta\xi_m)^2 \ln \left[1 - \left(\frac{\delta\xi}{\delta\xi_m} \right)^2 \right], \tag{11.25}$$

where $\delta\xi = \xi - \xi_{eq}$ is the deviation from the equilibrium value, with a maximum extension $\delta\xi_m$ and K_ξ is the stiffness constant. Here $\xi = s_{ij}$ for stretching, and $\xi = \theta_{ij}$ for bending, while ξ_{eq} is either the equilibrium bond length s_{eq} or angle θ_{eq}. When the bond distance and the bond angle are equal to s_{eq} and θ_{eq}, respectively, the pair bonding energy takes the U_0 value. In this approximation the stretching energy does not depend on the orientation of the bond vector $\boldsymbol{s}_{ij}$, but only on its length s_{ij}. The algorithm for simulating the polymerization process [Berardi et al., 2004c] generalizes the method proposed by

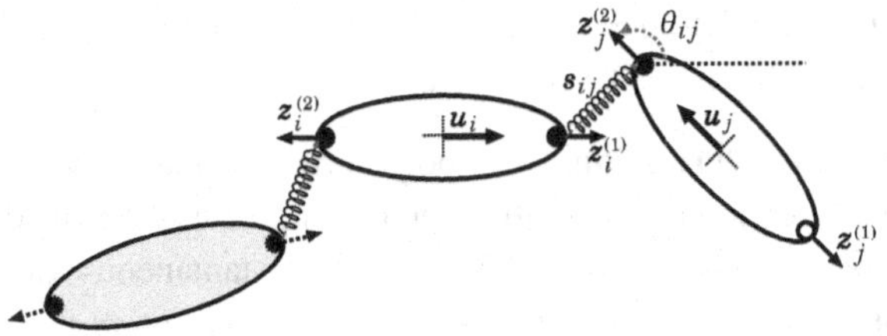

Figure 11.29 Schematic drawing of a bonded monomer i with orientation $\boldsymbol{u}_i$ endowed with two reactive sites (1) and (2) with orientations $z_i^{(1)} = -z_i^{(2)}$, and positions $(0,0,\sigma_e/2)$, and $(0,0,-\sigma_e/2)$. The site-site vector $\boldsymbol{s}_{ij}$ and the bending angle θ_{ij} between the bonded $z_i^{(1)}$ and $z_j^{(2)}$ sites are also shown. Bonded sites are represented as ● and active propagating sites, like $z_j^{(1)}$, as ○, while the grey particle on the left represents the remaining part of the chain [Berardi et al., 2004c].

Kurdikar et al. [1995] to mimic chain-reaction polymerization (free radical mechanism). Every reaction site can undergo polymerization along a preferential unit vector, $z_i^{(1)} = \boldsymbol{u}_i$ for site 1 and $z_i^{(2)} = -\boldsymbol{u}_i$ for site 2. The polymerization starts with the random generation of a certain number N_r of radical initiators, modelled for simplicity as particles like the monomers but carrying a single reacting site (see Fig. 11.28). The initiation step is followed by a modified MC evolution where, besides the translational and orientational moves (see Chapter 8), moves involving the reaction steps leading to chain growth, are also attempted. After a successful polymerization event, the other, non-bonded, site of the terminal monomeric unit becomes reactive. The iteration of these steps is continued for a chosen number of MC cycles until the polymerization is quenched. The simulation then continues with the reaction step switched off and the chains (Fig. 11.29) are left to evolve and relax for a certain number of conventional MC cycles and a production run during which averages of observable properties are computed. The approach described does not involve a chain-transfer mechanism, thus the number of growing polymer chains N_c equals that of initiators N_r and, in addition, the propagation reaction is irreversible. The details of the chain-growth process are the following: given the ith radical with an active terminal site, the monomer most likely to form a new chemical bond is determined by examining its neighbour list [Allen and Tildesley, 2017]. Only the set of monomers $\{j\}$ with site-site distances s_{ij} and bond angles θ_{ij} falling within acceptable ranges Δs_m and $\Delta\theta_m$ from the optimal bond distance s_p and angle θ_p are considered (see Fig. 11.28). The monomer chosen for reacting is the one presenting the smallest deviation function $\Phi = \left(\frac{|\theta_{ij}-\theta_p|}{\Delta\theta_m} + \frac{|s_{ij}-s_p|}{\Delta s_m}\right)$. In Fig. 11.28 we show how only monomer 2 can react with the propagating radical 0 as both reaction criteria are satisfied. If no monomer is found, the chain propagation does not take place and the propagating radical remains reactive. The methodology has been applied by Berardi et al. [2004c], using (N,V,T) MC, to systems of $N = 4096$ GB(3,5,1,3) monomers at density $\rho^* = 0.30$, in a cubic box with PBC (Section 8.2.1) and considering various temperatures in the isotropic and nematic state and different mole fractions of radical initiators. In all cases the monomeric sample can be equilibrated in the nematic or in the isotropic phase, before simulating the chain growth process, allowing a preparation protocol similar to different experiments to be followed.

LC Polymer chain characterization. Once polymerization has taken place it is important to try and describe, with some quantitative metric, the chains obtained. A few standard observables used for polymers [Flory, 1953] are the instantaneous monomer conversion $C_{\text{mon}} = N_{\text{pol}}/N$ and the *number-average degree of polymerization* $\bar{x}_n = N_{pol}/N_c$, where $N_{\text{pol}} = \sum_x xN(x)$ is the number of reacted monomers, and the length density $N(x)$ counts the number of chains formed by x monomeric units in the sample, so that $N_c = \Sigma_x N(x)$. For all values of N_c studied the chains become on average longer (i.e. with larger $\bar{x}_n$) as temperature decreases and order increases. This is consistent with a higher probability of successful MC reactive moves at temperatures where the orientational order is higher, favouring chain growth, since aligned monomers are more likely to have an orientation appropriate for reaction (Fig. 11.28b). Furthermore, the system with the lowest number of propagating radicals exhibits the longest chains, due to the higher concentration of monomers in the neighbourhood of the propagating sites during the polymerization process.

A classical descriptor for chains being extended or folded is the average *end-to-end distance* $\langle r_{ee}\rangle = \sum_{m=1}^{M}\sum_{k=1}^{N_c} r_{ee}^{(m,k)}/(MN_c)$, where $r_{ee}^{(m,k)}$ is the distance between the tips of the first and last monomeric units in the kth chain for the mth MC configuration.

A measure of *polymer deformability* can be obtained from the mean square deviation from equilibrium values of bond lengths s_{eq} and bending angles θ_{eq} namely, $\Delta s = \langle (s_{ij} - s_{eq})^2\rangle^{1/2}$, and $\Delta\theta = \langle(\theta_{ij} - \theta_{eq})^2\rangle^{1/2}$. The values obtained for our model systems show that bond length fluctuations Δs slightly decrease with temperature, as one would expect, depending on the value of the elastic constant K_s. Concerning the description of the anisotropic properties of the samples we can introduce a global second-rank orientational order parameter $\langle P_2^{(d)}\rangle$ relative to all N bonded and non-bonded monomers and referred to the phase director $\boldsymbol{d}^{(J)}$, computed in the standard fashion by diagonalization of the $\mathbf{Q}^{(J)} = \langle 3\boldsymbol{u}_i^{(J)} \otimes \boldsymbol{u}_i^{(J)} - \mathbf{1}\rangle_i/2$ ordering matrix (Section 3.5)

$$\langle P_2^{(d)}\rangle = \frac{1}{MN}\sum_{J=1}^{M}\sum_{i=1}^{N} P_2(\boldsymbol{u}_i^{(J)}\cdot\boldsymbol{d}^{(J)}), \tag{11.26}$$

with $\boldsymbol{u}_i^{(J)}$ the orientation of the ith monomer in the Jth configuration,

An additional order parameter $\langle P_2^{(e)}\rangle$, describing the average alignment of each monomeric unit $\boldsymbol{u}^{(J,k)}$ with respect to the kth chain axis $\hat{\boldsymbol{e}}^{(J,k)}$ in configuration J can be introduced as

$$\langle P_2^{(e)}\rangle = \frac{1}{MN_c}\sum_{J=1}^{M}\sum_{k=1}^{N_c}\frac{1}{x_k}\sum_{l=1}^{x_k} P_2\big(\boldsymbol{u}_l^{(J,k)}\cdot\hat{\boldsymbol{e}}^{(J,k)}\big), \tag{11.27}$$

where each chain axis $\hat{\boldsymbol{e}}^{(J,k)}$ is determined as the eigenvector of the chain inertia tensor (see Eq. 4.24) corresponding to the lowest eigenvalue. For a well-ordered linear chain the two order parameters are very similar.

The persistence length of orientation along the chain can be obtained from the decay of the orientational pair correlations between monomers n units away along the polymeric

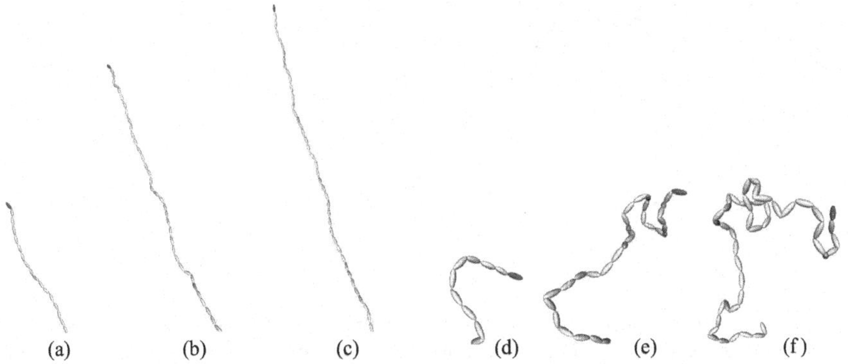

Figure 11.30 Sequences of snapshots of a growing polymeric chain at $T^* = 2.8$ (nematic: a, b, c) and $T^* = 3.8$ (isotropic: d, e, f) for a system of N = 4096 monomers with $N_r = 30$ initiators (and chains), after 20 (a, d), 60 (b, e) and 100 (c, f) MC kcycles of the polymerization reaction. The terminal, reactive monomer is represented as a dark grey ellipsoid. To help the visualization, chains have been unwound from the MC periodic boundary conditions [Berardi et al., 2004c].

strand: $\langle C_2(n)\rangle = \frac{1}{MN_c}\sum_{J=1}^{M}\sum_{k=1}^{N_c}\left[\frac{1}{N_k-n}\sum_{l=1}^{N_k-n} P_2\left(\boldsymbol{u}_l^{(J,k)}\cdot\boldsymbol{u}_{l+n}^{(J,k)}\right)\right]$, where the quantity in square brackets is relative to monomers n units apart in the kth chain from the Jth MC configuration. The correlation function profiles $\langle C_2(n)\rangle$ at the various temperatures go from 1 for $n = 0$ to an asymptotic value $\langle P_2^{(d)}\rangle^2$ for $n \to \infty$ (long-range). $\langle P_2^{(d)}\rangle$ essentially corresponds to the order parameter of the monomeric particles, and can be of use to judge the influence of the phase anisotropy on polymeric chains ordering. The first portion of the $\langle C_2(n)\rangle$ profiles gives instead information on the short-range orientational order around a monomeric unit. Examples of these properties can be found in Berardi et al. [2004c] and Micheletti et al. [2005].

Simulations show that the chain conformations are indeed influenced by the orientational order: at the lowest nematic temperature ($T^* = 2.8$) chains are fairly elongated, then at $T^* = 3.4$ they become less straight, while in the isotropic phase ($T^* = 3.8$) chains are disordered but still not quite random coil. This is reasonable because the polymer is soluble in the monomers solution and there is no evidence of segregation. These features can be more easily seen if we unwind the MC periodic boundaries and show only one polymeric strand. Fig. 11.30 shows a sequence of three snapshots of a single growing chain for the $T^* = 2.8$, and $T^* = 3.8$ samples after 20, 60 and 100 MC kcycles where the differences are quite evident. Consistent with experimental work [Lub et al., 1998], a polymerization reaction performed in the LC monomer solution results in the formation of more ordered, thus more birefringent, materials.

Although polymer chains generated from the nematic phase are generally straighter, it is important to see if they can occasionally form sharp U-turns (Fig. 11.31), the so-called *hairpins* [de Jeu and Ostrovskii, 2012]. These have been observed experimentally in some main chain polyesters [Li et al., 1993] and polyethers [Hardouin et al., 1995]. To detect

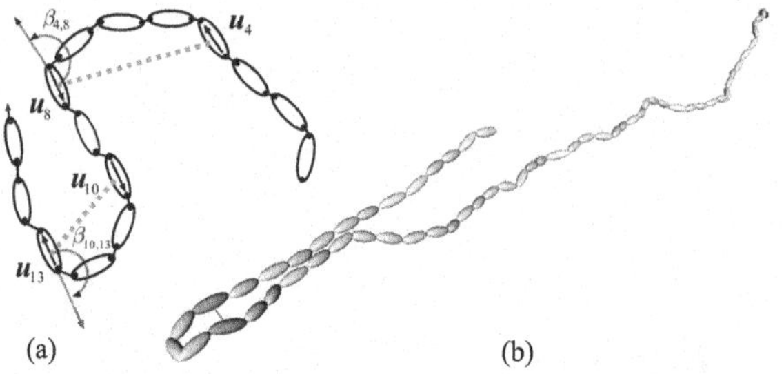

Figure 11.31 (a) Sketch of a polymeric chain showing two hairpins involving four and five monomer units, respectively. The particles at the start and end of the hairpins are shaded and joined with a dotted line. (b) Snapshot of a 64 units chain showing a hairpin in the nematic phase for a $N_c = 30$ system [Berardi et al., 2004c].

hairpins from simulated configurations, we need an operational definition. One such definition could be the following [Berardi et al., 2004c]: the shortest chain portion whose first and last monomeric units i, j exhibit a value of $(\boldsymbol{u}_i \cdot \boldsymbol{u}_j) = \cos\beta_{ij} \leq \cos\theta_{hp}$, where θ_{hp}, the hairpin threshold angle, has been chosen as $\theta_{hp} = 160°$ (see Fig. 11.31a) and two additional constraints can be enforced to avoid overlapping hairpins and considering wide bends: neither i or j must be within another pin, and the maximal hairpin length should be of six units, so that any U-turn involving more than six monomeric units is considered as a wide turn, rather than a proper hairpin. At low temperatures inside the nematic ($T^* = 2.8$, $\langle P_2^{(d)} \rangle = 0.81$) only 5%–10% of chains have a hairpin (Fig. 11.31b), but this becomes 20%–40% by increasing the temperature to $T^* = 3.4$ ($\langle P_2^{(d)} \rangle = 0.67$) still in the nematic, since the lower orientational order allows larger bending fluctuations. Finally, the number of hairpins slightly decreases in the isotropic phase, since the high chain flexibility favours chains with larger curvature radii. Interestingly, this is similar to what was observed experimentally [Hardouin et al., 1995] and it is understandable since hairpins only determine a local perturbation in the system, without affecting the overall orientational ordering of the system. In the simulations, the average length of such hairpins, L_{hp}, is of four monomeric units.

11.10 Liquid Crystal Elastomers

Liquid crystal elastomers (LCEs) [Warner and Terentjev, 2003; White and Broer, 2015] are weakly cross-linked polymeric networks containing LC units, either belonging to the polymer backbone (main-chain LCEs) or attached to it (side-chain LCEs). Some of their fascinating properties and applications have been briefly introduced in Section 1.16. LCE networks, sketched in Fig. 1.60, are characterized by a pronounced coupling of strain and mesogenic alignment [de Gennes, 1975]. They can respond to external stimuli such as temperature changes, UV light or external (mechanical or electrical) fields by large (up to several 100%) elastic (reversible) deformations. LCE properties depend critically on their preparation strategy. The fabrication process of an LCE sample from single monomers normally consists of two polymerization stages, the first generating the polymer strands with only a weak cross link and the second cross-linking them to the desired extent ($\lesssim$10%). If a suitable aligning field (mechanical or otherwise) is applied after the first polymerization step, a monodomain transparent sample with a uniform director can be obtained [Küpfer and Finkelmann, 1991]. The LCE obtained is stable and can also be heated from the N to the I state and back. In a stress-free sample, a polydomain is instead obtained, with locally aligned micron-sized regions with director $\boldsymbol{d}_L$, resulting in pronounced light scattering and an opaque appearance [Küpfer and Finkelmann, 1991; Brommel et al., 2012]. Even though most modelling work on these systems has been performed at continuum level [Warner and Terentjev, 2003], off-lattice particle-based models [Lyulin et al., 1998; Skačej and Zannoni, 2011, 2012, 2014] have the great advantage of being suitable for the introduction of some essential features of LCEs. In particular, molecular models take into account the shape and interaction anisotropy of the monomers, allowing the study of smectic as well as nematic LCEs, modelling main-chain or side-chain LCEs, varying the amount of cross-linking and, importantly, including protocols for the fabrication. This last aspect, often neglected, allows us to begin understanding, for example, the difference in properties between LCEs produced by polymerization in the isotropic phase and then cooled down (the so-called *isotropic genesis*) from those where polymerization takes place in an oriented liquid crystal state to start with. Since the properties of LCEs depend on the preparation procedure, the simulation methodology introduced in the last section appears particularly suitable also for the modelling of LCEs. A practical problem, however, is in equilibrating the systems and avoiding jamming and in this case, it is essential to employ the GB soft-core version [Berardi et al., 2009, 2011] which reduces the steepness of the repulsive part of the GB potential, replacing the standard inverse 12 power dependence on particle separation with a linear one (Section 5.6.5). The possibility of limited particle-particle 'interpenetration', that the potential allows offers a simplified description of a flexible molecular system, where two colliding particles can deform and yield to some extent rather than just harshly repelling each other. In addition, it facilitates MC equilibration, without significantly affecting the phase behaviour of the molecular system as a whole. The total potential energy for the LCE model system of N particles is thus,

$$U^{\text{LCE}} = \sum_{i=1}^{N} \sum_{j=i+1}^{N} \left(U_{ij}^{\text{GBS}} + w_{ij}[U^{\text{FENE}}(s_{ij}) + U^{\text{FENE}}(\theta_{ij})] \right), \tag{11.28}$$

where U_{ij}^{GBS} is the soft-core GB introduced in Eq. 5.69 and $U^{\text{FENE}}(s_{ij})$, $U^{\text{FENE}}(\theta_{ij})$ are the stretching and bending contributions already described (Eq. 11.25). All monomers are endowed with two reactive bonding sites at the particle ends that, upon polymerization, provide the main-chain strands. Monomers are also decorated with an additional equatorial reactive site, as necessary for cross linking. Here we introduce the methodology for LCE modelling considering two examples of LCEs based on the soft-core GB(3,5,1,3) pair potential, covering (i) the mechanical properties [Skačej and Zannoni, 2011] and actuation by an electric field in monodomain LCEs [Skačej and Zannoni, 2012] and (ii) the poly-to-monodomain transition and supersoft elasticity and elastic moduli [Skačej and Zannoni, 2014].

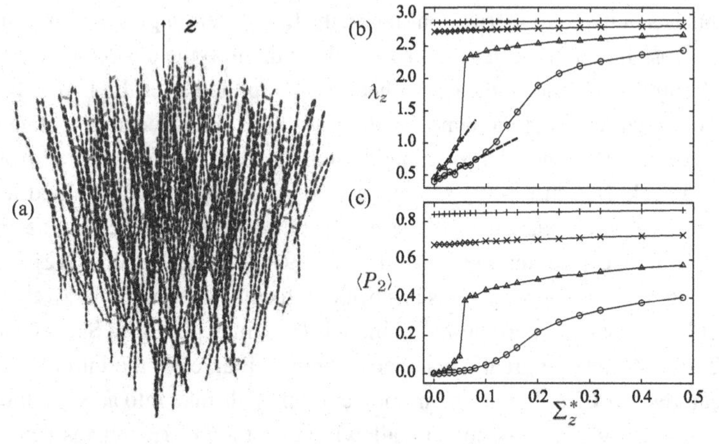

Figure 11.32 (a) Snapshot of a swollen monodomain LCE (swelling molecules not shown) with $\boldsymbol{d} \parallel z$. Change of (b) sample length λ_z and of (c) order parameter $\langle P_2 \rangle$ upon application of a reduced stress Σ_z^* along z ($\Sigma_z^* = 0.1 \approx 1.1$ MPa engineering stress) from MC runs at $T^* = 3.0\,(+)$, $5.0\,(\times)$,$7.0\,(\Delta)$ and $9.0\,(\circ)$. The dashed lines in (b) are the linear fits used to estimate the LCE elastic moduli (770 MPa for $T^* = 3.0$ and 310 MPa for $T^* = 5.0$). The FENE parameters are $s_e = 0.15\sigma_s$, $\delta s_m = 0.25\sigma_s$, $K_s = 1000\varepsilon_0/\sigma_s^2 \approx 5.52\,\mathrm{Nm}^{-1}$, $\theta_e = 0°$, $\delta\theta_m = 150°$ and $K_\theta = 7.6 \times 10^{-4}\varepsilon_0\mathrm{deg}^{-2} \approx 3.44 \times 10^{-21}$ Jrad^{-2} [Skačej and Zannoni, 2011].

11.10.1 Liquid Crystal Elastomer Monodomains: Actuation Properties

We start, discussing the preparation of a main-chain monodomain LCE sample, inspired by the Küpfer and Finkelmann [1991] cross linking procedure, where the polymer strands are first prepared and aligned by mechanical stress, then weakly cross linked to a desired extent (some 10%). To prepare this simulated uniformly aligned sample the monomers are first assembled into a regular square array and linked linearly to their nearest neighbours forming parallel polymer strands. Additional randomly placed monomers representing cross-links are then added (see Fig. 11.32). Only up to one bond per monomer pair is possible, while the average distance between cross-links along a polymer strand is ≈ 7.5 monomers [Skačej and Zannoni, 2011, 2012]. If a swollen LCE is required, as we assume for the present examples, and as is often used in real experiments [Urayama et al., 2006; Wu et al., 2020], the sample is also uniformly filled with additional GB monomers to a certain swollen/dry volume ratio ϕ_{SW}. The sample is then compressed almost to the maximum packing density, yielding the cubic reference sample used as a starting point for the simulated experiments, with PBC applied to mimic a bulk-like sample. Upon applying this procedure, a well-defined director is imprinted into the system, resulting in an oriented monodomain sample at low temperatures. After this preparation stage, the sample shape is allowed to change as the system evolves through isostress MC simulations (Section 8.3.2), first performed at zero stress, as in the experiments of Urayama et al. [2005, 2006] and Fukunaga et al. [2009], dealing with unconstrained samples. In more detail, the following attempted MC evolution steps are performed: (i) purely translational and (ii) purely rotational moves of a single particle, (iii) bonded pair rotations, dealing only with doubly linked GB particles in the polymer strands, excluding cross links. The selected particle pair is rotated by a random angle about the end-to-end vector of the particle pair, and (iv) changing sample shape, attempting a random variation of two sides of the sample box chosen at random, while the third box side is determined by the constant-volume constraint (note that also real LCE deformations maintain sample volume). This move allows the initially cubic sample to be deformed as a whole and gradually turn into an orthogonal slab. The polymer strands are deformed affinely, i.e. applying the macroscopic box deformation uniformly to the microscopic level. The trial configurations violating the maximum-length or maximum-angle constraints of the FENE potential are rejected. Note that all the above trial move types are reversible and unbiased. As a result of this zero external stress simulations, the shape of the sample changes as a result of the internal stress resulting from the fabrication process and of the temperature from the reference cube of sides ($\lambda_x^{(R)} = \lambda_y^{(R)} = \lambda_z^{(R)} = \lambda^{(R)}$) to rather different ones. After this sample fabrication, virtual experiments can be performed and we report two examples.

(i) Stress-strain virtual experiment [Skačej and Zannoni, 2011]. This aims to study the monodomain LCE mechanical properties monitoring how the sample length along the director $\boldsymbol{d} \parallel z$, is varied by the application of a stretching/compression stress along the z-axis, over the xy facet area of the reference sample (Fig. 11.32). The engineering stress Σ_z is expressed in reduced units: $\Sigma_z^* \equiv \Sigma_z \sigma_s^3/\varepsilon_0$, with σ_s the GB monomer width (and unit length, $\sigma_0 = \sigma_s$). A constant volume resize move, deforming the sample box along z by $\Delta\lambda_z = (\lambda_z)_{\text{new}} - (\lambda_z)_{\text{old}}$ is accepted with a probability $\min\{1, \exp[-N\varepsilon_0(\Delta U^* - \Sigma^*\Delta\lambda_z/\rho^*)/(k_B T)]\}$ ($\min\{a,b\}$ returns the lowest between a and b) [Raos and Allegra, 2000; Pasini et al., 2005a]. The same experiment can also provide a connection with ordering showing how the stress increases the orientational order $\langle P_2 \rangle$ (Fig. 11.32c) as experimentally found [Warner and Terentjev, 2003].

(ii) Actuation. Let us now consider the actuation (sample deformation) produced by a transversal electric field. This is particularly appealing, even though not easy to implement experimentally, since rather high field strengths are required to induce deformations and this can, among other practical problems, lead to dielectric breakdown. However, electromechanical actuation has been demonstrated, by Urayama and collaborators, in unconstrained samples swollen by an excess of nematic solvent upon application of an external field perpendicular to the director and using optical and infrared spectroscopy to detect changes in orientational ordering [Urayama et al., 2005, 2006; Fukunaga et al., 2009]. These experiments were accompanied by various analyses using continuum theory [Fukunaga et al., 2008; Corbett and Warner, 2009]. An insight at molecular level into the actuation phenomena in these LCEs can be obtained by a direct simulation of the switching experiments just mentioned, extending and applying the GB modelling as proposed by Skačej and Zannoni [2012]. Then an external electric field is applied along the x-axis, transversally with respect to the director (Fig. 11.33). The coupling energy of an external electric field $\boldsymbol{E}$ with a GB particle with long axis $\boldsymbol{u}_i$ is modelled as we did in Section 10.8.2, adding to U^{LCE} in Eq. 11.28 the term $U_N^{\text{F}} = -\epsilon\xi \sum_{i=1}^{N} (\boldsymbol{u}_i \cdot \hat{\boldsymbol{E}})^2$ where $\hat{\boldsymbol{E}} = \boldsymbol{E}/|\mathbf{E}|$ and ξ is the dimensionless

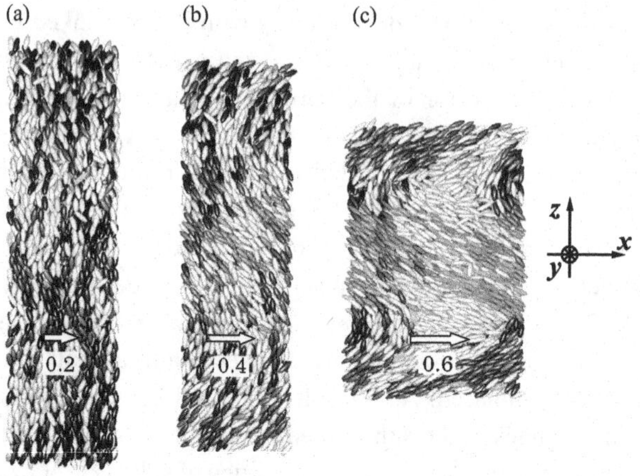

Figure 11.33 Actuation by application of a transversal electric field $\boldsymbol{E} \parallel \boldsymbol{x}$ (arrows) to a monodomain LCE with $\boldsymbol{d} \parallel z$. MC snapshots with applied field strength parameter $\xi =$ (a) 0.2, (b) 0.4, (c) 0.6 of a swollen sample with $N = 8000$ particles and a swelling ratio $\phi_{\mathrm{SW}} \approx 2.1$. Particle orientation and type are grey-coded with dark and light areas corresponding to clockwise and counterclockwise rotations of the polymer network, while all swelling monomers are shown in white regardless of their orientation [Skačej and Zannoni, 2012].

field coupling strength $\xi = \varepsilon_0 \Delta\varepsilon E^2 \mathcal{V}/(2\epsilon)$, with $\varepsilon_0, \varepsilon$ dielectric constants of vacuum and material, $\Delta\varepsilon$ the molecular dielectric constant anisotropy and $\mathcal{V}$ the effective volume occupied by a mesogen. Using typical values for a nematic (e.g. 5CB at $T \approx 300$ K), $\xi = 0.1$ approximately corresponds to 60 V/μm. The mechanical analogue of such an experiment would be the stretching perpendicular to the director. Fig. 11.33 shows three representative snapshots for increasing values of ξ. In the snapshots, director rotation (accompanied by macroscopic sample deformation) is seen above a switching threshold estimated as $\xi \approx 0.35 \pm 0.05$ (or ≈ 110 MV/m) in Fig. 11.33b,c. To meet the periodic boundary conditions assumed in simulation, domains of opposite director rotation and shear appear in the sample. These domains are similar to the 'stripe domains' observed after applying an external stress perpendicular to the nematic director [Verwey et al., 1996; Finkelmann et al., 1997; Zubarev et al., 1999; Conti et al., 2002].

11.10.2 Liquid Crystal Elastomer Supersoft Elasticity

Nematic polydomain elastomers (NPD) cross-linked while in the isotropic phase (I-PDE) exhibit a stress-strain curve (Fig. 1.61b) with a peculiar plateau corresponding to supersoft elastic deformations that require vanishingly little work to stretch the sample, while samples cross-linked in the aligned nematic phase (N-PDE) exhibit a Hooke-like response [Uchida, 2000; Biggins et al., 2009, 2012; Urayama et al., 2009; Takebe and Urayama, 2020]. Since the exceptionally soft elasticity of LCEs is exhibited by materials prepared polymerized in the isotropic state, it is useful to describe the simulated fabrication procedure leading to a

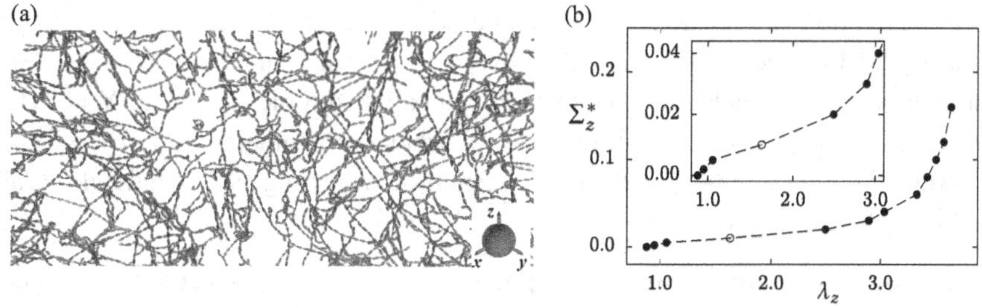

Figure 11.34 (a) Snapshot of a simulated isotropic generation LCE (swelling solvent monomers not shown). (b) Reduced stress-strain plot at $T^* = 3.0$, with low-stress behavior also shown as the inset. Results from isostress MC, $N = 216,000$ particles [Skačej and Zannoni, 2014].

polydomain LCE system in some detail. We recall that, as before, monomers are decorated with two terminal reactive bonding sites that, upon polymerization, provide the main-chain strands with an additional reactive site on the equator allowing for cross-linking. Different to the previous case, we aim here to obtain an isotropic polydomain with only local order. In the proposed polymerization protocol, the polymer chains are first grown at low density along a randomly oriented straight line, each starting from random coordinates. In each growth step, chains are terminated with a probability of 0.125, yielding polydisperse strands $\approx 8 \pm 7.5$ monomers long. Once this first polymerization stage is completed, reticulation is performed by transversally connecting the active head/tail site of each terminal monomer to the nearest available equatorial site belonging to another chain (Fig. 11.34a). The sample is then soaked with swelling monomers ($\phi_{SW} \approx 2.1$) to mimic experimental systems [Urayama et al., 2009] and isotropically compressed in an MC run, at a temperature inside the nematic phase, up to a density $\rho^* = N\sigma_s^3/V \approx 0.287$. This produces a highly interconnected and, as long as the sample size is sufficiently large, almost isotropic polymer network. In any case, it is worth discussing the issue of local and global order quantitatively.

LCE polydomain order parameters. Different to a monodomain system where $\boldsymbol{d}$ and $\langle P_2 \rangle$ are uniform across the sample, polydomain systems are characterized by a static or quasi-static distribution of ordered local domains L of particles (here bonded and swelling monomers) with a local director $\boldsymbol{d}_L$ and order parameter $\langle P_2 \rangle_L = \left[3\langle(\boldsymbol{u}_i \cdot \boldsymbol{d}_L)^2\rangle_{i \in L} - 1\right]/2$. One can also consider the overall ('global') order parameter $\langle P_2 \rangle_G$, calculated over the whole sample, together with a global director $\boldsymbol{d}_G$. In the case of uniaxial ordering, one has $\langle P_2 \rangle_G \approx \langle P_2 \rangle_L \langle P_2 \rangle_d$ (following from the spherical harmonics addition theorem, Eq. F.15), where $\langle P_2 \rangle_d = \left[3\left\langle (\boldsymbol{d}_L \cdot \boldsymbol{d}_G)^2 \right\rangle_d - 1\right]/2$ is an order parameter of the local director $\boldsymbol{d}_L$ with respect to the global one, $\boldsymbol{d}_G$, the average $\langle \ldots \rangle_d$ being performed over domains. Hence, $\langle P_2 \rangle_d$ measures the overall anisotropy of the local director distribution in space and an isotropic polydomain sample, having $\langle P_2 \rangle_d \approx 0$ will exhibit global disorder, i.e. $\langle P_2 \rangle_G \to 0$, even if the local order $\langle P_2 \rangle_L \neq 0$. On the contrary, in monodomain systems $\langle P_2 \rangle_d \to 1$, and there is only little difference between $\langle P_2 \rangle_G$ and $\langle P_2 \rangle_L$ [Skačej and Zannoni, 2011]. Experimentally, $\langle P_2 \rangle_G$ could be measured, e.g. from the dielectric or magnetic susceptibility anisotropy of the sample, while $\langle P_2 \rangle_L$ is accessible, e.g. by magnetic resonance (NMR or ESR [Zannoni et al., 1981]). $\langle P_2 \rangle_G$ can then be used as a test that the preparation protocol is leading to an effectively globally isotropic sample. A $N = 216,000$ sample (Fig. 11.34a) appears an almost fully isotropic polydomain, with $\langle P_2 \rangle_G \approx 0.15$. This 'residual' order coming from the preparation process (acting as a virtual field) is along a director $\boldsymbol{d}_G$ here along $\boldsymbol{y}$.

Stress-Strain Behaviour. To explore the possibility of achieving super soft elastic deformations, stress-strain virtual experiments with the $N = 216,000$ sample were carried out in the nematic at $T^* = 3.0$. An external stress Σ_z^* is applied by pulling along the z-axis (i.e. perpendicular to the residual $\boldsymbol{d}_G$). The stress-strain curve obtained is shown in Fig. 11.34b. Qualitatively, the curve is similar to the experimental ones for I-PDE nematics [Giamberini et al., 2005; Tokita et al., 2006; Urayama et al., 2009; Higaki et al., 2012]: it shows an initial region of small elongation when low stress is applied, then a very soft region where a major elongation of more than 200% is obtained with a small increase in stress and a third region where the LCE stiffens and higher stress is required for further elongation. In the first low-stress region, the large sample deforms only slightly ($\approx 20\%$), with Young's elastic modulus (estimated from the curve slope) around $E_Y \approx 200$ kPa. Then after the weak residual alignment has been overcome at $\lambda_z \approx 1.0$ and $\Sigma^* \approx 0.005$ (corresponding to ≈ 55 kPa), the slope is significantly reduced and the curve exhibits a slightly tilted plateau with $E_Y \approx 100$ kPa (inset of Fig. 11.34b), until a deformation of $\approx 200\%$ is reached. Then the elastomer becomes significantly stiffer, and Young's modulus reaches several MPa towards the end of the simulated curve.

In essence, the polydomain-monodomain transition appears characterized by a simultaneous domain growth and rotation mechanism, assisted by the release of the elastic energy stored during the fabrication in the isotropic phase, that leads to an isotropic distribution of locally ordered domains with conflicting local directors when cooled down to the nematic phase.

12
Atomistic Simulations

Soft crystals exist undoubtedly, there may also be flowing crystals as far as I am concerned, but liquid crystals? Never!

G. Tamman, Annual Meeting of the German Chemical Society, 1905

The underlying physical laws necessary for the mathematical theory of a large part of physics and the whole of chemistry are thus completely known, and the difficulty is only that the exact application of these laws leads to equations much too complicated to be soluble.

P. A. M. Dirac, Proc. Roy. Soc. A, 1929

One of the continuing scandals in the physical sciences is that it remains in general impossible to predict the structure of even the simplest crystalline solids from a knowledge of their chemical composition. Solids such as crystalline water (ice) are still thought to lie beyond mortals' ken.

J. Maddox, 1988

12.1 Introduction

Atomistic simulations consist of methodologies to calculate material (in our case mainly LC) properties starting from molecular structures with chemical (and thus atomic level) detail. The task could be dealt with, in principle, using either of the MC or MD techniques described in previous chapters. In practice, however, generating new equilibrium configurations from an existing one in systems of atomistic complexity is much easier with MD and this is essentially the only approach used for the purpose, and the one we shall discuss here. Computer simulations at atomistic level are often called realistic as they offer, at least in principle, the possibility of reproducing or even predicting in full the properties of LC phases. Unfortunately, mesogenic molecules normally contain a number of atoms so large as to make fully atomistic simulations very demanding. Correspondingly, the number of LC systems that have been simulated providing proof of true realism, e.g. the reproduction of transition temperatures within a few degrees, as well as of relevant observables, such as order parameters and their temperature dependence in good agreement with experiment, is still very limited. The source of the difficulty depends on at least three issues: force fields (FFs), sample sizes and equilibration times. Atomistic molecular dynamics (AMD) simulations performed until 2000, say, [Wilson, 1999], have particularly suffered by all these aspects: FFs were not particularly optimized, sample sizes were of at most of a few hundred molecules and timescales were of the order of 1ns or less, well below the reorientation time of a molecule in LCs, thus inevitably leading to a memory of the (arbitrary) initial configurations chosen. In particular, only in very few cases were ordered LC phases generated by cooling down an initial isotropic configuration, as would be recommended nowadays to confirm equilibration.

The most commonly available force fields, described in Chapter 5, were not developed to reproduce LC properties, but rather simple liquids or biological systems like phospholipids or proteins, where the typical chemical structures and the relevant observables are rather different. Only relatively recently some recommendation on the most appropriate choice among the many force fields available [Cheung et al., 2002; Boyd and Wilson, 2018] has been made. Since 2000, samples with a number of molecules of the order $N \approx 10^3$ [Cook and Wilson, 2001b; Wang et al., 2001; Tiberio et al., 2009] or more [Palermo et al., 2015] have been simulated and the results validated against experiment. An effort towards FF optimization has also been made [Cacelli et al., 2005, 2007; Tiberio et al., 2009] and is continuing [Boyd and Wilson, 2015, 2018].

Notwithstanding the many open issues, there is a class of problems and observations, very important for the understanding of LCs, that completely depends on an atomistic level description. The foremost example is probably the prediction of phase transition temperatures, and particularly the nematic-isotropic one, T_{NI}. This is a crucial element in the design of viable materials for LC devices that have to exist and operate in a certain temperature range. The task is now tackled empirically [Thiemann and Vill, 1997; Johnson and Jurs, 1999] and is far from trivial, since even small changes of structure can dramatically alter T_{NI} (see Fig. 2.22). The other outstanding problem that can only be tackled with atomistic resolution is the prediction of spectroscopic observables (e.g. from proton or deuterium NMR).

Here we shall show a few examples of atomistic MD applications to LCs focussing, as in previous chapters, on the specific methodological aspects relating to anisotropic fluids and necessarily referring to original papers for the very specific details.

12.2 Quinquephenyl and the Rigid Rod Approximation

In total contrast with the complexity of real liquid crystal molecules, it is striking that a vast number of theories (Chapter 7) and computer simulations of LCs resort to the drastic simplification of representing mesogenic molecules as simple rigid objects like spherocylinders or ellipsoids (Chapter 11). Indeed, these minimalist models are still the main tools for our understanding of these complex materials and their generic behaviour. Atomistic simulations can offer a way of testing, as far as possible, how well these approximate treatments work, at least for molecules that are not too removed from the assumptions of the theory, either because of their complexity or flexibility or for being chemically specific (H bond or similar). To compare theory and experiment, an ideal, simple, rigid and symmetric molecule would be highly desirable. As vigorously pointed out by Dingemans et al.

[2006], speaking of *p*-quinquephenyl (P5 or PPPPP), 'If ever there were a calamitic mesogen that corresponded to the approximations used to derive S, the rod-like thermotropic LC PPPPP ... is one among them.' Studying such a molecule without the major approximations introduced in simple statistical mechanics treatments, can help to clarify the importance of flexible chains or polar groups, here absent, or of the aspect ratio in determining nematic behaviour. Unfortunately, P5 is a difficult system to study experimentally, in view of its high transition temperature (see Table 1.2) and of the possibility of chemical decomposition [Sherrel and Crellin, 1979], and there is still a lack of general consensus even on basic things like the sequence of phases observed when cooling from isotropic down to crystal and the value of T_{NI} [Irvine et al., 1984; Wunderlich, 1999; Dingemans et al., 2001, 2006; Rodrigues et al., 2013]. A realistic computer simulation of P5 is also not trivial [Cacelli et al., 2003; Olivier et al., 2014], in view of the number of atoms in the molecule, the need to consider internal rotations of the rings and more generally of optimizing the intramolecular and intermolecular contributions to the FF. It is worth mentioning, however, that an additional reason for interest in an atomistic study is that *p*-quinquephenyl, and more generally polyphenyl derivatives, are, and have been for a long time [Baker et al., 1993], model organic semiconductor molecules for transistors [Hlawacek et al., 2011] and OLED [Grimsdale et al., 2009] devices.

Let us now discuss an AMD simulation of *p*-quinquephenyl [Olivier et al., 2014], its validation and the prediction of LC properties, as well as the assessment of its actual rigidity in relation with classical liquid crystal theories.

Force field and simulations. We already discussed the basic elements of atomistic FFs in Section 5.4. Quantum chemistry (QC) techniques (at MP2//cc-pVDZ level of theory [Cramer, 2004]) have been employed for P5 to obtain partial atomic charges with the ESP scheme [Besler et al., 1990], and inter-ring torsional potentials, used to tune the standard general AMBER FFs (GAFFs) [Wang et al., 2004]. As for P5 rigidity, we can see in Fig. 5.10 that the various QC methods employed produce consistent qualitative results for the internal torsions, with a tilt between the rings of ≈ 40 degrees and energy barriers of the order of ≈ 2 kcal/mol, suggesting the possibility of significant internal rotations at the high temperatures required to melt the material. We also saw in Section 5.4.2 that relying only on QC calculations in vacuum is not sufficient to predict fluid properties. Even for a molecule as chemically simple as P5 (only C and H atoms and a conjugated aromatic structure) we need to empirically adjust the LJ well depth for carbon so as to match (with the help of some test MD runs) the experimental density at one temperature in the nematic (see Fig. 5.11). An optimal value $\varepsilon_C = 0.105$ kcal/mol, together with the standard OPLS parameters $r_C = 1.908$ Å for carbon and $\varepsilon_H = 0.015$ kcal/mol, $r_H = 1.459$ Å for hydrogen [Berendsen et al., 1984] were used in the (N, P, T) MD runs ($P = 1$ atm) carried out with the code NAMD 2.8 [Phillips et al., 2005]. Periodic boundary conditions, Berendsen et al. [1984] barostat and thermostat and a smooth particle mesh Ewald (PME) method for electrostatic interactions [Essmann et al., 1995] were all employed. A sample of $N = 1000$ P5 molecules in an orthorhombic box was studied, starting from the isotropic phase at $T = 750$ K, and gradually cooling down to $T = 600$ K, with equilibration times of at least 50 ns and production times of 40 to 70 ns each for calculating properties.

Figure 12.1 Typical MD snapshots inside the isotropic ($T = 750$ K), nematic ($T = 690$ K), smectic A ($T = 650$ K), smectic A_X ($T = 630$ K), different temperatures, taken from the cooling-down temperature sequence studied for P5 [Olivier et al., 2014].

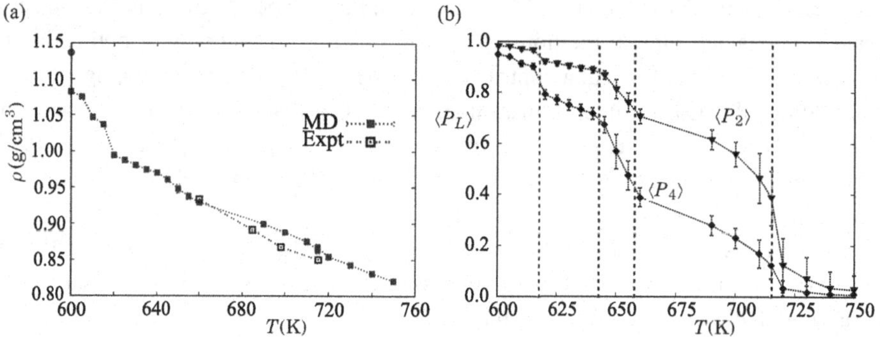

Figure 12.2 (a) Simulated (■) and experimental (□) [Irvine et al., 1984] densities ρ for P5 at various temperatures T. (b) Second- and fourth-rank order parameters $\langle P_2 \rangle$ and $\langle P_4 \rangle$ vs temperature for P5 as obtained from MD [Olivier et al., 2014].

Quinquephenyl phases. In Fig. 12.1 we show typical equilibrium snapshots of the molecular organizations obtained in various phases, using a grey scale for the orientations. The images show at once the disordered isotropic state (right), and on cooling down, the onset of ordered phases that we shall now discuss.

Density. A first important validation of the simulations is obtained comparing the simulated mass density and its temperature dependence with experiment. This is plotted in Fig. 12.2a, as obtained using AMD when cooling down from the isotropic phase, together with experimental data [Irvine et al., 1984]. Note that even if the carbon LJ well depth was optimized against experimental density at a single temperature (see Fig. 5.11), we see good agreement over the whole available range [Irvine et al., 1984]. Four phase transitions can be identified at $T \approx 715, 657, 642, 617$ K from changes of slope of the curve.

Orientational order. The characterization of an LC phase inevitably deals with the investigation of its orientational order parameters with respect to the director $\boldsymbol{d}$. Here the director $\boldsymbol{d}(t_k)$, at successive time steps t_k of the MD trajectory is determined as described in Section 3.5, i.e. as the eigenvector corresponding to the largest eigenvalue of the ordering matrix

Q

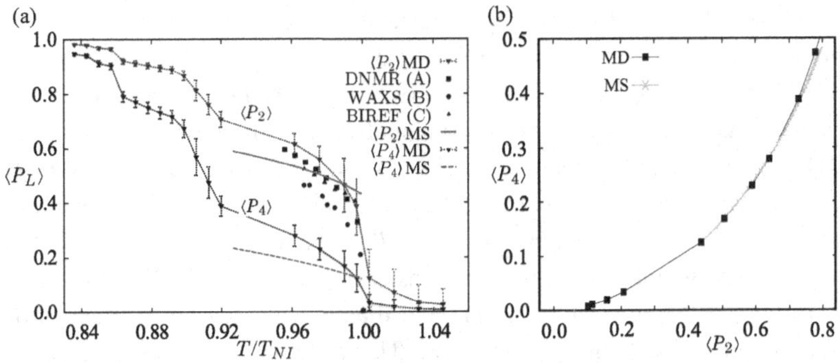

Figure 12.3 (a) P5 order parameters $\langle P_2\rangle$ and $\langle P_4\rangle$ for the inertia tensor axis of P5 (····▾····) as a function of T/T_{NI} compared with experimental data from DNMR (A) [Dingemans et al., 2006], WAXS (B) and birefringence BIREF (C) [Kuiper et al., 2011] and with Maier–Saupe (MS) MFT predictions for $\langle P_2\rangle$ (—) and $\langle P_4\rangle$ (- - -). (b) $\langle P_4\rangle$ vs $\langle P_2\rangle$ from MD simulation (■) and MS theory (line) [Olivier et al., 2014].

(Eq. 3.47) [Berardi et al., 2004a; Tiberio et al., 2009]. Given that molecules change their conformation over time, the reference axis $\boldsymbol{u}_i$ for molecule i can be taken along the eigenvector of its inertia tensor $\mathbf{I}_i(t_k)$ (Eq. 4.24) corresponding to the lowest eigenvalue: $I_{i,min}(t_k)$. The instantaneous sample ordering matrix $\mathbf{Q}(t_k)$ can then be obtained, using $\boldsymbol{u}_i(t_k)$ and averaging over all molecules (Eq. 3.47). The scalar order parameter $P_2(t_k) = -2\lambda_0(t_k)$ at time t_k is gathered from the middle of the three eigenvalues $\lambda_{min}(t_k) < \lambda_0(t_k) < \lambda_{max}(t_k)$ of $\mathbf{Q}(t_k)$ [Eppenga and Frenkel, 1984] and averaged over a sufficiently long equilibrium trajectory, yielding the order parameter: $\langle P_2\rangle = \langle P_2(t_k)\rangle_{t_k} = -2\langle\lambda_0(t_k)\rangle$, where the angular brackets indicating a time average over the trajectory. $\langle P_2\rangle$ and higher-order parameters can also be obtained employing the instantaneous director at time t_k, $\boldsymbol{d}(t_k)$, using the Legendre polynomial definitions (Eq. A.48), e.g.

$$\langle P_2\rangle = \frac{1}{2NM}\sum_{k=1}^{M}\sum_{i=1}^{N}\Big[3\,(\boldsymbol{d}(t_k)\cdot\boldsymbol{u}_i(t_k))^2 - 1\Big], \tag{12.1a}$$

$$\langle P_4\rangle = \frac{1}{8NM}\sum_{k=1}^{M}\sum_{i=1}^{N}\Big[35\,(\boldsymbol{d}(t_k)\cdot\boldsymbol{u}_i(t_k))^4 - 30\,(\boldsymbol{d}(t_k)\cdot\boldsymbol{u}_i(t_k))^2 + 3\Big], \tag{12.1b}$$

where $(\boldsymbol{d}(t)\cdot\boldsymbol{u}_i(t)) = \cos\beta_i(t)$. These order parameters (Fig. 12.2b) allow a first assessment of the phase transitions observed. We shall see later in this section that the phase contiguous to the isotropic one has $\langle P_2\rangle > 0$ but is devoid of positional order, confirming its classification as nematic. Estimating T_{NI} from the simulations is not trivial, because of the finite and relatively small sample size that gives a residual order ($\gtrsim\sqrt{N}$) even in the isotropic phase and the relative small number of temperatures simulated. Considering the highest temperature at which $\langle P_2\rangle > 0.15$, we have 715 K $\leq T_{NI} <$ 720 K. Some refinement can be obtained examining not only the average value of the order, but also the moments of the histogram of all the instantaneous values of $\langle P_2\rangle$ observed during the simulation, as we did for lattice models (Section 10.2.1). In particular, the change of sign of the third moment, indicative of a first-order transition like the NI one (see Fig. 10.3) gives $T_{NI} = 715$ K. Comparing with experiment we see that this overestimates T_{NI} by ≈ 17 K ($\lesssim 3\%$), even if it should be said that the various experimental results differ from one another by up to 7 K [Olivier et al., 2014]. In Fig. 12.3 we show $\langle P_2 \rangle$ and $\langle P_4 \rangle$ versus T/T_{NI} and compare them with available experimental data for $\langle P_2 \rangle$ while, as far as we are aware, experimental results are not (yet) available for $\langle P_4 \rangle$. We note that the agreement between simulation and experiment for $\langle P_2 \rangle$ is good, considering deuterium NMR [Dingemans et al., 2006], birefringence [Kuiper et al., 2011] and diamagnetic susceptivity anisotropy data [Sherrel and Crellin, 1979] (not shown here). The wide angle X-ray scattering (WAXS) data [Kuiper et al., 2011] are more at variance with the simulations, but also with other experimental results. Fitting the $\langle P_2 \rangle$ results into the nematic to the Haller [1975] empirical equation (see Eqs. 3.32 and 11.9), gives a value of the pseudo-critical exponent $\beta_H = 0.18$. Regarding the temperature dependence of $\langle P_2 \rangle$ and $\langle P_4 \rangle$ we also note that, even though $\langle P_4 \rangle$ and $\langle P_2 \rangle$ in Fig. 12.3a do not follow the curves predicted by the Maier–Saupe (MS) MFT very well (Section 7.2 and Fig. 7.3), the simulated results for $\langle P_4 \rangle$ versus $\langle P_2 \rangle$ are in excellent agreement with the MS prediction (Fig. 12.3b). However, this good agreement only indicates that the orientational distribution has, at each temperature, a second-rank exponential character (i.e. that $P(\cos\beta) \propto \exp[a_2 P_2(\cos\beta)]$), even if the effective field strength a_2 is not simply proportional to $\langle P_2 \rangle$ as would be predicted by MS theory (Chapter 7). We note that this should also apply to Onsager theories, since it is well known that their more complex density functional expression for $P(\cos\beta)$ (Eq. 7.68) can also be well represented by expanding its exponent in a Legendre polynomial series truncated at the second term (see Eq. 7.81) [Isihara, 1951; Lasher, 1970].

Smectic phases and positional order. Cooling down from the nematic, other orientationally ordered phases appear. It is apparent from the snapshots in Fig. 12.1 that these have a layered structure, with molecules essentially orthogonal to the layers, suggesting that they could be smectics (thus, e.g. S_A or S_B) or crystalline. Various transitions have indeed been observed experimentally, but the situation is rather confusing. The phases are characterized by their positional order parameters τ_n (see Eq. 3.14)

$$\tau_n = \int_0^{\ell_z} dz P(z) \cos(q_n z) = \left\langle \cos\left(\frac{2\pi n z}{\ell_z}\right)\right\rangle, \quad n \geq 1, \tag{12.2}$$

that represent the expansion coefficients in an orthogonal Fourier basis of the probability $P(z)$ of finding a molecule at a position z along the normal to the layers (see Eq. 3.11) of thickness ℓ_z, assuming the origin of the director frame to be such that $P(z) = P(-z)$. In Fig. 4.11b we have already shown the first three positional order parameters of P5, obtained from MD trajectories with a fit of the translationally invariant two-particle density autocorrelation $g(z_{12})$ (Eq. 4.108b), obtained from the number density of molecules at distance z_{12} divided by the average density. The fit gives the simulated layer spacing as $\ell_z \approx 24$ Å, similar to what was reported by Kuiper et al. [2011] and to the length of P5, indicating a non-interdigitated organization. The positional order parameter τ_1 grows very quickly as T decreases below the nematic temperature range and into the smectic

(Fig. 4.11b). The two low temperature phases are thus crystalline or smectic with high orientational ($\langle P_2 \rangle > 0.9$) and positional order. We note also significant jumps in τ_2, τ_3 at the low temperature phase changes, showing a sharpening of the layers.

Spatial and space-orientational distributions. We now take a closer look at the local structure of the phases. We have already shown in Fig. 4.2a the radial distribution, $g(r) = \langle \delta(r - r_{ij}) \rangle_{ij} / (4\pi r^2 \rho)$, of P5 at a few temperatures. In particular, we see that $g(r)$ is liquid-like for the I and N phases: no sharp peaks are present at medium-to-long range, with a quick tailing to the asymptotic value of 1, as expected for a disordered phase. A close up of $g(r)$ at a temperature below and one above T_{NI} (inset of Fig. 4.2a) shows that they are very similar, reinforcing the concept that the immediate surroundings of a molecule are very similar in both phases as suggested from the weak first-order character of the NI transition [Luckhurst and Zannoni, 1977]. The radial distribution also confirms the high structuring present in the two low temperature phases: the crystalline and the lower temperature smectic, that we shall tentatively call S_{AX}, as well as some difference between the two.

For P5, or oligophenyls in general, the relative orientation of the phenyl rings is of great importance as face-to-face orientation should favour the possibility of charge hopping from one molecule to the other, thus hopefully improving its semiconducting properties. A phenyl centre-phenyl centre radial distribution $g_{PP}(r)$ (Fig. 4.2b) shows that the first neighbour peak is located at ≈ 5.5 Å, rather than at ≈ 3.5 Å, as would be typically expected for face-to-face, π-π packing, corresponding to a somewhat skewed local organization. This closest approach distance is similar to the value of 5.0 Å in the experimental crystal structure at room temperature [Baker et al., 1993]. We also see that the crystal phase has a local hexagonal structure, as shown by the splitting of the second peak [Berardi et al., 1993], while this feature is missing in the two upright smectic phases, that we can then classify as of type S_A, rather than S_B.

The space correlation between two molecules is expected to change as the angles β_{ij} between their c.o.m. separation vector $\boldsymbol{r}_{ij}$ and the director $\boldsymbol{d}$ varies. This can be expressed by the anisotropic radial distribution $g(r, \cos\beta_r)$, introduced in Section 4.9.1 $g(r, \cos\beta_r) = \left\langle \delta(r - r_{ij})\, \delta(\cos\beta_r - \hat{\boldsymbol{r}}_{ij} \cdot \boldsymbol{d}) \right\rangle_{ij} / (4\pi r^2 \rho)$, where $\hat{\boldsymbol{r}}_{ij} = \boldsymbol{r}_{ij}/r_{ij}$, and the average is over particle pairs i, j. In Fig. 12.4 we show contour maps of $g(r, \cos\beta_r)$ at three temperatures. The S_A phase at 650 K shows only a liquid-like correlation transversal to $\boldsymbol{d}$, similar to what we saw for GB(3,5,1,3) particles in Fig. 4.10. Cooling to 630 K shows differences: an increase of these in-layer peaks but also the appearance at $r \approx 25$ Å, of a peak parallel to the director corresponding to the adjacent layer. Upon reaching the crystal, the structuring increases as expected, with a sharpening of the peaks and disappearance of appreciable correlations at intermediate angles β_r.

Quinquephenyl dynamics. We still have to consider if the low temperature layered phases are smectic or crystalline, and for this it is useful to examine their fluidity, monitoring the translational diffusion coefficients along the director, $D_{\parallel}$, and perpendicular to it, $D_{\perp}$ from the mean square displacements in the respective directions, using the classical Einstein formula (see Eq. 6.51). We see from Fig. 12.5a that the three high temperature phases are

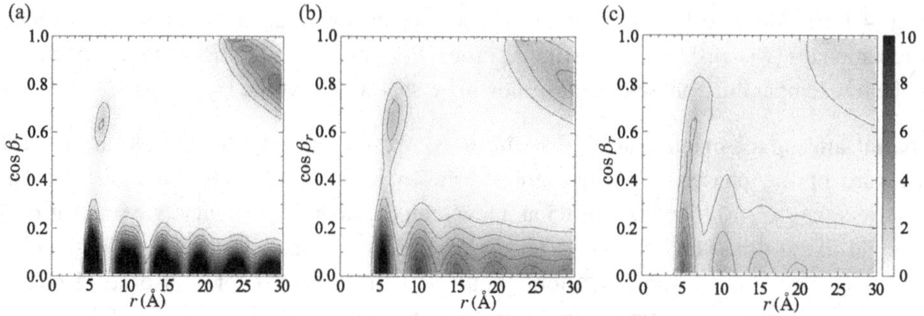

Figure 12.4 Map of the angular-radial distribution function $g(r, \cos\beta_r)$ for P5 at temperatures (a) $T = 610$ K (crystal), (b) $T = 630$ K (S_A^X) and (c) $T = 635$ K (S_A), with β_r is the angle between the director and the interparticle vector $\boldsymbol{r}$. The palette on the far right indicates the value of $g(r, \cos\beta_r)$ with contour lines drawn at integer values [Olivier et al., 2014].

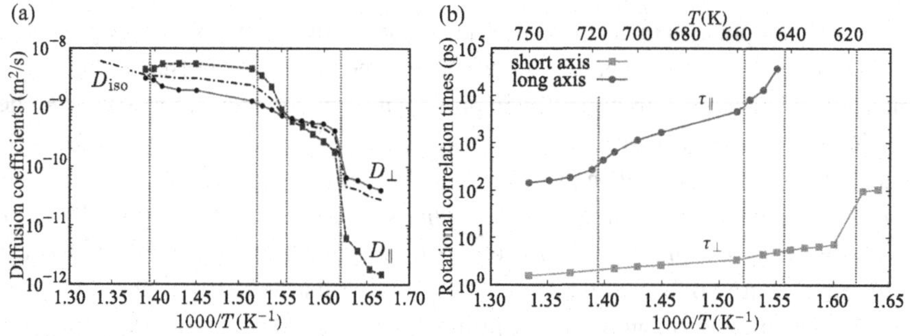

Figure 12.5 (a) Arrhenius plots for the simulated translational diffusion coefficients $D_\parallel$, $D_\perp$ and D_{iso} of a P5 molecule moving parallel and perpendicular to the director and (b) the correlation times for rotation around the long axis $\tau_\parallel$ (spinning) and of the long axis itself $\tau_\perp$ (tumbling) as a function of inverse temperature $1000/T$ (K^{-1}). The dotted vertical lines indicate phase transitions [Olivier et al., 2014].

clearly fluid with translational diffusion coefficients of the order of 10^{-9} m^2/s, comparable with those of other nematics [Miyajima, 2001].

We also see that both in the N and the S_A phase the diffusion along the director is faster than the one perpendicular to it. This is similar to what was found for GB fluids (see Fig. 6.3) but is surprising given the idealized picture of a smectic as a stack of 2D, somehow independent fluids, as we have discussed in Section 6.7. For P5 this is unexpected since molecular flexibility as enforced by flexible tails, here missing, is also believed to play an important role in favouring interlayer diffusion [Mukherjee et al., 2013]. The translational diffusion becomes easier within the layers than across them in the low temperature smectic phase S_{A_X}, a feature typical of smectic phases with high interlayer energy barriers, such as, e.g. tilted ones [Cifelli et al., 2006, 2012]. In any case, the S_{A_X} phase retains some fluidity in both directions, qualifying it as a highly ordered smectic phase, rather than a conformationally disordered crystal [Wunderlich, 1999]. In the crystalline phase the dynamics is slowed down by three orders of magnitude compared to the nematic one and a very slow diffusion is possible only inside a layer (the values of $D_{||} \approx 10^{-12}$ m^2/s reported correspond to a MSD of a few Ångstrom on the simulation timescale). It should anyway be stressed that the aim of the P5 study was to reproduce its LC phases, not the room temperature crystal polymorphs, notoriously nearly impossible to obtain by a cooling down process from the isotropic melt [Oganov, 2010] also because of the MD timescales.

Switching to reorientation dynamics, this can be described with the single-molecule correlation functions for the long axis $\phi_{||}(t) = \langle z(0) \cdot z(t) \rangle$ and for the short axis $\phi_{\perp}(t) = \langle \boldsymbol{x}(0) \cdot \boldsymbol{x}(t) \rangle$ (see the molecular frame in Fig. 5.10). Fig. 12.5b shows, through their decay times to 1/e of the initial value, i.e. the spinning and tumbling times, that rotations around the long axis are fast (of the order of picoseconds) all the way from isotropic to smectic phases. Long-axis reorientation times $\tau_{||}$ start instead from being two orders of magnitude slower in the isotropic phase with a further progressive increase of tumbling time that becomes longer than 100 ns in the smectic.

Is quinquephenyl a rigid rod? The molecular shape anisotropy, expressed by the length to breadth aspect ratio (L/B), is a most important parameter in understanding LC phase behaviour, such as the onset of nematic and smectic phases, and the change in transition temperatures induced by small chemical modifications. Many of these effects have been extensively studied with the help of hard [Allen et al., 1993] and attractive-repulsive rigid models [Zannoni, 2001b]. A deformable molecular model was also proposed [Muccioli and Zannoni, 2006] to bridge between limiting descriptions that assume mesogens to be either completely rigid or fully flexible. Among the shape effects on hard spherocylinders, a progressive increase of the length to breadth ratio is known to induce the nematic phase, widening it, and finally stabilizing a smectic phase [Bolhuis and Frenkel, 1997]. To be more quantitative we consider, like in Berardi et al. [2004a] and Tiberio et al. [2009], the minimal rectangular box containing the molecule rotated to its inertial frame and we take its side lengths L_x, L_y, L_z as molecular size indicators. This molecule-in-a-box approach provides an upper limit to the actual molecular dimensions, difficult to define unambiguously for a non-rigid molecule and was used for other flexible mesogens [Berardi et al., 2004a; Tiberio et al., 2009] and to sexithiophene [Pizzirusso et al., 2011]. The sources of the conformation changes leading to shape polydispersity in P5 appear to be quite different and involve mainly two mechanisms. One is the change in the effective width of the molecule due to internal torsions around the phenyl-phenyl bonds defining the long molecular axis, while the length remains constant. The other mechanism is the presence of some overall bending of the molecule, as shown by the distribution of aspect ratios (Fig. 12.6a) and of bending angles θ (Fig. 12.6b) at various temperatures. van der Schoot [1995] showed, generalizing Onsager theory, that even a small flexibility stabilizes the nematic phase. Also, simulations of mono- and polydisperse hard spherocylinders, have shown that intrinsic polydispersity goes in the direction of widening the nematic phase of P5 (which should be absent for aspect ratios lower than 4.7, as we see in Fig. 12.7) at the expense of the smectic [Bolhuis and Frenkel, 1997]. Polydispersity is also indicated as an important factor for obtaining smectic

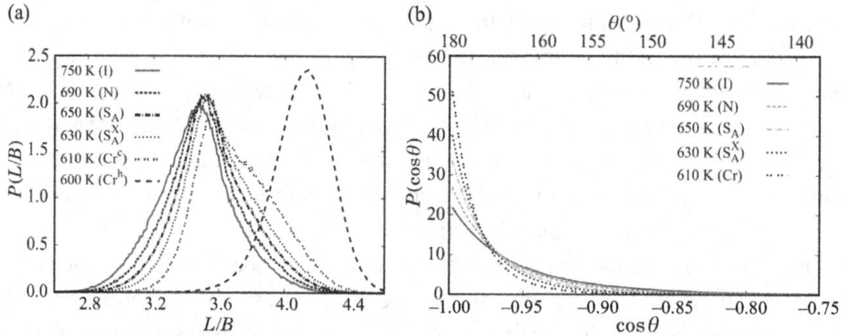

Figure 12.6 (a) Distribution of observed length (L) to breadth (B) aspect ratios $P(L/B)$ of P5. (b) Observed distribution of bending angle represented as $P(\cos\theta)$ vs $\cos\theta$ and θ (°) for P5. Various temperatures shown [Olivier et al., 2014].

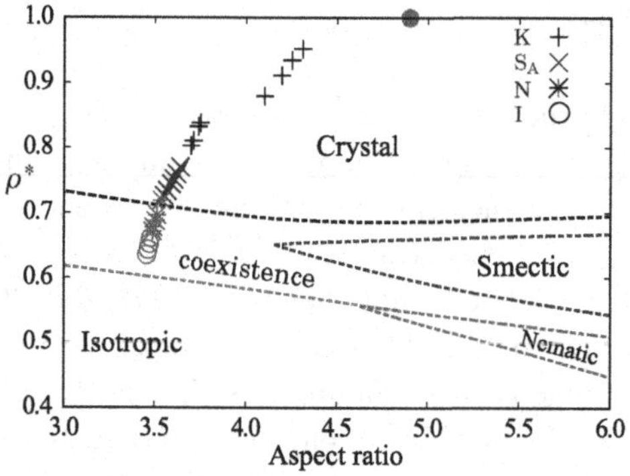

Figure 12.7 Comparison of atomistic simulation results for crystal (K), smectic A (S_A), nematic (N) and isotropic (I) (symbols) with the MC phase diagram for hard spherocylinders (HSCs) by Bolhuis and Frenkel [1997]. The aspect ratio is defined as length over breadth: L/B, corresponding to $(l + d)/d$ in the usual notation for HSCs (see Fig. 11.8), while ρ^* is the reduced density expressed as fraction of the close-packed one ρ_{cp}. The density of the experimental crystal cell at room temperature, $\rho_{cp} = 1.292$ g/cm^3, is used for scaling the atomistic data. The aspect ratio is 4.9 for the crystal, $\rho^* = 1$ (●) [Olivier et al., 2014].

and nematic phases for colloidal systems with relatively low aspect ratio, with smectic phases expected only if the polydispersity is below $\approx 18\%$ [Kuijk et al., 2012]. The other important factor is the presence of anisotropic attractive forces for P5, missing in the hard-particle, that should make it easier for rod-like molecules to order, even for relatively small aspect ratios.

It would seem that a combination of anisotropic dispersive and repulsive interactions is essential to give an LC, rather than a crystal phase, for these relatively small aspect ratios, and moreover, that polydispersity in the aspect ratio provided by internal torsions and bending, even for an apparently rigid molecule like P5, is important to yield a nematic, rather than a smectic phase.

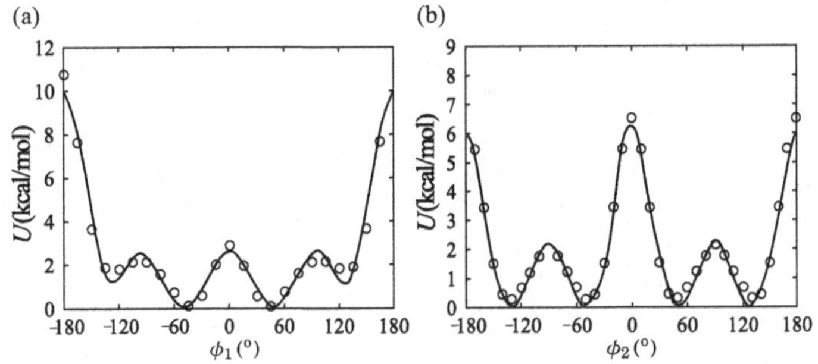

Figure 12.8 The symmetrized and non-optimized MP2/STO–3G level torsional energies (○) for the conformational angles (a) ϕ_1 (°), and (b) ϕ_2 (°) of the cinnamate series (see Fig. 2.22), and the Fourier series approximations (—) using the first six even expansion terms [Berardi et al., 2004a].

12.3 Odd-Even Effect in Cinnamates

Tackling the problem of understanding the large alternation in transition temperatures in the series of cinnamates of Gray and Harrison [1971] (see Fig. 2.22) with atomistic MD requires the reliable prediction of T_{NI} for similar molecular structures, a task for a long time judged not feasible [Yakovenko et al., 1998]. The phase transitons of the first three homologues of the mesogenic series have been studied [Berardi et al., 2004a] with the MD code ORAC [Procacci et al., 1997]. The stretching and bending force constants $K_r^{t_it_j}$ and $K_\theta^{t_it_jt_k}$ were taken from the AMBER FF [Cornell et al., 1995] and LJ parameters from the OPLS all atom FF [Jorgensen and Tirado-Rives, 1988]. The bond lengths and angles r_{eq} and θ_{eq} were derived from *ab initio* calculations at the MP2/3–21G* level performed with the Gaussian 98 package [Frisch et al., 2002]. The partial charges e_i on each atom were obtained fitting the MP2 electrostatic potential with the CHELPG method [Breneman and Wiberg, 1990]. The torsional parameters for the N-phenyl, and O-phenyl bonds (cf. torsions ϕ_1 and ϕ_2 of Fig. 2.22b) were obtained from quantum chemistry MP2/STO–3G energy profiles and approximated with a six terms Fourier-cosine series (see Fig. 12.8). (N, P, T) MD PBC simulations were run for small samples of $N = 98$ molecules of the $n = 0, 1, 2$ homologues (4214, 4508 and 4802 atomic centres, respectively) with temperature and pressure controlled using a velocity scaling thermostat [Fincham and Heyes, 1985] and an isotropic barostat [Parrinello and Rahman, 1981]. Electrostatic interactions were calculated using a variant of the Ewald technique (Section 5.4.4), the smooth particle mesh Ewald (SPME) method [Darden et al., 1993]. Integration of the equations of motion used a multiple time step scheme [Tuckerman et al., 1991a] (see Section 9.3.4) with equilibration lasting to 40 ns for each homologue.

The order parameter $\langle P_2 \rangle$ (Eq. 12.1a), calculated for the molecular axis $\boldsymbol{u}$ attached to the rigid part of each molecule (see Fig. 2.22) shows (Fig. 12.9) that all systems present ordered and disordered phases in the temperature range studied, with $\langle P_2 \rangle$ going to 0 within statistical error, i.e. $\langle P_2 \rangle \approx 1/\sqrt{N}$, at the NI transition. Even if the transitions appear continuous, given the small sample size, the large difference in T_{NI} is apparent, with the odd, $n = 1$,

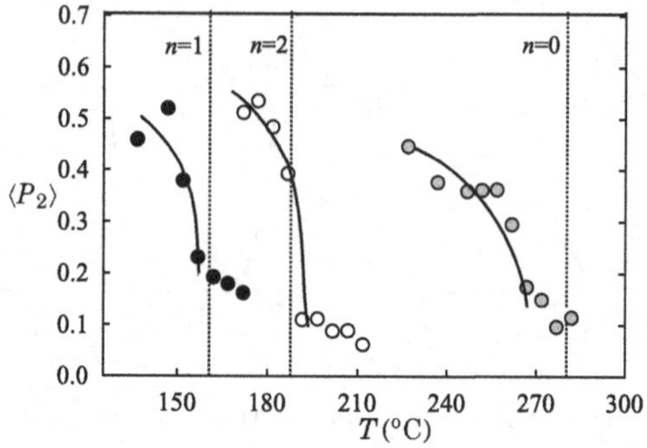

Figure 12.9 The MD order parameter $\langle P_2 \rangle$ vs temperature and experimental T_{NI} (dotted lines) for the $n = 0, 1, 2$ mesogens of the cinnamate series [Berardi et al., 2004a].

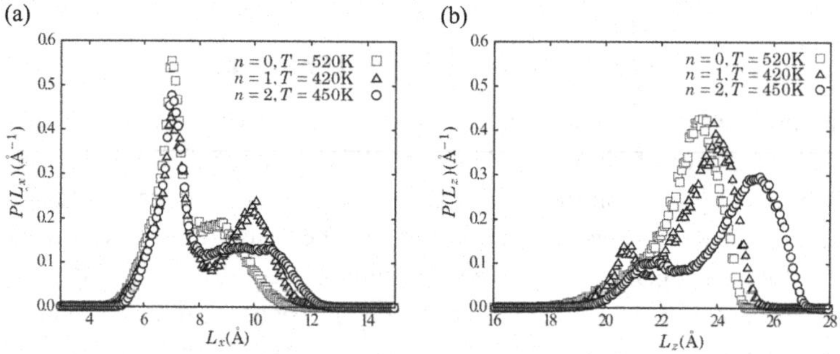

Figure 12.10 Distribution functions of the molecular dimensions (a) L_x(Å), and (b) L_z(Å) for the three cinnamate homologues at three similar reduced temperatures T/T_{NI} in the N phase: $n = 0$ ($T = 520$ K, $T/T_{NI} = 0.96$, $\langle P_2 \rangle = 0.36$); $n = 1$ $T = 420$ K, $T/T_{NI} = 0.98$, $\langle P_2 \rangle = 0.52$), and $n = 2$ ($T = 450$ K, $T/T_{NI} = 0.97$, $\langle P_2 \rangle = 0.54$) [Berardi et al., 2004a].

term definitely less stable compared to the two even ones. This good agreement validates the FF adopted [Berardi et al., 2004a] and leads us to examine the origin of the odd-even effect. The anisometry of each structure at a certain time t can be evaluated, as we did for P5, from the dimensions $L_x(t)$, $L_y(t)$, $L_z(t)$ of the smallest rectangular box containing each molecule, evaluated from the *end-to-end* atomic distances along the axes of its inertia tensor principal frame. From the peak positions of their histograms, shown in Fig. 12.10, we see that the second CH_2 group has the effect of increasing the average molecular length $\langle L_z \rangle$, while the first one has rather the net effect of increasing $\langle L_x \rangle$ and $\langle L_y \rangle$, so that the $n = 1$ homologue is wider than the even homologues $n = 0$ and 2, and thus has lower anisotropy $\langle \Delta L \rangle$. Very concisely, the odd-even effect seems due to the absence of a substantial increase in the molecular length when passing from an even to an odd number of -CH_2- groups see Fig. 12.10). The importance of non-fully stretched conformations cannot be neglected, as they are responsible for the decrease of the transition temperatures along the even and odd homologues series.

12.4 Cyano-biphenyls

Cyano-biphenyls are arguably the most important and well-studied family of liquid crystals, both from the experimental and computational sides. Much work has also been invested on specific force fields [Amovilli et al., 2002; Cacelli et al., 2005; De Gaetani et al., 2006; Cacelli et al., 2007]. Here we discuss an nCB FF at united-atom (UA) resolution, which has the advantage of reducing the number of centres and of allowing larger integration time steps, thus increasing the affordable system size and the time window that the MD simulation can span [Tiberio et al., 2009]. The FF tuning procedure and the optimized set of LJ parameters, as well as other details, are reported in full in the supplementary material of Tiberio et al. [2009].

Simulation conditions. Each of the simulated samples contained at least $N = 250$ UA molecules in a cubic box (sides in the 4.5–5.0 nm range) with PBC, corresponding to 4500, 4750, 5000, 5250 and 5500 atomic centres for 4, 5, 6, 7 and 8CB [Tiberio et al., 2009]. Larger samples of 750 and 3000 molecules have also been studied for 8CB [Palermo et al., 2013] to investigate its smectic phase. Every sample was equilibrated with (N, P, T) MD ($P = 1$ atm, $T = 275–330$ K) and electrostatic interactions evaluated with the particle mesh Ewald method [Essmann et al., 1995]. Trajectories were generated with the NAMD [Phillips et al., 2005] code, using time steps of 2, 8 and 16 fs for bonded, short-range and long-range non-bonded interactions, respectively, employing the Berendsen et al. [1984] barostat and thermostat. For each nCB sample the starting configuration was an isotropic one at 340 K which was then cooled, with equilibration lasting at least 20 ns and production runs of 20–45 ns in the isotropic and of 40–95 ns in the nematic phase. We also mention that UA FFs are known to speed up rotational dynamics [Budzien et al., 2002; Tiberio et al., 2009]. Even though this cooling procedure is considerably more demanding than starting from a crystal state and heating up the sample towards more disordered states, the spontaneous onset of ordered configurations from disordered ones gives more confidence in the reliability of the results [Peláez and Wilson, 2006, 2007]. We now briefly review the MD simulation results for cyano-biphenyls [Tiberio et al., 2009; Palermo et al., 2013].

Density. As we already mentioned for quinquephenyl, a first important validation of the simulated results comes from comparing mass densities with experimental ones. As we see from Fig. 12.11 the density is in good agreement with experimental values [Würflinger and Sandmann, 2001], with a deviation of less than $\pm 3\%$ in the temperature range.

Orientational order parameters. For a uniaxial LC phase with a director $\boldsymbol{d}$ and assuming for now effective molecular uniaxiality, with a rod axis $\boldsymbol{u}$, the orientational distribution reduces to $P(\cos\beta)$, with $\cos\beta = \boldsymbol{u} \cdot \boldsymbol{d}$. To calculate $P(\cos\beta)$ from the simulated trajectories, the necessary step is to determine the director at successive times t of the MD trajectory. As we have seen in this chapter for P5, this can be done by setting up and diagonalizing an ordering matrix at time t, $\mathbf{Q}(t)$, Eq. 3.47, and averaging its eigenvalues as already described. An alternative way, also discussed for P5, proceeds determining the instantaneous director $\boldsymbol{d}(t)$ from the eigenvectors of $\mathbf{Q}(t)$ and calculating the order parameters $\langle P_2(\boldsymbol{d} \cdot \boldsymbol{u}_i)\rangle_i$ and $\langle P_4(\boldsymbol{d} \cdot \boldsymbol{u}_i)\rangle_i$ (as in Eqs. 12.1a and 12.1b). The NI transition can be located from the

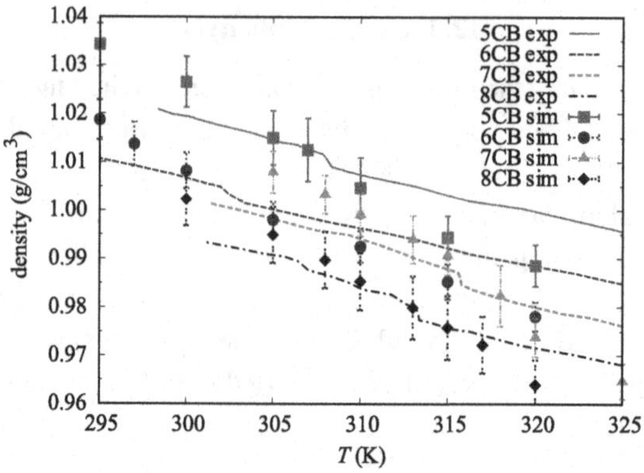

Figure 12.11 Comparison between simulated (symbols) and experimental (lines) mass densities [Würflinger and Sandmann, 2001] for nCB homologues [Tiberio et al., 2009].

temperature variation of $\langle P_2 \rangle$. For the shorter homologue, 4CB, which experimentally does not present a reversible LC behaviour, the simulations also do not indicate a nematic phase, with $\langle P_2 \rangle$ that remains essentially isotropic across the temperature range. Referring to the $n = 5 - 8$ homologues, we see instead from Fig. 12.12 that, on cooling, a jump from isotropic values of the order parameters ($\langle P_2 \rangle < 0.15$) to those typical of a nematic phase ($\langle P_2 \rangle > 0.3$) is apparent for all the homologues. Near T_{NI} the order parameters switch between the ordered and the disordered state; while for $T < T_{NI}$ the fluctuations become much smaller and $\langle P_2 \rangle$ steadily increases with decreasing temperature. We note that observing the onset of crystallization cooling from isotropic is hardly possible from MD since nCB mesophases are known to be easily supercooled. In Fig. 12.12 we compare the simulated $\langle P_2 \rangle$ and $\langle P_4 \rangle$ with several sets of experimental data, in particular birefringence [Horn, 1978; Dunmur et al., 2001; Chirtoc et al., 2004] and Raman data [Picken, 2001]. We see that experimental results, even using the same technique, show differences, due also to the method used to extract $\langle P_2 \rangle$ from raw data. Polarized Raman measurements are also of value as the technique is among the very few accessing $\langle P_4 \rangle$. We note that the agreement between simulation and experiment is good, particularly considering the spread between the various experimental data sets. A precise determination of the transition temperature from simulated data is a non-trivial task, as we have seen even for lattice models (Section 10.2.1), due to the limited number of state points available, the first-order transition coexistence effects causing large fluctuations of the order parameters and also to the small but non-zero-value of the order in the isotropic phase (cf. Fig. 12.12). T_{NI} can be assigned for these nCB by inspection of the histogram of $P_2(t)$ values at each temperature, like that for 8CB [Palermo et al., 2013] shown as an example in Fig. 12.13. We note the non-Gaussian shape of the distribution when approaching the transition from both the isotropic and from the nematic side. This is reflected in the large standard deviations of $\langle P_2 \rangle$ (see Fig. 12.12) and is in agreement with the predictions of Landau theory (Section 2.7.2) for a weakly first-order transition as already observed for the LL model in Section 10.2.1 [Fabbri and Zannoni, 1986] and for Gay–Berne models [Berardi et al., 1993]. The simulated and experimental values of T_{NI}

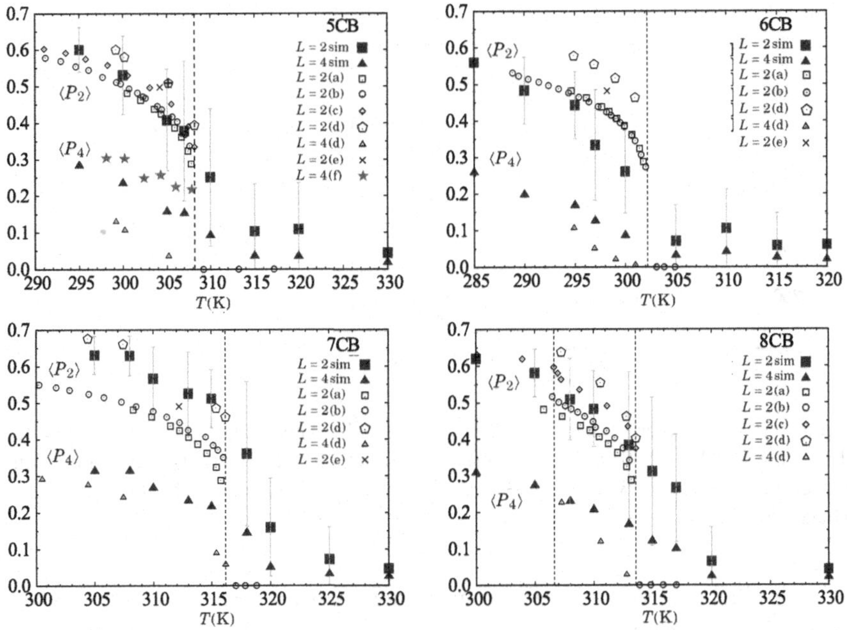

Figure 12.12 Simulated order parameters $\langle P_2 \rangle$ (■) and $\langle P_4 \rangle$ (▲) for the inertia tensor axis of 5CB, 6CB, 7CB, 8CB as a function of temperature T [Tiberio et al., 2009]. Experimental values from various techniques are also shown: refractive index anisotropy: (a, □) [Chirtoc et al., 2004], (b, ○) [Dunmur et al., 2001] and (c, ◇) [Horn, 1978]; Raman data from (d) Picken [2001] for $\langle P_2 \rangle$ (⬠) and $\langle P_4 \rangle$ (△) and (f) from Sanchez-Castillo et al. [2010] for $\langle P_4 \rangle$ (⋆); NMR $\langle P_2 \rangle$ values for 5CB aromatic rings [Fung et al., 1986] (e, ×).

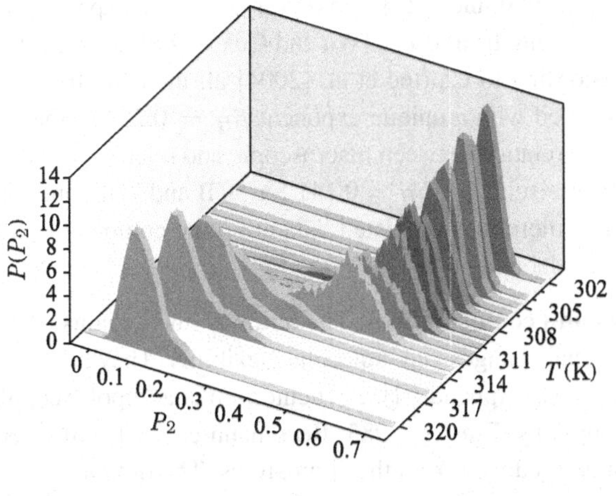

Figure 12.13 Histograms of $P(P_2)$, the frequency of occurrence of instantaneous P_2 values for 8CB at a series of temperatures T (K). The location of T_{NI} is apparent from the discontinuity in peak positions and the change of skewness [Palermo et al., 2013].

(a)

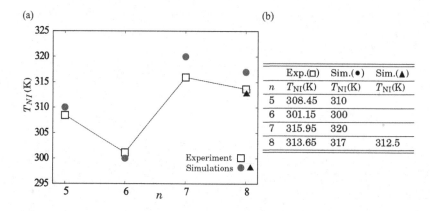

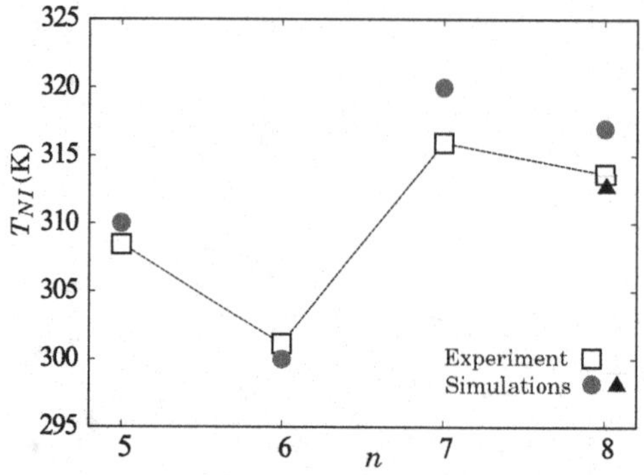

(b)

	Exp.(□)	Sim.(•)	Sim.(▲)
n	T_{NI}(K)	T_{NI}(K)	T_{NI}(K)
5	308.45	310	
6	301.15	300	
7	315.95	320	
8	313.65	317	312.5

Figure 12.14 (a) Experimental [Hird, 2001] (□) and simulated with $N = 250$ (●) [Tiberio et al., 2009] or $N = 750$ (▲) molecules [Palermo et al., 2013] nematic-isotropic transition temperatures for the nCB series. Note that $T_{NI}^{\mathrm{odd}} > T_{NI}^{\mathrm{even}}$. (b) Numerical values of T_{NI} in tabular from.

shown in Fig. 12.14 are in excellent agreement (the worst-case difference is less than 4 K), and the small odd-even effect in the n CB series is well reproduced, demonstrating the level of realism that can be reached with atomistic simulations (see also [Cacelli et al., 2007; Zhang et al., 2011]). The temperature dependence of $\langle P_2 \rangle$ in the nematic is in fair agreement with experiments, as we can see from the plots in Fig. 12.12. These results also allow a fair comparison of simulated $\langle P_4 \rangle$ values with experiments. Fitting the nematic order parameter to the empirical Haller [1975] equation (Eq. 11.9) in the range $-10 < T - T_{NI} < 1$ gives a value of the pseudo-critical exponent $\beta_H = 0.226 \pm 0.04$ for 5CB, 0.241 ± 0.05 for 6CB, 0.191 ± 0.02 for 7CB and 0.231 ± 0.04 for 8CB. The experimental values are fairly scattered, with β_H ranging from 0.172 [Wu and Cox, 1988] to 0.19 for 5CB [Horn, 1978]. Yet at variance, according to Chirtoc et al. [2004] all their birefringence data for 5, 6, 7 and 8CB could be fitted with a unique exponent $\beta_H = 0.25$. In the case of diamagnetic anisotropy, where the relation between macroscopic and microscopic data is more straightforward still different results, i.e. $\beta_H = 0.141$ for 5CB and 7CB, are obtained [Stannarius, 1998a]. We see once more that simulated data are quite comparable with the best sets of experimental data available.

NMR observables and conformations. The terminal alkyl chains of the cyano-biphenyls provide conformational changes and molecular flexibility. They have been studied by various NMR techniques (see Appendix I), e.g. deuterium quadrupolar couplings, and nucleus-nucleus dipolar couplings [Emsley, 1985; Beckmann et al., 1986]. It is thus important to see how these can be predicted from the simulations. The dipolar term in the nuclear spin Hamiltonian in the presence of the strong magnetic field of the NMR spectrometer, stems from the orientational average of the nuclear magnetic dipole interaction: $\mathbf{I}_a \cdot \mathbf{D}_{ab} \cdot \mathbf{I}_b$ with a, b two spin-endowed nuclei on the same molecule (see Section 4.10.5 and Appendix I). This average vanishes in an isotropic phase, while in a uniaxial solvent it gives rise to measurable splittings, expressed in terms of the residual dipolar couplings (RDCs).

Residual dipolar couplings. D_{ab}:

$$D_{ab} = -\frac{\hbar\mu_0}{8\pi^2}\gamma_a\gamma_b\left\langle\frac{3\,(\hat{\boldsymbol{r}}_{ab}\cdot\hat{\boldsymbol{H}})^2-1}{2r_{ab}^3}\right\rangle, \tag{12.3}$$

where γ_a and γ_b are the magnetogyric ratios of nuclei a and b. The RDC represents the component along the field of the average dipolar tensor, (see Eq. C.10) which may be expressed as a function of the internuclear distance r_{ab} and of the orientation of the unit vector $\hat{\boldsymbol{r}}_{ab}$ with respect to the magnetic field $\boldsymbol{H}$. For cyano-biphenyls $\boldsymbol{H}$ aligns the nematic director $\boldsymbol{d}$ parallel to it and we can write:

$$D_{ab} = -\frac{\hbar\mu_0}{8\pi^2}\gamma_a\gamma_b\left\langle\frac{P_2(\hat{\boldsymbol{r}}_{ab}\cdot\boldsymbol{d})}{r_{ab}^3}\right\rangle, \tag{12.4}$$

where $(\hat{\boldsymbol{r}}_{ab}\cdot\boldsymbol{d}) = \cos\beta_{ab}$ and β_{ab} is the angle between the phase director and the internuclear vector. If the two nuclei belong to a rigid fragment, r_{ab} can be considered constant, except for small vibrational corrections, and the latter equation can be employed to derive the order parameter of the internuclear vector and of the fragment. For flexible molecules like nCB, the r_{ab} distances and the angles β_{ab} may vary due to conformational changes if the nuclei a, b reside on different mobile fragments connected by internal rotation axes [Zannoni, 1985]. The average observable dipolar couplings thus contain indirect information on the conformational distribution and in this case various methods have been proposed to extract it. One is the use of approximate models for the internal potential [Emsley et al., 1993], another the maximum entropy inversion of the data [Berardi et al., 1998b]. However, the best way of testing the validity of the conformational information obtained from the simulations is probably to compare calculated and measured NMR RDC, rather than order parameters. In fact, RDC are the raw (and unbiased) output of an NMR experiment in an ordered solvent, while any elaboration of them leading to order parameters and molecular conformations contains some additional assumptions and approximations. In the literature a number of experimental measurements of ^{13}C-^{13}C, ^{13}C-^{1}H and ^{1}H-^{1}H dipolar couplings have been reported [Emsley, 1985; Beckmann et al., 1986; Fung et al., 1986; Emsley et al., 1987]. While the former are directly accessible also in a UA molecular description, to compute couplings involving ^{1}H we have to assign the hydrogen positions from the position of the UA centres. This is easily accomplished assuming a perfect planar trigonal and tetrahedral geometry for sp^2 and sp^3 carbons, and a fixed C-H bond distance of 1.08 and 1.09 Å, respectively. Table 12.1 compares simulated and experimental RDCs for the hydrogenated aromatic carbons as measured by Fung et al. [1986]. We see that the results are in semi-quantitative agreement with experimental figures. The systematically higher values of the simulated RDCs can be ascribed to the different temperature ($T = T_{NI}^{sim} - 5$K in the simulation and $T = T_{NI} - 4$K in the experiment), which implies a higher nematic ordering and consequently higher absolute values of the simulated RDCs. In Fig. 12.15 we present instead the RDC for the vicinal carbon-hydrogen contacts along the alkyl chaini. We note that the sign of the couplings, that could not be determined experimentally, was assigned following geometrical considerations [Fung et al., 1986] or calculated with the assumption of a potential of mean torque [Emsley et al., 1987], while it can be directly obtained from the simulations (see Eq. 12.4, with β is the angle between the C-H vector and the director).

Table 12.1. *C-H dipolar couplings between aromatic carbons and directly bonded hydrogens for nCB* (kHz). *Experimental values were recorded at* $T = T_{NI} - 4$ K *[Fung et al., 1986; Emsley et al., 1987] while simulated values [Tiberio et al., 2009] at* $T = T_{NI} - 4$ K *for 8CB (313 K) and* $T = T_{NI}^{sim} - 5$K *for 5–7CB (305, 295, 315* K), *using Eq. 12.4*

	5CB		6CB		7CB		8CB	
Carbon	Expt	Sim	Expt	Sim	Expt	Sim	Expt	Sim
3′	0.81	0.98	0.84	1.07	0.91	1.26	–	0.97
2′	1.08	1.08	1.02	1.17	1.16	1.39	–	1.07
2	1.02	1.13	1.03	1.26	1.21	1.44	–	1.16
3	1.03	1.09	1.06	1.21	1.15	1.39	–	1.12

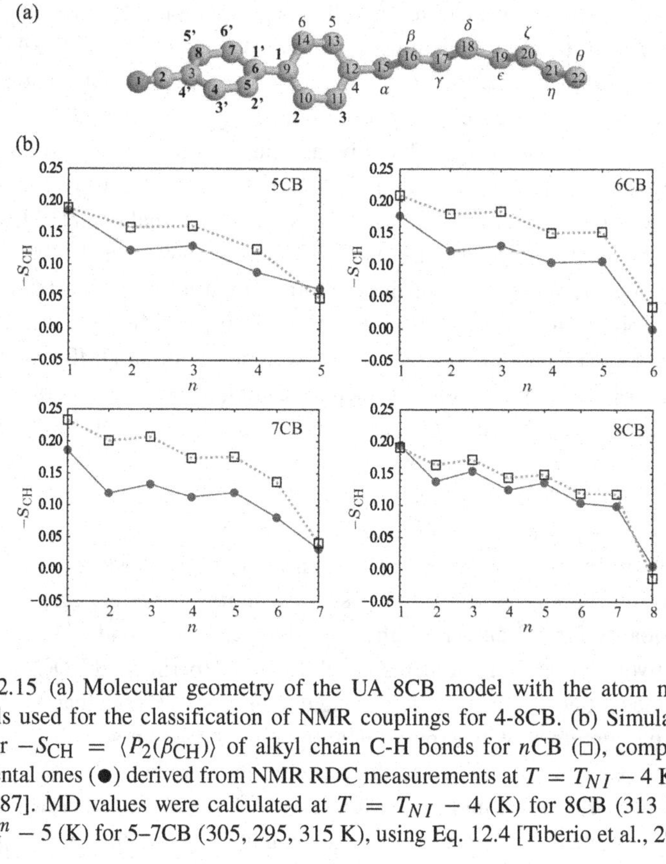

Figure 12.15 (a) Molecular geometry of the UA 8CB model with the atom numbering and labels used for the classification of NMR couplings for 4-8CB. (b) Simulated order parameter $-S_{CH} = \langle P_2(\beta_{CH}) \rangle$ of alkyl chain C-H bonds for nCB (□), compared with experimental ones (●) derived from NMR RDC measurements at $T = T_{NI} - 4$ K [Emsley et al., 1987]. MD values were calculated at $T = T_{NI} - 4$ (K) for 8CB (313 K) and at $T = T_{NI}^{sim} - 5$ (K) for 5–7CB (305, 295, 315 K), using Eq. 12.4 [Tiberio et al., 2009].

For this particular observable the approximation of a rigid inter-nuclear vector is acceptable and the calculation of the order parameters S_{CH} is straightforward. On the whole the behaviour along the chain is well reproduced, particularly for 8CB, where the simulated and experimental $T - T_{NI}$ coincide. In summary, even if hydrogens are not explicitly

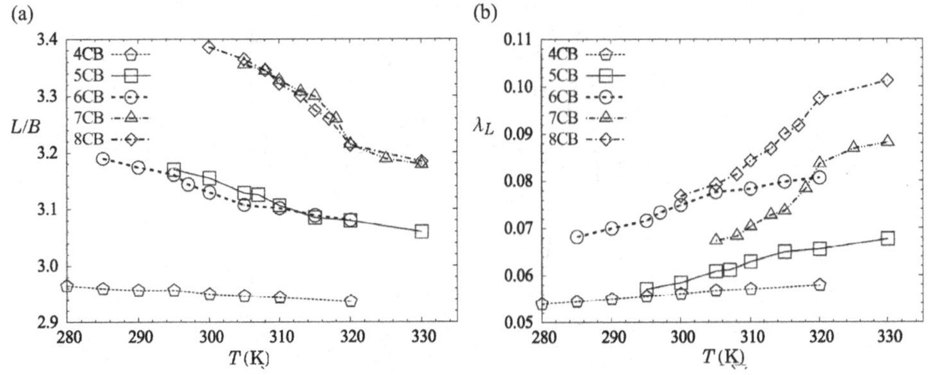

Figure 12.16 (a) Aspect ratio L/B of molecular dimensions and (b) shape biaxiality $\lambda_L = \sqrt{3/2}\left(L_y - L_x\right) / \left(2L_z - L_x - L_y\right)$ as a function of temperature T (K), and its probability at the first simulated temperature below the T_{NI}: 4CB (⬠), 5CB (□), 6CB (○), 7CB (△), 8CB (◇) [Tiberio et al., 2009].

considered in the UA FF, the MD simulations are in good agreement with the available NMR data, without the introduction of any fitting parameter. Moreover, simulations have a clear potential as a complement to NMR techniques with oriented solvents towards resolving spectral assignment issues, as shown in some specific examples [Weber et al., 2012, 2015].

Molecular shape. As we saw in Section 12.3 the molecular shape descriptors, expressed in terms of the relative ratios, the anisotropy and the biaxiality of the molecular dimensions, are important parameters in understanding LC phase behaviour and the change in T_{NI} induced by small chemical modifications. To estimate the effective shape of the nCBs, we consider, as in Section 12.3, the edges L_x, L_y, L_z of the minimal rectangular boxes containing each molecule in its inertial frame. This represents an approximation to the actual, time-dependent, molecular dimensions that are difficult to define unambiguously for a flexible molecule. In Fig. 12.16 we show the temperature dependence of the ratio between average molecular length $L = \langle L_z \rangle$ and molecular breadth $B = \langle L_x + L_y \rangle/2$, respectively. We note that the aspect ratio L/B, always below 3 for 4CB, does not grow linearly with n, as we might have expected, but it increases only when n goes from even to odd (e.g. from 4 to 5 and from 6 to 7), as demonstrated by the similar aspect ratios for the couples 5CB, 6CB and 7CB, 8CB in Fig. 12.16a. With the additional help of the shape biaxiality $\lambda_L = \sqrt{3/2}(\langle L_y \rangle - \langle L_x \rangle)/(2\langle L_z \rangle - \langle L_x \rangle - \langle L_y \rangle)$ (Fig. 12.16b), we can see that the $n = 5, 6$ and $7, 8$ terms have approximately the same aspect ratio, but that $n = 5$ and 7 are considerably more biaxial ($L_y > L_x$). This increase in λ_L tends to counteract the effect of the increase in molecular length, leading to the even-odd profile of T_{NI} (Fig. 12.14).

Spatial and space-orientational distributions. We now take a closer look at the molecular organization of this series of compounds. To do this, the 'default' observable is the radial distribution function $g(r) = \langle \delta(r - r_{ij}) \rangle_{ij} / (4\pi r^2 \rho)$, shown in Fig. 12.17, where $\boldsymbol{r}_{ij}$ is the vector connecting the chosen reference centres on molecules i and j and ρ is number density. In the N phase of nCB, $g(r)$ is liquid-like, in the sense that no sharp peaks are

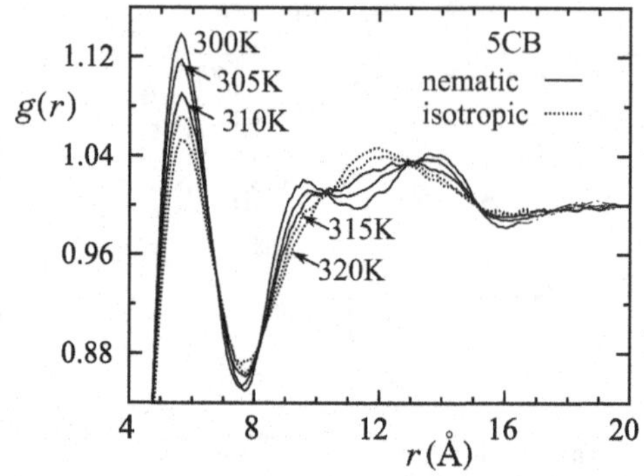

Figure 12.17 Temperature dependence of c.o.m. $g(r)$ for 5CB [Tiberio et al., 2009].

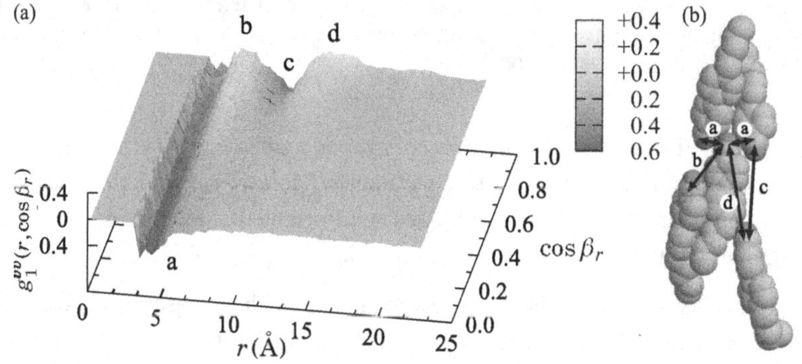

Figure 12.18 (a) The angular-radial distribution function $g_1^{vv}(r, \cos\beta_r)$ for 5CB at $T = 295$ K. Here $\boldsymbol{r} \equiv \boldsymbol{r}_{\mathrm{CN-CN}}$ is the vector connecting the CN point dipole $\boldsymbol{v}$ on a molecule to the same on another, while β_r is the angle between $\boldsymbol{r}$ and the director (Eq. 12.5). The range of $\cos\beta_r$ goes from 0 ($\boldsymbol{r} \perp \boldsymbol{d}$) to 1 ($\boldsymbol{r} \parallel \boldsymbol{d}$). (b) Snapshot of few near neighbours with an indication of relevant distances: a, b, c, d [Tiberio et al., 2009].

present at medium-long range and that it quickly goes to the asymptotic value of 1, as expected for a fluid [Tiberio et al., 2009]. It is also useful to introduce more specific site-site anisotropic correlations of rank L, e.g. $L = 1$ between a chosen vector $\boldsymbol{v}$ belonging to two different molecules: $g_1^{vv}(r, \cos\beta_r)$, or $L = 2$ between some second-rank tensor, e.g. electric quadrupoles $\boldsymbol{\Theta}$: $g_2^{\boldsymbol{\Theta\Theta}}(r, \cos\beta_r)$. In particular for cyano-biphenyls, it is interesting to consider the correlation between the strongly polar CN groups on two different molecules:

$$g_1^{vv}(r, \cos\beta_r) = \left\langle \delta(r - r_{ij})\delta(\cos\beta_r - \cos\beta_{r_{ij}})(\boldsymbol{v}_i \cdot \boldsymbol{v}_j) \right\rangle_{ij} \tag{12.5}$$

where $\boldsymbol{v}_i$, $\boldsymbol{v}_j$ are unit vectors along the CN bond on molecules i and j, while $\boldsymbol{r}$ is the separation vector between the CN point dipoles of the two molecules and $\cos\beta_r = (\hat{\boldsymbol{r}} \cdot \boldsymbol{d})$. This almost coincides with choosing the para axis of the biphenyl, since the *ab initio* electrostatic charges are concentrated on this portion of the molecule. In Fig. 12.18a we show this correlation for 5CB in the N phase, together with the configuration of neighbours, indicating

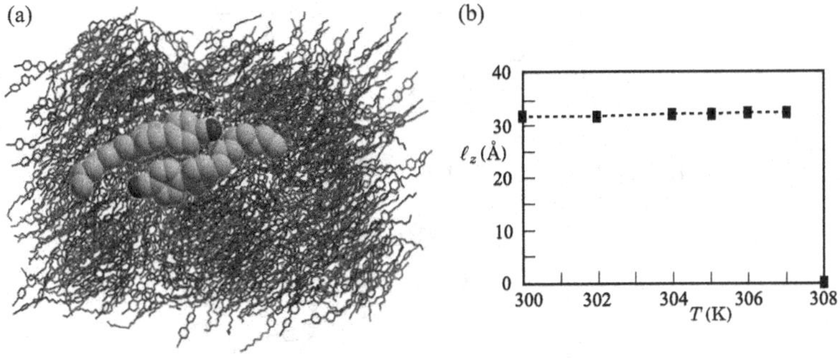

Figure 12.19 (a) MD results for layer interdigitation in the smectic A phase of 8CB. (b) Simulated layer spacing ℓ_z at various temperatures [Palermo et al., 2013].

the formation of antiparallel dimer-like pairs. We note that the formation of antiparallel 5CB dimers is favoured mainly for side-side and head-tail pairs (a in Fig. 12.18b), while parallel dimers arise only when aligned also with the director (b, d). The g_1 gives information of the local polar ordering and shows that a pronounced short-range polar structuring is present. The sign of g_1 of the neighbour is distance-dependent and is determined by an alternation of prevalently antiparallel ($g_1 < 0$) and parallel ($g_1 > 0$) orientations. The dipole correlation between two molecules is anisotropic and varies when $\boldsymbol{r}$ is at different angles β_r with respect to the director (Fig. 12.18).

8CB Smectic phase. 8CB is the first in the homologous cyano-biphenyl series to show a smectic phase and is arguably the most studied smectic system ever, both theoretically and experimentally. It is then particularly interesting to investigate to what extent the FF can reliably predict its smectic as well as nematic features. Samples with $N = 750$ and $N = 3000$ molecules of 8CB (16,500 UA centres) in a deformable sample box of size ≈ 90 Å with PBC were simulated [Palermo et al., 2013] with N, P, T ($P = 1$ atm) using NAMD. An isotropic Parrinello–Rahman–Nosé barostat (Section 9.6) and a simple velocity scaling thermostat have been used to control pressure and temperature and PME employed for electrostatics. The smectic was obtained by gradual cooling from an isotropic ($T = 360$ K) disordered configuration, with post-equilibration production runs of up to 400 ns. We have already seen the excellent agreement of T_{NI} (Fig. 12.13) and the agreement is found to be good (307 ± 2 K compared to 306.8 K experimental) also for the T_{S_AN} transition temperature. The layer thickness obtained from MD is $\ell_z = 32.4 \pm 0.2$ Å, while X-ray experiments give $\ell_z = 31.6$ Å [Leadbetter et al., 1979], $\ell_z = 31.432$ Å [Krentsel et al., 1997] or $\ell_z = 31$ Å [Urban et al., 2005]. As we see in Fig. 12.19a this layer spacing is sensibly lower than twice the molecular length of 8CB and corresponds to a strongly interdigitated structure (much more so than what we saw for 5CB in the nematic, in Fig. 12.18). The layer spacing is also found to be essentially constant with temperature, (Fig. 12.19b) in accord with the experimental results of Urban et al. [2005], confirming the validity of the UA FF described in [Tiberio et al., 2009] for the family of n-alkyl cyano-biphenyls. In a sense the FF optimization corresponds to devising a computational chemistry synthesis of *in silico*

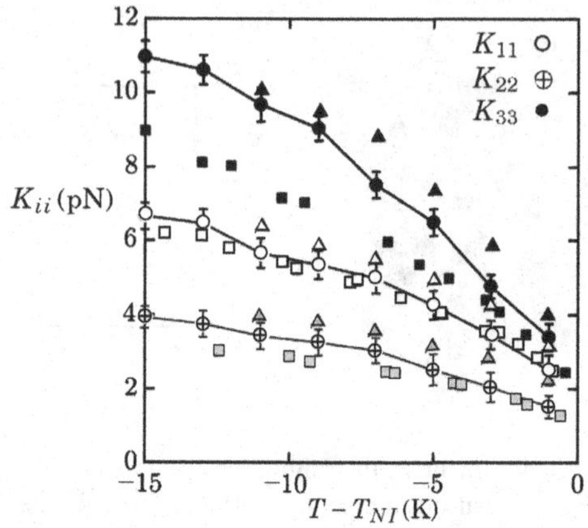

Figure 12.20 Temperature dependence of splay, twist and bend bulk elastic constants K_{11} (○), K_{22} (⊕), K_{33} (●) for 5CB evaluated with AMD and a free energy perturbation technique [Sidky et al., 2018] compared to experimental data from Madhusudana and Pratibha [1982] (squares) and Chen et al. [1989] (triangles).

nCB mesogens, that can then be used to study other properties and model devices. We shall see in Section 12.5, applications to the investigation of surface anchoring in thin films. For other bulk properties it is important to mention at least:

Elastic constants. The Frank elastic constants (Section 1.2.3) for cyano-biphenyls and some of their mixtures have been determined by Sidky et al. [2018], Shi et al. [2020] using the FF by Tiberio et al. [2009] just described and the free energy perturbation method already mentioned for rod-like (Section 11.5.1) and disc-like (Section 11.5.2) Gay–Berne systems. Fig. 12.20 shows a good agreement with experiment, particularly for the splay and twist constants. Note that no information on elastic constants was used when setting up the FF.

12.5 Surface Anchoring

The alignment of organic molecules at an interface (anchoring) [Allara, 2005] is a key issue for LC physics and display technology [De Luca et al., 2008; Ricci et al., 2010], as well as for applications to organic transistors and photovoltaic cells where it is known to strongly affect charge transport [Vilan et al., 2010; Beljonne et al., 2011]. The description of anchoring is, however, essentially empirical, and methods for predicting the molecular structuring and orientation at a support surface have been surprisingly scarce. Indeed, the commonly employed approaches [Yokoyama, 1988; Jérôme, 1991] typically amount to a continuum description of the surface free energy per unit area when the LC director is at an angle β from the surface normal, while its equilibrium value (*easy axis*) corresponds to an angle $\beta_{\rm A}$, as in the celebrated Rapini and Papoular [1969] expression:

$$W(\beta) = w_0^{\rm A} - \frac{1}{2} w_2^{\rm A} \sin^2 (\beta - \beta_{\rm A}), \tag{12.6}$$

where w_2^A is called the anchoring strength and, for planar anchoring, $\beta_A = \frac{\pi}{2}$. More general formulations including higher-rank angular terms [Teixeira and Sluckin, 1992], or an in-plane azimuthal angle ϕ [Fukuda et al., 2007] have been proposed, but the problem of determining the anchoring strength coefficients from realistic microscopic interactions still remains, even if various approximate approaches, based on specific assumptions on the type of interaction model, exist. For instance, dispersive, electrostatic or excluded volume mechanisms [de Gennes, 1974; Jérôme, 1991; Evangelista and Ponti, 1995; Andrienko and Allen, 2002] have been proposed to date, albeit with limited success. To further complicate the problem, the alignment induced by the surface not only depends on the chemical nature of the LC-substrate interactions, but also on the corrugation of the surface itself and its roughness and defects that may cause elastic distortions of the director [Berreman, 1972; Fukuda et al., 2007]. In real experiments the chemical and morphological aspects are entangled and it is extremely difficult to assess their relative importance. On the positive side, important progress has been made in characterizing experimentally the surface-induced order and structuring of LCs close to a surface [Carbone and Rosenblatt, 2005; De Luca et al., 2008; Carbone et al., 2009; Lee et al., 2009; Nazarenko et al., 2010; Voitchovsky et al., 2010], making the need for a theoretical description even more pressing. Atomistic MD simulations can provide a viable bottom-up approach and here we briefly examine examples for a nematic on different hard and soft surfaces, emphasizing the methodologies that can be applied to other systems as well. In particular, the alignment orientation, the order and molecular organization of nano-thick nematic and isotropic films of 5CB at the interface with (i) atomically flat, H-terminated (001) crystalline silicon surfaces [Pizzirusso et al., 2012a], (ii) crystalline and glassy silica of different roughness [Roscioni et al., 2013], (iii) some lapped and disordered polymers [Palermo et al., 2017] and (iv) self-assembled monolayers [Roscioni et al., 2017] will be examined.

12.5.1 5CB on Crystalline Silicon

The atomistic simulations of 5CB on a crystalline silicon slab [Pizzirusso et al., 2012a] was arguably the first realistic one for a nematic thin film deposited on a solid slab, showing that a quantitative description of anchoring can be obtained from molecular and surface properties. The deposited film is in contact with air (vacuum strictly) on the free surface. When choosing silicon as a substrate, e.g. cleaving a slice from a bulk crystal, the first choice to make is on how to terminate the unavoidable dangling bonds. For many practical applications this is done by exposing the slab to oxygen at an appropriate pressure and temperature to yield a so-called SiO_x. Unfortunately, x is not in a stoichiometric ratio, also because of surface restructuring reactions, and is often only approximately known. This ill-defined composition is not really suitable for an atomistic simulation and a better alternative for choosing a chemically well-defined and atomically flat silicon surface is an hydrogen-terminated one [Irene, 2008; Bellec et al., 2009; Vilan et al., 2010]. This is also of importance as a micro- and nano-fabrication platform, in particular for opto-electronic applications such as non-linear optics, thin-film displays, lithography and molecular electronics.

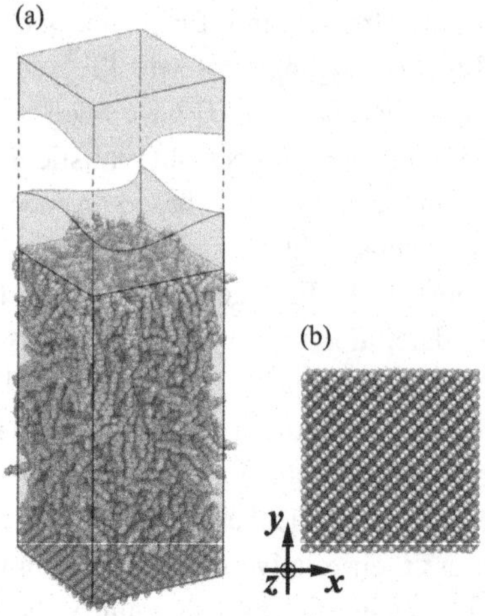

Figure 12.21 (a) The simulated sample cell replicated with PBC showing the 5CB film and the vacuum above. (b) The 'laboratory' axes frame with a snapshot of the hydrogenated silicon (001) surface showing the (1, 1) nanogrooves [Pizzirusso et al., 2012a].

Simulation details. In the study of Pizzirusso et al. [2012a] two 5CB films with $N = 1000$ or 2000 molecules, corresponding to a thickness of ≈ 12 nm and ≈ 24 nm, respectively, were prepared over the hydrogenated Si(001):H flat surface of a silicon slab of dimensions 59.73 Å × 59.73 Å and thickness 18 Å. The support slab was obtained cleaving a silicon crystal along the (001) crystal direction, and saturating completely the dangling bonds with hydrogens (see Fig. 12.21b). The Si surface was modelled at atomistic level, with force constants for the Si-Si and Si-H covalent bonds taken from the COSMIC force field for a polyalkylsilanes [Szabó et al., 1999] whereas the equilibrium bond distances and valence angles have been assumed to be those typical for a silicon diamond-like lattice [Ashcroft and Mermin, 1976]. The atomic electrostatic point charges were calculated by QC techniques as described by Pizzirusso et al. [2012a] and further symmetrized for H and Si atoms with equivalent local environment. Thus, all H atoms were assigned the same average charge (−0.138e), and four types of Si atoms were defined, i.e. those only bound to other Si atoms, with zero atomic charge and those bonded to one (+0.138e), two (+0.276e) or three (+0.414e) hydrogens. After equilibration, the solid slab was 'frozen' during the anchoring simulation. (N, V, T) MD simulations were run with the NAMD code [Phillips et al., 2005]. A general problem arising with thin film relates to the use of the very efficient variants of Ewald summations for electrostatic calculations, since these require an infinite periodic crystal and thus periodic boundary conditions (PBCs) in all directions. To do this the sample, i.e. the elementary cell, is extended to ≈ 300 Å (more than ten times the film thickness) in the z-direction (i.e. across the film) to accommodate ample empty space above the whole surface (Fig. 12.21a). The typical MD production runs were ≈ 70 ns for the smaller system and ≈ 40 ns for the larger one.

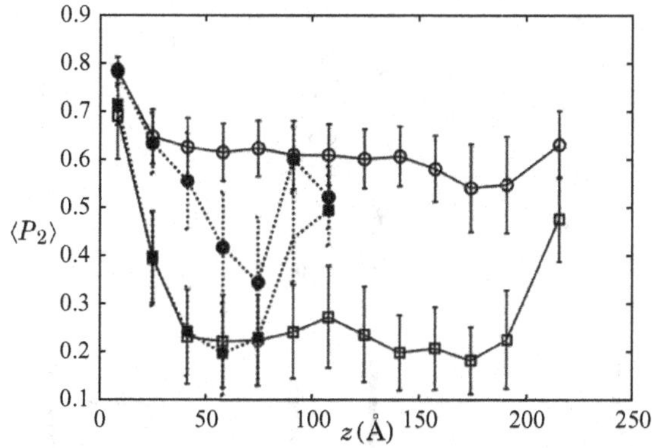

Figure 12.22 Local order parameter $\langle P_2 \rangle$ of 5 CB vs distance z (Å) from the silicon Si(001):H support surface for the thin sample with $N = 1000$: $T = 300$ K (●) and $T = 315$ K (■), and the thicker one with $N = 2000$ 5CB molecules: 300 K (○) and 315 K) (□) [Pizzirusso et al., 2012a].

Order across the film. A fundamental question about the effect of surfaces on the molecular organization, is whether they increase or disrupt the orientational order [Jérôme, 1991]. In Fig. 12.22 we see the variation of the local orientational order parameter $\langle P_2 \rangle$ as a function of the distance z from the Si(001):H surface. We see that at $T = 315$ K where 5CB in the bulk is isotropic, the surface still induces a high nematic order (≈ 0.7) in the LC overlayer, as predicted by theoretical models [Sheng, 1982]. The order is lost in a few nanometres and almost vanishes (within numerical error) in the centre of the film, then increases again approaching the vacuum interface ($\langle P_2 \rangle \approx 0.5$) . Thus, both interfaces induce a local increase of the order parameter. In the nematic, at $T = 300$ K, we observe a quite different behaviour, depending on film thickness. In the case of the 24 nm film, we see again that the nematic order close to the silicon surface is higher ($\langle P_2 \rangle \approx 0.8$) than the equilibrium bulk phase value ($\langle P_2 \rangle \approx 0.5$) [Tiberio et al., 2009] while the corresponding value in the middle of the film is considerably lower for the thinner sample (0.3) and higher for the thicker one (0.6). We note that the order at the solid interface is essentially the same at the two temperatures and in the two phases (nematic and isotropic), both for the thin and thick films, indicating that their ordering is dominated by surface effects. This finding is consistent with theoretical arguments on the so-called 'subsurface deformation' at flat planar surfaces [Rajteri et al., 1996; Barbero et al., 1998] induced by dispersive interactions.

Director across the film. The way the director and the order change across the sample for a hybrid geometry, with planar alignment on a surface and perpendicular (homeotropic) anchoring at the other one, has been in itself the object of many theoretical [Barbero and Barberi, 1983; Palffy-Muhoray et al., 1994; Chiccoli et al., 2003] and experimental [Carbone et al., 2009; Chiccoli et al., 2019] investigations. In particular, three possibilities have been outlined: (i) a uniform director orientation across the sample, imposed by the surface with strongest anchoring, (ii) a biaxial structure with director exchange, where each

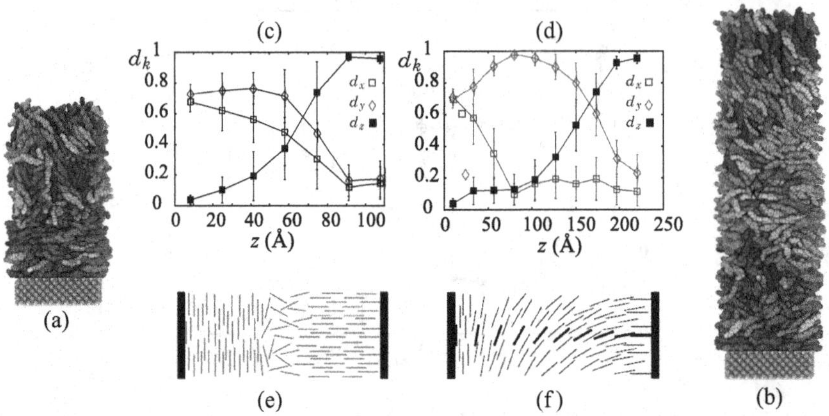

Figure 12.23 Snapshots of the molecular organization of a 5CB film on a Si(001):H surface at $T = 300$ K, viewed from the x-axis, for the $N = 1000$ (a) and $N = 2000$ (b) samples of different thickness studied. We also show the director components $d_k(k = x, y, z)$ across the film as a function of the distance z from the surface for the thin (c) and thick (d) film [Pizzirusso et al., 2012a]. The organization is compared with a sketch of those predicted by continuum theory for the case of a very thin film with abrupt changes between the planar and homeotropic director orientations (e) or a continuous bend deformation (f) [Barbero and Barberi, 1983; Palffy-Muhoray et al., 1994; Chiccoli et al., 2003].

surface imposes its local orientation with a discontinuous transition from one region to the other with order disruption at their boundary (Fig. 12.23e), and (iii) a bent structure where the director changes regularly between the two perpendicular orientations (Fig. 12.23f). Barbero and Barberi [1983] showed, within the one-constant approximation $K_{11} = K_{22} = K_{33} = K$, that in a hybrid film with anchoring strengths w_2^{A} and w_2^{B} and $w_2^{\mathrm{B}} < w_2^{\mathrm{A}}$, the bent-director configuration is only favoured if the film is thicker than a critical length, $d_{\mathrm{c}} = K\left(1/w_2^{\mathrm{B}} - 1/w_2^{\mathrm{A}}\right)$. In Fig. 12.23c, d we show the 5CB director components across both films. For the thinner film (Fig. 12.23a), we note that d_x and d_y at $T = 300$ K decrease with z, but remain very similar, suggesting that the change from planar to homeotropic does not involve an azimuthal reorientation of the director. Instead, the director switches from planar to homeotropic alignment between 4 and 7 nm, from the silicon surface, i.e. on a very short interval, comparable to a few molecular lengths. Thus, for the thin film the director variation resembles a discontinuous, rather than a continuous, director reorientation, and we see some evidence of order disruption in the middle of the 5CB slab, with the decrease of $\langle P_2 \rangle$ noticed earlier on, as expected for a perfectly discontinuous rearrangement [Palffy-Muhoray et al., 1994] (see Fig. 12.23e). For the thicker film (Fig. 12.23f) we see instead that the director changes continuously from planar homogeneous along the (110) direction $\boldsymbol{d} \approx (\frac{1}{\sqrt{2}}, \frac{1}{\sqrt{2}}, 0)$, to homeotropic $\boldsymbol{d} \approx (0, 0, 1)$ at the vacuum interface, in two stages: first a twist in the xy-plane from $d_x = d_y$, $d_z = 0$ to $d_x \approx 0$, $d_y \approx 1$), followed by a bend up to $\boldsymbol{d} \approx (0, 0, 1)$ Thus, the 5CB molecules adjacent to the Si(001):H surface align along the hydrogen rows or the grooves, then twist continuously towards the direction perpendicular to the grooves in the middle of the nematic slab, and finally bend upwards until reaching a homeotropic arrangement close to the vacuum (see also Figs. 12.23a,b).

Anchoring on Si(001):H. Atomistic simulations offer a unique opportunity of trying to understand the surface alignment mechanism. In Fig. 12.24a we show a contour plot of $P(z, \cos\beta)$, i.e. the probability of finding a molecule at a given angle and distance from the silicon surface, calculated for the ≈ 24 nm samples at $T = 300$ K, with a sub-nanometre resolution. The distribution shows a sharp peak (in black) at ≈ 3 Å from the surface and at $\cos\beta = 0$, corresponding to the planar anchoring of the 5CB molecules. We see that, after the first layer, the distribution becomes asymmetric, showing that 5CB molecules arrange with a slightly tilted orientation ($\approx 15°$) from the surface, corresponding to the CN-group preferentially pointing towards the silicon ($\cos\beta < 0$). The distribution changes completely at the vacuum interface, where we see (Fig. 12.24a) two peaks at $\cos\beta \approx \pm 1$, both corresponding to a perpendicular alignment, but shifted along z with respect to each other. The outermost one is for CN directed inside the film and thus with methylene chains outside. The other, at $\cos\beta = +1$, corresponds to molecules with CN pointing outwards in the position suitable for dimerization with the interfacial ones.

The anchoring coefficients in the Rapini–Papoular expression (Fig. 12.24b) are obtained by deriving first an effective work function from the positional orientational distribution $P(z_i, \cos\beta)$ obtained as a histogram from the simulated configurations (see Fig. 12.24a) by a Boltzmann inversion, scaled by the number of molecules $N(z)$ per unit surface area A as

$$W(z_i, \cos\beta) = -k_B T \ln P(z_i, \cos\beta)\, N(z_i)/A + W_0, \tag{12.7}$$

and then fitting the parameters in the Rapini–Papoular expression at the surface A by minimizing the mean square deviation

$$\chi^2_{\mathrm{RP}} = \sum_{z=z_1}^{z=z_2} \sum_{\cos\beta=-1}^{\cos\beta=+1} P(z, \cos\beta) \left\{ W(z, \cos\beta) - \left[w_0^{\mathrm{A}} - \frac{1}{2} w_2^{\mathrm{A}} \sin^2\left(\beta - \beta_{\mathrm{A}}^{d}(z)\right)\right]\right\}^2 . \tag{12.8}$$

We see from Fig. 12.24b that the anchoring coefficient w_2^{A} varies strongly with distance from the solid surface, different to the simple Rapini–Papoular expression Eq. 12.6, indicating that the definition of anchoring energy requires attention at the nanoscale, in comparison with traditional experimental techniques probing much larger distances. In particular, consideration should be given to the fact that the first layer of molecules can be strongly adherent to the surface itself [Voitchovsky et al., 2010], thus in a way renormalizing its properties. Different experiments can explore a solid-nematic interface at different depths more or less close to the surface and this, added to the difficulty of obtaining surfaces with exactly the same physical and chemical properties after a series of chemical treatments or mechanical rubbing, might explain the huge spread in anchoring energies found in the literature. The value at the overlayer is quite strong compared to published experimental results, even though on different substrates [Blinov et al., 1989].

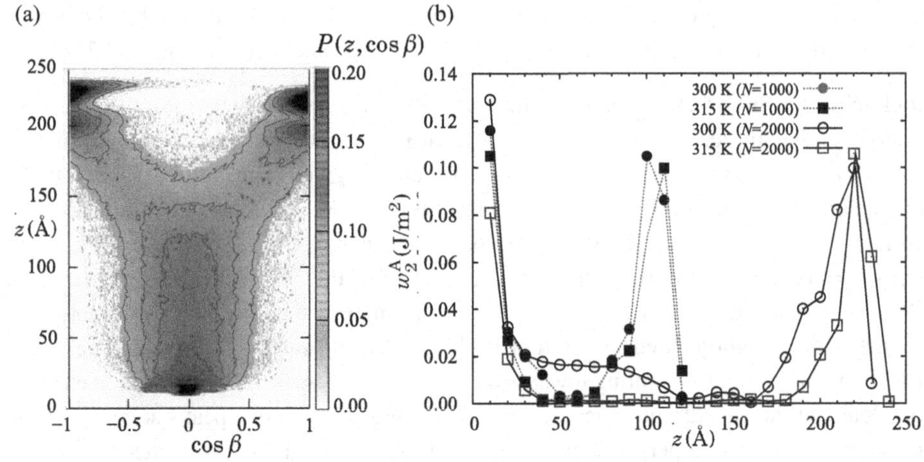

Figure 12.24 (a) The orientational distribution $P(z, \cos\beta)$ for 5CB at 300K and (b) the Rapini-like anchoring coefficient w_2^A of 5CB at various distances z from the silicon plane. Part (b) shows the results for the thin (filled symbols) and thick samples (empty symbols), at $T = 300$ K (circles) and $T = 315$ K (squares) [Pizzirusso et al., 2012a].

In summary, the 5CB films adopt a hybrid configuration, with a change of the director orientation from planar uniform at the silicon to perpendicular at the free surface, as shown by the snapshots (a , b) in Fig. 12.23. This variation is nearly discontinuous for the ≈ 12 nm thick film, and continuous with twist and bend for the thicker (≈ 24 nm) one. These findings usefully complement experimental results at nanometric resolution [De Luca et al., 2008; Carbone et al., 2009; Lee et al., 2009; Voitchovsky et al., 2010] and possibly open the way to new ones, ideally on the same substrate.

12.5.2 5CB on Quartz or Glass: Morphology Effects

One very interesting question related to anchoring is to what extent, for a given type of mesogen, are the chemical nature of the substrate and/or its morphology (crystalline versus glassy, for instance) determining the anchoring orientation and its strength. In many ways silica (SiO_2) is an ideal substrate to investigate this, since both its crystalline form (quartz, or actually a certain type of quartz, since there are various morphologies) and its disordered solid (glass) are easily available and have been investigated. In this case study we shall briefly describe the work of Roscioni et al. [2013] that addressed this issue looking at 5CB where, as we have just discussed, a reliable FF is available, on cristobalite quartz and on glass with a controlled roughness.

Quartz. The (cubic) cristobalite crystal structure was optimized at the molecular mechanics level with the General Utility Lattice Program (GULP) [Gale and Rohl, 2003]. The pair potential was assumed to be a sum of Coulomb and Buckingham (Section 5.3) terms: $U_{ij} = e_i e_j / r + A_{ij} \exp\left(-B_{ij} r\right) - C_{ij}/r^6$, with the silica parametrization of Du and

Cormack [2005]. A 63.5 Å × 63.5 Å slab, approximately 60 Å thick and exposing the (001) surface was obtained from the optimized bulk.

Bulk amorphous silica. Amorphous silica glass was prepared following the procedure of Della Valle and Andersen [1992]. (N, V, T) MD simulations were performed with LAMMPS starting from a cristobalite sample of 4608 SiO_2 units (13824 atoms) with a density corresponding to the experimental one of vitreous silica $(2.2\ g/cm^3)$. The sample was heated to 4000 K, melting it, then cooled, at a rate of 10 K/ps, to 300 K, yielding an amorphous glass. A silica slab $(58.7\ Å \times 58.7\ Å \times 60\ Å)$ was obtained eliminating PBC in the z-direction
(cf. Fig. 12.21a) and SiO_2 units were then randomly removed on the exposed surface obtaining a controlled roughness of the otherwise atomically flat surfaces. A further annealing and energy minimization followed the final slab organization. The solvent-accessible surface (SAS) roughness can be defined as the root mean square (RMS) deviation from the mean of the SAS along the normal to the surface, which in turn can be computed choosing as the virtual solvent a rigid sphere of a certain diameter, e.g. using the 3V rolling-sphere algorithm by Voss and Gerstein [2010]. A spherical probe with a diameter of 1.6 Å was used in the calculations, while Si and O atoms were represented as hard spheres with radii of 2.10 Å and 1.52 Å respectively (Blue Obelisk – Open Babel library [Guha et al., 2006]). Two silica surfaces were prepared: a 'smooth' one (surface 1) with an RMS of 1.5 Å, and a 'rough' (surface 2) with an RMS of 3.2 Å. The morphology and surface topography of the final system is shown in Fig. 12.25. Silicon and oxygen atoms belonging to the silica slab were kept frozen in their initial positions during the simulation.

LC film. The LC sample consisted in all cases of 2000 5CB molecules ($\approx$22 nm thick film) deposited on top of one of the anhydrous silica slabs. 5CB is described as before by the united-atom FF developed by Tiberio et al. [2009].

LC-silica interaction. The 5CB-SiO_2 interaction was described by a long-range Coulomb plus a shorter range LJ potential term [Cruz-Chu et al., 2006]. MD simulations were performed with NAMD under (N, V, T) conditions: at $T = 300$ K and $T = 320$ K for the nematic and isotropic phases of 5CB, respectively.

Results for 5CB on silica. The effects of surface roughness on the orientational order across the film show clearly (Fig. 12.26) that anchoring also depends on slab morphology and that each interface influences the orientational ordering across the 5CB films. As in the previous case of 5CB on silicon, the interface with the vacuum induces the formation of a highly oriented molecular double layer with $\langle P_2 \rangle$ higher than in the central 'bulk-like' region of the film. Similarly, the interface with cristobalite induces a strong ordering in the 5CB overlayer and an increase of $\langle P_2 \rangle$ with respect to that in the middle of the film. This surface order seems to be only weakly dependent on temperature, being similar both for the nematic and the isotropic temperatures studied. Notwithstanding the same chemical composition, the amorphous silica surfaces have opposite effects on the 5CB order, with $\langle P_2 \rangle$ decreasing rather than increasing, at the silica interface in agreement with expectations [Jérôme, 1991]. The effect seems to be due to the disordered and to some extent irregular

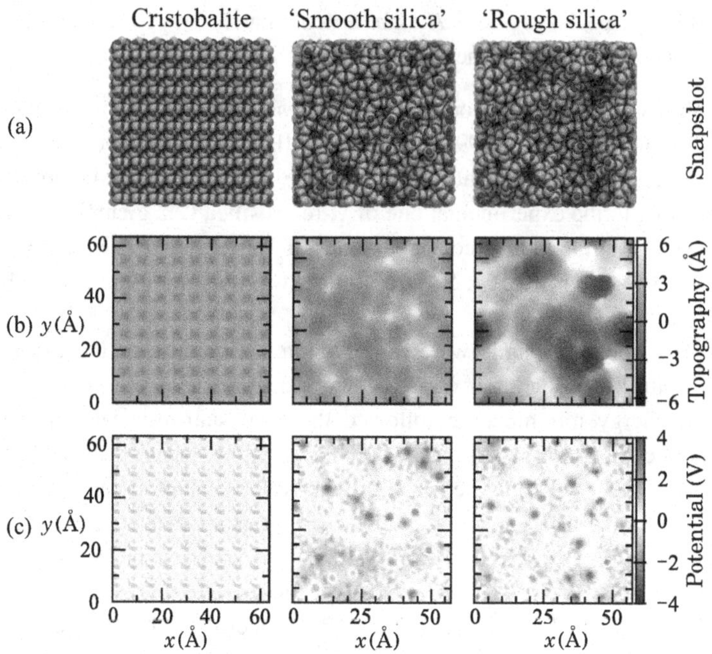

Figure 12.25 Silica support surfaces. (a) Top view of cristobalite: 001 crystal facet (rms roughness $\sigma_R = 0.7$ Å, a, left), smooth amorphous surface ($\sigma_R = 1.5$ Å, a, center panel) and rough amorphous surface ($\sigma_R = 3.2$ Å, a, right panel). The corresponding surface topographies are shown as the (b) solvent-accessible surfaces (SAS) and (c) the grey-coded electrostatic potential maps computed on the SAS [Roscioni et al., 2013].

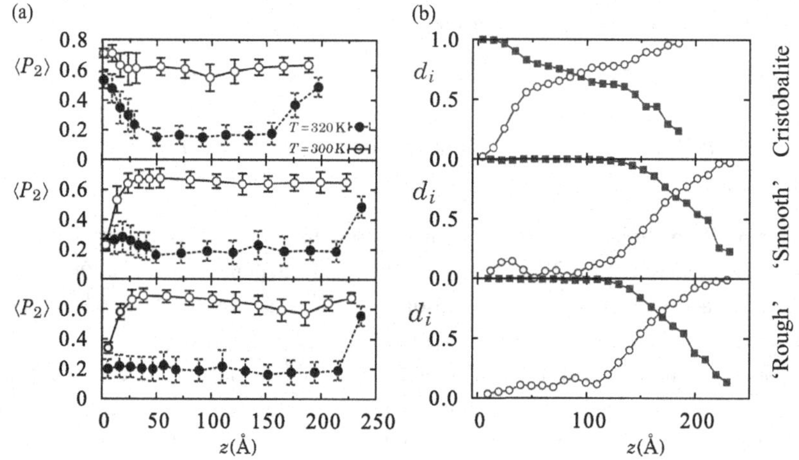

Figure 12.26 (a) Scalar order parameter of 5CB thin films at $T = 300$ K (○) and $T = 320$ K (●) on crystalline cristobalite and 'smooth' and 'rough' glassy silica computed at distance z from the surface. (b) In-plane and out-of-plane director components $d_{xy} = \sqrt{[n_x^2 + n_y^2]}$ (■) and d_z (□) across the film on the same three surfaces and at $T = 300$ K [Roscioni et al., 2013].

nature of these surfaces, which favours different local orientations of 5CB molecules with a disruptive effect on the nematic order for the first few nanometres above the support surface. After this region a bulk-like value of $\langle P_2 \rangle$, similar to that measured for the cristobalite sample, is recovered. The decrease of order observed with increasing roughness agrees with experimental observations on SiO [Monkade et al., 1997].

In summary, both silicon and silica crystalline surfaces enhance local orientational order at the interface while amorphous silica reduces it, as found experimentally, but enhances the long-range director alignment.

12.5.3 5CB on Aligned and Disordered Polymers

In very many practical applications the alignment of LC thin films at a polymer support surface is achieved by a mechanical process of rubbing [Chatelain, 1943; Stöhr and Samant, 1999] or buffing [Geary et al., 1987; Hayashi and Matsumoto, 1994; Brown et al., 1998b]. During the production process of these aligning layers, the polymer-coated surfaces are softly rubbed in one direction by a rotating cylinder covered with a suitable cloth, to achieve a uniform orientation of the director. In typical cases the LCs at the surface of the rubbed polymer film layer are aligned parallel to the rubbing direction. In these systems, the suggested explanation is that sub-micrometric grooves are generated on the film surface along the rubbing producing alignment. Another hypothesis proposed is that 'the oriented state of the polymer chains, and not scratching or grooving of the surface, is necessary to produce alignment' [Geary et al., 1987]. However, LC alignment along the rubbing direction is not universal. For instance, this is the case for polymethyl methacrylate (PMMA), while polystyrene (PS) surprisingly induces a planar alignment, but perpendicular to the rubbing direction. To try to verify if this is reproduced by simulations and clarify the origin of the aligning mechanism, a thin film of 5CB, sandwiched between two PMMA and PS slabs was studied [Palermo et al., 2017]. We briefly describe the simulation, as usual focussing on the approach.

Polymer modelling. Each polymer was built by linking 50 monomers in the reacted form (Fig. 12.27, top), obtaining isotactic polymer chains. The relatively short chain length was chosen for practical reasons of feasibility. The macromolecules were modelled with full atomistic detail using a customized AMBER-OPLS molecular mechanics FF [Cornell et al., 1995; Jorgensen et al., 1996] previously employed in the study of the electrostatic interaction between pentacene and PS or PMMA [Martinelli et al., 2009]. Charges were calculated with QC (Hartree-Fock AM1 using Gaussian 03 and ESP at the DFT level). The validation of the polymer FF, relied on in the extensive investigations by Soldera and Metatla [2005, 2006] and Soldera [2012], who showed that experimental densities and bulk moduli for PS and PMMA are well reproduced with these OPLS FF parameters.

Polymer surfaces. Four disordered or oriented confining polymer slabs (≈ 5 nm thick) were prepared as follows.

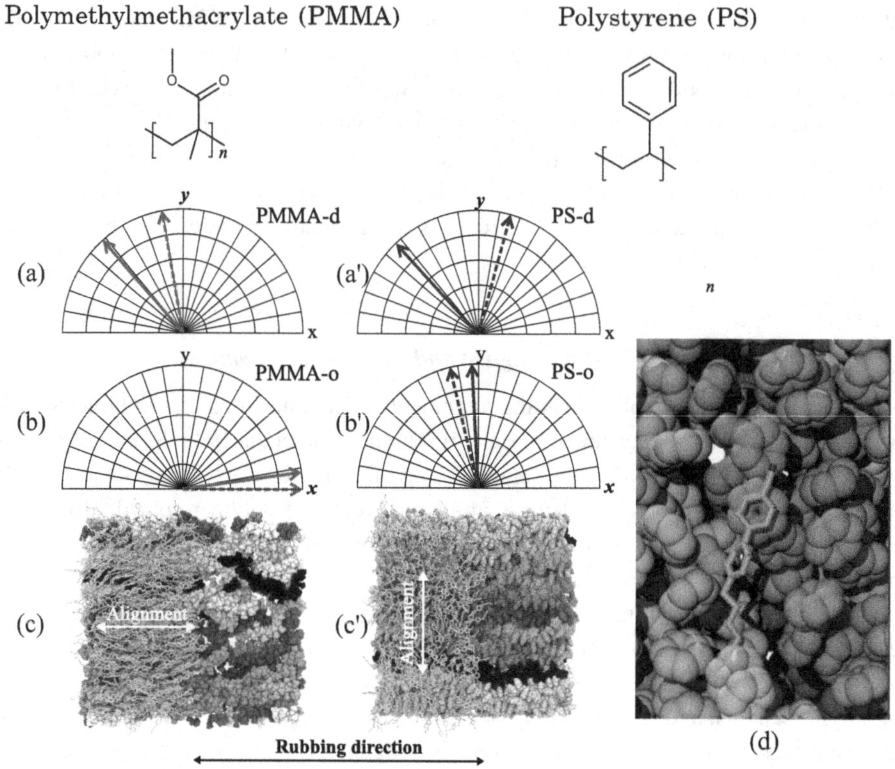

Figure 12.27 Chemical structure of the PMMA and PS polymer substrates making the top and bottom slabs trapping the 5CB film. The director ***d*** orientation of 5CB at the top (- - -) and bottom (—) interfaces with PMMA (a, b) and PS(a′, b′) are shown for disordered (a, a′) or rubbed (b, b′) slabs. The rubbed surfaces align uniformly ***d*** along or perpendicularly to the rubbing direction as also shown by the MD snapshots (c) and (c′). The close up in (d) shows the alignment of 5CB molecules along PS nanochannels [Palermo et al., 2017].

(i) Amorphous PS consisting of 60 chains (of 50 monomers) arranged in a disordered fashion (PS-d) with surface area $\approx$ 10 nm $\times$ 10 nm.

(ii) Ordered PS, consisting of 60 elongated chains arranged in a parallel fashion (PS-o), with surface area $\approx$ 10.3 nm $\times$ 10.3 nm.

(iii) Amorphous PMMA surface consisting of 72 chains arranged in a disordered fashion (PMMA-d), with surface area $\approx$ 10 nm $\times$ 10 nm.

(iv) Ordered PMMA surface, consisting of 72 elongated chains arranged parallel to each other (PMMA-o), with surface area of $\approx$ 11.1 nm $\times$ 9.9 nm. The amorphous polymers were prepared in the melt, at 700 K, a temperature well above the experimental glass transition temperature, first applying a sufficiently high pressure (P = 1000 atm) to achieve a density similar to the experimental one, equilibrating at P = 1 atm, still at

700 K, then at 300 K. Ordered slabs were built, at first, replicating regularly in space a single elongated polymeric chain. The elongation of the chains was achieved by applying two equal stretching forces (1 kcal $mol^{-1}Å^{-1}$), in opposite directions, to the terminal carbon atoms of each chain, during the packing process performed as for the amorphous samples. After compression, all samples were equilibrated for about 50 ns at 300 K until they reached a constant equilibrium thickness. The vertical pressure was set to 1 atm, keeping the horizontal section of the simulation cell fixed, with the Langevin piston method [Feller et al., 1995], similar to the extended Lagrangian formulation by Andersen (Section 9.6), but where the deterministic equations of motion for the piston degree of freedom are replaced by a Langevin equation.

Liquid crystal film. The LC was 5CB, modelled with the Tiberio et al. [2009] FF as used in previous examples. The LC samples consisted of 1000 or 3000 5CB molecules, the last corresponding to a film thickness of $\approx$ 12 nm with ample space left along z (see Fig. 12.21) so as to be able to implement PBC.

Simulations. Atomistic (N,P,T) MD simulations were run with the NAMD package [Phillips et al., 2005] at $T = 300$ K, $P = 1$ atm. Equilibration times were over 50 ns and production times over 100 ns.

5CB surface alignment. It can be seen from Fig. 12.27 that for both the PMMA-d and PS-d cells, the director at the bottom and top surfaces are unrelated, consistent with fact that no easy surface axis is present. More interestingly, if we compare the effects of the two ordered surfaces, we see that the director is oriented along the chain alignment direction at the two interfaces for PMMA-o, while it is pointing perpendicular to it for PS-o. The surface alignment of 5CB at the interfaces, might be ascribed to the π-π interaction between the 5CB aromatic moieties and the phenyl PS side groups. In practice, the 5CB alignes along nanogrooves created by the PS phenyls that essentially stick out of the surface. Experimental confirmation of the alignment can be found in various works [Ishihara et al., 1989; Oh-e et al., 2002; Lee et al., 2003].

We can also look at the order across the film for the various cells, like we did for the silicon and silica surfaces, recalling that, in that case, one of the two surfaces was free, while here the film is between two solid slabs. The plots for $\langle P_2 \rangle$ across the film are shown in Fig. 12.28. It can be seen that the PMMA-o cell is the only one featuring order parameter similar to those found for the bulk nematic phase (see Fig. 12.12-top left), while all other systems are more disordered. Here all samples present slightly lower values of $\langle P_2 \rangle$ close to the surfaces, consistently with the fact that even the ordered polymer surfaces PS-o and PMMA-o are not really endowed with the regularity of a crystal. An additional disrupting effect is probably due to the observed diffusion of 5CB within the polymer film. We also see that the $\langle P_2 \rangle$ profiles are not symmetric, particularly for PS-d and PS-o. This is reasonable, since for these systems the two polymeric surfaces are not identical and lower values are found in proximity of the roughest surface [Palermo et al., 2017].

In conclusion, the results show that, at least for the studied polymers, both chains stretching and the chemical nature of the polymer contribute to the orientational order of the LC phase, and that surface microgrooves created by rubbing are not necessary to reproduce the

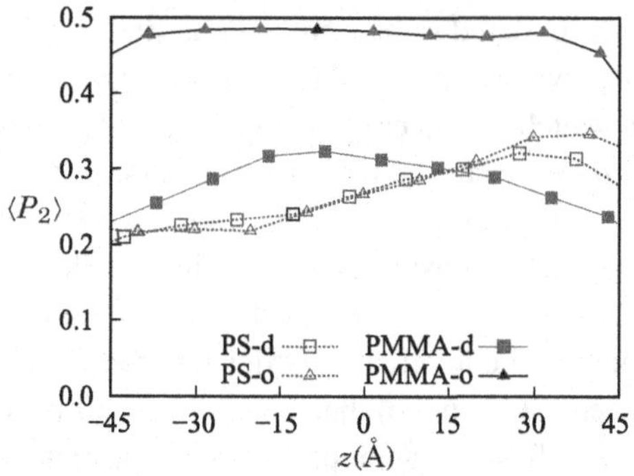

Figure 12.28 Orientational order parameter $\langle P_2 \rangle$ of 5CB across the simulation box for samples composed of $N_M = 3000$ 5CB molecules at $T = 300$ K between two ordered and disordered PMMA (filled symbols) and PS (empty symbols) [Palermo et al., 2017].

experimental results for the alignment of 5CB on PS and PMMA surfaces. This essentially supports the suggestion by Geary et al. [1987] that chains ordering, rather than scratching or grooving, is the key element for alignment at a polymer surface.

12.5.4 Self-assembled Monolayers on Silica

The various hard solid surfaces we have discussed up to now essentially yield in-plane orientation of a nematic (5CB). Inducing perpendicular, homeotropic alignment is also very important and has traditionally been obtained by coating glass solid support surfaces with lecithins or some self-assembled monolayers (SAMs). The approach is to a large extent empirical, but atomistic MD can now be used to study at least some of these orienting soft surfaces. Here we describe the MD simulation of two prototypical SAM-forming silanes: octadecyltrichlorosilane (OTS) and 1H, 1H, 2H, 2H- perfluorodecyl-trichlorosilane (FDTS), chemisorbed on glass (an atomically flat amorphous SiO_2) [Roscioni et al., 2016]. An outline of the procedure is as follows. A slab of amorphous silica (90 Å $\times$ 75 Å with a thickness of 67 Å, corresponding to $\approx$ 30,000 atoms) is initially fabricated as discussed in Section 12.5.2 [Roscioni et al., 2013]. The resulting top surface is hydrated capping unsaturated oxygen and silicon atoms on the surface with H or OH groups, respectively. The hydrated surface is then equilibrated at 300 K in (N, V, T) conditions ($\approx$ 20 ns). After this step the surface presents an irregular distribution of silanol SiOH groups, similar to that of real amorphous silica surfaces, with the surface roughness and inhomogeneities that represent important features in governing alignment. The next stage in order to construct OTS and FDTS SAMs is performed grafting silane molecules to the silanol (SiOH) groups on the surface. The method is somehow inspired by the experimental formation of monodentate, non-cross-linked SAMs [Wang et al., 2005] from the reaction between a trichlorosilane derivative and the hydroxylated surface of amorphous silica, followed by hydroxylation of the remaining chlorine atoms [Tripp and Hair, 1995]. As a result, the alkylsilane groups are fully hydroxylated and bonded to the silica surface by single Si_{surf} -$OSi(OH)_2R$ bonds,

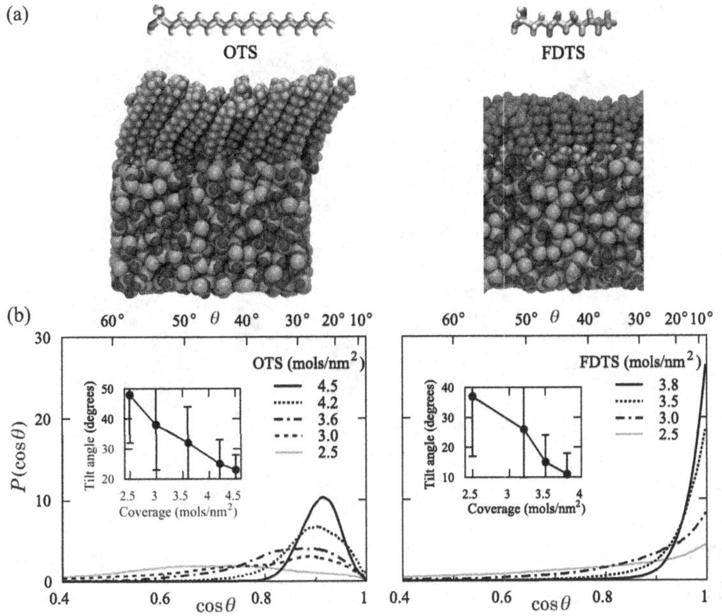

Figure 12.29 (a) MD snapshots showing the tilt angle of OTS alkylsilane (left) and of the perfluoroalkyl silane FDTS (right) on silica glass. (b) Distribution of tilt angles $P(\cos\theta)$ vs θ at different surface coverages, showing the different arrangement of the two SAMs. The insets show the variation of tilt angle with coverage [Mityashin et al., 2014; Roscioni et al., 2016].

with R being the alkyl or fluoroalkyl chain. The computational approach does not reproduce these steps but only the final product. In practice, the SAM is composed of molecular ions, each with the formal charge of the replaced -OH group ($-0.525e$ in the CLAYFF force field [Cygan et al., 2004]), distributed evenly over the three oxygens, so as to maintain the charge neutrality of the slab. Even if the surface Si atom and the oxygen of the $OSi(OH)_2R$ SAM residue are not covalently bound, the electrostatic and LJ interactions are sufficiently strong to keep the molecules adsorbed, while allowing enough mobility for the self-organization during the MD simulations as a function of surface coverage. The calculated morphologies, tilt angle, film thickness and lattice parameters agree with experiments, demonstrating the accuracy of the methodology. In particular, it is found that OTS molecules show a coverage-dependent tilt, while FDTS molecules are always vertically oriented, regardless of the coverage (Fig. 12.29). It is worth noting that these MD simulations also provide the detailed atomistic configurations needed for the quantum chemical calculation of electronic and semiconductor properties of the two SAM-coated surfaces [Mityashin et al., 2014].

12.5.5 5CB on Self-assembled Monolayers and Homeotropic Alignment

Having prepared a SAM-coated silica surface similar to the ones experimentally used to induce alignment perpendicular to the surface we can try to examine if this type of surface

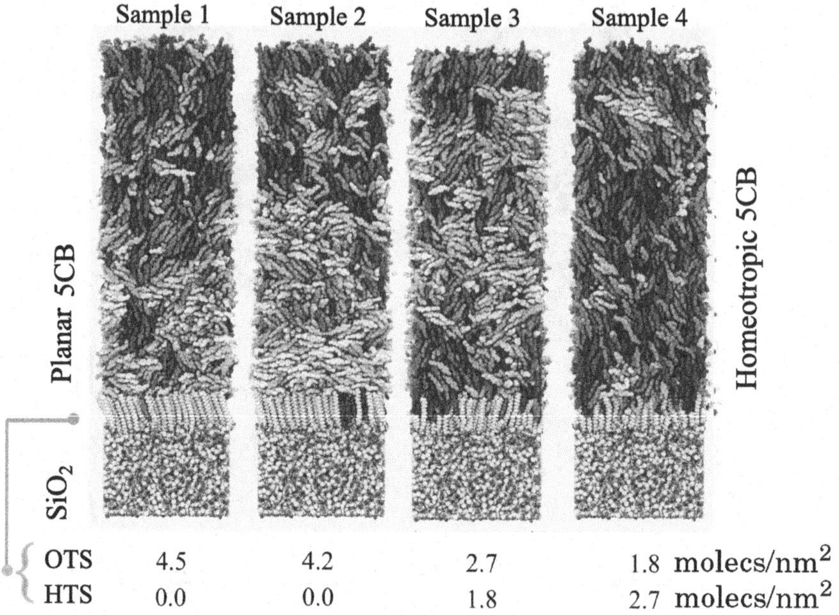

Figure 12.30 Lateral view of nematic 5CB films on various SiO_2/SAM surfaces at $T = 300$ K. 5CB molecules are grey coded according to their orientation with respect to the surface: from black, perpendicular, to white, parallel, to the support surface. The composition of the SAMs and coverage densities (molecules/nm^2) of samples 1–4 are also shown [Roscioni et al., 2017].

actually produces 'homeotropic alignment' of an LC deposited on it. The family of SAMs typically used experimentally to induce perpendicular alignment is of the alkyl type like OTS, perhaps surprisingly, given the tilted alignment of the overlayer as seen in Fig. 12.29. To understand this Roscioni et al. [2017] studied a film of 4100 5CB molecules ($\approx$20 nm thick) over four SAMs systems. Each of these monolayers consists of 300 molecules of mixed 'long' and 'short' alkylsilanes: TS and hexyltrichlorosilane (HTS) in various proportions, coating an amorphous silica slab with surface 9.0 nm $\times$ 7.5 nm. All MD calculations were carried out with NAMD in (N, V, T) conditions at 300 and 320 K, corresponding to 5CB being in the N or I phase, respectively. Other technical details are given in Roscioni et al. [2017]. Snapshots of the films showing the molecular alignment of the 5CB and the SAM are reported in Fig. 12.30. In all cases, the 5CB is aligned perpendicular to the vacuum interface as usual. More specifically the different SAM coatings are:

(i) OTS with density of 4.5 molecs/nm^2 (essentially full packing). In this case, the 5CB alignment over the SAM is planar, not homeotropic.
(ii) Again, pure OTS but with a density of 4.2 molecs/nm^2, thus with vacancies that allow some 5CB insertion. In this case, a few 5CB molecules can infiltrate in the vacancies, assuming a strict homeotropic alignment. A predominantly planar alignment is instead found above the SAM interface and maintained up to the middle of the sample. It is very interesting to compare the tilt of OTS molecules in the two cases: we see from

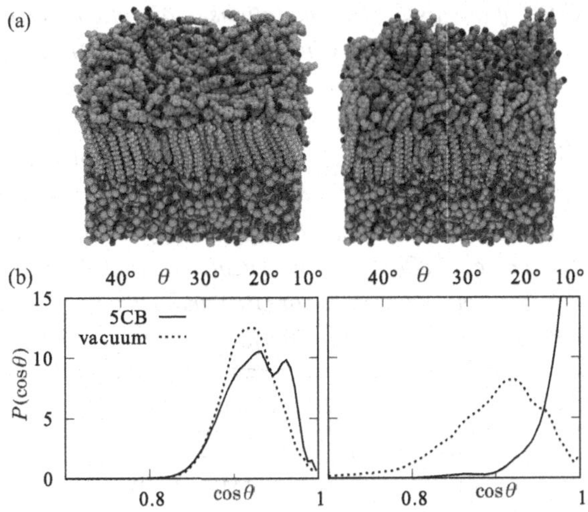

Figure 12.31 (a) Snapshots and (b) plots of tilt-angle distributions of a SAM of OTS in vacuum (Fig. 12.29) and in the presence of the 5CB film for the full coverage sample 1 (left) and for the less dense sample 2 (right) described in Fig. 12.30 [Roscioni et al., 2017].

Fig. 12.31 that, for the tightly packed SAM in sample 1, the tilt of OTS molecules ($\approx 23°$) is very similar to that of the SAM exposed to a vacuum in Fig. 12.29a [Roscioni et al., 2016]. In sample 2 the OTS molecules in the vacuum-exposed film are again tilted in the same way, although with a wide distribution due to the increased free space available to the OTS chains. However, when the 5CB film is present, the penetration of 5CB molecules in the SAM has the effect of aligning the OTS molecules vertically. This configuration is also evident in Fig. 12.30. The remaining two soft surfaces are:

(iii) A mixed long- and short-chains SAM with 60% OTS and 40% of HTS and total density 4.5 molecs /nm^2.

(iv) A SAM with inverted proportions of short and long chains: 40 % OTS and 60% of HTS, total density 4.5 molecs/nm^2. For samples 3 and 4, we observe at last an essentially uniform homeotropic alignment of 5CB across the film, which seems to be of better quality for the sample with the higher concentration of short HTS chains (sample 4 in Fig. 12.30). The rationale is that the 5CB molecules stacking out of the SAM, particularly in the HTS region, provide the template for homeotropic alignment across the film.

In summary, MD simulations clarify that not all the SAMs induce homeotropic alignment and provide an indication of the conditions required to produce this alignment effect.

12.6 Discotics and Columnar Phases

Atomistic simulations of discotic systems are not very abundant, partly because of the considerable complexity (number of constituent atoms in particular) of these mesogens

Table 12.2. *A small list of atomistic MD simulations of discotic mesogens: (a) hexakis(pentyloxy) triphenylene (HAT5) [Cinacchi et al., 2004]; (b) metal - porphyrazine complex (ZnP4) [Cristinziano and Lelj, 2007]; (c) various hexa substituted hexabenzocoronens HBC-(R)$_6$ with R (i) linear alkyl chains,* C_n, *with* $n = 10, 12, 14, 16$ *(cf. Table 1.6), (ii) branched side chains,* C_{10-6}, *and (iii) dodecylphenyl-substituted side chains,* PhC_{12} *[Kirkpatrick et al., 2007, 2008]; (d) tetra* C_{12}-C_8 *alkoxy-substituted metal free phthalocyanine (PHT) [Olivier et al., 2009]; (e) triangular and semitriangular polyaromatics (TSP) [Feng et al., 2009]; (f) carbazole macrocycle (CM) [Vehoff et al., 2010]; (g) a substitute perylene electron donor (PE) and benzoperylene diimido diester electron acceptor (BP) columnar-columnar interface [Idé et al., 2014].*

System	Phases	Simulation details and comments	Ref.
HAT5	D_h	MD, (N, P, T), $N = 80$, $P = 0.1$ MPa, $T = 375$ K, 3 ns Start: 6 columns, UA, AMBER-OPLS	(a)
ZnP4	I, D_h^t	UA, GROMACS, $N = 64$, 40 ns, start: I Tilt wrt column axis = 28.5°	(b)
HBC−(R)$_6$	D_r	$N = 160$, $T = 300$ K, $P = 0.1$ MPa, 100 ns Start:16 columns of 10 HBC-(R)$_6$ Charge mobility	(c)
PHT	D_r, D_h	MD, (N, P, T), $N = 80$, $P = 1$ atm, NAMD, time = 65 ns $T = 300$ K (D_r), $T = 425$ K (D_h) Charge transport	(d)
TSP	D_h^h	Start:16 columns of 60 TSPs with helical structure. Charge mobility	(e)
CM	D	PBC, GROMACS OPLS Start: 16 columns of 48 CMs	(f)
PE + BP	D_h	A stack of 5 × 5PE and 4 × 4BP columns of 30 PE or BP on SiO_2 modelling a planar interface organic solar cell MD (N, P, T), NAMD, UA. AMBER, time > 50 ns	(g)

(see Section 1.10) and possibly because of the more limited technological applications, as of now. It is also worth recalling that very few discotic mesogens have a nematic phase and that it is difficult to obtain a well-ordered columnar organization in a cooling down sequence. Indeed, most of the simulations to date have started from a set of columns and used the MD runs to test their stability and adjust the organization of the molecules inside the columns. There are however a number of interesting simulations and in Table 12.2 we list a few representative ones with an indication of their main features. We have omitted some of the older, even if pioneering, ones when lasting less than a few nanoseconds, i.e. times shorter that what is, with hindsight, considered necessary for equilibration. Even if the intent here is not to provide a review it is important to point out that various of the most recent simulations are devoted to obtaining detailed molecular configurations to be used as input for quantum-type calculations of properties of interest for organic electronic applications such as organic field-effect transistors, organic light-emitting diodes and organic photovoltaics [Brédas and Marder, 2016; Da Como et al., 2016].

12.7 Lyotropics

The examples of atomistic simulations shown up to now concern thermotropic LCs, but essentially the same MD methods can be used, at least in principle, for the lyotropic phases shown in Chapter 1, even if these mixtures of at least two (and very often more) components including amphiphilic molecules and solvents are extremely demanding in computational resources. To help in this direction, united-atom FFs that, like we saw for thermotropic LCs, combine each carbon with its bonded hydrogen atoms, are very often used [Kukol, 2009]. By reducing the number of sites representing a molecule (e.g. for DPPC from 130 atoms to 50 UAs) and allowing longer integration time steps, UAs usually speed up the simulations 3–5 fold compared with the all-atoms counterpart, while maintaining essential atomistic details. In any case, there is ample scope for simulations of lyotropics, since many of the phases described in Sections 1.11 and 2.14 are still unexplored with realistic AMD. One exception is that of micelles and bilayer membranes, and indeed, some of the pioneering work was performed on lyotropics [Egberts and Berendsen, 1988], even if much of the research has concerned biophysical or biomedical aspects [Marrink et al., 2019] rather than LC properties. Chromonics systems (Section 1.12) are another exception and have also been studied in some detail [Chami and Wilson, 2010; Thind et al., 2018]. Here we focus on a few examples dealing with micelles and lipid bilayers formed by surfactants and water, discussing some of the methodologies of general applicability.

12.7.1 Micellar Phases: SDS

We have already introduced the self-organization of surfactants in water and the various types of aggregates it yields, in Section 1.11.1. Some of these systems have now been simulated at atomistic level. In particular, sodium dodecyl sulphate (SDS) is one of the most studied anionic amphiphiles, both experimentally and with computer simulations. We recall that at concentrations, low but above the critical micellar concentration (*CMC*) of ≈ 8 mM [Wennerström and Lindman, 1979], SDS forms in water spherical micelles of ≈ 5 nm in diameter that, increasing the SDS concentrations, become much larger until they transform into cylindrical aggregates with lengths of the order of $\sim$100–1000 nm. The transition is significantly affected by the presence of salts in the solution [Zana and Kaler, 2007]. The properties of SDS solutions have been studied extensively by MD simulations with atomistic [Sammalkorpi et al., 2007], coarse-grained [Marrink et al., 2019] and hybrid particle-field/MD [Schafer et al., 2020] techniques. Two main approaches can be pursued for this and other surfactants.

(i) Single micelle structure. A first type of problem concerns the shape of the SDS aggregates in water and the amount of water penetration inside the aggregate [Palazzesi et al., 2011; Chun et al., 2015; Prior and Oganesyan, 2017]. In this type of work surfactant molecules are initially prearranged in space in a form similar to that of the target aggregate, whose stability is tested running the MD simulations. For instance, MacKerrel et al. [1995]; Bruce et al. [2002], Palazzesi et al. [2011], Chun et al. [2015] and Prior and Oganesyan [2017] have employed 60 SDS molecules (N60SDS), radially disposed to model the initial configuration of spherical micelles. Taking as an example the work of Palazzesi et al. [2011] a molecular mechanics energy minimization is first used to relax undesired contacts. Ad hoc programs, like the PACKMOL software [Martinez et al., 2009] or similar ones can also be used for this preliminary stage. Having prepared the initial configuration, equilibration and production runs can follow, employing a certain FF. Various FFs for lipids are available and the development and performance of the most important ones, such as the CHARMM, AMBER, GROMOS, OPLS, and coarse-grained MARTINI [Marrink et al., 2019] families have been compared [Leonard et al., 2019]. Palazzesi et al. [2011] adopted the general AMBER FF for SDS. Sodium ions are added in order to neutralize the system and the micelle is solvated with *explicit water* molecules interacting with the TIP3P potential (see Section 5.11.1). Examining the results, we note (Fig. 12.32) that many of the alkyl chains in the SDS micelle are at least partially water exposed. This is rather different from the classical picture where the hydrophobic chains are essentially dry (see Fig. 1.11.1). The radii, defined as the distance from the micelle centre of mass to the sulphur atom of the N60SDS micelle is calculated to be 1.97 ± 0.04 nm [Palazzesi et al., 2011] or (≈ 1.94 nm [Tang et al., 2014]). The shape of the micelles is not exactly spherical, and its instantaneous fluctuations can be quantified by the eccentricity parameter of its inertia tensor $\mathbf{I}$ (cf. Eq. 4.24): $e(t) = 1 - (I_{min}(t)/I(t))$, where I_{min} is the smallest eigenvalue and $I = \text{Tr}\mathbf{I}/3$, which reduces to $e = 0$ for a perfect sphere [Salaniwal et al., 2001]. The time averaged eccentricity of the N60SDS micelle is $\langle e(t) \rangle \approx 0.12$ (Fig. 12.32a). Prior and Oganesyan [2017] also reported MD simulations of N60SDS micelles as well as of cationic micelles of dodecyltrimethylammonium chloride (DTAC), with the inclusion of a nitroxide spin probe to directly simulate EPR spectra from the MD trajectories. We shall not go into the EPR spin probe part of the work, but only on the micelle organization. In particular, we report the segmental order along the SDS chain

$$S_{\text{CH}(n)} = \left\langle P_2 \left(\boldsymbol{r}_{\text{CH}(n)} \cdot \boldsymbol{d}_m \right) \right\rangle, \tag{12.9}$$

where $\boldsymbol{r}_{\text{CH}(n)}$ is the C-H internuclear vector at position n along the chain (see Fig. 12.32b) and $\boldsymbol{d}_m$ is the local director, taken as the normal to the micelle surface. The simulation results obtained for the CH bonds along the chain are compared in Fig. 12.32c with ^{13}C [Ellena et al., 1987] and ^{2}H spin-lattice and spin-spin relaxation NMR data [Söderman et al., 1988]. We report the modulus of the order since the NMR experiments yield the square of $S_{\text{CH}(n)}$. The MD values along the SDS alkyl chain appear only in qualitative agreement with the NMR ones. This might in principle be due to some approximations about the model required in the experimental relaxation data analysis or it could suggest that the GAFF force field employed produces overly ordered alkyl chains, as was also found in lipid bilayers [Dickson et al., 2012].

(ii) Surfactants self-organization. A second type of problem deals with the actual process of aggregation of SDS (or other) amphiphiles in water and can be particularly demanding in computational resources. In this case, the simulations are started from a random solution of surfactant molecules in water, observing the formation of surfactant aggregates brought

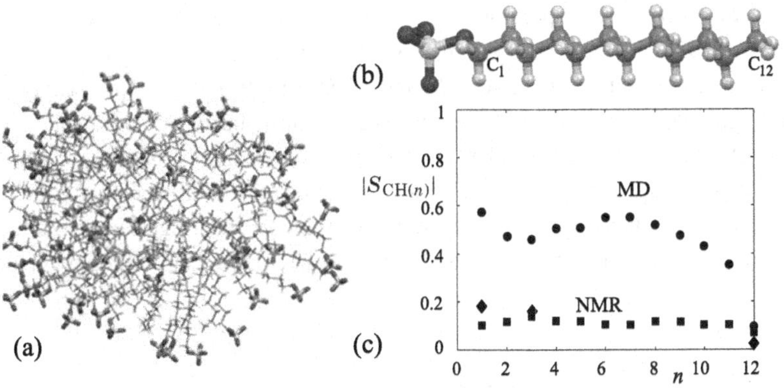

Figure 12.32 (a) Structure of a N60SDS micelle after a 20 ns MD equilibration run [Palazzesi et al., 2011]. (b) Ball and stick representation of dodecyl sulphate. (c) Order parameters $|S_{CH(n)}|$ of the CH bond at position n in a N60SDS micelle at 310 K. Comparison between MD [Prior and Oganesyan, 2017] (●) and experimental [Ellena et al., 1987] (■) and [Söderman et al., 1988] (♦) NMR results.

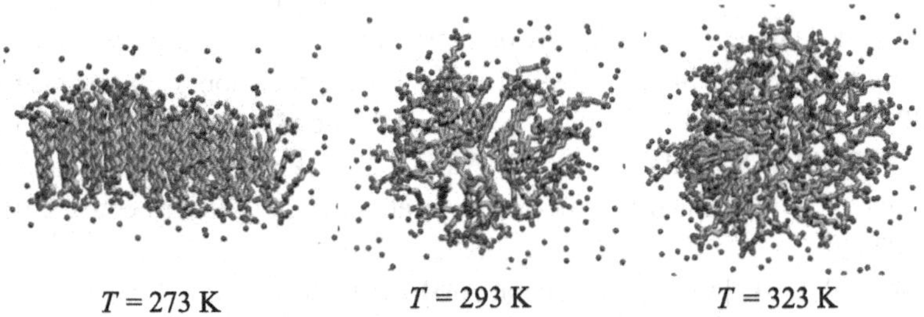

Figure 12.33 MD snapshots of SDS self-organized structures: a crystalline aggregate ($T = 273$ K) and quasi-spherical micelles at $T = 293$ K and a larger one at $T = 323$ K. Here the spheres represent sodium atoms, while the surrounding micelles and the explicit water molecules have been omitted to facilitate visualization of the aggregate. The snapshots correspond to the largest aggregate in the system [Sammalkorpi et al., 2007].

about by a change in temperature or concentration. Simulations of SDS micelles in water have been studied using four popular empirical force fields, namely, CHARMM, OPLS, AMBER and GROMOS united atoms, and some additional specialized force fields with both (N, V, T) and (N, P, T) ensembles. In particular, Sammalkorpi et al. [2007] examined the formation of SDS aggregates in explicit SPC water (Section 5.11.1) with extensive MD simulations using an ad hoc tuned parametrization of SDS within the UA GROMACS 3.3 [van der Spoel et al., 2005] and long runs of 200 ns (see Fig. 12.33). Specifically at low temperatures, the surfactants form crystalline aggregates, whereas at elevated temperatures, they form micelles. The dependence of aggregation on surfactant concentration (200 mM, 800 mM and 1 M) is also examined. The concentration limit (1 M) was investigated by simulating the dynamics of 200 SDS molecules at a temperature $T = 253$ K (well below the critical micellization temperature T_{CMT}) and 323 K (well above T_{CMT}) over 200 ns. The largest micelle in the 1 M system is slightly deformed towards an elliptical form, whereas the micelles with size between 40 and 80 SDS molecules are approximately spherical. The aggregates in the 200 mM system appear significantly less spherical.

It is worth noticing that as simulations become more realistic, also moving to the self-assembly of larger aggregates like cylindrical micelles, new challenges and discrepancies between results obtained with current models may emerge, leading in turn to the development of new more accurate FFs or parameterizations of existing ones [Taddese et al., 2020].

12.7.2 Lipid Bilayers: DPPC

Lipid bilayers are, as we briefly mentioned in Section 1.11.2, a fundamental component of a cell membrane, separating its interior and exterior environments and providing a barrier to the free migration of ions, proteins, etc. in or out of the cell [van Meer et al., 2008]. Simulations of lipid bilayer membranes formed by phospholipids have been abundant since the nineties [Venable et al., 1993], even though those pioneering simulations did unavoidably deal with sub-nanosecond times and relatively small particle numbers. Nowadays time windows of the order of 100–1000 ns or more are easily accessible using high performance computing (HPC), in particular using clusters of processors with graphic processing units (GPUs) and the GPU accelerated version of MD codes, like NAMD [Phillips et al., 2020] and GROMACS [Pall et al., 2020]. Various specific FFs for lipids have been produced, updated, validated and compared [Poger et al., 2016; Leonard et al., 2019; Yu and Klauda, 2020].

Dickson et al. [2012] tuned the General AMBER force field (GAFF) to allow the accurate tensionless simulation of a number of different lipid types using AMBER. Specifically tested systems include DLPC, DMPC, DPPC, DOPC, POPC and POPE phospholipid bilayers.

Pluhackova et al. [2016] also simulated various phospholipid bilayers (DMPC, POPC and POPE) comparing four atomistic lipid FFs, namely, the united-atoms GROMOS54a7 and the all-atoms ones CHARMM36, Slipids [Jämbeck and Lyubartsev, 2012] and Lipid14, the AMBER lipid FF [Dickson et al., 2014]. In particular, the membrane melting transition temperature for the investigated FFs was studied from heating simulations, starting from a membrane in its gel phase of a DPPC bilayer.

The additional possibility of combining AA and UA models is available in various FFs, e.g. the C36U lipid FF employs CHARMM36 AA parameters for headgroups and OPLS-UA Lennard–Jones (LJ) parameters for the acyl tails FF [Klauda et al., 2010; Lee et al., 2014]. Yu and Klauda [2020] presented an updated version of this FF (C36UAr), tuned using bulk liquid properties (density, heat of vapourization, isothermal compressibility and diffusion coefficient) of hydrocarbons model compounds for the LJ parameters, and dihedrals fitted to either quantum chemical (QC) or potential of mean force calculations.

The quantity of literature on the topic of both model and biological membranes as well as FFs is huge, and we certainly do not aim to provide a full review [Marrink et al., 2019] that would anyway be out of place here. Rather, we mainly focus on a well-studied case, that of DPPC, that we take as an example for the methodologies that can be used for many other membrane systems.

Lipid bilayer structures. Phosphatidylcholine (PC) aggregates in water show a rich phase behaviour [Koynova and Caffrey, 1998; Mouritsen and Bagatolli, 2016]. In Fig. 1.50 we showed a sketch of the main lipid bilayer phases including L_β, $L_{\beta'}$, $P_{\beta'}$, L_α. The various phases have different molecular organization and fluidity, with the gel phases (the ones with a subscript β) having a lateral translational diffusion coefficient in the membrane plane of the order of 10^{-9} cm^2 s^{-1}, i.e. some two orders of magnitude less than that of the fluid phase. Chains can be on average straight or tilted (indicated by a primed subscript). In the $L_{\beta'}$ gel phase the lipid tails are stretched (mainly all trans), tilted with respect to the bilayer normal and ordered in a hexagonal array. Bilayers are embedded in water and stabilized by the interactions between the water molecules and the amphiphilic phospholipids, particularly their zwitterionic polar heads and, possibly because of these strong interactions, it is found that water molecules located at the membrane interface region (the hydration layer) have quite different properties from the water in the bulk [Marrink et al., 1993; Gurtovenko et al., 2004]. The L_α is fluid-like and is also called the liquid crystal phase, since an L_α multilayer strongly resembles a smectic A type structure. Under physiological conditions the membrane lipids are in this L_α phase, an important prerequisite for their proper functioning. The set of structures observed is, however, not the same for all lipid systems. For instance, the ripple $P_{\beta'}$ phase[1] occurs for saturated phosphatidyl cholines but appears to be missing for the unsaturated ones. Moreover, unsaturated lipids have significantly lower melting temperature and thus behave more fluid-like around room temperature. DPPC and DMPC bilayers in water do form the rippled bilayer phase $P_{\beta'}$ between the gel phase $L_{\beta'}$ and the fluid liquid crystalline phase in the temperature range between the so-called pretransitional temperature T_P and the main transition temperature T_M (see the calorimetry plot in Fig. 2.31).

Atomistic simulations allow us to examine the molecular organization beyond the oversimplified sketches (Fig. 1.50) and various of the phase organizations and phase transitions (Section 2.14.3) have actually been simulated at UA [Lindahl and Edholm, 2000; Scott, 2002] or AA (Fig. 12.34) level. Let us now discuss in some more detail the simulations of the L_α and the $P_{\beta'}$ phases.

Liquid crystal L_α phase. A system of $N = 256$ fully hydrated DPPC molecules in the L_α phase was simulated by Lindahl and Edholm [2000] at a temperature $T = 323$ K, separately maintained for lipids and water by using a Berendsen thermostat (cf. Section 9.5.1). The starting configuration was a bilayer structure and, after equilibration, 5 ns trajectories were studied. The (x, y) coordinates of this structure were scaled to produce five systems with fixed areas per lipid A_{lip} of 0.605, 0.620, 0.635, 0.650 and 0.665nm^2. The bilayers were hydrated with 23 SPC waters per lipid, bringing the systems to 30,464 atoms each. The normal pressure of the system was kept to 1 atm by scaling the box z-coordinate

[1] Note that the notation is often confusing, e.g. the $L_{\beta'}$ and $P_{\beta'}$ are called L_β and P_β in Losada-Perez et al. [2014] and Khakbaz and Klauda [2018].

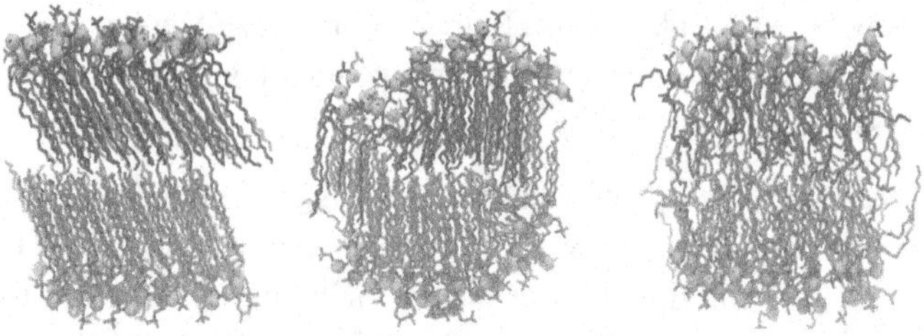

Figure 12.34 Snapshots of lipid bilayers obtained by cooling from the L_α to the $P_{\beta'}$ then to $L_{\beta'}$ phase using AA MD, with NAMD and CHARM36 FF, $N = 72$ lipids, $t = 300$ ns for each of 27 replicates. Note the interdigitation in the minor arm region [Khakbaz and Klauda, 2018].

with a time constant of 1 ps. All simulations were carried out using the UA GROMOS FF, charges from QC calculations and dihedrals by Ryckaert and Bellemans [1975, 1978] were used to describe the hydrocarbon chains. We recall that in the Ryckaert–Bellemans' (RB) model each hydrogenated carbon of the chain is represented by a UA, all C-C bonds are rigid and of length $l_{C-C} = 1.53$ Å, while the angles between adjacent bonds are fixed at $109°28'$ (the tetrahedral sp^3 geometry). The potential energy associated with the relative rotation of the two parts of a chain adjacent to a C-C bond is a function of the dihedral angle ϕ: $U^{RB}(\phi)/k_B = [1.116 + 1.462 \cos\phi - 1.578 \cos^2\phi - 0.368 \cos^3\phi + 3.156 \cos^4\phi - 3.788 \cos^5\phi] 10^3$ K, with the special values (in kcal/mol), $U^{RB}(0) = 0$ (*trans*), $U^{RB}(\pm\pi/3) = 2.95$, $U^{RB}(\pm 2\pi/3) = 0.70$ and $U^{RB}(\pm\pi) = 10.7$. The LJ parameters for two acyl CH_2 (or CH_3) belonging to two different molecules or separated by at least three centres are $\epsilon_{CC}/k_B = 72$ K, $\sigma_{CC} = 3.923$ Å. Lindahl and Edholm [2000] proposed various methods to calculate bilayer properties. In particular, they discussed how to obtain the energetic and entropic contributions to the total surface tension obtained from the virial expression (see Eq. 4.152). From the average surface tension γ obtained at various fixed area A they obtained the bilayer area compressibility from the slope: $\kappa_A = A\partial\gamma/\partial A$, finding $\kappa_A \approx 0.3$ N/m.

As we have seen for thermotropics, the most appropriate way of ensuring proper convergence to equilibrium of an ordered structure would be to obtain the same organization starting from a more ordered or from a disordered configuration. Simulation of the spontaneous aggregation of phospholipids into bilayers is of course difficult, and most studies have started from pre-assembled bilayers. However, simulations from a random mixture of lipids and water are not impossible and can provide a revealing view of the self-assembly mechanism and of the times involved. In pioneering work Marrink et al. [2001] studied systems containing 64 DPPC lipids and 3000 water molecules with PBC, using GROMACS and the GROMOS UA FF at $T = 323$ K, well above the main phase transition at 315 K and $P = 1$ atm in vertical and lateral directions, corresponding to a stress-free bilayer. The total simulation time exceeded 500 ns. Starting from a random water-lipid mixture, they found a rapid separation into lipid and aqueous domains on a timescale $t_{sep} \approx 200$ ps. This was

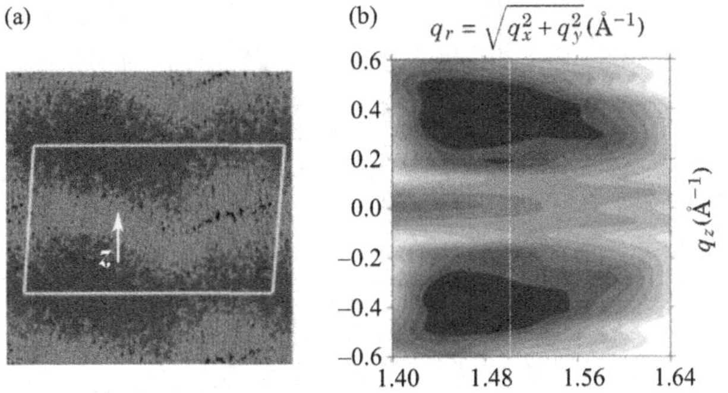

Figure 12.35 (a) Snapshot of the DPPC ripple structure and (b) WAXS intensity calculated from the positions of the atoms. The intensity is plotted as contours on a grey scale, using logarithmic procession of the levels. In the snapshot, water appears black and the lines show the projection of the unit cell on the x-z-plane while the white arrow indicates the z-axis [de Vries et al., 2005].

followed, on a timescale $t_{\text{bil}} \approx 5$ ns, by the formation of a bilayer-like phase still containing defects like water pores, which eventually yields a defect-free bilayer. The last step takes a much longer time of ≈15 ns, which is determined by the energy of the transition state. It is interesting that a bilayer with a matching number of lipids in the two sheets was obtained. Atomistic MD simulations have also been used to study the spontaneous aggregation of a concentrated solution of DPPC molecules in water to follow the spontaneous formation of small vesicles [de Vries et al., 2004]. Starting from a random solution of DPPC modelled with a UA FF in SPC water, an oblong-shaped vesicle with a long axis of 15 nm and short axes of 10 nm was formed spontaneously in less than 90 ns at 323 K.

The ripple $P_{\beta'}$ phase. The ripple phase can be experimentally obtained either from heating up the $L_{\beta'}$ phase with tilted chains or cooling down the L_α phase [Tenchov et al., 1989]. For the $P_{\beta'}$ phase, which is the more structurally complex one, X-ray studies ([Sengupta et al., 2003] and references therein) show it consists of two domains of different lengths and orientation (the minor and major arms), connected by a kink region, with the organization of the lipids in the longer domain being a splayed gel and in the minor arm a gel-like and fully interdigitated one.

de Vries et al. [2005] simulated the formation of the $P_{\beta'}$ phase (Fig. 12.35a) by sudden cooling down from a L_α bilayer. They generated simulated wide angle X-ray scattering (WAXS) data by calculating the real part of the intensity, $I(\boldsymbol{q})$, from the positions of all lipid atoms of a representative configuration of the system by using the formula (see Appendix J and Eq. 3.98b):

$$I(\boldsymbol{q}) \propto \sum_i \sum_j Z_i Z_j \cos\left(-2\pi \left[\boldsymbol{q} \cdot \left(\boldsymbol{r}_j - \boldsymbol{r}_i\right)\right]\right), \tag{12.10}$$

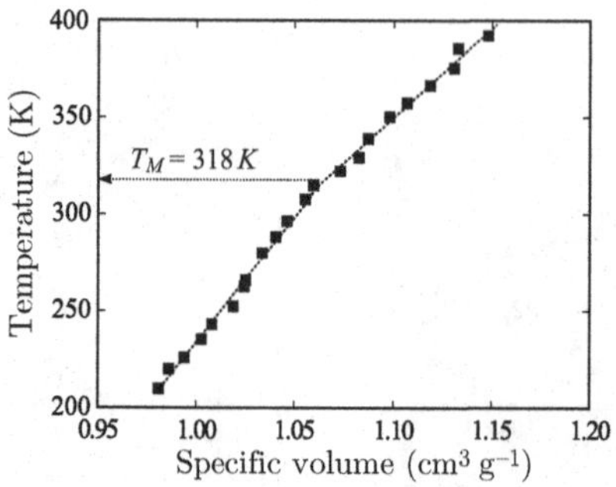

Figure 12.36 The main (L_α-$P_{\beta'}$) phase transition temperature of the DPPC bilayers, located at $T_M \approx 318$ K from the change in slope of the temperature-specific volume curve obtained from MD data [Youssefian et al., 2017].

where $\boldsymbol{q}$ is the scattering vector in the 3D reciprocal space, $(\boldsymbol{r}_j - \boldsymbol{r}_i)$ is the separation vector between the positions of atoms i and j, and Z_i and Z_j are the number of electrons associated with atoms i and j, respectively. The z-axis in real space was chosen to be the ripple stacking direction, with the transform from real space to reciprocal space performed on the fly on a grid spaced 0.1 nm^{-1} in x, y, z-directions. The data were mapped from the 3D representation on a 2D grid by projecting the $I(\boldsymbol{q})$ on q_z and on the plane perpendicular to $q_z, q_r = (q_x^2 + q_y^2)^{1/2}$ as shown in Fig. 12.35b. The ripple phase of DPPC and DMPC was also been studied more recently at AA level by Khakbaz and Klauda [2018]. They found that $P_{\beta'}$ major arm has a structure similar to that of the $L_{\beta'}$ (Fig. 12.34), while the thinner minor arm has interdigitated chains and the transition region between these two domains has large-chain splay and disorder. In the concave part of the kink region between the domains the lipids are disordered. At lower temperatures, their MD simulations predict the formation of the $L_{\beta'}$ phase with tilted fatty acid chains (Fig. 12.34a).

Main ($P_{\beta'}$-L_α) transition temperature. In simulations, the melting transition from gel to fluid for membrane bilayers has typically been located monitoring the change of specific volume or area per lipid or thermal conductivity [Youssefian et al., 2017]. The simulation of a system of 72 DPPC and 2560 water molecules using the consistent force field (CFF91) [Sun, 1995] was carried out after an initial equilibration leading to a bilayer with $A_{\text{lip}} = 62.5$ Å^2. The phase transformations were studied by heating the system to 400 K, then cooling it down all the way to 200 K at 0.2 K/ps, controlling the temperature and pressure by a Nosé thermostat and Berendsen barostat. The phase transition temperature was obtained from the temperature where the slope of the specific volume versus temperature curve changes (Fig. 12.36). This gives an estimate for the main transition temperature of DPPC at about 318 K which is in excellent agreement with the experimental result of 315 K [Biltonen and Lichtenberg, 1993; Pennington et al., 2016]. The paper also studies in detail the thermal conductivity of the bilayer as a function of temperature.

Yu and Klauda [2020] performed (N, P, T) MD simulations of bilayers of $N = 72$ DPPC molecules using NAMD and the already mentioned C36UAr FF, with a modified TIP3P water model [Jorgensen et al., 1983]. Up to 200 ns simulations were produced to get static equilibrium properties. Other longer simulations were run with the OpenMM platform[2] to calculate trasport properties. The pressure was set to be 1 bar allowing the square section cell (L_x constrained to be equal to L_y) to vary independently with respect to L_z. The observable area per lipid (A_{lip}) for such a simulation can be simply calculated as the area of the sample cell in the x-y-plane divided by the number of lipids per sheet: $A_{\text{lip}} = (2L_x L_y)/n_{\text{lip}}$. It is found that A_{lip} reproduces the experimental values for various lipid heads. For DPPC at $T = 323.15$ K, $A_{\text{lip}} = 63.3$ Å^2 for (C36Ar) and 63.0 Å^2 from experiment [Kucerka et al., 2008]. The variation in surface area per lipid with temperature can be used to locate the transition from the L to the P′ and then to the L′ phase. A sharp decrease in the surface area per lipid occurred at 308.15 K suggesting a transition from the L_α to the $P_{\beta'}$ or the $L_{\beta'}$ phase, while the DPPC bilayer is still in the L phase at 313.15 K, with some discrepancy from experiment. Another important property, the area compressibility κ_A, can be evaluated from the mean square deviation $\langle \delta A^2 \rangle$ in the total surface area $\langle A \rangle$: $\kappa_A = (k_B T \langle A \rangle)/\langle \delta A^2 \rangle$. For DPPC it is found that the simulated $\kappa_A = 170 \pm 20$ dyn/cm (C36Ar) and 230 dyn/cm from experiment. It is worth noting that κ_A is much higher in the more rigid ripple than in the LC phase [Khakbaz and Klauda, 2018].

Dickson et al. [2012] calculated the volume per lipid, V_{lip}, from that of the simulation box V_{box} according to: $V_{\text{lip}} = (V_{\text{box}} - n_W V_W)/n_{\text{lip}}$, where n_W is the number of water molecules of volume $V_W \approx 30.53$ A^3 for TIP3P water molecules [Rosso and Gould, 2008]. The V_{lip} converges fast [Anezo et al., 2003] and shows lower fluctuation than A_{lip}, and as such, could provide a better metric for validating simulation results by comparing with experimental structural parameter [Costigan et al., 2000; Nagle and Tristram-Nagle, 2000]. The V_{lip} values obtained with the GAFF lipid FF show agreement within 2% from experimental ones.

Order parameters along the acyl chain. Deuterium order parameters have been measured by NMR for a variety of different lipids and lipid mixtures for 30 years. A very direct method is to use deuterium quadrupole coupling, even if this requires the synthesis of deuterated analogues. From the residual quadrupole splitting $\Delta\nu$, the order parameter S_{CD} of the C-D bond can be calculated (see Sections 3.10.6 and 10.8.2 and Appendix I) according to

$$\Delta\nu = \frac{3}{4}\left(e^2 q Q/h\right) S_{\text{CD}}. \tag{12.11}$$

The deuteron quadrupole splitting constant $\left(e^2 q Q/h\right)$ was found to be 170 KHz for paraffin chains [Burnett and Muller, 1971]. The average orientation of the CD bond is essentially perpendicular to the bilayer normal; hence it is reasonable to assume that S_{CD} is negative.

The order parameter at different positions of the DPPC alkyl chains was calculated by MD [Hofsäss et al., 2003] and compared with those measured by deuterium NMR (of course after deuteration) [Seelig and Seelig, 1974, 1975; Douliez et al., 1996]. Such an

[2] http://docs.openmm.org/latest/userguide/application.html

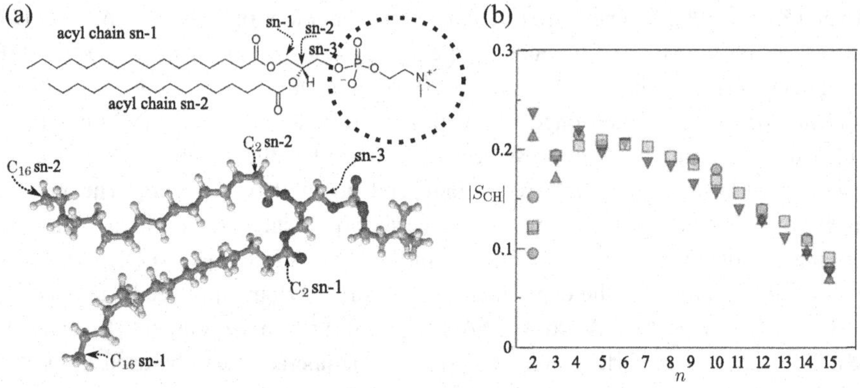

Figure 12.37 (a) The chemical formula and a ball and stick representation of DPPC with the definition of the two acyl chains sn-1 and sn-2 and of the carbon atom numbering. (b) Modulus of order parameter, $|S_{CH}|$ of the n lipid chain C-H bond for a MD simulation of a $N = 72$ DPPC bilayer system using NAMD averaged over $\approx$200 ns and comparison with experiment (▲ sn-1), (● sn-2) at $T = 323.15$ K both for the simulation [Yu and Klauda, 2020] and experiments: from DNMR quadrupolar splitting of selectively deuterated positions in the chains [Seelig and Seelig, 1974, 1975; Klauda et al., 2008].

order parameter may be defined for every CH_2 group in the chains as we did in Eq. 12.9, where the local director $\boldsymbol{d}_m$ is now taken as the normal to the membrane bilayer. When using UA models in the simulations the CH bond has to be located from the positions of three successive CH_2 groups assuming tetrahedral geometry of the CH_2 groups. The brackets in Eq. 12.9 indicate averaging over the two bonds in each CH_2 group, all the lipids and time. The two chains of a phospholipid (see Fig. 12.37) are not equivalent, and the sn-2 chain, being attached to the middle carbon of the glycerol backbone, is expected to be positioned slightly closer to the membrane surface than the sn-1 chain. Thus, the order parameters, especially in the upper part of the chain, are slightly larger (more negative) in the sn-2 chain compared to sn-1 chain in fair agreement with experiment. For large systems, undulations complicate things a bit as shown by a slight difference between simulated order parameters in a 64-lipid and a 1024-lipid DPPC system. A typical feature for the order parameter profiles of pure phospholipid bilayers is that there is a plateau region in the upper and middle part of the chains in which the order parameters vary only slightly. Then there is a drop towards 0 at the end of the chain.

In summary, also for these lyotropic liquid crystal systems atomistic simulations are now approaching the quality of real experimental data for the major force fields, even if some differences exist.

12.8 Some Comments and an Outlook

In many ways simulations, particularly atomistic ones, represent a theoretician's dream of what an experiment should be. A certain potential is chosen to describe the interaction between the particles (atoms, molecules, ...) of a system of interest and comparison is made between the essentially exact results produced by the simulations and those produced by approximate theories or experiments, as appropriate. We have the possibility of predicting properties not readily accessible, or perhaps not available at all, from experiment, controlling at will external conditions such as temperature and pressure. As we have seen we can calculate not only thermodynamic observables such as energy, heat capacity, etc. but also positional and orientational distributions and multiparticle correlation functions that allow us to predict the outcome of experiments yet to be performed. In many cases we can also evaluate separately the contributions provided by different terms in the intermolecular potential, e.g. switching on and off some terms (charges or whatever) in the FF. A major problem is that only a limited number of particles can be studied. For atomistic simulations this currently goes from a few hundred to a few tens of thousands of molecules, depending on their complexity and on the computer resources available. With the predictable rise in computer power the sample sizes are certainly due for an increase, although it is difficult to foresee many order-of-magnitude increments in the number of particles. The situation is more complicated for systems near a phase change, since close to a transition finite size effects might be particularly important and, as in real experiments, a long time may be required to reach equilibrium. These considerations can be important for liquid crystals, where very often we are not far from a phase transition.

In general terms, the predictive capability of atomistic simulations has been demonstrated. Even though this approach is very demanding in terms of resources, it provides a test bench for microscopic and phenomenological theories of nematics based on a certain FF. More importantly, atomistic MD opens exciting possibilities for the molecular design of novel mesogens with the prediction of their properties, even in advance of their chemical synthesis.

Computer simulations do not cover, as yet, all the many liquid crystal systems introduced in Chapter 1 at the atomistic or even molecular resolution level, but the number has drastically increased and should increase further. For instance, systems like active LCs, until now treated at a mesoscopic level with the tools of hydrodynamics equations (see, e.g., [Marchetti et al., 2013; Turiv et al., 2020]) should become amenable to investigation at molecular (see, e.g., [de Souza et al., 2022]) or atomistic level.

On closing, a few points perhaps worth considering are:

(i) **Computer power.** The increasing success of realistic atomistic simulations shows that the outlook for predictive modelling has to be clearly optimistic if computer performance, that has already increased by a factor of the order of 10^4 in the last 20 years [Strohmaier et al., 2015], will continue to evolve from current petascale to hexascale and beyond.

(ii) **Force fields.** An increase in computer performance, while important, is not sufficient in itself. FF development, if not an art, is still the result of previous theoretical knowledge, quantum chemistry and experience combined with a variety of experimental data. It should thus benefit greatly from developments in machine learning and some applications have already started to appear [Jackson et al., 2019; Bedolla et al., 2021].

(iii) **Improvements in algorithms.** Improved methods are much in demand, in particular, better techniques for time integration of Newton's equation of motion, that would allow increasing the elementary time step [Leimkuhler and Matthews, 2015] and thus the overall time window for observation of a system.
(iv) **Multiscale integration.** We have discussed models at different scales as separate, independent approaches but a coarse-graining procedure that allows going in a systematic way from the atomistic to a molecular level where molecules are represented, e.g. by suitably connected sets of GB beads, would be highly desirable in perspective and starts to be feasible at least for certain classes of organic functional materials employed in organic electronics [Roscioni and Zannoni, 2016; Ricci et al., 2019].
(v) **Modelling experiments and devices.** While we have been concerned with performing simulations of equilibrium phases, guided by a free energy minimization principle, many practical fabrication techniques involve non-equilibrium processes, e.g. evaporation, vapour deposition, casting, and so on. A huge amount of work remains to be done in this area, both from the methodological and predictive point of views. Hopefully this book will be a helpful companion also for these new developments.

Appendix A

A Modicum of Linear Algebra

A.1 Vectors and All That

Here we wish to provide a brief reminder of basic linear algebra as required for the developments described in the main text. Detailed and much more general descriptions can be found in many specialized books, e.g. Shapira [2019]. One of our aims is to see how a function can be expanded in terms of a suitable set of linearly independent functions (a basis set). This is quite similar to expanding a vector in a basis set of orthogonal vectors, so we shall start from this example. A familiar way of explicitly writing down an ordinary 3D vector $\boldsymbol{a}$ is by listing its components (real numbers for a vector in $\mathbb{R}^3$ space or complex numbers in $\mathbb{C}^3$ space) along three chosen orthogonal directions $\boldsymbol{x}$, $\boldsymbol{y}$, $\boldsymbol{z}$ (or $\boldsymbol{e}_1$, $\boldsymbol{e}_2$, $\boldsymbol{e}_3$, also often called $\boldsymbol{i}, \boldsymbol{j}, \boldsymbol{k}$). Thus, we write, e.g.

$$\boldsymbol{a} = a_1\boldsymbol{e}_1 + a_2\boldsymbol{e}_2 + a_3\boldsymbol{e}_3 = \sum_{i=1}^{3} a_i\, \boldsymbol{e}_i \tag{A.1}$$

and write the components of $\boldsymbol{a}$ as the column vector

$$|\, a\rangle \equiv \boldsymbol{a} = \begin{pmatrix} a_1 \\ a_2 \\ a_3 \end{pmatrix} \tag{A.2}$$

or as its *transpose complex conjugate*, the row vector

$$\langle a \,|\equiv \boldsymbol{a}^{\dagger} \equiv \boldsymbol{a}^{T*} = (a_1^*, a_2^*, a_3^*). \tag{A.3}$$

The quantities $\langle a\,|, |\,a\rangle$ are called, using Dirac [1958] terminology, a *bra* and a *ket*, respectively, elements of *dual* spaces. We have considered for generality the case of a vector with complex elements by taking the *adjoint*, i.e. the complex conjugate (indicated by *) of the transpose (superscript T) when going from a ket to a bra and vice versa. The notation can be extended to vectors in n dimensions considering components a_i, $i = 1, \ldots, n$. We shall assume that what we say holds for this more general n-dimensional space from now on, even though we shall continue to use mainly 3D vectors ($n = 3$) as explicit examples.

Scalar product. The *scalar* or *dot* or *inner* product of two vectors $\boldsymbol{a}$, $\boldsymbol{b}$ can be written as

$$\boldsymbol{a} \cdot \boldsymbol{b} \equiv \sum_{i=1}^{n} a_i^* b_i \equiv \langle a \mid b\rangle. \tag{A.4}$$

The result of combining a bra and a ket in this way (with paired angular brackets) gives a quantity invariant by rotation (a *scalar*). The *modulus* or 'length' or *norm* of a vector is the quantity $a = ||\boldsymbol{a}|| = \sqrt{\langle a \mid a \rangle}$. A vector is said to be normalized if it has unit length and in that case is often called a *versor*. When it is necessary to stress that a vector is of unit length, we use a *cap* (or hat) and write, e.g. $\boldsymbol{a} = a\hat{\boldsymbol{a}}$. However, to keep notation simpler, we avoid writing the hat systematically if the vector has been explicitly defined of unit length (e.g. we write $\boldsymbol{d}$ and not $\hat{\boldsymbol{d}}$ for the director). The scalar product has the following defining properties [Birkhoff and Mac Lane, 1997; Knowles, 1998]:

$$\langle a \mid a \rangle \geq 0, \tag{A.5a}$$

$$\langle a \mid b \rangle = \langle b \mid a \rangle^*, \tag{A.5b}$$

$$\langle a \mid kb \rangle = k\langle a \mid b \rangle, \tag{A.5c}$$

$$\langle a \mid k_1 b + k_2 c \rangle = k_1\langle a \mid b \rangle + k_2\langle a \mid c \rangle, \tag{A.5d}$$

with k, k_1, k_2 real or complex numbers, $|a\rangle, |b\rangle, |c\rangle$ vectors and with the scalar product in Eq. A.4 and with Eq. A.5a being 0 only if $|a\rangle = \mathbf{0}$, i.e. has all its components equal to 0. An n-dimensional vector space endowed with this scalar product is called a Euclidean space $\mathbb{E}^n$. The *orthonormality* of the unit vectors $\boldsymbol{e}_1, \boldsymbol{e}_2, \ldots, \boldsymbol{e}_n$ (e.g. the Cartesian axis $\boldsymbol{x}$, $\boldsymbol{y}$, $\boldsymbol{z}$ in three dimensions) can be written concisely as $\boldsymbol{e}_i \cdot \boldsymbol{e}_j \equiv \langle e_i \mid e_j \rangle = \delta_{i,j}$, where $\delta_{i,j}$ is the *Kronecker delta* symbol, defined as

$$\delta_{i,j} \equiv \delta(i-j) = \begin{cases} 1, & \text{if } i = j, \text{ with } i, j \text{ integers} \\ 0, & \text{otherwise} \end{cases} \tag{A.6}$$

corresponding to the elements of the identity matrix $\mathbf{1}$. The notation $\delta(i-j)$ is less frequent, but convenient at times. The set of orthonormal vectors $\{\boldsymbol{e}_i\}$ is called a *basis set* and Eq. A.1 is called the *expansion* of the vector $\boldsymbol{a}$ in that basis set. Taking a scalar product with $|e_i\rangle$ on both sides, we can find $a_i = \langle a|e_i\rangle$, i.e. $\sum_{j=1}^{n} \delta_{i,j} a_j = a_i$.

Vector product. The inner product generates a scalar starting from two vectors, but we can also combine two vectors to obtain another vector. For example, in three dimensions we can introduce the *vector* (or *wedge* or *cross*) product

$$\boldsymbol{c} = \boldsymbol{a} \times \boldsymbol{b} = -\boldsymbol{b} \times \boldsymbol{a} = \begin{vmatrix} \boldsymbol{e}_1 & \boldsymbol{e}_2 & \boldsymbol{e}_3 \\ a_1 & a_2 & a_3 \\ b_1 & b_2 & b_3 \end{vmatrix}, \tag{A.7}$$

where the last equation is written as a determinant. The components of the vector product of $\boldsymbol{a}$, $\boldsymbol{b}$ can also be conveniently written as

$$c_\alpha = \sum_{i,j} \varepsilon_{\alpha ij} a_i b_j \equiv \varepsilon_{\alpha\beta\gamma} a_\beta b_\gamma, \tag{A.8}$$

introducing the Levi-Civita *permutation symbol* (or *alternator*) $\varepsilon_{\alpha\beta\gamma}$:

$$\varepsilon_{\alpha\beta\gamma} = \begin{cases} +1, & \text{if } (\alpha\beta\gamma) = (123),\ (231),\ (312) \\ -1, & \text{if } (\alpha\beta\gamma) = (132),\ (213),\ (321) \\ 0, & \text{otherwise.} \end{cases} \tag{A.9}$$

In particular, if any two of its indices are equal, $\varepsilon_{\alpha\beta\gamma}$ is 0. We use here, and unless otherwise specified in the rest of the book, the convention of implying a summation over repeated Greek subscripts. Another way of writing the antisymmetric Levi-Civita symbol is as a determinant:

$$\varepsilon_{\alpha\beta\gamma} = \begin{vmatrix} \delta_{\alpha,1} & \delta_{\alpha,2} & \delta_{\alpha,3} \\ \delta_{\beta,1} & \delta_{\beta,2} & \delta_{\beta,3} \\ \delta_{\gamma,1} & \delta_{\gamma,2} & \delta_{\gamma,3} \end{vmatrix} \tag{A.10}$$

or as *pseudoscalar product* $\varepsilon_{\alpha\beta\gamma} = \boldsymbol{e}_\alpha \cdot (\boldsymbol{e}_\beta \times \boldsymbol{e}_\gamma)$. Note that changing the sign of all vectors, i.e. performing an *inversion* operation, changes the sign of a pseudoscalar product. A term of this type is often used in introducing molecular chirality (e.g. Section 5.10). Some useful identities for vector products are

$$\boldsymbol{a} \cdot (\boldsymbol{b} \times \boldsymbol{c}) = \boldsymbol{b} \cdot (\boldsymbol{c} \times \boldsymbol{a}) = \boldsymbol{c} \cdot (\boldsymbol{a} \times \boldsymbol{b}), \tag{A.11a}$$

$$\boldsymbol{a} \times (\boldsymbol{b} \times \boldsymbol{c}) = \boldsymbol{b}(\boldsymbol{a} \cdot \boldsymbol{c}) - \boldsymbol{c}(\boldsymbol{a} \cdot \boldsymbol{b}), \tag{A.11b}$$

$$(\boldsymbol{a} \times \boldsymbol{b}) \cdot (\boldsymbol{c} \times \boldsymbol{d}) = (\boldsymbol{a} \cdot \boldsymbol{c})(\boldsymbol{b} \cdot \boldsymbol{d}) - (\boldsymbol{a} \cdot \boldsymbol{d})(\boldsymbol{b} \cdot \boldsymbol{c}). \tag{A.11c}$$

Note that the volume of a parallelepiped with edges given by the three non-coplanar real vectors $\boldsymbol{a}$, $\boldsymbol{b}$, $\boldsymbol{c}$ is the absolute value of the pseudoscalar product

$$V = |\boldsymbol{a} \cdot (\boldsymbol{b} \times \boldsymbol{c})| = |\varepsilon_{\alpha\beta\gamma} a_\alpha b_\beta c_\gamma| = \begin{vmatrix} a_1 & a_2 & a_3 \\ b_1 & b_2 & b_3 \\ c_1 & c_2 & c_3 \end{vmatrix}. \tag{A.12}$$

A few useful properties of the Levi-Civita symbol are (again the sum over repeated Greek subscripts is implied)

$$\varepsilon_{ijk}\, \delta_{i,j} = 0, \tag{A.13a}$$

$$\varepsilon_{\alpha\beta m} \varepsilon_{\alpha\beta n} = 2\, \delta_{m,n}, \tag{A.13b}$$

$$\varepsilon_{\alpha\beta\gamma} \varepsilon_{\alpha\beta\gamma} = 6, \tag{A.13c}$$

$$\varepsilon_{\alpha ij} \varepsilon_{\alpha lm} = \delta_{i,l}\delta_{j,m} - \delta_{i,m}\delta_{j,l}, \tag{A.13d}$$

$$\varepsilon_{ijk}\varepsilon_{lmn} = \begin{vmatrix} \delta_{i,l} & \delta_{i,m} & \delta_{i,n} \\ \delta_{j,l} & \delta_{j,m} & \delta_{j,n} \\ \delta_{k,l} & \delta_{k,m} & \delta_{k,n} \end{vmatrix} = \delta_{i,l}\left(\delta_{j,m}\delta_{k,n} - \delta_{j,n}\delta_{k,m}\right) - \delta_{i,m}\left(\delta_{j,l}\delta_{k,n} - \delta_{j,n}\delta_{k,l}\right) + \delta_{i,n}\left(\delta_{j,l}\delta_{k,m} - \delta_{j,m}\delta_{k,l}\right). \tag{A.13e}$$

Direct product. We can also combine two vectors to obtain a matrix, as the *direct product* or *outer product* or *dyadic product* or *tensor product* of their elements, indicated with various notations as:

$$\boldsymbol{a} \otimes \boldsymbol{b} \equiv \boldsymbol{ab} \equiv |a\rangle\langle b|, \tag{A.14}$$

with elements $(\boldsymbol{a} \otimes \boldsymbol{b})_{ij} = a_i b_j^*$ in complex numbers space, so we have in general $\boldsymbol{a} \otimes \boldsymbol{b} \neq \boldsymbol{b} \otimes \boldsymbol{a}$. In three dimensions

$$\boldsymbol{a} \otimes \boldsymbol{b} = \begin{pmatrix} a_1 b_1^* & a_1 b_2^* & a_1 b_3^* \\ a_2 b_1^* & a_2 b_2^* & a_2 b_3^* \\ a_3 b_1^* & a_3 b_2^* & a_3 b_3^* \end{pmatrix}. \tag{A.15}$$

Resolution of the identity. The *identity matrix* **1** can be written (or 'resolved') as a sum of n direct products of the unit vectors along the orthogonal axis (each $\mid e_i\rangle\langle e_i \mid$ is also called a *projector*). In n dimensions:

$$\mathbf{1} = \sum_{i=1}^{n} \mid e_i\rangle\langle e_i \mid, \tag{A.16}$$

is particularly useful. For instance, we can use the identity in Eq. A.16 to express a vector property, e.g. a dipole moment $\boldsymbol{\mu}$, as: $|\boldsymbol{\mu}\rangle = \mathbf{1}|\boldsymbol{\mu}\rangle = \sum_{i=1}^{n} \mid e_i\rangle\langle e_i|\mu\rangle$, giving the vector components along the frame axes as $\mu_i = \boldsymbol{\mu} \cdot \boldsymbol{e}_i = \langle \mu|e_i\rangle = \langle e_i|\mu\rangle$. Thus, the expansion coefficients of a vector in an orthonormal basis are just the projections (scalar products) of the vector on the basis vectors, generalizing the result of Eq. A.1 to n dimensions.

Gradient. We can introduce the *gradient* (nabla) operator $\langle\nabla|$ as a vector whose components are the partial derivatives along the three Cartesian directions. Writing in different common notations, we have for the gradient of a scalar differentiable function $f = f(x, y, z)$

$$\langle\nabla| f \equiv \nabla f \equiv \operatorname{grad} f \equiv \langle x|\frac{\partial f}{\partial x} + \langle y|\frac{\partial f}{\partial y} + \langle z|\frac{\partial f}{\partial z} = \sum_i \boldsymbol{e}_i \nabla_i f, \tag{A.17}$$

where $\langle x|, \langle y|, \langle z|$ and $\boldsymbol{e}_i$ are orthogonal unit vectors. Some useful relations for the combination of the gradient operator with a vector:

Divergence of a vector is the dot product of the gradient with that vector

$$\nabla \cdot \boldsymbol{a} \equiv \operatorname{div}\boldsymbol{a} = \sum_i \nabla_i a_i = \nabla_\alpha a_\alpha = \frac{\partial a_x}{\partial x} + \frac{\partial a_y}{\partial y} + \frac{\partial a_z}{\partial z}. \tag{A.18}$$

Curl of a vector is the vector product of the gradient and a vector

$$\nabla \times \boldsymbol{a} \equiv \operatorname{curl} a \equiv \operatorname{rot}\boldsymbol{a} = \boldsymbol{e}_\alpha \varepsilon_{\alpha\beta\gamma} \nabla_\beta a_\gamma \tag{A.19a}$$

$$= \boldsymbol{i}\Big(\frac{\partial a_z}{\partial y} - \frac{\partial a_y}{\partial z}\Big) + \boldsymbol{j}\Big(\frac{\partial a_x}{\partial z} - \frac{\partial a_z}{\partial x}\Big) + \boldsymbol{k}\Big(\frac{\partial a_y}{\partial x} - \frac{\partial a_x}{\partial y}\Big). \tag{A.19b}$$

and also

$$\boldsymbol{a} \cdot \nabla \times \boldsymbol{a} = \frac{\partial a_y}{\partial x} - \frac{\partial a_x}{\partial y}, \tag{A.20}$$

$$|\boldsymbol{a} \times \nabla \times \boldsymbol{a}|^2 = \frac{\partial^2 a_x}{\partial z^2} + \frac{\partial^2 a_y}{\partial z^2}. \tag{A.21}$$

Note that when the vector $\boldsymbol{a}$ is the director $\boldsymbol{d}$, Eqs. A.20 and A.21 appear in the Frank elastic energy expression (Eq. 1.8) as the splay and bend terms. The gradient of a vector is a matrix: $(\nabla\boldsymbol{a})_{ij} = \nabla_i a_j$ or $\nabla\boldsymbol{a} = \nabla_\alpha a_\beta \boldsymbol{e}_\alpha \otimes \boldsymbol{e}_\beta$ and its norm or scalar contraction (see Section B.5)

$$||\nabla\boldsymbol{a}|| \equiv \nabla\boldsymbol{a} : \nabla\boldsymbol{a} = \sum_{ij} \nabla_i a_j \nabla_i a_j = \frac{\partial a_\beta}{\partial e_\alpha}\frac{\partial a_\beta}{\partial e_\alpha}. \tag{A.22}$$

Using the gradient vector ∇ with components $\nabla_i = \partial/\partial e_i$ we have

$$[\text{curl}\, \boldsymbol{v}]_i = [\nabla \times \boldsymbol{v}]_i = \varepsilon_{i\beta\gamma} \nabla_\beta v_\gamma, \tag{A.23}$$

implying again summation over repeated Greek subscripts, and

$$[\nabla \times (\nabla \times \boldsymbol{a})]_i = \varepsilon_{i\beta\gamma} \nabla_\beta \varepsilon_{\gamma\beta'\gamma'} \nabla_{\beta'} a_{\gamma'} = \varepsilon_{\gamma i\beta} \varepsilon_{\gamma\beta'\gamma'} \nabla_\beta \nabla_{\beta'} a_{\gamma'}, \tag{A.24a}$$

$$= [\delta_{i,\beta'} \delta_{\beta,\gamma'} - \delta_{i,\gamma'} \delta_{\beta,\beta'}] \nabla_\beta \nabla_{\beta'} a_{\gamma'} = \nabla_\beta \nabla_i a_\beta - \nabla_\beta \nabla_\beta a_i, \tag{A.24b}$$

$$= \nabla_i \nabla_\beta a_\beta - \nabla_\beta \nabla_\beta a_i = \left[\nabla(\nabla \cdot \boldsymbol{a}) - \nabla^2 \boldsymbol{a}\right]_i, \tag{A.24c}$$

where $\nabla^2 = \nabla \cdot \nabla = \text{div grad} = (\partial^2/\partial x^2,\, \partial^2/\partial y^2,\, \partial^2/\partial z^2)$ is the *Laplacian*. Other useful identities are

$$(\nabla \times \boldsymbol{d})^2 = (\boldsymbol{d} \cdot \nabla \times \boldsymbol{d})^2 + (\boldsymbol{d} \times \nabla \times \boldsymbol{d})^2, \tag{A.25}$$

$$\nabla \cdot [(\boldsymbol{a} \cdot \nabla)\boldsymbol{a} - (\nabla \cdot \boldsymbol{a})\boldsymbol{a}] = \nabla_\alpha a_\beta \nabla_\beta a_\alpha - \nabla_\alpha \nabla_\beta a_\beta a_\alpha, \tag{A.26a}$$

$$= ([\nabla \otimes a]_{\alpha\alpha})^2 - (\nabla_\alpha a_\alpha)^2 = \text{Tr}(\nabla \boldsymbol{a})^2 - (\nabla \cdot \boldsymbol{a})^2. \tag{A.26b}$$

An identity [Stewart, 2004] for the scalar contraction (Eq. A.22):

$$\|\nabla \boldsymbol{a}\|^2 = (\nabla \cdot \boldsymbol{a})^2 + (\nabla \times \boldsymbol{a})^2 + \nabla \cdot [(\boldsymbol{a} \cdot \nabla)\boldsymbol{a} - (\nabla \cdot \boldsymbol{a})\boldsymbol{a}] \tag{A.27}$$

can be used to conveniently rewrite the Frank elastic energy in the one-constant approximation. Considering a unit vector, e.g. the director of a uniform monodomain liquid crystal: $\boldsymbol{d} = (0, 0, 1)$, $\boldsymbol{d} \cdot \boldsymbol{d} = 1$ and some small deformations,

$$\nabla(\boldsymbol{d} \cdot \boldsymbol{d}) = 2\mathbf{e}_\alpha d_\beta \nabla_\alpha d_\beta = 0. \tag{A.28}$$

A.2 Orthogonal Functions and Basis Sets

The relations we have just recalled apply to an n-dimensional space defined by n orthogonal unit vectors $\{\boldsymbol{e}_i\}$. What is even more useful is that quite similar relations can be written for the expansion or decomposition of functions, by defining a suitable space of basis functions, orthogonal to each according to a certain scalar product. For example, given a function of a real variable x, we have the analogue of Eq. A.1

$$f(x) = \sum_{i=1}^{n_f} f_i \, \phi_i(x), \tag{A.29}$$

where f_i, $\phi_i(x)$ can have real or complex values and with the functions $\phi(x)$ playing the role of unit vectors, with n_f that is normally $n_f = \infty$. A condition to be able to write Eq. A.29 is that we can define the orthogonality of the basis functions and thus, what we have to do first is to define a generalization of the scalar product for two functions, e.g.

$$\langle f(x) \mid g(x) \rangle \equiv \int_a^b \mathrm{d}x \, \mu(x) f(x)^* g(x), \tag{A.30}$$

where $\mathrm{d}x\,\mu(x)$ is the volume element or *measure* in the space of the variable x, with $a \le x \le b$. Thus, we define the scalar product of two functions as their *overlap integral*, i.e. the integral of the product of one function by the complex conjugate of the other. Note that as long as the integral on the right-hand side (RHS) exists and is finite, this definition satisfies the basic properties of a scalar product, Eqs. A.5a–A.5d. The space of all these quadratically integrable functions is called a *Hilbert space*. In particular, we call two functions $\phi_m(x)$, $\phi_n(x)$ of a certain set orthogonal if their scalar product is 0:

$$\langle \phi_m | \phi_n \rangle = \frac{1}{k_n} \int_a^b \mathrm{d}x\, \mu(x)\, \phi_m^*(x)\, \phi_n(x) = \delta_{m,n}, \tag{A.31}$$

where k_n is a normalization constant ($k_n = 1$ for orthonormal functions). If the set is complete, then these scalar products are just the matrix elements of the identity, that we can resolve, like in Eq. A.16 as

$$\hat{\mathbf{1}} = \sum_k |\phi_k\rangle\langle\phi_k|. \tag{A.32}$$

It is also convenient to consider [Dennery and Krzywicki, 1969] the set of values of a function $f(x)$ as the components of an abstract vector $|f\rangle$ with respect to a basis of vectors $|x\rangle$, with a continuous 'index' x labelling these vectors ($a \le x \le b$). In this sense we can write $f(x) \equiv \langle x | f \rangle$ and rewrite Eq. A.31 as

$$\frac{1}{k_n} \int_a^b \mathrm{d}x\, \mu(x)\, \langle \phi_m | x \rangle \langle x | \phi_n \rangle = \delta_{m,n}. \tag{A.33}$$

Thus, the identity operator can also be written as

$$\hat{\mathbf{1}} = \frac{1}{k_n} \int_a^b \mathrm{d}x\, \mu(x)\, |x\rangle\langle x|. \tag{A.34}$$

Using these generalized definitions, we can have basis sets for functions as we have for vectors. The coefficients f_i in Eq. A.29 are found to be

$$f_n = \frac{\langle f(x) \mid \phi_n(x) \rangle}{\langle \phi_n(x) \mid \phi_n(x) \rangle} = \frac{1}{k_n} \int_a^b \mathrm{d}x\, \mu(x) f(x)^* \phi_n(x). \tag{A.35}$$

It is clear that listing all the infinite number of coefficients completely defines the function $f(x)$. Once we have specified the basis set in a certain space defined for a variable x we could represent any function $f(x)$ by a string of coefficients f_n. In a number of conditions this digital representation will be quite a useful way of storing information about a function. This is particularly true if the use of a limited number of coefficients allows a satisfactory representation of the function itself. For instance, we could represent a sound signal by expanding it in harmonics and obtain in this way a useful digital representation. In a more general situation, we shall have a function of many variables and the integral in Eq. A.35

will really be a multiple integral. A typical example is that of a function of the polar angles α, β where we have a two-fold integral and $\mathrm{d}x\,\mu(x)$ becomes $\mathrm{d}\alpha\mathrm{d}\beta \sin\beta$.

A.2.1 Orthogonalization

Gram–Schmidt orthogonalization. Let us start from m non-orthogonal functions $g_i(x)$ which we take to be linearly independent, in the sense that no linear combination of these functions is equal to 0:

$$\sum_{i=1}^{m} c_i g_i(x) \neq 0, \tag{A.36}$$

for any set of (not all 0) real coefficients $c_1, c_2, \ldots, c_m$ and proceed by building our orthogonal combinations one after the other. If we arbitrarily start from $\phi_1(x) = g_1(x)$ then, clearly, the function

$$\phi_2(x) = g_2(x) - \frac{\langle g_2(x)|\phi_1(x)\rangle}{\langle \phi_1(x)|\phi_1(x)\rangle}\phi_1(x), \tag{A.37}$$

where we have removed the 'projection' of g_2 on ϕ_1, is orthogonal to ϕ_1:

$$\langle\phi_1|\phi_2\rangle = \langle\phi_1|g_2\rangle - \frac{\langle g_2|\phi_1\rangle}{\langle\phi_1|\phi_1\rangle}\langle\phi_1|\phi_1\rangle = 0. \tag{A.38}$$

We can then use ϕ_1 and ϕ_2 to obtain ϕ_3 and so on, writing:

$$\phi_{k+1}(x) = g_k(x) - \sum_{i=1}^{k} \frac{\langle g_k(x)|\phi_i(x)\rangle}{\langle\phi_i(x)|\phi_i(x)\rangle}\phi_i(x). \tag{A.39}$$

Let us consider, as an example, the set of powers $\{g_i(x)\} = \{x^0, x^1, x^2, \ldots, x^n\}$, with x real and $-1 \leq x \leq 1$. Then, using

$$\int_{-1}^{1} \mathrm{d}x\, x^m x^n = \begin{cases} 0 & \text{if } n+m \text{ is odd} \\ 2/(m+n+1) & \text{if } n+m \text{ is even,} \end{cases} \tag{A.40}$$

we find the orthogonal set

$$\phi_0(x) = 1, \tag{A.41a}$$

$$\phi_1(x) = x, \tag{A.41b}$$

$$\phi_2(x) = x^2 - \frac{1}{3}, \tag{A.41c}$$

$$\phi_{2k}(x) = x^{2k} - \sum_{j=0,2.}^{2k-2} \frac{(2j+1)}{(2k+j+1)}\phi_j(x), \tag{A.41d}$$

$$\phi_{2k+1}(x) = x^{2k+1} - \sum_{j=1,3.}^{2k-1} \frac{(2j+1)}{(2k+j+2)}\phi_j(x). \tag{A.41e}$$

We can still choose the normalization of these orthogonalized functions. If the choice is to normalize ϕ_n to $2/(2n+1)$, these correspond to the Legendre polynomials that we have used in Chapter 3 and elsewhere to write down expressions for the orientational order parameters. Note that the Gram–Schmidt procedure allows the generation of an infinite set of orthogonal functions.

Löwdin orthogonalization. There are various other procedures to generate a set of orthonormal functions $\phi_i(x)$ from m linearly independent non-orthogonal functions $g_i(x)$. The scalar products $S_{ij} = \langle g_i(x)|g_j(x)\rangle = S^*_{ji}$, can be used to form an *overlap matrix* **S** that has a non-zero determinant. **S** is *Hermitian* or *self-adjoint* (i.e. $\mathbf{S} = \mathbf{S}^{T*}$) and can be diagonalized with a unitary eigenvector matrix **U** to the diagonal eigenvalues matrix **s** with principal values s_i

$$\mathbf{SU} = \mathbf{Us}. \tag{A.42}$$

Each column of **U**, e.g. $\boldsymbol{u}^{(i)} \equiv (U_{1i}, U_{2i}, \ldots, U_{ni})$, corresponds to one eigenvector, e.g. $\mathbf{S}\,\boldsymbol{u}^{(i)} = s_i\,\boldsymbol{u}^{(i)}$. The linear combinations $\phi_i(x) = \sum_{j=1}^m U_{ji}\,g_j(x)$ are orthogonal:

$$\langle\phi_i(x)|\phi_j(x)\rangle = \sum_{k,l=1}^{m} U_{ki}\,\langle g_k(x)|g_l(x)\rangle U_{lj} = \sum_{k,l=1}^{m} U_{ki}\,S_{kl}U_{lj} = s_j\delta_{i,j} \tag{A.43}$$

and the linear combinations $\psi_i(x) = \frac{1}{\sqrt{s_i}}\sum_{j=1}^m U_{ji}\,g_j(x)$, are orthonormal. The Löwdin method has the advantage of treating all starting functions in an unbiased, symmetric way and thus of defining orthogonal functions that are as similar as possible to the original ones. This is particularly convenient in quantum chemistry where it is used, e.g. on atomic orbitals. Note, however, that only a finite basis set can be treated in this way.

We now list a few useful examples of basis sets.

Harmonics. The harmonic functions $\{\cos(nx)\}$, for $0 \leq x \leq 2\pi$ and integer n form a basis set with the orthogonality relation

$$\int_0^{2\pi} dx\,\cos(mx)\cos(nx) = \pi(\delta_{m,0}\delta_{n,0} + \delta_{m,n}) \tag{A.44}$$

and similarly, the set: $\{\sin(nx)\}$, for $0 \leq x \leq 2\pi$ and integer n:

$$\int_0^{2\pi} dx\,\sin(mx)\sin(nx) = \pi(\delta_{m,0}\delta_{n,0} + \delta_{m,n}) \tag{A.45}$$

and in complex exponential form:

$$\int_0^{2\pi} dx\,e^{-imx}\,e^{inx} = 2\pi\,\delta_{m,n}. \tag{A.46}$$

Legendre polynomials. A Legendre polynomial of rank L, $P_L(x)$, is a polynomial of degree L of x which contains only odd or even powers of x when L is, respectively, odd or even and with L 0s in the range $-1 \le x \le 1$. The $P_L(x)$ can be generated with the Rodrigues formula:

$$P_L(x) = \frac{1}{2^L L!}\frac{\mathrm{d}^L}{\mathrm{d}x^L}(x^2-1)^L = \frac{1}{2^L}\sum_{k=0}^{L}(-1)^k \frac{(2L-2k)!}{k!\,(L-k)!\,(L-2k)!}x^{L-2k}. \tag{A.47}$$

The explicit form of the first few Legendre polynomials is:

$$P_0(x) = 1, \tag{A.48a}$$
$$P_1(x) = x, \tag{A.48b}$$
$$P_2(x) = (3x^2-1)/2, \tag{A.48c}$$
$$P_3(x) = (5x^3-3x)/2, \tag{A.48d}$$
$$P_4(x) = (35x^4-30x^2+3)/8, \tag{A.48e}$$
$$P_5(x) = (63x^5-70x^3+15x)/8, \tag{A.48f}$$
$$P_6(x) = (231x^6-315x^4+105x^2-5)/16, \tag{A.48g}$$

where $x \equiv \cos\beta$. At $x = 1$ (i.e. $\beta = 0$), $P_{2L}(1) = 1$. At $x = 0$ (i.e. $\beta = \pi/2$) we have [Tricomi, 1948], $P_{2L}(0) = [(-1)^L(2L)!\,]/[2^{2L}(L!\,)^2]$. It is also useful to write the inverse relations

$$1 = P_0(x), \tag{A.49a}$$
$$x = P_1(x), \tag{A.49b}$$
$$x^2 = \frac{1}{3}[P_0(x)+2P_2(x)], \tag{A.49c}$$
$$x^3 = \frac{1}{5}[3P_1(x)+2P_3(x)], \tag{A.49d}$$
$$x^4 = \frac{1}{35}[7P_0(x)+20P_2(x)+8P_4(x)], \tag{A.49e}$$
$$x^5 = \frac{1}{63}[27P_1(x)+28P_3(x)+8P_5(x)], \tag{A.49f}$$
$$x^6 = \frac{1}{231}[33P_0(x)+110P_2(x)+72P_4(x)+16P_6(x)]. \tag{A.49g}$$

The Legendre polynomials $\{P_L(x)\}$, for $x = \cos\beta$, $-1 \le x \le 1$ and L integers with $L \ge 0$ form a complete orthogonal basis set:

$$\int_{-1}^{1}\mathrm{d}x\,P_L(x)P_{L'}(x) = \int_0^{\pi}\mathrm{d}\beta\,\sin\beta\,P_L(\cos\beta)P_{L'}(\cos\beta) = \frac{2}{(2L+1)}\delta_{L,L'}. \tag{A.50}$$

They satisfy the recurrence relation

$$LP_L(x) = (2L-1)xP_{L-1}(x) - (L-1)P_{L-2}(x) \tag{A.51}$$

and their derivative can be written as

$$\frac{\mathrm{d}}{\mathrm{d}x}P_L(x) = (2L-1)P_{L-1}(x) + (2L-5)P_{L-3}(x) + \cdots \begin{cases} +1 & (L \text{ odd}) \\ +3P_1(x) & (L \text{ even}) \end{cases}. \tag{A.52}$$

Spherical harmonics. Spherical harmonics are familiar in quantum mechanics and chemical physics since they represent the angular part of the atomic orbital of a hydrogenoid atom (see, e.g., [Atkins, 1983]). The rank $L = 0, 1, 2, \ldots$ of a spherical harmonic $Y_{L,m}$ corresponds to the angular momentum for $s, p, d, \ldots$ orbitals, and m to the angular momentum projection quantum number. Explicit expressions can be derived from the definition

$$Y_{L,m}(\alpha, \beta) = (-)^m \left(\frac{(2L+1)(L-m)!}{4\pi(L+m)!} \right)^{\frac{1}{2}} \mathrm{e}^{im\alpha} (1 - \cos^2 \beta)^{m/2} \frac{\mathrm{d}^m P_L(\cos \beta)}{\mathrm{d}(\cos \beta)^m}, \quad \text{(A.53)}$$

with $L = 0, 1, 2, \ldots, m = -L, -L+1, \ldots, L-1, L$. The first few $Y_{L,m}(\alpha, \beta)$ are:

$$Y_{0,0}(\alpha, \beta) = \sqrt{1/(4\pi)}, \quad \text{(A.54a)}$$

$$Y_{1,0}(\alpha, \beta) = \sqrt{3/(4\pi)} \cos \beta, \quad \text{(A.54b)}$$

$$Y_{1,\pm 1}(\alpha, \beta) = \mp \sqrt{3/(8\pi)} \sin \beta \, \mathrm{e}^{\pm i\alpha}, \quad \text{(A.54c)}$$

$$Y_{2,0}(\alpha, \beta) = \sqrt{5/(4\pi)} \left(\frac{3}{2} \cos^2 \beta - \frac{1}{2} \right), \quad \text{(A.54d)}$$

$$Y_{2,\pm 1}(\alpha, \beta) = \mp \sqrt{15/(8\pi)} \sin \beta \cos \beta \, \mathrm{e}^{\pm i\alpha}, \quad \text{(A.54e)}$$

$$Y_{2,\pm 2}(\alpha, \beta) = \sqrt{15/(32\pi)} \sin^2 \beta \, \mathrm{e}^{\pm i2\alpha}. \quad \text{(A.54f)}$$

The functions $\{Y_{L,m}(\alpha, \beta)\}$, with $0 \le \beta \le \pi$ and $0 \le \alpha \le 2\pi$ form a basis set, with the orthogonality relation

$$\int_0^{2\pi} \mathrm{d}\alpha \int_0^{\pi} \mathrm{d}\beta \sin \beta \, Y_{L,m}(\alpha, \beta) Y^*_{L',m'}(\alpha, \beta) = \delta_{L,L'} \, \delta_{m,m'}. \quad \text{(A.55)}$$

A generalization of spherical harmonics to the space of Euler angles (α, β, γ) leads to Wigner rotation matrices $\mathcal{D}^L_{m,n}(\alpha, \beta, \gamma)$ that are particularly important for discussing orientational distributions and order parameters and will be treated in detail in Appendix F.

Appendix B

Tensors and Rotations

B.1 Scalars and Vectors

Physical properties often need to be classified according to their number of independent components and their behaviour under rotations. We are familiar with some of these, e.g. scalars and vectors, but here we shall try to give a more systematic classification in terms of tensors. To do this we briefly review the behaviour under rotation of scalars and vectors and then introduce tensors of rank n as quantities that transform under rotation as the nth power of a vector.

Scalars. These are one-component quantities that are invariant under rotation. Some examples are the number density ρ, i.e. the number of particles per unit volume of a given sample, or the mass m and the volume V of an object.

Vectors. The most familiar examples are those of vectors in two and three dimensions. In general, using the Dirac notation introduced in Appendix A, the rotated vector will be obtained starting from the identity $|v\rangle = |v\rangle$, where we insert on the right-hand side the identity matrix written in terms of the unit vectors $|e_j\rangle$ along the axis of the chosen coordinate system (cf. Eq. A.16)

$$|v\rangle = \sum_{j=1}^{3} |e_j\rangle\langle e_j|v\rangle. \tag{B.1}$$

The components of the vector along the ith direction of the new, rotated frame (that we indicate with a prime), are obtained multiplying on the left by $\langle e'_i|$

$$\langle e'_i|v\rangle = \sum_{j=1}^{3} \langle e'_i|e_j\rangle\langle e_j|v\rangle. \tag{B.2}$$

If we write this using the more familiar subscript notation for the components of a vector we have

$$v'_i = \sum_{j=1}^{3} R_{ij} v_j, \tag{B.3}$$

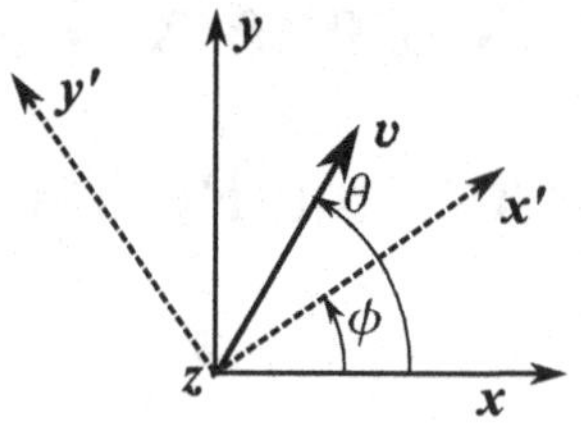

Figure B.1 A vector $\boldsymbol{v}$ at an angle $\theta - \phi$ with respect to coordinate frame $(\boldsymbol{x}', \boldsymbol{y}')$ rotated by an angle ϕ around $z = z'$ with respect to frame $(\boldsymbol{x}, \boldsymbol{y})$.

where we have introduced the Cartesian rotation matrix $\mathbf{R}$ whose elements $R_{ij} = \langle e_i' | e_j \rangle = \ell_{i,j}$ are the *direction cosines* $\ell_{i,j}$ of the new frame axis with respect to the old ones.

Cartesian rotation matrix. We can now write the rotation matrix elements in terms of angles of rotation for a vector $\boldsymbol{v} = (v_p \cos\theta, v_p \sin\theta, v_z)$, $v_p = \sqrt{v^2 - v_z^2}$. We rotate the coordinate frame of an angle ϕ around the z-axis (Fig. B.1) and find the components of the vector in the new (primed) frame, $\boldsymbol{v}' = (v_{x'}, v_{y'}, v_z)$ as

$$v_{x'} = v_p \cos(\theta - \phi) = v_p[\cos\phi\cos\theta + \sin\phi\sin\theta], \tag{B.4a}$$

$$v_{y'} = v_p \sin(\theta - \phi) = v_p[-\sin\phi\cos\theta + \cos\phi\sin\theta], \tag{B.4b}$$

$$v_{z'} = v_z. \tag{B.4c}$$

We then find, substituting $v_x = v_p \cos\theta$, $v_y = v_p \sin\theta$, that $v_x' = +v_x \cos\phi + v_y \sin\phi$: $v_y' = -v_x \sin\phi + v_y \cos\phi$ and $v_z' = v_z$. Thus, we write the 3D Cartesian matrix $\mathbf{R}_z(\phi)$ for a frame rotation of ϕ around the z-axis as

$$\mathbf{R}_z(\phi) = \begin{pmatrix} \cos\phi & \sin\phi & 0 \\ -\sin\phi & \cos\phi & 0 \\ 0 & 0 & 1 \end{pmatrix}. \tag{B.5}$$

Similarly, we get the matrices for a rotation of the frame around the x- or y-axis

$$\mathbf{R}_x(\phi) = \begin{pmatrix} 1 & 0 & 0 \\ 0 & \cos\phi & \sin\phi \\ 0 & -\sin\phi & \cos\phi \end{pmatrix} \quad \mathbf{R}_y(\phi) = \begin{pmatrix} \cos\phi & 0 & \sin\phi \\ 0 & 1 & 0 \\ -\sin\phi & 0 & \cos\phi \end{pmatrix}. \tag{B.6}$$

More complex rotations can be obtained by performing rotations about the three axes in a suitable sequence. The matrix for the composite rotation is the product of the individual rotation matrices taken in the proper order. Indeed, rotations in three dimensions do not commute, different to rotations in a plane, i.e. in two dimensions. The most general rotation needed to transform from a right-handed laboratory frame to an arbitrarily rotated coordinate system (e.g. from a laboratory to a molecular fixed system) can be written as a product of three rotations. In the Rose [1957] and Brink and Satchler [1968] convention, these are:

(i) a rotation of α around the z laboratory axis, (ii) a rotation of β around the new y-axis, and (iii) a rotation of γ around the new z-axis.[1] Thus,

$$\mathbf{R}(\alpha, \beta, \gamma) = \mathbf{R}_{z''}(\gamma)\mathbf{R}_{y'}(\beta)\mathbf{R}_z(\alpha), \tag{B.7}$$

where xyz indicate a laboratory frame and $x'y'z'$, $x''y''z''$ the axis frame after the first, or respectively, the second transformation. The three angles α, β, γ are called *Euler angles.* The explicit result is

$$\mathbf{R}(\alpha, \beta, \gamma) = \begin{pmatrix} c\alpha\, c\beta\, c\gamma - s\alpha\, s\gamma & s\alpha\, c\beta\, c\gamma + c\alpha\, s\gamma & -s\beta\, c\gamma \\ -c\alpha\, c\beta\, s\gamma - s\alpha\, c\gamma & -s\alpha\, c\beta\, s\gamma + c\alpha\, c\gamma & s\beta\, s\gamma \\ c\alpha\, s\beta & s\alpha\, s\beta & c\beta \end{pmatrix}, \tag{B.8}$$

where we have used the shorthand $s\alpha$, $c\alpha$ for $\sin\alpha$, $\cos\alpha$, etc. $\mathbf{R}$ is an orthogonal matrix, in the sense that its inverse is the transpose of $\mathbf{R}$. This is a consequence of the fact that rotations do not change the length of a vector. Thus, $\langle v'|v'\rangle = \langle v|\mathbf{R}^T\mathbf{R}|v\rangle = \langle v|v\rangle$, so that: $\mathbf{R}^T\mathbf{R} = \mathbf{R}\mathbf{R}^T = \mathbf{1}$, expressing the fact that for a rotation matrix $\mathbf{R}^{-1} = \mathbf{R}^T$, i.e. $\mathbf{R}$ is *orthogonal* (its inverse is just its transpose).

B.2 Tensors of Rank 2

The *direct* or *outer* or *tensor product* of two vectors $\boldsymbol{u}, \boldsymbol{w}$ can be represented by a matrix $\mathbf{A}$ with elements $A_{ij} \equiv (vw)_{ij} = v_i w_j$, often indicated with different notations: $\mathbf{A} = \boldsymbol{v} \otimes \boldsymbol{w} = |v\rangle\langle w| = \boldsymbol{vw}$. Using Eq. B.3 the matrix elements will transform under rotation as $(v')_i(w')_k = \sum_{jl} R_{ij} R_{kl} v_j w_l$ or, if we write the rotation for the matrix $\mathbf{A}$,

$$A'_{ik} = \sum_{jl} R_{ij} A_{jl} R^T_{lk}, \tag{B.9}$$

i.e. in matrix form, $\mathbf{A}' = \mathbf{R}\,\mathbf{A}\,\mathbf{R}^T$. This expression for the rotation of a matrix is valid for any matrix, not only the direct product ones. Thus, a matrix transforms as the second direct power of a vector and in this sense it is a tensor of rank 2. The coordinate frame where a matrix is diagonal, if it exists, is called its *principal frame* and the non-vanishing elements are the *principal values* or *eigenvalues*. The operation bringing a matrix to its diagonal form is called *diagonalization*

$$\mathbf{A}\,\mathbf{U} = \mathbf{U}\,\mathbf{a} = \mathbf{U}\,\mathrm{diag}\,(a_1, a_2, a_3), \tag{B.10}$$

where the notation $\mathrm{diag}\,(a_1, a_2, \ldots, a_n)$ indicates the diagonal matrix whose entries are a_k. Here $a_1, \ldots, a_n$ are the eigenvalues of $\mathbf{A}$: $\sum_j A_{ij} U_{jk} = \sum_{j'} U_{ij'} a_{j'k} \delta_{j',k} = a_k U_{ik}$ or, more compactly, $\mathbf{A}\,\boldsymbol{u}^{(k)} = a_k \boldsymbol{u}^{(k)}$, where the eigenvector $\boldsymbol{u}^{(k)}$ corresponding to the kth eigenvalue,

[1] Another popular convention is that of Goldstein [1980], also used by Evans [1977] and Allen and Tildesley [2017] where the second rotation is around the x- instead of the y-axis. The angles in the Goldstein (G) and Rose (R) conventions are related by $\alpha_G = \alpha_R + \pi/2$, $\beta_G = \beta_R$, $\gamma_G = \gamma_R - \pi/2$.

a_k is the kth column of the eigenvector matrix $\mathbf{U}$. The eigenvalues are the root of the secular equation, or characteristic polynomial:

$$\det(\lambda\mathbf{I} - \mathbf{A}) = \lambda^3 - I_1\lambda^2 + I_2\lambda - I_3 = 0, \tag{B.11}$$

where I_1, I_2, I_3 are the three scalar invariants of the 3×3 matrix $\mathbf{A}$

$$I_1 = A_{11} + A_{22} + A_{33} = a_1 + a_2 + a_3 = \mathrm{Tr}\mathbf{A}, \tag{B.12a}$$

$$\begin{aligned} I_2 &= A_{11}A_{22} + A_{11}A_{33} + A_{22}A_{33} - A_{12}^2 - A_{13}^2 - A_{23}^2 \\ &= a_1a_2 + a_1a_3 + a_2a_3 = \frac{1}{2}\left[\mathrm{Tr}(\mathbf{A})^2 - \mathrm{Tr}\left(\mathbf{A}^2\right)\right], \end{aligned} \tag{B.12b}$$

$$\begin{aligned} I_3 &= 2A_{12}A_{13}A_{23} + A_{11}A_{22}A_{33} - A_{13}^2A_{22} - A_{23}^2A_{11} - A_{12}^2A_{33} \\ &= a_1a_2a_3 = \det(\mathbf{A}) = \frac{1}{3}\left[\mathrm{Tr}(\mathbf{A}^3) - \frac{3}{2}\mathrm{Tr}(\mathbf{A})\,\mathrm{Tr}(\mathbf{A}^2) + \frac{1}{2}\mathrm{Tr}(\mathbf{A})^3\right]. \end{aligned} \tag{B.12c}$$

Note that the invariants, often used to describe the shape of a tensor, can be evaluated from the eigenvalues, but also directly from the matrix elements. For a traceless matrix, like the ordering matrix $\mathbf{Q}$ defined in Eq. 3.50, the invariants are just

$$I_1 = 0, I_2 = -\frac{1}{2}\mathrm{Tr}\left(\mathbf{Q}^2\right), I_3 = \frac{1}{3}\mathrm{Tr}(\mathbf{Q}^3). \tag{B.13}$$

B.3 Tensors of Rank *n*

The direct product is the device that allows us to construct higher-rank tensors from lower-rank ones. We can generalize, calling *tensor of rank n* a quantity that transforms as the nth direct power of a vector. We have

$$[v_1'v_2'\cdots v_n']_{i_1i_2\ldots i_n} = \sum_{j_1,\ldots,j_n} R_{i_1j_1}R_{i_2j_2}\cdots R_{i_nj_n}[v_1v_2\cdots v_n]_{j_1j_2\cdots j_n}. \tag{B.14}$$

In particular, just as the direct product of two vectors is a tensor of rank 2, the direct product of four vectors is a tensor of rank 4. The invariants of a fourth-rank tensor $\mathbf{T}$, with components T_{ijkl} can be written as [Betten, 1987]

$$I_1 = -S_1, \quad I_2 = \left(S_1^2 - S_2\right)/2!, \tag{B.15a}$$

$$I_3 = -\left(S_1^3 - 3S_1S_2 + 2S_3\right)/3!, \tag{B.15b}$$

$$I_4 = \left(S_1^4 + 8S_1S_3 - 6S_2S_1^2 + 3S_2^2 - 6S_4\right)/4!, \tag{B.15c}$$

$$I_5 = -\left(S_1 - 30S_1S_4 + 15S_1S_2^2 - 20S_2S_3 - 10S_2S_1^3 + 20S_3S_1^2 + 24S_5)/5!\right), \tag{B.15d}$$

$$\begin{aligned} I_6 = \Big(&S_1^6 + 144S_1S_5 - 120S_1S_2S_3 - 15S_2S_1^4 + 90S_2S_4 \\ &+ 405S_3S_1^3 - 15S_2^3 - 90S_4S_1^2 + 40S_3^2 + 45S_2^2S_1^2 - 120S_6\Big)/6!, \end{aligned} \tag{B.15e}$$

where the traces are $S_m \equiv \mathrm{Tr}\mathbf{T}^m = T_{\alpha_1\beta_1\alpha_2\beta_2}T_{\alpha_2\beta_2\alpha_3\beta_3}\ldots T_{\alpha_m\beta_m\alpha_1\beta_1}$.

B.4 Spherical Tensors

We can think of the transformation Eq. B.9, as a single-matrix transformation from the nine components of $\mathbf{A}$ to the nine components of $\mathbf{A}'$:

$$A'_{ij} = \sum_{ijkl} T_{ijkl} A_{kl}, \tag{B.16}$$

where comparing with Eq. B.9, $T_{ijkl} \equiv R_{ik}R_{jl}$. The 9×9 transformation matrix mixes in general all the nine components; however, it can be brought to block diagonal form if we take suitable combinations of the Cartesian elements. The 3^n-dimensional representation of the rotation group realized in this way can be decomposed into irreducible spherical tensors, so that the spherical components of rank L, $A^{L,m}$, transform only among themselves under rotation. The combinations of Cartesian tensor components of a physical property A transforming according to the representation $\mathbf{D}^{(L)}$ are called *spherical tensor components* of rank L and are denoted by $A^{L,m}$,

$$\begin{pmatrix} A'^{0,0} \\ A'^{1,-1} \\ A'^{1,0} \\ A'^{1,1} \\ A'^{2,-2} \\ A'^{2,-1} \\ A'^{2,0} \\ A'^{2,1} \\ A'^{2,2} \end{pmatrix} = \begin{pmatrix} 1 & 0 & 0 & 0 & 0 & 0 & 0 & 0 & 0 \\ 0 & \mathscr{D}^1_{-1,-1} & \mathscr{D}^1_{-1,0} & \mathscr{D}^1_{-1,1} & 0 & 0 & 0 & 0 & 0 \\ 0 & \mathscr{D}^1_{0,-1} & \mathscr{D}^1_{00} & \mathscr{D}^1_{0,1} & 0 & 0 & 0 & 0 & 0 \\ 0 & \mathscr{D}^1_{1,-1} & \mathscr{D}^1_{1,0} & \mathscr{D}^1_{1,1} & 0 & 0 & 0 & 0 & 0 \\ 0 & 0 & 0 & 0 & \mathscr{D}^2_{-2,-2} & \mathscr{D}^2_{-2,-1} & \mathscr{D}^2_{-2,0} & \mathscr{D}^2_{-2,1} & \mathscr{D}^2_{-2,2} \\ 0 & 0 & 0 & 0 & \mathscr{D}^2_{-1,-2} & \mathscr{D}^2_{-1,-1} & \mathscr{D}^2_{-1,0} & \mathscr{D}^2_{-1,1} & \mathscr{D}^2_{-1,2} \\ 0 & 0 & 0 & 0 & \mathscr{D}^2_{0,-2} & \mathscr{D}^2_{0,-1} & \mathscr{D}^2_{0,0} & \mathscr{D}^2_{0,1} & \mathscr{D}^2_{0,2} \\ 0 & 0 & 0 & 0 & \mathscr{D}^2_{1,-2} & \mathscr{D}^2_{1,-1} & \mathscr{D}^2_{1,0} & \mathscr{D}^2_{1,1} & \mathscr{D}^2_{1,2} \\ 0 & 0 & 0 & 0 & \mathscr{D}^2_{2,-2} & \mathscr{D}^2_{2,-1} & \mathscr{D}^2_{2,0} & \mathscr{D}^2_{2,1} & \mathscr{D}^2_{2,2} \end{pmatrix} \begin{pmatrix} A^{0,0} \\ A^{1,-1} \\ A^{1,0} \\ A^{1,1} \\ A^{2,-2} \\ A^{2,-1} \\ A^{2,0} \\ A^{2,1} \\ A^{2,2} \end{pmatrix}. \tag{B.17}$$

The matrix elements $\mathscr{D}^L_{mn}(\Omega)$ will be functions of the rotation $\Omega = \alpha, \beta, \gamma$ connecting the initial to the final coordinate frame. They are called Wigner rotation matrix (elements) and will be discussed in detail in Appendix F. Spherical tensors are particularly convenient in order to write the components of a tensor property in one frame in terms of the components of the same property in another. For instance, the components of $\mathbf{A}$ measured in the laboratory frame can be related to those in the molecular frame

$$A^{L,m}_{\text{LAB}} = \sum_{n=-L}^{L} \mathscr{D}^{L*}_{m,n}(\alpha, \beta, \gamma) A^{L,n}_{\text{MOL}}, \tag{B.18}$$

and vice versa

$$A_{\mathrm{MOL}}^{L,m} = \sum_{n=-L}^{L} \mathscr{D}_{n,m}^{L}(\alpha,\beta,\gamma) A_{\mathrm{LAB}}^{L,n}. \tag{B.19}$$

These equations illustrate the main reason for the usefulness of irreducible tensors in problems involving rotations, i.e. that their transformation properties are very simple. The set of $(2L+1)$ components, $\mathbf{A}^{(L)}$, corresponds to an irreducible tensor of rank L. If we consider in particular a tensor $\mathbf{A} = \boldsymbol{u} \otimes \boldsymbol{w}$, i.e. the direct product of two possibly different vectors, the explicit irreducible components of $\mathbf{A}$ in terms of the Cartesian components of the vectors $\boldsymbol{v}$, $\boldsymbol{w}$ or of their polar analogues, e.g. $u_{\pm} = \mp(u_x \pm i u_y)$, are

$$A^{0,0} = -\frac{1}{\sqrt{3}}(u_x w_x + u_y w_y + u_z w_z) = -\frac{1}{\sqrt{3}}\left[u_z w_z + \frac{1}{2}(u_+ w_- + u_- w_+)\right], \tag{B.20a}$$

$$A^{1,0} = -i\frac{1}{\sqrt{2}}(u_y w_x - u_x w_y) = -\frac{1}{2} - \frac{1}{\sqrt{2}}[u_+ w_- - u_- w_+], \tag{B.20b}$$

$$A^{1,\pm 1} = \frac{1}{2}[u_z w_x - u_x w_z \pm i(u_z w_y - u_y w_z)] = -\frac{1}{2}\left[u_{\pm} w_z - u_z w_{\pm}\right], \tag{B.20c}$$

$$A^{2,0} = \sqrt{\frac{2}{3}}\left[u_z w_z - \frac{1}{2}(u_x w_x + u_y w_y)\right] = \sqrt{\frac{2}{3}}\left[u_z w_z - \frac{1}{4}(u_+ w_- + u_- w_+)\right], \tag{B.20d}$$

$$A^{2,\pm 1} = \mp\frac{1}{2}[u_x w_z + u_z w_x \pm i(u_y w_z + u_z w_y)] = \mp\frac{1}{2}\left(u_{\pm} w_z + u_z w_{\pm}\right), \tag{B.20e}$$

$$A^{2,\pm 2} = \frac{1}{2}[u_x w_x - u_y w_y \pm i(u_x w_y + u_y w_x)] = \frac{1}{2} u_{\pm} w_{\pm}. \tag{B.20f}$$

More generally, a 3×3 matrix $\mathbf{A}$ can be decomposed as

$$A^{0,0} = -\frac{1}{\sqrt{3}}\mathrm{Tr}\mathbf{A}, \tag{B.21a}$$

$$A^{1,0} = -i\frac{1}{\sqrt{2}}(A_{yx} - A_{xy}), \tag{B.21b}$$

$$A^{1,\pm 1} = \frac{1}{2}[A_{zx} - A_{xz} \pm i(A_{zy} - A_{yz})], \tag{B.21c}$$

$$A^{2,0} = \sqrt{\frac{2}{3}}\left[A_{zz} - \frac{1}{2}(A_{xx} + A_{yy})\right], \tag{B.21d}$$

$$A^{2,\pm 1} = \mp\frac{1}{2}[A_{xz} + A_{zx} \pm i(A_{yz} + A_{zy})], \tag{B.21e}$$

$$A^{2,\pm 2} = \frac{1}{2}[A_{xx} - A_{yy} \pm i(A_{xy} + A_{yx})]. \tag{B.21f}$$

A generic second-rank Cartesian tensor can be written in terms of spherical tensors of rank $0, 1, 2$.

$$A_{xx} = -\frac{1}{\sqrt{3}}A^{0,0} - \frac{1}{\sqrt{6}}A^{2,0} + \frac{1}{2}(A^{2,2} + A^{2,-2}), \tag{B.22a}$$

$$A_{xy} = \frac{-i}{2}(A^{2,2} - A^{2,-2}) - \frac{i}{\sqrt{2}}A^{1,0}, \tag{B.22b}$$

$$A_{xz} = \frac{1}{2}(A^{2,-1} - A^{2,1}) - \frac{1}{2}(A^{1,-1} + A^{1,1}), \tag{B.22c}$$

$$A_{yx} = \frac{-i}{2}(A^{2,2} - A^{2,-2}) + \frac{i}{\sqrt{2}}A^{1,0}, \tag{B.22d}$$

$$A_{yy} = -\frac{1}{\sqrt{3}}A^{0,0} - \frac{1}{\sqrt{6}}A^{2,0} - \frac{1}{2}(A^{2,2} + A^{2,-2}), \tag{B.22e}$$

$$A_{yz} = \frac{i}{2}(A^{2,1} + A^{2,-1}) + \frac{i}{2}(A^{1,1} - A^{1,-1}), \tag{B.22f}$$

$$A_{zx} = \frac{1}{2}(A^{2,-1} - A^{2,1}) + \frac{1}{2}(A^{1,-1} + A^{1,1}), \tag{B.22g}$$

$$A_{zy} = \frac{i}{2}(A^{2,1} + A^{2,-1}) - \frac{i}{2}(A^{1,1} - A^{1,-1}), \tag{B.22h}$$

$$A_{zz} = -\frac{1}{\sqrt{3}}A^{0,0} + \sqrt{\frac{2}{3}}A^{2,0}. \tag{B.22i}$$

The first-rank terms vanish if the Cartesian tensor is symmetric, as is the case for many physical properties, and in this limit the explicit relations become:

$$A_{xx} = -\frac{1}{\sqrt{3}}A^{0,0} - \frac{1}{\sqrt{6}}A^{2,0} + \frac{1}{2}(A^{2,2} + A^{2,-2}), \tag{B.23a}$$

$$A_{xy} = \frac{-i}{2}(A^{2,2} - A^{2,-2}), \tag{B.23b}$$

$$A_{xz} = \frac{1}{2}(A^{2,-1} - A^{2,1}), \tag{B.23c}$$

$$A_{yy} = -\frac{1}{\sqrt{3}}A^{0,0} - \frac{1}{\sqrt{6}}A^{2,0} - \frac{1}{2}(A^{2,2} + A^{2,-2}), \tag{B.23d}$$

$$A_{yz} = \frac{i}{2}(A^{2,1} + A^{2,-1}), \tag{B.23e}$$

$$A_{zz} = -\frac{1}{\sqrt{3}}A^{0,0} + \sqrt{\frac{2}{3}}A^{2,0}. \tag{B.23f}$$

As mentioned previously a vector is a Cartesian tensor of rank 1. Its Cartesian components can be written as the components of a rank 1 spherical tensor:

$$v^{1,0} = v_z, \tag{B.24a}$$

$$v^{1,\pm 1} = \mp\frac{1}{\sqrt{2}}(v_x \pm i v_y). \tag{B.24b}$$

The quantities $\mathscr{D}^L_{m,n}$, combinations of Cartesian rotation matrix elements, are the Wigner rotation matrices of rank L to be discussed in Appendix F.

B.5 Tensor Contraction

The direct product operation allows the generation of tensors of higher rank. On the contrary, the *contraction* or inner product operation generates a tensor of lower rank or even a scalar from two tensors by summing over the product of tensor components with the same index. For rank 1 tensors (vectors) this is the familiar scalar product (cf. Appendix A). Thus, if $\boldsymbol{v}$, $\boldsymbol{w}$ are tensors of rank 1, $\mathbf{V}$, $\mathbf{W}$ are tensors of rank 2 and $\mathcal{V}$, $\mathcal{W}$ tensors of rank 3 we can write the effect of the contraction as

$$\boldsymbol{v} \cdot \mathbf{w} = \sum_i v_i w_i^*, \tag{B.25a}$$

$$\mathbf{V{:}W} = \sum_{ij} V_{ij} W_{ij}^* = \mathrm{Tr}(\mathbf{V}\mathbf{W}^{T*}), \tag{B.25b}$$

$$\mathcal{V}\vdots\mathcal{W} = \sum_{ijk} \mathcal{V}_{ijk} \mathcal{W}_{ijk}^*, \tag{B.25c}$$

while partial contractions could be

$$\mathbf{V} \cdot \mathbf{v} \equiv \mathbf{V}\mathbf{v} = \sum_j V_{ij} v_j, \tag{B.26a}$$

$$\mathcal{V}{:}\mathbf{W} = \sum_{ij} \mathcal{V}_{ijk} W_{jk}^*. \tag{B.26b}$$

For vectors and matrices self-contractions are also written as

$$||\boldsymbol{v}|| = v^2 = |\boldsymbol{v}|^2 = \boldsymbol{v} \cdot \boldsymbol{v} = \sum_i v_i v_i^*, \tag{B.27a}$$

$$||\mathbf{V}|| = \mathbf{V{:}V} = \sum_{ij} V_{ij} V_{ij}^* = \mathrm{Tr}(\mathbf{V}\mathbf{V}^{T*}). \tag{B.27b}$$

For the dyadic product of two vectors whose components commute

$$||\boldsymbol{ab}|| = |\boldsymbol{ab}|^2 = \sum_{ij} a_i b_j a_i^* b_j^* = |\boldsymbol{a}|^2 |\boldsymbol{b}|^2 = \mathrm{Tr}([\boldsymbol{a} \otimes \boldsymbol{b}][\boldsymbol{a} \otimes \boldsymbol{b}]^{T*}). \tag{B.28}$$

More generally we could use the symbol $\odot$ for the total contraction operation of two tensors of any rank *n:*

$$\mathbf{V} \odot \mathbf{W} = \sum_{i_1 \ldots i_n} V_{i_1 \ldots i_n} W_{i_1 \ldots i_n}^*. \tag{B.29}$$

The Cartesian contraction operation in Eq. B.29 can be written in terms of spherical tensors as

$$\mathbf{V} \odot \mathbf{W} = \sum_{L,m} V^{L,m} W^{L,m*} = \sum_{L,m} (-1)^{L+m} V^{L,m} W^{L,-m}, \quad L \text{ integer}. \tag{B.30}$$

Appendix C

Taylor Series

For a scalar function $f(x)$ of a single real variable x which is continuous and differentiable n times in a point x_0, the value of f at a point $x + h_x$ near x can be written as a Taylor series:

$$f(x+h_x) = f(x) + \left(\frac{\mathrm{d}f}{\mathrm{d}x}\right)_x h_x + \frac{1}{2!}\left(\frac{\mathrm{d}^2 f}{\mathrm{d}x^2}\right)_x h_x^2 + \cdots + \frac{1}{n!}\left(\frac{\mathrm{d}^n f}{\mathrm{d}x^n}\right)_x h_x^n + \mathcal{O}(h_x^{n+1}), \tag{C.1}$$

where the derivatives are evaluated at the original point x and $\mathcal{O}(h_x^{n+1})$ indicates the order of magnitude of the error in truncating the series at the nth power of the small increment h_x. The aim here is to find a similar expansion for a scalar function $f(\boldsymbol{r})$ depending on a vector $\boldsymbol{r} \equiv (x, y, z)$, in order to find an approximate value of the function when the vector changes slightly, say from $\boldsymbol{r}$ to $\boldsymbol{r} + \boldsymbol{h}$. Taylor expanding each Cartesian component in turn, we have

$$\begin{aligned} f(\boldsymbol{r}+\boldsymbol{h}) &= f(x+h_x, y+h_y, z+h_z) \\ &= f(x,y,z) + \left(\frac{\partial f}{\partial x}\right)_{\boldsymbol{r}} h_x + \left(\frac{\partial f}{\partial y}\right)_{\boldsymbol{r}} h_y + \left(\frac{\partial f}{\partial z}\right)_{\boldsymbol{r}} h_z + \cdots, \end{aligned} \tag{C.2}$$

where terms like $\left(\frac{\partial f}{\partial x}\right)_{\boldsymbol{r}}$ are partial derivatives evaluated at the starting point. Using the gradient ∇ and introducing the displacement vector $|h\rangle = |x\rangle h_x + |y\rangle h_y + |z\rangle h_z$, we can rewrite Eq. C.2 as

$$f(\boldsymbol{r}+\boldsymbol{h}) = f(\boldsymbol{r}) + \langle \nabla f(\boldsymbol{r}) | h \rangle . \tag{C.3}$$

The second-order term in the expansion can be written in a compact way with the help of the matrix of second derivatives (*Hessian* matrix) constructed from the direct product of the gradient vector:

$$|\nabla\rangle\langle\nabla| f \equiv \nabla\nabla f = \begin{pmatrix} \frac{\partial^2 f}{\partial x^2} & \frac{\partial^2 f}{\partial x \partial y} & \frac{\partial^2 f}{\partial x \partial z} \\ \frac{\partial^2 f}{\partial y \partial x} & \frac{\partial^2 f}{\partial y^2} & \frac{\partial^2 f}{\partial y \partial z} \\ \frac{\partial^2 f}{\partial z \partial x} & \frac{\partial^2 f}{\partial z \partial y} & \frac{\partial^2 f}{\partial z^2} \end{pmatrix}. \tag{C.4}$$

Thus, we can write the Taylor expansion for the function $f(\boldsymbol{r})$ taken at a position $\boldsymbol{r}+\boldsymbol{h}$ differing from $\boldsymbol{r}$ by a small increment vector $\boldsymbol{h}$ as

$$f(\boldsymbol{r}+\boldsymbol{h}) = f(\boldsymbol{r}) + \langle\nabla f(\boldsymbol{r})|\boldsymbol{h}\rangle + \frac{1}{2}\langle\boldsymbol{h}|\nabla\rangle\langle\nabla f(\boldsymbol{r})|\boldsymbol{h}\rangle + \cdots, \tag{C.5a}$$

$$= f(\boldsymbol{r}) + \nabla f(\boldsymbol{r})\cdot\boldsymbol{h} + \frac{1}{2}\boldsymbol{h}\cdot\nabla\nabla f(\boldsymbol{r})\cdot\boldsymbol{h} + \cdots, \tag{C.5b}$$

$$= f(\boldsymbol{r}) + \nabla f(\boldsymbol{r})\cdot\boldsymbol{h} + \frac{1}{2}\nabla\nabla f(\boldsymbol{r}) : \boldsymbol{hh} + \cdots, \tag{C.5c}$$

where we have introduced the second-rank tensor $\boldsymbol{hh}$ (see Appendix B) and used the contraction operation indicated by the double dot (:). The last equation is useful because it hints that the terms of order (n) in the general expansion of the scalar $f(\boldsymbol{r}+\boldsymbol{h})$ can be generated by contracting the n-rank tensor constructed from the n direct power of the increment vector $\boldsymbol{h}$ with the n-rank tensor constructed from the direct product of the derivatives evaluated at the original point. Thus,

$$f(\boldsymbol{r}+\boldsymbol{h}) = f(\boldsymbol{r}) + \nabla f(\boldsymbol{r})\cdot\boldsymbol{h} + \frac{1}{2!}\nabla\nabla f(\boldsymbol{r}) : \boldsymbol{hh} + \frac{1}{3!}\nabla\nabla\nabla f(\boldsymbol{r}) \vdots\, \boldsymbol{hhh} + \cdots. \tag{C.6}$$

An important example of application of these formulas is the expansion of the function $f(\boldsymbol{r}+\boldsymbol{h}) = 1/|\boldsymbol{r}+\boldsymbol{h}|$ appearing in the Coulomb interactions:

$$\frac{1}{|\boldsymbol{r}+\boldsymbol{h}|} = \frac{1}{r} + \nabla\frac{1}{r}\cdot\boldsymbol{h} + \frac{1}{2!}\nabla\nabla\frac{1}{r} : \boldsymbol{hh} + \frac{1}{3!}\nabla\nabla\nabla\frac{1}{r} \vdots\, \boldsymbol{hhh} + \cdots + \frac{1}{n!}\mathbf{T}^{(n)}(\boldsymbol{r}) \odot \boldsymbol{h}^{\otimes n} + \cdots, \tag{C.7}$$

where $r \equiv |\boldsymbol{r}| = \sqrt{x^2+y^2+z^2}$ and we have indicated with $\odot$ the general contraction operation, consisting of generating a scalar by summing over all components in n dimensions. We have also used the direct power notation $\boldsymbol{h}^{\otimes n}$ for the direct product of n vectors $\boldsymbol{h}$. With this notation we can write

$$\frac{1}{|\boldsymbol{r}+\boldsymbol{h}|} = \sum_n \frac{1}{n!}\mathbf{T}^{(n)}(\mathbf{r}) \odot \boldsymbol{h}^{\otimes n}. \tag{C.8}$$

As we see, the calculation relies on the evaluation, that can be done once and for all, of the tensors $\mathbf{T}^{(n)}(\boldsymbol{r}) = \nabla^{\otimes(n)}(1/r)$. The first few are

$$\mathbf{T}^{(0)} = \frac{1}{r}, \tag{C.9a}$$

$$\mathbf{T}^{(1)}(\boldsymbol{r}) = \nabla\frac{1}{r} = -\frac{\boldsymbol{r}}{r^3}, \tag{C.9b}$$

$$\mathbf{T}^{(2)}(\boldsymbol{r}) = \nabla\nabla\frac{1}{r} = \frac{3(\boldsymbol{rr} - r^2\mathbf{1})}{r^5}, \tag{C.9c}$$

$$\mathbf{T}^{(3)}(\boldsymbol{r}) \equiv \nabla\nabla\nabla\frac{1}{r} = \frac{3}{r^7}[5\boldsymbol{rrr} - (r^2\boldsymbol{r}\mathbf{1} + \boldsymbol{r}r^2\mathbf{1} + \boldsymbol{r}\mathbf{1}r^2)], \tag{C.9d}$$

with a dependence on distance r going as $(1/r)^{n+1}$. The second-rank tensor $\mathbf{T}^{(2)}(\boldsymbol{r})$, also called the dipolar tensor, is particularly important and it is convenient to write it down explicitly in matrix form:

$$\mathbf{T}^{(2)}(\boldsymbol{r}) = -\frac{1}{r^5}\begin{pmatrix} 3x^2 - r^2 & 3xy & 3xz \\ 3xy & 3y^2 - r^2 & 3yz \\ 3xz & 3yz & 3z^2 - r^2 \end{pmatrix}, \qquad \text{(C.10)}$$

with components $T^{(2)}_{ab}(\boldsymbol{r}) = \nabla_a \nabla_b (1/r)$. Note that the dipolar tensor is traceless: $\mathrm{Tr}\left(\mathbf{T}^{(2)}(\boldsymbol{r})\right) = 0$. Thus, when we take the trace of the product of the tensor $\mathbf{T}^{(2)}(\boldsymbol{r})$ with another tensor $\mathbf{A}$, we can add an arbitrary constant to the diagonal of that tensor without changing the result:

$$\mathrm{Tr}\big(\mathbf{T}^{(2)}(\boldsymbol{r})\,(\mathbf{A} + \lambda \mathbf{1})\big) = \mathrm{Tr}\big(\mathbf{T}^{(2)}(\boldsymbol{r})\,\mathbf{A}\big). \qquad \text{(C.11)}$$

For example, the constant λ could be minus the trace of the matrix $\mathbf{A}$ itself to make the matrix a traceless one. We also give the explicit Cartesian components of $\mathbf{T}^{(3)}(\boldsymbol{r})$:

$$T^{(3)}_{abc}(\boldsymbol{r}) = \frac{3}{r^7}\big[5r_a r_b r_c - r^2(r_a\delta_{b,c} + r_b\delta_{a,c} + r_c\delta_{a,b})\big]. \qquad \text{(C.12)}$$

Appendix D

The Dirac Delta Function

Definition. A one-dimensional delta 'function' [Hoskins, 2009] can be defined as a generalized function or distribution [Strichartz, 1994] that is 0 everywhere except in an infinitesimal domain around 0, where it goes to infinity:

$$\delta(x) = \begin{cases} 0 & \text{for } x \neq 0 \\ \infty & \text{for } x = 0, \end{cases} \tag{D.1}$$

in such a way that it has a finite area underneath: $\int_{-\infty}^{\infty} \mathrm{d}x\, \delta(x) = 1$. For every well-behaved function $f(x)$ (also called *test function*) and x_0 real we have

$$\int_{-\infty}^{\infty} \mathrm{d}x\, \delta(x - x_0) f(x) = f(x_0), \tag{D.2}$$

and as a special case, $\int_{-\infty}^{\infty} \mathrm{d}x\, \delta(x) f(x) = f(0)$. Clearly, the range of integration does not need to be from $-\infty$ to ∞ but has just to include the singular point x_0, i.e. the location of the delta peak. The Dirac delta function can be written in terms of a sequence of functions $\delta_n(x - x_0)$ with x_0 a real number [Lighthill, 1958]:

$$\lim_{n\to\infty} \int_{-\infty}^{\infty} \mathrm{d}x\, \delta_n(x - x_0) f(x) = f(x_0). \tag{D.3}$$

Such a sequence, of which we can have very many examples, is called a *delta sequence* and we write, symbolically,

$$\lim_{n\to\infty} \delta_n(x - x_0) = \delta(x - x_0), \qquad x \in \mathbb{R}. \tag{D.4}$$

For instance, some useful delta sequences are:

$$\delta_n(x) = \begin{cases} 0, & x < -\frac{1}{2n} \\ n, & -\frac{1}{2n} < x < \frac{1}{2n} \\ 0, & x > \frac{1}{2n} \end{cases}, \tag{D.5a}$$

$$\delta_n(x) = \frac{n}{\sqrt{\pi}} \exp(-n^2 x^2), \tag{D.5b}$$

$$\delta_n(x - x_0) = \frac{1}{2\pi} \sum_{k=-n}^{n} \mathrm{e}^{ik(x - x_0)} = \frac{\sin\left[(n + \frac{1}{2})(x - x_0)\right]}{2\pi \sin\left[\frac{1}{2}(x - x_0)\right]}. \tag{D.5c}$$

General properties.

$$\delta(-x) = \delta(x), \tag{D.6a}$$

$$\delta(ax) = \frac{1}{|a|}\delta(x), \tag{D.6b}$$

$$x\,\delta(x) = 0, \tag{D.6c}$$

$$\delta(x^2 - x_0^2) = \frac{1}{2|x_0|}[\delta(x - x_0) + \delta(x + x_0)], \tag{D.6d}$$

$$\delta(x_1 - x_2) = \int_{-\infty}^{\infty} \mathrm{d}x\, \delta(x - x_1)\,\delta(x - x_2). \tag{D.6e}$$

For a continuous differentiable function $f(x)$ with n real roots at x_i, where $f(x_i) = 0$, and $|f'(x_i)| \neq 0$ for all x_i, we have, Taylor expanding around each root that $f(x) = f(x_i) + f'(x_i)(x - x_i) + \cdots$. Thus, using Eq. D.6b we have

$$\delta\left(f(x)\right) = \sum_{i=1}^{n} \frac{\delta(x - x_i)}{|f'(x_i)|}. \tag{D.7}$$

The first derivative of a delta function can be introduced, integrating by parts, as:

$$\int_{-\infty}^{\infty} \mathrm{d}x f(x)\delta'(x) = -f'(0), \tag{D.8}$$

and generalizing to the nth derivative of the delta: $\int_{-\infty}^{\infty} \mathrm{d}x f(x)\delta^{(n)}(x) = (-1)^n f^{(n)}(0)$.

Delta function in more than one dimension. The definition of a delta function can be readily generalized to more than one dimension:

- In 2D, for Cartesian coordinates:

$$\delta(\boldsymbol{r} - \boldsymbol{r}_0) = \delta(x - x_0)\,\delta(y - y_0), \tag{D.9}$$

  and using polar coordinates, $\delta(\boldsymbol{r} - \boldsymbol{r}_0) = \frac{1}{r}\delta(r - r_0))\delta(\phi - \phi_0)$.
- In 3D we have, similarly, $\delta(\boldsymbol{r} - \boldsymbol{r}_0) = \delta(x - x_0)\,\delta(y - y_0)\,\delta(z - z_0)$, and

$$\int_V \mathrm{d}x\mathrm{d}y\mathrm{d}z f(\boldsymbol{r})\,\delta(\boldsymbol{r} - \boldsymbol{r}_0) = \begin{cases} f(\boldsymbol{r}_0), & \text{if } \boldsymbol{r}_0 \text{ is inside V} \\ 0, & \text{if } \boldsymbol{r}_0 \text{ is outside V,} \end{cases} \tag{D.10}$$

  with $\int_V \mathrm{d}x\mathrm{d}y\mathrm{d}z\,\delta(\boldsymbol{r}) = 1$. Using polar coordinates,

$$\delta(\boldsymbol{r} - \boldsymbol{r}_0) = \frac{1}{r^2}\delta\left(r - r_0\right)\delta\left(\cos\theta - \cos\theta_0\right)\delta\left(\phi - \phi_0\right) \tag{D.11a}$$

$$= \frac{1}{r^2 \sin\theta}\delta(r - r_0)\delta(\theta - \theta_0)\delta(\phi - \phi_0). \tag{D.11b}$$

Analytic representations as limits. A few useful representations are:

- Delta function as the limit of certain oscillating functions, e.g.

$$\delta(x - x_0) = \lim_{t\to\infty} \frac{\sin[t(x - x_0)]}{\pi(x - x_0)}, \tag{D.12}$$

and

$$\delta(x-x_0) = \lim_{t\to\infty} \frac{\sin^2(t(x-x_0))}{\pi t(x-x_0)^2}. \tag{D.13}$$

- Delta function as a Gaussian of vanishing width σ in 1D or 3D

$$\delta(x-x_0) = \lim_{\sigma\to 0^+} \frac{1}{\sigma\sqrt{2\pi}} \mathrm{e}^{-(x-x_0)^2/(2\sigma^2)}, \tag{D.14}$$

$$\delta(\boldsymbol{r}-\boldsymbol{r}_0) = \lim_{\sigma\to 0^+} \left(\frac{1}{2\pi\sigma^2}\right)^{\frac{3}{2}} \mathrm{e}^{-|\boldsymbol{r}-\boldsymbol{r}_0|^2/(2\sigma^2)}. \tag{D.15}$$

- Delta function as a Lorentzian of vanishing width σ

$$\delta(x) = \frac{1}{\pi} \lim_{\sigma\to 0^+} \frac{\sigma}{x^2+\sigma^2}. \tag{D.16}$$

- Defining a *rectangular box* or *pulse function* with base d and height h:

$$\Pi_{d,h}(x) = \begin{cases} h, & \text{for } -\frac{d}{2} < x < +\frac{d}{2} \\ 0, & \text{for } +\frac{d}{2} < x < -\frac{d}{2} \end{cases}, \tag{D.17}$$

and taking $h = 1/d$, so that the area under the function is 1, we can write

$$\delta(x) = \lim_{d\to 0} \Pi_{d,\frac{1}{d}}(x)\,\mathrm{d}x. \tag{D.18}$$

Delta function as a derivative of the step or sign functions.

$$\delta(x-x_0) = \frac{\mathrm{d}}{\mathrm{d}x} H(x-x_0), \tag{D.19}$$

where the step or Heaviside function $H(x)$ (also called theta function $\Theta(x)$), which is 0 when $x<0$, $x = 1/2$ when $x = 0$ and 1 when $x > 0$ [Abramowitz and Stegun, 1965], or

$$H(x) = \frac{1}{2}\left[1 + \frac{x}{|x|}\right]. \tag{D.20}$$

$H(x)$ is then the primitive of a delta function. A useful step-like function is the *sign function* $\mathrm{sgn}(x) = -1$ for $x < 0$, 0 for $x = 0$, 1 for $x > 0$, or $\mathrm{sgn}(x) = H(x) - H(-x)$. In terms of this,

$$\delta(x) = \frac{1}{2}\frac{\mathrm{d}}{\mathrm{d}x}\mathrm{sgn}(x). \tag{D.21}$$

Integral representations. The Fourier form is particularly useful

$$\delta(x-x_0) = \frac{1}{2\pi}\int_{-\infty}^{\infty} \mathrm{d}q\ \mathrm{e}^{iq(x-x_0)}. \tag{D.22}$$

$$\delta(\boldsymbol{r}-\boldsymbol{r}_0) = \frac{1}{(2\pi)^3}\int_{-\infty}^{\infty} \mathrm{d}\boldsymbol{q}\ \mathrm{e}^{i\boldsymbol{q}\cdot(\boldsymbol{r}-\boldsymbol{r}_0)}. \tag{D.23}$$

Orthogonal expansions. Given the complete orthonormal basis set of functions $\{\psi_i(x)\}$ over the variable x in the interval $a \le x \le b$, we have

$$\delta_{m,n} = \int_a^b \mathrm{d}x\, \psi_m^*(x)\, \psi_n(x) = \langle m|n\rangle, \tag{D.24}$$

which can be considered an expansion of the Kronecker delta $\delta_{m,n}$, with m, n discrete, integer numbers. We can write $\langle x|\psi_n\rangle = \psi_n(x)$ as an integral mapping from the continuous variable x to the discrete one n [Dennery and Krzywicki, 1969] using Eq. A.34. Then using an alternative, more suggestive, notation for the Kronecker delta and inserting the representation of the identity in Eq. A.32, we have

$$\delta(m-n) \equiv \delta_{m,n} = \langle\psi_m|\psi_n\rangle = \int_a^b \mathrm{d}x \langle\psi_m|x\rangle\langle x|\psi_n\rangle. \tag{D.25}$$

Similarly, inserting the representation of the identity Eq. A.34

$$\delta(x-x') = \langle x|x'\rangle = \sum_n \langle x|\psi_n\rangle\langle\psi_n|x'\rangle = \sum_n \psi_n^*(x)\,\psi_n(x'). \tag{D.26}$$

For instance, referring to the respective spaces where the functions are orthogonal (cf. Appendix A) we can express the delta function in terms of:

- Fourier complex exponentials (see Eq. A.46)

$$\delta\left(x-x'\right) = \frac{1}{2\pi}\sum_{n=-\infty}^{\infty} \mathrm{e}^{in(x-x')}. \tag{D.27}$$

- Fourier cosines (see Eq. A.44)

$$\delta(x) = \frac{1}{2\pi} + \frac{1}{\pi}\sum_{n=1}^{\infty} \cos nx, \qquad -\pi \le x \le \pi. \tag{D.28}$$

- Legendre polynomials (see Eq. A.50)

$$\delta\left(x-x'\right) = \sum_{L=0}^{\infty} \frac{2L+1}{2} P_L(x)\, P_L\left(x'\right). \tag{D.29}$$

- Spherical harmonics (see Eq. A.55)

$$\delta\left(\cos\theta_1 - \cos\theta_2\right)\,\delta\left(\phi_1-\phi_2\right) = \sum_{L=0}^{\infty}\sum_{m=-L}^{L} Y_{L,m}\left(\theta_1,\phi_1\right) Y_{L,m}^*(\theta_2,\phi_2). \tag{D.30}$$

- A generic orthogonal basis set. A representation of the delta function in a generic complete orthogonal basis set $\psi_{\boldsymbol{\lambda}}(\boldsymbol{\Phi})$ for the variables $\boldsymbol{\Phi} = (\phi_1,\phi_2,\ldots,\phi_n)$ labelled by a set of parameters $\boldsymbol{\lambda} = (\lambda_1,\lambda_2,\ldots,\lambda_m)$,

$$\delta(\boldsymbol{\Phi}-\boldsymbol{\Phi}') = \sum_{\boldsymbol{\lambda}} \frac{1}{k_{\boldsymbol{\lambda}}}\psi_{\boldsymbol{\lambda}}(\boldsymbol{\Phi})\psi_{\boldsymbol{\lambda}}^*(\boldsymbol{\Phi}'), \tag{D.31}$$

  and

$$\int_V \mathrm{d}\boldsymbol{\Phi}\, \psi_{\boldsymbol{\lambda}}(\boldsymbol{\Phi})\psi_{\boldsymbol{\lambda}'}^*(\boldsymbol{\Phi}) = k_{\boldsymbol{\lambda}}\delta(\boldsymbol{\lambda}'-\boldsymbol{\lambda}). \tag{D.32}$$

Appendix E

Fourier Series and Transforms

E.1 Fourier Series

We give here some of the basic equations and properties of Fourier series and Fourier transforms [Arfken and Weber, 1995; Bracewell, 2000], focussing on those most used in the main text. Let us consider a function $f(x)$ defined for real arguments x, $x \in \mathbb{R}^1$, which is either periodic or defined over a finite interval $-d/2 \leq x \leq d/2$ endowed with periodic boundary conditions, so that $f(x+d) = f(x)$ and d becomes the *period*. Let us also assume that $f(x)$ is *piecewise regular*: i.e. single-valued and continuous, except possibly at a finite number of jump discontinuities, and that it has only a finite number of maxima and minima. Then $f(x)$ can be represented by a complex Fourier series:

$$f(x) = \sum_{n=-\infty}^{\infty} f_n \, \mathrm{e}^{inqx}, \tag{E.1}$$

where n is an integer and qx is dimensionless. In our example, two common occurring cases are that x is a length and $q \equiv 2\pi/d$, or that x is a time t and $q = 2\pi/\tau = \omega$ an angular frequency, and $d = \tau$ is now the time period. Multiplying both sides for $\exp(-inqx)$, integrating over x and using the orthogonality integral:

$$\int_c^{c+d} \mathrm{d}x \, \mathrm{e}^{-imx} \, \mathrm{e}^{inx} = d \, \delta_{m,n}, \tag{E.2}$$

where, e.g. we can have $d = 2\pi$, we find the expansion coefficients as

$$f_n = \frac{1}{d} \int_c^{c+d} \mathrm{d}x f(x) \, \mathrm{e}^{-inqx}, \tag{E.3}$$

where c is a real constant (common choices are $c = -d/2$ or $c = 0$). The series can also be written in real form, using the Euler identity $\mathrm{e}^{ix} = \cos x + i \sin x$. Then

$$f(x) = a_0 + \sum_{n=1}^{\infty} a_n \cos(nqx) + \sum_{n=1}^{\infty} b_n \sin(nqx), \tag{E.4}$$

where $a_0 = \frac{1}{d} \int_c^{c+d} \mathrm{d}x f(x)$, $a_n = \frac{2}{d} \int_c^{c+d} \mathrm{d}x f(x) \cos(nqx)$ and $b_n = \frac{1}{d} \int_c^{c+d} \mathrm{d}x f(x) \sin(nqx)$. The equivalence between the complex and real form is established by $f_0 = a_0$, $f_n = \frac{1}{2}(a_n - ib_n)$ and $f_{-n} = \frac{1}{2}(a_n + ib_n)$. As an explicit application, we have used Fourier series in Chapter 3 to expand $P(z)$, the distribution of molecular positions in a smectic with layer spacing ℓ_z using the basis set $\left\{\cos\frac{n2\pi z}{\ell_z}\right\}$, for $0 \leq z \leq \ell_z$ and integer n. The orthogonality relation in Eq. A.44 becomes with a change of variable

$$\int_0^d \mathrm{d}x \cos(m2\pi z/\ell_z)\,\cos(n2\pi z/\ell_z) = \ell_z(\delta_{m,0}\delta_{n,0} + \delta_{m,n})/2. \tag{E.5}$$

We can now examine a few useful Fourier expansions.

- **Rectangular wave** of width w, height h with period d (*box wave function*) [Menzel, 1960]

$$f(x) = \begin{cases} h, & \text{for } c < x < c \leq c + w, \\ 0, & \text{for } c + w < x < c + d, \\ f(x+d) \end{cases} \tag{E.6a}$$

$$= \frac{hw}{d} + \frac{2h}{\pi}\sum_{n=1}^{\infty}\frac{1}{n}\sin\frac{n\pi w}{d}\cos\frac{2n\pi}{d}\left(x - c - \frac{w}{2}\right). \tag{E.6b}$$

- **Triangular wave** of height h, defined in the interval $-d \leq x \leq d$:

$$f(x) = \begin{cases} h + hx/d, & \text{for } -d \leq x < 0, \\ h - hx/d, & \text{for } 0 \leq x \leq d, \end{cases} \tag{E.7a}$$

$$= \frac{h}{2} + h\sum_{\substack{n=-\infty \\ n\neq 0}}^{N}\frac{1-\cos(n\pi)}{n^2\pi^2}\,\mathrm{e}^{in\pi/d}. \tag{E.7b}$$

- **Gaussian** function (non-normalized) $g(x)$,

$$g(x) = \mathrm{e}^{-x^2/\sigma^2} \tag{E.8}$$

truncated to $-d \leq x \leq d$. The coefficients of its Fourier series can be written in terms of the erf function [Abramowitz and Stegun, 1965]

$$\int_{-\frac{d}{2}}^{\frac{d}{2}} \mathrm{d}x\, \mathrm{e}^{-x^2/\sigma^2}\cos\left(\frac{n2\pi x}{d}\right)$$
$$= \frac{\sqrt{\pi}\sigma}{2}\,\mathrm{e}^{-(\pi^2n^2\sigma^2)/d^2}\left(\mathrm{erf}\left(\frac{d}{2\sigma} - \frac{i\pi n\sigma}{d}\right) + \mathrm{erf}\left(\frac{d}{2\sigma} + \frac{i\pi n\sigma}{d}\right)\right). \tag{E.9}$$

A much simpler expression is obtained assuming d is large enough:

$$\int_{-\infty}^{\infty} \mathrm{d}x\, \mathrm{e}^{-x^2/\sigma^2}\cos\left(\frac{n2\pi x}{d}\right) = \sqrt{\pi}\sigma e^{-[n\pi\sigma/d]^2}, \tag{E.10}$$

which gives the Fourier series expansion as:

$$g(x) = \frac{\sqrt{\pi}\sigma}{d}\sum_{n=-\infty}^{\infty}\mathrm{e}^{-[n\pi\sigma/d]^2}\cos\left(\frac{n2\pi x}{d}\right). \tag{E.11}$$

Table E.1. *Some relevant real Fourier transforms*

$f(t)$	$\tilde{f}(\omega) = \sqrt{\frac{2}{\pi}} \int_0^\infty \mathrm{d}t \cos(\omega t) f(t)$	$\omega > 0$
$\begin{cases} 1, & \text{for } 0 < t \leq a \\ 0, & \text{otherwise} \end{cases}$	$\sqrt{\frac{2}{\pi}} \frac{\sin(a\omega)}{\omega}$	
$\frac{1}{a^2+t^2}$	$\sqrt{\frac{\pi}{2}} \frac{\mathrm{e}^{-a\omega}}{a}$	$\mathrm{Re}(a) > 0$
e^{-at}	$\sqrt{\frac{2}{\pi}} \frac{a}{a^2+\omega^2}$	$\mathrm{Re}(a) > 0$
e^{-at^2}	$\sqrt{\frac{1}{2a}} \mathrm{e}^{-\omega^2/(4a)}$	$\mathrm{Re}(a) > 0$

E.2 Fourier Transforms

We can introduce the exponential Fourier transform [Abramowitz and Stegun, 1965; NIST, 2016][1]

$$\tilde{f}(q) = \mathcal{F}[f(x)] = \frac{1}{\sqrt{2\pi}} \int_{-\infty}^{\infty} \mathrm{d}x \exp(+iqx) f(x). \tag{E.12}$$

As before, the exponent qx has to be dimensionless, so if q and x represent physical quantities, their units must be inverse to each other. For instance, we could have time and frequency as conjugated spaces, with $q = \omega = 2\pi\nu$, ω the angular frequency and ν the frequency. Another common example is with x a linear coordinate and $q = 2\pi/\lambda$ a wavevector modulus. The Fourier transform operator $\mathcal{F}[\ldots]$ can be inverted:

$$f(x) = \mathcal{F}^{-1}[\tilde{f}(q)] = \frac{1}{\sqrt{2\pi}} \int_{-\infty}^{\infty} \mathrm{d}q \exp(-iqx) \tilde{f}(q) \tag{E.13}$$

or, more generally, in n dimensions:

$$\tilde{f}(\boldsymbol{q}) = \mathcal{F}[f(\boldsymbol{r})] = \frac{1}{(2\pi)^{n/2}} \int_{-\infty}^{\infty} \mathrm{d}\boldsymbol{r} \exp(+i\boldsymbol{q}\cdot\boldsymbol{r}) f(\boldsymbol{r}), \tag{E.14}$$

$$f(\boldsymbol{r}) = \mathcal{F}^{-1}[\tilde{f}(\boldsymbol{q})] = \frac{1}{(2\pi)^{n/2}} \int_{-\infty}^{\infty} \mathrm{d}\boldsymbol{r} \exp(-i\boldsymbol{q}\cdot\boldsymbol{r}) \tilde{f}(\boldsymbol{q}). \tag{E.15}$$

Clearly it is assumed that the integrals exist, and this is true for rapidly decreasing functions, i.e. functions belonging to the so-called Schwartz space, consisting of smooth functions whose derivatives (including the function itself) decay at infinity faster than any power [Reed and Simon, 1975]. It is often convenient to introduce also the Fourier cosine transform $\tilde{f}(\omega) = \sqrt{\frac{2}{\pi}} \int_0^\infty \mathrm{d}t \cos(\omega t) f(t)$, of which a few examples are shown in Table E.1.

[1] Note that, different to the common mathematical convention, we have chosen the plus sign in the phase factor to comply with the crystallographic convention [Morelhao, 2016].

E.2.1 Convolution Theorem

Given the correlation function (also called convolution integral or *faltung*) between the functions $f(t)$ and $g(t)$, i.e.

$$C_{fg}(t) = \int_{-\infty}^{\infty} \mathrm{d}t' f(t')g^*(t-t') = C_{gf}(t) \equiv f(t) * g(t), \tag{E.16}$$

where we have introduced the 'star' convolution operator. The convolution, as defined, is linear and commutative, but also associative and distributive: $f*(g*h) = (f*g)*h$ and $f*(g+h) = f*g + f*h$. The Fourier transform of C_{fg}, i.e. the spectral density j_{fg}, is the product of the Fourier transforms of $f(t)$ and $g(t)$:

$$j_{fg}(\omega) = \tilde{C}_{fg} = \int_{-\infty}^{\infty} \mathrm{d}t\,[f(t)*g(t)]\exp(+i\omega t) = \tilde{f}(\omega)\tilde{g}^*(\omega). \tag{E.17}$$

In fact,

$$\begin{aligned}&\int_{-\infty}^{\infty} \mathrm{d}t\,\mathrm{e}^{+\omega t}\int_{-\infty}^{\infty} \mathrm{d}t' f(t')g^*(t-t')\\ &= \left\{\int_{-\infty}^{\infty} \mathrm{d}t' f(t')\mathrm{e}^{+i\omega t'}\right\}\left\{\int_{-\infty}^{\infty} \mathrm{d}t\;\mathrm{e}^{+i\omega(t-t')}g^*(t-t')\right\} = \widetilde{f}(\omega)\widetilde{g}^*(\omega),\end{aligned} \tag{E.18}$$

and in the case of autocorrelations

$$j_{ff}(\omega) = \widetilde{f}(\omega)\widetilde{f}^*(\omega) = \int_{-\infty}^{\infty} \mathrm{d}t\,\mathrm{e}^{+i\omega t}\int_{-\infty}^{\infty} \mathrm{d}t' f(t')f^*(t-t'). \tag{E.19}$$

We can also write $C_{fg}(t) = \int_{-\infty}^{\infty} \mathrm{d}\omega\,\mathrm{e}^{+i\omega t}\widetilde{f}(\omega)\widetilde{g}^*(\omega) = \mathscr{F}^{-1}[\widetilde{f}(\omega)\widetilde{g}^*(\omega)]$. For a Gaussian the convolution can be performed analytically. The calculation of Fourier transforms can be very fast using so-called Fast Fourier transform (FFT). This method was proposed and used to calculate time correlations from the trajectories in time obtained in MD simulations [Futrelle and McGinty, 1971; Kestemont and VanCraen, 1976]. An important result is the Parseval formula

$$\int_{-\infty}^{\infty} \mathrm{d}\omega|\tilde{f}(\omega)|^2 = \int_{-\infty}^{\infty} \mathrm{d}t\,|f(t)|^2. \tag{E.20}$$

E.2.2 Laplace Transform

We also introduce the definition of a Laplace transform:

$$\mathscr{L}(f(t);s) = \int_0^{\infty} \mathrm{d}t\,\mathrm{e}^{-st} f(t), \tag{E.21}$$

where s can have a real and imaginary part. The special case with $s = i\omega$, i.e. purely imaginary:

$$\mathscr{L}(f(t);i\omega) = \int_0^{\infty} \mathrm{d}t\,\mathrm{e}^{-i\omega t} f(t), \tag{E.22}$$

is called a Fourier–Laplace transform, and is often useful in applications (see, e.g., Section 6.9).

Appendix F

Wigner Rotation Matrices and Angular Momentum

F.1 Wigner Matrices

The operator $\hat{\mathscr{D}}(\Omega) \equiv \hat{\mathscr{D}}(\alpha, \beta, \gamma)$ that performs a rotation of Euler angles $\Omega = (\alpha, \beta, \gamma)$ of a coordinate frame can be written, using the convention of Rose [1957], as a sequence of three rotations: the first of an angle α around the laboratory Z-axis, the second of an angle β around the new y'-axis and the third of an angle γ around the resulting, body-fixed, z''-axis:

$$\hat{\mathscr{D}}(\alpha, \beta, \gamma) = \mathrm{e}^{-i\gamma \hat{\mathbf{J}}_{z''}} \, \mathrm{e}^{-i\beta \hat{\mathbf{J}}_{y'}} \, \mathrm{e}^{-i\alpha \hat{\mathbf{J}}_{Z}}, \tag{F.1}$$

where $\hat{\mathbf{J}}_Z$, $\hat{\mathbf{J}}_{y'}$, $\hat{\mathbf{J}}_{z''}$ are angular momentum operators. Using the commutation properties: $\left[\hat{\mathbf{J}}_a, \hat{\mathbf{J}}_b\right] = i\hbar \sum_c \varepsilon_{abc} \hat{\mathbf{J}}_c$, where ε_{abc} is the Levi-Civita permutation symbol (Eq. A.9), it is possible to rewrite this product of operators in a common laboratory frame, with axis X, Y, Z as:

$$\hat{\mathscr{D}}(\alpha, \beta, \gamma) = \mathrm{e}^{-i\alpha \hat{\mathbf{J}}_Z} \, \mathrm{e}^{-i\beta \hat{\mathbf{J}}_Y} \, \mathrm{e}^{-i\gamma \hat{\mathbf{J}}_Z}. \tag{F.2}$$

The matrix elements $\mathscr{D}^L_{m,n}(\alpha, \beta, \gamma)$ of the rotation operator in a basis $|Lm\rangle$, where the angular momentum $\hat{\mathbf{J}}^2$ and its projection, $\hat{\mathbf{J}}_Z$, are diagonal, are

$$\mathscr{D}^L_{m,n}(\alpha, \beta, \gamma) = \langle Lm \mid \mathrm{e}^{-i\alpha \hat{\mathbf{J}}_Z} \, \mathrm{e}^{-i\beta \hat{\mathbf{J}}_Y} \, \mathrm{e}^{-i\gamma \hat{\mathbf{J}}_Z} \mid Ln \rangle \tag{F.3}$$

and are called Wigner rotation matrices [Wigner, 1959] or Wigner functions or generalized spherical harmonics of rank L. The Euler angles (α, β, γ) determine the rotations which carry the original ('laboratory') coordinate system into the rotated ('molecular') one. From Eq. F.3 we can express $\mathscr{D}^L_{m,n}(\alpha, \beta, \gamma)$ as

$$\mathscr{D}^L_{m,n}(\alpha, \beta, \gamma) = \mathrm{e}^{-im\alpha} d^L_{m,n}(\beta) \, \mathrm{e}^{-in\gamma}, \tag{F.4}$$

where the real quantities,

$$d^L_{m,n}(\beta) \equiv \langle Lm \mid \mathrm{e}^{-i\beta \hat{\mathbf{J}}_Y} \mid Ln \rangle, \tag{F.5}$$

are called reduced or small Wigner matrices and will be discussed and tabulated at the end of this appendix. The functions $\mathscr{D}^L_{m,n}(\alpha, \beta, \gamma)$ constitute a complete orthogonal set spanning the space of the Euler angles α, β, γ. Some properties of Wigner matrices are:

(i) **Complex conjugation.**

$$\mathscr{D}^{L*}_{m,n}(\alpha,\beta,\gamma) = (-1)^{m-n}\mathscr{D}^{L}_{-m,-n}(\alpha,\beta,\gamma). \tag{F.6}$$

(ii) **Inverse and unitarity.** The Wigner rotation matrices are unitary, i.e. their inverse corresponds to the complex conjugate of their transpose matrix. Thus, $\mathscr{D}^{-1} = \mathscr{D}^{T*}$ and for the matrix elements, i.e. the individual functions,

$$\mathscr{D}^{L*}_{m,n}(\alpha,\beta,\gamma) = (-1)^{m-n}\mathscr{D}^{L}_{-m,-n}(\alpha,\beta,\gamma) = \mathscr{D}^{L}_{n,m}(-\gamma,\,-\beta,\,-\alpha). \tag{F.7}$$

Taking matrix elements of the operator relations $\hat{\mathscr{D}}(\Omega)\hat{\mathscr{D}}^{-1}(\Omega) = \hat{\mathscr{D}}^{-1}(\Omega)\hat{\mathscr{D}}(\Omega) = \hat{\mathbf{1}}$ in the angular momentum basis, we have:

$$\sum_{n=-L}^{L} \mathscr{D}^{L}_{m,n}(\Omega)\mathscr{D}^{L*}_{m',n}(\Omega) = \delta_{m,m'}. \tag{F.8}$$

(iii) **Orthogonality.**

$$\int \mathrm{d}\alpha \mathrm{d}\beta \sin\beta \mathrm{d}\gamma\, \mathscr{D}^{L*}_{m,n}(\alpha,\beta,\gamma)\mathscr{D}^{L'}_{m',n'}(\alpha,\beta,\gamma) = \frac{8\pi^2}{(2L+1)}\delta_{m,m'}\delta_{n,n'}\delta_{L,L'}. \tag{F.9}$$

(iv) **Completeness and delta function representation.** Wigner rotation matrices constitute a complete basis set in the space of Euler angles, and they can be used (see Eq. D.31) to write the angular delta function in that space as

$$\delta(\Omega - \Omega') = \sum_{L=0}^{\infty}\sum_{m=-L}^{L}\sum_{n=-L}^{L} \frac{(2L+1)}{8\pi^2}\mathscr{D}^{L}_{m,n}(\Omega)\mathscr{D}^{L^*}_{m,n}(\Omega'), \tag{F.10}$$

with $\Omega = (\alpha,\beta,\gamma)$, $0 \le \alpha \le 2\pi$, $0 \le \beta \le \pi$, $0 \le \gamma \le 2\pi$ and $\mathrm{d}\Omega = \mathrm{d}\alpha\mathrm{d}\beta\sin\beta\mathrm{d}\gamma$.

(v) **Special cases** Wigner matrices reduce to spherical harmonics $Y_{L,m}$ or to Legendre polynomials P_L when one or two of their subscripts are 0,

$$\mathscr{D}^{L}_{m,0}(\alpha,\beta,\gamma) = \left(\frac{4\pi}{2L+1}\right)^{1/2} Y^{*}_{L,m}(\beta,\alpha) \equiv C^{*}_{Lm}(\beta,\alpha), \tag{F.11}$$

$$\mathscr{D}^{L}_{0,0}(\alpha,\beta,\gamma) = d^{L}_{0,0}(\beta) = P_L(\cos\beta), \tag{F.12}$$

where we have also introduced the definition of the modified spherical harmonics used, e.g. by Brink and Satchler [1968], Stone [1996] and Luckhurst [2015]. Thus, the properties of spherical harmonics and Legendre polynomials can be obtained as particular cases of the ones for Wigner matrices.

(vi) **Closure.** Rotation matrices form a group, so that a product of two rotation matrices is another rotation matrix (closure). Alternatively, a rotation matrix of a set of Euler angles $(\Omega_{\mathrm{CA}}) \equiv (\alpha,\beta,\gamma)$ from frame A to C, i.e. $\mathbf{D}(\Omega_{\mathrm{CA}})$, can always be written as a rotation of $(\Omega_{\mathrm{BA}}) \equiv (\alpha_1,\beta_1,\gamma_1)$ from A to an intermediate frame B followed by a rotation of $(\Omega_{\mathrm{CB}}) \equiv (\alpha_2,\beta_2,\gamma_2)$ from this frame to C. Thus,

$$\mathbf{D}(\Omega_{\mathrm{CA}}) = \mathbf{D}(\Omega_{\mathrm{BA}})\mathbf{D}(\Omega_{\mathrm{CB}}). \tag{F.13}$$

Note that rotations are applied in reverse order (i.e. the first on the left) and that the order in which they are taken is important since rotations in three dimensions, in contrast to rotations in a plane, do not commute. Taking matrix elements of Eq. F.13 we have that the product of two successive rotations of Euler angles $(\alpha_1, \beta_1, \gamma_1)$ and $(\alpha_2, \beta_2, \gamma_2)$ to give (α, β, γ) is

$$\mathcal{D}^L_{m,n}(\alpha, \beta, \gamma) = \sum_{q=-L}^{L} \mathcal{D}^L_{m,q}(\alpha_1, \beta_1, \gamma_1)\mathcal{D}^L_{q,n}(\alpha_2, \beta_2, \gamma_2). \quad \text{(F.14)}$$

We can use this closure property to obtain the very useful *spherical harmonics addition theorem*. Quite often we need to rewrite a Legendre polynomial depending on the relative orientation between two axes, 1 and 2: $P_L(\cos\beta_{12})$ in terms of orientations of 1 and 2 with respect to a certain coordinate frame (e.g. a laboratory frame L). If we consider two vectors $\boldsymbol{u}_1$, $\boldsymbol{u}_2$ as the z-axis of two frames 1 and 2, then we have

$$P_L(\boldsymbol{u}_1 \cdot \boldsymbol{u}_2) = P_L(\cos\beta_{21}) \equiv \mathcal{D}^L_{0,0}(\Omega_{21}) = \sum_{q=-L}^{L} \mathcal{D}^{L*}_{q,0}(\Omega_{1L})\mathcal{D}^L_{q,0}(\Omega_{2\mathrm{L}}), \quad \text{(F.15)}$$

where we have employed the unitarity of the Wigner rotation matrices introduced earlier. Two important special cases are the decomposition of a first- and second-rank Legendre polynomial in terms of trigonometric functions as

$$P_1(\cos\beta_{21}) = \cos\beta_{21} = \cos\beta_1\cos\beta_2 + \sin\beta_1\sin\beta_2\cos(\alpha_2 - \alpha_1) \quad \text{(F.16a)}$$

$$P_2(\cos\beta_{21}) = P_2(\cos\beta_1)P_2(\cos\beta_2) + 3\sin\beta_1\cos\beta_1\sin\beta_2\cos\beta_2\cos(\alpha_1 - \alpha_2),$$
$$+\frac{3}{4}\sin^2\beta_1\sin^2\beta_2\cos(2\alpha_1 - 2\alpha_2), \quad \text{(F.16b)}$$

which can be used, e.g. to write functions of the relative orientation between the axis of two molecules 1 and 2 in terms of the orientations of each of the two molecules with respect to a common laboratory frame.

F.1.1 Clebsch–Gordan Coefficients and Wigner Matrices Coupling

The product of two Wigner matrices of the same argument can be written as

$$\mathcal{D}^{L'}_{m',n'}\mathcal{D}^{L''}_{m'',n''} = \sum_{L=|L''-L'|}^{L'+L''} C(L', L'', L; m', m'')C(L', L'', L; n', n'')\mathcal{D}^L_{m'+m'',n'+n''}, \quad \text{(F.17)}$$

where $C(a, b, c; d, e)$ are *Clebsch–Gordan coefficients* [Rose, 1957]. These coefficients, that play an essential role in a variety of problems involving addition of angular momenta and tensor manipulation, can be defined as the coupling coefficients arising when we combine two states with angular momentum L' and L'' to yield a state $|Lm\rangle$. Writing the state $|Lm\rangle$ in the basis set $|L'm'\rangle|L''m''\rangle$ by inserting the identity resolved in the same product basis (cf. Appendix A)

$$|Lm\rangle = \sum_{m',m''} \left[|L'm'\rangle|L''m''\rangle\langle L'm'|\langle L''m''|\right] |Lm\rangle, \tag{F.18a}$$

$$\equiv \sum_{m',m''} C(L',L'',L;m',m'',m)\,|L'm'\rangle|L''m''\rangle. \tag{F.18b}$$

The matrix element $C(L',L'',L;m',m'',m) \equiv [\langle L'm'|\langle L''m''|)]|Lm\rangle$ is a Clebsch–Gordan coefficient and L',L'',L can take non-negative integer or semi-integer values. Since $m = m' + m''$, the notation $C(L',L'',L;m',m'') \equiv C(L',L'',L;m',m'',m)$ is normally used for conciseness. The angular momenta L', L'',L have to form a triangle $\Delta(L'L''L)$, in the sense that allowed values obey

$$\Delta\left(L',L'',L\right) : \begin{cases} L' + L'' - L \geqq 0 \\ L' - L'' + L \geqq 0 \\ -L' + L'' + L \geqq 0, \end{cases} \tag{F.19}$$

where $(L' + L'' + L)$ is an integer. Coefficients formed with combinations of angular momenta not satisfying this triangular relation equal 0. The Clebsch–Gordan coefficients can be calculated explicitly from closed expressions due to Wigner and Racah [Rose, 1957]. A tabulation, together with symmetry relations and a comparison of the many different notations available, is given in Pasini and Zannoni [1984b]. Routines for their calculations are now available in the major computer algebra languages (e.g. Maple© and Mathematica©). Recalling that $\mathscr{D}^L_{0,0}(0\beta 0) = P_L(\cos\beta)$ we immediately find from Eq. F.17 the coupling relation for Legendre polynomials of the same argument:

$$P_{L'}(\cos\beta)\,P_{L''}(\cos\beta) = \sum_{L=|L'-L''|}^{L'+L''} C(L',L'',L;0,0)^2\,P_L(\cos\beta). \tag{F.20}$$

Note that the coupling of even rank polynomials only yields even rank P_L since the Clebsch–Gordan coefficient $C(L',L'',L;0,0)$ is 0 unless $(L' + L'' + L)$ is even. An equivalent expression can be obtained in terms of the so-called Wigner 3-j symbol [Brink and Satchler, 1968]

$$\begin{pmatrix} L' & L'' & L \\ m' & m'' & -m \end{pmatrix} \equiv \frac{(-1)^{L'-L''+m}}{\sqrt{2L+1}} C(L',L'',L;m',m'')\,\delta_{-m,m_1+m_2}. \tag{F.21}$$

The 3-j is invariant under cyclic permutations of the columns, and is multiplied by a phase factor $(-1)^{L'+L''+L}$ under non-cyclic ones:

$$\begin{pmatrix} L' & L'' & L \\ m' & m'' & m \end{pmatrix} = \begin{pmatrix} L'' & L & L' \\ m'' & m & m' \end{pmatrix} = \begin{pmatrix} L & L' & L'' \\ m & m' & m'' \end{pmatrix},$$
$$= (-1)^{L'+L''+L} \begin{pmatrix} L' & L & L'' \\ m' & m & m'' \end{pmatrix} = (-1)^{L'+L''+L} \begin{pmatrix} L' & L'' & L \\ -m' & -m'' & -m \end{pmatrix}. \tag{F.22}$$

It is also convenient to introduce the *Racah coefficient* that appears when coupling three angular momenta or, for our purposes, in the simplification of sums of products of Clebsch–Gordan coefficients

$$W(L_1,L_2,L_3,L_4;L_5,L_6) = [(2L_5+1)(2L_6+1)]^{-1/2} \times \sum_{m_1,m_2} C(L_1,L_2,L_5;m_1,m_2)\,C(L_5,L_4,L_3;m_1+m_2,0) \times C(L_2,L_4,L_6;m_2,0)\,C(L_1,L_6,L_3;m_1,m_2). \tag{F.23}$$

We list some useful Clebsch–Gordan coefficients relations.

Orthogonality.

$$\sum_{m'} C(L',L'',L;m',m'')\,C(L',L'',L;m',m'') = \delta_{L,L'}, \tag{F.24}$$

$$\sum_{L} C(L',L'',L;m',m'')\,C(L',L'',L;n',n'') = \delta_{m',n'}\,\delta_{m'',n''}. \tag{F.25}$$

Symmetries.

$$C(L',L'',L;m',m'') = (-1)^{L'+L''-L}C(L',L'',L;-m',-m''), \tag{F.26a}$$

$$= (-1)^{L'+L''-L}C(L'',L',L;m'',m'), \tag{F.26b}$$

$$= (-1)^{L'-m'}\sqrt{\frac{2L+1}{2L''+1}}C(L',L,L'';m',-m'-m''), \tag{F.26c}$$

$$= (-1)^{L''+m''}\sqrt{\frac{2L+1}{2L'+1}}C(L,L'',L';-m'-m'',m''). \tag{F.26d}$$

Explicit expressions for some special Clebsch–Gordan coefficients.

$$C(L,L',0;m,-m) = (-1)^{L-m}\delta_{L,L'}/\sqrt{2L+1}, \tag{F.27}$$

$$C(1,1,2;m,-m) = (1/2)^{|m|}\sqrt{(2/3)}, \tag{F.28}$$

$$C(2,2,2;m,-m) = (-1)^m C(2,2,2;0,m) = (-1)^m(m^2-2)/\sqrt{14}, \tag{F.29}$$

$$C(2,2,L;0,0) = (-12)^{\frac{L}{2}}\sqrt{(2L+1)(4-L)!}\;(\delta_{L,0}+\delta_{L,2}+\delta_{L,4})\,/\,\sqrt{(5+L)!}. \tag{F.30}$$

Clebsch–Gordan coefficients are also enter in other useful expressions:

Wigner matrices decomposition. A Wigner rotation matrix can be written as a linear combination of products of Wigner functions of lower rank.

$$\mathcal{D}^{L}_{m,n} = \sum C(L',L'',L;m',m'')\,C(L',L'',L;n',n'')\,\mathcal{D}^{L'}_{m',n'}\,\mathcal{D}^{L''}_{m'',n''}\delta_{m'+m'',m}\delta_{n'+n'',n},$$

$$= \sum C(L',L'',L;m',m-m')\,C(L',L'',L;n',n-n')\,\mathcal{D}^{L'}_{m',n'}\,\mathcal{D}^{L''}_{m-m',n-n'}, \tag{F.31}$$

where the sum extends to all indices not appearing on the left-hand side.

Integral of three Wigner rotation matrices. The integral, often called *Gaunt formula,* can be expressed in terms of Clebsch–Gordan coefficients:

$$\int \mathrm{d}\alpha\mathrm{d}\beta\sin\beta\mathrm{d}\gamma\,\mathcal{D}^{L''*}_{m'',n''}(\alpha,\beta,\gamma)\mathcal{D}^{L'}_{m',n'}(\alpha,\beta,\gamma)\mathcal{D}^{L}_{m,n}(\alpha,\beta,\gamma)$$

$$= \frac{8\pi^2\delta_{m+m',m''}\delta_{n+n',n''}}{(2L''+1)}\,C(L,L',L'';m,m')C(L,L',L'';n,n'). \tag{F.32}$$

F.2 Effect of Angular Momentum Operators on the Wigner Matrices

We recall the classical definition of angular momentum for a particle at position $\boldsymbol{r}$ and moving with a velocity $\boldsymbol{v}$ arising uniquely from its rotation [Goldstein, 1980]:

$$\boldsymbol{J} = m\boldsymbol{r} \times \boldsymbol{v} = m[\boldsymbol{r} \times \dot{\boldsymbol{\Omega}} \times \boldsymbol{r}] = m[\dot{\boldsymbol{\Omega}}\, r^2 - \boldsymbol{r}\, \dot{\Omega}_\alpha r_\alpha] = \boldsymbol{r} \times \boldsymbol{p} \tag{F.33}$$

and the classical quantum correspondence $\boldsymbol{p} = i\nabla$, in units of $\hbar = 1$,

$$\hat{\mathbf{J}} = -i(\boldsymbol{r} \times \nabla) = \begin{vmatrix} \boldsymbol{e}_1 & \boldsymbol{e}_2 & \boldsymbol{e}_3 \\ x & y & z \\ p_x & p_y & p_z \end{vmatrix}. \tag{F.34}$$

The Wigner matrices are eigenfunctions of the $\hat{\mathbf{J}}^2$angular momentum operator:

$$\hat{\mathbf{J}}^2 \mathscr{D}^L_{m,n} = L(L+1)\, \mathscr{D}^L_{m,n}, \tag{F.35}$$

as well as of the body-fixed angular momentum projection operator

$$\hat{\mathbf{J}}_z \mathscr{D}^L_{m,n} = -n\, \mathscr{D}^L_{m,n}, \tag{F.36}$$

and of the laboratory-fixed angular momentum projection operator

$$\hat{\mathbf{J}}_Z \mathscr{D}^L_{m,n} = -m\, \mathscr{D}^L_{m,n}. \tag{F.37}$$

The $\hat{\mathbf{J}}^2$ angular momentum operator is explicitly, in terms of Euler angles,

$$\hat{\mathbf{J}}^2 = -\frac{\partial^2}{\partial^2 \beta} - \cot\beta \frac{\partial}{\partial \beta} - \frac{1}{\sin^2\beta}\left(\frac{\partial^2}{\partial^2\alpha} + \frac{\partial^2}{\partial^2\gamma} - 2\cos\beta \frac{\partial^2}{\partial\alpha\partial\gamma}\right), \tag{F.38}$$

while the laboratory-fixed angular momentum component operators are

$$\hat{\mathbf{J}}_X = -i\left(-\cos\alpha \cot\beta \frac{\partial}{\partial\alpha} - \sin\alpha \frac{\partial}{\partial\beta} + \frac{\cos\alpha}{\sin\beta}\frac{\partial}{\partial\gamma}\right), \tag{F.39a}$$

$$\hat{\mathbf{J}}_Y = -i\left(-\sin\alpha \cot\beta \frac{\partial}{\partial\alpha} + \cos\alpha \frac{\partial}{\partial\beta} + \frac{\sin\alpha}{\sin\beta}\frac{\partial}{\partial\gamma}\right), \tag{F.39b}$$

$$\hat{\mathbf{J}}_Z = -i\frac{\partial}{\partial\alpha}. \tag{F.39c}$$

We can also define step up and down (or ladder) operators $\hat{\mathbf{J}}_+$, $\hat{\mathbf{J}}_-$ whose action on the Wigner rotation matrices is

$$\hat{\mathbf{J}}_+ \mathscr{D}^{L*}_{m,n} = (\hat{\mathbf{J}}_x + i\hat{\mathbf{J}}_y)\mathscr{D}^{L*}_{m,n} = \sqrt{(L-n)(L+n+1)}\, \mathscr{D}^{L*}_{m,n+1}, \tag{F.40}$$

$$\hat{\mathbf{J}}_- \mathscr{D}^{L*}_{m,n} = (\hat{\mathbf{J}}_x - i\hat{\mathbf{J}}_y)\mathscr{D}^{L*}_{m,n} = \sqrt{(L+n)(L-n+1)}\, \mathscr{D}^{L*}_{m,n-1}. \tag{F.41}$$

We now wish to give explicit expressions for the Wigner rotation matrices of rank $L = 0,1,2,4$ and thus implicitly for the most often used order parameters, which correspond to their orientational averages (see Chapter 3). From Eq. F.3 we see that what we really need are expressions for the small matrices $d^L_{m,n}(\beta)$. Some tabulations can be found in the literature [Brink and Satchler, 1968; Zannoni, 1979]. In general, the $d^L_{m,n}$ are real and explicit expressions that can be obtained from available formulas due to Wigner and Racah [Rose, 1957]. We also have a few useful relations:

$$d^L_{m,n}(-\beta) = (-1)^{m-n} d^L_{n,m}(\beta) = d^L_{-m,-n}(\beta), \tag{F.42a}$$
$$d^L_{m,n}(\beta) = d^L_{-n,-m}(\beta) = (-1)^{m-n} d^L_{-m,-n}(\beta), \tag{F.42b}$$
$$d^L_{m,n}(\beta) = (-1)^{L+m} d^L_{m,-n}(\pi - \beta). \tag{F.42c}$$

Using $d^L_{m,n}(0) = \delta_{m,n}$ (obvious from Eq. F.5) and Eq. F.42c we find

$$d^L_{m,n}(\pi) = (-1)^{L+m} \delta_{m,-n}, \tag{F.43a}$$
$$d^L_{m,n}(2\pi) = (-1)^{2L} \delta_{m,n}. \tag{F.43b}$$

Another useful relation is [St. Pierre and Steele, 1975]

$$d^L_{m,0}\left(\frac{\pi}{2}\right) = \frac{\cos[(L-m)(\pi/2)][(L-m)!\,(L+m)!\,]^{1/2}}{2^L[(L+m)/2]!\,[(L-m)/2]!}. \tag{F.44}$$

In particular, $d^L_{0,0}\left(\frac{\pi}{2}\right) = P_L(\cos\frac{\pi}{2}) = P_L(0) = [\cos(\pi L/2)(L)!\,]/[2^L((L/2)!\,)^2]$. We now report explicit expressions for the small Wigner matrices $d^L_{m,n}(\beta)$ of rank $L = 0, 1, 2, 4$, using the shorthand $c \equiv \cos\left(\frac{\beta}{2}\right)$ and $s \equiv \sin\left(\frac{\beta}{2}\right)$,

$L = 0$

$$d^0_{0,0} = 1. \tag{F.45}$$

$L = 1$

$$d^1_{1,1} = d^1_{-1,-1} = c^2, \tag{F.46a}$$
$$d^1_{1,0} = -d^1_{-1,0} = -d^1_{0,1} = d^1_{0,-1} = -\sqrt{2}sc, \tag{F.46b}$$
$$d^1_{1,-1} = d^1_{-1,1} = s^2, \tag{F.46c}$$
$$d^1_{0,0} = -1 + 2c^2. \tag{F.46d}$$

$L = 2$

$$d^2_{2,2} = d^2_{-2,-2} = c^4, \tag{F.47a}$$
$$d^2_{2,1} = -d^2_{-2,-1} = -d^2_{1,2} = d^2_{-1,-2} = -2c^3 s, \tag{F.47b}$$
$$d^2_{2,0} = d^2_{-2,0} = d^2_{0,2} = d^2_{0,-2} = \sqrt{6}c^2 - \sqrt{6}c^4, \tag{F.47c}$$
$$d^2_{2,-1} = -d^2_{-2,1} = -d^2_{-1,2} = d^2_{1,-2} = -2s^3 c, \tag{F.47d}$$
$$d^2_{2,-2} = d^2_{-2,2} = s^4, \tag{F.47e}$$
$$d^2_{1,1} = d^2_{-1,-1} = -3c^2 + 4c^4, \tag{F.47f}$$
$$d^2_{1,0} = -d^2_{-1,0} = -d^2_{0,1} = d^2_{0,-1} = \sqrt{6}cs - 2\sqrt{6}c^3 s, \tag{F.47g}$$
$$d^2_{1,-1} = d^2_{-1,1} = 3s^2 - 4s^4, \tag{F.47h}$$
$$d^2_{0,0} = 1 - 6c^2 + 6c^4. \tag{F.47i}$$

$L = 4$

$$d^4_{4,4} = d^4_{-4,-4} = c^8, \tag{F.48a}$$
$$d^4_{4,3} = -d^4_{-4,-3} = -d^4_{3,4} = d^4_{-3,-4} = -2\sqrt{2}c^7 s, \tag{F.48b}$$
$$d^4_{4,2} = d^4_{-4,-2} = d^4_{2,4} = d^4_{-2,-4} = 2\sqrt{7}c^6 s^2, \tag{F.48c}$$
$$d^4_{4,1} = -d^4_{-4,-1} = -d^4_{1,4} = d^4_{-1,-4} = -2\sqrt{14}c^5 s^3, \tag{F.48d}$$
$$d^4_{4,0} = d^4_{-4,0} = d^4_{0,4} = d^4_{0,-4} = \sqrt{70}c^4 s^4, \tag{F.48e}$$
$$d^4_{4,-1} = -d^4_{-4,1} = -d^4_{-1,4} = d^4_{1,-4} = -2\sqrt{14}c^3 s^5, \tag{F.48f}$$
$$d^4_{4,-2} = d^4_{-4,2} = d^4_{-2,4} = d^4_{2,-4} = 2\sqrt{7}c^2 s^6, \tag{F.48g}$$
$$d^4_{4,-3} = -d^4_{-4,3} = d^4_{3,-4} = -d^4_{-3,4} = -2\sqrt{2}cs^7, \tag{F.48h}$$
$$d^4_{4,-4} = d^4_{-4,4} = s^8, \tag{F.48i}$$
$$d^4_{3,3} = d^4_{-3,-3} = c^6(c^2 - 7s^2), \tag{F.48j}$$
$$d^4_{3,2} = -d^4_{-3,-2} = -d^4_{2,3} = d^4_{-2,-3} = -\sqrt{14}c^5 s(c^2 - 3s^2), \tag{F.48k}$$
$$d^4_{3,1} = d^4_{-3,-1} = d^4_{1,3} = d^4_{-1,-3} = \sqrt{7}c^4 s^2(3c^2 - 5s^2), \tag{F.48l}$$
$$d^4_{3,0} = -d^4_{-3,0} = -d^4_{0,3} = d^4_{0,-3} = -2\sqrt{35}c^3 s^3(c^2 - s^2), \tag{F.48m}$$
$$d^4_{3,-1} = d^4_{-3,1} = d^4_{-1,3} = d^4_{1,-3} = \sqrt{7}c^2 s^4(5c^2 - 3s^2), \tag{F.48n}$$
$$d^4_{3,-2} = -d^4_{-3,2} = -d^4_{-2,3} = d^4_{2,-3} = -\sqrt{14}cs^5(3c^2 - s^2), \tag{F.48o}$$
$$d^4_{3,-3} = d^4_{-3,3} = s^6(7c^2 - s^2), \tag{F.48p}$$
$$d^4_{2,2} = d^4_{-2,-2} = c^4(c^4 - 12c^2 s^2 + 15s^4), \tag{F.48q}$$
$$d^4_{2,1} = -d^4_{-2,-1} = -d^4_{1,2} = d^4_{-1,-2} = -\sqrt{2}c^3 s(3c^4 - 15c^2 s^2 + 10s^4), \tag{F.48r}$$
$$d^4_{2,0} = d^4_{-2,0} = d^4_{0,2} = d^4_{0,-2} = \sqrt{10}c^2 s^2(3c^4 - 8c^2 s^2 + 3s^4), \tag{F.48s}$$
$$d^4_{2,-1} = -d^4_{-2,1} = -d^4_{-1,2} = d^4_{1,-2} = -\sqrt{2}cs^3(10c^4 - 15c^2 s^2 + 3s^4), \tag{F.48t}$$
$$d^4_{2,-2} = d^4_{-2,2} = s^4(15c^4 - 12c^2 s^2 + s^4), \tag{F.48u}$$
$$d^4_{1,1} = d^4_{-1,-1} = c^2(c^6 - 15c^4 s^2 + 30c^2 s^4 - 10s^6), \tag{F.48v}$$
$$d^4_{1,0} = -d^4_{-1,0} = -d^4_{0,1} = d^4_{0,-1} = -2\sqrt{5}cs(c^6 - 6c^4 s^2 + 6c^2 s^4 - s^6), \tag{F.48w}$$
$$d^4_{1,-1} = d^4_{-1,1} = s^2(10c^6 - 30c^4 s^2 + 15c^2 s^4 - s^6), \tag{F.48x}$$
$$d^4_{0,0} = c^8 - 16c^6 s^2 + 36c^4 s^4 - 16c^2 s^6 + s^8. \tag{F.48y}$$

Appendix G

Molecular and Mesophase Symmetry

G.1 Symmetry and Order Parameters

Here we wish to provide some tools for simplification of single and pair distributions and, in particular, for the identification of the relevant order parameters in the presence of some symmetry of the mesophase and its constituent molecules [Zannoni, 1979c]. There are cases, like that of cylindrical symmetry, where this is quite intuitive. However, a more formal procedure is required for determining the independent order parameters. We shall define the symmetry group of a phase as the group of transformations of the laboratory system that leaves the single-particle distribution as well as the higher ones invariant. Similarly, we can define an effective symmetry for the molecule in terms of the group of molecular transformations leaving the single-particle distribution (see Chapter 3) unchanged. The same considerations hold, of course, for the purely orientational distribution $P(\Omega_{\text{ML}}) \equiv P(\Omega)$ which is of primary concern to us. In the language of group theory, we would say that $P(\Omega_{\text{ML}})$ belongs to the totally symmetric representation of the group of the molecule and of the mesophase. Therefore, one way of applying symmetry is to project the distribution onto the totally symmetric representation of these groups [Lax, 1974]. In practice, this symmetrization can be performed applying a projection operator $\hat{\mathscr{P}}$ that sums over all possible transformed functions generated by the action of the various symmetry operations $\hat{\mathscr{O}}_s$ [Zannoni, 1979c]. For a generic function f and a group of n_s discrete transformations

$$\hat{\mathscr{P}}[f] = \frac{1}{n_s} \sum_{s=1}^{n_s} \hat{\mathscr{O}}_s f. \tag{G.1}$$

For a continuous group like the rotation group in three dimensions, $SO(3)$, the projection operator is

$$\hat{\mathscr{P}}[f] = \frac{1}{\int \mathrm{d}\Omega} \int \mathrm{d}\Omega \, \hat{\mathscr{O}}[\Omega] f[\Omega]. \tag{G.2}$$

A convenient way for the symmetry simplification of order parameters starts from their definition (cf. Chapter 3):

$$\langle \mathscr{D}^L_{m,n} \rangle = \int \mathrm{d}\Omega_{\text{ML}} \, P(\Omega_{\text{ML}}) \mathscr{D}^L_{m,n}(\Omega_{\text{ML}}), \tag{G.3}$$

where $\Omega_{\rm ML} = (\alpha, \beta, \gamma)$ is the rotation angle from the laboratory to the molecular frame and $P(\Omega_{\rm ML})$ is the single-particle distribution studied in Chapter 3. A symmetry transformation $\hat{\mathcal{O}}$ does not alter $\int d\Omega_{\rm ML}$ and also leaves $P(\Omega_{\rm ML})$ invariant. Thus, the effect of the operator $\hat{\mathcal{O}}$ on the order parameters can be obtained from its action on the Wigner matrices. Since these are a representation of the rotation operator $\mathcal{D}^L(\Omega_{\rm ML})$ in an angular momentum basis set $\mid Ln\rangle$, i.e. $\mathcal{D}^L_{m,n}(\Omega_{\rm ML}) = \langle Lm \mid \hat{\mathcal{D}}(\Omega_{\rm ML}) \mid Ln\rangle$ (Appendix F), to investigate the effect of a molecular symmetry operation $\hat{\mathcal{O}}_{\rm M}$ we need to determine the matrix elements $\langle Lm \mid \hat{\mathcal{D}}(\Omega_{\rm ML})\hat{\mathcal{O}}_{\rm M} \mid Ln\rangle$. Note that successive rotations $\hat{\mathcal{D}}$ are applied in reverse order, that is from left to right (cf. Eq. F.13). Similarly, the application of a laboratory frame transformation, such as a symmetry operation of the mesophase $\hat{\mathcal{O}}_{\rm L}$, yields $\langle Lm \mid \hat{\mathcal{O}}_{\rm L}\hat{\mathcal{D}}(\Omega_{\rm ML}) \mid Ln\rangle$. Every symmetry operation for molecule or phase adds a relation that order parameters have to satisfy, i.e. indicating explicitly the orientational averages: $\langle \mathcal{D}^L_{m,n}\rangle_{\Omega_{\rm MM'}} = \big\langle \langle Lm \mid \hat{\mathcal{D}}(\Omega_{\rm ML})\hat{\mathcal{O}}_{\rm M} \mid Ln\rangle \big\rangle_{\Omega_{\rm MM'}}$ for the molecule and $\langle \mathcal{D}^L_{m,n}\rangle_{\Omega_{\rm LL'}} = \big\langle \langle Lm \mid \hat{\mathcal{O}}_{\rm L}\hat{\mathcal{D}}(\Omega_{\rm ML}) \mid Ln\rangle \big\rangle_{\Omega_{\rm LL'}}$ for the phase symmetries. Any point group operation can be written as a certain combination of rotation and inversion operations. Let us consider each of these in turn. Thus, supposing that the symmetry operation of the mesophase $\hat{\mathcal{O}}_{\rm L}$ is a rotation, then it will transform the original laboratory frame L into another frame L′ and give

$$\langle Lm \mid \hat{\mathcal{O}}_L\hat{\mathcal{D}}(\Omega_{\rm ML}) \mid Ln\rangle = \langle Lm \mid \hat{\mathcal{D}}(\Omega_{\rm L'L})\hat{\mathcal{D}}(\Omega_{\rm ML'}) \mid Ln\rangle, \tag{G.4a}$$

$$= \sum_{q=-L}^{L} \mathcal{D}^L_{m,q}(\Omega_{\rm L'L})\mathcal{D}^L_{q,n}(\Omega_{\rm ML'}). \tag{G.4b}$$

Quite analogously, if the molecule has a rotation $\hat{\mathcal{O}}_{\rm M}$ among its symmetry operations, we find

$$\langle Lm \mid \hat{\mathcal{D}}(\Omega_{\rm ML})\hat{\mathcal{O}}_{\rm M} \mid Ln\rangle = \langle Lm \mid \hat{\mathcal{D}}(\Omega_{\rm ML})\hat{\mathcal{D}}(\Omega_{\rm M'M}) \mid Ln\rangle, \tag{G.5a}$$

$$= \sum_{q=-L}^{L} \mathcal{D}^L_{m,q}(\Omega_{\rm ML})\mathcal{D}^L_{q,n}(\Omega_{\rm M'M}). \tag{G.5b}$$

A particularly interesting case is that of a $\hat{C}_k$ symmetry axis operation, corresponding to a rotation of $2\pi/k$ around the molecular z-axis:

$$\langle Lm \mid \hat{\mathcal{D}}(\Omega_{\rm ML})\hat{C}_k(z)_{\rm M} \mid Ln\rangle = \langle Lm \mid \hat{\mathcal{D}}(\Omega_{\rm ML})\hat{\mathcal{D}}(0,0,2\pi/k) \mid Ln\rangle, \tag{G.6a}$$

$$= \sum_{q=-L}^{L} \mathcal{D}^L_{m,q}(\Omega_{\rm ML})\, d_{q,n}(0)\, {\rm e}^{-in2\pi/k}, \tag{G.6b}$$

$$= \mathcal{D}^L_{m,n}(\Omega_{\rm ML})\, {\rm e}^{-in2\pi/k}, \tag{G.6c}$$

where we have used $d^L_{q,n}(0) = \delta_{q,n}$. Thus, if we consider order parameters of rank $L = 2$, as is most often the case in experiments (see Chapter 3), then $-2 \le n \le 2$ and the presence of a C_k-axis, for $k \geq 3$, requires $n = 0$ and is, in this respect, equivalent to that of a true cylindrical symmetry axis C_∞, yielding for the order parameter

$$\langle \mathscr{D}^2_{m,n} \rangle = \delta_{n,0} \langle \mathscr{D}^2_{m,0} \rangle \,. \tag{G.7}$$

Similarly, when looking at rank L properties, a C_k-axis, $k \geq L+1$ would be enough to yield effective uniaxial symmetry. It is worth noting that analogous considerations do not allow us to say, if we measure only second-rank tensor properties, that the director of a monodomain liquid crystal is truly an axis of cylindrical symmetry or just a C_k-axis, with $k \geq 3$, a fact already pointed out by Landau [1965]. If a symmetry operation $\hat{\mathscr{O}}$ contains the inversion (or *parity*) operation, $\hat{\mathscr{I}}$, which corresponds to a transformation $(x, y, z) \rightarrow (-x, -y, -z)$ or, in polar coordinates $(r, \beta, \alpha) \longrightarrow (r, \pi - \beta, \pi + \alpha)$, it is clear that to investigate its effect we also need the matrix elements $\langle Lm \mid \hat{\mathscr{I}} \mid Ln \rangle$. These are obtained recalling that $\hat{\mathscr{I}} \mid Lm \rangle = (-1)^L \mid Lm \rangle$, thus $\langle Lm \mid \hat{\mathscr{I}} \mid Ln \rangle = (-1)^L \delta_{m,n}$ and

$$\langle Lm \mid \hat{\mathscr{I}} \hat{\mathscr{D}}(\Omega_{\rm ML}) \mid Ln \rangle = \langle Lm \mid \hat{\mathscr{D}}(\Omega_{\rm ML}) \hat{\mathscr{I}} \mid Ln \rangle = (-1)^L \mathscr{D}^L_{m,n}(\Omega_{\rm ML}). \tag{G.8}$$

The inversion and rotation operations commute, then the inversion has the same effect in either a laboratory or a molecular frame, even if in a composite operation including inversion other operations have to be taken in the proper order. As a simple example, let us suppose that the mesophase has a plane of symmetry perpendicular to the z-axis, that is $\hat{\mathscr{O}}_{\rm L} = \hat{\sigma}_h = \hat{\sigma}(xy)_{\rm L}$. Since we can write $\hat{\sigma}(xy)_{\rm L} = \hat{\mathscr{I}} \hat{\mathscr{D}}(\pi 00)$ we find

$$\langle Lm \mid \hat{\sigma}(xy)_{\rm L} \hat{\mathscr{D}}(\Omega_{\rm ML}) \mid Ln \rangle$$

$$= \sum_{q,p=-L}^{L} \langle Lm \mid \hat{\mathscr{I}} \mid Lq \rangle \langle Lq \mid \hat{\mathscr{D}}(\pi 00) \mid Lp \rangle \langle Lp \mid \hat{\mathscr{D}}(\Omega_{\rm ML}) \mid Ln \rangle, \tag{G.9a}$$

$$= (-1)^L \delta_{m,q} \exp(-iq\pi) \delta_{q,p} \mathscr{D}^L_{p,n}(\Omega_{\rm ML}) = (-1)^{L+m} \mathscr{D}^L_{m,n}(\Omega_{\rm ML}). \tag{G.9b}$$

A symmetry plane $\hat{\sigma}(xy)$ in the molecule would give instead

$$\langle Lm \mid \hat{\mathscr{D}}(\Omega_{\rm ML}) \hat{\sigma}(xy)_{\rm M} \mid Ln \rangle = (-1)^{L+n} \mathscr{D}^L_{m,n}(\Omega_{\rm ML}), \tag{G.10}$$

since the inversion operation commutes with rotations, but the rotation in the molecular and lab frame do not commute. Proceeding in a similar way we obtain the results reported in Table G.1 for the effect of various molecular symmetry operations on the Wigner functions. The effect of mesophase (laboratory frame) symmetries can also be easily obtained from the table remembering that the order of operations is modified so that, for example, the first subscript, m, in $\mathscr{D}^L_{m,n}$ is affected instead of n. Knowing the effect of symmetry transformations on the Wigner matrices a substitution in Eq. G.3 gives the corresponding symmetry relations for the order parameters. Thus, in the example we have just seen, a $\hat{\sigma}(xy)$ plane in the laboratory frame gives $\langle \mathscr{D}^L_{m,n} \rangle = (-1)^{L+m} \langle \mathscr{D}^L_{m,n'} \rangle$, while a $\hat{\sigma}(xy)$ operation in the molecule yields $\langle \mathscr{D}^L_{m,n} \rangle = (-1)^{L+n} \langle \mathscr{D}^L_{m,n'} \rangle$. We also have, from the reality of $P(\Omega_{\rm ML})$ and Eq. F.6 for the Wigner functions, $\langle \mathscr{D}^{L*}_{m,n} \rangle = (-1)^{m-n} \langle \mathscr{D}^L_{-m,-n} \rangle$.

Table G.1. *Effect of various molecular point group symmetry operations [Cotton, 1990] $\hat{\mathscr{O}}_M$ on the Wigner matrices [Zannoni, 1979c]*

Operator $\hat{\mathscr{O}}_M$	$\langle Lm\vert\hat{\mathscr{O}}_M\hat{\mathscr{D}}\vert Ln\rangle$	Symmetry operation description
$\hat{\mathscr{I}} = \hat{\mathscr{S}}_2$	$(-1)^L\mathscr{D}^L_{m,n}$	Inversion
$\hat{\sigma}(xy)_M$	$(-1)^{L+n}\mathscr{D}^L_{m,n}$	Plane perpendicular to z
$\hat{\sigma}(xz)_M$	$(-1)^n\mathscr{D}^L_{m,-n}$	Plane perpendicular to y
$\hat{\sigma}(yz)_M$	$\mathscr{D}^L_{m,-n}$	Plane perpendicular to x
$\hat{\sigma}(\phi)_M$	$(-1)^n\exp(-i2n\phi)\mathscr{D}^L_{m,-n}$	Plane making an angle ϕ with (zx)
$\hat{C}_2(z)_M$	$(-1)^n\mathscr{D}^L_{m,n}$	Rotation of π about z-axis
$\hat{C}_k(z)_M$	$\mathscr{D}^L_{m,n}\exp(-i2n\pi/k)$	Rotation of $2\pi/k$ about z-axis
$\hat{C}_\gamma(z)_M$	$\mathscr{D}^L_{m,n}\exp(-in\gamma)$	Rotation of γ around z-axis
$\hat{C}_2(x)_M$	$(-1)^L\mathscr{D}^L_{m,-n}$	Rotation of π about x-axis
$\hat{C}_k(x)_M$	$(-1)^{L-n}\mathscr{D}^L_{m,-n}\exp(-in2\pi/k)$	Rotation of $2\pi/k$ about x-axis
$\hat{C}_\gamma(x)_M$	$(-1)^{L-n}\mathscr{D}^L_{m,-n}\exp(-in\gamma)$	Rotation of γ around x-axis
$\hat{C}_2(y)_M$	$(-1)^{L-n}\mathscr{D}^L_{m,-n}$	Rotation of π about y-axis
$\hat{C}_\beta(y)_M$	$\sum_q\mathscr{D}^L_{m,q}d^L_{q,n}(\beta)$	Rotation of β around y-axis
$\hat{C}_2(\phi)_M$	$(-1)^L\mathscr{D}^L_{m,n}\exp(-i2n\phi)$	Rotation of π about an axis perpendicular to z and at ϕ from x
$[\hat{\mathscr{S}}_k]_M$	$(-1)^{L+n}\mathscr{D}^L_{m,n}\exp(-in2\pi/k)$	A k-fold roto-reflection axis.

From all these relations we can find the relevant order parameters for various molecular and phase symmetries (see Table 3.1).

G.2 Rotationally Invariant Pairwise Functions

In a number of situations, we have to consider pairwise rather than single particle properties, functions of the positions $\boldsymbol{r}_1$, $\boldsymbol{r}_2$ and orientations $\Omega_{1L} = \Omega_1$, $\Omega_{2L} = \Omega_2$ of two rigid particles, $f(\boldsymbol{X}_1, \boldsymbol{X}_2) = f(\boldsymbol{r}_1\Omega_{1L}, \boldsymbol{r}_2\Omega_{2L})$, that have to be invariant for an arbitrary translation and rotation of the coordinate system. Examples are the angular-radial pair correlation $G(\boldsymbol{r}_1\Omega_{1L}, \boldsymbol{r}_2\Omega_{2L})$ of a macroscopically isotropic fluid (Section 4.6) and the pair potential $U(\boldsymbol{r}_1\Omega_{1L}, \boldsymbol{r}_2\Omega_{2L})$ between two rigid molecules (Chapter 5).

We first impose that the function should be translationally invariant by an arbitrary shift $\boldsymbol{R}$: $f(\boldsymbol{r}_1\Omega_{1L}, \boldsymbol{r}_2\Omega_{2L}) = f(\boldsymbol{R} + \boldsymbol{r}_1, \Omega_{1L}, \boldsymbol{R} + \boldsymbol{r}_2, \Omega_{2L})$. In general, such an invariant function will be $f(\boldsymbol{r}, \Omega_{1L}, \Omega_{2L}) = f(r, \Omega_{rL}, \Omega_{1L}, \Omega_{2L})$, where $\boldsymbol{r} \equiv \boldsymbol{r}_2 - \boldsymbol{r}_1$ is the inter-centre vector with orientation Ω_{rL}. General rotationally invariant combinations of $f(r, \Omega_{rL}, \Omega_{1L}, \Omega_{2L})$ can be constructed following Blum and Torruella [1972], Stone [1978, 1979] and Zannoni [1979c]. Starting from an arbitrary laboratory frame L′ we can then expand our pair function as

$$f(r,\Omega_{1L'},\Omega_{2L'},\Omega_{rL'}) = \sum_{\lambda} f^{m_1m_2m;n_1n_2}_{L_1L_2.L}(r)\, \mathscr{D}^{L_1*}_{m_1,n_1}(\Omega_{1L'})\mathscr{D}^{L_2*}_{m_2,n_2}(\Omega_{2L'}\mathscr{D}^{L}_{m,0}(\Omega_{rL'}), \tag{G.11}$$

where the sum extends to the set λ of all the integer angular momenta L_1, L_2, L and their components m_1, m_2, m, n_1, n_2 appearing on the RHS and not on the LHS. Note that, according to the convention of Rose [1957] for the Euler angles that we have adopted, the coefficients m_1, m_2, m refer to rotations around the laboratory z-axis while n_1, n_2 refer to rotations around the molecular z-axis. Thus, the last subscript is 0 because only the two angles α_r, β_r are needed to specify the intermolecular vector orientation Ω_{rL}. We now perform an arbitrary rotation of the lab frame from L′ to L. Using the closure relation of Wigner matrices (cf. Eqs. F.13 and F.14) the rotations from the arbitrary auxiliary frame L′ can be rewritten, for each term, starting from the lab frame L. For instance

$$\mathscr{D}^{L_1*}_{m_1,n_1}(\Omega_{1L'}) = \sum_{q_1} \mathscr{D}^{L_1*}_{m_1,q_1}(\Omega_{LL'})\mathscr{D}^{L_1*}_{q_1,n_1}(\Omega_{1L}) = \sum_{q_1} \mathscr{D}^{L_1}_{q_1,m_1}(\Omega_{L'L})\mathscr{D}^{L_1*}_{q_1,n_1}(\Omega_{1L}). \tag{G.12}$$

Substituting the three expressions on the RHS of Eq. G.11 gives

$$\begin{aligned}&\mathscr{D}^{L}_{m,0}(\Omega_{rL'})\mathscr{D}^{L_2*}_{m_2,n_2}(\Omega_{2L'})\mathscr{D}^{L_1*}_{m_1,n_1}(\Omega_{1L'})\\ &= \sum_{q_1,q_2,q} \mathscr{D}^{L*}_{q,m}(\Omega_{L'L})\mathscr{D}^{L_2}_{q_2,m_2}(\Omega_{L'L})\mathscr{D}^{L_1}_{q_1,m_1}(\Omega_{L'L})\mathscr{D}^{L_1*}_{q_1,n_1}(\Omega_{1L})\mathscr{D}^{L_2*}_{q_2,n_2}(\Omega_{2L})\mathscr{D}^{L}_{q,0}(\Omega_{rL}).\end{aligned} \tag{G.13}$$

We then integrate on all possible orientations $(\Omega_{L'L})$ (equivalent to $\hat{\mathscr{P}}_{O(3)}$) using the Gaunt formula, Eq. F.32, and obtain the rotationally invariant combinations

$$\begin{aligned}&\int d\Omega_{L'L}\mathscr{D}^{L*}_{m,0}(\Omega_{rL'})\mathscr{D}^{L_2*}_{m_2,n_2}(\Omega_{2L'})\mathscr{D}^{L_1}_{m_1,n_1}(\Omega_{1L'})\\ &= \sum_{q_1,q_2,q} \mathscr{D}^{L}_{q,0}(\Omega_{rL})\mathscr{D}^{L_2*}_{q_2,n_2}(\Omega_{2L})\mathscr{D}^{L_1*}_{q_1,n_1}(\Omega_{1L})\\ &\quad \int d\Omega_{L'L}\mathscr{D}^{L*}_{q,m}(\Omega_{L'L})\mathscr{D}^{L_2}_{q_2,m_2}(\Omega_{L'L})\mathscr{D}^{L_1}_{q_1,m_1}(\Omega_{L'L}),\\ &= \frac{8\pi^2}{(2L+1)}C(L_1,L_2,L;m_1,m_2)\\ &\quad \sum_{q_1,q_2} C(L_1,L_2,L;q_1,q_2)\mathscr{D}^{L_1*}_{q_1,n_1}(\Omega_{1L})\mathscr{D}^{L_2*}_{q_2,n_2}(\Omega_{2L})\mathscr{D}^{L}_{q_1+q_2,0}(\Omega_{rL}).\end{aligned} \tag{G.14}$$

Thus, a scalar pair function can be written as an expansion like

$$\begin{aligned}&f(r,\Omega_{1L},\Omega_{2L},\Omega_{rL})\\ &= \sum_{\substack{L_1,L_2,L,\\ n_1,n_2}} \left\{ \sum_{m_1,m_2} \frac{8\pi^2}{(2L+1)} f^{m_1m_2m;n_1n_2}_{L_1L_2.L}(r)\, C(L_1,L_2,L;m_1,m_2)\right\}\\ &\quad \times \sum_{q_1,q_2} C(L_1,L_2,L;q_1,q_2)\mathscr{D}^{L_1*}_{q_1,n_1}(\Omega_{1L})\mathscr{D}^{L_2*}_{q_2,n_2}(\Omega_{2L})\mathscr{D}^{L}_{q_1+q_2,0}(\Omega_{rL}),\\ &= \sum_{\substack{L_1,L_2,L\\ n_1,n_2}} f^{n_1n_2}_{L_1,L_2,L}(r)\, \Phi^{n_1,n_2}_{L_1,L_2L}(\Omega_{1L},\Omega_{2L},\Omega_{rL}),\end{aligned} \tag{G.15}$$

where L_1, L_2, L are not arbitrary any more, but have to satisfy the triangular condition (Eq. F.19) enforced by the Clebsch–Gordan coefficients and we have introduced the rotationally invariant combinations

$$\Phi^{n_1,n_2}_{L_1,L_2 L}(\Omega_{1L}, \Omega_{2L}, \Omega_{rL}) \equiv \sum_{q_1,q_2} C(L_1, L_2, L; q_1, q_2) \mathscr{D}^{L_1 *}_{q_1,n_1}(\Omega_{1L}) \mathscr{D}^{L_2 *}_{q_2,n_2}(\Omega_{2L'}) \mathscr{D}^{L}_{q_1+q_2,0}(\Omega_{rL}), \tag{G.16}$$

equivalent to those of Blum [1972].[1] Since the expansion is valid in an arbitrary frame, we can in particular adopt the intermolecular (IM) frame, with z-axis along the inter-centre axis and $\Omega_{rL} = (000)$. In the IM frame, $\mathscr{D}^{L}_{q_1+q_2,0}(\Omega_{rL}) = \mathscr{D}^{L}_{q_1+q_2,0}(000) = \delta_{q_1,-q_2}$ and

$$\begin{aligned} f_{\rm IM}(r, \Omega_{1L}, \Omega_{2L}) &= \sum_{\substack{L_1,L_2,L \\ q_1;n_1,n_2}} f^{q_1,-q_1;n_1 n_2}_{L_1,L_2,L}(r) C(L_1, L_2, L; q_1, -q_1) \mathscr{D}^{L_1 *}_{q_1,n_1}(\Omega_{1L}) \mathscr{D}^{L_2 *}_{-q_1,n_2}(\Omega_{2L'}), \\ &\equiv \sum_{\substack{L_1,L_2,L \\ q_1;n_1,n_2}} [f_{\rm IM}]^{q_1;n_1,n_2}_{L_1,L_2}(r) \mathscr{D}^{L_1 *}_{m_1,n_1}(\Omega_{1L}) \mathscr{D}^{L_2 *}_{-m_1,n_2}(\Omega_{2L}). \end{aligned} \tag{G.17}$$

Thus, we can switch from a space-fixed to a molecule-fixed expansion writing the expansion coefficients of one representation in terms of those of the other:

$$[f_{\rm IM}]^{q_1;n_1,n_2}_{L_1,L_2}(r) = \sum_{L} C(L_1, L_2, L; q_1, -q_1) f^{n_1,n_2,m_1}_{L_1,L_2,L}(r). \tag{G.18}$$

The rotational invariant combinations $\Phi^{n_1,n_2}_{L_1,L_2 L}(\Omega_{1L}, \Omega_{2L}, \Omega_{rL})$ defined in Eq. G.16 correspond, apart from a constant, to the rotational invariants of Stone [1978], where a phase factor is introduced so as to make $S^{00}_{L_1,L_2,L}$ always real:

$$S^{n_1,n_2}_{L_1,L_2,L}(\Omega_{1L}, \Omega_{2L}, \Omega_{rL}) = \frac{(i)^{L_1-L_2+L}}{\sqrt{2L+1}} \sum_{q_1,q_2} C(L_1, L_2, L; q_1, q_2) \mathscr{D}^{L_1 *}_{q_1,n_1}(\Omega_{1L}) \mathscr{D}^{L_2 *}_{q_2,n_2}(\Omega_{2L}) \mathscr{D}^{L}_{q_1+q_2,0}(\Omega_{rL}), \tag{G.19}$$

$$= (i)^{L_1-L_2-L} \sum_{q_1,q_2} \begin{pmatrix} L_1 & L_2 & L \\ q_1 & q_2 & -q_1-q_2 \end{pmatrix} \mathscr{D}^{L_1 *}_{q_1,n_1}(\Omega_{1L}) \mathscr{D}^{L_2 *}_{q_2,n_2}(\Omega_{2L}) \mathscr{D}^{L *}_{-q_1-q_2,0}(\Omega_{rL}), \tag{G.20}$$

where the last expression is written in terms of 3-j coefficients, as in Stone [1978]. The definition of Blum and Torruella [1972] is written using Edmonds, [1960] convention for rotations instead of the one of Rose [1957] that we follow. The conversion between the two is [Brink and Satchler, 1968]

$$\underset{\text{Rose}}{\mathscr{D}^{L}_{m,n}(\alpha, \beta, \gamma)} \Longleftrightarrow \underset{\text{Edmonds}}{\mathscr{D}^{L}_{m,n}(-\alpha, -\beta, -\gamma) = (-1)^{m-n} \mathscr{D}^{L *}_{m,n}(\alpha\beta\gamma)}. \tag{G.21}$$

[1] Note, however, that Blum [1972] adopts the convention of Edmonds [1960] for the Euler angles which describes rotations from MOL to LAB instead, as we do, from LAB to MOL. Brink and Satchler [1968].

Thus, expression G.19 gives

$$\underset{\text{Stone}}{S^{n_1,n_2}_{L_1,L_2,L}(\Omega_{1\mathrm{L}},\Omega_{2\mathrm{L}},\Omega_{r\mathrm{L}})} \Longleftrightarrow \underset{\text{Blum and Torruella}}{(i)^{L_1-L_2-L}(-1)^{n_1+n_2}\Phi^{L_1L_2L}_{n_1n_2}(\Omega_{1\mathrm{L}},\Omega_{2\mathrm{L}},\Omega_{r\mathrm{L}})}, \tag{G.22}$$

which can be useful when comparing with literature data using either convention. The rotational invariants form an orthogonal basis set of functions in the angular space $\{\Omega_{1\mathrm{L}},\Omega_{2\mathrm{L}},\Omega_{r\mathrm{L}}\}$:

$$\int d\Omega_{1\mathrm{L}}d\Omega_{2\mathrm{L}}d\Omega_{r\mathrm{L}}S^{n_1,n_2*}_{L_1,L_2,L}(\Omega_{1\mathrm{L}},\Omega_{2\mathrm{L}},\Omega_{r\mathrm{L}})\,S^{n'_1,n'_2}_{L'_1,L'_2,L'}(\Omega_{1\mathrm{L}},\Omega_{2\mathrm{L}},\Omega_{r\mathrm{L}})$$
$$= \frac{256\pi^5}{(2L_1+1)(2L_2+1)(2L+1)}\delta_{L_1L'_1}\delta_{L_2,L'_2}\delta_{L,L'}\delta_{n_1,n'_1}\delta_{n_2,n'_2}. \tag{G.23}$$

We shall generally adopt in this book the Stone definition, since it is more popular and well-described in the literature, and briefly review some properties. Using this definition and the symmetry properties of Clebsch–Gordan coefficients (cf. Eq. F.26b), the complex conjugate of the rotational invariant is

$$S^{n_1,n_2}_{L_1,L_2,L}(\Omega_{1\mathrm{L}},\Omega_{2\mathrm{L}},\Omega_{r\mathrm{L}})^* = (-1)^{n_1+n_2}S^{n_1,n_2}_{L_1,L_2,L}(\Omega_{1\mathrm{L}},\Omega_{2\mathrm{L}},\Omega_{r\mathrm{L}}). \tag{G.24}$$

For uniaxial molecules we only need the subset

$$S_{L_1,L_2,L}(\Omega_{1\mathrm{L}},\Omega_{2\mathrm{L}},\Omega_{r\mathrm{L}}) \equiv S^{0,0}_{L_1,L_2,L}(\Omega_{1\mathrm{L}},\Omega_{2\mathrm{L}},\Omega_{r\mathrm{L}}),$$
$$= \frac{i^{L_2-L_1+L}}{\sqrt{2L+1}}\sum_{q_1,q_2} C(L_1,L_2,L;q_1,q_2)\mathscr{D}^{L_1*}_{q_1,0}(\Omega_{1\mathrm{L}})\mathscr{D}^{L_2*}_{q_2,0}(\Omega_{2\mathrm{L}})\,\mathscr{D}^{L}_{q_1+q_2,0}(\Omega_{r\mathrm{L}}), \tag{G.25}$$

which are real functions. Recalling the effect of the inversion operation on the Wigner matrices (Table G.1) we also have $\hat{\mathscr{I}}S^{n_1,n_2}_{L_1,L_2,L} = (-1)^{L_1+L_2+L}S^{n_1,n_2}_{L_1,L_2,L}$. Thus, we see that in the description of a centrosymmetric phase like a nematic or non-chiral isotropic, only S functions with L_1+L_2+L even parity need to be considered. On the contrary chiral phases, e.g. cholesterics, will also need combinations with odd L_1+L_2+L. Explicit expressions of the first few rotational invariants for **uniaxial molecules** in Cartesian coordinates, and using the unit vectors $\hat{z}_1$, $\hat{z}_2$, $\hat{\boldsymbol{r}}$ along the z-axis of the two molecules and of the separation vector $\boldsymbol{r} \equiv \boldsymbol{r}_2 - \boldsymbol{r}_1$ can be written, extending the list reported by Stone [1978, 1996], as follows:

$$S_{0,0,0} = 1, \tag{G.26}$$
$$S_{1,1,0} = -(\hat{z}_1\cdot\hat{z}_2)\sqrt{3}, \tag{G.27}$$
$$S_{1,0,1} = -(\hat{z}_1\cdot\hat{\boldsymbol{r}})/\sqrt{3}, \tag{G.28}$$
$$S_{0,1,1} = (\hat{z}_2\cdot\hat{\boldsymbol{r}})/\sqrt{3}, \tag{G.29}$$
$$S_{1,1,2} = \left[(\hat{z}_1\cdot\hat{z}_2) - 3(\hat{z}_1\cdot\hat{\boldsymbol{r}})(\hat{z}_2\cdot\hat{\boldsymbol{r}})\right]/\sqrt{30}, \tag{G.30}$$
$$S_{1,2,1} = \left[(\hat{z}_1\cdot\hat{\boldsymbol{r}}) - 3(\hat{z}_1\cdot\hat{z}_2)(\hat{z}_2\cdot\hat{\boldsymbol{r}})\right]/\sqrt{30}, \tag{G.31}$$
$$S_{2,1,1} = -\left[(\hat{z}_2\cdot\hat{\boldsymbol{r}}) - 3(\hat{z}_1\cdot\hat{z}_2)(\hat{z}_1\cdot\hat{\boldsymbol{r}})\right]/\sqrt{30}, \tag{G.32}$$
$$S_{2,2,0} = \left[-1 + 3(\hat{z}_1\cdot\hat{z}_2)^2\right]/\sqrt{20}, \tag{G.33}$$
$$S_{2,0,2} = \left[-1 + 3(\hat{z}_1\cdot\hat{\boldsymbol{r}})^2\right]/\sqrt{20}, \tag{G.34}$$

$$S_{0,2,2} = \left[-1 + 3(\hat{z}_2 \cdot \hat{\boldsymbol{r}})^2\right] / \sqrt{20}, \tag{G.35}$$

$$S_{2,2,2} = \left[2 - 3(\hat{z}_1 \cdot \hat{z}_2)^2 - 3(\hat{z}_1 \cdot \hat{\boldsymbol{r}})^2 - 3(\hat{z}_2 \cdot \hat{\boldsymbol{r}})^2 + 9(\hat{z}_1 \cdot \hat{z}_2)(\hat{z}_1 \cdot \hat{\boldsymbol{r}})(\hat{z}_2 \cdot \hat{\boldsymbol{r}})\right]\sqrt{70}, \tag{G.36}$$

$$S_{1,2,3} = \sqrt{3}\,\left[(\hat{z}_1 \cdot \hat{\boldsymbol{r}}) + 2(\hat{z}_1 \cdot \hat{z}_2)(\hat{z}_2 \cdot \hat{\boldsymbol{r}}) - 5(\hat{z}_1 \cdot \hat{\boldsymbol{r}})(\hat{z}_2 \cdot \hat{\boldsymbol{r}})^2\right] / \sqrt{140}, \tag{G.37}$$

$$S_{2,3,1} = -\sqrt{3}\left[(\hat{z}_2 \cdot \hat{\boldsymbol{r}}) + 2(\hat{z}_1 \cdot \hat{\boldsymbol{r}})(\hat{z}_1 \cdot \hat{z}_2) - 5(\hat{z}_1 \cdot \hat{z}_2)^2(\hat{z}_2 \cdot \hat{\boldsymbol{r}})\right] / \sqrt{140}, \tag{G.38}$$

$$S_{2,1,3} = -\sqrt{3}\left[(\hat{z}_2 \cdot \hat{\boldsymbol{r}}) + 2(\hat{z}_1 \cdot \hat{z}_2)(\hat{z}_1 \cdot \hat{\boldsymbol{r}}) - 5(\hat{z}_1 \cdot \hat{\boldsymbol{r}})^2(\hat{z}_2 \cdot \hat{\boldsymbol{r}})\right] / \sqrt{140}, \tag{G.39}$$

$$S_{1,3,2} = \sqrt{3}\left[(\hat{z}_1 \cdot \hat{z}_2) + 2(\hat{z}_1 \cdot \hat{\boldsymbol{r}})(\hat{z}_2 \cdot \hat{\boldsymbol{r}}) - 5(\hat{z}_1 \cdot \hat{z}_2)(\hat{z}_2 \cdot \hat{\boldsymbol{r}})^2\right] / \sqrt{140}, \tag{G.40}$$

$$S_{3,1,2} = \sqrt{3}\left[(\hat{z}_1 \cdot \hat{z}_2) + 2(\hat{z}_1 \cdot \hat{\boldsymbol{r}})(\hat{z}_2 \cdot \hat{\boldsymbol{r}}) - 5(\hat{z}_1 \cdot \hat{z}_2)(\hat{z}_1 \cdot \hat{\boldsymbol{r}})^2\right] / \sqrt{140}, \tag{G.41}$$

$$S_{3,2,1} = \sqrt{3}\left[(\hat{z}_1 \cdot \hat{\boldsymbol{r}}) + 2(\hat{z}_1 \cdot \hat{z}_2)(\hat{z}_2 \cdot \hat{\boldsymbol{r}}) - 5(\hat{z}_1 \cdot \hat{z}_2)^2(\hat{z}_1 \cdot \hat{\boldsymbol{r}})\right] / \sqrt{140}, \tag{G.42}$$

$$\begin{aligned} S_{1,3,4} = \frac{1}{12\sqrt{7}}\big[&-3(\hat{z}_1 \cdot \hat{z}_2) + 15(\hat{z}_1 \cdot \hat{\boldsymbol{r}})(\hat{z}_2 \cdot \hat{\boldsymbol{r}}) + 15(\hat{z}_1 \cdot \hat{z}_2)(\hat{z}_2 \cdot \hat{\boldsymbol{r}})^2 \\ &-35(\hat{z}_1 \cdot \hat{\boldsymbol{r}})(\hat{z}_2 \cdot \hat{\boldsymbol{r}})^3\big], \end{aligned} \tag{G.43}$$

$$\begin{aligned} S_{3,1,4} = \frac{1}{12\sqrt{7}}\big[&-3(\hat{z}_1 \cdot \hat{z}_2) + 15(\hat{z}_1 \cdot \hat{z}_2)(\hat{z}_1 \cdot \hat{\boldsymbol{r}})^2 + 15(\hat{z}_1 \cdot \hat{\boldsymbol{r}})(\hat{z}_2 \cdot \hat{\boldsymbol{r}}) \\ &-35(\hat{z}_1 \cdot \hat{\boldsymbol{r}})^3(\hat{z}_2 \cdot \hat{\boldsymbol{r}})\big], \end{aligned} \tag{G.44}$$

$$\begin{aligned} S_{2,2,4} = \frac{1}{4\sqrt{70}}\big[&1 + 2(\hat{z}_1 \cdot \hat{z}_2)^2 - 5(\hat{z}_1 \cdot \hat{\boldsymbol{r}})^2 - 5(\hat{z}_2 \cdot \hat{\boldsymbol{r}})^2 - 20(\hat{z}_1 \cdot \hat{z}_2)(\hat{z}_1 \cdot \hat{\boldsymbol{r}})(\hat{z}_2 \cdot \hat{\boldsymbol{r}}) \\ &+35(\hat{z}_1 \cdot \hat{\boldsymbol{r}})^2(\hat{z}_2 \cdot \hat{\boldsymbol{r}})^2\big], \end{aligned} \tag{G.45}$$

$$\begin{aligned} S_{4,2,2} = \frac{1}{4\sqrt{70}}\big[&1 - 5(\hat{z}_1 \cdot \hat{z}_2)^2 - 5(\hat{z}_1 \cdot \hat{\boldsymbol{r}})^2 + 2(\hat{z}_2 \cdot \hat{\boldsymbol{r}})^2 - 20(\hat{z}_1 \cdot \hat{z}_2)(\hat{z}_1 \cdot \hat{\boldsymbol{r}})(\hat{z}_2 \cdot \hat{\boldsymbol{r}}) \\ &+35(\hat{z}_1 \cdot \hat{z}_2)^2(\hat{z}_1 \cdot \hat{\boldsymbol{r}})^2\big], \end{aligned} \tag{G.46}$$

$$\begin{aligned} S_{2,4,2} = \frac{1}{4\sqrt{70}}\big[&1 - 5(\hat{z}_1 \cdot \hat{z}_2)^2 + 2(\hat{z}_1 \cdot \hat{\boldsymbol{r}})^2 - 5(\hat{z}_2 \cdot \hat{\boldsymbol{r}})^2 - 20(\hat{z}_1 \cdot \hat{z}_2)(\hat{z}_1 \cdot \hat{\boldsymbol{r}})(\hat{z}_2 \cdot \hat{\boldsymbol{r}}) \\ &+35(\hat{z}_1 \cdot \hat{z}_2)^2(\hat{z}_2 \cdot \hat{\boldsymbol{r}})^2\big], \end{aligned} \tag{G.47}$$

$$S_{3,3,0} = \frac{1}{2\sqrt{7}}\left[3(\hat{z}_1 \cdot \hat{z}_2) - 5(\hat{z}_1 \cdot \hat{z}_2)^3\right], \tag{G.48}$$

$$S_{3,0,3} = \frac{1}{2\sqrt{7}}\left[3(\hat{z}_1 \cdot \hat{\boldsymbol{r}}) - 5(\hat{z}_1 \cdot \hat{\boldsymbol{r}})^3\right], \tag{G.49}$$

$$S_{0,3,3} = -\frac{1}{2\sqrt{7}}\left[3(\hat{z}_2 \cdot \hat{\boldsymbol{r}}) - 5(\hat{z}_2 \cdot \hat{\boldsymbol{r}})^3\right], \tag{G.50}$$

$$\begin{aligned} S_{3,3,2} = +\frac{1}{4\sqrt{105}}\big[&-21(\hat{z}_1 \cdot \hat{z}_2) + 30(\hat{z}_1 \cdot \hat{z}_2)(\hat{z}_1 \cdot \hat{\boldsymbol{r}})^2 + 30(\hat{z}_1 \cdot \hat{z}_2)(\hat{z}_2 \cdot \hat{\boldsymbol{r}})^2) \\ &+25(\hat{z}_1 \cdot \hat{z}_2)^3 + 3(\hat{z}_1 \cdot \hat{\boldsymbol{r}})(\hat{z}_2 \cdot \hat{\boldsymbol{r}}) - 75(\hat{z}_1 \cdot \hat{z}_2)^2(\hat{z}_1 \cdot \hat{\boldsymbol{r}})(\hat{z}_2 \cdot \hat{\boldsymbol{r}})\big], \end{aligned} \tag{G.51}$$

$$\begin{aligned} S_{3,2,3} = \frac{1}{4\sqrt{105}}\big[&-21(\hat{z}_1 \cdot \hat{\boldsymbol{r}}) + 25(\hat{z}_1 \cdot \hat{\boldsymbol{r}})^3 + 30(\hat{z}_1 \cdot \hat{z}_2)^2(\hat{z}_1 \cdot \hat{\boldsymbol{r}}) + 3(\hat{z}_1 \cdot \hat{z}_2)(\hat{z}_2 \cdot \hat{\boldsymbol{r}}) \\ &-75(\hat{z}_1 \cdot \hat{z}_2)(\hat{z}_1 \cdot \hat{\boldsymbol{r}})^2(\hat{z}_2 \cdot \hat{\boldsymbol{r}}) + 30(\hat{z}_1 \cdot \hat{\boldsymbol{r}})(\hat{z}_2 \cdot \hat{\boldsymbol{r}})^2\big], \end{aligned} \tag{G.52}$$

$$\begin{aligned} S_{2,3,3} = +\frac{1}{4\sqrt{105}}\big[&-3(\hat{z}_1 \cdot \hat{z}_2)(\hat{z}_1 \cdot \hat{\boldsymbol{r}}) + 21(\hat{z}_2 \cdot \hat{\boldsymbol{r}}) - 30(\hat{z}_1 \cdot \hat{\boldsymbol{r}})^2(\hat{z}_2 \cdot \hat{\boldsymbol{r}}) \\ &-30(\hat{z}_1 \cdot \hat{z}_2)^2(\hat{z}_2 \cdot \hat{\boldsymbol{r}}) + 75(\hat{z}_1 \cdot \hat{z}_2)(\hat{z}_1 \cdot \hat{\boldsymbol{r}})(\hat{z}_2 \cdot \hat{\boldsymbol{r}})^2 - 25(\hat{z}_2 \cdot \hat{\boldsymbol{r}})^3\big], \end{aligned} \tag{G.53}$$

$$S_{4,4,0} = \frac{1}{24}[3 - 30\,(\hat{z}_1 \cdot \hat{z}_2)^2 + 35\,(\hat{z}_1 \cdot \hat{z}_2)^4], \tag{G.54}$$

$$S_{4,0,4} = \frac{1}{24}[3 - 30\,(\hat{z}_1 \cdot \hat{\boldsymbol{r}})^2 + 35\,(\hat{z}_1 \cdot \hat{\boldsymbol{r}})^4], \tag{G.55}$$

$$S_{0,4,4} = \frac{1}{24}[\,3 - 30\,(\hat{z}_2 \cdot \hat{\boldsymbol{r}})^2 + 35\,(\hat{z}_2 \cdot \hat{\boldsymbol{r}})^4], \tag{G.56}$$

$$\begin{aligned} S_{4,4,2} = \frac{\sqrt{5}}{12\sqrt{77}}[&-6 + 51\,(\hat{z}_1 \cdot \hat{z}_2)^2 + 9\,(\hat{z}_1 \cdot \hat{\boldsymbol{r}})^2 + 9\,(\hat{z}_2 \cdot \hat{\boldsymbol{r}})^2 \\ &-27\,(\hat{z}_1 \cdot \hat{z}_2)(\hat{z}_1 \cdot \hat{\boldsymbol{r}})(\hat{z}_2 \cdot \hat{\boldsymbol{r}}) - 63\,(\hat{z}_1 \cdot \hat{z}_2)^2(\hat{z}_1 \cdot \hat{\boldsymbol{r}})^2 - 63\,(\hat{z}_1 \cdot \hat{z}_2)^2(\hat{z}_2 \cdot \hat{\boldsymbol{r}})^2 \\ &+147\,(\hat{z}_1 \cdot \hat{z}_2)^3(\hat{z}_1 \cdot \hat{\boldsymbol{r}})(\hat{z}_2 \cdot \hat{\boldsymbol{r}}) - 49\,(\hat{z}_1 \cdot \hat{z}_2)^4], \end{aligned} \tag{G.57}$$

$$\begin{aligned} S_{4,2,4} = \frac{\sqrt{5}}{12\sqrt{77}}[&-6 + 9\,(\hat{z}_1 \cdot \hat{z}_2)^2 + 51\,(\hat{z}_1 \cdot \hat{\boldsymbol{r}})^2 + 9\,(\hat{z}_2 \cdot \hat{\boldsymbol{r}})^2 \\ &-27\,(\hat{z}_1 \cdot \hat{z}_2)(\hat{z}_1 \cdot \hat{\boldsymbol{r}})(\hat{z}_2 \cdot \hat{\boldsymbol{r}}) - 63\,(\hat{z}_1 \cdot \hat{z}_2)^2(\hat{z}_1 \cdot \hat{\boldsymbol{r}})^2 \\ &-63\,(\hat{z}_1 \cdot \hat{\boldsymbol{r}})^2(\hat{z}_2 \cdot \hat{\boldsymbol{r}})^2 + 147\,(\hat{z}_1 \cdot \hat{z}_2)(\hat{z}_1 \cdot \hat{\boldsymbol{r}})^3(\hat{z}_2 \cdot \hat{\boldsymbol{r}}) - 49\,(\hat{z}_1 \cdot \hat{\boldsymbol{r}})^4], \end{aligned} \tag{G.58}$$

$$\begin{aligned} S_{2,4,4} = \frac{\sqrt{5}}{12\sqrt{77}}[&-6 + 9\,(\hat{z}_1 \cdot \hat{z}_2)^2 + 9\,(\hat{z}_1 \cdot \hat{\boldsymbol{r}})^2 + 51\,(\hat{z}_2 \cdot \hat{\boldsymbol{r}})^2 \\ &-27\,(\hat{z}_1 \cdot \hat{z}_2)(\hat{z}_1 \cdot \hat{\boldsymbol{r}})(\hat{z}_2 \cdot \hat{\boldsymbol{r}}) - 63\,(\hat{z}_1 \cdot \hat{z}_2)^2(\hat{z}_2 \cdot \hat{\boldsymbol{r}})^2 \\ &-63\,(\hat{z}_1 \cdot \hat{\boldsymbol{r}})^2(\hat{z}_2 \cdot \hat{\boldsymbol{r}})^2 + 147\,(\hat{z}_1 \cdot \hat{z}_2)(\hat{z}_1 \cdot \hat{\boldsymbol{r}})(\hat{z}_2 \cdot \hat{\boldsymbol{r}})^3 - 49\,(\hat{z}_2 \cdot \hat{\boldsymbol{r}})^4], \end{aligned} \tag{G.59}$$

$$\begin{aligned} S_{4,4,4} = \frac{1}{24\sqrt{2002}}[&109 - 545\,(\hat{z}_1 \cdot \hat{z}_2)^2 - 545\,(\hat{z}_1 \cdot \hat{\boldsymbol{r}})^2 - 545\,(\hat{z}_2 \cdot \hat{\boldsymbol{r}})^2 \\ &+3700\,(\hat{z}_1 \cdot \hat{z}_2)(\hat{z}_1 \cdot \hat{\boldsymbol{r}})(\hat{z}_2 \cdot \hat{\boldsymbol{r}}) + 490\,(\hat{z}_1 \cdot \hat{z}_2)^4 + 490\,(\hat{z}_1 \cdot \hat{\boldsymbol{r}})^4 \\ &+490\,(\hat{z}_2 \cdot \hat{\boldsymbol{r}})^4 + 875\,(\hat{z}_1 \cdot \hat{z}_2)^2(\hat{z}_1 \cdot \hat{\boldsymbol{r}})^2 + 875\,(\hat{z}_1 \cdot \hat{z}_2)^2(\hat{z}_2 \cdot \hat{\boldsymbol{r}})^2 \\ &+875\,(\hat{z}_1 \cdot \hat{\boldsymbol{r}})^2(\hat{z}_2 \cdot \hat{\boldsymbol{r}})^2 - 4900\,(\hat{z}_1 \cdot \hat{z}_2)(\hat{z}_1 \cdot \hat{\boldsymbol{r}})(\hat{z}_2 \cdot \hat{\boldsymbol{r}})^3 \\ &-4900\,(\hat{z}_1 \cdot \hat{z}_2)(\hat{z}_1 \cdot \hat{\boldsymbol{r}})^3(\hat{z}_2 \cdot \hat{\boldsymbol{r}}) - 4900\,(\hat{z}_1 \cdot \hat{z}_2)^3(\hat{z}_1 \cdot \hat{\boldsymbol{r}})(\hat{z}_2 \cdot \hat{\boldsymbol{r}}) \\ &+8575\,(\hat{z}_1 \cdot \hat{z}_2)^2(\hat{z}_1 \cdot \hat{\boldsymbol{r}})^2(\hat{z}_2 \cdot \hat{\boldsymbol{r}})^2]. \end{aligned} \tag{G.60}$$

Higher invariants can be generated, e.g. using the coupling formula of two rotational invariants, obtained from the coupling formulae for Wigner rotation matrices [Price et al., 1984].

We also report a small tabulation for **chiral systems** ($L_1 + L_2 + L$ odd)

$$S_{1,1,1} = \frac{1}{\sqrt{6}}[\hat{z}_1 \cdot \hat{z}_2 \times \hat{\boldsymbol{r}}], \tag{G.61}$$

$$S_{1,2,2} = \frac{\sqrt{3}}{\sqrt{10}}[\hat{z}_1 \cdot \hat{z}_2 \times \hat{\boldsymbol{r}}](\hat{z}_2 \cdot \hat{\boldsymbol{r}}), \tag{G.62}$$

$$S_{2,1,2} = -\frac{\sqrt{3}}{\sqrt{10}}[\hat{z}_1 \cdot \hat{z}_2 \times \hat{\boldsymbol{r}}](\hat{z}_1 \cdot \hat{\boldsymbol{r}}), \tag{G.63}$$

$$S_{2,2,1} = -\frac{\sqrt{3}}{\sqrt{10}}[\hat{z}_1 \cdot \hat{z}_2 \times \hat{\boldsymbol{r}}](\hat{z}_1 \cdot \hat{z}_2), \tag{G.64}$$

$$S_{2,2,3} = \frac{3}{\sqrt{280}}[\hat{z}_1 \cdot \hat{z}_2 \times \hat{\boldsymbol{r}}][(\hat{z}_1 \cdot \hat{z}_2) - 5(\hat{z}_1 \cdot \hat{\boldsymbol{r}})(\hat{z}_2 \cdot \hat{\boldsymbol{r}})], \tag{G.65}$$

$$S_{2,3,2} = \frac{3}{\sqrt{280}}[\hat{z}_1 \cdot \hat{z}_2 \times \hat{\boldsymbol{r}}][(\hat{z}_1 \cdot \hat{\boldsymbol{r}}) - 5(\hat{z}_1 \cdot \hat{z}_2)(\hat{z}_2 \cdot \hat{\boldsymbol{r}})], \tag{G.66}$$

$$S_{3,2,2} = -\frac{3}{\sqrt{280}}[\hat{z}_1 \cdot \hat{z}_2 \times \hat{\boldsymbol{r}}][(\hat{z}_2 \cdot \hat{\boldsymbol{r}}) - 5(\hat{z}_1 \cdot \hat{z}_2)(\hat{z}_1 \cdot \hat{\boldsymbol{r}})], \tag{G.67}$$

$$S_{2,4,3} = \frac{\sqrt{5}}{\sqrt{112}}[\hat{z}_1 \cdot \hat{z}_2 \times \hat{\boldsymbol{r}}][(\hat{z}_1 \cdot \hat{z}_2) + 2(\hat{z}_1 \cdot \hat{\boldsymbol{r}})(\hat{z}_2 \cdot \hat{\boldsymbol{r}}) - 7(\hat{z}_1 \cdot \hat{z}_2)(\hat{z}_2 \cdot \hat{\boldsymbol{r}})^2], \tag{G.68}$$

$$S_{2,3,4} = \frac{\sqrt{5}}{\sqrt{112}}[\hat{z}_1 \cdot \hat{z}_2 \times \hat{\boldsymbol{r}}][(\hat{z}_1 \cdot \hat{\boldsymbol{r}}) + 2(\hat{z}_1 \cdot \hat{z}_2)(\hat{z}_2 \cdot \hat{\boldsymbol{r}}) - 7(\hat{z}_1 \cdot \hat{\boldsymbol{r}})(\hat{z}_2 \cdot \hat{\boldsymbol{r}})^2], \tag{G.69}$$

$$S_{3,2,4} = -\frac{\sqrt{5}}{\sqrt{112}}[\hat{z}_1 \cdot \hat{z}_2 \times \hat{\boldsymbol{r}}][(\hat{z}_2 \cdot \hat{\boldsymbol{r}}) + 2(\hat{z}_1 \cdot \hat{z}_2)(\hat{z}_1 \cdot \hat{\boldsymbol{r}}) - 7(\hat{z}_1 \cdot \hat{\boldsymbol{r}})^2(\hat{z}_2 \cdot \hat{\boldsymbol{r}})], \tag{G.70}$$

$$S_{3,4,2} = -\frac{\sqrt{5}}{\sqrt{112}}[\hat{z}_1 \cdot \hat{z}_2 \times \hat{\boldsymbol{r}}][(\hat{z}_2 \cdot \hat{\boldsymbol{r}}) + 2(\hat{z}_1 \cdot \hat{z}_2)(\hat{z}_1 \cdot \hat{\boldsymbol{r}}) - 7(\hat{z}_1 \cdot \hat{z}_2)^2(\hat{z}_2 \cdot \hat{\boldsymbol{r}})], \tag{G.71}$$

$$S_{4,2,3} = \frac{\sqrt{5}}{\sqrt{112}}[\hat{z}_1 \cdot \hat{z}_2 \times \hat{\boldsymbol{r}}][(\hat{z}_1 \cdot \hat{z}_2) + 2(\hat{z}_1 \cdot \hat{\boldsymbol{r}})(\hat{z}_2 \cdot \hat{\boldsymbol{r}}) - 7(\hat{z}_1 \cdot \hat{\boldsymbol{r}})^2(\hat{z}_1 \cdot \hat{z}_2)], \tag{G.72}$$

$$S_{4,3,2} = \frac{\sqrt{5}}{\sqrt{112}}[\hat{z}_1 \cdot \hat{z}_2 \times \hat{\boldsymbol{r}}][(\hat{z}_1 \cdot \hat{\boldsymbol{r}}) + 2(\hat{z}_1 \cdot \hat{z}_2)(\hat{z}_2 \cdot \hat{\boldsymbol{r}}) - 7(\hat{z}_1 \cdot \hat{\boldsymbol{r}})(\hat{z}_1 \cdot \hat{z}_2)^2], \tag{G.73}$$

$$S_{1,3,3} = -\frac{\sqrt{3}}{\sqrt{112}}[\hat{z}_1 \cdot \hat{z}_2 \times \hat{\boldsymbol{r}}][1 - 5(\hat{z}_2 \cdot \hat{\boldsymbol{r}})^2], \tag{G.74}$$

$$S_{3,1,3} = -\frac{\sqrt{3}}{\sqrt{112}}[\hat{z}_1 \cdot \hat{z}_2 \times \hat{\boldsymbol{r}}][1 - 5(\hat{z}_1 \cdot \hat{\boldsymbol{r}})^2], \tag{G.75}$$

$$S_{3,3,1} = -\frac{\sqrt{3}}{\sqrt{112}}[\hat{z}_1 \cdot \hat{z}_2 \times \hat{\boldsymbol{r}}][1 - 5(\hat{z}_1 \cdot \hat{z}_2)^2], \tag{G.76}$$

$$\begin{aligned} S_{3,3,3} = {} & \frac{1}{\sqrt{168}}[\hat{z}_1 \cdot \hat{z}_2 \times \hat{\boldsymbol{r}}][2 - 5(\hat{z}_1 \cdot \hat{\boldsymbol{r}})^2 - 5(\hat{z}_1 \cdot \hat{z}_2)^2 - 5(\hat{z}_2 \cdot \hat{\boldsymbol{r}})^2 \\ & +25(\hat{z}_1 \cdot \hat{\boldsymbol{r}})(\hat{z}_1 \cdot \hat{z}_2)(\hat{z}_2 \cdot \hat{\boldsymbol{r}}))], \end{aligned} \tag{G.77}$$

$$S_{1,4,4} = \frac{1}{\sqrt{80}}[\hat{z}_1 \cdot \hat{z}_2 \times \hat{\boldsymbol{r}}](\hat{z}_2 \cdot \hat{\boldsymbol{r}})(2 + 35(\hat{z}_2 \cdot \hat{\boldsymbol{r}})^3], \tag{G.78}$$

$$S_{4,1,4} = -\frac{1}{\sqrt{80}}[\hat{z}_1 \cdot \hat{z}_2 \times \hat{\boldsymbol{r}}](\hat{z}_1 \cdot \hat{\boldsymbol{r}})[2 + 35(\hat{z}_1 \cdot \hat{\boldsymbol{r}})^3], \tag{G.79}$$

$$S_{4,4,1} = -\frac{1}{\sqrt{80}}[\hat{z}_1 \cdot \hat{z}_2 \times \hat{\boldsymbol{r}}](\hat{z}_1 \cdot \hat{z}_2)[2 + 35(\hat{z}_1 \cdot \hat{z}_2)^3], \tag{G.80}$$

$$\begin{aligned} S_{4,4,3} = {} & -\frac{1}{\sqrt{997920}}[\hat{z}_1 \cdot \hat{z}_2 \times \hat{\boldsymbol{r}}][759(\hat{z}_1 \cdot \hat{z}_2) - 225(\hat{z}_1 \cdot \hat{\boldsymbol{r}})(\hat{z}_2 \cdot \hat{\boldsymbol{r}}) \\ & -1050(\hat{z}_1 \cdot \hat{z}_2)(\hat{z}_1 \cdot \hat{\boldsymbol{r}})^2 - 1050(\hat{z}_1 \cdot \hat{z}_2)(\hat{z}_2 \cdot \hat{\boldsymbol{r}})^2 - 1225(\hat{z}_1 \cdot \hat{z}_2)^3 \\ & +1470(\hat{z}_1 \cdot \hat{z}_2)^4 + 3675(\hat{z}_1 \cdot \hat{z}_2)^2(\hat{z}_1 \cdot \hat{\boldsymbol{r}})(\hat{z}_2 \cdot \hat{\boldsymbol{r}})], \end{aligned} \tag{G.81}$$

$$\begin{aligned} S_{4,3,4} = {} & -\frac{1}{\sqrt{997920}}[\hat{z}_1 \cdot \hat{z}_2 \times \hat{\boldsymbol{r}}][\ 759(\hat{z}_1 \cdot \hat{\boldsymbol{r}}) - 225(\hat{z}_1 \cdot \hat{z}_2)(\hat{z}_2 \cdot \hat{\boldsymbol{r}}) \\ & -1050(\hat{z}_1 \cdot \hat{z}_2)^2(\hat{z}_1 \cdot \hat{\boldsymbol{r}}) - 1050(\hat{z}_1 \cdot \hat{\boldsymbol{r}})(\hat{z}_2 \cdot \hat{\boldsymbol{r}})^2 - 1225(\hat{z}_1 \cdot \hat{\boldsymbol{r}})^3 \\ & +1470(\hat{z}_1 \cdot \hat{\boldsymbol{r}})^4 + 3675(\hat{z}_1 \cdot \hat{z}_2)(\hat{z}_1 \cdot \hat{\boldsymbol{r}})^2(\hat{z}_2 \cdot \hat{\boldsymbol{r}})], \end{aligned} \tag{G.82}$$

$$S_{3,4,4} = \frac{1}{\sqrt{997920}}[\hat{z}_1 \cdot \hat{z}_2 \times \hat{r}][759(\hat{z}_2 \cdot \hat{r}) - 225(\hat{z}_1 \cdot \hat{z}_2)(\hat{z}_1 \cdot \hat{r}) \\ -1050(\hat{z}_1 \cdot \hat{r})^2(\hat{z}_2 \cdot \hat{r}) - 1050(\hat{z}_1 \cdot \hat{z}_2)^2(\hat{z}_2 \cdot \hat{r}) - 1225(\hat{z}_2 \cdot \hat{r})^3 \\ +1470(\hat{z}_2 \cdot \hat{r})^4 + 3675(\hat{z}_1 \cdot \hat{z}_2)(\hat{z}_1 \cdot \hat{r})(\hat{z}_2 \cdot \hat{r})^2]. \tag{G.83}$$

All the invariants reported are for uniaxial particles. In what follows we also list a small selection of explicit expressions for biaxial particles [Berardi and Zannoni, 2000]. In this case, the molecular frame requires to specify the three axes for the two molecules: $\hat{\boldsymbol{x}}_1, \hat{\boldsymbol{y}}_1, \hat{\boldsymbol{z}}_1$ and $\hat{\boldsymbol{x}}_2, \hat{\boldsymbol{y}}_2, \hat{\boldsymbol{z}}_2$.

First-rank biaxial rotational invariants.

$$S^{1,0}_{1,0,1} = \frac{1}{\sqrt{6}}[+(\hat{x}_1 \cdot \hat{r}) - i\,(\hat{y}_1 \cdot \hat{r})], \tag{G.84}$$

$$S^{0,1}_{0,1,1} = \frac{1}{\sqrt{6}}[-(\hat{x}_2 \cdot \hat{r}) + i\,(\hat{y}_2 \cdot \hat{r})], \tag{G.85}$$

$$S^{1,0}_{1,1,0} = \frac{1}{\sqrt{6}}[+(\hat{x}_1 \cdot \hat{z}_2) - i\,(\hat{y}_1 \cdot \hat{z}_2)], \tag{G.86}$$

$$S^{0,1}_{1,1,0} = \frac{1}{\sqrt{6}}[+(\hat{z}_1 \cdot \hat{x}_2) - i\,(\hat{z}_1 \cdot \hat{y}_2)], \tag{G.87}$$

$$S^{-1,0}_{1,0,1} = \frac{1}{\sqrt{6}}[-(\hat{x}_1 \cdot \hat{r}) - i\,(\hat{y}_1 \cdot \hat{r})], \tag{G.88}$$

$$S^{0,-1}_{0,1,1} = \frac{1}{\sqrt{6}}[+(\hat{x}_2 \cdot \hat{r}) + i\,(\hat{y}_2 \cdot \hat{r})], \tag{G.89}$$

$$S^{-1,0}_{1,1,0} = \frac{1}{\sqrt{6}}[-(\hat{x}_1 \cdot \hat{z}_2) - i\,(\hat{y}_1 \cdot \hat{z}_2)], \tag{G.90}$$

$$S^{0,-1}_{1,1,0} = \frac{1}{\sqrt{6}}[-(\hat{z}_1 \cdot \hat{x}_2) - i\,(\hat{z}_1 \cdot \hat{y}_2)], \tag{G.91}$$

$$S^{1,\pm 1}_{1,1,0} = \frac{1}{2\sqrt{3}}\{\mp(\hat{x}_1 \cdot \hat{x}_2) + (\hat{y}_1 \cdot \hat{y}_2) + i\,[(\hat{x}_1 \cdot \hat{y}_2) + (\hat{y}_1 \cdot \hat{x}_2)]\}, \tag{G.92}$$

$$S^{-1,\pm 1}_{1,1,0} = \frac{1}{2\sqrt{3}}\{\pm(\hat{x}_1 \cdot \hat{x}_2) + (\hat{y}_1 \cdot \hat{y}_2) - i\,[(\hat{x}_1 \cdot \hat{y}_2) - (\hat{y}_1 \cdot \hat{x}_2)]\}. \tag{G.93}$$

Second-Rank biaxial rotational invariants.

$$S^{0,\mp 2}_{0,2,2} = -\frac{\sqrt{3}}{\sqrt{40}}[\mp i\,(\hat{x}_2 \cdot \hat{r}) + (\hat{y}_2 \cdot \hat{r})]^2, \tag{G.94}$$

$$S^{0,\mp 1}_{0,2,2} = \frac{\sqrt{3}}{\sqrt{10}}[\pm(\hat{x}_2 \cdot \hat{r}) + i\,(\hat{y}_2 \cdot \hat{r})]\,(\hat{z}_2 \cdot \hat{r}), \tag{G.95}$$

$$S^{\mp 2,0}_{2,0,2} = -\frac{\sqrt{3}}{\sqrt{40}}[\mp i\,(\hat{x}_1 \cdot \hat{r}) + (\hat{y}_1 \cdot \hat{r})]^2, \tag{G.96}$$

$$S^{\mp 1,0}_{2,0,2} = \frac{\sqrt{3}}{\sqrt{10}}[\pm(\hat{x}_1 \cdot \hat{r}) + i\,(\hat{y}_1 \cdot \hat{r})]\,(\hat{z}_1 \cdot \hat{r}), \tag{G.97}$$

$$S_{2,2,0}^{-2,\mp 2} = \frac{1}{4\sqrt{5}}\{\mp(\hat{x}_1\cdot\hat{x}_2) + (\hat{y}_1\cdot\hat{y}_2) - i\,[(\hat{x}_1\cdot\hat{y}_2) + (\hat{y}_1\cdot\hat{x}_2)]\}^2, \tag{G.98}$$

$$S_{2,2,0}^{-2,\mp 1} = \frac{1}{2\sqrt{5}}\{-(\hat{x}_1\cdot\hat{y}_2) \mp (\hat{y}_1\cdot\hat{x}_2) \pm i\,[(\hat{x}_1\cdot\hat{x}_2) \mp (\hat{y}_1\cdot\hat{y}_2)]\}$$
$$[-i\,(\hat{x}_1\cdot\hat{z}_2) + (\hat{y}_1\cdot\hat{z}_2)], \tag{G.99}$$

$$S_{2,2,0}^{-2,0} = \frac{\sqrt{3}}{\sqrt{40}}\{-(\hat{x}_1\cdot\hat{x}_2)^2 - (\hat{x}_1\cdot\hat{y}_2)^2 + (\hat{y}_1\cdot\hat{x}_2)^2 + (\hat{y}_1\cdot\hat{y}_2)^2$$
$$-2i\,[(\hat{x}_1\cdot\hat{x}_2)(\hat{y}_1\cdot\hat{x}_2) + (\hat{x}_1\cdot\hat{y}_2)(\hat{y}_1\cdot\hat{y}_2)]\}, \tag{G.100}$$

$$S_{2,2,0}^{-1,-2} = \frac{1}{2\sqrt{5}}\{-(\hat{x}_1\cdot\hat{y}_2) - (\hat{y}_1\cdot\hat{x}_2) + i\,[(\hat{x}_1\cdot\hat{x}_2) - (\hat{y}_1\cdot\hat{y}_2)]\}$$
$$[-i\,(\hat{z}_1\cdot\hat{x}_2) + (\hat{z}_1\cdot\hat{y}_2)], \tag{G.101}$$

$$S_{2,2,0}^{-1,-1} = \frac{1}{2\sqrt{5}}\{(\hat{x}_1\cdot\hat{x}_2) - (\hat{y}_1\cdot\hat{y}_2) + i\,[(\hat{x}_1\cdot\hat{y}_2) + (\hat{y}_1\cdot\hat{x}_2)]\}[1 + 2(\hat{z}_1\cdot\hat{z}_2)], \tag{G.102}$$

$$S_{2,2,0}^{-1,0} = \frac{\sqrt{3}}{\sqrt{10}}\Big\{(\hat{x}_1\cdot\hat{z}_2) - (\hat{z}_1\cdot\hat{x}_2)[(\hat{x}_1\cdot\hat{x}_2) - (\hat{y}_1\cdot\hat{y}_2)] - (\hat{z}_1\cdot\hat{y}_2)[(\hat{x}_1\cdot\hat{y}_2) + (\hat{y}_1\cdot\hat{x}_2)]$$
$$+i\,[(\hat{y}_1\cdot\hat{z}_2) - (\hat{z}_1\cdot\hat{x}_2)[(\hat{x}_1\cdot\hat{y}_2) + (\hat{y}_1\cdot\hat{x}_2)]$$
$$+(\hat{z}_1\cdot\hat{y}_2)[(\hat{x}_1\cdot\hat{x}_2) - (\hat{y}_1\cdot\hat{y}_2)]\Big\}, \tag{G.103}$$

$$S_{2,2,0}^{-1,1} = -\frac{1}{2\sqrt{5}}\Big\{(\hat{x}_1\cdot\hat{x}_2) + (\hat{y}_1\cdot\hat{y}_2) + 2(\hat{x}_1\cdot\hat{z}_2)(\hat{z}_1\cdot\hat{x}_2) + 2(\hat{y}_1\cdot\hat{z}_2)(\hat{z}_1\cdot\hat{y}_2)$$
$$-i\,[+(\hat{x}_1\cdot\hat{y}_2) - (\hat{y}_1\cdot\hat{x}_2) - 2(\hat{y}_1\cdot\hat{z}_2)(\hat{z}_1\cdot\hat{x}_2) + 2(\hat{x}_1\cdot\hat{z}_2)(\hat{z}_1\cdot\hat{y}_2)]\Big\}, \tag{G.104}$$

$$S_{2,2,0}^{-1,2} = \frac{1}{2\sqrt{5}}\{(\hat{x}_1\cdot\hat{x}_2) + (\hat{y}_1\cdot\hat{y}_2) - i\,((\hat{x}_1\cdot\hat{y}_2) - (\hat{y}_1\cdot\hat{x}_2)))((\hat{z}_1\cdot\hat{x}_2) - i\,(\hat{z}_1\cdot\hat{y}_2)\}, \tag{G.105}$$

$$S_{2,2,0}^{0,\mp 2} = \frac{\sqrt{3}}{\sqrt{40}}\Big\{-(\hat{x}_1\cdot\hat{x}_2)^2 + (\hat{x}_1\cdot\hat{y}_2)^2 - (\hat{y}_1\cdot\hat{x}_2)^2 + (\hat{y}_1\cdot\hat{y}_2)^2$$
$$\mp 2i\,[(\hat{x}_1\cdot\hat{x}_2)(\hat{x}_1\cdot\hat{y}_2) + (\hat{y}_1\cdot\hat{x}_2)(\hat{y}_1\cdot\hat{y}_2)]\Big\}, \tag{G.106}$$

$$S_{2,2,0}^{0,\mp 1} = \frac{\sqrt{3}}{\sqrt{10}}\Big\{\mp(\hat{z}_1\cdot\hat{x}_2) \mp (\hat{x}_1\cdot\hat{x}_2)(\hat{x}_1\cdot\hat{z}_2) \mp (\hat{x}_1\cdot\hat{z}_2)(\hat{y}_1\cdot\hat{y}_2)$$
$$\pm(\hat{x}_1\cdot\hat{y}_2)(\hat{y}_1\cdot\hat{z}_2) \mp (\hat{y}_1\cdot\hat{x}_2)(\hat{y}_1\cdot\hat{z}_2) - i\,[(\hat{z}_1\cdot\hat{y}_2) + (\hat{x}_1\cdot\hat{y}_2)(\hat{x}_1\cdot\hat{z}_2)$$
$$+(\hat{x}_1\cdot\hat{x}_2)(\hat{y}_1\cdot\hat{z}_2) - (\hat{x}_1\cdot\hat{z}_2)(\hat{y}_1\cdot\hat{x}_2) + (\hat{y}_1\cdot\hat{y}_2)(\hat{y}_1\cdot\hat{z}_2)]\Big\}, \tag{G.107}$$

$$S_{2,2,0}^{1,-2} = \frac{1}{2\sqrt{5}}\{(\hat{x}_1\cdot\hat{y}_2) - (\hat{y}_1\cdot\hat{x}_2) - i\,[(\hat{x}_1\cdot\hat{x}_2) + (\hat{y}_1\cdot\hat{y}_2)]\}[-i\,(\hat{z}_1\cdot\hat{x}_2) + (\hat{z}_1\cdot\hat{y}_2)], \tag{G.108}$$

$$S_{2,2,0}^{1,-1} = \frac{1}{2\sqrt{5}}\{(\hat{x}_1\cdot\hat{x}_2) + (\hat{y}_1\cdot\hat{y}_2) + i\,((\hat{x}_1\cdot\hat{y}_2) - (\hat{y}_1\cdot\hat{x}_2))\}[1 - 2(\hat{z}_1\cdot\hat{z}_2)], \tag{G.109}$$

$$S_{2,2,0}^{1,0} = \frac{\sqrt{3}}{\sqrt{10}}\{-(\hat{x}_1\cdot\hat{z}_2) - (\hat{y}_1\cdot\hat{y}_2)(\hat{z}_1\cdot\hat{x}_2) + (\hat{x}_1\cdot\hat{x}_2)(\hat{z}_1\cdot\hat{x}_2) + (\hat{x}_1\cdot\hat{y}_2)(\hat{z}_1\cdot\hat{y}_2)$$
$$+(\hat{y}_1\cdot\hat{x}_2)(\hat{z}_1\cdot\hat{y}_2) + i\,[+(\hat{y}_1\cdot\hat{z}_2) + (\hat{x}_1\cdot\hat{x}_2)(\hat{z}_1\cdot\hat{y}_2) - (\hat{x}_1\cdot\hat{y}_2)(\hat{z}_1\cdot\hat{x}_2)$$
$$-(\hat{y}_1\cdot\hat{x}_2)(\hat{z}_1\cdot\hat{x}_2) - (\hat{y}_1\cdot\hat{y}_2)(\hat{z}_1\cdot\hat{y}_2)]\}, \tag{G.110}$$

$$S_{2,2,0}^{1,1} = \frac{1}{2\sqrt{5}}\{(\hat{x}_1\cdot\hat{x}_2) - (\hat{y}_1\cdot\hat{y}_2) - i\,[(\hat{x}_1\cdot\hat{y}_2) + (\hat{y}_1\cdot\hat{x}_2)]\}[1 + 2(\hat{z}_1\cdot\hat{z}_2)], \tag{G.111}$$

$$S_{2,2,0}^{1,2} = \frac{1}{2\sqrt{5}}\{(\hat{x}_1\cdot\hat{y}_2) + (\hat{y}_1\cdot\hat{x}_2) + i\,((\hat{x}_1\cdot\hat{x}_2) - (\hat{y}_1\cdot\hat{y}_2)))(i\,(\hat{z}_1\cdot\hat{x}_2) + (\hat{z}_1\cdot\hat{y}_2)), \tag{G.112}$$

$$S^{2,\mp 2}_{2,2,0} = \frac{1}{4\sqrt{5}}\{\pm(\hat{x}_1\cdot\hat{x}_2)+(\hat{y}_1\cdot\hat{y}_2)+i\,((\hat{x}_1\cdot\hat{y}_2)\mp(\hat{y}_1\cdot\hat{x}_2))\}^2, \tag{G.113}$$

$$S^{2,-1}_{2,2,0} = \frac{1}{2\sqrt{5}}\{(\hat{x}_1\cdot\hat{x}_2)+(\hat{y}_1\cdot\hat{y}_2)+i\,((\hat{x}_1\cdot\hat{y}_2)-(\hat{y}_1\cdot\hat{x}_2))((\hat{x}_1\cdot\hat{z}_2)-i\,(\hat{y}_1\cdot\hat{z}_2)), \tag{G.114}$$

$$S^{2,0}_{2,2,0} = \frac{\sqrt{3}}{40^{1/2}}\{-(\hat{x}_1\cdot\hat{x}_2)^2-(\hat{x}_1\cdot\hat{y}_2)^2+(\hat{y}_1\cdot\hat{x}_2)^2+(\hat{y}_1\cdot\hat{y}_2)^2$$
$$+2i\,[(\hat{x}_1\cdot\hat{x}_2)(\hat{y}_1\cdot\hat{x}_2)+(\hat{x}_1\cdot\hat{y}_2)(\hat{y}_1\cdot\hat{y}_2)]\}, \tag{G.115}$$

$$S^{2,1}_{2,2,0} = \frac{1}{2\sqrt{5}}\{(\hat{x}_1\cdot\hat{y}_2)+(\hat{y}_1\cdot\hat{x}_2)+i\,[(\hat{x}_1\cdot\hat{x}_2)-(\hat{y}_1\cdot\hat{y}_2)]\}[i\,(\hat{x}_1\cdot\hat{z}_2)+(\hat{y}_1\cdot\hat{z}_2)]. \tag{G.116}$$

The invariants can be used, e.g., to expand intermolecular potentials or pair correlation functions. In general, the rotationally invariant expansion of an arbitrary pairwise function will be

$$f(r,\Omega_{1L},\Omega_{2L},\Omega_{rL}) = \sum_{\substack{L_1,L_2,L\\k_1,k_2}} f^{k_1,k_2}_{L_1,L_2,L}(r)S^{k_1,k_2}_{L_1,L_2,L}(\Omega_{1L},\Omega_{2L},\Omega_{rL}). \tag{G.117}$$

Having dealt with rotational invariance, other symmetries of the pairwise function f can be implemented, when present. We have already seen in Section G.1 how to implement point group symmetries for particles and for phases. In addition, for identical particles, the permutation symmetry $S^{n_1,n_2}_{L_1,L_2,L}(\Omega_{1\mathrm{L}},\Omega_{2\mathrm{L}},\Omega_{r\mathrm{L}}) = S^{n_2,n_1}_{L_2,L_1,L}(\Omega_{2\mathrm{L}},\Omega_{1\mathrm{L}},\Omega_{\mathrm{L}r})$. A pair function for non-chiral molecules should also be invariant under inversion.

Appendix H

Quaternions and Rotations

Introducing quaternions. A vector in two-dimensions can be represented by a two components complex number, $z = (a, b) = a + i\, b$, with $i^2 = -1$. In turn, a complex number u with unit modulus: $uu^* = u^*u = 1$, can be seen as a rotation operator for 2D vectors. This is more transparent using the Euler exponential representation

$$u(\theta) \;=\; \mathrm{e}^{i\theta} = (\cos\theta + i\,\sin\theta). \tag{H.1}$$

This can be considered as an operator acting on an arbitrary 2D vector $\boldsymbol{v}$ written as a complex number $v = (v_1 + i v_2)$. In fact,

$$v' = u(\theta)v = (\cos\theta + i\,\sin\theta)(v_1 + i v_2) = (v_1\cos\theta - v_2\sin\theta) + i(v_1\sin\theta + v_2\cos\theta), \tag{H.2}$$

which corresponds to a counterclockwise rotation of θ of the vector. Thus, a unitary complex number could also be written as an orthogonal rotation matrix

$$u(\theta) = \mathbf{R}(\theta) = \begin{pmatrix} \cos\theta & -\sin\theta \\ \sin\theta & \cos\theta \end{pmatrix}. \tag{H.3}$$

To extend a similar formalism to 3D vectors, one can introduce a generalization of complex numbers, the quaternions $\mathbf{q}$, first introduced by Hamilton. They can be defined as an ordered set of four real numbers q_0, q_1, q_2, q_3 representing the components in a certain (quaternion) basis set $\{1, \boldsymbol{e}_1, \boldsymbol{e}_2, \boldsymbol{e}_3\}$:

$$\mathbf{q} = q_0 + q_1\boldsymbol{e}_1 + q_2\boldsymbol{e}_2 + q_3\boldsymbol{e}_3 = q_0 + \boldsymbol{q}\cdot\boldsymbol{e}, \tag{H.4}$$

where q_0 is called the scalar part and $\boldsymbol{q} \equiv (q_1, q_2, q_3)$ is a 3D vector, called the *vector part* of $\mathbf{q}$.[1] The basis set components $\boldsymbol{e}_1$, $\boldsymbol{e}_2$, $\boldsymbol{e}_3$ obey the combination rule

$$\boldsymbol{e}_i\boldsymbol{e}_j \;=\; -\delta_{i,j} + \varepsilon_{ijk}\boldsymbol{e}_k, \tag{H.5}$$

where $\delta_{i,j}$ is a Kronecker delta and ε_{ijk} a Levi–Civita symbol (see Eq. A.9). The quaternion has a complex conjugate $\mathbf{q}^*$ and a norm (or 'length') $\bar{q}$:

$$\sum_{i=0}^{3} q_i q_i^* = \bar{q}^2. \tag{H.6}$$

[1] Note that we use bold roman, e.g. $\mathbf{a}$, for quaternions and bold italic, e.g. $\boldsymbol{a}$, for vectors.

A convenient representation of quaternions can be realized in terms of the Pauli spin angular momentum matrices $\boldsymbol{\sigma}_1, \boldsymbol{\sigma}_2, \boldsymbol{\sigma}_3$. Indeed, defining the basis set components $\boldsymbol{e}_i$ as

$$\boldsymbol{e}_1 = -i\boldsymbol{\sigma}_1,\ \boldsymbol{e}_2 = -i\boldsymbol{\sigma}_2,\ \boldsymbol{e}_3 = -i\boldsymbol{\sigma}_3, \tag{H.7}$$

we see that the combination law Eq. H.5 is obeyed. It is easily verified using Eqs. H.4 and H.5 or directly from the Pauli matrices, that the product of two quaternions $\mathbf{q}$ (Eq. H.4) and $\mathbf{q}'$, i.e.

$$\mathbf{q}' = q_0' + q_1'\boldsymbol{e}_1 + q_2'\boldsymbol{e}_2 + q_3'\boldsymbol{e}_3, \tag{H.8}$$

is also a quaternion, that can be written as

$$\mathbf{q}\mathbf{q}' = (q_0 q_0' - \boldsymbol{q}\cdot\boldsymbol{q}') + [q_0\boldsymbol{q}' + q_0'\boldsymbol{q} + (\boldsymbol{q}\times\boldsymbol{q}')]\cdot\boldsymbol{e}, \tag{H.9}$$

and in particular, $\mathbf{q}\mathbf{q}^* = \bar{q}^2$, while

$$\mathbf{q}'\mathbf{q} = (q_0' q_0 - \boldsymbol{q}'\cdot\boldsymbol{q}) + [q_0'\boldsymbol{q} + q_0\boldsymbol{q}' + (\boldsymbol{q}'\times\boldsymbol{q})]\cdot\boldsymbol{e}. \tag{H.10}$$

The product is non-commutative, i.e. $\mathbf{q}\mathbf{q}' \neq \mathbf{q}'\mathbf{q}$, unless the vector product in Eqs. H.9 and H.10 is 0. The inverse of a unitary quaternion is $\mathbf{q}^{-1} = (q_0, -q_1, -q_2, -q_3)$:

$$\mathbf{q}\mathbf{q}^{-1} = q_0^2 + q_1^2 + q_2^2 + q_3^2 = 1. \tag{H.11}$$

Quaternions form a group and constitute a non-commutative algebra which, according to Frobenius theorem [Bahturin, 1993], is the only one, besides real and complex numbers, where every non-null element has an inverse, which is $\mathbf{q}^{-1} = \mathbf{q}^*/\mathbf{q}\mathbf{q}^*$. Just as a complex number can be written as a 2×2 matrix, a quaternion can also be written as a 4×4 matrix. Indeed, since a quaternion is a four-component column matrix and the product of two quaternions is also a quaternion, we can write the product in 4×4 matrix and vector form as

$$\mathbf{q}'' = \mathbf{q}\mathbf{q}' = \begin{pmatrix} +q_0 & -q_1 & -q_2 & -q_3 \\ +q_1 & +q_0 & -q_3 & +q_2 \\ +q_2 & +q_3 & +q_0 & -q_1 \\ +q_3 & -q_2 & +q_1 & +q_0 \end{pmatrix} \begin{pmatrix} q_0' \\ q_1' \\ q_2' \\ q_3' \end{pmatrix}. \tag{H.12}$$

Quaternions and rotations. We now wish to discuss quaternions as a means of representing the orientation of a rigid particle and their relation to Euler angles [Rose, 1957]. The three-dimensional rotation group $SO(3)$ is the group of transformations which leave the quantity $v_x^2 + v_y^2 + v_z^2$, i.e. the length of a vector $\boldsymbol{v}$, invariant. Any such transformation can be expressed in a quaternion form as [Girard, 1984] $\boldsymbol{v}' = \mathbf{u}\,\boldsymbol{v}\,\mathbf{u}^*$, where $\mathbf{u}$ is a unitary quaternion. In particular, to rotate the vector $\boldsymbol{v}$ of an angle θ around an axis $\boldsymbol{n}$, we can write $\mathbf{u} = \mathbf{u}_{\boldsymbol{n}}(\theta) = \hat{\mathbf{R}}_{\boldsymbol{n}}(\theta)$. To obtain an explicit expression for the unitary quaternion, which acts as rotation operator, we recall that the operator for a right-handed rotation of an angle θ around an axis $\boldsymbol{n}$ is $\hat{\mathbf{R}}_{\boldsymbol{n}}(\theta) = \exp\left(-i\theta\,\boldsymbol{n}\cdot\hat{\mathbf{J}}\right)$, where $\hat{\mathbf{J}}$ is the angular momentum operator. Choosing an angular momentum (1/2), $\hat{\mathbf{J}} = \hat{\boldsymbol{\sigma}}/2$, (also called *spinor*) representation and using the Pauli basis set, Eq. H.7, the rotation operator can be rewritten in the quaternion basis $\{\boldsymbol{e}\}$ as

$$\hat{\mathbf{R}}_n(\theta) = \exp\left(\frac{\theta}{2}\boldsymbol{n}\cdot\boldsymbol{e}\right). \tag{H.13}$$

Using the generalized Euler formula $\exp(\lambda\boldsymbol{n}\cdot\boldsymbol{e}) = \cos\lambda + (\boldsymbol{n}\cdot\boldsymbol{e})\sin\lambda$, which can be demonstrated by a Taylor series expansion of both sides, we can also write $\mathbf{u} = \cos\frac{\theta}{2} + \boldsymbol{n}\cdot\boldsymbol{e}\sin\frac{\theta}{2}$. Thus, the unitary quaternion, $\mathbf{u}$, with components

$$u_0 = \cos\frac{\theta}{2},\ u_1 = \sin\frac{\theta}{2}n_X,\ u_2 = \sin\frac{\theta}{2}n_Y,\ u_3 = \sin\frac{\theta}{2}n_Z \tag{H.14}$$

performs a rotation of an angle θ around the axis $\boldsymbol{n} = (n_x, n_y, n_z)$. Note that any 3D vector $\boldsymbol{a} = (a_x, a_y, a_z)$ can be considered as an 'axial' quaternion $\mathbf{a} = (0, a_x, a_y, a_z)$, but only if $\boldsymbol{a}\cdot\boldsymbol{a} = 1$ does the quaternion $\mathbf{a}$ corresponds to a rotation operator. To obtain an alternative explicit representation of quaternions in terms of Euler angles (instead of a rotation around a certain vector), we can consider the rotation operator, written in terms of laboratory-fixed angular momentum operators: $\hat{\mathcal{D}}(\alpha,\beta,\gamma) = \exp(-i\alpha\hat{\mathbf{J}}_Z)\exp(-i\beta\hat{\mathbf{J}}_Y)\exp(-i\gamma\hat{\mathbf{J}}_Z)$ (see Appendix F). In the $\mathbf{J} = \hat{\boldsymbol{\sigma}}/2$ representation

$$\hat{\mathcal{D}}(\alpha,\beta,\gamma) = \mathrm{e}^{-i\frac{\alpha}{2}\hat{\sigma}_Z}\,\mathrm{e}^{-i\frac{\beta}{2}\hat{\sigma}_Y}\,\mathrm{e}^{-i\frac{\gamma}{2}\hat{\sigma}_Z} = \mathrm{e}^{\frac{\alpha}{2}e_Z}\,\mathrm{e}^{\frac{\beta}{2}e_Y}\,\mathrm{e}^{\frac{\gamma}{2}e_Z}. \tag{H.15}$$

Repeated use of the generalized Euler formula gives $\mathcal{D}(\alpha,\beta,\gamma) = u_0 + \boldsymbol{u}\cdot\boldsymbol{e}$, with

$$u_0 = \cos\frac{\beta}{2}\cos\frac{\gamma+\alpha}{2}, \tag{H.16a}$$

$$u_1 = \sin\frac{\beta}{2}\sin\frac{\gamma-\alpha}{2}, \tag{H.16b}$$

$$u_2 = \sin\frac{\beta}{2}\cos\frac{\gamma-\alpha}{2}, \tag{H.16c}$$

$$u_3 = \cos\frac{\beta}{2}\sin\frac{\gamma+\alpha}{2}. \tag{H.16d}$$

The rank $\frac{1}{2}$ Wigner rotation matrix has components

$$\mathcal{D}^{\frac{1}{2}}_{\frac{1}{2},\frac{1}{2}} = u_0 + iu_3, \qquad \mathcal{D}^{\frac{1}{2}}_{\frac{1}{2},-\frac{1}{2}} = -u_2 - iu_1,$$
$$\mathcal{D}^{\frac{1}{2}}_{-\frac{1}{2},\frac{1}{2}} = u_2 - iu_1, \qquad \mathcal{D}^{\frac{1}{2}}_{-\frac{1}{2},-\frac{1}{2}} = u_0 - iu_3. \tag{H.17}$$

These rank $\frac{1}{2}$ functions can now be coupled, using Eq. F.31, to build Wigner rotation matrices of rank 1 [Rose, 1957; Zannoni and Guerra, 1981][2]

$$\mathcal{D}^1_{m,n} = \sum_{q,p} C\left(\frac{1}{2},\frac{1}{2},1;q,m-q\right)C\left(\frac{1}{2},\frac{1}{2},1;p,m-p\right)\mathcal{D}^{\frac{1}{2}}_{q,p}\mathcal{D}^{\frac{1}{2}}_{m-q,n-p}. \tag{H.18}$$

[2] Notice that the matrices $\mathcal{D}^{\frac{1}{2}}_{m,n}$ and $\mathcal{D}^{1}_{m,n}$ listed here correspond to $\mathcal{D}^{\frac{1}{2}*}_{-n,-m}$ and $\mathcal{D}^{1*}_{-n,-m}$ in the appendix of [Zannoni and Guerra, 1981].

The components of the rank 1 Wigner rotation matrix are

$$\mathscr{D}^1_{-1,-1} = u_0^2 - u_3^2 + 2iu_0u_3 \quad \text{(H.19a)}$$

$$\mathscr{D}^1_{-1,0} = \sqrt{2}(u_0u_2 - u_1u_3 + iu_2u_3 + iu_0u_1) \quad \text{(H.19b)}$$

$$\mathscr{D}^1_{-1,1} = u_2^2 - u_1^2 + 2iu_1u_2 \quad \text{(H.19c)}$$

$$\mathscr{D}^1_{0,-1} = -\sqrt{2}(u_0u_2 + u_1u_3 + iu_2u_3 - iu_0u_1) \quad \text{(H.19d)}$$

$$\mathscr{D}^1_{0,0} = u_0^2 + u_3^2 - u_2^2 - u_1^2 \quad \text{(H.19e)}$$

$$\mathscr{D}^1_{0,1} = \sqrt{2}(u_0u_2 + u_1u_3 - iu_2u_3 + iu_0u_1) \quad \text{(H.19f)}$$

$$\mathscr{D}^1_{1,-1} = u_2^2 - u_1^2 - 2iu_1u_2 \quad \text{(H.19g)}$$

$$\mathscr{D}^1_{1,0} = \sqrt{2}(u_0u_2 + u_1u_3 - iu_2u_3 + iu_0u_1) \quad \text{(H.19h)}$$

$$\mathscr{D}^1_{1,1} = u_0^2 - u_3^2 - 2iu_0u_3 \quad \text{(H.19i)}$$

Higher-rank matrices of integer rank can be obtained from these by systematic application of the decomposition scheme (Eq. F.31)

$$\mathscr{D}^L_{m,n} = \sum_{q,p} C(L-1,1,L;q,m-q)\,C(L-1,1,L;p,n-p)\,\mathscr{D}^{L-1}_{q,p}\,\mathscr{D}^1_{m-q,n-p}. \quad \text{(H.20)}$$

In particular, for the second-rank Wigner matrix components, we find

$$\mathscr{D}^2_{-2,-2} = (u_0 + iu_3)^4 \quad \text{(H.21a)}$$

$$\mathscr{D}^2_{-2,-1} = 2(u_2 + iu_1)(u_0 + iu_3)^3 \quad \text{(H.21b)}$$

$$\mathscr{D}^2_{-2,0} = \sqrt{6}(u_0 + iu_3)^2(u_2 + iu_1)^2 \quad \text{(H.21c)}$$

$$\mathscr{D}^2_{-2,1} = 2(u_0 + iu_3)(u_2 + iu_1)^3 \quad \text{(H.21d)}$$

$$\mathscr{D}^2_{-2,2} = (u_2 + iu_1)^4 \quad \text{(H.21e)}$$

$$\mathscr{D}^2_{-1,-2} = -2(u_2 - iu_1)(u_0 + iu_3)^3 \quad \text{(H.21f)}$$

$$\mathscr{D}^2_{-1,-1} = (u_0^2 - 3u_1^2 + u_3^2 - 3u_2^2)(u_0 + iu_3)^2 \quad \text{(H.21g)}$$

$$\mathscr{D}^2_{-2,-2} = (u_0 + iu_3)^4 \quad \text{(H.21h)}$$

$$\mathscr{D}^2_{-2,-1} = 2(u_2 + iu_1)(u_0 + iu_3)^3 \quad \text{(H.21i)}$$

$$\mathscr{D}^2_{-2,0} = \sqrt{6}(u_0 + iu_3)^2(u_2 + iu_1)^2 \quad \text{(H.21j)}$$

$$\mathscr{D}^2_{-2,1} = 2(u_0 + iu_3)(u_2 + iu_1)^3 \quad \text{(H.21k)}$$

$$\mathscr{D}^2_{-2,2} = (u_2 + iu_1)^4 \quad \text{(H.21l)}$$

$$\mathscr{D}^2_{-1,-2} = -2(u_2 - iu_1)(u_0 + iu_3)^3 \quad \text{(H.21m)}$$

$$\mathscr{D}^2_{-1,-1} = (u_0^2 - 3u_1^2 + u_3^2 - 3u_2^2)(u_0 + iu_3)^2 \quad \text{(H.21n)}$$

$$\mathscr{D}^2_{-1,0} = \sqrt{6}(u_0^2 - u_1^2 - u_2^2 + u_3^2)(u_0 + iu_3)(u_2 + iu_1) \quad \text{(H.21o)}$$

$$\mathscr{D}^2_{-1,1} = (3u_3^2 - u_2^2 - u_1^2 + 3u_0^2)(u_2 + iu_1)^2 \quad \text{(H.21p)}$$

$$\mathscr{D}^2_{-1,2} = 2(-iu_3 + u_0)(u_2 + iu_1)^3 \quad \text{(H.21q)}$$

$$\mathscr{D}^2_{0,-2} = \sqrt{6}(u_0 + iu_3)^2(u_2 - iu_1)^2 \quad \text{(H.21r)}$$

$$\mathscr{D}^2_{0,-1} = -\sqrt{6}(u_0^2 - u_1^2 - u_2^2 + u_3^2)(u_0 + iu_3)(u_2 - iu_1) \tag{H.21s}$$

$$\mathscr{D}^2_{0,0} = u_0^4 + u_3^4 - 4u_0^2u_1^2 - 4u_0^2u_2^2 + 2u_0^2u_3^2 - 4u_3^2u_1^2 - 4u_3^2 \\ u_2^2 + 2u_1^2u_2^2 + u_1^4 + u_2^4 \tag{H.21t}$$

$$\mathscr{D}^2_{0,1} = \sqrt{6}(u_0^2 - u_1^2 - u_2^2 + u_3^2)(-iu_3 + u_0)(u_2 + iu_1) \tag{H.21u}$$

$$\mathscr{D}^2_{0,2} = \sqrt{6}(-iu_3 + u_0)^2(u_2 + iu_1)^2 \tag{H.21v}$$

with the rest of 25 functions obtainable simply from Eq.F.7.

Let us now express the Cartesian rotation matrix $\mathbf{R}$ connecting a vector $\boldsymbol{v}$ in the rotated (primed) and original frames, i.e. $\boldsymbol{v}' = \mathbf{R}\boldsymbol{v}$. Inverting Eqs. H.16

$$\cos\beta = u_0^2 + u_3^2 - u_1^2 - u_2^2, \tag{H.22a}$$

$$\sin\beta = 2\sqrt{(u_0^2 + u_3^2)(u_1^2 + u_2^2)}, \tag{H.22b}$$

$$\sin\alpha = (u_0u_1 + u_2u_3)/\sqrt{(u_0^2 + u_3^2)(u_1^2 + u_2^2)}, \tag{H.22c}$$

$$\cos\alpha = (u_0u_2 - u_1u_3)/\sqrt{(u_0^2 + u_3^2)(u_1^2 + u_2^2)}, \tag{H.22d}$$

$$\sin\gamma = (u_2u_3 - u_1u_0)/\sqrt{(u_0^2 + u_3^2)(u_1^2 + u_2^2)}, \tag{H.22e}$$

$$\cos\gamma = (u_0u_2 + u_1u_3)/\sqrt{(u_0^2 + u_3^2)(u_1^2 + u_2^2)}, \tag{H.22f}$$

and we could substitute these in the expression for the Cartesian rotation matrix $\mathbf{R}(\alpha, \beta, \gamma)$, Eq. B.8. However, more elegantly, we can also obtain $\mathbf{R}$ from $\mathscr{D}^1$ recalling that $\mathscr{D}^1$ is the transpose of $\mathbf{R}$ when this is written in spherical coordinates, so that $\mathbf{R} = \mathbf{U}^{T*}(\mathscr{D}^1)^T\mathbf{U}$, where $\mathbf{U}$ is the matrix converting Cartesian to spherical vectors:

$$\mathbf{U} = \sqrt{\frac{1}{2}}\begin{pmatrix} -1 & -i & 0 \\ 0 & 0 & \sqrt{2} \\ 1 & -i & 0 \end{pmatrix}. \tag{H.23}$$

The Cartesian rotation matrix $\mathbf{R}$ in terms of quaternions, is $\mathbf{R} = \mathbf{R}(\mathbf{u})$,

$$\mathbf{R} = \begin{pmatrix} u_0^2 - u_3^2 + u_1^2 - u_2^2 & 2(u_0u_3 + u_1u_2) & -2(u_0u_2 - u_1u_3) \\ 2(u_1u_2 - u_0u_3) & u_o^2 - u_3^2 + u_2^2 - u_1^2 & 2(u_0u_1 + u_2u_3) \\ 2(u_0u_2 + u_1u_3) & -2(u_0u_1 - u_2u_3) & u_o^2 + u_3^2 - u_1^2 - u_2^2 \end{pmatrix}. \tag{H.24}$$

Note that writing $\mathbf{R}$ in this quaternion form avoids evaluating trigonometric functions.

Appendix I

Nuclear Magnetic Resonance

Even if we do not have the intention nor the space required to discuss the details and subtleties of Nuclear Magnetic Resonance (NMR) [Abragam, 1961] or its recent developments [Levitt, 2001; Bakhmutov, 2015], we wish to recall at least the essentials as far as liquid crystals studies are concerned. Doing this is important, as NMR, when applicable, is one of the techniques where the interpretation of the experimental data in terms of molecular parameters is relatively more straightforward and reliable. Moreover, NMR observables can be successfully obtained from atomistic MD simulations, as shown by Tiberio et al., [2009] and Pizzirusso et al. [2012b, 2014]. To start with, we recall that in an NMR experiment molecules are studied through those of their atomic nuclei that are endowed with a nuclear spin I_i (e.g. ^{1}H, ^{13}C, ^{15}N, ^{19}F, ^{31}P, which have $I = 1/2$, or deuterium (D) ^{2}H and ^{14}N which have $I = 1$). The sample to be studied is exposed to a strong static external magnetic field $\boldsymbol{H}_0$. $\boldsymbol{H}_0 = (0, 0, H_Z)$ that interacts with the nuclear spins, in particular, lifting the degeneracy between spin up and down energy levels. For instance, the two spin levels of a proton ^{1}H with spin $I = 1/2$ will have a $\Delta\mathscr{E} = \mathscr{E}_{+1/2} - \mathscr{E}_{-1/2}$ corresponding to a resonance frequency $\nu_0 \approx 600$ MHz in a magnetic field of $\approx$14.1 Tesla.

More generally, the positions and intensities of the absorption lines in an NMR spectrum can be obtained by solving the Schrödinger equation for an effective Hamiltonian (the *spin Hamiltonian*) containing interactions between nuclear magnetic moments and external fields, dipolar and indirect interactions between nuclear magnetic moments and possibly electrostatic interactions involving nuclear spins. The spin Hamiltonian $\hat{\mathscr{H}}$ representing this is

$$\hat{\mathscr{H}} = \hat{\mathscr{H}}_{\mathrm{Z}} + \hat{\mathscr{H}}_{\mathrm{J}} + \hat{\mathscr{H}}_{\mathrm{Q}} + \hat{\mathscr{H}}_{\mathrm{D}}. \tag{I.1}$$

The first, Zeeman, term $\hat{\mathscr{H}}_{\mathrm{Z}}$ represents the interaction between the nuclear magnetic dipole moments, $\gamma_i h \mathbf{I}_i$, and the magnetic field at the ith nucleus, $\boldsymbol{H}_i$. Here $\gamma_i = g_i \mu_N / \hbar$ is the so-called nuclear gyromagnetic ratio expressed in radians, g_i the nuclear g factor of nucleus i, and the nuclear magneton $\mu_N = 5.051 \times 10^{-27}$ JT^{-1}. $\boldsymbol{H}_i$ differs from the value of $\boldsymbol{H}_0$, the applied field, according to the shielding effect of the surrounding electrons. The shielding depends on the electronic distribution around each nucleus which is different in the different direction and is described by a second-rank *chemical shift* tensor $\boldsymbol{\sigma}$, so that

$$\hat{\mathscr{H}}_Z = -\gamma h H_Z \sum_i \hat{\mathbf{I}}_{i,Z}\,(1 - [\sigma_i]_{ZZ}), \tag{I.2a}$$

$$= -\gamma h H_Z \sum_i \hat{\mathbf{I}}_{i,Z} + \gamma h H_Z \sum_i \hat{\mathbf{I}}_{i,Z}\,[\sigma_i]_{ZZ}, \tag{I.2b}$$

where $\hat{\mathbf{I}}_Z = \sum_i \hat{\mathbf{I}}_{i,Z}$ and $\hat{\mathbf{I}}_{i,Z}$ is the nuclear spin projection operator, quantized along the magnetic field $\boldsymbol{H}_Z$ of the spectrometer, that is assumed to define the Z-axis. In brief, the various contributions are: the 'indirect' spin-spin coupling

$$\hat{\mathscr{H}}_J = \sum_{i \le j} \hat{\mathbf{I}}_i\,[\mathbf{J}_{ij}]\,\hat{\mathbf{I}}_j \tag{I.3}$$

and the quadrupolar nuclear term

$$\hat{\mathscr{H}}_Q = \sum_i \frac{eQ_i}{2I_i(2I_i - 1)\hbar} \hat{\mathbf{I}}_i\,\mathbf{V}_i\,\hat{\mathbf{I}}_i, \tag{I.4}$$

where Q_i and $\mathbf{V}_i$ are the quadrupole moment and the electric field gradient tensor at the site of nucleus i and e the proton charge. They are different from 0 for nuclei that have $I > 1/2$ since these have a non-spherical nuclear charge distribution.

$\hat{\mathscr{H}}_D$ corresponds to the effective spin Hamiltonian for the direct interaction between the two magnetic dipole moments of nuclei with spin $\hat{\mathbf{I}}_i$ and $\hat{\mathbf{I}}_j$, that can be written as

$$\hat{\mathscr{H}}_D = \sum_{i \le j} \hat{\mathbf{I}}_i[\mathbf{T}_{ij}]\hat{\mathbf{I}}_j, \tag{I.5}$$

where we have enclosed in square brackets the tensors $[\mathbf{T}_{ij}]$ and $[\mathbf{J}_{ij}]$ in Eq. I.3 involving a pair of nuclei to avoid interpreting ij as tensor components. $[\mathbf{T}_{ij}]$ is the dipolar coupling tensor between a pair of nuclei i and j with gyromagnetic ratios γ_i and γ_j, and can be written explicitly as

$$[\mathbf{T}_{ij}] = -\frac{h\,\gamma_i\,\gamma_j}{8\pi^2\,r_{ij}^5}\left(3\boldsymbol{r}_{ij} \otimes \boldsymbol{r}_{ij} - r_{ij}^2\,\mathbf{1}\right), \tag{I.6}$$

where $\mathbf{1}$ is the 3×3 identity matrix, and $\boldsymbol{r}_{ij} = \boldsymbol{r}_j - \boldsymbol{r}_i$ is the vector of length r_{ij} joining the two nuclei i and j, with $\boldsymbol{r}_i, \boldsymbol{r}_j$ the position of nuclei with respect to the laboratory reference frame. $[\mathbf{T}_{ij}]$ has the same mathematical form (see Eq. C.10) as the electrostatic dipole-dipole interactions tensor discussed in Chapter 5. The spin Hamiltonian can be given a matrix representation in a basis set of nuclear spin angular momentum eigenfunctions $|I_j, m_j\rangle$. The matrix elements of the various spin Hamiltonians can be obtained recalling their action on the basis functions, analogous to that of angular momentum operators

$$\hat{\mathbf{I}}_j^2 |I_j, m_j\rangle = I_j(I_j + 1)\,|I_j, m_j\rangle, \tag{I.7a}$$

$$\hat{\mathbf{I}}_{j,Z} |I_j, m_j\rangle = m_j\,|I_j, m_j\rangle. \tag{I.7b}$$

Note that in NMR the spin interactions with the external magnetic field applied (Zeeman interactions) are normally much greater than internal NMR interactions, that can accordingly be treated as perturbations (the usual *high-field approximation*) [Emsley and Lindon,

1975] and only the terms of the Hamiltonian commuting with $\hat{\mathbf{I}}_{j,Z}$ (called *secular*) are retained. The energy levels of a system of N coupled equal spins in a high magnetic field are characterized by the total magnetic Zeeman quantum number M:

$$\hat{\mathscr{H}}_Z|I,M\rangle = M|I,M\rangle, \tag{I.8}$$

with $M = \sum_j m_j$. The transition between the eigenstates of the spin Hamiltonian can be performed with the step up (or raising) and step down (or lowering) ladder operators $\hat{\mathbf{I}}_{j,+} \equiv \hat{\mathbf{I}}_{j,X} + i\hat{\mathbf{I}}_{j,Y}$ and $\hat{\mathbf{I}}_{j,-} \equiv \hat{\mathbf{I}}_{j,X} - i\hat{\mathbf{I}}_{j,Y}$ that applied to a state $|I,m\rangle$ effect the transition to the next higher or lower spin level:

$$\hat{\mathbf{I}}_{j,+}\,|I_j,m_j\rangle = \sqrt{(I_j - m_j)(I_j + m_j + 1)}\,|I_j,m_j + 1\rangle, \tag{I.9a}$$
$$\hat{\mathbf{I}}_{j,-}\,|I_j,m_j\rangle = \sqrt{(I_j + m_j)(I_j - m_j + 1)}\,|I_j,m_j - 1\rangle. \tag{I.9b}$$

For a spectrometer operating at a modest field (proton resonance frequency of 100 MHz), the Zeeman energy is $\approx 10^8$ Hz, three orders of magnitude larger than the dipolar and quadrupolar contributions. while the indirect spin-spin couplings are no bigger than 10^2 Hz [Dong, 2016]. Thus, the off-diagonal matrix elements of the spin Hamiltonian can be neglected to a good approximation and

$$\begin{aligned}\hat{\mathscr{H}} = &-\frac{H_0}{2\pi}\sum_j \gamma_j\left(1 - [\sigma_j]_{ZZ}\right)\hat{\mathbf{I}}_{j,Z} + \sum_{i<j} J_{ij}^{\mathrm{iso}}\hat{\mathbf{I}}_i\cdot\hat{\mathbf{I}}_j \\ &+\sum_{i<j}\left(2[T_{ij}]_{ZZ} + [J_{ij}]_{ZZ} - \frac{1}{3}\mathrm{Tr}\,[\mathbf{J}_{ij}]\right)\left[I_{i,Z}I_{j,Z} - \frac{1}{4}\left\{\hat{\mathbf{I}}_{i,+}\hat{\mathbf{I}}_{j,-} + \hat{\mathbf{I}}_{i,-}\hat{\mathbf{I}}_{j,+}\right\}\right] \\ &+\sum_j \frac{[q_j]_{ZZ}}{4I_j\left(2I_j - 1\right)}\left(3\hat{\mathbf{I}}_{j,Z}^2 - I_j(I_j + 1)\right).\end{aligned} \tag{I.10}$$

In practice, one-quantum transitions can be performed by a weak magnetic field $\boldsymbol{H}_1(\omega)$ oscillating in a direction perpendicular to $\boldsymbol{H}_0$

$$\hat{\mathscr{H}}_1 = -\gamma h\sum_j \hat{\mathbf{I}}_{j,X}(1 - \sigma_{j,XX})H_{1X}\cos\omega t, \tag{I.11}$$

with an angular frequency $\omega = 2\pi v$ in the radio frequency range to effect transitions between the spin levels. Note that $\hat{\mathbf{I}}_{jX} = \frac{1}{2}\left(\hat{\mathbf{I}}_{j+} + \hat{\mathbf{I}}_{j-}\right)$. This in turn means that these are the quantities that we can extract from the spectra, i.e. our observables. The transition probability between eigenstates $|I,m\rangle$ of the spin Hamiltonian with energy $\mathscr{E}_M$, $\mathscr{E}_{M'}$ is given, using quantum perturbation theory [Schatz and Ratner, 1993], by

$$W_{M,M'} \approx -\gamma^2h^2H_1^2\,|\langle M|\hat{\mathbf{I}}_X|M'\rangle|^2\delta\left(\omega_{MM'} - \omega\right), \tag{I.12a}$$
$$= -\gamma^2h^2H_1^2\,|\langle M|\hat{\mathbf{I}}_X|M'\rangle|^2\int_{-\infty}^{\infty} \mathrm{d}t\ \mathrm{e}^{-i(\omega_{MM'}-\omega)t}, \tag{I.12b}$$

where $\omega_{MM'} \equiv (\mathscr{E}_M - \mathscr{E}_{M'})/\hbar$, and we have used the Fourier integral representation of the delta function. The dipolar couplings can involve nuclei on the same molecule (*intramolecular*) or on different ones.

In low viscosity liquids and liquid crystals (although not in solids or gels) intermolecular contributions are largely averaged out by rapid intermolecular motions and only intramolecular contributions will be relevant. The effect of molecular reorientations occurring on a frequency scale faster than the spectral width $\Delta\omega$ will yield an orientationally averaged spectrum, rather than a superposition of individual spectra coming from each orientation, as the experimental observable. The splittings in each group are due to chemical shifts and couplings between the spins.

Allowed one-quantum transitions correspond to ($\Delta M = \pm 1$). Multiple quantum transitions are possible in different, specially designed, experiments using suitable sequences of pulses and Fourier transform techniques [Pines, 1988]. A problem is the number of transitions and the number of lines in the NMR spectrum of a system of coupled nuclei (e.g. protons). The number of energy levels with magnetic quantum number M is $\binom{N}{N/2+M}$, and if one is restricted to transitions between neighbouring Ms is $\binom{2N}{N+1}$ which is an upper bound to the number of one-quantum transitions. For $N = 4$ this number is 56, for $N = 8$ it is 11,440, and for $N = 12$ it is 2,496,144 [Pines, 1988]. This increase in the number of lines is practically forbidding the HNMR investigations of pure liquid crystals, unless some of the protons are eliminated by isotopic substitution with deuterium. However, it has been demonstrated since the pioneering work of Saupe [Saupe and Englert, 1963] that small molecules dissolved in liquid crystals can be studied and completely resolved NMR spectra of the partly oriented solute molecules can be obtained. The signal of the nematic solvent will normally just provide a broad background (see, for example, Fig. 3.19b). The number of interacting nuclei of the solute must, of course, be kept low so that comparatively few but strong lines result. It then becomes possible to investigate average orientation of the molecule and/or obtain structural information.

Appendix J
X-ray Diffraction

The Bragg derivation is simple but is convincing only because it reproduces the correct result

C. Kittel, Introduction to Solid State Physics, 2005

X-ray diffraction basics. X-ray diffraction [Guinier, 1994; Fiori and Spinozzi, 2010; Als-Nielsen and McMorrow, 2011; Morelhao, 2016] offers an important structural tool for hard and soft materials in the determination of:

(i) the type of molecular organization (crystalline, amorphous, fibrous),

(ii) the relevant morphology details (e.g. if the material is a liquid crystal, what type of liquid crystals) and

(iii) the relevant intermolecular or interatomic distances, that can typically be of the order of Angstroms. The use of electromagnetic radiation of wavelength similar to the size of desired structural details (see Table J.1) allows us to exploit interference as a tool to obtain information about the very small distances of interest. Let us consider the amplitude of the electric field of a radiation at a certain point in space $\boldsymbol{r}$ at time t:

$$\boldsymbol{E}(\boldsymbol{r},t) = \hat{\boldsymbol{e}} E_{in}\, \mathrm{e}^{i\boldsymbol{k}\cdot\boldsymbol{r}-i\omega t}, \tag{J.1}$$

where $\boldsymbol{k}$ is the wavevector along the propagation direction, with modulus $k = 2\pi/\lambda$, λ the wavelength, ω the angular frequency of the wave and $\hat{\boldsymbol{e}}$ the polarization. In the case of X-rays, the polarization will depend on the source (unpolarized for the common X-ray tube, linearly polarized from synchrotron sources) but we shall not consider it here. Like any other radiation, X-rays can be absorbed or scattered when impinging on a material. However, hard X-rays, with λ of a few Ångstrom (say λ <8 Å or $\hbar\omega$ >1500 eV) are far away from the typical atomic absorption edge for carbon (which is at $\lambda \approx 280$ eV) and they can go through an organic material basically without changing their direction (the refractive index at these wavelengths is $\approx$1). They interact with all the electrons of each atom as if they were free and make them oscillate, thus behaving as a dipolar antenna that irradiates a scattered spherical wave with the same wavelength as the incident wave (elastic scattering). The property measured in an ordinary scattering experiment is, however, not the electric field but rather an intensity, $I \propto EE^*$, thus the unchanged frequency phase factor will simplify and, from now on, we can avoid carrying it on. The waves scattered from the various electrons

Table J.1. *Some important regions of the electromagnetic spectrum in terms of indicative wavelength (λ), wavenumbers ($\tilde{\nu}$), energy (E).*

	Units	IR	VIS	UV	Soft X-ray	X-ray	Hard X-ray
λ	Å	10^4	5×10^3	10^3	10	1	10^{-1}
λ	nm	10^3	500	10^2	1	0.1	10^{-2}
$\tilde{\nu}$	cm^{-1}	10^4	2×10^4	10^5	10^7	10^8	10^9
E	eV	1.24	2.48	12.4	1.24×10^3	12.4×10^3	124×10^3
E	kcal/mol	28.59	57.19	285.94	28.59×10^3	28.59×10^4	28.59×10^5

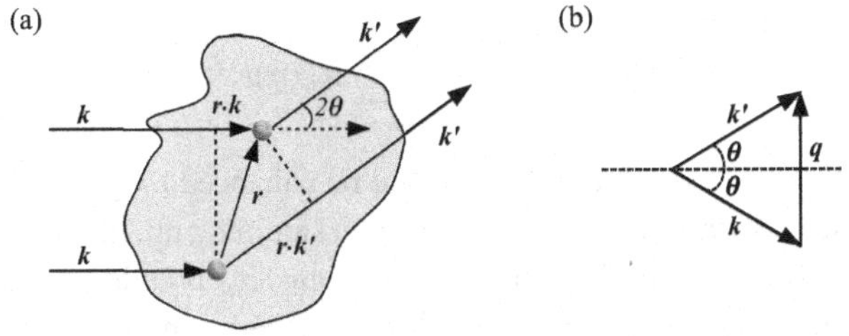

Figure J.1 (a) Phase delay $\boldsymbol{r} \cdot (\boldsymbol{k}' - \boldsymbol{k})$ between two X photons scattered by a pair of centres separated by a distance vector $\boldsymbol{r}$. (b) The scattering vector $\boldsymbol{q}$.

are coherent and thus can interfere. This physical phenomenon is the fundamental principle of many X-ray experiments, and we shall assume experimental conditions to be such that this coherent scattering is the only relevant one, while the incoherent inelastic contributions (Compton scattering), where the wavelength changes are instead negligible. Another key assumption made in discussing the scattering of X-rays from a molecular system is the Born or single-scattering approximation. This assumes that the amplitude of the scattered wave is very small compared to the incident wave so that when it encounters another electron it is not scattered a second time. If this is the case, only the primary incident wave generates scattering from the electrons. The electric field of radiation scattered from the various centres of electronic density travels through the sample and arrives at the detector. Since the X-rays interact with the electrons, we can also assume as a first approximation the scattering factor, b, of a certain centre to be proportional to the number of electrons (if the scattering centre is an atom, its atomic number), or more properly, to the local electron density $\rho(\boldsymbol{r})$

$$E(\boldsymbol{r}) \propto E_{in} \frac{e^{i\boldsymbol{k}\cdot\boldsymbol{r}}}{r} \rho(\boldsymbol{r}) = E_{in}(\boldsymbol{r})\rho(\boldsymbol{r}). \tag{J.2}$$

It is convenient to determine the travelled path difference between two scatterers considering one, arbitrary, centre at the origin and another at a vector distance $\boldsymbol{r}$ (see Fig. J.1). If the scattering is coherent and there is no phase change on scattering, the phase difference between the two waves scattered from a centre at the origin and one at $\boldsymbol{r}$ and arriving at the same detection point will depend only on the difference in travelled path length between the two rays: $\boldsymbol{r} \cdot \boldsymbol{k}' - \boldsymbol{r} \cdot \boldsymbol{k} = \boldsymbol{r} \cdot \boldsymbol{q}$, where the vector $\boldsymbol{q} = \boldsymbol{k}' - \boldsymbol{k}$ is called the *scattering vector*. In more detail, the total field from the two scattering centres will be

$$E_{tot}(\boldsymbol{r}) = E_1(\boldsymbol{r}) + E_2(\boldsymbol{r}) = E_{in} \frac{\mathrm{e}^{i\boldsymbol{k} \cdot \boldsymbol{r}}}{r} [1 + \mathrm{e}^{i\boldsymbol{q} \cdot \boldsymbol{r}}]. \tag{J.3}$$

Since the process is elastic, $k = k'$ and from Fig. J.2 and simple trigonometry we have $q^2/4 = k^2 - k^2 \cos^2 \theta$, thus

$$q = 2k \sin \theta = \frac{4\pi}{\lambda} \sin \theta. \tag{J.4}$$

The amplitude of the total electric field arriving at the same detection point will be the sum over all N scatterers contributions,

$$E_{tot}(\boldsymbol{q}) = E_{in} \frac{\mathrm{e}^{i\boldsymbol{k} \cdot \boldsymbol{r}}}{r} \sum_{m=1}^{N} b_m(\boldsymbol{q}) e^{i\boldsymbol{q} \cdot \boldsymbol{r}_m}, \tag{J.5}$$

and at each detection point (or at each $\boldsymbol{q}$) it could be enhanced (*constructive interference*), reduced or cancelled (*destructive interference)* when summing up, because of the phase difference. If we consider the scattering centres to be atoms, b_{at} is called the *atomic scattering factor*. It is clear that it is higher for atoms (or ions) with a higher number of electrons and that only atoms with a sufficient number of electrons will be visible (protons being the less visible!). However, the factor b_{at} decreases with q, contributing to the difficulty of observing higher reflections [Guinier, 1994]. The corresponding scattered intensity is

$$I(\boldsymbol{q}) \propto \langle |E_{tot}(\boldsymbol{q})|^2 \rangle = \frac{E_{in}^2}{r^2} \sum_{m=1}^{N} \sum_{m'=1}^{N} b_{at,m}(q)\, b_{at,m'}(q) \langle e^{i\boldsymbol{q} \cdot \boldsymbol{r}_{mm'}} \rangle, \tag{J.6a}$$

$$\approx \frac{E_{in}^2}{r^2} \sum_{m \neq m'=1}^{N} b_{at,m}(q)\, b_{at,m'}(q) \sum_{L} i^L (2L+1) \big\langle j_L(q r_{mm'}) \mathscr{D}_{00}^{L}(\hat{\boldsymbol{q}} \cdot \hat{\boldsymbol{r}}_{mm'}) \big\rangle, \tag{J.6b}$$

where to get the last equation we have used the Rayleigh plane-wave expansion

$$\mathrm{e}^{i\boldsymbol{q} \cdot \boldsymbol{r}} = \sum_{L} i^L (2L+1) j_L(qr) \mathscr{D}_{00}^{L}(\hat{\boldsymbol{q}} \cdot \hat{\boldsymbol{r}}), \tag{J.7}$$

with $j_L(qr)$ a spherical Bessel function [Abramowitz and Stegun, 1965]. In many cases it is convenient to consider not a discrete sum like in Eq. J.6b, but a continuous distribution of electronic density in space. Considering the contribution from all electronic centres, the total scattered electric field can be written as

$$E_{tot}(\boldsymbol{q}) \propto \int \mathrm{d}\boldsymbol{r}\, \rho(\boldsymbol{r})\, \mathrm{e}^{i\boldsymbol{q} \cdot \boldsymbol{r}} \propto \mathscr{F}[\rho(\boldsymbol{r})], \tag{J.8}$$

which is a 3D space Fourier transform of the electronic density $\rho(\boldsymbol{r})$ (see Appendix E). Taking an average over the distribution of electrons and its fluctuations in time caused by any atomic or molecular motion:

$$I(\boldsymbol{q}) \propto \left\langle |E_{tot}(\boldsymbol{q})|^2 \right\rangle. \tag{J.9}$$

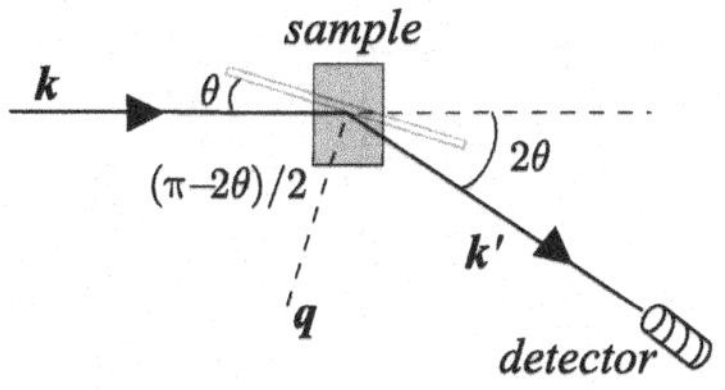

Figure J.2 Macroscopic geometry of scattering experiment. The X-ray beam $\boldsymbol{k}$ impinges on the sample and the scattered beam $\boldsymbol{k}'$ is detected at the angle 2θ. Note that from a formal point of view the scattering is equivalent to a reflection from the virtual Bragg plane R orthogonal to the scattering vector $\boldsymbol{q}$, even if no actual physical reflection is involved.

We can now consider a few simple examples.

1D lattice. For a 1D regular lattice of N identical atoms placed along a line with a spacing a between them, we have the electron density $\rho(\boldsymbol{r}) = \sum_m \rho(\boldsymbol{r} - \boldsymbol{r}_m) \approx \rho_{el} \sum_m \delta(\boldsymbol{r} - \boldsymbol{r}_m)$, where the electron density ρ_{el} is to a first approximation the atomic number Z and where $\boldsymbol{r}_m$ is the position of the mth centre. We assume (Fig. J.3) $\boldsymbol{r}_m = (0, 0, (m-1)a)$, finding the field scattered from the mth atom as

$$E_m = \frac{E_{in}}{r} b_{at}(q)\, \mathrm{e}^{i\boldsymbol{k}\cdot\boldsymbol{r} + im\boldsymbol{q}\cdot\boldsymbol{a}} \tag{J.10}$$

and the total scattered field from the row of N centres

$$E_{tot}(\boldsymbol{r}, \boldsymbol{q}) = \frac{E_{in}}{r} b_{at}(q)\, \mathrm{e}^{i\boldsymbol{k}\cdot\boldsymbol{r}} \sum_{m=0}^{N-1} (\mathrm{e}^{i\boldsymbol{q}\cdot\boldsymbol{a}})^m = \frac{E_{in}}{r} b_{at}(q)\, \mathrm{e}^{i\boldsymbol{k}\cdot\boldsymbol{r}} \frac{(\mathrm{e}^{iN\boldsymbol{q}\cdot\boldsymbol{a}} - 1)}{(\mathrm{e}^{i\boldsymbol{q}\cdot\boldsymbol{a}} - 1)}, \tag{J.11}$$

which has been written in closed form using: $\sum_{m=m_1}^{m_2} r^m = (r^{m_1} - r^{m_2+1})/(1-r)$, the geometric sum expression, where $|r| \neq 1$ and m, m_1, m_2 are integers. The intensity scattered by a row of centres is then

$$I_{tot}(q) = \frac{|E_{in}|^2}{r^2} b_{at}^2 \frac{\sin^2(N(\boldsymbol{q}\cdot\boldsymbol{a})/2)}{\sin^2(\boldsymbol{q}\cdot\boldsymbol{a}/2)}, \tag{J.12}$$

with $q = (4\pi/\lambda)\sin\theta$ (cf. Eq. J.4). As we see from Fig. J.4, as N increases, i.e. as the size of periodic linear cluster increases, the position of the peaks remains the same, at $\boldsymbol{q}\cdot\boldsymbol{a} = qa = n2\pi$, n integer, but the peaks become rapidly sharper. Note that the peaks occur at the Bragg pseudo-reflection condition $2a\sin\theta = n\lambda$ or, in other words, when $\boldsymbol{q}$ matches a vector of the reciprocal lattice. For this 1D lattice the reciprocal lattice can only be along the same direction, with a unit vector of length $2\pi/a$.

3D lattice. Let us now consider a regular 3D lattice, built from a repeated unit cell defined by the primitive vectors $\boldsymbol{a}_1$, $\boldsymbol{a}_2$, $\boldsymbol{a}_3$. Any site of the lattice can be written as $\boldsymbol{v} = (m_1\boldsymbol{a}_1 + m_2\boldsymbol{a}_2 + m_3\boldsymbol{a}_3)$, with integer m_i. We can always associate to the real space lattice a reciprocal one (see, e.g., [Lax, 1974; Kittel, 2005]. More explicitly, any point in the reciprocal lattice will be $\xi_{hkl} = 2\pi\,(h\boldsymbol{b}_1 + k\boldsymbol{b}_2 + l\boldsymbol{b}_3)$, with h, k, l integers and the reciprocal vectors $\boldsymbol{b}_i$, with $\boldsymbol{b}_i = \boldsymbol{a}_j \times \boldsymbol{a}_k / v_0$, and the orthonormality relation $\boldsymbol{a}_i \cdot \boldsymbol{b}_j = \delta_{i,j}$,

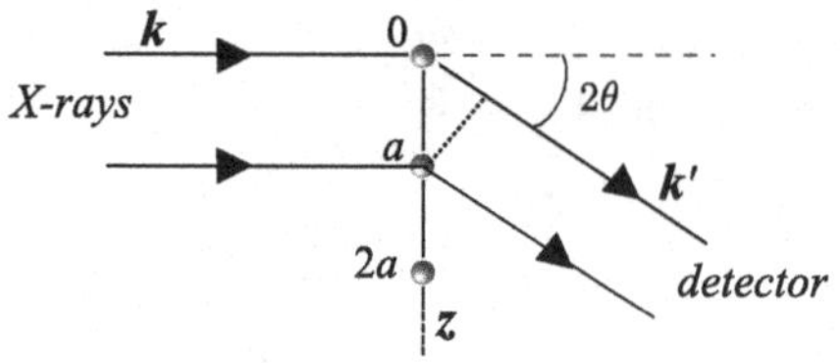

Figure J.3 Scattering from a row of atoms with electron density $\rho(\boldsymbol{r})$.

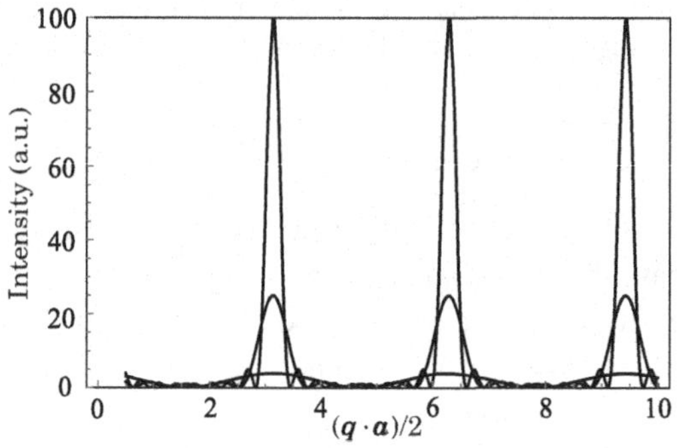

Figure J.4 Intensity scattered from 2, 5, 10 identical centres regularly spaced on a line. Note that the scattered peaks all have the same intensities, since a constant scattering factor has been assumed. In a real case the peaks will decrease with q.

with i, j, k an even permutation of 1,2,3, while $v_0 = \boldsymbol{a}_1 \cdot \boldsymbol{a}_2 \times \boldsymbol{a}_3$ is the volume of the primitive cell in the real lattice. In general, the three lattice vectors $\boldsymbol{a}_1, \boldsymbol{a}_2, \boldsymbol{a}_3$ will not be perpendicular to each other (e.g. for lattices tilted in one (*monoclinic*) or more (*triclinic*) directions), and in this case also the different reciprocal lattice vectors will also be non-orthogonal between themselves. However, the real and reciprocal lattice vectors are always perpendicular to each other. In the special case of an *orthorhombic* lattice, where $\boldsymbol{a}_1 \perp \boldsymbol{a}_2 \perp \boldsymbol{a}_3$ can be taken along x, y, z, the wavevector can be expressed as $\boldsymbol{k} = 2\pi\left[\frac{n_x}{\ell_x}\boldsymbol{x} + \frac{n_y}{\ell_y}\boldsymbol{y} + \frac{n_z}{\ell_z}\boldsymbol{z}\right]$, where ℓ_i are the lattice spacings in the three directions and n_i are integers. The scattering from a 3D lattice with N atoms along each of the three directions will be

$$E_{tot}(\boldsymbol{r}, \boldsymbol{q}) = E_{in}\,\rho_{el}\frac{e^{i\boldsymbol{k}\cdot\boldsymbol{r}}}{r}\sum_{m_1,m_2,m_3=0}^{N-1} e^{i\boldsymbol{q}\cdot(m_1\boldsymbol{a}_1+m_2\boldsymbol{a}_2+m_3\boldsymbol{a}_3)}, \tag{J.13a}$$

$$= E_{in}\,\rho_{el}\frac{e^{i\boldsymbol{k}\cdot\boldsymbol{r}}}{r}\frac{(e^{iN\boldsymbol{q}\cdot\boldsymbol{a}_1}-1)}{(e^{i\boldsymbol{q}\cdot\boldsymbol{a}_1}-1)}\frac{(e^{iN\boldsymbol{q}\cdot\boldsymbol{a}_2}-1)}{(e^{i\boldsymbol{q}\cdot\boldsymbol{a}_2}-1)}\frac{(e^{iN\boldsymbol{q}\cdot\boldsymbol{a}_3}-1)}{(e^{i\boldsymbol{q}\cdot\boldsymbol{a}_3}-1)}, \tag{J.13b}$$

which has been summed explicitly using the geometric sum. Diffraction occurs when the *Laue condition* $\boldsymbol{q} = \boldsymbol{\xi}$, where $\boldsymbol{\xi}$ is a reciprocal lattice vector, is satisfied, since this leads to $\exp(i\boldsymbol{q}\cdot\boldsymbol{\xi}) = 1$. The results can be visualized in terms of reflections from a plane. Indeed, any vector of reciprocal lattice is perpendicular to a plane in real space. If this plane in real

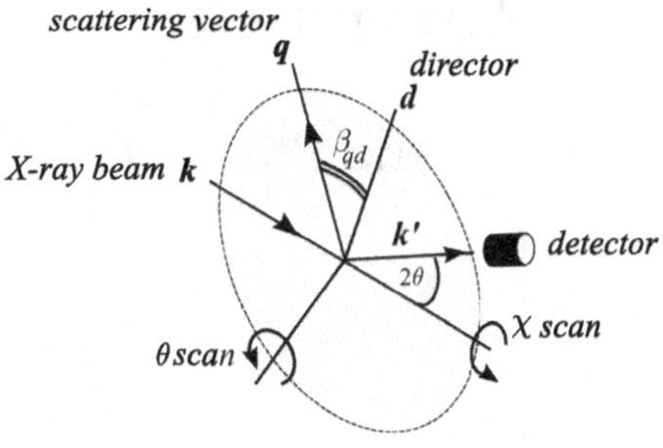

Figure J.5 The geometry of an X-ray experiment showing the main types of scanning experiments used to explore the structure of a liquid crystal with director $\boldsymbol{d}$ [Kumar, 2001]. In the $\theta - 2\theta$ scans the incident and scattered beams direction $\boldsymbol{k}$, $\boldsymbol{k}'$ are varied so that the magnitude q of the scattering vector $\boldsymbol{q}$ is changed while keeping the same 2θ. Thus, $I_{2\theta}(q)$ is recorded. In a θ scan instead the detector position is changed to receive the intensity at various angles and obtain $I(2\theta)$.

space goes through three points $m_1\boldsymbol{a}_1$, $m_2\boldsymbol{a}_2$, $m_3\boldsymbol{a}_3$ it can be labelled by the Miller indices (h, k, l) with, apart from an integer factor, $h = 1/m_1$, $k = 1/m_2$, $l = 1/m_3$. These are the components of the reciprocal vector and

$$|\boldsymbol{\xi}_{hkl}| = 2\pi/\ell_{hkl}, \tag{J.14}$$

with ℓ_{hkl} the distance between the Miller planes (h, k, l). At the diffraction condition $\boldsymbol{q} = \boldsymbol{\xi}$, e.g. if we have an ordered LC with the director along the z laboratory axis and we look in the meridian direction, at $\boldsymbol{q} = q(001)$ a smectic A with layers perpendicular to z, will give a reflection at $q = 2\pi/\ell_z$ (and possibly multiples of that). Using Eq. J.4 we also have, more generally, $2\ell_{hkl} \sin\theta = \lambda$.

In Fig. J.5 we show the geometry of the various experiments typically performed on liquid crystals [Kumar, 2001].

Appendix K

Stochastic Processes

The Russian mathematician Andrey Markov introduced the stochastic processes now named after him when studying the alternation of vowels and consonants in the poem 'Onegin' by Pushkin [Hayes, 2013]. This seems a problem rather far from practical applications, but we could just think that instead of vowels and consonants we have 'sun' or 'rain' in a sequence of daily weather forecasts. Let us thus imagine a sequence of such readings and that the probabilities of having a sunny or, respectively, a rainy day are p_s and p_r. If we assume no other weather condition (fog or whatever), this is a 'two states' process, so necessarily $p_s + p_r = 1$. The probability p_s can be approximated with the observed frequency of occurrence of a sunny day over a sufficiently long time (its 'equilibrium probability') and will change from place to place etc. We can now ask what is the probability that a sunny day will be followed by another sunny one or rather by rain. Strictly, the fact that the next day will be sunny depends not only on the present condition, but also on the sequence of previous ones. Moreover, in different places we can expect different treble or quartet sequences being more common. The simplest and roughest assumption we could make would be that the weather of each day is independent on the previous one, so there would be the same probability of having sun or rain, irrespective of the type of the previous one. This is rather extreme, and the proposal of Markov is instead to assume that the new observation depends just on the last one (what we now call a *Markov process*): Thus, in matrix form:

$$\begin{pmatrix} p_s^{(1)} \\ p_r^{(1)} \end{pmatrix} = \begin{pmatrix} p_s^{(0)} \\ p_r^{(0)} \end{pmatrix} \begin{pmatrix} \Pi_{ss} & \Pi_{sr} \\ \Pi_{rs} & \Pi_{rr} \end{pmatrix}, \tag{K.1}$$

or $\boldsymbol{p}^{(1)} = \boldsymbol{p}^{(0)}\boldsymbol{\Pi}$, where we have added a superscript to indicate the position in the chain of events. The matrix $\boldsymbol{\Pi}$ is called a *transition matrix* and, generalizing to a system that can be found in m distinct states, rather than just two, this will be $m \times m$. It is easy to see that its elements are all non-negative ($\Pi_{ij} \geq 0$), since Π_{ij} represents the probability of jumping form a state i to another state j, which cannot be negative. Clearly, the probability of hopping from the given state to another one should be 1:

$$\sum_{j=1}^{m} \Pi_{ij} = 1. \tag{K.2}$$

A square matrix with these properties is called a right *stochastic* matrix and has some special mathematical properties:

(i) If $\mathbf{\Pi}$ is stochastic, then $\mathbf{\Pi}^n$ is also stochastic.
(ii) One of its eigenvectors is always formed of all 1s.
(iii) For any eigenvalue λ_i of $\mathbf{\Pi}$, $|\lambda_i| \leq 1$.
(iv) Since $\mathbf{\Pi}$ is real, its eigenvalues are either real or they are complex conjugate in pairs so to give a real trace.

The probability vector after two jumps will be $\boldsymbol{p}^{(2)} = \boldsymbol{p}^{(1)}\mathbf{\Pi} = \boldsymbol{p}^{(0)}\mathbf{\Pi}^2$ and in general, after n jumps,

$$\boldsymbol{p}^{(n)} = \boldsymbol{p}^{(n-1)}\mathbf{\Pi} = \boldsymbol{p}^{(n-2)}\mathbf{\Pi}\,\mathbf{\Pi} = \cdots = \boldsymbol{p}^{(0)}\mathbf{\Pi}^{n-1}\mathbf{\Pi} = \boldsymbol{p}^{(0)}\mathbf{\Pi}^n. \tag{K.3}$$

Thus, the transition matrix $\mathbf{\Pi}^{(n)}$ for proceeding n steps is just the nth power of $\mathbf{\Pi}$ and the transition probability for going from site i to site j after n steps will be in the form of a recurrence relation

$$(\Pi^n)_{ij} = \sum_k (\Pi^{n-1})_{ik}\Pi_{kj}, \tag{K.4}$$

often called the Chapman–Kolmogorov equation. A property of Markov processes that is crucial for its application in the Monte Carlo method (Chapter 8) is the asymptotic behaviour of the transition matrix after a very large number of steps. Indeed, if every state can be reached from any other state with a certain sequence of steps (the configuration space is connected and the system is *ergodic*), then the limit

$$\lim_{n\to\infty} (\Pi^n)_{ij} = w_j \tag{K.5}$$

exists for every pair i, j and is independent of the starting state i. The asymptotic frequency factors w_j represent the occupation probability (the weight) of the jth state. The asymptotic weights have to be positive: $w_j > 0$, and obey the normalization condition: $\sum_{j=1}^m w_j = 1$. Taking the limit for n going to infinity on both sides of Eq. K.4 we have,

$$\boldsymbol{w}\,\mathbf{\Pi} = \boldsymbol{w}, \tag{K.6}$$

showing that the equilibrium distribution is an eigenvector of the transition matrix, corresponding to the eigenvalue 1. This ensures that the transition matrix does not modify the distribution when equilibrium is reached.

Appendix L

Simulating Polarized Optical Microscopy Textures

A propagating beam of light can be represented by a vector or a matrix. Two particularly useful vector representations are the Jones (or Fresnel–Maxwell) [Jones, 1948] and the Stokes [Azzam, 1978] ones, in terms of 2d and 4D vectors, respectively. The electric (or similarly the magnetic) field of a monochromatic radiation with frequency $\nu = 2\pi\omega$, propagating in vacuum or in an isotropic medium in a direction that we take as the z-axis, can be represented by a 2D vector field of the form

$$\boldsymbol{E}(z,t) = \boldsymbol{e} E_0 \exp\left\{ i[\omega t - \frac{n}{2\pi\lambda_0} z] \right\}, \tag{L.1}$$

where λ_0 is the radiation wavelength in vacuum, n the refractive index of the medium, E_0 the field amplitude and the Jones vector, $\boldsymbol{e} = (e_1, e_2)$, represents the direction of polarization of the radiation in the xy-plane. Any optical element in the path of the radiation will modulate some features of the incoming beam $\boldsymbol{E}^{in}(z, t)$, giving another vector $\boldsymbol{E}^{out}(z, t)$. Thus, the optical element introduced can be represented by a 2×2 matrix. While the Jones formalism is attractively simple, many experiments measure intensities, rather than electric field components and the Stokes representation that employs only measurable intensities is often more convenient. The Stokes vector $\boldsymbol{s} = (s_0, s_1, s_2, s_3)$ of a radiation has four components: s_0, the total intensity, that can be taken as 1, $s_1 = I_{45} - I_{135}$, $s_2 = I_+ - I_-$ and $s_3 = I_0 - I_{90}$. Here I_0, I_{90}, I_{45}, I_{135}, I_+ and I_- are the output intensities when the light is passed through perfect linear polarizers set at x, y, 45°, 135° and through right and left circular polarizers, respectively. It is important to note that the Stokes vector includes the description of partially or completely depolarized light, different to Jones vectors and that $s_0^2 \geq s_1^2 + s_2^2 + s_3^2$, with the equality holding only for totally polarized light. In this last case, the components of the Stokes vector, s_i, can be calculated from the Jones vector $\begin{pmatrix} e_1 \\ e_2 \end{pmatrix}$ and its conjugate transpose row vector (e_1^*, e_2^*), as follows: $s_i = \frac{1}{2}(e_1^*, e_2^*)\boldsymbol{\sigma}_i \begin{pmatrix} e_1 \\ e_2 \end{pmatrix}$, where $\boldsymbol{\sigma}_0$ is the unit matrix, while $\boldsymbol{\sigma}_i$, $i = 1, 2$ or 3 correspond to the 2×2 Pauli spin matrices [Rose, 1957]:

$$\boldsymbol{\sigma}_0 = \begin{pmatrix} 1 & 0 \\ 0 & 1 \end{pmatrix}, \boldsymbol{\sigma}_1 = \begin{pmatrix} 0 & 1 \\ 1 & 0 \end{pmatrix}, \boldsymbol{\sigma}_2 = \begin{pmatrix} 0 & -i \\ i & 0 \end{pmatrix}, \boldsymbol{\sigma}_3 = \begin{pmatrix} 1 & 0 \\ 0 & -1 \end{pmatrix}. \tag{L.2}$$

Clearly, if the incoming beam of light can be represented by a vector and the outgoing beam of light by another vector, any optical device operating on the incoming beam to produce the output one in a linear regime will be phenomenologically represented by a 4×4 matrix, called the Müller matrix, which converts the Stokes vector of the incoming radiation to that of the transmitted radiation. The images of polarized optical microscopy (POM) experiments can be obtained from the sample configurations generated from a molecular or atomistic simulation assuming that each elementary volume element (*voxel*) is described by a Müller matrix. In simulations of lattice systems, where a single spin represents a tightly packed cluster of real molecules, a voxel is usually taken to correspond to a spin [Berggren et al., 1994a]. The light beam crossing a sequence of sites across the sample is then retarded according to the matrix resulting from the product of the Müller matrices $\mathbf{M}^{(j)}$ for each voxel encountered along the light path. The effect of a uniaxial LC element on the light corresponds to a simple linear retarder, and the Müller matrix $\mathbf{M}^{(j)}$ introduced to represent the effect of the jth voxel along the light path can be written explicitly as [Schellman, 1998]

$$\mathbf{M}^{(j)} = \begin{pmatrix} 1 & 0 & 0 & 0 \\ 0 & \sin^2 2\alpha_j + \cos^2 2\alpha_j \cos\delta_j & -\cos 2\alpha_j \sin\delta_j & \sin 2\alpha_j \cos 2\alpha_j (1-\cos\delta_j) \\ 0 & \cos 2\alpha_j \sin\delta_j & \cos\delta_j & -\sin 2\alpha_j \sin\delta_j \\ 0 & \sin 2\alpha_j \cos 2\alpha_j (1-\cos\delta_j) & \sin 2\alpha_j \sin\delta_j & \cos^2 2\alpha_j + \sin^2 2\alpha_j \cos\delta_j \end{pmatrix}, \tag{L.3}$$

where α_j is the angle between the projection of the jth voxel director on the plane perpendicular to the beam direction (z, say) of the light and the x-axis and δ_j is the phase difference for voxel j [Xu et al., 1992] $\delta_j = \frac{2\pi h n_o}{\lambda}\left(\frac{n_e}{n_{e,j}} - 1\right)$, where h is the thickness of each layer, n_o, n_e are the ordinary and extraordinary refractive indices of the liquid crystal. The effective refractive index, $n_{e,j}$ is obtained as $n_{e,j} = \sqrt{n_o^2 + (n_e^2 - n_o^2)\cos^2\beta_j}$ and β_j is the angle between the director of the jth voxel and the light propagation direction. The resulting Stokes vector of the polarized and retarded light beam is given by [Xu et al., 1992; Berggren et al., 1994a; Schellman, 1998]

$$\mathbf{s}^{out} = \mathbf{P}^{out} \prod_j \mathbf{M}^{(j)} \mathbf{P}^{in} \mathbf{s}^{in}. \tag{L.4}$$

The input unpolarized light is represented by the vector $\mathbf{s}^{in} = (1,0,0,0)^T$ and the Müller matrices, $\mathbf{P}^{in}$ and $\mathbf{P}^{out}$, correspond to the polarizer and the analyzer, respectively set at angles in, out (degrees). Commonly used linear polarizers and analyzers are [Schellman, 1998]

$$\mathbf{P}^{(0,90)} = \frac{1}{2}\begin{pmatrix} 1 & 0 & 0 & \pm 1 \\ 0 & 0 & 0 & 0 \\ 0 & 0 & 0 & 0 \\ \pm 1 & 0 & 0 & 1 \end{pmatrix}, \quad \mathbf{P}^{(45,135)} = \frac{1}{2}\begin{pmatrix} 1 & \pm 1 & 0 & 0 \\ \pm 1 & 1 & 0 & 0 \\ 0 & 0 & 0 & 0 \\ 0 & 0 & 0 & 0 \end{pmatrix}, \tag{L.5}$$

while left or right circular polarizers can be represented by the matrix

$$\mathbf{P}^{(\pm 1)} = \frac{1}{2}\begin{pmatrix} 1 & 0 & \pm 1 & 0 \\ 0 & 0 & 0 & 0 \\ \pm 1 & 0 & 1 & 0 \\ 0 & 0 & 0 & 0 \end{pmatrix}. \tag{L.6}$$

To obtain a more accurate image of the polarized light from the simulated configurations an average is performed over N_c equilibrated configurations, and the intensity is then calculated as a projection in the plane perpendicular to the light propagation

$$\boldsymbol{s}^{out} = \mathbf{P}^{out} \frac{1}{N_c} \sum_{k=1}^{N_c} \left(\prod_j \mathbf{M}^{(j,k)} \right)_k \mathbf{P}^{in} \boldsymbol{s}^{in}. \tag{L.7}$$

The output intensity is proportional to the first element in the obtained Stokes vector $\boldsymbol{s}^{out}$, i.e. $I^{out} \propto [s^{out}]_0$. An image comparable to those experimentally observed by POM is rendered coding different intensities in a greyscale from black (no light), to white (full intensity). Typical values of the parameters used in the calculations in Chapter 10 are: layer thickness: $\ell = 5.3$ μm, wavelength: $\lambda = 545$ nm and refractive indices $n_0 = 1.50$, $n_e = 1.66$.

Appendix M

Units and Conversion Factors

Table M.1. *Some useful conversion factors between units of measurement. Here* au *stands for atomic unit (often used in quantum chemistry) and* esu *for the electrostatic unit of charge,* m_e *and e are the electron mass and proton charge, respectively.*

Quantity	Units
Length	1 au = 1 Bohr = $a_0 = 5.291772 \times 10^{-11}$ m
Mass	1 au = $m_e = 9.1095 \times 10^{-31}$ kg
Time	1 au = $\hbar E_h^{-1} = 2.4189 \times 10^{-17}$ s
Force	1 N = 1 J/m
Energy	1 au = 1 Hartee = $1E_h$ = 627.509 kcal mol^{-1} = 27.2135 eV = 4.3597×10^{-18} J
	1 eV = 23.0621 kcalmol^{-1}
	1 J = 1 kg m^2/s^2 = 10^7 erg
	1 zJ = 10^{-21} J $\approx$ 0.15 kcal/mol
	1 $k_B T_{293.15\,\mathrm{K}} \approx$ 0.6073 kcal/mol
Pressure	1 Pa = 1 N/m^2 = 10^{-5} bar = (1/101325) atm
Charge	1 au = 1 e = 1.6022×10^{-19} C = 4.8032×10^{-10} esu
	1 esu = 1 (erg cm)$^{1/2}$ = 3.33564×10^{-10} C
Electric dipole	1 au = 2.5418×10^{-18} esu cm = 2.5418 D
	1 D $\equiv$ 1 Debye = $1.0 \times 10^{-18} esucm = 3.336 \times 10^{-30} Cm$
Electric quadrupole	1au = 1 ea_0^2 = 4.4866×10^{-40} C m^2 = 1.34504×10^{-26} esu cm^2 = 1.34504 B
	1 B $\equiv$ 1 Buckingham = 1 DÅ = 1×10^{-26} esu cm^2
Electric potential	1 au = 9.076814×10^{-2} esu/cm = 271139 V
Electric field	1 au = 1.71527×10^7 esu/cm^2 = 5.14221×10^{11} V/m
Electric field gradient	1 au = 3.241390×10^{15} esu/cm^3 = 9.717362×10^{21} Vm^2
Polarizability	1 au = $1a_0^3$ = 0.14818×10^{-24} cm^3 = 0.16488×10^{-40} C^{2}m^{2}J^{-1}

Appendix N

Acronyms and Symbols

N.1 Abbreviations and Acronyms

1D	One dimension/one-dimensional
2D	Two dimensions/two-dimensional
3D	Three dimensions/three-dimensional
5CB	4-n-pentyl-4$'$-cyano-biphenyl
8CB	4-n-octyl-4$'$-cyano-biphenyl
aka	Also known as
a.u.	Arbitrary unit
au	Atomic unit
AA	All atoms
AC	Alternating current
AF	Antiferromagnetic
AFM	Atomic force microscopy
AMBER	Assisted model building and energy refinement (a force field)
AMD	Atomistic molecular dynamics
AO	Atomic orbital
APAPA	Anisylidene-p aminophenylacetate
B_n	Banana phase of type n ($n = 1, 2, \ldots, 7$)
BC	Boundary condition
BBC	Bipolar boundary condition
BCC	Body-centred cubic
BI	Boltzmann Inversion (method)
BP	Blue phase
nCBmCBn	n-cyano-biphenyl dimer with an m-alkyl spacer
CD	Circular dichroism
CG	Coarse-grained
CF	Correlation function
CFF	Consistent force field
CHARMM	Chemistry at Harvard molecular mechanics
CHELPG	Charges from electrostatic potentials using a grid-based method
CMC	Critical micellar concentration
CMC	Cluster Monte Carlo

CNT	Carbon nanotube
c.o.m.	Centre of mass
COMPASS	Condensed phase optimized molecular potentials for atomistic simulation studies
CPU	Central processing unit
CTAB	Cetyltrimethylammonium bromide
CUDA	Compute unified device architecture
D_h^d	Hexagonal columnar phase with inside column disorder
D_h^o	Hexagonal columnar phase with order inside columns
D_h^h	Hexagonal columnar phase with helicoidally ordered columns
D_r^d	Rectangular columnar phase with disorder in columns
D_r^o	Rectangular columnar phase with order in columns
D^t, D^{to}, D^{td}	Tilted columnar phase
DC	Direct current, a static field at zero frequency
DCM	4-(dicyanomethylene)-2-methyl-6-(4-dimethyl-amino styryl)-4-H-pyran
DFT	Density Functional Theory
DHPC	Dihexanol phosphatidylcoline
DHS	Dipolar hard sphere
DMANS	*p*-dimethylamino-*p*-nitro-stilbene
DMPC	Dimyristoil-phosphatidyl choline
DNA	Deoxyribonucleic acid
DNMR	Deuterium Nuclear Magnetic Resonance
DOPC	1,2-dioleoyl-sn-glycero-3-phosphocholine
DPD	Dissipative particle dynamics
DPH	1,6, diphenyl-hexatriene
DPL	Dipalmitoyl lecithin (or DPPC)
DPO	1,6, diphenyl-octatetraene
DPPC	Dipalmitoyl-phosphatidyl choline (DPL) or 1,2-dipalmitoyl-sn-glycero-3-phosphocholine
DSC	Differential scanning calorimetry
DSS	Dipolar soft spheres
DTBC	Disordered tangential boundary condition
DTH	2,2′-bithienyl
E7	A nematic mixture
E63	A nematic mixture
EBBA	4-ethoxybenzylidene-4′-*n*-butylaniline
Eq.	Equation
EFG	Electric field gradient
EPR	Electron paramagnetic resonance (also called ESR)
ESP	Electrostatic potential
ESR	Electron spin resonance (also called EPR)
F	A generic fluid phase
FCC	Face-centred cubic
FD	Fluorescence depolarization
FDTS	1H, 1H, 2H, 2H-perfluorodecyl-trichlorosilane

FENE	Finitely extendable non-linear elastic (potential)
Fig.	Figure
FF	Force field
FFT	Fast Fourier transform
FSC	Four site cluster
GA	Genetic algorithm
GAFF	General AMBER force field
GB	Gay–Berne
GHRL	Gruhn–Hess–Romano–Luckhurst
GPU	Graphics processing unit
GROMACS	Groningen machine for chemical simulation
GROMOS	Groningen molecular simulation force field and simulation package
GULP	General utility lattice program
H	Hexagonal phase
HAB	4,4′-diheptyl-azobenzene
HBA5PE	4-*n*-hexyloxy-benzoic acid-[4′-*n*-pentyl-phenylester]
HAN	Hybrid aligned nematic
HAT5	Hexakis (pentyloxy) triphenylene
HBC	Hexabenzocoronene
HCP	Hexagonal close packed
HF	Hartree–Fock
HGO	Hard Gaussian overlap model
HHTT	2,3,6,7,10,11-hexahexylthio triphenylene
HJL	Humphries–James–Luckhurst Mean Field Theory
HNMR	Proton nuclear magnetic resonance
HE	Hard ellipsoid
HPP	Hard polyhedra particle
HS	Hard sphere
HTP	Helical twisting power
HTS	Hexyltrichlorosilane
IF	Isotropic-ferroelectric
IN	Isotropic-nematic
IM	Intermolecular
IR	Infrared spectroscopy
I52	4-ethyl-2-fluoro-4′-[2-(trans-4-*n*-pentylcyclohexyl)-ethyl]-biphenyl *n*-propyl-cyclohexyl-ethyl-6-fluoro n-butyl-biphenyl
ITO	Indium-tin oxide
K	A solid crystal
KL	Potassium laurate
KMC	Kinetic Monte Carlo (method)
L	A liquid phase
L	Generic lamellar phase
L_α	Lamellar ‘liquid crystal’ phase with fluid like chains
$L_{\alpha'}$	Fluid lamellar phase with tilted mesogen chains
L_β	Gel lamellar phase with homeotropic mesogen chains

$L_{\beta'}$	Gel lamellar phase with tilted mesogen chains
LAB	Laboratory frame
LC	Liquid crystal
LCD	Liquid crystal display
LCE	Liquid crystal elastomer
LCP	Liquid crystal polymer
LD	Linear dichroism
LED	Light-emitting diode
LG	Liquid-gas
LHS	Left-hand side
LJ	Lennard Jones (potential)
LL	Lebwohl–Lasher
LSC	Light scattering
LZNS	Luckhurst–Zannoni–Nordio–Segre
mol	Mole
molecs	Molecules
M	'Middle' (hexagonal) chromonic phase
MBA5PE	4-methoxy-benzoic acid [4′-*n*-pentyl-phenylester]
MBBA	4-methoxybenzylidine-4″-*n*-butylaniline
MBBAHPE	4-*n*-butyl-benzoic acid- [4′-*n*-hexyloxy-phenylester]
MBCA	4-methoxybenzylidene-4′-cyanoaniline
MC	Monte Carlo (method)
MD	Molecular Dynamics (method)
ME	Maximum entropy
MEH-PPV	*poly* [2-methoxy-5-(2-ethylhexyloxy)-1,4-phenylene vinylene]
MESP	Molecular electrostatic potential
MF	Mean force
MFT	Mean Field (or Molecular Field) Theory
MIC	Minimum image convention
MM	Molecular mechanics
MOG	1-mono palmitoleoyl-rac-glycerol
MOL	Molecular (or particle) fixed frame
MO	Molecular orbital
MP2	Møller–Plesset perturbation treatment in quantum chemistry
MRI	Magnetic resonance imaging
MS	Maier–Saupe (Molecular Field Theory)
MSD	Mean square displacement
MTS	Multiple timescale
nCB	4-cyano-4′-*n*-alkyl-biphenyl (when n is an integer, it corresponds to the chain length)
N	Nematic
N*	Chiral nematic (cholesteric)
N_B	Biaxial nematic
N_C	Uniaxial lyotropic nematic formed by cylindrical micelles
N_D	Discotic nematic, e.g. formed by discotic micelles
N_{TB}	Twist-bend or heliconical phase

N(r)	Reentrant nematic
NEXAFS	Near-edge X-ray absorption fine structure
NMR	Nuclear magnetic resonance
NPD	Nematic polydomain elastomer
NI	Nematic-isotropic
N, P, T	Constant number of particles (N), pressure (P), temperature (T)
NSC	Neutron scattering
N, V, E	Constant number of particles (N), volume (V), energy (E)
N, V, T	Constant number of particles (N), volume (V), temperature (T)
ODF	Orientational distribution function
OH-MBBA	*o*-hydroxy-*p*-methoxybenzylidene-*p*′-butylaniline
OLED	Organic light-emitting diode
OPLS	Optimized potential for liquid simulation
OPV	Organic photovoltaic
OSC	Organic semiconductor
OTS	Octadecyltrichlorosilane
$P_{\beta'}$	Ripple lamellar phase
P3HT	Poly(3-hexylthiophene-2,5-diyl)
P5	*p*-quinquephenyl
PAA	4,4′-dimethoxy azoxybenzene
PBC	Periodic boundary condition
PBLG	poly(γ-benzyl-α, L-glutamate)
PBTTT	Poly[2,5-bis(3-alkylthiophen-2-yl)thieno[3,2-b]thiophene]
PCBM	[6,6]-phenyl C61 butyric acid methyl ester
PCH5	4′-*n*-pentyl-4-cyano phenyl cyclohexyl
PCH7	4′-*n*-heptyl-4-cyano phenyl cyclohexyl
PDLC	Polymer dispersed liquid crystal
PES	Potential energy surface
PK	Plastic crystal
PL	Parsons–Lee
PLPC	Palmitoyl lysophosphatidyl choline
PME	Particle mesh Ewald (method for calculating electrostatic interactions)
PNLC	Polymer network liquid crystal
POE	Polyoxyethylene
POM	Polarizing optical microscopy
PPV	Poly[phenylene vinylene]
PR-TRMC	Pulse radiolysis (technique to determine charge mobility)
PSLC	Polymer stabilized liquid crystal
QC	Quantum chemistry
QD	Quantum dot
QM	Quantum mechanics
RBC	Radial boundary condition
RDC	Residual dipolar coupling
RDF	Radial distribution function
RESPA	Reference system propagator algorithm

RHS	Right-hand side
RIS	Rotational isomeric state model
RM	Reactive monomer
RMS	Root mean square
RT	Random tiling
S	Solid of generic morphology (amorphous or crystal)
SAS	Solvent accessible surface
SDS	Sodium dodecyl sulphate
Sm	Generic smectic phase
S_A	Smectic A
S_{A_1}	Monolayer smectic A
S_{A_2}	Smectic A with bilayer structure
S_{A_d}	Smectic A with interdigitated partial bilayer structure
$S_{\tilde{A}}$	Smectic A with polar stripe structure (antiphase)
S_B	Smectic B
S_{B_K}	Crystal smectic B
S_{B_H}	Hexatic smectic B
S_C	Smectic C
S_{CSPF}	Sinclinic ferroelectric smectic C
S_{CSPA}	Sinclinic antiferroelectric smectic C
S_{CAPF}	Anticlinic ferroelectric smectic C
S_{CAPA}	Anticlinic antiferroelectric smectic C
$S_C{}^*$	Chiral smectic C
S_E	Smectic E
$SO(3)$	Three-dimensional rotation group
SPC	Simple point charge
SPME	Smooth particle mesh Ewald
STO	Slater-type orbital
StME	Staggered mesh Ewald
SWNT	Single-wall carbon nanotube
T6	α-sexithiophene
TBBA	Terephtalydene-bis-(4-n-butylaniline)
TBC	Toroidal boundary condition
TCP	Landau tricritical point
TCP	Topologically close packed
TEM	Transmission electron microscopy
TEOS	Tetraethyl orthosilicate
TGB	Twist grain boundary phase
TIP3P	Transferable intermolecular potential with 3 points
TMV	Tobacco mosaic virus
TN	Twisted nematic (display)
TOF	Time of flight (technique to determine charge mobility)
TSC	Two-site cluster
UA	United atoms (force field)
UFF	Universal force field

UV	Ultraviolet
vdW	van der Waals
VMD	Visual Molecular Dynamics
WAXD	Wide angle X-ray diffraction
WRT	With respect to
wt	Weight
w/w	Weight per weight concentration
XRD	X-ray diffraction
ZLI-1167	A nematic mixture

N.2 Symbols for Mathematical and Physical Quantities

When, occasionally, the same symbol is used for different quantities each is indicated on a separate line.

$\langle .. \rangle_X$	Average over variable X (X omitted if it is the only variable)
$\{\mathcal{H}, ...\}$	Classical Poisson brackets
$[\hat{A}, \hat{B}]$	Commutator, $\hat{A}\hat{B} - \hat{B}\hat{A}$
α, β, γ	Euler angles
α, α'	Heat capacity critical exponents
$\boldsymbol{a} \otimes \boldsymbol{b}$	Direct or tensor product of two vectors $\boldsymbol{a}, \boldsymbol{b}$
$\boldsymbol{a} \times \boldsymbol{b}$	Vector product of two vectors $\boldsymbol{a}, \boldsymbol{b}$
$\boldsymbol{\alpha}$	Molecular polarizability tensor
A	Surface area
Å	Ångstrom (10^{-10} m)
$A^{L,m}$	Spherical components of a rank L tensor
$\mathcal{A}$	Helmoltz (constant volume) free energy
β, β'	Order parameter critical exponents
β_n	n-th cluster integral
β_H	Haller order parameter pseudo critical exponents
β_T	Inverse temperature $\beta_T \equiv (1/k_B T)$
$\boldsymbol{\beta}$	Hyperpolarizability tensor
B_2	Second virial coefficient
B_n	nth virial coefficients
C	Concentration
C^*	Chiral dopant concentration
C_2	Symmetry axis for a rotation of π
$c(\boldsymbol{r}_{12}, \Omega_1, \Omega_2)$	Direct correlation function for two rigid molecules
$C_{AB}(t)$	Time correlation function between properties A and B
$\mathscr{C}_{AB}(t)$	Normalized correlation function
C_P	Heat capacity at constant pressure
C_V	Heat capacity at constant volume
$C(J_1, J_2, J_3; m_1, m_2)$	Clebsch–Gordan angular momentum coupling coefficient

$\boldsymbol{\chi}^{(e)} \equiv \boldsymbol{\chi}_{PP}$	Dielectric susceptibility tensor
$\boldsymbol{\chi}^{(m)}$	Magnetic susceptivity tensor
$\delta_{i,j}$ or $\delta(i-j)$	Kronecker delta
$\boldsymbol{d}$	Director
$d_a,\ a = x, y, z$	Direction cosines or director components
D	Diffusion coefficient
$D(x)$	Dawson function
D_{2h}	Point symmetry group for a parallelepiped
$\mathbf{D}$	Rotational diffusion tensor
$\mathbf{D}_T$	Translational diffusion tensor
$\mathscr{D}^L_{m,n}(\Omega)$	Component m, n of a Wigner matrix of rank L
$\boldsymbol{\varepsilon}$	Dielectric permittivity tensor
ϵ, ϵ_0	Energy terms in a Lennard–Jones and Gay–Berne potential
ε_{ijk}	Levi-Civita antisymmetric symbol
$\dot{\varepsilon}$	Shear rate
e_i	Electric charge on a particle or atom i
$\boldsymbol{e}$	Polarization direction of an electromagnetic radiation
$\boldsymbol{e}_i$	One of the unit vectors of an orthogonal basis set
E_Y	Young's (tensile) modulus
$\mathbf{E}$	Electric field vector
$\mathscr{E}_m$	mth energy level of a quantum system
$\mathrm{erf}(x)$	Error function
$\mathrm{erfc}(x)$	Complementary error function $(1 - \mathrm{erf}(x))$
η	Dielectric coupling strength between mesogen and electric field
η	Shear viscosity
η	Generic order parameter
ϕ	Volume fraction
$\phi_{\mu\mu}(t)$	Dipole-dipole single molecule time correlation function
$\phi^{LL'}_{mn;m'n'}(t)$	Wigner rotation matrix time correlation function
$\phi^{LL}_{mn}(t)$	Wigner rotation matrix time correlation for uniaxial molecules in a uniaxial medium
$\varphi^{LL'}_{mn;m'n'}(t)$	Real part of $\phi^{L,L'}_{mn;m'n'}(t)$
$\boldsymbol{\Phi} \equiv (\phi_1, \ldots, \phi_{N-1})$	The set of internal torsional angles
Φ_{ij} or $\Phi(x_i.x_j)$	Mayer function
$\dot{f}(t)$	Time derivative of a time dependent function
$f(\boldsymbol{q})$	Scattering factor
$\boldsymbol{f}$	Force vector
$F(t)$	Probability for a fluorescent molecule of being still excited at a time t after excitation event
$F[f(x)]$	Functional of $f(x)$

$\boldsymbol{F}$	Generic electric or magnetic field
$\widetilde{f}$	Fourier transform of a function f
$\mathscr{F}[f]$	Fourier transform operator acting on a function f
γ	Shape factor for convex bodies
γ	Surface tension
γ_1	Director rotational viscosity
γ_{AB}	Surface tension between two surfaces A, B
γ_i	Magnetogyric (or gyromagnetic) ratio
$\hat{\boldsymbol{\Gamma}}$	Time evolution operator (propagator)
$\mathcal{G}$	Gibbs (constant pressure) free energy
$\mathcal{G}_E$	Free energy contribution of an applied electric field
$\mathcal{G}_B$	Free energy contribution of an applied magnetic field
$\mathcal{G}_{el}$	Frank elastic energy
$G_i(\Omega_i, \boldsymbol{r})$	Gaussian function
$G_2(r)$	Second-rank angular-radial correlation
$G_6^H(r)$	Hexatic order radial correlation function
$G(\boldsymbol{r}_{12}, \Omega_1, \Omega_2)$	Positional-orientational pair distribution
$g(\boldsymbol{r}, \Omega_1, \Omega_2) \equiv g(r, \Omega_1, \Omega_2, \Omega_r)$	Reduced radial-orientational distribution
$g(r, \Omega_r)$	Radial-angular distribution for centres of mass
$g^{\mathrm{AB}}(r, \Omega_r)$	Radial-angular distribution for sites A, B
$g(r)$ or $g_0(r_{12})$	Radial distribution for centres of mass
$g^{\mathrm{AB}}(r)$ or $g_0^{\mathrm{AB}}(r_{12})$	Radial distribution for sites A, B
g_2	Second-rank Kirkwood coefficient
h	Planck constant
$\boldsymbol{h}_{a,i}$	Position vector of atom a in the molecule i frame
$h(\boldsymbol{r}_{12}, \Omega_1, \Omega_2)$	Total pair correlation for rigid molecules
$\boldsymbol{H}$	Magnetic field
$\mathcal{H}$	Enthalpy
$\mathscr{H}$	Classical Hamiltonian ($\mathscr{H} \equiv U + K$), i.e. potential plus kinetic energy
$\hat{\mathscr{H}}$	Hamiltonian operator
$\hat{\mathscr{H}}_{\mathrm{D}}$	Dipolar spin Hamiltonian
$\hat{\mathscr{H}}_{\mathrm{Z}}$	Zeeman spin Hamiltonian
$\hat{\mathscr{H}}_{\mathrm{J}}$	Indirect spin Hamiltonian
$\mathbf{I}$	Inertia tensor
I_{ab}	Cartesian components of inertia tensor
$\mathfrak{I}(\omega)$	Absorption intensity at frequency ω (aka spectrum)
$I^F(t)$	Time dependent fluorescence intensity
$I^F_{ab}(t)$	Time dependent fluorescence intensity for incident and output polarizers along $\boldsymbol{a}$ and $\boldsymbol{b}$
I_i	Nuclear spin of a certain isotope i
$\hat{\mathscr{I}}$	Inversion operator
$\mathbf{1}$	Identity matrix

$j_L(x)$	Spherical Bessel function of rank L
$\hat{\mathbf{J}}$	Angular momentum operator
$\boldsymbol{J}$	Classical angular momentum vector
$\boldsymbol{J}_Q$	Heat current
κ	Aspect ratio of a uniaxial ellipsoid
κ'	Potential well anisotropy of a uniaxial GB ellipsoid
k_B	Boltzmann constant
κ_G	Gaussian curvature
k_n	n-th cumulant of a distribution
κ_T	Isothermal compressibility
$\mathbf{K}$	Orientation factors matrix
$\mathscr{K}$	Kerr constant
K	Kinetic energy
$K(\tilde{\boldsymbol{p}})$	Global kinetic energy
K_R	Rotational kinetic energy
K_T	Translational kinetic energy
K_a	Elastic constant
K_{11}	Splay elastic constant
K_{22}	Twist elastic constant
K_{33}	Bend elastic constant
$\boldsymbol{\lambda}$	Thermal conductivity tensor
λ	Molecular biaxiality
λ	Electromagnetic radiation wavelength
λ_n	Nominal or engineering strain: dimensionless deformation per unit length
Λ	De Broglie thermal wavelength
ℓ_m	Typical molecular length
ℓ_z	Layer spacing in smectics
$\mathscr{L}[f]$	Laplace transform of a function f
$\hat{\mathscr{L}}$	Liouville operator
μ_0	Vacuum permeability
μ	Charge carrier mobility
$\boldsymbol{\mu}^{(a)}$	Absorption transition moment
$\boldsymbol{\mu}^{(e)}$	Emission transition moment
$\boldsymbol{\mu}$	Molecular dipole moment
$\boldsymbol{\mu}_M$	Magnetic dipole moment vector
$\boldsymbol{\mu}$	Electric dipole moment vector
m_i	Mass of particle i
m_n	nth moment of a distribution
$M(a,b,x)$	Kummer hypergeometric confluent function
$\boldsymbol{M}$	Magnetization (total magnetic dipole moment)
$\mathbf{M}$	4×4 Müller matrix in optics
ν	Frequency ($\sec^{-1}$)
$\mathbf{n}$	Refractive index tensor
n_d	Spacial dimensionality, $n_d = 1,2,3$

n_o	Spin dimensionality, $n_o = 1, 2, 3$
$n_\perp = n_o$	Ordinary refractive index of a uniaxial material
$n_{\|\|} = n_e$	Extraordinary refractive index of a uniaxial material
N	Number of particles in a sample
$\boldsymbol{N}$	Torque (moment of the force) vector
ω	Angular frequency (rad $\sec^{-1}$)
Ω	Orientation, e.g. Euler angles $\Omega = (\alpha, \beta, \gamma)$
Ω_{BA}	Rotation from coordinate frame A to B
$\Omega_i \equiv \Omega_{i\mathrm{L}} \equiv (M_i\mathrm{L})$	Orientation of ith particle frame with respect to the lab
$\dot{\boldsymbol{\Omega}}$	Angular velocity
$\hat{\mathscr{O}}_s$	Symmetry operator
$\psi_6(\mathbf{r})$	Hexatic bond order parameter
$\mathbf{\Pi}$	Markov transition matrix
Π_{ij}	Transition probability from state i to j in a Markov chain
Π	Osmotic pressure
p	Pitch or repeat distance of a helical structure (e.g. cholesteric or $\mathrm{N_{TB}}$)
$\boldsymbol{p}$	Linear momentum
$\boldsymbol{p}_T$	Total linear momentum of a system
$\langle P_2 \rangle$	Second-rank orientational order parameter
$\langle P_4 \rangle$	Fourth-rank orientational order parameter
$P_{abs}(0)$	Probability of absorption of a photon
$P_{em}(t)$	Emission probability from a fluorescent molecule
P	Pressure
P_C	Pressure at critical point
$P_T \equiv P_{GIS}$	Pressure at gas-isotropic liquid-solid triple point
P^*	Pressure in dimensionless units
$\boldsymbol{P}$	Overall electric dipole moment
$\boldsymbol{P}_E$	Electric polarization (overall dipole moment per volume)
$\hat{\mathscr{P}}$	Projection operator
$\mathscr{P}$	Shorthand for $P(\Omega_0 \| \Omega t)$
$P(\Omega_0 \| \Omega t)$	Orientational conditional probability
$P(\boldsymbol{X}_0 \| \boldsymbol{X}, t)$	Conditional probability
$P(\boldsymbol{X}_0; \boldsymbol{X}, t)$	Joint probability
$\hat{P}(\Omega_0 \| \Omega t)$	Symmetrized orientational conditional probability
$P_L^m(x)$	Associated Legendre functions
$P_L(x)$	Legendre polynomial of rank L
$\bar{q}$	Quaternion modulus
q^{T}	Total charge of a molecule (ion)
$\mathbf{q}$	Quaternion

$\mathbf{q}_k$	Nuclear quadrupole tensor at a nucleus k in NMR
$\mathbf{Q}$	Ordering matrix
$\mathcal{Q}(N, V, T)$	Full canonical partition function
$Q(N, V, T)$	Configurational canonical partition function
ρ	Density
ρ^*	Dimensionless density
$\varrho(\boldsymbol{r}, \boldsymbol{p})$	Probability density in phase space (spherical particles)
$\varrho(\boldsymbol{r}, \boldsymbol{\Omega}, \boldsymbol{p}, \dot{\boldsymbol{\Omega}})$	Probability density in phase space (rigid anisotropic particles)
$r(t)$	Fluorescence polarization anisotropy ratio
$\boldsymbol{r}_{a,i}$	Position vector of atom a of molecule i in the laboratory
$\mathbb{R}^n$	Real n-dimensional space
R	Gas constant
$\mathcal{R}$	Curvature of a body
$\mathbf{R}$	Cartesian rotation matrix in 3D
$\mathrm{R}^\bullet$	Radical: chemical species R with an unpaired electron
$\mathscr{R}^L_{m_1, m_2}(\alpha, \beta)$	Wigner matrices adapted for biaxial $\mathscr{D}_{2h}$ symmetry
σ	Shear stress
σ	Lennard–Jones spherical particle 'diameter'
σ_h	Hard spherical particle diameter
σ_s	Attractive range in square well potentials
σ_λ	Ratio of attractive to repulsive range diameter for a square well potential
$\boldsymbol{\sigma}(\omega)$	Frequency dependent electrical conductivity
$\boldsymbol{\sigma}$	Stress tensor
$\boldsymbol{\sigma}_i$	2×2 Pauli matrix
s	Topological charge of a structure (e.g. of a defect)
Σ_n	Nominal or engineering stress: applied force per initial unit cross section
$\mathcal{S}$	Entropy
$\mathcal{S}$	Surface of a body
S	Tsvetkov orientational order parameter, or $\langle P_2 \rangle$
$\mathbf{S}$	Saupe ordering matrix for biaxial molecules in a uniaxial phase
$\mathbf{S}_i$	Electron spin of an atom i
$S^{k_1,k_2}_{L_1,L_2,L} \equiv S^{k_1,k_2}_{L_1,L_2,L}(\Omega_{1\mathrm{L}}, \Omega_{2\mathrm{L}}, \Omega_{r\mathrm{L}})$	Stone rotational invariant
$S_{L_1,L_2,L}(\Omega_{1\mathrm{L}}, \Omega_{2\mathrm{L}}, \Omega_{r\mathrm{L}})$	Stone rotational invariant for uniaxial particles
$\overline{S^{k_1,k_2}_{L_1,L_2,L}}(r)$	Stone invariant angular-radial pair correlation
$S(\boldsymbol{q})$	Structure factor

τ	First positional order parameter in smectics
τ_F	Fluorescence lifetime
T^*	Dimensionless temperature, e.g. $T^* = k_B T/\epsilon$, where ϵ is an energy scale parameter
$T_{\alpha\beta}$	Transition temperature between α and β phases
T_{AI}	Smectic A-isotropic transition temperature
T_{AN}	Smectic A-nematic transition temperature
T_B	Boiling point temperature
T_B	Boyle temperature
T_C	Critical point temperature
T_{GIN}	Gas-isotropic-nematic triple point
T_g	Glass transition temperature
T_{KN}	Solid crystal-nematic transition temperature
T_M	Main (gel-fluid) transition temperature in lipid bilayers
T_{NI}	Nematic-isotropic transition temperature
T_{NSm}	Nematic-generic smectic transition temperature
T_{NI^*}	Pretransitional effects divergence temperature
$T_T \equiv T_{GIS}$	Gas-isotropic liquid-solid triple point temperature
T_{SN}	Solid-nematic transition temperature
T_{GNSm}	Gas-nematic-smectic triple point temperature
$T_r \equiv (T - T_C)/T_C$	Reduced temperature in terms of a critical temperature T_C
T_{YI}	Lyotropic-gel phase transition temperature
$[\mathbf{T}_{ij}]$	Dipolar coupling tensor between two nuclei i and j
$\mathcal{U}$	Average internal energy
$\mathscr{U}$	Dimensionless internal energy, $\mathscr{U} \equiv U/k_B T$
U	Potential energy
U_{ij}^{e}	Electrostatic pair potential
U_{ij}^{HE}	Hard ellipsoids pair potential
U_{ij}^{HP}	Hard particles pair potential
U_{ij}^{HPH}	Hard polyhedra pair potential
U_{ij}^{HSC}	Hard spherocylinders pair potential
$U_{ij}^{\mu\mu}$	Dipole-dipole pair potential
$U_{ij}^{\alpha\alpha}$	Dispersive pair potential
$U_{ij}^{\Theta\Theta}$	Quadrupole-quadrupole pair potential
U_{ij}^{LJ}	Lennard–Jones pair potential
$\boldsymbol{v}$	Linear velocity
$\mathcal{V}$	Volume of a particle or a molecule
V	Volume
V	Voltage
V^{e}	Electrostatic potential

V_C	Volume at a critical point
$V_T = V_{GIS}$	Gas-liquid-solid triple point volume
V^*	Volume in dimensionless units, e.g. $V^* = V/V_c$
$W(J_a, J_b, J_c; J_d, J_e)$	Racah angular momentum coupling coefficient
x	Molar fraction
$\tilde{\boldsymbol{X}}$	Set of generalized coordinates $(\boldsymbol{X}_1, \boldsymbol{X}_2, \ldots, \boldsymbol{X}_N)$ for N particles
$\boldsymbol{X}_i$	Generalized coordinates for the ith particle
Ξ	Grand canonical partition function
$\widetilde{\mathcal{X}} \equiv (\widetilde{q}, \widetilde{p})$	Phase space point
$\mathrm{d}\widetilde{\boldsymbol{X}}$	A shorthand for $(\mathrm{d}\boldsymbol{X}_1, \mathrm{d}\mathbf{X}_2, \ldots, \mathrm{d}\boldsymbol{X}_N)$
ξ	Asphericity factor for a convex particle
ξ	Generic field susceptibility
ξ	Typical size of an aligned domain or correlation length
ξ_2	Ornstein-Zernike correlation length
ξ_I	Typical size of ordered domains in the isotropic phase
$Y_{Lm}(\alpha, \beta)$	mth component of a spherical harmonic of rank L
z	Compressibility factor
$\mathcal{Z}_{N,V,T}$	Phase integral
Z	Configurational partition funtion
$Z_{N,V,T}$	Canonical configurational integral

References

Abascal, J. L. F. and Lago, S. 1985. A unified treatment of the equation of state of hard linear bodies. *J. Mol. Liq.*, **30**, 133–137.

Abascal, J. L. F. and Vega, C. 2005. A general purpose model for the condensed phases of water: TIP4P/2005. *J. Chem. Phys.*, **123**, 234505.

Abragam, A. 1961. *The Principles of Nuclear Magnetism*. Oxford: Oxford University Press.

Abraham, F. F. 1986. Computational statistical mechanics methodology, applications and supercomputing. *Adv. Phys.*, **35**, 1–111.

Abraham, M., Hess, B., van der Spoel, D., Lindahl, E. and development team, GROMACS. 2014. *GROMACS User Manual Version 5. 0. 4.*

Abramowitz, M. and Stegun, I. A. (eds.). 1965. *Handbook of Mathematical Functions*. New York: Dover.

Acharya, B. R., Primak, A. and Kumar, S. 2004. Biaxial nematic phase in bent-core thermotropic mesogens. *Phys. Rev. Lett.*, **92**, 145506.

Adam, C. J., Clark, S. J., Ackland, G. J. and Crain, J. 1997. Conformation-dependent dipoles of liquid crystal molecules and fragments from first principles. *Phys. Rev. E*, **55**, 5641–5649.

Adams, D. J., Luckhurst, G. R. and Phippen, R. W. 1987. Computer simulation studies of anisotropic systems. XVII. The Gay-Berne model nematogen. *Mol. Phys.*, **61**, 1575–1580.

Adlem, K., Čopič, M., Luckhurst, G. R., et al. 2013. Chemically induced twist-bend nematic liquid crystals, liquid crystal dimers, and negative elastic constants. *Phys. Rev. E*, **88**, 022503.

Afghah, S., Selinger, R. L. B. and Selinger, J. V. 2018. Visualising the crossover between 3D and 2D topological defects in nematic liquid crystals. *Liq. Cryst.*, **45**, 1–11.

Agarwal, U. and Escobedo, F. A. 2011. Mesophase behaviour of polyhedral particles. *Nat. Mater.*, **10**, 230–235.

Alder, B. J. and Wainwright, T. E. 1957. Phase transition for a hard sphere system. *J. Chem. Phys.*, **27**, 1208–1209.

Allara, D. L. 2005. A perspective on surfaces and interfaces. *Nature*, **437**, 638–639.

Allen, M. P. 1984. A molecular dynamics simulation study of octopoles in the field of a planar surface. *Mol. Phys.*, **52**, 717–732.

Allen, M. P. 2019. Molecular simulation of liquid crystals. *Mol. Phys.*, **117**, 2391–2417.

Allen, M. P. and Masters, A. 1993. Computer simulation of a twisted nematic liquid crystal. *Mol. Phys.*, **79**, 277–289.

Allen, M. P. and Tildesley, D. J. 2017. *Computer Simulation of Liquids*. 2nd ed. Oxford: Oxford University Press.

Allen, M. P. and Warren, M. A. 1997. Simulation of structure and dynamics near the isotropic-nematic transition. *Phys. Rev. Lett.*, **78**, 1291–1294.

Allen, M. P., Frenkel, D. and Talbot, J. 1989. Molecular dynamics simulation using hard particles. *Computer Phys. Rep.*, **9**, 301–353.

Allen, M. P., Evans, G. T., Frenkel, D. and Mulder, B. M. 1993. Hard convex body fluids. *Adv. Chem. Phys.*, **86**, 1–166.

Allen, M. P., Camp, P. J., Mason, C. P., Evans, G. T. and Masters, A. J. 1996a. Viscosities of isotropic hard particle fluids. *J. Chem. Phys.*, **105**, 11175–11182.

Allen, M. P., Warren, M. A., Wilson, M. R., Sauron, A. and Smith, W. 1996b. Molecular dynamics calculation of elastic constants in Gay-Berne nematic liquid crystals. *J. Chem. Phys.*, **105**, 2850–2858.

Allender, D. W. and Doane, J. W. 1978. Biaxial order parameters in liquid crystals: Their meaning and determination with nuclear quadrupole resonance. *Phys. Rev. A*, **17**, 1177–1180.

Allinger, N. L. 1977. Conformational analysis. 130. MM2. Hydrocarbon force field utilizing V_1 and V_2 torsional terms. *J. Amer. Chem. Soc.*, **99**, 8127–8134.

Allinger, N. L. and Lii, J. H. 1987. Benzene, aromatic rings, Van der Waals molecules, and crystals of aromatic-molecules in molecular mechanics (MM3). *J. Comput. Chem.*, **8**, 1146–1153.

Allinger, N. L., Chen, K. S. and Lii, J. H. 1996. An improved force field (MM4) for saturated hydrocarbons. *J. Comput. Chem.*, **17**, 642–668.

Allinger, N. L., Yuh, Y. H. and Lii, J. H. 1989. Molecular mechanics. The MM3 force field for hydrocarbons. 1. *J. Amer. Chem. Soc.*, **111**, 8551–8566.

Allinger, N. L., Tribble, M. T., Miller, M. A. and Wertz, D. H. 1971. Conformational analysis. 69. Improved force field for calculation of structures and energies of hydrocarbons. *J. Amer. Chem. Soc.*, **93**, 1637–1648.

Aloe, R., Chidichimo, G. and Golemme, A. 1991. Molecular reorientation in PDLC films monitored by 2H-NMR: electric field induced reorientation mechanism and optical properties. *Mol. Cryst. Liq. Cryst.*, **203**, 9–24.

Als-Nielsen, J. and McMorrow, D. 2011. *Elements of Modern X-ray Physics*. 2nd ed. Chichester: Wiley.

Altmann, S. L. 1977. *Induced Representations in Crystals and Molecules: Point, Space and Nonrigid Molecule Groups*. London: Academic Press.

Altmann, S. L. 2005. *Rotations, Quaternions, and Double Groups*. New York: Dover.

Ambrozic, M., Formoso, P., Golemme, A. and Žumer, S. 1997. Anchoring and droplet deformation in polymer dispersed liquid crystals: NMR study in an electric field. *Phys. Rev. E*, **56**, 1825–1832.

Amovilli, C., Cacelli, I., Campanile, S. and Prampolini, G. 2002. Calculation of the intermolecular energy of large molecules by a fragmentation scheme: application to the 4-n-pentyl-4′-cyanobiphenyl (5CB) dimer. *J. Chem. Phys.*, **117**, 3003–3012.

Andersen, H. C. 1980. Molecular dynamics at constant pressure and/or temperature. *J. Chem. Phys.*, **72**, 2384–2393.

Andersen, H. C. 1983. RATTLE-a velocity version of the SHAKE algorithm for molecular dynamics calculations. *J. Comput. Phys.*, **52**, 24–34.

Andersen, H. C., Chandler, D. and Weeks, J. D. 1976. Roles of repulsive and attractive forces in liquids: The equilibrium theory of classical fluids. *Adv. Chem. Phys.*, **34**, 105–155.

Anderson, J. C., Leaver, K. D., Rawlings, R. D. and Alexander, J. M. 1990. *Materials Science*. London: Chapman and Hall.

Anderson, P. W. 1954. A mathematical model for the narrowing of spectral lines by exchange or motion. *J. Phys. Soc. Japan*, **9**, 316–339.

Anderson, P. W. 1981. Some general thoughts about broken symmetry. In Boccara, N. (ed.), *Symmetries and Broken Symmetries in Condensed Matter Physics*, Paris: IDSET, pp. 11–20.

Andrienko, D. and Allen, M. P. 2002. Theory and simulation of the nematic zenithal anchoring coefficient. *Phys. Rev. E*, **65**, 021704.

Andrienko, D., Germano, G. and Allen, M. P. 2001. Computer simulation of topological defects around a colloidal particle or droplet dispersed in a nematic host. *Phys. Rev. E*, **63**, 041701.

Anezo, C., de Vries, A. H., Holtje, H. D., Tieleman, D. P. and Marrink, S. J. 2003. Methodological issues in lipid bilayer simulations. *J. Phys. Chem. B*, **107**, 9424–9433.

Anisimov, M. A. 1987. Universality of the critical dynamics and the nature of the nematic-isotropic phase transition. *Mol. Cryst. Liq. Cryst.*, **146**, 435–461.

Arcioni, A., Tarroni, R. and Zannoni, C. 1988. Fluorescence Depolarization in liquid crystals. In Samorì, B. and Thulstrup, E. (eds.), *Polarized Spectroscopy of Ordered Systems*. Dordrecht: Kluwer, pp. 421–453.

Arcioni, A., Bertinelli, F., Tarroni, R. and Zannoni, C. 1987. Time resolved Fluorescence Depolarization in a nematic liquid crystal. *Mol. Phys.*, **61**, 1161–1181.

Arcioni, A., Bertinelli, F., Tarroni, R. and Zannoni, C. 1990. A Fluorescence Depolarization study of the order and dynamics of 1,8-diphenyloctatetraene in a nematic liquid crystal. *Chem. Phys.*, **143**, 259–270.

Arfken, G. B. and Weber, H. J. 1995. *Mathematical Methods for Physicists*. San Diego, CA: Academic Press.

Arnold, A. and Holm, C. 2005. MMM1D: a method for calculating electrostatic interactions in 1D periodic geometries. *J. Chem. Phys.*, **123**, 144103.

Arunan, E., Desiraju, G. R., Klein, R. A., et al. 2011. Definition of the hydrogen bond (IUPAC Recommendations 2011). *Pure Appl. Chem.*, **83**, 1637–1641.

Ashcroft, N. W. and Mermin, N. D. 1976. *Solid State Physics*. New York: Harcourt.

Atkins, P. W. 1978. *Physical Chemistry*. Oxford: Oxford University Press.

Atkins, P. W. 1983. *Molecular Quantum Mechanics*. Oxford: Oxford University Press.

Attard, G. S., Beckmann, P. A., Emsley, J. W., Luckhurst, G. R. and Turner, D. L. 1982. Pretransitional behavior in nematic liquid crystals. A Nuclear Magnetic Resonance study. *Mol. Phys.*, **45**, 1125–1129.

Attard, G. S., Glyde, J. C. and Goltner, C. G. 1995. Liquid crystalline phases as templates for the synthesis of mesoporous silica. *Nature*, **378**, 366–368.

Atwood, J. L., Barbour, L. J., Hardie, M. J. and Raston, C. 2001. Metal sulfonatocalix[4,5]arene complexes: bi-layers, capsules, spheres, tubular arrays and beyond. *Coord. Chem. Rev.*, **222**, 3–32.

Avogadro. An Open-Source Molecular Builder and Visualization Tool, http://avogadro.openmolecules.net

Axenov, K. V. and Laschat, S. 2011. Thermotropic ionic liquid crystals. *Materials*, **4**, 206–259.

Ayton, G. and Patey, G. N. 1996. Ferroelectric order in model discotic nematic liquid crystals. *Phys. Rev. Lett.*, **76**, 239–242.

Azzam, R. M. A. 1978. Propagation of partially polarized light through anisotropic media with or without depolarization: a differential 4×4 matrix calculus. *JOSA*, **68**, 1756–1767.

Bacchiocchi, C. and Zannoni, C. 1998. Directional energy transfer in columnar liquid crystals: a computer simulation study. *Phys. Rev. E*, **58**, 3237–3244.

Bacchiocchi, C., Miglioli, I., Arcioni, A., et al. 2009. Order and dynamics inside H-PDLC nanodroplets. An ESR spin probe study. *J. Phys. Chem. B*, **113**, 5391–5402.

Baerends, E. J., Ziegler, T., Atkins, A. J., et al. 2021. ADF Manual. Amsterdam Modeling Suite 2021.1. Amsterdam: SCM.

Baggioli, A., Casalegno, M., Raos, G., et al. 2019. Atomistic simulation of phase transitions and charge mobility for the organic semiconductor Ph-BTBT-C10. *Chem. Mater.*, **31**, 7092–7103.

Bahr, C. 1994. Influence of dimensionality and surface ordering on phase transitions: studies of freely-suspended liquid crystal films. *Int. J. Mod. Phys. B*, **8**, 3051–3082.

Bahturin, Y. 1993. *Basic Structures of Modern Algebra*. Dordrecht: Kluwer.

Bailly-Reyre, A. and Diep, H. T. 2015. Phase transition of mobile Potts model for liquid crystals. *Physics Procedia*, **75**, 557–564.

Baker, G. A. and Gammel, J. L. 1970. *The Padé Approximant in Theoretical Physics*. New York: Academic Press.

Baker, K. N., Fratini, A. V., Resch, T., et al. 1993. Crystal-structures, phase transitions and energy calculations of poly(p-phenylene) oligomers. *Polymer*, **34**, 1571–1587.

Bakhmutov, V. I. 2015. *NMR Spectroscopy in Liquids and Solids*. Boca Raton, FL: CRC Press.

Balescu, R. 1975. *Equilibrium and NonEquilibrium Statistical Mechanics*. New York: Wiley.

Ball, J. M. 2017. Mathematics and liquid crystals. *Mol. Cryst. Liq. Cryst.*, **647**, 1–27.

Ball, J. M. and Majumdar, A. 2010. Nematic liquid crystals: from Maier-Saupe to a continuum theory. *Mol. Cryst. Liq. Cryst.*, **525**, 1–11.

BALLview. The BALL Website www.ballview.org/.

Bannister, N., Skelton, J., Kociok-Köhn, G., et al. 2019. Lattice vibrations of γ- and β-coronene from Raman microscopy and theory. *Phys. Rev. Mater.*, **3**, 125601.

Barbero, G. and Barberi, R. 1983. Critical thickness of a hybrid aligned nematic liquid crystal cell. *J. de Physique*, **44**, 609–616.

Barbero, G. and Evangelista, L. R. 2006. *Adsorption Phenomena and Anchoring Energy in Nematic Liquid Crystals*. London: Taylor & Francis.

Barbero, G., Ferrero, C., Gunzel, T., Skačej, G. and Žumer, S. 1998. Surface-induced nematic ordering and the localization of a twisted distortion in a nematic cell. *Phys. Rev. E*, **58**, 8024–8027.

Barboy, B. and Gelbart, W. M. 1979. Series representation of the equation of state for hard particle fluids. *J. Chem. Phys.*, **71**, 3053–3062.

Barker, J. A. and Henderson, D. 1976. What is 'liquid'? Understanding the states of matter. *Rev. Mod. Phys.*, **48**, 587–671.

Barker, J. A. and Watts, R. O. 1969. Structure of water; a Monte Carlo calculation. *Chem. Phys. Lett.*, **3**, 144–145.

Barois, P. 1992. Phase transitions in liquid crystals: introduction to phase transition theories. In Martellucci, S. and Chester, A. (eds.), *Phase Transitions in Liquid Crystals*. New York: Plenum Press, pp. 41–66.

Barois, P. 1999. Phase transition theories. In Demus, D., Goodby, J., Gray, G. W., Spiess, H. W. and Vill, V. (eds.), *Physical Properties of Liquid Crystals*. Weinheim: Wiley-VCH, pp. 179–207.

Barojas, J., Levesque, D. and Quentrec, B. 1973. Simulation of diatomic homonuclear liquids. *Phys. Rev. A*, **7**, 1092.

Barry, E., Hensel, Z., Dogic, Z., Shribak, M. and Oldenbourg, R. 2006. Entropy-driven formation of a chiral liquid crystalline phase of helical filaments. *Phys. Rev. Lett.*, **96**, 018305.

Bates, M. A. 1998. Planar anchoring at the smectic A-isotropic interface. A molecular dynamics simulation study. *Chem. Phys. Lett.*, **288**, 209–215.

Bates, M. A. and Luckhurst, G. R. 1996. Computer simulation studies of anisotropic systems. XXVI. Monte Carlo investigations of a Gay-Berne discotic at constant pressure. *J. Chem. Phys.*, **104**, 6696–6709.

Bates, M. A. and Luckhurst, G. R. 1998. Computer simulation studies of anisotropic systems XXIX. Quadrupolar Gay-Berne discs and chemically induced liquid crystal phases. *Liq. Cryst.*, **24**, 229–241.

Bates, M. A. and Luckhurst, G. R. 1999a. Computer simulation studies of anisotropic systems. XXX. The phase behavior and structure of a Gay-Berne mesogen. *J. Chem. Phys.*, **110**, 7087–7108.

Bates, M. A. and Luckhurst, G. R. 1999b. Computer simulation of liquid crystal phases formed by Gay-Berne mesogens. In Mingos, D. M. P. (ed.), *Liquid Crystals I. Struct. Bond.* Berlin: Springer, pp. 65–138.

Bates, M. A. and Zannoni, C. 1997. A molecular dynamics simulation study of the nematic-isotropic interface of a Gay–Berne liquid crystal. *Chem. Phys. Lett.*, **280**, 40–45.

Bates, M. A., Skačej, G. and Zannoni, C. 2010. Defects and ordering in nematic coatings on uniaxial and biaxial colloids. *Soft Matter*, **6**, 655–663.

Bauman, D., Zieba, A. and Mykowska, E. 2008. Orientational behaviour of some homologues of 4-n-pentylphenylthio-4′-n-alkoxybenzoate doped with dichroic dye. *Opto-Electronics Review*, **16**, 244–250.

Bautista-Carbajal, G. and Odriozola, G. 2014. Phase diagram of two-dimensional hard ellipses. *J. Chem. Phys.*, **140**, 204502.

Bayly, C. I., Cieplak, P., Cornell, W. D. and Kollman, P. A. 1993. A well-behaved electrostatic potential based method using charge restraints for deriving atomic charges-the RESP model. *J. Phys. Chem.*, **97**, 10269–10280.

Bechhoefer, J., Simon, A. J., Libchaber, A. and Oswald, P. 1989. Destabilization of a flat nematic-isotropic interface. *Phys. Rev. A*, **40**, 2042–2056.

Beckmann, P. A., Emsley, J. W., Luckhurst, G. R. and Turner, D. L. 1986. Nuclear spin-lattice relaxation rates in liquid crystals. Results for deuterons in specifically deuterated 4-n-pentyl-4′-cyanobiphenyl in both nematic and isotropic phases. *Mol. Phys.*, **59**, 97–125.

Bedolla, E., Padierna, L. C. and Castaneda-Priego, R. 2021. Machine learning for condensed matter physics. *J. Phys. Cond. Matter*, **33**, 053001.

Beljonne, D., Cornil, J., Muccioli, L., et al. 2011. Electronic processes at organic-organic interfaces: insight from modeling and implications for opto-electronic devices. *Chem. Mater.*, **23**, 591–609.

Bellec, A., Riedel, D., Dujardin, G., et al. 2009. Electronic properties of the n-doped hydrogenated silicon (100) surface and dehydrogenated structures at 5 K. *Phys. Rev. B*, **80**, 245434.

Belli, S., Dussi, S., Dijkstra, M. and van Roij, R. 2014. Density functional theory for chiral nematic liquid crystals. *Phys. Rev. E*, **90**, 020503.

Bellini, T., Cerbino, R. and Zanchetta, G. 2012. DNA-based soft phases. *Topics. Curr. Chem.*, **318**, 225–79.

Bellini, T., Radzihovsky, L., Toner, J. and Clark, N. A. 2001. Universality and scaling in the disordering of a smectic liquid crystal. *Science*, **294**, 1074–1079.

Ben-Reuven, A. and Gershon, N. D. 1969. Light scattering by orientational fluctuations in liquids. *J. Chem. Phys.*, **51**, 893–902.

Benning, S., Kitzerow, H. S., Bock, H. and Achard, M. F. 2000. Fluorescent columnar liquid crystalline 3, 4, 9, 10-tetra-(n-alkoxycarbonyl)-perylenes. *Liq. Cryst.*, **27**, 901–906.

Berardi, R. and Zannoni, C. 2000. Do thermotropic biaxial nematics exist? A Monte Carlo study of biaxial Gay-Berne particles. *J. Chem. Phys.*, **113**, 5971–5979.

Berardi, R. and Zannoni, C. 2015. Computer simulations of biaxial nematics. In Luckhurst, G. R. and Sluckin, T. J. (eds.), *Biaxial Nematic Liquid Crystals. Theory, Simulation and Experiment.* Chichester: Wiley, pp. 153–184.

Berardi, R., Emerson, A. P. J. and Zannoni, C. 1993. Monte Carlo investigations of a Gay-Berne liquid crystal. *J. Chem. Soc. Faraday Trans.*, **89**, 4069–4078.

Berardi, R., Cecchini, M. and Zannoni, C. 2003a. A Monte Carlo study of the chiral columnar organizations of dissymmetric discotic mesogens. *J. Chem. Phys.*, **119**, 9933–9946.

Berardi, R., Fava, C. and Zannoni, C. 1998a. A Gay-Berne potential for dissimilar biaxial particles. *Chem. Phys. Lett.*, **297**, 8–14.

Berardi, R., Fava, C. and Zannoni, C. 1995. A generalized Gay-Berne intermolecular potential for biaxial particles. *Chem. Phys. Lett.*, **236**, 462–468.

Berardi, R., Fehervari, M. and Zannoni, C. 1999. A Monte Carlo simulation study of associated liquid crystals. *Mol. Phys.*, **97**, 1173–1184.

Berardi, R., Muccioli, L. and Zannoni, C. 2004a. Can nematic transitions be predicted by atomistic simulations? A computational study of the odd-even effect. *Chem. Phys. Chem.*, 104–111.

Berardi, R., Muccioli, L. and Zannoni, C. 2008a. Field response and switching times in biaxial nematics. *J. Chem. Phys.*, **128**, 024905.

Berardi, R., Orlandi, S. and Zannoni, C. 1996a. Antiphase structures in polar smectic liquid crystals and their molecular origin. *Chem. Phys. Lett.*, **261**, 357–362.

Berardi, R., Orlandi, S. and Zannoni, C. 2003b. Molecular dipoles and tilted smectic formation: a Monte Carlo study. *Phys. Rev. E*, **67**, 041708.

Berardi, R., Ricci, M. and Zannoni, C. 2001. Ferroelectric nematic and smectic liquid crystals from tapered molecules. *ChemPhysChem*, **2**, 443–447.

Berardi, R., Ricci, M. and Zannoni, C. 2004b. Ferroelectric and structured phases from polar tapered mesogens. *Ferroelectrics*, **309**, 3–12.

Berardi, R., Spinozzi, F. and Zannoni, C. 1992. Maximum Entropy internal order approach to the study of intermolecular rotations in liquid crystals. *J. Chem. Soc. Faraday Trans.*, **88**, 1863–1873.

Berardi, R., Spinozzi, F. and Zannoni, C. 1994. The rotational-conformational distribution of $2,2'$-bithienyl in liquid crystals. *Liq. Cryst.*, **16**, 381–397.

Berardi, R., Spinozzi, F. and Zannoni, C. 1996b. The conformations of alkyl chains in fluids. A maximum entropy approach. *Chem. Phys. Lett.*, **260**, 633–638.

Berardi, R., Spinozzi, F. and Zannoni, C. 1996c. A new maximum entropy conformational analysis of biphenyl in liquid crystal solution. *Mol. Cryst. Liq. Cryst. A*, **290**, 245–253.

Berardi, R., Spinozzi, F. and Zannoni, C. 1998b. A multitechnique maximum entropy approach to the determination of the orientation and conformation of flexible molecules in solution. *J. Chem. Phys.*, **109**, 3742–3759.

Berardi, R., Kuball, H. G., Memmer, R. and Zannoni, C. 1998c. Chiral induction in nematics. A computer simulation study. *J. Chem. Soc. Faraday Trans.*, **94**, 1229–1234.

Berardi, R., Lintuvuori, J. S., Wilson, M. R. and Zannoni, C. 2011. Phase diagram of the uniaxial and biaxial soft-core Gay-Berne model. *J. Chem. Phys.*, **135**, 134119.

Berardi, R., Zannoni, C., Lintuvuori, J. S. and Wilson, M. R. 2009. A soft-core Gay-Berne model for the simulation of liquid crystals by Hamiltonian replica exchange. *J. Chem. Phys.*, **131**, 174107.

Berardi, R., Micheletti, D., Muccioli, L., Ricci, M. and Zannoni, C. 2004c. A computer simulation study of the influence of a liquid crystal medium on polymerization. *J. Chem. Phys.*, **121**, 9123–9130.

Berardi, R., Muccioli, L., Orlandi, S., Ricci, M. and Zannoni, C. 2008b. Computer simulations of biaxial nematics. *J. Phys. Condens. Matter*, **20**, 4631011–16.

Berardi, R., Muccioli, L., Orlandi, S., Ricci, M. and Zannoni, C. 2004d. Mimicking electrostatic interactions with a set of effective charges: a genetic algorithm. *Chem. Phys. Lett.*, **389**, 373–378.

Berardi, R., Orlandi, S., Photinos, D. J., Vanakaras, A. G. and Zannoni, C. 2002. Dipole strength effects on the polymorphism in smectic A mesophases. *Phys. Chem. Chem. Phys.*, **4**, 770–777.

Berardi, R., Cainelli, G., Galletti, P., et al. 2005. Can the π-facial selectivity of solvation be predicted by atomistic simulation? *J. Amer. Chem. Soc.*, **127**, 10699–10706.

Berendsen, H. J. C. 2007. *Simulating the Physical World. Hierarchical Modeling from Quantum Mechanics to Fluid Dynamics*. Cambridge: Cambridge University Press.

Berendsen, H. J. C., Grigera, J. R. and Straatsma, T. P. 1987. The missing term in effective pair potentials. *J. Phys. Chem.*, **91**, 6269–6271.

Berendsen, H. J. C., van der Spoel, D. and van Drunen, R. 1995. GROMACS: a message-passing parallel molecular dynamics implementation. *Comput. Phys. Commun.*, **91**, 43–56.

Berendsen, H. J. C., Postma, J. P. M, van Gunsteren, W. F. and Hermans, J. 1981. Interaction models of water in relation to protein hydration. In Pullman, B. (ed.), *Intermolecular Forces*. Dordrecht: Reidel, pp. 331–342.

Berendsen, H. J. C., Postma, J. P. M., van Gunsteren, W. F., Di Nola, A. and Haak, J. R. 1984. Molecular dynamics with coupling to an external bath. *J. Chem. Phys.*, **81**, 3684–3690.

Berggren, E. and Zannoni, C. 1995. Rotational diffusion of biaxial probes in biaxial liquid crystal phases. *Mol. Phys.*, **85**, 299–333.

Berggren, E., Tarroni, R. and Zannoni, C. 1993. Rotational diffusion of uniaxial probes in biaxial liquid crystal phases. *J. Chem. Phys.*, **99**, 6180–6200.

Berggren, E., Zannoni, C., Chiccoli, C., Pasini, P. and Semeria, F. 1992. Monte Carlo study of the molecular organization in model nematic droplets. Field effects. *Chem. Phys. Lett.*, **197**, 224–230.

Berggren, E., Zannoni, C., Chiccoli, C., Pasini, P. and Semeria, F. 1994a. Computer simulations of nematic droplets with bipolar boundary-conditions. *Phys. Rev. E*, **50**, 2929–2939.

Berggren, E., Zannoni, C., Chiccoli, C., Pasini, P. and Semeria, F. 1994b. Monte Carlo study of the effect of an applied-field on the molecular-organization of polymer dispersed liquid crystal droplets. *Phys. Rev. E*, **49**, 614–622.

Berggren, E., Zannoni, C., Chiccoli, C., Pasini, P. and Semeria, F. 1995. A Monte Carlo simulation of a Twisted Nematic liquid crystal display. *Int. J. Mod. Phys. C*, **6**, 135–141.

Berlman, I. B. 1971. *Handbook of Fluorescence Spectra of Aromatic Molecules*. 2nd ed. New York: Academic Press.

Bernassau, J. M., Black, E. P. and Grant, D. M. 1982. Molecular motion in anisotropic medium. I. The effect of the dipolar interaction on nuclear spin relaxation. *J. Chem. Phys.*, **76**, 253–256.

Berne, B. J. 1971. Time-dependent properties of condensed media. In Henderson, D. (ed.), *Physical Chemistry. An Advanced Treatise. Liquid State*, vol. 8B. New York: Academic Press, pp. 539–716.

Berne, B. J. and Pechukas, P. 1972. Gaussian model potentials for molecular-interactions. *J. Chem. Phys.*, **56**, 4213–4216.

Berne, B. J. and Pecora, R. 2000. *Dynamic Light Scattering*. New York: Dover.

Bernstein, J. 2007. *Polymorphism in Molecular Crystals*. Oxford: Oxford University Press.

Berreman, D. W. 1972. Solid surface shape and alignment of an adjacent nematic liquid crystal. *Phys. Rev. Lett.*, **28**, 1683–1686.

Besler, B. H., Merz, Jr, K. M. and Kollman, P. A. 1990. Atomic charges derived from semiempirical methods. *J. Comput. Chem.*, **11**, 431–439.

Betten, J. 1987. Irreducible invariants of 4th-order tensors. *Math. Model.*, **8**, 29–33.

Bhethanabotla, V. R. and Steele, W. 1987. A comparison of hard-body models for axially-symmetric molecules. *Mol. Phys.*, **60**, 249–251.

Bi, C. K., Yang, L., Duan, Y. L. and Shi, Y. 2019. A survey on visualization of tensor field. *J. Visualization*, **22**, 641–660.

Biggins, J. S., Warner, M. and Bhattacharya, K. 2009. Supersoft elasticity in polydomain nematic elastomers. *Phys. Rev. Lett.*, **103**, 037802.

Biggins, J. S., Warner, M. and Bhattacharya, K. 2012. Elasticity of polydomain liquid crystal elastomers. *J. Mech. Phys. Solids*, **60**, 573–590.

Biltonen, R. L. and Lichtenberg, D. 1993. The use of differential scanning calorimetry as a tool to characterize liposome preparations. *Chem. Phys. Lipids*, **64**, 129–142.

Binder, K. 1976. Monte Carlo investigations of phase transitions and critical phenomena. In Domb, C. and Green, M. S. (eds.), *Phase Transitions and Critical Phenomena*, vol. 5B. London: Academic Press, pp. 1–105.

Binder, K. 1987. Theory of first order phase transitions. *Rep. Progr. Phys.*, **50**, 783–859.

Binder, K. (ed.). 1995. *Monte Carlo and Molecular Dynamics Simulations in Polymer Science*. Oxford: Oxford University Press.

Binnemans, K. 2005. Ionic liquid crystals. *Chem. Rev.*, **105**, 4148–4204.

Bird, R. B., Armstrong, R. C. and Hassager, D. 1971. *Dynamics of Polymeric Liquids*. New York: Wiley.

Birgeneau, R. J., Garland, C. W., Kasting, G. B. and Ocko, B. M. 1981. Critical behavior near the nematic-smectic-A transition in butyloxybenzylidene octylaniline (4O.8). *Phys. Rev. A*, **24**, 2624–2634.

Birkhoff, G. and Mac Lane, S. 1997. *A Survey of Modern Algebra*. Wellesley, MA: A. K. Peters.

Biscari, P., Calderer, M. C. and Terentjev, E. M. 2007. Landau-de Gennes theory of isotropic-nematic-smectic liquid crystal transitions. *Phys. Rev. E*, **75**, 051707.

Biscarini, F., Zannoni, C., Chiccoli, C. and Pasini, P. 1991. Head tail asymmetry and ferroelectricity in uniaxial liquid crystals. Model calculations. *Mol. Phys.*, **73**, 439–461.

Biscarini, F., Chiccoli, C., Pasini, P., Semeria, F. and Zannoni, C. 1995. Phase diagram and orientational order in a biaxial lattice model – a Monte Carlo study. *Phys. Rev. Lett.*, **75**, 1803–1806.

Bisi, F., Luckhurst, G. R. and Virga, E. G. 2008. Dominant biaxial quadrupolar contribution to the nematic potential of mean torque. *Phys. Rev. E*, **78**, 021710.

Bisi, F., Romano, S. and Virga, E. G. 2007. Uniaxial rebound at the nematic biaxial transition. *Phys. Rev. E*, **75**, 0417051.

Bisoyi, H. K. and Kumar, S. 2010. Discotic nematic liquid crystals: science and technology. *Chem. Soc. Rev.*, **39**, 264–85.

Blachnik, N., Kneppe, H. and Schneider, F. 2000. Cotton-Mouton constants and pretransitional phenomena in the isotropic phase of liquid crystals. *Liq. Cryst.*, **27**, 1219–1227.

Blinc, R., O'Reilly, D. E., Peterson, E. M., Lahajnar, G. and Levstek, I. 1968. Proton spin-lattice relaxation study of the liquid crystal transition in p-anisaldazine. *Solid State Comm.*, **6**, 839–841.

Blinc, R., Burgar, M., Luzar, M., et al. 1974. Anisotropy of self-diffusion in smectic-A and smectic-C phases. *Phys. Rev. Lett.*, **33**, 1192–1195.

Blinov, L. M. 1998. On the way to polar achiral liquid crystals. *Liq. Cryst.*, **24**, 143–152.

Blinov, L. M. 2011. *Structure and Properties of Liquid Crystals*. Berlin: Springer.

Blinov, L. M., Kabayenkov, A. Y. and Sonin, A. A. 1989. Experimental studies of the anchoring energy of nematic liquid crystals. Invited Lecture. *Liq. Cryst.*, **5**, 645–661.

Blum, L. 1972. Invariant expansion. 2. Ornstein-Zernike equation for nonspherical molecules and an extended solution to mean spherical model. *J. Chem. Phys.*, **57**, 1862–1869.

Blum, L. and Torruella, A. J. 1972. Invariant expansion for 2-body correlations – Thermodynamic functions, scattering, and Ornstein-Zernike equation. *J. Chem. Phys.*, **56**, 303–310.

Blumstein, A. 1978. *Mesomorphic Order in Polymers and Polymerization in Liquid Crystalline Media*. Washington, DC: American Chemical Society.

Blumstein, A. 1985. *Polymeric Liquid Crystals*. New York: Plenum Press.

Blunk, D., Bongartz, N., Stubenrauch, C. and Gartner, V. 2009. Syntheses, amphitropic liquid crystallinity, and surface activity of new inositol-based amphiphiles. *Langmuir*, **25**, 7872–7878.

Boden, N. 1990. Self-assembly and self-organisation in fluids. *Chem. Britain*, **26**, 344–348.

Boden, N., Bushby, R. J. and Hardy, C. 1985. New mesophases formed by water soluble discoidal amphiphiles. *J. de Physique Lett.*, **46**, 325–328.

Boden, N., Bushby, R. J., Ferris, L., Hardy, C. and Sixl, F. 1986. Designing new lyotropic amphiphilic mesogens to optimize the stability of nematic phases. *Liq. Cryst.*, **1**, 109–125.

Bolhuis, P. and Frenkel, D. 1997. Tracing the phase boundaries of hard spherocylinders. *J. Chem. Phys.*, **106**, 666–687.

Bondi, A. 1964. van der Waals volumes and radii. *J. Phys. Chem.*, **68**, 441–451.

Bonomi, M., Branduardi, D., Bussi, G., et al. 2009. PLUMED: a portable plugin for free energy calculations with molecular dynamics. *Comput. Phys. Comm.*, **180**, 1961–1972.

Borshch, V., Kim, Y. K., Xiang, J., et al. 2013. Nematic twist-bend phase with nanoscale modulation of molecular orientation. *Nat. Commun.*, **4**, 2635.

Bose, T. K. and Saha, J. 2012. Origin of tilted-phase generation in systems of ellipsoidal molecules with dipolar interactions. *Phys. Rev. E*, **86**, 050701.

Boselli, L., Lopez, H., Zhang, W., et al. 2020. Classification and biological identity of complex nano shapes. *Commun. Mater.*, **1**, 35.

Böttcher, C. J. F. and Bordewijk, P. 1978. *Theory of Electric Polarization. vol. 2. Dielectrics in Time Dependent Fields*. Amsterdam: Elsevier.

Böttcher, C. J. F., van Belle, O. C., Bordewijk, P. and Rip, A. 1973. *Theory of Electric Polarization. vol. 1, Dielectrics in Static Fields*. 2nd ed. Amsterdam: Elsevier.

Böttcher, T., Elliott, H. L. and Clardy, J. 2016. Dynamics of snake-like swarming behavior of vibrio alginolyticus. *Biophys. J.*, **110**, 981–92.

Boublik, T. 1975. Hard convex body equation of state. *J. Chem. Phys.*, **63**, 4084–4084.

Boublik, T. and Nezbeda, I. 1986. PVT behaviour of hard body fluids. Theory and experiment. *Collect. Czechoslov. Chem. Commun.*, **51**, 2301–2432.

Bower, D. I. 1981. Orientation distribution functions for uniaxially oriented polymers. *J. Polym. Sci. B*, **19**, 93–107.

Bower, D. I. 1982. Orientation distribution functions for biaxially oriented polymers. *Polymer*, **23**, 1251–1255.

Bowers, K. J., Chow, D. E., Xu, H., et al. 2006. Scalable algorithms for Molecular Dynamics simulations on commodity clusters. *Proceedings of the ACM/IEEE Conference on Supercomputing (SC06)*.

Boyd, N. J. and Wilson, M. R. 2015. Optimization of the GAFF force field to describe liquid crystal molecules: the path to a dramatic improvement in transition temperature predictions. *Phys. Chem. Chem. Phys.*, **17**, 24851–24865.

Boyd, N. J. and Wilson, M. R. 2018. Validating an optimized GAFF force field for liquid crystals: T_{NI} predictions for bent-core mesogens and the first atomistic predictions of a dark conglomerate phase. *Phys. Chem. Chem. Phys.*, **20**, 1485–1496.

Bracewell, R. N. 2000. *The Fourier Transform and Its Applications*. 3rd ed. Boston, MA: McGraw Hill.

Braun, B., Hohla, M. and Kohler, J. 1999. Liquid crystal modeling: Electrostatic and van der Waals interaction energies for molecular building blocks from benzene to cholesterol. *Int. J. Mod. Phys. C*, **10**, 455–468.

Brédas, J.-L. and Marder, S. R. (eds.). 2016. *The WSPC Reference on Organic Electronics: Organic Semiconductors*. Singapore: World Scientific.

Breneman, C. M. and Wiberg, K. B. 1990. Determining atom-centered monopoles from molecular electrostatic potentials. The need for high sampling density in formamide conformational analysis. *J. Comput. Chem.*, **11**, 361–373.

Briels, W. J. 1980. An expansion of the intermolecular energy in a complete set of symmetry-adapted functions-convergence of the series for methane-methane and adamantane-adamantane interactions. *J. Chem. Phys.*, **73**, 1850–1861.

Brink, D. M. and Satchler, G. R. 1968. *Angular Momentum*. Oxford: Oxford University Press.

Brochard, F. 1977. Nematic fluids: some easy demonstration experiments. *Contemp. Phys.*, **18**, 247–264.

Brock, J. D., Birgeneau, R. J., Litster, J. D. and Aharony, A. 1989. Hexatic ordering in liquid crystal films. *Contemp. Phys.*, **30**, 321–335.

Brock, J. D., Aharony, A., Birgeneau, R. J., et al. 1986. Orientational and positional order in a tilted hexatic liquid crystal phase. *Phys. Rev. Lett.*, **57**, 98–101.

Broer, D. J., Finkelmann, H. and Kondo, K. 1988. In-situ photopolymerization of an oriented liquid crystalline acrylate. *Makromol. Chem.*, **189**, 185–194.

Broer, D. J, Hikmet, R. A. M. and Challa, G. 1989. In-situ photopolymerization of oriented liquid crystalline acrylates, 4. Influence of a lateral methyl substituent on monomer and oriented polymer network properties of a mesogenic diacrylate. *Macromolec. Chem Phys.*, **190**, 3201–3215.

Brommel, F., Kramer, D. and Finkelmann, H. 2012. Preparation of liquid crystalline elastomers. *Adv. Polym. Sci.*, **250**, 1–48.

Brommel, F., Stille, W., Finkelmann, H. and Hoffmann, A. 2011. Molecular dynamics and biaxiality of nematic polymers and elastomers. *Soft Matter*, **7**, 2387–2401.

Brooks, B. R., Bruccoleri, R. E., Olafson, D. J., et al. 1983. CHARMM: a program for macromolecular energy, minimization, and dynamics calculations. *J. Comput. Chem.*, **4**, 187–217.

Brooks, B. R., Brooks, III, C. L., Mackerell, A. D., et al. 2009. CHARMM: the biomolecular simulation program. *J. Comput. Chem.*, **30**, 1545–1614.

Brown, G. H. and Wolken, J. J. 1979. *Liquid Crystals and Biological Structures*. New York: Academic Press.

Brown, J. T., Allen, M. P., Martin del Rio, E. and de Miguel, E. 1998a. Effects of elongation on the phase behavior of the Gay-Berne fluid. *Phys. Rev. E*, **57**, 6685–6699.

Brown, K. R., Bonnell, D. A. and Sun, S. T. 1998b. Atomic force microscopy of mechanically rubbed and optically buffed polyimide films. *Liq. Cryst.*, **25**, 597–601.

Brown, R. G. and Ciftan, M. 1996. High-precision evaluation of the static exponents of the classical Heisenberg ferromagnet. *Phys. Rev. Lett.*, **76**, 1352–1355.

Bruce, C. D., Berkowitz, M. L., Perera, L. and Forbes, M. D. E. 2002. Molecular dynamics simulation of sodium dodecyl sulfate micelle in water: micellar structural characteristics and counterion distribution. *J. Phys. Chem. B*, **106**, 3788–3793.

Bucak, S., Cenker, C., Nasir, I., Olsson, U. and Zackrisson, M. 2009. Peptide nanotube nematic phase. *Langmuir*, **25**, 4262–4265.

Buckingham, A. D. 1967a. Angular correlation in liquids. *Faraday Discuss.*, **43**, 205–211.

Buckingham, A. D. 1967b. Permanent and induced molecular moments and long-range intermolecular forces. *Adv. Chem. Phys.*, **12**, 107–142.

Buckingham, R. A. 1938. The classical equation of state of gaseous helium, neon and argon. *Proc. Roy. Soc. (London) A*, **168**, 264–283.

Budzien, J., Raphael, C., Ediger, M. D. and de Pablo, J. J. 2002. Segmental dynamics in a blend of alkanes: Nuclear Magnetic Resonance experiments and molecular dynamics simulation. *J. Chem. Phys.*, **116**, 8209–8217.

Buining, P. A., Philipse, A. P. and Lekkerkerker, H. N. W. 1994. Phase-behavior of aqueous dispersions of colloidal boehmite rods. *Langmuir*, **10**, 2106–2114.

Bunning, T. J., Natarajan, L. V., Tondiglia, V. P. and Sutherland, R. L. 2000. Holographic polymer-dispersed liquid crystals (H-PDLCS). *Annu. Rev. Mater. Sci.*, **30**, 83–115.

Burnell, E. E. and de Lange, C. A. 1998. Prediction from molecular shape of solute orientational order in liquid crystals. *Chem. Rev.*, **98**, 2359–2387.

Burnell, E. E. and de Lange, C. A. 2003. *NMR of Ordered Liquids*. Dordrecht: Springer.

Burnett, L. J. and Muller, B. H. 1971. Deuteron quadrupole coupling constants in three solid deuterated paraffin hydrocarbons: C2D6, C4D10, C6D14. *J. Chem. Phys.*, **55**, 5829–5831.

Buscaglia, M., Lombardo, G., Cavalli, L., Barberi, R. and Bellini, T. 2010. Elastic anisotropy at a glance: the optical signature of disclination lines. *Soft Matter*, **6**, 5434–5442.

Bushby, R. J. and Lozman, O. R. 2002. Discotic liquid crystals 25 years on. *Curr. Opin. Colloid Interface Sci.*, **7**, 343–354.

Busselez, R., Cerclier, C. V., Ndao, M., et al. 2014. Discotic columnar liquid crystal studied in the bulk and nanoconfined states by molecular dynamics simulation. *J. Chem. Phys.*, **141**, 15.

Cacelli, I., Prampolini, G. and Tani, A. 2005. Atomistic simulation of a nematogen using a force field derived from quantum chemical calculations. *J. Phys. Chem. B*, **109**, 3531–3538.

Cacelli, I., De Gaetani, L., Prampolini, G. and Tani, A. 2007. Liquid crystal properties of the *n*-alkyl-cyanobiphenyl series from atomistic simulations with *ab initio* derived force fields. *J. Phys. Chem. B*, **111**, 2130–2137.

Cacelli, I., Cinacchi, G., Geloni, C., Prampolini, G. and Tani, A. 2003. Computer simulation of *p*-phenyls with interaction potentials from *ab-initio* calculations. *Mol. Cryst. Liq. Cryst.*, **395**, 171–182.

Callan-Jones, A. C., Pelcovits, R. A., Slavin, V. A., et al. 2006. Simulation and visualization of topological defects in nematic liquid crystals. *Phys. Rev. E*, **74**, 061701.

Camacho-Lopez, M., Finkelmann, H., Palffy-Muhoray, P. and Shelley, M. 2004. Fast liquid crystal elastomer swims into the dark. *Nat. Mater.*, **3**, 307–310.

Camp, P. J., Mason, C. P., Allen, M. P., Khare, A. A. and Kofke, D. A. 1996. The isotropic-nematic phase transition in uniaxial hard ellipsoid fluids: coexistence data and the approach to the Onsager limit. *J. Chem. Phys.*, **105**, 2837–2849.

Campostrini, M., Hasenbusch, M., Pelissetto, A., Rossi, P. and Vicari, E. 2001. Critical behavior of the three-dimensional XY universality class. *Phys. Rev. B*, **63**, 144520.

Campostrini, M., Hasenbusch, M., Pelissetto, A., Rossi, P. and Vicari, E. 2002. Critical exponents and equation of state of the three-dimensional Heisenberg universality class. *Phys. Rev. B*, **65**, 214503.

Caneda-Guzman, E., Moreno-Razo, J. A., Diaz-Herrera, E. and Sambriski, E. J. 2014. Molecular aspect ratio and anchoring strength effects in a confined Gay-Berne liquid crystal. *Mol. Phys.*, **112**, 1149–1159.

Cao, W., Munoz, A., Palffy-Muhoray, P. and Taheri, B. 2002. Lasing in a three-dimensional photonic crystal of the liquid crystal blue phase II. *Nat. Mater.*, **1**, 111–113.

Caprion, D., Bellier-Castella, L. and Ryckaert, J.-P. 2003. Influence of shape and energy anisotropies on the phase diagram of discotic molecules. *Phys. Rev. E*, **67**, 8.

Carbone, G. and Rosenblatt, C. 2005. Polar anchoring strength of a tilted nematic: confirmation of the dual easy axis model. *Phys. Rev. Lett.*, **94**, 057802.

Carbone, G., Lombardo, G., Barberi, R., Muševič, I. and Tkalec, U. 2009. Mechanically induced biaxial transition in a nanoconfined nematic liquid crystal with a topological defect. *Phys. Rev. Lett.*, **103**, 167801.

Care, C. M. and Cleaver, D. J. 2005. Computer simulations of liquid crystals. *Rep. Progr. Phys.*, **1**, 2665–2700.

Carnahan, N. F. and Starling, K. E. 1969. Equation of state for nonattracting rigid spheres. *J. Chem. Phys.*, **51**, 635–636.

Caspar, D. L. D. 1964. Assembly and stability of the tobacco mosaic virus particle. In Anfinsen, C. B., Anson, M. L. and Edsall, J. T. (eds.), *Advances in Protein Chemistry*, vol. 18. New York: Academic Press, pp. 37–121.

Catalano, D., Forte, C., Veracini, C. A. and Zannoni, C. 1983. Orientational ordering of some non-cylindrically symmetric solutes in nematic solvents. *Israel J. Chemistry*, **23**, 283–289.

Catalano, D., Di Bari, L., Veracini, C. A., Shilstone, G. N. and Zannoni, C. 1991. A maximum-entropy analysis of the problem of the rotameric distribution for substituted biphenyls studied by proton Nuclear Magnetic Resonance spectroscopy in nematic liquid crystals. *J. Chem. Phys.*, **94**, 3928–3935.

Cerutti, D. S., Duke, R. E., Darden, T. A. and Lybrand, T. P. 2009. Staggered Mesh Ewald: an extension of the Smooth Particle-Mesh Ewald method adding great versatility. *J. Chem. Theory Comput.*, **5**, 2322–2338.

Cestari, M., Diez-Berart, S., Dunmur, D. A., et al. 2011. Phase behavior and properties of the liquid crystal dimer 1″,7″-bis(4-cyanobiphenyl-4′-yl) heptane: a twist-bend nematic liquid crystal. *Phys. Rev. E*, **84**, 031704.

Chaikin, P. M. and Lubensky, T. C. 1995. *Principles of Condensed Matter Physics*. Cambridge: Cambridge University Press.

Chaikin, P. M., Donev, A., Man, W., Stillinger, F. H. and Torquato, S. 2006. Some observations on the random packing of hard ellipsoids. *Ind. Eng. Chem. Res.*, **45**, 6960–6965.

Chakrabarti, D. and Bagchi, B. 2009. Dynamics of thermotropic liquid crystals across the isotropic-nematic transition and their similarity with glassy relaxation in supercooled liquids. *Adv. Chem. Phys.*, **141**, 249–319.

Chakrabarty, S., Chakrabarti, D. and Bagchi, B. 2006. Power law relaxation and glassy dynamics in Lebwohl-Lasher model near the isotropic-nematic phase transition. *Phys. Rev. E*, **73**, 061706.

Chalam, M. K., Gubbins, K. E., de Miguel, E. and Rull, L. F. 1991. A molecular simulation of a liquid crystal model: bulk and confined fluid. *Molec. Simul.*, **7**, 357–385.

Chami, F. and Wilson, M. R. 2010. Molecular order in a chromonic liquid crystal: a molecular simulation study of the anionic azo dye Sunset Yellow. *J. Amer. Chem. Soc.*, **132**, 7794–7802.

Chan, H., Demortiere, A., Vukovic, L., Kral, P. and Petit, C. 2012. Colloidal nanocube supercrystals stabilized by multipolar Coulombic coupling. *ACS Nano*, **6**, 4203–4213.

Chandler, D., Weeks, J. D. and Andersen, H. C. 1983. van der Waals picture of liquids, solids, and phase transformations. *Science*, **220**, 787–794.

Chandrasekhar, S. 1982. Liquid crystals of disklike molecules. In Brown, G. H. (ed.), *Advances in Liquid Crystals*, vol. 5. New York: Academic Press, pp. 47–78.

Chandrasekhar, S. 1988. Recent developments in the physics of liquid crystals. *Contemp. Phys.*, **29**, 527–558.

Chandrasekhar, S. 1992. *Liquid Crystals*. 2nd ed. Cambridge: Cambridge University Press.

Chandrasekhar, S. 1993. Discotic liquid crystals. A brief review. *Liq. Cryst.*, **14**, 3–14.

Chang, S. S. 1983. Heat-capacity and thermodynamic properties of para-terphenyl: study of order-disorder transition by automated high-resolution adiabatic calorimetry. *J. Chem. Phys.*, **79**, 6229–6236.

Chao, C. Y., Maclennan, J. E., Pang, J. Z., Hui, S. W. and Ho, J. T. 1998. Phase behavior of liquid crystal films exhibiting the surface smectic-L phase. *Phys. Rev. E*, **57**, 6757–6760.

Chapman, D. 1975. Phase transitions and fluidity characteristics of lipids and cell-membranes. *Q.Rev. Biophys.*, **8**, 185–235.

Charvolin, J. and Hendrikx, Y. 1985. Amphiphilic molecules in lyotropic liquid crystals and micellar phases. In Emsley, J. W. (ed.), *Nuclear Magnetic Resonance of Liquid Crystals*. Dordrecht: Reidel, pp. 449–472.

Chatelain, P. 1943. Orientation of liquid crystals. *Bull. Soc. Franc. Miner*, **66**, 105.

Chen, D., Porada, J. H., Hooper, J. B., et al. 2013. Chiral heliconical ground state of nanoscale pitch in a nematic liquid crystal of achiral molecular dimers. *Proc. Nat. Acad. Sci. USA*, **110**, 15931–15936.

Chen, G. P., Takezoe, H. and Fukuda, A. 1989. Determination of $K_i(i = 1-3)$ and $\mu_j(j = 2-6)$ in 5CB by observing the angular dependence of Rayleigh line spectral widths. *Liq. Cryst.*, **5**, 341–347.

Chen, K., Ferrenberg, A. M. and Landau, D. P. 1993. Static critical behavior of three-dimensional classical Heisenberg models: A high-resolution Monte Carlo study. *Phys. Rev. B*, **48**, 3249–3256.

Chen, X., Korblova, E., Dong, D., et al. 2020. First-principles experimental demonstration of ferroelectricity in a thermotropic nematic liquid crystal: polar domains and striking electro-optics. *Proc. Nat. Acad. Sci. USA*, **117**, 14021–14031.

Chen, Y., Ma, P. and Gui, S. 2014. Cubic and hexagonal liquid crystals as drug delivery systems. *Biomed. Res. Int.*, **2014**, 815981.

Cheung, D. L., Clark, S. J. and Wilson, M. R. 2002. Parametrization and validation for a force field for liquid-crystal forming molecules. *Phys. Rev. E*, **65**, 051709.

Chiarelli, P., Faetti, S. and Fronzoni, L. 1983. Structural transition at the free-surface of the nematic liquid crystals MBBA and EBBA. *J. de Physique*, **44**, 1061–1067.

Chiccoli, C., Pasini, P. and Zannoni, C. 1987. A Monte Carlo simulation of the inhomogeneous Lebwohl-Lasher lattice model. *Liq. Cryst.*, **2**, 39–54.

Chiccoli, C., Pasini, P. and Zannoni, C. 1988a. Can Monte Carlo detect the absence of ordering in a model liquid crystal? *Liq. Cryst.*, **3**, 363–368.

Chiccoli, C., Pasini, P. and Zannoni, C. 1988b. A Monte Carlo investigation of the planar Lebwohl-Lasher lattice model. *Physica A*, **148**, 298–311.

Chiccoli, C., Pasini, P. and Zannoni, C. 1999a. Hybridly aligned liquid crystal films. A Monte Carlo study of molecular organization and thermodynamics. *Mol. Cryst. Liq. Cryst.*, **336**, 123–131.

Chiccoli, C., Pasini, P. and Zannoni, C. 2010. Elastic anisotropy and anchoring effects on the textures of nematic films with random planar surface alignment. *Mol. Cryst. Liq. Cryst.*, **516**, 1–11.

Chiccoli, C., Lavrentovich, O. D., Pasini, P. and Zannoni, C. 1997. Monte Carlo simulations of stable point defects in hybrid nematic films. *Phys. Rev. Lett.*, **79**, 4401–4404.

Chiccoli, C., Pasini, P., Biscarini, F. and Zannoni, C. 1988c. The P_4 model and its orientational phase transition. *Mol. Phys.*, **65**, 1505–1524.

Chiccoli, C., Pasini, P., Guzzetti, S. and Zannoni, C. 1998. A Monte Carlo simulation of an in-plane switching liquid crystal display. *Int. J. Mod. Phys. C*, **9**, 409–419.

Chiccoli, C., Pasini, P., Semeria, F. and Zannoni, C. 1990. A computer simulation of nematic droplets with radial boundary conditions. *Phys. Lett. A*, **150**, 311–314.

Chiccoli, C., Pasini, P., Semeria, F. and Zannoni, C. 1992. Computer simulations of nematic droplets with toroidal boundary-conditions. *Mol. Cryst. Liq. Cryst.*, **221**, 19–28.

Chiccoli, C., Pasini, P., Semeria, F. and Zannoni, C. 1993. An application of Cluster Monte Carlo method to the Heisenberg model. *Int. J. Mod. Phys. C*, **4**, 1041–1048.

Chiccoli, C., Pasini, P., Sarlah, A., Zannoni, C. and Žumer, S. 2003. Structures and transitions in thin hybrid nematic films: a Monte Carlo study. *Phys. Rev. E*, **67**, 050703(R).

Chiccoli, C., Pasini, P., Semeria, F., Sluckin, T. J. and Zannoni, C. 1995. Monte Carlo simulation of the hedgehog defect core in spin systems. *J. de Physique II*, **5**, 427–436.

Chiccoli, C., Pasini, P., Skačej, G., Zannoni, C. and Žumer, S. 1999b. NMR spectra from Monte Carlo simulations of polymer dispersed liquid crystals. *Phys. Rev. E*, **60**, 4219–4225.

Chiccoli, C., Pasini, P., Skačej, G., Zannoni, C. and Žumer, S. 2000. Dynamical and field effects in polymer-dispersed liquid crystals: Monte Carlo simulations of NMR spectra. *Phys. Rev. E*, **62**, 3766–3774.

Chiccoli, C., Pasini, P., Škacej, G., Žumer, S. and Zannoni, C. 2005. Lattice spin models of polymer-dispersed liquid crystals. In Pasini, P., Zannoni, C. and Žumer, S. (eds.), *Computer Simulations of Liquid Crystals and Polymers*. Dordrecht: Kluwer, pp. 1–25.

Chiccoli, C., Pasini, P., Skačej, G., Zannoni, C. and Žumer, S. 2013. Chirality transfer from helical nanostructures to nematics: a Monte Carlo study. *Mol. Cryst. Liq. Cryst.*, **576**, 151–156.

Chiccoli, C., Feruli, I., Lavrentovich, O. D., et al. 2002. Topological defects in schlieren textures of biaxial and uniaxial nematics. *Phys. Rev. E*, **66**, 0307011.

Chiccoli, C., Pasini, P., Zannoni, C., et al. 2019. From point to filament defects in hybrid nematic films. *Sci. Rep.*, **9**, 17941.

Chilaya, G. S. and Lisetski, L. N. 1986. Cholesteric liquid crystals – physical-properties and molecular-statistical theories. *Mol. Cryst. Liq. Cryst.*, **140**, 243–286.

Chirlian, L. E. and Francl, M. M. 1987. Atomic charges derived from electrostatic potentials: a detailed study. *J. Comput. Chem.*, **8**, 894–905.

Chirtoc, I., Chirtoc, M., Glorieux, C. and Thoen, J. 2004. Determination of the order parameter and its critical exponent for nCB ($n = 5$–8) liquid crystals from refractive index data. *Liq. Cryst.*, **31**, 229–240.

Chmielewski, A. G. 1986. Viscosities coefficients of some nematic liquid crystals. *Mol. Cryst. Liq. Cryst.*, **132**, 339–352.

Chu, K. C. and McMillan, W. L. 1977. Unified Landau theory for nematic, smectic A, and smectic C phases of liquid crystals. *Phys. Rev. A*, **15**, 1181–1187.

Chuang, T. J. and Eisenthal, K. B. 1972. Theory of Fluorescence Depolarization by anisotropic rotational diffusion. *J. Chem. Phys.*, **57**, 5094–5097.

Chun, B. J., Choi, J. I. and Jang, S. S. 2015. Molecular dynamics simulation study of sodium dodecyl sulfate micelle: water penetration and sodium dodecyl sulfate dissociation. *Colloids Surf. A*, **474**, 36–43.

Ciccotti, G. and Ryckaert, J.-P. 1986. Molecular dynamics simulation of rigid molecules. *Computer Phys. Reports*, **4**, 345–392.

Cifelli, M., Cinacchi, G. and De Gaetani, L. 2006. Smectic order parameters from diffusion data. *J. Chem. Phys.*, **125**, 164912.

Cifelli, M., Domenici, V., Dvinskikh, S. V., Veracini, C. A. and Zimmermann, H. 2012. Translational self-diffusion in the smectic phases of ferroelectric liquid crystals: an overview. *Phase Transit.*, **85**, 861–871.

Cinacchi, G. and Schmid, F. 2002. Density functional for anisotropic fluids. *J. Phys. Cond. Matter*, **14**, 12223–12234.

Cinacchi, G., Colle, R. and Tani, A. 2004. Atomistic molecular dynamics simulation of hexakis(pentyloxy)triphenylene: structure and translational dynamics of its columnar state. *J. Phys. Chem. B*, **108**, 7969–7977.

Cinacchi, G., De Gaetani, L. and Tani, A. 2005. Numerical study of a calamitic liquid crystal model: phase behavior and structure. *Phys. Rev. E*, **71**, 031703.

Cladis, P. E. and Kleman, M. 1972. Non-singular disclinations of strength $s = +1$ in nematics. *J. de Physique*, **33**, 591–598.

Cladis, P. E., Bogardus, R. K., Daniels, W. B. and Taylor, G. N. 1977. High-pressure investigation of reentrant nematic-bilayer-smectic-A transition. *Phys. Rev. Lett.*, **39**, 720–723.

Clark, M. G. 1976. Algebraic derivation of the free energy of a distorted nematic liquid crystal. *Mol. Phys.*, **31**, 1287–1289.

Clark, S. J. 2001. Measurements and calculation of dipole moments, quadrupole moments and polarizabilities of mesogenic molecules. In Dunmur, D. A., Fukuda, A. and Luckhurst, G. R. (eds.), *Physical Properties of Liquid Crystals: Nematics*, vol. 25. London: INSPEC, IEE, pp. 113–124.

Clark, S. J., Adam, C. J., Ackland, G. J., White, J. and Crain, J. 1997. Properties of liquid crystal molecules from first principles computer simulation. *Liq. Cryst.*, **22**, 469–475.

Cleaver, D. J. and Allen, M. P. 1991. Computer simulations of the elastic properties of liquid crystals. *Phys. Rev. A*, **43**, 1918–1931.

Cleaver, D. J., Care, C. M., Allen, M. P. and Neal, M. P. 1996. Extension and generalization of the Gay-Berne potential. *Phys. Rev. E*, **54**, 559–567.

Coates, D. and Gray, G. W. 1973. Synthesis of a cholesterogen with Hydrogen-Deuterium asymmetry. *Mol. Cryst. Liq. Cryst.*, **24**, 163–177.

Cognard, J. 1984. The anisotropy of the surface tension of polar liquids: the case of liquid crystals. *J. Adhes.*, **17**, 123–134.

Cole, K. S. and Cole, R. H. 1941. Dispersion and absorption in dielectrics I. Alternating current characteristics. *J. Chem. Phys.*, **9**, 341–351.

Coleman, D. A., Fernsler, J., Chattham, N., et al. 2003. Polarization modulated smectic liquid crystal phases. *Science*, **301**, 1204–1211.

Coles, H. J. and Jennings, B. R. 1978. Optical and electrical Kerr Effect in 4-n-pentyl-4′-cyanobiphenyl. *Mol. Phys.*, **36**, 1661–1673.

Coles, H. J. and Pivnenko, M. N. 2005. Liquid crystal 'Blue phases' with a wide temperature range. *Nature*, **436**, 997–1000.

Collings, P. J. 1990. *Liquid Crystals: Nature's Delicate Phase of Matter*. Bristol: Adam Hilger.

Collings, P. J. and Hird, M. 1997. *Introduction to Liquid Crystals*. London: Taylor & Francis.

Collyer, A. A. 1990. Lyotropic liquid crystal polymers for engineering applications. *Mater. Sci. Tech.*, **6**, 981–992.

Cometti, G., Dalcanale, E., Duvosel, A. and Levelut, A. M. 1992. A new, conformationally mobile macrocyclic core for bowl-shaped columnar liquid crystals. *Liq. Cryst.*, **11**, 93.

Conradi, M., Ravnik, M., Bele, M., et al. 2009. Janus nematic colloids. *Soft Matter*, **5**, 3905–3912.

Conti, S., DeSimone, A. and Dolzmann, G. 2002. Semisoft elasticity and director reorientation in stretched sheets of nematic elastomer. *Phys. Rev. E*, **66**, 061710.

Cook, M. J. and Wilson, M. R. 2001a. Development of an all-atom force field for the simulation of liquid crystal molecules in condensed phases (LCFF). *Mol. Cryst. Liq. Cryst.*, **357**, 149–165.

Cook, M. J. and Wilson, M. R. 2001b. The first thousand-molecule simulation of a mesogen at the fully atomistic level. *Mol. Cryst. Liq. Cryst.*, **363**, 181–193.

Corbett, D. and Warner, M. 2009. Deformation and rotations of free nematic elastomers in response to electric fields. *Soft Matter*, **5**, 1433–1439.

Cornell, W. D., Cieplak, P., Bayly, C. I., et al. 1995. A second generation force field for the simulation of proteins, nucleic acids, and organic molecules. *J. Amer. Chem. Soc.*, **117**, 5179–5197.

Costigan, S. C., Booth, P. J. and Templer, R. H. 2000. Estimations of lipid bilayer geometry in fluid lamellar phases. *Biochim. Biophys. Acta* **1468**, 41–54.

Cotton, F. A. 1990. *Chemical Applications of Group Theory*. 3rd ed. New York: Wiley.

Coveney, P. V. and Wan, S. 2016. On the calculation of equilibrium thermodynamic properties from molecular dynamics. *Phys. Chem. Chem. Phys.*, **18**, 30236–30240.

Cox, J. S. G., Woodard, G. D. and McCrone, W. C. 1971. Solid state chemistry of cromolyn sodium (disodium cromoglycate). *J. Pharmaceut. Sci.*, **60**, 1458–1465.

Cramer, C. J. 2004. *Essentials of Computational Chemistry. Theories and Models*. New York: Wiley.

Crawford, G. P. and Žumer, S. 1995. Saddle-splay elasticity in nematic liquid crystals. *Int. J. Mod. Phys. B*, **9**, 2469–2514.

Crawford, G. P. and Žumer, S. (eds.). 1996. *Liquid Crystals in Complex Geometries Formed by Polymer and Porous Networks*. London: Taylor & Francis.

Crawford, G. P., Ondris Crawford, R., Žumer, S. and Doane, J. W. 1993. Anchoring and orientational wetting transitions of confined liquid crystals. *Phys. Rev. Lett.*, **70**, 1838–1841.

Cristinziano, P. L. and Lelj, F. 2007. Atomistic simulation of discotic liquid crystals: Transition from isotropic to columnar phase example. *J. Chem. Phys.*, **127**, 134506.

Crooker, P. P. 1983. The cholesteric blue phase - a progress report. *Mol. Cryst. Liq. Cryst.*, **98**, 31–45.

Crooker, P. P. 2001. Blue Phases. In Kitzerow, H. S. and Bahr, C. (eds.), *Chirality in Liquid Crystals*. Berlin: Springer, pp. 186–222.

Croxton, C. A. 1975. *Introduction to Liquid State Physics*. London: Wiley.

Cruz-Chu, E. R., Aksimentiev, A. and Schulten, K. 2006. Water-silica force field for simulating nanodevices. *J. Phys. Chem. B*, **110**, 21497–21508.

Cuetos, A., Ilnytskyi, J. M. and Wilson, M. R. 2002. Rotational viscosities of Gay-Berne mesogens. *Mol. Phys.*, **100**, 3839–3845.

Cuetos, A., Dennison, M., Masters, A. and Patti, A. 2017. Phase behaviour of hard board-like particles. *Soft Matter*, **13**, 4720–4732.

Cummins, P. G., Dunmur, D. A. and Laidler, D. A. 1975. Dielectric properties of nematic $4,4'$-n-pentylcyanobiphenyl. *Mol. Cryst. Liq. Cryst.*, **30**, 109–121.

Cygan, R. T., Liang, J.-J. and Kalinichev, A. G. 2004. Molecular models of hydroxide, oxyhydroxide, and clay phases and the development of a general force field. *J. Phys. Chem. B*, **108**, 1255–1266.

Da Como, E., De Angelis, F., Snaith, H. and Walker, A. B. 2016. *Unconventional Thin Film Photovoltaics*. London: Royal Society of Chemistry.

Damasceno, P. F., Engel, M. and Glotzer, S. C. 2012. Predictive self assembly of polyhedra into complex structures. *Science*, **337**, 453–457.

Darden, T., York, D. and Pedersen, L. 1993. Particle Mesh Ewald: an $N \log(N)$ method for Ewald sums in large systems. *J. Chem. Phys.*, **98**, 10089–10092.

Dash, D. and Wu, X. L. 1997. A shear-induced instability in freely suspended smectic-A liquid crystal films. *Phys. Rev. Lett.*, **79**, 1483–1486.

Davidson, P. and Gabriel, J.-C. P. 2005. Mineral liquid crystals. *Curr. Opin. Colloid Interface Sci.*, **9**, 377–383.

De Gaetani, L., Prampolini, G. and Tani, A. 2006. Modeling a liquid crystal dynamics by atomistic simulation with an *ab initio* derived Force Field. *J. Phys. Chem. B*, **110**, 2847–2854.

de Gennes, P. G. 1971. Short range order effects in the isotropic phase of nematics and cholesterics. *Mol. Cryst. Liq. Cryst.*, **12**, 193–214.

de Gennes, P. G. 1972. An analogy between superconductors and smectics A. *Solid State Comm.*, **10**, 753–756.

de Gennes, P. G. 1974. *The Physics of Liquid Crystals*. Oxford: Oxford University Press.

de Gennes, P. G. 1975. Réflexions sur un type de polymères nématiques (Some reflections about a type of nematic liquid crystal polymers). *C. R. Acad. Sc. Paris*, **281**, 101–103.

de Gennes, P. G. 1977. Polymeric liquid crystals: Frank elasticity and light scattering. *Mol. Cryst. Liq. Cryst.*, **34**, 177–182.

de Gennes, P. G. and Pincus, P. A. 1970. Pair correlations in a ferromagnetic colloid. *Physik Kondens. Mater.*, **11**, 189–198.

de Gennes, P. G. and Prost, J. 1993. *The Physics of Liquid Crystals*. Oxford: Oxford University Press.

de Gennes, P. G., Hebert, M. and Kant, R. 1997. Artificial muscles based on nematic gels. *Macromol. Symp.*, **113**, 39–49.

de Jeu, W. H. 1973. First and second-order nematic-smectic A phase transitions in the series of di-n-alkyl azoxybenzenes. *Solid State Comm.*, **13**, 1521–1523.

de Jeu, W. H. 1981. Physical properties of liquid crystalline materials in relation to their applications. *Mol. Cryst. Liq. Cryst.*, **63**, 83–109.

de Jeu, W. H. (ed.). 2012. *Liquid Crystal Elastomers: Materials and Applications*. Adv. Polym. Sci., vol. 250. Heidelberg: Springer.

de Jeu, W. H. and Ostrovskii, B. I. 2012. Order and disorder in Liquid-Crystalline Elastomers. In de Jeu, W. H. (ed.), *Liquid Crystal Elastomers: Materials and Applications*. Adv. Polym. Sci., vol. 250, Heidelberg: Springer, 187–234.

de Jeu, W. H., Ostrovskii, B. I. and Shalaginov, A. N. 2003. Structure and fluctuations of smectic membranes. *Rev. Mod. Phys.*, **75**, 181–235.

de Leeuw, S. W., Perram, J. W. and Smith, E. R. 1980a. Simulation of electrostatic systems in periodic boundary conditions. I. Lattice sums and dielectric constants. *Proc. Roy. Soc. London A*, **373**, 27–56.

de Leeuw, S. W., Perram, J. W. and Smith, E. R. 1980b. Simulation of electrostatic systems in periodic boundary conditions. II. Equivalence of boundary conditions. *Proc. Roy. Soc. London A*, **373**, 57–66.

De Luca, A., Barna, V., Atherton, T. J., et al. 2008. Optical nanotomography of anisotropic fluids. *Nat. Phys.*, **4**, 869–872.

De Matteis, G. and Romano, S. 2009. Mesogenic lattice models with partly antinematic interactions producing uniaxial nematic phases. *Phys. Rev. E*, **80**, 031702.

De Matteis, G., Romano, S. and Virga, E. G. 2005. Bifurcation analysis and computer simulation of biaxial liquid crystals. *Phys. Rev. E*, **72**, 041706.

De Michele, C., Schilling, R. and Sciortino, F. 2007. Dynamics of uniaxial hard ellipsoids. *Phys. Rev. Lett.*, **98**.

de Miguel, E. 2002. Reexamining the phase diagram of the Gay-Berne fluid. *Mol. Phys.*, **100**, 2449–2459.

de Miguel, E. and Martin Del Rio, E. 1999. Simulation of nematic free surfaces. *Int. J. Mod. Phys. C*, **10**, 431–433.

de Miguel, E. and Vega, C. 2002. The global phase diagram of the Gay-Berne model. *J. Chem. Phys.*, **117**, 6313–6322.

de Miguel, E., Rull, L. F. and Gubbins, K. E. 1992. Dynamics of the Gay-Berne fluid. *Phys. Rev. A*, **45**, 3813–3822.

de Miguel, E., Martin Del Rio, E., Brown, J. T. and Allen, M. P. 1996. Effect of the attractive interactions on the phase behavior of the Gay-Berne liquid crystal model. *J. Chem. Phys.*, **105**, 4234–4249.

de Miguel, E., Rull, L. F., Chalam, M. K. and Gubbins, K. E. 1990. Liquid-vapor coexistence of the Gay-Berne fluid by Gibbs-Ensemble simulation. *Mol. Phys.*, **71**, 1223–1231.

de Miguel, E., Rull, L. F., Chalam, M. K. and Gubbins, K. E. 1991. Liquid crystal phase diagram of the Gay-Berne fluid. *Mol. Phys.*, **74**, 405–424.

De Raedt, H. and De Raedt, B. 1983. Applications of the generalized Trotter formula. *Phys. Rev. A*, **28**, 3575.

De Santo, P. 2016. *SICL conference*. 3rd Italy-Brazil Workshop on Liquid Crystals, Portonovo, 19–21 June.

de Souza, R. F., Zaccheroni, S., Ricci, M. and Zannoni, C. 2022. Dynamic self-assembly of active particles in liquid crystals. *J. Mol. Liq.*, **352**, 118692.

de Vries, A. H., Mark, A. E. and Marrink, S. J. 2004. Molecular dynamics simulation of the spontaneous formation of a small DPPC vesicle in water in atomistic detail. *J. Amer. Chem. Soc.*, **126**, 4488–4489.

de Vries, A. H., Yefimov, S., Mark, A. E. and Marrink, S. J. 2005. Molecular structure of the lecithin ripple phase. *Proc. Nat. Acad. Sci. USA*, **102**, 5392–5396.

Deamer, D. W. 2010. From 'banghasomes' to liposomes: a memoir of Alec Bangham, 1921–2010. *FASEB J.*, **24**, 1308–10.

Delafuente, M. R., Jubindo, M. A. P., Zubia, J., Iglesias, T. P. and Seoane, A. 1994. Low and high-frequency Dielectric-Spectroscopy on a liquid crystal with the phase sequence N*-S_A-Sc*. *Liq. Cryst.*, **16**, 1051–1063.

Della Valle, R. G. and Andersen, H. C. 1992. Molecular dynamics simulation of silica liquid and glass. *J. Chem. Phys.*, **97**, 2682–2689.

Demus, D. 1989. One hundred years of liquid crystal chemistry: thermotropic liquid crystals with conventional and unconventional molecular structure. *Liq. Cryst.*, **5**, 75–110.

Demus, D., Goodby, J., Gray, G. W., Spiess, H. W. and Vill, V. (eds.). 1998. *Handbook of Liquid Crystals. Low Molecular Weight Liquid Crystals I*. Weinheim: Wiley-VCH.

den Boer, W. 2005. *Active Matrix Liquid Crystal Displays. Fundamentals and Applications*. Amsterdam: Elsevier.

Denham, J. Y., Luckhurst, G. R., Zannoni, C. and Lewis, J. W. 1980. Computer-simulation studies of anisotropic systems.3. Two-dimensional nematic liquid crystals. *Mol. Cryst. Liq. Cryst.*, **60**, 185–205.

Dennery, P. and Krzywicki, A. 1969. *Mathematics for Physicists*. New York: Harper & Row.

Dewar, M. J. S., Zoebisch, E. G., Healy, E. F. and Stewart, J. J. P. 1985. Development and use of quantum mechanical molecular models. 76. AM1: a new general purpose quantum mechanical molecular model. *J. Amer. Chem. Soc.*, **107**, 3902–3909.

Dickson, C. J., Rosso, L., Betz, R. M., Walker, R. C. and Gould, I. R. 2012. GAFFlipid: a General AMBER force field for the accurate molecular dynamics simulation of phospholipid. *Soft Matter*, **8**, 9617–9627.

Dickson, C. J., Madej, B. D., Skjevik, A. A., et al. 2014. Lipid14: the AMBER Lipid force field. *J. Chem. Theory Comput.*, **10**, 865–879.

Dierking, I. 2003. *Textures of Liquid Crystals*. New York: Wiley.

Dill, K. A. and Bromberg, S. 2011. *Molecular Driving Forces: Statistical Thermodynamics in Chemistry, Physics, Biology, and Nanoscience*. 2nd ed. New York: Garland Science.

Dingemans, T. J., Murthy, N. S. and Samulski, E. T. 2001. Javelin-, hockey stick-, and boomerang-shaped liquid crystals. Structural variations on *p*-quinquephenyl. *J. Phys. Chem. B*, **105**, 8845–8860.

Dingemans, T. J., Madsen, L. A., Zafiropoulos, N. A., Lin, W. B. and Samulski, E. T. 2006. Uniaxial and biaxial nematic liquid crystals. *Phil. Trans. Roy. Soc. A*, **364**, 2681–2696.

Dirac, P. A. M. 1929. Quantum mechanics of many-electron systems. *Proc. Roy. Soc. A* **123**, 714–733.

Dirac, P. A. M. 1958. *The Principles of Quantum Mechanics*. 4th ed. Oxford: Clarendon Press.

Diroll, B. T., Greybush, N. J., Kagan, C. R. and Murray, C. B. 2015. Smectic nanorod superlattices assembled on liquid subphases: structure, orientation, defects, and optical polarization. *Chem. Mater.*, **27**, 2998–3008.

Doane, J. W. 1985a. Determination of biaxial structures in lyotropic materials by DNMR. In Emsley, J. W. (ed.), *Nuclear Magnetic Resonance of Liquid Crystals*. Dordrecht: Reidel, pp. 413–420.

Doane, J. W. 1985b. Phase biaxiality in cholesteric and blue phases. In Emsley, J. W. (ed.), *Nuclear Magnetic Resonance of Liquid Crystals*. Dordrecht: Reidel, pp. 421–429.

Doane, J. W. 1990. Polymer dispersed liquid crystal displays. In Bahadur, B. (ed.), *Liquid Crystal Applications and Uses*, vol. 1. Singapore: Word Scientific, pp. 362–396.

Dogic, Z. and Fraden, S. 1997. Smectic phase in a colloidal suspension of semiflexible virus particles. *Phys. Rev. Lett.*, **78**, 2417–2420.

Dogic, Z. and Fraden, S. 2000. Cholesteric phase in virus suspensions. *Langmuir*, **16**, 7820–7824.

Dogic, Z. and Fraden, S. 2006. Ordered phases of filamentous viruses. *Curr. Opin. Colloid Interface Sci.*, **11**, 47–55.

Doi, M. 2003. OCTA (Open computational Tool for advanced material technology). *Macromol. Symp.*, **195**, 101–107.

Domenici, V., Geppi, M., Veracini, C. A. and Zakharov, A. V. 2005. Molecular dynamics in the smectic A and C* phases in a long-chain ferroelectric liquid crystal: H-2 NMR, dielectric properties, and a theoretical treatment. *J. Phys. Chem. B*, **109**, 18369–18377.

Donald, A. M. and Windle, A. H. 1992. *Liquid Crystalline Polymers*. Cambridge: Cambridge University Press.

Donev, A., Stillinger, F. H., Chaikin, P. M. and Torquato, S. 2004a. Unusually dense crystal packings of ellipsoids. *Phys. Rev. Lett.*, **92**, 255506.

Donev, A., Cisse, I., Sachs, D., et al. 2004b. Improving the density of jammed disordered packings using ellipsoids. *Science*, **303**, 990–993.

Dong, R. Y. 1997. *Nuclear Magnetic Resonance of Liquid Crystals*. New York: Springer.

Dong, R. Y. 2016. Recent NMR Studies of Thermotropic Liquid Crystals. *Ann. Reports NMR Spectroscopy*, **87**, 41–174.

Dong, R. Y. and Shen, X. 1996. Rotational diffusion of asymmetric molecules in liquid crystals: a global analysis of deuteron relaxation data. *J. Chem. Phys.*, **105**, 2106–2111.

Dong, R. Y., Emsley, J. W. and Hamilton, K. 1989. Orientational order and dynamics of molecules in the nematic phase of 4-trans-(4-trans-*n*-propylcyclohexyl)cyclohexanenitrile. *Liq. Cryst.*, **5**, 1019–1031.

Donnio, B. and Bruce, D. W. 1999. Metallomesogens. *Struct. Bond.*, **95**, 194–247.

Donnio, B., Wermter, H. and Finkelmann, H. 2000. Simple and versatile synthetic route for the preparation of main-chain, liquid crystalline elastomers. *Macromolecules*, **33**, 7724–7729.

Doucet, J., Levelut, A. M. and Lambert, M. 1973. Long and short range order in the crystalline and smectic B phases of terephthal-bis-butylaniline (TBBA). *Mol. Cryst. Liq. Cryst.*, **24**, 317–329.

Douliez, J. P., Bechinger, B., Davis, J. H. and Dufourc, E. J. 1996. C-C bond order parameters from ^{2}H and ^{13}C solid-state NMR. *J. Phys. Chem.*, **100**, 17083–17086.

Dozov, I. 2001. On the spontaneous symmetry breaking in the mesophases of achiral banana-shaped molecules. *Europhys. Lett.*, **56**, 247–253.

Dozov, I. and Penchev, I. I. 1980. Effect of the rotational depolarization in fluorescent measurements of the nematic order parameters. *J. Lumin.*, **22**, 69–78.

Dozov, I., Kirov, N. and Fontana, M. P. 1984. Determination of reorientational correlation functions in ordered fluids: IR absorption spectroscopy. *J. Chem. Phys.*, **81**, 2585–2590.

Dozov, I., Kirov, N. and Petroff, B. 1987. Molecular biaxiality and reorientational correlation functions in nematic phases. Theory. *Phys. Rev. A*, **36**, 2870–2878.

Dreher, R., Meier, G. and Saupe, A. 1971. Selective reflection by cholesteric liquid crystals. *Mol. Cryst. Liq. Cryst.*, **13**, 17–26.

Drzaic, P. 1988. A new director alignment for droplets of nematic liquid crystal with low bend-to-splay ratio. *Mol. Cryst. Liq. Cryst.*, **154**, 289–306.

Drzaic, P. and Drzaic, P. S. 2011. Putting liquid crystal droplets to work: a short history of polymer dispersed liquid crystals. *Liq. Cryst.*, **33**, 1281–1296.

Drzaic, P. S. 1995. *Liquid Crystal Dispersions*. Singapore: World Scientific.

Du, J. and Cormack, A. N. 2005. Molecular dynamics simulation of the structure and hydroxylation of silica glass surfaces. *J. Amer. Ceramic Soc.*, **88**, 2532–2539.

Dubois, E., Perzynski, R., Boue, F. and Cabuil, V. 2000. Liquid-gas transitions in charged colloidal dispersions: small-angle neutron scattering coupled with phase diagrams of magnetic fluids. *Langmuir*, **16**, 5617–5625.

Dubois-Violette, E. and Pansu, B. 1988. Frustration and related topology of blue phases. *Mol. Cryst. Liq. Cryst.*, **165**, 151–182.

Dunmur, D. A. 2001. Measurements of bulk elastic constants of nematics. In Dunmur, D. A., Fukuda, A. and Luckhurst, G. R. (eds.), *Physical Properties of Liquid Crystals: Nematics*. London: INSPEC-IEE, pp. 216–229.

Dunmur, D. A., Fukuda, A. and Luckhurst, G. R. (eds.). 2001. *Physical Properties of Liquid Crystals: Nematics*. London: INSPEC-IEE.

Dussi, S., Belli, S., van Roij, R. and Dijkstra, M. 2015. Cholesterics of colloidal helices: predicting the macroscopic pitch from the particle shape and thermodynamic state. *J. Chem. Phys.*, **142**, 074905.

Dvinskikh, S. V. and Furó, I. 2001. Anisotropic self-diffusion in the nematic phase of a thermotropic liquid crystal by ^{1}H-spin-echo Nuclear Magnetic Resonance. *J. Chem. Phys.*, **115**, 1946–1950.

Dvinskikh, S. V. and Furó, I. 2012. Anisotropic self-diffusion in nematic, smectic-A, and reentrant nematic phases. *Phys. Rev. E*, **86**, 031704.

Eargle, J., Wright, D. and Luthey-Schulten, Z. 2006. Multiple alignment of protein structures and sequences for VMD. *Bioinformatics*, **22**, 504–506.

Eastman, P., Friedrichs, M. S., Chodera, J. D., et al. 2013. OpenMM 4: a reusable, extensible, hardware independent library for high performance molecular simulation. *J. Chem. Theory Comput.*, **9**, 461–469.

Ebbens, S., Tu, M. H., Howse, J. R. and Golestanian, R. 2012. Size dependence of the propulsion velocity for catalytic Janus-sphere swimmers. *Phys. Rev. E*, **85**, 020401.

Edmonds, A. R. 1960. *Angular Momentum in Quantum Mechanics*. 2nd ed. Princeton, NJ: Princeton University Press.

Edwardes, D. 1892. Steady motion of a viscous liquid in which an ellipsoid is constrained to rotate about a principal axis. *Q. J. Math*, **26**, 68.

Edwards, D. J., Jones, J. W., Lozman, O., et al. 2008. Chromonic liquid crystal formation by Edicol Sunset Yellow. *J. Phys. Chem. B*, **112**, 14628–14636.

Egberts, E. and Berendsen, H. J. C. 1988. Molecular-dynamics simulation of a smectic liquid crystal with atomic detail. *J. Chem. Phys.*, **89**, 3718–3732.

Eggers, D. F., Gregory, N. W., Halsey, G. D. and Rabinovitch, B. S. 1964. *Physical Chemistry*. New York: Wiley.

Eichhorn, H., Bruce, D. W. and Wöhrle, D. 1998. Amphitropic mesomorphic phthalocyanines – a new approach to highly ordered layers. *Adv. Mater.*, **10**, 419–422.

Ellena, J. F., Dominey, R. N. and Cafiso, D. S. 1987. Molecular dynamics in sodium dodecyl sulfate micelles elucidated using carbon-13 and proton spin-lattice relaxation, carbon-13 spin-spin relaxation, and proton nuclear Overhauser effect spectroscopy. *J. Phys. Chem.*, **91**, 131–137.

Emerson, A. P. J., Luckhurst, G. R. and Whatling, S. G. 1994. Computer-simulation studies of anisotropic systems 23. The Gay-Berne discogen. *Mol. Phys.*, **82**, 113.

Emerson, A. P. J., Faetti, S. and Zannoni, C. 1997. Monte Carlo simulation of the nematic-vapour interface for a Gay-Berne liquid crystal. *Chem. Phys. Lett.*, **271**, 241–246.

Emsley, J. W. (ed.). 1985. *Nuclear Magnetic Resonance of Liquid Crystals*. Dordrecht: Reidel.

Emsley, J. W. and Lindon, J. C. 1975. *NMR Spectroscopy Using Liquid Crystal Solvents*. Oxford: Pergamon Press.

Emsley, J. W., Luckhurst, G. R. and Stockley, C. P. 1981. The deuterium and proton-(deuterium) NMR-spectra of the partially deuteriated nematic liquid crystal 4-n-pentyl-4′-cyanobiphenyl. *Mol. Phys.*, **44**, 565–580.

Emsley, J. W., Fung, B. M., Heaton, N. J. and Luckhurst, G. R. 1987. The potential of mean torque for flexible mesogenic molecules. Determination of the interaction parameters from carbon-hydrogen dipolar couplings for 4-n-alkyl-4′-cyanobiphenyls. *J. Chem. Phys.*, **87**, 3099–3103.

Emsley, J. W., Luckhurst, G. R., Palke, W. E. and Tildesley, D. J. 1992. Computer simulation studies of the dependence on density of the orientational order in nematic liquid crystals. *Liq. Cryst.*, **11**, 519–530.

Emsley, J. W., Wallington, I. D., Catalano, D., et al. 1993. Comparison of the Maximum-Entropy and Additive Potential methods for obtaining rotational potentials from the NMR spectra of samples dissolved in liquid crystalline solvents – the case of 4-nitro-1-(β,β,β-trifluoroethoxy)benzene. *J. Phys. Chem.*, **97**, 6518–6523.

Endo, N., Matsumoto, T., Kikuchi, H. and Kimura, M. 2016. Study of polymer-stabilised blue phase liquid crystal on a single substrate. *Liq. Cryst.*, **43**, 66–76.

Ennari, J., Hamara, J. and Sundholm, F. 1997. Vibrational spectra as experimental probes for molecular models of ion-conducting polyether systems. *Polymer*, **38**, 3733–3744.

Eppenga, R. and Frenkel, D. 1984. Monte Carlo study of the isotropic and nematic phases of infinitely thin hard platelets. *Mol. Phys.*, **52**, 1303–1334.

Erdélyi, A., Magnus, W., Oberhettinger, F. and Tricomi, F. G. 1953. *Bateman Manuscript Project. Higher Transcendental Functions*. vol. 1. New York: McGraw-Hill.

Ericksen, J. L. 1966. Inequalities in liquid crystal theory. *Phys. Fluids*, **9**, 1205.

Ernst, R. R., Bodenhausen, G. and Wokaun, A. 1991. *Principles of Nuclear Magnetic Resonance in One and Two Dimensions*. Oxford: Oxford University Press.

Erpenbeck, J. J. and Wood, W. W. 1977. Molecular dynamics techniques for hard-core systems. In Berne, B J. (ed.), *Statistical Mechanics B: Time Dependent Processes*. New York: Plenum Press, pp. 1–40.

Escobedo, F. A. 2014. Engineering entropy in soft matter: the bad, the ugly and the good. *Soft Matter*, **10**, 8388–8400.

Essmann, U., Perera, L., Berkowitz, M. L., et al. 1995. A Smooth Particle mesh Ewald method. *J. Chem. Phys.*, **103**, 8577–8593.

Evangelista, L. R. and Ponti, S. 1995. Intrinsic part of the surface energy for nematics in a pseudo-molecular approach: comparison with experimental results. *Phys. Lett. A*, **197**, 55–62.

Evans, D. J. 1977. Representation of orientation space. *Mol. Phys.*, **34**, 317–325.

Evans, D. J. and Murad, S. 1977. Singularity free algorithm for molecular dynamics simulation of rigid polyatomics. *Mol. Phys.*, **34**, 327–331.

Evans, D. J. and Murad, S. 1989. Thermal conductivity in molecular fluids. *Mol. Phys.*, **68**, 1219–1223.

Everaers, R. and Ejtehadi, M. R. 2003. Interaction potentials for soft and hard ellipsoids. *Phys. Rev. E*, **67**, 041710.

Fabbri, U. and Zannoni, C. 1986. Monte Carlo investigation of the Lebwohl-Lasher lattice model in the vicinity of its orientational phase transition. *Mol. Phys.*, **58**, 763–788.

Faber, T. E. 1980. A continuum theory of disorder in nematic liquid crystals. 4. Application to lattice models. *Proc. Roy. Soc. London A*, **370**, 509–521.

Fano, U. 1957. Description of states in quantum mechanics by density matrix and operator techniques. *Rev. Mod. Phys.*, **29**, 74–93.

Fatuzzo, E. and Mason, P. R. 1967. A theory of dielectric relaxation in polar liquids. *Proc. Phys. Soc.*, **90**, 741–750.

Favre-Nicolin, V. and Cerny, R. 2004. A better FOX: using flexible modelling and maximum likelihood to improve direct-space *ab initio* structure determination from powder diffraction. *Z. Kristallogr.*, **219**, 847–856.

Favre-Nicolin, V. and Cerny, R. 2007. FOX: a friendly tool to solve nonmolecular structures from powder diffraction. *Z. Kristallogr.*, **222**, 105–113.

Feller, S. E., Zhang, Y., Pastor, R. W. and Brooks, B. R. 1995. Constant pressure molecular dynamics simulation: the Langevin piston method. *J. Chem. Phys.*, **103**, 4613–4621.

Feng, X. L., Marcon, V., Pisula, W., et al. 2009. Towards high charge-carrier mobilities by rational design of the shape and periphery of discotics. *Nat. Mater.*, **8**, 421–426.

Ferrarini, A., Luckhurst, G. R., Nordio, P. L. and Roskilly, S. J. 1994. Prediction of the transitional properties of liquid crystal dimers. A molecular field calculation based on the surface tensor parametrization. *J. Chem. Phys.*, **100**, 1460–1469.

Ferrarini, A., Moro, G. J., Nordio, P. L. and Luckhurst, G. R. 1992. A shape model for molecular ordering in nematics. *Mol. Phys.*, **77**, 1–15.

Ferrario, M. and Ryckaert, J.-P. 1985. Constant pressure-constant temperature molecular dynamics for rigid and partially rigid molecular systems. *Mol. Phys.*, **54**, 587–603.

Ferrenberg, A. M. and Swendsen, R. H. 1988. New Monte Carlo technique for studying phase-transitions. *Phys. Rev. Lett.*, **61**, 2635–2638.

Feynman, R. P., Leighton, R. and Sands, M. 1963. *The Feynman Lectures on Physics. Mechanics, Radiation, Heat.* Reading: Addison-Wesley.

Figueirinhas, J. L., Cruz, C., Filip, D., et al. 2005. Deuterium NMR investigation of the biaxial nematic phase in an organosiloxane tetrapode. *Phys. Rev. Lett.*, **94**, 107802.

Figuereido Neto, A. and Salinas, S. R. A. 2005. *The Physics of Lyotropic Liquid Crystals. Phase Transitions and Structural Properties*. Oxford: Oxford University Press.

Fincham, D. and Heyes, D. M. 1985. Recent advances in molecular dynamics computer simulation. *Adv. Chem. Phys.*, **63**, 493–575.

Findenegg, G. H., Jähnert, S., Akcakayiran, D. and Schreiber, A. 2008. Freezing and melting of water confined in silica nanopores. *ChemPhysChem*, **9**, 2651–2659.

Finkelmann, H. 1982. Synthesis, structure and properties of liquid crystalline side chain polymers. In Ciferri, A., Krigbaum, W. R. and Meyer, R. (eds.), *Polymer Liquid Crystals*. New York: Academic Press, pp. 35–62.

Finkelmann, H. and Rehage, G. 1984. Liquid crystal side-chain polymers. *Adv. Polym. Sci.*, **60-1**, 97–172.

Finkelmann, H., Kundler, I., Terentjev, E. M. and Warner, M. 1997. Critical stripe-domain instability of nematic elastomers. *J. Phys. II France*, **7**, 1059–1069.

Finkenzeller, U., Geelhaar, T., Weber, G. and Pohl, L. 1989. Liquid crystalline reference compounds. *Liq. Cryst.*, **5**, 313–321.

Fiore, A., Mastria, R., Lupo, M. G., et al. 2009. Tetrapod-shaped colloidal nanocrystals of II-VI semiconductors prepared by seeded growth. *J. Amer. Chem. Soc.*, **131**, 2274–2282.

Fiori, F. and Spinozzi, F. 2010. X-Rays and Neutrons Scattering. In Rustichelli, F., Skrzypek, J. and Albertini, G. (ed.), *Innovative Technological Materials : Structural Properties by Neutron Scattering, Synchrotron Radiation and Modeling*. Heidelberg: Springer, pp. 21–38.

Fisher, M. E. 1972. Phase transitions, symmetry and dimensionality. In Conn, G. K. T. and Fowler, G. N. (eds.), *Essays in Physics*, vol. 4, pp. 43–89. New York: Academic Press, pp. 43–89.

Fisz, J. J. 1987. Symmetry simplifications in the description of molecular order and reorientational dynamics in uniaxial molecular systems. 1. Symmetry constraints on the joint probability-distribution function. *Chem. Phys.*, **114**, 165–185.

Flory, P. J. 1953. *Principles of Polymer Chemistry*. Ithaca, NY: Cornell University Press.

Flory, P. J. 1956. Phase equilibria in solutions of rod-like particles. *Proc. Roy. Soc. A*, **234**, 73–89.

Flory, P. J. 1969. *Statistical Mechanics of Chain Molecules*. New York: Wiley.

Flury, B. N. and Constantine, G. 1985. Algorithm AS 211: the FG diagonalization algorithm. *J. Roy. Stat. Soc C*, **34**, 177–183.

Flury, B. N. and Gautschi, W. 1986. An algorithm for simultaneous orthogonal transformation of several positive definite symmetric matrices to nearly diagonal form. *SIAM J. Sci. Stat. Comput.*, **7**, 169–184.

Flyvbjerg, H. and Petersen, H. G. 1989. Error estimates on averages of correlated data. *J. Chem. Phys.*, **91**, 461–466.

Foley, J. D. and Van Dam, A. 1982. *Fundamentals of Interactive Computer Graphics*. Reading: Addison-Wesley.

Fong, C., Le, T. and Drummond, C. J. 2012. Lyotropic liquid crystal engineering-ordered nanostructured small molecule amphiphile self-assembly materials by design. *Chem. Soc. Rev.*, **41**, 1297–1322.

Fontana, M. P., Rosi, B., Kirov, N. and Dozov, I. 1986. Molecular orientational motions in liquid crystals: a study by Raman and infrared band-shape analysis. *Phys. Rev. A*, **33**, 4132–4142.

Fontes, E., Heiney, P. A. and de Jeu, W. H. 1988. Liquid crystalline and helical order in a discotic mesophase. *Phys. Rev. Lett.*, **61**, 1202–1205.

Forster, D. 1974. Microscopic theory of flow alignment in nematic liquid crystals. *Phys. Rev. Lett.*, **32**, 1161–1164.

Forster, D. 1975. *Hydrodynamic Fluctuations, Broken Symmetry, and Correlation Functions*. New York: Addison-Wesley.

Forsyth, P. A., Marcelja, S., Mitchell, D. J. and Ninham, B. W. 1978. Ordering in colloidal systems. *Adv. Colloid Interface Sci.*, **9**, 37–60.

Fraden, S., Maret, G. and Caspar, D. L. D. 1993. Angular-correlations and the isotropic-nematic phase-transition in suspensions of Tobacco Mosaic-Virus. *Phys. Rev. E*, **48**, 2816–2837.

Francescangeli, O., Stanic, V., Gobbi, L., et al. 2003. Structure of self-assembled liposome-DNA-metal complexes. *Phys. Rev. E*, **67**, 011904.

Francescangeli, O., Stanic, V., Torgova, S. I., et al. 2009. Ferroelectric response and induced biaxiality in the nematic phase of a bent-core mesogen. *Adv. Funct. Mater.*, **19**, 2592–2600.

Frank, F. C. 1958. Liquid crystals. On the theory of liquid crystals. *Faraday Discuss.*, **25**, 19–28.

Freed, J. H. 1964. Anisotropic rotational diffusion and Electron Spin Resonance linewidths. *J. Chem. Phys.*, **41**, 2077–2083.

Freed, J. H., Nayeem, A. and Rananavare, S. B. 1994. ESR and slow motions in liquid crystals. In Luckhurst, G. R. and Veracini, C. A. (eds.), *The Molecular Dynamics of Liquid Crystals*, vol. 431. Dordrecht: Kluwer, pp. 365–402.

Freiser, M. J. 1970. Ordered states of a nematic liquid. *Phys. Rev. Lett.*, **24**, 1041–1043.

Frenkel, D. 1986. Free energy computation and first-order phase transitions. In Ciccotti, G. and Hoover, W. G. (eds.), *Molecular Dynamics Simulation of Statistical-Mechanical Systems. Proceedings of the International School of Physics 'Enrico Fermi', Varenna.* Amsterdam: North-Holland, pp. 151–188.

Frenkel, D. 1987. Computer-simulation of hard-core models for liquid crystals. *Mol. Phys.*, **60**, 1–20.

Frenkel, D. and Mulder, B. M. 1985. The hard ellipsoid-of-revolution fluid I. Monte Carlo simulations. *Mol. Phys.*, **55**, 1171–1192.

Frenkel, D. and Smit, B. 2002. *Understanding Molecular Simulations. From Algorithms to Applications.* San Diego, CA: Academic Press.

Frezza, E., Ferrarini, A., Kolli, H. B., Giacometti, A. and Cinacchi, G. 2013. The isotropic-to-nematic phase transition in hard helices: theory and simulation. *J. Chem. Phys.*, **138**, 164906.

Friedli, S. and Velenik, Y. 2017. *Statistical Mechanics of Lattice Systems: A Concrete Mathematical Introduction.* Cambridge: Cambridge University Press.

Friedman, H. L. 1985. *A Course in Statistical Mechanics*. New York: Prentice Hall.

Frisch, M. J., Trucks, G. W., Schlegel, H. B., Scuseria, G. E., Robb, M. A., Cheeseman, J. R., Zakrzewski, V. G., Montgomery, J. A., Stratmann, R. E., Burant, J. C., Dapprich, S., Millam, J. M., Daniels, A. D., Kudin, K. N., Strain, M. C., Farkas, O., Tomasi, J., Barone, V., Cossi, M., Cammi, R., Mennucci, B., Pomelli, C., Adamo, C., Clifford, S., Ochterski, J., Petersson, G. A., Ayala, P. Y., Cui, Q., Morokuma, K., Salvador, P., Dannenberg, J. J., Malick, D. K., Rabuck, A. D., Raghavachari, K., Foresman, J. B., Cioslowski, J., Ortiz, J. V., Baboul, A. G., Stefanov, B. B., Liu, G., Liashenko, A., Piskorz, P., Komaromi, I., Gomperts, R., Martin, R. L., Fox, D. J., Keith, T., Al-Laham, M. A., Peng, C. Y., Nanayakkara, A., Challacombe, M., Gill, P. M. W., Johnson, B., Chen, W., Wong, M. W., Andres, J. L., Gonzalez, C., Head-Gordon, M., Replogle, E. S. and Pople, J. A. 2002. *Gaussian 98 (Revision A. 11. 3)*. Pittsburgh, PA: Gaussian, Inc.

Frisch, M. J., Trucks, G. W., Schlegel, H. B., Scuseria, G. E., Robb, M. A., Cheeseman, J. R., Scalmani, G., Barone, V., Mennucci, B., Petersson, G. A., Nakatsuji, H., Caricato, M., Li, X., Hratchian, H. P., Izmaylov, A. F., Bloino, J., Zheng, G., Sonnenberg, J. L., Hada, M., Ehara, M., Toyota, K., Fukuda, R., Hasegawa, J., Ishida, M., Nakajima, T., Honda, Y., Kitao, O., Nakai, H., Vreven, T., Montgomery, Jr., J. A., Peralta, J. E., Ogliaro, F., Bearpark, M., Heyd, J. J., Brothers, E., Kudin, K. N., Staroverov, V. N., Kobayashi, R., Normand, J., Raghavachari, K., Rendell, A., Burant, J. C., Iyengar, S. S., Tomasi, J., Cossi, M., Rega, N., Millam, J. M., Klene, M., Knox, J. E., Cross, J. B., Bakken, V., Adamo, C., Jaramillo, J., Gomperts, R., Stratmann, R. E., Yazyev, O., Austin, A. J., Cammi, R., Pomelli, C., Ochterski, J. W., Martin, R. L., Morokuma, K., Zakrzewski, V. G., Voth, G. A., Salvador, P., Dannenberg, J. J., Dapprich, S., Daniels, A. D., Farkas, O., Foresman, J. B. and Ortiz, J. V. 2009. *Gaussian 09 Revision D. 01.* Wallingford, CT: Gaussian, Inc.

Fukuda, J., Yoneya, M. and Yokoyama, H. 2002. Defect structure of a nematic liquid crystal around a spherical particle: adaptive mesh refinement approach. *Phys. Rev. E*, **65**, 041709.

Fukuda, J., Yoneya, M. and Yokoyama, H. 2007. Surface-groove-induced azimuthal anchoring of a nematic liquid crystal: Berreman's model reexamined. *Phys. Rev. Lett.*, **99**, 139902.

Fukunaga, A., Urayama, K., Koelsch, P. and Takigawa, T. 2009. Electrically driven director-rotation of swollen nematic elastomers as revealed by polarized Fourier transform infrared spectroscopy. *Phys. Rev. E*, **79**, 051702.

Fukunaga, A., Urayama, K., Takigawa, T., DeSimone, A. and Teresi, L. 2008. Dynamics of electro-opto-mechanical effects in swollen nematic elastomers. *Macromolecules*, **41**, 9389–9396.

Fukunishi, H., Watanabe, O. and Takada, S. 2002. On the Hamiltonian replica exchange method for efficient sampling of biomolecular systems: application to protein structure prediction. *J. Chem. Phys.*, **116**, 9058–9067.

Fuller, G. J., Luckhurst, G. R. and Zannoni, C. 1985. Computer simulation studies of anisotropic systems.11. 2nd-rank and 4th-rank molecular-interactions. *Chem. Phys.*, **92**, 105–115.

Fung, B. M., Afzal, J., Foss, T. L. and Chau, M. H. 1986. Nematic ordering of 4-n-alkyl-4′-cyanobiphenyls studied by carbon-13 NMR with off-magic angle spinning. *J. Chem. Phys.*, **85**, 4808–4814.

Futrelle, R. P. and McGinty, D. J. 1971. Calculation of spectra and correlation-functions from molecular dynamics data using Fast Fourier Transform. *Chem. Phys. Lett.*, **12**, 285–287.

Gabriel, A. T., Meyer, T. and Germano, G. 2008. Molecular graphics of convex body fluids. *J. Chem. Theory Comput.*, **4**, 468–476.

Gale, J. D. and Rohl, A. L. 2003. The General Utility Lattice Program (GULP). *Molec. Simul.*, **29**, 291–341.

Gallo, P., Arnann-Winkel, K., Angell, C. A., et al. 2016. Water: a tale of two liquids. *Chem. Rev.*, **116**, 7463–7500.

Gamez, F. and Caro, C. 2015. The second virial coefficient for anisotropic square-well fluids. *J. Mol. Liq.*, **208**, 21–26.

Ganzke, D., Wróbel, S. and Haase, W. 2004. Dielectric studies of bicyclohexylcarbonitrile nematogens with large negative dielectric anisotropy. *Mol. Cryst. Liq. Cryst.*, **409**, 323–333.

Garland, C. W. 2001. Calorimetric studies. In Kumar, S. (ed.), *Liquid Crystals. Experimental Study of Physical Properties and Phase Transitions*. Cambridge: Cambridge University Press, pp. 240–294.

Garland, C. W. and Nounesis, G. 1994. Critical-behavior at nematic-smectic A phase transitions. *Phys. Rev. E*, **49**, 2964–2971.

Garland, C. W., Meichle, M., Ocko, B. M., et al. 1983. Critical behavior at the nematic-smectic-A transition in butyloxybenzylidene heptylaniline (4O.7). *Phys. Rev. A*, **27**, 3234–3240.

Gasparoux, H. 1980. Carbonaceous mesophase and disk-like molecules. In Heppke, G. and Helfrich, W. (eds.), *Liquid Crystals of One- and Two-Dimensional Order*. Berlin: Springer, pp. 373–382.

Gavezzotti, A. 1997. *Theoretical Aspects and Computer Modeling of the Molecular Solid State*. New York: Wiley.

Gay, J. G. and Berne, B. J. 1981. Modification of the overlap potential to mimic a linear site-site potential. *J. Chem. Phys.*, **74**, 3316–3319.

Gear, C. V. 1971. *Numerical Initial Value Problems in Ordinary Differential Equation*. Englewood Cliffs, NJ: Prentice-Hall.

Gearba, R. I., Lehmann, M., Levin, J., et al. 2003. Tailoring discotic mesophases: columnar order enforced with hydrogen bonds. *Adv. Mater.*, **15**, 1614–1618.

Geary, J. M., Goodby, J. W., Kmetz, A. R. and Patel, J. S. 1987. The mechanism of polymer alignment of liquid crystal materials. *J. Appl. Phys.*, **62**, 4100–4108.

Gelbart, W. M. and Ben-Shaul, A. 1996. The 'new' science of 'complex fluids'. *J. Phys. Chem.*, **100**, 13169–13189.

Gelbart, W. M. and Gelbart, A. 1977. Effective one-body potentials for orientationally anisotropic fluids. *Mol. Phys.*, **33**, 1387–1398.

Gell-Mann, M. 1956. The interpretation of the new particles as displaced charge multiplets. *Il Nuovo Cimento*, **4**, 848–868.

Giamberini, M., Cerruti, P., Ambrogi, V., et al. 2005. Liquid crystalline elastomers based on diglycidyl terminated rigid monomers and aliphatic acids. Part 2. Mechanical characterization. *Polymer*, **46**, 9113–9125.

Gibbs, J. W. 1902. *Elementary Principles in Statistical Mechanics*. New York: Charles Scribners Sons.

Gilli, G. and Gilli, P. 2009. *The Nature of the Hydrogen Bond: Outline of a Comprehensive Hydrogen Bond Theory*. Oxford: Oxford University Press.

Giordano, M., Leporini, D., Martinelli, M., et al. 1982. Electron resonance investigation of a cholesteric mesophase induced by a chiral probe. *J. Chem. Soc. Faraday Trans.2*, **78**, 307–316.

Girard, P. R. 1984. The quaternion group and modern physics. *Eur. J. Phys.*, **5**, 25.

Glarum, S. H. 1960. Dielectric relaxation of polar liquids. *J. Chem. Phys.*, **33**, 1371–1375.

Glasser, L. 2002. Equations of state and phase diagrams. *J. Chem. Educ.*, **79**, 874.

Gleim, W. and Finkelmann, H. 1989. Side chain liquid crystal elastomers. In McArdle, C. B. (ed.), *Side Chain Liquid Crystal Polymers*. Glasgow: Blackie and Son, pp. 287–308.

Glotzer, S. C., Solomon, M. J. and Kotov, N. A. 2004. Self-assembly: from nanoscale to microscale colloids. *AIChE Journal*, **50**, 2978–2985.

Goldberg, D. E. 1989. *Genetic Algorithms in Search, Optimization, and Machine Learning*. Reading: Addison-Wesley.

Goldfarb, D., Belsky, I., Luz, Z. and Zimmermann, H. 1983a. Axial-biaxial phase transition in discotic liquid crystals, studied by deuterium NMR. *J. Chem. Phys.*, **79**, 6203–6213.

Goldfarb, D., Poupko, R., Luz, Z. and Zimmermann, H. 1983b. Deuterium NMR of biaxial discotic liquid crystals. *J. Chem. Phys.*, **79**, 4035–4047.

Goldstein, A. N., Echer, C. M. and Alivisatos, A. P. 1992. Melting in semiconductor nanocrystals. *Science*, **256**, 1425–1427.

Goldstein, H. 1980. *Classical Mechanics*. Reading: Addison-Wesley.

Goldstein, H., Poole Jr, C. P. and Safko, J. L. 2001. *Classical Mechanics*. 3rd ed. Reading: Addison-Wesley.

Golemme, A., Žumer, S., Allender, D. W. and Doane, J. W. 1988a. Continuous nematic-isotropic transition in submicron-size liquid crystal droplets. *Phys. Rev. Lett.*, **61**, 2937–2940.

Golemme, A., Žumer, S., Doane, J. W. and Neubert, M. E. 1988b. Deuterium NMR of polymer dispersed liquid crystals. *Phys. Rev. A*, **37**, 559–569.

Goodby, J. W. and Gray, G. W. 1979. Classification of smectic polymorphic phases. *Mol. Cryst. Liq. Cryst.*, **49**, 217–223.

Goodby, J. W. and Gray, G. W. 1999. Guide to nomenclature and classification of liquid crystals. In Demus, D., Goodby, J., Gray, G. W., Spiess, H. W. and Vill, V. (eds.), *Physical Properties of Liquid Crystals*. Weinheim: Wiley-VCH, pp. 17–24.

Goodby, J. W., Waugh, M. A., Stein, S. M., et al. 1989. Characterization of a new helical smectic liquid crystal. *Nature*, **337**, 449–452.

Goossens, K., Nockemann, P., Driesen, K., et al. 2008. Imidazolium ionic liquid crystals with pendant mesogenic groups. *Chem. Mater.*, **20**, 157–168.

Goossens, W. J. A. 1971. Molecular theory of cholesteric phase and of twisting power of optically active molecules in a nematic liquid crystal. *Mol. Cryst. Liq. Cryst.*, **12**, 237–244.

Goossens, W. J. A. 1987. The smectic A-smectic C phase transition – a molecular statistical theory. *Europhys. Lett.*, **3**, 341–346.

Gordon, R. G. 1968. Correlation functions for molecular motion. *Adv. Mag. Res.*, **3**, 1–42.

Gordon, R. G. and Messenger, T. 1972. Magnetic resonance line shapes in slowly tumbling molecules. In Muus, L. T. and Atkins, P. W. (eds.), *Electron Spin Relaxation in Liquids*. New York: Plenum Press, pp. 341–382.

Gorkunov, M. V., Osipov, M. A., Lagerwall, J. and Giesselmann, F. 2007. Order-disorder molecular model of the smectic-A-smectic-C phase transition in materials with conventional and anomalously weak layer contraction. *Phys. Rev. E*, **76**, 051706.

Gottlob, A. P. and Hasenbusch, M. 1993. Critical behavior of the 3D XY-model – a Monte Carlo study. *Physica A*, **201**, 593–613.

Govers, E. and Vertogen, G. 1984. Elastic continuum theory of biaxial nematics. *Phys. Rev. A*, **30**, 1998–2000.

Gowda, A. and Kumar, S. 2018. Recent advances in discotic liquid crystal-assisted nanoparticles. *Materials*, **11**, 382.

Grabert, H. 1982. *Projection Operator Techniques in Nonequilibrium Statistical Mechanics*. Berlin: Springer-Verlag.

Graham, C., Imrie, D. A. and Raab, R. E. 1998. Measurement of the electric quadrupole moments of CO_2, CO, N_2, Cl_2 and BF_3. *Mol. Phys.*, **93**, 49–56.

Gramsbergen, E. F., Longa, L. and de Jeu, W. H. 1986a. Landau theory of the nematic-isotropic phase transition. *Phys. Rep.*, **135**, 195–257.

Gramsbergen E., Hoving H., de Jeu W. H., Praefcke K. and Kohne B., 1986b. X-ray investigation of discotic mesophases of alkylthio substituted triphenylenes. *Liq. Cryst.*, **1**, 397–400.

Gray, C. G. and Gubbins, K. E. 1984. *Theory of Molecular Fluids. vol. 1. Fundamentals*. Oxford: Clarendon Press.

Gray, G. W. 1962. *Molecular Structure and the Properties of Liquid Crystals*. London: Academic Press.

Gray, G. W. 1979. Liquid crystals and molecular structure: nematics and cholesterics. In Luckhurst, G. R. and Gray, G. W. (eds.), *The Molecular Physics of Liquid Crystals*. London: Academic Press, pp. 1–29.

Gray, G. W. 1987. *Thermotropic Liquid Crystals*. New York: Wiley.

Gray, G. W. and Goodby, J. 1984. *Smectic Liquid Crystals. Textures and Structures*. Glasgow: Leonard Hill.

Gray, G. W. and Harrison, K. J. 1971. Molecular theories and structure. Some effects of molecular structural change on liquid crystalline properties. *Symp. Faraday Soc.*, **5**, 54–67.

Gray, G. W., Harrison, K. J. and Nash, J. A. 1973. New family of nematic liquid crystals for displays. *Electron. Lett.*, **9**, 130–131.

Gray, G. W., Hird, M., Lacey, D. and Toyne, K. J. 1989. The synthesis and transition-temperatures of some $4,4''$-dialkyl-$1,1'$-$4',1''$-terphenyl and $4,4''$-alkoxyalkyl$-1,1'$-$4',1''$-terphenyl with $2,3$-difluoro or $2',3'$-difluoro substituents and of their biphenyl analogs. *J. Chem. Soc. Perkin Trans. 2*, 2041–2053.

Greer, A. L. 2000. Too hot to melt. *Nature*, **404**, 134–5.

Grimsdale, A. C., Chan, K. L., Martin, R. E., Jokisz, P. G. and Holmes, A. B. 2009. Synthesis of light-emitting conjugated polymers for applications in electroluminescent devices. *Chem. Rev.*, **109**, 897–1091.

Group d'Etude des Cristaux Liquides, Orsay. 1969. Dynamics of fluctuations in nematic liquid crystals. *J. Chem. Phys.*, **51**, 816–822.

Gruhn, T. and Hess, S. 1996. Monte Carlo simulation of the director field of a nematic liquid crystal with three elastic coefficients. *Z. Naturforsch. A*, **51**, 1–9.

Guggenheim, E. A. 1945. The principle of corresponding states. *J. Chem. Phys.*, **13**, 253–261.

Guha, R., Howard, M. T., Hutchison, G. R., et al. 2006. The Blue Obelisk – interoperability in chemical informatics. *J. Chem. Inf. Model.*, **46**, 991–998.

Guillon, D. 1999. Columnar order in thermotropic mesophases. *Struct. Bond.*, **95**, 41–82.

Guillon, D. 2000. Molecular engineering for ferroelectricity in liquid crystals. *Adv. Chem. Phys.*, **113**, 1–49.

Guinier, A. 1994. *X-Ray Diffraction in Crystals, Imperfect Crystals and Amorphous Bodies*. New York: Dover.

Gurtovenko, A. A., Patra, M., Karttunen, M. and Vattulainen, I. 2004. Cationic DMPC/DMTAP lipid bilayers: molecular dynamics study. *Biophys. J.*, **86**, 3461–3472.

Hagen, M. H. J., Meijer, E. J., Mooij, G. C. A. M., Frenkel, D. and Lekkerkerker, H. N. W. 1993. Does fullerene C60 have a liquid phase? *Nature*, **365**, 425–431.

Hahn, O., Delle Site, L. and Kremer, K. 2001. Simulation of polymer melts: from spherical to ellipsoidal beads. *Macromol. Theory Simul.*, **10**, 288–303.

Hait, D. and Head-Gordon, M. 2018. How accurate is Density Functional Theory at predicting dipole moments? An assessment using a new database of 200 benchmark values. *J. Chem. Theory Comput.*, **14**, 1969–1981.

Haller, I. 1975. Thermodynamic and static properties of liquid crystals. *Progr. Solid State Chem.*, **10**, 103–118.

Halperin, B. I. and Nelson, D. R. 1978. Theory of two-dimensional melting. *Phys. Rev. Lett.*, **41**, 121.

Hamaker, H. C. 1937. The London-van der Waals attraction between spherical particles. *Physica*, **4**, 1058–1072.

Hammersley, J. M. and Handscomb, D. C. 1965. *Monte Carlo Methods*. London: Methuen.

Hansen, J.-P. 1977. Correlation functions and their relationship with experiments. In Dupuy, J. and Dianoux, A. J. (eds.), *Microscopic Structure and Dynamics of Liquids*. New York: Plenum Press, pp. 1–68.

Hansen, J.-P. and McDonald, I. R. 2006. *Theory of Simple Liquids*. 3rd ed. Amsterdam: Academic Press.

Hanson, H., Dekker, A. J. and Van der Woude, F. 1977. Composition and temperature-dependence of pitch in cholesteric binary-mixtures. *Mol. Cryst. Liq. Cryst.*, **42**, 1025–1042.

Hanwell, M. D., Curtis, D. E., Lonie, D. C., et al. 2012. Avogadro: an advanced semantic chemical editor, visualization, and analysis platform. *J. Cheminformatics*, **4**, 17.

Harasima, A. 1958. Molecular theory of surface tension. *Adv. Chem. Phys.*, **1**, 203–237.

Hardouin, F., Sigaud, G., Achard, M. F., et al. 1995. SANS study of a semiflexible main chain liquid crystalline polyether. *Macromolecules*, **28**, 5427–5433.

Harris, F. E. 2014. *Mathematics for Physical Science and Engineering: Symbolic Computing Applications in Maple and Mathematica*. Amsterdam: Elsevier.

Harvey, M. J., Giupponi, G. and De Fabritiis, G. 2009. ACEMD: Accelerating biomolecular dynamics in the microseconds time scale. *J. Chem. Theory Comput.*, **5**, 1632–1639.

Hasenbusch, M., Pinn, K. and Vinti, S. 1999. Critical exponents of the three-dimensional Ising universality class from finite-size scaling with standard and improved actions. *Phys. Rev. B*, **59**, 11471–11483.

Hashim, R., Sugimura, A., Minamikawa, H. and Heidelberg, T. 2012. Nature-like synthetic alkyl branched-chain glycolipids: a review on chemical structure and self-assembly properties. *Liq. Cryst.*, **39**, 1–17.

Hayashi, Y. and Matsumoto, K. 1994. X-ray photoelectron-spectroscopy analysis of buffed polyimide film. *Nippon Kagaku Kaishi.*, **1994**, 490–492.

Hayes, B. 2013. First links in the Markov chain. *Am. Sci.*, **101**, 252.

Haynes, W. M., Lide, D. R. and Bruno, T. J. (eds.). 2014. *CRC Handbook of Chemistry and Physics*. 93rd ed. Boca Raton, FL: CRC Press.

Headen, T. F., Howard, C. A., Skipper, N. T., et al. 2010. Structure of π-π interactions in aromatic liquids. *J. Amer. Chem. Soc.*, **132**, 5735–5742.

Hedin, F., El Hage, K. and Meuwly, M. 2016. A toolkit to fit nonbonded parameters from and for condensed phase simulations. *J. Chem. Inf. Model.*, **56**, 1479–89.

Hegmann, T., Qi, H. and Marx, V. M. 2007. Nanoparticles in liquid crystals: synthesis, self-assembly, defect formation and potential applications. *J. Inorg. Organomet. Polym. Mater.*, **17**, 483–508.

Heinz, H., Paul, W. and Binder, K. 2005. Calculation of local pressure tensors in systems with many-body interactions. *Phys. Rev. E*, **72**, 066704.

Hess, B., Bekker, H., Berendsen, H. J. C. and Fraaije, J. G. E. M. 1997. LINCS: a linear constraint solver for molecular simulations. *J. Comput. Chem.*, **18**, 1463–1472.

Hess, B., Kutzner, C., van der Spoel, D. and Lindahl, E. 2008. GROMACS 4: algorithms for highly efficient, load-balanced, and scalable molecular simulation. *J. Chem. Theory Comput.*, **4**, 435–447.

Higaki, H., Urayama, K. and Takigawa, T. 2012. Memory and development of textures of polydomain nematic elastomers. *Macromol. Chem. Phys.*, **213**, 1907–1912.

Hilbers, C. W. and MacLean, C. 1972. NMR of molecules oriented in electric fields. In Diehl, P., Fluck, E. and Kosfeld, R. (eds.), *NMR Basic Principles and Progress*. Berlin: Springer-Verlag, pp. 1–52.

Hird, M. 2001. Relationship between molecular structure and transition temperatures for calamitic structures in nematics. In Dunmur, D. A., Fukuda, A. and Luckhurst, G. R. (eds.), *Physical Properties of Liquid Crystals: Nematics*. London: INSPEC-IEE, pp. 3–16.

Hird, M. 2007. Fluorinated liquid crystals - properties and applications. *Chem. Soc. Rev.*, **36**, 2070–2095.

Hirschmann, H. and Reiffenrath, V. 1998. TN, STN displays. In Demus, D., Goodby, J., Gray, G. W., Spiess, H. W. and Vill, V. (eds.), *Handbook of Liquid Crystals. Low Molecular Weight Liquid Crystals I*, vol. 2A. Weinheim: Wiley-VCH, pp. 199–229.

Hlawacek, G., Khokhar, F. S., van Gastel, R., Poelsema, B. and Teichert, C. 2011. Smooth growth of organic semiconductor films on graphene for high-efficiency electronics. *Nanolett.*, **11**, 333–337.

Hobdell, J. and Windle, A. 1995. Topological point-defects in liquid crystalline polymers. *Liq. Cryst.*, **19**, 401–407.

Hockney, R. W. 1989. *Computer Simulation Using Particles*. New York: McGraw-Hill.

Hofsäss, C., Lindahl, E. and Edholm, O. 2003. Molecular dynamics simulations of phospholipid bilayers with cholesterol. *Biophys. J.*, **84**, 2192–2206.

Hohenberg, P. C. 1967. Existence of long-range order in one and two dimensions. *Phys. Rev.*, **158**, 383–386.

Holm, C. and Janke, W. 1993. Critical exponents of the classical three-dimensional Heisenberg model: a single-cluster Monte Carlo study. *Phys. Rev. B*, **48**, 936–950.

Holm, C. and Janke, W. 1997. Critical exponents of the classical Heisenberg ferromagnet. *Phys. Rev. Lett.*, **78**, 2265–2265.

Homer, J. and Mohammadi, M. S. 1987. Polyatomic London dispersion forces. *J. Chem. Soc., Faraday Trans. II*, **83**, 1957–1974.

Hong, Q., Wu, T. X. and Wu, S. T. 2003. Optical wave propagation in a cholesteric liquid crystal using the finite element method. *Liq. Cryst.*, **30**, 367–375.

Hoover, W. G. 1985. Canonical dynamics: equilibrium phase-space distributions. **31**, 1695–1697.

Hoover, W. G., Ashurst, W. T. and Olness, R. J. 1974. 2-dimensional computer studies of crystal stability and fluid viscosity. *J. Chem. Phys.*, **60**, 4043–4047.

Horn, R. G. 1978. Refractive-indexes and order parameters of two liquid crystals. *J. de Physique*, **39**, 105–109.

Hornreich, R. M. 1985. Landau theory of the isotropic-nematic critical point. *Phys. Lett. A*, **109**, 232–234.

Horton, J. C., Donald, A. M. and Hill, A. 1990. Coexistence of 2 liquid crystalline phases in poly(gamma-benzyl-alpha,l-glutamate) solutions. *Nature*, **346**, 44–45.

Hoskins, R. F:. 2009. *Delta Functions: An Introduction to Generalised Functions*. 2nd ed. Chichester: Horwood.

Houssa, M., Oualid, A. and Rull, L. F. 1998a. Reaction field and Ewald summation study of mesophase formation in dipolar Gay-Berne model. *Mol. Phys.*, **94**, 439–446.

Houssa, M., Rull, L. F. and McGrother, S. C. 1998b. Effect of dipolar interactions on the phase behavior of the Gay-Berne liquid crystal model. *J. Chem. Phys.*, **109**, 9529–9542.

Houssa, M., Rull, L. F. and McGrother, S. C. 1999. Dipolar Gay-Berne liquid crystals: a Monte Carlo study. *Int. J. Mod. Phys. C*, **10**, 391–401.

Houssa, M., Rull, L. F. and Romero-Enrique, J. M. 2009. Bilayered smectic phase polymorphism in the dipolar Gay-Berne liquid crystal model. *J. Chem. Phys.*, **130**, 154504.

Huang, C.-C., Baus, M. and Ryckaert, J.-P. 2015. On the calculation of the absolute grand potential of confined smectic-A phases. *Mol. Phys.*, **113**, 2643–2655.

Huang, C. C., Ramachandran, S. and Ryckaert, J.-P. 2014. Calculation of the absolute free energy of a smectic-A phase. *Phys. Rev. E*, **90**, 12.

Hudson, S. A. and Maitlis, P. M. 1993. Calamitic metallomesogens: metal containing liquid crystals with rodlike shapes. *Chem. Rev.*, **93**, 861–885.

Hudson, S. D. and Thomas, E. L. 1989. Frank elastic-constant anisotropy measured from transmission-electron-microscope images of disclinations. *Phys. Rev. Lett.*, **62**, 1993–1996.

Hughes, J. R., Luckhurst, G. R., Praefcke, K., Singer, D. and Tearle, W. M. 2003. Chemically-induced discotic liquid crystals. Structural studies with NMR spectroscopy. *Mol. Cryst. Liq. Cryst.*, **396**, 187–225.

Hughes, Z. E., Stimson, L. M., Slim, H., et al. 2008. An investigation of soft-core potentials for the simulation of mesogenic molecules and molecules composed of rigid and flexible segments. *Comp. Phys. Comm.*, **178**, 724–731.

Humpert, A. and Allen, M. P. 2015a. Elastic constants and dynamics in nematic liquid crystals. *Mol. Phys.*, **113**, 2680–2692.

Humpert, A. and Allen, M. P. 2015b. Propagating director bend fluctuations in nematic liquid crystals. *Phys. Rev. Lett.*, **114**, 028301.

Humphrey, W., Dalke, A. and Schulten, K. 1996. VMD – Visual molecular dynamics. *J. Mol. Graph.*, **14**, 33–38.

Humphries, R. L., James, P. G. and Luckhurst, G. R. 1972. Molecular field treatment of nematic liquid crystals. *J. Chem. Soc. Faraday Trans.*, **68**, 1031–1044.

Huntress, W. T. 1970. The study of anisotropic rotation of molecules in liquids by NMR quadrupolar relaxation. *Adv. Magn. Reson*, **4**, 1–37.

Hurd, A. J., Fraden, S., Lonberg, F. and Meyer, R. B. 1985. Field-induced transient periodic structures in nematic liquid crystals – the splay Frederiks transition. *J. de Physique*, **46**, 905–917.

Ibrahim, I. H. and Haase, W. 1976. Order parameter temperature-dependence of some nematic liquids related to magnetic and optical anisotropies. *Z. Naturforschung. A*, **31**, 1644–1650.

Idé, J., Méreau, R., Ducasse, L., et al. 2014. Charge dissociation at interfaces between discotic liquid crystals: the surprising role of column mismatch. *J. Amer. Chem. Soc.*, **136**, 2911–2920.

Irene, E. A. 2008. *Surfaces, Interfaces, and Thin Films for Microlectronics*. Hoboken, NJ: Wiley.

Irikura, K. K. 2019. Glossary of common terms and abbreviations in Quantum Chemistry. In Johnson III, R. D. (ed.), *NIST Computational Chemistry Comparison and Benchmark Database, NIST Standard Reference Database N.101 Rel.20*. https://cccbdb.nist.gov/glossary.asp

Irrgang, M. E., Engel, M., Schultz, A. J., Kofke, D. A. and Glotzer, S. C. 2017. Virial coefficients and equations of state for hard polyhedron fluids. *Langmuir*, **33**, 11788–11796.

Irvine, P. A., Wu, D. C. and Flory, P. J. 1984. Liquid crystalline transitions in homologous para-phenylenes and their mixtures. 1. Experimental results. *J. Chem. Soc. Faraday Trans. I*, **80**, 1795–1806.

Irving, J. H. and Kirkwood, J. G. 1950. The statistical mechanical theory of transport processes. IV. The equations of hydrodynamics. *J. Chem. Phys.*, **18**, 817–829.

Ishihara, S., Wakemoto, H., Nakazima, K. and Matsuo, Y. 1989. The effect of rubbed polymer-films on the liquid crystal alignment. *Liq. Cryst.*, **4**, 669–675.

Ishikawa, K., Yoshikawa, K. and Okada, N. 1988. Size effect on the ferroelectric phase transition in $PbTiO_3$ ultrafine particles. *Phys. Rev. B*, **37**, 5852.

Isihara, A. 1951. Theory of anisotropic colloidal solutions. *J. Chem. Phys.*, **19**, 1142–1147.

Israelachvili, J. 1992. *Intermolecular and Surface Forces*. New York: Academic Press.

Israelachvili, J. 2011. *Intermolecular and Surface forces*. 3rd ed. Waltham: Academic Press.

Jackson, N. E., Webb, M. A. and de Pablo, J. J. 2019. Recent advances in machine learning towards multiscale soft materials design. *Curr. Opin. Chem. Eng.*, **23**, 106–114.

Jacobsen, J. P. and Pedersen, E. J. 1981. ^{1}H and ^{2}H NMR spectra of pyridine and pyridine-N-oxide in liquid crystalline phase. *J. Magn. Res.*, **44**, 101–108.

Jadzyn, J. and Kedziora, P. 2006. Anisotropy of static electric permittivity and conductivity in some nematics and smectics A. *Mol. Cryst. Liq. Cryst.*, **145**, 17–23.

Jákli, A., Bailey, C. and Harden, J. 2007. Physical properties of banana liquid crystals. In Ramamoorthy, A. (ed.), *Thermotropic Liquid Crystals: Recent Advances*. Dordrecht: Springer, pp. 59–84.

Jákli, A., Lavrentovich, O. D. and Selinger, J. V. 2018. Physics of liquid crystals of bent-shaped molecules. *Rev. Mod. Phys.*, **90**, 045004.

Jämbeck, J. P. M. and Lyubartsev, A. P. 2012. Derivation and systematic validation of a refined all-atom force field for phosphatidylcholine lipids. *J. Phys. Chem. B*, **116**, 3164–3179.

Jang, W. G., Glaser, M. A., Park, C. S., Kim, K. H., Lansac, Y., and Clark, N. A. 2001. Evidence from infrared dichroism, X-ray diffraction, and atomistic computer simulation for a "zigzag" molecular shape in tilted smectic liquid crystal phases. *Phys. Rev. E*, **64**, 051712.

Jansen, H. J. F., Vertogen, G. and Ypma, J. G. J. 1977. Monte Carlo calculation of nematic-isotropic phase transition. *Mol. Cryst. Liq. Cryst.*, **38**, 445–453.

Jasz, A., Rak, A., Ladjanszki, I. and Cserey, G. 2020. Classical molecular dynamics on graphics processing unit architectures. *Wiley Interdiscip. Rev. Comput. Mol. Sci.*, **10**, e1444.

Jaynes, E. T. 1957a. Information theory and statistical mechanics I. *Phys. Rev.*, **106**, 620–630.

Jaynes, E. T. 1957b. Information theory and statistical mechanics II. *Phys. Rev.*, **108**, 171–190.

Jen, S., Clark, N. A., Pershan, P. S. and B., Priestley E. 1973. Raman-Scattering from a nematic liquid crystal – orientational statistics. *Phys. Rev. Lett.*, **31**, 1552–1556.

Jenkins, F. A. and White, H. E. 2001. *Fundamentals of Optics*. 4th ed. New York: McGraw-Hill.

Jenz, F., Osipov, M. A., Jagiella, S. and Giesselmann, F. 2016. Orientational distribution functions and order parameters in 'de Vries'-type smectics: a simulation study. *J. Chem. Phys.*, **145**, 134901.

Jepsen, D. W. and Friedman, H. L. 1963. Cluster expansion methods for systems of polar molecules: some solvents and dielectric properties. *J. Chem. Phys.*, **38**, 846–864.

Jérôme, B. 1991. Surface effects and anchoring in liquid crystals. *Rep. Progr. Phys.*, **54**, 391–451.

Jiang, S. and Granick, S. 2012. *Janus Particle Synthesis, Self-Assembly and Applications*. Cambridge: Royal Society of Chemistry.

Jmol. An Open-Source Java Viewer for Chemical Structures In 3D. www.jmol.org

John, B. S., Juhlin, C. and Escobedo, F. A. 2008. Phase behavior of colloidal hard perfect tetragonal parallelepipeds. *J. Chem. Phys.*, **128**, 044909.

Johnson, R. D. III. 2019. *NIST Computational Chemistry Comparison and Benchmark Database*. NIST Standard Reference Database N. 101, Rel. 20, August 2019. http://cccbdb.nist.gov/

Johnson, S. R. and Jurs, P. C. 1999. Prediction of the clearing temperatures of a series of liquid crystals from molecular structure. *Chem. Mater.*, **11**, 1007–1023.

Jones, R. C. 1948. A new calculus for the treatment of optical systems. VII. Properties of the N-matrices. *JOSA*, **38**, 671–685.

Jönsson, B., Nilsson, P. G., Lindman, B., Guldbrand, L. and Wennerström, H. (eds.). 1984. *Principles of Phase Equilibria in Surfactant-Water Systems*. Surfactants in Solution. New York: Plenum Press.

Jorgensen, W. L. and Jenson, C. 1998. Temperature dependence of TIP3P, SPC, and TIP4P water from NPT Monte Carlo simulations: seeking temperatures of maximum density. *J. Comput. Chem.*, **19**, 1179–1186.

Jorgensen, W. L. and Tirado-Rives, J. 1988. The OPLS potential functions for proteins, energy minimizations for crystals of cyclic peptides and crambin. *J. Amer. Chem. Soc.*, **110**, 1657–1666.

Jorgensen, W. L., Maxwell, D. S. and Tirado-Rives, J. 1996. Development and testing of the OPLS all-atom force field on conformational energetics and properties of organic liquids. *J. Amer. Chem. Soc.*, **118**, 11225–11236.

Jorgensen, W. L., Chandrasekhar, J., Madura, J. D., Impey, R. W. and Klein, M. L. 1983. Comparison of simple potential functions for simulating liquid water. *J. Chem. Phys.*, **79**, 926–935.

Joshi, A. A., Whitmer, J. K., Guzman, O., Abbott, N. L. and de Pablo, J. J. 2014. Measuring liquid crystal elastic constants with free energy perturbations. *Soft Matter*, **10**, 882–893.

Jozefowicz, W. and Longa, L. 2007. Frustration in smectic layers of polar Gay-Berne systems. *Phys. Rev. E*, **76**, 011701.

Juszynska, E., Jasiurkowska, M., Massalska-Arodz, M., Takajo, D. and Inaba, A. 2011. Phase transition and structure studies of a liquid crystalline Schiff-base compound (4O.8). *Mol. Cryst. Liq. Cryst.*, **540**, 127–134.

Kaafarani, B. R. 2011. Discotic liquid crystals for opto-electronic applications. *Chem. Mater.*, **23**, 378–396.

Kabadi, V. N. and Steele, W. A. 1985. Molecular dynamics of fluids – the Gaussian Overlap model. *Ber. Bunsen-Ges. Phys. Chem.*, **89**, 2–9.

Kadanoff, L. P. 1966. Scaling laws for Ising models near T_C. *Physics*, **2**, 263–272.

Kalkura, A. N., Shashidhar, R., Venkatesh, G. and Weissflog, W. 1982. High-pressure studies on polymorphic liquid crystals. *Mol. Cryst. Liq. Cryst.*, **84**, 275–284.

Kamberaj, H., Low, R. J. and Neal, M. P. 2005. Time reversible and symplectic integrators for molecular dynamics simulations of rigid molecules. *J. Chem. Phys.*, **122**, 224114.

Kamien, R. D. 1996. Liquids with chiral bond order. *J. de Physique II*, **6**, 461–475.

Kapernaum, N. and Giesselmann, F. 2008. Simple experimental assessment of smectic translational order parameters. *Phys. Rev. E*, **78**, 062701.

Kaplan, J. I. and Drauglis, E. 1971. On the statistical theory of the nematic mesophase. *Chem. Phys. Lett.*, **9**, 645–645.

Karahaliou, P. K., Vanakaras, A. G. and Photinos, D. J. 2002. Tilt order parameters, polarity, and inversion phenomena in smectic liquid crystals. *Phys. Rev. E*, **65**, 031712.

Karahaliou, P. K., Vanakaras, A. G. and Photinos, D. J. 2009. Symmetries and alignment of biaxial nematic liquid crystals. *J Chem. Phys.*, **131**, 124516.

Karat, P. P. and Madhusudana, N. V. 1976. Elastic and optical-properties of some $4'$-n-alkyl-4-cyanobiphenyls. *Mol. Cryst. Liq. Cryst.*, **36**, 51–64.

Karjalainen, J., Lintuvuori, J., Telkki, V. V., Lantto, P. and Vaara, J. 2013. Constant-pressure simulations of Gay-Berne liquid crystalline phases in cylindrical nanocavities. *Phys. Chem. Chem. Phys.*, **15**, 14047–14057.

Kaszynski, P., Pakhomov, S. and Tesh, K. F. 2001. Carborane-containing liquid crystals: synthesis and structural, conformational, thermal, and spectroscopic characterization of diheptyl and diheptynyl derivatives of p-carboranes. *Inorg. Chem.*, **40**, 6622–6631.

Kato, T. 1998. Hydrogen-bonded systems. In Demus, D., Goodby, J., Gray, G. W., Spiess, H. W. and Vill, V. (eds.), *Handbook of Liquid Crystals. Low Molecular Weight Liquid Crystals II*, vol. 2B. Weinheim: Wiley-VCH, pp. 969–979.

Katriel, J., Kventsel, G. F., Luckhurst, G. R. and Sluckin, T. J. 1986. Free energies in the Landau and Molecular Field approaches. *Liq. Cryst.*, **1**, 337–355.

Kats, E. I. and Monastyrsky, M. I. 1984. Ordering in discotic liquid crystals. *J. de Physique*, **45**, 709–714.

Keddie, J. L, Jones, R. A. L. and Cory, R. A. 1994. Size-dependent depression of the glass transition temperature in polymer films. *EPL*, **27**, 59–64.

Keith, C., Lehmann, A., Baumeister, U., Prehm, M. and Tschierske, C. 2010. Nematic phases of bent-core mesogens. *Soft Matter*, **6**, 1704–1721.

Kestemont, E. and VanCraen, J. 1976. Computation of correlation functions in molecular dynamics experiments. *J. Comput. Phys.*, **22**, 451–458.

Kestenbach, H.-J., Loos, J. and Petermann, J. 1999. Transcrystallization at the interface of polyethylene single-polymer composites. *Mater. Res.*, **2**, 261–269.

Khakbaz, P. and Klauda, J. B. 2018. Investigation of phase transitions of saturated phosphocholine lipid bilayers via molecular dynamics simulations. *Biochim. Biophys. Acta* **1860**, 1489–1501.

Khare, R. S., de Pablo, J. J. and Yethiraj, A. 1996. Rheology of confined polymer melts. *Macromolecules*, **29**, 7910–7918.

Khokhlov, A. R. 1991. Theories based on the Onsager approach. In Ciferri, A. (ed.), *Liquid Crystallinity in Polymers. Principles and Fundamental Properties*. New York: VCH, pp. 97–129.

Khoo, I. C. 2007. *Liquid Crystals*. 2nd ed. New York: Wiley.

Kielich, S. 1972. General molecular theory and electric field effects in isotropic dielectrics. In Davies, M. (ed.), *Dielectric and Related Molecular Processes*, vol. 1. London: The Chemical Society, pp. 192–387.

Kihara, T. 1963. Convex molecules in gaseous and crystalline systems. *Adv. Chem. Phys.*, **5**, 147–188.

Kihara, T. 1967. Intermolecular forces for polyatomic molecules. *Progr. Theor. Phys. Suppl.*, **40**, 177–206.

Kikuchi, H. 2008. Liquid crystalline blue phases. *Struct. Bond.*, **128**, 99–117.

Kikuchi, H., Yokota, M., Hisakado, Y., Yang, H. and Kajiyama, T. 2002. Polymer-stabilized liquid crystal blue phases. *Nat. Mater.*, **1**, 64–68.

Kilian, A. 1993. Computer simulations of nematic droplets. *Liq. Cryst.*, **14**, 1189–1198.

Kilian, A, and Hess, S. 1989. Derivation and application of an algorithm for the numerical calculation of the local orientation of nematic liquid crystals. *Z. Naturfors. A*, **44**, 693–703.

Kim, D. G., Kim, Y. H., Shin, T. J., et al. 2017. Highly anisotropic thermal conductivity of discotic nematic liquid crystalline films with homeotropic alignment. *Chem. Comm.*, **53**, 8227–8230.

Kim, K. H. and Song, J. K. 2009. Technical evolution of liquid crystal displays. *NPG Asia Materials*, **1**, 29–36.

Kim, S., Thiessen, P. A., Bolton, E. E., et al. 2016. PubChem substance and compound databases. *Nucleic Acids Res.*, **44**, D1202–D1213.

Kipnis, A. Ya., Yavelow, B. E. and Rowlinson, J. S. 1996. *van der Waals and Molecular Sciences*. Oxford: Clarendon Press.

Kirkpatrick, J., Marcon, V., Nelson, J., Kremer, K. and Andrienko, D. 2007. Charge mobility of discotic mesophases: a multiscale quantum and classical study. *Phys. Rev. Lett.*, **98**.

Kirkpatrick, J., Marcon, V., Kremer, K., Nelson, J. and Andrienko, D. 2008. Columnar mesophases of hexabenzo-coronene derivatives. II. Charge carrier mobility. *J. Chem. Phys.*, **129**, 094506.

Kirov, N., Dozov, I. and Fontana, M. P. 1985. Determination of orientational correlation functions in ordered fluids: Raman scattering. *J. Chem. Phys.*, **83**, 5267–5276.

Kittel, C. 2005. *Introduction to Solid State Physics*. New York: Wiley.

Kitzerow, H. S. 1994. Polymer dispersed liquid crystals – from the nematic curvilinear aligned phase to ferroelectric-films. *Liq. Cryst.*, **16**, 1–31.

Kitzerow, H. S. and Bahr, C. (eds.). 2001. *Chirality in Liquid Crystals*. Berlin: Springer.

Klauda, J. B., Eldho, N. V., Gawrisch, K., Brooks, B. R. and Pastor, R. W. 2008. Collective and noncollective models of NMR relaxation in lipid vesicles and multilayers. *J. Phys. Chem. B*, **112**, 5924–5929.

Klauda, J. B., Venable, R. M., Freites, J. A., et al. 2010. Update of the CHARMM All-Atom Additive force field for lipids: validation on six lipid types. *J. Phys. Chem. B*, **114**, 7830–7843.

Kleman, M. 1982. *Points, Lines and Walls: In Liquid Crystals, Magnetic Systems and Various Ordered Media*. New York: Wiley.

Kleman, M. 1991. Defects and textures in liquid crystalline polymers. In Ciferri, A. (ed.), *Liquid Crystallinity in Polymers. Principles and Fundamental Properties*. New York: VCH, pp. 365–394.

Kleman, M. and Lavrentovich, O. D. 2003. *Soft Matter Physics*. Berlin: Springer.

Kleman, M. and Lavrentovich, O. D. 2006. Topological point defects in nematic liquid crystals. *Philos. Mag.*, **86**, 4117–4137.

Knotts IV, T. A., Rathore, N., Schwartz, D. C. and de Pablo, J. J. 2007. A coarse grain model for DNA. *J. Chem. Phys.*, **126**, 084901.

Knowles, J. K. 1998. *Linear Vector Spaces and Cartesian Tensors*. Oxford: Oxford University Press.

Knuth, D. E. 1998. *The Art of Computer Programming, Vol. 2, Seminumerical Algorithms*. 3rd ed., Boston, MA.: Addison-Wesley.

Ko, S. W., Huang, S. H., Fuh, A. Y. G. and Lin, T. H. 2009. Fabrications of liquid crystal polarization converters and their applications. In Khoo, I. C. (ed.), *SPIE, Liquid Crystals XIII*, vol. 7414. Bellingham, WA: SPIE, 001–006.

Kobashi, J., Yoshida, H. and Ozaki, M. 2016. Planar optics with patterned chiral liquid crystals. *Nat. Photonics*, **10**, 389–392.

Kohlrausch, R. 1854. Theorie des elektrischen rückstandes in der Leidener flasche. *Annalen der Physik*, **167**, 179–214.

Kolli, H. B., Frezza, E., Cinacchi, G., et al. 2014a. Communication: from rods to helices: evidence of a screw-like nematic phase. *J. Chem. Phys.*, **140**, 081101.

Kolli, H. B., Frezza, E., Cinacchi, G., et al. 2014b. Self-assembly of hard helices: a rich and unconventional polymorphism. *Soft Matter*, **10**, 8171–8187.

Kosterlitz, J. M. and Thouless, D. J. 1973. Ordering, metastability and phase transitions in two-dimensional systems. *J. Phys. C*, **6**, 1181–1203.

Kovshev, E. I., Blinov, L. M. and Titov, V. V. 1977. Thermotropic liquid crystals and their application. *Russian Chem. Rev.*, **46**, 395–419.

Koynova, R. and Caffrey, M. 1998. Phases and phase transitions of the phosphatidylcholines. *Biochim. Biophys. Acta*, **1376**, 91–145.

Kralj, S., Žumer, S. and Allender, D. W. 1991. Nematic-isotropic phase-transition in a liquid crystal droplet. *Phys. Rev. A*, **43**, 2943–2954.

Krentsel, T. A., Lavrentovich, O. D. and Kumar, S. 1997. In-situ X-ray measurements of light-controlled layer spacing in a smectic-A liquid crystal. *Mol. Cryst. Liq.Cryst. A*, **304**, 463–469.

Kubo, R. 1986. Brownian-motion and nonequilibrium statistical-mechanics. *Science*, **233**, 330–334.

Kucerka, N., Nagle, J. F., Sachs, J. N., et al. 2008. Lipid bilayer structure determined by the simultaneous analysis of neutron and X-ray scattering data. *Biophys. J.*, **95**, 2356–2367.

Kuijk, A., Byelov, D. V., Petukhov, A. V., van Blaaderen, A. and Imhof, A. 2012. Phase behavior of colloidal silica rods. *Faraday Discuss.*, **159**, 181–199.

Kuiper, S., Norder, B., Jager, W. F., et al. 2011. Elucidation of the orientational order and the phase diagram of *p*-quinquephenyl. *J. Phys. Chem. B*, **115**, 1416–1421.

Kukol, A. 2009. Lipid models for united-atom molecular dynamics simulations of proteins. *J. Chem. Theory Comput.*, **5**, 615–626.

Kulkarni, C. V. 2012. Lipid crystallization: from self assembly to hierarchical and biological ordering. *Nanoscale*, **4**, 5779–91.

Kumar, S. 2001. Structure: X-ray diffraction studies of liquid crystals. In Kumar, S. (ed.), *Liquid Crystals: Experimental Study of Physical Properties and Phase Transitions*. Cambridge: Cambridge University Press, pp. 65–94.

Kumar, S. 2002. Discotic liquid crystals for solar cells. *Curr. Sci.*, **82**, 256–257.

Kumar, S. 2004. Recent developments in the chemistry of triphenylene-based discotic liquid crystals. *Liq. Cryst.*, **31**, 1037–1059.

Kumar, S. 2006. Self-organization of disk-like molecules: chemical aspects. *Chem. Soc. Rev.*, **35**, 83–109.

Küpfer, J. and Finkelmann, H. 1991. Nematic liquid single crystal elastomers. *Macromol. Chem. Rapid Commun.*, **12**, 717–726.

Kurdikar, D. L., Boots, H. M. J. and Peppas, N. A. 1995. Network formation by chain polymerization of liquid crystalline monomer: a first off-lattice Monte Carlo study. *Macromolecules*, **28**, 5632–5637.

Kurik, M. V. and Lavrentovich, O. D. 1988. Defects in liquid crystals: homotopy theory and experimental studies. *Physics-Uspekhi*, **31**, 196–224.

Kushick, J. and Berne, B. J. 1973a. Role of attractive forces in self-diffusion in dense Lennard-Jones fluids. *J. Chem. Phys.*, **59**, 3732–3736.

Kushick, J. and Berne, B. J. 1973b. Methods for experimentally determining angular velocity relaxation in liquids. *J. Chem. Phys.*, **59**, 4486–4490.

Kushick, J. and Berne, B. J. 1977. Molecular dynamics methods: continuous potentials. In Berne, B. J. (ed.), *Statistical mechanics B: Time Dependent Processes*. New York: Plenum Press, pp. 41–64.

Kventsel, G. F., Luckhurst, G. R. and Zewdie, H. B. 1985. A molecular field theory of smectic A liquid crystals – a simpler alternative to the McMillan theory. *Mol. Phys.*, **56**, 589–610.

Kwak, C. H. and Kim, G. Y. 2016. Generalised descriptions on orientational order parameters and mean field theory for uniaxial and biaxial nematic liquid crystals. *Liq. Cryst.*, **43**, 32–48.

Lagardere, L., Jolly, L. H., Lipparini, F., et al. 2018. Tinker-HP: a massively parallel molecular dynamics package for multiscale simulations of large complex systems with advanced point dipole polarizable force fields. *Chemical Science*, **9**, 956–972.

Lagarias, J. C. (ed.). 2011. *The Kepler Conjecture: the Hales-Ferguson Proof by Thomas Hales, Samuel Ferguson*. New York: Springer.

Lagerwall, J. and Giesselmann, F. 2006. Current topics in smectic liquid crystal research. *ChemPhysChem*, **7**, 20–45.

Lamarra, M., Muccioli, L., Orlandi, S. and Zannoni, C. 2012. Temperature dependence of charge mobility in model discotic liquid crystals. *Phys. Chem. Chem. Phys.*, **14**, 5368–5375.

Landau, D. P. 1976. Finite-size behavior of the simple-cubic Ising lattice. *Phys. Rev. B*, **14**, 255–262.

Landau, D. P. and Binder, K. 2000. *A Guide to Monte Carlo Simulations in Statistical Physics*. Cambridge: Cambridge University Press.

Landau, E. M. and Rosenbusch, J. P. 1996. Lipidic cubic phases: a novel concept for the crystallization of membrane proteins. *Proc. Nat. Acad. Sci. USA*, **93**, 14532–14535.

Landau, E. M., Rummel, G., Cowan-Jacob, S. W. and Rosenbusch, J. P. 1997. Crystallization of a polar protein and small molecules from the aqueous compartment of lipidic cubic phases. *J. Phys. Chem. B*, **101**, 1935–1937.

Landau, L. D. 1965. On the theory of phase transitions (I: JETP, 7, 1, 1937; II:JETP, 7, 627, 1937). *In Collected Papers of L. D. Landau*. New York: Gordon and Breach, pp. 193–216.

Landau, L. D. and Lifshitz, E. M. 1958. *Quantum Mechanics. Non-Relativistic Theory*. Reading: Addison-Wesley.

Landau, L. D. and Lifshitz, E. M. 1980. *Statistical Physics. Part 1*. Oxford: Pergamon Press.

Landau, L. D. and Lifshitz, E. M. 1993. *Mechanics*. 3rd ed. Oxford: Butterworth-Heinemann.

Langevin, D. 1972. Analyse spectrale de la lumière diffusée par la surface libre d'un cristale liquide nématique. Mesure de la tension superficielle et des coefficients de viscosité. *J. de Physique*, **33**, 249–256.

Langevin, D. and Bouchiat, M. A. 1973. Molecular order and surface-tension for nematic-isotropic interface of MBBA, deduced from light reflectivity and light-scattering measurements. *Mol. Cryst. Liq. Cryst.*, **22**, 317–331.

Langner, M., Praefcke, K., Kruerke, D. and Heppke, G. 1995. Chiral radial pentaynes exhibiting cholesteric discotic phases. *J. Mater. Chem.*, **5**, 693–699.

Lasher, G. 1970. Nematic ordering of hard rods derived from a scaled particle treatment. *J. Chem. Phys.*, **53**, 4141–4146.

Lasher, G. 1972. Monte Carlo results for a discrete-lattice model of nematic ordering. *Phys. Rev. A*, **5**, 1350–1354.

Lau, M. H. and Dasgupta, C. 1989. Numerical investigation of the role of topological defects in the 3-dimensional Heisenberg transition. *Phys. Rev. B*, **39**, 7212–7222.

LaViolette, R. A. and Stillinger, F. H. 1985. Consequences of the balance between the repulsive and attractive forces in dense, nonassociated liquids. *J. Chem. Phys.*, **82**, 3335–3343.

Lavrentovich, O. D. and Nastishin, Y. A. 1990. Defects in degenerate hybrid aligned nematic liquid crystals. *Europhys. Lett.*, **12**, 135–141.

Lavrentovich, O. D. and Pergamenshchik, V. M. 1995. Patterns in thin liquid crystal films and the divergence (surface like) elasticity. *Int. J. Mod. Phys. B*, **9**, 2389–2437.

Lavrentovich, O. D., Pasini, P., Zannoni, C. and Žumer, S. (eds.). 2001. *Defects in Liquid Crystals: Computer Simulations, Theory and Experiments*. Dordrecht: Kluwer.

Lax, M. 1974. *Symmetry Principles in Solid State and Molecular Physics*. New York: Wiley.

Le Fèvre, R. J. W. 1964. *Dipole Moments: Their Measurement and Application in Chemistry*. 3rd ed. London: Methuen.

Le Fèvre, R. J. W. 1965. Molecular refractivity and polarizability. In *Advances in Physical Organic Chemistry*, vol. 3. London: Academic Press, pp. 1–90.

Le Fèvre, R. J. W. and Murthy, D. S. N. 1969. Molecular susceptibility. Diamagnetic anisotropies of some polynuclear aromatic hydrocarbons. *Austral. J. Chem.*, **22**, 1415.

Le Fèvre, R. J. W. and Radom, L. 1967. Molecular polarisability. Carbon-carbon bond polarisabilities in relation to bond lengths. *J. Chem. Soc. B*, 1295–1298.

Le Fèvre, R. J. W., Radom, L. and Ritchie, G. L. D. 1967. Molecular polarisability. Anisotropic polarisabilities of anthracene and several halogenated anthracenes. *J. Chem. Soc. B*, 595.

Leach, A. R. 2001. *Molecular Modelling: Principles and Applications*. 2nd ed. Harlow: Prentice Hall.

Leadbetter, A. J. 1979. Structural studies of nematic, smectic A and smectic C phases. In Luckhurst, G. R. and Gray, G. W. (eds.), *The Molecular Physics of Liquid Crystals*. London: Academic Press, pp. 285–316.

Leadbetter, A. J. and Norris, E. K. 1979. Distribution functions in three liquid crystals from X-ray diffraction measurements. *Mol. Phys.*, **38**, 669–686.

Leadbetter, A. J., Frost, J. C., Gaughan, J. P., Gray, G. W. and Mosley, A. 1979. The structure of smectic A phases of compounds with cyano end groups. *J. de Physique*, **40**, 375–380.

Lebowitz, J. L. and Perram, J. W. 1983. Correlation functions for nematic liquid crystals. *Mol. Phys.*, **50**, 1207–1214.

Lebowitz, J. L., Percus, J. K. and Verlet, L. 1967. Ensemble dependence of fluctuations with application to machine computations. *Phys. Rev.*, **153**, 250–254.

Lebwohl, P. A. and Lasher, G. 1972. Nematic liquid crystal order. A Monte Carlo calculation. *Phys. Rev. A*, **6**, 426–429.

Lee, J.-H., Liu, D. N. and Wu, S.-T. 2008. *Introduction to Flat Panel Displays*. Chichester: Wiley.

Lee, J. H., Atherton, T. J., Barna, V., et al. 2009. Direct measurement of surface-induced orientational order parameter profile above the nematic-isotropic phase transition temperature. *Phys. Rev. Lett.*, **102**, 167801.

Lee, S., Tran, A., Allsopp, M., et al. 2014. CHARMM36 United Atom chain model for lipids and surfactants. *J. Phys. Chem. B*, **118**, 547–556.

Lee, S. D. 1987. A numerical investigation of nematic ordering based on a simple hard-rod model. *J. Chem. Phys.*, **87**, 4972–4974.

Lee, S. D. 1988. The Onsager type theory for nematic ordering of finite-length hard ellipsoids. *J. Chem. Phys.*, **89**, 7036–7037.

Lee, S. W., Chae, B., Kim, H. C., Lee, B., et al. 2003. New clues to the factors governing the perpendicular alignment of liquid crystals on rubbed polystyrene film surfaces. *Langmuir*, **19**, 8735–8743.

Leenhouts, F., de Jeu, W. H. and Dekker, A. J. 1979. Physical properties of nematic Schiff bases. *J. de Physique*, **40**, 989–995.

Lehmann, O. 1889. Über fliessende krystalle. *Z. Phys. Chem.*, **4**, 462–472.

Lehninger, A. L., Nelson, D. L. and Cox, M. M. 2005. *Principles of Biochemistry*. New York: W. H. Freeman.

Leimkuhler, B. and Matthews, C. 2015. *Molecular dynamics with Deterministic and Stochastic Numerical Methods*. Heidelberg: Springer.

Leimkuhler, B., Margul, D. T. and Tuckerman, M. E. 2013. Stochastic, resonance-free multiple time-step algorithm for molecular dynamics with very large time steps. *Mol. Phys.*, **111**, 3579–3594.

Lekkerkerker, H. N. W. and Anderson, V. J. 2002. Insight into phase transition kinetics from colloid science. *Nature*, **416**, 811–815.

Lekkerkerker, H. N. W. and Vroege, G. J. 2013. Liquid crystal phase transitions in suspensions of mineral colloids: new life from old roots. *Phil. Trans. Roy. Soc. A*, **371**, 20120263.

Lelidis, I. and Durand, G. 1993. Electric-field-induced isotropic-nematic phase-transition. *Phys. Rev. E*, **48**, 3822–3824.

Lelidis, I. and Durand, G. 1994. Electrically induced isotropic-nematic smectic-A phase-transitions in thermotropic liquid crystals. *Phys. Rev. Lett.*, **73**, 672–675.

Leonard, A. N., Wang, E., Monje-Galvan, V. and Klauda, J. B. 2019. Developing and testing of lipid force fields with applications to modeling cellular membranes. *Chem. Rev.*, **119**, 6227–6269.

LeSar, R. 2013. *Introduction to Computational Materials Science: Fundamentals to Applications*. Cambridge: Cambridge University Press.

Levelut, A. M., Malthête, J. and Collet, A. 1986. X-Ray structural study of the mesophases of some cone-shaped molecules. *J. de Physique*, **47**, 351–357.

Levelut, A. M., Tarento, R. J., Hardouin, F., Achard, M. F. and Sigaud, G. 1981. Number of S_A phases. *Phys. Rev. A*, **24**, 2180.

Levesque, D. 2017. New solid phase of dipolar systems. *Condens. Matter Phys.*, **20**, 1–8.

Levesque, D. and Weis, J. J. 1994. Orientational and structural order in strongly interacting dipolar hard spheres. *Phys. Rev. E*, **49**, 5131–5140.

Levesque, D., Weis, J. J. and Zarragoicoechea, G. J. 1993. Monte Carlo simulation study of mesophase formation in dipolar spherocylinders. *Phys. Rev. E*, **47**, 496–505.

Levine, R. D. and Tribus, M. 1978. *Maximum Entropy Formalism*. Boston, MA: MIT Press.

Levitt, M. H. 2001. *Spin Dynamics. Basics of Nuclear Magnetic Resonance*. Chichester: Wiley.

Lewars, E. G. 2016. *Computational Chemistry: Introduction to the Theory and Applications of Molecular and Quantum Mechanics*. New York: Springer.

Li, L. S., Walda, J., Manna, L. and Alivisatos, A. P. 2002. Semiconductor nanorod liquid crystals. *Nano Lett.*, **2**, 557–560.

Li, L. S., Marjanska, M., Park, G. H., Pines, A. and Alivisatos, A. P. 2004. Isotropic-liquid crystalline phase diagram of a CdSe nanorod solution. *J. Chem. Phys.*, **120**, 1149–52.

Li, M. H., Brûlet, A., Davidson, P., Keller, P. and Cotton, J. P. 1993. Observation of hairpin defects in a nematic main-chain polyester. *Phys. Rev. Lett.*, **70**, 2297–2300.

Li, X., Hill, R. M., Scriven, L. E. and Davis, H. T. 1996. Liquid crystals in ternary polyoxyethylene trisiloxane surfactant-silicone oil-H_2O system. *MRS Online Proceeding Library Archive*, **425**, 173–178.

Liang, C. X., Yan, L. Q., Hill, J. R., et al. 1995. Force-field studies of cholesterol and cholesteryl acetate crystals and cholesterol cholesterol intermolecular interactions. *J. Comput. Chem.*, **16**, 883–897.

Lighthill, M. J. 1958. *Introduction to Fourier Analysis and Generalised Functions*. Cambridge: Cambridge: University Press.

Limbach, H. J., Arnold, A., Mann, B. A. and Holm, C. 2006. ESPResSo an extensible simulation package for research on soft matter systems. *Comput. Phys. Comm.*, **174**, 704–727.

Lin, Z. X. and van Gunsteren, W. F. 2015. On the use of a weak-coupling thermostat in replica-exchange molecular dynamics simulations. *J. Chem. Phys.*, **143**.

Lindahl, E. and Edholm, O. 2000. Spatial and energetic-entropic decomposition of surface tension in lipid bilayers from molecular dynamics simulations. *J. Chem. Phys.*, **113**, 3882–3893.

Linden, C. D. and Fox, C. F. 1975. Membrane physical state and function. *Acc. Chem. Res.*, **8**, 321–327.

Link, D. R., Natale, G., Shao, R., et al. 1997. Spontaneous formation of macroscopic chiral domains in a fluid smectic phase of achiral molecules. *Science*, **278**, 1924–1927.

Lintuvuori, J. S. and Wilson, M. R. 2008. A new anisotropic soft-core model for the simulation of liquid crystal mesophases. *J. Chem. Phys.*, **128**, 0449061.

Lipkin, M. D. and Oxtoby, D. W. 1983. A systematic density functional-approach to the mean field theory of smectics. *J. Chem. Phys.*, **79**, 1939–1941.

Liu, P., Kim, B., Friesner, R. A. and Berne, B. J. 2005. Replica exchange with solute tempering: a method for sampling biological systems in explicit water. *Proc. Nat. Acad. Sci. USA*, **102**, 13749–13754.

Lo Nostro, P., Ninham, B. W., Fratoni, L., et al. 2003. Effect of water structure on the formation of coagels from ascorbyl-alkanoates. *Langmuir*, **19**, 3222–3228.

Longa, L., Cholewiak, G., Trebin, R. and Luckhurst, G. R. 2001. Representation of pair correlations in nematics. *Eur. Phys. J. E*, **4**, 51–57.

Losada-Perez, P., Jimenez-Monroy, K. L., van Grinsven, B., Leys, et al. 2014. Phase transitions in lipid vesicles detected by a complementary set of methods: heat-transfer measurements, adiabatic scanning calorimetry, and dissipation-mode quartz crystal microbalance. *Phys. Stat. Solid A*, **211**, 1377–1388.

Lub, J., Broer, D. J., Martinez Antonio, M. E. and Mol, G. N. 1998. The formation of a liquid crystalline main chain polymer by means of photopolymerization. *Liq. Cryst.*, **24**, 375–379.

Lubensky, T. C. and Priest, R. G. 1974. Critical exponents for a symmetric traceless tensor field theory model. *Phys. Lett. A*, **48**, 103–104.

Lubensky, T. C. and Prost, J. 1992. Orientational order and vesicle shape. *J. de Physique II*, **2**, 371–382.

Lubensky, T. C. and Stark, H. 1996. Theory of a critical point in the Blue-Phase-III-isotropic phase diagram. *Phys. Rev. E*, **53**, 714–720.

Luckhurst, G. R. 1985. Molecular field theories of liquid crystals: systems composed of uniaxial, biaxial or flexible molecules. In Emsley, J. W. (ed.), *Nuclear Magnetic Resonance of Liquid Crystals*. Dordrecht: Reidel, pp. 53–84.

Luckhurst, G. R. 1988. Pretransitional behavior in liquid crystals: the roles of Nuclear Magnetic Resonance spectroscopy and molecular field theory. *J. Chem. Soc. Faraday Trans.*, **84**, 961–986.

Luckhurst, G. R. 2001. Biaxial nematic liquid crystals: fact or fiction? *Thin Solid Films*, **393**, 40–52.

Luckhurst, G. R. 2004. Liquid crystals – A missing phase found at last? *Nature*, **430**, 413–414.

Luckhurst, G. R. 2015. Biaxial nematics: order parameters and distribution functions. In Luckhurst, G. R. and Sluckin, T. J. (eds.), *Biaxial Nematic Liquid Crystals. Theory, Simulation and Experiment*. Chichester: Wiley, pp. 25–53.

Luckhurst, G. R. and Gray, G. W. (eds.). 1979. *The Molecular Physics of Liquid Crystals*. London: Academic Press.

Luckhurst, G. R. and Romano, S. 1980a. Computer simulation studies of anisotropic systems. II. Uniaxial and biaxial nematics formed by noncylindrically symmetric molecules. *Mol. Phys.*, **40**, 129–139.

Luckhurst, G. R. and Romano, S. 1980b. Computer-simulation studies of anisotropic systems. 4. The effect of translational freedom. *Proc. Roy. Soc. (London) A*, **373**, 111–130.

Luckhurst, G. R. and Romano, S. 1999. Computer simulation study of a nematogenic lattice model based on an elastic energy mapping of the pair potential. *Liq. Cryst.*, **26**, 871–884.

Luckhurst, G. R. and Sanson, A. 1972. Angular dependent linewidths for a spin probe dissolved in a liquid crystal. *Mol. Phys.*, **24**, 1297–1311.

Luckhurst, G. R. and Satoh, K. 2003. Computer simulation of the field-induced alignment of the smectic A phase of the Gay-Berne mesogen GB(4.4,20.0,1,1). *Mol. Cryst. Liq. Cryst.*, **394**, 153–169.

Luckhurst, G. R. and Sluckin, T. J. (eds.). 2015. *Biaxial Nematic Liquid Crystals. Theory, Simulation and Experiment*. Chichester: Wiley.

Luckhurst, G. R. and Yeates, R. N. 1976. Negative order parameters for nematic liquid crystals. *Mol. Cryst. Liq. Cryst.*, **34**, 57–61.

Luckhurst, G. R. and Zannoni, C. 1975. A theory of dielectric relaxation in anisotropic systems. *Proc. Roy. Soc. A*, **343**, 389–398.

Luckhurst, G. R. and Zannoni, C. 1977. Why is the Maier-Saupe theory of nematic liquid crystals so successful? *Nature*, **267**, 412–414.

Luckhurst, G. R., Simpson, P. and Zannoni, C. 1987. Computer simulation studies of anisotropic systems.16. The smectic E-smectic B transition. *Liq. Cryst.*, **2**, 313–334.

Luckhurst, G. R., Stephens, R. A. and Phippen, R. W. 1990. Computer simulation studies of anisotropic systems. XIX. Mesophases formed by the Gay-Berne model mesogen. *Liq. Cryst.*, **8**, 451–464.

Luckhurst, G. R., Zannoni, C., Nordio, P. L. and Segre, U. 1975. Molecular field theory for uniaxial nematic liquid crystals formed by non-cylindrically symmetric molecules. *Mol. Phys.*, **30**, 1345–1358.

Luo, Z., Xu, D. and Wu, S.-T. 2014. Emerging quantum-dots-enhanced LCDs. *J. Display Tech.*, **10**, 526–539.

Luz, Z., Goldfarb, D. and Zimmermann, H. 1985. Discotic liquid crystals and their characterization by deuterium NMR. Emsley, J. W. (ed.), *Nuclear Magnetic Resonance of Liquid Crystals*. Dordrecht: Reidel, pp. 343–420.

Luzzati, V, and Tardieu, A. 1974. Lipid phases: structure and structural transitions. *Annu. Rev. Phys. Chem.*, **25**, 79–94.

Luzzati, V., Gulik-Krzywicki, T. and Tardieu, A. 1968. Polymorphism of lecithins. *Nature*, **218**, 1031–1034.

Luzzati, V., Mustacchi, H. and Skoulios, A. 1957. Structure of the liquid crystal phases of the soap-water system: middle soap and neat soap. *Nature*, **180**, 600–601.

Lydon, J. 1998. Chromonic liquid crystal phases. *Curr. Opin. Colloid Interface Sci.*, **3**, 458–466.

Lydon, J. 2004. Chromonic mesophases. *Curr. Opin. Colloid Interface Sci.*, **8**, 480–490.

Lynden-Bell, R. M. 1980. Are models necessary to describe molecular reorientation of symmetrical molecules? *Chem. Phys. Lett.*, **70**, 477–480.

Lyulin, A., Al-Barwani, M., Allen, M. P., et al. 1998. Molecular dynamics simulation of main chain liquid crystalline polymers. *Macromolecules*, **31**, 4626–4634.

Ma, S.-K. 1976. *Modern Theory of Critical Phenomena*. Reading: Benjamin.

Ma, S.-K. 1985. *Statistical Mechanics*. Singapore: World Scientific.

Mabrey-Gaud, S. 1981. Differential Scanning Calorimetry of liposomes. Knight, C. G. (ed.), *Liposomes: From Physical Structure to Therapeutic Applications*. Amsterdam: Elsevier – North Holland, pp. 105–138.

MacKerrel, A. D., Wirkeiwicz-Kuczera, J. and Karplus, M. 1995. An all-atom empirical energy function for the simulation of nucleic acids. *J. Amer. Chem. Soc.*, **117**, 11946–11975.

Madhusudana, N. V. and Pratibha, R. 1982. High strength defects in nematic liquid crystals. *Curr. Sci.*, **51.**, 877–881.

Madsen, L. A., Dingemans, T. J., Nakata, M. and Samulski, E. T. 2004. Thermotropic biaxial nematic liquid crystals. *Phys. Rev. Lett.*, **92**, 145505.

Maddox, J. 1988. Crystals from first principles. *Nature*, **335**, 201.

Maeda, Y., Shankar Rao, D. S., Krishna Prasad, S., Chandrasekhar, S. and Kumar, S. 2001. Phase behaviour of the discotic mesogen 2,3,6,7,10, 11-hexahexylthiotriphenylene (HHTT) under hydrostatic pressure. *Liq. Cryst.*, **28**, 1679–1690.

Maeda, Y., Shankar Rao, D. S., Krishna Prasad, S., Chandrasekar, S. and Kumar, S. 2003. Phase behaviour of the discotic mesogen 2,3,6,7,10,11-hexahexyl thiotriphenylene (HHTT) under pressure. *Mol. Cryst. Liq. Cryst.*, **397**, 429–442.

Mahanty, J. and Ninham, B. W. 1976. *Dispersion Forces*. London: Academic Press.

Maier, W. and Saupe, A. 1958. Eine einfache molekulare theorie des nematischen kristallinflüssigen zustandes. *Z. Naturforsch. A*, **13**, 564–566.

Maier, W. and Saupe, A. 1959. Eine einfache molekularstatistische theorie der nematischen kristallinflüssigen phase. Teil I. *Z. Naturforsch. A*, **14**, 882–889.

Maier, W. and Saupe, A. 1960. Eine einfache molekularstatistische theorie der nematischen kristallinflüssigen phase. Teil II. *Z. Naturforsch. A*, **15**, 287–292.

Malikova, N., Pastoriza-Santos, I., Schierhorn, M., Kotov, N. A. and Liz-Marzan, L. M. 2002. Layer-by-layer assembled mixed spherical and planar gold nanoparticles: control of interparticle interactions. *Langmuir*, **18**, 3694–3697.

Malthête, J., Collet, A. and Levelut, A. M. 1989. Mesogens containing the DOBOB group. *Liq. Cryst.*, **5**, 123–131.

Man, W. N., Donev, A., Stillinger, F. H., et al. 2005. Experiments on random packings of ellipsoids. *Phys. Rev. Lett.*, **94**, 198001.

Mandle, R. J. and Goodby, J. W. 2018. A nanohelicoidal nematic liquid crystal formed by a non-linear duplexed hexamer. *Angew. Chem. Intern. Ed.*, **57**, 7096–7100.

Mandle, R. J., Davis, E. J., Archbold, C. T., et al. 2015. Apolar bimesogens and the incidence of the Twist-Bend nematic phase. *Chem. Eur. J.*, **21**, 8158–8167.

Manoharan, V. N. 2015. Colloids. Colloidal matter: packing, geometry, and entropy. *Science*, **349**, 1253751.

Mansoori, G. A., Carnahan, N. F., Starling, K. E. and Leland, T. W. 1971. Equilibrium thermodynamic properties of the mixture of hard spheres. *J. Chem. Phys.*, **54**, 1523–1525.

Marchetti, M. C., Joanny, J. F., Ramaswamy, S., et al. 2013. Hydrodynamics of soft active matter. *Rev. Mod. Phys.*, **85**, 1143–1189.

Margola, T., Satoh, K. and Saielli, G. 2018. Comparison of the mesomorphic behaviour of 1:1 and 1:2 mixtures of charged Gay-Berne GB(4.4,20.0,1,1) and Lennard-Jones particles. *Crystals*, **8**, 371.

Marguta, R. G., Martin del Rio, E. and de Miguel, E. 2006. Revisiting McMillan's theory of the smectic A phase. *J. Phys. Condens. Matter*, **18**, 10335–10351.

Marrink, S. J., Berkowitz, M. and Berendsen, H. J. C. 1993. Molecular dynamics simulation of a membrane water interface – the ordering of water and its relation to the hydration force. *Langmuir*, **9**, 3122–3131.

Marrink, S. J., Lindahl, E., Edholm, O. and Mark, A. E. 2001. Simulation of the spontaneous aggregation of phospholipids into bilayers. *J. Amer. Chem. Soc.*, **123**, 8638–8639.

Marrink, S. J., Corradi, V., Souza, P. C. T., et al. 2019. Computational modeling of realistic cell membranes. *Chem. Rev.*, **119**, 6184–6226.

Martin, A. J., Meier, G. and Saupe, A. 1971. Extended Debye theory for dielectric relaxations in nematic liquid crystals. *Symp. Faraday Soc.*, **5**, 119–133.

Martin del Rio, E. and de Miguel, E. 1997. Computer simulation study of the free surfaces of a liquid crystal model. *Phys. Rev. E*, **55**, 2916.

Martin del Rio, E., de Miguel, E. and Rull, L. F. 1995. Computer simulation of the liquid-vapor interface in liquid crystals. *Physica A*, **213**, 138–147.

Martinelli, N. G., Savini, M., Muccioli, L., et al. 2009. Modeling polymer dielectric/pentacene interfaces: on the role of electrostatic energy disorder on charge carrier mobility. *Adv. Funct. Mater.*, **19**, 3254–3261.

Martinez, L., Andrade, R., Birgin, E. G. and Martinez, J. M. 2009. PACKMOL: a package for building initial configurations for molecular dynamics simulations. *J. Comput. Chem.*, **30**, 2157–2164.

Martins, A. F., Ferreira, J. B., Volino, F., Blumstein, A. and Blumstein, R. B. 1983. NMR study of some thermotropic nematic polyesters with mesogenic elements and flexible spacers in the main chain. *Macromolecules*, **16**, 279–287.

Martonosi, M. 1974. Thermal-analysis of sarcoplasmic-reticulum membranes. *FEBS Lett.*, **47**, 327–329.

Maruani, J. and Serre, J. (eds.). 1983. *Symmetries and Properties of Non-Rigid Molecules. A Comprehensive Survey*. Amsterdam: Elsevier.

Maruani, J. and Toro-Labbe, A. 1983. Symmetry analysis and conformational dependence of molecular properties in nonrigid systems. In Maruani, J. and Serre, J. (eds.), *Symmetries and Properties of Non-Rigid Molecules. A Comprehensive Survey*. Amsterdam: Elsevier, pp. 291–314.

Marynissen, H., Thoen, J. and Van Dael, W. 1983. Heat-capacity and enthalpy behavior near phase transitions in some alkylcyanobiphenyls. *Mol. Cryst. Liq. Cryst.*, **97**, 149–161.

Matthey, T., Cickovski, T., Hampton, S. S., et al. 2004. ProtoMol: an object-oriented framework for prototyping novel algorithms for molecular dynamics. *ACM Trans. Math. Softw.*, **30**, 237–265.

Mayo, S. L., Olafson, B. D. and Goddard, W. A. 1990. DREIDING – a generic force-field for molecular simulations. *J. Phys. Chem.*, **94**, 8897–8909.

Mazenko, G. 2000. *Equilibrium Statistical Mechanics*. New York: Wiley.

McArdle, C. B. 1989. *Side Chain Liquid Crystal Polymers*. Glasgow: Blackie.

McBain, J. W. and Sierichs, W. C. 1948. The solubility of sodium and potassium soaps and the phase diagrams of aqueous potassium soaps. *J. Amer. Oil Chem. Soc.*, **25**, 221–225.

McBride, C. and Lomba, E. 2007. Hard biaxial ellipsoids revisited: numerical results. *Fluid Phase Equilibria*, **255**, 37–45.

McCammon, J. A. and Harvey, S. C. 1987. *Dynamics of Proteins and Nucleic Acids*. Cambridge: Cambridge University Press.

McLaughlin, E., Shakespeare, M. A. and Ubbelohde, A. R. 1964. Pre-freezing phenomena in relation to liquid crystal formation. *Trans. Faraday Soc.*, **60**, 25–32.

McMillan, W. L. 1971. Simple molecular model for the smectic-A phase of liquid crystals. *Phys. Rev. A*, **4**, 1238–1246.

McMillan, W. L. 1972. X-Ray scattering from liquid crystals. I. Cholesteryl nonanoate and myristate. *Phys. Rev. A*, **6**, 936–947.

McMillan, W. L. 1973. Simple molecular theory of the smectic C phase. *Phys. Rev. A*, **8**, 1921–1929.

Mederos, L., Velasco, E. and Martinez-Raton, Y. 2014. Hard-body models of bulk liquid crystals. *J. Phys. Cond. Matter*, **26**, 463101.

Mei, S. and Zhang, P. W. 2015. On a molecular based Q-tensor model for liquid crystals with density variations. *Multiscale Model. Simul.*, **13**, 977–1000.

Memmer, R. 1998. Computer simulation of chiral liquid crystal phases – VII. The chiral Gay-Berne discogen. *Ber. Bunsenges. Phys. Chem.*, **102**, 1002–1010.

Memmer, R. 2002. Liquid crystal phases of achiral banana-shaped molecules: a computer simulation study. *Liq. Cryst.*, **29**, 483–496.

Memmer, R., Kuball, H. G. and Schönhofer, A. 1993. Computer-simulation of chiral liquid crystal phases.I. The polymorphism of the chiral Gay-Berne fluid. *Liq. Cryst.*, **15**, 345–360.

Memmer, R., Kuball, H. G. and Schönhofer, A. 1996. Computer simulation of chiral liquid crystal phases.III. A cholesteric phase formed by chiral Gay-Berne atropisomers. *Mol. Phys.*, **89**, 1633–1649.

Menzel, D. H. 1960. *Fundamental Formulas of Physics*. vol. 1. New York: Dover.

Mercury. Crystal Structure Visualisation, Exploration and Analysis Made Easy. www.ccdc.cam.ac.uk/products/mercury/

Merkel, K., Kocot, A., Vij, J. K., et al. 2004. Thermotropic biaxial nematic phase in liquid crystalline organo-siloxane tetrapodes. *Phys. Rev. Lett.*, **93**, 237801.

Mermin, N. D. 1979. The topological theory of defects in ordered media. *Rev. Mod. Phys.*, **51**, 591–648.

Mermin, N. D. 1990. *Boojums All The Way Through*. Cambridge: Cambridge University Press.

Mermin, N. D. and Wagner, H. 1966. Absence of ferromagnetism or antiferromagnetism in one- or two-dimensional isotropic Heisenberg models. *Phys. Rev. Lett.*, **17**, 1133–1136.

Metropolis, N., Rosenbluth, A. W., Rosenbluth, M. N., Teller, A. H. and Teller, E. 1953. Equation of state calculations by fast computing machines. *J. Chem. Phys.*, **21**, 1087–1092.

Meyer, R. B. 1973. Existence of even indexed disclinations in nematic liquid crystals. *Philos. Mag.*, **27**, 405–424.

Micheletti, D., Muccioli, L., Berardi, R., Ricci, M. and Zannoni, C. 2005. Effect of nanoconfinement on liquid crystal polymer chains. *J. Chem. Phys.*, **123**, 224705.

Michl, J. and Thulstrup, E. W. 1986. *Spectroscopy with Polarized Light*. New York: Wiley.

Miesowicz, M. 1946. The three coefficients of viscosity of anisotropic liquids. *Nature*, **158**, 27–27.

Millett, F. S., and Dailey, B. P. 1972. NMR determination of some deuterium quadrupole coupling constants in nematic solutions. *J. Chem. Phys.*, **56**, 3249–3256.

Mills, S. J., Care, C. M., Neal, M. P. and Cleaver, D. J. 1998. Computer simulation of an unconfined liquid crystal film. *Phys. Rev. E*, **58**, 3284–3294.

Minkin, V. I., Osipov, O. A., Zhdanov, Y. A., Hazzard, B. J. and Vaughan, W. E. 1970. *Dipole Moments in Organic Chemistry*. New York: Plenum.

Mitov, M. 2012. *Sensitive Matter*. Cambridge, MA: Harvard University Press.

Mitov, M. 2017. Cholesteric liquid crystals in living matter. *Soft Matter*, **13**, 4176–4209.

Mityashin, A., Roscioni, O. M., Muccioli, L., et al. 2014. Multiscale modeling of the electrostatic impact of self-assembled mono layers used as gate dielectric treatment in organic thin-film transistors. *ACS Appl. Mater. Interfaces*, **6**, 15372–15378.

Miyajima, S. 2001. Meaurement of translational diffusion in nematics. In Dunmur, D. A., Fukuda, A. and Luckhurst, G. R. (eds.), *Physical Properties of Liquid Crystals: Nematics*. London: INSPEC-IEE, pp. 457–463.

Mohanty, S., Chou, S. H., Brostrom, M. and Aguilera, J. 2006. Predictive modeling of self assembly of chromonics materials. *Molec. Simul.*, **32**, 1179–1185.

Moin, P. 2010. *Fundamentals of Engineering Numerical Analysis*. 2nd ed. New York: Cambridge University Press.

Moll, A, Hildebrandt, A, Lenhof, H. P. and Kohlbacher, O. 2006a. BALLView: a tool for research and education in molecular modeling. *Bioinformatics*, **22**, 365–366.

Moll, A., Hildebrandt, A., Lenhof, H.-P. and Kohlbacher, O. 2006b. BALLView: an object-oriented molecular visualization and modeling framework. *J. Comput. Aided Mol. Des.*, **19**, 791–800.

Monkade, M., Martinot Lagarde, P., Durand, G. and Granjean, C. 1997. SiO evaporated films topography and nematic liquid crystal orientation. *J. de Physique II*, **7**, 1577–1596.

Monson, P. A. and Gubbins, K. E. 1983. Equilibrium properties of the Gaussian Overlap fluid – Monte Carlo simulation and thermodynamic Perturbation Theory. *J. Phys. Chem.*, **87**, 2852–2858.

Morelhao, S. L. 2016. *Computer Simulation Tools for X-ray Analysis. Scattering and Diffraction Methods*. Cham: Springer.

Mori, H. and Nakanishi, H. 1988. On the stability of topologically non-trivial point defects. *J. Phys. Soc. Japan*, **57**, 1281–1286.

Mori, H., Gartland, E. C., Kelly, J. R. and Bos, P. J. 1999. Multidimensional director modeling using the Q tensor representation in a liquid crystal cell and its application to the π cell with patterned electrodes. *Jap. J. Appl. Phys.*, **38**, 135–146.

Morozov, I. V., Kazennov, A. M., Bystryi, R. G., et al. 2011. Molecular dynamics simulations of the relaxation processes in the condensed matter on GPUs. *Comput. Phys. Comm.*, **182**, 1974–1978.

Mottram, N. J. and Newton, C. J. P. 2014. Introduction to Q-tensor theory. *arXiv:1409.3542*.

Mottram, N. J. and Newton, C. J. P. 2016. Liquid crystal theory and modelling. Chen, J., Cranton, W. and Fihn, M. (eds.), *Handbook of Visual Display Technology, Vols 1-4*. Berlin: Springer, pp. 2021–2052.

Mouritsen, O. and Bagatolli, L. A. 2016. *Life – as a Matter of Fat: The Emerging Science of Lipidomics*. 2nd ed. Berlin: Springer.

Movahed, H. B., Hidalgo, R. C. and Sullivan, D. E. 2006. Phase transitions of semiflexible hard-sphere chain liquids. *Phys. Rev. E*, **73**, 032701.

Muccioli, L. and Zannoni, C. 2006. A deformable Gay-Berne model for the simulation of liquid crystals and soft materials. *Chem. Phys. Lett.*, **423**, 1–6.

Muccioli, L., D'Avino, G., Berardi, R., et al. 2014. Supramolecular organization of functional organic materials in the bulk and at organic/organic interfaces: a modeling and computer simulation approach. In Beljonne, D. and Cornil, J. (eds.), *Topics in Current Chemistry. Multiscale Modelling of Organic and Hybrid Photovoltaics*, vol. 352. Berlin: Springer, pp. 39–102.

Mukherjee, B., Peter, C. and Kremer, K. 2013. Dual translocation pathways in smectic liquid crystals facilitated by molecular flexibility. *Phys. Rev. E*, **88**, 010502.

Mukherjee, B., Delle Site, L., Kremer, K. and Peter, C. 2012. Derivation of coarse grained models for multiscale simulation of liquid crystalline phase transitions. *J. Phys. Chem. B*, **116**, 8474–8484.

Mukherjee, P. K. 1998. The puzzle of the nematic-isotropic phase transition. *J. Phys. Cond. Matter*, **10**, 9191.

Mukherjee, P. K. and Saha, M. 1997. Critical exponents for the Landau-de Gennes model of the nematic-isotropic phase transition. *Mol. Cryst. Liq. Cryst.*, **307**, 103–110.

Mundoor, H., Park, S., Senyuk, B., Wensink, H. H. and Smalyukh, I. I. 2018. Hybrid molecular-colloidal liquid crystals. *Science*, **360**, 768–771.

Muševič, I. 2017. *Liquid Crystal Colloids*. Cham: Springer.

Nagle, J. F. and Tristram-Nagle, S. 2000. Structure of lipid bilayers. *Bioch. Biophys. Acta.*, **1469**, 159–195.

Nakata, M., Zanchetta, G., Chapman, B. D., et al. 2007. End-to-end stacking and liquid crystal condensation of 6 to 20 base pair DNA duplexes. *Science*, **318**, 1276–9.

Nazarenko, V. G., Boiko, O. P., Park, H. S., et al. 2010. Surface alignment and anchoring transitions in nematic lyotropic chromonic liquid crystals. *Phys. Rev. Lett.*, **105**, 017801.

Neal, M. P. and Parker, A. J. 1998. Computer simulations using a longitudinal quadrupolar Gay-Berne model: effect of the quadrupole magnitude on the formation of the smectic phase. *Chem. Phys. Lett.*, **294**, 277–284.

Neal, M. P. and Parker, A. J. 1999. Computer simulations using a quadrupolar Gay-Berne model. *Mol. Cryst. Liq. Cryst. A*, **330**, 1809–1816.

Nehring, J. and Saupe, A. 1971. Elastic theory of uniaxial liquid crystals. *J. Chem. Phys.*, **54**, 337–343.

Nehring, J. and Saupe, A. 1972. Calculation of elastic-constants of nematic liquid crystals. *J. Chem. Phys.*, **56**, 5527–5528.

Nelson, D. R. 1977. Recent developments in phase-transitions and critical phenomena. *Nature*, **269**, 379–383.

Nelson, D. R. 2002. Toward a tetravalent chemistry of colloids. *Nano Lett.*, **2**, 1125–1129.

Ness, D. and Niehaus, J. 2011. Semiconductor nanoparticles. A review. *The Strem Chemiker*, **25**, 39–47.

Neubert, M. E. 2001a. Characterization of mesophase types and transitions. In Kumar, S. (ed.), *Liquid Crystals: Experimental Study of Physical Properties and Phase Transitions*. Cambridge: Cambridge University Press, pp. 29–64.

Neubert, M. E. 2001b. Chemical structure-property relationships. In Kumar, S. (ed.), *Liquid Crystals: Experimental Study of Physical Properties and Phase Transitions*. Cambridge: Cambridge University Press, pp. 393–476.

Newman, M. E. J. and Barkema, G. T. 1999. *Monte Carlo Methods in Statistical Physics*. Oxford: Clarendon Press.

Newton, I. 1686. *Philosophiae Naturalis Principia Mathematica*. London: J. Streater.

Nicastro, A. J. and Keyes, P. H. 1984. Electric-field-induced critical phenomena at the nematic-isotropic transition and the nematic-isotropic critical point. *Phys. Rev. A*, **30**, 3156.

Ninham, B. W. and Lo Nostro, P. 2010. *Molecular Forces and Self Assembly: In Colloid, Nano Sciences and Biology*. Cambridge: Cambridge University Press.

Niori, T., Sekine, T., Watanabe, J., Furukawa, T. and Takezoe, H. 1996. Distinct ferroelectric smectic liquid crystals consisting of banana shaped achiral molecules. *J. Mater. Chem.*, **6**, 1231–1233.

NIST. 2016. *Digital Library of Mathematical Functions*. http://dlmf.nist.gov/

NIST. 2017. Electronic Book Section. http://webbook.nist.gov/chemistry/

NIST Center for Neutron Research. 2016. www.ncnr.nist.gov/

Nobili, M. and Durand, G. 1992. Disorientation-induced disordering at a nematic-liquid crystal solid interface. *Phys. Rev. A*, **46**, R6174–R6177.

Nordio, P. L. 1976. General magnetic resonance theory. In Berliner, L. J. (ed.), *Spin Labeling. Theory and Applications*. New York: Academic Press, pp. 5–52.

Nordio, P. L. and Busolin, P. 1971. Electron Spin Resonance line shapes in partially ordered systems. *J. Chem. Phys.*, **55**, 5485–5490.

Nordio, P. L. and Segre, U. 1975. ESR linewidth in oriented solvents with two-angle-dependent distribution function. *Chem. Phys.*, **11**, 57–62.

Nordio, P. L. and Segre, U. 1977. Magnetic relaxation from first-rank interactions. *J. Magn. Res.*, **27**, 465–473.

Nordio, P. L. and Segre, U. 1979. Rotational diffusion. In Luckhurst, G. R. and Gray, G. (eds.), *The Molecular Physics of Liquid Crystals*. London: Academic Press, pp. 411–426.

Nordio, P. L., Rigatti, G. and Segre, U. 1973. Dielectric relaxation theory in nematic liquids. *Mol. Phys.*, **25**, 129–136.

Nosé, S. 1984. A molecular dynamics method for simulations in the canonical ensemble. *Mol. Phys.*, **52**, 255–268.

Nosé, S. and Klein, M. L. 1983. Constant pressure molecular dynamics for molecular systems. *Mol. Phys.*, **50**, 1055–1076.

Nounesis, G., Garland, C. W. and Shashidhar, R. 1991. Crossover from three-dimensional XY to tricritical behavior for the nematic-smectic-A1 phase transition. *Phys. Rev. A*, **43**, 1849–1856.

O'Connor, C. J. 1982. Magnetochemistry. Advances in theory and experimentation. *Progr. Inorg. Chem.*, 203–283.

Odriozola, G. 2012. Revisiting the phase diagram of hard ellipsoids. *J. Chem. Phys.*, **136**, 134505.

Oganov, A. R. 2010. *Modern Methods of Crystal Structure Prediction*. Berlin: Wiley-VCH.

Oh-e, M., Hong, S. C. and Shen, Y. R. 2002. Orientations of phenyl sidegroups and liquid crystal molecules on a rubbed polystyrene surface. *Appl. Phys. Lett.*, **80**, 784–786.

Ohm, C., Brehmer, M. and Zentel, R. 2012. Applications of liquid crystalline elastomers. In de Jeu, W.H. (ed.), *Liquid Crystal Elastomers: Materials and Applications*. Berlin: Springer, pp. 49–93.

Ohzono, T., Katoh, K., Wang, C. G., et al. 2017. Uncovering different states of topological defects in schlieren textures of a nematic liquid crystal. *Sci. Rep.*, **7**, 16814.

Olivier, Y., Muccioli, L. and Zannoni, C. 2014. Quinquephenyl: the simplest rigid-rod-like nematic liquid crystal, or is it? An atomistic simulation. *ChemPhysChem*, **15**, 1345–1355.

Olivier, Y., Muccioli, L., Lemaur, V., et al. 2009. Theoretical characterization of the structural and hole transport dynamics in liquid crystalline phthalocyanine stacks. *J. Phys. Chem. B*, **113**, 14102–14111.

Ondris-Crawford, R., Boyko, E. P., Wagner, B. G., et al. 1991. Microscope textures of nematic droplets in polymer dispersed liquid crystals. *J. Appl. Phys.*, **69**, 6380–6386.

O'Neill, M. and Kelly, S. M. 2003. Liquid crystals for charge transport, luminescence, and photonics. *Adv. Mater.*, **15**, 1135–1146.

Onsager, L. 1944. Crystal statistics. I. A two-dimensional model with an order-disorder transition. *Phys. Rev.*, **65**, 117–149.

Onsager, L. 1949. The effects of shape on the interaction of colloidal particles. *Ann. New York. Acad. Sci.*, **51**, 627–659.

Orlandi, S., Berardi, R., Steltzer, J. and Zannoni, C. 2006. A Monte Carlo study of the mesophases formed by polar bent-shaped molecules. *J. Chem. Phys.*, **124**, 124907.

Orlandi, S., Muccioli, L., Ricci, M. and Zannoni, C. 2007. Core charge distribution and self assembly of columnar phases: the case of triphenylenes and azatriphenylenes. *Chemistry Central J.*, **1**, 15–28.

Orlandi, S., Benini, E., Miglioli, I., et al. 2016. Doping liquid crystals with nanoparticles. A computer simulation of the effects of nanoparticle shape. *Phys. Chem. Chem. Phys.*, **18**, 2428–2441.

Orr, R. and Pethrick, R. A. 2011. Viscosities coefficients of nematic liquid crystals: I. Oscillating plate viscometer measurements and rotational viscosity measurements: K15. *Liq. Cryst.*, **38**, 1169–1181.

Ortiz, C., Ober, C. K. and Kramer, E. J. 1998. Stress relaxation of a main-chain, smectic, polydomain liquid crystalline elastomer. *Polymer*, **39**, 3713–3718.

Orville-Thomas, W. J. 1974. *Internal Rotations in Molecules*. New York: Wiley.

Oseen, C. W. 1933. The theory of liquid crystals. *Trans. Faraday Soc.*, **29**, 883–899.

Ostrovskii, B. I. 1993. Structure and phase transitions in smectic-A liquid crystals with polar and sterical asymmetry. *Liq. Cryst.*, **14**, 131–157.

Oswald, P. and Pieranski, P. 2005. *Nematic and Cholesteric Liquid Crystals. Concepts and Physical Properties Illustrated by Experiment*. Boca Raton, FL: Taylor & Francis.

Oswald, P. and Pieranski, P. 2006. *Smectic and Columnar Liquid Crystals: Concepts and Physical Properties Illustrated by Experiments*. Boca Raton: Taylor & Francis.

Ouellette, R. J. and Rawn, J. D. 2015. *Principles of Organic Chemistry*. Amsterdam: Elsevier.

OVITO. The Open Visualization Tool. www.ovito.org

Oyarzun, B., van Westen, T. and Vlugt, T. J. H. 2013. The phase behavior of linear and partially flexible hard-sphere chain fluids and the solubility of hard spheres in hard-sphere chain fluids. *J. Chem. Phys.*, **138**, 204905.

Oyarzun, B., van Westen, T. and Vlugt, T. J. 2015. Isotropic-nematic phase equilibria of hard-sphere chain fluids. Pure components and binary mixtures. *J. Chem. Phys.*, **142**, 064903.

Ozel, T., Ashley, M. J., Bourret, G. R., et al. 2015. Solution-dispersible metal nanorings with deliberately controllable compositions and architectural parameters for tunable plasmonic response. *Nano Lett.*, **15**, 5273–5278.

Paci, E. and Marchi, M. 1996. Constant-pressure molecular dynamics techniques applied to complex molecular systems and solvated proteins. *J. Phys. Chem.*, **100**, 4314–4322.

Pajak, G. and Osipov, M. A. 2013. Unified molecular field theory of nematic, smectic-A, and smectic-C phases. *Phys. Rev. E*, **88**, 012507.

Palazzesi, F., Calvaresi, M. and Zerbetto, F. 2011. A molecular dynamics investigation of structure and dynamics of SDS and SDBS micelles. *Soft Matter*, **7**, 9148–9156.

Palermo, M. F., Muccioli, L. and Zannoni, C. 2015. Molecular organization in freely suspended nano-thick 8CB smectic films. An atomistic simulation. *Phys. Chem. Chem. Phys.*, **17**, 26149–26159.

Palermo, M. F., Bazzanini, F., Muccioli, L. and Zannoni, C. 2017. Is the alignment of nematics on a polymer slab always along the rubbing direction? A molecular dynamics study. *Liq. Cryst.*, **44**, 1764–1774.

Palermo, M. F., Pizzirusso, A., Muccioli, L. and Zannoni, C. 2013. An atomistic description of the nematic and smectic phases of 4-n-octyl-4′ cyanobiphenyl (8CB). *J. Chem. Phys.*, **138**, 204901.

Palffy-Muhoray, P., Gartland, E. C. and Kelly, J. R. 1994. A new configurational transition in inhomogeneous nematics. *Liq. Cryst.*, **16**, 713–718.

Pall, S., Zhmurov, A., Bauer, P., et al. 2020. Heterogeneous parallelization and acceleration of molecular dynamics simulations in GROMACS. *J. Chem. Phys.*, **153**.

Palma, M., Levin, J., Lemaur, V., et al. 2006. Self-organization and nanoscale electronic properties of azatriphenylene-based architectures: a scanning probe microscopy study. *Adv. Mater.*, **18**, 3313–3317.

Pana, A., Pasuk, I., Micutz, M. and Circu, V. 2016. Nematic ionic liquid crystals based on pyridinium salts derived from 4-hydroxypyridine. *CrystEngComm*, **18**, 5066–5069.

Paolini, G. V., Ciccotti, G. and Ferrario, M. 1993. Simulation of site site soft-core liquid crystal models. *Mol. Phys.*, **80**, 297–312.

Parrinello, M. and Rahman, A. 1981. Polymorphic transitions in single crystals: a new molecular dynamics method. *J. App. Phys.*, **52**, 7182–7190.

Parsegian, V. A. 2006. *Van der Waals Forces: A Handbook for Biologists, Chemists, Engineers, and Physicists*. New York: Cambridge University Press.

Parsons, J. D. 1979. Nematic ordering in a system of rods. *Phys. Rev. A*, **19**, 1225–1230.

Parthasarathi, S., Rao, D. S. S., Palakurthy, N. B., Yelamaggad, C. V. and Prasad, S. K. 2017. Effect of pressure on dielectric and Frank elastic constants of a material exhibiting the twist bend nematic phase. *J. Phys. Chem. B*, **121**, 896–903.

Pasechnik, S. V., Chigrinov, V. G. and Shmeliova, D. V. 2009. *Liquid Crystals: Viscous and Elastic Properties*. Weinheim: Wiley-VCH.

Pasini, P. and Zannoni, C. 1984a. Orientational correlation-functions in ordered fluids – the short-time expansion. *Mol. Phys.*, **52**, 749–756.

Pasini, P. and Zannoni, C. 1984b. *Tables of Clebsch-Gordan coefficients for integer angular momentum J=0-6*. Report TC-83/19. INFN.

Pasini, P. and Zannoni, C. (eds.). 2000. *Advances in the Computer Simulations of Liquid Crystals*. Dordrecht: Kluwer.

Pasini, P., Chiccoli, C. and Zannoni, C. 2000. Liquid crystal lattice models II. Confined systems. In Pasini, P. and Zannoni, C. (eds.), *Advances in the Computer Simulations of Liquid Crystals*. Dordrecht: Kluwer, pp. 121–138.

Pasini, P., Semeria, F. and Zannoni, C. 1991. Symbolic computation of orientational correlation-function moments. *J. Symb. Comput.*, **12**, 221–231.

Pasini, P., Skačej, G. and Zannoni, C. 2005a. A microscopic lattice model for liquid crystal elastomers. *Chem. Phys. Lett.*, **413**, 463–467.

Pasini, P., Zannoni, C. and Žumer, S. (eds.). 2005b. *Computer Simulations of Liquid Crystals and Polymers*. Dordrecht: Kluwer.

Patey, G. N. and Valleau, J. P. 1974. Dipolar hard spheres – Monte Carlo study. *J. Chem. Phys.*, **61**, 534–540.

Pathria, R. K. 1972. *Statistical Mechanics*. Oxford: Pergamon Press.

Pathria, R. K. and Beale, P. D. 2011. *Statistical Mechanics*. 3rd ed. Amsterdam: Elsevier.

Pauling, L. 1960. *The Nature of the Chemical Bond and the Structure of Molecules and Crystals*. 3rd ed. Ithaca, NY: Cornell University Press.

Peczak, P. and Landau, D. P. 1989. Monte Carlo study of finite-size effects at a weakly 1st-order phase-transition. *Phys. Rev. B*, **39**, 11932–11942.

Peierls, R. 1936. On Ising's model of ferromagnetism. *Math. Proc. Cambridge Phil. Soc.*, **32**, 477–481.

Peláez, J. and Wilson, M. 2006. Atomistic simulations of a thermotropic biaxial liquid crystal. *Phys. Rev. Lett.*, **97**, 267801.

Peláez, J. and Wilson, M. 2007. Molecular orientational and dipolar correlation in the liquid crystal mixture E7: a molecular dynamics simulation at a fully atomistic level. *Phys. Chem. Chem. Phys.*, **9**, 2968–2975.

Pelzl, G., Diele, S. and Weissflog, W. 1999. Banana shaped compounds – a new field of liquid crystals. *Adv. Mater.*, **11**, 707–724.

Pennington, E. R., Day, C., Parker, J. M., Barker, M. and Kennedy, A. 2016. Thermodynamics of interaction between carbohydrates and unilamellar dipalmitoyl phosphatidylcholine membranes. *J. Therm. Anal. Calorim.*, **123**, 2611–2617.

Peroukidis, S. D., Karahaliou, P. K., Vanakaras, A. G. and Photinos, D. J. 2009. Biaxial nematics: symmetries, order domains and field-induced phase transitions. *Liq. Cryst.*, **36**, 727–737.

Perram, J. W., Petersen, H. G. and De Leeuw, S. W. 1988. An algorithm for the simulation of condensed matter which grows as the 3/2 power of the number of particles. *Mol. Phys.*, **65**, 875–893.

Perram, J. W., Rasmussen, J., Praestgaard, E. and Lebowitz, J. L. 1996. Ellipsoid contact potential: theory and relation to overlap potentials. *Phys. Rev. E*, **54**, 6565–6572.

Perrin, C. L. and Nielson, J. B. 1997. 'Strong' hydrogen bonds in chemistry and biology. *Annu. Rev. Phys. Chem.*, **48**, 511–544.

Perrin, F. 1934. Mouvement brownien d'un ellipsoide – I. Dispersion dielectrique pour des molecules ellipsoidales. *J. Phys. Radium*, **5**, 497–511.

Pershan, P. S., Aeppli, G., Litster, J. D. and Birgeneau, R. J. 1981. High-resolution X-Ray study of the smectic-A-smectic-B phase transition and the smectic-B phase in butyloxybenzylidene octylaniline. *Mol. Cryst. Liq. Cryst.*, **67**, 861–870.

Pestov, S. and Vill, V. 2005. Liquid Crystals. In Martienssen, W. and Warlimont, H. (eds.), *Springer Handbook of Condensed Matter and Materials Data*. Berlin: Springer, pp. 941–978.

Pettersen, E.F, Goddard, T. D., Huang, C. C., et al. 2004. UCSF Chimera. A visualization system for exploratory research and analysis. *J. Comput. Chem.*, **25**, 1605–1612.

Phillips, J. C., Braun, R., Wang, W., et al. 2005. Scalable molecular dynamics with NAMD. *J. Comput. Chem.*, **26**, 1781–1802.

Phillips, J. C., Hardy, D. J., Maia, J. D. C., et al. 2020. Scalable molecular dynamics on CPU and GPU architectures with NAMD. *J. Chem. Phys.*, **153**, 044130.

Photinos, D. J., Samulski, E. T. and Toriumi, H. 1990. Alkyl chains in a nematic field. 1. A treatment of conformer shape. *J. Phys. Chem.*, **94**, 4688–4694.

Pick, R. M. and Yvinec, M. 1983. Symmetry properties in plastic crystals. In Maruani, J. and Serre, J. (eds.), *Symmetries and Properties of Non-Rigid Molecules. A Comprehensive Survey*. Amsterdam: Elsevier, pp. 439–460.

Picken, S. J. 2001. Measurements and values for selected order parameters. In Dunmur, D. A., Fukuda, A. and Luckhurst, G. R. (eds.), *Physical Properties of Liquid Crystals, vol. 1: Nematics*. London: INSPEC-IEE, pp. 89–102.

Pikin, S. A. 1991. *Structural Transformations in Liquid Crystals*. New York: Gordon and Breach.

Pimentel, G. C. and McClellan, A. L. 1960. *The Hydrogen Bond*. San Francisco, CA: Freeman.

Pindak, R. and Ho, J. T. 1976. Cholesteric pitch of cholesteryl decanoate near smectic-A transition. *Phys. Lett. A*, **59**, 277–278.

Pindak, R. S., Huang, C. C. and Ho, J. T. 1974. Divergence of cholesteric pitch near a smectic-A transition. *Phys. Rev. Lett.*, **32**, 43–46.

Pines, A. 1988. Lectures on pulsed NMR. In Maraviglia, B. (ed.), *Physics of NMR Spectroscopy in Biology and Medicine*. Proceedings of the International School of Physics 'Enrico Fermi', Varenna. Amsterdam: North-Holland, pp. 43–120.

Pines, A. and Chang, J. J. 1974. Effect of phase-transitions on C-13 Nuclear Magnetic-Resonance spectra in para-azoxydianisole, a nematic liquid crystal. *J. Amer. Chem. Soc.*, **96**, 5590–5591.

Pippard, A. B. 1966. *Classical Thermodynamics*. Cambridge: Cambridge University Press.

Pisula, W., Tomovic, Z., Simpson, C., et al. 2005. Relationship between core size, side chain length, and the supramolecular organization of polycyclic aromatic hydrocarbons. *Chem. Mater.*, **17**, 4296–4303.

Pizzirusso, A., Savini, M., Muccioli, L. and Zannoni, C. 2011. An atomistic simulation of the liquid crystalline phases of sexithiophene. *J. Mater. Chem.*, **21**, 125–133.

Pizzirusso, A., Berardi, R., Muccioli, L., Ricci, M. and Zannoni, C. 2012a. Predicting surface anchoring: molecular organization across a thin film of 5CB liquid crystal on silicon. *Chem. Sci.*, **3**, 573–579.

Pizzirusso, A., Di Cicco, M. B., Tiberio, G., et al. 2012b. Alignment of small organic solutes in a nematic solvent: the effect of electrostatic interactions. *J. Phys. Chem. B*, **116**, 3760–3771.

Pizzirusso, A., Di Pietro, M. E., De Luca, G., et al. 2014. Order and conformation of biphenyl in cyanobiphenyl liquid crystals: a combined atomistic molecular dynamics and ^{1}H NMR study. *ChemPhysChem*, **15**, 1356–1367.

Plimpton, S. 1995. Fast parallel algorithms for short-range molecular dynamics. *J. Comput. Phys.*, **117**, 1–19.

Pluhackova, K., Kirsch, S. A., Han, J., et al. 2016. A critical comparison of biomembrane force fields: structure and dynamics of model DMPC, POPC, and POPE Bilayers. *J. Phys. Chem. B*, **120**, 3888–3903.

Poger, D., Caron, B. and Mark, A. E. 2016. Validating lipid force fields against experimental data: progress, challenges and perspectives. *Biochim. Biophys. Acta* **1858**, 1556–1565.

Poggi, Y., Filippini, J. C. and Aleonard, R. 1976. Free energy as a function of order parameter in nematic liquid crystals. *Phys. Lett. A*, **57**, 53–56.

Pohl, L. and Finkenzeller, U. 1990. Physical properties of liquid crystals. In Bahadur, B. (ed.), *Liquid Crystals. Applications and Uses*, vol. 1. Singapore: World Scientific, pp. 139–170.

Polnaszek, C. F., Bruno, G. V. and Freed, J. H. 1973. ESR line shapes in slow-motional region – anisotropic liquids. *J. Chem. Phys.*, **58**, 3185–3199.

Portugall, M., Ringsdorf, H. and Zentel, R. 1982. Synthesis and phase behaviour of liquid crystalline polyacrylates. *Makromol. Chem.*, **183**, 2311–2321.

Pottel, H., Herreman, W., Van der Meer, B. W. and Ameloot, M. 1986. On the significance of the fourth-rank orientational order parameter of fluorophores in membranes. *Chem. Phys.*, **102**, 37–44.

Praefcke, K. 2001. Relationship between molecular structure and transition temperatures for organic materials of a disc-like molecular shape in nematics. Dunmur, D. A., Fukuda, A. and Luckhurst, G. R. (eds.), *Physical Properties of Liquid Crystals, vol. 1: Nematics*. London: INSPEC-IEE, pp. 17–35.

Praefcke, K., Singer, D., Kohne, B., et al. 1991. Charge-transfer induced nematic columnar phase in low-molecular-weight disk-like systems. *Liq. Cryst.*, **10**, 147–159.

Praefcke, K., Singer, D., Langner, M., et al. 1992. Further low mass liquid crystal systems with nematic columnar phase. *Mol. Cryst. Liq. Cryst.*, **215**, 121–126.

Preeti, G. S., Murthy, K. P. N., Sastry, V. S. S., et al. 2011. Does the isotropic-biaxial nematic transition always exist? A new topology for the biaxial nematic phase diagram. *Soft Matter*, **7**, 11483–11487.

Press, W., Teukolsky, S. A., Vetterling, W. T. and Flannery, B. P. 1992. *Numerical Recipes*. 2nd ed. Cambridge: Cambridge University Press.

Price, S. L. and Stone, A. J. 1983. A distributed multipole analysis of the charge-densities of the azabenzene molecules. *Chem. Phys. Lett.*, **98**, 419–423.

Price, S. L., Stone, A. J. and Alderton, M. 1984. Explicit formulas for the electrostatic energy, forces and torques between a pair of molecules of arbitrary symmetry. *Mol. Phys.*, **52**, 987–1001.

Priestley, E. B., Wojtowicz, P. J. and Sheng, P. 1975. *Introduction to Liquid Crystals*. New York: Plenum Press.

Priezjev, N. V. and Pelcovits, R. A. 2001. Cluster Monte Carlo simulations of the nematic-isotropic transition. *Phys. Rev. E*, **63**, 062702.

Prior, C. and Oganesyan, V. S. 2017. Prediction of EPR spectra of lyotropic liquid crystals using a combination of molecular dynamics simulations and the model-free approach. *Chem. Eur. J.*, **23**, 13192–13204.

Procacci, P. and Marchi, M. 2000. Multiple time steps algorithms for the atomistic simulations of complex molecular systems. In Pasini, P. and Zannoni, C. (eds.), *Advances in the Computer Simulations of Liquid Crystals*. Dordrecht: Kluwer, pp. 333–388.

Procacci, P., Paci, E., Darden, T. A. and Marchi, M. 1997. ORAC: a molecular dynamics program to simulate complex molecular systems with realistic electrostatic interactions. *J. Comput. Chem.*, **18**, 1848–1862.

Pryde, J. A. 1969. *The Liquid State*. London: Hutchinson.

Pulvirenti, E. and Tsagkarogiannis, D. 2012. Cluster expansion in the canonical ensemble. *Commun. Math. Phys.*, **316**, 289–306.

Qi, H. and Hegmann, T. 2008. Impact of nanoscale particles and carbon nanotubes on current and future generations of liquid crystal displays. *J. Mater. Chem.*, **18**, 3288.

QMGA. Qt-Based Molecular Graphics Application.
http://qmga.sourceforge.net

Querciagrossa, L., Berardi, R. and Zannoni, C. 2018. Can off-centre mesogen dipoles extend the biaxial nematic range? *Soft Matter*, **14**, 2245–2253.

Querciagrossa, L., Ricci, M., Berardi, R. and Zannoni, C. 2013. Mesogen polarity effects on biaxial nematics. Centrally located dipoles. *Phys. Chem. Chem. Phys.*, **15**, 19065–19072.

Querciagrossa, L., Ricci, M., Berardi, R. and Zannoni, C. 2017. Can multi-biaxial mesogenic mixtures favour biaxial nematics? A computer simulation study. *Phys. Chem. Chem. Phys.*, **19**, 2383–2391.

Rackers, J. A., Wang, Z., Lu, C., et al. 2018. Tinker 8: software tools for molecular design. *J. Chem. Theory Comput.*, **14**, 5273–5289.

Radu, M., Pfleiderer, P. and Schilling, T. 2009. Solid-solid phase transition in hard ellipsoids. *J. Chem. Phys.*, **131**, 164513.

Rahman, A. and Stillinger, F. H. 1971. Molecular dynamics study of liquid water. *J. Chem. Phys.*, **55**, 3336–3359.

Rahman, M. D., Mohd Said, S. and Balamurugan, S. 2015. Blue Phase liquid crystal: strategies for phase stabilization and device development. *Sci. Technol. Adv. Mater.*, **16**, 033501.

Rai, P. K., Pinnick, R. A., Parra-Vasquez, A. N. G., et al. 2006. Isotropic-nematic phase transition of single-walled carbon nanotubes in strong acids. *J. Amer. Chem. Soc.*, **128**, 591–595.

Raimondi, M. E. and Seddon, J. M. 1999. Liquid crystal templating of porous materials. *Liq. Cryst.*, **26**, 305–339.

Rajteri, M., Barbero, G., Galatola, P., Oldano, C. and Faetti, S. 1996. van der Waals induced distortions in nematic liquid crystals close to a surface. *Phys. Rev. E*, **53**, 6093–6100.

Raos, G. and Allegra, G. 2000. Mesoscopic bead-and-spring model of hard spherical particles in a rubber matrix. I. Hydrodynamic reinforcement. *J. Chem. Phys.*, **113**, 7554–7563.

Rapaport, D. C. 2004. *The Art of Molecular dynamics Simulation*. 2nd ed. Cambridge: Cambridge University Press.

Rapini, A. and Papoular, M. 1969. Distorsion d'une lamelle nématique sous champ magnétique conditions d'ancrage aux parois. *J. de Physique Colloq.*, **30**, 54–56.

Rappé, A., Casewit, J., Colwell, K., Goddard, W. and Skiff, W. 1992. UFF, a full periodic table force field for molecular mechanics and molecular dynamics simulations. *J. Amer. Chem. Soc.*, **114**, 10024–10035.

Rappé, A. K. and Casewit, C. J. 1997. *Molecular Mechanics across Chemistry*. Sausalito, CA: University Science Books.

Rastogi, S., Höhne, G. W. H. and Keller, A. 1999. Unusual pressure-induced phase behavior in crystalline poly (4-methylpentene-1): calorimetric and spectroscopic results and further implications. *Macromolecules*, **32**, 8897–8909.

Ravi, P., Murad, S., Hanley, H. J. M. and Evans, D. J. 1992. The thermal-conductivity coefficient of polyatomic-molecules – benzene. *Fluid Phase Equilibria*, **76**, 249–257.

Reddy, R. A. and Tschierske, C. 2006. Bent-core liquid crystals: polar order, superstructural chirality and spontaneous desymmetrisation in soft matter systems. *J. Mater. Chem.*, **16**, 907–961.

Reed, M. and Simon, B. 1975. *Methods of Modern Mathematical Physics. vol.2 Fourier Analysis*. New York: Academic Press.

Reed, M. A. 1993. Quantum dots. *Sci. Am.*, **268**, 118–123.

Remler, D. K. and Haymet, A. D. J. 1986. Phase transitions in nematic liquid crystals – a mean field theory of the isotropic, uniaxial, and biaxial phases. *J. Phys. Chem.*, **90**, 5426–5430.

Renkes, G. D. 1981. Symmetry groups and representations of Hamiltonians for several coupled degrees of freedom: application to non-rigid molecular vibrations. *Chem. Phys.*, **57**, 261–278.

Renn, S. R. and Lubensky, T. C. 1988. Abrikosov dislocation lattice in a model of the cholesteric to smectic-A transition. *Phys. Rev. A*, **38**, 2132–2147.

Ricci, M., Berardi, R. and Zannoni, C. 2008. Columnar liquid crystals formed by bowl-shaped mesogens. A Monte Carlo study. *Soft Matter*, **4**, 2030–2038.

Ricci, M., Berardi, R. and Zannoni, C. 2015. On the field-induced switching of molecular organization in a biaxial nematic cell and its relaxation. *J. Chem. Phys.*, **143**, 084705.

Ricci, M., Roscioni, O. M., Querciagrossa, L. and Zannoni, C. 2019. MOLC. A reversible coarse grained approach using anisotropic beads for the modelling of organic functional materials. *Phys. Chem. Chem. Phys.*, **21**, 26195–26211.

Ricci, M., Mazzeo, M., Berardi, R., Pasini, P. and Zannoni, C. 2010. A molecular level simulation of a Twisted Nematic cell. *Faraday Discuss.*, **144**, 171–185.

Rigby, M. 1989. Hard Gaussian overlap fluids. *Mol. Phys.*, **68**, 687–697.

Rivera, B. O., van Westen, T. and Vlugt, T. J. H. 2016. Liquid-crystal phase equilibria of Lennard-Jones chains. *Mol. Phys.*, **114**, 895–908.

Rodrigues, A. S. M. C., Rocha, M. A. A. and Santos, L. M. N. B. F. 2013. Isomerization effect on the heat capacities and phase behavior of oligophenyls isomers series. *J. Chem. Thermodyn.*, **63**, 78–83.

Romanelli, M. J. 1960. Runge-Kutta methods for the solution of ordinary equations. In Ralston, A. and Wilf, H. S. (eds.), *Mathematical Methods for Digital Computers*. New York: Wiley, pp. 110–120.

Romano, S. 1998. Elastic constants and pair potentials for nematogenic lattice models. *Int. J. Mod. Phys. B*, **12**, 2305–2323.

Romano, S. 2004a. Computer simulation study of a biaxial nematogenic lattice model associated with a three-dimensional lattice and involving dispersion interactions. *Physica A*, **339**, 511–530.

Romano, S. 2004b. Mean-Field and computer simulation study of a biaxial nematogenic lattice model mimicking shape amphiphilicity. *Phys. Lett. A*, **333**, 110–119.

Roscioni, O. M. and Zannoni, C. 2016. Molecular dynamics simulation and its applications to thin-film devices. In Da Como, E., De Angelis, F., Snaith, H. and Walker, A. B. (eds.), *Unconventional Thin Film Photovoltaics*. London: Royal Society of Chemistry, pp. 391–419.

Roscioni, O. M., Muccioli, L. and Zannoni, C. 2017. Predicting the conditions for homeotropic anchoring of liquid crystals at a soft surface. 4-n-pentyl-4′-cyanobiphenyl on alkylsilane self-assembled monolayers. *ACS Appl. Mater. Interfaces*, **9**, 11993–12002.

Roscioni, O. M., Muccioli, L., Mityashin, A., Cornil, J. and Zannoni, C. 2016. Structural characterization of alkylsilane and fluoroalkylsilane self-assembled monolayers on SiO_2 by molecular dynamics simulations. *J. Phys. Chem. C*, **120**, 14652–14662.

Roscioni, O. M., Muccioli, L., Della Valle, R. G., et al. 2013. Predicting the anchoring of liquid crystals at a solid surface: 5-cyanobiphenyl on cristobalite and glassy silica surfaces of increasing roughness. *Langmuir*, **29**, 8950–8958.

Rose, M. E. 1957. *Elementary Theory of Angular Momentum*. New York: Wiley.

Rosen, M. E., Rucker, S. P., Schmidt, C. and Pines, A. 1993. 2-dimensional proton NMR-Studies of the conformations and orientations of n-alkanes in a liquid crystal solvent. *J. Phys. Chem.*, **97**, 3858–3866.

Rosenbluth, M. N. and Rosenbluth, A. W. 1954. Further results on Monte Carlo equations of state. *J. Chem. Phys.*, **22**, 881–884.

Rosso, L. and Gould, I. R. 2008. Structure and dynamics of phospholipid bilayers using recently developed general all-atom force fields. *J. Comput. Chem.*, **29**, 24–37.

Rosso, R. 2007. Orientational order parameters in biaxial nematics. Polymorphic notation. *Liq. Cryst.*, **34**, 737–748.

Roussel, O., Kestemont, G., Tant, J., et al. 2003. Discotic liquid crystals as electron carrier materials. *Mol. Cryst. Liq. Cryst.*, **396**, 35–39.

Rowley, C. 2015. https://en.wikibooks.org/wiki/Molecular_Simulation/Radial_Distribution_Functions

Ruessink, B. H., Barnhoorn, J., Bulthuis, J. and Maclean, C. 1988. Electric-field NMR of pretransitional effects in liquid crystals – a solute study. *Liq. Cryst.*, **3**, 31–41.

Rühle, V., Junghans, C., Lukyanov, A., Kremer, K. and Andrienko, D. 2009. Versatile object-oriented toolkit for coarse-graining applications. *J. Chem. Theory Comput.*, **5**, 3211–3223.

Rull, L. F. and Romero-Enrique, J. M. 2017. Computer simulation study of the nematic – vapour interface in the Gay-Berne model. *Mol. Phys.*, **115**, 1–11.

Rull, L. F., Romero-Enrique, J. M. and Fernandez-Nieves, A. 2012. Computer simulations of nematic drops: coupling between drop shape and nematic order. *J. Chem. Phys.*, **137**, 0345051–7.

Russel, W. B., Saville, D. A. and Schowalter, W. R. 1989. *Colloidal Dispersions*. Cambridge: Cambridge University Press.

Rust, B., Burrus, W. R. and Schneeberger, C. 1966. A simple algorithm for computing the generalized inverse of a matrix. *Commun. ACM*, **9**, 381–385.

Ryckaert, J.-P. and Bellemans, A. 1975. Molecular dynamics of liquid normal-butane near its boiling-point. *Chem. Phys. Lett.*, **30**, 123–125.

Ryckaert, J.-P. and Bellemans, A. 1978. Molecular dynamics of liquid alkanes. *Faraday Discuss.*, **66**, 95–106.

Ryckaert, J.-P., Ciccotti, G. and Berendsen, H. J. C. 1977. Numerical integration of the cartesian equations of motion of a system with constraints: Molecular dynamics of n-alkanes. *J. Comput. Phys.*, **23**, 327–341.

Sacanna, S., Rossi, L., Kuipers, B. W. M. and Philipse, A. P. 2006. Fluorescent monodisperse silica ellipsoids for optical rotational diffusion studies. *Langmuir*, **22**, 1822–1827.

Sackmann, H. 1989. Smectic liquid crystals. A historical review. *Liq. Cryst.*, **5**, 43–55.

Sage, I. 1992. Thermochromic liquid crystals in devices. In Bahadur, B. (ed.), *Liquid Crystals. Applications and Uses*, vol. 3. Singapore: World Scientific, pp. 301–343.

Sage, I. 2011. Thermochromic liquid crystals. *Liq. Cryst.*, **38**, 1551–1561.

Saielli, G. and Satoh, K. 2019. A coarse-grained model of ionic liquid crystals: the effect of stoichiometry on the stability of the ionic nematic phase. *Phys. Chem. Chem. Phys.*, **21**, 20327–20337.

Saielli, G., Margola, T. and Satoh, K. 2017. Tuning Coulombic interactions to stabilize nematic and smectic ionic liquid crystal phases in mixtures of charged soft ellipsoids and spheres. *Soft Matter*, **13**, 5204–5213.

Salaniwal, S., Cui, S. T., Cochran, H. D. and Cummings, P. T. 2001. Molecular simulation of a dichain surfactant/water/carbon dioxide system. 1. Structural properties of aggregates. *Langmuir*, **17**, 1773–1783.

Salikolimi, K., Sudhakar, A. A. and Ishida, Y. 2020. Functional ionic liquid crystals. *Langmuir*, **36**, 11702–11731.

Sambasivarao, S. V. and Acevedo, O. 2009. Development of OPLS-AA forcel field parameters for 68 unique ionic liquids. *J. Chem. Theory Comput.*, **5**, 1038–1050.

Sammalkorpi, M., Karttunen, M. and Haataja, M. 2007. Structural properties of ionic detergent aggregates: a large-scale molecular dynamics study of sodium dodecyl sulfate. *J. Phys. Chem. B*, **111**, 11722–11733.

Samulski, E. T. 1985. Macromolecular structure and liquid crystallinity. *Faraday Discuss.*, **79**, 7–20.

Sanchez-Castillo, A., Osipov, M. A. and Giesselmann, F. 2010. Orientational order parameters in liquid crystals: a comparative study of X-ray diffraction and polarized Raman spectroscopy results. *Phys. Rev. E*, **81**, 021707.

Sanders, C. R. and Landis, G. C. 1995. Reconstitution of membrane-proteins into lipid-rich bilayered mixed micelles for NMR studies. *Biochemistry*, **34**, 4030–4040.

Sanders, C. R. and Schwonek, J. P. 1992. Characterization of magnetically orientable bilayers in mixtures of dihexanoylphosphatidylcholine and dimyristoylphosphatidylcholine by solid-state NMR. *Biochemistry*, **31**, 8898–8905.

Santoro, P. A., Sampaio, A. R., da Luz, H. L. F. and Palangana, A. J. 2006. Temperature dependence of refractive indices near uniaxial – biaxial nematic phase transition. *Phys. Lett. A*, **353**, 512–515.

Santos, A. 2016. *A Concise Course on the Theory of Classical Liquids: Basics and Selected Topics*. Heidelberg: Springer.

Sarman, S. 1994. Molecular dynamics of heat-flow in nematic liquid crystals. *J. Chem. Phys.*, **101**, 480–489.

Sarman, S. and Evans, D. J. 1993. Statistical-mechanics of viscous-flow in nematic fluids. *J. Chem. Phys.*, **99**, 9021–9036.

Sarman, S. and Laaksonen, A. 2011. The heat conductivity of liquid crystal phases of a soft ellipsoid string-fluid evaluated by molecular dynamics simulation. *Phys. Chem. Chem. Phys.*, **13**, 5915–5925.

Sarman, S. and Laaksonen, A. 2013. Thermomechanical coupling, heat conduction and director rotation in cholesteric liquid crystals studied by molecular dynamics simulation. *Phys. Chem. Chem. Phys.*, **15**, 3442–3453.

Satoh, K. 2008. Influence of dipolar interaction on the molecular dynamics of the dipolar Gay-Berne model GB(3,5,1,2). *Mol. Cryst. Liq. Cryst.*, **480**, 202–218.

Satoh, K., Mita, S. and Kondo, S. 1996a. Monte Carlo simulations on mesophase formation using dipolar Gay-Berne model. *Liq. Cryst.*, **20**, 757–763.

Satoh, K., Mita, S. and Kondo, S. 1996b. Monte Carlo simulations using the dipolar Gay-Berne model: effect of terminal dipole moment on mesophase formation. *Chem. Phys. Lett.*, **255**, 99–104.

Saupe, A. 1966. The average orientation of solute molecules in nematic liquid crystals by proton NMR measurements and orientation dependent intermolecular forces. In Brown, G. H., Dienes, G. J. and Labes, M. M. (eds.), *Liquid Crystals*. New York: Gordon and Breach, pp. 207–221.

Saupe, A. 1974. Statistical theories of nematic liquid crystals. *Ber. Bunsen-Ges. Phys. Chem.*, **78**, 848–855.

Saupe, A. and Englert, G. 1963. High-resolution Nuclear Magnetic Resonance spectra of orientated molecules. *Phys. Rev. Lett.*, **11**, 462–466.

Sawamura, M., Kawai, K., Matsuo, Y., et al. 2002. Stacking of conical molecules with a fullerene apex into polar columns in crystals and liquid crystals. *Nature*, **419**, 702–705.

Sayle, R. 2000. *RASMOL program manual*. www.rasmol.com

Sayle, R. A. and Milner-White, E. J. 1995. RASMOL: biomolecular graphics for all. *Trends Biochem. Sci.*, **20**, 374–376.

Scalfani, V. F., Williams, A. J., Tkachenko, V., et al. 2016. Programmatic conversion of crystal structures into 3D printable files using JMOL. *J. Cheminform.*, **8**, 66.

Schafer, K., Kolli, H. B., Christensen, M. K., et al. 2020. Supramolecular packing drives morphological transitions of charged surfactant micelles. *Angew. Chem. Intern. Ed.*, **59**, 18591–18598.

Schatz, G. C. and Ratner, M. A. 1993. *Quantum Mechanics in Chemistry*. Englewoods Cliffs, NJ: Prentice Hall.

Schellman, J. A. 1998. Polarization modulation spectroscopy. Samorì, B. and Thulstrup, E. W. (eds.), *Polarized Spectroscopy of Ordered Systems*. Dordrecht: Kluwer, pp. 231–274.

Schlick, T. 2002. *Molecular Modeling and Simulation*. Berlin: Springer.

Schmid, F. and Phuong, N. H. 2002. Spatial order in liquid crystals: computer simulations of systems of ellipsoids. In Mecke, K. and Stoyan, D. (eds.), *Morphology of Condensed Matter: Physics and Geometry of Spatially Complex Systems*. Lecture Notes in Physics, vol. 600. Berlin: Springer-Verlag, pp. 172–186.

Schmidt-Rohr, K., Nanz, D., Emsley, L. and Pines, A. 1994. NMR measurement of resolved heteronuclear dipole couplings in liquid crystals and lipids. *J. Phys. Chem.*, **98**, 6668–6670.

Schopohl, N. and Sluckin, T. J. 1987. Defect core structure in nematic liquid crystals. *Phys. Rev. Lett.*, **59**, 2582–2584.

Schreiber, A., Ketelsen, I. and Findenegg, G. H. 2001. Melting and freezing of water in ordered mesoporous silica materials. *Phys. Chem. Chem. Phys.*, **3**, 1185–1195.

Schrödinger, L. L. C. 2010. *The PyMOL Molecular Graphics System, Version 1. 3r1.*

Schultz, A. J. and Kofke, D. A. 2014. Fifth to eleventh virial coefficients of hard spheres. *Phys. Rev. E*, **90**, 023301.

Schultz, T., Özarslan, E. and Hotz, I. 2017. *Modeling, Analysis, and Visualization of Anisotropy*. Cham: Springer.

Schweizer, W. B. and Dunitz, J. D. 2006. Quantum mechanical calculations for benzene dimer energies: present problems and future challenges. *J. Chem. Theory Comput.*, **2**, 288–291.

Scott, H. L. 2002. Modeling the lipid component of membranes. *Curr. Opin. Struct. Biol.*, **12**, 495–502.

Scott, W. R. P., Hünenberger, P. H., Tironi, I. G., et al. 1999. The GROMOS biomolecular simulation program package. *J. Phys. Chem. A*, **103**, 3596–3607.

Sebastian, N., Robles-Hernandez, B., Diez-Berart, S., et al. 2017. Distinctive dielectric properties of nematic liquid crystal dimers. *Liq. Cryst.*, **44**, 177–190.

Seddon, J. M. 1990. Structure of the inverted hexagonal (HII) Phase, and non-lamellar phase transitions of lipids. *Biochim. Biophys. Acta*, **1031**, 1–69.

Seddon, J. M. 1998. Structural studies of liquid crystals by X-ray diffraction. In Demus, D., Goodby, J., Gray, G. W., Spiess, H. W. and Vill, V. (eds.), *Handbook of Liquid Crystals. Low Molecular Weight Liquid Crystals II*, vol. 2B. Weinheim: Wiley-VCH, pp. 635–679.

Seddon, J. M. and Templer, R. H. 1993. Cubic phases of self-assembled amphiphilic aggregates. *Phil. Trans. Roy. Soc. London A*, **344**, 377–401.

Seelig, A. and Seelig, J. 1974. Dynamic structure of fatty acyl chains in a phospholipid bilayer measured by deuterium magnetic resonance. *Biochemistry*, **13**, 4839–4845.

Seelig, A. and Seelig, J. 1975. Bilayers of dipalmitoyl-3-sn-phosphatidylcholine: conformational differences between the fatty acyl chains. *Biochim. Biophys. Acta*, **406**, 1–5.

Seeman, N. C. 2020. DNA nanotechnology at 40. *Nano Lett.*, **20**, 1477–1478.

Selinger, J. V. and Nelson, D. R. 1988. Theory of hexatic-to-hexatic transitions. *Phys. Rev. Lett.*, **61**, 416–419.

Selinger, J. V. and Nelson, D. R. 1989. Theory of transitions among tilted hexatic phases in liquid crystals. *Phys. Rev. A*, **39**, 3135–3147.

Semenza, P. 2007. Can anything catch TFT LCDs? *Nat. Photonics*, **1**, 267–268.

Senftle, T. P., Hong, S., Islam, M. M., et al. 2016. The ReaxFF reactive force-field: development, applications and future directions. *NPJ Comput. Mater*, **2**, 15011.

Sengupta, K., Raghunathan, V. A. and Katsaras, J. 2003. Structure of the ripple phase of phospholipid multibilayers. *Phys. Rev. E*, **68**, 031710.

Sepelj, M., Lesac, A., Baumeister, U., et al. 2007. Intercalated liquid crystalline phases formed by symmetric dimers with an α,ω-diiminoalkylene spacer. *J. Mater. Chem.*, **17**, 1154–1165.

Sergeyev, S., Pisula, W. and Geerts, Y. H. 2007. Discotic liquid crystals: a new generation of organic semiconductors. *Chem. Soc. Rev.*, **36**, 1902–1929.

Serrano, J. L. and Sierra, T. 1996. Low molecular weight calamitic metallomesogens. In Serrano, J. L. (ed.), *Metallomesogens: Synthesis, Properties, and Applications*. New York: VCH, pp. 43–130.

Shapira, Y. 2019. *Linear Algebra and Group Theory for Physicists and Engineers*. Cham: Birkhäuser.

Shashidhar, R, and Venkatesh, G. 1979. High pressure studies on $4'$-n-alkyl-4-cyanobiphenyls. *J. de Physique Colloq. C3*, **40**, 396–400.

Shekhar, R., Whitmer, J. K., Malshe, R., et al. 2012. Isotropic-nematic phase transition in the Lebwohl-Lasher model from density of states simulations. *J. Chem. Phys.*, **136**, 234503.

Sheng, P. 1982. Boundary-layer phase transition in nematic liquid crystals. *Phys. Rev. A*, **26**, 1610–1617.

Sheng, P. and Wojtowicz, P. J. 1976. Constant-coupling theory of nematic liquid crystals. *Phys. Rev. A*, **14**, 1883–1894.

Sherrel, P. and Crellin, D. 1979. Susceptibilities and order parameters of nematic liquid crystals. *J. de Physique, Colloq., C3*, **40**, 211–216.

Sherrill, C. D., Sumpter, B. G., Sinnokrot, Mutasem O., et al. 2009. Assessment of standard force field models against high-quality *ab initio* potential curves for prototypes of π-π, CH-π and SH-π interactions. *J. Comput. Chem.*, **30**, 2187–2193.

Shi, J., Sidky, H. and Whitmer, J. K. 2020. Automated determination of n-cyanobiphenyl and n-cyanobiphenyl binary mixtures elastic constants in the nematic phase from molecular simulation. *Mol. Syst. Des. Eng.*, **5**, 1131–1136.

Shibaev, V. P. and Bobrovsky, A. Yu. 2017. Liquid crystalline polymers: development trends and photocontrollable materials. *Russian Chem. Rev.*, **86**, 1024–1072.

Sidky, H. and Whitmer, J. K. 2016. Elastic properties of common Gay-Berne nematogens from density of states (DOS) simulations. *Liq. Cryst.*, **43**, 2285–2299.

Sidky, H., de Pablo, J. J. and Whitmer, J. K. 2018. In silico measurement of elastic moduli of nematic liquid crystals. *Phys. Rev. Lett.*, **120**, 107801.

Simova, P., Kirov, N., Fontana, M. P. and Ratajczak, H. 1988. *Atlas of Vibrational Spectra of Liquid Crystals.* Singapore: World Scientific.

Sinanoglu, O. 1967. Intermolecular forces in liquids. *Adv. Chem. Phys.*, **12**, 283–328.

Singer, S. J. and Nicolson, G. L. 1972. Fluid mosaic model of structure of cell-membranes. *Science*, **175**, 720–731.

Singh, G. S. and Kumar, B. 1996. Geometry of hard ellipsoidal fluids and their virial coefficients. *J. Chem. Phys.*, **105**, 2429–2435.

Singh, G. S. and Kumar, B. 2001. Molecular fluids and liquid crystals in convex-body coordinate systems. *Annals Phys.*, **294**, 24–47.

Singh, S. 2000. Phase transitions in liquid crystals. *Phys. Reports*, **324**, 108–269.

Singh, S. 2019. Impact of dispersion of nanoscale particles on the properties of nematic liquid crystals. *Crystals*, **9**, 475.

Singh, U. C. and Kollman, P. A. 1984. An approach to computing electrostatic charges for molecules. *J. Comput. Chem.*, **5**, 129–145.

Skačej, G. and Zannoni, C. 2008. Controlling surface defect valence in colloids. *Phys. Rev. Lett.*, **100**, 197802.

Skačej, G. and Zannoni, C. 2011. Main-chain swollen liquid crystal elastomers: a molecular simulation study. *Soft Matter*, **7**, 9983–9991.

Skačej, G. and Zannoni, C. 2012. Molecular simulations elucidate electric field actuation in swollen liquid crystal elastomers. *Proc. Nat. Acad. Sci. USA*, **109**, 10193–10198.

Skačej, G. and Zannoni, C. 2014. Molecular simulations shed light on supersoft elasticity in polydomain liquid crystal elastomers. *Macromolecules*, **47**, 8824–8832.

Skačej, G. and Zannoni, C. 2021. The nematic-isotropic transition of the Lebwohl-Lasher model revisited. *Phil. Trans. Roy. Soc. A*, **379**, 20200117.

Sluckin, T. J. and Poniewierski, A. 1985. Novel surface phase-transition in nematic liquid crystals – wetting and the Kosterlitz-Thouless transition. *Phys. Rev. Lett.*, **55**, 2907–2910.

Sluckin, T. J., Dunmur, D. A. and Stegemeyer, H. 2004. *Crystals that Flow: Classic Papers from the History of Liquid Crystals.* London: Taylor & Francis.

Smith, Y. 2002. DL POLY: application to molecular simulation. *Mol. Sim.*, **28**, 385–471.

Smondyrev, A. M., Loriot, G. B. and Pelcovits, R. A. 1995. Viscosities of the Gay-Berne nematic liquid crystal. *Phys. Rev. Lett.*, **75**, 2340–2343.

Söderman, O., Carlström, G., Olsson, U. and Wong, T. C. 1988. Nuclear Magnetic Resonance relaxation in micelles. Deuterium relaxation at three field strengths of three positions on the alkyl chain of sodium dodecyl sulphate. *J. Chem. Soc., Faraday Trans. I*, **84**, 4475–4486.

Soldera, A. 2012. Atomistic simulations of vinyl polymers. *Molec. Simul.*, **38**, 762–771.

Soldera, A. and Metatla, N. 2005. Glass transition phenomena observed in stereoregular PMMAs using molecular modeling. *Composites Part A Appl.*, **36**, 521–530.

Soldera, A. and Metatla, N. 2006. Glass transition of polymers: atomistic simulation versus experiments. *Phys. Rev. E*, **74**, 061803.

Soler-Illia, G. J. A. A., Sanchez, C., Lebeau, B. and Patarin, J. 2002. Chemical strategies to design textured materials: from microporous and mesoporous oxides to nanonetworks and hierarchical structures. *Chem. Rev.*, **102**, 4093–4138.

Song, L. and Deng, Z. X. 2017. Valency control and functional synergy in DNA-bonded nanomolecules. *ChemNanoMat*, **3**, 698–712.

Song, W. H. and Windle, A. H. 2005. Isotropic-nematic phase transition of dispersions of multiwall carbon nanotubes. *Macromolecules*, **38**, 6181–6188.

Song, W. H., Kinloch, I. A. and Windle, A. H. 2003a. Nematic liquid crystallinity of multiwall carbon nanotubes. *Science*, **302**, 1363–1363.

Song, W. H., Tu, H. J., Goldbeck-Wood, G. and Windle, A. H. 2003b. Elastic constant anisotropy and disclination interaction in nematic polymers II. Effect of disclination interaction. *Liq. Cryst.*, **30**, 775–784.

Song, W. H., Tu, H. J., Goldbeck-Wood, G. and Windle, A. H. 2005. Effect of the elastic constant anisotropy on disclination interaction in the nematic polymers. *J. Phys. Chem. B*, **109**, 19234–19241.

Sonnet, A., Kilian, A. and Hess, S. 1995. Alignment tensor versus director: description of defects in nematic liquid crystals. *Phys. Rev. E*, **52**, 718–722.

Sonnet, A. M., Virga, E. G. and Durand, G. E. 2003. Dielectric shape dispersion and biaxial transitions in nematic liquid crystals. *Phys. Rev. E*, **67**, 061701.

Southern, C. D. and Gleeson, H. F. 2007. Using the full Raman depolarisation in the determination of the order parameters in liquid crystal systems. *Eur. Phys. J. E*, **24**, 119–127.

Spencer, T. 2000. Universality, phase transitions and statistical mechanics. In Alon, N., Bourgain, J., Connes, A., Gromov, M., and Milman, V. (eds.), *Visions in Mathematics. GAFA 2000 Special Volume, Part II*. Basel: Birkhäuser Verlag, pp. 839–858.

St. Pierre, A. G. and Steele, W. A. 1975. Cross-correlation functions for angular-momentum and orientation. *J. Chem. Phys.*, **62**, 2286–2300.

St. Pierre, A. G. and Steele, W. A. 1981. Some exact results for rotational correlation-functions at short times. *Mol. Phys.*, **43**, 123–140.

Stanley, H. E. 1971. *Introduction to Phase Transitions and Critical Phenomena*. Oxford: Oxford University Press.

Stannarius, R. 1998a. Diamagnetic properties of nematic liquid crystals. In Demus, D., Goodby, J., Gray, G. W., Spiess, H. W. and Vill, V. (eds.), *Handbook of Liquid Crystals*, vol. 2A. Weinheim: Wiley-VCH, pp. 113–127.

Stannarius, R. 1998b. Elastic properties of nematic liquid crystals. In Demus, D., Goodby, J., Gray, G. W., Spiess, H. W. and Vill, V. (eds.), *Handbook of Liquid Crystals. Low Molecular Weight Liquid Crystals I*, vol. 2A. Weinheim: Wiley-VCH, pp. 60–90.

Stannarius, R. and Cramer, C. 1997. Computer simulation of the liquid-vapor interface in liquid crystals. *Liq. Cryst.*, **23**, 371–375.

Steed, J. M., Dixon, T. A. and Klemperer, W. 1979. Molecular beam studies of benzene dimer, hexafluorobenzene dimer, and benzene – hexafluorobenzene. *J. Chem. Phys.*, **70**, 4940–4946.

Steele, W. 1976. The rotation of molecules in dense phases. *Adv. Chem. Phys.*, **34**, 1–104.

Steele, W. 1983. Symmetry constraints on the rotational time-correlations functions of rigid and non-rigid molecules. In Maruani, J. and Serre, J. (eds.), *Symmetries and Properties of Non-Rigid Molecules. A Comprehensive Survey*. Amsterdam: Elsevier, pp. 427–438.

Steele, W. A. 1963. Statistical mechanics of nonspherical molecules. *J. Chem. Phys.*, **39**, 3197–3208.

Steele, W. A. 1980. Symmetry constraints on the configurational properties of non-linear molecules tetrahedra. *Mol. Phys.*, **39**, 1411–1422.

Steinhardt, P. J., Nelson, D. R. and Ronchetti, M. 1983. Bond-orientational order in liquids and glasses. *Phys. Rev. B*, **28**, 784–805.

Stelzer, J., Berardi, R. and Zannoni, C. 1999. Flexoelectric effects in liquid crystals formed by pear-shaped molecules. A computer simulation study. *Chem. Phys. Lett.*, **299**, 9–16.

Stelzer, J., Longa, L. and Trebin, H. R. 1995. Molecular dynamics simulations of a Gay-Berne nematic liquid crystal. Elastic properties from direct correlation-functions. *J. Chem. Phys.*, **103**, 3098–3107.

Stelzer, J., Bates, M. A., Longa, L. and Luckhurst, G. R. 1997. Computer simulation studies of anisotropic systems. 27. The direct pair correlation function of the Gay-Berne discotic nematic and estimates of its elastic constants. *J. Chem. Phys.*, **107**, 7483–7492.

Stephen, M. J. and Straley, J. P. 1974. Physics of Liquid Crystals. *Rev. Mod. Phys.*, **46**, 617–704.

Stephenson, J. 1971. Critical phenomena: static aspects. In Henderson, D. (ed.), *Liquid State. Physical Chemistry. An Advanced Treatise*, vol. VIIIB. New York: Academic Press, pp. 717–795.

Sternberg, U., Witter, R. and Ulrich, A. S. 2007. All-atom molecular dynamics simulations using orientational constraints from anisotropic NMR samples. *J. Biomol. NMR*, **38**, 23–39.

Stewart, I. W. 2004. *The Static and Dynamical Continuum Theory of Liquid Crystals: a Mathematical Introduction*. London: Taylor & Francis.

Stewart, I. W. 2015. Continuum theory of biaxial nematic liquid crystals. In Luckhurst, G. R. and Sluckin, T. J. (eds.), *Biaxial Nematic Liquid Crystals. Theory, Simulation and Experiment*. Chichester: Wiley pp. 185–203.

Stinson, T. W. and Litster, J. D. 1970. Pretransitional phenomena in the isotropic phase of a nematic liquid crystal. *Phys. Rev. Lett.*, **25**, 503–506.

Stockmayer, W. H. 1941. Second virial coefficients of polar gases. *J. Chem. Phys.*, **9**, 398–402.

Stöhr, J. and Samant, M. G. 1999. Liquid crystal alignment by rubbed polymer surfaces: A microscopic bond orientation model. *J. Electron Spectros. Relat. Phenomena*, **98**, 189–207.

Stone, A. 1996. *The Theory of Intermolecular Forces*. Oxford: Oxford University Press.

Stone, A. J. 1978. Description of bimolecular potentials, forces and torques – S and V function expansions. *Mol. Phys.*, **36**, 241–256.

Stone, A. J. 1979. Intermolecular forces. In Luckhurst, G. R. and Gray, G. W. (eds.), *The Molecular Physics of Liquid Crystals*. London: Academic Press, pp. 31–50.

Stone, A. J. 1981. Distributed multipole analysis, or how to describe a molecular charge-distribution. *Chem. Phys. Lett.*, **83**, 233–239.

Stone, A. J. 1985. Distributed polarizabilities. *Mol. Phys.*, **56**, 1065–1082.

Straley, J. P. 1973. Gas of long rods as a model for lyotropic liquid crystals. *Mol. Cryst. Liq. Cryst.*, **22**, 333–357.

Straley, J. P. 1974. Ordered phases of a liquid of biaxial particles. *Phys. Rev. A*, **10**, 1881–1887.

Strichartz, R. S. 1994. *A Guide to Distribution Theory and Fourier Transforms*. Boca Raton, FL: CRC Press.

Strohmaier, E., Meuer, H. W., Dongarra, J. and Simon, H. D. 2015. The TOP500 list and progress in high-performance computing. *Computer*, **48**, 42–49.

Stukowski, A. 2010. Visualization and analysis of atomistic simulation data with OVITO – the Open Visualization Tool. *Modelling Simul. Mater. Sci. Eng.*, **18**, 015012.

Sugimura, A. and Luckhurst, G. R. 2016. Deuterium NMR investigations of field-induced director alignment in nematic liquid crystals. *Progr. NMR Spectr.*, **94–95**, 37–74.

Sugita, Y. and Okamoto, Y. 1999. Replica-exchange molecular dynamics method for protein folding. *Chem. Phys. Lett.*, **314**, 141–151.

Sun, H. 1995. *Ab-Initio* calculations and force-field development for computer-simulation of polysilanes. *Macromolecules*, **28**, 701–712.

Suurkuusk, J., Lentz, B. R., Barenholz, Y., Biltonen, R. L. and Thompson, T. E. 1976. Calorimetric and fluorescent-probe study of gel-liquid crystalline phase transition in small, single-lamellar dipalmitoylphosphatidylcholine vesicles. *Biochemistry*, **15**, 1393–1401.

Swager, T. M. and Xu, B. 1994. Liquid crystalline calixarenes. *J. Incl. Phenom. Mol. Recogn. Chem.*, **19**, 389.

Sweet, J. R. and Steele, W. A. 1967. Statistical mechanics of linear molecules. I. Potential energy functions. *J. Chem. Phys.*, **47**, 3022–3028.

Swendsen, R. H. 1991. Acceleration methods for Monte Carlo computer simulations. *Comput. Phys. Commun.*, **65**, 281–288.

Swendsen, R. H. 2012. *An Introduction to Statistical Mechanics and Thermodynamics*. Oxford: Oxford University Press.

Swift, J. 1976. Fluctuations near nematic-smectic C phase transition. *Phys. Rev. A*, **14**, 2274–2277.

Szabo, A. and Ostlund, N. S. 1996. *Modern Quantum Chemistry*. New York: Dover.

Szabó, M. J., Szilagyi, R. K., Unaleroglu, C. and Bencze, L. 1999. DTMM and COSMIC molecular mechanics parameters for alkylsilanes. *J. Mol. Struct.-Theochem*, **490**, 219–232.

Taddese, T., Anderson, R. L., Bray, D. J. and Warren, P. B. 2020. Recent advances in particle-based simulation of surfactants. *Curr. Opin. Colloid Interface Sci.*, **48**, 137–148.

Takebe, A. and Urayama, K. 2020. Supersoft elasticity and slow dynamics of isotropic-genesis polydomain liquid crystal elastomers investigated by loading- and strain-rate-controlled tests. *Phys. Rev. E*, **102**, 012701.

Takezoe, H. and Eremin, A. 2017. *Bent-Shaped Liquid Crystals : Structures and Physical Properties*. 1st ed. Boca Raton, FL: CRC Press.

Tamman, G. 1905. In Annual Meeting of the German Chemical Society. University of Karlsruhe.

Tang, X. M., Koenig, P. H. and Larson, R. G. 2014. Molecular dynamics simulations of sodium dodecyl sulfate micelles in water-the effect of the force field. *J. Phys. Chem. B*, **118**, 3864–3880.

Tao, P., Wu, X. W. and Brooks, B. R. 2012. Maintain rigid structures in Verlet based Cartesian molecular dynamics simulations. *J. Chem. Phys.*, **137**.

Tarroni, R. and Zannoni, C. 1991. On the rotational diffusion of asymmetric molecules in liquid crystals. *J. Chem. Phys.*, **95**, 4550–4564.

Tarumi, K., Finkenzeller, U. and Schuler, B. 1992. Dynamic behaviour of twisted nematic liquid crystals. *Jpn. J. Appl. Phys.*, **31**, 2829–2836.

te Velde, G., Bickelhaupt, F. M., Baerends, E. J., et al. 2001. Chemistry with ADF. *J. Comput. Chem.*, **22**, 931–967.

Teixeira, P. I. C. and Sluckin, T. J. 1992. Microscopic theory of anchoring transitions at the surfaces of pure liquid crystals and their mixtures. I. The Fowler approximation. *J. Chem. Phys.*, **97**, 1498–1509.

Teixeira, P. I. C., Osipov, M. A. and Luckhurst, G. R. 2006. Simple model for biaxial smectic-A liquid crystal phases. *Phys. Rev. E*, **73**, 061708.

Tenchov, B. G., Yao, H. and Hatta, I. 1989. Time-resolved X-ray diffraction and calorimetric studies at low scan rates: I. Fully hydrated dipalmitoylphosphatidylcholine (DPPC) and DPPC/water/ethanol phases. *Biophys. J.*, **56**, 757.

Ter Beek, L. C., Zimmerman, D. S. and Burnell, E. E. 1991. The conformation of 2,2′-dithiophene in nematic solvents determined by ^{1}H-NMR. *Mol. Phys.*, **74**, 1027–1035.

Thiem, H., Strohriegl, P., Shkunov, M. and McCulloch, I. 2005. Photopolymerization of reactive mesogens. *Macromolec. Chem. and Phys.*, **206**, 2153–2159.

Thiemann, T. and Vill, V. 1997. Development of an incremental system for the prediction of the nematic-isotropic phase transition temperature of liquid crystals with two aromatic rings. *Liq. Cryst.*, **22**, 519–523.

Thind, R., Walker, M. and Wilson, M. R. 2018. Molecular simulation studies of cyanine-based chromonic mesogens: spontaneous symmetry breaking to form chiral aggregates and the formation of a novel lamellar structure. *Adv. Theory Simul.*, **1**, 1800088.

Thoen, J. 1988. Adiabatic scanning calorimetric results for the blue phases of cholesteryl nonanoate. *Phys. Rev. A*, **37**, 1754–1759.

Thoen, J. 1995. Thermal investigations of phase transitions in thermotropic liquid crystals. *Int. J. Mod. Phys. B*, **9**, 2157–2218.

Thompson, D'Arcy W. 1917. *On Growth and Form*. Cambridge: Cambridge University Press.

Thompson, I. R., Coe, M. K., Walker, A. B., et al. 2018. Microscopic origins of charge transport in triphenylene systems. *Phys. Rev. Materials*, **2**, 064601.

Tiberio, G., Muccioli, L., Berardi, R. and Zannoni, C. 2009. Towards *in silico* liquid crystals. Realistic transition temperatures and physical properties for *n*-cyanobiphenyls via molecular dynamics simulations. *ChemPhysChem*, **10**, 125–136.

Tiddy, G. J. T. 1980. Surfactant-water liquid crystal phases. *Phys. Rep.*, **57**, 1–46.

Tikhonov, A. N. and Arsenin, V. I. 1977. *Solutions of Ill-Posed Problems*. Washington, WA: V.H. Winston.

Tilley, R. D. 2000. *Colour and Optical Properties of Materials*. New York: Wiley.

Tjandra, N. and Bax, A. 1997. Direct measurement of distances and angles in biomolecules by NMR in a dilute liquid crystalline medium. *Science*, **278**, 1111–1114.

Tjipto-Margo, B. and Evans, G. T. 1990. The Onsager theory of the isotropic-nematic liquid crystal transition. Incorporation of the higher virial coefficients. *J. Chem. Phys.*, **93**, 4254–4265.

Tjipto-Margo, B. and Sullivan, D. E. 1988. Molecular interactions and interface properties of nematic liquid crystals. *J. Chem. Phys.*, **88**, 6620–6630.

Tokita, M., Tagawa, H., Niwano, H., Sada, K. and Watanabe, J. 2006. Temperature-induced reversible distortion along director axis observed for monodomain nematic elastomer of cross-linked main-chain polyester. *Jpn. J. Appl. Phys*, **45**, 1729–1733.

Tolédano, J.-C. and Tolédano, P. 1987. *The Landau Theory of Phase Transitions*. Singapore: World Scientific.

Torquato, S. and Stillinger, F. H. 2010. Jammed hard-particle packings: from Kepler to Bernal and beyond. *Rev. Mod. Phys.*, **82**, 2633–2672.

Torras, N., Zinoviev, K. E., Esteve, J. and Sanchez-Ferrer, A. 2013. Liquid crystalline elastomer micropillar array for haptic actuation. *J. Mater. Chem. C*, **1**, 5183–5190.

Tournilhac, F., Blinov, L. M., Simon, J. and Yablonsky, S. V. 1992. Ferroelectric liquid crystals from achiral molecules. *Nature*, **359**, 621–623.

Toxvaerd, S. 1982. A new algorithm for molecular dynamics calculations. *J. Comput. Phys.*, **47**, 444–451.

Toxvaerd, S. 1983. Energy-conservation in molecular dynamics. *J. Comput. Physics*, **52**, 214–216.

Trebin, H-R. 1982. The topology of non-uniform media in condensed matter physics. *Adv. Phys.*, **31**, 195–254.

Tricomi, F. G. 1948. *Serie orthogonali di funzioni*. Torino: Istituto Editoriale Gheroni.

Tripp, C. P. and Hair, M. L. 1995. Reaction of methylsilanols with hydrated silica surfaces: the hydrolysis of trichloro-, dichloro-, and monochloromethylsilanes and the effects of curing. *Langmuir*, **11**, 149–155.

Tschierske, C. 2001. Non-conventional soft matter. *Ann. Rep. Progr. Chem. C*, **97**, 191–267.

Tschierske, C. 2002. Liquid crystals stack up. *Nature*, **419**, 681–683.

Tschierske, C. and Photinos, D. J. 2010. Biaxial nematic phases. *J. Mater. Chem.*, **20**, 4263–4294.

Tsvetkov, V. F. 1939. *Acta Physicochim. URSS*, **10**, 555–561.

Tsykalo, A. L. 1991. *Thermophysical Properties of Liquid Crystals*. New York: Gordon & Breach.

Tsykalo, A. L. and Bagmet, A. D. 1978. Molecular dynamics study of nematic liquid crystals. *Mol. Cryst. Liq. Cryst.*, **46**, 111–119.

Tuchband, M. R.., Shuai, M., Graber, K. A., et al. 2017. Double-helical tiled chain structure of the Twist-Bend liquid crystal phase in CB7CB. *arXiv:1703.10787*.

Tuckerman, M. E. 2010. *Statistical Mechanics: Theory and Molecular Simulation*. Oxford: Oxford University Press.

Tuckerman, M. E., Berne, B. J. and Martyna, G. J. 1991a. Molecular dynamics algorithm for multiple time scales: systems with long-range forces. *J. Chem. Phys.*, **94**, 6811–6815.

Tuckerman, M. E., Berne, B. J. and Martyna, G. J. 1992. Reversible multiple time scale molecular dynamics. *J. Chem. Phys.*, **97**, 1990–2001.

Tuckerman, M. E., Berne, B. J. and Rossi, A. 1991b. Molecular dynamics algorithm for multiple time scales. Systems with disparate masses. *J. Chem. Phys.*, **94**, 1465–1469.

Turiv, T., Koizumi, R., Thijssen, K., et al. 2020. Polar jets of swimming bacteria condensed by a patterned liquid crystal. *Nat. Phys.*, **16**, 481.

Turzi, S. S. 2011. On the Cartesian definition of orientational order parameters. *J. Math. Phys.*, **52**, 053517.

Uchida, N. 2000. Soft and nonsoft structural transitions in disordered nematic networks. *Phys. Rev. E*, **62**, 5119–5136.

UCSF Chimera Package. www.cgl.ucsf.edu/chimera

Ulmius, J., Wennerström, H., Lindblom, G. and Arvidson, G. 1977. Deuteron Nuclear Magnetic Resonance studies of phase equilibriums in a lecithin-water system. *Biochemistry*, **16**, 5742–5745.

Ungar, G., Percec, V. and Zuber, M. 1992. Liquid-crystalline polyethers based on conformational isomerism.20. nematic-nematic transition in polyethers and copolyethers based on 1-(4-hydroxyphenyl)-2-(2-r-4-hydroxyphenyl)ethane with R = Fluoro, Chloro, and methyl and flexible spacers containing an odd number of methylene units. *Macromolecules*, **25**, 75–80.

Unsöld, A. 1927. Quantum theory of the hydrogen molecular ion and the Born-Landé repulsive forces. *Z. Phys.*, **43**, 563–574.

Urayama, K., Honda, S. and Takigawa, T. 2005. Electrooptical effects with anisotropic deformation in nematic gels. *Macromolecules*, **38**, 3574–3576.

Urayama, K., Honda, S. and Takigawa, T. 2006. Deformation coupled to director rotation in swollen nematic elastomers under electric fields. *Macromolecules*, **39**, 1943–1949.

Urayama, K., Kohmon, E., Kojima, M. and Takigawa, T. 2009. Polydomain-Monodomain transition of randomly disordered nematic elastomers with different cross-linking histories. *Macromolecules*, **42**, 4084–4089.

Urban, S., Przedmojski, J. and Czub, J. 2005. X-ray studies of the layer thickness in smectic phases. *Liq. Cryst.*, **32**, 619–624.

Urbanski, M. and Lagerwall, J. P. F. 2017. Why organically functionalized nanoparticles increase the electrical conductivity of nematic liquid crystal dispersions. *J. Mater. Chem. C*, **5**, 8802–8809.

Uzunov, D. I. 2010. *Introduction to the Theory of Critical Phenomena: Mean Field, Fluctuations and Renormalization*. 2nd ed. Singapore: World Scientific.

Vaidya, D., Kofke, D. A., Tang, S. and Evans, G. T. 1994. Self-diffusion in the nematic-A-phase and smectic-A-phase of an aligned fluid of hard spherocylinders. *Mol. Phys.*, **83**, 101–112.

Valleau, J. P. and Torrie, G. M. 1977. A guide to Monte Carlo for statistical mechanics: 2. Byways. In Berne, B. J. (ed.), *Statistical Mechanics. Part A: Equilibrium Techniques*. New York: Plenum, pp. 169–194.

van Bruggen, M. P. B., van der Kooij, F. M. and Lekkerkerker, H. N. W. 1996. Liquid crystal phase transitions in dispersions of rod-like colloidal particles. *J. Phys. Cond. Matter*, **8**, 9451–9456.

Van der Est, A. J., Kok, M. Y. and Burnell, E. E. 1987. Size and shape effects on the orientation of rigid molecules in nematic liquid crystals. *Mol. Phys.*, **60**, 397–413.

Van der Haegen, R., Debruyne, J., Luyckx, R. and Lekkerkerker, H. N. W. 1980. 4 particle cluster approximation for the Maier-Saupe model of the isotropic-nematic phase transition. *J. Chem. Phys.*, **73**, 2469–2473.

van der Kooij, F. M. and Lekkerkerker, H. N. W. 1998. Formation of nematic liquid crystals in suspensions of hard colloidal platelets. *J. Phys. Chem. B*, **102**, 7829–7832.

van der Kooij, F. M., Kassapidou, K. and Lekkerkerker, H. N. W. 2000. Liquid crystal phase transitions in suspension of polydisperse plate-like particles. *Nature*, **406**, 868–871.

Van der Meer, B. W. and Vertogen, G. 1979. Elastic-constants as key for a molecular-model of cholesterics. *Phys. Lett. A*, **71**, 486–488.

Van der Meer, B. W., Vertogen, G., Dekker, A. J. and Ypma, J. G. J. 1976. Molecular-statistical theory of temperature-dependent pitch in cholesteric liquid crystals. *J. Chem. Phys.*, **65**, 3935–3943.

van der Schoot, P. 1995. Phase ordering of marginally flexible linear micelles. *J. de Physique II*, **5**, 243.

van der Spoel, B., Lindahl, E., Hess, B., et al. 2005. GROMACS: fast, flexible and free. *J. Comput. Chem.*, **26**, 1701–1718.

van Duin, A. C. T., Dasgupta, S., Lorant, F. and Goddard, W. A. 2001. ReaxFF: a Reactive force field for hydrocarbons. *J. Phys. Chem. A*, **105**, 9396–9409.

van Gunsteren, W. F. and Berendsen, H. J. C. 1987. *Groningen Molecular Simulation (GROMOS) Library Manual.* Biomos, Groningen.

Van Kampen, N. G. 1961. A simplified cluster expansion for the classical real gas. *Physica*, **27**, 783–792.

van Meer, G., Voelker, D. R. and Feigenson, G. W. 2008. Membrane lipids: where they are and how they behave. *Nat. Rev. Mol. Cell Biol.*, **9**, 112–124.

Van Roie, B., Denolf, K., Pitsi, G. and Thoen, J. 2005. Characterization of the smectic-A-hexatic-B transition in 65OBC by adiabatic scanning calorimetry. *Eur. Phys. J. E*, **16**, 361–364.

van Westen, T., Oyarzun, B., Vlugt, T. J. H. and Gross, J. 2013. The isotropic-nematic phase transition of tangent hard-sphere chain fluids. Pure components. *J. Chem. Phys.*, **139**, 034505.

Vanakaras, A. G. and Photinos, D. J. 2008. Thermotropic biaxial nematic liquid crystals: spontaneous or field stabilized? *J. Chem. Phys.*, **128**, 1545121.

Vanzo, D., Ricci, M., Berardi, R. and Zannoni, C. 2012. Shape, chirality and internal order of freely suspended nematic nanodroplets. *Soft Matter*, **8**, 11790–11800.

Vanzo, D., Ricci, M., Berardi, R., and Zannoni, C. 2016. Wetting behaviour and contact angles anisotropy of nematic nanodroplets on flat surfaces. *Soft Matter*, **12**, 1610–1620.

Varetto, U. *Molekel molecular visualization*. Lugano, CH: Swiss National Supercomputing Centre. http://molekel.cscs.ch

Varga, S. and Szalai, I. 2000. Modified Parsons-Lee theory for fluids of linear fused hard sphere chains. *Mol. Phys.*, **98**, 693–698.

Varga, S., Szalai, I., Liszi, J. and Jackson, G. 2002. A study of orientational ordering in a fluid of dipolar Gay-Berne molecules using density-functional theory. *J. Chem. Phys.*, **116**, 9107–9119.

Vause, C. A. 1986. Connection between the isotropic-nematic Landau point and the paranematic-nematic critical-point. *Phys. Lett. A*, **114**, 485–490.

Vega, L., de Miguel, E., Rull, L. F., Jackson, G. and Mclure, I. A. 1992. Phase equilibria and critical-behavior of square-well fluid of variable width by Gibbs Ensemble Monte Carlo simulation. *J. Chem. Phys.*, **96**, 2296–2305.

Vehoff, T., Baumeier, B. and Andrienko, D. 2010. Charge transport in columnar mesophases of carbazole macrocycles. *J. Chem. Phys.*, **133**, 134901.

Venable, R. M., Zhang, Y. H., Hardy, B. J. and Pastor, R. W. 1993. Molecular-dynamics simulations of a lipid bilayer and of hexadecane – an investigation of membrane fluidity. *Science*, **262**, 223–226.

Verlet, L. 1967. Computer 'experiments' on classical fluids. I. Thermodynamical properties of Lennard-Jones molecules. *Phys. Rev.*, **159**, 98–103.

Versmold, H. 1977. Symmetries of molecular-reorientation processes in liquids. *Mol. Phys.*, **33**, 1051–1061.

Vertogen, G. and de Jeu, W. H. 1988. *Thermotropic Liquid Crystals: Fundamentals*. Berlin: Springer.

Verwey, G. C., Warner, M. and Terentjev, E. M. 1996. Elastic instability and stripe domains in liquid crystalline elastomers. *J. Phys. II France*, **6**, 1273–1290.

Vieillard-Baron, J. 1972. Phase transitions of the classical hard ellipse system. *J. Chem. Phys.*, **56**, 4729–4744.

Vilan, A., Yaffe, O., Biller, A., et al. 2010. Molecules on Si: electronics with chemistry. *Adv. Mater.*, **22**, 140–159.

Virga, E. G. 1994. *Variational Theories for Liquid Crystals*. London: Chapman & Hall.

Visser, J. 1972. On Hamaker constants: a comparison between Hamaker constants and Lifshitz-van der Waals constants. *Adv. Colloid Interface Sci.*, **3**, 331–363.

Vita, F., Adamo, F. C. and Francescangeli, O. 2018. Polar order in bent-core nematics: an overview. *J. Mol. Liq.*, **267**, 564–573.

Vliegenthart, G. A. and Lekkerkerker, H. N. W. 2000. Predicting the gas-liquid critical point from the second virial coefficient. *J. Chem. Phys.*, **112**, 5364–5369.

Voets, G., Martin, H. and Van Dael, W. 1989. Calorimetric investigation of phase transitions in cholesteryl oleate. *Liq. Cryst.*, **5**, 871–875.

Vogel, N., Retsch, M., Fustin, C. A., del Campo, A. and Jonas, U. 2015. Advances in Colloidal assembly. The design of structure and hierarchy in two and three dimensions. *Chem. Rev.*, **115**, 6265–6311.

Voitchovsky, K., Kuna, J. J., Contera, S. A., Tosatti, E. and Stellacci, F. 2010. Direct mapping of the solid-liquid adhesion energy with subnanometre resolution. *Nat. Nanotechnol.*, **5**, 401–405.

Vold, M. J. 1957. The van der Waals interaction of anisometric colloidal particles. *Proc. Indian Acad. Sci. A*, **46**, 152–166.

Vold, R. R. 1985. Nuclear spin relaxation. In Emsley, J. W. (ed.), *Nuclear Magnetic Resonance of Liquid Crystals*. Dordrecht: Reidel, pp. 253–288.

Vold, R. R. and Prosser, R. S. 1996. Magnetically oriented phospholipid bilayered micelles for structural studies of polypeptides. Does the ideal bicelle exist? *J. Mag. Res. B*, **113**, 267–271.

Voss, N. R. and Gerstein, M. 2010. 3V: cavity, channel and cleft volume calculator and extractor. *Nucleic Acids Res.*, **38**, W555–W562.

Vroege, G. J. 2013. Biaxial phases in mineral liquid crystals. *Liq. Cryst.*, **41**, 342–352.

Vroege, G, J. and Lekkerkerker, H. N. W. 1992. Phase transitions in lyotropic colloidal and polymer liquid crystals. *Rep. Progr. Phys.*, **55**, 1241–1309.

Vuillermot, P. A. and Romerio, M. 1973a. Exact solution of Maier-Saupe model of unidimensional nematic liquid crystal. *Helv. Phys. Acta*, **46**, 467–468.

Vuillermot, P. A. and Romerio, M. V. 1973b. Exact solution of Maier-Saupe model for a nematic liquid crystal on a one-dimensional lattice. *J. Phys. C*, **6**, 2922–2930.

Wang, J., Wolf, R. M., Caldwell, J. W., Kollman, P. A. and Case, D. A. 2004. Development and testing of a general AMBER force field. *J. Comput. Chem.*, **25**, 1157–1174.

Wang, L. Y. and Li, Y. D. 2007. Controlled synthesis and luminescence of lanthanide doped NaYF4 nanocrystals. *Chem. Mater.*, **19**, 727–734.

Wang, M., Liechti, K. M., Wang, Q. and White, J. M. 2005. Self-assembled silane monolayers: fabrication with nanoscale uniformity. *Langmuir*, **21**, 1848–1857.

Wang, S. P., Chen, A. F. T. and Schwartz, M. 1988. Rotational diffusion of tribromobenzene in solution. *Mol. Phys.*, **65**, 689–693.

Wang, X. L., In, M., Blanc, C., Nobili, M. and Stocco, A. 2015. Enhanced active motion of Janus colloids at the water surface. *Soft Matter*, **11**, 7376–7384.

Wang, Z. Q., Lupo, J. A., Patnaik, S. and Pachter, R. 2001. Large scale molecular dynamics simulations of a 4-*n*-pentyl-4′-cyanobiphenyl (5CB) liquid crystalline model in the bulk and as a droplet. *Comput. and Theor. Polym. Sci.*, **11**, 375–387.

Warman, J. M. and Van de Craats, A. M. 2003. Charge mobility in discotic materials studied by PR-TRMC. *Mol. Cryst. Liq. Cryst.*, **396**, 41–72.

Warner, M. and Terentjev, E. M. 2003. *Liquid Crystal Elastomers*. Oxford: Oxford University Press.

Wassmer, K. H., Ohmes, E., Portugall, M., Ringsdorf, H. and Kothe, G. 1985. Molecular order and dynamics of liquid crystal side-chain polymers – an Electron Spin Resonance study employing rigid nitroxide spin probes. *J. Amer. Chem. Soc.*, **107**, 1511–1519.

Weber, A. C. J., Burnell, E. E., Meerts, W. L., et al. 2015. Communication: molecular dynamics and ^{1}H NMR of *n*-hexane in liquid crystals. *J. Chem. Phys.*, **143**.

Weber, A. C., Pizzirusso, A., Muccioli, L., et al. 2012. Efficient analysis of highly complex Nuclear Magnetic Resonance spectra of flexible solutes in ordered liquids by using molecular dynamics. *J. Chem. Phys.*, **136**, 174506.

Weeks, J. D. and Broughton, J. Q. 1983. van der Waals theory of melting in two and three dimensions. *J. Chem. Phys.*, **78**, 4197–4205.

Wegdam, G. H., Evans, G. J. and Evans, M. 1977. The properties of some derivative autocorrelation functions computed with the atom-atom potential. *Mol. Phys.*, **33**, 1805–1811.

Wei, D. and Patey, G. N. 1992. Orientational order in simple dipolar liquids: computer simulation of a ferroelectric nematic phase. *Phys. Rev. Lett.*, **62**, 2043–2045.

Weiner, P. K. and Kollman, P. A. 1981. AMBER-Assisted Model-Building with Energy Refinement. A general program for modeling molecules and their interactions. *J. Comput. Chem.*, **2**, 287–303.

Weiner, S. J., Kollmann, P. A., Case, D. A., et al. 1984. A new force field for molecular mechanical simulation of nucleic acids and proteins. *J. Amer. Chem. Soc.*, **106**, 765–784.

Weis, J. J. 2005. The ferroelectric transition of dipolar hard spheres. *J. Chem. Phys.*, **123**, 044503.

Weis, J. J. and Levesque, D. 2006. Orientational order in high density dipolar hard sphere fluids. *J. Chem. Phys.*, **125**, 34504.

Weis, J. J., Levesque, D. and Zarragoicoechea, G. J. 1992. Orientational order in simple dipolar liquid crystal models. *Phys. Rev. Lett.*, **69**, 913–916.

Wen, X., Garland, C. W. and Heppke, G. 1991. Calorimetric investigation of nematic-smectic A_2 and smectic A_2-smectic C_2 transitions. *Phys. Rev. A*, **44**, 5064–5068.

Wennerström, H. and Lindman, B. 1979. Micelles. Physical chemistry of surfactant association. *Phys. Rep.*, **52**, 1–86.

Wermter, H. and Finkelmann, H. 2001. Liquid crystalline elastomers as artificial muscles. *e-Polymers*, **1**, 1–13.

Wermuth, C. G., Aldous, D., Raboisson, P. and Rognan, D. (eds.). 2015. *The Practice of Medicinal Chemistry*. 4th ed. Amsterdam: Elsevier.

Westin, C.-F., Maier, S. E., Mamata, H., et al. 2002. Processing and visualization for diffusion tensor MRI. *Medical Image Analysis*, **6**, 93–108.

White, T. J. and Broer, D. J. 2015. Programmable and adaptive mechanics with liquid crystal polymer networks and elastomers. *Nat. Mater.*, **14**, 1087–1098.

Whittle, M. and Masters, A. J. 1991. Liquid crystal formation in a system of fused hard-spheres. *Mol. Phys.*, **72**, 247–265.

Widom, B. 1996. Theory of phase equilibrium. *J. Phys. Chem.*, **100**, 13190–13199.

Wigner, E. P. 1959. *Group Theory and its Application to the Quantum Mechanics of Atomic Spectra*. New York: Academic Press.

Williams, G. 1994. Dielectric relaxation behaviour of liquid crystals. In Luckhurst, G. R. and Veracini, C. A. (eds.), *The Molecular Dynamics of Liquid Crystals*, Dordrecht: Kluwer, pp. 431–450.

Williams, G. and Watts, D. C. 1970. Non-symmetrical dielectric relaxation behaviour arising from a simple empirical decay function. *Trans. Faraday Soc.*, **66**, 80–85.

Wilson, M. R. 1999. Atomistic simulations of liquid crystals. In Mingos, D. M. P. (ed.), *Structure and Bonding: Liquid Crystals I*, vol. 94. Heidelberg: Springer-Verlag, pp. 41–64.

Wilson, M. R. and Allen, M. P. 1993. A computer simulation study of liquid crystal formation in a semi-flexible system of linked hard spheres. *Mol. Phys.*, **80**, 277.

Wittgenstein, L. 1922. *Tractatus Logico–Philosophicus*. London: Kegan Paul, Trench, Trubner & Co.

Wohrle, T., Wurzbach, I., Kirres, J., et al. 2016. Discotic liquid crystals. *Chem. Rev.*, **116**, 1139–1241.

Wolarz, E. and Bauman, D. 2006. Polarized fluorescence studies of orientational order in some nematic liquid crystals doped with stilbene dye. *Mol. Cryst. Liq. Cryst.*, **197**, 1–13.

Wood, W. W. and Jacobson, J. D. 1957. Preliminary results from a recalculation of the Monte Carlo equation of state of hard spheres. *J. Chem. Phys.*, **27**, 1207–1208.

Woodcock, L. V. 1971. Isothermal molecular dynamics calculations for liquid salts. *Chem. Phys. Lett.*, **10**, 257–261.

Woodcock, L. V. 1997. Entropy difference between the face-centred cubic and hexagonal close-packed crystal structures. *Nature*, *385*, 141–143.

Wu, F. 1982. The Potts model. *Rev. Mod. Phys.*, **54**, 235–268.

Wu, S. T. and Cox, R. J. 1988. Optical and electro-optic properties of cyanotolanes and cyanostilbenes – potential infrared liquid crystals. *J. Appl. Phys.*, **64**, 821–826.

Wu, Y. H., Yang, Y., Qian, X. J., et al. 2020. Liquid-crystalline soft actuators with switchable thermal reprogrammability. *Angew. Chem. Intern. Ed.*, **59**, 4778–4784.

Wunderlich, B. 1999. A classification of molecules, phases, and transitions as recognized by thermal analysis. *Thermochim. Acta*, **340-341**, 37–52.

Würflinger, A. and Sandmann, M. 2001. Equations of state for nematics. Dunmur, D. A., Fukuda, A. and Luckhurst, G. R. (eds.), *Physical Properties of Liquid Crystals: Nematics*. EMIS Datareview Series, vol. 25. London: INSPEC, IEE, pp. 151–161.

Xu, B. and Swager, T. M. 1993. Rigid bowlic liquid crystals based on tungsten-oxo calix [4] arenes: host-guest effects and head-to-tail organization. *J. Amer. Chem. Soc.*, **115**, 1159–1160.

Xu, F., Kitzerow, H. S. and Crooker, P. P. 1992. Electric-field effects on nematic droplets with negative dielectric anisotropy. *Phys. Rev. A*, **46**, 6535–6540.

Xue, X., Chandler, G., Zhang, X., et al. 2015. Oriented liquid crystalline polymer semiconductor films with large ordered domains. *ACS Appl. Mater. Interfaces*, **7**, 26726–26734.

Yakovenko, S. Y., Muravski, A. A., Eikelschulte, F. and Geiger, A. 1998. Temperature dependence of the properties of simulated PCH5. *Liq. Cryst.*, **24**, 657–671.

Yang, C. 1961. *An Approach to the Ising Problem Using a Large Scale Fast Digital Computer.* Report. IBM T. J. Watson Center, Yorktown Heights, New York.

Yang, D. K. and Wu, S. T. 2006. *Fundamentals of Liquid Crystal Devices.* Chichester: Wiley.

Yang, D. K., Huang, X. Y. and Zhu, Y. M. 1997. Bistable cholesteric reflective displays: materials and drive schemes. *Ann. Rev. Mater. Sci.*, **27**, 117–146.

Yang, L. J., Tan, C. H., Hsieh, M. J., et al. 2006. New-generation AMBER united-atom force field. *J. Phys. Chem. B*, **110**, 13166–13176.

Yashonath, S. and Rao, C. N. R. 1985. Molecular design and computer simulations of novel mesophases. *Mol. Phys.*, **54**, 245–251.

Yildirim, A., Eroglu, E. and Yilmaz, S. 2011. Investigation of anisotropic thermal conductivity of uniaxial and biaxial Gay-Berne particles with molecular dynamics simulation. *Molec. Simul.*, **37**, 1179–1185.

Yokoyama, H. 1988. Surface anchoring of nematic liquid crystals. *Mol. Cryst. Liq. Cryst.*, **165**, 265–316.

Yoshida, H. 1990. Construction of higher order symplectic integrators. *Phys. Lett. A*, **150**, 262–268.

Young, C. Y., Pindak, R., Clark, N. A. and Meyer, R. B. 1978. Light-scattering study of 2-dimensional molecular-orientation fluctuations in a freely suspended ferroelectric liquid crystal film. *Phys. Rev. Lett.*, **40**, 773–776.

Young, M. J., Lei, W., Nounesis, G., Garland, C. W. and Birgeneau, R. J. 1994. X-ray-diffraction study of the smectic-$\tilde{A}$ fluid antiphase and its transitions to smectic-A_1 and smectic-A_2 phases. *Phys. Rev. E*, **50**, 368–376.

Youssefian, S., Rahbar, N., Lambert, C. R. and Van Dessel, S. 2017. Variation of thermal conductivity of DPPC lipid bilayer membranes around the phase transition temperature. *J. Roy. Soc. Interface*, **14**, 20170127.

Yu, L. J. and Saupe, A. 1980. Observation of a biaxial nematic phase in potassium laurate-1-decanol-water mixtures. *Phys. Rev. Lett.*, **45**, 1000–1003.

Yu, Y. L. and Klauda, J. B. 2020. Update of the CHARMM36 United Atom chain model for hydrocarbons and phospholipids. *J. Phys. Chem. B*, **124**, 6797–6812.

Zana, R. and Kaler, E. W. 2007. *Giant Micelles: Properties and Applications.* Boca Raton, FL: CRC Press.

Zannoni, C. 1975. *On the Molecular Theories of Liquid Crystals.* Ph.D. Thesis, University of Southampton.

Zannoni, C. 1979a. Computer simulations. In Luckhurst, G. R. and Gray, G. W. (eds.), *The Molecular Physics of Liquid Crystals.* London: Academic Press, pp. 191–220.

Zannoni, C. 1979b. Mean field theory of a model anisotropic potential of rank higher than two. *Mol. Cryst. Liq. Cryst. Letters*, **49**, 247–253.

Zannoni, C. 1979c. Order parameters and orientational distributions in liquid crystals. In Luckhurst, G. R. and Gray, G. W. (eds.), *The Molecular Physics of Liquid Crystals.* London: Academic Press, pp. 51–83.

Zannoni, C. 1979d. A theory of time dependent Fluorescence Depolarization in liquid crystals. *Mol. Phys.*, **38**, 1813–1827.

Zannoni, C. 1981. A theory of Fluorescence Depolarization in membranes. *Mol. Phys.*, **42**, 1303–1320.

Zannoni, C. 1985. An internal order parameter formalism for non-rigid molecules. In Emsley, J. W. (ed.), *Nuclear Magnetic Resonance of Liquid Crystals.* Dordrecht: Reidel, pp. 35–52.

Zannoni, C. 1986. A Cluster Monte Carlo method for the simulation of anisotropic systems. *J. Chem. Phys.*, **84**, 424–433.

Zannoni, C. 1988. Order parameters and orientational distributions in liquid crystals. In Samorì, B. and Thulstrup, E. W. (eds.), *Polarized Spectroscopy of Ordered Systems.* Dordrecht: Kluwer, pp. 57–83.

Zannoni, C. 2000. Liquid crystal observables. Static and dynamic properties. In Pasini, P. and Zannoni, C. (eds.), *Advances in the Computer Simulations of Liquid Crystals.* Dordrecht: Kluwer, pp. 17–50.

Zannoni, C. 2001a. Molecular design and computer simulations of novel mesophases. *J. Mater. Chem.*, **11**, 2637–2646.

Zannoni, C. 2001b. Results of generic model simulations. In Dunmur, D. A., Fukuda, A. and Luckhurst, G. R. (eds.), *Physical Properties of Liquid Crystals, vol. 1: Nematics.* London: INSPEC-IEE, pp. 624–634.

Zannoni, C. 2020. *Molecular dipoles, quadrupoles, and polarizabilities for mesogenic molecules.* Unpublished calculations performed with Quantum Chemistry suite ADF 2019.303 (rel. 21/02/2020) using DFT meta GGA-TPSS-D3BJ-TZP (triple zeta polarized basis set).

Zannoni, C. and Guerra, M. 1981. Molecular dynamics of a model anisotropic system. *Mol. Phys.*, **44**, 143–154.

Zannoni, C., Arcioni, A. and Cavatorta, P. 1983. Fluorescence Depolarization in liquid crystals and membrane bilayers. *Chem. Phys. Lipids*, **32**, 179–250.

Zannoni, C., Pedulli, G. F., Masotti, L. and Spisni, A. 1981. The polyliquid crystalline EPR spectra of nitroxide spin probes and their interpretation. *J. Mag. Res.*, **43**, 141–153.

Zasadzinski, J. A. N. and Meyer, R. B. 1986. Molecular imaging of Tobacco Mosaic-Virus lyotropic nematic phases. *Phys. Rev. Lett.*, **56**, 636–638.

Zemansky, M. W. and Dittman, R. 1997. *Heat and Thermodynamics: An Intermediate Textbook*. 7th ed. New York: McGraw-Hill.

Zewdie, H. 1998. Computer simulation studies of liquid crystals: a new Corner potential for cylindrically symmetric particles. *J. Chem. Phys.*, **108**, 2117–2133.

Zhang, J., Domenici, V., Veracini, C. A. and Dong, R. Y. 2006a. Deuterium NMR of the TGBA* phase in chiral liquid crystals. *J. Phys. Chem. B*, **110**, 15193–15197.

Zhang, J. G., Su, J. Y. and Guo, H. X. 2011. An atomistic simulation for 4-cyano-4′-pentylbiphenyl and its homologue with a reoptimized force field. *J. Phys. Chem. B*, **115**, 2214–2227.

Zhang, S., Kinloch, I. A, and Windle, A. H. 2006b. Mesogenicity drives fractionation in lyotropic aqueous suspensions of multiwall carbon nanotubes. *Nano Lett.*, **6**, 568–572.

Zhang, Z. P., Mouritsen, O. G. and Zuckermann, M. J. 1992. Weak 1st-order orientational transition in the Lebwohl-Lasher model for liquid crystals. *Phys. Rev. Lett.*, **69**, 2803–2806.

Zhang, Z. P., Zuckermann, M. J. and Mouritsen, O. G. 1993. Phase transition and director fluctuations in the 3-dimensional Lebwohl-Lasher model of liquid crystals. *Mol. Phys.*, **80**, 1195–1221.

Zhao, J. G., Gulan, U., Horie, T., et al. 2019. Advances in biological liquid crystals. *Small*, **15**, 1900019.

Zheng, Q., Durben, D. J., Wolf, G. H. and Angell, C. A. 1991. Liquids at large negative pressures: water at the homogeneous nucleation limit. *Science*, **254**, 829–832.

Zheng, X. and Palffy-Muhoray, P. 2007. Distance of closest approach of two arbitrary hard ellipses in two dimensions. *Phys. Rev. E*, **75**, 061709.

Zhengmin, S. and Kleman, M. 1984. Measurement of the 3 elastic constants and the shear viscosity γ_1 in a main-chain nematic polymer. *Mol. Cryst. Liq. Cryst.*, **111**, 321–328.

Zhou, S., Nastishin, Y. A., Omelchenko, M. M., et al. 2012. Elasticity of lyotropic chromonic liquid crystals probed by director reorientation in a magnetic field. *Phys. Rev. Lett.*, **109**, 037801.

Zhou, X., Hu, X., Zhou, S., et al. 2017. Ultrathin 2D GeSe2 rhombic flakes with high anisotropy realized by van der Waals epitaxy. *Adv. Funct. Mater.*, **27**, 1703858.

Zhu, C. H., Tuchband, M. R., Young, A., et al. 2016. Resonant carbon K-edge soft X-ray scattering from lattice-free heliconical molecular ordering: soft dilative elasticity of the Twist-Bend liquid crystal phase. *Phys. Rev. Lett.*, **116**, 147803.

Zhu, X., Lopes, P. E. M. and MacKerell, A. D. 2012. Recent developments and applications of the CHARMM force fields. *Wiley Interdiscip. Rev. Comput. Mol. Sci.*, **2**, 167–185.

Zhuang, X. W., Marrucci, L. and Shen, Y. R. 1994. Surface-monolayer-induced bulk alignment of liquid crystals. *Phys. Rev. Lett.*, **73**, 1513–1516.

Zimmerman, D. and Burnell, E. 1990. Size and shape effects on the orientation of solutes in nematic liquid crystals. *Mol. Phys.*, **69**, 1059–1071.

Zimmermann, H., Poupko, R., Luz, Z. and Billard, J. 1985. Pyramidic mesophases. *Z. Naturforsch. A*,, **40**, 149–160.

Zubarev, E. R., Kuptsov, S. A., Yuranova, T. I., Talroze, R. V. and Finkelmann, H. 1999. Monodomain liquid crystalline networks: reorientation mechanism from uniform to stripe domains. *Liq. Cryst.*, **26**, 1531–1540.

Zwanzig, R. 2001. *Nonequilibrium Statistical Mechanics*. Oxford: Oxford University Press.

Zwanzig, R. and Ailawadi, N. K. 1969. Statistical error due to finite time averaging in computer experiments. *Phys. Rev.*, **182**, 280–283.

Index

Printed in the United States
by Baker & Taylor Publisher Services